Earth Surface Processes, Landforms and Sediment Deposits

Earth surface processes, landforms, and sediment deposits are intimately related – involving erosion of rocks, generation of sediment, and transport and deposition of sediment through various Earth surface environments. These processes, and the deposits and landforms that they generate, have a fundamental bearing on engineering, environmental and public safety issues; on recovery of economic resources; and on our understanding of Earth history.

This unique textbook brings together the traditional disciplines of sedimentology and geomorphology to explain Earth processes, landforms, and sediment deposits in a comprehensive and integrated way. Progressing from small-scale to large-scale phenomena, with frequent reference to environmental and economic applications, this text presents a rigorous treatment that assumes only a modest background knowledge of mathematics and science. It is the ideal resource for a two-semester course in sedimentology, stratigraphy, geomorphology, and Earth surface processes from the intermediate undergraduate to beginning graduate level. The book is accompanied by a website hosting illustrations and material on field and laboratory methods for measuring, describing, and analyzing Earth surface processes, landforms, and sediments (www.cambridge.org/earthsurfaceprocesses).

JOHN BRIDGE was awarded a Ph.D. from St. Andrews University, Scotland, in 1973. He now holds the position of Professor of Geological Sciences at the State University of New York at Binghamton where his research interests focus on fluvial sedimentology and Earth surface processes. Professor Bridge has served on the editorial boards of several sedimentology journals and he is also the author/editor of two other books.

ROBERT DEMICCO was awarded a Ph.D. from the Johns Hopkins University in 1981. He is now Professor of Geology at State University of New York at Binghamton where his research interests focus on carbonate sedimentology and geochemical modeling of the carbonate system. Professor Demicco is the author/editor of two other books.

The authors have taught numerous courses in sedimentology, stratigraphy, and Earth surface processes at their institution and this book is the culmination of years of experience in training students in these fields.

Earth Surface Processes, Landforms and Sediment Deposits

John Bridge

and

Robert Demicco

CAMBRIDGE UNIVERSITY PRESS

Cambridge, New York, Melbourne, Madrid, Cape Town, Singapore, São Paulo

Cambridge University Press
The Edinburgh Building, Cambridge CB2 8RU, UK

Published in the United States of America by Cambridge University Press, New York

www.cambridge.org
Information on this title: www.cambridge.org/9780521857802

First published 2008

Printed in the United Kingdom at the University Press, Cambridge

A catalog record for this publication is available from the British Library

ISBN 978-0-521-85780-2 hardback

Contents

Acknowledgments	*page* vii
Figure credits	viii

PART 1   Introduction

1

1 Definitions, rationale, and scope of the book	3
2 Overview of the Earth	9

PART 2   Production of sediment at the Earth's surface

43

3 Weathering of rocks, production of terrigenous sediment, and soils	45
4 Biogenic and chemogenic sediment production	85

PART 3   Fundamentals of fluid flow, sediment transport, erosion, and deposition

119

5 Unidirectional turbulent water flow, sediment transport, erosion, and deposition	121
6 Air flow, sediment transport, erosion, and deposition	195
7 Multidirectional water flow, sediment transport, erosion, and deposition	213
8 Movement of sediment by gravity	255
9 Generation and movement of volcaniclastic sediment	278
10 Ice flow, sediment transport, erosion, and deposition	292
11 Biogenic and chemogenic depositional structures	311
12 Post-depositional deformation of soft sediment	352

PART 4   Environments of erosion and deposition

363

13 Rivers, alluvial plains, and fans	365
14 Lakes	462
15 Coasts and shallow seas	473
16 Arid environments	563
17 Glacial and periglacial environments	595
18 Deep seas and oceans	630

PART 5   Sediment into rock: diagenesis

671

19 Diagenesis	673

**PART 6   Long-term, large-scale processes: mountains
and sedimentary basins**   701

20 **Tectonic, climatic, and eustatic controls on long-term, large-scale erosion
      and deposition**   703

References   742
Index   795

The plate section falls between pages 24 and 25

Acknowledgments

This book is based on the insights we have gained from our teachers and from many years of our own study, research, teaching, and discussions with many colleagues and our students. We thank all of them. J. S. B. acknowledges in particular the important influence of his teachers, Percy Allen and John Allen at Reading University, and Ken Walton at St Andrews University. Long-term academic and personal interactions with Mike Leeder, Jim Best, and Rudy Slingerland have been a source of inspiration. Heather has been a wonderful source of comfort and encouragement. R. V. D. acknowledges the early influences of Greg Horne and Peter Patton at Wesleyan University, and William Dupre, now at the University of Houston. R. V. D.'s long-term academic and personal interactions with Lawrence Hardie at Johns Hopkins University and his cohort of students there, including Ron Spencer, Tim Lowenstein, and Raymond Mitchell, have been particularly important. Karen has been an unending source of encouragement and inspiration.

We have been very privileged to have various chapters in this book reviewed by experts in the field, and want to extend our sincere thanks to these reviewers for their time and effort: Gail Ashley, Jeff Barker, Dave Evans, Lawrence Hardie, Nick Harris, Steve Hasiotis, Gary Kocurek, Paul Komar, Lee Kump, Nick Lancaster, Mike Leeder, Don Lowe, Tim Lowenstein, Henry Posamentier, J. Fred Read, Greg Retallack, Bruce Simonson, Rudy Slingerland, Dorrick Stow, and Robert (Bo) Tye. We especially thank Jim Best for his review of the entire book. We also want to absolve these reviewers of any responsibility for deficiencies in this book. Some of the reviewers also provided us with photos, and additional photos were provided by Chris Ammon, Peter Ashmore, Jaco Baas, Henk Berendsen, Jan Boelhouwers, Rob Brander, Ann Dittmer, Guy Gelfenbaum, Bob Ginsburg, John Grotzinger, Lawrie Hardie, Mitch Harris, Paul Hoffman, Dave Hyndman, Richard Kesel, Peter Knuepfer, Suzanne Leclair, Tim Lowenstein, Joe Mason, Ray Mitchell, Art Palmer, Arjan Reesink, Kenji Satake, Bruce Simonson, Joe Smoot, Norman Smith, Elizabeth Turner, Janrik Van Den Berg, Goncalo Vieira, and Brent Waters. Anne Hull and Dave Tuttle are thanked for their assistance in preparing some of the diagrams.

Figure credits

The publishers and individuals listed below are gratefully acknowledged for giving their permission to use figures and illustrations in journals and books for which they own the copyright. In most cases, the figures have been redrawn. The original source (author and copyright date) is cited in the figure captions and details of the publication can be found in the references. Every effort has been made to obtain permission to use copyrighted materials, and sincere apologies are rendered for any errors or omissions. The authors would welcome these being brought to their attention.

Copyright owner	Figure number
Allen, J. R. L.	5.9A, 5.22, 7.8A, 7.8B, 7.18A, 7.25A, 7.25B, 7.25C, 7.28, 8.12A, 8.12B, 8.15A, 12.4, 15.18B, 15.26D, 15.34
American Association for the Advancement of Science	
Science	2.28, 7.35, 9.13, 11.28D
American Association of Petroleum Geologists	
AAPG Bulletin	13.3A,13.46E, 13.51, 15.59D, A27
AAPG Computer Applications in Geology 3	13.47A, 13.74
AAPG Memoir 33	15.59A, 15.59B, 15.60A, 15.60B
AAPG Memoir 77	4.10A, 4.10B, 4.10C, 4.10F, 4.10G, 4.10H, 4.10I, 4.10J, 4.10K, 4.10L, 4.13C, 4.13D, 4.23A, 19.4D, 19.9B, 19.9C, 19.9D, 19.9E, 19.10D, 19.14D, 19.14H
American Geological Association	
Image Bank	1.1F
American Geophysical Union	
Geophysical Monograph	4.19A, 4.19B, 4.20
Journal of Geophysical Research	4.7A, 4.7B, 5.26B, 15.3B, 15.5B
Reviews of Geophysics	19.3A, 19.3B, 19.3C
Water Resources Research	5.31B, 13.5B, 13.25
American Journal of Science	19.13, Table 4.3
American Society of Civil Engineers	
Journal of Hydraulic Engineering	5.25
Annual Review of Earth and Planetary Sciences	6.10A, Table 2.7
Aqua Publications	5.17A, 5.17B
Berner, E. K. and Berner, R. A	3.9A, 3.9B, 3.9C, 3.9D, Table 3.3
Blackwell Publishing	
An Introduction to Marine Ecology	15.50A, 15.51
Carbonate Sedimentology	15.1A, 15.1B, 15.1C, 18.11A, 18.11B, 18.11C
Earth Surface Processes	7.12A, 12.13A, 18.14, 20.22
Fossil Invertebrates	4.9, 7.12B, 11.8
Glacial Science and Environmental Change	17.9A, 17.9B

IAS Special Publications	5.64, 5.74B, 13.50, 18.12A, 18.12B, 20.19A
Marine Geochemistry	18.15, 18.19B
River Flows and Channel Form	13.8
Sedimentary Environments: Processes, Facies and Stratigraphy	9.3A, 9.7, 10.7D, 17.21A, 17.21B, 18.19A, 18.20A, 18.27, Table 9.1
Sedimentary Petrology: An Introduction to the Origin of Sedimentary Rocks	20.5, A23, A26
Sedimentology and Sedimentary Basins: From Turbulence to Tectonics	2.15A, 2.15B, 2.15C, 5.46, 6.4A, 7.11, 7.15, 7.18B, 13.3B, 13.68, 15.3A, 15.3C, 15.3D, 15.10A, 16.1, 16.8, 18.18B, 18.26D, 18.26E, 20.2, 20.10B, A30
Sequence Stratigraphy	20.3B
Soils of the Past, 2nd edn.	Table 3.7
Blackwell Publishing Journals	
Basin Research	20.6, 20.15A, 20.15B, 20.15C, 20.15D, 20.16A, 20.16B, 20.16C, 20.16D
Sedimentology	4.6, 5.34A, 5.34B, 5.50B, 5.50C, 5.50D, 5.62, 5.65, 5.68A, 5.68B, 5.68D, 5.69A, 5.69B, 5.72, 5.73, 6.12D, 7.31A, 7.31B, 7.31C, 8.11, 9.11A, 13.20, 13.35, 13.59A, 13.71, 13.72, 15.2B, 16.2B, 16.13, 20.8, 20.9, 20.21A, A16, A17, A19
Boersma, J. R.	7.21A
Bromley, R. G.	11.2A, 11.2B, 11.3A, 11.5, 11.7A, 11.7B, 11.7C, 11.7D
Bryant, E.	7.33, 7.34A, 7.34B, 7.36
Cambridge University Press	
Geodynamics, 2nd edn.	2.2A, Table 2.1
Cambridge University Press Journals	
Journal of Fluid Mechanics	8.9D
Canadian Society of Petroleum Geologists	
Memoir 5	5.49
Memoir 16	7.29A, 15.35D
Cengage Learning Services Ltd	6.1
Colorado State University	13.5A
Columbia University Press	
Sedimentographica, 2nd edn.	11.11E, 11.11F, 12.2B
Elsevier	
Climates Throughout Geologic History	2.25
Developments in Sedimentology 12	4.11A, 15.38A, Table 19.2
Developments in Sedimentology 20	11.20B, 11.25A, 11.25B, 16.22A, 16.22E
Developments in Sedimentology 30	5.35, 5.48B, 5.60, 5.63A, 5.63B, 5.63C, 5.74C, 6.1, 7.21B, 12.10, 13.22
Developments in Sedimentology 50	16.2A, 16.2C
Academic Press – *Global Physical Climatology*	2.5A, 2.5B, 2.10A, 2.10B, 2.11A, 2.11B, 2.11C, 2.13A, 2.13B, 2.14, Table 2.4
Pergamon Press – *Descriptive Physical Oceanography*	18.17A, 18.18A, 2.23
Pergamon Press – *The Physics of Glaciers*	10.2A, 10.2B
Elsevier Journals	
Earth Science Reviews	10.7C, 10.9
Geomorphology	13.17B
Marine Geology	20.3A

Manson Publishing Limited
    A Colour Atlas of Carbonate Sediments and   19.10F, 19.14B, 19.14C, 19.14E, 19.14F, 19.14G
        Rocks Under the Microscope
Maryland Geological Survey
    Report of Investigations 36   11.4
McGraw Hill
    Bringing Fossils to Life   2.29
NASA   1.1B, 1.1E, 6.3A, 15.49B, 16.2D, 16.5B, 16.20A, 16.23A,
     17.4C, 18.16, 18.17B

Nature Publishing Group
    Nature   6.12E, 8.5D, 17.23C
Palm Beach County Environmental Resources   1.4D
Management
Parabolic Press
    An Album of Fluid Motion   5.1, 7.3B, 7.8C
Pearson
    Atlas of Sedimentary Rocks Under the   3.20A, 3.20B, 3.20C, 3.20D, 3.20E, 3.20F, 3.21A, 3.21B,
        Microscope   3.21C, 3.21D, 3.21E, 4.10D, 4.10E, 4.11C, 4.11F, 4.13A,
     4.13B, 4.23B, 4.23C, 19.10E

    Marine Ecology   11.6B
    Surface Processes and Landforms   3.23A, 3.23B, 3.23E, 17.14, 17.22A, 17.22B, 17.24A,
     17.24B
    Prentice-Hall – *Beach Processes and*   7.2A, 7.2B, 7.4, 7.10, 7.27, 15.17
        Sedimentation
    Prentice-Hall – *Geochemistry of Natural Waters*   3.7, 3.8, 3.13A, 3.13B
    Prentice-Hall – *Glacial Geomorphology and*   17.23A, 17.23B, 17.23D, 17.25, 17.26A
        Geology
    Prentice-Hall – *Nature and Properties of Soils,*   3.16B
        13th edn.
    Prentice-Hall – *Origin of Sedimentary Rocks*   3.18A, 3.18B
    Prentice-Hall – *Principles and Applications of*   3.2, 3.10A, 3.10B
        Geochemistry
Princeton University Press
    Phanerozoic Diversity Patterns   11.16
Ritter, D. F., Kochel, R. C., and Miller, J. R.   8.16A, 8.16B, 8.17
Routledge
    Geomorphology of Desert Dunes   6.6, 6.8, 6.11A, 6.11B, 6.12A, 6.12B, 16.5A, 16.6
Science Museum/Science and Society Picture   9.5
    Library
Senckenbergiana Maritima   15.18A
Slingerland, R.   20.7
Smithsonian Contributions to Earth Science 33   16.26, 16.27
Society of Economic Paleontologists and
Mineralogists
    Hydrologic Models of Sedimentary Aquifers   13.75A
    JSP/JSR   4.7C, 4.8A, 4.8B, 5.58, 5.74A, 5.74D, 6.12C, 7.14A, 7.14B,
     7.19A, 8.6, 11.23A, 12.5, 13.59B, 15.11A, 15.26A, 15.26B,
     15.35A, 15.35B, 16.10A, 16.10B, 16.14A, 16.14B, 16.14C
    Palaios   11.23E, 11.23F

SEPM Contributions in Sedimentology and Paleotology 8	11.6A, 15.58A, 16.24, 18.10A, 18.10B, 20.20A, 20.20B
SEPM Short Course 9	5.55
SEPM Short Course 16	17.12A, 17.12B, 17.12C
SEPM Slide Set 4	15.6A
SEPM Special Publication 28	15.44B
SEPM Special Publication 33	15.9A
SEPM Special Publication 52	20.11A
SEPM Special Publication 59	20.17
SEPM Special Publication 61	7.29C
SEPM Special Publication 67	11.23B, 11.23C, 11.23C
SEPM Special Publication 83	11.9, 15.34C, 15.35E, 15.35F
SEPM Special Publication 84	18.3, 18.4B, 18.6, 18.8, 18.21A, 18.21B, 18.22, 18.23B, 18.24B, 18.25C
Springer	
Atlas and Glossary of Primary Sedimentary Structures	11.11A, 11.21D, 12.1A, 12.9A, 12.11B, 12.12, 15.49A
Depositional Sedimentary Environments: With Special Reference to Terrigenous Clastics	12.2A, 12.3, 12.6B
The Geology of Continental Margins	15.36A, 15.36B, 20.13B
Hot Brines and Recent Metal Deposits in the Red Sea: A Geochemical and Geophysical Account	4.21A, 4.21B
Lakes: Chemistry Geology and Physics	Table 3.1
Microfacies of Carbonate Rocks: Analysis, Interpretation and Application	19.8
The Persian Gulf: Holocene Carbonate Sedimentation and Diagenesis in a Shallow Epicontinental Sea	16.20B
Sand and Sandstone	3.21F, A14, A15, A24, A25, Table 3.10
Volcanism	9.2, 9.4A, 9.4B, 9.4C, 9.4D, 9.4E, 9.6, 9.10D, 9.10E, 9.11B
Kluwer Academic Press – *Atlas of Deep Water Environments:*	18.26A, 18.26B, 18.26C, 18.29A, 18.29B
Kluwer Academic Press – *Diagenesis, a Quantitative Perspective:*	19.5A, 19.5B
Kluwer Academic Press – *Encyclopedia of Sediments and Sedimentary Rocks*	4.14, 4.15, 4.18A, 4.18B, 4.22B
Kluwer Academic Press – *Offshore Tidal Sands*	15.34A
Springer Journals	
Mathematical Geology	13.47B, 13.75B
Steur, H.	19.10C
Thompson Brooks/Cole	
Natural Hazards and Disasters, 2005 Hurricane Edition	7.7, 8.3A, 8.7C, 8.15B
USDA Soil Survey	3.15A, 3.15B, 3.15C, 3.16A
USGS	
Alaska Volcano Observatory	9.11C
Cascades Volcano Observatory	9.3B, 9.3C, 9.8a, 9.8b, 9.9, 9.10A, 9.10B, 9.10C, 9.12
Professional Papers	5.12, 13.24, Table 3.5
University of Chicago Press	
Journal of Geology	13.12

University of Washington Press
    Glacier Ice (Revised Edition)   10.3B, 10.3C, 10.4B, 10.4C, 10.5B, 10.8B, 17.2A, 17.2B,
                                                 17.2C

Vallis Press
    Submarine Channels: Processes and Architecture   18.5

W. H. Freeman and Company
    Earth, 2nd edn.   2.2B, 2.2C, 2.2D, 2.3A, 2.3B, 2.3C, 2.3D
    Understanding Earth, 4th edn.   18.1

Wolman, M. G.   5.31, 13.77C

W. W. Norton Company
    Evolution of Sedimentary Rocks   Table 3.6

Young, R.   1.3

Part 1
Introduction

1 Definitions, rationale, and scope of the book

Definition of terms

Earth surface processes include weathering; sediment production by weathering and biochemical or chemical precipitation; erosion, transport, and deposition of sediment under the influence of gravity, flowing water, air, and ice; earthquakes and Earth surface motions; volcanic eruptions and movement of volcanic ejecta. Study of the landforms (morphology) of the Earth's surface, including the processes responsible for such landforms, is called *geomorphology*. The study of sediment, specifically the nature and origin of unconsolidated sediments and consolidated sedimentary rocks, is called *sedimentology*. *Stratigraphy* strictly means the description of (sedimentary) strata, which, by definition, is another aspect of sedimentology.

Importance of Earth surface processes

Earth surface processes are important for scientific, engineering, environmental, and economic reasons, as explained below.

Shaping of the Earth's surface

Many of the Earth surface processes responsible for landforms involve the formation, erosion, transport, and deposition of sediment (Figure 1.1). It is impossible to understand the formation of depositional landforms (such as deltas and beaches for example) without knowledge of the sediments that compose them. Furthermore, the nature of sedimentary deposits is likely to be related to the processes of sediment production and erosion in the source area. Therefore, the disciplines of geomorphology and sedimentology are intimately related.

Interpretation of ancient surface processes, landforms, and sedimentary deposits

The only rational way of interpreting the origin of ancient sedimentary deposits (Figure 1.2) is to compare them with modern sedimentary deposits, or with theoretical models based on knowledge of modern sedimentary processes. Then, the sedimentary record can be interpreted in terms of past Earth surface processes and landforms, leading to reconstruction of the geography, tectonics, and climate of the past (i.e., Earth history). Evolution of life on Earth and the habitats and lifestyles of past organisms are also reconstructed from the fossil evidence preserved in sedimentary deposits.

Knowledge of Earth surface processes, landforms, and sedimentary deposits is being used increasingly to interpret the present and past surface environments on other planets (e.g., Mars), and such interpretation requires particular care in view of the different physical, chemical, and biological conditions on other planets.

Engineering, environmental, and public-safety issues

Many Earth surface processes involve engineering, environmental, and public-safety issues, including landslides and debris flows; river floods; storm surges; riverbank and beach erosion; sedimentation in navigation channels; ground movements, landslides, and tsunamis associated with earthquakes; eruptions of volcanic ash and lava, and their effects on climate, river floods, and sediment gravity flows (Figure 1.3). Earth surface processes must be understood when constructing anything on the Earth's surface: buildings, roads, railways, canals, dams, coastal barrages

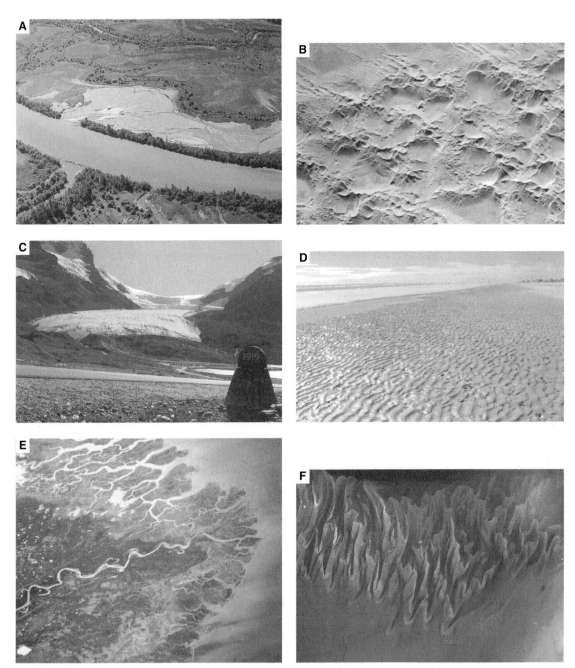

FIGURE 1.1. Depositional landforms. (A) Channels, levees, and crevasse splays on the Columbia River, BC, Canada. Photo courtesy of Henk Berendsen. (B) Issaouame sand sea in eastern Algeria. From NASA Earth from Space. (C) Athabasca Glacier, Canada, showing marginal deposits and the position of the terminus in 1919. Photo courtesy of Henk Berendsen. (D) Beach with ridge and runnel, Galveston Island, Texas. Photo courtesy of Robert Tye. (E) The Indus River delta. From NASA Earth from Space. (F) Carbonate-sand tidal sand shoals from the Bahamas. From Earth Science World Image Bank. (See Plate 1 for color versions.)

(Figure 1.4). Earth surface processes must also be understood when repairing human damage to Earth's surface environment: e.g., restoration of natural ecosystems; remediation of polluted water and air.

FIGURE 1.2. A large outcrop of sedimentary rocks from the Miocene Siwalik Group, northwest Pakistan. (See Plate 2 for color version.)

Economic resources in sedimentary deposits

Knowledge of sedimentary deposits is essential for the exploration, development, and management of economic resources that they contain: water, oil, and gas in their pore spaces; stone, sand, and gravel for building; clay for pottery and bricks, and as a source of aluminum and iron; limestone for cement; evaporite minerals; coal and lignite; placer minerals such as gold (Figure 1.5).

Rationale and scope of the book

It is clear that Earth surface processes, landforms, and sedimentary deposits are intimately related. They have a fundamental bearing on understanding Earth history; modern engineering, environmental, and public-safety issues; and exploitation and management of economic resources in sedimentary deposits. Geomorphologists, sedimentologists, environmental

FIGURE 1.3. (A) Tsunami damage in Banda Aceh, Sumatra, 2004. Photo courtesy of Guy Gelfenbaum of the US Geological Survey (http://walrus.wr.usgs.gov/tsunami/sumatra05/). (B) Sand deposits in New Orleans associated with levee breaches during Hurricane Katrina in 2005. Photo courtesy of Suzanne Leclair. (C) Beach erosion, Dauphin Islands, Alabama, following Hurricane Katrina in 2005. Photo courtesy of Robert Young. (D) A slump and landslide, northern Austria. Photo courtesy of Henry Posamentier. (See Plate 3 for color versions.)

FIGURE 1.4. (A), (B) River engineering in the Netherlands: levees and groins. Photos courtesy of Henk Berendsen. (C) The storm surge barrier across the Oosterschelde Estuary, Netherlands. Photo courtesy of Robert Hoeksema. (D) Beach replenishment in Palm Beach County, Florida. Photo from http://www.pbcgov.com/erm/enhancement/nourishment.asp. (See Plate 4 for color versions.)

FIGURE 1.5. (A) An oil well and outcrop that is an analog for a subsurface hydrocarbon reservoir in Patagonia, Argentina. (B) A coal mine near Osnabrück, northwest Germany. (C) An iron-ore mine in Paleoproterozoic Hamersley Group iron formations, Western Australia. (See plate 5 for color versions.)

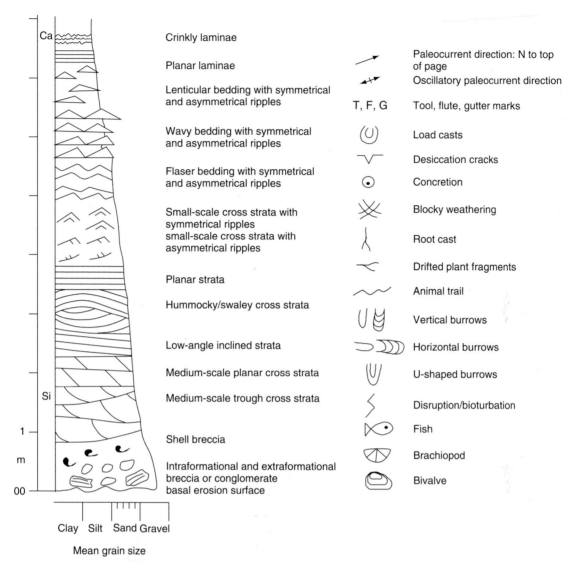

FIGURE 1.6 Typical (recommended) legend used for sedimentological logs. The leftmost column can be used for describing sediment composition (e.g., siliciclastic, calcareous). The grain-size scale is normally subdivided (e.g., very-fine, fine, medium, coarse, very-coarse sand). Additional symbols can be used as necessary. Symbols used should ideally look something like the features they represent.

scientists, and engineers traditionally study different aspects of Earth surface processes, landforms, and sedimentary deposits, commonly in different academic departments of institutions of learning. A more integrated view should enhance our appreciation of these topics. In particular, it is being realized that, in order to understand the nature of sedimentary deposits in subsiding basins, it is necessary to understand the generation of sediment in the source area and the nature of erosion and transport through the drainage network to the place of deposition. These processes are dependent upon tectonism and climate in the uplands and in the depositional basins, which are most likely all interrelated.

This book links Earth surface processes, landforms, and sedimentary deposits in a way that has not been attempted before. The book is divided into six parts. The first, introductory, part includes an overview of the Earth's lithosphere, hydrosphere, atmosphere, and biosphere, and their evolution through time, in order to set the stage for the main part of the book. Part 2 deals with the physical, chemical, and biological processes at and near the Earth's surface that cause the breakdown of rocks (weathering) into sediments (including soils), modify the chemical composition of water and air, form sediments by chemical or biochemical precipitation, and produce characteristic landforms. Part 3 is concerned with the physical processes of fluid flow, sediment transport, erosion, and deposition associated with the movement over the Earth's surface of water, air, ice, and sediment–fluid mixtures. Such movements are typically due to the effects of gravity, wind, volcanic eruptions, and earthquakes. Chemical and biological processes of erosion and deposition are also considered. These different types of processes are commonly associated with distinctive landforms (e.g., ripples, dunes) and types of sedimentary strata (called *lithofacies*, as manifested in stratal thickness, orientation, grain size and shape, and sedimentary structures). Part 4 considers environments of erosion and deposition on the Earth's surface. An *environment* is a part of the Earth's surface where erosion and deposition are proceeding that has a distinctive association of physical, chemical, and biological landforms and processes, and hence sediment deposits (*lithofacies associations*). Examples of environments are rivers on floodplains, glaciated regions, deserts, lakes, coasts, continental shelves, coral reefs, and ocean basins. In some texts, certain processes (e.g., ice flow) are treated together with the environments (e.g., glaciated areas) in which they are generally active. However, we have deliberately not done this, because it is common for several different processes to occur in a given environment (e.g., glaciated environments may include processes of ice flow, river flow, wind, wind waves, and sediment gravity flows). Parts 5 and 6 deal with the largest-scale (in terms of space and time) processes and landforms on the Earth. Part 5 concerns the transformation of loose sediment into sedimentary rock (*diagenesis*) as sedimentary strata are buried beneath the Earth's surface. Part 6 is about the accumulation of large-scale sequences of sedimentary strata in sedimentary basins, and the associated formation and erosion of mountains. These processes are controlled by large-scale tectonics, and by changes in climate and sea level. Thus, the discussion of the various Earth surface processes, associated landforms, and sedimentary deposits

in Parts 3–6 progresses from relatively small-scale and short-term to large-scale and long-term. Throughout the book, there is frequent reference to engineering, environmental, and economic issues. An appendix (located on the book's web page) contains information on methods for studying Earth surface processes, landforms, and sediments in the field, in the laboratory, and using theoretical models. Figure 1.6 is a legend for the sedimentological logs which appear throughout the text.

The subject areas of this book are developing and changing rapidly, and the quantity (if not the quality) of published literature in these areas is growing exponentially. The task of critically examining and summarizing this literature and keeping up to date has been challenging, and is a never-ending process. Obviously, most of the content of this book is not based on our own original research. The main reference books that we have used (and are therefore considered important) are listed at the beginning of each chapter, and it will be noticed that we have declined to make subjective comments about the relative merits of these books. Citations throughout the text are kept to a minimum in the interests of flow and clarity. We had to be very selective in choosing the published work that is cited in the list of references. We apologize in advance if we have missed or misrepresented important work of others, and encourage any offended parties to complain to us about this.

The book is designed for all levels of undergraduate students and beginning graduate students with interests in Earth surface processes, physical geography (geomorphology), sedimentology, stratigraphy, Earth history, environmental science, and engineering geology. Modest but realistic background knowledge of mathematics, physics, chemistry, and biology is assumed. There is enough material in this book to support the traditional undergraduate courses in Earth surface processes (geomorphology) and in sedimentology–stratigraphy. We have tended to lean towards more in-depth information rather than less, on the premise that it is easier to skip some information in a book than to look elsewhere for it.

2 Overview of the Earth

Introduction: spheres of the Earth

This chapter is an overview of the nature of the Earth's lithosphere, atmosphere, hydrosphere, and biosphere, and the interactions among them at the Earth's surface today. The long-term evolution of these "spheres" throughout Earth's history is also discussed briefly. The material in this chapter forms the basis of the more detailed information in the remainder of the book.

The Earth's shape closely approximates an oblate spheroid with a polar radius of 6,357 km and an equatorial radius of 6,378 km. A sphere with the same volume as the Earth has a radius of 6,370 km. The center of the Earth comprises a spherical core with a radius of approximately 3,470 km. Outside the core are several more or less concentric spheres. From inside out, these spheres are the mantle, lithosphere (composed of the uppermost mantle and overlying crust), hydrosphere (mostly the liquid ocean), and gaseous atmosphere (Figure 2.1). In addition, the biosphere is concentrated at the boundaries between the crust, ocean, and atmosphere. The biosphere extends down into the underlying crust for as much as 5 km, and to some as yet unknown depth into the crust beneath the oceans. The biosphere also extends up into the atmosphere. Coincidently, the surface of the Earth discussed in this book corresponds more or less to the distribution of the biosphere. The masses of the various spheres of the Earth are given in Table 2.1.

All of the spheres of the Earth are in motion. Fluid motions in the outer core produce the Earth's magnetic field. Convection cells comprising creeping motions in the mantle are manifest at the Earth's surface most noticeably as volcanoes and earthquakes, but also as the relative motions of a dozen or so rigid, tectonic *plates* (Figure 2.2). The plates extend laterally for hundreds to thousands of kilometers, and are up to 200 km thick. The plates are composed of the crust and upper portions of the mantle that together comprise the lithosphere (Figure 2.1). The velocities of convection currents in the mantle, and lateral velocities of lithospheric plates, are on the order of a few tens of millimeters per year. The atmosphere and hydrosphere are fluid, and spatially variable solar heating of these fluids ultimately produces fluid motions that range in scale from millimeters up to tens of thousands of kilometers. Currents of air (wind) in the atmosphere may have velocities on the order of tens of meters per second, whereas currents and oscillatory motions of liquid water typically range up to meters per second. The more important large-scale circulation patterns of the atmosphere and hydrosphere are outlined briefly below. Motions of the atmosphere and hydrosphere, and how these motions interact with the loose granular material (sediment) at the Earth's surface, are major topics of this book.

The lithosphere, hydrosphere, atmosphere, and biosphere interact physically as they move relative to one another, and interact chemically by exchanging materials that are stored for varying time spans. The most obvious manifestations of these chemical interactions are chemogenic and biochemogenic particles that are precipitated out of the atmosphere and hydrosphere and accumulate as sediment at the surface of the Earth. The hydrosphere, atmosphere, and biosphere are the main arteries of material exchange on the Earth, whereas the lithosphere and hydrosphere act as the main storage regions. The mantle is also important in interchange of materials in the Earth.

The long-term evolutionary histories of the lithosphere, atmosphere, hydrosphere, and biosphere are intimately interlinked as materials are interchanged among the different spheres. Changes in the storage of material among the various spheres have led to changes in the composition of the atmosphere, hydrosphere, and lithosphere throughout Earth's history.

Materials such as fossil fuels may be sequestered in the crust for up to hundreds of millions of years. The lithosphere is subject to recycling as lithospheric plates are subducted back into the mantle. Some of the subducted material may be transferred back into the crust and atmosphere as igneous magmas and associated volcanic gases. Some material may be completely removed from the Earth as gases escape into space.

Much of the material in this chapter is covered in standard introductory geoscience text books such as Grotzinger et al. (2007), and Turcotte and Schubert (2002) provide advanced treatment of the lithosphere dynamics. Additional information on the oceans can be found in Chester (2003), Open University (1989), Pernetta (1994), Pinet (2006), Summerhayes and Thorpe (1996), Thurman and Trujilo (2004), and Weaver and Thomson (1987), whereas Hartmann (1994) provides a comprehensive treatment of climatology and atmospheric circulation. The biosphere is covered in Campbell et al. (1999); early life on Earth

is covered in Schopf (1983) and Bengtson (1999). Earth science textbooks that emphasize interactions among the spheres include Berner and Berner (1996), Ernst (2000), and Kump et al. (2004). Stanley (2005) provides a comprehensive history of the lithosphere, atmosphere, hydrosphere, and biosphere.

The lithosphere

Composition

The core of the Earth probably consists mostly of iron, nickel, and sulfur. The elemental compositions of the mantle, continental crust, and oceanic crust are given in Table 2.2. The upper 660 km of the mantle is mostly composed of the mineral olivine. Various high-pressure polymorphs of olivine with closer packing structures predominate below this level. The crust of the Earth is composed of various aluminosilicate minerals, the most important of which are (1) the plagioclase feldspar solid-solution series ($CaAl_2Si_2O_8 \leftrightarrow NaAlSi_3O_8$); (2) the olivine solid-solution series ($Fe_2SiO_4 \leftrightarrow Mg_2SiO_4$); (3) pyroxene (($Mg,Fe,Ca)Si_2O_6$); (4) amphibole (hydrous Na, Ca, Fe, Mg aluminosilicates); (5) biotite mica ($K(Mg,Fe)_3(AlSi_3O_{10})(OH)_2$); (6) quartz ($SiO_2$); (7) potassium feldspar ($KAlSi_3O_8$); and (8) muscovite mica ($KAl_2(AlSi_3O_{10})(OH)_2$). Continental crust and oceanic crust have slightly different elemental and mineralogical compositions (Table 2.2). Approximately 95% of the mass of the continental crust is composed of granitic igneous and metamorphic rocks, mostly made up of Na-rich plagioclase, pyroxene, amphibole, quartz, biotite, muscovite, and K-feldspar. The ocean crust is

TABLE 2.1. Masses of the "spheres" of the Earth (Faure, 1991)

	Mass ($\times 10^{20}$ kg)
Core	18,830
Mantle	40,430
Crust	236
Hydrosphere	15
Atmosphere	0.0053
Biosphere	0.00003

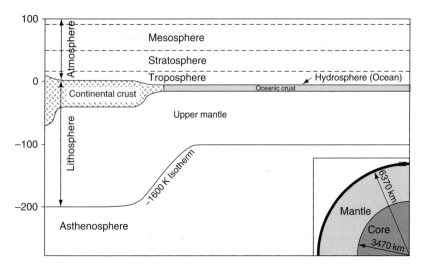

FIGURE 2.1. The spheres of the Earth. Inset on the lower right is a section of the Earth showing the core and mantle. The outer rind comprises the lithosphere, hydrosphere, and atmosphere, shown in the main panel.

dominantly composed of gabbros and basalts, igneous rocks mostly made up of Ca-rich plagioclase feldspars and pyroxenes.

TABLE 2.2. Elemental composition of Earth's rock shells in percentage by mass (Faure, 1991)

	Mantle	Continental crust	Ocean crust
O	44.3	45.5	44.1
Si	21.5	26.8	23.1
Al	1.9	8.4	8.1
Fe	6.3	7.1	7.8
Ca	2.2	5.3	8.9
Mg	23.0	3.2	4.5
Na	0.4	2.3	1.0
K	0.1	0.9	0.3
Ti	0.1	0.5	0.8
P	0.0	0.1	0.08
Everything else	<0.2	<0.1	<1.3

The outermost crustal rocks are generally covered by loose, granular material (sediment) that originates from chemical and physical breakdown of pre-existing rocks. Sediments are unconsolidated, whereas sedimentary rocks are their consolidated counterparts. Sediments and sedimentary rocks comprise only approximately 5% of the mass of the crust, but cover approximately 75% of the land surface area and perhaps 95% of the sea floor. Sedimentary rocks up to 15 km thick accumulate in local depressions called sedimentary basins (see the section on plate tectonics below). However, sediments and sedimentary rocks also occur as a widespread veneer (less than 1 km thick) over many continental interiors and oceanic basin plains. The proportions of the various kinds of common sedimentary rocks on the continents are given in Table 2.3.

Physical properties

Important physical properties of the lithosphere include density, pressure, temperature, and rheological

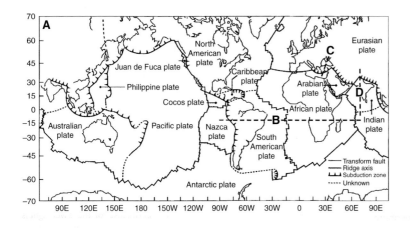

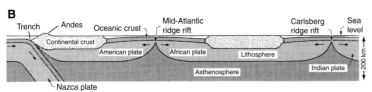

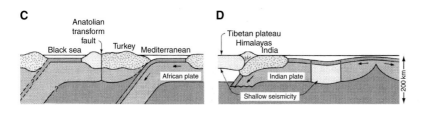

FIGURE 2.2. Elements of plate tectonics. (A) The distribution of major tectonic plates of the lithosphere. Spreading ridges (accretionary boundaries), subduction zones (convergent boundaries), and transform boundaries are indicated. Modified from Turcotte and Schubert (2002). The positions of schematic cross sections through the lithospheric plates and underlying asthenosphere (B), (C), and (D) are indicated. The cross sections emphasize the plate movements. Modified from Press and Siever (1978).

TABLE 2.3. Volumetric abundances of sedimentary rocks in Earth's continental crust

Rock type	Approximate percentage
Shale and mudstone	60
Sandstone	20
Carbonate rocks (limestone and dolostone)	15
Saline deposits	5
Conglomerate	<1
Volcaniclastic rocks	<1
Ironstones	<1
Phosphatic rocks	<1

(deformation and flow) properties such as elasticity, flexural rigidity, viscosity, and viscoelasticity. All of these properties vary spatially within the lithosphere. The average density of the lithosphere is $2,750 \, \text{kg m}^{-3}$. Lithostatic pressure increases linearly with depth as density $\times$ gravitational acceleration per unit of length: that is $2,750 \times 9.81 \approx 27,000$ pascals per meter of depth. However, dynamic pressures also exist in the lithosphere due to relative motions within and between plates. The average geothermal gradient in the upper part of the lithosphere is about $30\,^{\circ}\text{C}$ per kilometer of depth, but is higher in volcanically active areas, and lower in old continental areas. The geothermal gradient is not constant with depth, however, and must decrease downwards.

Rocks that are put under compressive or tensile stress undergo a change in length (called strain) in the direction of action of the stress. If the rock responds to the stress with no time lag, and the original length of the rock is regained immediately after removal of the stress, the rock is said to behave elastically. Parameters that describe elasticity (e.g., Young's modulus, Poisson's ratio, and flexural rigidity) are defined in Turcotte and Schubert (2002). Fractures in rocks occur when elastic deformation exceeds the yield strength of the rock. However, other (viscoelastic) rocks deform both elastically and by viscous (plastic) deformation. In this case, the viscous part of the strain resulting from an applied stress is irreversible, and the strain may be time-dependent. For example, melting of ice following the last glacial maximum removed a load on the lithosphere. The elastic response was immediate uplift: however, the viscoelastic response is still going on today. Molten rocks deform entirely viscously. The flow of molten rock is normally laminar (see Chapter 5) in view of their high viscosity (tens to thousands of Pa s). Elastic deformation is typical of cold, shallow rocks under small confining pressures, and some geologists assume that most of the lithosphere behaves elastically. Viscoelastic behavior is more likely as temperature and confining pressure increase with depth. A viscoelastic model for lithosphere rheology is more general, since it allows for purely viscous deformation.

Plate tectonics

Plate tectonics is a theory that describes the geometry and movement of the rigid plates that comprise the Earth's lithosphere (Figure 2.2). These plates are in motion with relative velocities on the order of a few tens of millimeters per year. Plate tectonics is responsible for major topographic features of the Earth's surface such as mountains and oceans. Most earthquakes and volcanic eruptions occur at or near plate boundaries, as do major mountain belts. Ancient mountain belts record the former positions of plate boundaries.

Plates are continually being created and destroyed (Figure 2.2). Adjacent plates move away from each other at mid-ocean ridges (a process known as *sea-floor spreading*), and hot, mantle-derived basaltic magmas ascend to fill the gap between the diverging plates. Mid-ocean ridges are sites of shallow earthquakes and vigorous hydrothermal circulation where seawater descends through fractures, reacts with basalts at high temperatures, and returns to the sea floor either as diffuse flow or concentrated in spectacular mineralized sea-floor geysers (*black smokers*). As the plates move away from the mid-ocean ridges, they cool, thermally contract, and subside. However, they also thicken by accretion of solidifying magma at their base and sediment on the top. Mid-ocean-ridge plate boundaries are commonly referred to as *accreting plate boundaries* or *divergent plate boundaries*.

Plates are destroyed where they converge with and are overridden by other plates, and descend into the mantle by a process called *subduction*. Oceanic trenches mark where oceanic plates begin their descent (Figure 2.2). These types of plate boundaries are referred to as *convergent plate boundaries*. Major faults separate the descending plate from the overriding plate, and these faults are the sites of many earthquakes, including most great earthquakes such as the Chile Earthquake of 1960, the Alaskan Earthquake of 1964, and the Sumatra Earthquake of 2004. The epicenters of earthquakes delineate the lithospheric plate as it

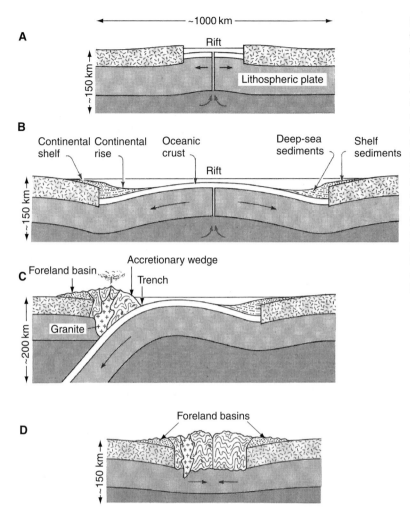

FIGURE 2.3. Sedimentary basins and the Wilson cycle. Modified from Press and Siever (1978). (A) A newly formed ocean basin such as the Red Sea. (B) A mature ocean basin such as the Atlantic with thick accumulations of sediment developed beneath continental shelf, continental slope, and continental rise. (C) Subduction of ocean crust with resultant deformation, metamorphism, and igneous intrusion of old continental shelf–slope–rise deposits (an accretionary wedge). Surface volcanicity and sediment accumulation in marginal foreland basins occur as well. (D) Final closure of an ocean basin by continent–continent collision. Sediment shed off mountain belt accumulates in marginal foreland basins.

descends into the mantle. These planar zones of earthquakes at convergent boundaries are known as *Wadati–Benioff zones*. Lines of active volcanoes parallel to the ocean trench occur in nearly every plate overriding a descending slab. The lines of volcanoes are a few hundreds of kilometers away from the trench (Figure 2.2). Volcanic island arcs, such as the Aleutians and Japan, form where the overriding plate is ocean floor. Where trenches lie next to continents, the volcanoes develop as part of a mountain range, such as the Andes Mountains of South America (Figure 2.2).

Interactions among plates generate an assortment of depressions (up to 15 km deep) of the Earth's surface that may become filled with sediments, thus forming *sedimentary basins* (Figure 2.3). Most sedimentary basins can be put into a plate-tectonic framework known as the *Wilson Cycle* in honor of J. Tuzo

Wilson, an early pioneer in studying the relationship between mountain building and plate tectonics. According to the Wilson cycle, ocean basins are initiated due to the rifting of continents (Figure 2.3). The East African Rift System and the Red Sea are modern-day examples of the initial stages. As sea-floor spreading continues and the ocean basin widens, the crust at the site of the old rupture cools and sinks. *Continental-margin* deposits bury the old rupture. Continental margins consist of a shallow (up to 200 m deep) *continental shelf*, that is inclined at less than one degree oceanward. At the *shelf–slope break*, the continental shelf abruptly gives way to the *continental slope*, inclined seaward at ∼4 degrees, that gradually decreases in slope into a deep-water *continental rise*. Modern continental-shelf–slope–rise regions comprise the largest sedimentary accumulations on the Earth. Eventually, subduction is initiated and either island arcs develop or, as shown in Figure 2.3, Andean-type mountains develop. The former continental margin overlying the subduction zone is deformed and augmented as material from the top of the downgoing oceanic plate is scraped off and incorporated into a so-called *accretionary wedge*. Intrusive and extrusive igneous rocks are introduced into the sedimentary pile by partial melting of the descending ocean crust and sediment, and contact and regional metamorphism occur. The deformation associated with the subduction and igneous activity produces mountains and additional associated sedimentary basins. Finally, the ocean is completely destroyed when the continental shelf and continental rise on the approaching plate are welded into a new mountain chain similar to the Himalayan Mountains of Asia. Sedimentary basins develop throughout the process of continent–continent collision. The origin and evolution of various basins within this plate-tectonics framework are discussed in Chapter 20.

The stratigraphic record

One of the great scientific achievements of the nineteenth and early twentieth centuries was the mapping and establishment of the relative age of all of the different rocks that cover the Earth, resulting in the *geological column* or *stratigraphic column*. The stratigraphic column was assembled on the basis of information from (1) the fossil content of sediments and sedimentary rocks; (2) application of a few common-sense rules based on the fact that sediments and sedimentary rocks were deposited layer upon layer on gently inclined surfaces; and (3) the recognition that discordances in the geological record represent time gaps (see the Appendix).

In the mid to late twentieth century, mapping and relative age dating of rocks below the surface of the continents and ocean floors were facilitated by seismic and other geophysical surveying techniques, improved land-based drilling, and ocean-going platforms capable of supporting ocean-floor drilling.

One of the most important developments of all of twentieth-century science was the development of radioactive-isotopic dating techniques for rocks (see the Appendix). These allowed the relative time scale based on fossils to be fixed absolutely in time. Moreover, rocks that did not contain fossils could now be dated within the limits of accuracy of the various radiometric dating techniques. This combination of relative and absolute age dating resulted in the *geological time scale* (Figure 2.4).

Another relative-age dating technique that was developed in the latter half of the twentieth century is based on the fact that the magnetic field of the Earth changes polarity at intervals of 10^5–10^6 years. The time scale of change was established using radiometric dating of continental and ocean-floor basalts. A chronology of past geomagnetic polarity reversals has now been established for the last 70 Myr or so. The best-known use of this polarity-reversal scale was in interpretation of the so-called *magnetic stripes* parallel to mid-ocean ridges, and that led to the theories of sea-floor spreading and plate tectonics. Other methods of relative and absolute dating are described in the Appendix.

The atmosphere

Composition

The atmosphere has three important components: (1) a relatively constant mixture of gases known as dry air; (2) variable amounts of water vapor; and (3) suspended particulate matter. The main gases comprising dry air are molecular nitrogen (N_2), 78.08% by volume; molecular oxygen (O_2), 20.95%; and argon (Ar), 0.934%. Many minor gases are found in the atmosphere (Table 2.4), including carbon dioxide, ozone, and methane. The amount of water vapor dissolved in the atmosphere is highly variable and strongly dependent on temperature, with warm air holding more water vapor than cool air. Water vapor, carbon dioxide, ozone, and methane are important gases influencing the temperature of the atmosphere, because they absorb different wavelengths of electromagnetic energy. The suspended particulate matter in the atmosphere includes carbon and other particles from combustion (commonly lumped as soot), volcanic ash, wind-blown dust, and 2–20-μm-diameter grains of sea-salt aerosols. Sea-salt aerosols form from droplets of seawater blown off the surface of the oceans that evaporate and leave behind particles of halite and gypsum suspended in the atmosphere.

Physical properties

Solids and liquids have a constant density (kg m^{-3}) at constant temperature because they are incompressible. Gases, on the other hand, do not have a constant density at a given temperature because they are compressible. Instead, a so-called ideal gas follows the ideal-gas law:

$$PV = nRT \qquad (2.1)$$

where P is pressure (N m^{-2}), V is volume (m^3), n is the number of moles of gas, R is the universal gas constant (8.131 43 J K^{-1} mol^{-1}), and T is the temperature in degrees Kelvin. Most real gases, including air, more or less follow the ideal-gas law. In other words, for a given number of moles of gas at a constant temperature, the pressure times the volume is constant. The density of air increases as temperature decreases and increases as pressure increases. At 25 °C, the density of air is ~1.3 kg m^{-3} at the Earth's surface.

Temperature and pressure are commonly used to describe the physical state of the atmosphere. The atmosphere is divided into layers based primarily on vertical variation of its temperature structure (Figure 2.5A). Temperature decreases from the Earth's surface to about −70 °C at the top of the troposphere (the *tropopause*). This temperature decrease is determined primarily by a balance between radiative cooling and convection of heat from the Earth's surface (Hartmann, 1994). Above the tropopause, temperature steadily rises into the overlying stratosphere, reaching approximately 0 °C at the *stratopause*, the boundary between the stratosphere and the mesosphere at an altitude of approximately 50 km.

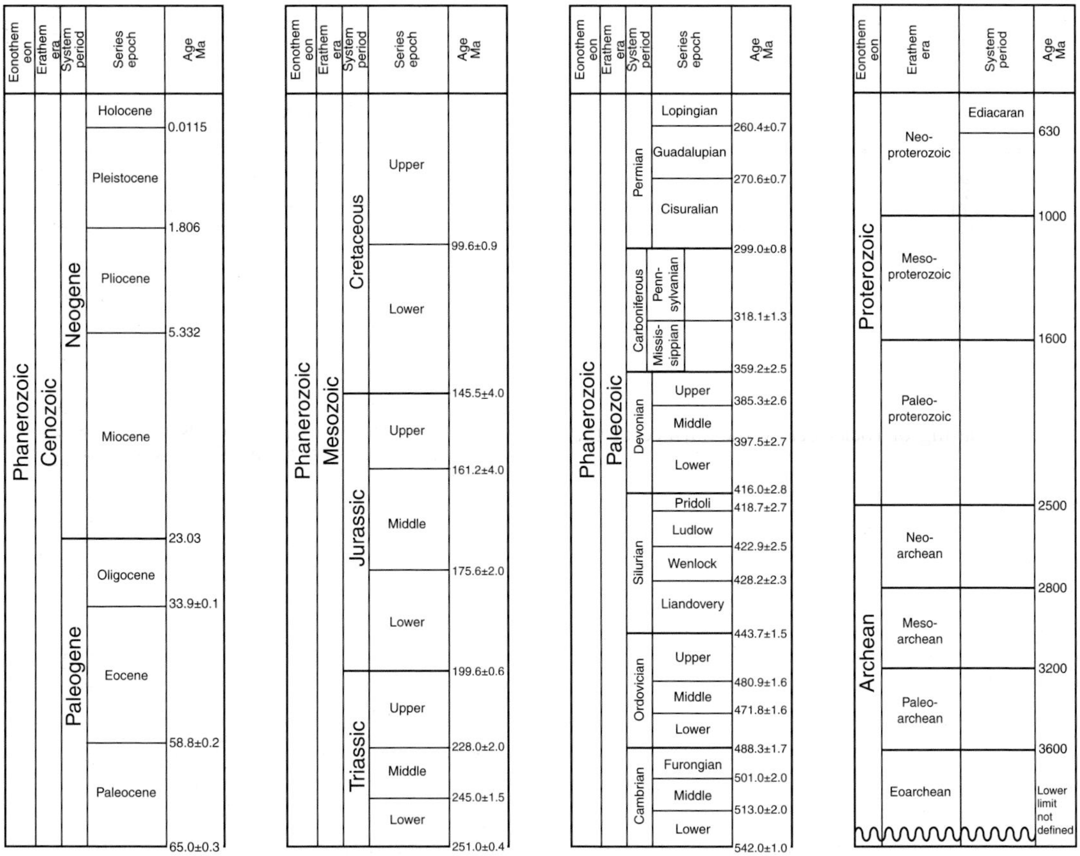

TABLE 2.4. The composition of the atmosphere (Hartmann, 1994)

Component	Proportion
N$_2$	78.08%
O$_2$	20.95%
Argon	0.934%
CO$_2$	~380 ppm and rising
Ne	~20 ppm
Kr	1.14 ppm
He	5.24 ppm
Methane	~2 ppm and rising
H	500 ppb
N$_2$O	310 ppb

FIGURE 2.4. The geological time scale (stratigraphic chart), modified from the 2004 International Commission on Stratigraphy.

A

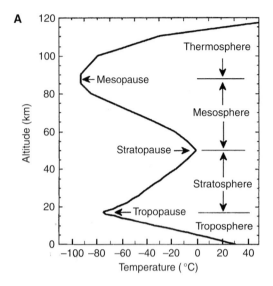

B

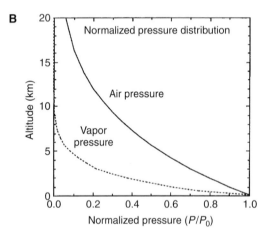

FIGURE 2.5. (A) Thermal structure and subdivisions of the atmosphere. From Hartmann (1994). (B) Air-pressure and water-vapor-pressure distributions in the atmosphere shown as "normalized" (measured pressure/standard pressure at sea level). From Hartmann (1994).

This rise in temperature is primarily due to the absorption of ultraviolet electromagnetic radiation in the *ozone layer*, a slightly increased concentration of O_3 that accumulates at the stratopause. The temperature steadily drops through the mesosphere until the outermost region of the atmosphere, the thermosphere. Although the SI unit of pressure is the pascal ($N\,m^{-2}$), the unit of pressure used in meteorology is the *bar*, where 1 bar $= 10^6$ dynes cm$^{-2} = 100$ kilopascals. At sea level, the average pressure of the atmosphere (one standard atmosphere – P_0) is 1,013.25 millibars $= 101.325$ kilopascals $= 760$ mm of mercury. Most of the mass of the atmosphere is concentrated

in the troposphere, the lower 15–18 km of the atmosphere (Figure 2.5B).

The average surface air temperature and air pressure of the Earth for January and July are given in Figures 2.6 and 2.7, and the average annual precipitation is given in Figure 2.8. The average temperature of the Earth's surface is 15 °C, with extremes of −89 °C at Vostok, Antarctica, and 58 °C at Al Aziziyah in the Libyan Desert. Spatial variations in the surface temperature are due to latitudinal variations in the heating of the land and ocean surfaces by solar radiation, to the distribution of land and sea, and to the elevation of the land. Equatorial regions have the highest surface air temperatures, and these vary little throughout the year (Figure 2.6). The South Pole stays at a fairly constant low temperature, whereas the North Pole, while still cold in summer, is considerably warmer than it is in winter. The greatest summer-to-winter temperature changes occur in the middle of continents, particularly Siberia, northwest Canada, the Tibetan Plateau, and the eastern Sahara Desert. The southern hemisphere has much less land relative to water, and exhibits smaller temperature variations from summer to winter. The distribution of land and water is responsible for the interhemispheric differences noted above. Water has a much higher heat capacity than rock, so the oceans change temperature much more slowly than land areas do.

The atmospheric pressure distribution is related to air temperature (pressure increasing as temperature decreases), but also to the existence of regions of ascending or descending air. For example, the subtropical high-pressure zones are related to descending air (Figure 2.7). Spatial variations in air temperature and pressure over the Earth result in air movements (wind). Surface winds tend to blow from high-pressure regions to low-pressure regions, but conservation of mass demands a return flow in the upper atmosphere. These pressure-driven air flows are strongly influenced by the Coriolis force associated with the rotation of the Earth, as discussed below.

Precipitation over the Earth results from cooling of air to the point of condensation of water vapor. Regions of high precipitation are associated with ascending (cooling) air that has moved over extensive water bodies. Ascending air is associated with regional zones of low pressure, and with motion over topographically high areas (Figure 2.8). Descending air is associated with regional high-pressure zones and the lee sides of mountains, leading to lack of precipitation. Continental interiors distant from surface-water bodies also tend to lack precipitation.

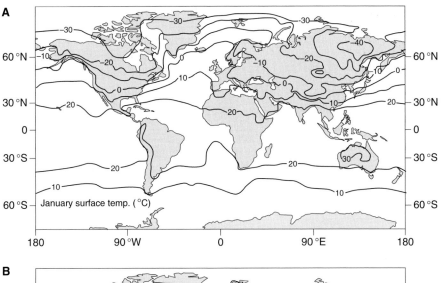

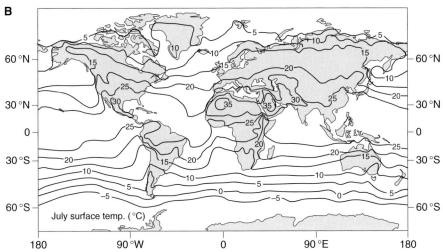

FIGURE 2.6. Surface air temperatures for January (A) and July (B). Data from Vose et al. (1992).

Earth rotation effects on atmospheric and oceanic circulation

The large-scale circulations of the atmosphere and ocean are interconnected, and both are affected strongly by the rotation of the Earth. The Earth rotates once on its axis about every 24 hours (~86,400 seconds). The angular velocity (Ω) is the number of radians the Earth turns through (360 angular degrees = 2π radians = 6.2832: radians are dimensionless) per second, or $2\pi/86,400\,\text{s} \sim 7.23 \times 10^{-5}\,\text{s}^{-1}$.

The Coriolis force, named after the eighteenth-century scientist who pointed out its importance on Earth, affects bodies that are moving over a rotating body (such as the Earth), and acts at right angles to the direction a body is moving in. The magnitude of the Coriolis force is

$$F_c = m2U\Omega \qquad (2.2)$$

where m is the mass of the object, Ω is the angular velocity, and U is the velocity of the body. Bodies moving over a counter-clockwise-rotating northern hemisphere of the Earth are deflected to the right. Objects moving over the clockwise-rotating southern hemisphere of the Earth are deflected to the left. Since the sense of the Coriolis deflection must reverse between hemispheres it is dependent on latitude:

$$F_c = m2U\Omega \sin \phi \qquad (2.3)$$

where ϕ is latitude. Thus the Coriolis force is maximum (deflecting bodies to the right) at the North Pole, directed to the right of motion in the northern hemisphere, decreases to 0 at the equator (Figure 2.9A), directed to

A

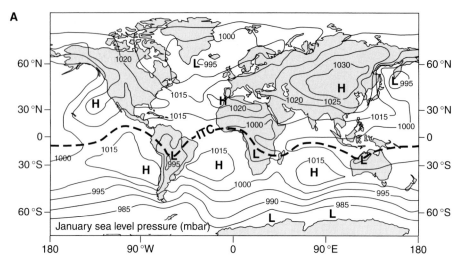

January sea level pressure (mbar)

B

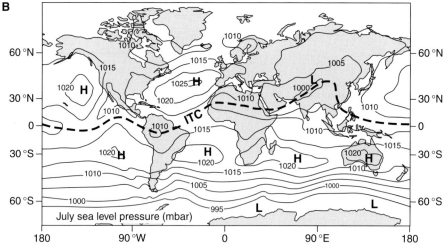

July sea level pressure (mbar)

FIGURE 2.7. Sea-level atmospheric pressure (measured or calculated in millibars) for January (A) and July (B). The location of the Intertropical Convergence (ITC) zone is shown by a heavy dashed line. Data from Vose et al. (1992). Locations of standing high (H) and low (L) pressure are indicated.

the left in the southern hemisphere, and is maximum (deflecting bodies to the left) at the South Pole. The Coriolis force is important in large-scale atmospheric and oceanic flow, giving rise to circulating motions such as cyclones, hurricanes, and surface ocean gyres.

The centrifugal force is given by

$$F_{centrifugal} = m\Omega^2 r \qquad (2.4)$$

where m is the mass of a rotating body, Ω is the angular velocity, and r is the radial distance of the body from the axis of rotation. The centrifugal force on a 1-kg object decreases from ~0.034 N at the equator to 0 at the pole (Figure 2.9B). Since the gravitational force on this object is ~10 N, the radially directed centrifugal force is four orders of magnitude less than gravity. The centrifugal force is directed

radially outward from the axis of rotation, and is responsible for the bulges in the Earth and ocean normal to the axis of Earth's rotation.

Atmospheric circulation, weather, and climate

Weather is the day-to-day state of the atmosphere as expressed by familiar parameters such as air temperature, atmospheric pressure, wind speed and direction, percentage cloud cover, how much and what kind of precipitation has fallen over the past 24 hours, relative humidity, and dew point. These properties can be highly variable from place to place and change from minute to minute over the course of a day. Most of the weather at the surface of the Earth is due to atmospheric circulation in the troposphere and lowermost

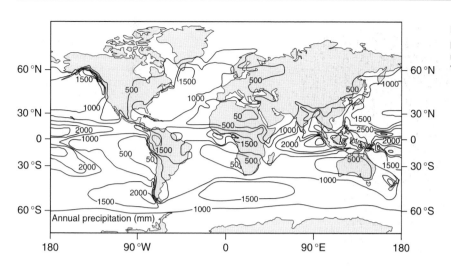

FIGURE 2.8. Annual global precipitation (in millimeters). Data from Vose et al. (1992).

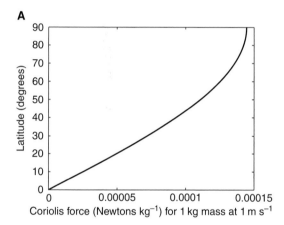

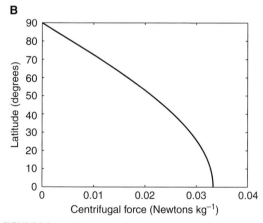

FIGURE 2.9. (A) The value of the Coriolis force on a 1-kg mass moving at 1 m s^{-1} from the equator (0 degrees) to the North Pole (90 degrees). (B) The value of the centrifugal force acting on a 1-kg mass from the equator to the North Pole.

stratosphere. So-called weather systems are commonly associated with zones of low pressure and ascending air that are called depressions, cyclones, hurricanes, and typhoons. These weather systems are commonly associated with mixing of different air masses along boundaries called fronts. Warm fronts occur at the leading edge of warm air masses that override cold air, and cold fronts occur at the leading edge of cold air masses that flow beneath warmer air. Thunderstorms and tornados are commonly associated with the fronts that separate air masses in low-pressure weather systems.

Climate is the long-term (many years) average of weather, and is commonly expressed using properties such as average annual temperature, average annual precipitation, atmospheric pressure, and wind direction. Global annual variation in these properties (e.g., Figures 2.6–2.8) results in the *general circulation* of the atmosphere. Moreover, the general circulation of the atmosphere is important in setting the oceans in motion. Global ocean circulation in turn modifies atmospheric circulation, such that atmospheric and oceanic circulations are intimately related. Weather and climate have an important effect on Earth surface processes. The spatial and temporal variation of climate over the Earth is briefly reviewed here (see also Hartmann, 1994; Leeder, 1999; and Kump *et al.*, 2004).

The long-term average air circulation in the troposphere can be resolved into east–west components (Figure 2.10) and north–south components (Figure 2.11) (Hartmann, 1994). The east–west component is known as the *zonal mean wind*. Zonal winds in the troposphere are westerly (meaning from west to east) at most latitudes between 30 and 70 degrees, but are easterly (from

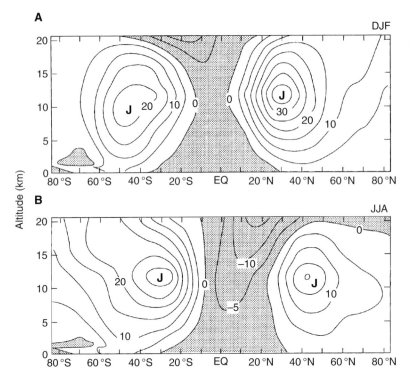

FIGURE 2.10. East–west components of wind speed (m s^{-1}) in the troposphere (zonal mean winds): (A) for December, January, and February; and (B) for June, July, and August. From Hartmann (1994). White areas denote east-directed winds; shaded areas denote west-directed winds. Locations of subtropical jets are indicated by J.

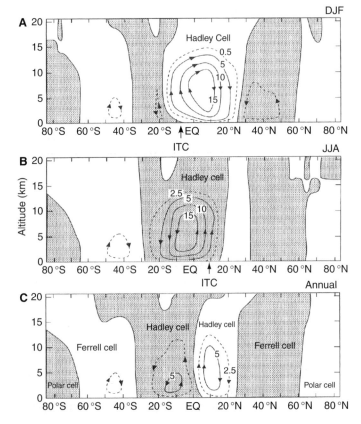

FIGURE 2.11. North–south (meridional) components of atmospheric circulation showing streamlines of mass transport, in 10^{10} kg s^{-1}. From Hartmann (1994). Clear areas are where transport at the top of the troposphere is to the north, shaded areas where transport at the top of the troposphere is to the south: (A) December, January, and February; (B) June, July, and August; and (C) annual average. Summer and winter are dominated by one Hadley cell, with the Intertropical Convergence (ITC) zone indicated. The annual average comprises two Hadley cells and other cells as idealized in Figure 2.12.

east to west) in the tropics. The most prominent features are the two subtropical jet streams centered about 12 km above the surface of the Earth. These shift position slightly with the seasons, and the stronger jet is located in the winter hemisphere. The northern-hemisphere winter jet reaches speeds in excess of $30\,\mathrm{m\,s^{-1}}$ and aids west-to-east air travel.

The north–south components of tropospheric circulation that compose the *mean meridional circulation* are much weaker than the zonal mean wind and have a significant vertical component as well (Figure 2.11). Maximum average meridional winds are only about $1\,\mathrm{m\,s^{-1}}$, with vertical speeds two orders of magnitude smaller. For these reasons, the mean meridional circulation is depicted in Figure 2.11 as streamlines of mass transported rather than speed. During the northern-hemisphere winter, the meridional circulation is dominated by a single circulation cell that rises approximately 10–15 degrees south of the equator, flows toward the north, and then sinks back toward the Earth's surface at about 20–30 degrees north with flow back toward the equator along the surface. During the southern-hemisphere winter, the meridional circulation is dominated by a single circulation cell that rises approximately 10–15 degrees north of the equator, flows toward the south, and then sinks back at about 20–30 degrees south with flow back toward the equator along the surface. These cells are commonly called Hadley cells in honor of George Hadley, who proposed them to explain the trade winds. The *intertropical convergence zone* (ITC) is located beneath the rising limb of the main Hadley cell (Figure 2.11) and can usually be easily picked out on satellite and Apollo photographs of the Earth as a dense band of clouds over the oceans near the equator. It is also the major equatorial precipitation band on Figure 2.7. The meridional circulation comprises two Hadley cells of nearly equal strength on either side of the equator during spring and fall, and when circulation is averaged over the year (Figure 2.11C). Two weaker cells, termed Ferrel cells, occur centered over latitudes 45 °N and 45 °S. Polar circulation cells are centered over 75 °N and 75 °S.

The meridional circulation is commonly simplified, with ascending air at the equator and 60 °N and 60 °S associated with latitudinal belts of standing lower atmospheric pressure, and descending air around 30 °N and 30 °S associated with a latitudinal belt of standing higher atmospheric pressure (Figure 2.12). The equatorial belt of low pressure is commonly referred to as *the Doldrums* because sailing ships

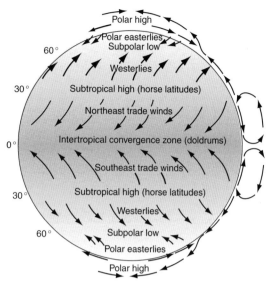

FIGURE 2.12. A schematic depiction of the global wind field. From Kump et al. (2004).

would become becalmed in this area. However, there would be nearly daily rainfall over the open oceans at the intertropical convergence zone, so dehydration would not be a problem faced by the crews. However, in the subtropical high-pressure zones, the descending air mass would warm and evaporate moisture, leading to arid conditions (Figure 2.7). Sailors becalmed in these areas would throw livestock overboard, hence the old expression *Horse Latitudes* for these zones. Many of the world's deserts are located beneath this subtropical belt of high pressure. At about 60 °N and 60 °S is the low-pressure *subpolar-front* area. During late fall, winter, and spring, low-pressure weather systems (cyclones) commonly track along the subpolar front. Also, hurricanes and typhoons are generated in equatorial regions during summer and fall months.

Winds blow from areas of high pressure toward areas of low pressure, so on a non-rotating Earth winds in the northern hemisphere would blow to the south between 30 and 0 degrees, to the north between 30 and 60 degrees, and again to the south between 90 and 60 degrees. The simple pattern of average pressure and circulation is complicated by: (1) the Coriolis force; (2) the distribution of land and sea; and (3) seasonal variations in temperature and pressure. On the rotating Earth, the winds blowing from higher-pressure areas to lower-pressure areas are deflected by the Coriolis force to the right of their track in the northern hemisphere, and to the left of their track in the southern hemisphere.

A

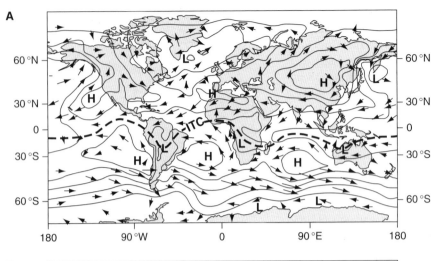

FIGURE 2.13. Average global wind fields for northern-hemisphere winter (A) and summer (B). The pressure field is from Figure 2.7, with wind vectors modified from Hartmann (1994). The length of the longest arrow indicates a wind speed of $\sim$10 m s^{-1}.

B

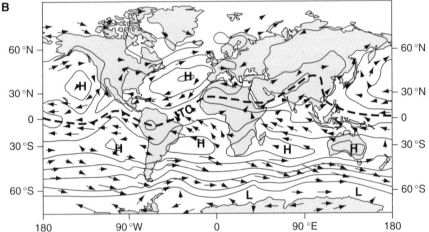

The Coriolis force does not exist on the equator but does become important a few degrees to the north and south (Figure 2.9A). Using this simplified picture, the atmospheric circulation of the Earth comprises various surface winds of the Earth flowing from high- to low-pressure areas. These are the trade winds, the prevailing westerlies, and the prevailing easterlies, as shown in Figure 2.12.

On the real Earth, however, the unequal distribution of land and sea disrupts the zonal symmetry of atmospheric pressure and circulation depicted in Figure 2.12. Pressure gradients in the atmosphere and ocean are nearly perfectly balanced by the Coriolis force. Thus, the isobars of pressure on Figure 2.13 also outline the main circulation at the Earth's surface. In the northern hemisphere, winds circulate clockwise around high-pressure systems and counter-clockwise around low-pressure systems. Also,

the steeper the pressure gradient (the more closely spaced the isobars) the greater the average wind speed. The real wind pattern on Earth is therefore quite a bit more complicated than the simple scheme shown in Figure 2.12.

Circulation in the southern hemisphere does not change as much from winter to summer as does that in the northern hemisphere. The southern hemisphere has a deep low-pressure trough at about 60 °S and a series of high-pressure zones over the Indian Ocean, the central South Atlantic, the eastern Pacific and Australia. During January, the ITC is confined to the southern hemisphere and passes through northern South America, southern Africa, and northern Australia. In contrast, circulation in the northern hemisphere changes profoundly from winter to summer. During the winter, cells of very low pressure are centered over the Aleutians and southeastern Greenland.

A very large high-pressure cell develops over the Tibetan Plateau during January, with smaller high-pressure systems over the eastern North Atlantic and eastern North Pacific. During the northern-hemisphere summer, large high-pressure cells develop over the North Atlantic and North Pacific Oceans while a deep low-pressure cell develops over northern India and Pakistan. The ITC moves up over northern Africa and over southern Asia at this time of year. The large July high-pressure system over the North Atlantic is known as the Bermuda High, and is responsible for the humid hot weather over eastern North America in the northern-hemisphere summer. The alternation of the low pressure over the Tibetan Plateau during summer with high pressure in winter is responsible for the alternating wet–dry "monsoon" climate of India and adjacent Southeast Asia. High pressure over Tibet results in winds over India blowing from the west and northwest and dry conditions prevail. Low pressure over Tibet results in winds over India and adjacent southeast Asia blowing to the west and northwest and delivering moisture-laden air from the Indian Ocean. Smaller-scale versions of monsoonal circulation also affect sub-Saharan Africa.

Milankovitch orbital variations and climate

The daily amount of solar radiation reaching the top of the atmosphere of the Earth is slightly more in the southern hemisphere than in the northern hemisphere, and the southern hemisphere receives more solar insolation during its summer than the northern hemisphere does (Figure 2.14). The Earth's orbit through space changes slightly with time, principally because of gravitational forces from the Sun and Moon that result in torques on the equatorial bulge of the Earth, and the gravitational effects of other planets in the solar system, notably Jupiter. These orbital variations produce temporal changes in the amount of solar radiation received at various latitudes. In the early twentieth century, a Serbian astronomer, Milutan Milankovitch, calculated how orbital variations would influence the global distribution of solar radiation over time scales of 10^4–10^6 years. Three variations in the Earth's orbit combine to cause variations in the amount of solar radiation received by the Earth: (1) *eccentricity*; (2) *precession*; and (3) *obliquity*.

The eccentricity of the elliptical orbit of the Earth around the Sun (Figure 2.15) is the ratio of the distance between the foci to the length of the major axis of the

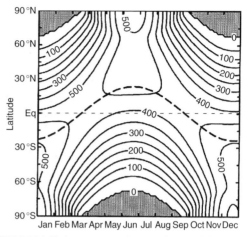

FIGURE 2.14. Daily average insolation at the top of the atmosphere as a function of latitude and time of year. The Sun is directly overhead at noon along the dashed line. From Hartmann (1994). The contour interval is $50\,\mathrm{W\,m^{-2}}$. Hemispheres receive differing amounts of insolation (note particularly the 500-W-m^{-2} contours).

ellipse. The Sun is centered on one of the foci. Earth's closest distance to the Sun is called the perihelion, whereas the aphelion is Earth's farthest distance from the Sun. Figure 2.16A shows how the eccentricity of the Earth calculated on the basis of orbital mechanics varies with time. The eccentricity over the past million years comprises two superimposed sine waves, one with a period of ~100,000 years and one with a period of ~400,000 years. The Earth's current eccentricity is ~0.016 72. The slight variability of the distance from the Sun changes the amount of solar energy reaching the upper part of the atmosphere by about 7% over the course of a year.

The precession and obliquity of the Earth's orbit do not change the total amount of solar insolation in the way the eccentricity does, but change its distribution over the surface of the Earth. Precession arises from the fact that the Earth's rotational axis is tilted relative to its equatorial plane by 23.5 degrees, giving rise to summer and winter seasons (Figure 2.15). However, the rotational axis changes its absolute direction in space by precessing, similarly to a spinning top. This precession of the Earth causes the location of the winter and summer solstices and the spring and autumnal equinoxes to change with time (Figure 2.15) by moving in a retrograde fashion around the orbit of the Earth (i.e., in the opposite direction to the Earth's track). Currently, the June solstice occurs where Earth is near aphelion and the December solstice occurs where Earth is near

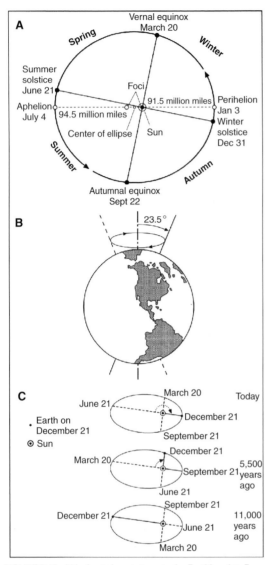

FIGURE 2.15. Milankovitch variations in the Earth's orbit. From Leeder (1999). (A) A view looking down on Earth's orbit around the Sun from a point in space above Earth's North Pole, showing the elliptical orbit and the positions of the perihelion and aphelion relative to the solstice and equinox points. (B) Precession of Earth's axis of rotation. (C) Effects of precession on the Earth's position in orbit relative to perihelion and aphelion.

perihelion. Variations in position of the equinoxes and solstices around the elliptical orbit of the Earth affect solar insolation (Figure 2.16B). The precession of the Earth's axis produces a complicated time series of changes of solar insolation at various latitudes with the main periods at ~20,000 and ~400,000 years (Figure 2.16B). Variations in high-latitude insolation due to the precession of the Earth's orbital points around the elliptical orbit can amount to changes on

the order of 15% in the amount of solar insolation received at high latitudes.

Obliquity is the shift up and down of the Earth's orbit relative to a fixed plane through the center of the Sun. Obliquity controls the annual mean equator-to-pole gradient of insolation. The main period of the obliquity is ~40,000 years (Figure 2.16C). Obliquity variations can also produce ~15% variations in summer insolation in high latitudes.

Figure 2.16D shows the calculated cumulative effects of the eccentricity, obliquity, and precession on the amount of solar insolation received at latitude 65 °N for the past 1 Myr. Milankovitch proposed that such variation in the amount of solar insolation reaching the high northern latitudes was responsible for the waxing and waning of ice sheets that covered large areas of the northern hemisphere in the Pleistocene. Today, only Greenland and Antarctica have large continental ice sheets on them. The Milankovitch theory of orbital variations controlling ice-sheet advances and retreats, and its role in producing changes in climate and Earth surface processes in general, is currently under vigorous debate.

The hydrosphere

Chemical properties of water

Water (H_2O) is a unique compound that exists in three states at the Earth's surface, solid ice, water vapor, and liquid. In addition, at the elevated temperatures and pressures that occur at depth in the Earth, water can exist as a supercritical fluid. Water is a polar molecule with a positive area (the two H protons) and a negative area (the O atom), and the polar nature of water accounts for its uniqueness. Two important chemical properties of water are its very high dissolving power and very high dielectric constant.

Water dissolves more substances in greater quantities than any other liquid. Pure liquid H_2O is hard to come by on Earth and virtually every element on Earth is found dissolved in seawater. Even rainwater invariably has material dissolved in it. As a result, all natural waters on and within the Earth's crust are *aqueous solutions* that have an important influence on Earth surface processes. Where precipitation falls on Earth materials, it reacts and generates new aqueous solutions that seep into the ground, only to react further in the subsurface. All solid chemical and biochemical sedimentary grains are extracted from Earth surface aqueous solutions.

PLATE 1. (Color version of Figure 1.1)

PLATE 2. (Color version of Figure 1.2)

PLATE 3. (Color version of Figure 1.3)

PLATE 4. (Color version of Figure 1.4)

PLATE 5. (Color version of Figure 1.5)

PLATE 6. (Color version of Figure 9.5)

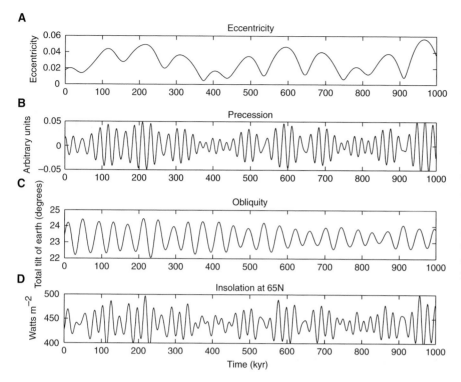

FIGURE 2.16. Effects of Milankovitch orbital variations at latitude 65° N for the past million years calculated by Berger and Loutre (1991). (A) Variations in the eccentricity of the Earth's orbit. (B) The effect of precession on variations of solar insolation at 65° N. (C) Variations in the obliquity of Earth's orbit. (D) Cumulative effects of (A), (B), and (C) on insolation reaching the top of the atmosphere at 65° N for the past million years.

In addition, aqueous solutions in the pores between grains take part in many chemical changes that affect sediment after deposition, particularly during burial. Such chemical changes, together with various physical changes, are referred to as *diagenetic* processes (Chapter 19).

Pure water has the highest *dielectric constant* of any liquid due to its polar nature. This means that inorganic compounds generally dissociate into ions in water. The solution remains electrically neutral but the material in solution is charged. Examples of dissociation reactions include

$$NaCl \leftrightarrow Na^+ + Cl^- \tag{2.5}$$

$$CaCO_3 \leftrightarrow Ca^{2+} + CO_3^{2-} \tag{2.6}$$

Water itself dissociates as

$$H_2O \leftrightarrow H^+ + OH^-. \tag{2.7}$$

Composition of the hydrosphere

The reservoirs of water in the hydrosphere, and fluxes of water between the reservoirs, are given in Table 2.5 and Figure 2.17. The chemical compositions of these reservoirs are given in Table 2.6. The range of chemical composition of water at or near the surface of the Earth is enormous, with about a dozen ions as common components (Table 2.6). The main cations (positively charged ions) are Ca^{2+}, Na^+, Mg^{2+}, and K^+, whereas the main anions (negatively charged ions) are HCO_3^-, Cl^-, SO_4^{2-}, CO_3^{2-}, and Br^-. The amount of dissolved silica is low and, where found, is commonly represented as neutral $H_4SiO_4^0$. In addition to these species, lesser amounts of dissolved Sr, Ba, Fe, Mn, Zn, Al, N (as nitrate), P (as orthophosphate), and Li are also found. A simple scheme to classify natural water is 0–1,000 ppm, *fresh* or potable; 1,000–10,000 ppm, *brackish*; 10,000–100,000 ppm, *saline* (seawater is roughly 35,000 ppm); and > 100,000 ppm, *brines*. Some waters from deep metamorphic and igneous terranes have 500,000 ppm! The amount of dissolved material affects the density of Earth surface waters (see below).

The composition of the water differs among the various reservoirs (Table 2.6), as explained below. Evaporation of water from the oceans and lakes, or evapotranspiration from land plants, dissolves pure water vapor into the atmosphere. Wherever water vapor condenses and falls as precipitation either to the surface of the oceans, and the ice sheets of the polar regions and high mountains, or to the land, it obtains small amounts of dissolved solid materials and atmospheric gases (see Chapter 3). Precipitation tends to nucleate on suspended particles in the air, including marine aerosols, natural dust particles, and soot particles from combustion of fossil fuels. More importantly, the carbon dioxide that dissolves into precipitation produces a weak acid (carbonic acid) that interacts with Earth surface materials, as described in detail in Chapter 3. The most obvious effects are chemical weathering of rocks, the generation of soil, and putting various dissolved solutes into shallow groundwaters (meteoric waters). Shallow groundwater has solute concentrations of up to a few hundred parts per million, and typical solute compositions of these waters are $Ca > Na$, Mg, K, $HCO_3 > Cl$, SO_4 (Table 2.6). The shallow meteoric groundwater lenses are a major source of the world's fresh drinking water and contribute *baseflow* to the world's river systems. Beneath the shallow meteoric lenses there is a deeper reservoir of very concentrated brines. The compositions of these brines are highly variable but, in general, they are commonly

TABLE 2.5. Hydrosphere reservoirs

	Mass ($\times 10^{15}$ kg)
Oceans	1,370,000
Ice sheets and glaciers	29,000
Shallow groundwater	4,200
Deep groundwater	5,300
Lakes	125
Soil moisture	65
Atmosphere	13
Rivers	1.7
Biosphere	0.6

TABLE 2.6. Dissolved inorganic solutes in various portions of the hydrosphere (in ppm – mg per kg)

	Rainwater	World river water (natural)	Shallow groundwater (typical ranges)	Seawater	Ca-rich deep groundwater
Cl^-	0.2	5.8	1.8–46	19,350	114,080
Na^+	0.6	5.2	2.3–362	10,760	23,300
SO_4^{2-}	1.6	8.3	0.9–2,170	2,710	0
Mg^{2+}	0.2	3.4	1.2–143	1,290	215
Ca^{2+}	0	13.4	3.1–416	411	46,300
K^+	0.6	1.3	0.6–7.2	399	1,640
HCO_3^-	3	52.0	21–374	142	100

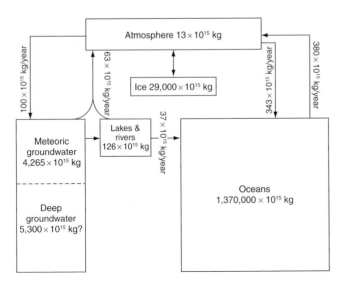

FIGURE 2.17. Earth's hydrologic cycle, showing reservoirs (excluding the biosphere) with masses and fluxes of materials among the reservoirs.

dominated by Na and Cl. Many have high Ca and low Mg and high metallic content from reactions with minerals during diagenesis. These concentrated waters are commonly referred to as *oil-field brines* and their origin and importance are taken up in Chapter 19. The high salinity of these brines indicates that they are left over after the deposition of saline deposits (so-called *residual evaporite brines*) or are formed by dissolving evaporites at depth.

Most of the water on Earth is found in the oceans. One cubic meter of ocean water weighs on average 1,028 kilograms, of which ~36 kilograms is dissolved inorganic solids (Table 2.6) and ~992 kilograms is water. There will also typically be between 0.01 and 10 grams (average 1 gram) of suspended inorganic matter, mostly clay minerals. In this cubic meter of seawater there will also be between 0.001 gram and 1 gram of mostly dead suspended organic matter and some living phytoplankton. The amount of organic matter is highly variable, especially in coastal areas where "blooms" of phytoplankton occur. There is also approximately the same mass of dissolved organic matter, including the metabolic wastes of living organisms and the decay products from dead organisms. Atmospheric gases are also dissolved in the seawater. In the cubic meter of seawater there can be up to 8 liters of dissolved O_2. When the temperature of water (including seawater) increases, the amount of dissolved oxygen and other gases decreases. There are only a few regions of the deep ocean that have no dissolved oxygen in them.

Physical properties of the hydrosphere

Water has the highest heat capacity of all solids and liquids (except ammonia, NH_3), and absorbs a great deal of heat before it changes temperature. The calorie is defined as the amount of heat required to raise 1 gram of water at one atmosphere pressure from 14.5 to 15.5 °C. Water has a high latent heat of fusion: at a constant temperature of 0 °C, 80 calories per gram must be added to melt ice or subtracted to freeze water. Water also has a very high latent heat of evaporation: to change 1 gram of liquid water at 100 °C to 1 gram of water vapor at 100 °C requires the addition of 540 calories. Latent heat of evaporation released back into the atmosphere upon the condensation of water vapor is one of the most important ways in which the atmosphere is heated. It is well known that ice is much less dense than an equal mass of liquid water. However, the density of liquid water also has an anomalous dependence on temperature; water is most dense at

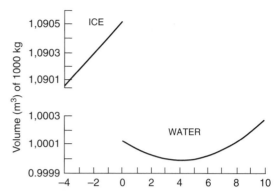

FIGURE 2.18. Changes in the density of ice and liquid water between −4 and 10 °C.

3.98 °C (Figure 2.18). Indeed, the SI unit of mass, the kilogram, is defined by the mass of matter contained in liquid water at 3.98 °C, where $1 m^3 = 1,000 kg$. The anomalous density of pure water is important in the dynamics and ecology of temperate lakes (see Chapter 14).

The range of surface ocean temperatures (Figure 2.19) is closely correlated with air temperatures (Figure 2.6). Ocean surface temperatures range from 0 to nearly 30 °C from pole to equator, and tongues of relatively cooler ocean water extend out from the southwest coasts of Africa and South America. Ocean temperature changes (along with many other ocean properties) are routinely monitored by satellites, and spectacular false-color images of real-time ocean surface temperatures are easily obtained on the Internet. Cross-sectional plots of the temperature distribution in the western portion of the Atlantic Ocean and the central portion of the Pacific Ocean (Figure 2.20) show lenses of warm water a few hundred meters thick centered on the equator that overlie masses of cold water at depth that are continuous with surface water at the poles. The upper 200 meters or so of the tropical oceans are warm, but beneath this warm *mixed layer* is a zone of thickness 500 meters or so known as the *thermocline*, where the temperature changes abruptly (Figure 2.20). Beneath this is a huge mass of cold seawater at approximately 4 °C: the *deep zone*. In summer in mid latitudes, a seasonal thermocline develops. The mixed layer is up to 500 meters thick and is underlain by the thermocline, which again gives way to uniform 4-°C deep water at about a kilometer depth. At high latitudes, the water is nearly isothermal. The relatively thin layer of warm water that sits over the tropical and temperate oceans is due to solar radiation. However, most of the volume of the oceans is filled with cold water that

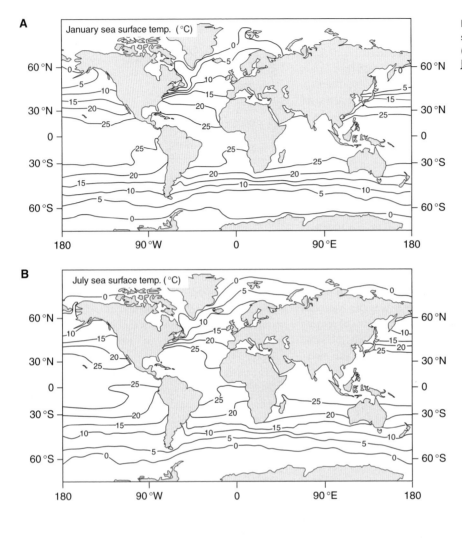

FIGURE 2.19. Sea-surface temperatures (°C) for January (A) and July (B)

originates at the poles. Details of the temperature variation in lakes and in the ocean are given in Chapters 14 and 15.

Sea-surface salinities (Figure 2.21) are much less variable than sea-surface temperatures. Cross-sections of the salinity distribution in the western portion of the Atlantic Ocean and the central portion of the Pacific Ocean (Figure 2.22) are more difficult to interpret than the temperature plots, but indicate deep ocean circulation that is further described in Chapter 18. In terms of salinity, 99.95% of world seawater is between 30 and 40 parts per thousand (ppt), 90% of world seawater is between 34.33 and 35.10 ppt, and 50% of global seawater is between 34.61 and 34.79 ppt. Salinity decreases where major rivers flow into the oceans, and increases in isolated basins in desert areas such as the Arabian Gulf. Also, formation of sea ice affects the salinity of surface ocean water by removing freshwater. However,

throughout the range in salinity, the ratios of the ions dissolved in seawater are remarkably uniform, implying that the oceans are well mixed. The salinity of lakes is discussed in Chapters 14 and 16.

The density of seawater is controlled by temperature (T), salinity (S), and depth (pressure) (Figure 2.23), and ranges from $1,021 \, \text{kg m}^{-3}$ to $1,070 \, \text{kg m}^{-3}$. Water is about as compressible as granite but the depth effects on density do become important in the deep oceans. For example, seawater at $0 \, °\text{C}$ and 35 ppt salinity has a density of $1,028 \, \text{kg m}^{-3}$ at the surface, but at a depth of 4,000 meters seawater at the same temperature and salinity has a density of $1,048.5 \, \text{kg m}^{-3}$ (a 2% increase due to compression). The change in density with salinity is nearly uniform over the entire T–S range of Figure 2.23. However, the change in density with temperature is much more nonlinear (Figure 2.23). The addition of salt lowers the freezing

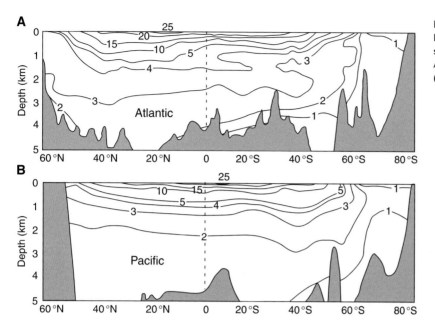

FIGURE 2.20.
Depth–temperature (°C) cross sections through the western Atlantic (A) and central Pacific (B) Oceans.

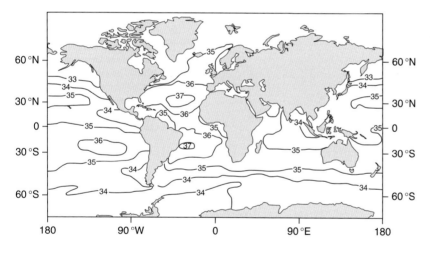

FIGURE 2.21. Mean annual salinity of ocean surface water (parts per thousand). Data from NOAA.

temperature of water, which is why people in temperate latitudes salt icy pavements and people everywhere put a slush of salt water and ice in their ice-cream makers. For the salinity range of the oceans, seawater will freeze at its maximum density, in contrast to freshwater that freezes at lower than its maximum density. Furthermore, freezing of ice out of seawater increases the salinity of the remaining seawater. Such cold and saline seawater has an important effect on the deep, thermohaline circulation of the oceans (Chapter 18). Rheological properties of water and ice, such as viscosity, are discussed in Chapters 5 and 10.

Circulation in the hydrosphere

The *hydrological cycle* (Figure 2.17) shows how water is moved from reservoir to reservoir on Earth. Each of the reservoirs has its own circulation. Atmospheric circulation (described above) carries water vapor around in the atmosphere and deposits it as precipitation (Figure 2.7). Freshwater flow in rivers (Chapter 13) and storage in lakes (Chapter 14) are part of the circulation in the shallower aquifers, the subject of groundwater hydrology (Bear, 1972; Freeze and Cherry, 1979). Whereas a great deal is known

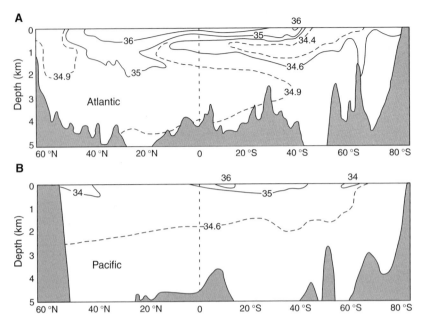

FIGURE 2.22. Depth–salinity (parts per thousand) cross sections through the western Atlantic (A) and central Pacific (B) Oceans.

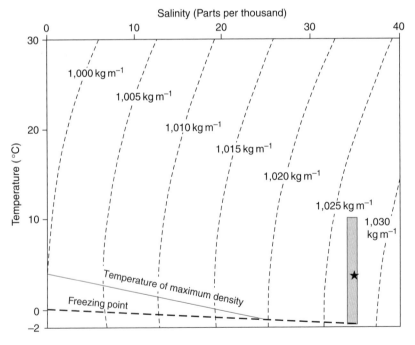

FIGURE 2.23. The density of seawater as a function of temperature and salinity. From Pickard and Emery (1990). The shaded box encompasses the temperature and salinity of 90% of the world's ocean; the star indicates the mean ocean temperature and salinity.

about the circulation in the shallower aquifers, the origin and circulation of deep brines is much less well known (Chapter 19). Circulation at coasts and adjacent continental shelves is covered in Chapter 15. The circulation in the deep ocean is described in Chapter 18.

The biosphere

The three domains of life

Approximately 1.85 million species have been described and given binomial Latin names by biologists (e.g., *Homo sapiens*). All organisms on Earth are made up of cells that share an amazing number of similarities in their chemical processes. Organisms on Earth are organized into a hierarchical structure prosaically referred to as the *tree of life*, and elucidating this hierarchy is the work of a branch of biology known as systematics. One of the biggest revolutions in biological systematics has been the use of universal cellular chemical features such as nucleic acids that reside in the ribosomes (subcellular systems common to all cells) to uncover the evolutionary relationships between species. This revolution is now in full swing, so the descriptions below are subject to change.

Life on Earth is divided into three domains: (1) bacteria; (2) archaea; and (3) eukaryotes. The bacteria and archaea stand apart from the eukaryotes as prokaryotic organisms. There are only about 5,000 known species of prokaryotes, but they are certainly significantly undercounted.

Archaea and bacteria (prokaryotes)

Prokaryotes are mostly unicellular, but can be organized into aggregates of nearly identical cells, and some aggregates may possess two to three specialized cells. Prokaryotic cells tend to be small (1–5 microns (μm) in diameter) and are usually spheroidal, rod-like, or helical (or, in the parlance of microbiologists, *coccoid*, *bacilli*, and *spiro*). Prokaryotes have one double strand of DNA in the form of a ring with little protein associated with it. Prokaryotic DNA may be concentrated in a nucleoid region but prokaryotes do not have a membrane-bound nucleus. Prokaryotic cells have ribosomes (subcellular organelles where protein synthesis occurs), and some have enfolded membrane systems. Prokaryotes have a great diversity of distinctive cell walls with different architectures exterior to their cell membranes or sandwiched between an internal and an external cell membrane. Prokaryotic cells have no mechanism to bring particles into their cells: instead, they produce enzymes that they release into the environment exterior to the cell to break down complex molecules into products that can then pass though the cell membrane into the cell. Prokaryotes divide, grow, and have limited motility, but what they lack in behavior they make up for in metabolic diversity (see below). General references on how prokaryotes affect Earth surface processes include Westbroek and De Jong (1983), Crick (1986), Mann (2001), Dove *et al.* (2003), and Baeuerlein (2004).

The prokaryotic bacteria and archaea are ubiquitous, being the oldest forms of life on Earth, and thrive in many environments that are inimical to eukaryotes. In terms of metabolic impact and sheer numbers, prokaryotes dominate the biosphere. Oxygen is one of the main controls on prokaryotic ecosystems: i.e., aerobes need oxygen to survive, obligate anaerobes are killed by oxygen, and facultative anaerobes can either use oxygen or not. Archaea in particular are prokaryotes that live in the most extreme environments on Earth and have therefore attracted the attention of the so-called astrobiologists, who look for signs of life on other planets. Three important groups of archaea are (1) methanogens (see below); (2) extreme halophiles (salt lovers) that live in brines of salinity up to ten times seawater concentrations; and (3) extreme thermophiles (heat lovers) that live in 60–80-°C hot springs. Bacteria include such familiar research organisms as *E. coli* and the pathogens that induce bacterial infections. A geologically important group of the bacteria is constituted by the *cyanobacteria* and their importance is further outlined below.

Metabolism essentially comprises oxidation–reduction reactions. The biggest difference between prokaryotes and eukaryotes is their metabolic diversity in carrying out these reactions. What the prokaryotes lack in complex behavior they make up for in metabolic diversity. Many geochemical reactions would be too sluggish, or kinetically blocked, were it not for the enzymatic catalysts provided by prokaryotes. It is the metabolic diversity of prokaryotes that drives many important "biogeochemical" cycles. A simple classification of prokaryotic metabolic types (Table 2.7) is based on (1) how the organism obtains the carbon required to synthesize organic compounds; (2) how the organism obtains the energy required to synthesize organic compounds; (3) what substances donate the electrons; and (4) what substances accept the electrons.

Most bacteria are *heterotrophs* that consume organic molecules to get their carbon for organic synthesis, and get their energy by "oxidizing" some of those organic molecules. Aerobic respiration (essentially the opposite of photosynthesis) is practiced by all eukaryotes and aerobic decomposing bacteria:

$$CH_2O + O_2 \rightarrow CO_2 + H_2O \qquad (2.8)$$

Mitochondria are aerobic bacteria that live symbiotically in eukaryotic cells. However, free elemental oxygen, O_2, need not be the electron acceptor that oxidizes the organic matter. Other electron acceptors include NO_3^-, Mn^{4+}, Fe^{3+}, and SO_4^{2-}. Representative equations for some of these reactions (Jorgenson, 2000) are the following.

TABLE 2.7. Metabolic types of prokaryotes (Nealson, 1997)

General type	Carbon source	Energy source	Electron donor	Electron acceptor
Heterotrophs	Organic C	Organic C		
Aerobes			Organic C	O_2
Denitrifiers			Organic C	NO_3^-
Mn reducers			Organic C	Mn^{4+}
Fe reducers			Organic C	Fe^{3+}
Sulfate reducers			Organic C	SO_4^{2-}
Sulfur reducers			Organic C	S^0
Methanogens			Organic C/H_2	CO_2
Syntrophs			Organic C	Organic C
Acetogens			Organic C/H_2	CO_2
Fermentors			Organic C	Organic C
Phototrophs	CO_2	Light		
Cyanobacteria			H_2O	
Photosynthetic bacteria			S compounds, H_2, organic C	
Lithotrophs	CO_2/organic C	Inorganic redox		
H_2 oxidizers			H_2	O_2, NO_3^-, Mn^{4+} Fe^{3+}, SO_4^{2-}
Fe oxidizers			Fe^{2+}	O_2, NO_3^-
S oxidizers			H_2S, S^0, $S_2O_3^{2-}$	O_2, NO_3^-
N oxidizers			NH_3, NO_2^-	O_2
CH_4 oxidizers			CH_4	O_2

Denitrification:

$$5CH_2O + 4NO_3^- \rightarrow 2N_2 + 4HCO_3^- + CO_2 + 3H_2O \tag{2.9}$$

Mn^{4+} reduction:

$$CH_2O + 3CO_2 + H_2O + 2MnO_2 \rightarrow 2Mn^{2+} + 4HCO_3^- \tag{2.10}$$

Fe^{3+} reduction:

$$CH_2O + 7CO_2 + 4Fe(OH)_3 \rightarrow 4Fe^{2+} + 8HCO_3^- + 3H_2O \tag{2.11}$$

Sulfate reduction:

$$2CH_2O + SO_4^{2-} \rightarrow H_2S + 2HCO_3^- \tag{2.12}$$

Methanogens are archaea that derive their energy by using hydrogen to reduce CO_2 according to the reaction

$$CO_2 + 4H_2 \rightarrow CH_4 + 2H_2O \tag{2.13}$$

These important reactions take place in subaqueous fine-grained sediments everywhere (usually in the order listed above) and many of these reactions are involved in the precipitation of minerals in sediments. Prokaryotes are also the only organisms that can take inorganic N_2 from the atmosphere and "fix" it into organic compounds such as ammonia (NH_3).

Phototrophs derive their energy from sunlight and their carbon from either CO_2 (*photoautotrophs*) or organic carbon (*photoheterotrophs*). Prokaryotes are the only organisms that have the metabolic pathways to perform photosynthesis, and there are several different photosynthetic pathways in prokaryotes. The most important group of photoautotrophs is the *cyanobacteria*, a group that uses a pigment known as *chlorophyll a* and produces O_2 as a metabolic byproduct. Chloroplasts, the organelles where photosynthesis takes place in green algae and higher plants, are evolved from endosymbiotic cyanobacteria. A tree can be viewed as a platform that cyanobacteria have coerced eukaryotes to build for them to get them more light and more CO_2. Photosynthesis can be represented by the equation

$$CO_2 + H_2O \rightarrow CH_2O + O_2 \tag{2.14}$$

Not all prokaryotic photosynthesis produces O_2 as a byproduct (e.g., photosynthetic bacteria in Table 2.7).

Lithotrophs can use either inorganic or organic carbon to synthesize their organic molecules, but derive their energy by using enzymes to catalyze redox chemical reactions between inorganic molecules in the environment. The extreme thermophile *Sulfolobus* derives metabolic energy by oxidizing sulfur. Many minerals, especially iron minerals, are precipitated in and around colonies of chemoautotrophs as a result of the chemical reactions they utilize. The most famous lithotrophs are the symbiotic prokaryotes that live in the tissues of gutless tube worms and clams that cluster around the hot spring vents known as black smokers at mid-ocean ridges. The field of *geobiology* in part concerns study of how prokaryotes might induce the precipitation of various minerals. This is a growing area of study, with some enthusiastic researchers suggesting that *all* low-temperature mineral reactions on Earth are either mediated or directly controlled by prokaryotes.

It is hard to understate the importance of bacteria in supporting eukaryotic life on Earth. They are ubiquitous in soils and sediments (Chapters 3 and 11). Prokaryotes are also noted for their ability to colonize stable sedimentary surfaces. In fact, many benthic invertebrates feed on microbial bottom mats, keeping them closely cropped (Chapter 11). If bottom-dwelling invertebrates are excluded from an area, the mat can flourish and become many millimeters thick, as is the case in the intertidal and supratidal zones of tidal flats in evaporitic settings (see Chapters 11, 15, and 16). Wherever light can penetrate, the microbial bottom mats are dominated by cyanobacteria. Thick cyanobacterial mats develop a surface-parallel zonation with various heterotrophic bacteria occurring in layers beneath the cyanobacteria. These mats may be so impermeable to gases that O_2 is rapidly depleted within them and anaerobic respiration, particularly sulfate reduction, takes over. Strong redox gradients within the mats induce precipitation of many minerals. In addition, prokaryotes, and in particular cyanobacteria, secrete thick slimy sheaths of mucopolysaccharides around their coccoid or filamentous colonies that act like flypaper to trap sedimentary particles on the mats. Some cyanobacteria, the *Oscillatoracea*, are motile and phototaxic and can slide out of their sheaths and recolonize the new sediment surface. The importance of microbial mats to sedimentation is further discussed in Chapters 11, 15, and 16 and this is a burgeoning area of research.

Eukaryotes

Some *eukaryotes* are unicellular but most are multicellular with complex specialized organ systems. Eukaryotic cells are generally 10–100 μm in diameter and a eukaryotic cell has several thousand times more DNA than a prokaryotic cell, with the DNA organized in linear molecules surrounded by proteins and complexly folded to form chromosomes. The number and morphology of chromosomes is characteristic of each species. The chromosomes in a eukaryotic cell are contained in a nucleus with its own membrane. In addition to ribosomes, eukaryotic cells carry a host of smaller so-called *organelles*, specialized bodies suspended in the cells' cytoplasm with specialized functions. *Mitochondria* and *chloroplasts* are organelles that have their own bacteria-like DNA strands and evolved from bacterial symbionts to be part of eukaryotic cells. In addition, eukaryotic cells possess a system of fibers and microtubules of cytoplasm that comprise a *cytoskeleton* that supports the cell and can be molded into pseudopods for locomotion or feeding in unicellular eukaryotes such as amoebas. Eukaryotes have complex behaviors but limited metabolic diversity.

Most of the 1.85 million or so species that have been described are eukaryotic organisms. Ecologists and biologists reckon that the actual number of species on the Earth is more like 4 to 10 million. Most of the unnamed species would come from five inconspicuous groups: bacteria, archaea, nematodes, insects, and mites. A conservative estimate is that there are on the order of 3 million eukaryotic species on Earth today. In contrast, only about 275,000 fossil species have been described and named, the overwhelming majority of which are eukaryotes. Most fossils are preserved in Phanerozoic sedimentary rocks deposited in marine environments, particularly nearshore settings, but fossils of terrestrial organisms such as plants and tetrapod vertebrates are also found in the deposits of lakes, river channels, and floodplains. Fossils comprise (1) the hard exo-skeletons of protists and multicellular invertebrates; (2) endo-skeletal hard parts of vertebrate fish and tetrapods; (3) plant fragments; (4) impressions of hard parts (molds and casts); (5) rare impressions of soft-bodied invertebrates; and (6) tracks, trails, and other traces recording animal behaviors. Preserved hard parts, and their casts and molds, are known as *body fossils*, whereas tracks and trails are called *trace fossils* or *ichnofossils* (see Chapter 11).

It might seem odd that there are so few fossil species. Fossilized species cover some 600 million years of Earth history, yet the number of fossilized species is only 15% that of the currently extant species and a much tinier fraction of all the species that have ever existed on Earth. There are several reasons for this apparent paradox. First, of the 1.85 million known species, approximately 1.4 million species (~75%) are from terrestrial groups: 1 million species of insects, spiders and mites; 280,000 species of land plants; 100,000 species of fungi; and 25,000 species of reptiles, birds, and mammals. The preservation potential of fossils in terrestrial environments is not high. Second, many of the 250,000 extant marine invertebrates are soft-bodied worm-like animals that stand little chance of preservation. Third, marine and terrestrial ecosystems are very efficient at recycling both organic matter and inorganic remains. Scavenging animals and decomposing bacteria (Table 2.7) and fungi are particularly adept at recycling soft-bodied organisms and the soft-body portions of organisms with endo- or exoskeletons. The soft parts are, after all, mostly water. Finally, most organisms are transported *post mortem*, with the transport further degrading soft and hard parts. *Taphonomy* is the study of how these processes affect the preservation of fossils and of how the fossil record is related to the living assemblage of organisms.

It is impossible to describe here in detail the hierarchical groupings of eukaryotes. Table 2.8 lists seven of the more important highest-order divisions of the eukaryotes known as *kingdoms* and some of the important groups found in those kingdoms. The multicellular animals (metazoans) are divided into about thirty groups known as *phyla* that represent distinctive basic body plans of organisms. Fourteen of the most important metazoan phyla are shown, together with one version of their kinships, in Figure 2.24. Green algae and the closely allied higher plants have very complicated divisions and subdivisions. Many of the groups of plants are known only from fossils. Three groups of plants have taken turns dominating terrestrial ecosystems: (1) large tree-like ferns and their allies dominated Paleozoic forests (see Figure 2.4 for the geological time scale); (2) conifer trees dominated Mesozoic forests until the Late Cretaceous, when they were replaced by (3) the modern flowering plants. Grasses apparently developed in the Late Mesozoic and spread worldwide in the Neogene. Most of the higher plants are terrestrial. However, a few important species have developed

TABLE 2.8. Important eukaryote "kingdoms" (the classification is still unsettled)

Animals	Further subdivided in Figure 2.24
Fungi	Important decomposers
Stramenopiles	Diatoms, brown algae, cocolithophores
Alveolates	Dinoflagellates, ciliates
Rhodophyta	Coralline red algae
Chlorophyta and *higher plants*	Green algae, vascular and non-vascular land plants
Other *protists* (kingdoms unsettled)	Foraminifera, radiolarians, amoebas

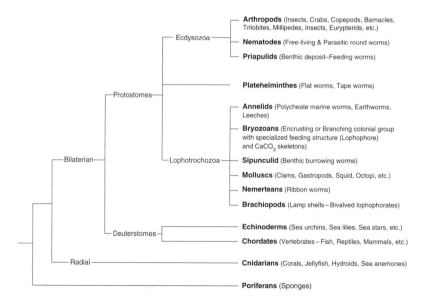

FIGURE 2.24. Fourteen important phyla of animals and one version of their phylogenetic relationships.

tolerance to salt water, including mangroves and salt-marsh grasses such as *Spartina*. Eel grass and turtle grass are two species of flowering plants that have evolved to live entirely immersed in shallow seas.

Ecology is the study of the interrelationships among organisms and the physical and chemical environment. An *ecosystem* is a three-dimensional block of lithosphere, hydrosphere, and atmosphere that contains specific organisms. Ecosystems are broadly divided into marine (ocean) ecosystems and terrestrial (land) ecosystems. The ocean ecosystem is commonly subdivided into the water column or *pelagic* realm and the sediment–water interface or *benthic* realm. These two realms are, in turn, traditionally broken up into marine ecosystems based on water depth. The pelagic realm is broken up into the (1) epipelagic zone, 0–200 m deep; (2) mesopelagic zone, 200–1,000 m deep; (3) bathypelagic zone, 1,000–2,000 m deep; (4) abyssal pelagic zone, 2,000–6,000 m deep; and (5) hadal pelagic zone, >6,000 m deep. Similarly, the benthic realm is divided into the (1) intertidal littoral (or coastal) zone; (2) sublittoral (subtidal) zone, 0–200 m deep; (3) bathyal zone, 200–2,000 m deep; (4) abyssal zone, 2,000–6,000 m deep; and (5) hadal zone, >6,000 m deep. The single biggest factor controlling the distribution of marine organisms is depth of light penetration. In the open ocean, the upper epipelagic zone more or less corresponds to the photic zone (~100 meters deep in the clearest ocean waters). Photosynthesis in the open ocean is principally carried out by microscopic unicellular organisms such as diatoms, coccolithophores, dinoflagellates, and cyanobacteria. Dinoflagellates are the most important unicellular photosynthetic group in coastal waters. Almost all metazoans living beneath the photic zone must be predators and/or scavengers or otherwise derive their nutrition from table scraps that rain out of the productive surface waters. Lithotrophic bacteria and archaea can make do without organic input, and some rare metazoans survive at deep-sea volcanic vents by hosting colonies of lithotrophs, but these organisms are the exceptions. The depth of light penetration can be as much as 100 meters in clear tropical shelf seas away from major rivers, but the depth of light penetration can be considerably less than 100 m (perhaps a few tens of meters) over other continental shelves where suspended sediment concentrations are high.

The intertidal zone is commonly divided into ecosystems based on substrate type. Divisions include mud (tidal) flats, sand beaches, and rocky shorelines. Estuaries are commonly brackish and have circulation set up by the density contrasts between freshwater and seawater. Lagoons without freshwater input can become hypersaline in evaporative conditions. Depth of light penetration is crucial for classifying benthic ecosystems of continental shelves below the tide line. Wherever light in sufficient strength reaches the bottom, multicellular algae, cyanobacteria, and flowering plants can flourish. Thus, animals that graze on these photosynthesizers are restricted to the *shallow photobenthic* zone. Waters above the shallow photobenthic zone also have abundant photosynthetic suspended organisms and small organisms that feed upon them. Thus, filter feeding is most effective in the photobenthic zone. Below the photic zone on murky continental shelves, only benthonic, lithotrophic, prokaryotic mats are available, together with the table scraps from the overlying water. The photobenthic zone can extend across clear tropical shelves but not over most of the continental shelves of the Earth. The benthic zone of continental shelves can be further subdivided: (1) below storm wave base (>100–200 m); (2) between storm and fair-weather wave base; and (3) above fair-weather wave base (<10–20 m). The aphotic–photic benthic zone boundary need not coincide with the wave-base boundaries, and all of these boundaries can change position seasonally and with rises and falls in sea level. Finally, the erodibility and composition of bottom material (i.e., gravel, sand or mud) influences the stability of the sea bed, burrowing activity, food supply, and sediment transport on the bed and in the water column.

The eukaryotic organisms that live in marine ecosystems are classified on the basis of their method of obtaining food (*feeding strategy*) and their locomotion. Common feeding strategies include (1) photosynthesis (algae, photosynthetic protists, flowering plants, cyanobacterial mats); (2) predation and scavenging; (3) deposit feeding – eating the sea floor for its contained organic matter and prokaryotes; (4) filter feeding – filtering suspended living and dead organic particles; and (5) grazing on microbial mats or multicellular algae and plants. Pelagic organisms are considered *nekton* (adjective nektonic) if they have substantial powers of locomotion and *plankton* (adjective planktonic) if they are more or less at the mercy of the currents. Benthonic organisms are either sedentary or mobile and either live in the bottom, where they comprise *infauna* (adjective infaunal), or on the bottom, where they comprise *epifauna* (adjective epifaunal). Sedentary epifauna can simply lie on the sea floor or can encrust over hard rock at the sea floor. Some epifauna are capable of cementing the bottom to produce a hard rock bottom for themselves. The trophic (feeding) interrelationships among marine organisms are complicated and commonly described as a *food web*. In addition, most benthic invertebrates have a juvenile larval stage that spends weeks to months in the surface plankton. The distribution of benthic organisms is strongly influenced by the bottom materials, water currents, and sediment transport.

There are ~250,000 marine invertebrate species and perhaps another 25,000 species of marine vertebrates, mostly bony fishes. In contrast, there are some 1.4 million species of terrestrial organisms: 1 million species of insects, spiders, and mites; 280,000 species of land plants; 100,000 species of fungi; and 25,000 species of amphibians, reptiles, birds, and mammals. The reason for this disparity in diversity between marine and terrestrial species stems from the fact that environmental conditions in the oceans vary gradually in time and space whereas terrestrial environmental conditions change rapidly over short distances and time frames. Thus, there is a greater diversity of ecosystems and habitats in terrestrial environments. Terrestrial ecosystems (except deep lakes – Chapters 14 and 16) are all built around photosynthesis carried out principally by multicellular plants, most with vascular tissues, that live at the lithosphere–atmosphere boundary. Whereas the largest marine ecosystems are three-dimensional, terrestrial ecosystems are essentially two-dimensional and the distribution of plants is closely linked to amount of solar insolation, climate (temperature and rainfall), and soil (Chapter 3). Trophic relationships in terrestrial ecosystems tend to be simple food chains rather than complicated food webs as in marine ecosystems. Filter feeding and deposit feeding (two common marine feeding strategies) are rare in terrestrial ecosystems. Instead, herbivores ingest the macrovegetation characteristic of terrestrial environments. Herbivores include small invertebrates such as insects and large vertebrates. Herbivores are fed upon by carnivores and scavengers. Unlike in the oceans, where there may be four or five levels of carnivores of increasing size, there are only one or two levels of carnivores in terrestrial ecosystems. There are eight major terrestrial ecosystems: (1) tropical rainforests; (2) seasonal wet–dry grassland known as savanna; (3) arid desert with limited vegetation; (4) seasonal wet–dry, shrub-dominated coastal chaparral; (5) temperate grasslands also known as veldts, steppes, prairies, or pampas; (6) temperate deciduous forests; (7) boreal coniferous forests; and (8) arctic and high-mountain grasslands called tundra underlain by permanently frozen soils (permafrost). In addition, Antarctica, Greenland, portions of Iceland, and many high mountain areas are covered by glacial ice.

Evolution of the spheres through time

The lithosphere, atmosphere, hydrosphere, and biosphere have all changed (evolved) with time, and changes in one of the spheres directly affect the others. The long-term records of the evolution of the Earth's outer spheres are archived in the sediments and sedimentary rocks that form the outer skin of the lithosphere. It is impossible to cover this topic comprehensively here, so only the points that will recur in this book are outlined. For thorough discussion of paleogeographic reconstructions of continents, mountain belts, and ocean basins, and evolution of the biosphere, the reader is referred to Stanley (2005).

Evolution of the lithosphere

Most scientists these days subscribe to the *nebula hypothesis* for the origin of the Sun and planets of our solar system. The nebula hypothesis was originally proposed by the German philosopher Immanuel Kant in 1755, but it has evolved considerably recently as a result of data collected using spacecraft and modern telescopes. The nebula was a diffuse, roughly spherical, slowly rotating cloud of gas and dust. The gases were mostly hydrogen and helium (two elements that make up most of the Sun), and the dust was chemically similar to material that makes up the Earth. According to the hypothesis, the nebula contracted due to gravitational attraction between the bits of matter (because of their mass), and this caused the rotation rate to increase and flattening of the nebula into a disk shape. Most of the matter drifted towards the center of the nebula, eventually forming the Sun. Compression of this matter under its own weight caused a tremendous increase in temperature in the central part of the nebula, which eventually became hot enough for hydrogen-fusion reactions. When the Sun ignited, light materials such as hydrogen, helium, and water were vaporized in the inner solar system and blasted outward by the so-called solar wind, a stream of ionized particles ejected from the Sun. The disk-shaped nebula then started to cool and many of the gases condensed into liquid or solid. In the hottest, inner regions of the nebula, silicates and iron can condense at thousands of degrees centigrade, whereas water ice, methane ice, and ammonia ice condensed in cooler, outer portions of the nebula. Gravitational attraction caused the dust and condensed matter to stick together (accrete) into kilometer-sized chunks called planetesimals, and these in turn collided and stuck together to form the nine planets. The inner planets (Mercury, Venus, Earth, and Mars) are relatively small and made of dense minerals and rocks containing abundant iron and silicates. The lighter elements in the outer portions of the solar system became the giant gas planets (Jupiter, Saturn, Uranus, and Neptune). Meteorites, asteroids, and comets are thought to be remnants of the planetesimal stage, and their age (determined by radiometric dating) suggests that the inner planets began to accrete about 4.56 billion years ago. Planetary accretion could have happened in less than 100 million years.

Violent impacts from planetesimals were common on the early Earth, and the kinetic energy of these impacting masses was converted to heat. The Moon is thought to have formed from a particularly large impact of a Mars-sized object that struck the Earth a glancing blow about 4.5 billion years ago. This impact ejected a large amount of debris into space that ultimately accreted to form the Moon. The Earth's rotation rate and axis of rotation were changed by this major impact, and an enormous amount of heat was generated. Dating of lunar surface materials indicates that most of the Moon's craters range between 4.4 and 3.8 billion years in age. Many large impacts with the Earth during this time generated much heat, and additional heat was generated by radioactive decay of certain elements in the Earth (e.g., uranium). As the oldest rocks found on Earth are about 4 billion years old, most of the direct evidence of this early bombardment phase has long since vanished from Earth due to subsequent erosion.

Perhaps 50% of the Earth was either molten or in a plastic state during bombardment of the early Earth. This molten and plastic state allowed relatively heavy material to drift towards the center of the Earth to form the core and lighter material to move towards the surface to form the mantle and crust (i.e., the Earth became concentrically zoned, or differentiated). The core is made mainly of iron and other heavy elements, whereas the mantle and crust are made mainly of silicates (compounds of silicon, oxygen, aluminum, iron, magnesium, calcium, sodium, and potassium). The rising molten material brought heat to the surface and radiated it out into space. Thus, the Earth cooled down and became mostly solid, apart from the liquid outer core and isolated patches of hot molten material in the mantle and crust. It appears that the formation and movement of lithospheric plates over the surface of the Earth, associated with earthquakes and volcanic eruptions, and giving rise to continents, mountains, and ocean basins, were in existence by at least 3 billion years ago. During this differentiation stage, the lightest material escaped from the solid Earth as gas and liquid to form the oceans and atmosphere (discussed below). Although the oceans and atmosphere have evolved through time, Earth surface processes resulting from the interaction of the atmosphere, hydrosphere, and surface of the Earth must have been in existence by about 3.85 Ga.

Most Archean (Figure 2.4) rocks preserved on the Earth occur in linear greenstone belts of chlorite-grade metamorphosed mafic and ultramafic volcanic rocks that commonly exhibit pillow textures, and low-grade metamorphosed sedimentary rocks (sandstones, mudstones, cherts, and iron-rich volcanogenic sediments). The greenstone belts occur between 40–50-km-diameter masses of high-grade metamorphic feldspathic gneisses. This assemblage of rocks is thought to represent ancient island arcs associated with subduction zones. The pillows of the volcanic rocks and the sediments indicate that substantial amounts of liquid water and weathering acted to produce the sedimentary particles. The greenstone belts and associated felsic blocks were sutured together and had produced substantial continental blocks by the end of the Archean. The oldest such continent known (>3 billion years old) is preserved in South Africa and is notable in that two large sedimentary basins on that continental block (the Witwatersrand and Pongola basins) contain sedimentary sequences 2.7–3.1 billion years old. The gold-bearing Witwatersrand deposits are reckoned to be braided stream deposits, and the gold is believed to have accumulated as detrital placer deposits. The Pongola basin contains quartz sandstones and mudstones reckoned to be tidal-flat and other shallow marine deposits, as well as glaciogenic sedimentary rocks.

The Proterozoic was substantially like the Phanerozoic in that sediments accumulated in basins associated with continental drift and the Wilson cycle (Figure 2.3). However, the sedimentary rocks of the Proterozoic differ in a few respects from their Phanerozoic counterparts. First, from about 2.5 to 1.8 billion years ago there was widespread deposition of iron-rich sedimentary rocks on Earth. The iron was deposited in deep basins as well as on continental shelves and is interbedded with an equal volume of chert. In the basinal deposits, these rocks commonly comprise striking millimeter-to-centimeter alternations of red-colored chert and dark-colored iron minerals that have earned these rocks the moniker *banded iron formations* (BIFs). The chert-rich iron formations deposited in shallower settings tend not to be so rhythmically layered but instead comprise cross-bedded sandstones of chert and iron peloids. These iron formations are thought to be related to the buildup of atmospheric oxygen (see below). There are also two well-documented periods of widespread glacial deposition: one just at the beginning of the Proterozoic (the so-called *Huronian* glacial event) and one just before the Ediacaran period (the so-called *Cryogenian*). From the end of the Proterozoic to the present, the overall position of continents and oceans on the Earth is fairly well known, as are the locations and timing of mountain-building episodes.

Evolution of the atmosphere

It is reckoned that the asteroids and planetesimals that accreted to form the Earth were too small to carry atmospheres, and that the Earth's atmosphere and ocean were emitted from the Earth itself. Such degassing would have been facilitated if the Earth had experienced substantial melting during accretion. The degassing from the interior continues to this day, with volcanic gases mostly comprising water vapor, carbon dioxide, carbon monoxide, HCl, HF, and SO_2. Most researchers agree that the Earth's original atmosphere probably contained quite a bit more carbon dioxide than is present today and very little free oxygen. Estimates of the partial pressure of CO_2 in the early Earth's atmosphere range from 1 to about 80 atmospheres (it is currently at about $10^{-3.5}$ atmospheres, or 380 ppm). Throughout the history of the Earth, carbon dioxide has been removed from the atmosphere by the weathering processes described in Chapter 3 and sequestered in carbonate rocks and fossil fuels. Cyanobacteria invented O_2-producing photosynthesis circa 3.0 Ga, and, starting at about 2.4 Ga, there has been a buildup of oxygen in the atmosphere. The history of the chemical composition of the Earth's atmosphere and the effect of that composition on climate is an area of active research. Water vapor, carbon dioxide, methane, and certain other gases are transparent to solar radiation, but absorb the radiation that is given off by the surface of the Earth, leading to a warmer atmosphere of Earth than would be the case if these gases were not in the atmosphere. This is the so-called *greenhouse effect*. The anthropogenic increases in concentration of these trace atmospheric gases (especially

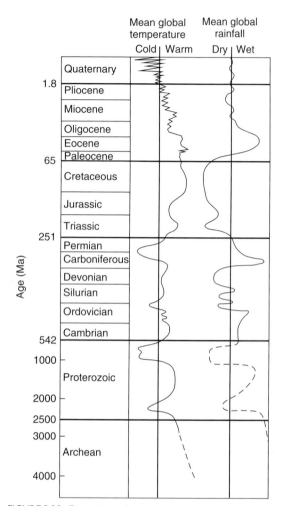

FIGURE 2.25. Estimations of mean global temperature and mean global rainfall throughout the Earth's history. From Frakes (1979). These estimates are based on sedimentary deposits and the fossils they contain.

CO_2) since the industrial revolution began are well known, but the contribution of these increases to bringing about the observed $\sim$1 °C rise in global average air temperature since the 1960s is still being debated (although many scientists are convinced of a causal relationship).

The climatic history of the Earth includes about half a dozen major episodes of continental-scale glaciation (Figure 2.25). The best known of these are the (1) Paleoproterozoic (Huronian glaciation); (2) Neoproterozoic (Snowball Earth glaciation); (3) Late Ordovician; (4) Permo-Carboniferous (Gondwana glaciation); and (5) Neogene glaciation, which the Earth is still experiencing. Al Fischer (1982) termed the Phanerozoic episodes of glaciation *icehouse* conditions

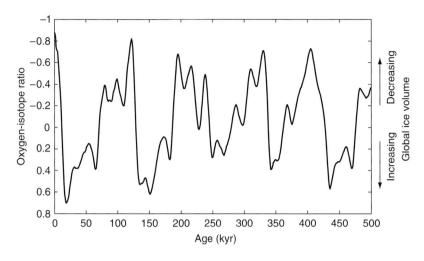

FIGURE 2.26. Estimates of global ice volume for the past half million years based on oxygen-isotope measurements of foraminiferan shells from two Indian Ocean cores. From Imbrie and Imbrie (1979) and Imbrie *et al.* (1984, 1993).

and the intervening warmer periods *greenhouse* conditions. The Cretaceous period saw an archetypical greenhouse Earth, with cold-blooded reptilian species existing in what were then polar and subpolar latitudes. Greenhouse conditions continued into the Paleogene to a thermal maximum at the *Early Eocene Climatic Optimum* (EECO). Since then, the climate of the Earth has cooled and entered an ice age as ice sheets began to develop in Antarctica during the late Eocene, and later developed in the northern hemisphere in the late Miocene (Zachos *et al.*, 2001) (Chapter 17).

The oxygen isotopic composition of many marine planktonic foraminifera shells depends on the oxygen isotopic composition of the seawater and the temperature of formation of the shell and closely follows inorganic fractionation. Many foraminifera, however, deviate from the inorganic isotopic fractionation law in ways that are species-specific (the so-called vital effect). The oxygen isotopic composition of planktonic foraminifera recovered from deep-sea cores has been studied intensively for the past 30 or so years with the purpose of recovering information on the paleoclimatic conditions of Earth's ocean surface waters. Reconstruction of global ice volumes based on oxygen isotopes of foraminifera (Figure 2.26) shows a distinctive pattern of decreased ice volume every 100,000 years or so followed by stepwise cooling to a maximum ice volume, followed by abrupt warming. These variations in ice volume are almost, but not quite, in tune with the Milankovitch orbital variations described above. The Last Glacial Maximum (LGM) occurred approximately 18,000 years ago. At that time, world sea levels were lower by about 130 m because a

substantial volume of seawater was sequestered in the ice sheets (Chapter 17). The LGM ended abruptly about 15,000 years ago (Figure 2.26). There was, however, a distinct global cooling episode about 11,000 years ago in an event known as the Younger Dryas (named after a particular plant pollen). The so-called medieval warm period lasted from 900 A.D. to 1300 A.D. and was the time that the Vikings colonized Iceland, Greenland, and North America. This period of equitable climate in Europe was followed by the so-called little ice age, which lasted into the 1800s, and during which all of Europe's mountain glaciers advanced down their valleys, and canals in the Netherlands and Long Island Sound routinely froze over. Following the little ice age, and continuing to this day, there has been a general global warming trend, shown in the instrument record and recently in detailed satellite records of ocean surface temperatures.

Evolution of the hydrosphere

The ocean originated together with the atmosphere by degassing of the early Earth. When the Earth had cooled enough, the water vapor in the atmosphere condensed and the earliest oceans formed. As today, SO_2 and CO_2 from the atmosphere dissolved in this rain produced acid rain for chemical weathering. However, there was also a substantial amount of HCl in the original atmosphere. Reactions between this HCl rain and lithosphere rocks probably resulted in an episode of intense chemical weathering and provided an instant chloride ocean, with only minor additions of chloride to the oceans since then. The salinity and exact composition of the original oceans are not known. However, it is likely that the major ion chemistry (Na^+, Mg^{2+}, Ca^{2+}, K^+, Cl^-, and SO_4^{2-}) of seawater changed throughout the Phanerozoic (and perhaps the Proterozoic), particularly the Mg^{2+}, Ca^{2+}, and SO_4^{2-}. This is suggested by a number of observations. The mineralogy of marine, chemically precipitated $CaCO_3$ sediment (ooids – Chapter 4) and cements has varied cyclically over the Late Proterozoic and Phanerozoic between predominantly aragonite ($> \sim 540$ Ma, ~ 340–180 Ma, and ~ 30–0 Ma) and predominantly calcite (~ 540–340 Ma and ~ 180–30 Ma) (Sandberg, 1983) (Chapter 4). The most likely control of these changes was variations in the Mg^{2+}/Ca^{2+} ratio of seawater (Füchtbauer and Hardie, 1980; Hardie, 1996; Morse et al., 1997), which was recently corroborated with measurements from ancient echinoderms (Dickson, 2002). Certain so-called hypercalcifying organisms (e.g., corals and sponges) responded with aragonitic species being more abundant during aragonite seas and declining in abundance during calcite seas, and vice versa for calcitic species (Stanley and Hardie, 1998). Stanley et al. (2002) showed that the Mg/Ca ratio of coral produced by high-magnesium calcite-secreting coralline algae grown in aquaria varied with the Mg^{2+}/Ca^{2+} ratio of the ambient "seawater," and similar changes occurred in the shells of echinoids, crabs, and shrimps, and the calcareous tubes secreted by marine serpulid worms (Ries, 2004).

Marine evaporites have been particularly useful in investigations of ancient seawater chemistry. Changes between so-called *aragonite seas* and *calcite seas* were mirrored by cyclic variations in the mineralogy of marine evaporites, with $MgSO_4$ evaporites during periods of aragonite precipitation, and KCl evaporites during periods of calcite precipitation (Hardie, 1996). Direct measurements of the ionic composition of evaporated seawater in fluid inclusions from marine evaporites of various ages (Figure 2.27) have been used to back-calculate the ionic composition of the original seawater (Lowenstein et al., 2001, 2003; Horita et al., 2002).

Ever since Fischer's (1982) paper on Earth's icehouse and greenhouse climate modes, there has been a growing consensus that global sea level, global climate, seawater chemistry (Lowenstein et al., 2001; Horita et al., 2002), atmospheric concentrations of carbon dioxide (Berner and Kothavala, 2001), nonskeletal carbonate mineralogy (Sandberg, 1983), evaporite mineralogy (Hardie, 1996), and perhaps evolution of shell-building organisms (Stanley and Hardie, 1998) have all varied through time more or less in lock step.

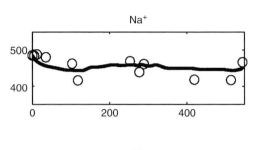

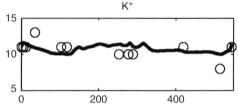

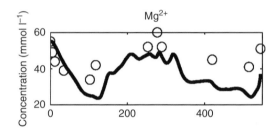

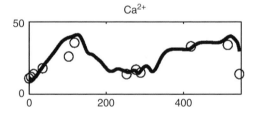

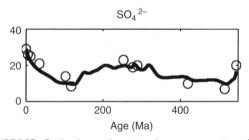

FIGURE 2.27. Circles denote the major-element composition of seawater for the past 550 million years estimated from measurements made on fluid inclusions from marine halites (Lowenstein et al., 2001, 2003; Demicco et al., 2005). The line shows the results of a simple model of seawater as a mixture of river water and high- and low-temperature alterations at mid-ocean ridges (Demicco et al., 2005).

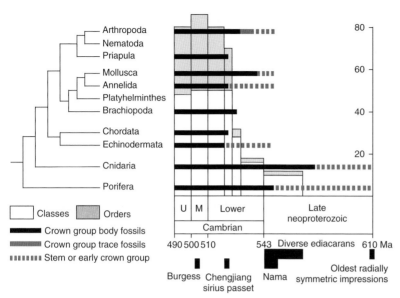

FIGURE 2.28. The "Cambrian explosion." From Knoll and Carroll (1999). "Crown" groups are the major phyla (many of which have a good fossil record). Burgess (Canada), Chengjiang (China), Sirius Passett (Siberia), and Nama (Namibia) are important fossil localities.

The ultimate driver of these global cycles is reckoned to be secular variation in the rates of ocean-floor spreading and concomitant increases and decreases in global volcanicity. This view of the ultimate controls on ocean chemistry has been challenged by Rowley (2002) and Holland (2005).

Evolution of the biosphere

Prokaryotic life arose on the Earth in the Archean (Figure 2.4). High-grade metamorphic rocks interpreted to be sedimentary iron deposits from Greenland contain ^{13}C-depleted graphite interpreted to be the signature of the oldest life on Earth at >3.85 Ga (Mojzsis, 1996). Filamentous microfossils from chert as old as 3.5 Ga have been claimed to have been recovered from Australia (Schopf, 1983). Both these claims for evidence of earliest life have been disputed, however. Undoubted prokaryotic microfossils occur in Mesoarchean rocks (refer to Figure 2.4 for stratigraphic nomenclature and absolute dates). One of the most important events in the history of life on Earth was the development of photosynthesis by cyanobacteria, probably in the Neoarchean. Prior to this development, the atmosphere had very low partial pressures of oxygen. The worldwide episode of iron deposition in banded iron formations (BIFs) that took place in the earliest Proterozoic between 2.5 and 1.8 Ga ago is usually interpreted as recording the buildup of atmospheric O_2 levels due to cyanobacterial photosynthesis. However, results of sulfur isotopic studies (Farquhar et al., 2000; Bekker

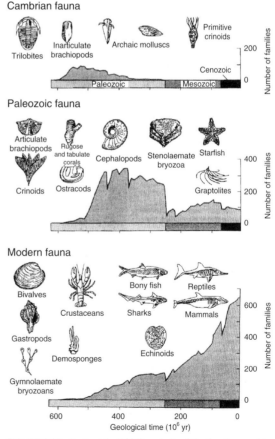

FIGURE 2.29. Sepkoski's (1981) analysis of the fossil record in terms of three distinct evolutionary fauna. From Prothero (2004).

et al., 2004) indicate that oxygen began to build up in the atmosphere at 2.4 Ga, with significant deposition of BIFs both before and after.

The earliest impressions of what are taken to be multi-cellular algae and organic-walled fossils of acritarchs (eukaryotic plankton similar to dinoflagellates) are also found in Paleoproterozoic rocks. Mesoproterozoic and Neoproterozoic carbonate rocks are dominated by stromatolites (Chapter 11), which are reckoned to record the interaction of sedimentary and chemical depositional processes with microbial benthic mats. Features interpreted to be the burrows of metazoan worm-like organisms (but this is disputed) are found in sandstones as old as 1.6 Ga from India. Only towards the end of the Proterozoic are fossils of a distinctive fauna, referred to as *Ediacaran*, found in deposits worldwide. The rapid development of most of the important shell-building groups is known as the *Cambrian Explosion* (Figure 2.28). Paleontologists tend to view the origin and diversification of the phyla outlined in the section on the biosphere as having been a relatively rapid process that took perhaps a few hundreds of millions of years prior to the Cambrian Explosion. However, biologists who use the genetic codes that govern cellular chemistry to tackle the origin and diversification of animals on Earth suggest that perhaps as much as a billion years of metazoan evolution took place prior to the Cambrian Explosion. These so-called molecular clocks measure times of divergence of metazoan phyla on the basis of measured differences in their ribosomal DNA and in other DNA sequences that control specific protein synthesis.

Sepkoski (1981) analyzed the history of families of preserved marine fossils and recognized three distinct evolutionary fauna: (1) Cambrian Fauna; (2) Paleozoic Fauna; and (3) Modern Fauna (Figure 2.29). The Cambrian Fauna comprised trilobites, inarticulate brachiopods, archaic molluscs, and primitive echinoderms, and, after a short period of dominance in the Early Paleozoic, gradually declined in abundance. The Paleozoic Fauna diversified rapidly in the Ordovician, after which there was a long period of relative stasis in the number of preserved fossil families. One of the most profound events in the history of life occurs within a few millions of years of the end of the Permian, when 60%–70% of marine families became extinct. The Paleozoic Fauna, comprising the rugose and tabulate corals, crinoids, articulate brachiopods, graptolites, starfish, and primitive bryozoans, was particularly hard hit by the end-Permian extinction and never recovered. There have been other mass extinctions that altered the course of evolution on the Earth in a less profound manner. The latest mass extinction, 65 million years ago at the Cretaceous–Paleogene boundary, was more or less coincident with the impact of a large extraterrestrial body. This impact partially or wholly caused the mass extinction, and there is debate over its effects. The Modern Fauna is characterized by bivalve and gastropod molluscs, crustaceans, echinoids, and marine vertebrates. There has been a steady rise in the diversity of the Modern Fauna since about the Ordovician, but it became dominant after the mass extinction at the end of the Cretaceous.

Part 2
Production of sediment at the Earth's surface

3 Weathering of rocks, production of terrigenous sediment, and soils

Introduction to weathering

Weathering is the physical, chemical, and biochemical breakdown of Earth materials at the interface between the lithosphere and the atmosphere, and can be seen today in the deterioration of buildings and gravestones. Weathering is the starting point of sediment production by producing a soil: a layer of loose, unconsolidated sediment of variable thickness over the land surface that hosts plant growth in all but the driest, coldest, and most saline areas. These sediment grains (also known as *clasts* or *detritus*; adjectives *clastic* and *detrital*) are called *terrigenous* because they originate from the land surface. Terrigenous sediment grains are transported from the soils of the source area by sediment gravity flows, rivers, wind, ice, and various marine currents (see Part 3). Weathering determines in large part the initial mineralogy, size, and shape of the sediment grains eroded and transported out of a source area. In certain cases, the mineral composition of the sediment grains can be traced back to source-area rocks (i.e., their *provenance*), especially if weathering has not been too destructive. However, the mineralogy, size, and shape of sediment grains can be greatly altered by weathering and physical abrasion during their transport.

Weathering also produces dissolved solutes that leave the source area in surface water and groundwater, and the composition of these solutes depends on the source rocks and the nature of weathering. The composition of surface water and groundwater entering lakes and the sea in turn determines the composition of chemogenic or biogenic sedimentary grains precipitated from lake water or seawater (e.g., calcium carbonate, silica). The chemical composition of groundwater is a key control of chemical reactions that take place in sediments during burial (such *diagenetic* alterations are further described in Chapter 19). Dissolution during weathering can leave void spaces of various sizes in the dissolved rocks and distinct landforms can be related to the nature of weathering.

Weathering has traditionally been divided into physical (or mechanical) weathering and chemical weathering, but it is becoming apparent that biochemical weathering is also important. In reality, physical, chemical, and biochemical weathering processes work together. There is no alteration of the chemical or mineral composition of Earth materials during physical weathering, only a physical breakage of rocks. Chemical and biochemical weathering, on the other hand, alter the chemical and mineral composition of the weathered Earth materials, produce new solid materials (e.g., clay minerals), and put dissolved solutes into the groundwater. Chemical and biochemical weathering involve interactions among minerals, aqueous solutions, and organisms, and are by far the most important. It is difficult to separate the effects of chemical and biochemical weathering. Physical weathering is described below first, followed by a short review of solution chemistry that sets up the ensuing section on chemical and biochemical weathering. Finally, soils and weathering landscapes are described. References on physical, chemical, and biochemical weathering include Garrels and MacKenzie (1971), Sposito (1989), Faure (1991), Berner and Berner (1996), Drever (1997), and Birkeland (1999).

Physical (or mechanical) weathering

Freeze–thaw

The main mechanism of physical weathering at the Earth's surface is the freeze–thaw cycle of water. Solid water (ice) is less dense and has about 9% greater volume than an equivalent mass of liquid **45**

FIGURE 3.1. Physical weathering. (A) A scree slope (talus) of angular limestone clasts, Dolomite Mountains, Italy. (B) Joints separating sheets of granite, Yosemite National Park, California, USA (courtesy of Peter Knuepfer).

water. The conversion of water to ice in a pre-existing confined space such as a joint, closed crack, or large pore can exert up to approximately one kilobar (100 kilopascals) of pressure. Freeze–thaw does not occur in the tropics and in permanently frozen polar areas but is restricted to temperate zones, at both mid latitudes and high altitudes. The splitting process in freeze–thaw produces fresh, angular fragments of rock that accumulate as *scree* (or *talus*) at the base of slopes. Scree deposits at the base of mountains in seasonally cold climates present many sharp edges to climbers (Figure 3.1A). Fractures are not essential for freeze–thaw. *Ice segregation* occurs where soil moisture is drawn to freezing sites in a porous medium by a temperature gradient. Ice can build up as layers and lenses at the freezing site, inducing doming and fracturing, particularly in permafrost regions (Murton *et al.*, 2006). Another common example of freeze–thaw is pavement potholes. Roads comprise a layer of relatively impermeable asphalt that is typically applied over a graded roadbed of permeable gravel. The

gravels can have variable water content and, if the water freezes, the asphalt is domed upwards and cracked. Vehicles break up the domes to produce potholes. This process is exacerbated by the growth of salt crystals in the road materials.

Pressure release

All rocks contain fractures. *Joints* are fractures along which no motion has occurred, whereas *faults* are fractures along which motion has occurred. Joints and faults occur due to brittle failure associated with lithostatic and/or dynamic stresses. Lithostatic pressures can cause rocks to expand in the direction of least compressive stress, producing joints parallel to the principal stress direction. Release of lithostatic pressure due to removal of overlying material allows rocks to expand upwards, producing joints parallel to the land surface. Dynamic stresses are needed to generate faults and folds in layered rocks, with associated fracture cleavage and slaty cleavage. Cooling of igneous rocks causes them to contract, producing joints in a range of directions. Joints are therefore commonly parallel and normal to layering, possibly with complex orientations, depending on the stress history. Open fracture patterns in rocks exist to a depth of a few kilometers.

Erosion of rocks from the Earth's surface and concomitant isostatic uplift cause the underlying rocks to expand, producing joints parallel and normal to the land surface. These new joint sets may cut across older fracture sets that also widen and become more numerous. The new set of joints parallel to the land surface produces sheets of rock that are typically meters thick near the surface (Figure 3.1B). The thickness of the sheets increases downwards. In quarries and mines, strain built up in the rocks may be explosively released as the stresses are reduced by removal of overlying and adjacent material. It is hard to consider these processes as weathering, but the formation of joints and faults in rocks contributes to their breakdown at the Earth's surface.

Other mechanisms

Several other mechanisms of physical weathering have been proposed: (1) breakdown of rocks due to differential thermal expansion and contraction of individual rock-forming minerals during diurnal temperature changes; (2) the growth of salt minerals in fractures and pores exerting increased pressure analogous to ice; and (3) the hydration of anhydrous salts (such as anhydrite to gypsum) within rocks. These mechanisms are commonly cited as physical weathering processes in arid regions where extremes of temperature and moisture occur. The efficacy of these processes is not clear. In particular, repeated heating and cooling of rock materials in laboratories does not lead to physical weathering since the thermal conductivity of rocks is very small and the effects can reach only a few millimeters into rocks. However, the extreme temperatures produced by forest fires can apparently produce significant spalling of surface rock slabs. Plants and animals also have physical effects during weathering, in addition to their biochemical effects described below. Plant roots commonly split rocks and buckle pavement slabs. Burrowing earthworms and moles break up materials and bring them to the surface where they can be weathered. In any event, the main contribution of physical weathering is the propagation and widening of cracks that increase the efficiency of chemical and biochemical weathering by increasing the surface area accessible to weathering waters.

Fundamentals of solution chemistry

Chemical and biochemical weathering reactions include dissolution, precipitation, and oxidation–reduction (redox) mediated by the aqueous solutions known as soil waters. In addition, there is a class of sedimentary grains produced by chemical and biochemical precipitation from groundwater, lake water, river water, and seawater (Chapter 4), and nearly all diagenetic chemical reactions involve an aqueous fluid (Chapter 19). The two most important classes of reactions that precipitate (or dissolve) minerals out of natural waters are (1) acid–base reactions that involve the exchange of protons (H^+ ions) and (2) oxidation–reduction reactions that involve the exchange of electrons (e^-). Many weathering reactions are acid–base reactions, and oxidation–reduction reactions are particularly important for the weathering of iron-bearing minerals. Two traditional approaches toward predicting how much of what minerals will precipitate (or dissolve) in aqueous solutions are equilibrium thermodynamics and kinetics. The equilibrium thermodynamics of aqueous solutions is discussed below. Kinetics are dealt with briefly in later sections. Useful references on solution chemistry include Stumm and Morgan (1981), Faure (1991), Drever (1997), and Walther (2005).

The chemical reactions that occur between minerals and solutions can be represented in a general way as

$$a\text{A} + b\text{B} \leftrightarrow c\text{C} + d\text{D} \tag{3.1}$$

in which a moles of substance A and b moles of substance B interact with the solution to form c moles of substance C and d moles of substance D. For reactions in solutions, A, B, C, and D are usually minerals, water, or dissolved ionic substances. Water does not explicitly appear in the reaction unless it is incorporated into one of the substances. Owing to the high dielectric constant of water, many of the common substances dissolved in the water exist as either positively charged *cations* or negatively charged *anions*. The substances A and B are called the *reactants* and the substances C and D are called the *products*. For aqueous solutions under all conditions, the *equilibrium constant K* is defined as

$$K = \frac{(a_\text{C})^c (a_\text{D})^d}{(a_\text{A})^a (a_\text{B})^b} \tag{3.2}$$

where a_A, a_B, a_C, and a_D are the *activities* of the species A, B, C, and D. For pure solid mineral phases at 25 °C and 1 atmosphere pressure (the so-called *standard state*), activities are defined as equal to 1. For pure gases, the standard state is 1 atmosphere of pressure.

However, for dissolved substances, activities are best thought of as *effective concentrations*.

The activity of the *i*th ionic substance in an aqueous solution is defined as

$$a_i = m_i \gamma_i \tag{3.3}$$

where m_i is the *molality* of species i (number of moles of i per kilogram of water) and γ_i is the *activity coefficient* of species i. Activity coefficients are usually <1, get closer to 1 as a solution becomes more dilute, and equal 1 at infinite dilution (pure water). The difference between molality and activity stems from a number of sources. Two of the more important sources are interactions between the ions and the polar water molecules, and interactions among the ions. Ions interact with the polar water molecules, and hydration shells of water molecules a few molecules thick surround each ion. At low concentrations, this effect is small, since only a small percentage of the water molecules is involved. However, in seawater *all* of the water molecules take part in hydration shells around the dissolved ions. These hydration shells "protect" the ions and thus lower their effective concentrations. In addition, the ions in a solution interact with each other and form complexes between positive and negative ions. Two positive ions and two negative ions can also become covalently bonded molecules in solution. These interactions in effect take some of the molecules out of the pool of "free" (uncomplexed) ions that are reacting. Ions form complexes with other ions in the solution, typically making many different neutral and charged complexes, and, in essence, lower the amount of an ion available to take part in chemical reactions. For example, a partial list of complexes of Ca^{2+} includes $CaHCO_3^+$, $CaCO_3^0$, $CaCl^+$, $CaCl_2^0$, $CaNO_3^+$, $Ca(NO_3)_2^0$, $CaOH^+$, and $Ca(OH)_2^0$ (Faure, 1991).

In very dilute solutions (1–10 ppm), molalities are essentially the same as activities. However, activities should be used instead of molalities in thermodynamic calculations at and above typical groundwater concentrations (a few tens to a few hundreds of ppm). This adds extra complexity to thermodynamic calculations involving aqueous solutions. Debye–Hückel theory is commonly used to calculate activity coefficients in dilute water, but the semi-empirical Pitzer theory must be used in seawater and in relatively concentrated brines (Pitzer, 1973). Computer programs for thermodynamic calculations in aqueous systems generally use Pitzer theory (Harvie *et al.*, 1984; Plummer *et al.*, 1988; Spencer *et al.*, 1990).

The formalism of thermodynamics postulates that during any chemical reaction (e.g.,

$aA + bB \leftrightarrow cC + dD$) there is a change in a quantity known as the *Gibbs free energy*. The change in the Gibbs free energy (ΔG) of a reaction is defined as

$$\Delta G = c\,\Delta G_C + d\,\Delta G_D - a\,\Delta G_A - b\,\Delta G_B \tag{3.4}$$

The change in the Gibbs free energy (ΔG) of a reaction equals the sum of the changes in the Gibbs free energies of the products minus the changes in the Gibbs free energies of the reactants. The changes in the Gibbs free energies of the substances are multiplied by the numbers of moles of the substances involved in the reaction. The changes in the Gibbs free energy of each of the substances are defined as follows:

$$\Delta G_A = \Delta G_{fA}^0 + RT \ln(a_A) \tag{3.5}$$

$$\Delta G_B = \Delta G_{fB}^0 + RT \ln(a_B) \tag{3.6}$$

$$\Delta G_C = \Delta G_{fC}^0 + RT \ln(a_C) \tag{3.7}$$

$$\Delta G_D = \Delta G_{fD}^0 + RT \ln(a_D) \tag{3.8}$$

where ΔG_{fA}^0 is known as the *Gibbs free energy of formation* of substance A in its standard state (pure solid, 1 molar aqueous species, or pure gas at 1 atmosphere). Likewise, ΔG_{fB}^0 is known as the Gibbs free energy of formation of substance B, and so forth. The Gibbs free energies of formation of most geological substances have been calculated and tabulated, and comprise thermodynamic data bases (Robie and Hemingway, 1995). Here R is the universal gas constant (8.3143 $J\,K^{-1}\,mol^{-1}$) and T is the temperature in Kelvin. Most of the reactions of interest at the Earth's surface occur at $\sim 25\,°C$, which is $\sim 298\,K$.

For the general reaction represented by Equation (3.1), substituting Equations (3.5)–(3.8) into Equation (3.4) yields

$$\Delta G = \Delta G_f^0 + RT \ln K \tag{3.9}$$

where

$$\Delta G_f^0 = c\,\Delta G_{fC}^0 + d\,\Delta G_{fD}^0 - a\,\Delta G_{fA}^0 - b\,\Delta G_{fB}^0 \tag{3.10}$$

and K is defined in Equation (3.2).

Equilibrium is a particular subset of the general conditions outlined above and is described qualitatively with a simple example. If salt is added to water, the salt will start to dissolve and, while salt dissolution continues, the solution is said to be *undersaturated*. However, there comes a point when no more salt will dissolve into the water. The solution is then said to be *saturated*, and the solution and the salt are in equilibrium. A conceptual model of what is happening between the solution and the "undissolved" salt is that there is a constant vigorous exchange of material, whereby individual Na^+ and Cl^- ions come off the salt only to be replaced by an equal number of Na^+ and Cl^- ions coming out of the solution onto the salt. If the temperature of the solution is increased, additional salt can be dissolved in the water and a new equilibrium is obtained. However, if the temperature of the system is lowered, the solution becomes *supersaturated* with Na^+ and Cl^- ions, and salt crystals will grow if there are suitable nucleation sites available.

At thermodynamic equilibrium, the change in the Gibbs free energy of a reaction is $\Delta G \equiv 0$, by definition. Under these special conditions, $K = K_{T,P}$, where the subscripts T and P indicate that equilibrium is for a fixed temperature and pressure and that the value of $K_{T,P}$ is different at different temperatures and pressures. Most of the reactions described in this chapter (and Chapter 4) are modeled as if they take place under so-called standard conditions of $25\,°C$ (298 K) and 1 atmosphere pressure. Tabulated data of Gibbs free energies of formation can be used to *calculate* equilibrium constants from Equation (3.9) as

$$-\Delta G_f^0 = RT \ln K_{T,P} \quad \text{or} \quad K_{T,P} = \exp\left(-\frac{\Delta G_f^0}{RT}\right)$$
(3.11)

At equilibrium

$$K_{T,P} = \frac{(a_C)^c (a_D)^d}{(a_A)^a (a_B)^b}$$
(3.12)

which is the so-called *law of mass action* and can be used in a number of ways. First, it is used to define two scales with which to classify the natural waters on the Earth: (1) the pH scale, used to define acid–base waters; and (2) the pe scale, used to define oxidizing and reducing waters. Second, the law of mass action can be used to calculate how much of a mineral will dissolve into water at equilibrium. The third use of the law of mass action, and one that is particularly important for weathering reactions, chemogenic sediments, and diagenetic studies, is using the equilibrium constant to calculate the state of saturation of a solution with respect to any mineral.

The pH, pe, and Eh scales for natural waters

The acid–base scale used to classify natural waters is based on the equilibrium dissociation of water at $25\,°C$ and 1 atmosphere pressure:

$$H_2O \leftrightarrow H^+ + OH^-$$
(3.13)

$$\Delta G_f^0 = \Delta G_{fH^+}^0 + \Delta G_{fOH^-}^0 - \Delta G_{fH_2O}^0$$
(3.14)

The appropriate values of the Gibbs free energy of formation (from Robie and Hemingway, 1995) are

$$\Delta G_f^0 = 0\ \frac{\text{joules}}{\text{K mole}} + \left(-157,300\ \frac{\text{joules}}{\text{K mole}}\right) - \left(-237,100\ \frac{\text{joules}}{\text{K mole}}\right)$$
(3.15)

so

$$\Delta G_f^0 = 79,800\ \frac{\text{joules}}{\text{K mole}}$$
(3.16)

Using this value of 79,800 $J\,K^{-1}\,mol^{-1}$ for ΔG_f^0 in Equation (3.11), and converting from natural logs to base-ten logs by multiplying the natural logs by 2.303, yields

$$K_{T,P} = 10^{-14}$$
(3.17)

so

$$10^{-14} = \frac{(a_{H^+})(a_{OH^-})}{a_{H_2O}}$$
(3.18)

By definition, the activity of a pure substance (in this case water) at the temperature and pressure of interest equals 1, and Equation (3.18) becomes

$$10^{-14} = (a_{H^+})(a_{OH^-})$$
(3.19)

This solution is so dilute that using molalities instead of activities is justified. Since the number of moles of H^+ equals the number of moles of OH^- (Equation (3.13)),

$$a_{H^+} = a_{OH^-} = 10^{-7}$$
(3.20)

These activities define the pH scale used to describe whether a substance is an acid (pH < 7; tends to give off protons) or a base (pH > 7; tends to accept protons, also called alkaline).

$$pH = -\log(a_{H^+})$$
(3.21)

In other words, pH is the negative of the base-ten logarithm of the activity of hydrogen ions in solution. Neutral pH is defined on the basis of equilibrium dissociation of water outlined in Equations (3.13)–(3.21) and is therefore $-(-7)$ or 7. Strong acids dissociate, putting large numbers of H^+ ions into solution, whereas weak acids tend to dissociate only partially and put fewer moles of H^+ ions, into solution. In one liter of pure water there are $\sim 10^{-7}$ moles of H^+ ions, so approximately 1 atom out of 2×10^8 is a H^+ ion. At a pH of 1, the activity coefficient of the H^+ ions is 0.8–0.9. Equations (3.22) and (3.3) then can be used to give

$$a_{H^+} = 10^{-1} = 0.1 = (m_{H^+} \gamma_{H^+}) = (0.8m_{H^+}) \qquad (3.22)$$

which shows that a pH of 1 implies that there are about 0.125 moles of H^+ in 1 kilogram of water or ~0.2% (one molecule in 500) of the solution is naked hydrogen ions. A low pH implies a high concentration of H^+. However, the equilibrium equation (3.20) fixes the number of OH^- ions. If the pH of a solution is 1, then

$$10^{-14} = (a_{H^+})(a_{OH^-}) = 10^{-13}(a_{OH^-}) \qquad (3.23)$$

Under these circumstances, $a_{OH^-} = 10^{-13}$. Since the solutions must maintain electrical neutrality, there must be negative ions other than OH^- to balance the H^+ ions. For example, in hydrochloric acid the anion is Cl^-.

At very high pH, an alkaline solution, the concentration of H^+ is very low. However, the equilibrium equation (3.19) fixes the number of OH^- ions. If the pH of a solution is 13, then

$$10^{-14} = (a_{H^+})(a_{OH^-}) = 10^{-13}(a_{OH^-}) \qquad (3.24)$$

$a_{OH^-} = 10^{-1}$ and ~0.2% of the solution is naked OH^- ions. Cations other than H^+ must be present in the water to maintain electrical neutrality. "Naked" H^+ and OH^- ions do not exist in the solutions; this is only a convenient way to model what are really quite complex chemical phenomena.

Oxidation–reduction reactions (or *redox* reactions) are also important in natural waters at the Earth's surface and in weathering and soil formation, and some of the most important redox reactions involve metals such as manganese and iron:

$$Fe^{3+} + e^- \leftrightarrow Fe^{2+} \qquad (3.25)$$

In this example Fe^{3+} (the oxidized substance) is reduced to Fe^{2+} (the reduced substance) by accepting an electron. Equivalently, reduced Fe^{2+} could be oxidized to Fe^{3+} by giving off an electron. Electrons do not exist free in the solution, but a useful thermodynamic formalism has arisen on the basis of defining the activity of the electron a_{e^-}. At equilibrium for reaction (3.25)

$$K_{T,P} = \left(\frac{a_{Fe^{2+}}}{a_{Fe^{3+}}}\right)\left(\frac{1}{a_{e^-}}\right) \qquad (3.26)$$

where a_{e^-} is the activity of the electron and

$$\log K_{T,P} = -\log(a_{e^-}) + \log\left(\frac{a_{Fe^{2+}}}{a_{Fe^{3+}}}\right) \qquad (3.27)$$

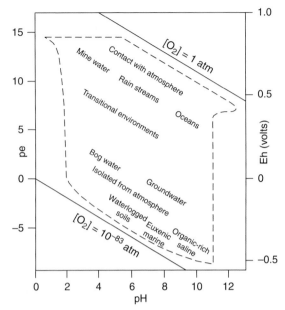

FIGURE 3.2. The pH–pe limits of natural waters. From Faure (1991). The Eh scale is also shown.

The pe scale is defined as the negative log of the activity of the electrons: $pe = -\log(a_{e^-})$,

$$pe = \log K_{T,P} + \log\left(\frac{a_{Fe^{3+}}}{a_{Fe^{2+}}}\right) \qquad (3.28)$$

The term $\log K_{T,P}$ is defined as pe^0 (explained below) so that

$$pe = pe^0 + \log\left(\frac{a_{Fe^{3+}}}{a_{Fe^{2+}}}\right) \qquad (3.29)$$

The pe of natural water ranges from around -12 to 16 (Figure 3.2). A low negative pe implies that the activity of the electrons is high and the setting is described as *reduced*. A high positive pe implies that the activity of the electons is low and the setting is described as *oxidized*. At pe of 0, $a_{e^-} = 10^0 = 1$.

The pe of a solution is an equilibrium property such that there needs to be an equilibrium balance between the oxidized and reduced substances. Under equilibrium conditions, pe^0 is related to the Gibbs free energy of formation. Direct application of Equation (3.11) leads to

$$\Delta G_f^0 = -RT \ln K_{T,P} \qquad (3.30)$$

Making the conversion from natural logs to base-ten logs yields

$$pe^0 = \frac{\Delta G_f^0}{2.303RT} \qquad (3.31)$$

The change in Gibbs free energy of formation of oxidation–reduction reactions is the sum of the Gibbs free energies of formation of the products minus the sum of the Gibbs free energies of formation of the reactants. The Gibbs free energy of formation of an electron is defined as 0. Thus, the change in Gibbs free energy of formation of the reaction in Equation (3.26) at equilibrium is

$$\Delta G_f^0 = \Delta G_{fFe^{2+}}^0 - 0 - \Delta G_{fFe^{3+}}^0 \qquad (3.32)$$

These types of reactions can be represented symbolically as

Reduced species $\leftrightarrow$ oxidized species $+ ne^-$ $\qquad (3.33)$

For this more general expression,

$$pe = pe^0 + \frac{1}{n} \log \left(\frac{\Pi(a_{oxidized})^{\nu_{oxidized}}}{\Pi(a_{reduced})^{\nu_{reduced}}} \right) \qquad (3.34)$$

The pe^0 is defined in Equation (3.31) and is calculated using tables of Gibbs free energies of formation.

Redox reactions in natural waters are envisioned as occurring simultaneously with reactions that either supply or use the electrons. Thus, redox reactions are most commonly treated as two parallel *half reactions*, one that supplies the electron (involving the reducing agent) and one that takes up the electron (involving the oxidizing agent). These reactions can be complex and involve a number of different electron donors and acceptors. Common electron acceptors (oxidizing agents) in natural waters include O_2 gas, Fe^{3+}, and complex ions such as NO_3^-, SO_4^{3-}, and PO_4^{3-}. Diatomic oxygen (O_2) is a very important electron acceptor. A pe of 0 is approximately equal to a partial pressure of oxygen of 10^{-4} atmospheres (approximately the limit of detection for dissolved oxygen in solution). Environments in which the partial pressure of oxygen is less than 10^{-4} atmospheres (and pe < 0) are commonly referred to as *anoxic*. Certain metal ions, particularly Fe^{2+} and Mn^{4+}, and S^- are important reducing agents. Fully balanced reaction equations can be quite complicated. For example, the oxidation of pyrite into iron oxyhydroxide written in a balanced equation is (Faure, 1991)

$$4FeS_2 + 15O_2 + 14H_2O \rightarrow 4Fe(OH)_3 + 8SO_4^{2-} + 16H \qquad (3.35)$$

The complementary half reactions of this reaction are

$$4FeS_2 + 4H_2O \rightarrow 4Fe(OH)_3 + 8SO_4^{2-} + 76H^- + 60e^-$$
reduction $\qquad (3.36)$

$$15O_2 + 60H^+ + 60e^- \rightarrow 30H_2O \qquad \text{oxidation} \qquad (3.37)$$

The Fe^{2+} and the S^- in the pyrite are both reducing agents, whereas the O_2 is the electron acceptor (oxidizing agent) and is reduced to O^{2-}. The equilibrium pe of these reactions can be calculated as described above. Also, this reaction has the transfer of protons in it in order to balance the charge. Thus, this reaction is dependent not only on the pe state of the water, but also on the pH of the water.

Many natural redox reactions involve acid–base (proton) reactions as well as the transfer of electrons. Two-dimensional diagrams with pH as the x axis and pe as the y axis are useful in considering surface biogeochemical processes, and Drever (1997) gives a thorough review of the calculation of these types of diagrams. The stability field of water plotted on such a diagram outlines the limits of natural waters on Earth (Figure 3.2). The stability fields of minerals and aqueous phases are commonly presented on these diagrams (see the example given below). These diagrams are somewhat limited because the equilibrium relations depicted by the various fields are commonly not obtained and, in many cases, the error bars on the underlying thermodynamic data are large. Also, these diagrams are most commonly drawn using another classification scale for oxidation–reduction reactions, known as Eh. The Eh scale developed on the basis of electro-chemical investigations that measured voltage potentials that developed between different reaction couples, so Eh has the units of volts. At 25 °C, Eh = 0.059pe. Finally, most of the redox reactions that occur at the Earth's surface involve the metabolism of prokaryotes as outlined in Chapter 2.

Testing for equilibrium

One of the most useful applications of the law of mass action (Equation (3.12)) is to test the saturation state of natural water with respect to any mineral. Consider the reaction

$$aA + bB \leftrightarrow cC + dD \qquad (3.38)$$

If the system is at equilibrium,

$$K_{T,P} = \frac{(a_C)^c (a_D)^d}{(a_A)^a (a_B)^b} \qquad (3.39)$$

If the reaction is not at equilibrium, the expression on the right-hand side of Equation (3.39) is called the ion activity product (IAP) and, by comparing the equilibrium constant with the IAP, the saturation state of the solution with respect to any mineral can be determined. For example, consider the precipitation of calcite from water:

$$Ca^{2+} + CO_3^{2-} \leftrightarrow CaCO_3 \qquad (3.40)$$

The ion activity product of Ca^{2+} and CO_3^{3-} is the product of their activities in the solution:

$$IAP = (a_{Ca^{2+}})(a_{CO_3^{2-}}) \qquad (3.41)$$

For seawater, the activity of Ca^{2+} is $10^{-2.62}$ and the activity of CO_3^{2-} is $10^{-5.05}$, so the IAP is $10^{-7.67}$. The equilibrium constant of this reaction is $K_{T,P} = 10^{-8.48}$. The test for saturation involves comparing the IAP of the dissolved components of the mineral in the water with the equilibrium constant for the reaction:

$$\frac{IAP}{K_{T,P}} \qquad (3.42)$$

If the ratio is approximately 1, the solution is nearly at equilibrium and the solution is saturated. If the ratio is $\ll 1$, then the solution is undersaturated. If the ratio is $\gg 1$, the solution is supersaturated, and the potential exists for the mineral to precipitate. For seawater, the ratio of the IAP of Ca^{2+} and CO_3^{2-} to the equilibrium constant of calcite is 6.45 and the solution is just saturated with respect to calcite. The state of saturation of seawater with respect to dolomite provides another example:

$$Ca^{2+} + Mg^{2+} + 2CO_3^{2-} \leftrightarrow CaMg(CO_3)_2 \quad K_{T,P} = 10^{-17}$$
$$(3.43)$$

For seawater, the activity of Ca^{2+} is $10^{-2.62}$, the activity of CO_3^{2-} is $10^{-5.05}$, and the activity of Mg^{2+} is $10^{-1.86}$, so the IAP is $10^{-14.58}$. The ratio $IAP/K_{T,P} \sim 263$ or $10^{2.42}$. Seawater is about two and one half orders of magnitude supersaturated with dolomite. However, dolomite does not precipitate from natural waters or laboratory waters at 25 °C and 1 atmosphere pressure because of the kinetics of the reaction. The equilibria predicted by classical thermodynamics are rarely obtained at the low temperatures and pressures that are commonly found at the Earth's surface, but thermodynamics does provide a convenient basis with which to understand the real reactions at the Earth's surface.

Chemical and biochemical weathering

Most minerals in igneous and metamorphic rocks formed at around 1000 °C and at many thousands of kilopascals pressure under reducing conditions. However, where they are exposed at the Earth's surface at approximately 15–25 °C, at 1 atmosphere pressure, and with abundant free oxygen dissolved in meteoric surface waters, igneous and metamorphic minerals readily break down chemically in various reactions known as chemical weathering. Chemical and biochemical weathering is the attack of oxygenated, acidic surface waters on Earth materials exposed at the lithosphere–atmosphere contact. The source of the water is meteoric precipitation.

Rainwater

Rainwater generally contains about 10 parts per million (ppm) total dissolved solids (TDS) (Table 3.1). The sources of solutes in rainwater include sea-salt aerosols,

TABLE 3.1. Chemical analyses of atmospheric precipitation in ppm (from Eugster and Hardie, 1978). By permission of Springer Science and Business Media.

	Snow; New Mexico	Snow; New Mexico	Rain; North Carolina	Rain; Baltimore	Rain; Rothamsted England	Rain; Northern Europe	Rain; Southeast Australia	Rain; San Diego	Rain; Colorado
H_4SiO_4	n.d.	0.1		0.1					
Ca^{2+}	0.3	0.2	0.65	0.3	1.7	1.42	1.20	0.67	3.41
Mg^{2+}	n.d.	n.d.	0.14	0.1	0.3	0.39	0.50		
Na^+	0.5	0.1	0.56	0.2	1.3	2.05	2.46	2.17	0.69
K^+	0.3	0.3	0.11	0.2	0.3	0.35	0.37	0.21	0.17
HCO_3^-	3.2	1.9		0.3					
SO_4^{2-}	n.d	n.d	2.18	1.7	1.8	2.19	Trace	1.66	2.37
Cl^-	0.2	0.2	0.57	0.6	2.7	3.47	4.43	3.31	0.28
NO_3^-	0.1	n.d.	0.62			0.27		3.13	2.63
Total	4.8	2.8	>2.83	3.5	>8.1	>10.14	>8.96		
pH	6.1	5.3	4.8–5.9	4.6	4.9	5.47			

n.d., no data.

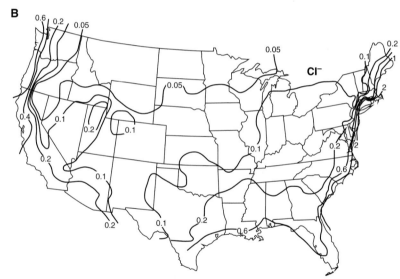

FIGURE 3.3. Average sodium (A) and chloride (B) concentrations ($mg\,l^{-1}$) in rainwater over the USA for 2003. Data from US Geological Survey National Atmospheric Deposition Program/National Trends Network.

mineral dust particles, and particulate industrial pollutants. Sea-salt aerosols arise where droplets of spray from breaking ocean waves are caught up in the atmosphere and evaporate, leaving microscopic particles of sea salts (principally halite and gypsum) suspended in the atmosphere. The contribution of sea-salt aerosols to precipitation is indicated by the highest concentrations of Na^+ and Cl^- in coastal areas (Figure 3.3). The deserts of the world are the principal sources of airborne dust, which includes small particles of silicates, carbonates, gypsum, and anhydrite. Additional sources of airborne dust are volcanic eruptions and meteorite debris. Industrial pollutants include suspended particles of soot and ash as well as dissolved nitrates and sulfates. The small particulates suspended in the

atmosphere form nuclei for the condensation of rain drops and snow flakes.

Atmospheric gases dissolve into rainwater as it falls. Nitrogen and oxygen are the principal dissolved gases in the atmosphere, and oxygen affects the oxidation–reduction potential (see below) of rainwater but not the pH of rainwater. Carbon dioxide, on the other hand, is most important for chemical weathering (even though it is only a small component of the atmosphere) because it does affect the pH of rainwater. The following reactions go on simultaneously at equilibrium:

$$CO_{2(g)} + H_2O_{(1)} \Leftrightarrow H_2CO_3{}^+{}_{(aq)} \qquad K_{T,P} = 10^{-1.47} \quad (3.44)$$

$$H_2CO_3{}^+{}_{(aq)} \Leftrightarrow H^+{}_{(aq)} + HCO^-{}_{(aq)} \qquad K_{T,P} = 10^{-6.35} \quad (3.45)$$

$$HCO_3^-_{(aq)} \Leftrightarrow H^+_{(aq)} + CO_3^{2-}_{(aq)} \quad K_{T,P} = 10^{-10.33} \tag{3.46}$$

$$H_2O_{(l)} \Leftrightarrow H^+_{(aq)} + OH^-_{(aq)} \quad K_{T,P} = 10^{-14} \tag{3.47}$$

where the subscripts identify the type of phase: $g =$ gas, $l =$ liquid, $aq =$ dissolved as an aqueous solute, and $s =$ solid (used in later equations). The hydrogen ions produced by these reactions comprise the acidity of precipitation. These equilibria, although complex, are quite well known, as are the values of the equilibrium constants for each reaction. For Equation (3.45) at equilibrium at 25 °C and 1 atmosphere pressure, application of the definition of the equilibrium constant yields

$$10^{-6.35} = \frac{(a_{H^+})(a_{HCO_3^-})}{a_{H_2CO_3}} \tag{3.48}$$

When $a_{HCO_3^-}$ equals a_{H2CO_3}, $a_{H^+} = 10^{-6.35}$, so the pH of this equilibrium equals 6.35. At pH less than 6.35 the activity of H_2CO_3 must be greater than the activity of HCO_3^-. A similar argument can be made for Equation (3.46):

$$10^{-10.33} = \frac{(a_{H^+})(a_{CO_3^{2-}})}{a_{HCO_3^-}} \tag{3.49}$$

Here again, when the activity of carbonate is equal to the activity of bicarbonate, the pH is fixed at 10.33. At pH below 10.33, the activity of $HCO_3^- >$ activity of CO_3^{2-}, whereas at pH above 10.33 $a_{CO_3}^{n-} > a_{HCO_3}^-$. These relationships are graphed in Figure 3.4, a so-called Bjerrum plot. The sum of activities of all ions dissolved in the water is $(H_2CO_3^0 + HCO_3^- + CO_3^{2-}) = 10^{-2}$ in this example. The sum of all the carbonate species in a natural water is known as the *carbonate alkalinity*. The Bjerrum plot is called a dominance plot because all three of the carbonate species exist at all times in any natural water, but which species is dominant depends on the pH. For example, CO_3^{2-} is important only in water where the pH is 10 or above: i.e., in *alkaline* systems.

Given the current partial pressure of carbon dioxide in the Earth's atmosphere, $pCO_2 \cong 10^{-3.4}$ atmospheres (approximately 380 parts per million by volume), and a temperature of 25 °C, Equations (3.44)–(3.47) predict the pH of rainwater to be approximately 5.6. The measured range of rainwater pH in areas away from pollution is 5.5–6.0, quite close to the calculated value (Figure 3.5A). Even though pCO_2 varies with time, these reactions guarantee that rainwater has always been acidic. *Acid rain* results from industrial combustion gases (including oxides of sulfur and nitrogen) that dissolve into the rainwater, locally lowering pH to values of 4–5. Sulfur emission deposited as sulfate is a problem in the northeastern USA (Figure 3.5B). Finally, volcanic gases containing HCl, SO_2, and SO_3 dissolve into meteoric waters during eruptions or in some hot springs and can make local waters with very low pH, pH < 0.

Macroscopic plants usually intercept some of the precipitation as it approaches the ground, with the rest directly falling onto the ground (Figure 3.6). The intercepted precipitation may drip off the leaves, flow down trunks and stems (*stemflow*), or evaporate back into the atmosphere. In addition, plants continually transpire from their leaves (i.e., *evapo-transpiration*) water vapor that they have sucked up from the ground by their roots. Plants and the ground surface have variable amounts of dry material deposited (so-called *dry deposition*) on them from the atmosphere (mostly soot, sea-salt aerosols, and desert-derived wind-blown dust), and some of these aerosols may dissolve into the rain, further altering the dissolved solids in the rainfall. Meteoric waters that reach the surface can directly *run off* into surface water bodies such as streams and lakes but most rainfall infiltrates into the ground.

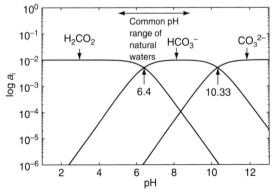

FIGURE 3.4. A dominance (Bjerrum) plot of carbonate species in solution at various values of pH, at temperature 25 °C and with the total activity of CO_2 fixed at 0.01.

Subsurface water

Shallow subsurface water occurs in (1) the unsaturated (or *vadose*) zone; (2) the saturated (or *phreatic*) zone; and (3) a *capillary fringe* separating the two (Figure 3.6). In the unsaturated zone, the infiltrated water co-exists with modified atmospheric gases in the pore spaces, and surface tension of the infiltrated

A

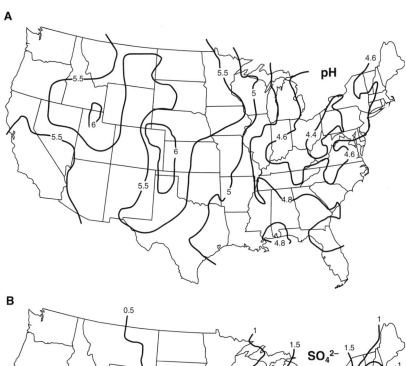

B

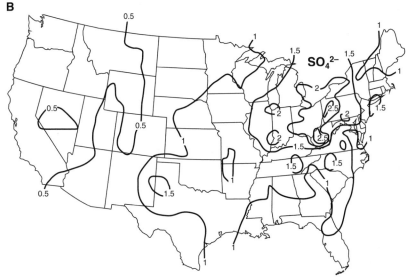

FIGURE 3.5. (A) The average pH of rainwater over the USA for 2003. (B) The average sulfate concentration (mg l^{-1}) in rainwater over the USA for 2003. Data from US Geological Survey National Atmospheric Deposition Program/National Trends Network.

water becomes important in differentially wetting particles. In the phreatic zone, all pore spaces are filled with groundwater. A very complicated ecosystem of bacteria, fungi, and invertebrates is found in the unsaturated zone intertwined with the roots of higher plants (see below). This biological activity results in significant further acidification of the infiltrated water from (1) an increase in concentration of soil gas CO_2; (2) organic acids derived from the breakdown of organic matter; and (3) organic acids released from many different organisms. The unsaturated zone contains variable amounts of organic matter, especially near the ground surface, and aerobic bacterial decay of this organic matter coupled with animal respiration commonly produces additional CO_2 gas, which can reach concentrations up to an order of magnitude more than that in the atmosphere. Water equilibrated with this higher pCO_2 has pH of 4–5. The organic acids that form from the degradation of organic matter are of high relative molecular mass, and are poorly defined compounds generally referred to as *humic* or *fulvic* acids. In addition, the root hairs of plants and hyphae of fungi produce specific organic acids that locally decrease the pH of unsaturated-zone water to 2. Acid mine runoff is a big environmental problem in some areas, where reduced sulfides from ore bodies or coal

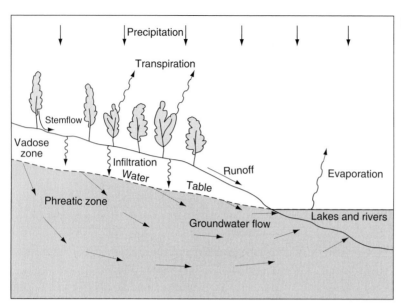

FIGURE 3.6. Near-surface hydrology.

seams are oxidized to sulfuric acid, producing local surface water with pH of 3–4:

$$4FeS_2 + 15O_2 + 8H_2O \rightarrow 2Fe_2O_3 + 8H_2SO_4 \qquad (3.50)$$

Most of the oxidation of the sulfides is catalyzed by bacteria.

Garrels and MacKenzie (1971) provided a classic description of chemical weathering as H^+ ions released by reactions described by Equations (3.44)–(3.47) attack and alter Earth materials, liberating the common metal cations Ca^{2+}, Na^+, Mg^{2+}, and K^+, which are dissolved into the groundwater, where they are electrically balanced by the HCO_3^- ions. Paul and Clark (1989), Sposito (1989), and Berner and Berner (1996), among others, suggested that the increased biological acidification of unsaturated-zone waters is responsible for most chemical weathering, and thus weathering should be considered a biochemical process. However, the chemical makeup of shallow groundwater in most geological terrains can be predicted closely by considering the mineralogical makeup of the surface materials and the chemical processes outlined below (Garrels and MacKenzie, 1971; Drever and Stillings, 1997). More importantly, HCO_3^- is the main anion found in groundwater and balances the positively charged anions. Thus the neutralization of organic acids in the unsaturated zone leaves virtually no evidence of its importance, and does not produce cations that are balanced by measurable organic anions. This is because the organic acids themselves

are usually degraded to H_2O and CO_2 in the lower portions of the unsaturated zone. There is, however, quite convincing evidence of the local importance of organic acids in weathering, as discussed later. The traditional geochemical way of describing weathering is followed here, as in Garrels and MacKenzie (1971). However, all the reactions described below could be written with an organic acid as the main proton contributor (Berner and Berner, 1996). Garrels and MacKenzie (1971) distinguished *congruent* and *incongruent* chemical weathering reactions. In *congruent* reactions, a natural salt dissolves into a solution with the same ionic composition, leaving a hole in the Earth materials where the mineral was, and putting ions into solution. In *incongruent* reactions a new amorphous substance or mineral with decreased solubility gradually replaces pre-existing minerals, and dissolved ions are also produced that are carried off in solution.

Congruent chemical weathering reactions

Congruent reactions may but need not be dependent on the pH of the soil water. For example, the congruent dissolution of halite,

$$NaCl_{(s)}^{\,halite} \Leftrightarrow Na^+_{(aq)} + Cl^-_{(aq)} \qquad (3.51)$$

is not dependent on pH. The solubility of halite is affected by other dissolved ions, but this effect is slight in the low concentrations of most natural waters.

Approximately 360 g of NaCl dissolves in 1 kg of H_2O at 25 °C (~350,000 ppm).

The congruent dissolution of gypsum,

$$CaSO_4 \cdot 2H_2O_{(s)}^{\text{gypsum}} \Leftrightarrow Ca^{2+}_{(aq)} + SO_4^{2-}_{(aq)} + 2H_2O_{(1)} \quad (3.52)$$

can be affected by pH in very acidic waters where the H^+ ion will combine with sulfate. In the normal pH range of shallow subsurface waters, gypsum is orders of magnitude less soluble than halite. The congruent dissolution of gypsum produces a solution with 2000 ppm solute at complete saturation, of which approximately 600 ppm is Ca^{2+} ions. Anhydrite ($CaSO_4$) has approximately the same solubility as gypsum.

Calcite dissolves congruently according to the reaction

$$CaCO_{3(s)}^{\text{calcite}} + H_2O_{(1)} + CO_{2(g)} \Leftrightarrow Ca^{2+}_{(aq)} + 2HCO_3^-_{(aq)} \quad (3.53)$$

For the pH range of shallow subsurface water, it is appropriate to write the reaction with the bicarbonate anion as a product. Under these conditions, the dissolution of calcite is dependent on pCO_2: the greater the pCO_2, the lower the pH and the greater the solubility of calcite (see Chapter 4). One of the carbonate ions in the bicarbonate comes from the calcite whereas the other comes from atmospheric or soil gas CO_2. Calcite is approximately an order of magnitude less soluble than gypsum, and at a pCO_2 of $10^{-3.5}$ atmospheres approximately 20 ppm Ca^{2+} is put into solution and approximately 100 ppm of bicarbonate ion is in solution. The congruent dissolution of dolomite ($CaMg(CO_3)_2$) is similar to that of calcite but also puts Mg^{2+} ions into solution. Congruent dissolution of carbonate rocks is responsible for the caves, caverns, and karst landscape features described below.

The congruent dissolution of silica is complicated. At normal shallow subsurface groundwater pH, the dissolution reaction of quartz,

$$SiO_{2(s)}^{\text{quartz}} + H_2O_{(1)} \Leftrightarrow H_4SiO_{4(aq)} \quad (3.54)$$

produces non-ionized silicic acid (H_4SiO_4). At 25 °C, approximately 6 ppm silicic acid is put into solution: quartz is not very soluble. Non-crystalline amorphous silica gels are considerably more soluble, putting up to 120 ppm silicic acid into solution. The solubility of quartz is significantly increased by increasing temperature. At 300 °C, approximately 600 ppm silicic acid can be dissolved in groundwater of normal pH ranges. Another way to increase significantly the solubility of

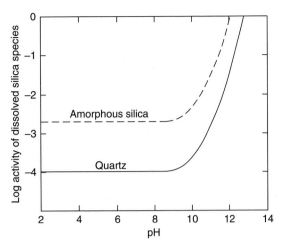

FIGURE 3.7. Solubilities of quartz and amorphous silica as functions of pH. From Drever (1997).

quartz is to raise the pH. At pH > 9 (at 25 °C) silicic acid dissociates:

$$H_4SiO_{4(aq)} \Leftrightarrow H^+_{(aq)} + H_3SiO_4^-_{(aq)} \quad (3.55)$$

At the elevated pH of alkaline water, this reaction is driven to the right and the solubility both of quartz and of amorphous silica is greatly enhanced (Figure 3.7).

Incongruent chemical weathering reactions

Incongruent weathering reactions produce new amorphous substances or minerals with lower solubilities than that of the weathered mineral. The most important incongruent weathering reactions involve the most common aluminosilicates: feldspars, amphiboles, pyroxenes, and micas. A typical reaction for albite feldspar would be

$$2NaAlSi_3O_{8(s)}^{\text{albite}} + 11H_2O_{(1)} + 2CO_{2(g)}$$
$$\Leftrightarrow Al_2Si_2O_5(OH)_{4(s)}^{\text{kaolinite}} + 2Na^+_{(aq)}$$
$$+ 2HCO_3^-_{(aq)} + 4H_4SiO_{4(aq)} \quad (3.56)$$

Albite placed in a carbonic acid solution produces a new solution in which the molar ratio of dissolved silicic acid to sodium is 2:1, as predicted in the reaction. Moreover, the feldspar grains have tiny amounts of amorphous aluminum oxyhydroxides along cleavage planes, twin planes, and crystal defects. In nature, however, it is common to find crystalline *clay minerals* such as kaolinite, illite, and smectite along cleavages in weathered feldspars (Figure 3.8). At a pCO_2 of $10^{-3.5}$ at 25 °C, equilibrium thermodynamic calculations indicate that this reaction would put 18 ppm Na^+, 48 ppm HCO_3^-, and 150 ppm H_4SiO_4 into solution. Although

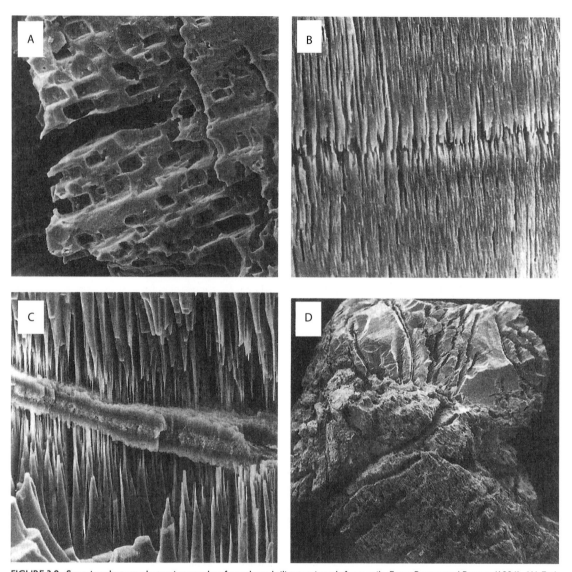

FIGURE 3.8. Scanning electron photomicrographs of weathered silicate minerals from soils. From Berner and Berner (1996). (A) Etch pits on the surface of plagioclase feldspar. (B) Etch pits developed on the surface of an amphibole following dislocations. (C) A clay-filled crack surrounded by dissolving amphibole. (D) A weathering pyroxene grain.

quartz is relatively insoluble, dissolution of feldspars puts a significant amount of dissolved silica in the form of neutral silicic acid ($H_4SiO_4^0$) into groundwater. Hydrogen ions are rapidly consumed in this reaction such that the groundwater becomes neutral to slightly alkaline with pH in the range 7.5–8.2. These reactions are also commonly referred to as *hydrolysis* because the clays have OH^- groups in them. Under typical CO_2 concentrations in subsurface gases, albite rapidly weathers with 2 moles of albite producing 1 mole of kaolinite.

Clay minerals comprise a diverse group of sheet silicates, the most common ones being kaolinite,

smectite, illite, and gibbsite. The structure and chemistry of clay minerals are further described below, as are some of the controls on which clays tend to form under different weathering conditions.

Under certain conditions, forsterite (the magnesium end member of the olivine solid solution) reacts congruently with acidic groundwater and generates high concentrations of both magnesium and silicic acid in solution:

$$Mg_2SiO_{4(s)}^{\text{forsterite}} + 4H^+_{(aq)} \Leftrightarrow 2Mg^{2+}_{(aq)} + H_4SiO_{4(aq)}$$

$$(3.57)$$

Approximately 1,400 ppm Mg^{2+} and 2,800 ppm H_4SiO_4 are put into solution at 25 °C and normal pH of shallow subsurface water. Large amounts of silicic acid are released from this reaction, and the highest concentrations of silicic acid in natural groundwaters are in the range 3,000–4,000 ppm in ultrabasic bedrock terrains. However, the silicic acid in solution commonly reacts incongruently with the dissolved magnesium ion and precipitates new magnesium silicates with lower solubility. In the case of forsterite weathering, a complex suite of magnesium silicates precipitates, including

talc	$Mg_6Si_8O_{20}(OH)_4$
chrysotile	$Mg_3Si_2O_5(OH)_4$
sepiolite	$Mg_2Si_3O_8 \cdot nH_2O$
brucite	$Mg(OH)_2$

These distinctive green minerals are the weathering products of basic rocks such as basalts and gabbros and comprise *serpentinites*, and this weathering process is sometimes referred to as *serpentinization*. Whether forsterite dissolves congruently or incongruently is complex. Forsterite apparently dissolves congruently until the dissolved concentrations of silicic acid and magnesium ions reach some critical value, at which point the new mineral phases are precipitated. The newly precipitated magnesium silicates are amorphous (i.e., not crystalline), although they develop an increasing degree of crystallinity with time.

Wherever transition metals such as Fe are involved in weathering environments, oxidation–reduction reactions are common and important. In the case of fayalite, the pure iron end member of the olivine solid solution,

$$Fe_2SiO_{4(s)}^{\text{ fayalite}} + 4H^+_{(aq)} \Leftrightarrow 2Fe^{2+}_{(aq)} + H_4SiO_{4(aq)} \tag{3.58}$$

with Fe^{2+} released. Unlike forsterite, iron silicates are not produced as fayalite is weathered. In most weathering environments, oxygen competes with the silicic acid and makes minerals orders of magnitude less soluble than any iron silicate. In the presence of O_2, ferrous iron (Fe^{2+}) oxidizes to ferric iron (Fe^{3+}). Once ferric iron is available, it is precipitated as oxides and hydroxides:

$$Fe^{3+}_{(aq)} + 3H_2O_{(l)} \Leftrightarrow Fe(OH)_3^{\text{(amorphous)}} + 3H^+_{(aq)} \tag{3.59}$$

The initially formed oxyhydroxides are amorphous. At a pH of around 2, the ferric iron content of a shallow subsurface groundwater equilibrated with amorphous iron oxyhydroxide is approximately 4,335 ppm at 25 °C. At a pH of 4, however, the ferric iron content in solution is much lower, approximately 0.004 ppm. As time

progresses, even less soluble iron oxyhydroxides such as goethite form:

$$Fe^{2+}_{(aq)} + 2H_2O_{(l)} \Leftrightarrow FeO(OH)_{(s)}^{\text{ goethite}} + 3H^+_{(aq)} \tag{3.60}$$

At a pH of around 2, the ferric iron content of shallow subsurface groundwater equilibrated with goethite is approximately 0.35 ppm at 25 °C. At a pH of 4, however, the ferric iron content in solution is much lower, approximately 0.35×10^{-6} ppm, whereas at a pH of around 6, 10^{-12} ppm ferric iron will be left in solution. The iron oxide that is ultimately stable is hematite:

$$2Fe^{3+}_{(aq)} + 3H_2O_{(l)} \Leftrightarrow Fe_2O_3(OH)_{(s)}^{\text{ hematite}} + 6H^+_{(aq)} \tag{3.61}$$

At a pH of 2, a solution equilibrated with this reaction has a ferric iron content of 0.000 56 ppm, whereas at a pH of 8, the ferric iron content of the solution would be $\sim 10^{-25}$ ppm. Therefore, if oxygen is present, as it is in most weathering environments, and if the pH is in the normal range of 4–8, iron is immobile and iron oxyhydroxides and iron oxides will build up and be concentrated in the weathering residue as other elements are dissolved away (Figure 3.9). These iron minerals give oxidized soils their yellow and red hues, and only a

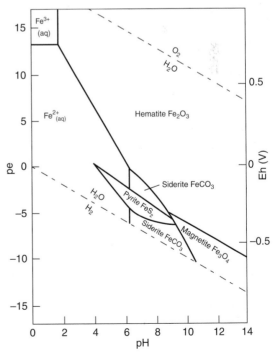

FIGURE 3.9. A pe–pH diagram showing the stability relations of some of the minerals and phases in the system $Fe-O-H_2O-S-CO_2$ at 25 °C, total $S = 10^{-6}$ and total $CO_2 = 10^0$. From Drever (1997).

few percent iron is necessary for color. However, at low pH (<2), appreciable amounts of iron can be in solution, even in fully oxygenated settings (Figure 3.9). For example, there are hot springs in New Zealand with extremely acidic pH of around 1 that have 10,000 ppm ferric iron in solution.

If there is no free oxygen available (pe < 0), iron can go into solution as ferrous iron (Fe^{2+}) and iron sulfides and iron carbonates can be precipitated (Figure 3.9). Although such anoxic conditions are not common in weathering profiles, they are common in the sediments at the bottoms of lakes, ponds, wetlands, bays, estuaries, and the ocean. In these environments, the reaction

$$Fe_2O_{3(s)}^{\text{hematite}} + 6H^+{}_{(aq)} + 2e^- \Leftrightarrow 2Fe^{2+}{}_{(aq)} + 3H_2O_{(1)} \tag{3.62}$$

becomes important. Since hydrogen ions are involved in this oxidation–reduction reaction, the pH must also be fixed. Where the sediments at the bottom of a water column have only a few percent organic matter, the oxygen in the upper few millimeters to meters of sediment is consumed by aerobic bacterial respiration. Where the free oxygen is gone, anaerobic bacteria with a variety of metabolisms can degrade the remaining organic matter (Chapter 2). The most common, and best-known, anaerobic bacteria are the sulfate-reducing bacteria, that reduce sulfate in the water to sulfide, in the process using the oxygen from the sulfate ion as the electron acceptor in their respiration:

$$Fe_2O_{3(s)}^{\text{hematite}} + 4SO_4^{2-}{}_{(aq)} + 8H^+{}_{(aq)} \Leftrightarrow 2FeS_{2(s)}^{\text{pyrite}}$$
$$+ \frac{15}{2}O^2{}_{(g)} + 4H_2O_{(1)} \tag{3.63}$$

Under anoxic conditions, up to 80 ppm of ferrous iron can be put into solution before the iron combines with the reduced sulfur to form amorphous iron sulfides (such as greigite and mackinawite) that eventually crystallize into pyrite and allied sulfides. These reactions account for the common occurrence of disseminated pyrite in mudstones.

Incongruent hydrolysis and oxidation–reduction chemical reactions can be written for most of the common aluminosilicate minerals that comprise the lithosphere: feldspars, amphiboles, pyroxenes, and micas. Where oxygen is available and pH is within a normal range, clay minerals and iron oxyhydroxides gradually replace these minerals as weathering proceeds, and released cations (Ca^{2+}, K^+, Mg^{2+}, Na^+) enter the groundwater. The chemical weathering reactions that affect granite can be summarized as follows:

potassium feldspar $\rightarrow$ clay minerals: K^+ and $H_4SiO_4^0$ in solution

sodium-rich plagioclase $\rightarrow$ clay minerals: Na^+, Ca^{2+}, and $H_4SiO_4^0$ in solution

muscovite $\rightarrow$ muscovite

quartz $\rightarrow$ quartz

biotite $\rightarrow$ clay minerals + iron oxyhydroxides: Mg^{2+} and $H_4SiO_4^0$ in solution

hornblende $\rightarrow$ clay minerals + iron oxyhydroxides: Mg^{2+} and $H_4SiO_4^0$ in solution

Feldspars and other aluminosilicate minerals can be released by weathering before they are totally altered. However, the incongruent weathering of feldspars accounts for the fact that, whereas feldspars are the most common minerals in igneous and metamorphic rocks, shales and mudstones (rocks composed of clay minerals) are the most common sedimentary rocks, comprising 60%–70% by volume of all sedimentary rocks. Quartz and muscovite are also common in sedimentary rocks because they are resistant to chemical weathering. However, under the most extreme tropical weathering conditions (high rainfall, high temperature, low slopes) even quartz and micas will eventually weather and the soils will end up being a thick accumulation of the mineral gibbsite ($Al(OH)_3$) or, more rarely, iron oxides. These extremely weathered residues may be unconsolidated or have indurated, pisolitic textures. Extremely weathered gibbsite accumulations are the main source of *bauxite* ores for aluminum, and weathered portions of iron formations provide iron-enriched, silica-depleted iron ores.

Rates of chemical reactions

Equilibrium thermodynamics predicts whether a reaction is possible for a given set of temperature, pressure, and compositional conditions, but tells nothing about the *rates* of reactions. Rates of reactions are the province of kinetics, a complicated subject beyond the scope of this book. The rates of chemical weathering reactions are assessed by empirical observations of soils and dissolution experiments in laboratories. Soil scientists developed a number of empirical schemes to describe relative mineral stability. One of the first such schemes was produced by Goldrich (1938), who looked at weathering of an Archean gneiss in southern Minnesota (Figure 3.10). In this case, there was a rapid loss of biotite and plagioclase feldspar, and a slower removal of quartz and potassium feldspar, but, in the most weathered places, potassium feldspar was gone and quartz, magnetite, and ilmenite remained (Faure 1991). These mineralogical changes were accompanied by changes in the chemistry of the residue, with rapid and complete removal of the calcium and sodium, and more gradual removal of the magnesium and potassium. The *weathering series* that Goldrich (1938) proposed amounts to Bowen's reaction series in reverse (Figure 3.11). Another relative-weatherability scheme from Berner and Berner (1996) is given in Table 3.2. Some aluminosilicates are notably resistant to weathering, including epidote, tourmaline, zircon, and rutile. The oldest known minerals on Earth are zircon crystals that are 4.4 billion years old. There have been many

experimental studies of rates and mechanisms of mineral dissolution (e.g., Wollast, 1976; Berner and Holdren, 1977, 1979; Berner *et al.*, 1985; Drever, 1994), which recently were reviewed by Kump *et al.* (2000), who compiled a data base of rate constants for

TABLE 3.2. Mineral susceptibility to chemical weathering according to Berner and Berner (1996) (decreasing from top to bottom)

Halite
Gypsum-anhydrite
Pyrite
Calcite
Dolomite
Volcanic glass
Olivine
Ca-plagioclase
Pyroxenes
Ca–Na plagioclase
Amphiboles
Na-plagioclase
Biotite
K-feldspar
Muscovite
Smectite
Quartz
Kaolinite
Gibbsite, hematite, goethite

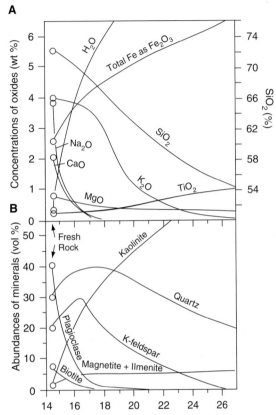

FIGURE 3.10. Faure's (1991) plots of data from Goldrich (1938), showing the effects of progressive weathering of granitic gneiss in Minnesota. Increased weathering is indicated by an increased Al_2O_3 content (the x-axis). (A) Variation in chemical composition reported as oxides; the scale on the right is for SiO_2, the scale on the left for remaining oxides. (B) Variation in mineral content with increased weathering.

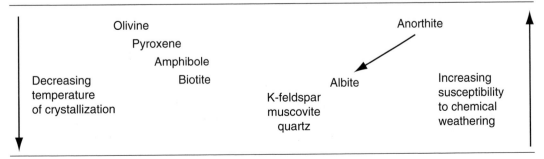

FIGURE 3.11. Bowen's reaction series and mineral susceptibility to chemical weathering.

TABLE 3.3. Lifetime of a theoretical 1-mm-diameter crystal at pH 5, at 25 °C, in dilute aqueous solution (from Kump et al., 2000)

Mineral	Lifetime (years)
Anorthite	10^2
Forsterite	$10^{3.5}$
Diopside	10^4
Albite	10^6
Microcline	10^6
Muscovite	$10^{6.5}$
Quartz	$10^{7.5}$

silicate dissolution. Their compilations show that the "Goldrich" series is generally valid (Table 3.3) but the rates of dissolution of silicate minerals in detail depend not only on the mineral composition, but also on the mechanism of dissolution, temperature of weathering solutions, presence of electrolytes and organic solutions, pH, supply of water, chemical affinity, surface area, and biological factors.

Chemistry of weathering-reaction water

Table 3.4 gives an average composition for rainwater, compositions of some typical meteoric groundwaters that have been through chemical and biochemical weathering, and the average composition of world river water. Whereas rainwater typically has ~10 ppm dissolved material (see Table 3.1), groundwater that has been through chemical weathering reactions has 100–1,000 ppm of dissolved solids. The composition of groundwater is dependent on the suite of minerals reacting, i.e., the composition of the bedrock (Table 3.5). Under natural conditions, the main cations are Na^+, Ca^{2+}, Mg^{2+}, and K^+. The main anion is usually HCO_3^-, with much lesser amounts of SO_4^{2-} and Cl^-. However, acid pollution in particular has increased the amount of sulfate in some groundwater and river water. There are thermodynamic models that can fairly accurately predict the composition of groundwater given the mineralogy of the weathered material.

Surface water tends to have lesser concentrations of solutes than groundwater, mainly due to the fact that most groundwater has completed the chemical weathering regime outlined above. Groundwater can become more concentrated the longer it remains in the subsurface. Overland flow does not completely react with Earth surface materials and generally has lower concentrations of dissolved solutes. Concentrations of solutes in river water depend on the relative contribution of shallow groundwater, deep groundwater, and overland flow. During floods, the percentage of groundwater is reduced and rivers have low concentrations of dissolved solutes. At low flow, the percentage of groundwater is increased and rivers tend to have higher concentrations of dissolved solutes. Observations of the Amazon watershed (Stallard and Edmond, 1981, 1983, 1987; Kump et al., 2000) led to the concept of two end members for the concentration of solutes in water issuing from a particular watershed: transport-limited and weathering-limited regimes. The thick, cation-depleted soils of the Amazon jungle lowland actually supply very small quantities of dissolved cations, because chemical weathering in such tropical areas is limited by the thick noneroding soil mantle. Soil water flows in these materials and does not interact with the deep bedrock so, although the potential for weathering is high, chemical weathering rates are low and streams are dilute (the transport-limited regime). Most of the dissolved cations in the Amazon instead come from the headwaters where, paradoxically, soils are poorly developed and the erosion rate is high. In these weathering-limited watersheds, precipitation percolates quickly to the bedrock–soil interface and reacts with the minerals most susceptible to dissolution, producing streams with high concentrations of dissolved solutes. It is the most readily weathered minerals that come to dominate the cations in the stream chemistry.

Clay minerals

A number of different clay minerals form during chemical weathering of aluminosilicates, and clay minerals also form or are altered during diagenesis (Chapter 19). Clay mineralogy is complex, with different crystal structures and chemistry leading to a large variety of clay mineral types such as kaolinite, smectite, illite, chlorite, "mixed-layer" clays, and saponite to name a few. The structure of clays is reviewed here briefly. The basic building block of the clay minerals is the silica tetrahedron SiO_4 (Figure 3.12A), and sheets are formed where infinite arrays of silica tetrahedrons share oxygens as shown in Figure 3.12B. All the oxygen atoms not involved in forming the base of the sheet point in the same direction. This produces a Si : O ratio of 2 : 5 over the sheet. The second basic building block of common clay minerals is an octahedron composed of an aluminum atom surrounded by six OH groups (Figure 3.12C). These octahedrons can also be joined in

TABLE 3.4. Chemical analyses of meteoric surface waters in the USA in ppm (from Livingstone, 1963)

	Rainwater	Granite, RI	Basalt, ID	Quartz sandstone, AR	Lithic sandstone, Pa	Arkose, Ma	Shale, NY	Ocala limestone, FL	Niagara dolomite, NY	Schist, MD	Glacial outwash, NY	World river water
Cations												
Ca^{2+}	0	6.5	54	50	44	27	227	39	35	3.1	75	15
Mg^{2+}	0.2	2.6	20	6.0	11	10	29	15	33	1.2	18	4.1
Na^+	0.6	5.9	51	2.4	60	2.3	12	7.5	28	3.3	17	6.3
K^+	0.6	0.8	7.2	3.0	4.1	0.8	2.7	1.3	1.3	0.8	1.3	2.3
SiO_2	0	20	44	12	14	14	5.5	25	18	14	7.7	13.1
Al^{3+}	–	0	0.05	0.2	0	0.1	0	0.1	0.2	0	–	–
Fe^{3+}	–	0.19	0.03	0.06	1.3	0.08	3.5	0.17	0.39	0.16	0.43	0.67
Mn^{2+}	–	0	0	0	0	0.10	0.13	0	0.03	0.02	0.37	–
Cu^{2+}	–	0	–	–	0	0	0.02	0	0	0	–	–
Zn^{2+}	–	0.07	–	–	0	0	0	0.02	0	0.18	–	–
Anions												
HCO_3^-	3	38	242	184	327	80	288	196	241	21	237	58.4
SO_4^{3-}	1.6	0.9	61	2.1	22	31	439	1.8	88	1.2	54	11.2
Cl^-	0.2	5.0	46	1.8	4.4	5.6	24	9.8	1.0	2.4	34	7.8
F^-	–	0.5	0.4	0	0.2	0.1	0	0.5	0.9	0	0.2	–
NO_3^-	0.1	1.5	6.0	6.5	2.0	18	0.9	0.8	1.2	0.4	0	1
PO_4^{3-}	–	0	0	0	0	0.1	0	0.1	0	0.1	–	–
TDS	4.8	82	532	268	490	189	1030	297	448	48	323	120
pH	5.6	7.6	7.7	7.4	7.7	7.8	7.6	8.0	8.2	8.3	8.1	

TDS, total dissolved soilds.

TABLE 3.5. Garrels and McKenzie's (1971) mass-balance calculation for the Sierra Nevada watershed (in this case, weathering reactions completely account for the groundwater composition)

	Na^+	Ca^{2+}	Mg^{2+}	K^+	HCO_3^-	SO_4^{2-}	Cl^-	SiO_2
Initial concentration in springs								
ppm	6	10	2	1.6	55	2	0.6	25
millimoles	1.34	0.78	0.29	0.28	3.28	0.10	0.14	2.73
Step 1								
Minus concentration in snow	1.10	0.68	0.22	0.20	3.10	0.00	0.00	2.70
Step 2								
Change kaolinite back into plagioclase	0.00	0.00	0.22	0.20	0.64	0.00	0.00	0.50

$$1.23Al_2Si_2O_5(OH)_4 + 1.10Na^+ + 0.68Ca^{2+} + 2.44HCO_3^- + 2.20SiO_2 \rightarrow$$
$$1.77Na_{0.62}Ca_{0.68}Al_{1.38}Si_{2.62}O_8 + 2.44CO_2 + 3.67H_2O$$

	Na^+	Ca^{2+}	Mg^{2+}	K^+	HCO_3^-	SO_4^{2-}	Cl^-	SiO_2
Step 3								
Change kaolinite back into biotite	0.00	0.00	0.00	0.13	0.13	0.00	0.00	0.35

$$0.037Al_2Si_2O_5(OH)_4 + 0.073K^+ + 0.22Mg^{2+} + 0.15SiO_2 + 0.51HCO_3^- \rightarrow$$
$$0.073KMg_3Al_3O_{10}(OH)_2 + 0.51CO_2 + 0.26H_2O$$

	Na^+	Ca^{2+}	Mg^{2+}	K^+	HCO_3^-	SO_4^{2-}	Cl^-	SiO_2
Step 4								
Change kaolinite back into K-feldspar	0.00	0.00	0.00	0.00	0.00	0.00	0.00	0.12

$$1.23Al_2Si_2O_5(OH)_4 + 0.13K^+ + 0.13HCO_3^- + 0.26SiO_2$$
$$\rightarrow 0.13KAlSi_3O_8 + 0.13CO_2 + 1.195H_2O$$

Row 1: chemical analysis (ppm and millimoles) of groundwaters from ephemeral springs at the base of a small drainage basin in the Sierra Nevada underlain by a uniform granodiorite: (1) approximately 90% plagioclase feldspar of approximately 62 mol% Na and 38 mol% Ca; (2) approximately 6% K-feldspar; and (3) approximately 4% biotite. The soils in the drainage basin contained kaolinite as the clay mineral.

Row 2: Step 1, subtract a chemical analysis of melted snowpack from the groundwater analysis, thereby eliminating the chemical contribution of the initial precipitation to the groundwater. All the Cl^- and SO_4^{2-} in the groundwater were derived from sea-salt aerosols.

Row 3: Step 2, use the Na^+, Ca^{2+}, and appropriate HCO_3^- and silica to reconstitute an appropriate mass of plagioclase feldspar from kaolinite.

Row 4: Step 3, use the Mg^{2+} and appropriate amounts of K^+, SiO_2, and HCO_3^- to reconstitute an appropriate mass of biotite from kaolinite.

Row 5: Step 4, use the remaining potassium to make K-feldspar from kaolinite. At the end of the exercise, only a small amount of silica is unaccounted for. The back-reactions produced 1.77 moles of plagioclase with the appropriate stoichiometry, 0.073 moles of biotite, and 0.13 moles of K-feldspar. These are the proportions of these minerals found in the original granodiorite within the error limits of this study. Note also that CO_2 and H_2O are produced during the back-reactions.

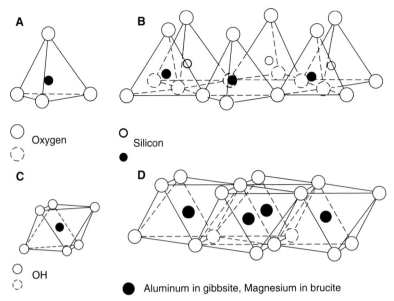

A

B

◯ Oxygen

◌

○ Silicon

●

C

D

◯ OH

◌

● Aluminum in gibbsite, Magnesium in brucite

FIGURE 3.12. Schematic diagrams of structural elements of clay minerals. (A) The Silica tetrahedron. (B) A sheet of tetrahedra with linked oxygen atoms. (C) An aluminum atom in an octahedron with six OH groups. (D) Sheet-linked octahedra – gibbsite or brucite.

a sheet (Figure 3.12D) to form the mineral *gibbsite*, $Al(OH)_3$. The mineral brucite, $Mg(OH)_2$, has the same structure as gibbsite but Mg^{2+} is the central cation in the octahedral layer instead of an aluminum ion.

The mineral *kaolinite* ($Al_2Si_2O_5(OH)_4$) belongs to a family of clays with a *two-layered structure* composed of alternating gibbsite and tetrahedral sheets (Figure 3.13A), with the apical oxygen atoms of the tetrahedral sheet taking up one third of the OH sites in the octahedral sheet. Sheets with this two-layer structure are held together by van der Waals bonds. Kaolinite has a number of polymorphs, including dickite, nacrite, and halloysite, that have slightly different crystal structures. So, although kaolinite is platy and can comprise fat books of amalgamated plates, halloysite forms tubular crystals. *Serpentine* is a two-layer clay where brucite octahedral layers replace the gibbsite octahedral layers.

Smectites have a three-layer structure with one octahedral layer sandwiched between two tetrahedral layers (Figure 3.13B). Smectites have two important properties: (1) ion-exchange capacity and (2) the ability to undergo expansion and contraction (swelling or bloating behavior). In smectites, Mg^{2+} and Fe^{2+} ions substitute into the octahedral layers for the Al ions, which have a charge of +3. The charge imbalance produced by this substitution of ions with charge +2 for a +3 ion is typically compensated for by Na^+ and Ca^{2+} ions that occupy *interlayer* sites (Figure 3.13B). These interlayer cations are *exchangeable*, which means that they can be moved in and out of the interlayer sites and can be exchanged with dissolved ions in ambient solutions. Thus smectites have a high ion-exchange capacity, similarly to the zeolites used in commercial water softeners. In addition, various numbers of water layers (up to ten) can pop in and out of the interlayer space, giving smectites their shrinking and swelling behavior. The amount of water in the structure responds to both temperature and the relative humidity (i.e., the activity) of water. Clay preparations must be dried or treated with ethylene glycol in order to identify with X-ray diffraction the clays capable of absorbing water into their structures. Finally, these low-temperature phases tend to have small crystals and not be ideal. The range in chemical composition of the smectites is illustrated by the chemical formula for montmorillonite, which is given by Deer *et al.* (1962) as

$$(\tfrac{1}{2}Ca,Na)_{0.7}(Al,Mg,Fe)_4[(Si,Al)_8O_{20}](OH)_4 \cdot nH_2O$$

Illite (also known as hydromica) is a potassium-rich three-layer clay with two tetrahedral sheets bounding an octahedral layer (Figure 3.13C). In illite, the potassium ions that make up the charge imbalance occur in interlayer positions. The chemical formula for illite is

$$K_{1-1.5}Al_4[Si_{6.5-7}Al_{1-1.5}O_{20}](OH)_4$$

Vermiculite is similar to illite, but Mg^{2+} ions occupy the interlayer sites. Chlorite,

$$(Mg,Al,Fe)_{12}[(Si,Al)_8O_{20}](OH)_{16}$$

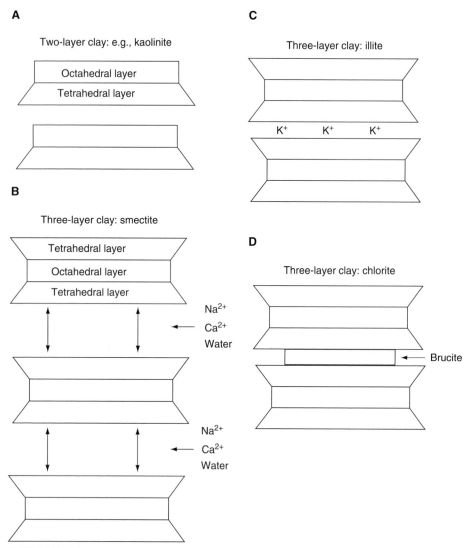

A

Two-layer clay: e.g., kaolinite

Octahedral layer
Tetrahedral layer

B

Three-layer clay: smectite

Tetrahedral layer
Octahedral layer
Tetrahedral layer

Na^{2+}
Ca^{2+}
Water

Na^{2+}
Ca^{2+}
Water

C

Three-layer clay: illite

K^+ K^+ K^+

D

Three-layer clay: chlorite

Brucite

FIGURE 3.13. Schematic representations of the structures and interlayer chemistry of various clay minerals: (A) two-layer clays such as kaolinite, (B) the three-layer clay smectite, (C) the three-layer clay illite, and (D) chlorite.

is similar to illite in that it has a three-layer structure, but an octahedral layer of brucite is sandwiched between two tetrahedral layers instead of cations (Figure 3.13D). Both illite and chlorite have low cation-exchange capability and little or no shrink–swell tendency. Finally, there is a large group of clay minerals called *mixed-layer* clays. These have a highly variable chemical composition and are composed of alternating layers of smectite, illite, and chlorite structural layers.

Clays that form at the low temperatures of weathering zones not only have variable chemical compositions, but also the ion substitutions mentioned above tend to deform the lattices of the clay minerals, such that they tend to have very tiny crystals (typically on the

order of 1 micron). Exactly which clay mineral forms in any particular weathering environment is still not entirely understood, but is apparently a complex function of rock composition, annual temperature, annual rainfall, drainage of the soil (sloping versus flat areas) time, vegetation, and, perhaps, soil microbes. Figure 3.14 (Drever, 1997) shows how the clay mineral compositions of the surface layers of soils from acid and basic igneous rocks from California vary with different amounts of precipitation. In general, illite is the most common clay and forms in most settings, whereas kaolinite and gibbsite tend to occur at low latitudes, and chlorite at high latitudes. Smectite mainly forms from volcanic glass. Hydraulic sorting and flocculation can

A

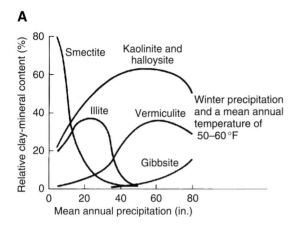

B

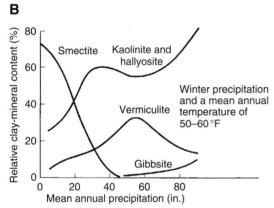

FIGURE 3.14. Clay minerals formed by weathering from acid (A) and basic (B) igneous rocks in California as a function of precipitation. From Drever (1997).

affect the compositional makeup of transported clays. Also, chemical reactions substantially alter clay minerals under increasing temperature and pressure (Chapter 19).

Soils

The weathered material at the Earth's surface is called soil, although some scientists will use this term only if the weathered material supports macroscopic plants. Modern soils are obviously important to humans, and have been studied and modified extensively (USDA Soil Survey Staff, 1993, 2003; Birkeland, 1999; Brady and Weil, 2002; Kaiser, 2004; Lal, 2004; McNeill and Winiwarter, 2004). Ancient soils (paleosols) are studied by Earth scientists because they are a direct record of Earth surface conditions (Retallack, 1997, 2001, 2003). The degree of development of soils and their properties depend on the parent material (mineralogy and degree of fracturing), the intensity of weathering (dependent on climate and vegetation), the rate of removal of

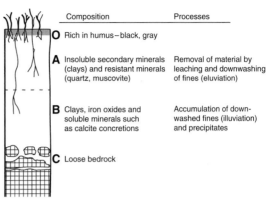

FIGURE 3.15. Horizons of a soil profile.

weathered material or addition of fresh material (dependent on factors such as slope, erosion, and deposition rate), and the length of time that weathering has acted on the Earth surface material. For example, steep, high mountain sides with poorly developed vegetation will not experience complete chemical weathering, and the incompletely weathered debris can easily be transported downslope by water flows and sediment gravity flows, thus leaving large exposures of bare rock. The lack of mature soil cover in these regions would be exacerbated in an arid climate. In contrast, lowland areas of low slope and abundant vegetation experience high rates of chemical weathering (especially in humid climates), and may also receive soil transported from higher regions.

Typical features of soils include: (1) layers differing in composition, texture, and structure, called *horizons*; (2) textural and chemical evidence of removal of soluble materials, precipitation of secondary minerals, and physical transport of clays and other fine grained colloidal particles (including iron oxyhydroxides) from percolating groundwater; (3) disruption of original structures by burrowing organisms, plant roots, change in moisture content, and growth of secondary minerals; and (4) characteristic coloration and mottling associated with chemical alteration of parent material and formation of new minerals. The thickness of well-developed (*mature*), horizonated soils is commonly decimeters to meters, and exceptionally tens of meters. These soil features are described below, and some of the rampant jargon developed by soil scientists for describing and classifying soils will be elucidated.

Soil horizons

Soils usually comprise several layers differing in composition, texture, and structure, called horizons (Figure 3.15

TABLE 3.6. Definition of soil horizons (Retallack, 2001, after USDA Soil Survey Staff)

O	Surface layer primarily composed of organic material with only a small amount of mineral matter. Black or dark brown in color. *Histic "epipedon"* has at least 12% original organic matter.
A	Uppermost horizon of soils lacking O horizon, composed of mineral matter with finely divided organic matter that produces a gray color. Minerals are clay, quartz, and other resistant minerals (depending on parent material) interpreted as evidence of chemical weathering, leaching, and elluviation of clays. Contains roots and burrows.
	Mollic "epipedon" usually has structure comprising granular clots of soil known as peds, dark color, and at least 1% organic matter – generally associated with grasslands.
	Umbric "epipedon" is similar to mollic epipedon, but has a platy to massive structure and is typically associated with forests.
	Ochric "epipedon" is light-colored and low in organic matter.
E	Underlies O or A horizon and is primarily composed of mineral grains resistant to weathering and too large to be moved as colloids. Light-colored because of relative lack of clay, sesquioxides (Fe_2O_3 and Al_2O_3), and organic material. Also known as *albic* horizon.
B	Underlies A or E horizon and is enriched in clays, oxides, and hydroxides of iron, manganese, and aluminum. These have been removed from overlying A and E horizons by (1) colloidal transport and redeposited as clay linings (cutans), and (2) solution transport and precipitated as void linings, void fillings, and concretions. Contains roots and burrows. Commonly strongly colored by iron oxides and hydroxides.
	Argillic horizon has more clay than A or E horizon or assumed parent material. Clay illuviation can be recognized by noting clay coatings of grains, clays lining or partly filling voids, and clay films on ped surfaces (illuviation cutans).
	Kandic horizon is similar to argillic, but the clay is kaolinite and there is little unweathered material.
	Calcic horizon is enriched in calcite or dolomite, is at least 0.15 m thick, and has at least 5% more carbonate than adjacent horizons. The carbonate is in the form of coatings, void linings and fillings, and nodules.
	Spodic horizon has concentration of organic matter and sesquioxides translocated from overlying E horizon.
	Oxic horizon is highly weathered, with hydrated oxides of iron (goethite) and aluminum (gibbsite).
C	Lowest, partially weathered horizon. Oxidation rinds on minerals. Possible parent material includes bedrock or sediments such as glacial outwash and till or alluvial or colluvial deposits.
K	Lowest, partially weathered horizon, cemented with calcium carbonate to form a hard pan. The upper part of this horizon is commonly laminated.
R	Unweathered material.

and Table 3.6). Horizons occur because weathering works its way downwards from the surface, such that the degree of weathering decreases downwards. Furthermore, some of the weathered material is moved downwards in water percolating towards the water table in solution (*leaching*) or as fine particles (*eluviation*), where it may accumulate by chemical precipitation or physical deposition (*illuviation*). Some of the details and terminology of the various horizons are given in Table 3.6. There are also subdivisions within horizons and gradational change between adjacent horizons (with additional terminology). Some of the descriptive terms require laboratory analysis in order to determine proportions of clay, calcium carbonate, organic matter, and so on. The definitions apply strictly to modern soils, but in many cases they can be applied to paleosols. However, diagenesis of paleosols may prevent strict adherence to the descriptive terminology used for modern soils.

The uppermost horizon is usually rich in organic matter and is designated O (Figures 3.15 and 3.16). In the underlying A horizon (Figure 3.15 and 3.16), chemical weathering is essentially complete, and (depending on the source rocks) the minerals left are quartz, muscovite, and clay. Most metal cations have been leached from the A horizon. Since soils commonly host plant

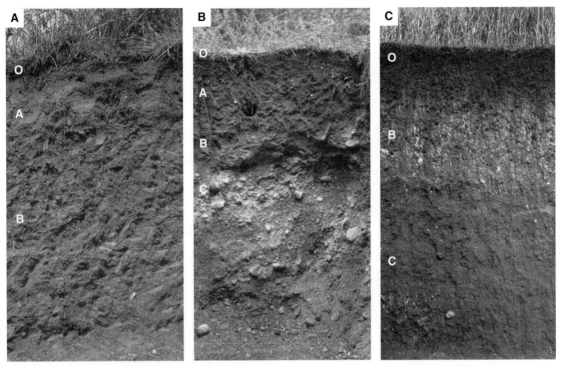

FIGURE 3.16. Examples of soil profiles (from USDA Soil Survey), see Table 3.6 for descriptions of labeled soil horizons and Table 3.8 for definitions of soil types: (A) ultisol, (B) inceptisol, and (C) mollisol – white blebs in B-horizon are calcium carbonate.

communities, partly to completely degraded plant remains (collectively referred to as *humus*) typically comprise up to a few tens of percent of the A horizon and give this horizon a darker color than those below. The humus content of the soil decreases with depth. In the B horizon, the weathering reactions are also more or less complete, but it is common to find iron oxyhydroxides that give a yellow or red coloration. Two mechanisms concentrate the oxyhydroxides in the B horizon. The Fe^{2+} ions have been dissolved by organic acids in the A horizon, and then leached into the B horizon, where they are precipitated. Decomposition of humus in the A horizon produces *chelating agents –* organic chemicals that bind to metals to produce soluble metal–organic complexes. These chelating agents bind iron and aluminum in particular, that are otherwise very insoluble, and promote their dissolution. Clays and colloidal particles of iron and aluminum are physically transported from the A horizon (eluviated) and deposited in the B horizon (illuviated). Ions (e.g., K^+, Mg^{2+}, Ca^{2+}, Na^+) released by chemical weathering and dissolved in the soil waters are picked up by plants. Phosphorus is another vital nutrient. Nitrates are also important nutrients but do not

generally come from weathering reactions. Rather, bacteria can take N_2 from soil gases and "fix" it as organic compounds such as ammonia. In the C horizon, the parent material is only partly decomposed and commonly the original structures of the rock or sediment are apparent. However, the minerals in the C horizon commonly exhibit oxidation rinds. The R horizon is the unweathered parent material.

Soil structure

Terms used to describe the structure of soil have arisen largely from the work of Brewer (1976; reviewed by Retallack, 2001), but other schemes are available (Bullock *et al.*, 1988; Fitzpatrick, 1984). Characteristic structural features of soils are *peds, cutans, glaebules, pedotubules*, void fillings, and orientation of clay minerals as seen under the microscope.

Peds are clumps of soil separated by planes of weakness such as shrinkage cracks, roots, and burrows. Peds vary in shape and size according to the parent material, the position in a soil profile, and the type of soil-forming processes. For example, platy peds are oriented parallel to the land surface and reflect soil modification

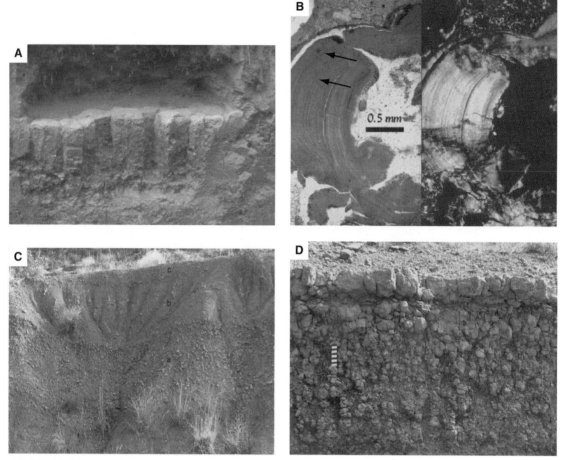

FIGURE 3.17. Structures of soils. (A) Blocky peds (from USDA Soil Survey Staff). (B) Well-developed clay cutan. From Brady and Weil (2002). Individual layers (arrows) are composed of oriented illuviated clays, seen under plane-polarized light (left) and crossed polarizers (right). (C) Nodular calcareous B horizon (a) overlain by non-calcareous mottled A horizon (b), overlain by stratified non-pedogenic deposits (c). The soil profile is 2–3 m thick. (D) Interconnected calcareous glaebules in a B horizon. The scale bar is 0.1 m long. Images (C) and (D) are from the Siwaliks of Pakistan, courtesy of Mike Zaleha.

of primary stratification. Prismatic, columnar peds (Figure 3.17A) are bounded by elongated vertical mudcracks (so-called prism cracks) and reflect shrinking and swelling of clays due to drying and wetting, primarily in B horizons. Blocky peds are also formed in this way, but indicate the presence of more roots and burrows. Granular and crumb-like peds are an indication of very active bioturbation in A horizons and eluviation of clays, sesquioxides, and organic matter. Peds range from angular to rounded, depending partly on the degree of movement of water around their edges.

Cutans are surfaces within a soil, such as the margins of peds or glaebules, that have been altered in some way. Clay cutans (*argillans*) are most common, and are formed either by washing clay into the cracks between peds or other open voids (illuviation cutans –

Figure 3.17B), or by compressive orientation of clays due to swelling (stress cutans). The clays of illuviated cutans can be oriented along surfaces by surface tension of the soil water. Stress cutans may have shiny, slickensided surfaces as a result of translation between adjacent peds. These slickensides may be inclined relative to the land surface and define anticlines and synclines (referred to as *pseudoanticlines*), the surface expression of which is polygonal surface ridges called *gilgae*. Such features occur in soils with abundant swelling clays (smectite) and in climates with extremes of wetting and drying. However, they may also be associated with expansion of a soil due to precipitation of secondary minerals as glaebules.

Glaebules are concentrations of mineral matter such as nodules and concretions (Figure 3.17).

Nodules and concretions have well-defined, smooth or rough boundaries, whereas color mottles are glaebules with diffuse boundaries. Glaebules are near-spherical, ellipsoidal, or tubular, and are millimeters to centimeters in diameter. They are commonly composed of calcium carbonate (micrite and calcite spar), siderite, pyrite, or amorphous iron oxide. Nodules have no internal structure, whereas concretions have concentric banding indicative of periodic addition of material of different composition or texture. *Septaria* are nodules or concretions with radiating and concentric cracks resulting from shrinkage or hydraulic fracturing.

Calcium carbonate glaebules occur when evaporation exceeds precipitation (i.e., in seasonally dry climates). Groundwater in the capillary fringe above the water table becomes saturated with respect to calcium carbonate during evaporation or evapotranspiration. Precipitation is initially in the form of micritic linings and fillings of pores and voids, such as roots and burrows. Subsequently, the calcium carbonate becomes replacive and displacive of the parent material. Displacive growth of calcium carbonate results in expansion of the soil, and, together with expansion associated with wetting and drying of expandable clays, may be responsible for the formation of slickensides, pseudoanticlines, and gilgae relief. As nodules and concretions grow in size and volumetric proportion in the soil, they may join together to form nodular layers (Figure 3.17). These nodular layers may then develop laminated tops and contain pisoliths (also see Chapter 11). Such well-developed soil carbonate layers are commonly called *caliche*, *calcrete*, or *kankar*. The formation of calcium carbonate coatings and small nodules takes on the order of 10^3 years, that of well-developed nodular horizons requires on the order of 10^4 years, and continuous nodular horizons with laminated crusts and pisoliths require on the order of 10^5 years to form (Table 3.7). However, there is much variability in the length of time needed to develop calcareous horizons, dependent upon groundwater composition (parent materials), type of vegetation, and climate. In seasonally wet and dry climates, calcium carbonate glaebules may co-precipitate with glaebules (nodules, concretions) of amorphous hydrated iron oxide. Calcium carbonate and iron oxide may precipitate alternately in the same concretion. The calcium carbonate precipitates during extremes of dryness and alkalinity, whereas the iron oxide precipitates under somewhat wetter and lower-pH conditions. Siderite and pyrite glaebules form under waterlogged (i.e., permanently

saturated) conditions with negative pe (reducing) and near neutral pH.

Duricrusts are hard layers within highly weathered soils that require very long periods of stable land surface for their formation. Duricrusts include *laterite*, *bauxite*, and *silcrete*. Laterite is rich in hematite, but also contains gibbsite and kaolinite. Laterite occurs beneath forests and woodlands in tropical areas of high rainfall, but with a warm dry season. Bauxite is similar to laterite but is composed mainly of gibbsite. The formation of bauxite requires more intense, continuous rainfall than does that of laterite; however, there is much uncertainty about the origin of laterite and bauxite. The silica that is precipitated as silcrete comes from intense chemical weathering of terrigenous material or from devitrification of volcanic glass. Silcretes can form under highly alkaline, evaporative conditions, such as in arid areas, where they may be associated with calcium carbonate and evaporites. They may also form around volcanic hot springs.

Color mottles are diffuse patches of contrasting color, such as red, yellow, brown, greenish gray, and bluish gray. Such colors are related mainly to the presence or absence of iron-bearing minerals and the redox state of such minerals. Very pale yellow colors normally indicate the absence of iron-bearing minerals, due for instance to leaching of clays or extreme chemical weathering of ferromagnesian minerals. Yellow and brown colors normally indicate the presence of oxidized iron minerals such as goethite and limonite. Strong red colors indicate the presence of hematite. Abundant hematite normally indicates diagenesis, as discussed below. Drab gray colors indicate the presence of reduced iron minerals and abundant organic matter.

Pedotubules are tubes that cannot be recognized specifically as root casts or burrows. There are various other kinds of spaces in soils, including vertical and horizontal cracks (e.g., desiccation cracks) and dissolution vugs. These voids may be filled with local sediment grains and/or different generations and compositions of precipitates, ranging from amorphous linings of clays and oxyhydroxides to crystalline cements.

Microfabrics of soils can be seen only under the microscope. The fine-grained material in peds (clays and amorphous iron oxide coatings) is referred to as *plasma*. The different types of microfabrics in soils are based on the proportion and nature of the plasma. Streaks of variably oriented, highly birefringent clay grains are classified into a number of different *sepic plasmic* fabrics (Brewer, 1976; Retallack, 2001). These fabrics are formed by pressures exerted during compaction and wetting of clays.

Organic features of soils

In addition to the obvious macroscopic plants, moles, voles, and earthworms, soils host intricate ecosystems dominated by microorganisms such as fungi and bacteria (Paul and Clark, 1989; Brady and Weil, 2002; Jahren, 2004; Pennisi, 2004; Wardle *et al.*, 2004; Young and Crawford, 2004). The role that macroscopic and microscopic organisms play in soil development is an active area of research and is briefly reviewed here. Roots and rhizomes can be preserved in soils as original organic material, but more commonly as molds or casts. Root casts are commonly composed of clay, calcium carbonate, or chert. Root casts can normally be distinguished from other tube-like soil structures by virtue of their downward and lateral branching into tubes of smaller diameter (see also Chapter 11). The various shapes and sizes of roots are related to the evolution of land plants and to environmental conditions (Retallack, 1997, 2001; Driese *et al.*, 2000; Driese and Mora, 2001). For example, as land plants developed in size and diversity through the Silurian and Devonian periods, their rhizome and root systems progressively increased in depth, size, and complexity. In modern plants, tabular root systems with shallow roots spreading out laterally are typical of high water tables. In contrast, deeply penetrating and highly branched roots are typical of low water tables. Plants in seasonally wet and dry climates may have both deep and shallow roots.

The zone around roots that is biochemically influenced by the plant and associated bacteria and fungi is called a *rhizosphere*. One element of the rhizosphere comprises various specific mutualistic relationships between plant roots and fungi known as *mychorrhizae*. Organic and carbonic acids in the rhizosphere release cations such as Mg^{2+}, K^+, Na^+, Ca^{2+}, Mn^{2+}, and Fe^{2+} from the parent soil material for use by the plant. However, the rhizosphere is not always acidic, and is commonly near neutral in terms of pH and pe. In some climates, there may be cycles of wet and acidic conditions and of dry and alkaline conditions. Such conditions may give rise to precipitation of minerals around roots known as *rhizoconcretions* or *rhizoliths*. Under dry (evaporation exceeds precipitation) and alkaline conditions, calcium carbonate may be precipitated, whereas hydrated iron oxides (e.g., goethite) may be

precipitated under wet and acidic conditions. In permanently waterlogged soils (e.g., marshes and swamps), conditions in the rhizosphere are reducing, and drab bluish and greenish gray haloes form around roots. Such gray and organic-rich soils are called *gley* soils. Mobilization of iron and manganese under these conditions may result in precipitation of nodules of siderite, pyrite, or iron–manganese hydroxide. When a root dies and rots, the interior of a rhizoconcretion once occupied by the root may be filled with illuviated clay or a crystalline cement such as calcite.

Other evidence of organic activity associated with soils and the overlying land surface is hard parts of animals and plants such as vertebrate bones (calcium phosphate), shells of molluscs such as bivalves and gastropods (calcium carbonate), organic material such as plant axes, leaves, pollen, and spores, mineralized organic material, and trace fossils such as burrows, tracks, trails, and coprolites (Behrensmeyer *et al.*, 1992; Retallack, 2001). Fossils in paleosols commonly occur where the organisms lived or died, and are therefore useful for reconstructing ancient terrestrial communities. Furthermore, physical and chemical features of the paleosols that indicate environmental conditions, such as climate, topography, and vegetation cover, may be relevant to interpreting the ecology of organisms preserved as fossils.

Bone and calcareous shells can be preserved under alkaline conditions (high pH), but not under acidic conditions. Siliceous material is stable under most pe and pH conditions, but dissolves under highly alkaline conditions (pH > 9). Organic matter is destroyed under oxidizing conditions (positive pe), in the presence of aerobic microbes such as bacteria and fungi, and under highly alkaline conditions, but can be preserved under reducing conditions. Some organic remains are preserved by mineralization (such as *phytoliths* of calcium carbonate or opaline silica, fungi preserved in coal or calcium carbonate, and insects encased in amber). Indirect evidence for the activities of microbes (e.g., bacteria, algae, fungi) is in the precipitation of characteristic mineral structures (e.g., calcite laminae in nodules and crusts, pyrite framboids), in their soil-binding effects, and in the decomposition of organic matter.

A variety of burrow types occurs in soils, attributable to a large range of invertebrates (worms, molluscs, insects, arthropods) and vertebrates (e.g., rodents) (see Chapter 11). Burrows are commonly filled with sediment derived from the surrounding soil, including fecal pellets made by the burrower. Burrows and burrow fills

TABLE 3.7. Stages in development of soils

1. Little evidence of soil formation, no horizons, little disruption of original fabric, some roots, no glaebules. *Very weakly developed.* Order of 10^2 years.
2. Incipient horizons, surface rooted zone, incipient chemical and physical alteration such as leaching, color mottling, precipitation of secondary minerals, clay translocation, and ped formation. Surface peaty (O) horizon up to decimeters thick. *Weakly developed.* Order of 10^3 years.
3. Well-defined horizons with chemical and physical alteration: peds, cutans, sepic plasmic fabric, glaebules. *Moderately developed.* Order of 10^4 years.
4. Thick, red, clayey B horizon, with well-developed nodular layers of calcium carbonate, or hydrated iron–aluminum oxides and organic matter (spodic horizons), or O horizons (meters to tens of meters thick). *Strongly developed.* Order of 10^5 years.
5. Exceptionally thick A (leached) and B (including spodic) horizons, or O horizon (tens of meters thick), or calcium carbonate horizons with laminated crusts and pisoliths. *Very strongly developed.* Order of 10^6 years.

may act as conduits for groundwater movement, and thereby contain concentrations of illuvial clay or chemical precipitates such as calcium carbonate. Trackways can yield information on the mode of locomotion and activities of the organism responsible. Reconstruction of dinosaur locomotion and life activities from trackways has received much attention recently (e.g., Lockley, 1991; Lockley and Hunt, 1995).

Stages in development of soils

The rate and degree of formation of these soil features are controlled by factors such as deposition rate, parent materials, groundwater composition, climate, and vegetation. Table 3.7 (adapted from Retallack, 2001, and Kraus and Brown, 1986) summarizes the stages in the development of soils, with approximate time spans required to reach these stages. It is generally accepted that well-developed soils represent long periods of low deposition rate (e.g., less than 1 mm yr^{-1} for more than the order of 10^3 years).

Lateral and vertical variation in thickness and type of soils

Soils commonly vary laterally in thickness, and in the nature and development of their various features. Such lateral variations may be associated with lateral variations in deposition rate, parent sediment type, vegetation, or soil moisture (related to height above the water table). Lateral variation of different types of soils beneath a land surface is called a *catena*. A common

type of catena in floodplains is associated with the decrease in elevation of the floodplain surface and in the grain size of floodplain sediment with distance from main channels (Chapter 13). Soils developed on the elevated, sandy, and well-drained sediments of levees and crevasse splays tend to have a relatively thick oxidized and leached zone (developed in the zone of aeration) underlain by a gleyed horizon (developed within the zone of intermittent and permanent saturation). The lower, muddy, and poorly drained deposits of floodbasins are more gleyed. In soils that undergo extremes of wetting and drying, calcium carbonate may be leached from the zone of aeration but accumulated in the capillary fringe above the groundwater table, at the fringe of the root zone, or where the maximum respiration of soil organisms takes place (Retallack, 2005). Poorly drained soils in low-lying areas may be more or less permanently saturated, allowing co-precipitation of nodular calcium carbonate and iron–manganese hydroxide throughout. These types of soils may also be associated with calcareous muds deposited in marshy settings. It has also been suggested that soils may become relatively more mature with distance from channel belts, because of an inferred decrease in deposition rate. Vertical variations in the nature of individual soils or groups of soils in floodplain deposits have been related to their lateral variability in catenas, to variations in floodplain deposition rate as channel belts move episodically over floodplains, and to changes in sediment type, deposition rate, and weathering processes due to tectonism, climate, or base-level change (see Chapters 13 and 20).

TABLE 3.8. Soil orders and their properties (from Faure 1991; Retallack 1997)

Entisol – early stage of soil development composed of A and C or A and R horizons – no B horizon development.

Inceptisol – young soil with some leaching in the A horizon but a weakly developed B horizon.

Andisol – like inceptisol but formed on volcanic ash.

Vertisol – abundant swelling clay (mainly smectite) subjected to seasonal extremes of wetting and drying (expansion and shrinkage), commonly has blocky to columnar ped structures due to mudcracks and prism cracks, cutans, slickensides, pseudoanticlines, and calcareous and iron oxide nodules ("glaebules").

Aridisol – desert soil, light color due to low prevalence of organic matter. Commonly with calcareous caliche layers within a meter of the surface or in-place-precipitated evaporite minerals such as gypsum. May contain silcrete or fericrete horizons – layers cemented by iron or silica minerals.

Mollisol – thick, organic-rich A horizon that is soft when dry, with abundant roots and burrows, and B horizon enriched in clay or calcium carbonate. Grassland soil.

Histosol – thick O horizon (peat) formed by accumulation of organic matter that fails to decompose. Uncompacted thickness of O horizon at least 0.4 m. Swamp soil.

Spodosol – composed of thick, well-differentiated A and E horizons with B horizon enriched in Al and Fe oxides and oxyhydroxides that form a hardpan. B horizon with little or no clay or calcium carbonate. Forest and woodland soil.

Alfisol – strongly leached acidic A horizon containing decomposing organic matter. Thick, well-differentiated horizons with B horizon enriched in clay, red sesquioxides. Forest and woodland soil.

Ultisol – similar to alfisol but more highly weathered, with well-developed, clay-rich B horizon. May contain laterite or bauxite horizons. Forest and woodland soil.

Oxisol – thick, well-differentiated clayey soil, highly oxidized and red lower horizons, clay in the B horizon decomposed to oxides of Al and Fe. May contain laterite or bauxite horizons. Rainforest soil.

Classification of soil types (pedotypes)

The soil classification of the US Soil Conservation Service is shown in Table 3.8 and Figure 3.18 (from Retallack, 1997). There are many other types of soil classification. The soil types (*pedotypes*) in Table 3.8 and Figure 3.18 are defined on the basis of criteria such as the degree of physical disruption, the abundance in various parts of the soil profile of secondary minerals such as clay, calcium carbonate, hydrated Fe and Al oxides (sesquioxides), and evaporite minerals, and the abundance of organic material. These features are related to time stage in development, source materials, groundwater level and composition, and climate. For example, in some *vertisols* the evidence of wetting and drying of clays, and the precipitation of both calcium carbonate and iron oxide glaebules, is an indication of extremes of temperature and moisture content in a seasonal climate. The depth to the carbonate concretionary layer increases with annual rainfall. These soils do not necessarily indicate semi-arid conditions, as is commonly believed. *Entisols* and *inceptisols* indicate immaturity of soils. *Aridisols* indicate an arid climate. *Histosols* occur in areas where the soil remains wet (precipitation exceeds evaporation throughout the year) so that surface organic matter is not oxidized. Such soils occur in a wide range of latitudes and climatic zones, but not in arid areas. *Ultisols* and *oxisols* are mature soils that occur mainly in wet tropical climates where chemical weathering is extreme. Sedimentary units with distinctive, observable features that resulted from soil modifications are known as *pedofacies* and may comprise multiple successive pedotypes.

Diagenesis of soils

Study of modern soils has aided greatly in the interpretation of *paleosols*. However, paleosols may have features that are not present in modern soils because terrestrial environments in the past may have been different from those in the present, especially in terms of the biosphere and atmosphere. Furthermore, diagenesis of soils following burial affects the original composition, texture, and structure of a soil. Diagenetic processes (Chapter 19) include alteration of organic matter, dehydration and recrystallization of hydrated iron oxides, cementation, recrystallization

FIGURE 3.18. Orders of soils of the US Conservation Service, with associated climate (Sun/cloud symbols), time of formation (longer as more sand has passed through the hourglass), vegetation, and types of horizons. From Retallack (1997).

of calcium carbonate, compaction, and illitization of smectite clays. In view of diagenesis of paleosols, an alternative scheme of classification may be appropriate for paleosols (e.g., Mack *et al.*, 1993; Mack and James, 1994). Another non-genetic classification is field designation of pedotypes.

Organic matter can be decomposed in an aerobic burial environment, especially in the presence of aerobic microbes. Gleying can also occur in a burial environment as anaerobic bacteria consume organic matter and reduce iron oxides and hydroxides. Burial reduction is likely if there is not other evidence for waterlogging of the original soil. The burial compaction and heating of peat layers gives rise to coal, and to the carbonization of pollen and spores. Decarboxylation of buried organic matter (kerogen), its maturation to oil, and ultimate cracking to natural gas occur at depths of burial from 1 to 5 km, and temperatures of 50–150 °C. Acidic solutions present during this process may lead to creation of secondary porosity (e.g., oversized vugs cutting across primary fabrics and containing deep-burial cements).

The bright red colors of paleosols are associated with hematite, which forms by dehydration and recrystallization of iron oxides such as goethite. Hematite forms at various stages of burial diagenesis, and may take on different magnetic orientations as revealed by stepwise demagnetization. Magnetization removed at temperatures less than 300 °C is oriented in the modern-day magnetic field. Magnetization that is stable at higher temperature is associated with hematite that formed during pedogenesis and early burial, and is thought to be oriented in the magnetic field prevailing at that time.

Common soil cements are calcite, dolomite, siderite, hematite, and silica. Some cements occur during pedogenesis, especially during the precipitation of void linings and fillings, and in the formation of glaebules and hardpans. Burial cements may be recognizable from their texture and composition. For example, deeper calcite cements tend to be relatively coarse-grained and occur as equant or drusy spar, or poikiloblastic. These cement textures are also those formed by the common process of recrystallization of calcium carbonate (neomorphism). The carbon isotopes and luminescence of calcite burial cements may be different from those of soil cements. Siderite and dolomite are more common as burial cements. The most widespread silica cements are likely to occur after burial during diagenesis of clays and pressure solution of quartz grains.

Compaction is most important in clays and peat. Evidence for compaction of clays is bedding-plane cleavage, slickensides, and differential compaction phenomena such as folds in vertical sandstone fills of tubes and cracks, breakage of shells and bones, and elliptical shapes of once-circular sections of burrows, nodules, and logs. Many clay-rich paleosols are illitic, whereas many soils contain illite or mixed-layer illite and smectite. Illitization of clay is a complicated process that occurs at depths of burial of 1.2–2.3 km (55–100 °C).

Interpretation of paleosols

The composition, texture, and structure of paleosols (i.e., pedofacies) can be compared with those of modern soils to yield information about past climate, vegetation, elevation of the land surface relative to the water table, parent materials, deposition rate on floodplains, and time of formation. However, interpretation of paleosols is complicated by virtue of the facts that several controlling factors may act together to produce a given soil feature, burial diagenesis may alter the appearance of a soil, and environmental conditions have evolved over time. There are, however, certain soil features or soil types that are useful indicators of atmospheric composition, paleoclimate, vegetation type, and topography relative to the groundwater table (Retallack, 1997, 2001).

Ratios of carbon and oxygen isotopes in calcareous paleosol concretions can indicate the nature of *vegetation* and *variations in atmospheric temperature* (Cerling, 1999). Ratios of stable carbon isotopes in paleosol carbonates can also indicate *carbon dioxide* levels in the paleoatmosphere, which can in turn be related to paleoclimate (Mora and Driese, 1993, 1999; Mora *et al.*, 1996; Cerling, 1999). However, great care must be taken in interpreting paleoatmospheric implications of stable-carbon-isotopic compositions, because such compositions are also influenced by depositional environment, soil temperature, soil parent material, and soil respiration rates, which particularly influence the pCO_2 of the soil gas. Air temperature is also indicated by periglacial soil features such as pingos and ice-wedge polygons (Chapter 17).

Mean annual rainfall can be related to the composition of paleosols. For example, the proportion of gibbsite, the kaolinite : smectite ratio, and the depth of calcic horizons (for mean annual rainfall < 1,000 mm) all increase as mean annual rainfall increases. Evaporites indicate very low rainfall relative to evaporation. Extreme seasonality of rainfall and temperature is indicated by the presence of vertic features, charcoal, and the co-occurrence of calcium carbonate and iron oxide glaebules, and of shallow and deep roots.

Rainforests have a tabular mat of large and small roots, and are commonly underlain by oxisols and ultisols. Forests and woodlands have deep roots of various sizes seated in alfisols, spodosols, and ultisols. Grasslands have abundant fine, near-surface roots and a few large deep roots, and are associated with mollisols or inceptisols. Desert scrub is associated with sparse and widely spaced, medium-to-large-sized deep roots, and the soils are typically aridisols, inceptisols or entisols. Marshes and swamps have fine to large roots in a tabular mat and are associated with histosols. The different types of vegetation in histosols can be deduced from the peat or coal macerals. The stages in development of soils must be borne in mind when making associations between vegetation type and soil type. Preserved plant remains of course give direct information on vegetation types and on climatic and drainage conditions.

Soils with deep water tables (uplands and well-drained lowlands) have deeply penetrating roots and burrows, and oxidized peds, cutans, and sepic plasmic fabrics. Intermittently waterlogged soils have shallow and deeply penetrating roots, mixed terrestrial and aquatic burrows, oxidized and reduced peds and cutans, sepic and asepic plasmic fabrics, and glaebules of iron oxide and calcium carbonate. Waterlogged lowland soils have a tabular root mat, mainly aquatic creatures, relict bedding because of reduced activity of organisms and lack of cracking, undulic and asepic plasmic fabric, peat, and siderite and pyrite glaebules and gleying due to an anaerobic environment.

Composition and texture of terrigenous grains

The texture (i.e., grain size and shape) and composition of terrigenous grains depend on (1) source rock; (2) chemical and physical alteration of source rocks by weathering in the source area and during transportation, and after deposition and burial (diagenesis); and (3) physical abrasion and breakage during sediment transport. Furthermore, many terrigenous grains have been through many cycles of weathering, erosion, transport, and deposition in different Earth surface environments. Under some circumstances, it is possible to link the textures and compositions of terrigenous grains to their original source rocks (their provenance) and to the processes of weathering, erosion, transport, deposition, and post-depositional alteration (diagenesis). This is a difficult task, as discussed below, and in books such as Folk (1965a), Pettijohn (1975), Blatt et al. (1980), and Blatt (1982). Photographic atlases of terrigenous sedimentary rocks include Adams et al. (1984) and Scholle (1979).

Texture

Chemical weathering completely dissolves some minerals and leaves others with weathering residue (e.g., clay minerals, iron oxyhydroxides) along planes that water can penetrate (crystal boundaries, cleavage and twin planes). The degree of alteration decreases inwards. This alteration process causes rounding of crystal edges and corners that are attacked chemically from two or three sides. Furthermore, growth of secondary minerals (clays) involves a volume increase that may

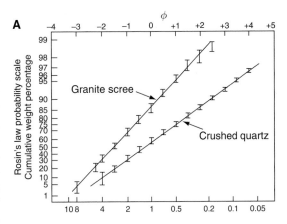

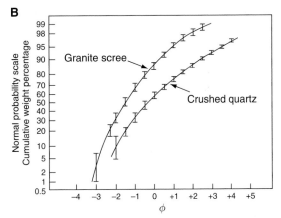

FIGURE 3.19. Comparisons of size distributions of crushed quartz and mineral grains taken from weathering profile developed on granite scree. From Blatt et al. (1980). (A) Cumulative weight percentages plotted against the probability scale of Rosin's law. (B) Cumulative weight percentages plotted against the normal probability scale. ϕ grain size scale is explained in the appendix.

lead to pushing whole crystals, or parts of them, outwards. Weathered residues from igneous rocks and gneisses follow the same size distribution (called Rosin's law) as that of mechanically crushed quartz (Figure 3.19) (Pettijohn, 1975; Blatt et al., 1980). Other processes that influence the shapes and sizes of terrigenous grains are discussed in Part 3, and methods of measuring and analyzing grain size and shape are described in the Appendix. Sediments and sedimentary rocks are commonly classified according to their mean grain size, and the relative proportions of clay, sand, and gravel (see the Appendix).

The chemical rounding process can also act on massive blocks of rock (typically basalts and granite) bounded by joint planes (Figure 3.20). Weathering fluid moves along joint planes and attacks the rock, initially producing a millimeter-scale rim of chemical

FIGURE 3.20. Spheroidal weathering of basalt.

alteration. Expansion of the secondary minerals allows water to penetrate a little further into the block, and weathering proceeds as before. This process leads to separation of thin layers of altered rock like the skins of an onion, and the degree of alteration of the skins increases outwards. The edges and corners of the joint blocks become rounded off, resulting in an unweathered core that is close to spheroidal in shape. Hence this mode of weathering is called *spheroidal weathering* or *exfoliation*.

Composition

The genetic information that can be interpreted from the composition of terrigenous grains in coarse-grained sedimentary rocks (sandstones and gravelstones) can be explained using two end-member compositions: (1) a large range of mineral types and rock fragments, and angular grain shapes; and (2) mainly well-rounded quartz and resistant accessory minerals, and a small amount of potassium feldspar. The first end member would be termed mineralogically *immature*, whereas the second would be termed *mature*. The immature mineralogy is likely to be similar to that of the source rocks, and would therefore be useful for provenance studies (see below, however). This mineralogy also suggests a minimal amount of chemical weathering, which could indicate any or all of the following: mainly physical weathering in a cold or dry environment; high altitude and slope; and rapid transport, deposition, and burial. It is not surprising, therefore, that this immature mineralogy might be associated with tectonically active areas with large rates of uplift of mountains and subsidence in adjacent sedimentary basins (e.g., Krynine, 1950). Indeed, the mineralogy of terrigenous grains has been related to specific plate tectonic settings (Dickinson, 1976, 1995). In addition, a lack of chemical

weathering should result in a lack of generation of new clay minerals.

The mature mineralogy suggests a large amount of chemical weathering that probably occurred over many sedimentary cycles. In other words, the most recent source rock was probably a terrigenous sedimentary rock. Evidence for recycled sedimentary rocks is sediment grains with old rounded cement attached to their perimeters. It is unlikely that the majority of mineral types in igneous or metamorphic rocks could be removed in one weathering, transport, and deposition cycle, unless chemical weathering was long-lived and extremely intense (e.g., in a wet, warm, lowland area). Such a weathering environment would produce a large amount of clay. Mature mineralogy does not carry the provenance information of immature mineralogy. However, the tectonic setting in which mature mineralogy might be found is a relatively stable, low-relief continental region, in which terrigenous sediment is repeatedly weathered, transported, deposited, then uplifted to start the cycle over again.

The composition of terrigenous sediments has also been related to specific depositional environments. For example, mature quartz sandstones are typically interpreted as beach or dune deposits (Folk, 1965a). It is generally not the case that composition is related to depositional environment. Modern beaches have a huge range of sediment compositions (carbonate shell fragments, volcanic glass, garnet), and only beaches that have a pure quartz sediment source are pure quartz.

Classification of terrigenous sandstones on the basis of composition

Indices of mineralogical maturity for the main constituents of terrigenous sedimentary rocks are quartz/feldspar or (quartz + chert)/(feldspar + rock fragments), and for accessory (heavy) minerals (zircon + tourmaline + rutile)/total accessory minerals. In fact, terrigenous sandstones are commonly classified on triangular diagrams that use quartz (or quartz + chert), feldspar, and (unstable) rock fragments as end members (Pettijohn *et al.*, 1973; Pettijohn, 1975; Appendix). Such classifications can be thought of as indicators of tectonic setting, insofar as tectonic settings strongly influence the nature of weathering, transport, deposition, burial, and reworking of sediment deposits. Some of these triangular classifications also contain a fourth parameter, the proportion of clay matrix. The boundary between matrix and framework grains is usually arbitrarily chosen at 40 microns (μm),

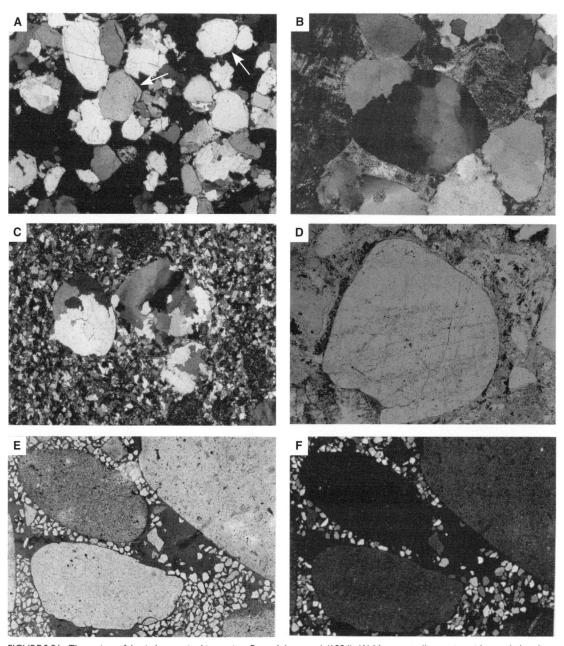

FIGURE 3.21. The variety of detrital quartz in thin section. From Adams *et al.* (1984). (A) Monocrystalline grains with rounded outlines demarcated by "dust rims" that have euhedral quartz overgrowths (arrows), seen under crossed polarizers. (B) Undulose extinction (crossed polarizers). (C) Three larger polycrystalline quartz grains surrounded by smaller quartz grains (crossed polarizers). (D) Inclusion-rich quartz (plane-polarized light). Large chert pebbles surrounded by smaller quartz grains seen under (E) plane-polarized light and (F) crossed polarizers.

although workers use boundaries from 30 to 62.5 μm. There is no hydraulic significance to this choice. Matrix is usually interpreted as fine-grained material *co-deposited* with the framework grains. However, most modern sands have very little matrix in them, strongly suggesting that matrix develops after deposition in most sedimentary rocks. Common mechanisms that produce "matrix" in sediments and sedimentary rocks include squashing mud pellets and soft rock fragments during compaction, bioturbation, and growth of clay

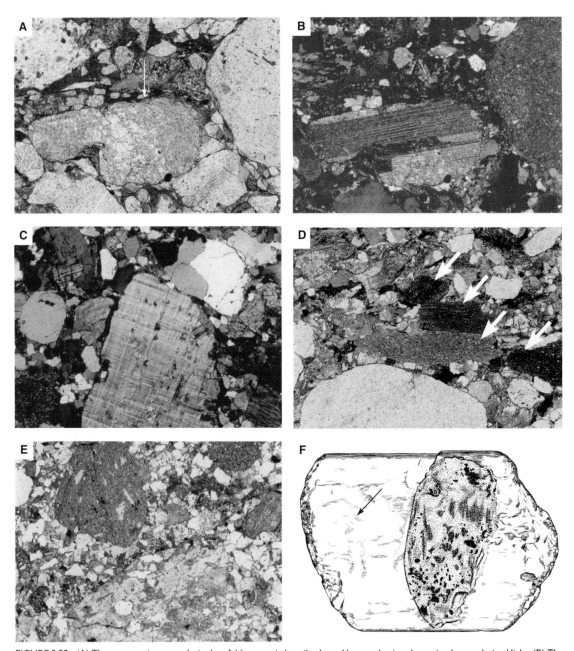

FIGURE 3.22. (A) The arrow points to a plagioclase feldspar grain heavily altered by weathering shown in plane-polarized light. (B) The same field of view as in (A) under crossed polarizers, showing twinning of plagioclase feldspar. (C) A large microcline potassium feldspar grain under crossed polarizers, showing characteristic "tartan plaid" twinning. (D) Shale chips/mud intraclasts (arrows). (E) Volcanic rock fragments. (F) Tourmaline crystal with detrital core and abraded overgrowth. Images (A)–(E) from Adams *et al.* (1984), and (F) from Pettijohn *et al.* (1972). By permission of Springer Science and Business Media.

minerals as cement (see the Appendix and Chapter 19). Classifications of sedimentary rocks that use the concept of matrix force interpretation of that material from the outset. Specific physical characteristics and genetic interpretations of terrigenous grain types of sand and gravel size are summarized below.

Terrigenous grain types

Quartz comprises ~65% of the average sandstone (Figure 3.21). Properties of quartz commonly observed under a petrographic microscope are the degree of crystal distortion (indicated by undulatory extinction;

TABLE 3.9. Heavy-mineral associations and provenance (from Pettijohn *et al.*, 1973), by permission of Springer Science and Business Media

Association	Source
Apatite, biotite, brookite, hornblende, monazite, muscovite, rutile, titanite, tourmaline (pink variety), zircon	Acid igneous rock
Cassiterite, dumortierite, fluorite, garnet, monazite, muscovite, topaz, tourmaline (blue variety), wolframite, xenotime	Granite pegmatite
Augite, chromite, diopside, hypersthene, ilmenite, magnetite, olivine, picotite, pleonaste	Basic igneous rock
Andalusite, chordrodite, corundum, garnet, phlogopite, staurolite, topaz, vesuvianite, wollastonite, zoisite	Contact metamorphic rocks
Andalusite, chloritoid, epidote, garnet, glaucophane, kyanite, sillimanite, staurolite, titanite, zoisite-clinozoisite	Regional metamorphic rocks
Barite, iron ores, leucoxene, rutile, tourmaline (rounded grains), zircon (rounded grains)	Reworked sedimentary rocks

Figure 3.21B), inclusions, color, grain shape, and whether it is monocrystalline or polycrystalline (Figure 3.21). Polycrystalline quartz grains can be bits of coarse-grained rocks such as metamorphic quartzites, granite, or cherts composed of microcrystalline quartz (Figure 3.21E, F). In general, these physical properties are not distinctive enough to relate to specific source rocks, but there are exceptions. Elongated grains with a bimodal size distribution in polycrystalline quartz grains are typical of metamorphic rocks. Milky-colored quartz grains are typically from hydrothermal veins, and the color is caused by abundant gas-filled or fluid-filled inclusions (Figure 3.21). Monocrystalline, euhedral beta-quartz with resorbed boundaries comes from volcanic rocks. Quartz grains with rounded rims of cement are from sedimentary rocks. Oxygen isotopes of quartz grains may indicate the original temperature of crystallization.

Feldspar comprises 10%–15% of the average sandstone (Figure 3.22). Compositional zoning and twinning are properties relating to source rocks (e.g., zoning in igneous rocks, absence of twinning in metamorphic rocks, perthite in subsolvus granites). Feldspar is not as resistant to abrasion during transport as quartz, because of the cleavage and twinning, but most feldspar is certainly able to survive one cycle of weathering, transport, and deposition. Feldspars in mature sediments tend to be mainly small amounts of potassium feldspar (especially microcline), and these would be well rounded and contain clay mineral alteration (Figure 3.22C). Feldspars can be very abundant in immature sediments where potassium feldspars can be relatively fresh. Most plagioclase in ancient sandstone is albite and exhibits characteristic twinning (Figure 3.22A and B), but calcic plagioclase is rare (a result of diagenesis – Chapter 19). Fresh, rounded feldspar indicates rounding by abrasion, or replacement of calcic plagioclase by albite, whereas an angular feldspar with a large amount of alteration (essentially a clay pseudomorph) was probably a fresh crystal upon deposition that was weathered after deposition.

Micas (biotite and muscovite) are common accessory minerals in sandstones, especially along bedding planes. Abundant micas normally indicate a metamorphic source rock (e.g., a schist). Biotite is commonly altered to and intergrown with the green clay mineral chlorite. Chlorite is responsible for the olive-green color of many sedimentary rocks, and hematite from the weathering of iron in biotite and chlorite imparts a reddish-gray color.

Quartz, feldspar, and mica are rock fragments, but this term is normally reserved for somewhat unstable grains composed of more than one mineral type. Rock (or *lithic*) fragments are particularly important in coarse-grained sandstones and gravelstones because they unambiguously indicate source rocks. However, the proportion of rock fragments of a given type of rock is not necessarily indicative of the distribution in the source area because (1) the proportion in the sediment depends on the relative grain sizes of the source rock and the sediment, and (2) some rock types are preferentially lost during weathering and transport. For example, coarse crystalline rocks such as granites and gneisses cannot occur in fine-grained sandstones, because the crystal size is larger than the sediment size. Fine-grained volcanic rock fragments could occur in fine-grained sandstones. However, fine-grained acidic volcanic rocks are difficult to distinguish from cherts and mudstones in thin section (Figure 3.22E). Limestones and sandstones cemented by calcium carbonate are uncommon rock fragments because they disintegrate quickly. Therefore, rock fragments can give a distorted view of the rock types in the source area. Rounded and angular mudstone fragments (Figure 3.22D) are quite common in sandstones and gravelstones, and, because they can disintegrate quickly during transport, they must have been derived locally (i.e., they are *intraformational*).

Heavy minerals are accessory terrigenous minerals (commonly less than 1% by volume) with density greater than 2,800 kg m^{-3}. Heavy minerals are usually concentrated prior to analysis. This involves disintegration of the sedimentary rock, and separation of heavy minerals from the light minerals using an organic liquid with the appropriate density (see the Appendix). Certain groups of heavy minerals can be distinguished using magnetic separation methods. Heavy minerals used to be identified from their optical properties observed in so-called "grain mounts," but this is difficult and a dying art that has been supplanted by the use of microprobes and SEMs with X-ray-dispersive equipment. Characteristic associations of heavy minerals can be linked to specific rock types (see Table 3.9 and the Appendix). However, zircon, tourmaline, rutile, and garnet are very stable and can survive many cycles of weathering, erosion, transport, burial, and exposure (Figure 3.22F).

Weathering and landforms

Rocks that are more resistant to weathering than adjacent rocks generally protrude above them. Resistance to weathering can be related to a high proportion of chemically and physically resistant minerals (such as quartzites and granites), or a lack of joint and fault planes where weathering fluids can penetrate. Features of differential chemical weathering, such as natural arches, towers, tors, and inselbergs (Figure 3.23), are commonly related to local variations in such planes of weakness.

Landforms associated with chemical weathering in humid climates with thick soils contrast with those in arid climates where slow rates of chemical weathering result in thin soils (Chapter 16). Rocks are more likely to be covered by smooth slopes of soil in humid climates than in arid climates, where bare rocks bordered

FIGURE 3.23. Landforms of differential weathering, Canyonlands National Park, Utah, USA. Upper photograph courtesy of Russ Finley; lower photograph courtesy of Gary Salamacha.

by thin soils are more common. In arid climates, limestones tend to form landscape features, whereas they tend to be dissolved away in humid climates.

Dissolution of limestone leaves distinctive features on and below the land surface, which are collectively referred to as *karst topography*, after the karst region of Serbia and Montenegro. Dissolution of limestone outcrops on the Earth's surface occurs preferentially along joints, leaving upstanding joint blocks (called *clints* and *grykes* in England), and in other places where solubility is locally high, leaving solution pits (Figure 3.24A). Solution hollows beneath the surface are also common,

FIGURE 3.24. Features of karst landscapes. (A) Solution-pitted surface (clint-and-gryke) and (B) a stream disappearing into a sinkhole, Yorkshire, England. From Easterbrook (1999), photography courtesy of Art Palmer. (C) Stalactites and stalagmites from Carlsbad Caverns, New Mexico, USA, and (D) flowstones from a cave in Virginia, USA (courtesy of Art Palmer). (E) Houses disappearing into a sinkhole in Winterpark, Florida, USA. From Easterbrook (1999).

and the largest ones are caves and conduits with intricate drainage networks (Figure 3.24). Dissolution commonly leaves insoluble residues of silt and clay at the base of the solution hollow. As subsurface water drips or flows into air-filled solution hollows, carbon dioxide is released into the air and calcium carbonate tufa is deposited on the roof, sides, and floor of the hollow.

Stalactites are columns of calcium carbonate that grow down from roofs of caves, and *stalagmites* grow up from the floor. Pillars and curtains of calcium carbonate form as stalagmites and stalactites meet (Figure 3.24). Spectacular patterns of tufa deposits occur in famous limestone caverns such as Mammoth Caves in Kentucky and Carlsbad Caverns in New Mexico. It is evident that many caves formed when they were full of water, because they are equally dissolved in all directions. Therefore, precipitation of tufa in air-filled caverns requires lowering of the permanent water table.

A *sinkhole* or *swallow hole* (or *cenote* in Mexico) is a limestone cavern that is open to the sky, typically tens of meters wide and deep, and has a downward tapering, funnel-like form. Some of these formed by collapse of cave roofs, but others were dissolved at the surface. Some sinkholes in the Floridan aquifer of Florida spontaneously appear, swallowing houses and cars as they develop (Figure 3.24). Streams may be diverted into sinkholes, leaving dry valleys downstream (Figure 3.24). Such dry valleys must have been eroded prior to the diversion of the water down the sinkhole, possibly when the water table was higher, or when water could not get into frozen ground.

Weathering, soils, and global climate

Weathering and soil production play important roles in regulating global climate (Berner and Berner, 1996; Kump *et al.*, 2000, 2004; Berner and Kothavala, 2001). Over the long-term carbon cycle ($>10^6$ years), silicate and carbonate weathering are a major sink of atmospheric CO_2 because the HCO_3^-, Ca^{2+}, and some Mg^{2+} produced by chemical and biochemical weathering are precipitated mainly in the oceans as various calcium carbonate minerals, and thus removed from the atmosphere. Raymo and Ruddiman (1992) made the controversial suggestion that tectonic uplift of the Tibetan Plateau caused a draw down of atmospheric CO_2 due to an increase in chemical weathering, that, in turn, led to late Cenozoic cooling. In this regard, the importance of plants to weathering rates is a large unknown factor controlling Phanerozoic global atmospheric CO_2. Over the short-term carbon cycle ($<10^6$ years), soils and their associated plants can be sources or sinks of atmospheric CO_2 (Lal, 2004). Burning of plants and disruption of soils adds CO_2 to the atmosphere as reduced carbon is oxidized, whereas increases of plant and organic soil humus biomass can remove CO_2 and temporarily store it. Unlike coal and oil, which are buried in sedimentary rocks and comprise long-term storage of carbon, plant biomass and soil carbon are temporary (short-term) storage sites.

One particular example of the transient nature of carbon storage in soils is provided by the permafrost soils (gelisols) of the Arctic (Stokstad, 2004). The frozen nature of these soils has guaranteed that plant biomass (reduced carbon) stored in them has been building up since tundra vegetation became established after deglaciation. This is because their frozen nature inhibits aerobic and anaerobic oxidation of the carbon by bacteria. However, as these soils begin to thaw due to global warming, they may provide a sudden large outpouring of CO_2 and methane as bacterial metabolism becomes permanent.

4 Biogenic and chemogenic sediment production

Introduction

Ancient biogenic (actually biochemogenic) and chemogenic sedimentary rocks include (1) limestones and dolostones (collectively referred to as carbonates); (2) cherts and related siliceous deposits; (3) saline deposits (commonly referred to as evaporites); (4) iron-rich sedimentary rocks; and (5) phosphorites (Pettijohn, 1975). Modern analogs help guide the interpretation of these ancient sediments. However, modern sediments generally differ from their ancient counterparts in terms of scale, details of mineralogy, and details of depositional environment and processes. Such differences stem from two sources. First, chemogenic and biogenic sedimentary rocks are very susceptible to changes in chemical composition and mineralogy during diagenesis (Chapter 19). Second, there have been changes in the compositions of the lithosphere, atmosphere, hydrosphere, and biosphere through geological time, for example (1) the outpourings of flood basalts in the Permian and Cretaceous, (2) the rise in concentration of oxygen in the Paleoproterozoic atmosphere, (3) variations in the amount of CO_2 in the atmosphere, (4) variation in the major-ion composition of seawater, and (5) the origin and evolution of shell-building organisms (see Chapter 2). Biogenic and chemogenic sediments are governed by these changes in the various spheres through geological time and are, in fact, the most important sources of information about these changes.

Two important features of modern chemogenic and biogenic sediments are that (1) they originate as either biochemical precipitates or physico-chemical precipitates in the depositional environment where they accumulate, and (2) the majority of the grains are subjected to physical transport and deposition, although some of the grains remain at the precipitation site. Examples of chemogenic and biogenic grains that remain at the precipitation site are coral skeletons that can comprise up to half the volume of modern and ancient reefs (Chapter 15), spring travertines and tufas (Chapters 11 and 16), and evaporite minerals that grow on the bottom of brine pools (Chapter 16). However, most chemogenic and biogenic sedimentary grains are moved about in their depositional settings by various currents and are subject to the same physical processes as those affecting terrigenous sediments (Figure 4.1). The transport may be minimal, as with salt crystals that nucleated at the surface of a shallow brine pond and settled to the bottom, or shells that are moved together into "coquinas" on the sea bed. However, the transport may be substantial, such as when sediments from shallow depositional environments are exported to adjacent deep-basin environments by sediment gravity flows.

In order for chemogenic and biogenic grains to be precipitated, the water of the depositional basin must be supersaturated with the various minerals that are formed. However, supersaturation on its own is commonly not enough to induce precipitation, especially for relatively insoluble minerals such as the carbonates. Kinetic barriers to mineral precipitation are commonly overcome by organisms, including bacteria, archaea, and eukaryotes, hence the term biochemogenic (shortened to biogenic) precipitate. Indeed, some organisms are able to precipitate shell material from water that is undersaturated with that particular mineral, although the shells begin to dissolve once free of their protective organic matrix. In order to accumulate deposits of a particular type of chemogenic or biogenic sediment, other sediment types must be overwhelmed or excluded from the depositional environment. Accumulation of carbonate sediment normally requires reduced siliciclastic sediment input and high carbonate productivity, typically in warm, shallow tropical water. Evaporites in arid settings and iron sulfides deposited around submarine hydrothermal vents at mid-ocean ridges are examples of chemogenic sediments that can completely overwhelm other sedimentation.

This chapter will deal mainly with modern carbonate sediments, which are the most common biogenic and chemogenic sediments. Carbonate minerals are described first, followed by an introduction to carbonate geochemistry, and then by description of the main carbonate grain types. The other main types of chemical sediments (cherts, evaporites, iron-rich sediments, and phosphorites) are briefly described in the same manner, with emphasis placed on mineralogy and grain types. Interpretation of ancient counterparts is also discussed in places.

Calcium carbonate mineralogy

Seven of the sixty or so naturally occurring carbonate minerals are important in Earth surface sediments (Table 4.1) (Lippman, 1973; Reeder, 1983). The main structural unit of all these minerals is the CO_3^{2-} group and all of them crystallize in the hexagonal system, except for aragonite, which is orthorhombic. The formulae given in Table 4.1 are for pure, ideal compositions. Natural dolomite, $CaMg(CO_3)_2$, is usually fairly close to ideal, but the other common carbonate minerals are rarely ideal and instead comprise various solid solutions. For example, various amounts of Ca^{2+} may substitute for Mg^{2+} in the crystal lattice of magnesite, and various amounts of Mg^{2+} may substitute for Ca^{2+} in the lattice of calcite. The formula for the solid solution is $Ca_{(1-x)}Mg_xCO_3$ (where x represents the mole fraction of magnesium ions). At the lower temperatures in sedimentary environments, the solid solution between calcite and magnesite is limited, with calcite that contains some magnesium being much more common than magnesite that contains some calcium. Most modern calcite, and almost all ancient calcite, has about 4 mol% magnesium substituting at

TABLE 4.1. Common carbonate minerals

Carbonate mineral	Formula
Calcite	$CaCO_3$
Magnesite	$MgCO_3$
Siderite	$FeCO_3$
Strontianite	$SrCO_3$
Dolomite	$CaMg(CO_3)_2$
Ankerite	$Ca(Fe,Mg)(CO_3)_2$
Aragonite	$CaCO_3$

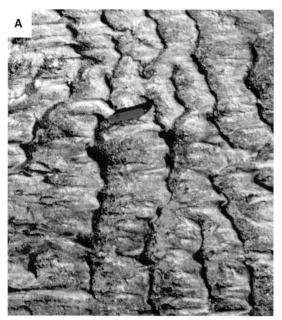

FIGURE 4.1. Bed-load transport of chemical sedimentary grains. (A) Asymmetrical ripples composed of gypsum crystal sand, Colorado River Delta, Baja California, Mexico. (B) Wave ripples in halite, Permian Salado Formation, New Mexico, USA (photograph courtesy of Tim Lowenstein).

random for calcium throughout the crystal structure. Technically, this mineral is referred to as *low-magnesium calcite* – $Ca_{0.96}Mg_{0.04}CO_3$, but commonly the modifier "low-magnesium" is omitted. This is in contrast to *high-magnesium calcite*, in which >4 mol% magnesium ions replace calcium ions. Organisms, particularly echinoderms and red algae, are capable of making up to 20-mol%-magnesium calcite with well-crystallized structure. However, inorganic precipitates of high-magnesium calcite with similar, or higher, mole percentages of Mg^{2+} are usually very fine-grained and, at low temperatures, this solid solution is limited to high-magnesium calcite with less than approximately 50 mol% Mg in the crystal lattice. This is apparently due to the relative sizes of the two ions being such that distortion of the calcite crystal lattice occurs as additional Mg^{2+} is added. When such non-biogenic, fine-grained, high-magnesium calcite was first found in evaporative tidal-flat settings in the 1960s, it was hailed as modern dolomite (Deffeyes *et al.*, 1965; Illing *et al.*, 1966; Kinsman, 1966; Shinn, 1968b; Patterson and Kinsman, 1982). However, in this high-magnesium calcite, the calcium and magnesium ions are not segregated into separate cation planes as in ordered, ideal dolomite. Therefore, this mineral is now usually referred to as *protodolomite* (Graf and Goldsmith, 1956).

Aragonite commonly has a few tens of thousands of ppm strontium in its crystal lattice. Nearly all ordered, natural dolomite is not quite ideal, but has a few percent Fe^{2+} substituting for Mg^{2+}, and is commonly known as ferroan dolomite – $Ca_{0.5}Mg_{0.046}Fe_{0.04}CO_3$. Dolomite with more than 4 mol% iron in its crystal lattice is commonly referred to as *ankerite*. Ferroan dolomite and ankerite can commonly be identified in the field by their tan color, which is produced by oxidation of the iron during weathering.

The identification of carbonate materials, including the solid solutions, is most confidently done with X-ray powder diffraction (Chave, 1954; Lowenstam, 1954; Goldsmith *et al.*, 1955; Tennant and Berger, 1957; Goldsmith and Graf, 1958; Lippmann, 1973; Reeder, 1983). Older techniques (see the Appendix) include use of organic compounds as stains to differentiate calcite from aragonite, and calcite from dolomite, in acetate peels and thin sections (Friedman, 1959; Evamy, 1962; Kummel and Raup, 1965; Dickson, 1966; Wolf *et al.*, 1967; Davies and Till, 1968).

The most common carbonate minerals in modern, shallow, tropical marine environments are aragonite and high-magnesium calcite (Bathurst, 1975; Lowenstam and Weiner, 1989; Flügel, 2004). Aragonite in modern marine environments principally occurs as skeletal fragments (green algae, corals, gastropods, and some bivalves; Figure 4.2), lime mud, ooids (see below), and cements (Chapter 19). Aragonite occurs as far back in time as the Ordovician, but is rare even in the Neogene and Paleogene. High-magnesium calcite occurs in modern settings as skeletal fragments (echinoderms, red algae, and bryozoans; Figure 4.2), lime mud, and cements. High-magnesium calcite is rare in sediment older than Paleogene. Freshwater carbonates are dominantly low-magnesium calcite, and some modern marine groups produce low-magnesium calcite skeletons (coccolithophores, brachiopods, and some foraminifera; Figure 4.2). Ancient carbonate rocks are most commonly composed of low-magnesium calcite and ordered dolomite, and the conversion of metastable aragonite and high-magnesium calcite to these minerals is one of the major processes of carbonate diagenesis (Chapter 19). However, it seems likely that, throughout the Phanerozoic, inorganic precipitates from seawater, such as ooids, some lime mud, and cements, have alternated between low-magnesium calcite (so-called *calcite seas*) and aragonite + high-magnesium calcite (so-called *aragonite seas* – such as today's ocean) (Figure 4.3). During times of calcite seas, such as in the Cretaceous, some invertebrate groups (such as the coccolithophores) that produce low-magnesium calcite skeletons produced much more robust and plentiful shells, whereas aragonite-secreting groups (such as Scleractinian corals) were less common. Conditions were reversed during times of aragonite seas, with certain aragonite-producing groups thriving and the shell-making abilities of calcite-producing groups (particularly the coccolithophores) being greatly reduced (Stanley and Hardie, 1998; Stanley *et al.*, 2002; Ries, 2004).

Calcium carbonate geochemistry

The equilibrium reactions of the carbonate system (with the appropriate equilibrium constants $K_{T,P}$ – see Chapter 3) are

$$CO_2 + H_2O \leftrightarrow H_2CO_3{}^0 \qquad K_{T,P} = 10^{-1.47} \qquad (4.1)$$

$$H_2CO_3{}^0 \leftrightarrow H^+ + HCO_3{}^- \qquad K_{T,P} = 10^{-6.35} \qquad (4.2)$$

$$HCO_3{}^- \leftrightarrow H^+ + CO_3{}^{2-} \qquad K_{T,P} = 10^{-10.33} \qquad (4.3)$$

$$Ca^{2+} + CO_3^{2-} \leftrightarrow CaCO_3 \qquad K_{T,P} = 10^{-8.48} \text{ for calcite}$$
$$K_{T,P} = 10^{-8.34} \text{ for aragonite}$$
$$(4.4)$$

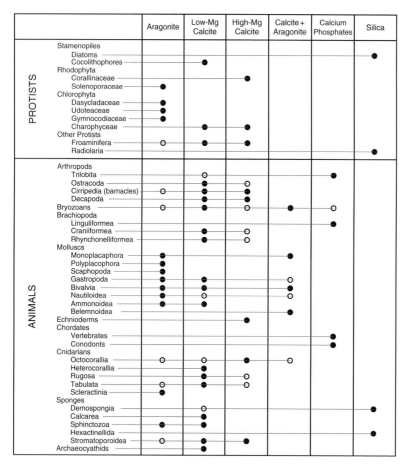

FIGURE 4.2. Mineralogy of skeletons of various sediment-producing organisms. Filled circles indicate common minerals, open circles less common minerals. Modified from Lowenstam and Weiner (1989) and Flügel (2004).

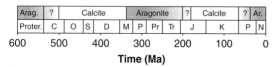

FIGURE 4.3. The distribution of calcite seas and aragonite seas for the past 600 million years. During calcite seas, ooids, mud, and marine cements are dominantly calcite with well-preserved microstructure, whereas during aragonite seas, ooids, mud, and marine cements undergo extensive alteration (except for modern examples), which is reckoned to indicate that these grains were dominantly aragonite or high-magnesium calcite when deposited.

When written as law-of-mass-action equations, this set of four equations contains six unknowns; $a(CO_2)$, $a(H_2CO_3{}^0)$, $a(H^+)$, $a(HCO_3{}^-)$, $a(CO_3{}^{2-})$, and $a(Ca^{2+})$. A fifth equation that comprises a charge balance can also be added:

$$m_{H^+} + 2m_{Ca^{2+}} = m_{HCO_3^-} + 2m_{CO_3{}^{2-}} \qquad (4.5)$$

There are now five equations and six unknowns. If one of the six variables is fixed, and the system is at equilibrium, all the remaining values are uniquely determined. The algebraic manipulation of the equilibrium equations is well covered in Drever (1997).

Most natural waters on the Earth's surface are very close to equilibrium with calcite: therefore, the equations for carbonate equilibria allow prediction of the circumstances under which carbonate minerals should be precipitated. The five principal ways that natural waters can become supersaturated with a carbonate mineral are (1) evaporation; (2) temperature change; (3) pressure change; (4) mixing of two or more undersaturated waters; and (5) loss of dissolved CO_2 (degassing), leading to the water becoming supersaturated. Evaporation raises the molalities of both Ca^{2+} and $CO_3{}^{2-}$, such that the IAP increases and supersaturation can be achieved. Changing the temperature or pressure changes the value of the equilibrium constants $(K_{T,P})$. For example, calcite and aragonite are less soluble in warm water than they are in cold water (in contrast to many other minerals such as quartz that are more soluble in warm water than in cold water). Calcite

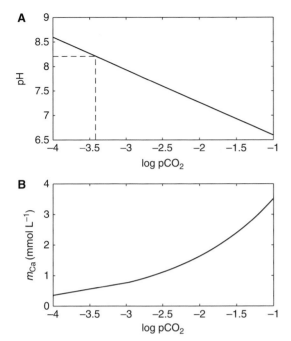

FIGURE 4.4. Equilibrium predictions for the $CO_2-H_2O-CaCO_3$ system. (A) The pH of pure water versus the logarithm of the partial pressure of CO_2 gas ($\log pCO_2$) in contact with that water at $T = 25\,°C$ and 1 atmosphere total pressure in equilibrium with calcite. The dotted line shows the current atmospheric pCO_2 of $\sim 10^{-3.4}$ atmospheres. (B) The molality of Ca^{2+} versus $\log pCO_2$ for pure water in equilibrium with calcite at $T = 25\,°C$ and 1 atmosphere total pressure.

and aragonite also become more soluble as the pressure increases. Thus, the warm surface water of the oceans is supersaturated with aragonite and calcite, but carbonate minerals dissolve in the deep oceans below the *compensation depths* (see Chapter 18), where pressure is high and temperature is low. Mixing two undersaturated waters can produce water that is supersaturated, and vice versa. Solutions can become saturated or undersaturated with carbonate by the removal or addition of CO_2. Figure 4.4A shows the pH of water versus the partial pressure of CO_2 (pCO_2) in a gas phase in contact with that water at $T = 25\,°C$ and 1 atmosphere total pressure in equilibrium with calcite. Figure 4.4B shows the molality of Ca^{2+} versus pCO_2 for the same system. If the partial pressure of CO_2 increases, the pH of the solution decreases, and the molality of Ca^{2+} increases, implying that additional calcite will dissolve (Figure 4.4). Figure 4.4 would be somewhat different for seawater because the activity coefficients of the various dissolved species for seawater will be different than for the $H_2O-CO_2-CaCO_3$ system (Drever, 1997).

At the current pCO_2 of $10^{-3.4}$ atmospheres, the predicted pH of surface water in equilibrium with calcite is about 8.2, very nearly the value for surface ocean water. However, as CO_2 levels in the atmosphere and shallow ocean increase, the ensuing increased solubility of carbonates may affect coral-reef ecosystems (Kleypas *et al.*, 1999). Likewise, if the partial pressure of CO_2 decreases, the pH of the solution increases, and the solution will precipitate calcite. Bacteria can readily add or subtract CO_2 from solution. Cyanobacterial photosynthesis decreases the concentration of dissolved CO_2 around the cells, whereas aerobic respiration increases it.

Geochemical calculations using equilibrium thermodynamics such as shown in Figure 4.4 serve as useful guides to some aspects of carbonate sedimentation, but are somewhat limited. Instead, the precipitation of carbonate minerals is characterized by non-ideal phases (such as the magnesium calcites) that precipitate in metastable equilibrium, by kinetics-controlled precipitation reactions and by biogenic precipitation. Dolomite is the phase that thermodynamics predicts should precipitate from seawater, yet ordered ideal dolomite has not been produced in a laboratory at temperatures below $\sim 100\,°C$. There is one reported case of the organic synthesis of low-temperature dolomite in a kidney stone removed from a Dalmatian dog. At temperatures $> 100\,°C$, calcite and aragonite are easily converted to dolomite if any magnesium is available (see Chapter 19). The carbonate minerals that precipitate out of natural waters at Earth surface temperatures are controlled by the Mg^{2+}/Ca^{2+} ratio and the temperature (Figure 4.5) (Füchtbauer and Hardie, 1980; Morse *et al.*, 1997). In the modern ocean the Mg^{2+}/Ca^{2+} ratio is ~ 5, and aragonite and magnesium calcite form (Figure 4.5A). The mole percentage of magnesium in calcite that precipitates from the ocean is controlled by temperature (Figure 4.5B). It follows from Figure 4.5A that the periods of calcite seas and aragonite seas (Figure 4.3) are controlled by the Mg^{2+}/Ca^{2+} ratio of seawater (Hardie, 1996; Lowenstein *et al.*, 2001; Dickson, 2002). The hypothesis that carbonate mineralogy is controlled by the Mg^{2+}/Ca^{2+} ratio is also supported by studies of the carbonate phases precipitated in saline lakes (Hardie, 2003a, b) (Chapter 16).

Precipitation of calcite versus aragonite from natural waters is an example of kinetic control of a reaction. The equilibrium thermodynamic approach provides useful guidelines for what reactions and minerals are possible, but it does not tell us which reactions are

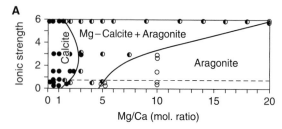

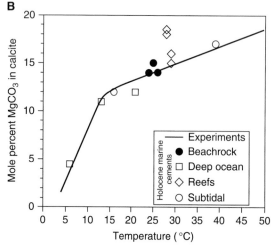

FIGURE 4.5. Controls on the mineralogy of modern carbonates. (A) Experimental data indicating the carbonate minerals that precipitate from $Mg-Ca-Na-Cl-HCO_3$ solutions as a function of total concentration (ionic strength) and the Mg^{2+}/Ca^{2+} ratio of the solution. Data from Füchtbauer and Hardie (1976, 1980). The dashed line shows the ionic strength of seawater. (B) An experimentally determined curve showing the mole percentages of magnesium in magnesium calcites expected to precipitate from seawater at various temperatures (Füchtbauer and Hardie, 1976, 1980). Points give values of mole percentage of magnesium versus temperature for modern high-magnesium calcite cements. Figure courtesy of L. A. Hardie.

likely and how quickly (or sluggishly) those reactions will proceed. This type of information about chemical reactions is the domain of kinetics. An in-depth description of kinetics is beyond the scope of this book, and interested readers should consult textbooks on geochemistry such as Stumm and Morgan (1981) and Drever (1997).

Rates and controls of production of modern carbonate sediments

Carbonate sediments accumulate today in a variety of depositional environments, the most important of which are (1) biogenic sediments from pelagic water of the oceans (see Chapter 18); (2) freshwater and saline

lakes (see Chapters 14 and 16); and (3) continental shelves and adjacent continental slopes and rises (see Chapters 15 and 16). Tropical shelf carbonates are of most interest to geologists since these are the best analogs for most carbonate sedimentary rocks.

Sediment-production rates on tropical carbonate shelves have been measured from standing crops of various carbonate-producing organisms, and calculated using the *alkalinity-titration* method. Stockman *et al.* (1967) pioneered measurements of standing-crop production of aragonite mud by a single genus of green algae, *Penicillus*. They measured the population density and turnover rate of populations of *Penicillus* at Rodriguez Key, a mudbank on the east side of the Florida Keys, and Cross Bank, a mudbank in Florida Bay on the west side of the Florida Keys. Multiplying the number of individual algae that grew per meter per year by the average mass of aragonite mud produced by an individual yielded production rates of 0.025 $kg\,m^{-2}\,yr^{-1}$ at Rodriguez Key and 0.003 $kg\,m^{-2}\,yr^{-1}$ in Florida Bay. Standing-crop production rates summed for all the biota at various mudbanks in the Florida reef tract and Florida Bay range from 0.20 to 0.75 $kg\,m^{-2}\,yr^{-1}$ (Bosence *et al.*, 1985; Bosence, 1989). There have been many standing-crop studies of coral reefs and individual corals (e.g., Chave *et al.*, 1972; Enos, 1991). These measurements suggest that reefs can produce carbonate at up to 10 $kg\,m^{-2}\,yr^{-1}$ (Stern *et al.*, 1977). Results from studies of individual corals such as *Montastrea annularis* (Figure 4.6) show that coral growth rates are proportional to water depth, because water depth is a primary control on the amount and wave length of light available for the symbiotic zooxanthellae algae that infuse the tissues of the corals and account for their high productivity (Schlager, 2005). Plots such as Figure 4.6 are appropriate for describing carbonate production on coral reefs along shelf margins and the open ocean, but their application to carbonate shelves has been questioned by Demicco and Hardie (2002). Results from standing-crop studies indicate that coral reefs produce much more carbonate sediment than back reef areas do.

Broecker and Takahashi (1966) calculated the carbonate production rate on the Great Bahama Banks northwest of Andros Island (Figure 4.7) with a geochemical technique that has since come to be known as the *alkalinity-titration* technique. As seawater from the open ocean flowed onto the shallow banks, the warming of the water coupled with degassing of CO_2 promoted precipitation of aragonite. Broecker and Takahashi used measurements of the pCO_2 of the

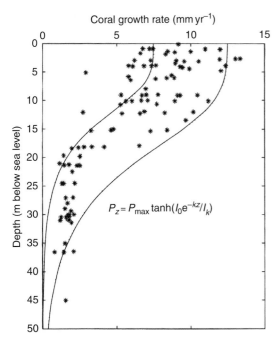

FIGURE 4.6. The growth rate of the Caribbean coral *Montastrea annularis* versus depth. Points are measured rates from Bosscher and Schlager (1992) and curves are growth rates (carbonate production) calculated according to the equation given, where P_z is the production at depth z, P_{max} is the maximum production at $z = 0$, I_0 is the surface light intensity, k is the extinction coefficient given by the Beer–Lambert law, and I_k is the light intensity at the base of the light saturation zone. See Bosscher and Schlager (1992) for further details.

seawater and its total carbon content (*carbonate alkalinity* – see Chapter 3) to calculate how much $CaCO_3$ had been lost from the bank water relative to seawater surrounding the bank (Figure 4.7). Morse *et al.* (1984) repeated these measurements and obtained essentially the same results. Broecker and Takahashi (1966) measured decay rates of radioactive isotopes (produced by open-air testing of nuclear weapons) to calculate the residence time of bank waters (Figure 4.7B). Calculated rates of carbonate production based on a uniform water depth of 4.5 meters range from 0.22 to 1.40 kg m^{-2} yr^{-1} (Figure 4.7A). Demicco and Hardie (2002) recalculated production rates on the basis of actual water depths in the same area (Figure 4.7C) and showed carbonate production to be highest at the shelf margins where the largest water depths, shortest residence times, and most rapid degassing and precipitation occur. Alkalinity-titration measurements of carbonate production on coral reefs at the margins of carbonate banks in different areas are in the range

5–10 kg m^{-2} yr^{-1} with a strong mode at 4 kg m^{-2} yr^{-1} (Smith and Kinsey, 1976). In contrast, alkalinity-titration calculations in back-reef lagoon areas within 5–10 km of the reef margin give values of about 1 kg m^{-2} yr^{-1} (Smith and Kinsey, 1976). In summary, these measurements and calculations show that carbonate production is greatest at the bank margins ($\sim$5–10 kg m^{-2} yr^{-1}), decreasing to around 1 kg m^{-2} yr^{-1} within 5–10 km of the platform margin, and to values of $\sim$0.5–0 kg m^{-2} yr^{-1} in the interior of shallow shelves. Carbonate production on shallow shelves such as the Great Bahama Banks is a function of water depth and the residence time of shelf water (Figure 4.8).

Carbonate grain types

The main types of chemogenic and biogenic grains are (1) skeletal grains, (2) carbonate mud, (3) fecal pellets, (4) intraclasts, (5) peloids, and (6) coated grains known as ooids and pisoids (Bathurst, 1975). Although some skeletal grains (e.g., crinoid ossicles) are single crystals, most skeletal carbonate grains are aggregates of smaller, commonly micron-sized crystals. Moreover, carbonate mud is commonly aggregated into gravel- and sand-sized grains ("mud masquerading as sand" according to Bob Ginsburg). Indeed, it is a hallmark of both chemogenic and terrigenous muds to be aggregated into larger grains, and these are some of the most common grains in chemogenic and biogenic sediments and sedimentary rocks. Descriptions and photomicrographs of carbonate grain types can be found in Bathurst (1975), Scholle (1978), Adams *et al.* (1984), Adams and MacKenzie (1998), Tucker (2001), and Scholle and Ulmer-Scholle (2003). Flügel (2004) gives detailed information and photomicrographs for identification of all types of carbonate grains, especially fossil fragments, in thin section.

Skeletal grains

Many groups of invertebrates, multicellular protozoans (algae), and unicellular protozoans make mineralized skeletons (Figure 4.9), and low-magnesium calcite, high-magnesium calcite, and aragonite are by far the dominant minerals used in skeleton construction (Figure 4.2). The groups that have dominated the production of skeletal carbonate have changed through geological time. An important example of this is the rise of planktonic coccolithophores (Figure 4.10A) and planktonic foraminiferans, which came to dominate global production of carbonate in Mesozoic

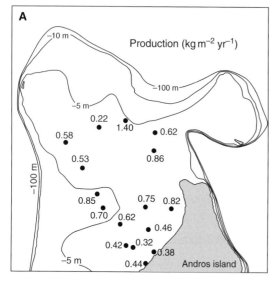

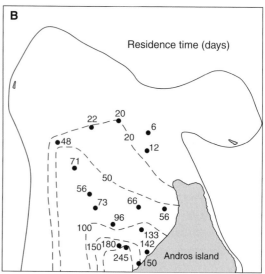

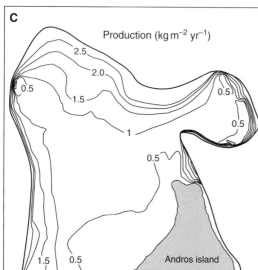

FIGURE 4.7. The carbonate-production rate on the northeastern Great Bahama Bank. (A) Bathymetry (meters) and carbonate-production rates (kg m^{-2} yr^{-1}) calculated at points shown by Broecker and Takahashi (1966) assuming a water depth of 4.5 m. (B) Residence times (days) of bank waters determined from radioisotope decays (Broecker and Takahashi, 1996), with the contour interval variable. (C) The carbonate-production rate on the northeastern Great Bahama Bank (kg m^{-2} yr^{-1}) using the water depths in (A) and residence times given in (B) (Demicco and Hardie, 2002).

oceans. Prior to the rise of these groups, carbonate sedimentation in the deep ocean was limited. These groups have had a substantial effect on altering global carbon budgets and ocean chemistry (Berner and Kothavala, 2001; Holland, 2005). The sediment production of some skeleton-building groups apparently responded to calcite seas and aragonite seas (Stanley and Hardie, 1998; Stanley *et al.*, 2002; Ries, 2004). Detailed understanding of the biochemical processes that control the growth of skeletal carbonate is currently a hot topic in carbonate biogeochemistry. Reviews can be found in Watabe and Wilbur (1976), Westbroek and de Jong (1982), Crick (1986),

Lowenstam and Weiner (1989), Mann (2001), Dove *et al.* (2003), and Baeuerlein (2004).

Skeletal grains range from boulder-sized coral heads to clay-sized fragments. Most of the identifiable skeletons and skeletal fragments of carbonate sediments and sedimentary rocks have been transported, and are in the sand to silt size range (Figure 4.10). The large array of skeletal shapes makes them difficult to identify when seen in randomly oriented cross sections. Skeletal green algae (Figure 4.10B) and red algae (Figure 4.10C) have been important sediment producers on shallow, tropical shelves throughout most of the Phanerozoic. Simply curved shells are created

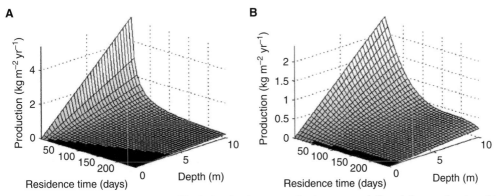

FIGURE 4.8. Maximum (A) and minimum (B) calculated carbonate-production rates on shallow shelves as a function of the water depth and residence time of shelf water (Broecker and Takahashi, 1966; Demicco and Hardie, 2002).

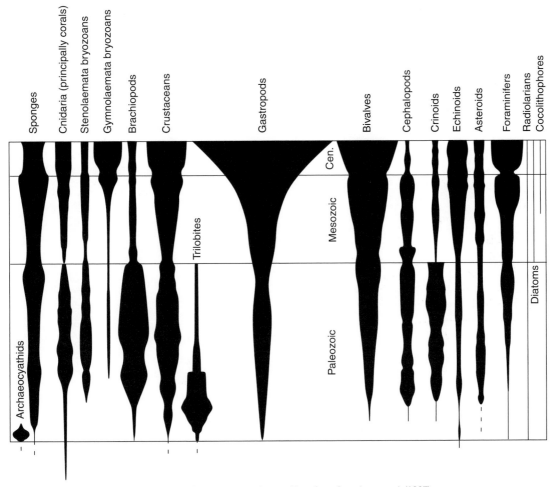

FIGURE 4.9. Relative abundance of various fossil groups with time. Data from Boardman et al. (1987).

by a number of groups including brachiopods (Figure 4.10D), bivalve molluscs (Figure 4.10E), and ostracods (Figure 4.10F). Bivalve mollusc and brachiopod shells in particular have complicated internal architecture. Gastropod shells comprise a series of whorls of expanding diameter about a central axis (Figure 4.10E). Some of the most common skeletal grains in Paleozoic carbonate sedimentary rocks are

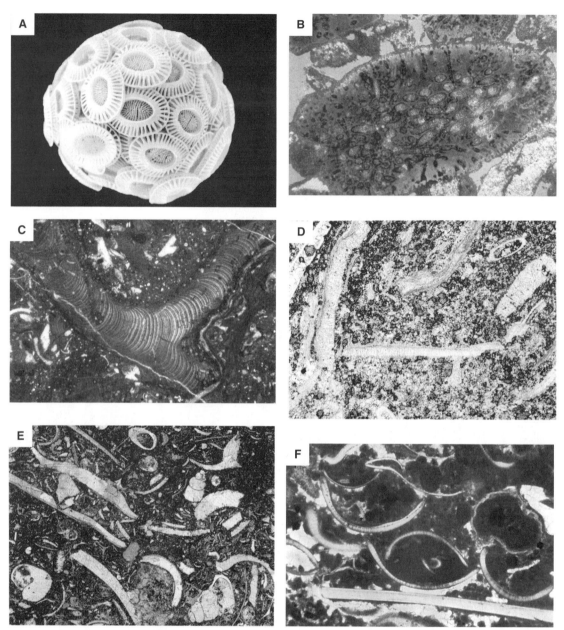

FIGURE 4.10. Skeletal grains. (A) A scanning electron photomicrograph of one of the most common coccolithophores, *Emiliania huxleyi*. Width of field ~15 microns. From Scholle and Ulmer-Scholle (2003). (B) A segment of the green alga *Halimeda*, a common sand grain on modern carbonate tropical shelves. Width of field ~2.5 mm. From Scholle and Ulmer-Scholle (2003). (C) Fragment of coralline red alga. Width of field ~5 mm. From Scholle and Ulmer-Scholle (2003). (D) Pieces of brachiopod shells, showing various wall structures. Width of field ~4 mm. From Adams *et al.* (1984). (E) Casts of mollusc fragments including bivalve and gastropod shells. Width of field ~8.5 mm. From Adams *et al.* (1984). (F) Ostracod shells above a straight trilobite fragment. Width of field ~4 mm. From Scholle and Ulmer-Scholle (2003). (G) Longitudinal sections through parts of two crinoid stems, each comprising nine or ten ossicles. Width of field ~10 mm. From Scholle and Ulmer-Scholle (2003). (H) An oblique cut through a branch of a bryozoan colony. Small partitions are where the individual zooids resided. Width of field ~12.5 mm. From Scholle and Ulmer-Scholle (2003). (I) A cross section through a trilobite carapace. Arrows point to characteristic "shepherd's crook" shapes. Width of field ~14.5 mm. From Scholle and Ulmer-Scholle (2003). (J) A cross section through a number of corallites of a modern aragonitic scleractinian coral. The epoxy resin in the pore space is stained blue. Width of field ~16 mm. From Scholle and Ulmer-Scholle (2003). (K) A grain mount of various foraminifera. Width of field ~5 mm. From Scholle and Ulmer-Scholle (2003). (L) Shell fragments (probably bivalve molluscs) that are preserved only as micrite envelopes. Width of field ~10 mm. From Scholle and Ulmer-Scholle (2003). All figures from Scholle and Ulmer-Scholle AAPG © (2003). Reprinted by permission of the AAPG, whose permission is required for further use.

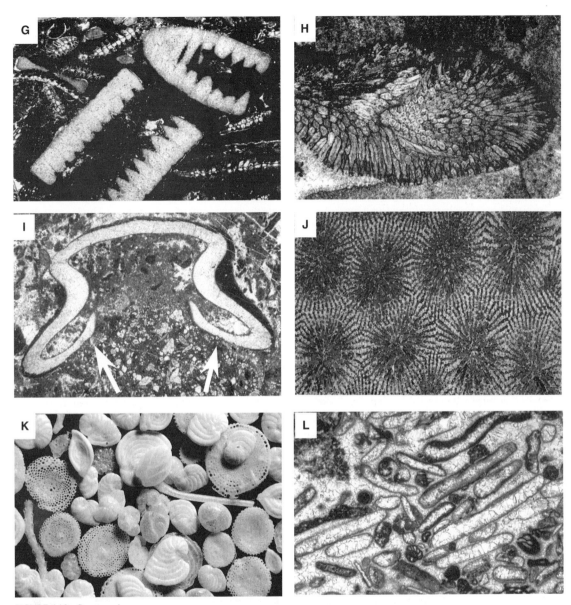

FIGURE 4.10. Continued.

crinoid stem pieces called *ossicles*, single cylindrical crystals of high-magnesium calcite with a central hole (Figure 4.10G). The shapes of crinoid ossicles have been likened to Polo Mints (UK) or Life Saver candies (USA). Two other common types of Paleozoic fossil are bryozoans (Figure 4.10H) and trilobites (Figure 4.10I). Three important groups of corals are the Paleozoic Tabulate and Rugose corals, and the Scleractinian corals (Figure 4.10J) that arose in the Triassic and comprise the modern coral fauna.

Benthic foraminifera (Figure 4.10K) can be important contributors of skeletal fragments in addition to the planktonic foraminifera.

Decomposition of soft parts of organisms, particularly muscles and ligaments, usually results in the disarticulation of skeletal elements in, for example, bivalve molluscs, brachiopods, and echinoderms. Transport also leads to disarticulation as well as to grain abrasion and breakage. Predators and scavengers abrade and break shells with claws, teeth, beaks, or, in the case of snails, the rasping radula. Shell material may also be modified in the guts of animals. Shells and shell fragments of dead organisms exposed on the sea floor are rapidly colonized by a host of organisms known as *epibionts*. The boring epibionts include cyanobacteria, fungi, bacteria, and sponges. The borings can extend well into the grains, given sufficient time. When these microboring epibionts die, their excavated galleyways become filled with carbonate mud (called *micrite* – see below). Thus, the process of shell boring and hole filling is referred to as *micritization*. Depending on the depth of boring, the grains could have so-called *micrite envelopes* or be almost completely *micritized* (see Figure 4.10L). Furthermore, micritization is not just limited to skeletal grains, as discussed below. It is common for shell fragments to go through a number of cycles of transport, deposition, epibiont colonization and micritization, burial, exposure, and transport again. Since aragonite and high-magnesium calcite are metastable, especially under burial conditions, skeletal fragments of these minerals tend to be replaced with low-magnesium calcite during carbonate diagenesis. These processes and their products are described in more detail in Chapter 19. Because of their unusual shapes, physical breakage, and chemical alteration, detailed identification of individual skeletal grains is a challenge; see Bathurst (1975) and Flügel (2004).

The odd shapes of skeletal fragments make it difficult to predict porosity, permeability, and packing (see the Appendix). Artificial packs of flattened shells have porosity up to 0.77 (Vinopal and Coogan, 1978), whereas natural unconsolidated carbonate sands and muddy sands have porosity of 0.4–0.7 (Enos and Sawatsky, 1981). Random two-dimensional slices through high-porosity shell packs rarely intersect any grain-to-grain contacts. Little is known about the hydrodynamic behavior of skeletal grains, although transported shells on beaches commonly lie in a stable concave-down position. There have been few studies of skeletal sands in laboratory flumes, although Allen (1982a, 1985) discussed the settling of bivalve mollusc shells in sediment suspensions. Needless to say, size, sorting, roundness, and sphericity have a distinctly different meaning in skeletal sands than they do in terrigenous sediments (see the Appendix).

Carbonate mud

Carbonate mud is the most voluminous carbonate sediment and the type of mud is related to the depositional environment. Detrital carbonate mud was termed *micrite* by Folk (1959), the word being a contraction of microcrystalline calcite. This term has broadened through time to refer to precipitated fine-grained cements as well as mud of detrital origin. We will use the explicit term carbonate mud. Common types of carbonate mud are (1) aragonite needles; (2) pelagic skeletons; (3) epibiont skeletons and fragments; (4) abraded skeletal debris; and (5) *tufa* and *periphyton* debris (so-called *automicrite*).

Aragonite-needle mud is the main component of the sediment that blankets the shallow interior of the Bahama Banks and Florida Bay (Chapter 15) and is mostly composed of ~4-μm-long, single crystals of aragonite with euhedral to rounded terminations (Figure 4.11A). The mud is reckoned to come from at least two sources: (1) codiacean green algae and (2) biochemical/chemical precipitates from supersaturated bank water. The green algae responsible for aragonite-needle mud include *Penicillus*, *Rhipoceophalus*, and *Udotea* (among others) (see Figure 15.46 later). These algae are lightly calcified, with aragonite needles embedded in organic sheaths around the algal filaments that are released upon death and decay of the organism. *Penicillus* alone could account for the total carbonate mud production in the Florida Bay and Reef Tract, according to Stockman *et al.* (1967).

Aragonite-needle-producing green algae are restricted to shallow water (the photic zone, <100 m maximum depth) in tropical areas, and are intolerant of high concentrations of suspended clay minerals. Since the same conditions are needed to initiate inorganic carbonate sedimentation on a shelf, direct inorganic chemical precipitation of aragonite needles from seawater must be considered a possibility. These environments are slightly supersaturated with aragonite (as well as calcite, dolomite, and magnesite), and the most compelling evidence of physico-chemical precipitation is the *whitings* seen on the Bahama Bank and elsewhere. Whitings are clouds of bank water that contain suspended aragonite needles. Whereas some whitings are clearly patches of aragonite mud stirred up from the bed by bottom-feeding fish (such as the famous bonefish), Shinn *et al.* (1989) have shown that whitings are indeed spontaneous precipitates. They form in the surface water, and propagate against prevailing current directions. The photosynthetic activities of phytoplanktonic coccolithophores or cyanobacteria may trigger whitings locally by supersaturating the waters by extracting CO_2. Samples of bank water will nucleate aragonite needles if

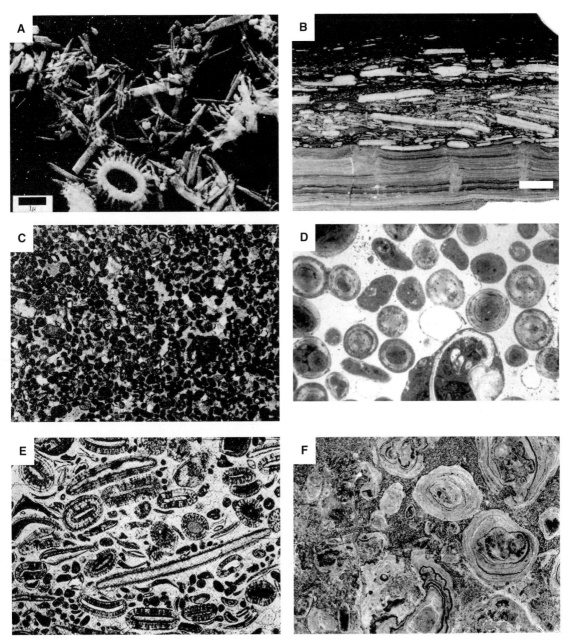

FIGURE 4.11. (A) A scanning electron photomicrograph of aragonite needle mud (and coccolith) from the Great Bahama Bank. Scale bar 1 micron. From Bathurst (1975). (B) A polished, etched slab of flat pebble intraclasts from a Silurian deposit of western Maryland, USA. Scale bar 2 cm. (C) Thin-section photomicrographs of peloids. Width of field ~3.5 mm. From Adams *et al.* (1984). (D) A thin-section photomicrograph of Bahamian aragonite ooids. Width of field ~6 mm. Courtesy of Robert Ginsburg. (E) Thin-section photomicrograph of radial calcite ooids. Width of field ~6.5 mm. From Cayeaux (1935). (F) A polished, etched slab of pisoids. Width of field ~6 cm. From Adams *et al.* (1984).

evaporated slightly. The Dead Sea and Great Salt Lake (Utah) also have whitings, with precipitation of aragonite needle mud at elevated salinities. Whitings are common in other lakes such as Lake Michigan, New York's Finger Lakes, and Lake Geneva, where calcite is precipitated rather than aragonite. The calcite grains that nucleate are equant aggregates of small rhombs (Chapter 14), and are thus very different from either aragonite-needle mud or the platy aluminosilicate clay minerals.

Questions about the origin of aragonite-needle mud are somewhat moot because most of that produced on the Bahama Banks does not stay where it was precipitated. For example, Traverse and Ginsburg (1966) showed that non-marine pollen grains ended up in the same areas as the thickest deposits of carbonate mud, indicating that the deposition of both grain types was largely controlled by physical rather than chemical factors.

Pelagic skeletal carbonate mud comes from (1) the circular plates (called coccoliths) that surround coccolithophorid cells, (2) foraminifera shells, and (3) pteropod shells. The controls on the geographic distribution of these pelagic organisms in surface ocean water, and the preservation of their skeletons in deep-ocean sediments, are described in Chapter 18. These skeletons are also intermixed with the aragonite-needle mud of shallow carbonate shelves. Coccolith plates and foraminifera shells are low-magnesium calcite, whereas pteropods, an aberrant group of pelagic snails, produce aragonite shells. Coccolithophorids are one of the most important photosynthetic groups in the surface waters of the oceans. A coccolithophore can be pictured as a ball (the eukaryote cell) surrounded by fancifully ornamented plates, the armoring plates of calcite known as coccoliths (Figure 4.10A). The plates are sloughed off throughout the lifetime of the organism. Coccolithophores arose in the Jurassic, but their heyday was in the Cretaceous, when chalks composed primarily of coccolith plates were deposited worldwide. Foraminifera are represented by about 60,000 species of amoeba-like eukaryotic protozoans. Benthic foraminifera comprise the vast majority of species but there are about 100 planktonic species. Many foraminifera secrete shells of calcite with a variety of shell sizes, and some modern and ancient benthic forms are up to millimeters across (Figure 4.10K). Benthic foraminifera are common contributors to carbonate muds. The planktonic foraminifera are mostly mud-sized, and are an important contributor to deep-sea sediment and some shelf muds (see Chapter 18).

Certain small melobesoid red algae are important epibionts on the turtle grass that flourishes on carbonate mudbanks on tropical shelves (Chapter 15). These red algae produce skeletons of high-magnesium calcite that disarticulate into mud-sized particles.

Carbonate mud arising from abrasion of skeletal fragments can range in size from clay to silt. All of the carbonate mud on temperate shelves is produced by abrasion of skeletal fragments.

Carbonate can precipitate around various photosynthetic organisms to form *tufa* or *periphyton*. Cyanobacteria take in CO_2 for photosynthesis, increasing the pH and the saturation of carbonate around their filaments, thus inducing precipitation of carbonate. This *periphyton* carbonate is mud and is released upon death of the filament. Periphyton carbonate mud is a substantial contributor of carbonate sediment in the Everglades in southern Florida. *Chara* is a freshwater green alga, commonly called stonewort. The fronds and stems of *Chara* and related genera (collectively known as charophytes) are usually encrusted by carbonate mud. This sediment is also precipitated from the water by the photosynthetic removal of CO_2 from the water adjacent to the fronds and stems. Charophytes can locally produce freshwater marls that form benches around the lake shore. Individual moss plants can also be coated with carbonate mud, especially around springs, in rivers, and beneath waterfalls. In marine settings, prokaryote-induced precipitation of carbonate mud may be particularly important in marginal-marine ponds (such as on Andros Island – Chapter 15). However, photosynthetic and non-photosynthetic bacteria are thought to induce precipitation of carbonate mud in caves (Macintyre, 1985; Reid *et al.*, 1990; Reitner 1993; Reitner and Neuweiler, 1995), and this mechanism is now invoked commonly to produce *in situ* carbonate mud (so-called *automicrite*) in mud mounds and in deep-sea settings (Reitner and Neuweiler, 1995; Flügel, 2004; Schlager, 2005).

Pellets

Carbonate mud beds host a variety of filter-feeding and deposit-feeding (mud-grubbing) organisms. Organisms with both feeding strategies commonly aggregate mud into fecal *pellets* (see Chapter 11, in particular Figures 11.1 and 11.7). Deposit-feeding organisms (both infaunal and epifaunal) ingest mud for its organic content (including bacteria) and excrete it as pellets either bound together by mucus or enclosed in a mucus sheath. Some of the bacteria metabolize the organic carbon contained in the sediment, but other bacteria are chemoautotrophs that derive energy from oxidation–reduction reactions of metals such as iron and manganese. Indeed, mud snails, given the choice between fresh fecal pellets and fecal pellets that are a few days old, will choose the older pellets, presumably because they host bacterial populations (Levinton, 1982). Filter-feeding organisms concentrate suspended inorganic particles, suspended living and dead plank-ton, and other organic material in their filtering appa-ratus from which they obtain nourishment. Many filter-feeding organisms ingest both organic and inor-ganic material, passing the inorganic part back to the environment. However, bivalves prescreen the contents of their filter-feeding apparatus and compress the inor-ganic material into *pseudo-fecal pellets*. Marsh-dwelling fiddler crabs (Figure 11.1A) use their mouth parts to work the muddy sediment into *mastication pellets*, which they leave strewn about their burrow entrances.

Populations of filter-feeding organisms can process tremendous volumes of water, extract the suspended organic and inorganic matter, and deposit the inor-ganic matter on the bed as sand-sized pellets. Verway (1952) reported that, in some areas of the Waddensee in the Netherlands, a few thousand common cockles occurred per square meter of sea bed, with each indi-vidual removing about 420 mg of suspended sediment per year from the water and depositing it as fecal pellets. In the western Waddensee, the population of the common mussel *Mytilus* was estimated to remove 175,000 metric tons of suspended sediment per year from the water and deposit it as fecal pellets. Along the southern Atlantic and Gulf Coasts of the USA, each *Callianassa* produces some 2,480 fecal pellets per day. Even with estimates of ten individuals per square meter, the *Callianassa* population in the shoreface zone of Sapelo Island (Georgia) (some 18 km^2) removes and pelletizes 12.3 metric tons of sediment per year (Pryor, 1975). In back-barrier lagoons, where the population of *Callianassa* can reach up to 500 individuals per square meter, they can pelletize and deposit up to 620 metric tons of suspended sediment per square kilo-meter per year (Pryor, 1975).

Fecal pellets are generally in the sand to silt size range and are ovoid with a long axis two to three times the length of the short axis. Some fecal pellets can have elaborate shapes, such as those of *Callianassa*, which has infolded flaps in its anus that give the fecal pellet an unusual cross section (Figure 11.7). Most fecal pellets are simple rods and ellipsoids, although some have rings that indicate peristaltic action of the guts of the organisms producing them. Fecal pellets are nor-mally soft but are made cohesive by the mucus of the animal's gut, which allows them to undergo transport. Fecal pellets tend to be transported together with finer-grained monomineralic sand grains, because the pellets are less dense. Bed forms commonly occur in pellet sands. In his 1969 SEPM presidential address, Bob Ginsburg pointed out that most of the aragonite-needle

mud on the interior of the Great Bahama Bank was in fact pellet sand. In addition, pellets, particularly those composed of aragonite mud, can become cemented internally and cemented together. Cemented clusters of hardened pellets and other rounded particles such as ooids on the Great Bahama Bank are known as *grapestone* particles, and are discussed below.

Intraclasts

Intraclasts are sand- and gravel-sized fragments of semiconsolidated aggregates of mud and sand that were eroded from the depositional environment in which they formed (Figure 4.11B). The term intraclast is used to indicate that these clasts originate within the depositional setting (intraformational) as opposed to being weathered pieces of carbonate rock transported in from outside the depositional environment (extra-formational or extraclasts). Mud beds can become con-solidated by compaction or desiccation and thereby become cohesive enough to be eroded and withstand significant transport. If the eroded mud was laminated, the intraclasts are commonly disk-shaped and are called *flat pebbles*. Intraclasts can be spheroidal if the eroded mud was massive. Eroded mud chips that were hardened by desiccation and disrupted by vertical des-iccation cracks and horizontal sheet cracks are very common in modern carbonate sediments and ancient carbonate rocks.

Carbonate sand and mud can also be cemented in the bed of the depositional setting by a number of early diagenetic processes (see Chapter 19). Beachrock is a common example of penecontemporaneously cemented sand, and occurs in both carbonate and ter-rigenous beach sediments when evaporation of sea-water produces aragonite and high-magnesium-calcite cements. The grapestones mentioned above are another common example of cemented carbonate sands. Early stages of cementation yield *firmgrounds*, whereas advanced cementation yields *hardgrounds*. Both firmgrounds and hardgrounds can be broken up and eroded into intraclasts. Cemented intraclasts may be referred to as *lithoclasts*, to distinguish them from uncemented intraclasts. However, it is commonly diffi-cult to decide whether a mudstone intraclast is from a semiconsolidated or cemented mud bed (see the discus-sion of hardgrounds in Chapter 19). Clasts of sand-sized material usually require some sort of internal cementation to make them cohesive enough to with-stand transport, and cements are commonly truncated along the erosional boundaries of lithoclasts. The term lithoclast is used only for extraformational clasts by some workers, but it is commonly difficult to distinguish extraformational and intraformational cemented grains.

Throughout the Florida–Bahamas carbonate-depositional area, the 120,000-year-old Pleistocene bedrock is thoroughly cemented, and skeletons and other carbonate grains that were aragonite and high-magnesium calcite when deposited have been converted (*stabilized*) to low-magnesium calcite due to exposure in freshwater (Chapter 19). In addition, a laminated soil crust a few tens of millimeters thick has developed on most of the Pleistocene surface during weathering. Clasts of the Pleistocene bedrock, with and without the soil crust, commonly fill karst pits in the Pleistocene surface, and overlie the surface. They are commonly encountered at the bottom of cores through the Holocene marine sediments, and some have been blackened, apparently by a number of organic processes, and by ancient forest and grassland fires (Shinn and Lidz, 1988). Such clasts were not formed within the depositional environment and are therefore *extraclasts*.

Peloids

It is not always possible to determine whether a sand-sized aggregate of carbonate mud is a pellet, an intraclast, or an extraclast (Figure 4.11C). In addition, pre-existing skeletal and other carbonate grains can be micritized, as discussed above. Carbonate mud precipitated as tufa or periphyton is also commonly aggregated into silt-sized aggregates (Macintyre, 1985; Chafetz, 1986). A useful non-genetic term for a silt- or sand-sized aggregate of carbonate mud is *peloid* (McKee and Gutschick, 1969). Proterozoic carbonates commonly contain peloids. Since there were presumably no macroinvertebrates to pelletize sediment until the latest Neoproterozoic, the term peloid is appropriate. Most workers now just apply the term peloid to all carbonate-mud pellets and do not concern themselves with the origin of the grain.

Ooids and pisoids

Ooids are sand-sized (0.0625–2-mm) spherical to ovoid chemogenic grains that possess a nucleus and either concentric internal fabric or a combination of radial and concentric fabrics (Figure 4.11D, E). *Pisoids* are similar to ooids but by definition are larger than 2 mm (Figure 4.11F). Oolite and pisolite are older terms applied to rocks consisting primarily of these grains.

In addition to these well-established terms, other terms of dubious usefulness have been suggested, such as spring peas, cave pearls (German *Holenperlen*), and even walnutoid (Esteban and Pray, 1983). An *oncoid* is a pisoid that is interpreted to be the result of the sediment-trapping and -binding abilities of a microbial mat, although some oncoids are clearly the result of encrusting of skeletal algae or other organisms (Flügel, 2004). Distinguishing pisoids from oncoids is difficult (Dunham, 1969b; Flügel, 2004). Reviews of ooids and pisoids and other so-called *coated grains* include Bathurst (1975), Simone (1980), Peryt (1983), and Flügel (2004).

Modern ooids and pisoids occur in a wide variety of settings, including marine shoals, shoals in saline lakes, caves, and calcareous soils. Ooids and pisoids are complicated, polygenetic grains, most of which are interpreted to be *in situ* chemical precipitates formed very near their depositional site. Modern marine ooids typically occur in sand shoals. On the Great Bahama Banks, the shoals occur on the shelf near the shelf–slope break as tidal ridges and wash-over fans (see Chapter 15). In the Persian Gulf, ooid sands comprise flood- and ebb-tidal deltas (Purser, 1973). These modern marine ooids have come to be known as *Bahamian-type ooids*, and commonly have a nucleus of either a peloid or a skeletal fragment surrounded by microns-thick concentric layers of tangentially oriented aragonite needles (Figure 4.11D). The origin of modern Bahamian-type ooids is still unknown, although everyone agrees that constant wave and current agitation in the shoals is a key factor in their formation. Sorby (1879) originally proposed that modern marine ooids were essentially snowball-like aggregates of chemically or biochemically precipitated aragonite-needle mud, accreted to the outside of the grains by adhering to a microbial community growing on the surface of the grain. The tangential arrangement of the aragonite needles is due to flattening during bed-load sediment transport. However, Cayeaux (1935) proposed that Bahamian-type ooids were direct chemical precipitates from seawater. Ooids dissolved in acid reveal a complex of microbial filaments, and Persian Gulf ooids treated with hydrogen peroxide to remove organic matter disarticulate into aragonite needles. These observations point to the importance of biofilms in the origin of ooids. The current consensus is that the organic molecules of the microbes act as templates for the direct precipitation of aragonite needles out of seawater. Other, rare ooids composed of radial fibers of high-magnesium calcite occur in Baffin Bay, Texas (a hypersaline lagoon) (Land *et al.*, 1979), on the Amazon shelf (Milliman and Barretto, 1975), and in deep waters off the Great Barrier Reef (Marshall and Davies, 1975). These ooids are generally thought to be chemical precipitates.

Ooids and pisoids also occur in a number of modern non-marine settings, including hot and cold springs, cave deposits, saline lakes, boilers, and tea kettles. These grains are commonly calcite with either a tangential fabric or a combination of tangential and radial fabrics, and most are thought to be chemical precipitates. Aragonite ooids occur in saline lakes with high Mg^{2+}/Ca^{2+} ratios, such as the Great Salt Lake, Utah. Great Salt Lake ooids are notable for having both radial and tangential fabrics. In all these cases, agitation is called on to allow for the even growth of the carbonate, and many of the ooids and pisoids of springs, lakes, and caves grow as freely moving grains, at least in the early stages of their formation. In the Johnson City area of upstate New York, people would put a marble in their tea kettles to intercept the calcite as it precipitated out of the boiling water in the tea kettle. Ooids and pisoids are also a common component of modern calcareous soils (Chapter 3), and, unlike the other modern examples, grow in place in the sediment. Calcareous soil pisoids can be distinguished from other types of pisoids by virtue of their lacking a nucleus, having inclusions of corroded soil material, and having mutually interfering, non-spherical shapes due to competitive *in situ* growth.

The interpretation of ancient ooids and pisoids is complicated by the fact that metastable aragonite and high-magnesium calcite are readily altered to low-magnesium calcite during diagenesis (in addition to micritization). Low-magnesium calcite ooids should not be as altered during diagenesis as aragonite and high-magnesium calcite ooids (although they can still be micritized). Many ancient ooids possess a radial fabric comprising crystallites of calcite growing out from a central core with concentric bands outlined by layers of included solid and fluid inclusions (Figure 4.11D). Other ancient ooids are heavily altered (further discussion is given in Chapter 19). Sandberg (1975, 1983), Wilkinson (1979), and Mackenzie and Pigott (1981) examined ancient ooids that are now all low-magnesium calcite in order to determine their original mineralogy. Sandberg (1983) in particular noticed that ooids he interpreted to have originally been aragonite (commonly heavily altered) were confined to certain time intervals (corresponding to aragonite seas), and were absent from time intervals with abundant

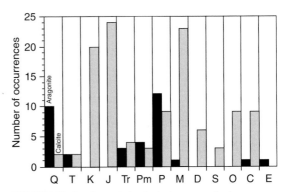

FIGURE 4.12. Secular variation in the interpreted primary mineralogy of marine ooids, showing Phanerozoic data from Sandberg (1983) and Ediacaran (latest Proterozoic) data from Singh (1987) (courtesy of L. A. Hardie).

radial, unaltered ooids that he interpreted to have had an original low-magnesium-calcite mineralogy (calcite seas – Figure 4.12). Hardie (1996) extended Sandberg's data on ooids with data from the Ediacaran (Singh, 1987) and added data on (1) marine cements that were also reckoned to have had an original aragonite mineralogy that had been converted to low-magnesium calcite during diagenesis and (2) evaporite mineralogy that also revealed a secular variation in primary mineralogy (see below).

Siliceous sediments and sedimentary rocks

Siliceous chemical sediments are similar to carbonates in that they include biogenic deposits (particularly of pelagic organisms), abiogenic spring deposits called sinter, silcretes (analogous to calcretes), and abiogenic non-marine fine-grained deposits. Although both silica and carbonates are important cements, silica minerals commonly replace diagenetically other sediment types, most notably carbonates, evaporites, pyroclastic deposits, and wood. Cementation and replacement of other sediment grains by silica minerals are mentioned here and further discussed in Chapter 19. References on siliceous sediments and sedimentary rocks include Pettijohn (1975), Hesse (1990a, 1990b), Chester (2003), Knauth (2003), and Maliva *et al.* (2005).

Silica mineralogy

The important silica minerals in chemogenic and biogenic sediments are opal-A, opal-CT, microcrystalline quartz, and chalcedony. Amorphous, hydrous opal ($SiO_2 \cdot nH_2O$), commonly called opal-A, contains

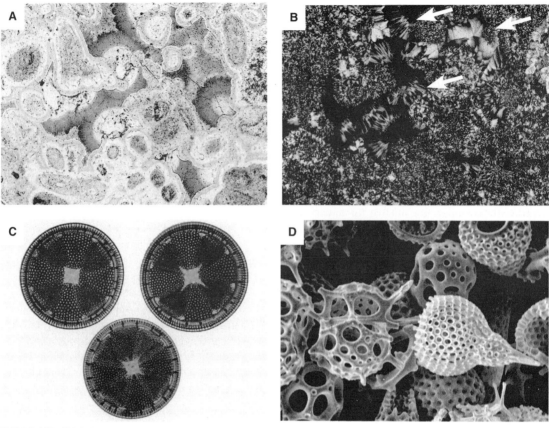

FIGURE 4.13. (A) A thin-section photomicrograph under plane-polarized light showing a variety of quartz types replacing limestone and cementing the replaced grains. (B) As in (A) under crossed-polarized light. Most of the thin section comprises microquartz, with the arrow pointing to fibrous chalcedony. Width of field in (A) and (B) ~1 mm. Both from Adams *et al.* (1984). (C) A grain mount of diatom skeletons. Width of field ~0.5 mm. From Scholle and Ulmer-Scholle (2003). (D) A scanning electron photomicrograph of radiolarian shells. Width of field ~375 microns. From Scholle and Ulmer-Scholle (2003). AAPG © 2003 reprinted by permission of the AAPG, whose permission is required for further use.

~10% water and is the most common silica phase used by modern organisms to make skeletons. Opal-CT is typically found as a diagenetic mineral, contains less water than opal-A, and has weak X-ray peaks of cristobalite and trydimite (polymorphs of quartz). Microcrystalline quartz (Figure 4.13A, B) comprises mosaics of α-quartz crystals that are up to a few tens of microns in size (submicron microcrystalline quartz is sometimes referred to as cryptocrystalline quartz). In microcrystalline mosaics, the quartz crystals commonly have *compromise boundaries*, which are most easily visualized as the boundaries that soap bubbles have in a froth. The boundaries are curvilinear planes, and in this geometry the crystals have minimized their surface free energies. Chalcedony is fibrous microcrystalline quartz, commonly found as radiating bundles of micron-diameter fibers filling voids (i.e., as cements, Figure 4.13A, B). The *C* axes of the quartz crystals

are usually not parallel to the elongate fibers. Chalcedony has quartz X-ray peaks: however, there is ~1% H_2O in the fibers and they have an anomalously high refractive index under a petrographic microscope.

Chert is a rock composed mainly of microcrystalline quartz. Chert has conchoidal fracture and is commonly white to gray, but small amounts of impurities can give it different hues. Hematite gives chert a red hue and such chert is called jasper. Black and dark-gray chert (called *flint*) was commonly used by early humans for arrowheads, axes, and so on. Chert can be nodular and bedded, as described below.

Silica geochemistry

River water typically contains less than 15 ppm dissolved silica, shallow meteoric groundwater typically has 10–50 ppm dissolved silica, and the ocean has between 0.5 and 10 ppm dissolved silica. Silica concentrations in the ocean are lowest in the surface layer, and fairly constant at around 10 ppm below the thermocline. Deep saline sodium chloride and calcium chloride groundwaters contain comparatively little dissolved silica (30–80 ppm), although some saline waters in shales that have dissolved organic acids contain up to 330 ppm dissolved silica.

At normal environmental pH, the dissolution–precipitation reaction of quartz,

$$SiO_{2(s)}{}^{quartz} + H_2O_{(1)} \leftrightarrow H_4SiO_4{}^0{}_{(aq)} \qquad (4.6)$$

produces non-ionized silicic acid ($H_4SiO_4{}^0$). At 25 °C, this reaction puts only approximately 6 ppm silicic acid into solution because quartz is not very soluble. Therefore, most of the silicic acid in river water and groundwater is reckoned to come from the incongruent dissolution of silicate minerals such as feldspars during weathering (Chapter 3). Non-crystalline amorphous silica gels are considerably more soluble, putting up to 120 ppm silicic acid into solution. The solubility of quartz is significantly affected by an increase in temperature, and at 300 °C approximately 600 ppm silicic acid can be dissolved in groundwater of normal pH. Where such hot water rises to the surface in hot springs, siliceous *sinter* is produced (this is analogous to carbonate travertines), and in the subsurface these waters also cause considerable silica mineral replacement and other ore-deposit mineralization. Another way to affect the solubility of quartz significantly is to raise the pH. At pH > 9 (at 25 °C) silicic acid dissociates:

$$H_4SiO_{4(aq)} \leftrightarrow H^+{}_{(aq)} + H_3SiO_4{}^-{}_{(aq)} \qquad (4.7)$$

With the elevated pH of alkaline water, this reaction is driven to the right and the solubility of both quartz and amorphous silica is greatly enhanced (see Figure 3.7). Alkaline lakes can contain up to 645 ppm dissolved silica as $H_3SiO_4{}^-{}_{(aq)}$.

Modern siliceous sediments

Modern siliceous sediments accumulate as biogenic marine oozes, biogenic freshwater-lake deposits, chemical precipitates in alkaline lakes, chemical precipitates in soils (silcrete), and chemical precipitates around subaqueous and subaerial hot springs.

The most important modern siliceous marine sediments are opal-A skeletons of planktonic diatoms (Figure 4.13C), a group that scavenges virtually all of the silica in the surface waters of the Earth. Diatoms arose in the Mesozoic, but have become particularly abundant over the past 30 million years and now dominate silica deposition in the oceans (Knauth, 2003; Katz *et al.*, 2004). The oceans are everywhere undersaturated with amorphous silica, so diatoms build their shells in defiance of thermodynamics, and perhaps 90% of dead diatoms' tests dissolve before they settle to the sea bed. However, appreciable thicknesses of diatom oozes can accumulate on the sea floor beneath areas with high productivity of diatoms, namely polar areas and areas of upwelling where there is a high flux of sinking diatoms, such as off the California coast. Other silica-secreting, single-celled eukaryotic plankton include the heterotrophic radiolarians (Figure 4.13D) and silicoflagellates. Radiolarian oozes are common along equatorial zones of upwelling (see Figure 18.19, in Chapter 18, for the distribution of siliceous biogenic marine sediments). Pore waters of siliceous oozes remain undersaturated with respect to amorphous silica for some depth below the surface, and diagenetic reorganization of silica is common in deep-sea siliceous sediments. This involves the slow conversion of opal-A to opal-CT, and eventually to microcrystalline quartz, via a complicated series of pathways involving quartz cementing and replacing the oozes (Hesse, 1990a). Indeed, Deep Sea Drilling Project (DSDP) drilling has encountered chert layers within otherwise unconsolidated Eocene siliceous oozes. The oldest sedimentary opal-A preserved in DSDP cores is Cretaceous in age.

Another important group of marine multicellular benthic organisms that have opaline silica skeletal elements is the glass sponges. Their skeletal elements are known as *spicules* and can be fused into massive skeletons, but are more commonly loose elements of sediment. Sponge spicules can comprise a few percent of the mud fraction on continental shelves, and spicules can accumulate as nearly pure deposits in some places.

Diatoms are the most important photosynthetic group in freshwater lakes (Chapter 14). Diatoms characteristically undergo explosive population growth in the spring and fall of the year in temperate lakes. This is when the lakes turn over, allowing nutrient-rich bottom waters to reach the surface at times when the temperature is high enough to support diatom growth. Most freshwater lakes are undersaturated with amorphous silica, so the diatom tests dissolve as they sink, and continue to dissolve in the bottom sediment.

The alkaline marginal-marine lagoons and lakes of the Coorong district of South Australia are sites of abiogenic opal-CT precipitation. The precipitation is within the sediment or just at the surface as opal gels that can comprise up to 6% by weight of the high-magnesium carbonate sediment (Peterson and von der Borch, 1965). The source of the silica is reckoned to be *in situ* dissolution under seasonal high-pH conditions of quartz, which reprecipitates as opal-CT where subsurface groundwater with lower pH contacts the surface brines. A chert found in Pleistocene sediments of an alkaline lake that occupied the Lake Magadi Basin (Eugster, 1969) comprises compact, well-indurated layers and nodules in otherwise unconsolidated lake sediments. The chert contains shrinkage cracks, complicated folds, and laterally grades into gels of hydrous sodium silicate minerals such as magadiite ($NaSi_7O_{13}(OH)_3 \cdot 3H_2O$). These gels are replaced in a piecemeal fashion by chert prior to compaction (Schubel and Simonson, 1990), and in places preserve laminae in the original magadiite. The deformation features are interpreted as recording the shrinkage of the gels as they lost sodium and dehydrated.

Silcretes are mostly known from southern Africa and Australia and apparently form under the same weathering conditions as calcretes and ferricretes (Summerfield, 1983; Hesse, 1990b, Bustillo, 2003). Silcretes comprise profiles up to 3 m thick with various amounts of silica cementation and replacement. Closely associated with silcretes are so-called groundwater silcretes, which are formed not by weathering, but by silica cementation and replacement at the water table.

Subaerial siliceous sinters exhibit many of the same features as calcareous travertines, forming mounds, terraces, and pools around hydrothermal vents and geysers. The sediments consist of light-colored laminated masses of opal-A that can be stromatolitic. Siliceous sinters have attracted the attention of astrobiologists because of their potential to preserve thermophilic archaea. Photographs and descriptions can be found in Walter (1976), Lowe *et al.* (2001), Lowe and Braunstein (2003), and Maliva *et al.* (2005). Submarine siliceous sinters are described below under iron deposits.

Ancient cherts

The Miocene Monterey Formation of California is a well-known siliceous sedimentary deposit and displays every gradation from pure unconsolidated diatomite through porcelain-like beds to beds of microcrystalline chert (Knauth, 2003). These deposits also show the stepwise replacement of opal-A with opal-CT, and ultimately microcrystalline α-quartz, that would later be seen by DSDP drilling of sea-floor diatom deposits (Chapter 18). Similar Cenozoic diatomaceous deposits

are preserved in the circum-Pacific area. Ancient diatom-rich clay deposits are commonly used as kitty litter. Paleozoic and Mesozoic bedded cherts include the Devonian white novaculite cherts of Arkansas and Texas (Folk and McBride, 1976), and bedded cherts of the Alps (Grunau, 1965; Pettijohn, 1975). Chemical etching of these bedded cherts reveals ghosts of radiolarians, and these cherts are now interpreted as recording times and places where radiolarians were abundant enough to accumulate in deep-sea settings similar to those of Cenozoic diatomaceous deposits. The Devonian Rhynie and Windyfield cherts of Scotland are interpreted as being subaerial siliceous sinter deposits that entombed some of the earliest land plants known. The plants grew in pools associated with the springs and were replaced and cemented by chert, preserving many fine details of the plants' anatomies (Trewin and Rice, 1992; Trewin, 1994).

Many ancient limestones contain nodules of chert (Pettijohn, 1975; Maliva and Siever, 1989; Hesse, 1990b; Knauth, 2003; Maliva *et al.*, 2005). The nodules range from simple ovoid shapes to quite complicated anastomozing forms. Primary stratification commonly passes through the boundaries of the nodules, and carbonate skeletal fragments are also replaced by silica within the nodules. Nodules can be found within single limestone layers, but they also commonly cut across bedding. Since these cherts cut across primary stratification and replace carbonate grains, they are diagenetic replacements of portions of the limestone (Chapter 19). They formed from carbonate-undersaturated, silica-saturated groundwater at some point after deposition of the limestone. In some cases, the replacement was prior to any major compaction, and sedimentary features are better preserved in the cherts than in adjacent compacted limestones. Chert nodules from Precambrian carbonates have provided most of the Precambrian microfossils. In addition to limestone replacements, chert replacements are common in many other sedimentary rocks. Chert commonly replaces plant fossils and volcaniclastic sediments.

Bedded cherts and silicified strata are common in Archaean greenstone belts. Also abundant in the early Proterozoic are the enigmatic iron-rich deposits. These iron deposits accumulated in deep basins as well as on continental shelves, and contain nearly equal volumes of chert and iron minerals. In the basinal deposits, these rocks commonly comprise striking millimeter- to centimeter-thick alternations of red-colored chert and dark-colored iron minerals that have earned these rocks the name *banded iron formations* (which are further described below). The chert-rich iron formations deposited in shallower-water settings tend not to be so rhythmically layered but are instead cross-bedded rocks composed of chert and iron peloids, some of which contain chert stromatolites. These chert-rich iron deposits are further described below.

Saline minerals (evaporites)

Saline deposits are purely chemogenic sediments and, because they commonly form from the evaporative concentration of an aqueous solution, are referred to as evaporites. Evaporite deposits require the following unique conditions to form: (1) an arid climate where the annual rate of evaporation exceeds inflow; (2) a hydrologically closed or restricted basin; and (3) a substantial inflow bringing solutes into the basin over a long period of time. Evaporite deposits are of major economic importance and are mined on all continents except Antarctica for raw materials for manufacture of inorganic chemicals. Important evaporite minerals include halite (NaCl), gypsum ($CaSO_4 \cdot 2H_2O$), sylvite (KCl), sodium carbonate salts, and sodium sulfate salts (Table 4.2). Modern surface brines and subsurface brines are mined for elements such as chlorine, bromine, and lithium. In addition, evaporites provide caps and seals for about half the world's known oil and gas reserves. References on evaporites include Sonnenfeld (1984), Handford et al. (1985), Hardie et al. (1985), Drever (1997), Hardie (1991, 2003a, b), and Warren (2006).

Evaporite mineralogy

Gypsum, anhydrite, and halite are the most common minerals in evaporites. Most of the other salts listed in Table 4.2 precipitate after the brines have become concentrated enough to precipitate halite. These so-called *bittern salts* are relatively rare but provide the most information on the compositions of the evaporating brines. Most of the 30 or so common evaporite minerals (Table 4.2) are composed of the dominant inorganic substances dissolved in natural waters: Ca^{2+}, Mg^{2+}, Na^+, K^+, HCO_3^- (or CO_3^{2-}), Cl^-, and SO_4^{2-}. Some of these minerals have the same compositions except for the amount of water in the structure, for example gypsum ($CaSO_4 \cdot 2H_2O$) and anhydrite ($CaSO_4$), mirabilite ($Na_2SO_4 \cdot 10H_2O$) and thenardite (Na_2SO_4), and kieserite ($MgSO_4 \cdot H_2O$), hexahydrite ($MgSO_4 \cdot 6H_2O$), and epsomite ($MgSO_4 \cdot 7H_2O$). Many of these saline minerals are highly soluble, so they precipitate only from very concentrated brines when most of the water

TABLE 4.2. Saline minerals

Carbonates	
Aragonite	$CaCO_3$
Calcite	$CaCO_3$
Mg-Calcite	$Ca_{1-x}Mg_xCO_3$
Dolomite	$CaCO_3 \cdot MgCO_3$
Nahcolite	$NaHCO_3$
Natron	$Na_2CO_3 \cdot 10H_2O$
Trona	$NaHCO_3 \cdot Na_2CO_3 \cdot 2H_2O$
Shortite	$2CaCO_3 \cdot Na_2CO_3$
Gaylussite	$CaCO_3 \cdot Na_2CO_3 \cdot 5H_2O$
Pirssonite	$CaCO_3 \cdot Na_2CO_3 \cdot 2H_2O$
Chlorides	
Halite	$NaCl$
Sylvite	KCl
Bischofite	$MgCl_2 \cdot 6H_2O$
Antarcticite	$CaCl_2 \cdot 6H_2O$
Carnallite	$KCl \cdot MgCl_2 \cdot 6H_2O$
Tachyhydrite	$CaCl_2 \cdot 2MgCl_2 \cdot 12H_2O$
Sulfates	
Gypsum	$CaSO_4 \cdot 2H_2O$
Anhydrite	$CaSO_4$
Thenardite	Na_2SO_4
Mirabilite	$Na_2SO_4 \cdot 10H_2O$
Bloedite	$Na_2SO_4 \cdot MgSO_4 \cdot 4H_2O$
Kieserite	$MgSO_4 \cdot H_2O$
Hexahydrite	$MgSO_4 \cdot 6H_2O$
Epsomite	$MgSO_4 \cdot 7H_2O$
Glauberite	$CaSO_4 \cdot Na_2SO_4$
Langbeinite	$2MgSO_4 \cdot K_2SO_4$
Schoenite	$MgSO_4 \cdot K_2SO_4 \cdot 6H_2O$
Polyhalite	$2CaSO_4 \cdot MgSO_4 \cdot K_2SO_4 \cdot 2H_2O$
Other	
Kainite	$MgSO_4 \cdot KCl \cdot 2.75H_2O$
Magadite	$Na_2Si_7O_{13} \cdot 3H_2O$

has been removed by evaporation. Such minerals are easily dissolved and recrystallized. Diagenetic alteration of evaporites is common (Hardie et al., 1985), as is further discussed in Chapter 19.

Chemical divides and the minerals precipitated from evaporating water

Although evaporative concentration of natural water is most important for the precipitation of the saline minerals, precipitation can also be induced by mixing of different waters, microbially mediated reactions, and

TABLE 4.3. Saline minerals and their parent brine types (modified after Hardie, 1984, Table 4). Reprinted by permission of the *American Journal of Science*.

Brine composition and source	Major saline minerals (more or less in the order formed)	Key indicator minerals
(1) Na–K–CO$_3$–Cl–SO$_4$ Non-marine, meteoric waters	Alkaline-earth carbonates, gaylussite, pirssonite, mirabolite, thenardite, trona, nahcolite, natron, halite, sylvite	Na$_2$CO$_3$ minerals
(2) Na–K–Mg–Cl–SO$_4$ Seawater, meteoric waters, volcanogenic waters, diagenetic waters	Alkaline-earth carbonates, gypsum, anhydrite, halite, mirabolite, thenardite, glauberite, polyhalite, epsomite, hexahydrite, kieserite, bloedite, kainite, carnallite, sylvite, bischofite	MgSO$_4$ or Na$_2$SO$_4$ minerals
(3) Na–K–Mg–Ca–Cl Non-marine waters, principally hydrothermal waters	Alkaline-earth carbonates, gypsum, anhydrite, halite, sylvite, carnalite, bischofite, tachyhydrite, antarcticite	KCl and CaCl$_2$ minerals in the absence of Na$_2$SO$_4$ and MgSO$_4$ minerals

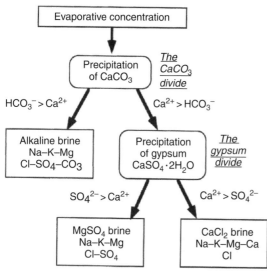

FIGURE 4.14. Chemical divides and the evolution of three major brine types (from Hardie, 2003a).

temperature changes. Lawrence Hardie (Hardie and Eugster, 1970; Eugster and Hardie, 1978) was the first to point out that the types and volumes of saline minerals formed during evaporative concentration, and their order of formation, are basically controlled by a series of *chemical divides* that force different types of evaporating waters to take fundamentally different evolutionary paths (Figure 4.14, Table 4.3). These paths are determined by the early formation of the

relatively insoluble carbonate minerals (low-magnesium calcite, high-magnesium calcite, or aragonite) and gypsum. The Mg^{2+}/ Ca^{2+} ratio in the brine controls whether the first-formed carbonate is low-magnesium calcite (Mg^{2+}/ Ca^{2+} < 2), high-magnesium calcite (2 > Mg^{2+}/ Ca^{2+} > 5), or aragonite (Mg^{2+}/ Ca^{2+} > 5). As the carbonate mineral is precipitated, either the Ca^{2+} is used up, or the HCO$_3^-$ is used up. If the Ca^{2+} is used up, an alkaline brine (pH > 10 – where the dominant carbonate species is CO$_3^{2-}$) forms with Na^{2+}, K$^+$, SO$_4^{2-}$, and Cl$^-$ as the other predominant dissolved species. Upon further evaporation, such an alkaline brine cannot produce gypsum (because all the Ca^{2+} has been used up) but instead produces a diagnostic suite of sodium carbonate minerals such as natron and trona. At very high concentrations, these brines can produce halite and sylvite (Table 4.3). Modern examples of this type of brine are found in Mono Lake, California, and Lake Magadi, Kenya. The most famous ancient example of this type of saline deposit is the Eocene Green River Formation of the western USA. If, on the other hand, the HCO$_3^-$ is used up during the initial formation of carbonates, the water will become enriched in Ca^{2+} and Mg^{2+}, and a neutral brine very depleted in carbonate is formed. The next chemical divide reached by a water depleted in HCO$_3^-$ occurs upon the precipitation of gypsum (CaSO$_4 \cdot 2$H$_2$O). As gypsum is precipitated, either the Ca^{2+} in the brine is used up, or the SO$_4^{2-}$ in the brine is used up. If the Ca^{2+} is depleted, a calcium-free brine rich in Na^+, K^+, Mg^+, Cl^-, and SO_4^{2-} will be the final product and the diagnostic minerals formed will include sodium sulfate and magnesium sulfate minerals. If, however, the sulfate is used up, the final brine will be rich in Na^+, K^+, Mg^{2+}, Ca^{2+}, and Cl^-. Such a calcium chloride brine will produce diagnostic potassium chloride and calcium chloride minerals such as sylvite, tachyhydrite, and antarcticite (Table 4.3). Any brine in which $Ca^{2+} > HCO_3^-$ will produce carbonate, gypsum, and halite as its first-formed and most abundant minerals.

Although the theory of chemical divides is important, a few problems complicate its application to saline-mineral deposition. Two problems arise from the thermodynamics and kinetics of the precipitation processes: (1) whether or not the evolving brine back-reacts with previously formed minerals (i.e., whether fractional crystallization prevails or not); and (2) whether or not metastable phases precipitate out of the evolving brine. In addition, the concentration of evaporating water can vary in space as well as in time. Concentration gradients are commonly found in evaporative lagoons (and salt evaporators), whereby concentrations can be maintained at halite saturation (for example) by continued inflow (and withdrawal of the concentrating brine). In natural systems, dense evaporated brines can leak out as dense underflows or as groundwater. Other problems are related to *recycling* of previously formed evaporite minerals and to diagenesis. It is quite common for dry evaporative environments to have ephemeral crusts of gypsum, and particularly halite, on the surface. Where undersaturated floodwater inundates these surfaces, it dissolves the mineral crusts and in effect becomes instant brine at or near saturation with gypsum or halite. In other words, this recycling process selectively adds solutes to the brines and thereby alters their chemistry.

Sources of evaporating water

Natural water that evaporates to form saline minerals includes (1) seawater; (2) meteoric, chemical-weathering water (including river flow and groundwater discharge); (3) deep saline groundwater; and (4) mixtures of the various types (Smoot and Lowenstein, 1991).

Seawater is the largest source of water for evaporite basins. Usiglio (1894) directly evaporated Mediterranean seawater and observed the mineral sequence. van 't Hoff and his students were the first to apply laboratory measurements of solubility to the problem of theoretically predicting the mineral sequence that should precipitate from evaporating seawater (Faure, 1991). The solubility data were used to assemble the complicated phase diagrams necessary to depict seawater evaporation. The final breakthrough in understanding evaporite-mineral precipitation came with the development of the EQUIL computer program (Figure 4.15) (Harvie et al., 1980, 1984). EQUIL used Pitzer theory to calculate activity coefficients at high ionic strengths and thermodynamics to calculate the mineral sequences.

The chemistry of some dilute surface water and groundwater is given in Table 3.3 in Chapter 3. Water with slightly different chemical composition commonly flows into the closed basins of the southwestern USA (Chapter 16), and the specific chemical composition of the inflow water depends on the type of bedrock weathered (Chapter 3). The evolution of these inflows as they undergo evaporative concentration, and the sequences of minerals they produce in the various closed basins, can be explained by the theory of chemical divides (Hardie and Eugster, 1970; Eugster and Hardie, 1978).

It has long been known that the sequence and masses of evaporite minerals in most ancient evaporite deposits (except late-Permian evaporites) do not match the sequence and masses predicted for modern seawater (Stewart, 1963; Borchert and Muir, 1964; Braitsch, 1971; Hardie, 1984, 1991). Most workers assumed that the composition of seawater had remained constant through time (e.g., Holland, 1972) and, because many evaporite basins had to be almost completely surrounded by land, it was reckoned that the compositions of ancient evaporites reflected recycling of salts, contamination of seawater by other water (particularly meteoric water or deep saline groundwater), and later diagenetic alteration of the evaporites (Hardie, 1984). Many evaporite basins are formed by tectonic processes such as rifting that allow the input of deep saline groundwater as well as meteoric groundwater. Currently, seawater contains $\sim 3.5\%$ by weight of dissolved inorganic solutes, and is principally a Na^+, Mg^{2+}, Ca^+, Cl^-, SO_4^{2-} water. Meteoric water is derived from the chemical and biochemical weathering reactions described in Chapter 3, and is commonly dilute, with Na^+, Ca^{2+}, and HCO_3^- as the main ions. However, the chemical composition varies depending on the type of bedrock weathered. Deep saline groundwater is a mixture of diagenetic water, hydrothermal reaction water, and volcanogenic water. Diagenetic water is formed by the low-temperature (below $\sim 200\,°C$) and low-pressure (below ~ 1 kilobar) interactions between sediment and pore water. These

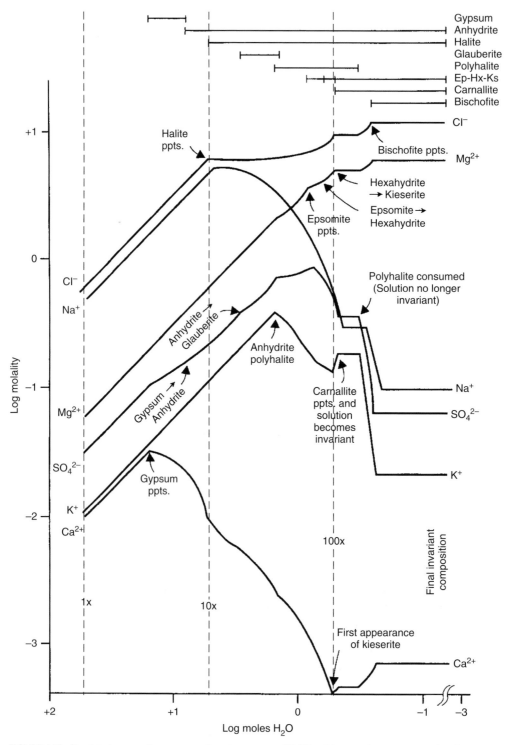

FIGURE 4.15. Simulated evaporative concentration of seawater at 25 °C and 1 atmosphere pressure using the computer program of Harvie *et al.* (1980), showing the evolution of brine chemistry, and, at the top, the sequence of minerals precipitated. The dashed vertical lines are at 1, 10, and 100 times the concentration of seawater. Modified from Hardie (2003a).

reactions (reviewed in Chapter 19) include the conversion of calcite and aragonite to dolomite, the conversion of calcium plagioclase to albite, and reactions among the clay minerals. Hydrothermal reaction water is produced by the circulation and interaction of hot groundwater with igneous, sedimentary, or low-grade-metamorphic rocks, and is distinctive in that it is commonly rich in Na^+ and Ca^{2+}, with Cl^- as the dominant anion. Moreover, at least part of the Ca^{2+} is balanced by Cl^-, leading these brines to be called *calcium chloride brines* by Hardie (1990).

Hardie (1996) showed that systematic changes in the late-stage (post-halite) evaporite minerals paralleled the calcite seas and aragonite seas postulated by Sandberg (1983). Evaporites formed from aragonite seas generally included $MgSO_4$ minerals in the late stages of evaporation, whereas late-stage evaporites from calcite seas are more likely to contain KCl minerals (Figure 4.16). Hardie (1996) therefore interpreted aragonite versus calcite seas as times when seawater was on different sides of the gypsum divide. During times of calcite seas, in terms of concentration $Ca^{2+} > 2HCO_3^- + SO_4^{2-}$ in seawater so that, upon gypsum precipitation, the concentration of sulfate declined and no sulfate minerals could be formed. During times of aragonite seas, in terms of concentration $Ca^{2+} < 2HCO_3^- + SO_4^{2-}$ in seawater so that, upon gypsum precipitation, the concentration of Ca^{2+} declined and magnesium sulfate minerals could be formed. The most reliable source of information on ancient brine composition has been from fluid inclusions, tiny pockets of brine that the crystals are growing from that are trapped along the irregular growth faces of the minerals. Lowenstein *et al.* (2001) and Horita *et al.* (2002) analyzed evaporated brines in fluid inclusions from ancient evaporites and used these data to back-calculate ancient seawater composition. Lowenstein *et al.* (2001) analyzed evaporites from several time periods in different geographic areas in order to get global seawater composition as opposed to a local mixed seawater composition. Results of these studies and Dickson (2002) confirmed that there have been measurable changes in the major-ion chemistry of seawater similar to those postulated by Hardie (1996, see Figure 2.30); see Figure 4.17.

Evaporite mineral textures and sedimentary structures

The textures and sedimentary structures of evaporite minerals are intimately related to the environment of

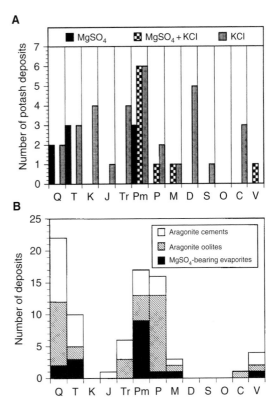

FIGURE 4.16. Evidence for secular changes in seawater composition. (A) Secular variation in three main types of marine evaporites, two that contain magnesium sulfate bitterns and one that doesn't (Zharkov, 1984; Hardie, 1990) (courtesy of L. A. Hardie). (B) Comparison of secular changes in ooid and cement mineralogy (Figure 4.12) and magnesium sulfate-bearing evaporites (A) above. From L. A. Hardie (1996). Periods of aragonite seas correspond to magnesium sulfate-bearing evaporites; periods of calcite seas correspond to KCl evaporites with no magnesium sulfate.

deposition. These features are therefore more fully discussed in Chapter 16 on arid environments, where evaporites are important components of the depositional environment.

Iron minerals and sedimentary grains

Iron is one of the most abundant elements of the Earth's crust (7%–8% by mass; see Table 2.2), and few sediments and sedimentary rocks are iron-free. Vanishingly small amounts of oxidized iron are transported dissolved in surface water, and small amounts of iron are transported as suspended *colloids*. Most subsurface water is anoxic, however, and has a few ppm of dissolved iron. Table 4.4 is a list of the names and chemical formulae for the most common iron

TABLE 4.4. Sedimentary iron minerals (underlined minerals are important components of iron-formation "facies")

Oxides	
Magnetite	Fe_3O_4
Hematite	Fe_2O_3
Goethite	$FeO \cdot OH$
Ilmenite	$FeTiO_3$
Sulfides	
Pyrite	FeS_2
Marcasite	FeS_2
Pyrrhotite	$Fe_{(1-x)}S$
Greigite	Fe_3S_4
Mackenawite	FeS
Carbonates	
Siderite	$FeCO_3$
Ankerite	$(Fe,Ca,Mg)(CO_3)_2$
Silicates	
Greenalite	$(OH)_{12}Fe^{II}_9Fe^{III}_2Si_8O_{22} \cdot 2H_2O$
Minnesotatite (iron "talc")	$Fe_3(OH)_2Si_4O_{10}$
Stilpnomelane	$(OH)_4(K,Na,Ca)_{0-1}(Fe,Mg,Al)_{7-8}$ $Si_8O_{23-24} \cdot 24H_2O$
Chamosite (iron "chlorite")	$(Mg,Fe,Al)_6(Si,Al)_4O_{14}(OH)_8$
Glauconite (iron "biotite")	$K_2(Mg,Fe)_2Al_6(Si_4O_{10})_3(OH)_{12}$

minerals that occur as chemical and biochemical precipitates in sedimentary rocks. In ancient sedimentary rocks, it is commonly difficult to sort out which iron minerals were formed by depositional processes, which are products of diagenetic alteration, and which formed via weathering upon exhumation and exposure. At certain times in Earth's history, chemogenic (or perhaps biochemogenic) iron-rich (>20% iron) sediments have been deposited in great abundance. The apparent restriction of the iron- and chert-rich sedimentary deposits to the Archean and Proterozoic, commonly referred to as *iron formations*, provides a strong indication of the evolution of the lithosphere, atmosphere, hydrosphere, and perhaps biosphere through geological time (further discussed below). In the Phanerozoic era, there are less-extensive iron-rich chemogenic or biochemogenic sediments, called *ironstones*, containing aluminosilicate minerals and less chert (Pettijohn, 1975; Simonson, 2003a, b). Ironstones include the Silurian "Clinton" ores of the eastern USA and the Jurassic "Minette" ores of western Europe. Ironstones have been mined since antiquity, but Proterozoic iron formations contain the vast majority of the world's iron reserves and host large iron-ore bodies that formed via secondary enrichment processes (Clout and Simonson, 2005). Archean iron formations in greenstone belts

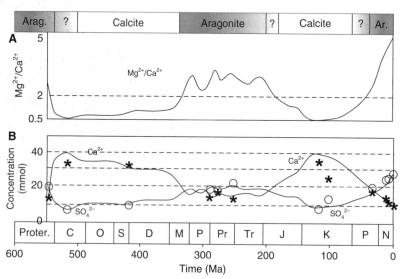

FIGURE 4.17. Secular changes in seawater. (A) Variation in the Mg^{2+}/Ca^{2+} ratio calculated by Demicco *et al.* (2005). At Mg^{2+}/Ca^{2+} ratios >2, high-magnesium calcite and aragonite precipitate, whereas below this value calcite precipitates. Compare this with the distribution of calcite seas and aragonite seas. (B) The Ca^{2+} molality (asterisks) and SO_4^{2-} molality (circles) back-calculated from brine compositions measured in fluid inclusions in marine halite (Lowenstein *et al.*, 2001; Demicco *et al.*, 2005). Curves are calculated changes in molalities of these ions in seawater (Demicco *et al.*, 2005). Where the curves cross, ancient seawater passes back and forth over the gypsum divide (Figure 4.14). When $Ca^{2+} > SO_4^{2-}$, all the sulfate is used up in the precipitation of gypsum and magnesium sulfate bitterns do not form. These times correspond to KCl evaporites and calcite seas. When $Ca^{2+} < SO_4^{2-}$, all the calcium is used up in the precipitation of gypsum and sulfate builds up in the brine, leading to magnesium sulfate bitterns.

also commonly host gold mineralization (Simonson, 2003a, b).

Modern environments of iron-mineral formation

Oxygen is abundant in most weathering environments, and the pH is in the range 4–8. Therefore, iron is quite immobile, and iron oxyhydroxides and iron oxides tend to accumulate in weathering profiles as more soluble elements are leached. In some subtropical areas such as South Africa and Western Australia, the iron concentrated in ultisols and oxisols forms a 1–2-m-thick calcrete-like accumulation in the B horizon, known as a ferricrete (a type of laterite). Modern ferricretes are composed of hematite and goethite pisoids and laminated crusts.

If there is no free oxygen available, iron can go into solution as ferrous iron (Fe^{2+}). A pe of 0 is approximately equal to a partial pressure of oxygen of 10^{-4} atmospheres (the limit of detection for dissolved oxygen in solution). Environments where the partial pressure of oxygen is less than 10^{-4} atmospheres (and the pe is <0) are commonly referred to as anoxic, and the reaction

$$Fe_2O_{3(s)}^{\ hematite} + 6H^+_{(aq)} + 2e^- \leftrightarrow 2Fe^{2+}_{(aq)} + 3H_2O_{(1)} \tag{4.8}$$

becomes important. Since hydrogen ions are involved in this redox (oxidation–reduction) reaction, the pH must also be fixed. Low pe is common in bogs, and in the sediments at the bottoms of lakes, ponds, wetlands, bays, estuaries, and the ocean. Groundwater is also commonly anoxic. One of the biggest problems affecting large groundwater-extraction wells is the deposition of iron oxyhydroxides in well screens as the extracted water comes into contact with the atmosphere. Crusts and cements of iron oxides are common at paleowater tables where reduced groundwater comes into contact with oxygenated soil gas. A similar reaction also occurs in bogs where highly reduced water with Fe^{2+} in solution comes into contact with oxygenated groundwater, resulting in precipitation of goethite and amorphous oxyhydroxides. Bacteria catalyze these reactions and induce the precipitation of iron. Bog iron ores were some of the first sources of iron used by early humans.

Where sediments beneath a water column contain even a few percent of organic matter, the dissolved oxygen in pore water of the upper few millimeters to meters of sediment is consumed by aerobic bacterial respiration. In deeper pore water from which the free

oxygen has been removed, anaerobic bacteria with a variety of metabolic pathways can degrade remaining organic matter and lower the pe even further (Chapter 2). Among the most common and best-known types of anaerobic bacteria are sulfate-reducing bacteria. These organisms reduce sulfate in the water to sulfide, using the oxygen cracked from the sulfate molecule as the electron acceptor in their respiration:

$$Fe_2O_{3(s)}^{\ hematite} + 4SO_{4\ (aq)}^{2-} + 8H^+_{(aq)}$$

$$\leftrightarrow 2FeS_{2(s)}^{\ pyrite} + \frac{15}{2}O_{2\ (g)} + 4H_2O_{(1)} \tag{4.9}$$

Under anoxic conditions, up to 80 ppm ferrous iron can be put into solution before the iron combines with the reduced sulfur to form small aggregates of amorphous iron sulfides (such as greigite and mackinawite) that eventually crystallize into pyrite and allied sulfides. These reactions account for the common occurrence of disseminated pyrite in gray and black mudstones. Iron carbonate (siderite – $FeCO_3$) may also form in this manner if there is a high pCO_2. The pyrite that is formed by diagenetic reorganization of the early-formed amorphous sulfides commonly occurs either as micron-diameter single crystals, or in small, raspberry-like particles known as *framboids* that may occur in clusters known as *polyframboids* (Ohfuji and Rickard, 2005; Wilken, 2003). The single crystals are usually subhedral to anhedral octahedrons and pyritohedrons, whereas framboids are usually less than 20 microns in size and are geometric aggregates of smaller crystals (Figure 4.18). Ordered, cell-like framboid aggregates have sparked a long-running debate as to whether they are biogenic or chemogenic features; see Wilken (2003, p. 702). Framboids and framboid-like aggregates without precise geometric packing have been produced abiogenically in laboratory settings (Fortin and Langley, 2005). Other references on bacterial mineralization include Lowenstam and Weiner (1989), Schulz and Zabel (2000), Mann (2001), Chester (2003), Dove et al. (2003), and Baeuerlein (2004).

Peloids composed of the iron silicates glauconite and chamosite are fairly common components of continental-shelf, slope, and rise sediments in depths from tens of meters to thousands of meters. In addition, chamosite ooids form in the shallows of Lake Chad. The fecal pellets that are most likely the precursors for these minerals apparently provide a *microreducing* environment. Recent experiments (Kim et al., 2004) have shown that, under the right conditions, bacteria

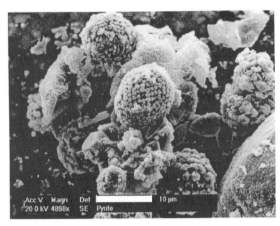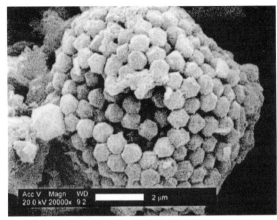

FIGURE 4.18. Scanning electron photomicrographs of pyrite framboids. From Wilken (2003). (A) Modern slope sediments from the Black Sea. Scale bar = 10 microns. (B) Cretaceous sediments. Scale bar = 2 microns.

can promote the reaction of smectites to illites, and it has long been suspected that bacteria may also be responsible for catalyzing the growth of glauconite and chamosite in modern marine and non-marine settings. Glauconite has a distinctive green color, and ancient glauconite-rich sands are known as greensands.

At low pH (<4), appreciable amounts of iron can be in solution, even in fully oxygenated settings. For example, there are hot springs in New Zealand with extremely acidic pH of around 1 that have 10,000 ppm ferric iron in solution. Acidic, anoxic, metal-rich waters vent from hot springs located on many mid-ocean ridges along ridge axes and just off the axes (see also Chapter 18). The deposits of these hot springs have potential economic importance, and provide the closest modern analogs to some iron formations and so-called Kuroko-type volcanogenic sulfide deposits (Ohmoto and Skinner, 1983). The fluids issuing from these springs range in temperature up to 400 °C and have pH of less than 4. This hot water has serpentinized basalts, and therefore has low SO_4^{2-} and Mg^{2+}, but elevated levels of Ca^{2+}, Fe^{2+}, and H_4SiO_4 as well as other metals such as Cu and Zn. Where this water issues from discrete conduits at the sea floor, it builds large (up to 45 m high and 30 m in diameter) *tufas* composed of metal sulfides, anhydrite, and opaline silica (Figure 4.19). As the hot (>250 °C) water mixes with the cold, oxygenated seawater, an opaque cloud of iron sulfide mud is precipitated, which is why these high-temperature vents are known as *black smokers*. In contrast, at lower temperatures (<250 °C), the vent discharge mixes with ambient seawater, and anhydrite, silica, and barite precipitate to produce a *white smoker*. The tufas that precipitate around black

and white smokers are commonly referred to as *chimneys*. Black-smoker tufa chimneys are principally composed of fine-grained crystals of pyrite, marcasite, chalcopyrite, and anhydrite, with lesser amounts of pyrrhotite and amorphous silica (Figure 4.19B). White-smoker-chimney tufas comprise pyrite, marcasite, sphalerite, amorphous silica, anhydrite, and barite (Hannington *et al.*, 1995). These hot-spring complexes also have substantial subsurface mineralization associated with them. The smokers have a lifespan of up to 10^4–10^5 years. At the end of a vent's lifetime, it can be surrounded by an apron of chimney breccia up to a few hundred meters in diameter that is mineralized in the subsurface (Figure 4.20). These vent-related deposits are analogous to iron-ore bodies known from rocks as old as 3.5 billion years (see the description of *Algoma-type* iron formations below).

Perhaps as much as 90% of the metals discharged by a typical black smoker is precipitated as metal sulfides in the hydrothermal plume and then carried away from the area. These iron sulfide chemogenic muds do not, however, settle out on the sea floor. Most of these mud particles are redissolved back into seawater, where they may form a source for the manganese–iron nodules that cover areas of the abyssal plains (see Chapter 18).

In the case of similar hydrothermal springs in the axial valleys of the Red Sea, the hydrothermal water has apparently interacted with evaporite deposits in the fault-block margins of the Red Sea, so the hydrothermal waters are not only hot (up to 40 °C), but also contain up to 250,000 ppm dissolved material, principally NaCl. The high salinity of this water makes it dense enough to accumulate in three or four deeper *pools* along the rift

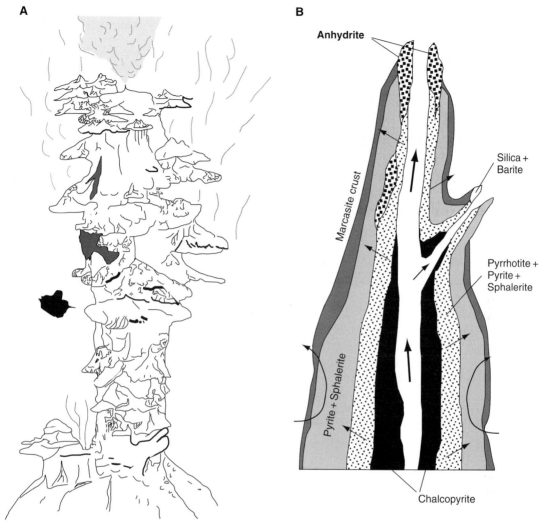

FIGURE 4.19. Black-smoker chimneys built around mid-ocean-ridge hydrothermal springs. From Hannington *et al.* (1995). (A) A sketch of a chimney ∼45 m tall, with the outline of submersible Alvin shown for scale. (B) A schematic diagram of the mineralogy of the chimney. Arrows indicate fluid-flow paths. From Hannington *et al.* (1995).

axis of the Red Sea. These hydrothermal brines contain appreciable amounts of iron and other dissolved metals, including manganese, zinc, and copper. Beneath the brine pools are spectacular laminated chemogenic, iron-rich muds interlaminated with pelagic sediments including pteropod, foraminiferal, and coccolith oozes together with amorphous opal, feldspar, and clay minerals (Figure 4.21) (Degens and Ross, 1969). The muds occur as four distinct layers up to 4.5 m thick, including (1) iron-rich smectite (clay); (2) amorphous iron oxyhydroxides and goethite; (3) manganosiderite and manganite; and (4) sulfides, including sphalerite, chalcopyrite, and pyrite. The iron-rich minerals occur mostly as 1–30-μm-diameter spherules. Nodular

anhydrite, morphologically similar to that found in modern saline mudflats, is also common. The exact precipitation mechanisms are not known, but precipitation most likely occurs along the interface between the brine and the cold, oxygenated bottom waters of the Red Sea. These deposits form substantial ore bodies, with one of the pools (known as the Atlantis II Deep) containing some 90 million metric tons (dry weight) of metal-rich sediments that average 45% iron, 2% zinc, and 0.5% Cu (Harrington *et al.*, 1995). Needless to say, this chemical sedimentary system has attracted a considerable amount of attention as an analog of large iron and sulfide deposits, including the iron formations of the Proterozoic (see below).

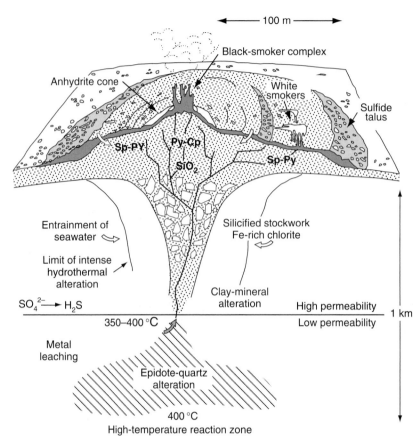

FIGURE 4.20. A schematic diagram of processes and products around mid-ocean-ridge hydrothermal vents. From Hannington *et al.* (1995). White arrows show flow paths: Sp = sphalerite, Py = pyrite, Cp = Chalcopyrite.

Ancient iron-bearing sedimentary rocks

Ancient iron-rich chemogenic deposits that are not closely associated with hydrothermal vent systems fall into two broad categories: (1) *iron formations* that are Archean and Proterozoic in age, contain chert, and are generally laminated; and (2) *ironstones* that are Phanerozoic in age, contain much less chert and more aluminum than iron formations, and are generally oolitic and bedded instead of laminated. The important iron minerals fall into four groups that have traditionally been used to define four facies of iron formation: oxide, silicate, carbonate, and sulfide (Table 4.4) (James, 1954). Most iron formations have been metamorphosed to some degree and many have been deformed. Less-altered iron formations can be divided into banded iron formations (BIFs) and granular iron formations (GIFs) (Simonson, 2003a, b; Clout and Simonson, 2005).

The BIFs are the most abundant of all chemogenic iron-rich sediments. They were deposited as chemogenic muds, most display millimeter- to centimeter-thick alternations of iron-rich and iron-poor chert layers, and they may contain oxides, silicates, carbonates, or sulfide minerals (Figure 4.22A). These BIFs can be up to hundreds of meters thick and persist laterally for hundreds of kilometers. Some iron formations exhibit significant facies changes laterally (e.g., Zajac, 1974) whereas others are remarkably uniform, to the point that individual millimeter-thick laminae can be correlated for over a hundred kilometers (Ewers and Morris, 1981). The BIFs can also exhibit cyclic variations in lamina thickness, in relative amounts of chert and iron minerals, and in the relative amounts of siliciclastic mudstone or volcaniclastic layers (Trendall and Blockley, 1970). Localized concretions of early chert cements, known as *chert pods*, are common. Differential compaction around the chert concretions indicates that the original chemogenic muds have been compacted by up to 70%–80% in some places.

Granular iron formations (GIFs) are rarer and composed mainly of sand-sized peloids (called *granules*) of chemogenic mud interpreted to be mostly reworked intraclasts of BIF material. Many GIFs

A

B

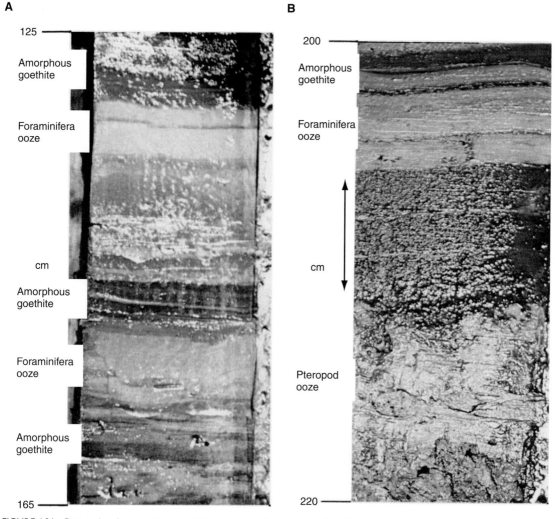

FIGURE 4.21. Cores taken from modern metal-rich sediments from brine pools at the bottom of the Red Sea. From Ross and Degens (1969). Sediments comprise interstratification of pelagic carbonate oozes (pteropods or foraminifera) with metallic chemical muds. Arrows in (B) outline zone of millimeter alternations of foraminiferal ooze and iron sulfides. By permission of Springer Science and Business Media.

contain an abundance of early chert, chalcedony, and quartz cement, and exhibit differential compaction features analogous to those of the chert pods in BIFs (Figure 4.22B). More rarely, GIFs contain layers of ooids, stromatolites, and flat pebbles of cherty BIF clasts. The iron-rich GIF sandstones are commonly cross stratified, and comprise oxide and silicate minerals. Layers of pure GIF thicker than a few meters are rare, and most are interstratified with BIFs. Shrinkage cracks, fenestrae, and vugs locally contain internal sediment with microfossils, and are filled with chalcedony or quartz cements. These features are interpreted to have formed due to post-depositional shrinkage as the gelatinous precursors dewatered (Simonson, 2003a,

b). Chert stromatolites and associated strata up to a meter thick in GIFs contain some of the best-preserved microfossils of the Proterozoic (Trendall and Morris, 1983; Simonson, 2003a, b).

Iron formations show a distinct change in character through time that is used to subdivide them into two main types: (1) an older *Algoma-type* and (2) a younger *Superior-type*. Algoma-type iron formations are almost always BIFs, are found worldwide in Archean greenstone belts, are one to two orders of magnitude smaller than Superior-type iron formations, and are associated with volcanic rocks, typically basaltic. In contrast, Superior-type iron formations are up to hundreds of meters thick, cover tens of thousands of square

FIGURE 4.22. Iron formations. (A) A banded iron formation (BIF) composed of gray iron sulfides and red chert. Variations in the relative proportions of chert and iron minerals and thicknesses in the layering give rise to "cycles." (B) A thin-section photomicrograph of a granular iron formation (GIF) where peloids ("granules") are chert and hematite that are cemented by chert. Width of field ~2.5 mm. From Simonson (2003a). In (C) and (D) are outcrop photographs of the Dales Gorge member of the Brockman Iron Formation, a Superior-type iron formation of the Hamersley Group, Western Australia.

kilometers, and commonly contain GIFs, although BIFs are still generally dominant (Figure 4.22C, D). Superior-type iron formations are associated mainly with sedimentary rocks such as shales, sandstones, carbonates, and some felsic volcaniclastics. Deposition of Superior-type iron formations happened worldwide for approximately 800 million years in the Neoarchean and Paleoproterozoic, then ceased at about 1.8 Ga, but with a short recurrence in the Neoproterozoic that is characterized by a hematite-dominated mineralogy and close association with glaciogenic deposits.

Some Algoma-type BIFs have been linked to submarine hydrothermal activity analogous to that which produces the deposits around black smokers, and all types of iron formations exhibit geochemical evidence of hydrothermal sources (Klein and Beukes, 1992; Simonson, 2003a, b). The closest modern analog to BIF is the metal-rich mud accumulating in the brine pools beneath the Red Sea, and it is now widely believed that hydrothermal water derived from alteration of the sea floor provided the iron and the silica so characteristic

of BIFs. In effect, large portions of deep ocean basins were filled with iron- and silica-rich hydrothermal fluids issuing from submarine hot springs. Much of the iron- and silica-rich precipitates that became iron formations probably formed as hydrous gels along *chemoclines* in those ocean basins. A chemocline is an abrupt change in the chemical properties of a water body analogous to the abrupt changes in temperature known as *thermoclines* (Chapter 14). As in the Red Sea, the exact details of precipitation are unknown, but data from isotopic studies suggest that microbes were involved, as they commonly are in redox reactions in modern environments (Beard *et al.*, 1999).

Superior-type deposits arose as continents began to form, suggesting that the change from Algoma-type to Superior-type iron formations is related to the development of large continental blocks. Chemoclines apparently were shallow enough towards the end of iron-formation deposition to cover shelf areas and to allow the sporadic deposition of cross-stratified GIFs. The demise of iron-formation deposition likely reflects the end of the time when the deep ocean served as a

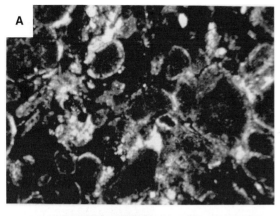

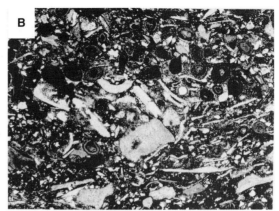

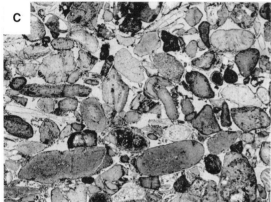

FIGURE 4.23. (A) Peloids of glauconite from Phanerozoic ironstone. Width of field ∼3 mm. From Scholle and Ulmer-Scholle (2003) AAPG © 2003 reprinted by permission of the AAPG, whose permission is required for further use. (B) Fossiliferous oolite ironstone from Phanerozoic ironstone. Larger fragments are echinoderm plates, and the coats on the ooids are chamosite and opaque iron oxides. Width of field ∼9 mm. From Adams et al. (1984). (C) Phosphatic pellets. Width of field ∼10 mm. From Adams et al. (1984).

large reservoir for dissolved iron. One idea holds that this transition is linked to the rise in concentration of oxygen in the atmosphere that took place during roughly 2.2–1.9 Ga (Holland, 1994), and that the deep oceans became too oxygenated to hold large quantities of dissolved iron. A newer theory suggests that iron solubility in the deep oceans was reduced as they became more sulfidic rather than oxic at this time (the so-called Canfield Ocean) (Anbar and Knoll, 2002). Steady cooling of the Earth with concomitantly decreased rates of hydrothermal activity may also have played a role. However, it is now accepted that black smokers have continued Algoma-type deposition in the world's oceans, apparently on a smaller scale, up to the present time. The chemogenic muds at the bottom of the Red Sea are richer in aluminum minerals than are Archean/Paleoproterozoic BIFs, and do not contain as much chert. The amount of chert is a key difference between modern and ancient oceanic hydrothermal deposition, as well as the difference between iron formations and ironstones. This probably reflects a transition from abiological silica deposition in the Archean and Proterozoic to the predominantly biologically controlled silica deposition of the Phanerozoic (Maliva et al., 2005).

Phanerozoic ironstones comprise a diverse group of iron-rich sedimentary rocks (Van Houten and Bhattacharyya, 1982; Van Houten, 1985; Cotter and Link, 1993). Ironstones contain detrital quartz, detrital carbonate grains, and phosphatic grains, but only minor amounts of chert. Ironstones have distinctive beds comprising hematite and chamosite ooids. The ooids develop around cores of quartz grains and fossil grains, particularly brachiopods and crinoid columnals (Figure 4.23B). The ooid-core fossil grains and other fossils around the ooids are partially to completely replaced by hematite. Much of the iron and phosphate is diagenetic, and occurs as cements and replacements. The ironstones contain shallow marine fossils, ooids and cross stratification, and were deposited in shelf settings where iron mineralization took place on or slightly below the sea floor, much as glauconite is formed today (Figure 4.23A). Most ironstones also underwent early replacement and cementation, which is possibly related to weathering alterations.

Phosphate minerals and sedimentary grains

Although $\sim$40 minerals are found in ancient phosphatic deposits, the most important mineral is usually referred to as carbonate fluorapatite (CFA). The chemical formula of CFA is given by Glenn and Garrison (2003) as $Ca_{10}[(PO_4)_{6-x}(CO_3)_x]F_{2+x}$ with many substitutions of cations and anions. Ancient and modern phosporite deposits are covered in Cook and Shergold (1986), Notholt *et al.* (1989), and Burnett and Riggs (1990). Modern phosphate-rich sediments are most common beneath eastern boundary currents, in areas where upwelling brings nutrient-rich water up to fertilize the surface waters. The resultant aerobic degradation of the rain of organic debris through the water column produces a so-called oxygen-minimum zone. The low pe of the pore waters allows PO_4^{3-} to be mobilized, and CFA is found as a diagenetic phase cementing and replacing pellets of clay minerals slightly beneath the sediment–water interface. The CFA also appears as ooids and pisoids, laminated crusts, and scattered nodules (Figure 4.23C). Microbes are suspected to be involved in the reactions. In addition, inarticulate brachiopod shells and vertebrate (usually fish) bones, teeth, scales, and armoring plates are also phosphatic. The latter are commonly found as *bone beds*. Phosphatic grains, phosphate-cemented hardgrounds, and concentrations of vertebrate remains are commonly interpreted as representing low sedimentation rates or complete cessation of sedimentation. Such layers are commonly interpreted as representing the deepest water reached during a marine transgression (see Chapter 20). These surfaces and thin beds are interpreted as representing conditions of poor oxygenation and slow sedimentation. In fact, glauconite is believed to accumulate under similar circumstances and is commonly associated with phosphatic grains.

For the assembly of large, ancient phosphorite deposits, Glenn and Garrison (2003) suggested a two-stage process. First is the generation of the primary phosphatic particles in a near-surface diagenetic setting. Second is reworking of the bottom to concentrate the particles. This is suggested by the fact that many ancient phosphate deposits are cross-stratified sands composed of phosphatic peloids and coated grains. Widespread deposition of phosphorites occurred on continental shelves during the Miocene. Miocene phosphorites in southeastern USA coastal-plain deposits are not associated with eastern boundary current upwelling, but are instead interpreted as recording complicated interactions among glacial eustatic sea-level changes, associated changes in climate, and Gulf Stream dynamics (Riggs, 1984).

Part 3
Fundamentals of fluid flow, sediment transport, erosion, and deposition

5 Unidirectional turbulent water flow, sediment transport, erosion, and deposition

Introduction

Unidirectional, turbulent water flow is probably the most important agent of erosion, sediment transport, and deposition on land. Together with sediment gravity flows, unidirectional water flows are the prime sculptors of the land surface, with air and ice flows of secondary importance. Water may flow in channels or as sheet-like flows over floodplains and hillslopes. However, unidirectional water flows also occur in the sea and lakes. This chapter deals with principles of unidirectional water flow, sediment transport and bed forms, erosion, and deposition. These principles are also relevant to unidirectional air flows and sediment gravity flows, and even periodically reversing water flows, as discussed in subsequent chapters. The extensive published literature on this subject is spread over the disciplines of Earth science, engineering, and physics. Useful reviews of various aspects of this subject are given in Allen (1982a, 1985), Best (1993, 1996), Bridge (2003), Leeder (1999), Middleton and Wilcock (1994), Nezu and Nakagawa (1993), Van Rijn (1990, 1993), and Yalin (1992).

Fundamentals of unidirectional water flow

Definition of physical properties of water and its motion

The physical properties of water that are of most concern here are density and viscosity, both of which depend on temperature. The flow of water is normally described in terms of depth, width, cross-sectional area, velocity, discharge, bed shear stress, Reynolds number, and Froude number. Methods of measurement of fluid flows are discussed in the appendix. Flow depth, width, velocity, and discharge are quite easy to measure, but bed shear stress (the most important quantity

controlling sediment transport) is not easily measured, and must be calculated. Reynolds and Froude numbers are important dimensionless numbers that define the character of the flow. All of these physical quantities are discussed below, and the units and standard symbols used to represent them are given in Table 5.1.

Flows in which water velocity does not change in time are called *steady*, whereas those in which velocity varies with time are called *unsteady*. Most water flows of interest to Earth scientists are *turbulent*, in which water particles travel in curved, swirling motions called *turbulent eddies* (Figure 5.1). The magnitude and direction of flow velocity at any point in a turbulent flow change over fractions of a second to seconds. Thus, natural turbulent flows are strictly unsteady, but this definition depends on the time scale over which the flow velocity is measured and averaged. For example, if the flow velocity is averaged over a period of one hour such that turbulent fluctuations in velocity are not considered, the flow may appear steady for days. In *laminar* flow, water particles follow straight paths (Figure 5.1), and the magnitude and direction of flow velocity change with time at a point in the flow only if the discharge changes with time. Most flows of groundwater, loose sediment, and lava are laminar.

Flows that do not vary along-stream in their velocity or cross-sectional area are called *uniform* flows: otherwise, they are *non-uniform*. Most natural flows are strictly non-uniform, but can be approximated as uniform over certain spatial scales, simplifying their analysis. A water flow in which the cross-sectional area, *a*, varies downstream is shown in Figure 5.2. Assuming that the discharge (volume of water flow per unit time, *Q*) is constant in space and time, and that water is incompressible (i.e., of constant density), an expression for the conservation of mass (and volume) is

$$Q = U_1 a_1 = U_2 a_2 \tag{5.1a}$$

TABLE 5.1. Physical properties of water and its motion, including units and mathematical symbols

Property	Units	Symbol
Mass	kg	m
Density	$kg\ m^{-3}$	ρ
Molecular viscosity	$kg\ m^{-1}\ s^{-1}$	μ
Kinematic viscosity	$m^2\ s^{-1}$	$\nu = \mu/\rho$
Depth	m	d
Cross-sectional area	m^2	a
Velocity	$m\ s^{-1}$	u, v, w, U, V, W
Discharge	$m^3\ s^{-1}$	Q
Bed shear stress	$N\ m^{-2}$ (Pa)	τ_o
Shear velocity	$m\ s^{-1}$	$U_* = \sqrt{\tau_o/\rho}$
Reynolds number	None	Re
Froude number	None	Fr

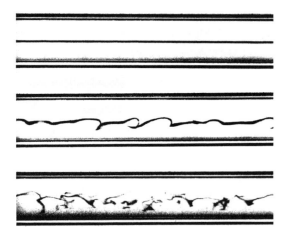

FIGURE 5.1. The distinction between turbulent flow and laminar flow as shown by flow visualization and path lines (dye inserted into the middle of a glass tube). The upper photo shows laminar flow, the middle photo shows transional flow, and the lower photo shows turbulent flow. From Van Dyke (1982).

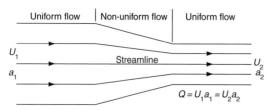

FIGURE 5.2. Definition of uniform and non-uniform flow, including streamlines.

$$\partial(Ua)/\partial x = 0 \qquad (5.1b)$$

where U is the mean flow velocity and x is the downstream direction. Equation (5.1) is known as the (one-dimensional) equation of continuity. Velocity must increase as the water flows into a smaller cross section (called convective acceleration), and must decrease as the cross section expands (convective deceleration). Non-uniform, spatially decelerating flows tend to be associated with deposition, whereas spatially accelerating flows tend to be associated with erosion, as shown below.

Water motion can be shown graphically using various kinds of flow lines. A *streamline* is an imaginary line parallel to the local mean flow direction. Streamlines are parallel in uniform flows, but not in non-uniform flows (Figure 5.2). Velocity increases as streamlines converge, and decreases as they diverge. A *skin-friction line* is a streamline that is very close to the solid boundary (bed) of a water flow. This line shows the direction of the near-bed velocity or bed shear stress. A *path line* is the trajectory of a single moving fluid particle, as shown by inserting a neutrally buoyant particle (a tracer) into a flow (Figure 5.1).

Forces acting on stationary and moving fluids

Force is defined, using Newton's second law of motion, as the rate of change of momentum (the product of mass and velocity), which is the same as the product of mass and acceleration (units are Newtons, where $N = kg\ m\ s^{-2}$). Forces may be classified as body forces or surface forces. Body forces act from a distance on every part of a substance, and depend on force fields such as the downward-directed force of the gravitational field (the gravity force). Surface forces act by direct contact between the surface of a substance and the medium applying the force. A force that is exerted tangentially to a surface is a *shear stress*, whereas a force normal to a surface is a *pressure*. Shear stresses and pressures are forces per unit area, and have units of $N\ m^{-2}$ or pascals. Pressures occur in fluids at rest (hydrostatic pressures) and when in motion (dynamic pressures). The gravity force moves water downslope, whereas shear stresses and pressures at the flow boundaries and within the flow resist this downslope movement. In addition, centrifugal forces are important in curved flows, and the Coriolis force associated with the Earth's rotation (see Chapter 2) influences large-scale flows.

Hydrostatic pressure exists at a point because gravity acts on the mass of water and air above that point. Therefore, hydrostatic pressures are expected to increase with distance below the water surface. The variation in hydrostatic pressure, p_h, with distance downward from the water–air interface, h, is given by

$$p_h = p_a + \rho hg \qquad (5.2)$$

where p_a is atmospheric pressure, ρ is the fluid density, and g is gravitational acceleration. This equation is derived by equating the pressure on any horizontal surface within the water to the weight per unit horizontal area of fluid and air above that surface. However, hydrostatic pressure at a point is the same in all directions, irrespective of the fact that it is caused by the downward-directed gravity force. The maximum hydrostatic force on the bank of a river channel or a dam is found where the water is deepest. Hydrostatic pressures exist in stationary and moving water, but dynamic pressures, shear stresses, and centrifugal forces exist only in moving fluids.

Dynamic pressure is exerted when a moving fluid experiences acceleration or deceleration, and is experienced when at the receiving end of a jet of water. Using Newton's second law of motion, force is rate of change of momentum, i.e.,

$$\text{Force} = \partial(mu)/\partial t = u\,\partial(mu)/\partial x = mu\,\partial u/\partial x \quad (5.3a)$$

$$\text{Dynamic pressure} = \rho u\,\partial u \qquad (5.3b)$$

The total dynamic pressure due to a change in velocity from u_1 to u_2 is obtained by integrating Equation (5.3b) over this velocity range, i.e.,

$$\text{Total dynamic pressure} = \rho(u_1 - u_2)^2/2 \qquad (5.4)$$

The *Bernoulli equation* is an important relationship between the velocity and pressure in a moving fluid. In developing this equation, three simplifying assumptions are made: that one has steady flow; that the fluid is incompressible; and that the fluid has no viscosity (i.e., it is *inviscid* or *ideal*). A small element of fluid (bounded at the sides by streamlines) in a non-uniform flow is shown in Figure 5.3. Since the flow is non-uniform, the velocity and cross-sectional area of the fluid element must be changing in the downstream direction. The rate of change of momentum due to this downstream variation in velocity must, by Newton's second law of motion, be equal to the sum of the forces acting on the fluid element. These forces are the gravity force and the pressure forces acting on the ends and sides of the fluid element. The resulting equation is

$$\rho u^2/2 + \rho gz + p = \text{constant} \qquad (5.5)$$

where z is the distance above an arbitrary datum. The first term in Equation (5.5) is the kinetic energy per unit volume, the second term is the potential energy per unit volume (because of gravity), and the third term, p, is the

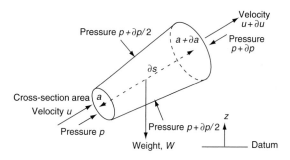

Forces on fluid element in direction of motion, s
Pressure on ends: $pa - (p + \partial p)(a + \partial a)$
Pressure on sides: $(p + \partial p/2)\partial a$
Weight of fluid: $-\rho(a + \partial a/2)\partial s\,g\,\partial z/\partial s$
Rate of change of momentum = $\rho au\,\partial u$

FIGURE 5.3. Forces on an element of a steadily moving, incompressible, and inviscid fluid, leading to Bernoulli's equation. Modified from Bridge (2003).

total fluid pressure comprising both hydrostatic and dynamic pressures. The constant is commonly referred to as the total energy per unit volume. This equation applies to any given streamline, and indicates that, if the velocity changes along the streamline, the elevation and/or the fluid pressure must also change. However, real fluids have *viscosity* (an internal frictional resistance to deformation, or relative motion of fluid elements), and *viscous shear stresses* exist in all real, viscous fluids where there is relative motion between different fluid elements, or between the fluid and solid objects in the flow. Energy must be expended to overcome these viscous shear stresses, such that the total energy per unit volume in the Bernoulli equation is not actually a constant. In fact, total energy decreases progressively in the flow direction as it is converted into low-grade thermal energy due to friction within the flow.

Laminar flow and turbulent flow: the Reynolds number

In order to understand viscosity and viscous shear stress, consider the movement of fluid past a solid boundary (Figure 5.4). Fluid in direct contact with the solid boundary does not move (the *no-slip* condition). At a small distance ∂y above the boundary, the fluid is moving at a small velocity ∂u. As the distance from the boundary increases, so does the flow velocity. A velocity gradient $\partial u/\partial y$ exists from the fixed boundary upwards. Because upper layers of fluid are moving faster than lower layers, they are shearing past each other: this is a *shear flow*. The magnitude of the viscous

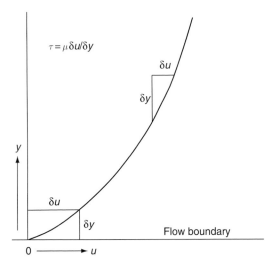

FIGURE 5.4. Definition of fluid viscosity, μ. Immobile fluid on the bed decelerates fluid above it because of viscosity, creating a velocity gradient, $\delta u/\delta y$, and shear stresses, τ, within the flow.

shear stress associated with a particular velocity gradient depends on how easy it is to deform or shear the fluid, as measured in terms of the fluid viscosity. In some fluids, specifically laminar flows of water moving at low velocity, there is a linear relationship between the shear stress, τ, and the velocity gradient, i.e.,

$$\tau = \mu \, \partial u/\partial y \qquad (5.6)$$

where μ is the *molecular viscosity*. In this case, the molecular viscosity is independent of the magnitude of the shear stress, and such a fluid is called a Newtonian fluid. The molecular viscosity of water is on the order of $10^{-3}\,\mathrm{kg\ m^{-1}\ s^{-1}}$, and varies with temperature and pressure. Molecular viscosity increases as temperature decreases.

Viscous shear stresses are important in laminar flows, so called because the water appears to move as a series of thin (molecular-scale) laminae. If dye is inserted at a point in a laminar flow, it will trace out a single straight line (Figure 5.1). In a laminar shear flow, the lower parts of the flow are decelerated relative to the upper parts, because of friction at the solid boundary. There must therefore be continuous transfer of momentum from the upper faster fluid to the lower slower fluid due to mutual acceleration and deceleration of fluid particles. This transfer (or diffusion) of momentum occurs as molecules of fluid are mutually decelerated and accelerated in moving between different levels in the flow. Equation (5.6) is strictly valid only for laminar flow, and can also be written as

$$\tau = (\mu/\rho)\partial(\rho u)/\partial y = \nu \, \partial(\rho u)/\partial y \qquad (5.7)$$

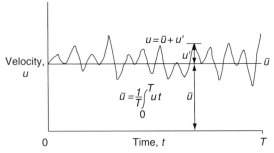

FIGURE 5.5. Definition of turbulent fluctuation in flow velocity. Modified from Bridge (2003).

Here, the shear stress can be thought of as the rate of vertical diffusion of fluid momentum between different levels in the flow, and is proportional to the vertical gradient of momentum per unit volume of fluid. The constant of proportionality, $\nu = \mu/\rho$, is called the *kinematic viscosity* or *diffusivity of momentum*, and is a measure of the degree of molecular mixing between laminae. The kinematic viscosity of water is on the order of $10^{-6}\,\mathrm{m^2\,s^{-1}}$.

Most natural water flows are turbulent, with temporal and spatial fluctuations in water velocity and direction (eddies) superimposed on the mean downstream motion (Figure 5.1). Figure 5.5 shows that the time-averaged, downstream flow velocity in a turbulent flow is given by

$$\bar{u} = \frac{1}{T}\int_0^T u \, dt \qquad (5.8)$$

The instantaneous velocities in the downstream, across-stream, and vertical directions can be written as

$$\begin{aligned} u &= \bar{u} + u' \\ w &= \bar{w} + w' \\ v &= \bar{v} + v' \end{aligned} \qquad (5.9)$$

The sums of the fluctuating velocity components (denoted by primes) are zero. Although $\bar{u}' = 0, \overline{u'u'} \neq 0$, and these *mean squares* are used to measure the magnitude of turbulence. For example

$$\mathrm{rms}(u') = \sqrt{\overline{u'u'}} \qquad (5.10)$$

is the root mean square (rms) of the velocity fluctuation, and is the standard deviation of the instantaneous velocities. Usually, the root mean square is made dimensionless using the time-averaged velocity at the point ($\mathrm{rms}(u')/\bar{u}$), or the shear velocity (defined later). Typically, $\mathrm{rms}(u')/\bar{u}$ in river flows is on the order of 10^{-1}. Many kinds of statistical analysis can be performed on velocity time series such as that shown in

Figure 5.5 (see the appendix). Such analysis gives information on the temporal and spatial scale (i.e., eddy size) of the velocity fluctuations.

Analysis of turbulence commonly involves simultaneous and repeated measurement of turbulent fluctuations of a velocity component, say u, at various (lag) distances apart, x. The (auto)correlation coefficients, $R(x)$, for the velocities (u', u'_x) at a given lag distance are given by

$$R(x) = \frac{\overline{u'u'_x}}{\text{rms}(u')\text{rms}(u'_x)} \tag{5.11}$$

Commonly, the velocity fluctuations at short lag distances have the largest correlation coefficients because they are related to the same turbulent eddy. The spatial scale (average size) of eddies is obtained by integrating the autocorrelation function (a plot of $R(x)$ versus x) with respect to the spatial lag, x. A similar type of analysis could be undertaken by repeated measurement of turbulent fluctuations in a velocity component at a given point, and then calculation of autocorrelation coefficients for velocity fluctuations at different time lags. The time scale (macroscale) of turbulence is calculated by integrating the autocorrelation function with respect to the time lag. The product of this turbulence macroscale and the mean flow velocity gives an estimate of the eddy length scale. An alternative method of analysis of turbulence is to obtain frequency spectra using power-spectral analysis (see the appendix). High-frequency fluctuations are associated with small eddies, and low frequencies are associated with large eddies. A typical eddy length scale is obtained by dividing the mean flow velocity at a point by the spectral wavenumber, given by $2\pi/\text{fluctuation period}$.

Shear stresses within turbulent flows (*turbulent shear stresses*) have much greater magnitudes than do the viscous shear stresses in laminar flows, and they depend on the nature of the turbulence, which makes turbulent flows difficult to analyze. Shear stresses in turbulent flows are not simply due to the transfer of momentum across shear planes on a molecular scale. Sizable parcels of fluid in turbulent eddies are being accelerated and decelerated as they move into different parts of the flow. Thus, dynamic pressure forces are involved in addition to viscous shear stresses. Shear stress within turbulent flow is given by

$$\tau = (\mu + \eta)\frac{d\bar{u}}{dy} \approx \eta\frac{d\bar{u}}{dy} = \varepsilon\frac{d(\rho\bar{u})}{dy} \tag{5.12}$$

where η is the eddy viscosity and $\varepsilon = \eta/\rho$ is the kinematic eddy viscosity. These apparent viscosities are

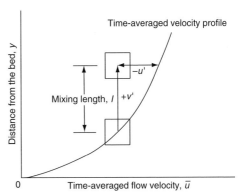

FIGURE 5.6. Definition of turbulent motions in a boundary layer, including the mixing length. Modified from Bridge (2003).

measures of the difficulty of accelerating or decelerating sizable parcels of fluid by virtue of their inertia or momentum. Eddy viscosity is several orders of magnitude greater than molecular viscosity, thus explaining why turbulent shear stresses are much greater than viscous shear stresses for a given velocity gradient. Unlike molecular viscosity, eddy viscosity is not constant, but depends on the nature of the turbulence, thus adding complexity to the analysis of turbulent shear stresses.

The increased shear stresses brought about by turbulence can be explained by considering the momentum exchange between fluid elements as they move within the flow. In Figure 5.6, relatively low-velocity fluid moves upwards with velocity v'. Its downstream velocity is deficient relative to that in its new surroundings by $-u'$. The relatively faster-moving fluid around it accelerates the upward-moving fluid and increases its downstream momentum. The change in downstream momentum per unit volume is $-\rho u'$. The rate of change of downstream momentum with time depends on the speed of the upward motion. Thus, the rate of change of downstream momentum of fluid moving vertically through a horizontal unit cross-sectional area is $-\rho u'v'$, which is a turbulent shear stress. The time-averaged shear stress in a turbulent flow can therefore be written as the sum of viscous shear stress and turbulent shear stress:

$$\tau = \mu\left(\frac{d\bar{u}}{dy}\right) - \rho\overline{u'v'} \tag{5.13}$$

$$-\rho u'v' = \rho\varepsilon\left(\frac{d\bar{u}}{dy}\right) \tag{5.14}$$

The viscous-shear-stress term in Equation (5.13) is commonly small enough to be ignored. Since the velocity fluctuations u' and v' are proportional to the time-averaged velocity at a point, turbulent shear stresses are proportional to the square of this mean velocity. By similar arguments, there are additional shear stresses due to velocity fluctuations across the flow and dynamic pressures in the flow direction. These are commonly referred to as *Reynolds stresses* or *apparent stresses*.

The shear stresses and dynamic pressures that arise from the motion of fluid in turbulent eddies (Reynolds stresses) are what distinguishes turbulent flows from laminar flows. The ratio of turbulent stresses to viscous stresses can be expressed in a general way as

$$\frac{\rho \bar{u}^2}{\mu \bar{u}/y} = \frac{\rho \bar{u} y}{\mu} \tag{5.15}$$

where y is a length normal to the flow direction. This important dimensionless number is called a *Reynolds number*. A Reynolds number can be defined for any physical situation where fluid particles are in relative motion, or fluids and solids are in relative motion. Reynolds numbers are used for defining which forces are most important in resisting motion. Physical systems with the same Reynolds number are called *dynamically similar*, irrespective of scale. For river flows, the Reynolds number is defined as

$$Re = \rho U d/\mu \tag{5.16}$$

where U is the time-averaged and depth-averaged flow velocity, and d is the mean flow depth. At low Reynolds numbers, flows are laminar, whereas they become turbulent as the Reynolds number increases. The transition from laminar to turbulent flows in open channels and pipes occurs between Reynolds numbers of 500 and 2,000.

A theoretical approach to predicting the transition from a laminar shear flow to a turbulent shear flow gives insight into the formation of turbulent eddies. Assume that a small wave-like disturbance affects a laminar shear flow (Figure 5.7). From the Bernoulli equation, regions where the streamlines are close together are regions of high velocity and low pressure, and vice versa. The wavy streamline is unstable because high pressure exists under the wave crest, tending to increase the wave amplitude. However, wave amplification is resisted by viscous shear stresses that dampen large velocity gradients and that make shear more uniform. Under conditions such that the dynamic pressure forces overcome the viscous shear forces, the wave will increase in amplitude, and the growing crest region will start to overtake the trough

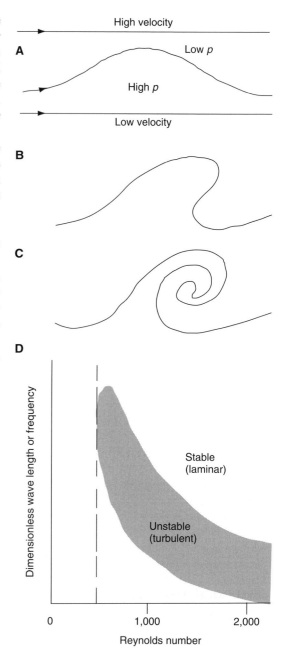

FIGURE 5.7. In (A)–(C), development of a Kelvin–Helmholtz instability (vortex) from a wave-like disturbance of the streamlines of a laminar shear flow. (D) Results of a theoretical stability analysis of a laminar boundary layer. Modified from Bridge (2003).

(Figure 5.7). The vortex thus formed (called a Kelvin–Helmholtz instability) will then become entrained within the flow as a distinct eddy. A *stability analysis* attempts to define mathematically the physical conditions for which a disturbance of a given wave length will be amplified (become unstable) or damped. One of the most successful stability theories, by Tollmein and Schlichting (Schlichting, 1979), fits experimental data well (Figure 5.7).

Fluid forces and equations of motion

Natural water flows normally vary spatially in three dimensions and vary in time due to turbulence and longer-term fluctuations in velocity. In order to describe and analyze turbulent water flows fully, it is necessary to determine flow velocities and fluid forces within a three-dimensional (3D) spatial coordinate system, such as the Cartesian system, where fluid motion and forces are related to three orthogonal directions (e.g., x, y, z). Using such a coordinate system, the equations governing fluid motion are derived by considering the conservation of mass (or volume for an incompressible fluid like water) and momentum as water passes through a unit volume within the flow. The 3D equation of conservation of mass (volume), also known as the continuity equation, is

$$\frac{\partial u}{\partial x} + \frac{\partial v}{\partial y} + \frac{\partial W}{\partial z} = 0 \tag{5.17}$$

where u, v, and w are the velocity components in the x, y, and z directions, respectively. A one-dimensional version of this equation averaged over the width and depth of a channel was given in Equation (5.1). The equations expressing conservation of momentum are based on Newton's second law of motion, and relate flow accelerations in time and space to the pressure and shear forces acting. These equations, known as the Navier–Stokes equations, are complicated and will not be reproduced here. Much simpler equations expressing the conservation of momentum can be obtained by ignoring viscosity and viscous shear stress (the Euler and Bernoulli equations). A simplified form of the Bernoulli equation is given in Equation (5.5). The continuity and momentum equations can be applied to turbulent water flows by replacing instantaneous velocity components with the sum of the average and fluctuating velocity components, and averaging over some time interval. The resulting conservation-of-momentum equations are commonly referred to as the Reynolds equations because they contain the turbulent Reynolds stresses mentioned previously. When applying the time-averaged Reynolds equations to turbulent flows such as in rivers, it is possible to ignore many of the Reynolds stresses. The Reynolds equation for a steady, two-dimensional turbulent flow is

$$\rho \left(u \frac{\partial \bar{u}}{\partial x} + v \frac{\partial \bar{u}}{\partial y} \right) = -\frac{\partial \bar{p}}{\partial x} + \frac{\partial}{\partial y} \left(\mu \frac{\partial \bar{u}}{\partial y} - \rho \overline{u'v'} \right) \tag{5.18}$$

where p is pressure and the overbars denote time-averaged values. Furthermore, it is common practice to express the continuity and momentum equations in even simpler form, for example by averaging over the flow depth and width. However, it is necessary to use the 3D Reynolds equations to describe some important 3D features of natural flows. The Reynolds equations must be solved numerically whatever forms are used.

It is now possible to solve numerically the full Navier–Stokes equations under specific boundary conditions, given recent advances in computer technology. This means that turbulent fluctuations in a water flow can be modeled. Direct numerical simulation (DNS) using the full Navier–Stokes equations requires many numerical grid points and long computing times to describe turbulent flows fully. Therefore, use of DNS has been limited to small-scale, low-Reynolds-number flows over short time intervals. Large-eddy simulation (LES) is another numerical technique that solves the Navier–Stokes equations only for the larger scales of turbulence (determined by the grid size used); LES can be used for simulating larger-scale and longer-term flows than are amenable to DNS.

Turbulent boundary layers in steady, uniform turbulent flows

Friction (or drag) at the solid boundary of a water flow causes a reduction of flow velocity near the boundary (Figure 5.8). The boundary zone where there is a velocity gradient and shear stresses is called a *boundary layer*. The zone where there is no velocity gradient is called the *free stream* or *external stream*. In rivers, the boundary layer extends to the water surface, such that there is no free stream. However, in the sea, the boundary layer of a unidirectional turbulent flow rarely reaches the water surface. Turbulent boundary layers are normally divided into at least three zones (Figure 5.8). The *viscous sublayer* is a thin layer (normally less than a millimeter thick) next to the boundary where viscous shear stress dominates and turbulence formation is suppressed. The flow behaves almost as if it were laminar (see below), and this region used to be called the laminar sublayer. The thickness of the viscous sublayer, δ_v, is given approximately by

$$\delta_v \approx 10\nu/U_* \quad \text{or} \quad 10 \approx \delta_v U_*/\nu = \delta_v^+ \tag{5.19}$$

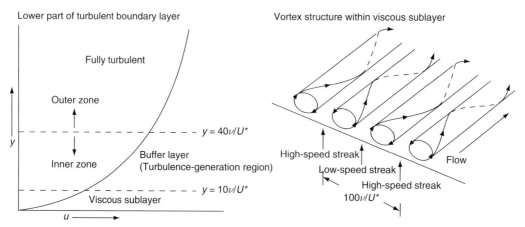

FIGURE 5.8. Main features of the structure of turbulent boundary layers. Modified from Bridge (2003).

where ν is the kinematic viscosity and U_* is the *shear velocity*, defined as $U_* = \sqrt{(\tau_0/\rho)}$, where τ_0 is the spatially averaged fluid shear stress at the bed and ρ is the fluid density. The shear velocity can be thought of as a measure of the flow velocity very close to the bed. The constant of 10 in the above equation has actually been taken to range from 5 to 12. As the flow in the viscous sublayer behaves as if it were laminar, the variation in velocity, u, with distance from the bed, y, is given using Equations (5.6) and (5.24) as

$$u = \tau_0 y/\mu \quad \text{and} \quad \frac{u}{U_*} = \frac{U_* y}{v} \qquad (5.20)$$

However, there are spiral motions in the viscous sublayer associated with Taylor–Gortler vortices that are oriented mainly parallel to the flow direction (Figure 5.8). These vortices give rise to a streaky structure that varies in time and space. Low-speed streaks occur between adjacent vortices where there is a common upward motion of fluid, and high-speed streaks occur where there is a common downward component of fluid. The mean across-stream (transverse) spacing of low-speed streaks, λ, in the viscous sublayer is given by

$$\lambda^+ = \lambda U_*/\nu \approx 100 \text{ (range } 60-200) \qquad (5.21)$$

This transverse spacing of the low-speed streaks increases if turbulence is suppressed by suspended sediment. The along-stream extent of the streaks is typically on the order of $600-1000\nu/U_*$. The flow-parallel vortices in the viscous sublayer are probably the trailing ends of horseshoe-shaped or hairpin-shaped vortex loops (Figure 5.9), structures that are considered to be fundamental in turbulent shear flows (Smith, 1996). The downstream ends of these vortex loops extend

above the viscous sublayer at angles from the bed of up to 45°. The viscous sublayer is important because bed grains immersed in it are protected from the large turbulent shear stresses above, and grains moving in it are concentrated into lanes beneath the low-speed streaks.

Water flow in the *buffer layer* is transitional between pseudo-laminar and fully turbulent. The buffer layer is three to seven times the thickness of the viscous sublayer (Figure 5.8). Intense, small-scale turbulent eddies are formed in the buffer layer and are advected into the fully turbulent region above. The *fully turbulent, outer region* occupies most of the boundary layer, and turbulent eddies generally increase in size with distance from the boundary in this region. The length and height of eddies viewed parallel to the flow direction may be comparable to the entire boundary-layer thickness (flow depth in rivers). The across-stream spacing of flow-parallel vortices (like those in the viscous sublayer) increases linearly with distance from the top of the viscous sublayer, and the diameters of the largest vortices scale with the flow depth. There is a large body of information on how the flow velocity, turbulence intensity, and length scale vary with distance from the boundary, and this information is critical to understanding how sediment is transported. The outer, fully turbulent region is commonly subdivided into a so-called *wall* region ($y/d < 0.2$) and an overlying *wake* region ($y/d > 0.2$), as discussed below.

Turbulence generation within turbulent boundary layers depends upon flow structures and events initiated within the viscous sublayer and buffer layer (Figure 5.9). Several low-speed streaks gradually rise from the viscous sublayer in response to a large

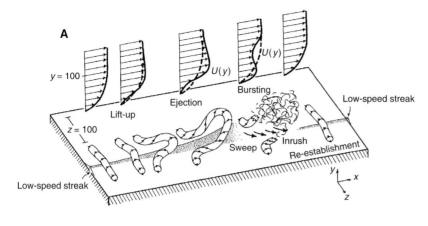

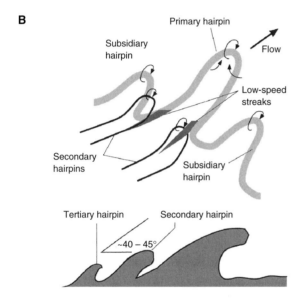

FIGURE 5.9. (A) The model of turbulent bursting according to Allen (1985), showing development of a hairpin vortex and associated low-speed streak, uplift and ejection of a low-speed streak, and downward movement of fluid as a sweep. Profiles of instantaneous velocity are also shown. (B) The model of hairpin vortices and low-speed streaks in the viscous sublayer and buffer layer according to Smith (1996) and Best (1996). The streamwise cross section shows ejection of viscous sublayer fluid (shaded) in hairpin vortices and downward movement of outer fluid (unshaded) in sweeps.

eddy (flow-transverse vortex) approaching the boundary from the outer region and creating a locally high-shear zone (Figure 5.9). The lifting low-speed streaks are parts of hairpin vortices. As the low-speed streaks lift up, they oscillate and rise rapidly into the buffer layer and then are ejected at high speed into the outer zone, where they mix with other eddies. High-speed fluid from the large eddy (called a sweep) then moves close to the boundary to take the place of the ejected fluid. This entire process is referred to as *bursting*.

Quadrant analysis is employed to distinguish the different kinds of turbulent events and their relative contributions to the total Reynolds shear stress at a point in the flow. The four quadrants are defined as (1) $u' > 0$, $v' > 0$, outward interaction; (2) $u' < 0$, $v' > 0$,

ejection; (3) $u' < 0$, $v' < 0$, inward interaction; and (4) $u' > 0$, $v' < 0$, sweep. Ejections and sweeps are most energetic and together account for about 70% of the turbulent shear stress in boundary layers.

Much of what is known of the bursting process is based on experiments using sediment-free fluids moving over smooth boundaries. The bursting process is also the dominant mechanism of turbulence production in natural water flows, but there are differences compared with the laboratory flows. For example, with rough boundaries the viscous sublayer streaks are much less conspicuous, and gradual lift-up is replaced by sudden violent ejections from between the roughness elements on the bed (e.g., large sediment grains). With very rough boundaries, such as those covered with dunes or boulders, turbulent eddies generated on

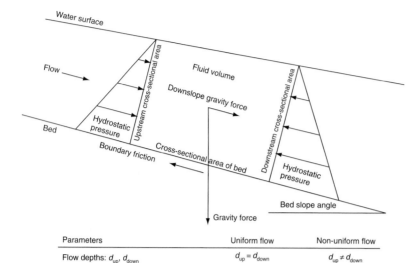

FIGURE 5.10. Time-averaged and depth-averaged forces (gravity, hydrostatic pressure, bed friction) acting on a fluid volume in straight, uniform, and non-uniform flows. Modified from Bridge (2003).

Parameters	Uniform flow	Non-uniform flow
Flow depths: d_{up}, d_{down}	$d_{up} = d_{down}$	$d_{up} \neq d_{down}$
Cross-sectional areas of flow: a_{up}, a_{down}	$a_{up} = a_{down}$	$a_{up} \neq a_{down}$
Mean hydrostatic pressure: P_{up}, P_{down}	$P_{up} = P_{down}$	$P_{up} \neq P_{down}$
Mean flow velocities: U_{up}, U_{down}	$U_{up} = U_{down}$	$U_{up} \neq U_{down}$
Bed shear stress: τ_0	$F = G \sin \alpha$	$F \neq G \sin \alpha$
Cross-sectional area of bed: a		
Boundary friction: $F = \tau_0 a$		
Gravity force: G		
Bed slope: $S = \sin \alpha$		

the lee side of the roughness elements are ejected into the flow so vigorously that they may reach the water surface as boils. These turbulent motions (and particularly ejections and sweeps) have a major influence on the mechanics of sediment transport and the generation of bed forms. Sediment in motion in turn affects the nature of the bursting process, as discussed below.

Vertical profiles of shear stress and velocity in steady, uniform boundary layers

Figure 5.10 illustrates the time-averaged forces acting on a volume of water in a straight, uniform, steady flow. The width of the flow is assumed to be much greater than the depth, such that friction at the sides of the flow can be ignored. Under these conditions, the balance of gravity, friction, and pressure forces (conservation of momentum) in the flow direction yields

$$\tau_0 = \rho g d S \tag{5.22}$$

in which τ_0 is the spatially averaged fluid shear stress at the bed, d is the mean flow depth, and S is the slope of the bed and water surface. This equation is actually the balance between the boundary friction and the gravity force, because the pressure forces cancel out. The same force-balancing procedure can be applied to a volume of fluid with its base within the flow and with thickness

$d - y$ (Figure 5.10). The mean shear stress along the base of this fluid volume is

$$\tau = \rho g (d - y) S \tag{5.23}$$

Combining Equations (5.22) and (5.23) gives

$$\tau = \tau_0 (1 - y/d) \tag{5.24}$$

indicating that the mean shear stress decreases linearly from a maximum at the bed to zero at the water surface. The bed shear stress is important because it controls the movement of sediment, as will be seen later.

Knowledge of the vertical variation in the time-averaged flow velocity (the so-called velocity profile) in turbulent boundary layers has enormous practical value, such as in the calculation of bed shear stress and for prediction of the suspended-sediment transport rate. However, the velocity profile is difficult to predict due to the fact that the eddy viscosity varies with the nature of the turbulence, which is in turn influenced by sediment in the flow. Furthermore, the geometry of a flow boundary composed of movable sediment is controlled by the nature of the flow, which is in turn controlled by the geometry of the flow boundary. Historically, velocity profiles in turbulent boundary layers were represented using empirical equations that were not generally applicable.

An important semi-theoretical approach to this velocity-profile problem is the Prandtl–Von Kármán mixing-length theory. In this theory, the turbulent fluctuations u' and v' are given by

$$u' = -l(d\bar{u}/dy) \quad \text{and} \quad v' = l(d\bar{u}/dy) \qquad (5.25)$$

in which l is the mixing length. The mixing length is the vertical distance over which the momentum of a fluid parcel is changed, and is analogous to the average eddy size at a point in the boundary layer (Figure 5.6). The turbulent shear stress at any point in the flow is then given, using Equations (5.13) and (5.25), as

$$\tau = -\rho \overline{u'v'} = \rho l^2 (d\bar{u}/dy)^2 \qquad (5.26)$$

In order to solve this equation, two assumptions were made. The first is that, close to the boundary, the shear stress is approximately equal to the bed shear stress ($\tau \approx \tau_0$). The second is that the mixing length increases linearly with distance from the bed, i.e., $l = \kappa y$, where κ is called Von Kármán's constant, which is normally taken as equal to 0.41 in clear-water flows. The resulting velocity-profile equation

$$\frac{\bar{u}}{U_*} = \frac{1}{\kappa} \ln\left(\frac{y}{y_0}\right); \quad \frac{d\bar{u}}{dy} = \frac{U_*}{\kappa y}; \quad U_* = \sqrt{(\tau_0/\rho)} \qquad (5.27)$$

is called the *law of the wall*, because the two assumptions made above necessarily restrict its application to the near-wall region of turbulent boundary layers ($y/d \leq 0.2$, but above the viscous sublayer). In this equation, y_0 is the value of y where the flow velocity is zero. In fact, the flow velocity is zero only when $y = 0$, and y_0 is actually an artifact of the solution of the equation. However, as will be seen below, the value of y_0 is actually related to the roughness of the bed, and is commonly called the *roughness height*. The law of the wall is very useful, particularly in the calculation of shear velocity (and hence bed shear stress) from measured velocity profiles.

In order to make the assumptions of the law of the wall less restrictive, the vertical variation in time-averaged shear stress is taken to be given by Equation (5.24) and the vertical variation in the mixing length is assumed to be

$$l = \kappa y (1 - y/d)^{0.5} \qquad (5.28)$$

The form of the law of the wall remains the same, but the law may now be applicable throughout the boundary layer. Also, the kinematic eddy viscosity, defined

in Equation (5.12), may be written, using Equations (5.24) and (5.27), as

$$\varepsilon = \kappa y U_* (1 - y/d) \qquad (5.29)$$

If the maximum velocity, $\bar{u}_{max}$, at the outer edge of the boundary layer, δ, is inserted into the law of the wall (Equation (5.27)), giving

$$\frac{\bar{u}_{max}}{U_*} = \frac{1}{\kappa} \ln\left(\frac{\delta}{y_0}\right) \qquad (5.30)$$

and Equation (5.27) is subtracted from Equation (5.30), the *velocity-defect law* is obtained:

$$\frac{\bar{u}_{max} - \bar{u}}{U_*} = \frac{1}{\kappa} \ln\left(\frac{\delta}{y}\right) = -\frac{1}{\kappa} \ln\left(\frac{y}{\delta}\right) \qquad (5.31)$$

More recently, it was proposed that the law of the wall and velocity-defect law under-predict the velocity in the outer (*wake*) region of turbulent boundary layers ($y/d > 0.2$), and that an additional *wake* term is required (Coles, 1956). Alternative empirical equations for the vertical variation in kinematic eddy viscosity and mixing length have also been utilized. However, the importance of the wake region has been challenged, on the basis that there are limitations in the experimental data used to suggest its existence. For example, velocity profiles in the outer, fully turbulent regions of experimental channels are particularly sensitive to the effects of flow non-uniformity and secondary currents. Recent, high-quality experimental data suggest that the law of the wall may be applicable to most of the boundary layer, and casts doubt on the need for a *wake* term (Bridge, 2003). Figure 5.11 shows vertical profiles of time-averaged flow velocity that are fitted well by the law of the wall and the velocity-defect law.

Hydraulically smooth, transitional, and rough boundaries

A *hydraulically smooth* boundary is defined as one where irregularities on the bed (e.g., sediment grains) are contained entirely within the viscous sublayer. The average height of bed irregularities above the bed (where $y = 0$) is commonly expressed using a so-called *equivalent sand roughness*, k_s. Although k_s was originally defined for flow boundaries covered with immobile sand grains (for which $k_s \approx D$, where D is a median or mean grain size), it is also used to refer to other types of boundary irregularity. For hydraulically smooth boundaries

$$k_s < \delta_v \approx 10\nu/U_* \quad \text{and} \quad k_s^+ = k_s U_*/\nu < 10 \qquad (5.32)$$

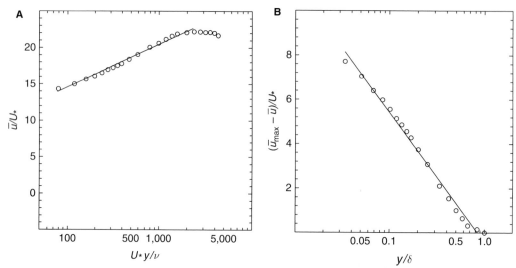

FIGURE 5.11. Vertical profiles of time-averaged flow velocity for clear-water flow over a planar sand bed. The straight line is the law of the wall in (A), and the velocity-defect law in (B), with $\kappa = 0.4$ in both cases. Modified from Bridge (2003).

although a value of 5 is commonly used instead of 10. The value of the roughness height y_0 for smooth boundaries is calculated by solving simultaneously the law of the wall (Equation (5.27)) and the velocity profile of the viscous sublayer (Equation (5.20)) at $y = \delta_v$, giving $y_0 = \nu/(9U_*)$. Substituting this value of y_0 back into the law of the wall gives a velocity-profile equation for the fully turbulent part of smooth boundary layers:

$$\frac{\bar{u}}{U_*} = \frac{1}{\kappa} \ln\left(\frac{yU_*}{\nu}\right) + \frac{2.2}{\kappa} \qquad (5.33)$$

This velocity profile is dependent on fluid viscosity, but is independent of the size of the bed irregularities.

With a *hydraulically rough* boundary, the bed irregularities poke through the viscous sublayer and the buffer layer (substantially disrupting them) and well into the fully turbulent region. In this case

$$k_s^+ = k_s U_*/\nu > 70 \qquad (5.34)$$

and, from experimental data, $y_0 = k_s/30.2$. Inserting this value of roughness height into the law of the wall results in

$$\frac{\bar{u}}{U_*} = \frac{1}{\kappa} \ln\left(\frac{y}{k_s}\right) + \frac{3.4}{\kappa} \qquad (5.35)$$

This velocity profile depends on the height of roughness elements, but is independent of fluid viscosity. This is because viscous shear stresses on the boundary are negligible compared with turbulent shear stresses. The dependence of the velocity profile on the height of roughness elements is related to the dynamic pressures exerted on the roughness elements. Dynamic pressures act on the upstream sides of the roughness elements and on their lee sides, in association with flow separation, vortex formation, and shedding. As the size of roughness elements increases, so does the size of the lee-side vortices and the frequency and intensity of their shedding into the flow above.

With *hydraulically transitional* boundaries, the bed irregularities protrude into the buffer layer, and

$$10 < k_s U_*/\nu < 70 \qquad (5.36)$$

In this case, y_0 is dependent on fluid viscosity and roughness height, and velocity profiles for hydraulically transitional flows are rather complicated.

Depth-averaged flow velocity and flow resistance

An important characteristic of turbulent water flows is that the shear stress is proportional to the square of the time-averaged flow velocity. One form of this relationship that relates the shear stress at the bed to the depth-averaged flow velocity, U, is

$$\tau_0 = \rho f U^2/8; \qquad \frac{U}{U_*} = \sqrt{\frac{8}{f}} \qquad (5.37)$$

in which f is a friction coefficient, the value of which depends on the roughness of the bed relative to the flow depth and on the Reynolds number (discussed below). This equation is a form of the general drag equation that is discussed later. Combining Equations (5.22) and (5.37) results in a relationship involving depth-averaged flow velocity, depth, slope, and bed roughness in steady, uniform flows:

$$U = \sqrt{8gdS/f} \tag{5.38}$$

This is called the d'Arcy–Weisbach (or Darcy–Weisbach) equation. The Chezy equation is similar and given by

$$U = C\sqrt{dS} \tag{5.39}$$

where $C = \sqrt{8g/f}$. A variation on these equations is the Manning equation:

$$U = 1.49d^{2/3}S^{1/2}/n \tag{5.40}$$

The Chezy and Manning equations are inferior to the Darcy–Weisbach equation because their resistance coefficients (C and n, respectively) are dimensional, as opposed to f, which is dimensionless. Thus the American Society of Civil Engineers recommended discontinuing use of the Manning equation in 1966. However, it is still in use by many engineers worldwide.

The depth-averaged flow velocity can also be expressed in terms of the bed shear stress, depth, and friction coefficient using the law of the wall, assuming that the law of the wall is applicable throughout the flow depth. Integration of the law of the wall over the flow depth leads to the result that the depth-averaged flow velocity occurs at a distance of $0.368d$ from the bed. Thus the depth-averaged versions of the law of the wall for hydraulically smooth and rough boundaries are

$$\frac{U}{U_*} = \sqrt{\frac{8}{f}} = \frac{1}{\kappa}\ln\left(\frac{dU_*}{\nu}\right) + \frac{1.2}{\kappa} \tag{5.41}$$

and

$$\frac{U}{U_*} = \sqrt{\frac{8}{f}} = \frac{1}{\kappa}\ln\left(\frac{d}{k_s}\right) + \frac{2.41}{\kappa} \tag{5.42}$$

respectively. The friction coefficient depends on the boundary Reynolds number for hydraulically smooth flows, but only on the boundary roughness for hydraulically rough flows. For rough flows, the friction coefficient increases as the *relative roughness* k_s/d increases. The equivalent sand roughness, k_s, is commonly equated with some measure of bed grain diameter (e.g., D_{50}, D_{65}, or D_{84}, where the subscript denotes the weight percentage of the grain-size distribution that is smaller than this grain size) for planar sediment beds with no sediment movement. For mobile-sediment beds, the equivalent sand roughness depends on at least the thickness of the bed-load layer and the geometry (e.g., height) of bed forms such as ripples and dunes. Both sand-bed rivers and gravel-bed rivers have bed load and bed forms at most flow stages. Values of f are approximately 0.02 and 0.1 for flat and dune-covered sand beds, respectively. Values of f for gravel-bed rivers typically range from 0.02 to 0.3. It is common practice to partition the friction coefficient into that part due to bed grains and that part due to bed forms such as gravel clusters, ripples, dunes, and bars. See Bridge (2003) for a review.

Spatial variation of turbulent fluctuations in straight, uniform flows

Time-averaged turbulent velocity fluctuations, rms (u') (in the streamwise direction) and rms (v') (in the vertical direction), decrease exponentially with distance from the boundary, and can be represented empirically in the fully turbulent region (Nezu and Nakagawa, 1993) by

$$\mathrm{rms}(u')/U_* = 2.3\exp(-y/d) \tag{5.43}$$

and

$$\mathrm{rms}(v')/U_* = 1.27\exp(-y/d) \tag{5.44}$$

Both components of turbulence intensity reach a maximum in the buffer layer and then decline to zero at the bed. The maximum value of the vertical turbulence intensity is close to 1, and is not affected much by bed roughness. The maximum value of the streamwise turbulence intensity is about 2.8 for smooth flows and 2.0 for rough flows (decreasing as k_s increases). The frequency distributions of the turbulent fluctuations u', v', $-\rho u'v'$, are approximately Gaussian or slightly positively skewed (the mode is smaller than the mean) above the viscous sublayer. The coefficient of variation (standard deviation/mean) of the near-bed shear-stress distribution is close to 0.4 for rough and transitional boundaries, but is larger for smooth boundaries.

Figure 5.12 illustrates Bagnold's (1966) concept of the anisotropy of vertical turbulence. Conservation of mass and momentum dictates that the mass (or volume) and momentum of upward-moving fluid must equal those for downward-moving fluid. With isotropic turbulence, the masses and vertical velocities of fluid moving in the upward and downward directions are all equal; however, this is not so for anisotropic turbulence. Bagnold (1966) thought that vertical turbulence would be most anisotropic in the buffer layer and become less anisotropic with distance above the

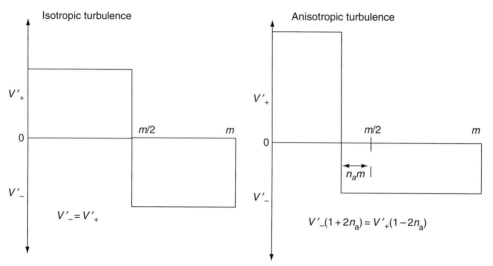

FIGURE 5.12. Definition of anisotropy of vertical turbulence according to Bagnold (1966). The coefficient of anisotropy is n_a.

A

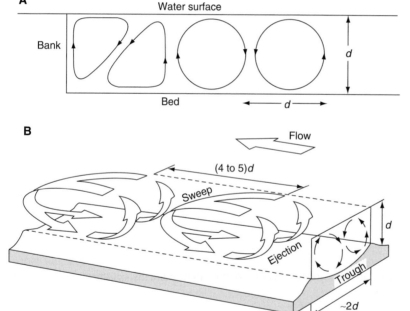

B

FIGURE 5.13. (A) Idealized pattern of secondary currents in partial cross section of a straight channel flow. From Bridge (2003). (B) A model of the three-dimensional, large-scale structure of open-channel flow over a mobile bed, as seen while moving at the mean flow velocity. From Shvidchenko and Pender (2001).

bed. A possible physical explanation for anisotropy of vertical turbulence intensity is the bursting phenomenon, whereby the maximum anisotropy would be associated with fluid ejections from within the buffer layer. Experimental data to test this theory are sparse.

Anisotropy between the v' and w' components of turbulence (i.e., those in the vertical plane normal to the mean flow direction) gives rise to streamwise vorticity (secondary currents) in straight, uniform flows. Such anisotropy is due to the geometry of the flow boundary (bed, banks, and water surface). Near a steep river bank, such secondary circulation takes the form of two vortices with common flows towards the corner between the bank and bed, near-bed flow away from the bank, and near-bank flow vertically upwards (Figure 5.13). Further away from the banks (greater than a flow depth), the secondary circulation takes the form of pairs of vortices with diameters approximately equal to the flow depth (Figure 5.13A). The velocity of the secondary flow is generally less than 5% of the streamwise flow, and the angle of deviation of flow from the downstream direction is generally less than a few degrees. These secondary currents cause the maximum flow velocity to be depressed below the water surface (Figure 5.13A). Bed shear stress and bed roughness are increased where there is downflow, and are decreased where there is upflow. In channels with movable beds, the bed areas experiencing converging and ascending secondary flow and low bed shear stress tend to be occupied by sediment ridges (e.g., sand ribbons). These secondary currents also affect the distribution of suspended sediment and the 3D form of transverse bed forms like dunes.

These depth-scale streamwise vortex pairs have been interpreted as parts of 3D turbulent structures that extend for four to five times the depth parallel to the flow and twice the depth transverse to the flow (Figure 5.13B). These structures are visible when they are observed while moving at the mean speed of the flow, and may be the transverse vortices associated with the bursting phenomenon discussed above (Figure 5.13B shows how sweeps and ejections may be related to the structures). Such depth-scale turbulent structures have important implications for the formation of dunes.

Turbulent boundary layers in steady, non-uniform flows and flow separation

Steady, non-uniform flow can be classified as *gradually varied flow* or *rapidly varied flow*. In gradually varied flow, variations of cross-sectional area and velocity of the flow are gradual enough in the along-stream direction that streamlines are almost parallel, and a hydrostatic pressure distribution can be assumed. Energy losses due to friction at the boundary are described in essentially the same way as for uniform flow. In rapidly varying flow, along-stream changes in cross-sectional area and flow velocity are rapid enough that streamlines are strongly curved, and the pressure distribution is no longer hydrostatic. Energy losses in such flows are commonly associated with rapid flow expansions, and those due to boundary friction are negligible in comparison. Turbulence characteristics in gradually varied flow are similar to those for uniform flow, but this is not so for rapidly varied flows.

An equation relating the along-stream variation in depth-averaged flow velocity U, depth d, bed slope S, and bed shear stress of a gradually varied channel flow can be obtained by analyzing the balance of forces acting (Figure 5.10). It is assumed that the channel is rectangular in cross section and that the width is constant and much greater than the depth, so that the effects of

bank friction can be ignored. In this case, the depth and hydrostatic pressure increase gradually downstream, so the mean velocity must decrease gradually downstream. A fluid element with height d, downstream length ∂x, and unit width is considered. The forces in the mean downstream x direction are the net hydrostatic pressure force, the downslope weight component, and the friction at the base of the element. The net hydrostatic pressure force per unit width is $-\rho g d \, \partial d$. The downslope weight component of the element per unit width is $\rho g d \, \partial x \, S$. The boundary friction force per unit width is $-\tau_0 \, \partial x$. The rate of change in momentum per unit width is $\rho d \, \partial x \, U \, \partial U / \partial x = \rho d U \, \partial U$. Since the rate of change of momentum equals the sum of the forces acting,

$$\rho d U \, \partial U = \rho g d \, \partial x \, S - \rho g d \, \partial d - \tau_0 \, \partial x \quad (5.45)$$

The one-dimensional continuity equation is

$$\frac{\partial (U d)}{\partial x} = U \frac{\partial d}{\partial x} + d \frac{\partial U}{\partial x} = 0 \quad (5.46)$$

Eliminating $\partial U / \partial x$ between Equations (5.45) and (5.46) results in

$$\left(1 - \frac{U^2}{g d}\right) \frac{\partial d}{\partial x} + \frac{\tau_o}{\rho g d} - S = 0 \quad (5.47)$$

The bed shear stress in the above equation may be written in terms of a uniform-flow equation such as the Darcy–Weisbach equation, resulting in an expression for the depth variation in a gradually varied flow. This equation is commonly used to predict the water-surface profiles in different types of gradually varied flow, and various techniques are available for solving it numerically. If a gradually varied flow expands and decelerates in the downstream direction, the water-surface slope decreases downstream, and this concave-upward water-surface profile is called a *backwater curve*. A flow that gradually contracts and increases in velocity downstream has a water-surface slope that increases in the downstream direction, and the profile is referred to as a *draw-down curve*.

In rapidly varied flows, abrupt changes in the flow direction, associated with abrupt changes in flow depth or curved channels with small radii of curvature, give rise to *flow separation*. Abrupt flow expansions give rise to upstream-directed hydrostatic pressures that are in excess of the downstream-directed dynamic fluid pressures (adverse pressure gradients), especially near the boundary where the flow velocity (and hence dynamic pressure) is relatively small (Figure 5.14). The result is separation of the boundary layer from the boundary, and reversal of the flow near the bed. The base of the

A

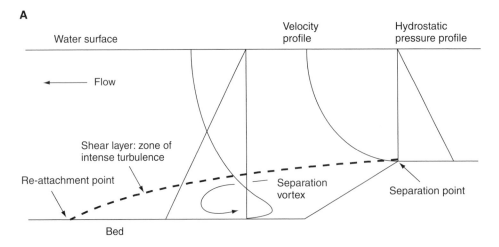

B

FIGURE 5.14. (A) Vertical profiles of time-averaged flow velocity and hydrostatic pressure associated with flow separation. (B) Flow separation on the lee side of a dune. Tracers show the flow pattern in the separation vortex. Flow is to the left. Modified from Bridge (2003).

separated boundary layer is associated with turbulence intensities two to three times greater than the average. Flow separation is important in the flow around sediment grains, the flow over ripples and dunes, and the flow around tight channel bends. Vertical profiles of flow velocity and eddy viscosity developed for uniform flows are generally not applicable to rapidly varied flows. Turbulence intensity generally decreases in accelerating flows, and increases in decelerating flows.

In curved flows, the centrifugal force, F_c, must be considered:

$$F_c = mU^2/r \tag{5.48}$$

Here r is the radius of curvature of the curved flow path. Figure 5.15 illustrates how the centrifugal force varies in a boundary-layer flow that is curved in the horizontal plane. Variation in the centrifugal force has important effects on flow patterns in curved river channels (see Chapter 13).

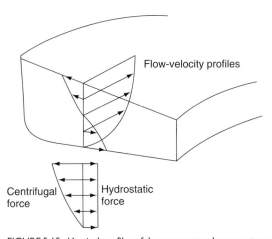

FIGURE 5.15. Vertical profiles of downstream and across-stream flow velocity (time-averaged) in steady, uniform curved-channel flow, and vertical variation in centrifugal and hydrostatic forces in across-stream section. Modified from Bridge (2003).

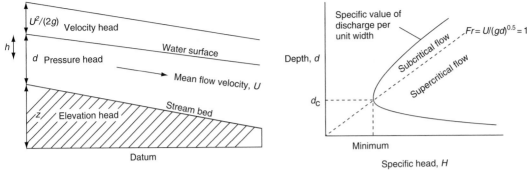

FIGURE 5.16. Definition of subcritical and supercritical flow. From Bridge (2003).

Subcritical and supercritical flow: the Froude number

Consider a steady flow in a wide channel in which the depth-averaged flow velocity, U, is constant across the width of the channel (Figure 5.16). Assume that the flow is either uniform or gradually varied, such that the pressure distribution can be assumed to be hydrostatic. The kinetic energy per unit weight of water is given by $mU^2/(2mg) = U^2/(2g)$. The potential energy per unit weight of water at any point in the flow is $z + d - y$ (Figure 5.16). The pressure energy per unit weight of water, due to hydrostatic pressure, is h. The total energy per unit weight of water, E, is thus

$$E = U^2/(2g) + d + z \qquad (5.49)$$

E is sometimes called the *total head*, and the terms on the right-hand side of Equation (5.49) are referred to as the *velocity head*, *pressure head*, and *elevation head*, respectively. Equation (5.49) is a form of the Bernoulli equation (5.5). If the streambed is used as the datum, $z = 0$, the specific head, H, is defined as

$$H = U^2/(2g) + d \qquad (5.50)$$

Since $Q = Udw$,

$$H = Q^2/(2gd^2W^2) + d \qquad (5.51)$$

If it is assumed that Q and w are constant along the length of the channel, the variation of depth with H can be examined (Figure 5.16). In general, there are two possible values of d for a given H. The smaller d is associated with larger U, and the flow is called *super-critical*. The larger d is associated with smaller U, and the flow is called *subcritical*. The critical depth is that at which H is at a minimum, and separates supercritical and subcritical flow. The critical depth, d_c, can be obtained by differentiating Equation (5.51) with respect to d and equating the result to zero, i.e.,

$$d_c = [Q^2/(gw^2)]^{1/3} \qquad (5.52)$$

which can be rewritten as

$$U/\sqrt{gd_c} = 1 \qquad (5.53)$$

The *Froude number*, $Fr = U/\sqrt{gd}$, thus determines whether a flow is subcritical ($Fr < 1$) or supercritical ($Fr > 1$). The geometries of the water surface relative to that of a movable-sediment bed for subcritical and supercritical flows are quite different, as will be seen subsequently. If there is a spatial transition between these two different types of flow, there is an abrupt step in the water surface called a *hydraulic jump* (see Figure 5.69 later).

Unsteady flows

The one-dimensional equation of continuity for unsteady, gradually varied flow in a wide channel of constant width can be written as

$$\frac{\partial d}{\partial t} + \frac{\partial (Ud)}{\partial x} = 0 \qquad (5.54)$$

and the conservation of momentum equation can be written as

$$\frac{\partial U}{\partial t} + U\frac{\partial U}{\partial x} + g\frac{\partial d}{\partial x} + \frac{\tau_0}{\rho d} - gS = 0 \qquad (5.55)$$

These equations, named after Saint-Venant, are used in water-routing problems in river systems. Exact integration of these equations is practically impossible, and they are usually solved numerically. A common type of water-routing problem, requiring solution of the unsteady, non-uniform flow equations is analysis of the movement of a flood wave. Flood waves are classified as long, free-surface waves, because their length is greater than 20 flow depths. If all of the terms in the equations of motion above are retained, the model for the flood wave is referred to as a *dynamic wave model*. In this case, it is necessary to express water discharge as a function of time at the upstream boundary of the system. However, it is common practice to ignore some of the terms in the equations of motion, depending on their relative magnitudes. For example, the first two terms in Equation (5.55), the acceleration terms, are commonly an order of magnitude less than the other terms. The acceleration terms are ignored in a *diffusive wave model*, and the momentum equation can be written as

$$g\frac{\partial d}{\partial x} + \frac{\tau_0}{\rho d} - gS = g\frac{\partial d}{\partial x} + \frac{fU^2}{8d} - gS = 0 \qquad (5.56)$$

Differentiating this equation with respect to x and eliminating $\partial U/\partial x$ using the continuity equation results in a convection–diffusion equation:

$$\frac{\partial d}{\partial t} + c\frac{\partial d}{\partial x} - K\frac{\partial^2 d}{\partial x^2} = 0 \qquad (5.57)$$

where the wave celerity (speed) $c = 1.5U$ and the dispersion coefficient $K = 4gd^2/(fU)$. When the wave height is small compared with the flow depth, c and K can be taken as constant. On introducing a moving coordinate system ($y = x - ct$), Equation (5.57) can be simplified to

$$\frac{\partial d}{\partial t} - K\frac{\partial^2 d}{\partial y^2} = 0 \qquad (5.58)$$

a solution of which is a moving Gaussian (normal) function. During wave propagation, the crest height decreases, and the wave length increases (Figure 5.17). When the flood-wave height is not small relative to the water depth, the propagation velocity is not constant. Usually the flow velocity under the wave crest is greater than that under the trough, so the crest of the wave propagates faster than the trough, and the wave front steepens.

In a *kinematic wave model*, the $\partial d/\partial x$ term is ignored in addition to the acceleration terms, so that water-level variations are solely described by the continuity equation. The continuity equation can be written as

$$\frac{\partial d}{\partial t} + c\frac{\partial d}{\partial x} = 0 \qquad (5.59)$$

where $c = U + d\,\partial U/\partial d$. Since $\partial U/\partial d = 4gS/(fU)$ (from the simplified momentum equation), $c = 1.5U$. Because the wave celerity is a function of the flow velocity (and hence of the water depth), the crest of the flood wave moves faster than the trough, so the wave front

becomes steeper. However, the wave height remains constant because of the absence of the diffusive term.

Fundamentals of sediment transport, erosion, and deposition

Grain properties

Description of sediment transport involves properties such as grain size, shape, and density, settling velocity, and sediment-transport rate. The units and standard symbols used to represent these properties are given in Table 5.2. A range of *grain size* is normally available for transport. Methods of measuring the size of unconsolidated sediment (discussed in the appendix) include direct measurement of individual grains (for gravel),

TABLE 5.2. *Physical properties of sediment and its motion, including units and mathematical symbols*

Property	Units	Symbol
Grain size	mm or m	D
Grain shape	None	
Grain density	kg m^{-3}	σ
Settling velocity	m s^{-1}	V_s
Sediment-transport rate	kg m^{-1} s^{-1}	i

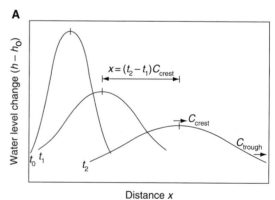

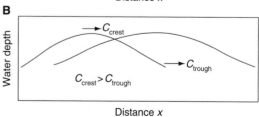

FIGURE 5.17. Flood-wave propagation according to a diffusive wave model (A) and a kinematic wave model (B). From Van Rijn (1990).

A

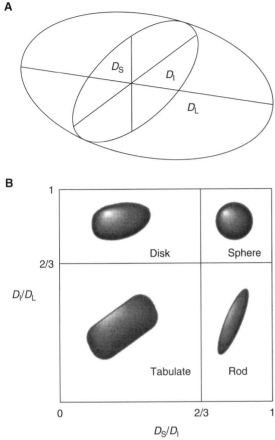

B

FIGURE 5.18. Definitions of grain shape. (A) A triaxial ellipsoid with three orthogonal axis lengths (D_S, D_I, D_L) indicated. The axes do not necessarily intersect in irregularly shaped grains. (B) Classification of grain shapes using ratios of axis lengths. From Bridge (2003).

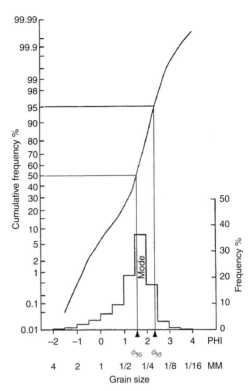

FIGURE 5.19. A histogram and cumulative-frequency curve of a grain-size distribution. The 50 and 95 phi percentiles are shown. The cumulative frequency is drawn using a Gaussian probability scale. From Bridge (2003).

sieving (for sand and gravel), use of settling tubes (for sand and mud), and laser-Doppler techniques. Grain size is difficult to define for non-spherical grains, and the definition used depends on the method of measurement. The shape of a non-spherical grain is normally approximated by a regular triaxial ellipsoid, allowing the definition of three orthogonal axis lengths (Figure 5.18). These three lengths are used in various combinations to define measures of grain size and shape (Figure 5.18). This type of size analysis is normally performed only on gravel grains for which the three axis lengths can be measured easily. In the case of sieving, the size of the sieve opening is assumed to be equivalent to the intermediate grain diameter (D_I in Figure 5.18).

The histogram of grain size is commonly plotted as weight (or number or volume) percentage versus the base-2 logarithm of grain size (Figure 5.19). The grain-size scale is in phi units, where phi is defined as

$$\phi = -\log_2[D \text{ (mm)}] \qquad (5.60)$$

The base 2 of logarithms is chosen such that the sizes of standard sieve openings (e.g., 0.5, 1, 2 mm, etc.) are whole numbers, halves, or quarters on the phi scale. The negative sign is more difficult to understand, since it results in the largest grain sizes being on the left-hand side of the graph. The cumulative frequency distribution scale is a Gaussian probability scale. If the grain-size distribution is log-Gaussian, the cumulative frequency curve will be a straight line. However, natural grain-size distributions are rarely log-Gaussian. Cumulative frequency distributions are also plotted using the logarithm of grain size (in millimeters) and an arithmetic scale for frequency. The median, mean, and standard deviation of grain-size distributions are represented by D_{50}, D_m, and D_σ, respectively. These parameters of the distribution can be calculated from combinations of percentiles (e.g., ϕ_{50}, ϕ_{95}; see Figure 5.19) determined from the cumulative curve, or, using raw data, by the method of moments (discussed in the appendix).

A sediment grain immersed in water experiences a greater hydrostatic pressure on its base than on its top: there is an upward-directed buoyancy force equal to the weight of water displaced by the grain (Archimedes' principle). If the weight of the grain exceeds the buoyancy force, the grain will sink. The *immersed weight* of the grain is the gravity force minus the buoyancy force, given by $Vg(\sigma - \rho)$, where V is the grain volume, g is the gravitational acceleration, σ is the sediment density, and ρ is the fluid density.

Drag on sediment grains

When a fluid flows over a bed of loose sediment, or a single sediment grain settles in still water, resistance to relative motion (*drag*) occurs due to viscous shear stress and dynamic pressure. Drag due to viscous shear stress is called *surface drag* or *skin-friction drag*, whereas drag due to dynamic pressure is called *form drag*. In turbulent flows, form drag is normally much more important than surface drag. Form drag is associated with individual grains and with bed forms such as ripples, dunes, and bars. A general equation for the drag force is

$$F_D = C_D a(\rho V_r^2/2) \tag{5.61}$$

in which C_D is a drag coefficient, a is the cross-sectional area exposed to the drag, and V_r is the relative velocity of the solid and fluid. The term in brackets is the dynamic pressure on the area of the solid that faces the fluid flow, but this term also accounts for the pressures on the lee side of the solid associated with flow separation and for viscous forces. Flow separation produces a pressure against the flow (a suction effect) that increases form drag. The drag coefficient takes into account whether or not flow separation occurs behind a solid body and the nature of such flow separation, including the size of the separation eddies. The drag coefficient therefore depends on the geometry of the solid body and on how streamlined it is. The drag coefficient is also dependent on an appropriate Reynolds number, because this determines the relative importance of surface (viscous) drag and form (pressure) drag.

Settling of grains in fluids

A sediment grain placed in a less-dense fluid will accelerate as it settles downwards under the influence of gravity, but its movement will be resisted by fluid drag. As the fluid drag on the accelerating grain

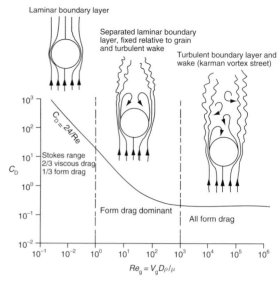

FIGURE 5.20. The drag coefficient C_D as a function of the Reynolds number, Re_g, for a single spherical grain settling in a still fluid. From Bridge (2003).

becomes equal to its immersed weight, the grain continues to fall at a constant velocity, referred to as the *terminal settling velocity*. The balance between the immersed weight and the drag force is given by

$$Vg(\sigma - \rho) = C_D a \rho V_s^2/2 \tag{5.62}$$

where V_s is the settling velocity. For the case of spheres, this equation reduces to

$$V_s^2 = 4Dg(\sigma - \rho)/(3\rho C_D) \tag{5.63}$$

The drag coefficient is related to the grain Reynolds number, $Re_g = \rho D V_s/\mu$, as shown in Figure 5.20. For $Re_g < 1$ (called the Stokes range), surface drag is dominant over form drag and a laminar boundary layer occurs next to the grain. For $1 < Re_g < 1,000$, form drag dominates over surface drag and a turbulent wake streams behind a separated laminar boundary layer that is fixed relative to the grain. For $1,000 < Re_g < 200,000$, there is a turbulent boundary layer around the grain and a turbulent wake. Separation vortices are alternately formed and detached from different sides of the grain, giving what is called a *Kármán vortex street*. Under these conditions, the drag coefficient is independent of the Reynolds number because surface drag is negligible relative to form drag.

In the Stokes range ($Re_g < 1$), it has been shown theoretically and experimentally that the drag on a smooth sphere settling in a fluid is

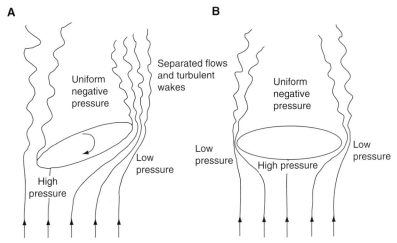

FIGURE 5.21. The pressure distribution around a disk or cylinder during settling through a still fluid for grain Reynolds number between 1 and 100. The grain in (A) would become oriented with its maximum projection area normal to the direction of settling, such that the pressure distribution is uniform (B). From Bridge (2003).

$$F_D = 6\pi(D/2)\mu V_s \qquad (5.64)$$

and, from the general drag equation,

$$F_D = C_D\pi(D/2)^2(\rho V_s^2/2) \qquad (5.65)$$

Combining these equations gives an expression for the drag coefficient,

$$C_D = 24/Re_g \qquad (5.66)$$

and the settling velocity,

$$V_s = (\sigma - \rho)gD^2/(18\mu) \qquad (5.67)$$

This equation, known as *Stokes' law of settling*, applies to sediment grains with diameters less than approximately 0.1 mm in water and less than 0.06 mm in air. Stokes' law is used to determine the grain size of fine-grained sediment from measured settling velocities (this is discussed in the appendix). Since Stokes' law does not apply to most sand and gravel grains, the drag coefficient and settling velocity for these grains must be determined experimentally. There are also empirical equations to relate settling velocities of sand and gravel grains to sediment and fluid parameters.

The relationship between the drag coefficient and the grain Reynolds number for non-spherical grains is similar to that for spherical grains (e.g., Figure 5.20), but the details differ. The orientation of a non-spherical grain may vary as it settles, and the grain might not follow a straight vertical path, depending on the grain's Reynolds number and moment of inertia. For $Re_g < 1$, there is no preferred orientation during settling, and the varying orientation of a grain has no effect on the drag because viscous forces are dominant. However, for $1 < Re_g < 100$, non-spherical particles tend to orient

themselves with their maximum projection area normal to the settling direction. This orientation results in a symmetrical distribution of dynamic pressure on the grain, as explained in Figure 5.21. For $Re_g > 100$, a non-spherical grain tends to oscillate or tumble about a horizontal axis that moves with the grain. Falling leaves commonly have this oscillatory settling behavior. Grains with a local concentration of mass (such as bivalve shells and seed cases) may rotate about vertical axes (spiraling) as they settle downwards. Rotating grains experience Magnus forces arising from the asymmetrical distribution of flow velocity and fluid pressure around the grain. Magnus forces tend to cause grains to deviate from their otherwise-straight paths, which is why spinning balls have curved paths (as known by players of soccer, baseball, cricket, pool, and snooker). The amount of rotation depends on the moment of inertia of the grain, which is represented by a stability number, $\sigma h/(D\rho)$, where σ is the sediment density, h is the thickness of a disk-like object, D is the mean grain diameter, and ρ is the fluid density (Figure 5.22).

Theoretical and empirical determinations of settling velocity discussed above are for single grains settling in fluids of "infinite" lateral extent, implying that there are no nearby flow boundaries or other grains that might interfere with their settling. However, with high concentrations of settling grains, the settling paths are complicated because of mutual interference between grains. In most cases, settling is hindered because of the upward movement of fluid displaced by downward-moving grains. Very small grains may even be swept upwards in this displaced fluid. Also, a suspension of very small grains increases the effective fluid viscosity

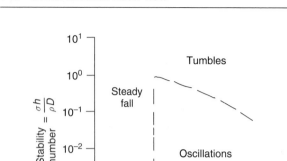

FIGURE 5.22. The mode of settling of discoidal grains as a function of the Reynolds number, Re, and stability number. Modified from Allen (1985).

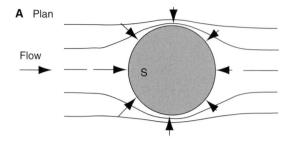

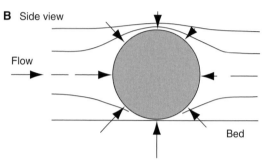

FIGURE 5.23. Streamlines and pressure distribution (relative magnitude indicated by arrow length) around a stationary sphere lying on the bed of a moving fluid at low Reynolds number: (A) plan view and (B) side view.

and density (Equations (5.81) and (5.82)), thus reducing the settling velocity of the larger grains. Clay minerals in close proximity to each other tend to flocculate, producing larger compound grains that settle faster than individual clay flakes. The settling velocity of a grain in a dense suspension, V_{ss}, can be expressed empirically as

$$V_{ss} = V_s(1 - C)^n \qquad (5.68)$$

where C is the volume concentration of grains and $n = 2.4$ for $Re_g > 100$. For smaller grain Reynolds numbers, n is larger and variable. At low Reynolds numbers, a flow boundary can affect the settling of a grain if it is closer to the grain than $100D$; however, this effect diminishes as the grain Reynolds number increases. Turbulence in natural flows also affects the settling behavior of sediment grains.

Drag and lift on bed grains

Stationary sediment grains on the bed of a unidirectional, turbulent flow are acted upon by viscous shear stress and dynamic pressure to varying degrees. Viscous (surface) drag is dominant over pressure (form) drag on hydraulically smooth boundaries. For hydraulically rough boundaries, pressure (form) drag is so dominant that viscous forces can be ignored.

The dynamic pressure at the upstream end of a stationary sediment grain on the bed can be calculated using the Bernoulli equation (5.5). Figure 5.23A shows that the spacing of streamlines near the upstream end of the grain is relatively large, indicating that the flow

velocity is relatively low and the fluid pressure is high. At the sides of the grain, the streamlines are close together, indicating relatively high flow velocity and low pressure. At point S (Figure 5.23A), a streamline attaches to the body and divides in a T shape. At this *stagnation point*, the velocity in the downstream direction is zero. Application of the Bernoulli equation (5.5), assuming that z does not vary along this streamline, gives the fluid pressure, p_s, at the stagnation point:

$$p_s = p_0 + \rho u_0^2/2 \qquad (5.69)$$

in which the terms on the right-hand side apply to the fluid well upstream of the stagnation point where the flow is uniform, p_0 is the hydrostatic pressure, and u_0 is the flow velocity. This equation indicates that the flow exerts a pressure at point S that is in excess of the hydrostatic pressure by an amount equal to the dynamic pressure.

However, the dynamic pressure varies in magnitude over the surface of the grain, and is greatly influenced by lee-side flow separation. Furthermore, the fluid pressure and shear forces on bed grains vary in time with turbulence. Since these time-varying fluid forces are difficult to predict at any point on the grain, it is normal to express the drag as a value averaged over the whole grain and over time, and taken to act through a specific point in the grain. The resultant direction of this mean drag is commonly resolved into components parallel to the bed (*drag*) and normal to the bed (*lift*). The mean drag on a sediment grain can be expressed using the general drag equation:

$$F_D = C_D a(\rho u^2/2) \tag{5.70}$$

where u is the mean velocity of the fluid at the level of effective mean drag. This velocity can be predicted using the law of the wall for hydraulically rough flows or the velocity profile for the viscous sublayer for hydraulically smooth flows. The level of effective drag depends on the shape of the grain and its position in the bed. The drag coefficient for single grains resting on the bed is assumed to be similar to that for settling grains. The area exposed to drag depends on the shape of the grain, and how it is packed in the bed relative to surrounding grains. This area will be proportional to the square of the grain size. Thus it can be shown that

$$F_D \propto \tau_0 D^2 \tag{5.71}$$

where the coefficient of proportionality depends on the grain shape and orientation, its relative protrusion above the mean bed level, and the grain Reynolds number.

Lift forces acting on bed grains can be both turbulent and non-turbulent. Non-turbulent lift forces are pressures caused by the asymmetrical distribution of flow velocity around bed grains. In the case of a stationary sediment grain resting on the bed (Figure 5.23B), the asymmetrical distribution of streamlines around the grain indicates that the fluid pressure will be less on top of the grain than near its base. Therefore, the grain will experience a net upward pressure (lift force). However, the flow velocity and pressure distribution around sediment grains resting on flow boundaries are much more complicated than shown in Figure 5.23B. Since turbulence originates very close to the bed of a stream, bed grains are undoubtedly affected by turbulent lift, and this is most likely to be dominant over non-turbulent lift (outside the viscous sublayer). Fluid lift is commonly expressed in a similar way to drag, but using a lift coefficient instead of a drag coefficient. The value of the lift coefficient is poorly known. Furthermore, the ratio of lift to drag is commonly and erroneously assumed to be constant.

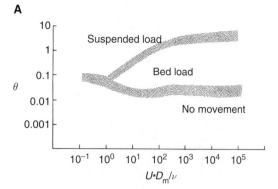

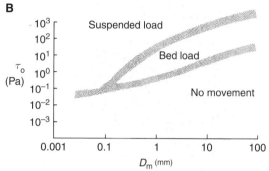

FIGURE 5.24. Thresholds of motion for bed load and suspended load in terms of the dimensionless bed shear stress and bed grain Reynolds number (A) or bed shear stress and mean bed-grain size (B). Boundaries between transport modes are zones rather than sharp lines. From Bridge (2003).

The threshold of transport of cohesionless sediment

If the drag and lift forces acting on stationary bed grains exceed the forces keeping the grains in place, the grains will be entrained by the flow. The forces keeping the grains in place are gravity and cohesive forces such as those associated with vegetation, cyanobacteria, cementation, and electrostatic clay-mineral interactions. The threshold of entrainment for cohesionless grains depends on the balance between the drag and gravity forces, F_D and F_G, respectively, as long as fluid lift is assumed to be a linear function of drag. Since

$$F_D \propto \tau_0 D^2 \qquad \text{and} \qquad F_G \propto (\sigma - \rho)gD^3 \tag{5.72}$$

the threshold of entrainment should depend on

$$F_D/F_G \propto \tau_0/(\sigma - \rho)gD = \Theta \tag{5.73}$$

in which Θ is referred to as the *dimensionless bed shear stress* (a measure of sediment mobility) and D is a grain size representative of all grains in the bed. Values of dimensionless bed shear stress at the threshold of motion, Θ_c, are typically between 0.03 and 0.06 (Figure 5.24). The exact value depends on the degree to which the grains are immersed in the viscous sublayer, the relative magnitude of the lift component of drag, the bed slope, and the relative sizes, shapes, and arrangements of the grains in the bed, as explained below.

In a classic theoretical and experimental study, Shields (1936) concluded that the two most important parameters controlling the threshold of sediment motion are the dimensionless bed shear stress Θ and the boundary grain Reynolds number, defined as

$$Re_b = U_* D / \nu \approx 10 D / \delta_v \tag{5.74}$$

in which δ_v is the thickness of the viscous sublayer. Thus, the boundary grain Reynolds number can be looked at as being proportional to the ratio of the grain diameter and the thickness of the viscous sublayer. For Re_b less than 10, the grains are substantially immersed in the viscous sublayer (i.e., we have hydraulically smooth flow), are subjected mainly to viscous forces, and are sheltered from the much greater drag forces of the turbulent flow above. This is the reason why the dimensionless bed shear stress must be relatively large to entrain these grains (Figure 5.24). However, this is also partly to do with increasing effects of cohesion with very small grains. If bed grains project high above the viscous sublayer ($Re_b > 70$–100; hydraulically rough flow), turbulent shear stresses and form drag are dominant, viscous forces are negligible, and the threshold of motion becomes independent of the Reynolds number (Figure 5.24). Since Shields' study, there have been numerous theoretical and experimental studies of the threshold of motion that have confirmed and extended his work. For fixed values of the fluid and sediment density, the threshold of entrainment can be expressed in terms of bed shear stress and grain size (Figure 5.24). The threshold of entrainment has also been expressed using the depth-averaged flow velocity and mean grain size. Since the bed shear stress is proportional to the square of the depth-averaged velocity and to the resistance coefficient (which depends on the boundary Reynolds number and the depth/bed-roughness height), the threshold of motion will depend on at least the flow depth in addition to the flow velocity and mean grain size. Thus, the threshold of entrainment cannot be represented adequately using the flow velocity and mean grain size.

The threshold of entrainment shown in Figure 5.24 is actually a zone that encompasses the range of experimentally determined values. This range of values is partly due to the fact that not all of the controlling

Instantaneous bed shear stress   Threshold bed shear stress for motion for single grains

No overlap of probability distributions

Increase in mean bed shear stress

Small overlap of probability distributions: movement of grains with lowest threshold bed shear stress

Increase in mean bed shear stress

Large overlap of probability distributions: movement of all grains except those with highest threshold shear stress

FIGURE 5.25. Probability distributions of the instantaneous bed shear stress and threshold of motion of individual grains on the bed. The degree of overlap of the distributions determines which grains will start to be entrained and the sediment-transport rate. Based on Grass (1970).

parameters (such as the bed slope and relative sizes, shapes, and arrangements of grains on the bed) are represented on plots of dimensionless bed shear stress and boundary grain Reynolds number. However, the main reason for the range of threshold values is the difficulty in actually defining the threshold of motion. Fluid drag and lift forces vary greatly in time and space in turbulent flow, and these forces act upon grains that vary in size, shape, and arrangement on the bed. Thus, a locally high instantaneous fluid force may dislodge a few of the most easily moved grains, but not produce general sustained movement. Indeed, such intermittent movement may be difficult to observe. Grass (1970) approached this problem by defining a frequency distribution of instantaneous turbulent bed shear stress and a frequency distribution of threshold bed shear stress for the individual bed grains that moved (Figure 5.25). As these two distributions start to overlap, the greatest shear stresses will start to entrain the most easily moved grains. The threshold of general motion can then be defined by some measure of the degree of overlap. When Shields defined the threshold of motion, he did not actually watch when grains started to move, as Grass did. He measured sediment transport rates at various values of dimensionless bed shear stress and deduced the threshold shear stress by linearly extrapolating the transport-rate–shear-stress curve to zero transport rate. This has been criticized on the grounds that the curve is not linear and is difficult to define at very small sediment-transport rates. Many other experimenters have defined the threshold of motion as that associated with a very small but finite sediment-transport rate; however, there has been little consistency in this definition.

The threshold of motion defined in Figure 5.24 applies only to flat beds with relatively small slopes. Theories that take the effects of bed slope into account are available. If bed waves such as ripples or dunes are present on the bed, the threshold of motion is complicated by non-negligible bed slope and the fact that the spatially averaged bed shear stress includes components of bed form drag (e.g., due to flow separation) that do not effectively act on bed grains. Thus, in order to define the threshold of motion on bed waves, it is necessary to define the local fluid drag and lift on their stoss sides, and to account for the upstream dipping bed surface.

The threshold of transport of grains of mixed sizes, shapes, and densities

The threshold of motion defined in Figure 5.24 applies only to the mean grain size of a population of grains that do not vary much in shape and density. Prediction of the threshold of entrainment of individual grains in a mixture of grains of varied size, shape, and density requires much more detailed analysis of the forces involved, including turbulent variations in fluid drag and lift. Approaches to this problem have been reviewed by Komar (1996) and Bridge (2003). A common semi-theoretical approach is to analyze the balance of time-averaged forces acting on a single grain as it begins to move out of position on a sloping bed. The many attempts at this type of force balance differ in terms of the simplifying assumptions involved in the specification of the fluid forces acting, the size of the grain relative to the other bed grains, the position of the grain in the bed, and the grain shape. A common empirical approach is to express the threshold bed shear stress as

$$\frac{\theta_{ci}}{\theta_{c50}} = \left(\frac{D_i}{D_{50}}\right)^{-n} \tag{5.75}$$

where θ_{ci} is the time-averaged dimensionless bed shear stress for entrainment for a given size fraction, θ_{c50} is the time-averaged dimensionless bed shear stress for

entrainment of the median grain size of the bed sediment, D_i is the diameter of size fraction i, D_{50} is the median diameter of the bed grains, and n ranges from 0.3 to nearly 1. The large ranges of values for n and θ_{c50} are due to the diversity of methods for defining the threshold of motion and to the fact that the equation results from trying to fit a straight line on log–log paper to parts of a family of curves. If $n = 1$, the threshold bed shear stress is the same for all grains, and grains of all sizes will start moving at the same time. This condition has been termed *equal mobility*, and claimed to exist under some natural conditions. However, it is not generally the case.

The approaches mentioned above predict that the larger-than-average grains are relatively easier to entrain than grains of all the same size, whereas the opposite is true for the smaller-than-average grains. This is because large grains protrude relatively higher into the flow where the average fluid velocity is greater, and can be rolled or slid relatively easily over a bed of smaller grains. Such grains may have dimensionless bed shear stress as low as 0.01 at the threshold of motion. In contrast, small grains sit low in the flow, may be hidden by the larger grains, and it is relatively difficult to roll or slide them over the larger grains surrounding them. However, these predictions are based on consideration of time-averaged fluid drag and lift. Experiments show that turbulence is very effective at lifting small grains out of their hiding places between immobile larger grains. Therefore, it is essential to consider turbulent variations in fluid drag and lift, because these cause different grain fractions to move at different times. The largest grains may actually move only under the influence of the largest instantaneous fluid drag or lift, and not be moved by the mean values. Furthermore, the largest instantaneous values of drag and lift do not occur simultaneously in turbulent flows (see the section on the bursting phenomenon), so the mean values of drag and lift cannot be assumed to act simultaneously as a grain is entrained into the flow. Since fluid turbulence has a dominating influence on sediment entrainment, and turbulent motions vary in time and space, it is not surprising that, when a particular size fraction starts to move on the bed, not all of the grains of that size fraction on the bed will be moving. This will be particularly true for the coarsest fractions that require the largest instantaneous turbulent stresses for movement. Some of the theoretical approaches to initiation of grain motion described above have been adapted for the case of turbulent variations in fluid drag and lift.

The threshold of transport of cohesive sediment (mud)

Cohesion in muds is the dominant control on resistance to entrainment; see reviews by Kuijper *et al.* (1989), Dade *et al.* (1992), and Black *et al.* (2002). Cohesion is partly due to electro-chemical forces, the magnitude of which depends on mineralogy, the size, shape, and spatial arrangement of the clay flakes, and the ionic properties of the pore water and eroding water. Another important type of cohesion is due to microbial organisms (bacteria, microphytobenthos). Resistance properties also depend on the state and history of consolidation. For example, there may be a flocculated structure, a pelleted structure due to desiccation and bioturbation, or fissility due to compaction. Compacted and dry clays are more difficult to erode than soft, wet clays. The nature of entrainment also depends on the state and history of consolidation, determining whether the clay is eroded as individual flakes, floccules, or fragments of consolidated mud called intraclasts. Also, the direct fluid stress may be augmented by impacts on the bed of grains already in motion.

Many experimental studies have been conducted to establish relationships between the critical bed shear stress for entrainment and some gross property of the clay such as the cohesive shear strength, liquid and plastic limits, percentage of clay, density, permeability, and porosity. These properties have usually been linked to critical shear stress one or two at a time using only one type of clay–water complex. There are no universal correlations for all types of clays in all aqueous environments. In general, the erosion rate of uncompacted cohesive mud increases with bed shear stress, once the critical shear strength has been exceeded.

Bed load

Sediment grains with diameters greater than approximately 0.1 mm (sand and gravel) generally move close to the bed (within about ten grain diameters) by rolling, sliding, and saltating (jumping). These *bed-load* grains move more slowly than the surrounding fluid because of their intermittent collisions with the bed and with each other. The continuing movement of bed-load grains requires the existence of an upward dispersive force (Bagnold, 1966, 1973) that must be exactly balanced by the immersed weight of the moving grains for steady bed-load transport. The two ways of dispersing grains normal to the flow boundary are by collisions between the grains and bed, and by fluid lift, as discussed below.

Saltation (jumping) is the dominant mode of bed-load transport, with rolling and sliding occurring only at low sediment-transport rates and between individual jumps. Rolling and sliding grains are frequently launched upwards as they are forced over protruding immobile grains, thus initiating jumps. Jumps may also occur due to elastic rebound of saltating grains when they impact on the bed, but only for relatively large grains at high transport rates. At low sediment-transport rates, saltation trajectories are unlikely to be interrupted by collisions with other moving grains. However, as sediment transport rates increase, jumps will be more commonly interrupted by collisions between moving grains.

Both turbulent and non-turbulent fluid lift forces act on bed-load grains. Non-turbulent lift forces are pressures arising from a net relative velocity between grains and surrounding fluid. They are caused by asymmetrical flow around near-bed grains (Figure 5.23), the reduced grain velocity relative to the velocity gradient in the surrounding fluid (shear drift), and the effects of grain rotation (Magnus lift). It is very difficult to specify their individual contributions to net lift, particularly when complex grain collisions occur within the bed load and when turbulent lift forces are dominant. Bed-load grains are undoubtedly affected by turbulent lift. Since near-bed turbulent fluctuations increase with bed shear stress (or shear velocity), grain jumps are expected to be modified more and more by turbulence as bed shear stress and sediment-transport rate increase. Turbulence affects different-sized grains in different ways. For example, at a low sediment-transport rate of gravel, most of the movement occurs during sweeps, and to a lesser extent during outward interactions. Ejections and inward interactions have little effect on coarse-sediment movement, because even the largest values of turbulent lift are much less than the immersed weight of the gravel grains. With smaller (lighter) grains or larger bed shear stresses, ejections have much more influence on bed-load movement.

The mean height of the bed-load (saltation) zone controls the roughness height, k_s. Furthermore, theories of bed-load transport require knowledge of the distance from the bed of effective fluid thrust on bed-load grains, which is related to saltation height. Many approaches to determining the mean height of the bed-load zone are semi-theoretical (empirical), predicting that saltation height increases with sediment size and some measure of the sediment-transport rate such as bed shear stress or shear velocity. Theoretical models of single saltation trajectories have also been used to determine saltation height. However, it is very difficult to model the influence of turbulent fluid motions on saltating grains, and the effects on saltation trajectories of grain impacts are not considered. Therefore, most theoretical models do not perform very well.

Suspended load

Relatively small grains (clay, silt, and very fine sand) travel within the flow, supported by fluid turbulence directed upwards against their tendency to settle. These *suspended-load* grains travel at approximately the same speed as the surrounding fluid because they are not being decelerated by intermittent collisions with the bed. Maintenance of a suspended-sediment load requires the existence of a net upward-directed fluid stress to balance the immersed weight of the suspended grains, which can arise only if the vertical component of turbulence affecting the grains is anisotropic (Bagnold, 1966). This anisotropy can be associated with the bursting phenomenon, whereby sediment is suspended mainly in vigorous ejections, and returns to the bed under the influence of gravity and downward-directed turbulent motions. This idea is supported by experimental observations of discrete clouds of suspended sediment associated with ejections, and of the wavy trajectories of suspended grains. However, sediment is also suspended in vortices generated by eddy shedding from the separated boundary layer in the trough regions of ripples and dunes. These vortices are much larger than an ejection from a viscous sublayer. In the case of dunes, these vortices and the sediment they carry can reach the water surface as boils or kolks.

Two criteria for the suspension of sediment are

$$\mathrm{rms}(v'_+) > V_s \tag{5.76}$$

and

$$\mathrm{rms}(v'_+) - \mathrm{rms}(v'_-) > V_s \tag{5.77}$$

in which $\mathrm{rms}(v')$ is the root mean square of the vertical turbulent velocities, the subscripts $+$ and $-$ indicate upward- and downward-directed turbulent fluctuations, respectively, and V_s is the settling velocity of the grains. The first criterion indicates that, as long as time-averaged upward-directed turbulent velocities are in excess of V_s, sediment can be moved upwards within the flow, even if the sediment eventually returns to the bed. The second criterion indicates that for a grain to *remain* in suspension the time-averaged upward-directed turbulent velocity experienced by the grain

must exceed the downward-directed velocity plus the settling velocity of the grain. The relationship between the time-averaged upward- and downward-directed velocities and the overall $\mathrm{rms}(v')$ depends on the coefficient of anisotropy, a (Figure 5.13). Using (sparse) experimental values of a, the suspension criteria (Equations (5.76) and (5.77)) near the bed become

$$a\,\mathrm{rms}(v') > V_s \quad \text{where } a \text{ is } 1.2-1.5 \tag{5.78}$$

$$b\,\mathrm{rms}(v') > V_s \quad \text{where } b \text{ is } 0.4-0.9 \tag{5.79}$$

Since $\mathrm{rms}(v')/U_* \approx 1$ near the bed (Equation (5.45)), the suspension criteria for near-bed conditions are

$$aU_* > V_s \quad \text{and} \quad bU_* > V_s \tag{5.80}$$

Also, since the shear velocity and the settling velocity can be related to the dimensionless bed shear stress and the grain Reynolds number, the threshold of suspension can be represented on the threshold of entrainment diagrams (Figure 5.24). Actually, sediment with a particular settling velocity (or grain size) may travel as both bed load and sediment load (called *intermittently suspended load*), because the bed shear stress varies in time and space. This is particularly true of the sand sizes greater than about 0.1 mm.

The *wash load* is commonly distinguished from *suspended bed-material load*. Wash load is very fine-grained and suspended even at low flow velocity. Furthermore, the volume concentration of wash load does not vary much with distance from the bed, whereas the volume concentration of suspended bed-material load decreases markedly with distance from the bed. Whereas the suspended bed-material load originates from the bed, the wash load can come from bank erosion and overland flow. The two different suspension criteria above effectively distinguish wash load, (5.77) and (5.79), from suspended bed-material load, (5.76) and (5.78).

Effect of sediment transport on flow characteristics

Sediment in motion modifies the density and viscosity of the fluid–sediment mixture and the turbulent flow characteristics of the transporting fluid. The density and viscosity of water–sediment mixtures are given by

$$\rho_s = \rho + (\sigma - \rho)C \tag{5.81}$$

$$\mu_s = \mu(1 + 2.5C + 6.25C^2 + 15.26C^3) \tag{5.82}$$

$$\nu_s = \mu_s/\rho_s \tag{5.83}$$

where μ and ν are the molecular and kinematic fluid viscosities of pure water and the subscript s refers to the fluid–sediment mixture. The Reynolds number decreases as grain concentration increases for a given flow velocity and depth because viscosity increases more than density as concentration increases. It might be expected, therefore, that the degree of turbulence decreases as the grain concentration increases. Turbulence modulation is not as simple as this, as will be seen below.

Bed load and suspended load cause reductions in eddy viscosity ε, mixing length l, and Von Kármán's κ, and increases in roughness height and near-bed velocity gradient. The main effect of sediment on turbulence is near the bed, where the sediment concentration is highest. Even weak bed-load transport affects near-bed turbulence, with κ being reduced from its clear-water value of 0.41 to around 0.3. Some workers deny that sediment motion affects the value of κ, and claim that turbulence modification in sediment-transporting flows is brought about by density stratification (associated with an upward decrease in suspended sediment concentration), which suppresses turbulent mixing. The effects of density stratification are represented by a *flux Richardson number*, Ri, one form of which is

$$Ri = \kappa g(\sigma/\rho - 1)V_s C/U_*^3 \tag{5.84}$$

The flux Richardson number has been used to modify the law of the wall, the velocity-defect–wake law, or an empirical expression for the vertical variation of eddy viscosity. However, such modifications to velocity-profile laws using the flux Richardson number do not agree well with data.

The effects of moving sediment on turbulence can be related to the size (or mass) of sediment grains relative to the size of turbulent eddies, bearing in mind that, as eddy size increases, turbulence intensity normally decreases and eddy time scale increases. Thus, if the grain size is small relative to the eddy size, the inertia (mass) of the grain is relatively small, the relative velocity of the grain and turbulent fluid is small, and the grain follows the motion of the fluid closely. Under these circumstances, some of the kinetic energy of the fluid is transferred to the grain, and turbulence intensity is decreased. Decrease in turbulence intensity is also related to concentration of this fine-grained sediment, because viscosity increases (the Reynolds number decreases) as concentration increases. For sediment grains that are large relative to the turbulent-eddy size, grain inertia (mass) is large, the relative velocity of the grains and fluid is large, and

the grains do not follow the eddy motion closely. The flow may separate in the lees of the grains, producing turbulent wakes of relatively small eddies. In this case, turbulence intensity is increased. The mixing length (eddy size) is relatively decreased because the proportion of small-scale eddies is increased by the relative velocity of grains and fluid and the increase in the development of turbulent wakes. This is particularly significant near the bed, where the concentration of relatively large grains is greatest. In this zone, grain collisions may also add to the increase in turbulence intensity. Also, relatively large grains increase the frequency of bursting but decrease the ejection velocities because of the relative velocity of the fluid and grains. As will be seen later, bed forms also have an important effect on turbulence intensity.

Sediment-transport rate (capacity)

The sediment-transport rate is the amount (weight, mass, or volume) of sediment that can be moved past a given width of flow in a given time. The sediment-transport rate controls the development of bed forms such as ripples and dunes, the nature of erosion and deposition, and the dispersal of physically transported pollutants. The sediment-transport rate has been investigated in great detail. Only the basics will be presented here.

The bed-load transport rate can be expressed as

$$i_b = Wu_b \tag{5.85}$$

in which i_b is the bed-load transport rate expressed as immersed weight passed per unit width, W is the immersed weight of bed-load grains over unit bed area, and u_b is the mean bed-load grain velocity. The contacts between bed-load grains and the bed result in resistance to forward motion. Therefore, the fluid must exert a mean downstream force on the grains to maintain their steady motion. For bed-load transport on a bed of slope angle β, the force balance parallel to the bed is

$$\tau_o + W\sin\beta = T + \tau_r \tag{5.86}$$

where τ_o is the bed shear stress applied to the moving grains and immobile bed (excluding form drag due to bed forms and banks), $W\sin\beta$ is the downslope weight component of the moving grains per unit bed area (with $\sin\beta$ positive if the bed slopes down flow), T is the shear resistance due to the moving bed load, and τ_r is the residual shear stress carried by the immobile bed. The bed-load shear resistance can be expressed as

$$T = (W\cos\beta - F_L)\tan\phi \tag{5.87}$$

where F_L is the net fluid lift on bed-load grains per unit bed area, taken as resulting dominantly from aniso-tropic turbulence, and given by

$$F_L = W(bU_*/V_s)^2 \tag{5.88}$$

where b is explained in Equation (5.79) and $\tan\phi$ is the dynamic friction coefficient. From experiments on shearing of grains in dry-grain flows and coaxial drums, where volumetric grain concentrations are approximately 0.5, $\tan\phi$ commonly lies between 0.4 and 0.75; $\tan\phi$ is not strongly dependent on shear rate, but decreases with increasing grain concentration. At low grain concentrations, $\tan\phi$ is in the range 0.75–1. The residual fluid stress τ_r equals the bed shear stress at the threshold of motion, τ_c, according to Bagnold (1956, 1966). Thus, once the applied fluid stress exceeds the value necessary for entrainment, grains will be entrained until the bed-load resistance caused by their motion reduces the applied fluid stress to the threshold value of τ_c. Therefore, the immersed weight of grains moving per unit bed area, W, is given, from Equations (5.86)–(5.88), as

$$W = \frac{\tau_0 - \tau_c}{\tan\phi[\cos\beta - (bU_*/V_s)^2] - \sin\beta} \tag{5.89}$$

As the threshold of suspension is reached, the term in square brackets approaches zero, indicating that the weight of bed-load grains becomes very large. This sit-uation could not arise in practice because high grain concentrations in the bed-load layer would occlude the bed from fluid drag and lift, thus limiting the weight of bed-load grains (e.g., Bagnold, 1966). Furthermore, high concentrations of near-bed grains modify turbu-lence characteristics above the bed. In many natural flows, the bed slope is small enough that $\sin\beta$ tends to zero, and the term in square brackets tends to 1, so

$$W = \frac{\tau_0 - \tau_c}{\tan\phi} \tag{5.90}$$

The mean velocity of the average bed-load grain is given by

$$u_b = u_D - u_R \tag{5.91}$$

where u_D is the time-averaged flow velocity at the level of effective drag on the grain and u_R is the relative velocity of the fluid and the grain. The relative velocity is that at which the fluid drag F_D plus the downslope weight component on the grain $F_G\sin\beta$ is in equili-brium with the bed-load shear resistance:

$$F_D + zF_G\sin\beta = F_G\tan\phi[\cos\beta - (bU_*/V_s)^2] \tag{5.92}$$

Using the general drag equation for F_D and the fact that F_G is equal to the fluid drag at a grain's terminal settling velocity results in

$$u_R = V_s\left\{\tan\phi[\cos\beta - (bU_*/V_s)^2] - \sin\beta\right\}^{0.5} \tag{5.93}$$

Finally, since $u_D/U_* \approx 10$ (Bridge, 2003), the mean velocity of bed-load grains is given by

$$u_b \approx 10U_* - V_s\left\{\tan\phi[\cos\beta - (bU_*/V_s)^2] - \sin\beta\right\}^{0.5} \tag{5.94}$$

and, for small bed slopes, by

$$u_b \approx 10U_* - V_g(\tan\phi)^{0.5} \tag{5.95}$$

Furthermore, since the grain velocity must be zero at the threshold of grain motion, this equation may also be written as

$$u_b \approx 10[U_* - U_{*c}(\tan\phi/\tan\phi_c)^{0.5}] \tag{5.96}$$

where $\tan\phi_c$ is the static friction coefficient, commonly about 0.6. At the threshold of suspension, the grain velocity approximately equals the fluid velocity and $u_b \approx 10U_*$, which is another suspension criterion.

Thus, the final approximate bed-load transport equation can be written, using Equations (5.85), (5.90), and (5.96), as

$$i_b \approx \frac{10}{\tan\phi}(U_* - U_{*c})(\tau_0 - \tau_c) \tag{5.97}$$

Bed-load transport equations of this form have been developed by many people and agree very well with experimental data for flows over plane beds (Bridge, 2003). For low sediment-transport rates (lower-stage plane beds), $\tan\phi$ approaches 1, but it is approxi-mately 0.6 for high sediment-transport rates (upper-stage plane beds). For beds covered with bed forms such as ripples and dunes, the bed-load transport equation can be applied in one of two ways. The most common method is to determine the average bed shear stress over a number of bed forms, and to remove the part of the total average stress that is due to form drag in the lee of the bed forms, because it is ineffective at moving bed-load grains. A second approach is to calculate the bed-load transport rate integrated over the back of a typical bed form from knowledge of the local fluid drag and lift in this zone. This is not done routinely because it is very difficult to predict the near-bed fluid motions on the backs of bed forms. As seen below, there are alternative ways of calculating bed-load transport rates associated with ripples and dunes.

The bed-load transport theory developed above is for mean flow and sediment characteristics, but it can be adapted for use with heterogeneous sediment and with temporal variations in bed shear stress. In that case, it is necessary to specify the proportions of the various grain fractions available in the bed for transport, and the thresholds of entrainment and suspension for each of the grain fractions. It is also necessary to specify the proportion of time that a particular bed shear stress acts upon the bed load, requiring a frequency distribution of bed shear stress. Such theories are essential for quantitative understanding of sediment sorting during erosion, transport, and deposition, as discussed below.

The *suspended-sediment transport rate* at a point in the flow is expressed as

$$i_s = u_s C \tag{5.98}$$

where u_s is the average speed of the sediment (approximately the same as the fluid velocity), and C is the volume concentration of suspended sediment. The units of i_s are volume of sediment transported past unit cross-sectional area per unit time. The vertical variation of the velocity of suspended sediment and fluid can be calculated using an appropriate velocity-profile law (e.g., the law of the wall). Calculation of the vertical variation of C in steady, uniform water flows is

traditionally based on the balance of downward settling of grains and their upward diffusion in turbulent eddies:

$$V_s C + \varepsilon_s \, dC/dy = 0 \tag{5.99}$$

where the first term is the rate of settling of a particular volume of grains per unit volume of fluid, the second term is the rate of turbulent diffusion of sediment per unit volume, and ε_s is the diffusivity of suspended sediment (equivalent to a kinematic eddy viscosity). This balance can also be written for individual grain fractions. In order to determine C at any height above the bed, y, it is necessary to calculate the vertical variation in ε_s. Assuming that $\varepsilon_s = B\varepsilon$, where B is close to 1, and that the law of the wall extends throughout the flow depth, results in the well-known Rouse equation

$$\frac{C}{C_a} = \left(\frac{d-y}{y} \cdot \frac{a}{d-a} \right)^z; \quad z = \frac{V_s}{B\kappa U_*} \tag{5.100}$$

where C_a is the value of C at $y = a$ and d is the flow depth. This equation predicts that C decreases continuously and smoothly with distance from the bed, as would be expected in view of the fact that turbulence intensity also decreases with distance from the bed. The distribution of suspended sediment becomes more uniform throughout the flow depth as the exponent z decreases, that is as the settling velocity (grain size) decreases and/or as the shear velocity (near-bed turbulence intensity) increases (Figure 5.26). For example, a

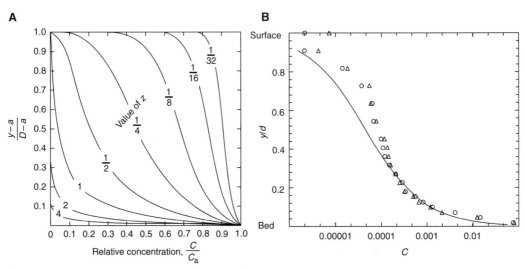

FIGURE 5.26. (A) Variation of the relative concentration of suspended sediment with distance from the bed as a function of z, according to the Rouse equation (5.100). The distribution of suspended sediment becomes more uniform throughout the flow depth as the exponent z decreases, that is as the settling velocity (grain size) decreases and/or the shear velocity (near-bed turbulence intensity) increases. (B) A comparison between the Rouse equation and data from an upper-stage plane bed. From Bennett *et al.* (1998). The predicted suspended-sediment concentration is less than that observed in the upper part of the flow.

decrease in water temperature causes an increase in fluid viscosity, which may in turn cause a decrease in settling velocity and an increase in suspended sediment concentration. The Rouse equation has been shown to agree with measured suspended-sediment concentrations in the lower part of streamflows but concentrations tend to be underestimated in the upper parts (Figure 5.26).

Some of the assumptions used in developing the Rouse equation have been criticized, unjustly in some cases. Perhaps the most serious concern with the Rouse equation is that the sediment diffusivity has rarely been calculated directly using quantitative observations of the motion of sediment in turbulent eddies. Thus, there is doubt about the vertical variation of ε_s and B. There are two main approaches to calculation of ε_s. The first, most common method involves assuming that the diffusion–settling balance (Equation (5.99)) is correct, such that ε_s can be calculated using measured values of V_s, C, and dC/dy. This led to the unreasonable conclusion that ε_s is generally greater than ε, such that $B > 1$. This approach also led to the idea that ε_s does not vary parabolically with y/d as ε does, but may approach a constant value above $y/d = 0.5$. A second, more desirable method for calculating ε_s is to use an approach similar to that for calculating ε, i.e.,

$$\varepsilon_s = \frac{-\overline{u'v'}}{d\overline{u}/dy} \tag{5.101}$$

where all values apply to sediment particles only. Experimental data on turbulent motions of suspended sediment are very rare, but Bennett et al. (1998) have shown that values of ε_s calculated using Equation (5.101) are generally similar to values of ε, suggesting that suspended sediment follows the mean fluid turbulence fairly closely. In contrast, values of ε_s calculated using the diffusion–settling balance tend to be significantly overestimated relative to ε in the upper half of the flow depth, because the sediment concentration is larger than expected. One reason for this discrepancy could be errors in the definition of ε_s using Equation (5.99), such that sediment grains that are suspended to the higher levels in the flow are not associated with mean turbulence characteristics, but with turbulent eddies with the greatest turbulence intensities and mixing lengths. Thus, the Rouse equation, which is based on average turbulence characteristics, may be applicable only to the lower parts of the flow.

Application of the Rouse equation in practice requires calculation of C_a and a. Normally, C_a is calculated at a position within or at the top of the bed-load layer. If bed waves are present, a can be taken as half the bed-wave height or equivalent roughness height. The mean volume concentration of grains in the bed-load layer, C_b, is the volume of grains per unit bed area divided by the mean thickness of the bed-load layer, y_b,

$$C_b = \frac{\tau_0 - \tau_c}{(\sigma - \rho)gy_b \, \tan \phi} \tag{5.102}$$

The mean grain concentration can be assumed to occur at a distance from the bed of half the thickness of the bed-load layer, or half the equivalent roughness height. In order to calculate the suspended-sediment transport rate integrated across the width of a flow, it is necessary to define vertical profiles of flow velocity and suspended-sediment concentration across the width.

The sediment-transport rate varies over a large range of time and space scales, associated for example with fluid turbulence, the migration of bed waves such as ripples, dunes, and bars, the break-up of armored or partially cemented beds, bank slumps, seasonal floods, flow diversions, and earthquake-induced changes in sediment supply. A characteristic feature of unsteady, non-uniform flows is that the sediment-transport rate commonly varies incongruently with temporal and spatial changes in water discharge: i.e., there is a lag or hysteresis. This may be due to variations in available sediment supply, or due to the lag of sediment and water flow behind a faster-moving flood wave. The wash load is never well correlated with water discharge, and commonly reaches a peak during rising discharge because of an overland flow source. The lag of the bed-load transport rate behind a spatial change in flow characteristics is very small (on the order of a few grain diameters); however, for suspended sediment this spatial lag is on the order of the flow depth.

Sorting and abrasion of sediment during transport in turbulent flow

The fact that bed load and suspended load travel at different speeds and in different parts of the flow leads to the possibility of *sorting*, or physical separation of sediment grains with different settling velocities (i.e., differing grain size, shape, and density), as shown in Figure 5.27. Figure 5.28 indicates *approximately* the range of grain sizes (for a given shape and density) that can be transported as bed load for a particular bed shear stress. Actually, Figure 5.28 represents the thresholds of entrainment and suspension for mean

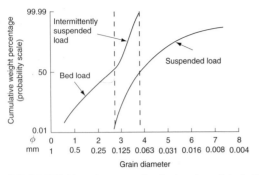

FIGURE 5.27. Typical grain-size distributions (cumulative) of bed load and suspended load sampled at a point in a river. The zone of overlap of the two distributions represents grain sizes that are intermittently suspended from the bed region. Modified from Bridge (2003).

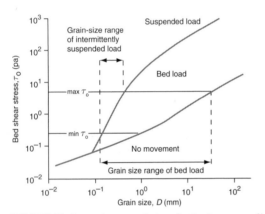

FIGURE 5.28. Approximate prediction of grain-size ranges of bed load and intermittently suspended load where the bed shear stress is allowed to vary between limits max τ_o and min τ_o. The mean grain size and range of sizes increase as the instantaneous bed shear stress increases. The prediction is only approximate because the thresholds of motion and suspension of individual grains in a mixture of different-sized grains are not necessarily exactly as shown. Threshold curves are based on sediment of constant grain size and density, and time-averaged bed shear stress. Modified from Bridge (2003).

flow conditions and for mean sediment size. The actual grain-size range in the bed load depends on what is available for transport, the threshold of entrainment or suspension for individual grain fractions in a distribution, and turbulent fluctuations in bed shear stress. Nevertheless, Figure 5.28 explains in a general way why intermittently suspended load exists, and why it requires temporal variations in bed shear stress. It is possible to calculate theoretically the range of sizes of grains of a given shape and density that will form the bed load and intermittently suspended load, given a distribution of available sediment. In general, the

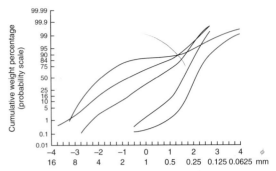

FIGURE 5.29. Typical grain-size distributions of bed material in rivers. Modified from Bridge (2003).

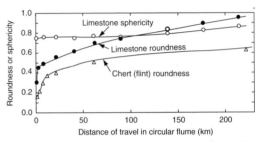

FIGURE 5.30. Rounding of broken limestone and chert (flint) fragments in a circular flume. The floor of the flume consisted of rounded flint pebbles. Initial weights of pebbles were about 40 g, and final weights were 15–30 g. The velocity of movement was 0.7 m s^{-1}. From Blatt et al. (1980).

mean and standard deviation of grain size in the bed load increase as the bed shear stress increases (Figure 5.29). The size distribution of suspended sediment is more difficult to specify. Although the coarse-grained end of the distribution is controlled by bed shear stress, the overall distribution is controlled by sediment availability. Calculation of the sorting of grains of differing size, shape, and density in *deposited* sediment requires consideration of changes in the rate of sediment transport, as discussed below.

Collisions between grains in transport, occurring mainly near the bed, result in *abrasion* of the grains. Results from experimental studies of abrasion of gravel during river transport (e.g., Figure 5.30) indicate the following changes with increasing duration and distance of transport: (1) a decrease in mean size of grains; (2) a rapid increase in roundness initially, then increase at a slower rate; and (3) a slow increase in sphericity at a constant rate. These changes increase in magnitude as the hardness of the grains decreases. In general, the downstream decrease in grain size of gravel is mainly due to hydraulic sorting, with abrasion playing a relatively minor role. Abrasion of sand in water is negligible (in contrast to that in air flows): thus rounded sand grains in fluvial deposits normally indicate a phase of air transport and/or that the source of the sediment was a pre-existing sedimentary rock.

Erosion and deposition of sediment

Erosion and deposition of sediment are due to changes in sediment-transport rate in space (mainly) and in time (a minor contribution), as expressed in the equation of conservation of sediment mass (the sediment continuity equation). A one-dimensional version of this equation is

$$-C_b\, \partial h/\partial t = \partial i_x/\partial x + \partial i_x/(u_{sx}\, \partial t) \tag{5.103}$$

where C_b is the volume concentration of sediment in the bed $(1 - \text{porosity})$, h is the bed height, t is time, i_x is the downstream sediment-transport rate by volume, x is the downstream distance, and u_{sx} is the downstream velocity of the sediment. In uniform flow, $\partial i_x/\partial x = 0$, but an increase in sediment-transport rate in time $(\partial i_x/(u_{sx}\, \partial t) > 0)$ would result in erosion. Such erosion would involve entrainment of a layer of bed grains not exceeding a few grain diameters thick: that is, erosion and deposition in uniform, unsteady flows are negligible. In contrast, rates of erosion and deposition in non-uniform flows can be substantial, depending on the magnitude of $\partial i_x/\partial x$. Erosion occurs in spatially accelerating flows where $\partial i_x/\partial x$ is positive, and deposition occurs in spatially decelerating flows where $\partial i_x/\partial x$ is negative.

It is common for relatively long periods of little net erosion or deposition to be punctuated by relatively short periods (e.g., during floods or storms at sea) when there is appreciable erosion followed by appreciable deposition or vice versa (i.e., the terms in Equation (5.103) vary over time). This is why sedimentary strata are normally bounded by surfaces that record erosion or non-deposition. If $\partial i_x/\partial x$ is negative and does not vary much in space, a layer of sediment that varies little in thickness laterally will be deposited. If $\partial i_x/\partial x$ is negative but changes markedly in space, sediment will be deposited as mounds or ridges (e.g., ripples, dunes, and bars). Similarly, a surface of erosion will tend to be planar if $\partial i_x/\partial x$ is positive and does not vary much in space, whereas it will tend to be strongly curved (e.g., trough-shaped or spoon-shaped) if $\partial i_x/\partial x$ is positive and changes locally. In reality, $\partial i_x/\partial x$ varies in space on various different scales, such that there are different scales of variation in the geometry of strata and erosion surfaces superimposed.

Deposition rates and erosion rates depend on the time interval and spatial scale for which they are calculated. Vertical deposition rates averaged over a few years and over bed waves such as ripples, dunes, and bars can be up to meters per year. However, many deposits will be completely or partially eroded before final burial, such that deposition rates averaged over progressively longer time spans tend to decrease. Deposition rates and erosion rates averaged over millions of years are typically on the order of 0.1 mm per year (Leeder, 1999). The deposits with the best *preservation potential* are those that were deposited in the topographically lowest areas (e.g., deepest parts of channels) and those that subsided rapidly.

Sediment sorting and orientation during erosion and deposition

Application of Equation (5.103) to individual grain fractions (differing in grain size, for example) leads to prediction of *sorting* in deposited sediment. Rational prediction of such sorting requires knowledge of (1) the sediment types available for transport; (2) the temporal and spatial variations in flow and sediment-transport rates of individual grain fractions; and (3) the flow and sediment transport over bed forms such as ripples, dunes, and bars. Such prediction has been attempted only rarely, using sediment-routing models as discussed below. In general, when erosion occurs, the coarsest and/or densest grains are most likely to remain on the bed. The large, immobile grains that are left on an eroded bed surface are commonly referred to as a *lag deposit*. The term *armoring* refers to the situation where there are sufficient large immobile (or only intermittently mobile) grains on an eroding bed to protect the potentially mobile grains underneath from entrainment, thus limiting the amount of erosion possible. Armored beds commonly occur on gravel beds during waning flow stages after river floods or storms at sea. The amount of erosion at a point is commonly also limited by generation during erosion of a progressively larger depression in the sediment bed, leading to expansion and deceleration of the flow, and thus cessation of erosion or relocation of the zone of erosion.

When deposition occurs, the coarsest and/or densest grain fractions in the bed load are preferentially deposited, but the character of the deposited sediment is closely related to that of the bed load from which it was derived. The deposits associated with peak flow conditions (high bed shear stress) are commonly coarser-grained than those associated with waning

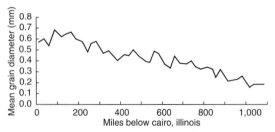

FIGURE 5.31. Downstream decrease in bed-material size along the lower Mississippi River. After Leopold *et al.* (1964).

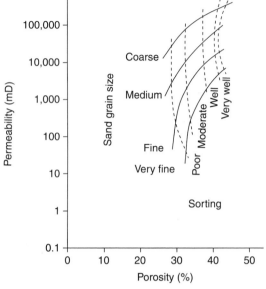

FIGURE 5.32. Porosity and permeability as functions of mean sand-grain size and sorting. Modified from Brayshaw *et al.* (1996).

and low flow conditions (low bed shear stress), because the mean grain size of sediment in transport and being deposited is directly related to bed shear stress.

The mean grain size of sediment in the bed surface and in transport near the bed commonly decreases exponentially in the direction of transport (Figure 5.31). This is mainly due to downstream decrease in the bed shear stress and turbulence intensity of the transporting and depositing flow, such that the coarsest bed-load grains are progressively lost in the downstream direction (i.e., size-selective deposition). Downstream reduction in size due to progressive abrasion associated with grain collisions is of minor importance, because only the larger and softer gravel grains suffer appreciable abrasion. Intermittently suspended grains may be deposited with bed-load grains, either as a relatively fine-grained surficial layer, or by infiltrating the pore spaces between the larger grains in the bed (thus reducing the degree of sorting). Most of the suspended load is not deposited with bed load, but accumulates in places where turbulence intensity is low (e.g., floodbasins, lakes, deep sea). Suspended sediment can be sorted by selective deposition according to settling velocity.

Heavy minerals (grain density >2800 kg m^{-3}: see the appendix) that are deposited from bed load and suspended load are commonly finer-grained than associated light minerals. Heavy minerals may be concentrated where a flow that is powerful enough to entrain and transport all grain fractions decelerates and deposits some of them. The relatively small grains of heavy minerals become protected from re-entrainment by the larger grains of light minerals that can continue to move above them, and some grains of heavy minerals may infiltrate into the pore spaces between the larger deposited grains. Subsequent erosion may remove the larger grains of light minerals but not the smaller grains of heavy minerals. Such conditions are found in association with various kinds of bed forms, as discussed in later sections.

Non-spherical grains tend to be *oriented* during bed-load transport immediately prior to deposition. On plane beds, ellipsoidal or rod-shaped grains tend to be oriented with their long axes approximately parallel to flow, and disk- and tabular-shaped pebbles tend to be oriented with their maximum projection planes dipping upstream (*imbrication*). However, these characteristic orientations are modified where bed forms are present, as explained below. Magnetic grains may be oriented during deposition with their poles aligned in the prevailing magnetic field, but only if the orienting effect of the transporting fluid is weak.

Sorting (by grain size, shape, and composition) and grain orientation within sediment deposits have a strong influence on *porosity and permeability* (see the appendix). The most important controls on porosity and permeability are grain size, sorting, and fabric. Grain size is overwhelmingly important for permeability. Permeability increases by a factor of eight for each increment in grain-size class (see the appendix and Figure 5.32). Permeability also increases as sorting improves (the standard deviation of grain size decreases). *Hydraulic conductivity* has been related to the grain-size distributions of unconsolidated sediments, and is particularly sensitive to the proportion of the finest sizes. Grain size does not have a big influence on porosity (they are theoretically independent), although some unconsolidated clays have very high porosity (up to 80%) because of loose clay-mineral packing. Sorting influences porosity, and well-sorted sediments are more porous than poorly sorted ones (Figure 5.32). However, grain size and sorting are not independent variables in most sediments (Figure 5.29). The relationships between fabric (grain packing and orientation) and porosity and permeability are difficult to quantify. Close packing reduces permeability and porosity, and grain orientation causes permeability to vary anisotropically. Porosity and permeability vary greatly with spatial variation in texture and fabric (i.e., with stratification), as discussed below.

Sorting, packing, and grain orientation (and hence permeability and porosity) in sedimentary rocks are strongly influenced by diagenesis (Chapter 19) as well as by processes of deposition. The amount of the finest sediment can be increased diagenetically by infiltration through the groundwater (e.g., illuviation), crushing of mudstone fragments, and chemical precipitation of clays. Grain shape, packing, and orientation can be changed by compaction, by crushing, and by dissolution. Composition of sediment can be changed by dissolution or precipitation. Porosity commonly increases with permeability in hydrocarbon reservoirs and aquifers (Figure 5.33), although the scatter in such plots is large.

Numerous attempts to relate the characteristics of grain-size distributions of deposited sediment to the mechanics of sediment transport and deposition have been made, and have largely failed because of: (1) poorly designed sediment-sampling strategies; (2) lack of understanding of the interaction between turbulent fluid flow and transport of sediment of mixed sizes, shapes, and densities; (3) failure to distinguish sorting during transport from sorting during deposition; and (4) failure to appreciate diagenetic effects on sediment textures. Some workers have estimated the minimum bed shear stress or mean flow velocity required to move sediment of a given size as bed load, or to cause its suspension. However, in most cases the sediment was well above its threshold of entrainment or suspension when deposited. Grain-size sorting can be approached rationally only by using sediment-routing models, as discussed below.

Sediment-routing models

The only rational way of analyzing and predicting bed-elevation changes and sediment sorting during erosion, transport, and deposition is by using sediment-routing models. Sediment-routing models are used to deal with problems such as bed degradation and armoring downstream of dams, and reservoir sedimentation. Sediment-routing models normally require treatment of unsteady, gradually varied flow acting upon movable, heterogeneous sediment beds. A realistic sediment-routing model requires specification of (1) sediment types available for transport; (2) the mean and turbulent fluctuating values of fluid force acting upon sediment grains, and how these vary in time and space; (3) the interaction between fluid forces and individual sediment fractions, resulting in their entrainment and transport as bed load or suspended load; (4) erosion and deposition as determined by a sediment continuity equation applied to the various grain fractions; (5) the concept of an active bed layer to which sediment can be added or from which it can be removed; and (6) an accounting scheme to record the nature of deposited sediment or armored, eroded surfaces. Many sediment-routing models do not have all of these desirable features, and contain many arbitrary correction factors. Figure 5.34 shows how a good sediment-routing model can successfully predict sediment-transport rates, patterns of erosion and deposition, and the grain-size distributions of the bed load and the sediment in the bed. When deposition occurs by overloading sediment at the upstream end of an experimental channel (Figure 5.34A), deposition initially occurs as a mound near the sediment feed point, producing a flow expansion (and reduction in bed shear stress and

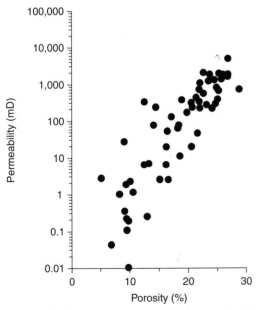

FIGURE 5.33. Porosity versus permeability for core plugs from a fluvial hydrocarbon reservoir, Triassic, southern England. From Brayshaw et al. (1996).

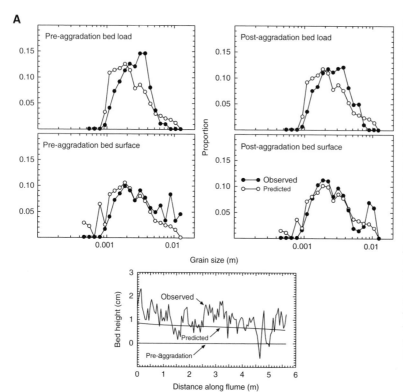

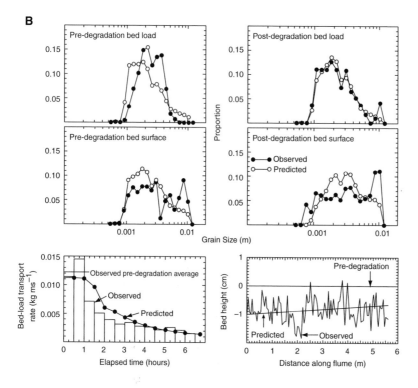

FIGURE 5.34. (A) A comparison of the MIDAS sediment-routing model with experimental data for the case of an aggrading bed. Bed-load and bed-surface grain-size distributions are given for pre-aggradation and post-aggradation equilibrium conditions. Changes in mean bed height following aggradation are shown on the lower diagram. The observed bed profile manifests low-relief bed waves. From Bennett and Bridge (1995). (B) A comparison of the MIDAS sediment-routing model with experimental data for the case of a degrading bed. Bed-load and bed-surface grain-size distributions are given for pre-degradation and post-degradation equilibrium conditions. The bed-load transport rate (lower left) decreases with time due to bed armoring. Changes in mean bed height following degradation are shown on the lower right. The observed bed profile manifests low-relief bed waves. From Bennett and Bridge (1995).

sediment-transport rate) downstream. This results in downstream progradation of deposition, producing a wedge-shaped deposit. As sediment is deposited at a point, the bed shear stress and sediment-transport rate increase locally. The deposited sediment is slightly coarser-grained than the sediment feed and the bed load, as a result of depositional sorting. Also, as local bed shear stress decreased in the flow direction, the coarsest sediment was deposited upstream, and the deposit became finer-grained downstream. In the case of erosion induced by sediment starvation at the upstream end of the experimental channel (Figure 5.34B), a scour zone moves downstream, and (with gravelly sediments) the bed becomes armored, thereby causing exponential decreases in bed-load transport rate and erosion rate at a point.

Bed forms and sedimentary structures

Bed forms are ubiquitous during sediment transport. Bed forms are important to sedimentary geologists because they are associated with characteristic types of sedimentary structures, and to engineers because they influence flow resistance and the sediment-transport rate. A *bed form* is defined as a single geometric element, such as a ripple or a dune (Allen, 1982a). A *bed configuration* is defined as the assemblage of bed forms of a given type that occurs in a particular bed area at a given time (e.g., a ripple bed configuration). An *equilibrium* or *steady-state* bed configuration is one in which the bed-form geometry, bed-form migration rate, and sediment-transport and flow characteristics are on average constant and adjusted to the steady supply of water and sediment. This does not mean, however, that individual bed forms in the bed configuration do not

vary in geometry and migration rate. Indeed, all bed forms have a finite *lifespan* (from creation to growth to decline to destruction) and *excursion* (distance of bed form migration during its lifespan). A *bed state* is the assemblage of all bed configurations of a given type. Some of the parameters used to describe bed forms and sedimentary structures are given in Table 5.3 (see also Figure 5.35).

Knowledge of bed forms and bed configurations has been aided greatly by laboratory flume studies, even though most flumes have severe scale limitations (see the appendix). Both equilibrium and disequilibrium bed configurations can be studied in flumes. Observations of bed states in natural environments have been made using sonic depth profilers (e.g., side-scan sonar, multibeam sonar) and underwater cameras. However, it is difficult to describe fully the geometry and migration rates of bed forms in natural environments, and to relate them to the formative flow and sediment-transport conditions. This is because: (1) most natural flows are unsteady and non-uniform; (2) bed configurations are commonly not in equilibrium with

TABLE 5.3. Physical properties of bed waves and cross strata, including units and mathematical symbols

Property	Units	Symbol
Bed-wave height	m or mm	H
Bed-wave length (flow parallel)	m	L
Bed-wave length (flow transverse)	m	
Bed-wave migration rate (celerity)	mm s^{-1}	c
Deposition rate	mm s^{-1}	r
Cross-set thickness	m or mm	s

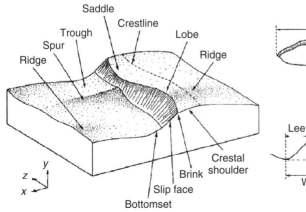

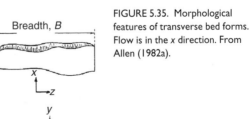

FIGURE 5.35. Morphological features of transverse bed forms. Flow is in the *x* direction. From Allen (1982a).

the flow and sediment transport at the time of observation; and (3) there are limits to the time periods and areas over which field observations are possible.

Bed states in cohesionless sediment

Bed forms in cohesionless sediment have been classified as *microforms*, *mesoforms*, and *macroforms* (e.g., Jackson, 1975). The occurrence and geometry of microforms (e.g., ripples) are controlled by near-bed flow characteristics such as the boundary Reynolds number or bed-grain size. Mesoform (e.g., dune) geometry is controlled by boundary-layer thickness or flow depth. Macroforms (bars) have lengths of the same order as the channel width, and heights comparable to the depth of the generating flow (American Society of Civil Engineers Task Force on Bed Forms, 1966). Macroforms are discussed in Chapter 13 because these relatively large-scale forms define the overall geometry of river channels. Microforms and mesoforms are discussed in this chapter, and comprise five main bed states in cohesionless sediment that are defined on the basis of their geometry relative to (uniform, steady) flow and sediment-transport conditions (Table 5.4). The characteristics of equilibrium bed forms in uniform, steady flows will be considered first, followed by those of disequilibrium bed forms in unsteady and non-uniform flows.

Ripples

Ripples form at a bed shear stress just above the threshold of bed-load motion if the flow is hydraulically smooth ($U_*D/\nu < 10$) and subcritical. The mean bed-grain size must be less than about 0.7 mm for the bed-grain Reynolds number to be less than 10 near the threshold of bed-load motion. Thus ripples cannot form in very coarse sands and gravels. Ripples also do not occur in cohesionless mud, because this fine-grained sediment is suspended as soon as it is entrained from the bed. However, ripples can form if the mud is in the form of sand-sized pellets that are transported as bed load.

The initiation and growth of ripples on a planar bed result from turbulent variation in near-bed flow velocity and sediment-transport rate; see the reviews by Allen (1982a) and Best (1992a, 1993, 1996). At the threshold of bed-grain motion, the grains move in bursts in patches of the bed affected by the most intense turbulent fluid motions. The grains are moved into a series of discontinuous ridges and depressions *parallel*

to the flow, with a relief of a few grain diameters, and a flow-transverse spacing of several millimeters to centimeters (i.e., *primary current lineations*: Figure 5.36). Primary current lineations reflect the streaky flow structure of the viscous sublayer. An exceptionally energetic turbulent fluid motion near the bed (e.g., a burst and sweep, or a group of them) may cause local erosion and re-deposition of a mound of sediment (a bed defect) several grains high with its crest *normal* to the flow. If the defect is sufficiently high, flow separation occurs on its lee side. Increased turbulence intensity in the zone of re-attachment causes further erosion and re-deposition further downstream as another defect. These defects develop progressively downstream, and grow in size, until the whole bed is covered with ripples of a specific mean size and shape (Figure 5.37; Table 5.4).

Ripples have asymmetrical profiles parallel to the flow direction, with the steeper lee side at the angle of repose (Figure 5.37). According to Allen (1982a), ripples are less than 0.6 m long and less than 0.04 m high (Table 5.4). Ripples are commonly 0.1–0.15 m long and 0.01–0.02 m high. The length and height of individual ripples in a configuration vary in space and time (even in steady, uniform flows). Most bed forms (not just ripples) commonly grow in height and length by amalgamation of smaller bed forms. For example, one bed form may catch up with the next bed form downstream, giving rise to a larger and higher combined bed form. In addition, relatively small, fast-moving bed forms commonly develop on the stoss sides of the larger ones. Such superimposed bed forms may migrate to the crest of the host bed form and leave the geometry of the host bed form more or less unchanged. However, these small superimposed bed forms may also combine to form a larger bed form, thereby causing the host bed form to become two separate bed forms. Thus, new bed forms are constantly being created while others are being modified or destroyed. Incipient bed forms on the stoss sides of larger forms are always smaller and more numerous than the larger, more developed forms, which is why frequency histograms of bed-wave height, length, and period are always asymmetrical (Paola and Borgman, 1991; Bridge, 1997) and commonly polymodal (Figure 5.38).

Ripple length is proportional to mean grain size (e.g., $L \approx 1{,}000D$; Figure 5.39 (Allen, 1982a; Yalin, 1977, 1992)) and independent of flow depth. This is related to the fact that the length of the flow-separation zone of an embryonic ripple (which is about half of the ripple length) is proportional to the ripple height,

TABLE 5.4. Bed-form geometry and migration rates, equilibrium flow conditions, flow resistance, sediment transport, and flow regime for selected bed states

	Ripples	Lower plane bed (+ bed-load sheets)	Dunes	Upper plane bed (+ bed-load sheets)	Antidunes
Geometry					
	$H < 40$ mm, mainly 10–20 mm	$H <$ few D	H typically 10^{-2}–10^1 m	$H <$ few mm	H typically tens of mm
	$L \propto D$ or U_*D/ν, normally <60 mm	$L \propto d$	$L \propto d$, typically 10^{-1}–10^2 m	$L \propto d$	$L = 2\pi d$
	$H/L < 0.1$	$H/L < 0.01$	$H/L < 0.06$	$H/L < 0.01$	$H/L < 0.06$
					symmetrical
	←————— asymmetrical in streamwise sections —————→				
	←——— lee side at angle of repose ———→			←——— sides less than angle of repose ———→	
	←——— straight, sinuous, linguoid or lunate in plan ———→				
Bed-wave migration rate					
	c is on the order of 10^{-1}–10^0 mm s^{-1}			$c \approx 10$ mm s^{-1}	
Equilibrium flow conditions					
	$U_*D/\nu < 10$ ($D < 0.7$ mm)	←— $U_*D/\nu > 10$ ($D > 0.7$ mm) ——	$U_*D/\nu > 10$ —→		
	←———— $\Theta < \sim0.5$ ————→			←———— $\Theta > \sim0.5$ ————→	
Flow resistance (f)					
	Moderate (0.04–0.08)	Low (0.02–0.03)	High (0.04–0.15)	Low (0.02–0.03)	Low (0.02–0.03)
Sediment transport					
	←———————— sediment-transport rate increases ————————→				
	←———————— suspended load/bed load increases ————————→				
Flow regime					
	←————— Lower —————			———— Upper ————→	
	←———— Subcritical ($Fr < 1$) —————			——— Supercritical ($Fr > 1$) ———→	

which is several grain diameters. As embryonic ripples grow by amalgamation, the relationship among ripple length, ripple height, and grain size is maintained. The mean height/length ratio of ripples is dependent on the dimensionless bed shear stress, or the bed-load transport rate (Figure 5.40). Incipient ripples tend to have straight crests, as can be seen in plan view. Fully developed ripples tend to have sinuous to linguoid (tongue-shaped) crestlines (Figure 5.37). Ripples formed in areas of limited sediment supply tend to have lunate (crescent-moon-shaped) crestlines. The crest heights of curved crested ripples vary transverse to flow, and flow-parallel spurs occur in the trough areas (Figures 5.35 and 5.37).

Lower-stage "plane" beds and associated bed forms

Lower-stage plane beds occur just above the threshold of bed-grain motion if the flow is subcritical and hydraulically transitional or rough ($U_*D/\nu > 10$). At these relatively low shear velocities, hydraulically transitional or rough flow occurs if the mean size of bed grains is greater than approximately 0.7 mm. The bed grains substantially disrupt the viscous sublayer, and flow separation occurs on the lee side of the bed grains. Near the threshold of bed-grain motion, bed defects up to a few grain diameters high develop and propagate downstream in much the same way as described for ripples (Leeder, 1980). However, in this case the defects do not grow progressively in height, but develop into low-relief bed waves called bed-load sheets (i.e., lower-stage plane beds are not truly planar: Figure 5.41). Bed-load sheets are also asymmetrical in along-stream profile, with the lee face at or near the angle of repose. Bed-load sheets are typically several bed-grain diameters thick, such that they are less than about 10 mm high in sediment with a mean grain size of 2 mm. The lengths of bed-load sheets are proportional to the flow depth (L/d is commonly 4–12), suggesting that these bed forms are incipient dunes. Therefore, bed-load sheets have much larger length/height ratios than ripples do.

FIGURE 5.36. Formation of flow-transverse "bed defects" (embryonic ripples) from a group of flow-parallel sediment ridges within the viscous sublayer, according to Best (1992a, 1996).

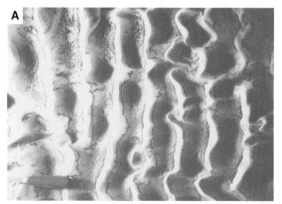

FIGURE 5.37. (A) Developing (straight-crested) ripples in a flume (from Jaco Baas). (B) Ripples with linguoid and sinuous crestlines on a sand bar.

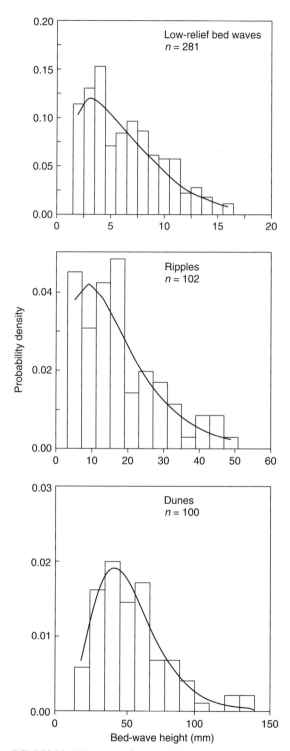

FIGURE 5.38. A frequency histogram of the height of n successive bed forms (low-relief bed waves, ripples, dunes) passing a point on the bed. From Bridge (1997). Histograms are asymmetrical and polymodal. A gamma density function is fitted.

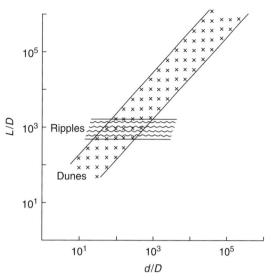

FIGURE 5.39. Experimental data envelopes for the length (L) of equilibrium ripples and dunes in terms of flow depth (d) and mean grain size (D). From Bridge (2003).

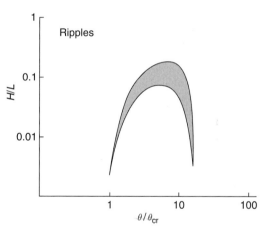

FIGURE 5.40. Variation in height/length ratio of equilibrium ripples with the dimensionless bed shear stress. The region of experimental data is stippled. From Bridge (2003).

Bed-load grains of different sizes are segregated by the turbulent flow on lower-stage plane beds. In sandy gravel, sand is moved out of the interstices of gravel that is either immobile or partially mobile (giving an armored bed). It is common for the coarsest gravel to be concentrated in the trough areas of the bed-load sheets as a partially armored bed, whereas finer gravel forms the crest areas (Figure 5.41). This is an indication that, at these low sediment-transport rates, the coarsest grain fractions are barely mobile. These mobile or

FIGURE 5.41. (A) Low-relief bed waves (bed-load sheets) on a gravel bar. Flow is from right to left. (B) Deposits of a bed-load sheet or low-height dune (80 mm of fining-up open-framework gravel above lens cap). Flow is from left to right. Modified from Bridge (2003).

immobile gravels from which sand has been removed are called *open-framework gravels*, and have very high permeability (thousands of darcies). These flow and sediment-transport conditions also occur on the backs of dunes and unit bars (discussed in Chapter 15), and appreciable thicknesses of open-framework gravel can be deposited on the lee sides of these bed forms, as will be seen below.

Pebble and cobble clusters are common on poorly-sorted lower-stage plane beds, particularly in the coarser-grained trough areas of bed-load sheets (Figures 5.41 and 5.42). Pebble or cobble clusters are groups of relatively large, temporarily immobile grains, with the largest grains forming a nucleus. In sandy fine gravels, a wedge of sand accumulates on the downstream side of the cluster, and a scoured zone (a current-crescent scour) occurs upstream and to the side of the cluster, in which heavy minerals may be concentrated. The clustered gravel grains

are *imbricated*, meaning that the plane containing the long and intermediate axes of discoidal or tabular pebbles dips upstream. However, such grains that accumulate on the steep lee side of the bed-load sheets dip downstream (*pseudo-imbrication*) (Figures 5.41 and 5.42).

Relatively small bed grains moving over lower-stage plane beds are also commonly segregated into distinct, flow-parallel lanes of sediment transport (e.g., *sand ribbons, longitudinal stripes*) that are adjacent to coarser-grained, partially armored bed areas (Figure 5.43). The flow-transverse spacing of sand ribbons is from one to four flow depths (most commonly two flow depths), suggesting that sand ribbons occur where the near-bed flow associated with pairs of depth-scale streamwise vortices converges and rises from the bed. This type of secondary circulation is related to across-stream variations in boundary roughness and turbulent stress, including bank effects. Sand ribbons may have rippled

FIGURE 5.42. A pebble cluster with imbricated grains. Flow is to the right. Photo courtesy of Jim Best.

FIGURE 5.43. (A) Sand ribbons on a gravel bed. The sand ribbons pass to the right into rippled sand. Scale divisions are 2 cm. The flow direction is towards the observer. From Bridge (2003).

FIGURE 5.44. Emergent sinuous-crested dunes from a Congaree River point bar. (A) The view looking upstream. (B) The view from river to floodplain, with flow from left to right. Dunes are about 0.2 m high and 4 m long.

surfaces if their mean grain diameter is less than approximately 0.7 mm.

The migration of bed-load sheets gives rise to planar stratification, which is a form of transcurrent stratification (Figure 5.41). The strata are recognizable because of vertical variation in grain size and fabric. The coarsest grains at the base of planar strata are associated with the trough areas of the bed-load sheets, whereas the finer grains above are associated with the main body of the migrating bed form. The lower parts of the strata may contain imbricated pebble clusters. However, pseudo-imbricated pebbles are associated with deposition on the steep front of a bed-load sheet. The thickness of preserved planar strata is less than the height of the formative bed-load sheets, and varies laterally (see below). Since bed-form height is proportional to grain size, the thickness of a stratum is also expected to be proportional to grain size (on the order

of millimeters for sand, and on the order of centimeters to decimeters for gravel).

Dunes

As the sediment-transport rate increases on rippled beds or lower-stage plane beds, dunes form if the mean grain size exceeds approximately 0.1 mm (Figure 5.44). Dunes occur only in subcritical, hydraulically transitional, and rough flows. Dunes are the most common bed waves in unidirectional flows with beds of sand and gravel, and cross stratification formed from the downstream migration of dunes is the most common sedimentary structure. Dunes include bed configurations that other authors have referred to as sand waves and megaripples (Allen, 1982a; Ashley, 1990; Best, 2005; Bridge, 2003; Carling, 1999).

Dunes have lengths greater than 0.6 m and heights greater than 0.04 m according to Allen (1982a) (Table 5.4). This definition is too restrictive, because it does not take into account smaller forms that are incipient or dying dunes. Adequate definition of dune geometry requires observation over time periods that allow recognition of the geometric evolution of the bed forms. Near the transition between plane beds and dunes, the equilibrium dune height may well be less than 0.04 m. Furthermore, because dune height and length are related to flow depth, dunes in shallow flows may be smaller than the limits given by Allen (1982a).

Although dunes are larger than ripples and bed-load sheets, all of these bed forms have similar asymmetrical streamwise profiles. However, not all dunes have lee sides at the angle of repose (Best, 2005), especially where smaller dunes are superimposed. Dunes that form near the transition from ripples or lower-stage plane beds are relatively long (their mean L/d is up to about 20) (Yalin, 1977; Allen, 1982a) and their crestlines are slightly sinuous in plan view. Ripples occur on the backs of these dunes if the sediment size is less than about 0.7 mm. In coarser sediment, bed-load sheets may occur on the backs of these dunes. As bed shear stress and sediment transport rate increase beyond the transitional stage, dunes become shorter relative to their height (Figure 5.45) and to the flow depth ($L \approx 6d$), and their crest lines become more sinuous to linguoid in plan view (Yalin, 1977; Allen, 1982a). The stoss sides (backs) of these types of dunes are *approximately* planar (low-relief bed waves are commonly present), and primary current lineations may occur in sand. The trough areas are relatively deep scours bounded laterally by flow-parallel spurs, and commonly contain ripples with diverse orientations

(Allen, 1982a). Lunate-crested dunes occur where the sediment supply is limited.

The mean height and length of dunes generally increase with the flow depth (Figures 5.39 and 5.46) (e.g., Allen, 1982a). The mean length of curved-crested, fully developed dunes is approximately six times the flow depth. The empirical data for mean dune-height/flow-depth ratios mainly fall within the range $3 < d/H < 20$. Empirical equations relating mean dune height and depth have low correlation coefficients, due primarily to two reasons. First, the mean height of dunes (relative to their length and flow depth) increases from near zero at the lower boundary of the dune-stability

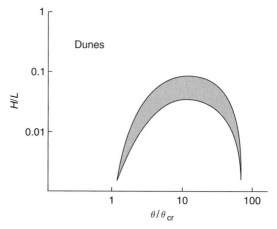

FIGURE 5.45. Variation in height/length ratio of equilibrium dunes with the dimensionless bed shear stress. The region of experimental data is stippled. From Bridge (2003).

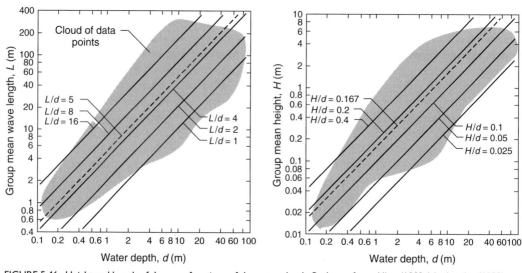

FIGURE 5.46. Height and length of dunes as functions of the water depth. Redrawn from Allen (1982a) by Leeder (1999).

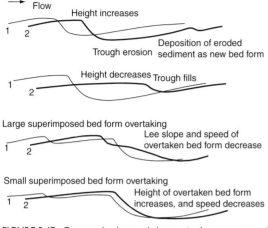

FIGURE 5.47. Commonly observed changes in the geometry and migration rate of individual dunes as seen in along-stream profiles. Thin lines indicate profiles at time 1, and thick lines indicate profiles at later time 2.

field (transition from ripples or lower-stage plane beds) to a maximum in the middle of the field and then to near zero at the transition to upper-stage plane beds (Figure 5.45). Second, when measurements of dune height and flow depth are made in natural rivers, it is not certain that the dunes were in equilibrium with the prevailing flow conditions (see later).

An individual dune varies in height, length, steepness, and migration rate along its crestline (i.e., across-stream), and during its lifespan (Figure 5.47). The individual dunes in a configuration also differ in geometry and migration rate. Therefore, the variability in dune geometry and migration rate measured on a series of along-stream transects reflects (1) the across-stream variability of 3D dunes; (2) variability within individual dunes; and (3) variability between individual dunes. The height of an individual dune may increase or decrease as it migrates. Increase in height of individual dunes is commonly accomplished by trough scouring, and/or by a relatively fast upstream dune overtaking and adding to the height of a downstream dune (Figure 5.47). With trough scouring, the eroded sediment may form a new small dune on the back of the downstream dune. However, trough scouring is not essential for spontaneous development of small dunes on the backs of larger dunes. With dune overtaking, the length of the downstream dune obviously decreases as it is overtaken, and the length of the combined dune becomes relatively large. In some cases of overtaking, the overtaking dune decreases in height and increases in speed as it reaches the crest of the dune being overtaken. In this case, the geometry of the overtaken dune

is not changed very much, except for a small increase in height (Figure 5.47). In other cases of overtaking, the overtaking dune is sufficiently high that its flow-separation zone causes the downstream dune to decelerate abruptly, and decrease in both height and lee-side slope (Figure 5.47). A decrease in speed of a dune as its height increases is expected if the sediment-transport rate is constant (see Equation (5.104)). However, this is not always the case, because the sediment-transport rate may vary in association with local scouring of individual dune troughs. A decrease in height of an individual dune commonly occurs as it climbs on to the back of a downstream dune, or is overtaken by an upstream dune. Rates of creation and destruction of individual dunes by these mechanisms, and their lifespans and excursions, are difficult to measure and not known well. Excursions are generally on the same order as dune length, and lifespans can be hours to days.

Rates of migration of individual dunes are extremely variable due to the temporal changes in geometry of individual dunes described above, and to the variability of lateral movement of the crests of 3D dunes. The migration rate, c, averaged in time over all dunes measured in a population is related to the average sediment-transport rate and dune height according to

$$c = i_c/(HC_b) = 2i_m/(HC_b) \qquad (5.104)$$

where i_c is the mean volumetric sediment-transport rate on dune crests, i_m is the mean volumetric sediment-transport rate over dunes that are approximately triangular in profile, H is the mean dune height, and C_b is the volume concentration of sediment in the bed (1 − porosity). Equation (5.104) is derived below. Since the sediment-transport rate is related to bed shear stress and dune height is related to flow depth (except where the system is transitional to plane beds or ripples), the mean migration rate of dunes is expected to increase with flow velocity for a given flow depth. The mean dune-migration rate in rivers commonly ranges from fractions of a millimeter per second to a few millimeters per second.

Natural unidirectional flows are inherently unsteady. The mean geometry of dunes changes in response to changing conditions of flow and sediment transport, but there is normally a *time lag* because of the volume of sediment that must be moved in order to effect a change in form. For example, the dune height and length may be smaller than their equilibrium values during a rapidly rising flow stage, but greater than their equilibrium values during a falling flow stage. These variations in dune geometry result in relatively low flow resistance and depth, but high velocity, during rising flow stages and relatively high flow resistance and depth and low velocity during falling flow stages. Therefore, plots of dune height and length, flow depth, and flow velocity versus discharge may be loop-shaped (Figure 5.48A). However, loop-shaped curves do not in themselves indicate the existence or the degree of dune lag. This is because the dune height and height/length ratio depend on the dimensionless bed shear stress rather than on the discharge. Therefore, a constant-discharge flow may have low velocity, large depth,

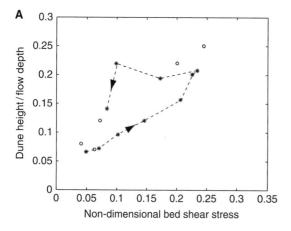

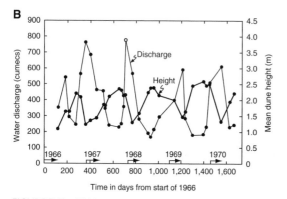

FIGURE 5.48. (A) Variation in mean dune height with bed shear stress during an unsteady-flow experiment lasting 20 hours. Water discharge and flow velocity were increased incrementally (from 0.063 to 0.09 cumecs, and from 0.7 to 1 m s^{-1}) over 10 hours, and then decreased incrementally by the same amounts for 10 hours. The mean sediment diameter was 1.5 mm, and the water depth was constant at 0.15 m. Arrows indicate the direction of change in dune height as time progressed. The dune height was lower than the equilibrium value (open circles) during the rising-flow stage, but approached or exceeded the equilibrium value during the falling-flow stage. Data from Lunt and Bridge (2006). (B) Time series of mean dune height and water discharge for the River Weser, Germany, indicating a phase lag of about π radians. From Allen (1982a).

and large dunes, or it may have high velocity, small depth, and small dunes.

Allen (1982a) defined the *relaxation time* as the time taken for dunes to equilibrate to a changed flow, and it is related to the amount of time taken to move sediment and form dunes with geometry appropriate to the changed flow conditions. The maximum relaxation time is the lifespan of a dune (Allen, 1982a). The *phase difference* is the time difference between the peaks and troughs of time series of two periodically varying parameters such as dune height and discharge. If these parameters are assumed to vary sinusoidally, the phase difference can be expressed in terms of radians. For example, if the discharge peak coincides with the lowest value of dune height, the phase difference would be π radians (Figure 5.48B). A phase difference does not mean that dune lag exists. For example, dune height might be at a minimum at high discharge because the bed configuration may be close to the transition to upper-stage plane beds. Dune height would then increase as discharge decreased and the dunes were in the middle of their existence field. In order to determine the degree of dune lag it is necessary to plot geometric characteristics of dunes against the appropriate controlling hydraulic parameters (e.g., bed shear stress, flow velocity, depth), not against discharge.

Bimodal or polymodal frequency distributions of dune height and length have been cited as evidence for dune lag, especially in rivers with high variability of discharge. However, polymodal distributions of dune height or length also occur with equilibrium dunes, as discussed above (Figure 5.38) (Bridge, 1997). The migration of relatively small dunes down the degraded, low-angle (less than 15°, say) lee sides of larger dunes is good evidence that the larger dune is not in equilibrium with the flow. Although the occurrence of relatively small dunes migrating over larger dunes is due to dune lag in some cases, such superposition could also possibly be an equilibrium phenomenon (Allen, 1982a). This question can be answered only if a time series of measurements of changing flow, sediment transport, and bed-form geometry is available (which is not normally the case). In some reported cases of co-existence of dunes of different sizes, the smaller superimposed forms are dunes but the larger host forms are unit bars (e.g., Figure 5.49). The length of the dunes is controlled by flow depth, but the length of unit bars is controlled by flow width.

Theoretical models for predicting changes in dune height and length as discharge changes are available

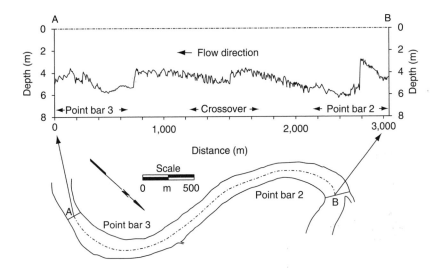

FIGURE 5.49. Depth sounding along the Congaree River, South Carolina, showing dunes superimposed on bars. From Levey (1978).

(Allen, 1982a; Bridge, 2003). Unfortunately, these models are very simplistic and there are many unjustified assumptions, so it is difficult to assess their usefulness.

Flow and sediment transport over ripples and dunes

Figure 5.50 illustrates the main features of flow and sediment transport along the length of ripples and dunes. The water-surface topography is out of phase with the bed topography in such subcritical flows. The flow accelerates over the stoss side of the bed wave and decelerates over the lee side (as demanded by the continuity equation). The boundary layer separates on the lee side of equilibrium ripples and dunes. Thus, there is significant along-stream variation in bed shear stress and in vertical profiles of time-averaged flow velocity and turbulence intensity (Figure 5.50). The near-bed flow is decelerated and reversed in the flow-separation zone, and the near-bed velocity gradient (and hence the bed shear stress) increases progressively from the flow-attachment zone towards the bed-form crest. The near-bed fluid pressure is lowest where the flow velocity is highest (as expected from Bernoulli's equation (5.5)) at the bed-form crest, and is highest where the velocity is lowest in the flow-separation zone. In the case of curved-crested bed waves, these parameters also vary transverse to flow. The spatial and temporal variations in flow and sediment transport over ripples and dunes are very difficult to measure in the field, as discussed below.

The *law of the wall* (Equation (5.28)) has been applied to the whole flow depth above ripples and dunes, supposedly yielding some measure of the spatially averaged bed shear stress and roughness height. Such a bed shear stress would include the resistance due to bed-form drag as well as that due to the mobile and immobile bed grains. Thus the equivalent sand roughness height is comparable to the bed-form height. The law of the wall has also been applied to the lowest parts of velocity profiles on the backs of bed forms to define local variations of bed shear stress and roughness height due to bed-grain and bed-load drag. Although the law of the wall is strictly valid for uniform flows, it applies close to the wall for slightly curved boundaries. Local bed shear stress and roughness height calculated in this way increase from the reattachment point to the crest of the bed form, but the roughness height is greater than that expected from bed-load and bed-grain drag. This is mainly because of the existence of small bed forms on the backs of the host bed forms.

Most of the data on *turbulence intensity* over bed forms come from measurements on fixed beds with no sediment transport, such that the fundamentally important interaction among turbulence, sediment transport, and the bed form cannot be addressed. Best (2005) reviews observations and numerical modeling of turbulent flow over dunes. Turbulence intensities and Reynolds stresses along the free-shear layer above the flow-separation zone and at reattachment are up to three times the magnitude of those in the unseparated boundary layer (Figure 5.50) (Allen, 1982a). Ejections and sweeps are the dominant turbulent structures in this region, and are related to eddy shedding from the free-shear layer. The effects of the flow-separation zone on turbulence intensity are not as marked for ripples as they are for dunes (Figure 5.51). Reynolds stresses are

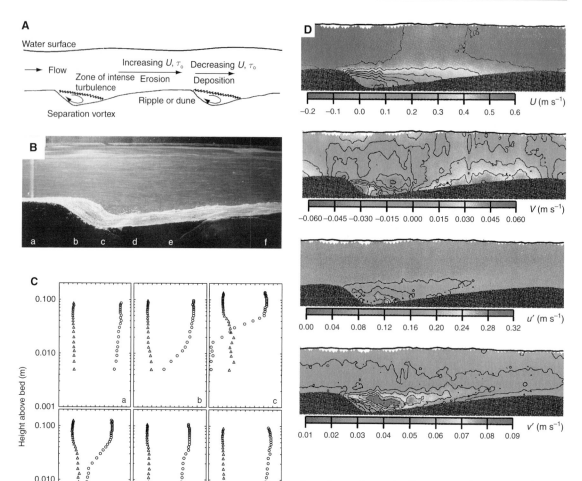

FIGURE 5.50. (A) The general character of flow and sediment transport over ripples and dunes. (B) Flow over an experimental fixed dune visualized with tracers, with positions of flow-measurement profiles a–f. (C) Measured vertical profiles of the mean downstream flow velocity, U, and the root mean square of downstream-velocity fluctuations, u'. Positions of profiles a–f are marked on (B). From Bennett and Best (1995). (D) Contour maps of mean flow and turbulence parameters over a fixed dune studied by Bennett and Best (1995). Flow is from left to right, a scale color bar is given for each map, and the vertical exaggeration is about 1.3; U and V are the mean downstream and vertical flow velocities, respectively, and u' and v' are root mean squares of turbulent fluctuations of these velocities.

relatively larger than the mean for dunes, and zones of high turbulence intensity can reach higher in the flow (Bennett and Best, 1996). *Macroturbulence* over dunes refers to large-scale eddies that are shed from the free-shear layer above the flow-separation zone and move up towards the water surface, where they may erupt as boils or kolks. These turbulent motions are responsible for entrainment of a significant amount of near-bed sediment into suspension, especially from the reattachment region (Figure 5.51). The increases in bed shear stress and bed-load transport rate on going from the flow-reattachment zone to the crest of the bed form result in erosion of this region. The abrupt decrease in bed shear stress as the flow separates results in deposition of bed load that periodically avalanches down the steep lee side. Some suspended sediment may settle through the zone of recirculating flow beneath the separated boundary layer. This and some of the avalanching bed load may be re-entrained and moved upstream in the bed-wave trough and lower lee slope.

When calculating a *sediment-transport rate* associated with ripples and dunes, it is necessary to use a bed shear stress that is effective at transporting the sediment. This bed shear stress may be considered to be some average value on the back of the bed form, outside the flow-separation zone. This bed shear stress

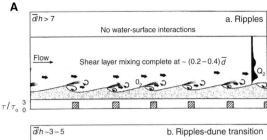

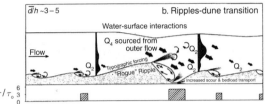

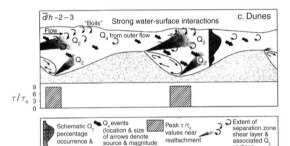

FIGURE 5.51. (A) Turbulent flow over ripples, dunes, and transitional ripples–dunes. From Bennett and Best (1996). In the case of dunes, vortices generated in the shear layer above the separation zone reach the water surface as boils. Here $\bar{d}/h$ is the mean flow depth/bed-form height. The τ ratio is the mean Reynolds stress near reattachment over the spatially averaged boundary Reynolds stress. Labels Q_2 and Q_4 refer to turbulent events from quadrant 2 (ejections) and quadrant 4 (sweeps), respectively.

FIGURE 5.51. (B) Boils on the water surface downstream from a dune crest. The smooth water is over the dune crest. Calamus River, Nebraska.

must be less than (commonly less than half of) the spatially averaged bed shear stress that includes the effect of the flow-separation zone. There have been many attempts at such "drag partitioning" (Yalin, 1977, 1992). An alternative to using this drag-partitioning approach is to calculate the local variation of sediment transport over bed forms, but this is very difficult because it is dependent on complicated spatial variations in turbulent lift and drag that are associated with flow separation and reattachment. Perhaps the best method for calculating the sediment-transport rate associated with bed forms uses knowledge of the bed-form height and migration rate, as discussed below.

The *variation in bed-load transport rate*, i_b, over the stoss side of a bed form can be calculated as a function of the local bed height, h, assuming that the bed form does not change geometry transverse to flow or as it migrates. Using the one-dimensional sediment continuity equation for steady flow

$$-C_b \, dh/dt = di_b/dx \tag{5.105}$$

and the fact that

$$-dh/dt = c \, dh/dx \tag{5.106}$$

where C_b is the volume concentration of grains in the bed, c is the bed-form celerity, and x is the downstream distance, results in

$$di_b/dx = C_b c \, dh/dx \tag{5.107}$$

This can be integrated to give

$$i_b - i_{b0} = C_b c(h - h_0) \tag{5.108}$$

where i_{b0} and h_0 are the bed-load transport rate and bed height in the bed-form trough. Since i_{bo} can be taken as zero, Equation (5.108) indicates that the bed-load transport rate increases linearly with distance above the trough, and reaches a maximum at the crest. If the stoss side was planar (straight in profile), the mean bed-load transport rate would occur at the position of half the bed-form height. This equation indicates that bed-form height increases with sediment-transport rate for a constant migration rate.

The *total bed-load transport rate* associated with bed-form migration can be estimated for the case of bed forms that are triangular in along-stream profile, and do not change geometry and migration rate transverse to flow or as they migrate. The volume of sediment per unit width in such a bed form is $C_b HL/2$. The time taken for this volume of sediment to move one bed-form length is L/c. Thus the volumetric sediment-transport rate per unit width is $C_b cH/2$. It can also be calculated easily that the transport rate at the crest, taken as equal to the rate of addition of sediment to the lee side, is twice the average transport rate. However, part of the bed-form migration may be associated with deposition of suspended sediment on the lee face, such that the actual bed-load transport rate may be overestimated. This type of relationship is used widely to estimate bed-load transport rates. Corrections may be made to account for the facts that bed forms are not triangular in profile, bed forms do not have invariant geometry and migration rate, and some suspended sediment may be involved in bed-form migration.

Grain sorting during sediment transport over ripples and dunes

At low sediment-supply and -transport rates, the trough areas of bed forms are relatively coarse-grained (and possibly armored), whereas the finer grains comprise the moving bed forms (Figure 5.41). The combination of relatively small bed shear stress, armored bed areas, and a limited supply of the transportable finer grains effectively limits the height of these bed forms. As the sediment-transport rate increases, the coarser grains can be transported to the bed-form crests, and some of the finer grains can be suspended. If sand supply is abundant, the pore spaces between gravel grains are occupied with sand, especially in bed-form trough areas. Gravel beds with sand filling interstices are relatively smooth, facilitating the movement of the large grains (e.g., *overpassing*). Overpassing happens also when isolated gravel grains move relatively quickly over sand beds. Separation of sand from gravel during their transport on the back of dunes and flow separation on the lee side result in open-framework gravel strata, as will be seen below.

Heavy minerals are normally finer-grained than the light minerals with which they are transported and deposited. Concentrations of heavy minerals occur commonly in scoured areas such as bed-form troughs and upstream of obstacles, and also near bed-form crests and in the depositional areas to the lee of obstacles. These are regions where relatively large grains of light minerals are entrained, leaving the smaller protected grains of heavy minerals to concentrate on the bed, where the small grains of heavy minerals can infiltrate between the pore spaces of the larger deposited grains, and where heavy-mineral concentrations can be re-entrained and deposited immediately downstream.

Cross strata formed by ripples and dunes

Cross strata accumulate on the lee sides of ripples and dunes, and are inclined at the angle of repose (30°–40°). The cross strata are defined by spatial variations in texture and composition that are due to (1) pre-sorting of sediment that arrives at the top of the lee side of the bed form; (2) sorting due to differential deposition of sediment on the lee side and associated grain flows; and (3) movement of sediment on the lee side by the water currents in the lee-side flow-separation zone (Allen, 1982a; Hunter and Kocurek, 1986; Bridge, 2003; Kleinhans, 2004, 2005; Reesink and Bridge, 2007). These processes are discussed in turn.

The texture, composition, and transport rate of bed load arriving at the top of the lee side of a bed form vary with time due to turbulence, the passage of smaller bed waves over the stoss side of the larger host bed form, and longer-term variations (related to floods, river diversions, or diurnal freezing and snowmelt). It is commonly suggested that relatively coarse-grained cross strata are deposited at high flow stage, whereas fine-grained cross strata are deposited at low flow stage (Figure 5.52A). However, this is not always the case, because deposited grain size is dependent on what is available to transport as well as flow strength.

Ripples, bed-load sheets, or small dunes may be superimposed on the backs of larger dunes during both steady and unsteady flows. Relatively large amounts of coarse-grained bed load are associated with the arrival at the lee side of a superimposed bed wave, whereas small amounts of relatively fine sediment are associated with arrival of the trough region (Figure 5.52B, C). If the superimposed bed form is relatively large, the flow in its separation zone causes the lee side of the host bed form to be reduced in slope. The fine-grained sediment deposited in the lee of the superimposed bed form accumulates on this low-angle slope (Figure 5.52C). If the superimposed bed form is relatively small (height less than about a fifth of the host bed-form height), the lee side of the host bed form remains at the angle of repose, and a discrete cross stratum is formed as the superimposed bed form advances over the host bed form (Figure 5.52B). In this case, the cross-sectional area (in the streamwise direction) of the cross stratum, A, is the same as cross-sectional area of the superimposed bed form. Therefore, if superimposed bed forms are assumed to be triangular in cross section, with height H_s and length L_s,

$$A = H_s L_s / 2 = Ht / \sin \alpha \qquad (5.109a)$$

with

$$t = H_s L_s \sin \alpha / (2H) \qquad (5.109b)$$

where H is the height of the host bed form, t is the thickness of the cross stratum, and α is the slope of the lee side of the host bed form (the residual angle of repose). For example, if H_s is 10 mm, L_s is 0.8 m, H is 0.1 m, and $\sin \alpha$ is 0.5, the cross-stratal thickness would be 20 mm.

Differential deposition and grain flow on the lee side of ripples and dunes also influence grain sorting in cross strata. Bed-load sediment (and some near-bed suspended sediment) that arrives at the brink point of the bed form is deposited on the lee side at a point that is controlled by (1) grain velocity; (2) height above the bed; (3) grain settling velocity; and (4) turbulence in the separated flow (Figure 5.53A). Grains travel further downstream from the brink point as grain velocity and height above the bed increase, and as the grain

A

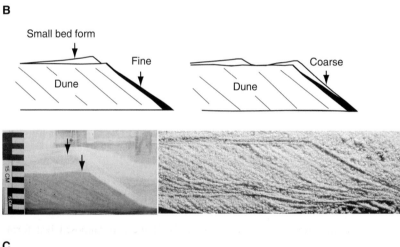

B

Small bed form

Fine

Dune

Coarse

Dune

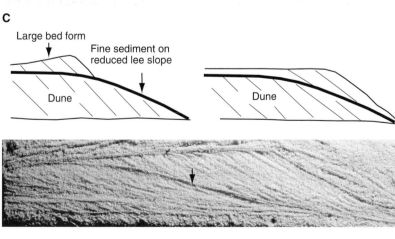

C

Large bed form

Fine sediment on reduced lee slope

Dune

Dune

FIGURE 5.52. Variation in texture of cross strata due to (A) long-term (diurnal?) variation in sediment supply, and passage of (B) small and (C) large superimposed bed waves. In (A) the light color is silt, and the dark color is sand. In the photo on the left of (B), dark bands are relatively fine-grained sand deposited during passage of the troughs of small bed forms superimposed on the dune. Superimposed bed forms (arrows) are forming cross strata of relatively coarse sand. The photo on the right of (B) is an epoxy-resin peel of the cross strata. A low-angle drape of fine sediment is shown by an arrow in the photo of (C).

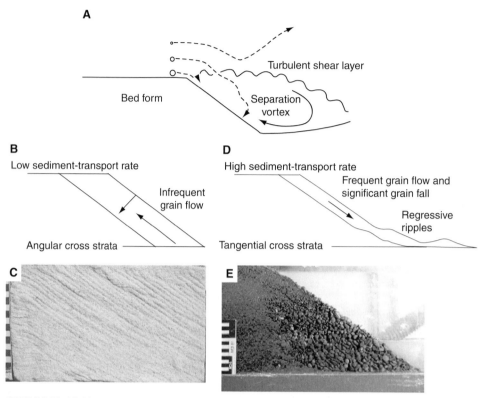

FIGURE 5.53. (A) Idealized paths of grainfall on the bed-form lee side for grains of various sizes and various distances from bed. (B) Grain-size sorting by intermittent grain flow on the bed-form lee side for a low sediment-transport rate. Arrows point to the direction of decreasing grain size. Angular cross strata are produced. (C) Epoxy-resin peel showing angular cross strata produced by intermittent grain flows. (D) Grain-size sorting due to frequent grainflow and significant grainfall for a high sediment-transport rate. The arrow points in the direction of decreasing grain size. Regressive ripples are produced by reverse flow in a separation vortex. Tangential cross strata are produced. (E) Segregation of gravel from sand by avalanching without forming cross strata.

settling velocity (and hence grain size) decreases. Therefore, the rate of deposition and grain size of deposited sediment decrease down the bed-form lee side. The downstream decrease in deposition rate on the lee side results in an increase in its slope, until it reaches the angle of initial yield. The angle of initial yield is commonly around 35°, and up to 10° larger than the residual angle after shearing. Avalanching and grain flow occur when the angle of initial yield is reached. Avalanching is intermittent at low sediment-transport rate, because of the time taken to build up the slope of the lee side. At high sediment-transport rate, the lee side is maintained at a high angle by the sediment deposition, and avalanching (or rolling of grains down the lee side) is very frequent. The size (cross-sectional area) of the wedge of sediment that avalanches depends on the downstream rate of decrease of deposition rate and on the distance traveled downstream by the sediment. For coarse-grained bed-load

sediment moving relatively slowly, most sediment is deposited close to the brink point, and the wedge is relatively small at the point of avalanching. For fine-grained bed load moving relatively fast, much of the sediment is transported well past the brink point, and the wedge would be larger.

Grain-size sorting during discrete avalanching (low sediment-transport rate) has been related to kinetic sieving, whereby fine grains settle between the spaces created between the larger grains during shearing and dilation (see Chapter 8). Thus large grains move towards the outer edge of the grain flow, where they can move downslope more quickly than finer grains. Therefore, the avalanche deposit becomes coarser-grained downslope and towards its outer edge (Figure 5.53B, C). If the largest grains are hindered in their movement to the base of the slope, due to frictional resistance, then the size sorting from basal surface to outer surface will be more marked. The frictional resistance depends on the relative sizes and shapes between the large and small grains. If there is a big difference in size between the largest and smallest grains, the largest grains can more easily move down the outer edge of the grain flow. The bed form must be high enough (its lee side must be long enough) for this process to be effective. The lack of settling of suspended sediment on the lower slope of the lee side results in an angular contact between the lee slope and the bed form trough (Figure 5.53B, C). Also, because the grain-flows are intermittent, it may be possible to distinguish relatively coarse-grained grain flow strata from relatively fine-grained grain-fall strata. However, these sorting processes are complicated by the entrainment of grains from the top of one grain flow by a subsequent grain flow (Kleinhans, 2004, 2005). At high sediment-transport rate, for which avalanching is almost continuous, much fine-grained sediment is deposited on the lower lee slope. This reduces the effectiveness of kinetic sieving, and grain-size sorting might not be obvious. Cross strata tend to be finer-grained at the base of the lee side than at the top. Furthermore, the buildup of fine sediment at the base of the lee side reduces the slope of the cross strata here (tangential bases: Figure 5.53D). Such basal fine sediments form significant barriers to fluid flow through the sediment.

The separated (reverse) flow over the lee side of a ripple or dune can reduce the velocity of the grain flows and move sediment (in some cases as *regressive ripples*: Figure 5.53D) up the lower lee slope, causing a decrease in slope here. Such redistribution of sediment does not strongly affect the thickness of cross strata, but does affect grain-size sorting.

Avalanching of sediments with a large range of grain size, such as sandy gravels, may lead to complete segregation of different grain sizes along the angle-of-repose slope, and to the absence of cross strata (Figure 5.53E). This is due to the fact that the large gravel grains moving over finer sand grains have much smaller pivoting angles (and less frictional resistance) than those of sand grains moving over gravel grains. Therefore, the largest grains readily move to the base of the lee slope. This means that cross strata in many sandy gravels must be related to superimposed bed forms or longer-term flow unsteadiness, rather than to avalanching.

The separation of sand from gravel associated with flow separation at the brink point of dunes (and unit bars: Chapter 13) is the main mechanism for forming cross-stratified, open-framework gravel (Figure 5.54), and the mechanism becomes more effective as bed-

form height increases. Infiltration of suspended sand into the gravel framework on the lee side is not significant for many sand–gravel mixtures. However, appreciable amounts of sand are deposited in the bed-form troughs. Variation in the mean grain size of individual cross strata in sandy gravels is commonly related to sorting within superimposed bed forms (Figures 5.52 and 5.54). Some relatively thick sandy or muddy layers in gravel cross strata may be related to long-term changes in sediment supply. The preservation of open-framework gravel is discussed below.

The *thickness of individual cross strata* is normally on the order of a millimeter or less for ripples, but varies from millimeters to centimeters for dunes. If strata are classified using McKee and Weir's (1953) arbitrary distinction between laminae and beds at a thickness of 10 mm, ripple migration gives rise to cross laminae whereas dune migration gives rise to cross laminae or cross beds, or a combination of the two. In view of the fact that superimposed bed forms and lee-side avalanches are 3D sediment bodies, an individual cross stratum is expected to vary laterally in thickness and texture, in addition to such variation from the top to bottom of the bed form.

The *three-dimensional shape of cross strata* depends on the shape of the surface on which the strata accumulate. Cross strata that accumulate on the lee side of straight-crested ripples and dunes do not vary in dip direction along the length of the crest, and appear straight in all sections parallel to the crest line (Figure 5.55). Cross strata that accumulate on the lee side of curved-crested ripples and dunes vary in dip direction along the length of the crest, and they appear curved in all cross sections (Figure 5.55). The basal surfaces of sets of cross strata (a set being produced by migration of a single bed wave: Figures 5.55 and 5.56) are also related to the shape and orientation of bed-form crestlines because these control the flow pattern in the erosive trough area of bed forms. Straight-crested bed forms have more or less straight flow-reattachment lines that generally sweep out planar surfaces of erosion as the bed forms migrate downstream. Thus, sets of cross strata produced by straight-crested bed forms rest upon planar basal surfaces, and are referred to as *planar cross stratification* (Figure 5.55). Curved-crested bed forms have curved flow-reattachment lines and spoon-shaped erosional regions that are bounded laterally by flow-parallel spurs (Figure 5.55). Thus, migration of curved-crested ripples and dunes generally sweeps out trough-shaped erosion surfaces upon which sets of cross strata accumulate, resulting in *trough*

FIGURE 5.54 Cross-stratified open-framework gravel (arrows) interbedded with sandy gravel. In (A), the 0.08-m-thick layer of open-framework gravel above the lens cover fines upwards. The basal coarsest layer contains imbricated pebble clusters. Above this, platy pebbles dip to the right (pseudo-imbrication), and were deposited on the steep lee side of a low dune. In (C), open-framework gravel is concentrated along the base of a 0.15-m-thick cross set, and also forms some cross strata. The variation in grain size of the individual cross strata is due mainly to superimposed bed forms, with the largest superimposed bed forms forming the gravelly cross strata.

cross stratification. Trough cross stratification is much more common than planar cross stratification, because straight-crested bed forms are rare.

A *cross set* is the set of cross strata produced by migration of a single bed form. Groups of sets formed by similar bed forms are called *cosets* (Figures 5.55 and 5.56). *Form sets* are single cross sets that have the shape of a bed form such as a ripple or a dune. They may, or might not, be preserved at the top of a coset of cross strata or planar strata. Also, the preserved bed form need not necessarily be related to the cross strata that it encloses.

The *thickness of cross sets* observed in cross section parallel to the mean flow direction depends on (1) the average rate of deposition relative to the rate of bed-form migration; (2) the sequence of bed forms of a given scour depth (related to the bed-form height) passing a given bed location; and (3) changes in the geometry and migration rate of individual bed forms as they migrate (Allen, 1982a; Paola and Borgman, 1991; Bridge, 1997; Leclair and Bridge, 2001). For bed forms of constant geometry (in time and space) migrating downstream at a constant rate with steady deposition, the thickness of sets of cross strata s is constant and given by

$$s = Lr/c = L \tan \delta \qquad (5.110)$$

where L is the bed-form length, r is the deposition rate, c is the bed-form speed (celerity), and δ is the angle of bed-form climb relative to the mean bed level (Figure 5.57). However, the geometry and migration

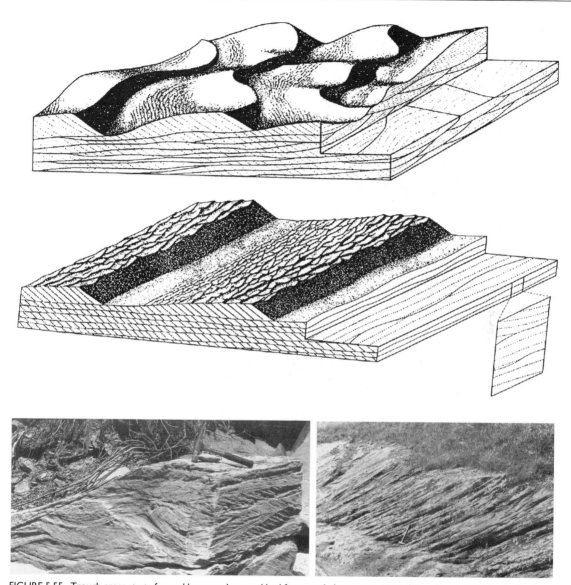

FIGURE 5.55. Trough cross strata formed by curved-crested bed forms and planar cross strata formed by straight-crested bed forms, from Harms *et al.* (1982), with photos of field examples of medium-scale trough cross strata (lower left) and planar cross strata (lower right).

rate are never constant in reality, and Equation (5.110) is not generally applicable. In order to account for the effect on cross-set thickness of the variability in height of bed waves passing a given bed location, Paola and Borgman (1991) derived the following equation for the probability density function of the set thickness, $p(s)$:

$$p(s) = ae^{-as}(e^{-as} + as - 1)/(1 - e^{-as})^2 \qquad (5.111)$$

where $1/a$ is the mean value of the exponential tail of the probability density function for topographic height

relative to a datum. The parameter a is inversely related to the breadth of the tail of the height distribution. The mean value of s is $1.64/a$, and the standard deviation is $1.45/a$. If a is derived from bed-form height data, its value must be increased by up to a factor of two, and the constants 1.64 and 1.45 may be up to half as much as the stated values (Leclair and Bridge, 2001). A simple adaptation of Paola and Borgman's (1991) theoretical model that applies to depositional conditions is

$$s_m = Lr/c + x/a \qquad (5.112)$$

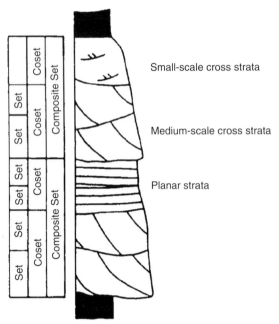

Small-scale cross strata

Medium-scale cross strata

Planar strata

FIGURE 5.56. Definition of sets of cross strata and planar strata. The strata within the sets may be beds or laminae, or both. From Bridge (2003).

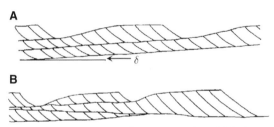

FIGURE 5.57. Preservation of cross strata where (A) bed waves climb and migrate with constant geometry, migration rate, and deposition rate (unrealistic); and (B) bed-wave geometry and migration rate vary with time (realistic). From Bridge (2003).

where x varies from 0.82 to 1.64, and increases from bed-load sheets to ripples to dunes (Bridge, 2003). The value of the parameter a decreases exponentially as the mean height of bed forms increases (Figure 5.58), and the mean bed-form height is approximately $5.3/a$. This means that, as bed forms increase in height, the second term on the right-hand side of Equation (5.112) increases in magnitude relative to the first term. For dunes, the second term is normally an order of magnitude more than the first term for natural deposition rates, and the deposition rate has negligible influence on the cross-set thickness (Figure 5.59). The deposition

rate has potentially more influence on set geometry for low-relief bed forms. Equation (5.112) can be used to estimate bed-form heights from measurements of the cross-set thickness and inclination of cross-set bases (Leclair and Bridge, 2001).

Cross sets (or planar laminae associated with migration of bed-load sheets on plane beds) result only from the bed forms with the largest heights. More of the smaller bed forms can form cross strata or planar strata as the deposition rate increases. Experimental data for dunes show that the ratio of cross-set thickness and formative dune height is less than 0.5 (Figure 5.59). The ratio of mean cross-set thickness and mean dune height is approximately 0.3 (Leclair and Bridge, 2001). This "*preservation ratio*" increases with deposition rate in the case of bed-load sheets and ripples, and may approach 1 (see below).

The frequency distribution of cross-set thicknesses compiled from many different sources is bimodal, with one mode between 0.02 m and 0.03 m and another mode around 0.3 m (Allen, 1982a). Representatives of the smaller mode are commonly referred to as *small-scale cross stratification* or *cross lamination* (because the individual cross strata are less than 10 mm thick) (McKee and Weir, 1953). Representatives of the larger mode are commonly referred to as *medium* or *large-scale cross stratification* or *cross bedding* (if the individual cross strata are thicker than 10 mm). The smaller-scale cross sets are commonly ascribed to the migration of ripples, whereas the thicker ones are ascribed to the migration of dunes. In fact, centimeter-thick cross sets can be formed by ripples or dunes, and some cross sets can be formed by migration of channel-scale bars, or delta-like accumulations. Different scales of cross set are commonly superimposed (this is called compound cross stratification by Allen, 1982a), and are normally ascribed to migration of relatively small-scale bed waves over the surfaces of larger-scale migrating bed waves. Great care must be exercised in interpreting the origin of cross sets seen in cores, because the crucial lateral variation in geometry and facies cannot be determined.

Climbing-ripple cross lamination (Figure 5.60) is formed where the average angle of climb of ripples equals or exceeds the angle of their stoss sides, indicating a very high deposition rate (see Equation (5.110)). In this special case, the stoss sides of ripples are not eroded during migration, and stoss-side laminae may be preserved (preservation ratio ≥ 1). These cross strata have been classified according to the development of stoss-side laminae (Allen, 1982a). Climbing-ripple

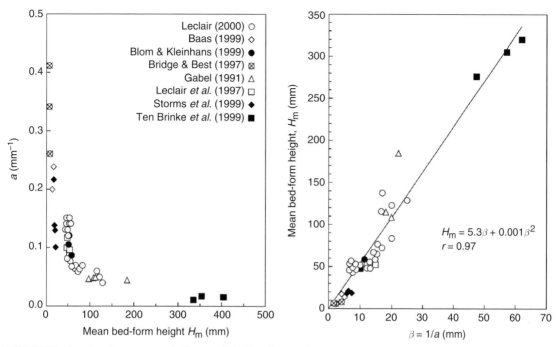

FIGURE 5.58. Variation of parameter *a* (in Equation (5.111)) as the mean height of bed waves increases. The Bridge and Best data are for bed-load sheets on an upper-stage plane bed. The Baas and Storms *et al.* data are for ripples. All other data are for dunes. From Leclair and Bridge (2001).

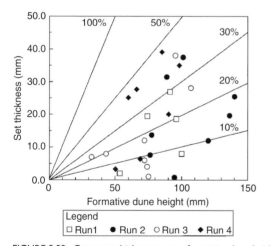

FIGURE 5.59. Cross-set thickness versus formative dune height. From Leclair *et al.* (1997). The aggradation rate increases from run 1 to run 4. The aggradation rate has negligible influence on the cross-set thickness in this example.

cross strata normally occur in trough cross sets, and individual cross strata have convex-upward sections in addition to concave-upward sections (Figure 5.60). This type of cross stratification also rarely occurs with dunes.

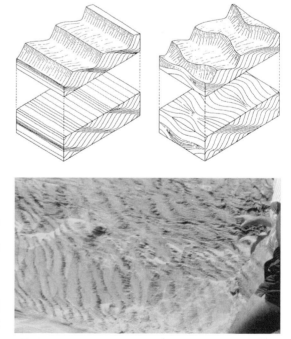

FIGURE 5.60. Climbing-ripple cross-stratification associated with straight-crested and curved-crested bed forms (from Allen (1982a)) and field example.

Transcurrent stratification is the name given to sets of more or less planar strata that are produced by the migration of low-relief bed waves (bed-load sheets) and/or those with a very small angle of climb (Allen, 1982a, pp. 357–360). Most transcurrent strata are laminae, being less than 10 mm thick (McKee and Weir, 1953). The bounding surfaces of the transcurrent planar strata are equivalent to those of cross sets, and inclined strata are rarely visible within the planar strata (Figure 5.61). The thickness of individual planar strata varies laterally for the same reasons as cross-set thickness varies laterally (see below). Transcurrent strata have been associated with migration of low-relief ripples and dunes, bed-load sheets on lower-stage and upper-stage plane beds, and antidunes.

The amount of open-framework gravel preserved in cross strata increases as the mean bed-form height increases. If the bed forms are relatively small (e.g., bed-load sheets), the preserved basal parts tend to contain sandy gravel associated with deposition of suspended sand on gravel in the bed-form troughs. Therefore, planar gravel strata tend to be composed mainly of sandy gravel. In contrast, relatively thick cross sets in gravel tend to have a high proportion of open-framework gravel.

As the geometry and migration rate of individual bed forms change as they migrate (*both in steady, uniform flows and in unsteady, non-uniform flows*), the resulting cross sets vary in thickness in the migration direction. Leclair (2002) related along-stream variation in the geometry of cross sets to the kinematics of individual dunes formed in a laboratory flume, building upon the earlier work on ripples by Allen (1973). Modes of dune migration such as trough scouring, overtaking, and climbing were recognized in the geometry of the cross sets (Figure 5.62). The characteristic concave-upward base of a cross set (in streamwise sections) is formed by dune trough scouring, migration, and eventual climbing and decrease in height of the dune. The upper boundary of a cross set is formed by the subsequent migration of several dunes. The mean length of cross sets in the along-stream direction is approximately half the mean dune length. This empirical result has important implications for interpretation of paleoflow depth, because the dune length is closely related to the flow depth. Results from experimental studies such as those of Allen (1973) and Leclair (2002) indicate that it is necessary to understand the geometry and migration of individual bed forms, and the interactions among individual bed forms, in order to understand and model the formation of cross strata sets in two or three dimensions.

A *reactivation surface* is an erosional surface that progressively cuts down into and reduces the thickness of a cross set. Such a surface can be produced by bed forms with time-varying geometry migrating under conditions of steady, uniform flow (Figures 5.52 and 5.62), or by bed-form migration under unsteady flows when bed-form migration ceased for a relatively long

FIGURE 5.61. Formation of planar laminae (transcurrent laminae) by migration of low-relief bed waves: (A) migration of bed waves with constant geometry, migration rate, and deposition rate; and (B) a more realistic case in which the bed-wave geometry and migration rate vary with time. From Bridge (2003).

FIGURE 5.62. Modes of experimental dune movement recognized in the geometry of cross strata. Modified from Leclair (2002). Flow is from right to left, and the width of the section is 0.5 m. The dune form at the top of the section is highlighted by a dark line. Trough erosion results in the base of the cross set declining at a relatively steep angle (arrow 1). At arrow 2, the lee side of a dune is preserved, because the dune was overtaken. Then, the overtaken dune decreased in height and its trough was filled with low-angle strata. Then the dune climbed up the back of a downstream dune and disappeared, leaving a cross-set base that is inclined upwards (arrow 3).

time (e.g., after a flood or waning tide) and subsequently resumed. For example, the reactivation surface shown in Figure 5.63B exhibits evidence of cessation of dune migration, low-stage migration of ripples along the lee face of the dune, and then resumption of dune migration as the flow stage increased.

Upper-stage plane beds

Ripples or dunes are transformed into upper-stage plane beds as bed shear stress and sediment-transport rate are increased, provided that the flow remains subcritical. Near the transition to upper-stage plane beds, the bed forms become lower relative to their length and hump-backed (the crest is displaced upstream to near the trough: Figure 5.64), and areas of plane bed develop. Upper-stage plane beds are not completely flat, because low-relief bed forms (millimeters high) occur (Figure 5.65; Table 5.4). These bed forms (called bed-load sheets here) are asymmetrical in streamwise profile, but the lee slope rarely approaches the angle of repose. Their mean length/depth ratio is similar to that of curved-crested dunes, suggesting that these low-relief bed forms are incipient dunes that cannot grow in height.

The suspended-sediment transport rate is significant over upper-stage plane beds, suggesting a possible reason for their existence. If an along-stream change in suspended-sediment transport rate lags behind the along-stream change in bed shear stress and turbulence intensity, deposition of suspended sediment would occur in the troughs of bed forms and erosion would

occur at the bed-form crests, thus reducing bed-form height. The high concentration of suspended sediment required for the existence of upper-stage plane beds explains why this bed state is most common in the finer sand sizes.

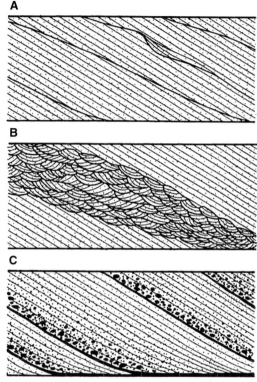

FIGURE 5.63. Types of reactivation surfaces in cross strata, according to Allen (1982a).

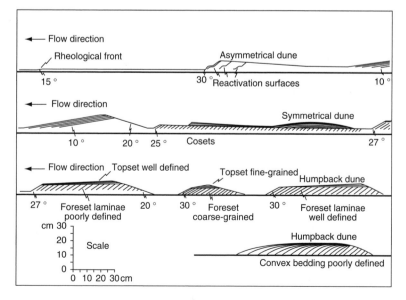

FIGURE 5.64. Types of experimental dunes and their internal structures formed near the transition to upper-stage plane beds. From Saunderson and Lockett (1983).

FIGURE 5.65. Experimental upper-stage plane beds with bed-load sheets. From Bridge and Best (1988). Flow, from right to left, is highlighted by Pliolite tracers. Trough of bed wave is shown by arrow. The white scale is 60 mm across.

Planar lamination associated with upper-stage plane beds

Deposition upon upper-stage plane beds gives rise to planar lamination (Allen, 1982a; Bridge, 2003). The laminae are defined by variations in grain size and composition (Figure 5.66): they coarsen upwards and/ or become finer upwards, and heavy-mineral concentrations occur in the coarser-grained parts. The laminae are fractions of a millimeter to millimeters in thickness, and vary in thickness laterally. The thickest laminae are commonly the coarsest. Individual laminae extend parallel to the flow direction for up to meters, but for lesser distances transverse to flow. Low-angle laminae occur rarely within the thicker planar laminae. Bounding surfaces of planar laminae normally have discontinuous ridges and troughs, with relief of a few grain diameters, that extend parallel to flow for decimeters and have transverse spacing of millimeters to a centimeter or so (i.e., *parting lineation* or *primary current lineation*: Figure 5.66). The long axes of ellipsoidal or rod-shaped grains are oriented approximately parallel to the parting lineations (Allen, 1982a).

Early explanations for the formation of planar laminae beneath upper-stage plane beds appealed mainly to the influence of turbulent variations in bed shear stress and sediment transport. Turbulence, and specifically the turbulent bursting process linked to large coherent structures in the boundary layer, cannot explain the thickness, lateral extent, and size grading within planar laminae. However, the migration of low-relief bed forms can explain planar laminae. The fine-grained bases of laminae are due to deposition from suspension in the troughs of the bed forms. Coarsening upwards from the lamina base is due to the migration of bed forms in which bed-load grain size decreases from the crest to the toe of the lee side. However,

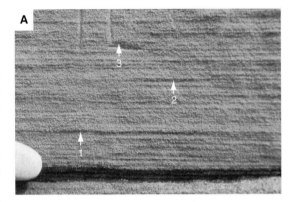

FIGURE 5.66. (A) Planar laminae in sandstone. A finger-tip is shown for scale. Coarser-grained parts are more protrusive than finer-grained parts. The sharp base to the relatively thick fining-upward lamina is shown by arrow 1. A laterally discontinuous lamina is indicated by arrow 2. Low-angle inclined laminae are shown by arrow 3. Photo courtesy of Jim Best. (B) Planar-laminated sandstone with primary-current lineation on the lamina surface. Note that the edges of laminae are not the lineations. The image is 30 cm across.

some of the largest bed forms have a lee side at the angle of repose (which is associated with lee-side flow separation). Coarse sediment accumulates at the base of this steep slope, and lies with sharp contact on the fine sediment of the bed-form trough. Fining-upward laminae result from preferential preservation of the lower parts of these large bed waves. The low-angle inclined laminae within the thicker planar laminae are related to turbulent fluctuations in sediment transport onto the depositional lee side of the bed waves. Parting lineations result from the streaky flow structure within the viscous sublayer, with the ridges occurring under the low-speed streaks. These lineations are a *potential* paleohydraulic indicator, because the mean transverse spacing of low-speed streaks, λ, is approximately $100\nu/U_*$ for clear water flows. It is necessary to determine what effect high sediment concentrations have on this spacing before this relationship can be used in practice.

Antidunes

The bed forms formed under supercritical ($Fr = U/\sqrt{g\,d} > 1$), unidirectional, turbulent water flows over cohesionless sediment beds include antidunes (manifested as transverse ribs in gravel), chutes-and-pools, and rhomboid ripples. Antidunes are low-amplitude sinusoidal waves that are in phase with water-surface waves (Figure 5.67). With supercritical flow, the shape of the water surface directly affects the shape of the bed. If a shallow-water gravity wave (i.e., one with wave length $>d/2$) formed on the water surface, a plane bed would experience deposition under the crest of the water-surface wave in response to flow expansion and deceleration, and erosion under the trough associated with flow convergence and acceleration. Subsequently, deposition would occur on the upstream side of the developing bed wave (due to flow deceleration) and erosion would occur on the downstream side (Figure 5.67).

The water-surface waves above incipient antidunes are symmetrical in profile and in phase with the bed waves. The length L_s and speed (celerity) c_s of shallow-water gravity waves are given by

$$L_s = 2\pi c_s^2/g \tag{5.113}$$

$$c_s = \sqrt{gd} \tag{5.114}$$

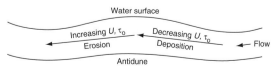

FIGURE 5.67. Generalized geometry, flow, and sediment transport over antidunes, at Froude number close to unity.

$$L_s = 2\pi d \tag{5.115}$$

If a shallow-water gravity wave moves upstream with speed equal to the mean velocity of the flow ($U = \sqrt{gd}$, or $Fr = 1$), the gravity wave (and bed wave) will be stationary relative to the boundary. An upstream-propagating gravity wave will theoretically move upstream or downstream relative to the bed if the Froude number is a little less than or greater than unity, respectively. To be correct, the *surface* flow velocity should be compared with the speed of the gravity wave to determine which way the waves migrate. If the surface flow velocity is greater than the mean flow velocity, stationary waves can occur at Froude numbers less than 1 (theoretically at $Fr > 0.84$) (Kennedy, 1963). The ability of antidunes to migrate upstream is the reason for their name, although they are not dunes, and they can migrate downstream also. Thus, the term antidune was poorly chosen, although it is probably here to stay.

Kennedy (1963) showed experimentally and theoretically that the minimum wavelength L of water-surface waves and two-dimensional antidunes is given by

$$L = 2\pi U^2/g \tag{5.116}$$

implying a Froude number of 1. An equation for determining the dominant wavelengths of antidunes is (Kennedy, 1963)

$$Fr^2 = U^2/(gd) = [2 + kd\tanh(kd)]/[(kd)^2 3kd\tanh(kd)] \tag{5.117}$$

where k is $2\pi/L$ and tanh is the hyperbolic tangent function.

The forms of the water surface and bed change constantly as the waves on the water surface and bed develop, migrate, and dissipate. It is common for water-surface waves and bed waves to increase in height with time. Antidunes grow by erosion of the trough areas and deposition on the crests. Although some antidunes stay more or less stationary as they grow, they more commonly migrate upstream at rates of centimeters per second. Downstream migration of antidunes is not common. As antidunes grow, their height remains less than (commonly half of) the height of the water-surface waves, such that the flow depth above antidune crests is greater than that over the troughs. Some antidunes decrease in height after growth, by erosion of the crest and deposition in the trough, returning to a nominally plane bed. In other

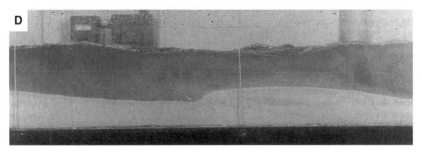

FIGURE 5.68. Kinematics of experimental antidunes. From Alexander *et al.* (2001). The flume section is 1.5 m wide, and flow is from right to left. (A) A standing water-surface wave and developing antidune. (B) An asymmetrical water-surface wave and antidune, with flow separation on the upstream side of the antidune prior to wave breaking. (C) A breaking wave. (D) An asymmetrical bed form migrating into an antidune trough after breaking.

cases, water-surface waves and antidunes increase in height and become asymmetrical, with their steeper sides facing upstream (Figure 5.68). As the degree of asymmetry increases, the water-surface wave travels upstream faster than the underlying bed wave, and the flow separates from the upstream slope of the underlying antidune. This flow separation causes a plume of sediment-laden fluid to rise steeply from the upstream trough of the antidune towards the water surface (Figure 5.68). The flow in the separation zone is highly turbulent and charged with suspended sediment. Rapid deposition occurs on the upstream face and crest of the antidune. At a critical steepness (water-wave height $0.142L$), the water-surface wave breaks upstream in a zone that is very turbulent and heavily charged with suspended sediment (Figure 5.68). Antidune crests are rapidly eroded, and deposition occurs in the trough areas after breaking. Such erosion and deposition may be in the form of asymmetrical bed forms (up to a few centimeters high, with a downstream face approaching the angle of repose) that migrate rapidly downstream. The ensuing bed topography is either planar or contains subdued bed waves (remnant antidunes and asymmetrical bed forms). Both types of bed form may form the nuclei of subsequent antidune growth. Breaking of water-surface

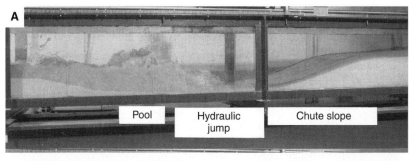

FIGURE 5.69. An experimental chute-and-pool showing upstream migration of a hydraulic jump (A) and its breaking on the chute slope (B). From Alexander *et al.* (2001).

waves may be an isolated event or several waves may break more or less simultaneously (leaving large areas of nearly plane bed).

Chutes-and-pools

Chutes-and-pools occur at higher Froude numbers than antidunes (the upper limit for formation of antidunes is $Fr = 1.77$), but there is a zone of overlap where antidunes are superimposed on chutes. Chutes are shallow, rapid flows over plane beds that slope downstream (at up to 30°) into pools where the flow depth abruptly increases and flow velocity decreases. The abrupt transition from rapid shallow flow to deep slower flow is a hydraulic jump (Figure 5.69). The accelerating flow within the chute slope is erosive, leading to upstream migration (at millimeters to centimeters per second) of the chute slope. As the sediment-charged flow of the chute passes into the hydraulic jump, it rises steeply upwards towards the water surface, in the same way as for breaking waves over antidunes. Suspended sediment and water in the lower part of this rising plume mix with the extremely turbulent, sediment-laden flow of the pool beneath. The rapidly decelerating flow immediately downstream of the hydraulic jump results in deposition from suspension on the upstream-facing slope of the pool. Rates of deposition can be up to millimeters per second. The rates of movement of the downstream and upstream slopes of the pool are variable and not entirely congruous. For example, it is

common for the hydraulic jump to move upstream relatively rapidly until it breaks on the chute slope (Figure 5.69). Exceptionally rapid deposition occurs downstream of the hydraulic jump as it moves upstream in this way. Following breaking of the hydraulic jump, the pool is filled rapidly by a downstream-migrating, asymmetrical bed wave, similarly to the case with breaking antidunes.

As the Froude number is reduced, chutes-and-pools change gradually into antidunes, beginning with periodic superposition of antidunes along the chutes. Then, the pools become shallower and shorter, and breaking antidunes cover most of the bed. With further reduction in Froude number, the length and height of the antidunes decrease, and wave breaking becomes less frequent.

Transverse ribs

Transverse ribs are ridges of gravel that are transverse to the flow direction (Figure 5.70) and are repetitive in the flow direction, with wavelength typically on the order of decimeters to meters. The ridges are more or less straight and extend normal to the flow for several wavelengths. Ribs normally comprise several imbricated gravel clasts, and the rib height is one to two clast diameters. The wavelength of ribs is commonly five to eight times the size of the gravel clasts forming the ribs.

Theories for the origin of transverse ribs are summarized by Allen (1982a). The most viable explanation

FIGURE 5.70. Transverse ribs on gravel bar. The photo shows imbricated pebble ribs with snow filling the troughs between ribs. Flow is towards the lower right of the photo. From Bridge (2003).

FIGURE 5.71. Rhomboid ripples. Flow is from bottom to top. From Bridge (2003).

is that transverse ribs represent the crests of antidunes. This explanation is consistent with the repetitive nature of the ribs and estimated flow conditions during rib formation.

Transverse ribs and their deposits have not been recognized in the fossil record, although gravel clusters have. It may be that some gravel clusters in the fossil record are transverse ribs. If this is the case, a lower-stage plane gravel bed with gravel clusters is difficult to distinguish from a plane gravel bed with transverse ribs. In both cases the gravel clasts are close to the threshold of grain motion, but in the latter case the spacing of the gravel clusters is controlled by water-surface waves.

Rhomboid ripples

Rhomboid ripples have a symmetrical, diamond shape in plan, are on the order of centimeters long and wide, and are a few grain diameters high (Figure 5.71). They occur in very shallow supercritical flows on bar tops, and may be superimposed on upper-stage plane beds or antidunes (Allen, 1982a). Rhomboid ripples are associated with two series of hydraulic jumps that are oblique to the mean flow direction (Allen, 1982a). The crests of the ripples correspond to the position of the hydraulic jump. The nature of associated sedimentary structures is unknown. It is possible that they could be responsible for some laminae associated with upper-stage plane beds or antidunes.

Cross stratification formed by antidunes

The sedimentary structures formed by antidunes, as observed in flow-parallel sections, are lenses with

trough-shaped, erosional bases that contain sets of dipping laminae (Alexander *et al.*, 2001) (Figure 5.72). The bases of sets are characteristically overlain by relatively fine sediment and concentrations of heavy minerals. The laminae within the sets are up to a few millimeters thick, and are distinguished by slight variations in grain size. Some laminae are indistinct, and some lenticular sets contain structureless deposits (Figure 5.72). Many laminae dip upstream (dips are of up to 20° but normally less than 10°) and decrease in dip near the base of the set to approach the set base asymptotically. Laminae that are concave upwards and more or less parallel to the trough-shaped base of the set are also common: these flatten out upwards, thus filling the trough. The thicker sets of this type most commonly contain poorly defined laminae. A few sets contain laminae dipping predominantly downstream at a relatively high angle (more than 20°). Rarely, laminae are convex upwards and define the shape of antidunes (Figure 5.72).

The trough-shaped sets overlap and truncate each other. In some cases, troughs are superimposed directly upon other troughs (giving the false impression of the laminated filling of a single trough). Such superposition of troughs is indicated by basal concentrations of heavy minerals and differences among the inclinations of the laminae within the different troughs (Figure 5.72). In most cases, superimposed troughs are offset from each other (Figure 5.72). The maximum thickness of sets is commonly two or three centimeters, and their length is commonly tens of centimeters.

Antidune forms may be preserved on the top of a bed. Each bed form contains a set of internal laminae that rests upon a concave-upward erosion surface (Figure 5.72). The basal laminae are concordant with the base of the set, but become flatter upwards and then

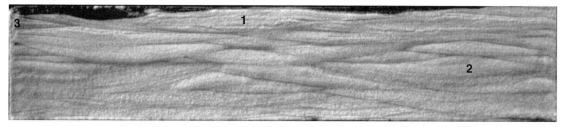

FIGURE 5.72. Internal structures of experimental, aggrading antidunes. From Alexander *et al.* (2001). The lightest colors are the coarsest sediments, and the dark layers are relatively fine-grained. Flow is from right to left. The panel is 1 m across and 0.2 m high. Below the antidune (1), convex-upward laminae overlie trough-shaped set bases. Troughs are commonly filled symmetrically with laminae (2). Upstream-dipping laminae occur at 3.

convex upwards and parallel to the form of the antidune. These laminae indicate growth of the antidune under a stationary wave. Other laminae within antidune forms dip upstream (Figure 5.72), indicating upstream migration.

Normal to the flow direction, sets are continuous for at least tens of centimeters, with relatively small changes in thickness (Figure 5.73). Within the sets, the laminae are either parallel to set boundaries or are inclined usually at less than 10°, but rarely up to 20°.

Sets of upstream-dipping laminae are formed during upstream migration of antidunes, although some of these are formed by the offset superposition of the trough bases of different antidunes. The trough-filling laminae are due to filling of antidune troughs either during gradual reduction in height of surface waves or following wave breaking. Trough fills with poorly defined laminae were deposited very rapidly following wave breaking. Concentrations of fine sand at the bases of troughs were deposited from suspension in the rapidly decelerating, sediment-charged flow in the troughs following wave breaking. The downstream-dipping laminae are due to rapid migration of asymmetrical bed forms with a steep leading edge, following wave breaking and smoothing of the bed. Rare, convex-upward laminae record growth of antidunes under stationary waves. The continuity of the lamina sets normal to the flow direction, but the variability in their thickness and in dips of internal laminae, indicates that individual antidunes extend across the flow but their heights vary. The origin of the individual laminae is not yet understood.

The preserved sets are related mainly to the largest antidunes that had the deepest trough scours, as has been determined for sets of cross strata formed by ripples and dunes. On the basis of somewhat sparse data, both the ratio of mean cross-set length to formative antidune length and the ratio of maximum set

thickness to formative antidune height are 0.4–0.6. Although more data are needed to refine these relationships, it appears that the geometry of lenticular lamina sets can be used to estimate antidune geometry, and hence paleohydraulic conditions. Up to this point, only the geometry of preserved antidune forms has been used to interpret paleoflow velocity and depth, using the well-known relationship of antidune length, depth, and flow velocity (Equations (5.115) and (5.116) and the fact that antidunes occur at Froude numbers around 1.) Sedimentary structures associated with antidunes superficially resemble hummocky cross strata and swaley cross strata formed by wave currents (see Chapter 7), and great care must be taken in distinguishing them.

Cross stratification associated with chutes-and-pools

Chutes-and-pools give rise to upstream-dipping laminae, commonly referred to as *backset* laminae (Allen, 1982a). In flow-parallel sections, such deposits consist of laterally adjacent lamina sets containing upstream-dipping laminae (with dips of up to 16°) or structureless sand (Figure 5.73). In cross-stream sections, the lamina sets extend for decimeters with minor variations in thickness, and the internal laminae may dip at up to 10°. These backset strata form during the upstream migration of chute-and-pool structures. The structureless sand is deposited by rapid deposition from suspension as the hydraulic jump moves upstream.

Porosity and permeability in stratified sands and gravels

The variation of sediment texture and fabric that defines stratification has a major influence on spatial

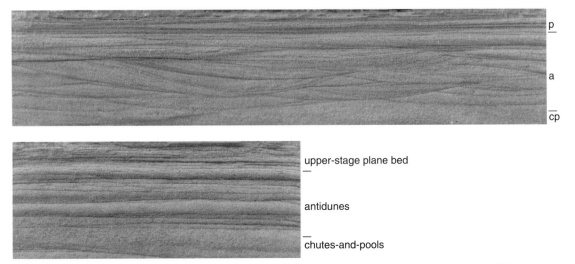

FIGURE 5.73. Internal structures associated with chutes-and-pools (cp), antidunes (a), and the upper-stage plane bed (p), resulting from a reduction in flow velocity during aggradation. From Alexander *et al.* (2001). The upper panel is parallel to the flow direction (right to left), and is 1 m long. The lower panel is normal to the flow direction, and is 0.5 m across.

variation of porosity and permeability, and therefore on the nature of flow in hydrocarbon reservoirs and aquifers; see Brayshaw *et al.* (1996) for a review. Measurement of porosity and permeability is discussed in the appendix. Permeability and porosity in cross-stratified sand and gravel are relatively large because the sediment is relatively coarse and poorly packed. The coarsest parts of cross strata have the highest permeability and porosity. The bases (bottomsets) of cross sets are commonly enriched in fine-grained material, and these are zones of reduced permeability or even barriers to flow. The bases of planar laminae formed by low-relief bed waves are analogous to bases of cross sets, and are similarly relatively fine-grained. The direction of maximum permeability is parallel to trough axes of trough cross sets, and the permeability is least in the direction perpendicular to the trough axes.

Hydraulic criteria for the existence of equilibrium bed states

Hydraulic criteria for the occurrence of equilibrium bed states are indicated in Table 5.4, and discussed above. The problem of establishing such criteria is similar to the problem of predicting sediment-transport rate, and important variables are expected to be the dimensionless bed shear stress and grain Reynolds number (and the Froude number for predicting antidunes). Approaches to this problem range from mainly

empirical to fully theoretical. The data used to test the various approaches are almost exclusively laboratory data, because of the uncertainties involved in describing bed configurations in nature, and in establishing whether or not they are in equilibrium with the flow. Laboratory data are necessarily limited in scale.

Plots of bed-state data with dimensionless bed shear stress and grain Reynolds number (or mean grain size) as axes separate quite well the existence fields of ripples, lower-stage plane beds, dunes, and upper-stage plane beds (Yalin, 1977; Allen, 1982a; Van den Berg and Van Gelder, 1993; Carling, 1999); see Figure 5.74A, B. A simplified version of this plot has bed shear stress and mean grain size as axes (Allen, 1982a; Leeder, 1982; Best, 1996). A related approach is to use stream power per unit bed area and mean grain size (Allen, 1982a); see Figure 5.74C. All of these diagrams show large amounts of overlap between the fields of ripples or dunes and upper-stage plane beds. This overlap is related to the bed-form drag associated with ripples and dunes. Thus, a given bed shear stress can be associated with a relatively high sediment-transport rate over an upper-stage plane bed (negligible bed-form drag) or a lower sediment-transport rate over a ripple or dune bed (appreciable bed-form drag). Antidunes cannot be predicted by any of these approaches because they depend on the Froude number. Best (1996) has associated the different bed states in plots of bed shear stress against mean grain size with fluid

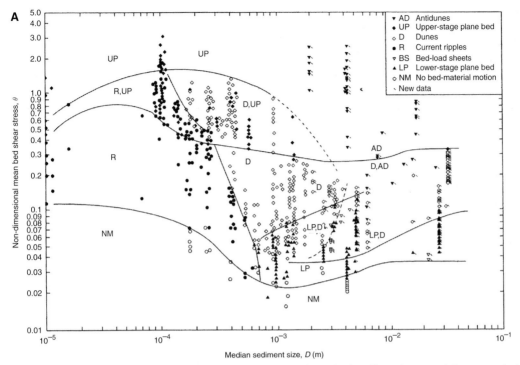

FIGURE 5.74. Equilibrium stability fields of the various subaqueous bed forms developed by uniform, steady flow over cohesionless silt, sand, and gravel beds. Most data come from straight laboratory channels, with some field data. Note the degree of overlap between stability fields, some of which is controlled by the differing nomenclature of individual experimenters. (A) As a function of dimensionless bed shear stress and median sediment diameter. From Carling (1999), modified from Allen (1982a), and including additional data from Carling and Shvidchenko (2002). (B) As a function of dimensionless bed shear stress and median sediment diameter, according to Van Den Berg and Van Gelder (1993). (C) As a function of stream power and mean grain size. Modified from Allen (1982a). (D) As a function of flow velocity, grain size, and flow depth. Modified from Southard and Boguchwal (1990).

flow and sediment-transport processes that he considers control the generation of the different bed states. Plots of flow velocity, depth, and mean grain size (Middleton and Southard, 1984; Southard and Boguchwal, 1990; Carling, 1999) (Figure 5.74D) effectively separate all of the bed states, although there is still some overlap of the existence fields. These parameters are not as easily related to sediment-transport rate as bed shear stress, grain Reynolds number, and stream power, but are much easier to measure. Allen (1982a) reviewed other criteria for the occurrence of equilibrium bed states.

The value of diagrams like Figure 5.74 is that they allow approximate quantitative prediction of changes in bed state (hence sedimentary structure) as flow conditions change over sediment beds of a given grain size; see Table 5.5 and the review in Allen (1982a). They also allow quantitative interpretation of changes in flow and bed state from data on sedimentary structures and mean grain size (Bridge, 2003), although it must

be remembered that bed states such as dunes might not be in equilibrium with the flow. Figure 5.75 illustrates some typical sedimentary sequences in which mean grain size and sedimentary structures vary vertically and laterally. They are interpreted in terms of temporal and spatial variations in flow strength (e.g., bed shear stress and/or flow velocity), sediment-transport rate, bed state, and deposition rate; see also Allen (1982a). Temporal variations in sediment transport rate ($\partial i/\partial t$) are associated with changes in the flow strength over a single flood, and produce vertical variations in mean grain size and sedimentary structures. Spatial decrease in sediment-transport rate ($\partial i/\partial x < 0$) is mainly responsible for the deposition and for downstream changes in mean grain size and sedimentary structures. The availability of various sediment sizes and the overall flow strength control the overall grain size of sediment in the sedimentary sequences. Sequences such as this will be incorporated within larger-scale sequences associated with larger topographic features such as bars (Chapter 13).

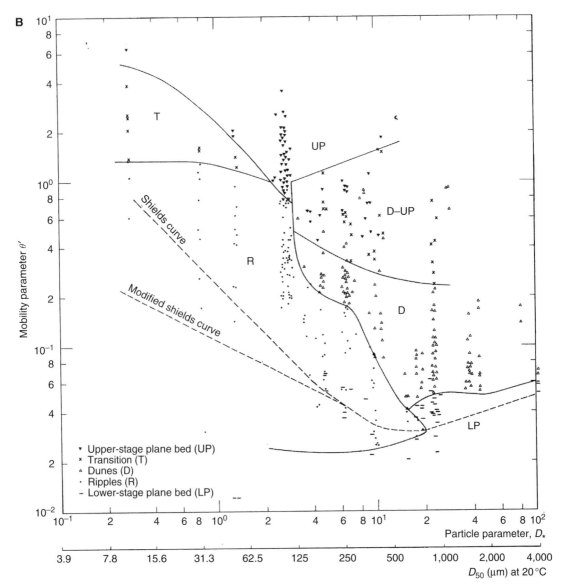

FIGURE 5.74. Continued.

Theories for the hydraulic stability and geometry of bed states

There are five main types of theory for the hydraulic stability of equilibrium bed states under unidirectional flow: stability analysis, shallow-water wave theory, kinematic wave theory, large-eddy theory, and flow-separation–reattachment theories. They are reviewed in Allen (1982a), McLean (1990), and Bridge (2003).

Stability analysis (the most common type of theory) starts by defining the uniform, steady flow and sediment transport over a plane bed. Fluid motion is

defined using the equations for conservation of momentum and conservation of mass (the continuity equation). Movement of sediment is defined using the conservation of mass (the sediment continuity equation) and a function that relates the sediment-transport rate to fluid-flow characteristics. It is then assumed that a sinusoidal disturbance of the planar bed (a bed defect) occurs, and the stability analysis attempts to predict whether a bed defect of a given wave length will become amplified (giving rise to bed forms) or will be damped (the bed remains plane). If a bed defect becomes amplified, those wave lengths with the fastest

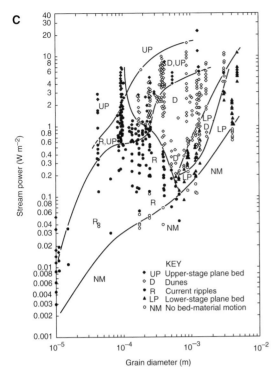

Stream power (W m^{-2})
Grain diameter (m)

KEY
UP Upper-stage plane bed
D   Dunes
R   Current ripples
LP Lower-stage plane bed
NM No bed-material motion

FIGURE 5.74. Continued.

and sediment-transport rate coincide with the bed-form crests or troughs. The spatial lag has been related to (1) the interaction between turbulent shear stress and bed topography; (2) the inertia of suspended load; and (3) the effects of local bed slope on bed-load transport rate. These are discussed by Allen (1982a), McLean (1990), and Bridge (2003). Although modern stability analyses account for the effects of bed topography and sediment transport on the lag distance required for bed-form growth, they do not consider the effects of high near-bed sediment concentration or streamwise pressure gradient on local turbulence structure and sediment-transport rate. Furthermore, the mechanisms that promote initial bed instability are not necessarily those that determine continued growth and existence of bed forms. In particular, the effects of flow separation downstream of bed-form crests are not considered, despite their control on the distribution of bed shear stress on the backs of bed forms. More information is needed on the interaction of turbulence structure, sediment transport, and bed-form dynamics before a comprehensive bed-form theory can be developed.

Wave theory of bed forms, discussed by Allen (1985), differs from stability analysis in that water-surface waves rather than bed waves are considered to be the main driving mechanism. It is well known that water-surface waves have a direct influence on the formation of antidunes under supercritical flows. Allen (1985) claims that water-surface waves also affect the bed in the case of subcritical flows. However, wave theory does not explain the wave lengths of ripples and all dunes, and ripples and dunes occur in closed conduits where there is no free surface, so water-surface waves are clearly not essential for the formation of subcritical bed waves.

Kinematic wave theory has been used to explain some types of bed form (straight-crested dunes and bars); see the review in Bridge (2003). A kinematic wave consists of a concentration of units (a bed form) through which the individual units (sediment grains) may move. A common example of a kinematic wave is traffic bunches on busy roads. A concentration of cars (e.g., behind a slow-moving vehicle) is a wave through which individual cars move (if they overtake). The individual cars move faster than the wave, but their velocity decreases as they become closer together. In the case of bed forms, the sediment grains move faster than the bed form. The transport rate of sediment grains over bed waves depends on how close they are. The sediment-transport rate increases with grain

rate of amplification are considered to become dominant. Thus, a stability analysis will normally predict the wave length of the resulting bed forms. Stability analyses have been successful in predicting the stability fields of antidunes, plane beds, and lower-flow-regime bed forms (ripples and dunes). The various attempts differ in the way the movements of water and sediment are treated, or in the simplifications to the equations of motion. Some treat the water as ideal or non-viscous (thereby ignoring boundary friction and flow resistance), whereas others consider the water to be a real viscous fluid. There is much variability in the sediment-transport functions used. Some consider the effects of local bed slope on sediment transport, whereas others do not. In most cases the initial bed disturbance is considered to be two-dimensional, but 3D disturbances have also been considered.

All stability analyses require introduction of a spatial lag of sediment-transport rate behind a spatial change in flow velocity and bed topography in order to allow amplification of the initial bed defect. If such a lag were not introduced, the bed defect would migrate downstream or upstream, but would not grow in height. This is because, in the absence of a lag, the maxima and minima in flow velocity, bed shear stress,

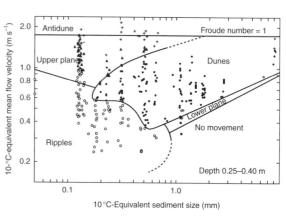

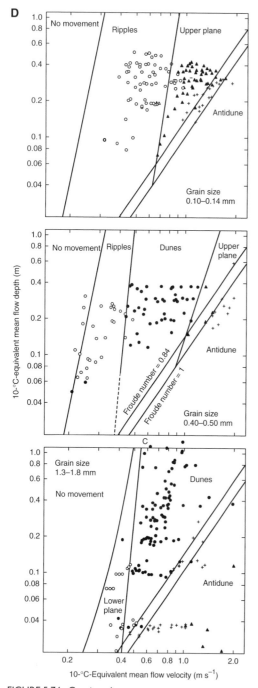

FIGURE 5.74. Continued.

waves does not consider the interactions among turbulent flow, sediment transport, and bed forms, and cannot explain the characteristic spacing and period of repetitive bed forms.

Large eddies superimposed on the mean flow have been invoked to explain the origin and geometry of dunes (Jackson, 1976; Yalin, 1977, 1992). The size of eddies is on the same scale as the flow depth, such that the length of the eddies and dunes is given by $L \approx 6d$. These eddies were linked to turbulent bursting in the bed region. Turbulent ejections were considered to occur preferentially from zones of decelerated flow (adverse pressure gradients) associated with the large eddies. Ejected fluid would ascend into the flow, interact with the large eddy, and subsequently return to the bed as a high-speed sweep. Erosion and dune–trough formation were associated with ejections, whereas deposition was associated with deceleration of sweeps. Turbulent boils on the water surface were related to these ejections of fluid that originated in dune troughs (Jackson, 1976). Several criticisms have been leveled at this theory, particularly the link between turbulent bursting and the origin of dunes (e.g., Allen, 1982a, 1985; Best, 1996): (1) turbulent motions affect sediment transport over all bed forms in unidirectional flows, yet the length of ripples and many dunes does not follow the scaling law above; (2) large eddies are shorter than $6d$, and are convected downstream at approximately the mean flow velocity, yet most bed forms have negligible speed relative to the water; (3) developing bed waves have a length much smaller than their equilibrium length; and (4) the boils that have been interpreted as turbulent ejections are probably eddies shed from the free-shear layer above the flow-separation zone.

concentration up to a point, but then decreases as the grains begin to jam against each other. These sediment jams may be bed forms. It is unknown why only two classes of bed form were singled out for explanation in terms of kinematic waves. The theory of kinematic

Flow-separation–reattachment theories have also been used to explain ripples and dunes. When flow separates on the lee side of a bed defect, it reattaches at a distance of about seven defect heights downstream of the crest. The flow between the crest and the reattachment zone is decelerated, and hence depositional. Downstream of the reattachment zone, the bed shear stress increases (and the bed is eroded) up to a maximum value about 15–40 defect heights downstream of the separation point. Deposition occurs downstream of the maximum in bed shear stress, forming another bed defect. This theory explains why ripples and dunes develop progressively downstream in turbulent flows, but does not explain the height and length of developed bed forms. The length of the defects is on the order of tens of defect heights. In order to understand why the length of ripples is proportional to grain size but the length of dunes is proportional to flow depth, it is necessary to understand what controls the limiting or equilibrium height of the growing bed defect.

TABLE 5.5. Sequences of bed states

Mean grain size	Sequence with increasing bed shear stress and/or flow velocity
$D < 0.1$ mm	Ripples to upper-stage plane beds or antidunes
0.1 mm $< D < 0.7$ mm	Ripples to dunes to upper-stage plane beds or antidunes
$D > 0.7$ mm	Lower-stage plane beds to dunes to upper-stage plane beds or antidunes

Erosional structures in cohesionless sediments

Although erosion associated with bed-wave migration and the formation of cross strata was mentioned above, specific structures associated with erosion have not yet been discussed. Erosion may be local, as in the

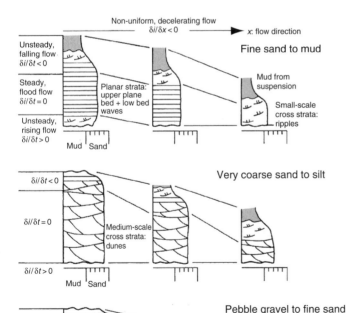

FIGURE 5.75. Typical sedimentary sequences produced by erosion and deposition over a single flood period, for three different grain-size ranges. Deposition, and downstream change in grain size and sedimentary structures, are caused by decreasing bed shear stress and sediment-transport rate (i) in the flow direction (x). Vertical changes in grain size and sedimentary structures are caused by changes in bed shear stress and sediment-transport rate with time (t) over the flood period. The thickness of flood-generated sedimentary sequences is typically centimeters to meters, and downflow extents are typically hundreds of meters to kilometers. Modified from Bridge (2003).

3 m

FIGURE 5.76. (A) A gutter cast on the base of a sandstone bed. Note shell impressions along the base of the cast. The scale bar is 5 cm long. (B) Current-crescent scour around a grounded ice block (now melted) on the surface of a gravel bar. Flow is to the right.

wide, called *gutter casts* (Figure 5.76). Allen (1982a) discusses these relatively small-scale erosional features at length.

Bed forms and sedimentary structures in cohesive sediments (muds)

The bed forms and sedimentary structures associated with cohesive mud depend on the state of consolidation of the mud. When soft mud is entrained, the clay and silt particles travel in suspension, and there are no bed forms like those in coarser, cohesionless sediments. However, various bed configurations may be eroded into the mud according to the flow strength, as reviewed by Allen (1982a). The sequence of bed configurations expected as the mean flow velocity increases is (1) straight longitudinal grooves and ridges, (2) meandering longitudinal grooves and ridges, (3) flute marks, and (4) transverse ridge marks (Figure 5.77). The longitudinal grooves and ridges are associated with the streaky flow structure of the viscous sublayer, and are analogous to parting lineations in cohesionless sediment. Flute marks are commonly the shape and size of spoons (minus handles), the blunt end facing upstream. They develop where locally increased near-bed turbulence results in local erosion and flow separation. The flow separation may also be associated with a burrow excavation or shell on the bed. If the flow is carrying sand in suspension, sand-blasting aids in the erosion and smoothes out the bed surface in the flute (that may otherwise contain longitudinal ridges and grooves). However, the flute will not form if sand is deposited in the separated flow. Transverse erosional marks are ripple-shaped and occur only in supercritical flows. They may contain longitudinal ridges and flutes on their surfaces.

Soft mud can also be molded by hard objects (tools) carried by the flow. *Tool marks* (Figure 5.78) are commonly long grooves of varying width and depth due to objects sliding and rolling along the bed and cutting down into the mud. Marks are also made by intermittent contacts of objects with the bed (e.g., bounce, skip, and prod marks). These and the current-formed marks are normally preserved as coarser sediment covers the mud more or less immediately following mark formation (e.g., *sole marks* on the base of sandstone strata). These marks are valuable indicators of paleocurrent direction. The objects responsible for the tool marks are only rarely found

case of migrating bed waves, or more widespread. Whatever the areal extent of surfaces of erosion, they are normally recognizable as sharp surfaces of truncation of underlying strata, and are overlain by the coarsest and/or densest grains from the underlying material that could not be entrained (i.e., a lag deposit). Only the sediment lying directly on the erosion surface comprises the lag deposit. Features indicative of local erosion include *current crescents*, which are horseshoe-shaped scours around the upstream margins of immobile obstacles to the flow (e.g., large clasts, tree trunks; Figure 5.76). Current crescents are commonly associated with pebbles or large shells on sand beds. The eroded material is deposited on the lee side as a ridge of sediment or *sand shadow*. Other local erosional features include elongated, flow-parallel channels, centimeters to decimeters deep and

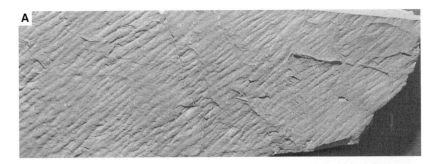

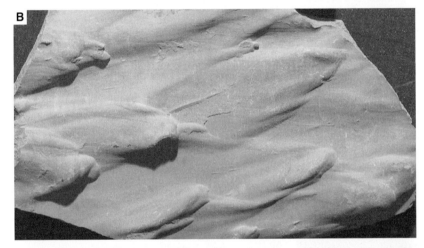

FIGURE 5.77. (A) Longitudinal ridge marks on the base of a sandstone bed, 30 cm across. (B) Flute marks on the base of a sandstone bed, 20 cm across. Paleoflow was from right to left. (C) Transverse erosional marks on the base of a sandstone bed. Paleoflow was from left to right. The sandstone filling the troughs has load structures (1), and these pass to the right into flute marks (2). Tool marks are also present (3). A head is shown for scale.

preserved at the end of the mark. These objects include shells, bits of wood, and shale chips. Tool marks and other erosional structures in muds are well illustrated in Pettijohn and Potter (1964) and Dzulynski and Walton (1965).

Deposition of mud from suspension commonly gives rise to a laminated deposit where the laminae drape over pre-existing bed topography (Figure 5.79). Such laminae (not to be confused with compactional

fissility: see Chapter 19) are probably due to relatively long-term changes in the velocity and turbulence intensity of the depositing fluid, for example associated with flood periods. However, as discussed below, not all laminated muds were deposited from suspension.

Consolidated mud is usually eroded as silt- to sand-sized pellets or tabular to discoidal chips of compacted, fissile mud, which may then travel as bed load. Cross

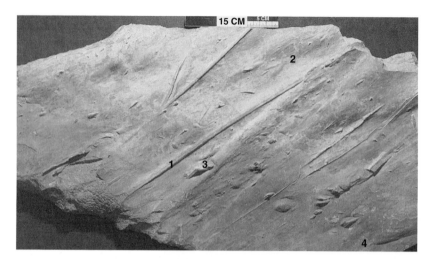

FIGURE 5.78. Assorted tool marks on the base of a sandstone bed, including groove marks (1), prod marks (2), and a twisted prod mark (3). There is a flute mark at (4). Prod and flute marks indicate that the paleoflow was from lower left to upper right.

FIGURE 5.79. Laminae in mud draping topography.

FIGURE 5.80. Intraformational mud chips eroded from desiccated mud.

strata in muddy sediments normally indicate that the mud was transported as sand-sized grains, although the original geometry of such grains may be difficult to discern in ancient deposits. Planar strata (laminae) in muds may also be associated with the transport of sand-sized mud grains, as low-relief bed waves. Gravel-sized chips of locally eroded mud commonly accumulate as erosional lags or are transported and deposited by strong currents as intraformational breccias and conglomerates (Figure 5.80). Erosional marks in consolidated mud include steep-walled gutter marks and potholes.

6 Air flow, sediment transport, erosion, and deposition

Introduction

Air flow is a significant agent of erosion, sediment transport, and deposition over subaerial surfaces that lack a protective cover of vegetation, such as deserts, sandy beaches, floodplains exposed after floods, and outwash plains of glacial meltwater streams. The mechanics of air flow, sediment transport, erosion, and deposition have been studied in the field and in wind tunnels, and are similar to those of unidirectional water flows (e.g., Allen, 1982a; Greeley and Iverson, 1985; Pye and Tsoar, 1990; Nickling, 1994; Lancaster, 1995, 2005). To avoid repetition of material in Chapter 5, emphasis will be placed on the differences between air flows and water flows and the associated sedimentary processes.

Mechanics of air flow

Air has much lower density than water ($1.3\,\text{kg m}^{-3}$ versus $1,000\,\text{kg m}^{-3}$), and much lower viscosity than water (1.78×10^{-5} N s m^{-3} versus 1.00×10^{-3} N s m^{-3}). The density of air increases as temperature decreases and increases as pressure increases because air is compressible. Air speeds regularly exceed water speeds by an order of magnitude. For example, a river in flood may have a mean flow velocity of $1\,\text{m s}^{-1}$, whereas a strong wind may be tens of m s^{-1}. This means that air flows are always turbulent, and resistance to motion is due to turbulent shear stress entirely. However, turbulent shear stresses in air flows are comparable to those in water flows because they are proportional to the product of fluid density and the square of flow velocity. Turbulent boundary layers of air flows have a viscous sublayer (thickness $3.5\nu/U*$), a buffer layer, and a fully turbulent region. The law of the wall applies to steady, uniform air flows. However, air flows over large objects such as sand dunes, trees, and buildings are clearly not uniform. It is difficult to define the mean bed level (where the flow velocity is zero) over such large roughness elements. As with water flows, the mean bed level may be taken as approximately $0.7k$, in which k is the mean height of the roughness elements. Flow patterns over aeolian dunes are similar to those over subaqueous dunes, and can be modeled in a similar way. However, it is very difficult to test such models because local, near-bed velocity and shear stress are difficult to measure in air flows. Air flows differ from water flows in having no free surface, in having density stratification associated with vertical variation in temperature, and in having strong convection currents associated with heating of the air by the surface over which it is moving. This means that turbulent boundary layers associated with air flow can be much more complicated than those of water flows.

Mechanics of sediment transport

Settling grains

The immersed weight of quartz-density sediment is 1.6 times greater in air than in water. The drag on a grain moving in air is much less than that in water for a given relative velocity, because of the low density and negligible viscosity of air. Therefore, the terminal settling velocity of grains in still air is much greater than that in water.

The threshold of entrainment of cohesionless sediment

The mechanism of entrainment of cohesionless bed sediment by air is similar to that in water, and the threshold of entrainment can be expressed using the dimensionless bed shear stress and the grain Reynolds number (Figure 6.1) (Allen, 1982a; Greeley and Iverson, 1985; **195**

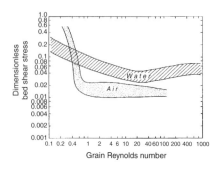 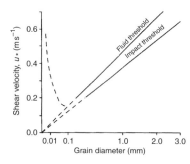

FIGURE 6.1. The threshold of entrainment of cohesionless sediment by air flow as a function of the dimensionless bed shear stress and boundary grain Reynolds number (from Allen (1982a)), and as a function of shear velocity and mean grain size (after Bagnold (1941)).

Pye and Tsoar, 1990; Lancaster, 1995). The threshold can also be expressed in simplified form using the shear velocity (or bed shear stress) and mean grain size (Figure 6.1). For grain Reynolds numbers less than 3.5 (grain diameters less than about 0.1 mm), the grains lie within the viscous sublayer, and the threshold value of the dimensionless bed shear stress starts to increase. This increase is also due to increasing effects of cohesion of finer grains. However, in addition to direct fluid stress, the threshold of entrainment in air flow is affected by already-entrained bed-load grains impacting the bed and disturbing stationary bed grains. Because of their momentum, these moving grains can exert a force great enough to move bed grains, as explained below. The threshold of entrainment accounting for grain impacts in addition to direct fluid shear stress is approximately 64% of the threshold due to fluid shear stress alone (for grain diameters greater than 0.1 mm). The threshold of motion is also greatly affected by the wetness of the sediment, because the surface tension of water in wet sediment greatly increases its resistance to motion. Organic films, clay minerals, and cements in bed sediments also greatly increase resistance to motion.

The threshold of motion of individual grain-size fractions in a mixture of different sizes (rather than of the mean size of well-sorted sediment) varies as for sediment entrainment by water. Relatively fine-grained sediment surrounded by coarse-grained sediment (i.e., hidden from the air flow) is more difficult to entrain than if it were surrounded by similar-sized grains, and the opposite is true for coarse-grained sediment that protrudes up above finer-grained sediment. Armor layers of relatively coarse grains can render finer grains immobile.

Mechanics of bed-load motion

The dominant mode of bed-load motion of sand is *saltation* (jumping). Saltation trajectories are caused and modified by turbulent lift and drag. However, saltation is also influenced by impacts of the moving grains with the bed, causing elastic rebound of saltating grains and/or splashing up of previously stationary bedgrains. The motion of splashed-up grains is referred to as *reptation*. Some of the larger grains (coarser sands and gravel) that cannot be splashed up into the flow are moved along the bed incrementally by the bombardment of the smaller saltating grains (a process known as *creep*). Creeping bed-load motion may constitute 20%–25% of the total bed-load transport.

The high speed and momentum of the saltating grains is the reason why they can exert such large forces on the bed. Such high speeds are attained as grains are lifted from the bed into the boundary layer to distances of hundreds or thousands of grain diameters (generally centimeters to decimeters) and accelerated by relatively fast-moving air. Grain speed is further increased as the grains return to the bed under the influence of gravity. The low viscosity and density of air ensures that the grains are not resisted greatly by drag during their upward and downward movements. In water flows, bed-load grains move relatively slowly within about ten grain diameters from the boundary, thus inhibiting large impact velocities with the bed. Nevertheless, the thickness of the bed-load layer is proportional to the bed shear stress in air flows, as it is for water flows.

Bed-load transport equations can be developed as for unidirectional water flow by analyzing the fluid stresses acting on bed grains. However, about 25% must be added to the calculated transport rate to account for the effects of grain bombardment on bed-grain movement. Furthermore, aeolian bed-load transport-rate equations must be able to account for limited availability of sediment, which could be related to the absence of a complete sediment cover on the bed, moisture content, or binding effects of vegetation. Bed-load transport equations are difficult to test because of

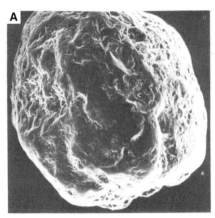 

FIGURE 6.2. (A) Surface texture of an aeolian sand grain, showing good rounding and upturned plates. (B) Faceted pebbles from the Namib Desert (photo courtesy of Gary Kocurek).

difficulty in measuring the effective bed shear stress and the inaccuracy of some bed-load-measuring traps.

Grain–bed and grain–grain collisions in air are effective at rounding off the edges of grains (Figure 6.2) and at disintegrating mechanically unstable grains. However, wind-blown sand is not always well rounded as is commonly believed, and most is subrounded to subangular (Lancaster, 1995). The surfaces of wind-blown sand grains commonly contain distinctive percussion marks associated with grain collisions (e.g., upturned plates, conchoidal fractures), but they also have patches of cement and pits caused by chemical weathering. Large bed grains (pebbles) that have been sandblasted and re-oriented several times have distinctive facets (Figure 6.2), and have been called *ventifacts* or *driekanters*. Wind carrying sediment is also effective at shaping rock outcrops (e.g., yardangs) by sandblasting, as discussed below.

Bed load transported by saltation under air flows is commonly fine- to medium-grained sand, and is very well to moderately well sorted. Bed load transported by creep is commonly coarse sand to granule size, and poorly sorted. Such coarse bed-load sediment may be bimodal due to the infiltration between the larger grains of silt and very fine sand carried in suspension by either wind or water.

Mechanics of transport of suspended sediment

The suspended load is composed mainly of silt and clay. The threshold of suspension can be calculated in the same way as with turbulent water flows, using V_s/U_* as a criterion (Chapter 5). According to Tsoar and Pye (1987), if $0.7 < V_s/U_* < 1.25$, saltation trajectories

are modified by turbulence, and this corresponds to intermittently suspended bed-material load in turbulent water flows. If $V_s/U_* < 0.7$, sediment is transported in suspension, and this corresponds to permanently suspended sediment in water flows (Chapter 5, Equation (5.80)).

If the air flow is stable (i.e., there are no density-related air currents), such that vertical air motions are due entirely to shear turbulence, the concentration distribution of suspended sediment can be calculated using an equation similar to that for water flows (Chapter 5, Equation (5.100)). Most of the suspended sediment is expected to be within a few meters of the ground. However, if the air flow is unstable, convection currents can carry sediment upwards to many kilometers from the ground, where flow velocities can reach many tens of $m\,s^{-1}$. Under these "dust-storm" conditions, suspended sediment can be carried for thousands of kilometers away from its source before being deposited (Figure 6.3). Silt and clay from the Sahara has been found accumulating on the mid-Atlantic ridge, in northern Europe, and in Florida. Aeolian silt and clay is commonly found in deep-sea cores and ice cores, and has been used as an indication of ancient periods of dryness. The atmosphere is commonly hazy because of the large amounts of dust that it contains.

Sediment that is transported and deposited in suspension is commonly bimodal in size, with one mode between 1 and 20 microns, and another between about 10 and 200 microns (Pye, 1987). The small-mode sediment is generally considered to be formed by attrition during transport due to grain–bed and grain–grain collisions. The large-mode sediment was most likely entrained from the bed, particularly from ephemeral

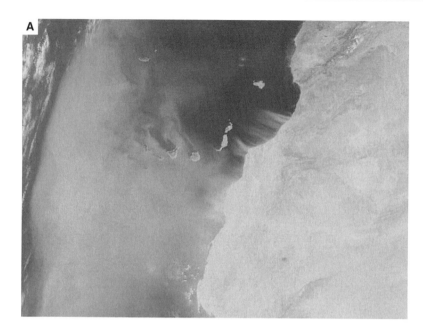

FIGURE 6.3. (A) Saharan dust blowing offshore over the Canary Islands. Note eddies to the lee side of the islands. Satellite photo from NASA (www.visibleearth.nasa.gov). (B) Pleistocene loess from Luochuan section, central Loess Plateau, China. Light-brown loess contrasts with reddish-brown paleosols. The inset is a close-up of unweathered loess. Photos courtesy of Joe Mason.

lake beds and river floodplains. A long transport path of suspended silt and clay in which the flow velocity and turbulence intensity decline in the direction of transport will lead to deposition of progressively finer-grained sediment with distance down flow. This leads to effective sorting out of different grain sizes.

A particular type of well-sorted silt with indistinct laminae, deposited from suspension in air flows, is called *loess*. Loess tends to be composed of 10–50-micron-sized grains of mainly quartz, but also substantial amounts of calcium carbonate, and produces very fertile soils (Pye, 1987). Leaching of wind-blown calcium carbonate dust may be the source of soil carbonate glaebules in some calcareous soils. High-latitude loess is spread widely over northern Asia, Europe, North America, and Argentina, and much of it is associated with deflation of glacial outwash plains during the last Ice Age (see Figure 17.17 later). Less extensive

A

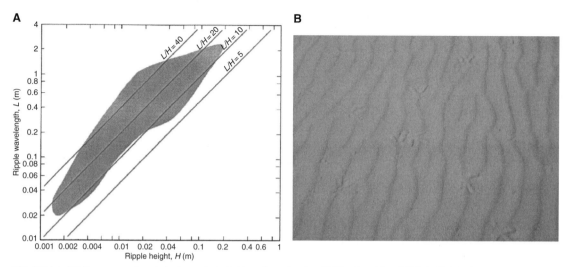

B

FIGURE 6.4. (A) Geometry of sand ripples and wind ridges. From Leeder (1999), after Allen (1982a). (B) Wind ripples with bird footprints. Flow is to the right.

and thinner accumulations of coarser-grained loess occur around low-latitude deserts.

Bed forms and sedimentary structures

Sand ripples and granule ridges

Sand ripples are virtually ubiquitous in sand seas, but granule ridges are not common. Sand ripples and granule ridges were originally thought to be different classes of bed form, but are now considered parts of a single, continuously varying population (Figure 6.4). Bedform wave length ranges from about 0.02 to 2 m (typically 0.05–0.2 m), height ranges from a few millimeters to about 0.1 m (typically 5–10 mm), and the length/height ratio is in the range 5–70 (mean about 20). These bed forms are asymmetrical in streamwise section, with the lee side approaching the angle of repose. Wave length increases with mean grain size and with wind speed and duration. The coarsest bed-load sediment is concentrated on the crests of the bed forms. The smaller bed forms, in very-fine- to medium-grained sands, have straight to sinuous crestlines in plan. The larger bed forms in coarse sand to granules have more curved crestlines in plan, and are less asymmetrical in section.

An early theory for the origin of sand ripples by Bagnold (1941) involves large grains creeping along the bed due to their being bombarded by smaller saltating grains. If a large grain is sheltered from bombardment by a following grain, it slows down and causes a "traffic jam" (cf. kinematic waves). Such "traffic jams" cause "bed defects" or incipient ripples, and the inclined lee sides of these incipient ripples are sheltered from bombardment by saltating grains. If there are limited numbers of saltating grains originating from the lee sides of these defects, there can be little bombardment at a downstream distance of a mean saltation length. Therefore, another accumulation of coarse grains will occur in this position. According to this theory, the mean length of ripples equals the mean saltation length (which is proportional to the bed shear stress). Bagnold's theory has been criticized because ripple spacing actually increases with time on an initially planar surface, and the equilibrium mean wave length does not equal the mean saltation length (Sharp, 1963; Allen, 1982a; Anderson, 1987).

In fact, the origin of wind ripples is much more similar to the origin of ripples under water than commonly thought, although turbulent structures in the viscous sublayer apparently do not play the same role (Anderson, 1987; Anderson and Bunas, 1993; Werner, 1995). As sediment is moved along by saltation and reptation (for smaller grains) and creeping (for larger grains), random bed defects arise, possibly related to the deceleration of a particularly large grain or due to a turbulent fluctuation in wind speed. Such a defect causes deceleration of other large grains on its upstream side and accumulation of relatively fine sediment on its lee side where the flow decelerates. As in the case of subaqueous ripples, the defect grows in length as its height increases, because the defect height controls the length of flow deceleration downstream. As the bed defect migrates downstream by upstream erosion and downstream deposition, the coarse sediment at the crest will become superimposed on the finer sediment to the lee. It is expected that the mean grain size of sediment in the bed load will increase with bed shear stress given an adequate supply of grains (as is the case for transport by water: Chapter 5). The height of the initial defect would obviously be greater for larger grains: however, it is not known what controls the limiting height (and hence length) of the ripple.

Dunes and draas

Aeolian dunes have heights up to several hundred meters and spacings of tens of meters to several kilometers, and vary greatly in their morphology. Aeolian dunes are commonly referred to using Arabic names (e.g., barchan, seif, aklé, rhourd) because of the locations where they were first described. Although most modern workers do not use this old terminology, these old names left the erroneous impression that aeolian dunes are quite different from subaqueous dunes, when in fact the forms and mechanics of formation of subaerial and subaqueous dunes are very similar (Figure 6.5). Aeolian dunes can be classified, following McKee (1979), Pye and Tsoar (1990), and Lancaster (1995), on the basis of their morphology, their orientation relative to the wind directions, and whether they can move freely or are fixed by vegetation or some other topographic feature. Free dunes include forms that have their crestlines more or less transverse to the flow direction, and in plan view these crestlines may be sinuous (crescentic dunes, aklé), lunate (barchan), or linguoid (parabolic). Some free dunes have their crestlines almost parallel (longitudinal) to the mean flow direction (linear or seif), and other free dunes that are subjected to several different flow directions are almost pyramidal in shape (rhourds, star dunes, reversing dunes). These free dunes can occur as single dunes, as dunes of a single type superimposed on larger dunes of the same type (compound dunes), or as dunes of a given type superimposed on dunes of a different type (complex dunes). Furthermore, spatial and temporal transitions between the various dune types are common. It is also common in a particular sand sea to see a bimodal distribution of dune sizes, and the largest (compound and complex) dunes (spacing >500 m) have been referred to as draas or megadunes (Wilson, 1972a, b; Lancaster, 1995). However, in a global sample, there is every gradation between the sizes of dunes and draas.

Many sand seas also contain areas of low dunes without slip faces, called zibar, and gently undulating areas with no obvious dunes (sand sheets). Dunes that are fixed by vegetation include shrub–coppice dunes (shadow dunes, nebkha) and elongated parabolic dunes (in some cases known as blowout dunes). Dunes that are fixed by topographic features (e.g., rock outcrops) include climbing or echo dunes (upstream of topographic obstacles) and lee-side or falling dunes (downstream of obstacles).

A persistent problem when classifying and interpreting the origin of dunes and draas is establishing whether or not they are in equilibrium with the currently prevailing wind regime (e.g., Lancaster, 1995). It is unusual for the formation and migration of dunes and draas to be observed, except in the case of small dunes. This is because of the lengths of time involved and the difficulty in making observations during the sand storms when these bed forms were most active. However, recently, the amount and direction of migration of large dunes has been established by examination of their internal structures using ground-penetrating radar and the timing of this migration (over periods of hundreds to thousands of years) has been established using optically stimulated luminescence dating (e.g., Lancaster et al., 2002; Bristow et al., 2005). The lifespan (maximum relaxation time: see Chapter 5) of a small dune that is on the order of 100 m long and migrates at about 10 m per year is on the order of decades. It is therefore possible that such dunes could be in equilibrium with wind regimes averaged over decades. However, large dunes that are kilometers long and migrate at rates on the order of 0.1 m per year would have lifespans on the order of tens of thousands of years. It is therefore unlikely that such bed forms are in equilibrium with present-day wind regimes. In the case of complex dunes, it is probable that the smaller superimposed dunes are in equilibrium with prevailing wind regimes, but that the larger host dunes are not. In the case of compound dunes, it is probable that the smaller superimposed forms are in equilibrium with the prevailing wind patterns, but it is not known whether or not the larger forms are in equilibrium. This dilemma also exists in the case of subaqueous dunes.

Another issue of concern when interpreting the origin of sand dunes is whether they were formed predominantly in a unidirectional air flow, or in a multidirectional flow regime. If the sand dunes were formed in a multidirectional flow regime, what was the velocity and duration of the wind in the various flow

FIGURE 6.5. (A) Compound transverse crescentic dunes, with linear dunes in the background. Flow is to the left. From Algodones, CA. (B) Transverse linguoid dunes. Flow is to the left. (C) A transverse lunate dune next to an alluvial fan (background). Flow is to the left. Images (B) and (C) from Great Sand Dunes National Park, San Luis Valley, CO. (D) Linear dunes with zibar and vegetation in interdune areas, from Algodones, CA. (E) Star dunes from Dumont, CA. (F) Linear dunes with zibar and shadow dunes in interdune areas. Flow is to the left. From Algodones, CA. (G) A blowout dune (foreground) from Denmark. Photos A, D, E, F, and G courtesy of Gary Kocurek.

directions? These are commonly not easy questions to answer, especially if the dunes are not in equilibrium with the modern wind regime. It is normally assumed that transverse dunes with crescentic, linguoid, and lunate crestlines in plan are formed under unidirectional wind regimes. However, longitudinal dunes have been interpreted as being formed in both unidirectional and bidirectional wind regimes. Rubin and

Hunter (1987) demonstrated experimentally that transverse, oblique, and longitudinal wind ripples can all be formed in a bidirectional flow regime by varying the angle between the two wind directions and their relative sand-transport rates. The wind used in the experiments was a relatively steady, unidirectional California sea breeze, the two wind directions were obtained by rotating the wind ripples on a board, and the different

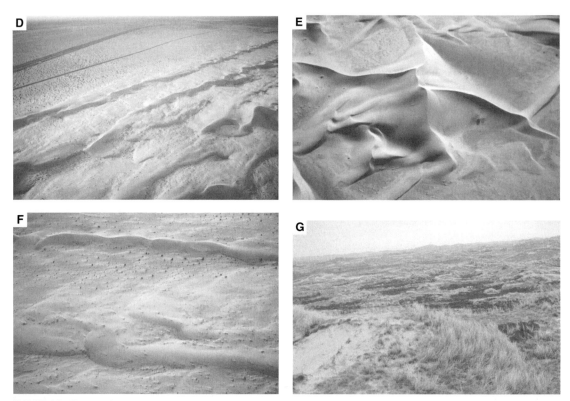

FIGURE 6.5. Continued.

sand-transport rates were obtained by varying the duration of a particular wind direction. The two wind directions were made to act on the ripples as they were forming, such that an equilibrium form under a single wind direction never existed. Whether or not these experiments are applicable to real dune formation is debatable.

Transverse dunes are asymmetrical in profile, with the stoss-side slope ranging from 2° to 10°, and the lee side at the angle of repose (Figure 6.5). Transverse dunes are commonly 50–1,000 m long and 2–60 m high. The mean height/length ratio is about 0.04 (range 0.02–0.08). The width of lunate dunes is up to about 300 m. Larger transverse forms up to several hundred meters high and kilometers long (called *draas* or *megadunes*) normally have smaller (typically transverse) dunes superimposed on them. Draas may, but need not, have downstream faces at the angle of repose. Crescentic dunes are the most common transverse dunes, and comprise around 40% of sand seas globally. Lunate dunes (barchans) occur where sand supply is limited, as is the case with lunate subaqueous bed forms, and they make up a small proportion of aeolian sand seas. Crescentic dunes pass laterally into lunate dunes at the margins of sand seas. If lunate dunes are subjected to bidirectional winds of unequal strength and/or duration, their downstream tips (wings) may be extended in the resultant flow direction, transforming the lunate dunes into more longitudinal-looking dunes (Figure 6.6). Linguoid (parabolic) dunes are common next to sandy beaches and in semi-arid regions. The trailing sides of these dunes may be stabilized by vegetation, while the central part actively migrates downwind, elongating the dune (Pye, 1993). The hollow upstream of the active dune is called a blowout, hence the term blowout dune.

The kinematics of most aeolian dunes are not known well. Transverse dunes migrate downwind by erosion of their stoss sides and deposition on their lee sides. Such migration is episodic due to changes in wind velocity and/or direction. Long-term migration rates for dunes up to 20 m high range from about 50 m per year for the smallest dunes to meters per year for the larger forms. The size and shape of dunes may change with changing wind speed and direction, but little is known of the details, or of how individual dunes change under steady-flow conditions. Dunes have been shown to develop from patches of sand by

A

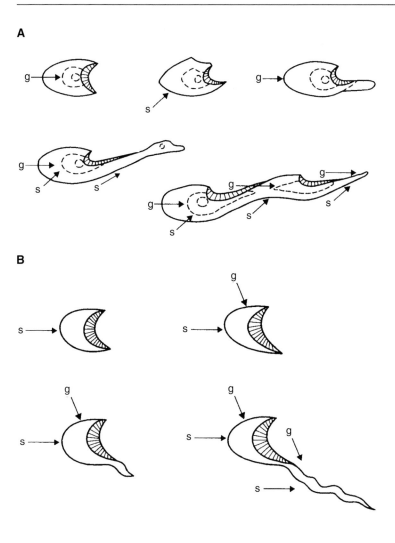

FIGURE 6.6. Transformation of barchan dunes to linear dunes according to (A) Bagnold (1954) and (B) Tsoar (1984); g = gentle winds, s = strong winds. From Lancaster (1995).

B

increasing height, length, and asymmetry (Kocurek *et al.*, 1992).

Transverse dunes generally form where there is low variability (<15°) in wind direction. It is likely, but not proven, that the lengths of transverse aeolian dunes are proportional to the thickness of the atmospheric boundary layer that existed when they formed; that is, on the order of hundreds or thousands of meters. Flow depth obviously does not severely limit the sizes of aeolian dunes as it does for subaqueous dunes. It may be that draas are large dunes formed under exceptional past flow conditions and that are in disequilibrium with modern flow conditions. As stated above, one reason for the lack of knowledge of the relationship between dune geometry and flow characteristics is that researchers find it difficult to make observations during the sand storms when dunes are active.

Longitudinal dunes are many kilometers long (commonly more than 20 km, hence the term *linear dunes*), meters to tens of meters high, and have spacing transverse to flow (width) of hundreds of meters to kilometers (Figure 6.5) (Lancaster, 1982, 1995; Tsoar, 1982, 1983, 1989). Their height/width ratio is 0.01–0.05. They comprise about 50% of all dunes in sand seas globally. Two varieties are commonly recognized: (1) straight dunes with rounded crests and partially vegetated sides; and (2) sinuous-crested dunes with sharp crests (looking sword-like in plan, hence the name *seif*). Crestlines of seif dunes are commonly discontinuous, or beaded. In cross section transverse to their elongation, these dunes are symmetrical to slightly asymmetrical. They commonly have a low-angle (4° to 14°) basal slope called a plinth or apron, and this side slope increases upwards to 20° and in places up to an angle of repose near the crest. The trough areas

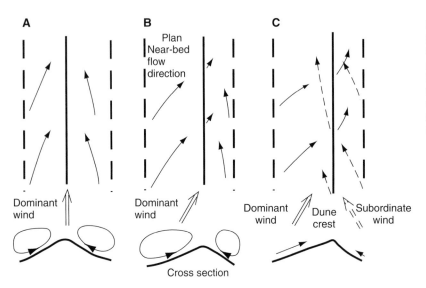

FIGURE 6.7. Hypothetical patterns of flow over linear dunes: (A) longitudinal vortices parallel to symmetrical dunes, (B) longitudinal vortices oblique to asymmetrical dunes, and (C) two wind-flow directions oblique to dunes, with unequal wind velocities.

between dunes are areas of deflation or sediment bypass and may be vegetated. Transverse dunes and wind ripples occur commonly on the flanks of large linear dunes, with their migration directions oriented at an oblique angle, or parallel, to the crest line.

Crestlines of linear dunes commonly shift position periodically, in some cases in a preferred direction. Rates of movement are on the order of meters per year. Cross-sectional asymmetry increases if the migration is persistently in one direction. The degree of cross-sectional asymmetry also changes seasonally as winds vary in speed and direction. The predominant direction of sand transport and direction is parallel to the length of the dunes, and the migration rate in this crest-parallel direction can be meters to tens of meters per year.

Several origins have been proposed for linear dunes: (1) transformation from lunate dunes (barchans) under two different wind directions (Figure 6.6); (2) formation due to longitudinal vortices associated with a single wind direction acting parallel, or nearly parallel, to dune crests; and (3) formation due to two different wind directions oblique to dune crests. An early explanation for the origin of longitudinal dunes (Bagnold, 1941; Tsoar, 1984) appealed to two different wind-flow directions of unequal strength at an acute angle to each other. One of the flow directions formed barchan dunes and the other one elongated the wings of the barchans. The long axis of the resulting beaded longitudinal dune bisected the angle between the two wind directions. Figure 6.6 shows that the theories of

Bagnold and Tsoar differ in the relative magnitudes and directions of the two different winds. The origin of longitudinal dunes has also been related to patterns of longitudinal vortices (Taylor–Gortler vortices) in which adjacent pairs have counter-rotating air motions (Figure 6.7) (Bagnold, 1941; Hanna, 1969; Wilson, 1972a, b; Cooke and Warren, 1973). The troughs are regions where near-bed flow diverges, and the crests are regions where near-bed flow converges. This theory has been challenged on the grounds that the expected widths of vortex pairs (2–6 km) are greater than the typical spacing of linear dunes (on the order of 1 km) and most linear dunes form in areas where two different wind directions occur (Pye and Tsoar, 1990; Lancaster, 1995; Bristow *et al.*, 2000, 2005). A theory for the formation of linear subaqueous bed forms (Huthnance, 1982a, b) predicts that bed forms will grow and remain stable if the angle between the crestline and the flow direction is approximately 20°. This theory predicts the mildly asymmetrical cross sections of linear dunes, their tendency to migrate laterally, and the orientation of superimposed dunes that move almost parallel to the steepest face and obliquely up the shallower face (Figure 6.7). Most modern workers think that linear dunes originate under two distinct wind directions of unequal strength acting at different times of the year (Figure 6.7) (Tsoar, 1983; Pye and Tsoar, 1990; Lancaster, 1995, 2005; Bristow *et al.*, 2000, 2005).

Pyramidal or star dunes are typically hundreds of meters across, and tens to hundreds of meters high

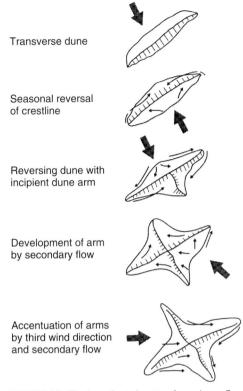

Transverse dune

Seasonal reversal
of crestline

Reversing dune with
incipient dune arm

Development of arm
by secondary flow

Accentuation of arms
by third wind direction
and secondary flow

FIGURE 6.8. The hypothetical origin of star dunes. From Lancaster (1995).

(Figure 6.5); see the reviews by Lancaster (1989, 1995). They represent about 8.5% of dune types in sand seas globally. They have a central peak from which radiate crestlines (typically three or four) that separate avalanche faces. These dunes have a low-angle plinth and tend to be surrounded by deflated areas. Multidirectional flow patterns are assumed to be responsible for pyramidal dunes (Figure 6.8), mostly in areas of multidirectional winds, but also in areas to the lee of land masses where flow separates periodically, giving erratic flow directions.

Zibar are low, nearly symmetrical features without avalanche faces that occur in areas between dunes. They are several hundred meters long and on the order of meters high. They might be incipient dunes associated with coarse sand.

Interdune areas may represent tens of percent of sand seas. Interdune areas may be deflationary (with erosional features, armored beds with faceted pebbles) or depositional. Depositional areas may be dry and contain sand sheets covered with wind ripples, zibar, and shadow dunes, or wet, with ephemeral or perennial lakes and evaporites (Chapter 18).

Internal structures of wind ripples, dunes, and draas

Observation of the internal structures of modern dunes and draas is very difficult because of the difficulty of digging large trenches in dry sand under challenging climatic conditions. Natural erosion can reveal large sections through dunes in some cases. However, ground-penetrating radar is being used increasingly to image the internal structure of dunes and draas (e.g., Schenk *et al.*, 1993; Bristow *et al.*, 1996, 2000, 2005). Interpretation of the origin of the observed stratification also poses problems, because the nature of the flow and sediment transport, bed-form geometry, and migration as the sedimentary structures were being produced are not normally known. The preservation potential of aeolian stratification has barely been investigated. This all means that much is to be learned about the generation of aeolian stratification, and interpretation of ancient aeolian strata must be treated with caution.

Sedimentary structures within sand ripples and granule ridges are mainly coarsening-upward planar strata (so-called transcurrent or translatent strata: Figure 6.9). Coarsening upwards is due to migration of the coarse-grained crest over the fine-grained trough. The rarity of cross strata is due to the relative lack of lee-side avalanching (Hunter, 1977; Rubin and Hunter, 1982). Strata associated with wind ripples occur within dunes, sand sheets, and interdune regions. The internal structure of zibar is low-angle ($<15°$) strata dipping in the flow direction, and formed by wind ripples migrating over their lee side (Nielson and Kocurek, 1986).

Periodic avalanching (grain flow) of sediment down the angle-of-repose lee side of aeolian dunes produces a lens of inclined sediment (a cross stratum) that thins towards its lateral and upslope edges, and coarsens outwards and downslope. The grain-size sorting is related to kinetic sieving. Relatively fine-grained sediment settling from suspension (grain fall) on the lee side of the dunes may be distinguished from the grain-flow deposits. However, at high sediment-transport rates, avalanching is more or less continuous, sorting by kinetic sieving is less effective, grain-fall sediment mixes more with grain-flow sediment, and cross strata are not as clearly size-sorted. Actually, voluminous grain-fall sediment settling on the lower lee slope may result in cross strata that coarsen upslope (Figure 6.10). It has been suggested (Hunter, 1977) that the alternation between relatively coarse grain-flow layers and

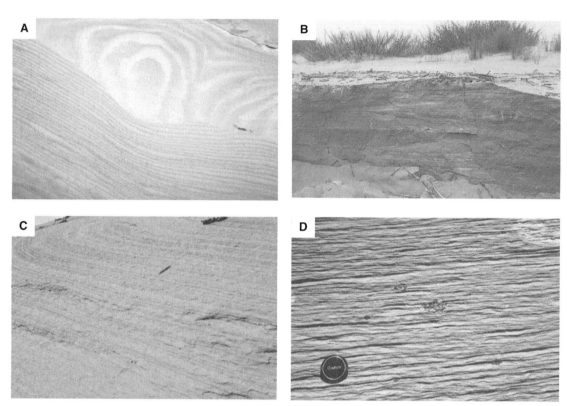

FIGURE 6.9. Internal structures of wind ripples and zibar. (A) Modern wind-ripple laminae in vertical and horizontal cross sections, from Padre Island. Dark parts of laminae are finer-grained than light parts. (B) Low-angle laminae associated with zibar on the backshore of a South Carolina beach. Images (C) and (D) show ancient wind-ripple laminae from Jurassic Page sandstone. In (C), the scale is a pen. Image (D) shows some ripple forms and low-angle cross strata. Photos (A), (C), and (D) courtesy of Gary Kocurek.

finer grain-fall layers is characteristic of aeolian cross strata. However, exactly the same processes and products occur with subaqueous dunes (Hunter, 1985); see Chapter 5. The influence of bed forms superimposed on dune stoss sides (e.g., wind ripples) on the formation of cross strata has not seriously been considered. However, it has been recognized that wind ripples may migrate over the lee side of the dune during and/ or after dune migration, and the associated wind-ripple lamination will be interstratified with the grain-flow and grain-fall deposits (Figure 6.10) (Kocurek, 1991, 1996). Rainfall on lee faces of dunes and subsequent slumping can produce soft-sediment deformation structures in aeolian cross strata.

The cross strata formed by transverse dunes dip downwind, and (by virtue of sinuous crestlines) occur in trough cross sets. Most observations of the cross strata formed by transverse dunes (e.g., McKee, 1966, 1979; Schenk *et al.*, 1993; Bristow *et al.*, 1996) find compound sets of cross strata that clearly indicate that the dune has grown by accretion of superimposed dunes as it migrated downwind (Figure 6.11). The

bases of the individual sets tend to be inclined downwind, and the sets also tend to thicken downwind. Only the smallest transverse dunes are composed of simple sets of cross strata. Individual sets also exhibit discordances related to the changing form and migration rate of the formative dune. Such changes might be related to changes in wind velocity and/or direction, but could also be due to normal dune kinematics under steady air flows. These kinds of features have exact parallels in subaqueous dunes. The various scales of bounding surfaces illustrated in Figure 6.11 were codified by Brookfield (1977) and by Kocurek (1988). Simple or compound cross sets may be bounded by interdune deposits. However, as seen below, such sets may be bounded by more regionally extensive surfaces related to changing water table and sand supply.

The internal structure of longitudinal (linear) dunes has traditionally been thought of as cross strata dipping away from the crest in opposite directions, therefore having a bimodal dip pattern (Figure 6.12A, B) (Bagnold, 1941; Tsoar, 1982). The cross strata vary in dip within different cross-cutting sets. Deposits at the

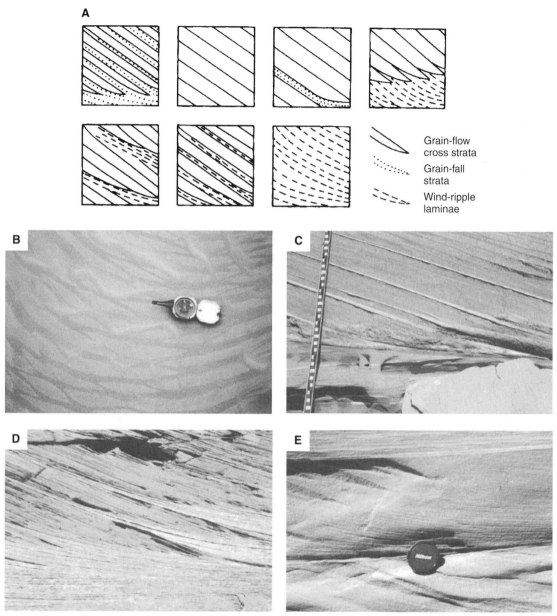

FIGURE 6.10. (A) Cross strata on dunes formed by grain flow, grain fall, and wind-ripple migration. After Kocurek (1991). (B) Modern grain-flow (toe) and grain-fall cross strata in horizontal section through a barchan dune, Padre Island. Grain-flow cross strata are lighter colored because they are drier than the grain-fall cross strata. (C) Grain-flow and grain-fall cross strata, and wind-ripple strata (light color) from the Jurassic Page sandstone. (D) Grain-flow cross strata in section parallel to flow, and basal wedges of wind-ripple strata (light color) associated with an oblique flow direction. (E) Inclined wind-ripple laminae at the base of a cross set with interstratified wedges of grain-flow cross strata (dark). Images (D) and (E) are from Jurassic Navajo Sandstone. All photos courtesy of Gary Kocurek.

base (plinth) of a longitudinal dune were assumed to be relatively fine-grained laminae at a relatively low angle, produced by grain fall and wind-ripple migration. Although such a bimodal pattern received limited support from field studies (e.g., McKee and Tibbitts, 1964), it could not be found readily in ancient deposits interpreted to be of aeolian origin. Other studies have shown that some of the inclined strata seen in cross sections through linear dunes are actually the end-on views of cross strata that dip downwind more or less parallel to dune faces, and must have been produced by smaller dunes migrating along the sides of the linear dune (Tsoar, 1982); see Figure 6.12C. It now appears that longitudinal dunes normally migrate laterally, producing cross strata dipping in the direction of lateral migration (Rubin and Hunter, 1985; Hesp *et al.*, 1989; Rubin, 1990; Bristow *et al.*, 2000, 2005) (Figure 6.12D, E, F). Some of these cross strata dip in a direction opposite to that of the dominant lateral migration, associated with the effects of a secondary wind direction (that produces crestline sinuosity; Figure 6.12E). If dunes are superimposed on the lee face of the linear dune, cross strata would be of the compound type, and the smaller-scale cross strata formed by superimposed dunes may dip approximately parallel to the crest line of the linear dune (Figure 6.12D, E), or down the lee slope of the linear dune (Figure 6.12F). The cross strata exhibit variations in set thickness and discordances that indicate changes in wind speed and direction (Figure 6.12F).

The internal structure of star dunes is poorly known, and is assumed to be cross strata dipping in different directions within trough cross sets (McKee, 1966, 1982). Basal parts of star dunes (plinths) are expected to be composed of fine-grained wind-ripple laminae or cross strata formed by small crescentic dunes. Needless to say, documented ancient examples are rare.

The internal structure of parabolic dunes is compound sets of trough cross strata with cross strata dipping mainly downstream. Vegetated coastal dunes adjacent to beaches tend to be blowout and star varieties, and have cross strata dipping in various directions. Basal parts may contain wind-ripple laminae

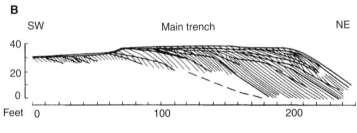

A

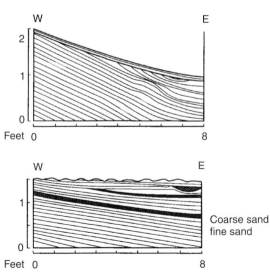

FIGURE 6.11. Cross strata within (A) barchan and (B) crescentic transverse dune. From McKee (1966). (C) Cross strata formed by transverse dunes, Navajo Sandstone, southern Utah. The wind direction is from left to right. The light-colored cross strata are formed by wind ripples. Photo courtesy of Gary Kocurek.

C

FIGURE 6.11. Continued.

A

B

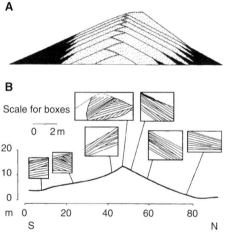

Scale for boxes

0   2m

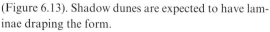

C

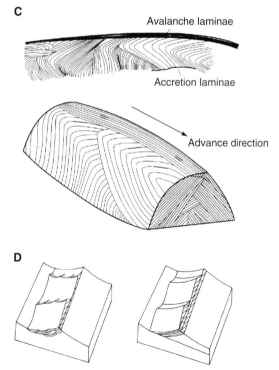

Avalanche laminae

Accretion laminae

Advance direction

FIGURE 6.12. Cross strata formed by linear dunes: (A) the hypothetical model of Bagnold (1954), and (B) modified from McKee and Tibbitts (1964) by Lancaster (1995). Note that exposures shown in boxes represent a very small, upper part of the dune. (C) A view of one side of a linear dune, and the model, modified from Tsoar (1982). (D) Hypothetical models of Rubin and Hunter (1985). (E) The model of Bristow et al. (2000) based on a GPR survey of a Namibian linear dune. (F) The GPR profile normal to the crestline of a Namibian linear dune, showing superimposed dunes and compound cross strata inclined from right to left. From Bristow et al. (2005).

D

(Figure 6.13). Shadow dunes are expected to have laminae draping the form.

The preservation of sets of aeolian cross strata has not been examined in the same detail as for subaqueous bedforms (Chapter 5). It is commonly implied that climbing of dunes of a fixed geometry is essential for preservation of cross strata (e.g., Rubin and Hunter, 1982; Rubin, 1987; Kocurek, 1988, 1996), and the importance of variability of dune height and migration rate is given short shrift (but see Kocurek

E

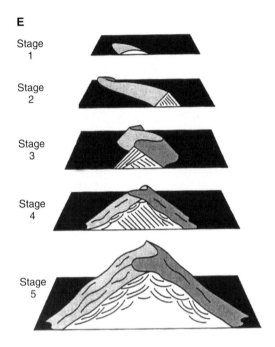

Stage
1

Stage
2

Stage
3

Stage
4

Stage
5

F

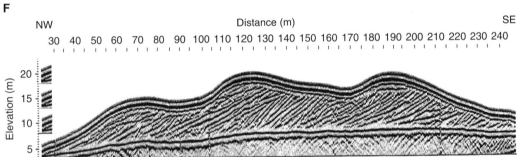

FIGURE 6.12. Continued.

(1996)). There is no theory for the relationship between variability of aeolian dune height and cross-set thickness as there is for subaqueous dunes (see Chapter 5). It is likely, as with subaqueous dunes, that it is only the lowest parts of the highest dunes that are preserved as cross strata. It is commonly stated that large cross-set thicknesses (up to tens of meters) are typical of wind-blown sands. However, similarly thick cross strata can be formed in deep subaqueous environments.

Dry interdune deposits are composed of wind-ripple laminae associated with sand sheets and zibar, and deposits of coppice dunes, whereas wet interdune deposits include desiccated (cracked) lacustrine muds and evaporites. Plenty of root casts and burrows are to be expected.

Regional (super) bounding surfaces of compound sets of aeolian strata are due to large-scale changes in sand seas, related to climate, eustasy (base level), or tectonism (Talbot, 1985; Kocurek, 1988). Such surfaces may be erosional, or may be overlain by interdune deposits containing soils, evaporites, and so on. These large-scale, long-term aspects of sand seas are discussed further in Chapter 16.

Erosional bed forms

Erosional bed forms range in scale from scours on the upstream sides of obstacles, to eroded protrusions of rock or sediment (yardangs) to deflated basins called pans (e.g., Greeley and Iversen, 1985; Goudie, 1999; Lancaster, 2005). Erosion of bedrock by wind is not

FIGURE 6.13. A view from backshore of trough cross strata and low-angle strata in a vegetated coastal dune. Wind ripples can be seen in the foreground.

FIGURE 6.14. Erosional structures. (A) Current-crescent scours and sand shadows around shells in wind-deflated backshore. (B) Sand shadows to the lee of boulders, and wind ripples in the backshore of a beach. (C) Yardangs cut into Pleistocene lacustrine deposits in the Dahkla area of Egypt. The wind direction is from left to right. Photo courtesy of Nick Lancaster.

significant in most areas (Lancaster, 2005). Current-crescent scours around shells and pebbles, and sand shadows on the downstream side, are common features of eroding sediment surfaces (Figure 6.14). The obstacle grain can become perched on a plinth above the general bed surface. Yardangs have the shape of an inverted boat hull, with the blunt end facing upwind and the tapered end facing downwind (Figure 6.14).

The blunt end facing upwind may be undercut. They can be 1–200 m high and a few meters to kilometers long. Their length/width ratio is commonly about 4. They are particularly common in fine-grained sedimentary rocks such as siltstones and volcanic ash. Yardangs are thought to be formed by abrasion (sandblasting), by deflation of weathered grains, and possibly by erosion by water. The streamlined shape apparently minimizes wind drag (Ward and Greeley, 1984).

Implications for humans

Strong winds associated with tornadoes and hurricanes have well-known harmful effects on life and property. The implications of wind transport of sediment can be good or bad. The widespread deposition of loess during the last Ice Age has had a very beneficial impact on cultivation. However, over-cultivation, over-grazing and deforestation can result in extensive soil erosion, especially when they occur in combination with droughts, as exemplified by the "dust bowl" of the Great Plains of the USA in the 1930s. This has stimulated the development of empirical equations for wind erosion of soil. Soil loss can be counteracted to some degree by increasing the resistance to erosion and decreasing the near-surface velocity of the wind. Increasing resistance to erosion can be accomplished by leaving plant stubble in the soil after harvesting, by ploughing soil to produce large clods, and by planting a vegetation cover that will not be cultivated. Decreasing the near-surface wind velocity is accomplished by increasing the boundary roughness using wind breaks, using tall cover vegetation, and by creating large furrows during ploughing.

Sand movement from deserts and coastal sand dunes can cause abrasion damage to crops and man-made structures, block roads and rivers, bury man-made structures, and increase flooding of back-beach areas during storms. Attempts to reduce such sand loss have involved planting or restoration of dune vegetation, erection of sand-trapping fences, surface stabilization with mulches (which are commonly environmentally unfriendly), and limitation of animal traffic (e.g., humans and their machines).

Excessive wind-blown dust in the atmosphere reduces air quality and can cause respiratory problems. Furthermore, it reduces solar insolation and affects radiation of heat within the atmosphere. For these reasons, sediment within the atmosphere is routinely monitored.

Ancient wind-deposited sands are relatively well sorted and, in the absence of excessive cementation, have relatively high porosity and permeability. Thus, aeolian sands are commonly important aquifers or hydrocarbon reservoirs (e.g., Permian of the North Sea region, and Jurassic of the southern and western USA).

7 Multidirectional water flow, sediment transport, erosion, and deposition

Introduction

Water currents that periodically reverse direction are associated with (1) water surface waves formed by the wind; (2) tidal waves caused by the gravitational attractions between the Earth, Moon, and Sun; and (3) tsunami waves caused by earthquakes, landslides, volcanic eruptions, and meteorite impacts. Temporal and spatial variation in the velocity and direction of these currents is complicated. These waves undergo fundamental changes as they move into shoaling water and towards land, further modifying the associated currents. These waves also normally act together in the sea, and it is common for unidirectional currents to be superimposed upon the wave currents. For example, the wind that acts on the water surface to produce wind waves also drags surface water in the general direction of the wind. Therefore, water currents associated with these various types of waves are truly multidirectional. Reviews of wind waves, tidal waves, and associated water currents and sediment transport are given by Defant (1958), J. R. L. Allen (1970, 1985), Komar (1976, 1998), Bowden (1983), Pond and Pickard (1983), Sleath (1984), Fredsoe and Deigaard (1992), Nielsen (1992), and P. A. Allen (1997). Reviews of tsunamis and their effects on the Earth include Iida and Iwasaki (1983), Mader (1988), Dawson (1994, 1999), Shiki *et al.* (2000), Bryant (2001), and Satake (2005). Wind waves and associated wind-drag currents will be discussed first, then tidal waves, and finally tsunami waves.

Definitions

The parameters and symbols used to represent wave geometry and movement are given in Table 7.1 and Figure 7.1. Water-surface waves can be classified using their period (the time taken for one wavelength to pass a point) and the forces responsible for forming them. *Wind waves* have periods ranging from a fraction of a second to 300 seconds (but mainly within the range 1–20 seconds) and are generated by the wind acting on the water surface. Wind waves are called *sea waves* when the wind is actively working on the water surface to generate waves, *swell waves* when they travel beyond the influence of the wind, and *surf waves* when they break on the shore. A further important classification of water-surface waves is based on their wavelength relative to undisturbed water depth:

$L/d < 2$	deep-water waves
$2 < L/d < 20$	intermediate-water waves
$L/d > 20$	shallow-water waves

Tidal waves have periods of about 12.5 hours (semidiurnal), 25 hours (diurnal), and longer. Tides are generated by gravitational forces and modified by the Coriolis force and friction. Since the wavelength of a semidiurnal tidal wave is half of the circumference of the Earth, and the mean oceanic depth is about 4 km, a tidal wave is a shallow-water wave. *Tsunami waves* result from isolated events such as submarine earthquakes, volcanic eruptions, and landslides. They have periods of minutes and lengths up to 500 km, and are shallow-water waves. Other kinds of water-surface waves include those formed when water is periodically driven against shorelines by the wind, temporarily increasing water level at the coast (e.g., storm surges at sea, or *seiches* in lakes: Chapter 14). Internal waves along density interfaces within water bodies related to spatial variations in salinity, temperature, or suspended sediment concentration are discussed in Chapters 14, 15, and 18.

Generation of wind waves

When sustained wind blows over the water surface, wind waves develop, with heights, wave lengths, and periods **213**

TABLE 7.1. Wave parameters

Parameter	Units	Symbol
Wave height	m	H
Water-surface elevation above a datum	m	h
Wave amplitude	m	A
Wave length	m	L
Wave-number ($=2\pi/\lambda$)	m^{-1}	k
Wave period	s	T
Radian frequency ($=2\pi/T$)	s^{-1}	σ
Wave celerity ($=\lambda/T$)	$m\ s^{-1}$	c
Undisturbed water depth	m	d
Orbital diameter of an Airy wave	m	d_o
Orbital velocity	$m\ s^{-1}$	u

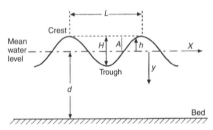

FIGURE 7.1. Definition of wave parameters

controlled by the wind velocity, duration of wind, and distance over which the wind blows (*fetch*). Exactly how wind waves are formed is not fully understood (Allen, 1985; Komar, 1998). Turbulent wind shear acting on the water surface drags water in the direction of the wind and also causes undulations in the water surface. According to the Bernouilli equation (Chapter 5), a wavy water surface causes differences in dynamic air pressure between the crest (low pressure) and the trough (high pressure), tending to increase the wave height. However, turbulent wind pressure acting normal to the water surface also influences wave geometry, and is necessary to create waves. The result is that a range of various wave sizes and speeds is formed in the generation area. In view of the incomplete understanding of wave generation, most attempts to predict wave parameters from wind velocity, duration, and fetch have been empirical (e.g., Carter, 1982; Komar, 1998). However, recent models for predicting wave parameters from wind data (e.g., NOAA's WAVEWATCH III: http://polar.ncep.noaa.gov/waves/) utilize theoretical models for wave generation and dispersion, and perform well

when the results are compared with data from buoys and satellite altimetry.

As sea waves leave the generation area, the waves with the longest periods travel the fastest (called *dispersion*: see Equation (7.5)). Also, viscous damping affects short-period waves preferentially. Therefore, complex sea waves become transformed into groups of waves with similar geometry and speed, and these regular swell waves can travel for thousands of kilometers across entire oceans. Waves generated by storms off Antarctica can reach the coast of Alaska. Changes that wind waves undergo as they move into shallow water and approach shorelines are described below.

Measurement and analysis of wind waves

Temporal variations in water-surface elevation associated with waves are recorded using a variety of techniques, involving hydrostatic pressure or acoustic sensors attached to the bed, electronic wave staffs on fixed platforms, buoys that measure water-surface slopes and accelerations, and reflection of radar or laser beams from satellites or aircraft. Wave speed and direction are calculated from pairs of measurements of water-surface elevation, and measured using current meters. Time series of wave parameters are likely to be complicated, because any one storm will generate a large variety of waves, and the waves arriving at a point may have been generated by different storms in different areas. It is normal to calculate the mean and variance of wave height, maximum wave height, and significant wave height (the mean height of the highest third of waves). There are many more complicated ways of analyzing time series (e.g., autocorrelation, spectral analysis: see the appendix). Simple statistical analysis of 40,000 observations of wave height (Komar, 1976) reveals that 45% of waves are less than 1.2 m high, 80% are less than 3.6 m high, and 90% are less than 6 m high. The largest wind wave ever recorded was about 34 m high. The reliability of direct observations of the largest waves can be questioned for obvious reasons. Indeed, larger waves have occurred, but most observers did not survive to report them. However, much more wave data has become available since the advent of remote sensing of the ocean surface. Most common wave periods are 1–10 seconds, and storm wave periods are typically 10–15 seconds.

Wave theories

Wave theories attempt to explain the relationship between the geometry and kinematics of simple

A

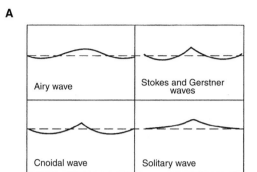

B

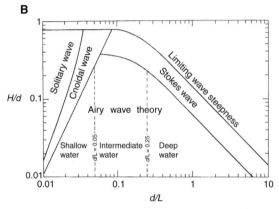

FIGURE 7.2. (A) Theoretical wave forms in profile. (B) Regions of applicability of wave theories. Modified from Komar (1998).

idealized waves and the water motions associated with them. These theoretical approaches involve simplifying the basic equations of motion (e.g., conservation of mass and momentum) in order to obtain analytical solutions. The more simplifications made, the simpler the theory. The four main theories and their ranges of applicability in terms of H/d and L/d are shown in Figure 7.2. The theories, in order of increasing mathematical complexity, are Airy, Stokes, Gerstner, and cnoidal. In addition to these analytical approaches to understanding wave mechanics, it is now possible to solve the equations of motion directly using numerical methods and fast, powerful computers. The Airy and Stokes wave theories are simple enough for this book, and give important insights into the nature of wave currents. These theories assume that the flow is non-viscous, irrotational, and two-dimensional, and that there are no superimposed unidirectional currents.

Airy's linear theory of oscillatory waves is called the small-amplitude theory, because it applies to waves with small amplitude (height) relative to depth (Figure 7.2). It is called a linear theory because it considers only the first-order terms in the series expansion

of the equations of motion. The idealized wave profile is sinusoidal, and given by

$$h = A \sin(kx - \sigma t) \tag{7.1}$$

where h is the water-surface elevation above or below the average, undisturbed water level at a given point in space and time, A is the wave amplitude (half of the wave height), $k = 2\pi/L$, where L is the wave length, x is the distance in the direction of wave motion, $\sigma = 2\pi/T$, where T is the wave period, and t is time (Table 7.1 and Figure 7.1). General equations for wave celerity (speed) and wavelength are

$$c^2 = (g/k)\tanh(kd) \tag{7.2}$$

$$c = (g/\sigma)\tanh(kd) \tag{7.3}$$

$$L = \left[gT^2/(2\pi)\right]\tanh(kd) \tag{7.4}$$

where tanh is hyperbolic tangent. For *deep-water waves*, kd is large, and $\tanh(kd)$ tends to 1 for large kd. Therefore, the celerity and wave length of deep-water waves are

$$c = \sqrt{(g/k)} = g/\sigma \tag{7.5}$$

$$L = gT^2/(2\pi) \tag{7.6}$$

Equation (7.5) is a wave-dispersion equation because it describes how waves with the largest wave length travel with the greatest speed, allowing waves of different lengths to be dispersed as they travel away from their generation region. For *shallow-water waves*, kd is small and $\tanh(kd)$ tends to kd, so

$$c = \sqrt{(gd)} \tag{7.7}$$

$$L = T\sqrt{(gd)} \tag{7.8}$$

With small-amplitude theory, water particles beneath a translating wave describe *closed orbits* (Figure 7.3), and the theory predicts the velocity of the water particles in these orbits. The fact that the water particles are not translated with the wave form can be appreciated by watching an object floating on the water surface as a wave passes. For *deep-water waves*, the orbits are circular, with diameter d_o equal to wave height H at the water surface. Since water particles complete one orbit in one wave period, the velocity of the orbital motions at the water surface u_s is $\pi H/T$. However, the orbit diameters decrease exponentially downwards for deep-water waves according to

$$d_o = H \exp(-ky) \tag{7.9}$$

A

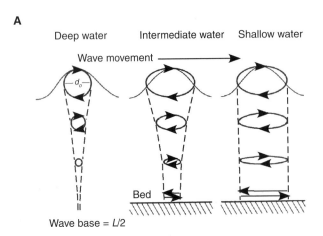

FIGURE 7.3. (A) Airy-wave theory: water-particle orbits for deep-, intermediate-, and shallow-water waves. (B) Water-particle trajectories for intermediate-water waves photographed over one wave period. Trajectories are almost circles at the upper surface, but are progressively flatter ellipses with decreasing long axis towards the bed. Trajectories are not completely closed. The wave amplitude is 0.04 times the wave length, and the depth is 0.22 times the wave length. From Van Dyke (1982).

B

where y is the distance below the undisturbed water level. At $y = L/2$, the orbital diameter is only 4% of its surface value, and the orbital velocity is considered to be negligible. This level is called *wave base*. Since $d > L/2$ for deep-water waves, the bed is always below wave base, and the bed does not interact with the orbital water motions. For intermediate-water waves, the orbits are ellipses with horizontal long axes that decrease in length with depth (Figure 7.3). For *shallow-water waves*, the orbits are ellipses that become flatter with depth but their long-axis lengths are constant (Figure 7.3).

The maximum near-bed horizontal velocity during an orbit is given as

$$u_{max} = \pi H / T \sinh(kd) \tag{7.10}$$

where sinh is hyperbolic sine. For deep-water waves, $\sinh(kd)$ increases as kd increases, such that u_{max} tends to zero, as expected. For shallow-water waves, $\sinh(kd)$ tends to kd as kd decreases. Therefore, the maximum near-bed orbital velocity is given as

$$u_{max} = Hc/(2d) = H\sqrt{(gd)}/(2d) \tag{7.11}$$

A shallow-water wave with a height of 0.4 m in water 5 m deep would have a maximum near-bed water velocity of about 0.3 m s^{-1}, easily enough to entrain fine-grained sand.

Directions of orbital water motion are related to the wave form (Figure 7.3). Water motion is in the direction of wave advance under the wave crest, but in the opposite direction under the trough. Upward components of water motion occur beneath the side of the wave facing the direction of wave advance, because the water level increases as the wave crest approaches. Downward components of water motion occur beneath the opposite side of the wave because the water level decreases as the trough approaches. As the wave form advances, so do the directions of water motion. In the case of intermediate-water waves, the near-bed water motions have a vertical component but are mainly horizontal and reversing in direction. Near-bed water motions for shallow-water waves are dominantly horizontal and reversing. The maximum near-bed horizontal water velocity in the direction of wave motion is underneath the wave crest, whereas that in the opposite direction is underneath the wave trough.

The implications of these near-bed flow patterns for sediment transport are given below.

Progressive waves have potential energy by virtue of the existence of crests and troughs, and kinetic energy by virtue of the motion of water beneath the waves. The total wave energy (kinetic and potential) per unit water-surface area is $0.5\rho g A^2$, and the wave power per unit width of wave crest is the product of wave energy per unit area and wave celerity.

Stokes' finite-amplitude theory takes into account second-order terms in the series expansion of the equations of motion, and the effects of finite wave height. This theory predicts that the orbits are not actually closed (Figure 7.2B) and there is a non-periodic drift of fluid in the direction of wave motion throughout the flow depth, due to the fact that orbital velocities under wave crests (in the direction of wave motion) are slightly greater than those under wave troughs (opposite to the direction of wave motion). As orbital diameters of deep-water waves decrease exponentially downwards, so do the drift velocities. This relationship is not applicable to waves in shallow water, where orbital diameters do not decrease downwards exponentially, and the effects of friction on net drift must be accounted for. Furthermore, if the waves are moving towards a solid boundary such as cliffs or a beach, continuity of water mass requires a return flow of equal mass. Theoretical flow-velocity distributions throughout the flow depth for Stokes non-periodic drift for various wave conditions (and including viscous effects; Longuet-Higgins (1953)) are shown in Figure 7.4. In all cases, there is a net drift at the bed in the direction of wave motion. However, there are many uncertainties about the veracity of these theoretical drift velocities. This is because prediction of drift velocities must also take into account the effects of drift

due to surface wind shear (e.g., geostrophic currents), and the coastal circulation system involving edge waves and rip currents (see below). Introduction of viscosity into Stokes' theory results in a bottom boundary layer, the thickness of which equals the square root of $\mu T /(\pi \rho)$. This oscillatory boundary layer is a few millimeters thick for large waves, and may be laminar or turbulent. The transition from laminar to turbulent occurs as the orbital diameter at the bed (and hence the water velocity) increases.

Another important result of Stokes' theory is a criterion for *wave breaking*. This occurs when the two faces of the wave make an angle of 120° at the crest, and the water velocity at the crest just equals the wave celerity. If the water velocity exceeds the wave speed, water overtakes the wave and the wave starts to break. The condition for breaking is

$$H/L = 0.142 \tanh(kd) \qquad (7.12)$$

which for shallow-water waves (small kd) leads to

$$H/L = 0.142(kd) \quad \text{or} \quad H = 0.89d \qquad (7.13)$$

Most waves would break when the wave height was less than this theoretical criterion. Measurements of H/d at wave-breaking range from 0.7 to 1.3, with larger values for steeper bed slopes (Paul Komar, personal communication).

Wind waves in shoaling water

Changes in wave form and speed

As wind waves move into progressively shallower water, wave celerity and wave length decrease (Equations (7.7) and (7.8)), wave height increases, the crests become more peaked while the troughs broaden (trochoidal shape), the waves become progressively more asymmetrical in profile, and they ultimately become oversteepened and break (Figure 7.5). In the absence of geostrophic currents and rip currents (see below), the magnitude and degree of asymmetry of near-bed currents (due to non-periodic drift) increase up to the point of breaking as the water depth decreases and wave height increases (Figure 7.5). The nature of wave breaking depends on the bed steepness (Figure 7.6). For low-angle beds, water gradually spills down the wave front and the wave height gradually decreases (a *spilling breaker*). As the bed steepness increases, a coherent mass of water from the crest plunges downwards, enclosing a tube or barrel (surfer

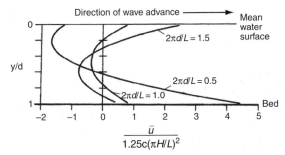

FIGURE 7.4. The theoretical vertical variation of the time-averaged drift velocity (dimensionless) in a laboratory channel for Stokes waves of various depth/wave length ratios. Modified from Longuet-Higgins (1953) by Komar (1998).

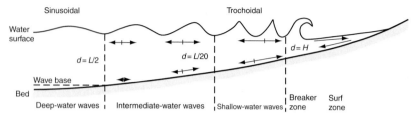

FIGURE 7.5. Waves and associated water currents in shoaling water. Waves change from long, low, symmetrical, and sinusoidal to short, high, asymmetrical, and trochoidal as the depth decreases. Arrows indicate the maximum orbital velocity under a wave trough or crest. Just above wave base, the maximum near-bed orbital velocity is small and of the same magnitude in either direction (Stokes drift is minimal). As waves change with decreasing depth, the near-bed orbital velocity increases, and the onshore velocity under the wave crest increases relative to the offshore velocity under the trough (due to Stokes drift). Near the water surface, the water velocity under wave crests and troughs increases onshore as waves increase in height. Differences between onshore- and offshore-directed flow velocities are due to Stokes drift. Arrows in the surf zone indicate swash and backwash.

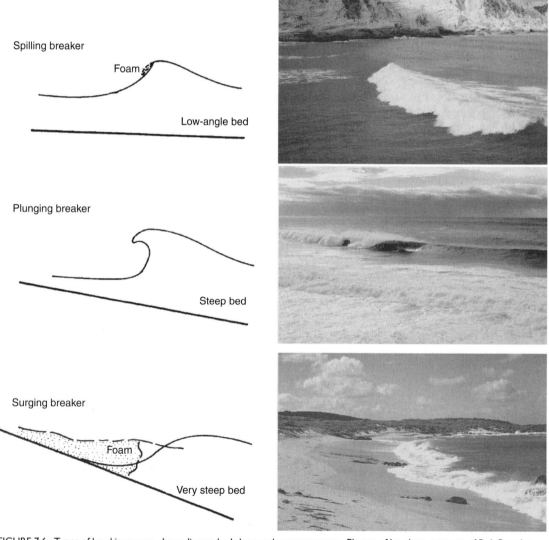

FIGURE 7.6. Types of breaking waves, depending on bed slope and wave steepness. Photos of breakers courtesy of Rob Brander.

term) of air (a *plunging breaker*). With very steep beds, the base of the wave surges up the beach before the wave can break (a *surging breaker*). The *surf zone* is landward of the *breaker zone*, the rush of water up the beach following wave breaking is called the *swash*, and the return flow is called the *backwash*. The water velocity in the surf zone following wave breaking is directly proportional to the maximum near-bed orbital velocity at the point of breaking. The relatively elevated water surface in the surf and swash zones (Figure 7.5) is called the *wave set-up*, and is due to the momentum flux of the breaking waves.

Wave reflection, refraction, and diffraction

Waves can be reflected, refracted, or diffracted as they approach a shoreline (Figure 7.7), and these phenomena can be treated in a similar way to light waves. Waves are reflected from cliffs and steep beaches, with the angle of reflection equal to the angle of incidence. The incident and reflected waves interact to give a pattern of distinct wave crests in different directions. If the angle of incidence is 0° (wave crests parallel to the shoreline), the incident and reflected waves interact to form *standing waves*. The profile of a standing Airy wave can be obtained by adding the wave-profile equations (e.g., Equation (7.1)) for waves of a given geometry moving in opposite directions, resulting in

$$h = 2A\sin(kx)\sin(\sigma t) \qquad (7.14)$$

The amplitude of the standing wave is double that of the individual waves. The streamline pattern for standing waves (Figure 7.8) shows that the near-bed flow direction reverses each wave period, but the maximum horizontal velocity occurs at the points where the water level does not change (*nodes*). The points that experience the maximum range of water level (*antinodes*) are associated with vertical (up and down) water motions only. Net drift associated with standing waves (Figure 7.8) gives rise to a convergence of near-bed flow within the boundary layer towards the nodes.

Wave refraction can be treated using Snell's law, in which the ratio of the sine of the angle between the wave crest and the topographic contours to the wave celerity is constant. Therefore, because the wave celerity decreases as depth decreases (Equation (7.7)), the angle that wave crests make with topographic contours also decreases with depth (Figure 7.7). This causes wave crests to become progressively closer to parallel to linear shorelines, to bend towards headlands, and to

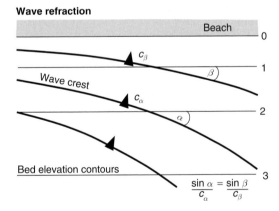

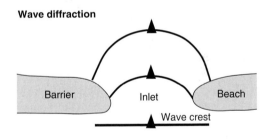

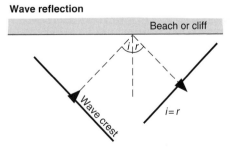

FIGURE 7.7. Explanation of wave reflection, refraction, and diffraction. A photo of wave-crest curvature by refraction in a New Zealand bay. From Hyndman and Hyndman (2006).

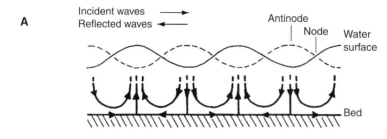

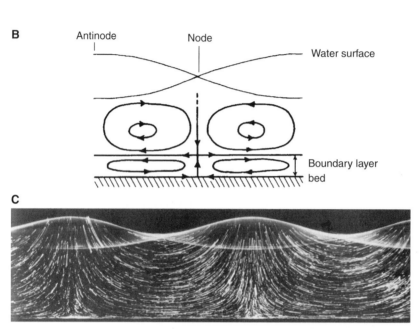

FIGURE 7.8. Standing waves:
(A) oscillatory currents and (B) net
drift (modified from Allen (1985)).
(C) Water-particle trajectories under
standing waves. From Van Dyke (1982).

bend away from submarine channels (Figure 7.7). Typical patterns of wave diffraction are shown in Figure 7.7.

Rip currents and nearshore circulation

Rip currents are localized unidirectional currents that originate in the breaker zone and flow offshore for tens to hundreds of meters (Figure 7.9). These currents are meters wide, and may be channelized near the breaker zone, where flow velocities may be on the order of several m s^{-1}. They spread laterally, decelerate, and dissipate offshore. The velocity and distance of off-shore water movement increase with incoming wave height. Rip currents are spaced regularly along the shoreline, with a mean spacing of tens to hundreds of meters. They occur where breaking waves have relatively low height. Breaking waves between rip currents have relatively large height. This difference in

breaking-wave height results in alongshore variation in water level in the surf zone (*wave set-up*), resulting in alongshore water flow towards the sources of the rip currents (Figure 7.9). This water flow in the surf zone follows a zig-zag path due to swash and backwash. Therefore, there is a pattern of water circulation in the near-shore zone that involves net shoreward drift associated with shoaling waves, alongshore drift in the surf zone, and offshore motion in rip currents (Shepard and Inman, 1950).

In the case of oblique wave approach, the pattern of near-shore circulation is modified (Figure 7.9). Rip currents are directed obliquely offshore, and the long-shore currents in the surf zone become unidirectional but spatially periodic in velocity. As the angle of wave approach increases, rip currents cease to exist, and wave-generated currents in the surf zone are dominantly alongshore. The velocity of these longshore currents increases with the angle of wave approach and the

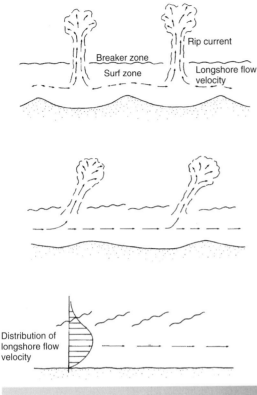

FIGURE 7.9. Nearshore water circulation involving net drift below incoming waves, longshore currents in the surf zone, and rip currents. The angle of wave approach increases from the top diagram to the bottom diagram. The length of arrows is proportional to the flow velocity. Modified from Komar (1998). The photo (courtesy of Rob Brander) shows rip currents (r) cutting through the breaker zone.

maximum orbital velocity at breaking (which increases with breaker height), and can reach several $m\,s^{-1}$. Net alongshore drift gives rise to shoreline features such as spits (discussed in Chapter 15).

This alongshore variation in wave set-up, nearshore water circulation, and specifically the regular spacing of rip currents probably owes its origin to standing *edge waves*, with their crestlines approximately normal to the shoreline (e.g., Bowen, 1969; Bowen and Inman, 1969; Huntley and Bowen, 1973; Guza and Davis, 1974; Guza and Bowen, 1975; Huntley *et al.*, 1981; Holman, 1983; Oltman-Shay and Guza, 1987). Progressive edge waves are due to refraction of oblique reflected waves back towards the shoreline, and this process requires a relatively steep bed slope and/or adjacent cliffs. The amplitude of these waves decreases exponentially offshore. The wave length of edge waves is proportional to the square of the wave period, and also increases slightly with bed slope. The period of edge waves can be the same as that of the incoming waves, but edge waves can also have a range of different lengths and periods. So-called subharmonic edge waves have periods that are twice those of the incident waves. It is quite possible that edge waves of the same period can be moving in opposite directions, thus producing standing edge waves. When the crest of a standing edge wave coincides with the crest of a breaking wave, the combined wave will have a large height, and the water level in the surf zone after breaking will be relatively high (Figure 7.10). The coincidence of this wave crest and the trough of the edge wave will give rise to low waves and low water level in the surf zone. Rip currents occur where incoming wave crests coincide with edge-wave troughs, and the spacing of rip currents equals the edge-wave length.

If rip currents form channels, the increased water depth over the channel will result in wave refraction (Figure 7.7), resulting in divergence of wave paths away from the channel and reinforcing the distribution of high and low waves due to edge waves. The near-shore circulation associated with edge waves also probably gives rise to regularly spaced ridges and troughs on beaches, called *beach cusps* (discussed in Chapter 15).

Wind shear and geostrophic currents

Wind acting on the water surface of the ocean and large lakes not only forms waves: it also drags surface water (up to a depth of about 100 m) in the general direction of the wind. The interaction of these wind-drag or wind-shear currents with wind-wave currents results in a complex current pattern, especially near coasts and on continental shelves. Therefore, it is difficult to understand sediment transport, erosion, and deposition by wind waves without also considering the associated wind-shear currents. Wind-shear currents in major oceans follow the major patterns of wind circulation (associated with polar easterlies, mid-latitude

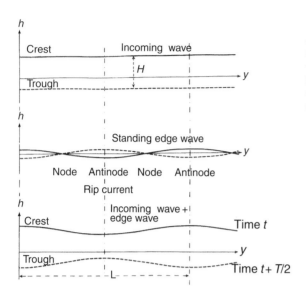

FIGURE 7.10. Addition of an incoming wave and a standing edge wave at the breaker zone, giving longshore variation in breaker height and wave set-up. The breaker height is greatest where the edge wave and incoming wave are in phase, and lowest where they are 180° out of phase. Rip currents occur where the breaker height and wave set-up are lowest. Modified from Komar (1998).

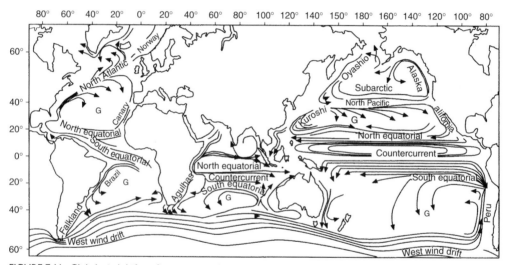

FIGURE 7.11. Global wind-drift surface currents, drawn as streamlines such that closer streamlines indicate stronger currents; G indicates subtropical gyre. From Leeder (1999).

westerlies, and tropical trade winds: Figure 7.11). These currents are deflected by the Coriolis force and by continents, giving rise to ocean-scale circulations that are clockwise in the northern hemisphere and anticlockwise in the southern hemisphere. An example is the Gulf Stream that transports warm Caribbean water to the shores of western Europe, having a major impact on the climate.

Wind-shear current velocities vary from a few centimeters per second to meters per second. Current velocity and direction depend on the velocity of the wind, the distance below the water surface, the depth of water, the latitude, and the proximity to land masses.

Current velocity decreases downwards in the water body (a surface boundary layer exists) because of the viscosity of water. Surface-water currents are deflected relative to the wind direction by the Coriolis force, which is given by

$$F_c = m2V\Omega \sin \phi \qquad (7.15)$$

where m is mass, V is the flow velocity, Ω is the angular velocity of the Earth's rotation (equal to 2π divided by 23 hours and 56 minutes, or 7.23×10^{-5} radians per second), and ϕ is latitude. Progressive deflection of water currents by the Coriolis force with increasing depth gives rise to so-called *Ekman spirals* (Figure 7.12). Ekman

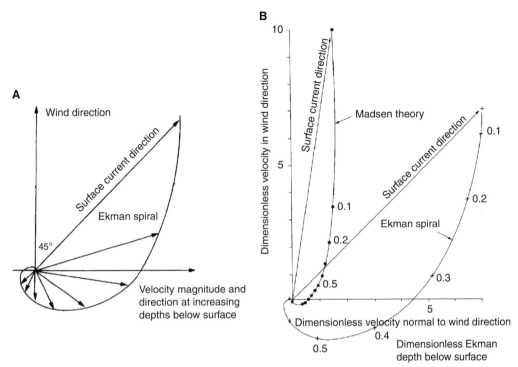

FIGURE 7.12. (A) A plan view of an Ekman spiral (for the northern hemisphere), showing how the magnitude and direction of velocity change with distance below the water surface. (B) A comparison of the variations of orthogonal velocity magnitudes with the distance below the water surface as predicted by Ekman and Madsen theories. Modified from Allen (1997).

spirals are predicted from the balance among wind shear, fluid friction, and the Coriolis force, assuming that there is no water-surface slope and no friction at the bed, and that the wind is steady and prolonged. This theoretical force balance gives expressions for the depth variation of flow velocity and direction, and for the depth at which currents become negligible (the *Ekman depth*, which is also the point at which there is a small velocity opposite to the surface velocity). Ekman theory predicts that the surface-water current should make an angle of 45° to the surface wind direction, and that the current velocity decreases exponentially downwards from the water surface.

Observations do not agree well with Ekman theory (Madsen, 1977; Keen and Glenn, 1994; Allen, 1997). The angle between the directions of the wind and surface currents is commonly less than 10° (Figure 7.12), and the surface-water currents are commonly about 3% of the wind velocity measured 10 m above the water surface, i.e.

$$V_s \approx 25u_* = 25(\rho_a/\rho_w)w_* \qquad (7.16)$$

where V_s is the surface current velocity, u_* is the water shear velocity, w_* is the wind shear velocity, and

$\rho_a/\rho_w = 0.035$ is the ratio of the densities of air and water. The wind shear velocity is given as

$$w_* = W\sqrt{C_d} \qquad (7.17)$$

where W is the wind velocity at 10 m above the water surface and C_d is a drag coefficient that depends on the wind velocity, and is on the order of 10^{-3}. The thickness of the Ekman layer is given by

$$L_E = 96w_*/\sin\phi \qquad (7.18)$$

Therefore, the velocity of a surface-water current and the thickness of the Ekman layer, for a wind velocity of 10 m s^{-1} at latitude 65°, are, using Equations (7.16)–(7.18) and $C_d = 0.0014$, 0.33 m s^{-1} and 39.6 m, respectively. In Madsen's analysis, the velocity decreases with depth more rapidly than predicted by Ekman (Figure 7.12). Actually, surface currents will normally be due to both wind shear and wave-induced Stokes drift, and the latter may be up to half of the velocity of the surface current. Also, bed friction greatly modifies Ekman spirals in shallow water (above the Ekman depth), as discussed below.

Wind-shear currents commonly result in changes in the elevation of the water surface, one example being

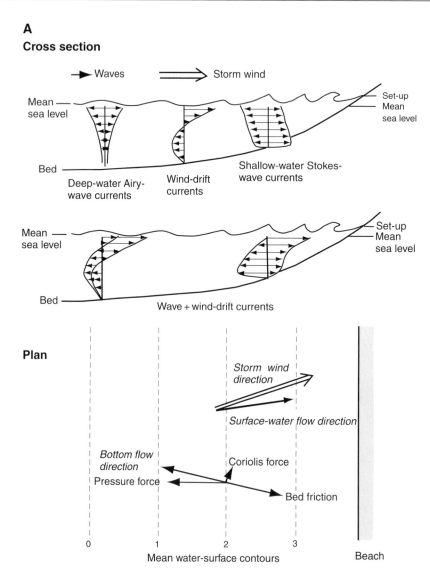

FIGURE 7.13. Idealized combined wave currents and wind-drift currents near a coastline in the northern hemisphere for (A) onshore wind, (B) offshore wind, and (C) alongshore wind. Cross-sections show wind-drift currents and the maximum horizontal component of wave-induced currents under wave troughs or crests, and combined currents. The length of arrows is proportional to the water velocity. Plans show water-surface contours and directions of wind-drift currents at various levels in the flow. The directions of wind-drift currents are determined by the direction and magnitude of the pressure (arising from the water-surface slope), friction, and Coriolis forces. Bottom currents are strongly influenced by bed friction.

when there is a component of water flow towards a land mass. A wind-shear current will obviously have an onshore component if the wind has an onshore component. However, an alongshore wind may also produce an onshore surface flow because of the Coriolis deflection of the surface water. The obstruction to flow causes a rise in water level at the coast, called a *wind set-up* or a *storm tide* (Figure 7.13), and an offshore-directed water-surface slope. The wind set-up at the coast may be calculated by balancing the pressure gradient arising from the sloping water surface, the wind shear stress, and the bed shear stress arising from water motion. This force balance is given as

$$\tau_w - \tau_b = \rho g(d + e)S \qquad (7.19)$$

where τ_w and τ_b are the wind shear stress and bed shear stress, respectively, ρ is the fluid density, g is the gravitational acceleration, $d + e$ is the sum of the water depth and super-elevation of the water surface, and S is the water-surface slope in the direction of the shear stresses. If the super-elevation e is small relative to the water depth, Equation (7.19) can be simplified to

$$\tau_w = C\rho g dS \qquad (7.20)$$

where C varies from 1 to 1.5. The set-up due to wind shear may be added to by the breaking of large waves (wave set-up), and by the effects of low atmospheric pressure (as in the center of a depression). The set-up may be several meters above normal mean sea level

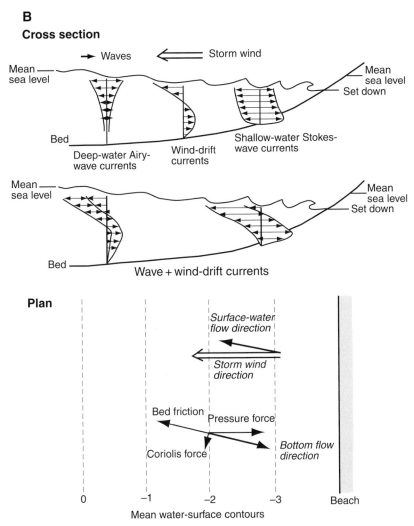

B

Cross section

Deep-water Airy-wave currents
Wind-drift currents
Shallow-water Stokes-wave currents

Wave + wind-drift currents

Plan

Surface-water flow direction

Storm wind direction

Bed friction
Pressure force

Coriolis force

Bottom flow direction

0 −1 −2 −3 Beach

Mean water-surface contours

FIGURE 7.13. Continued.

during major storms. Large set-ups combined with high tides can produce devastating flooding of coastal regions. If a wind-shear current has an offshore component, there is a *set-down* at the coast, and an onshore-directed water-surface slope.

Continuity of water mass requires that the onshore movement of water in wind-shear currents (plus wave-induced drift) is balanced by an offshore movement of the same mass of fluid. This offshore movement is in the form of a unidirectional bottom current that flows down the pressure gradient. During major storms, such bottom currents can reach velocities of meters per second, decreasing in velocity offshore (due to increasing depth and decreasing water-surface slope). Such strong currents can easily cut channels in the sea bed. Offshore-directed return flows that are normal to the

shoreline at the coast may be subsequently deflected by the Coriolis force, until the Coriolis force is exactly balanced by the pressure gradient (Figure 7.13), assuming that bed friction is ignored. Currents caused by pressure gradients and turned by the Coriolis force in the absence of bed friction are called *geostrophic currents*. Geostrophic means Earth-turned. However, the near-bed, offshore-directed flows must be influenced by bed friction as discussed below.

The influence of frictional resistance at the bed on wind-drift currents must be considered in shallow water, above the Ekman depth. A reduction in wind-shear flow velocity near the bed due to bed friction results in a reduction in the Coriolis force, and this results in a reduction in the angular deviation of the bottom flow from the surface flow. In very shallow

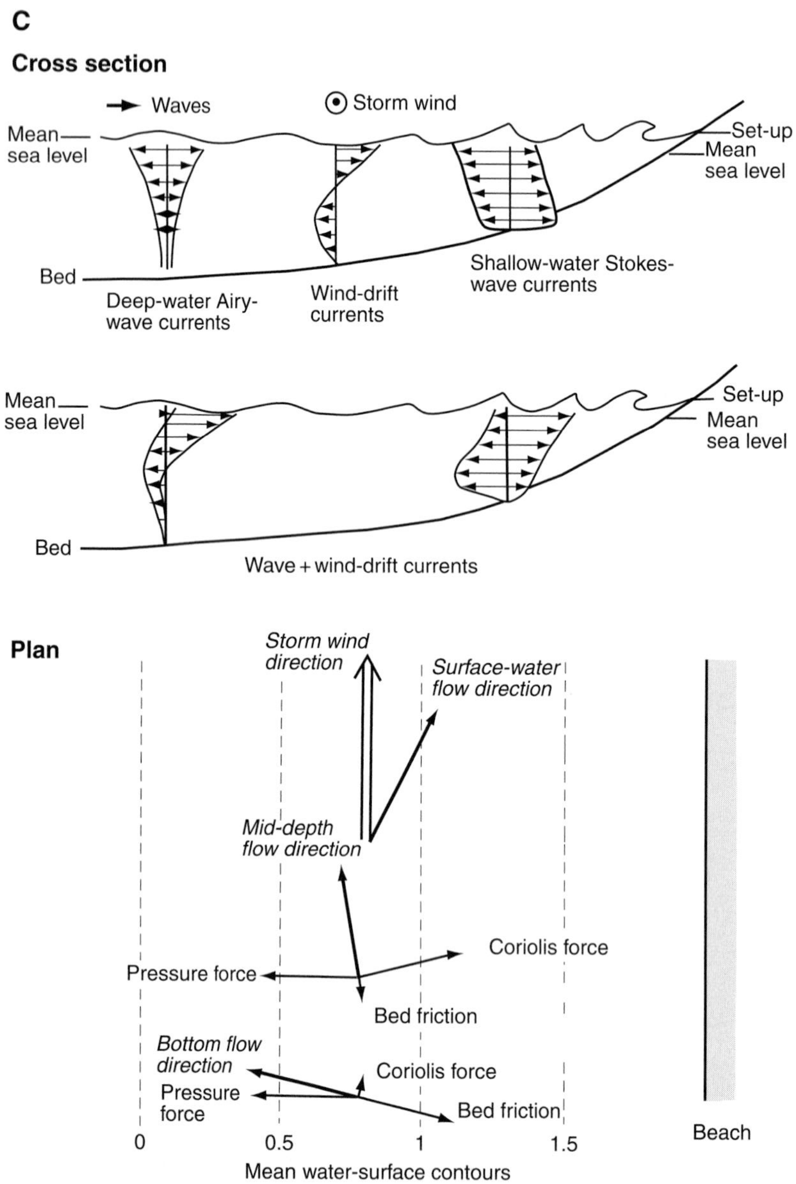

C

Cross section

FIGURE 7.13. Continued.

water (0.1 times the Ekman depth, say), the upper Ekman spiral and the bottom layer interact such that the net fluid transport is close to the wind direction. In the case of offshore-directed return flows, the lower regions should always be strongly influenced by bed friction, resulting in a bottom-flow direction very close to the water-surface slope, which is normal to the shoreline (Figure 7.13): compare with Swift *et al.* (1986) and Duke *et al.* (1991).

Wind-shear and geostrophic currents can greatly modify the currents associated with waves (oscillatory

motion, net drift, rip currents, longshore currents), producing combined, multidirectional currents. It is therefore probable that current patterns in shallow seas will be very complicated during major storms, when most erosional and depositional activity occurs. Figure 7.13 shows idealized, near-coastal, combined wind-shear and wave-induced currents for various wind directions for the northern hemisphere. The exact nature of these combined currents will depend on the relative magnitudes and directions of wind-shear and wave-induced currents, which are dependent upon wind magnitude and direction, swell-wave characteristics, sea-bed top-ography, and so on. In the case of an onshore wind-shear component, the bottom flow varies in velocity with wave period but is mainly offshore, and relatively weak onshore flow components occur only in shallow water. Surface flow is mainly onshore. Such offshore-directed, spatially decelerating bottom currents are expected to erode beaches and deposit bed-load sedi-ment offshore. In addition, high water levels at beaches may result in washovers (Chapter 15) and deposition landwards of the beach as washover fans. Wind-shear currents with an offshore component and a set-down at the coast have near-bed return currents directed onshore (Figure 7.13). This causes upwelling at the coast. When combined with wave-induced currents, near-bed cur-rents vary in magnitude with wave period, but are mainly directed onshore, with weak offshore currents only in shallow water. Surface currents are dominantly offshore. Such onshore accelerating bottom currents are expected to erode the sea bed and deposit bed-load sedi-ment on the beach, while suspended sediment is trans-ported and deposited offshore. It appears that oscillatory bottom currents occur near coasts only when wind-shear currents are negligible.

Thermohaline currents

Thermohaline currents result from horizontal pressure gradients associated with lateral variations in water temperature and/or salinity (and hence water density). These currents are discussed in Chapters 14, 16, and 18.

Sediment transport by wind waves and associated currents

The threshold of entrainment of cohesionless sediment

The threshold of entrainment of cohesionless sediment by oscillatory wave currents can be treated in essen-tially the same way as for unidirectional currents, even though wave currents are reversing and there is not a persistent turbulent boundary layer like that in unidir-ectional flows. The bed shear stress must be expressed in terms of the maximum near-bed orbital velocity using

$$\tau_o = f_w \rho u_{max}^2 / 2 \qquad (7.21)$$

which requires knowledge of the friction coefficient, f_w. The dimensionless bed shear stress and grain Reynolds number can then be used to predict the threshold of

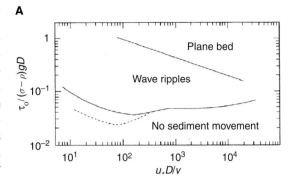

A

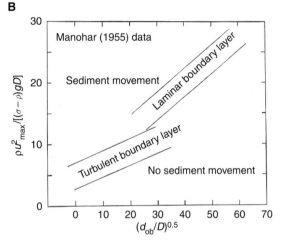

B

FIGURE 7.14. The threshold of sediment motion under waves based on (A) the dimensionless bed shear stress and grain Reynolds number (modified from Komar and Miller (1975a, b), and (B) the dimensionless maximum near-bed orbital velocity and dimensionless wave orbital diameter (modified from Komar and Miller (1973)). Envelopes of data (from Manohar) are shown for laminar and turbulent boundary layers. The boundary between laminar and turbulent boundary layers is approximately 0.5 mm at the threshold of motion.

motion, as for unidirectional flows (Figure 7.14A). An alternative approach (Komar and Miller, 1973, 1975a; Dingler, 1979; Allen, 1985) is to use a slightly different form of the dimensionless bed shear stress and the ratio of the near-bed orbital diameter, d_{ob}, and mean grain size:

$$\rho u_{max}^2 / [(\sigma - \rho)gD] \quad \text{and} \quad d_{ob}/D \qquad (7.22)$$

The form of this relationship depends on the magni-tude of d_{ob}/D, which controls whether the boundary layer is laminar (large values) or turbulent (small val-ues) (Figure 7.14B). The transition from laminar to turbulent boundary layer at the threshold of motion occurs at a grain diameter of approximately 0.5 mm. Since u_{max} is a function of the near-bed orbital diameter and wave period for Airy waves, the threshold of entrainment can be expressed as a function of u_{max}, D, and T for constant sediment density, fluid density, and viscosity. Large storm waves with periods between 10 and 15 seconds are capable of entraining sand in water depths of 100–200 m. However, Airy wave theory is approximate and inappropriate in shallow water. The threshold of motion calculated as above is for mean flow conditions and mean grain size only. Furthermore, it is likely that combined oscillatory and unidirectional currents will be operating at the threshold of grain motion in most natural situations.

Modes and rates of sediment transport

Sand and gravel are transported mainly as bed load, and silt and clay travel mainly in suspension. There have been many more studies of suspended-sediment transport than of bed-load transport by wave currents. Upward-directed water motions associated with gravity waves are known to suspend sediment. The near-bed concentration of suspended sediment increases with orbital velocity, and is dependent on the upward advection of suspended sediment associated with the vortices developed to the lee of ripple crests. Shear turbulence associated with superimposed unidirectional currents can also suspend sediment. However, the intensity of vertical turbulent motions and suspended sediment concentration is much greater in combined flows than in the component flows (Figure 7.15) (Kemp and Simons, 1982; Murray *et al.*, 1991; Osborne and Greenwood, 1993). The concentration of suspended sediment decreases exponentially upwards from the bed both in unidirectional currents and in wave-induced currents.

If there is symmetrical oscillatory water motion affecting the bed, as is the case just above wave base, no net bed-load transport occurs (Figure 7.16). Asymmetrical oscillatory motion near the bed, as is the case for intermediate- and shallow-water waves, produces net bed-load transport in the direction of wave motion as long as the offshore bed slope is not excessive (Figure 7.16). The addition of a unidirectional current to oscillatory currents has a multiplicative effect on the sediment-transport rate rather than an additive effect (e.g., Bagnold, 1963; Grant and Madsen, 1979; Vincent *et al.*, 1982; Green *et al.*, 1990; Murray *et al.*, 1991; Osborne and Greenwood, 1993); see Figures 7.15 and 7.16. This is because the sediment has already been put in motion by the oscillatory currents, and so the unidirectional current can more

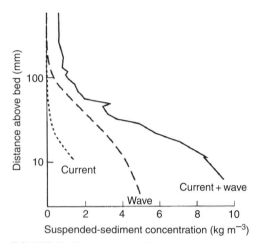

FIGURE 7.15. Experimental vertical variation of the time-averaged suspended sediment concentration for unidirectional current, wave currents, and combined unidirectional and wave currents. Modified from Leeder (1999).

easily transport the sediment. The sediment-transport rate is proportional to the cube of the resultant magnitude of u_{max}. The mean sediment-transport rate of longshore currents associated with oblique wave approach is proportional to the cube of the mean velocity of the longshore current (which is proportional to the maximum near-bed orbital velocity at wave breaking) and also increases with the angle of wave approach. The longshore sediment-transport rate is also proportional to the product of breaker height squared and mean longshore current velocity (Komar, 1998).

Sediment sorting during transport

Since u_{max} increases as water depth decreases and as wave height increases, so also should the mean grain size of the bed load and the suspended load (cf. Chapter 5). The mean grain size of bed load should therefore increase at any point on the sea bed during storms (compared with under fair-weather conditions), and increase from offshore towards the breaker zone. This is actually observed. However, relatively coarse-grained bed sediment occurs on parts of continental shelves where such sediment is not expected to be moved under present-day storm waves. This *relict* sediment was deposited during the last ice age when the sea level was much lower, and most of the continental shelf was a coastal plain traversed by rivers.

As the mean grain size of bed-load sediment increases, the standard deviation of grain size also

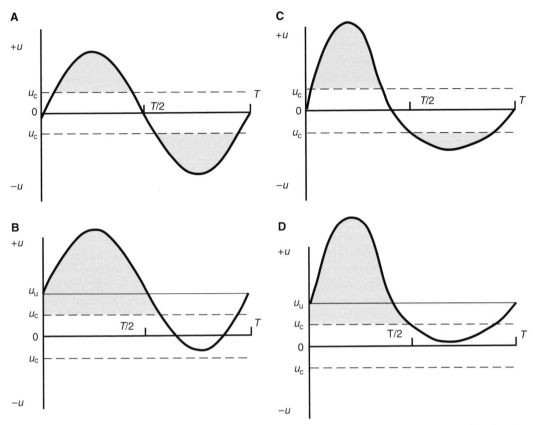

FIGURE 7.16. Hypothetical sediment transport by symmetrical and asymmetrical oscillatory wave currents and combined wave and unidirectional currents. (A) Variation over a wave period, T, of the horizontal component of the wave orbital velocity in the direction of wave motion, $+u$, and in the opposite direction, $-u$, for the case of a symmetrical, sinusoidal Airy wave. The threshold of sediment transport, u_c, is indicated, and the shaded areas represent periods when sediment transport occurs. (B) As for (A), but with a steady unidirectional current, u_u, added to the wave motion. (C) Orbital velocity variation and sediment transport for an asymmetrical Stokes wave. (D) As for (C), but with a steady unidirectional current, u_u, added to the wave motion.

increases, as is the case for bed sediment under unidirectional water flows. This suggests a common sorting process, but more details of the suspension threshold are needed to evaluate this. Beach sands are traditionally considered to be well sorted, and this is partly due to the fact that they are commonly fine- to medium-grained sands, and such sands are always well sorted (Chapter 5). However, the good sorting of beach sands may also be due to the constant reworking of beach sands by waves and to deflation by wind, both of which remove the fine-grained fraction in suspension. Alongshore variation in the mean grain size of beach sediment is expected in view of alongshore variation in the velocity of longshore currents. Shape sorting may also occur. The heavier grains of a given density and mean size (spheres) are more difficult to move than the lighter grains (flat grains such as disks), and will therefore require stronger currents to move them.

Bed forms and sedimentary structures due to wind waves and associated currents

Ripples form on a bed of cohesionless sand beneath wind waves at relatively low near-bed orbital velocities above the threshold of grain motion (Figure 7.14A). At larger near-bed orbital velocities, plane beds occur. The transition from wave ripples to plane beds occurs at a dimensionless bed shear stress of about 0.6 (Figure 7.14A), which is similar to the value for transition to upper-stage plane beds under unidirectional water flows. Plane beds beneath wind waves are probably dynamically similar to upper-stage plane beds. These bed configurations are modified in combined (oscillatory and unidirectional) flows, and new types appear. The origin, geometry, and dynamics of bed forms under wind waves and associated currents are described here, together with their associated sedimentary structures. These include ripples, hummocks and wave-modified dunes, and plane beds. Larger-scale bed forms are also associated with wind waves and associated currents, including longshore bars (ridge-and-runnel structures) and beach cusps at shorelines, and longitudinal ridges formed in shallow water. These larger-scale bed features are discussed in Chapter 15 on coasts and shallow seas.

Wave ripples

At the threshold of bed-grain movement, just above wave base, there is little asymmetry in near-bed orbital velocities. As sediment begins to move backwards and forwards on the bed, low ripples ($L/H > 10$) that are symmetrical in profile (so-called rolling-grain ripples) start to form (Figure 7.18). The length of rolling-grain ripples is less than the near-bed orbital diameter. At greater near-bed orbital velocities, ripples grow in height until flow separation occurs on alternate sides of the ripple (so-called vortex ripples: Bagnold (1946)). Vortex ripples have heights, H, of 0.003–0.25 m, lengths, L, of 0.009–2.5 m, and L/H ratios between 4 and 10 mainly (Allen, 1979a, 1982a, 1985) (Figures 7.17 and 7.18). The wave length of vortex ripples ranges from 0.65 to1.0 times d_{ob}, most commonly $0.8d_{ob}$, up to a limiting orbital diameter beyond which the ripple length remains approximately constant but increases with mean grain size: commonly around 0.04 m in very fine sand to 1 or 2 m for very coarse sand and gravel (Miller and Komar, 1980) (Figure 7.18A). The relationship between vortex ripple length and grain size exists because the near-bed orbital velocity (and bed shear stress) is proportional to the orbital diameter for a given wave period. Vortex ripples have a symmetrical trochoidal profile (in the absence of asymmetrical wave currents) and long, straight crestlines in plan, and thus are commonly referred to as two-dimensional (2D) wave ripples. Crestlines have characteristic bifurcations (Figure 7.17). Paleowave conditions have been interpreted from the geometry of vortex-wave ripples observed in ancient deposits (e.g., Komar, 1974; Allen, 1981; Clifton and Dingler, 1984; Diem, 1984).

At still larger near-bed orbital velocities, 2D vortex ripples change to 3D (dome-shaped; circular to elliptical in plan) wave ripples (Southard *et al.*, 1990) (Figure 7.17). Three-dimensional wave ripples formed under waves with large orbital diameters and periods may have heights of decimeters and lengths up to several meters, and such large bed forms have been called

hummocks. There is some disagreement over whether hummocks are large 3D wave ripples or a kind of wave-modified dune formed under combined flows (see below). At near-bed orbital velocities close to the transition to plane beds, L/H of wave ripples increases markedly (Figure 7.18A), and they become *postvortex (anorbital) wave ripples*.

Wave ripples formed in shoaling water are asymmetrical in profile, with the steep side facing the direction of shoaling (Figure 7.17), as a result of the net drift of the near-bed fluid. Such asymmetrical ripples may look superficially like current ripples, and care must be taken to distinguish them (Reineck and Wunderlich, 1968a). Two key differences between asymmetrical wave ripples and current ripples are that wave ripples have a smaller L/H ratio and are straight-crested. The L/H ratio of current ripples is commonly around 20, and current ripples never have long straight crestlines. *Interfering wave ripples* (Figure 7.17) occur where more than one set of waves (usually two) with different orientations occur in an area. Each set of waves simultaneously forms a distinctly oriented set of wave-ripple crests. The different wave sets may be associated with incident and reflected waves at a coastline. Interfering wave ripples are particularly common on tidal flats, where there are many bars that can reflect and refract waves. On tidal flats, wave ripples may also form interfering patterns with tidal current ripples.

Wave ripples will also become asymmetrical in profile under a combination of wind-wave currents and unidirectional currents. Addition of only a modest unidirectional current (0.1 m s^{-1}) to oscillatory currents will result in a degree of bed-form asymmetry that is difficult to distinguish from that due to purely unidirectional currents (Myrow and Southard, 1991; Dumas *et al.*, 2005) (Figure 7.19). It is therefore expected that combined flows will produce bed forms that are difficult to distinguish from those formed under unidirectional currents.

The pattern of flow and sediment transport over symmetrical and asymmetrical vortex wave ripples is rather complicated (Figure 7.20). As the crest of the wave passes, a flow-separation zone containing a high concentration of suspended sediment forms on the side of the wave ripple facing the direction of wave motion. As the trough passes, the same thing occurs on the other side of the ripple. In the intervening periods, the separation vortices and suspended sediment rise above the ripple (Bagnold, 1946; Longuet-Higgins, 1981), and this is an important mechanism for dispersing suspended sediment up from the bed. The growth of

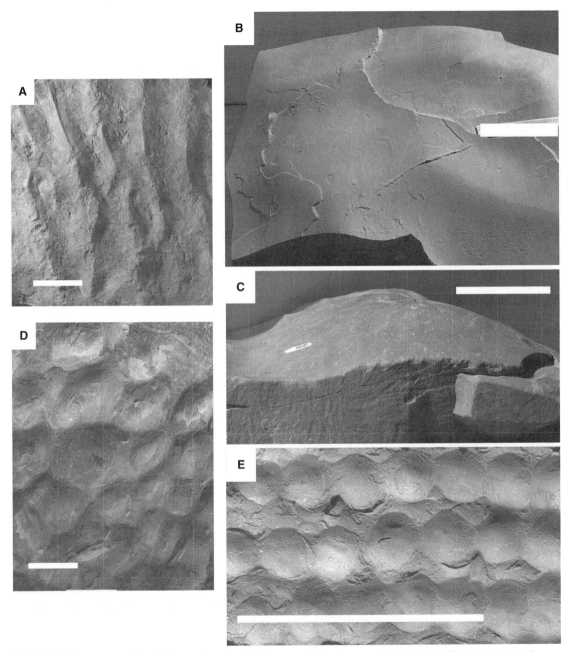

FIGURE 7.17. Ancient examples of (A) two-dimensional (2D) symmetrical vortex wave ripples, (B) and (C) three-dimensional hummocky wave ripples, and (D) and (E) interfering wave ripples. The white scale bars are 0.1 m long. Modern examples of (F) 2D symmetrical vortex wave ripples with a smaller set of wave ripples in troughs, (G) 2D asymmetrical vortex wave ripples with a set of superimposed symmetrical wave ripples, and (H) vortex wave ripples with double crests.

wave ripples is actually related to a non-periodic net drift of near-bed fluid and sediment from ripple troughs to crests, as confirmed in the visualization studies of Honji *et al.* (1980).

Symmetrical wave ripples generally do not migrate, because of the symmetry of the near-bed orbital velocities. However, natural wave currents are never completely symmetrical, and individual wave ripples may migrate for short distances either in the direction of wave motion or in the opposite direction. Since there is not expected to be any net sediment transport under symmetrical wave currents, erosion and deposition are

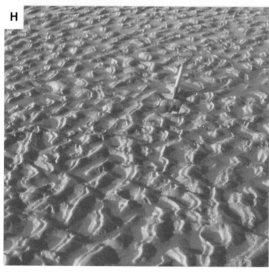

FIGURE 7.17. Continued.

sediment-transport rate changes in the net transport direction.

Sedimentary structures formed by wave ripples and associated currents

Sediment laminae are draped on both sides of vortex wave ripples, including in their trough regions (Boersma, 1970; de Raaf *et al.*, 1977; Allen, 1982a) (Figure 7.21). Since symmetrical wave ripples can change their geometry and move short distances, different sets of inclined laminae can be produced and superimposed on each other (Figure 7.21). In views parallel to straight wave crests, the laminae appear almost horizontal, planar, and parallel. Slightly asymmetrical wave ripples also have laminae draped along both sides, but the laminae are preferentially developed on the steeper side that faces the direction of net sediment transport (Figure 7.21). As the degree of ripple asymmetry increases, sets of cross laminae form only on the steepest face, and these cross strata look similar to those formed by current ripples in views parallel to the flow direction (Figure 7.21). Cosets of such wave-ripple cross laminae can form because wave-ripple height changes during migration, and because deposition is possible with asymmetrical wave currents. Such asymmetrical wave-ripple laminae are commonly of the climbing type (Figure 7.21) when the deposition rate is high. Quantitative relationships among wave-ripple height, deposition rate, and cross-set thickness have not yet been developed. It is probable that cross laminae formed by combined flows would be almost indistinguishable from those formed by unidirectional currents.

Sediment laminae are also draped on 3D wave ripples as the ripples grow and migrate under depositional conditions. If the ripples were symmetrical, such cross laminae would appear the same in cross sections in any direction (Figure 7.22). If 3D wave ripples became asymmetrical they could migrate laterally, and net deposition would be possible. For high rates of deposition, cross laminae formed both on crests and on troughs of ripples can be preserved, forming what is known as *hummocky cross stratification* (HCS) (Figure 7.22). This is the 3D wave-ripple equivalent of climbing-ripple cross stratification (Chapter 5, Figure 5.60). For lower deposition rates, the laminae formed on the crests of the ripples cannot be preserved, and trough laminae are preferentially preserved, giving rise to *swaley cross stratification* (SCS). The terms hummocky and swaley cross stratification have tended to be

not expected. Asymmetrical wave ripples migrate in the direction of the dominant current. Since there is net sediment transport in the case of asymmetrical wave currents and combined wave–unidirectional currents, there is the possibility of net erosion or deposition if the

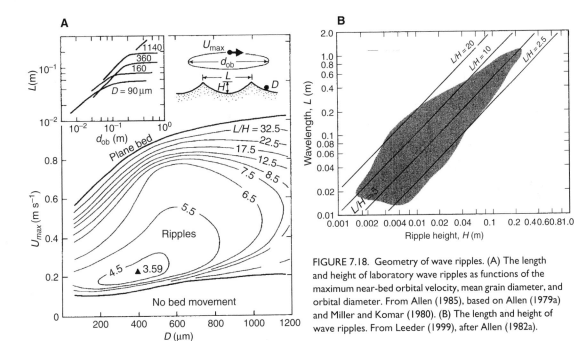

FIGURE 7.18. Geometry of wave ripples. (A) The length and height of laboratory wave ripples as functions of the maximum near-bed orbital velocity, mean grain diameter, and orbital diameter. From Allen (1985), based on Allen (1979a) and Miller and Komar (1980). (B) The length and height of wave ripples. From Leeder (1999), after Allen (1982a).

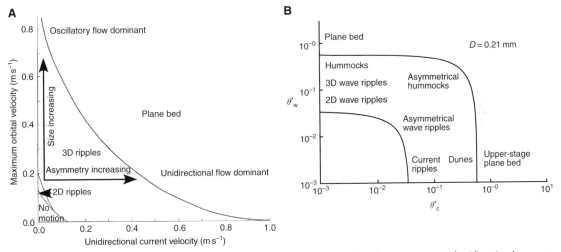

FIGURE 7.19. The equilibrium hydraulic stability of small-scale bed forms under combined wave currents and unidirectional currents, based on experimental data, in terms of (A) the water velocity, modified from Myrow and Southard (1991), and (B) the dimensionless bed-grain shear stress, courtesy of Maarten Kleinhans.

reserved for stratification associated with hummocks (and adjacent swales) that are meters across and decimeters high, with side slopes not exceeding 15°. However, cross strata formed by smaller 3D wave ripples are identical except for their scale (Craft and Bridge, 1987).

There has been much debate and disagreement about the origin of hummocky and swaley cross strata, because they have never been directly observed forming under natural conditions (e.g., Dott and Bourgeois, 1982; Harms et al., 1982; Hunter and Clifton, 1982; Swift et al., 1983; Walker, 1984; P. A. Allen, 1985;

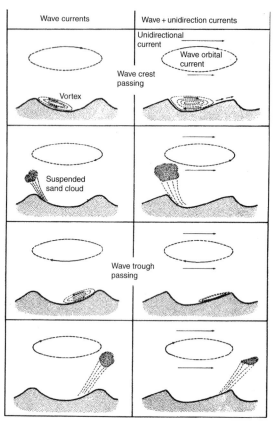

FIGURE 7.20. Flow and suspended-sediment transport over vortex wave ripples during a single wave period for the cases of wave currents only (left) and combined wave and unidirectional currents (right). Modified from Komar (1976).

Duke, 1985, 1990; Greenwood and Sherman, 1986; Leckie and Krystinik, 1989; Duke et al., 1991; Amos et al., 1996; Li and Amos, 1999). This is because HCS and SCS most likely form under very large waves during storms at sea when observation of bed conditions is difficult or impossible. Such conditions obviously cannot be reproduced in laboratory wave tanks, although Southard et al. (1990) and Dumas et al. (2005) have attempted to simulate waves with large orbital diameters and periods by using a flume in which the direction of unidirectional flow was periodically reversed. Hummocky bed forms have been observed on the sea bed following storms (e.g., Swift et al., 1983), but it is not known what the bed configuration looked like during the storm, and there are insufficient observations of the internal structure of the bed. Several questions that have been debated are the following. (1) How is it possible for decimeter- to meter-thick cosets of HCS and SCS to form without appreciable deposition rate, and hence net sediment transport? (2) Why does the

erosion surface at the base of cosets of HCS and SCS commonly show evidence of unidirectional currents, such as flute marks, gutter casts, and tool marks? (3) Do hummocks form only under combined flows? (4) What is the exact nature of the combined flows?

The erosion surface at the base of HCS and SCS units has been associated with offshore-directed unidirectional flows such as turbidity currents and geostrophic currents (e.g., Walker, 1984; Leckie and Krystinik, 1989; Duke, 1990). Ancient hummocky cross-stratified sandstone units typically pass laterally into thinner, finer-grained units composed mainly of wave-ripple cross lamination. This suggests that the magnitudes of bed shear stress and sediment-transport rate decrease in the direction of net sediment transport, which requires either asymmetrical wave currents and/ or combined currents. The unidirectional currents that formed the basal erosion surface could also have deposited the overlying sediment that was molded into hummocks by strong wave currents. However, it is rare to find evidence for strong unidirectional currents in HCS and SCS, such as stratal dips (including angle-of-repose cross strata) directed preferentially in the direction of thinning and fining of the deposit. It is known also that a unidirectional current with a velocity of as little as $0.1\,\mathrm{m\ s^{-1}}$ can cause wave ripples to become markedly asymmetrical. Therefore, HCS and SCS are probably formed by strong oscillatory wave currents, with only a minor unidirectional current superimposed (P. A. Allen, 1985; Nottvedt and Kreisa, 1987; Arnott and Southard, 1990; Myrow and Southard, 1991; Dumas et al., 2005). Hummocks are therefore not wave-modified dunes. As the magnitude of the unidirectional current increases relative to the magnitude of the oscillatory current, the hummocky bed form will become markedly asymmetrical and angle-of-repose cross strata will form. Such a combined flow would have a larger absolute magnitude, resulting in coarser-grained sediment (e.g., Nottvedt and Kreisa, 1987; Arnott and Southard, 1990; Cheel and Leckie, 1992; Bridge and Willis, 1994).

If HCS units are formed during storms that last on the order of days, a unit on the order of decimeters thick would require deposition rates on the order of $10^{-3}\,\mathrm{mm\ s^{-1}}$. Since the near-bed sediment concentration is very high during major storms at sea, a small change in a large sediment-transport rate with distance could yield such deposition rates. However, there has been no quantitative work on the relationship among hummock height, deposition rate, and set thickness.

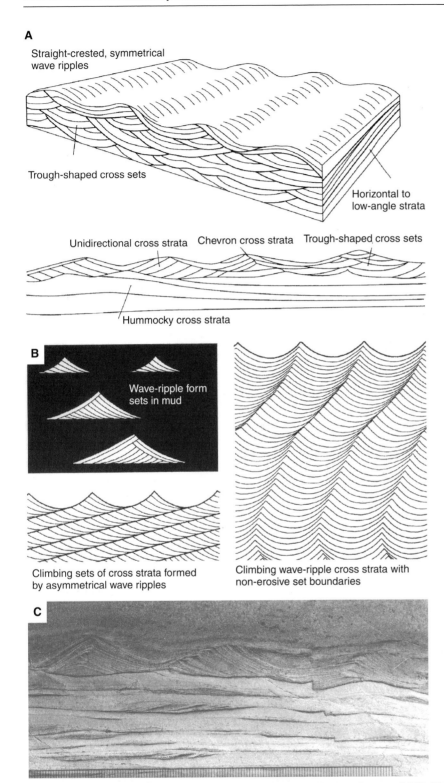

FIGURE 7.21. Types of cross strata formed by wave ripples. (A) Modified from Boersma (1970). (B) Modified from Allen (1982a). (C) A field example of cross strata formed by symmetrical and asymmetrical wave ripples (scale in millimeters).

FIGURE 7.22. Hummocky cross strata, (A) and (B), and swaley cross strata in orthogonal views, (C) and (D).

Plane beds and planar lamination

Plane beds form beneath waves with large near-bed orbital velocities, and in the surf zone of sandy beaches (Figure 7.23). Although these plane beds appear to be dynamically similar to the upper-stage plane beds of unidirectional turbulent flows, there are obviously some differences. Water flow reverses periodically under waves, and, as far as we know, low-relief bed forms have not been observed on wave-formed plane beds.

Deposition on wave-formed plane beds forms planar lamination that looks remarkably like that developed on upper-stage plane beds. However, it cannot have formed the same way. The individual laminae are most likely formed by the periodic deposition associated with the reversing wave currents as individual waves pass. However, if strong unidirectional currents are superimposed on wave currents, it is uncertain which current is responsible for the planar laminae. In the surf zone of sandy beaches, the planar laminae have characteristic concentrations of heavy minerals (Figure 7.23). This is the well-known *beach lamination* (Clifton, 1969; Reineck and Singh, 1980; Allen, 1982a). The character of sets of these laminae can be influenced

FIGURE 7.23. (A) A wave-formed plane bed on a beach and associated planar laminae. (B) The crest area of a longshore bar on a beach, showing rhomboid ripple marks, antidunes, and inclined strata associated with onshore migration of the steep lee face.

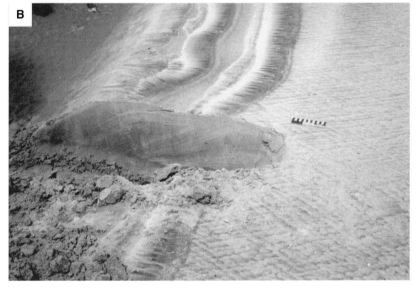

by tides, storms, or storm seasons. It is unknown whether low-relief bed forms play any role in the formation of laminae by wave currents.

Hydraulic criteria for the existence of equilibrium bed states under wind waves

The equilibrium hydraulic stability fields of bed states formed by purely oscillatory wave currents can be expressed in terms of the dimensionless bed shear stress and grain Reynolds number, as is the case for unidirectional water flows (e.g., Komar and Miller, 1975b; Miller and Komar, 1980) (Figure 7.14A). Simpler stability diagrams have been constructed using u_{max} and the mean grain size (Allen, 1979a, 1982a, 1985),

see Figure 7.18A, and u_{max}, mean grain size, and wave period (Harms *et al.*, 1982). However, these diagrams are based mainly on experimental data. More data from natural environments are needed, especially relating to larger-scale bed forms such as hummocks. There is also a pressing need for equivalent diagrams for a wide range of types and orientations of combined flows. Myrow and Southard's (1991) diagram (Figure 7.19) was a good start, but it is not general enough. However, Kleinhans (personal communication) has used a broad range of laboratory and field data to define the hydraulic stability fields of combined-flow bed forms in terms of dimensionless bed shear stress (Figure 7.19). Hydraulic stability diagrams can be used to predict sequences of bed forms

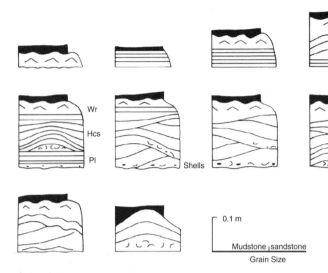

FIGURE 7.24. Typical vertical sequences of grain size and sedimentary structures due to waves. After Craft and Bridge (1987). Black is mudstone; white is sandstone. Bases of stratasets are sharp and erosional, and may contain flute marks, gutter casts, tool marks, and burrows. Burrows also occur in upper parts of stratasets. Top parts of stratasets may be eroded. Wr = wave-ripple lamination, Hcs = hummocky cross strata, Pl = planar strata.

and sedimentary structures formed under changing hydraulic conditions (e.g., Allen, 1985; Myrow and Southard, 1991) (Figure 7.24).

Generation of tides

Tides in the world's oceans are produced by the gravitational attraction between the Earth, Moon, and Sun. The tide-generating force is actually the resultant of the gravity force and the centrifugal force arising from co-rotation of the Earth, Moon, and Sun. The manifestation of the tide-generating force is deformation of the water surface of the Earth into an oblate spheroid, producing bulges in the water surface and concomitant hydrostatic pressure gradients. The so-called *equilibrium tide* is that where the hydrostatic pressures and tide-generating forces are balanced. The Moon has the dominant effect on ocean tides, and the Sun's effect is about half as strong because of its greater distance from the Earth. Figure 7.25A shows the Earth with bulges in the surface of the ocean (equivalent to high tides) on opposite sides of the Earth, caused mainly by the gravitational attraction of the Moon. The distance between the crests of these bulges (and the length of a tidal wave) is half of the local circumference of the Earth. Since the Earth rotates on its axis once every 24 hours, an observer on the surface of the Earth will experience two high tides and two low tides every 24 hours approximately (semidiurnal tides). However, the Moon moves around the Earth every month, so semidiurnal tides have a period of 12.5 hours rather than 12 hours. This is why the time of high tide is later by about one hour every day. The tidal wave amplitude actually differs for

successive semidiurnal tides (the *diurnal inequality*) because of the declination of the Earth's rotational axis (Figure 7.25B). As the Moon moves around the Earth, the gravitational attractions of the Sun and Moon will vary in direction throughout the month. When the Earth, Moon, and Sun are all in line (full and new moons), the amplitude of the tidal bulges is at its maximum, producing the highest tidal range (*spring tides*, occurring twice per month). When the Moon and Sun form a right angle with the Earth (Figure 7.25C), their gravitational attractions act against each other, and the tidal bulges have their minimum amplitude and tidal range (*neap tides*). This is a simple picture of tidal periods and amplitudes. There are many more tidal periods and amplitudes related to factors such as the elliptical orbits of the Earth and Moon, and changes in the declination of the rotational axes of the Earth and Moon through their orbits (Table 7.2). Figure 7.25C explains seasonal changes in the amplitudes of neap tides and spring tides, and inequalities of successive neap and spring tides.

Tides on a perfectly spherical Earth, with no friction between the water and the Earth, would have a maximum tidal range of less than 1 m. However, tidal ranges near coastal areas can be much greater than this, and may reach more than 15 m. Tidal waves in the open ocean are amplified as they approach land masses, due to the effects of (1) flow convergence, (2) wave reflection, and (3) Coriolis deflection. Tidal range may also be damped by friction between the water and the Earth's surface. The effects of these tide-modifying factors on tidal ranges and currents are discussed below.

TABLE 7.2. Most important tide-generating constituents (from Defant, 1958)

Type of tide	Symbol	Period (solar hours)	Relative amplitude	Constituent
Semidiurnal	M_2	12.42	100.00	Main lunar
	S_2	12.00	46.6	Main solar
	N_2	12.66	19.1	Monthly variation in Moon's distance from Earth
	K_2	11.97	12.7	Changes in declination of the Sun and Moon during orbits
Diurnal	K_1	23.93	58.4	Solar–lunar
	O_1	25.82	41.5	Main lunar
	P_1	24.07	19.3	Main solar
Long period	M_f	327.86	17.2	Moon's fortnightly

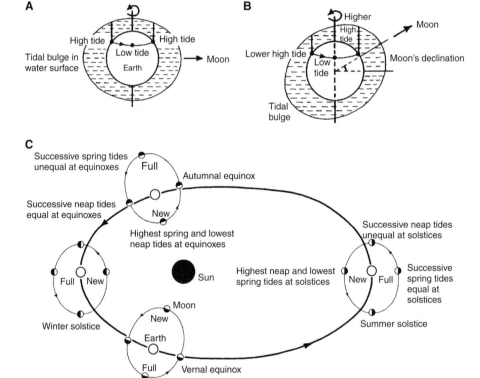

FIGURE 7.25. The origin of tides. (A) The bulge in Earth's ocean-water surface due to the position of the Moon explains the semidiurnal tidal period (two high tides and two low tides per day). (B) Diurnal inequality of semidiurnal tidal amplitudes due to Moon's declination. (C) The eccentric elliptical orbit of the Moon around Earth results in neap and spring tides (two of each per month), and inequality of successive neap-tidal amplitudes (at solstices) and spring-tidal amplitudes (at equinoxes). The eccentric elliptical orbit of Earth around the Sun results in changes in peak spring- and neap-tidal amplitudes. Modified from J. R. L. Allen (1985).

Tidal currents

Tidal waves are shallow-water waves ($L/d > 20$), and undergo a transformation similar to that undergone by wind waves as they move into shoaling water: the waves become asymmetrical in profile and their height increases. A breaking tidal wave is called a *tidal bore*. With asymmetrical tidal waves, water currents associated with the flood tide are stronger and of shorter duration than those associated with the ebb

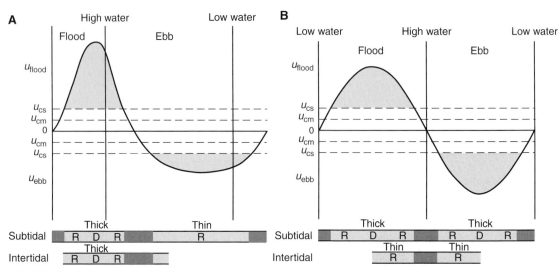

FIGURE 7.26. Idealized variation of tidal current velocity, u_{flood} and u_{ebb}, and sediment transport over one wave period for (A) an asymmetrical progressive tidal wave and (B) a symmetrical, standing tidal wave. Thresholds of sediment motion for sand, u_{cs}, and mud, u_{cm}, are indicated, and the light-shaded areas are periods of sand motion. Under the assumption that deposition occurs throughout the tidal cycle, and that peak tidal currents are strong enough to form dunes, the relative thickness, grain size, and internal structure of the deposits in subtidal and intertidal locations are suggested. Dark shading indicates mud, and light shading indicates sand; R indicates ripple migration and associated small-scale cross strata; D indicates dune migration and associated medium-scale cross strata.

tide. The maximum flood currents occur close to the time of high tide, and the maximum ebb currents occur near the time of low tide (Figure 7.26). This means that topographically high coastal regions (e.g., upper tidal flats) will experience maximum flood-current velocities, and topographically low coastal areas (e.g., channels) will experience maximum ebb-current velocities. The maximum flood and ebb currents may well follow different paths at the coast, and inequality of durations and magnitudes of flood and ebb currents at a point is to be expected. The transformation of a tidal wave as it moves into shoaling water is associated with flow convergence in the vertical plane. However, tidal waves also converge in the horizontal plane as they move through straits between land masses, and as they move up estuaries. This convergence also increases tidal ranges (see Figure 7.27).

A tidal wave moving into an enclosed basin or estuary will be reflected at its end, and the reflected wave may interact with the next incoming tidal wave to give *resonant amplification*. When two waves of equal period and height travel in opposite directions, a standing wave with double the amplitude of the individual waves develops (see above). Resonant amplification of standing waves occurs when the length of the basin or estuary is 1/4, 3/4, 5/4, and so on, of the length of the

standing wave, which is $T\sqrt{(gd)}$. The maximum tidal currents in standing waves occur at times of mid-tide level (Figure 7.26). However, it is likely that a tidal wave moving into an enclosed basin or estuary will also be modified by the effects of flow convergence, the Coriolis force, and bed friction. Therefore, a simple standing wave is unlikely, and the maximum tidal currents will not be associated with the mid-tide position. The Bay of Fundy in Newfoundland has the highest tide range in the world (spring tidal range >15 m), due to the effects of resonant amplification, flow convergence, and the Coriolis force. A quarter wave length of the standing wave with period 12.42 hours and mean depth 70 m is 290 km, which is very close to the length of the Bay of Fundy.

The Coriolis force deflects moving water to the right of its path in the northern hemisphere and to the left in the southern hemisphere. If a tidal wave in the northern hemisphere moves into an enclosed basin (e.g., the North Sea), a northward flow would be deflected towards the east side of the basin, and a southward flow would be deflected towards the west side of the basin. This flow deflection results in an amplification of tidal range that reaches a maximum at the coast, and a point in the center of the basin where the tidal range is zero (an *amphidromic point*). Another result is that the tidal wave moves around the basin in an anticlockwise

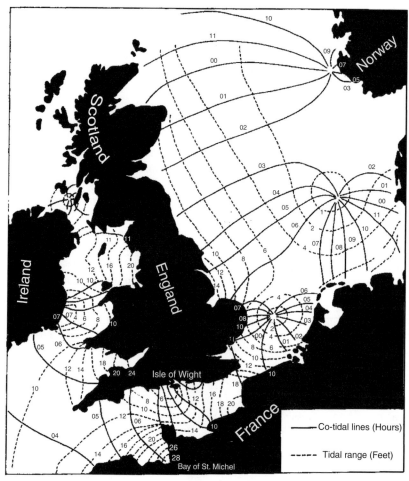

FIGURE 7.27. Tides around England. Co-tidal lines are positions of the tidal wave crest (high tide) throughout the daily cycle. In places, co-tidal lines indicate anticlockwise rotation of the tidal wave crest around amphidromic points, where the tidal range is zero. The tidal range (marked by dashed lines) increases away from amphidromic points and towards coastlines. The tidal range also increases into embayments and estuaries (e.g., the Severn), and towards the Straits of Dover, due to convergence of the tidal wave. From Komar (1998).

direction about an axis that is the amphidromic point (Figure 7.27). This means that tidal current directions vary around the basin rather than oscillating in a straight line parallel to a constant direction of tidal wave propagation. At a given point, the paths of water particles follow an elliptical path rather than oscillation about a line. In reality, tidal ellipses are not closed, are irregular in shape, and vary with depth at a point. Also, tidal currents in the North Sea are affected by flow convergence and wave reflection.

Bed friction causes damping of tidal waves, meaning reduction in wave height in the direction of wave propagation. Such damping modifies the nature of reflected tidal waves, because the reflected wave will be more damped than the incoming wave. Reduction in wave height in the direction of wave propagation can

also result in the displacement of amphidromic points to one side of an enclosed sea or gulf (P. A. Allen, 1997).

Maximum tidal current velocities increase with tidal range and with shoaling, as is the case for all shallow-water waves (Equation (7.11)). In the case of a simple shallow-water wave with a height (tidal range) of 2 m in water 10 m deep, the maximum near-bed velocity would be about $1\,\mathrm{m\,s^{-1}}$ using Equation (7.11). However, Equation (7.11) is not strictly applicable when flow is accelerated through channels or damped by friction over broad tidal flats. Nevertheless, it is common for maximum velocities of tidal currents to reach 1 or 2 meters per second in shallow enclosed seas and tidal channels. Tidal range (and associated current velocity) is commonly classified as *microtidal* (0–2 m), *mesotidal* (2–4 m), and *macrotidal* (>4 m). These terms are used to indicate the relative importance of tidal currents in shaping coastal environments (see Chapter 15). Wind-wave currents are also important in shallow marine and coastal areas, and it is normal for superposition of tidal currents and wind-wave currents to occur.

Sediment transport by tidal currents

Tidal currents commonly have sufficient velocity to move sand and gravel on the bed, and silt and clay in suspension. The asymmetry of tidal currents in shallow water, and the fact that maximum flood and ebb currents commonly follow different paths, result in characteristic sediment-transport paths for which there is net sediment transport in a given direction (Figure 7.28) (Allen, 1970; Johnson *et al.*, 1982; Harris *et al.*, 1995).

If the tidal range and maximum current velocity decrease in the direction of the sediment-transport path, the mean grain size of the sediment transported will decrease and deposition will occur. If the current velocity increases in the direction of net sediment transport, the mean grain size of transported sediment will increase and erosion will occur.

Since the maximum flood currents can affect topographically high areas, sediment can be deposited in these areas if the current velocity decreases in the direction of sediment transport. Such deposited sediment might not be removed by erosion by the weaker ebb currents. This is one reason for the preferential deposition of suspended mud in upper intertidal flats. Another reason may be that mud requires a greater bed shear stress to erode it than to deposit it. Deposition from a tidal current that is decelerating in time as well as space will give rise to a fining-upward

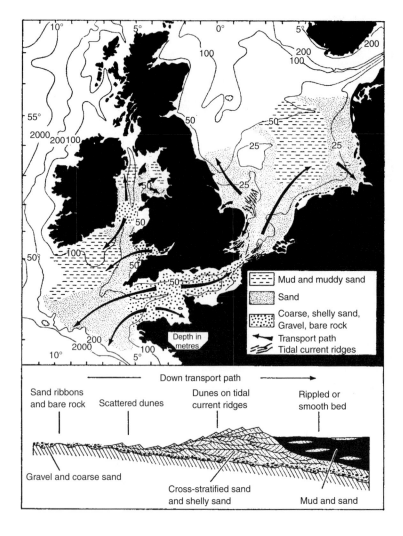

FIGURE 7.28. Sediment-transport paths associated with tidal currents around the North Sea. From Allen (1970).

unit, and deposition from a tidal current that is decelerating in space but accelerating in time will yield a coarsening-upward unit. If the deposition rate is on the order of millimeters per tidal cycle, a lamina could be formed during the flood tide, and another might be formed on the ebb tide (especially in areas below the low tide level). Such laminated deposits are called *laminites* (Figure 7.29A, B). The thickness of each lamina will be related to the deposition rate, the time over which it formed, and any subsequent erosion.

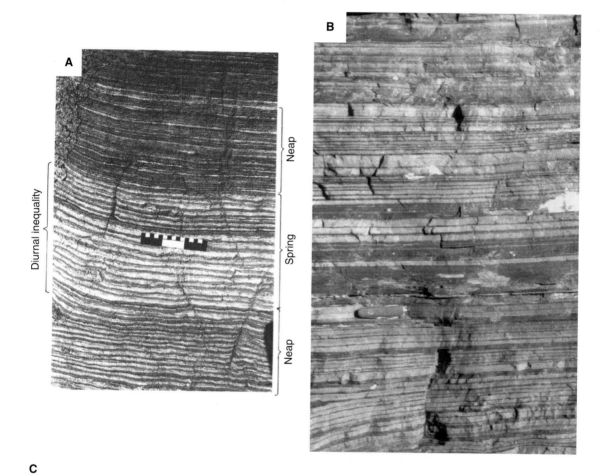

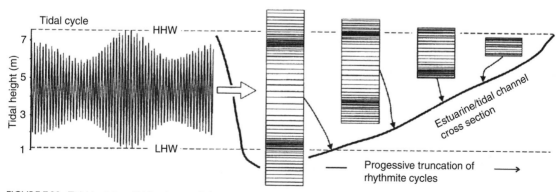

FIGURE 7.29. Tidal laminites. (A) Sandstone–shale couplets with variable sandstone/shale ratio, attributed to deposition over spring and neap tides. From Nio and Yang (1991). (B) Possible tidal laminites in limestones. (C) Decreasing preservation of tidal rhythmites as a function of elevation on a tidal flat. From Archer (1998).

Systematic variations in the thickness of laminae have been related to periodic variations in current strength and deposition rate over, say, spring–neap tidal cycles; see reviews by Nio and Yang (1991) and Archer (1998) and various papers in Smith *et al.* (1991) and Alexander *et al.* (1998). However, it is unlikely that thickness variations would follow periodicity in tidal range very closely in view of the effects of simultaneous non-periodic variation in wind-wave currents. Also, the relationship among tidal range, current velocity, and deposition rate might not be simple. Laminites formed in high intertidal regions may receive deposits only during the highest tides or major storms (Figure 7.29C). However, if 28 successive laminae exhibit a systematic thickening and thinning, it is difficult to avoid the interpretation of their having been formed during continuous deposition by semidiurnal tides over a spring–neap cycle.

Bed forms and sedimentary structures formed by tidal currents

Ripples and dunes are readily formed by tidal currents because the velocities required exist for on the order of hours (Figure 7.30). However, it is unlikely that the geometry of the larger bed forms is ever in equilibrium with the changing flow conditions. The common asymmetry of tidal currents means that different bed forms may exist in an area under flood and ebb conditions. Dunes commonly migrate in one dominant direction, although they may be modified by the flow in the opposite direction. Most tidal bed forms are influenced by simultaneous wave currents and by the effects of rising and falling water level. Larger-scale bed forms associated with tidal currents (on which ripples and dunes are superimposed) are tidal-current ridges, sand waves, and ebb- and flood-tidal deltas. These bed forms are discussed in Chapter 15 on coastal and shallow marine depositional environments.

Ripples

Ripples formed by tidal currents alone are essentially the same as those formed by unidirectional currents. However, tide-formed ripples are very commonly modified by wind-wave currents, and by the effects of falling water levels and reversing currents (Figure 7.30). Superposition of different patterns of tidal current ripples and wave ripples is common. As water flows off a rippled bed area during an ebbing tide, the crests of the ripples may be planed off. At any given time, the small-scale cross strata within a ripple may indicate flow in a direction different from that indicated by the ripple form. Superimposed sets of small-scale cross strata formed by tidal ripples (Figure 7.31) may indicate the occurrence of reversing flows if tidal currents were symmetrical (*herringbone cross stratification*). Such sets may also indicate flow in one direction if the tidal currents are strongly asymmetrical or if deposition from either asymmetrical or symmetrical tidal currents occurred only during the flood or ebb (in which case the formative flow could be misinterpreted as unidirectional).

In areas where ripples are formed by the maximum tidal currents, the ripples stop moving during the periods of slackening water flow (e.g., Figure 7.30). These ripples may then be draped with mud deposited from suspension. This mud is commonly pelleted by benthic invertebrates. If net deposition is proceeding, the alternations of rippled and cross-stratified sand and mud drapes formed during tidal cycles give rise to *heterolithic bedding* (Figure 7.31). Differences among varieties of heterolithic bedding depend on (1) the relative amounts of sand and mud in the deposits, depending on the current velocity and sediment supply during deposition; (2) the relative importance of tidal and wave currents during deposition; (3) whether the tidal currents are symmetrical or asymmetrical; (4) whether deposition occurs throughout the tidal cycle; and (5) whether the site of deposition was intertidal or subtidal.

Flaser bedding (Reineck and Wunderlich, 1968b; Reineck and Singh, 1980) occurs when mud drapes occur only in ripple troughs (Figure 7.31). *Wavy bedding* occurs when mud drapes over the whole of a continuous bed of rippled and cross-stratified sand. *Lenticular bedding* refers to isolated sand-ripple forms encased in mud. These types of heterolithic bedding are commonly associated with deposition on tidal flats. However, the basic alternation of sand and mud can occur in any environment where there is periodic variation in current strength. Specific features must be present for a tidal origin to be invoked, as explained below. In tidal settings, it is normally assumed that each mud drape in flaser and wavy bedding could be deposited during a single tidal cycle. However, the thicker mud layers in wavy and lenticular bedding could not be deposited during a single tidal cycle: there is not enough time given the expected suspended-sediment concentrations (McCave, 1970; Hawley, 1981). Such mud layers may be related to a sequence of tidal cycles or post-storm conditions. Mud layers should be examined closely to determine whether they are composed of sand-sized mud

FIGURE 7.30. Bed forms associated with tidal currents. (A) Asymmetrical wave ripples modified by tidal sediment transport towards the observer. (B) Wave-ripple marks superimposed on tidal dunes, some of which have crests truncated by tidal flows. (C) An ebb-oriented tidal dune with a ripple fan in the trough. (D) Flood-oriented tidal dunes modified by falling water level.

pellets (Demicco, 1983). If the mud was deposited as sand, greater thicknesses could be deposited during a single tidal cycle. Such pelleted mud would be expected to be cross stratified.

Variation in the thickness of sand–mud couplets in heterolithic bedding has been related to variation in tidal current strength and deposition rate associated with periodic variations in tidal range, such as neap–

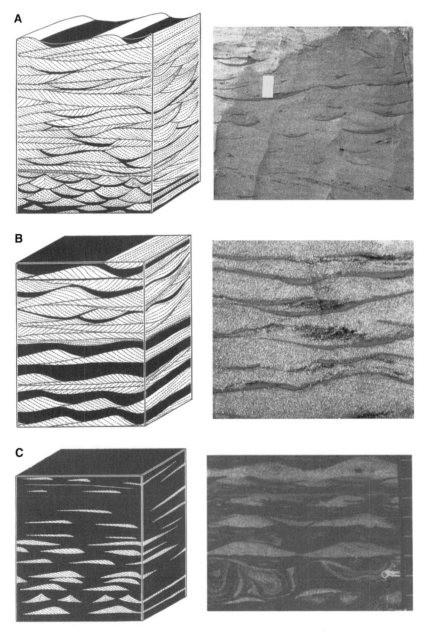

FIGURE 7.31. Flaser (A), wavy (B), and lenticular (C) bedding associated with current ripples, and symmetrical and asymmetrical wave ripples. From Reineck and Wunderlich (1968b). Mud is black, sand is white. Actual examples from North Sea tidal flats are shown to the right. Ripples are about 10 mm high.

spring cycles (as done for laminites). However, the relationship among tidal range, current speed, and deposition rate might not be straightforward, and variation in wind-wave currents may also have an influence (Terwindt, 1981; Nio and Yang, 1991).

Details of the spatial distribution of rippled sand and mud depend on the degree of asymmetry of the tidal currents, whether deposition occurred throughout the tidal cycle, and whether the depositional area is subjected to one (intertidal) or two (subtidal) slack water periods during a tidal cycle (Figures 7.26 and 7.31). If tidal currents are symmetrical and deposition occurs both during the ebb tide and during the flood tide, cross laminae in the sand would dip in opposing directions, and slackwater mud drapes would separate the ripples with opposing flow directions. If tidal currents are strongly asymmetrical, or deposition occurs only during the ebb or flood tide, the cross laminae would dip in one direction only, and the mud drapes would represent two slackwater periods and the subordinate tide (Figure 7.31). In subtidal settings, two slackwater mud drapes could occur during a single tidal cycle, but only one could occur in intertidal settings.

Dunes

Dunes formed by tidal currents may have straight crestlines (2D) or curved crestlines (3D); for reviews of dune geometry see Allen (1982a) and Dalrymple and Rhodes (1995). They are commonly asymmetrical in along-stream section, and the degree of asymmetry varies with the asymmetry of the tidal currents. Most dunes will undergo some change in shape (in along-stream section) in response to changes in flow direction. Frequency distributions of dune height and wave length are commonly polymodal, and change with time as tidal currents change. This polymodality is partly due to the delayed response of dune height and length to changing flow conditions: dune lag (Allen, 1982a). Wave ripples and tidal current ripples are commonly superimposed on the surfaces of dunes during periods of low tidal current velocity (Figure 7.30).

Migration of tidal dunes generates medium-scale cross strata. Superimposed sets of medium-scale cross strata formed by tidal dunes (Figure 7.32) may indicate reversing flows if tidal currents were near symmetrical (*herringbone cross stratification*). Superimposed sets may also indicate flow in one direction if the tidal currents are strongly asymmetrical or if deposition from either asymmetrical or symmetrical tidal currents occurred only during the flood or ebb. Reactivation surfaces are common in cross sets formed by tidal dunes. Some of these are due to dunes overtaking each other, as is the case with dunes migrating under unidirectional flows. Other types of reactivation surfaces are formed by slackening and reversal of tidal currents. As the tidal currents slacken, ripples may form and migrate on the dunes. As tidal current speed further declines, wave ripples may form, if waves are active, or mud drapes may be deposited (Figure 7.32).

Characteristic patterns of stratification are formed by dune migration under symmetrical and asymmetrical tidal currents for the case when deposition occurs throughout the tidal cycle, and where sand and mud are available for transport; e.g., Allen (1985) (Figures 7.26 and 7.32). For strongly asymmetrical tidal currents, the dune migrates and forms medium-scale cross strata under the dominant tide. As the tidal current slackens, current and/or wave ripples may form on the dune, and finally a mud drape may be deposited, depending on whether the setting is intertidal or subtidal. During the

subordinate tide, limited sand transport may occur as ripples, leaving a relatively thin deposit. Another mud drape may be deposited during the following slack-water period, depending on whether the setting is sub-tidal or intertidal. In the case of symmetrical tidal currents, dune migration is followed by ripple forma-tion and (perhaps) mud-drape deposition both during ebb tide and during flood tide (Figure 7.32). The char-acteristic deposit formed during a single tidal cycle is referred to as a *tidal bundle*. Systematic variation in the thickness of tidal bundles has been related to variations in tidal range associated with neap–spring cycles and even the diurnal inequality of the tide (Figure 7.32) (Visser, 1980; Boersma and Terwindt, 1981; Terwindt, 1981; Allen, 1982b; Nio *et al.*, 1983; Allen and Homewood; 1984; Dalrymple, 1984; De Mowbray and Visser, 1984; Nio and Yang, 1991). The basis of this interpretation is that the thickness of a medium-scale cross stratum is related to the rate of dune migration, which is in turn related to the tidal current velocity and tidal range. It seems improbable that such a linkage could exist. For example, the dune-migration rate varies with dune height as well as with the sediment-transport rate. Nevertheless, cases of systematic thinning and thickening of sequences of 28 tidal bundles have been recorded, and an alternative explanation is not likely.

Tsunami

Tsunami (from Japanese *tsun* meaning harbor and *ami* meaning wave) are water-surface waves generated by a sudden change in the elevation of the ocean floor (related to earthquakes and landslides) that displaces overlying ocean water vertically. Such movement affects the whole water column in addition to the water surface. Volcanic eruptions and meteorite impacts can also displace water and generate tsunami. Tsunami waves travel rapidly across the ocean, and their amplitude increases near coasts, commonly being on the order of meters. At the coast, the waves surge inland with water velocities on the order of meters per second and depths of meters. Tsunami-wave surges can reach elevations of tens of meters above mean sea level up to a few kilometers inland. Although tsunami waves are particularly common in the Pacific Ocean, they affect all oceans to some degree. Tsunami can have a devastating effect on life and prop-erty (Figure 1.3A), and human death tolls from major tsunami are typically tens of thousands. Warning sys-tems are in place, but not everywhere, and they are not always adequate. Tsunami waves are capable of

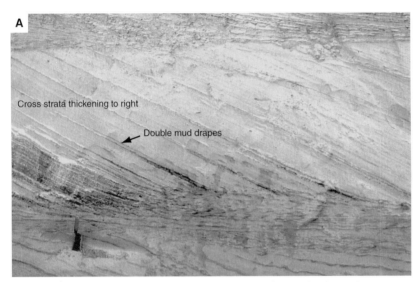

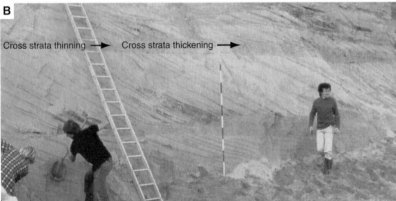

FIGURE 7.32. Cross strata due to tidal dunes: (A) and (B) tidal bundles with double mud drapes, and progressive thinning and thickening of cross strata associated with neap–spring tidal cycles, from subtidal channel deposits of the southwest Netherlands; and (C) tidal bundles and herringbone cross strata from subtidal channel deposits of the southwest Netherlands (photo from Janrik Van Den Berg).

performing significant erosion and deposition, particularly in coastal regions, and ancient sediments thought to be deposited by tsunami have been used to interpret past major earthquakes and meteorite impacts.

The tsunami that resulted from the Sumatra–Andaman earthquake of 26 December 2004 is undoubtedly the most damaging and best recorded in history up to this point (e.g., Lay *et al.*, 2005; Moore *et al.*, 2006). More than 283,000 people died, and millions more were rendered homeless. Numerous investigations of the nature of this earthquake and tsunami, and the impact of the tsunami on humans and the Earth's surface are under way, and these will be published in due course. Some of the preliminary information is included below.

Tsunami generation

Over the past 150 years, one to two tsunami have occurred per year on average, and more than 80% of these have been caused by submarine earthquakes. Most submarine earthquakes are not large enough to generate tsunami, and earthquakes with Richter magnitudes greater than 7 cause most tsunami. However, earthquakes of lower magnitude can generate tsunami if they initiate subaqueous landslides. Subduction zones are the prime sites for large-magnitude earthquakes, and these can be associated with ruptures of the ocean floor with vertical movements of meters and lengths of hundreds of kilometers, involving areas of up to 10^5 km^2 and volumes of rock of up to 10^3 km^3. Tsunami-wave amplitude (at the source and at some distant coast) increases with the volume of the displaced rock and with the magnitude of the earthquake (Bryant, 2001). The shapes of tsunami waves are quite variable at the point of generation, depending on the distribution of uplift and subsidence, and propagate at right angles to the rupture orientation (Figures 7.33 and 7.34).

The main shock of the Sumatra–Andaman earthquake of 26 December 2004 had a Richter magnitude of between 9.1 and 9.3, and was followed by many aftershocks with magnitudes in excess of 5 (Lay *et al.*, 2005). The earthquake was due to fault slip associated with the subduction zone separating the Indo-Australian and SE Eurasian plates. The fault slip occurred over a length of more than 1300 km, extending from NW Sumatra northwards to the Andaman Islands (Figure 7.35). The width of the faulted zone was up to about 240 km, and faulting occurred over a depth

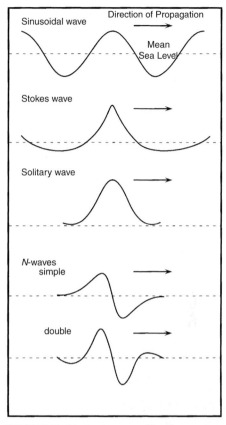

FIGURE 7.33. Idealized cross profiles of tsunami waves. From Bryant (2001).

of 30–45 km. The epicenter of the earthquake was just offshore and to the west of NW Sumatra (Figure 7.35), and the rupture propagated to the north, as recorded by aftershocks. The amplitude, propagation velocity, and duration of the fault slip varied in space and time. Slippage lasted for at least 500 seconds, and was up to 15 m near Banda Aceh.

Landslides can involve displacement of similar areas and volumes of rock or sediment to those displaced by earthquake ruptures. Earthquakes can, of course, initiate landslides, especially on the steep slopes of oceanic trenches and continental margins, but so can eruptive activity on the slopes of volcanoes. The character of tsunami waves generated by landslides depends on their volume, the flow depth, and the rate of movement. The lengths and periods of the initial waves (typically on the order of kilometers and minutes, respectively) are less than those generated by earthquakes (typically on the order of tens to hundreds of kilometers and minutes to tens of minutes, respectively), and the wave profile is typically N-shaped (Figure 7.33).

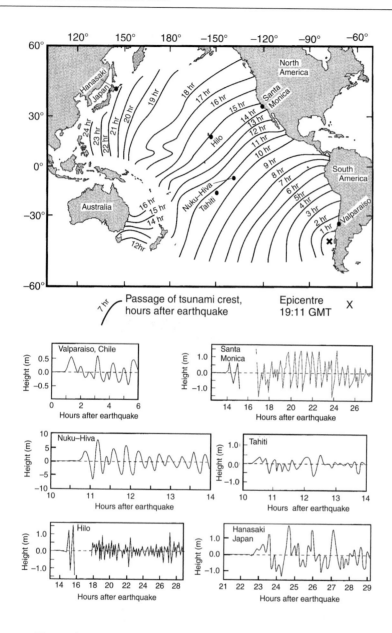

FIGURE 7.34. (A) Passage of a tsunami wave crest across the Pacific Ocean following the 22 May 1960 Chilean earthquake, and (B) water-level records at selected stations (daily tides removed). From Bryant (2001).

Tsunami waves associated with volcanic eruptions can be generated by earthquakes, landslides, submarine eruptions, subaerial flows of ash or lava into the sea, and collapse of volcanoes to form calderas. Spectacular examples include Krakatoa (1883), Pelee (1902), and Santorini (1650 BC). Ancient meteorite impacts have been inferred to generate tsunamis, and these inferences have been based on preserved sediment deposits (e.g., Bourgeois *et al.*, 1988) and computer simulations (Hills and Mader, 1997; Crawford and Mader, 1998). Figure 7.36 shows predictions of the height of tsunamis as a function of meteorite diameter, impact velocity, and distance from the impact.

Tsunami dynamics

The shapes, periods, and current velocities associated with tsunami waves are recorded by tide gauges, buoys and various other tethered instruments throughout the oceans, and satellite altimetry. The Deep Ocean Assessment and Reporting of Tsunamis (DART) Project is a U.S. National (NOAA) effort to maintain and improve the capability for the early detection and real-time reporting of tsunamis in the open ocean. The DART systems consist of an anchored sea-floor pressure recorder and a companion moored surface buoy for real-time communications. Data from the sea-floor

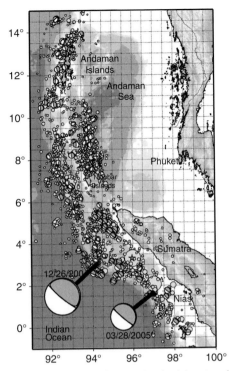

FIGURE 7.35. A map showing aftershock locations (circles with diameter proportional to seismic magnitude) for 13 weeks after the 26 December 2004 Sumatra–Andaman earthquake. From Lay *et al.* (2005). The epicenter is shown by a star within a large circle. The spatial distribution of aftershocks outlines the area of fault slip.

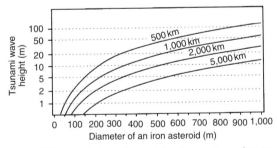

FIGURE 7.36. Predictions of the height of tsunamis as a function of meteorite diameter, impact velocity, and distance from the impact. From Bryant (2001).

recorder are transmitted acoustically to the surface buoy, and then relayed via satellite to ground stations.

Most waves generated by large earthquakes are sinusoidal waves with heights normally less than a meter, lengths of 10–500 km, and periods of 100–2000 s. Several waves are normally generated, and the position of the largest wave in the train is variable. These waves can be considered to be Airy shallow-water waves (wave length/water depth > 20) in water depths greater than a few tens of meters, because they have small amplitude and great length relative to the water depth. The celerity of Airy shallow-water waves decreases as the square root of depth (Equation (7.7)), and wave speeds (celerities) are typically $160\text{–}250\,\mathrm{m\,s^{-1}}$ in the open ocean, $30\text{–}85\,\mathrm{m\,s^{-1}}$ on continental shelves, and on the order of $10\,\mathrm{m\,s^{-1}}$ near the coast (Bryant, 2001). As such waves propagate away from the generation area, the long-period waves travel faster than do waves of shorter period (wave dispersion) such that the long waves reach coastlines first. Over long-distance travel, wave paths must be corrected for geometric spreading on a spherical surface.

Important transformations of Airy waves occur as they move into shoaling water and become more influenced by the bed, as discussed above. The waves become increasingly peaked (trochoidal), asymmetrical in profile, higher, and shorter. They change from Airy waves to Stokes waves. Some waves become solitary waves or N-shaped waves (Figure 7.33). Instead of water particles beneath the waves following closed circular orbits, the orbits become increasingly elliptical and unclosed. This results in net drift of near-bed fluid in the direction of wave propagation, but in the opposite direction nearer the water surface. The maximum near-bed orbital velocity is given approximately by Equation (7.11). Maximum velocities near coasts are typically $2\text{–}4\,\mathrm{m\,s^{-1}}$, which is similar to the velocity of large storm waves.

As shallow-water waves move over sea-bottom topography, especially that associated with coasts, they are refracted (due to variation of wave speed with depth). Wave diffraction occurs as waves pass through narrow spaces between land areas such as islands. Tsunami waves are always reflected at the coast, irrespective of bed slope and wave height. This means that edge waves should also be present. Resonant amplification of reflected waves, and of edge waves and incoming waves, means that wave heights will vary substantially along a coastline, and the highest wave at the coast may end up not being the largest incoming wave. Resonant amplification can double the height of an incoming wave.

The geometry and water motions of tsunami waves can be modeled numerically using the Navier–Stokes equations for low-height, long waves, considering also friction at the bed and the Coriolis force (e.g., Mader, 1974, 1988; Satake, 2005). The equations are solved using nonlinear, finite-difference methods using variable grid sizes and shapes. This requires detailed knowledge of the bed topography. Figure 7.37 shows one of

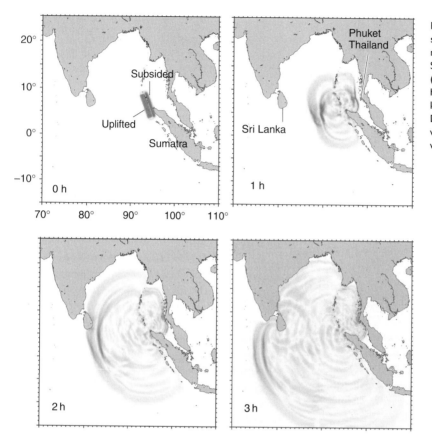

FIGURE 7.37. Numerical simulation of tsunami wave motions following the Sumatra–Andaman earthquake (from an animation available at http://staff.aist.go.jp/ kenji.satake/animation.gif). Dark areas are amplitude highs, with darkness increasing with wave amplitude.

several simulations of tsunami-wave motion for the Sumatra–Andaman tsunami. Accurate simulations can be produced within hours of tsunami initiation.

Tsunami waves tend to surge up the shore rather than to break. The surge of a tsunami wave can move inland for distances of hundreds of meters to kilometers, and reach elevations that exceed the maximum water level at the coast by a factor of two or more (meters to tens of meters). The distance inland reached by the run-up increases with the water level at the coast, and decreases with the flow resistance (bed roughness). With the 2004 Sumatra–Andaman tsunami, the surging water in Sumatra traveled inland for up to kilometers, and the water depth of the surge at Banda Aceh exceeded 20 m with run elevations in excess of 35 m (Moore *et al.*, 2006). If the depth of surging water is similar to the wave height at the coast, run-up velocities are typically meters per second, and return flows are probably faster. The water-surface slope is typically on the order of 0.001. Run-up height, length, and flow velocity can be predicted accurately by numerical solution of the shallow-water, long-wave equations.

Sediment transport, erosion, and deposition associated with tsunami

Much of what is known about water flow, sediment transport, erosion, and deposition associated with tsunami is inferred, because scientists have been reticent about making measurements during tsunami activity. Flow velocities of near-bed currents near the coast and during run-up on the land are expected to be meters per second, comparable to those of large storm waves. These flows are capable of moving gravel-sized sediment (even boulders up to meters across) on the bed, and sand-sized sediment and finer in suspension. The bed forms expected under these conditions are a near-plane bed or dunes with lengths of tens of meters and heights up to meters (Chapter 5), as long as there is sufficient time to form them. The time of action of these currents is on the order of minutes (compared with seconds for storm waves), but this might not be enough time to form dunes. Tsunami surges are also capable of causing great destruction to buildings, animals, and vegetation, due not only to the force of the moving water but also to the hard objects carried by the flow. Tree trunks are

FIGURE 7.38. A 2004-tsunami deposit from Banda Aceh, Sumatra. Photo courtesy of Guy Gelfenbaum of the US Geological Survey (http://walrus.wr.usgs.gov/tsunami/sumatra05/).

readily snapped off near their bases by tsunami waves, and debarked by "tools" carried by the flow.

Erosion and deposition of sediment requires along-stream increases or decreases in flow velocity and sediment-transport rate. The most common type of deposit attributed to tsunami in coastal settings is a sheet of sand up to about 0.5 meters thick that thins in the landward direction (e.g., Dawson *et al.*, 1991; Shiki *et al.*, 2000; Bryant, 2001) (Figure 7.38). The sand overlies coastal mud, soil, or peat with a sharp base. Erosion of underlying material is suggested by the presence of rip-up clasts. The deposit may contain wood, leaves, and marine fossils such as large shells, foraminifera, and diatoms. The mean grain size commonly decreases upwards and landwards. Large isolated clasts up to boulder size can occur in these deposits. There is little information about the internal structure of these deposits, although diffuse planar laminae have been recorded. A lack of internal structures suggests rapid deposition from suspension. A series of stacked fining-upward units has been interpreted as deposits of individual waves comprising one tsunami event. However, there is no reason to suppose that each wave will deposit a fining-upward layer. The sharp tops of some units have been interpreted as due either to backwash or to the passage of a later wave. Attempts have been made to interpret tsunami-wave characteristics from the thickness and grain size of tsunami sand layers, using theoretical models. However, such approaches are not well developed, and a comprehensive sediment-routing model (Chapter 5) has not been used for this purpose.

In the case of the 2004 Sumatra–Andaman tsunami at Banda Aceh, erosion occurred near the coast, but deposition occurred further landwards, as far as the inundation (Moore *et al.*, 2006). Buildings, soil layers, and trees were eroded. The deposit is generally 5–20 cm thick, but locally reaches 80 cm, and tends to fill low spots (Figure 7.38). The deposit is very coarse to medium sand, decreasing in grain size upwards and landwards. In some places, up to three distinct layers can be discerned, indicating deposition from discrete waves. The sand is either structureless or planar laminated. Larger clasts of soil, coral, and concrete also occurred. Further from shore, broken trees and contents of buildings form a larger part of the deposit.

Tsunami sand deposits are very difficult to distinguish from sand deposited by a storm-surge washover (Tuttle *et al.*, 2004; Chapter 15). Storm surges produce only one depositional unit at a time, but it is common for units deposited at different times to be superimposed. Although multiple fining-upward units in supposed tsunami deposits might be formed by different waves in one event, they could also be formed by different events. It is also difficult to date ancient deposits accurately enough to link them to a specific event, whether it is a tsunami or a major storm surge. Large lateral extent is commonly cited as evidence for tsunami deposits, but the lateral extents of storm-surge washovers can be comparable.

Despite the uncertainties in the unambiguous recognition of ancient tsunami deposits, such interpretations have been used to infer the nature of ancient earthquakes (e.g., Atwater and Hemphill-Haley, 1997; Shiki *et al.*, 2000; Atwater *et al.*, 2005; Bondevik *et al.*, 2005; Rhodes *et al.*, 2006).

Other coastal deposits attributed to tsunamis are dunes in sands and gravels with heights of meters and lengths of tens to hundreds of meters, and chenier-like ridges of shell-rich sand with similar dimensions to dunes. Mounds of unsorted sediment (diamictons) also occur, although their origin is uncertain. Isolated boulders or clusters of imbricated boulders occur at the base of cliffs or marking swash lines. Boulders that are 1–4 m across require flow velocities of 5–10 m s^{-1} to move them. It has also been claimed (Bryant, 2001) that many large-scale depositional features of coastlines such as overwash fans, barrier beaches, and flood-tidal deltas might owe their origin to tsunamis. Although tsunamis might influence the form of such features, the minutes of tsunami flow are surely not long enough to effect a great deal of deposition, and episodic deposition by storm waves over many years is more likely.

In deeper marine environments, deposits of sediment gravity flows (Chapter 8) have been associated with earthquakes, and related also to tsunamis (e.g., Pickering *et al.*, 1991). The sediment gravity flows might have been initiated either by earthquakes, or by the cyclic wave loading of sediment by the tsunami waves. Sediment-gravity-flow deposits have also been related to tsunamis supposedly generated by meteorite impacts (e.g., Bourgeois *et al.*, 1988). However, exceptional sediment-gravity-flow deposits are not necessarily initiated by tsunamis. We conclude that interpretation of ancient deposits as due to tsunamis is far from clear cut, since there are normally other possible origins.

Erosional features in bedrock that have been attributed to tsunamis are pits, grooves, flutes, and potholes that could be formed by cavitation, and impacts and abrasion by moving grains (Bryant, 2001). Loose-joint blocks can also be moved from cliffs by high dynamic pressures, and either fall to the base of the cliff or are transported by the flow. None of these processes is unique to tsunami (see the discussion of erosion of bedrock coastlines in Chapter 15). The occurrence of such features well above normal high water level has been used to justify a tsunami origin. However, these features could be formed by storm waves acting during a previous period of higher relative sea level. Furthermore, it is unlikely that major erosion of bedrock could occur during the minutes that water flows during a tsunami. Erosion of bedrock requires many years of pounding by water and sediment.

8 Movement of sediment by gravity

Introduction

Sediment can be moved down Earth-surface slopes under the influence of gravity either as a coherent mass (e.g., a slump or slide) or with the grains dispersed within air or water (sediment gravity flows such as grain flows, debris flows, mudflows, and turbidity currents). The interstitial air or water within a sediment gravity flow acts as a lubricant or a means of grain support rather than the driving force. These mass movements may be subaqueous or subaerial, and are distinguished on the basis of their coherence, grain size, geometry, and mode of movement (Figure 8.1). As will be seen below, a particular type of mass movement may be transformed into a different type. The initiation of motion of sediment due to gravity is discussed first, followed by the mechanics of motion and the character of sediment deposited from sediment gravity flows. Many Earth-surface slopes are intimately related to the mechanics of sediment gravity flows. Sediment gravity flows are capable of inflicting serious damage to life and property.

Initiation of sediment motion on slopes

Sediment motion is initiated when the downslope component of the gravity force on a sediment mass exceeds the forces resisting the motion (Figure 8.2). The resistance to motion (or *yield strength*) is associated with cohesion and friction within the sediment mass. This force inequity can be expressed mathematically as

$$mg \sin \alpha > \tau_s = \tau_c + (mg \cos \alpha - L)\tan \phi_c \qquad (8.1)$$

where m is the sediment mass per unit area of slope, g is the gravitational acceleration (9.81 m s^{-2}), α is the slope angle, τ_s is the total shear resistance (yield strength), τ_c is the shear resistance due to cohesion (enhanced by cements, clays, binding vegetation, and

sticky organic matter), the term in parentheses is the effective force normal to the Earth's surface (equal to the normal weight component $mg \cos \alpha$ minus any upward-directed force L, such as that due to an excess of pore-water pressure over normal hydrostatic pressure), and $\tan \phi_c$ is the static friction coefficient, where ϕ_c is the maximum angle to which loose sediment can be piled up before it starts to avalanche downslope, which is referred to as the angle of initial yield. The angle of initial yield commonly varies from 32° to 40°, and increases with increasing packing density and angularity of grains. After an avalanche, grains come to rest at their residual angle of repose, which is commonly 26° to 32°. The equation for shear resistance on the right-hand side of equation (8.1) is called the Coulomb equation. Equation (8.1) can be rewritten as

$$\tan \alpha > \frac{(\tau_c - L \tan \phi_c)}{\rho g y \cos \alpha} + \tan \phi_c \qquad (8.2)$$

where ρ is the material density and y is the thickness of the potential mass movement.

In order to initiate motion (cause *failure*) on a previously stable slope, the downslope component of the gravity force must be increased and/or the forces resisting motion must be decreased. The downslope component of the gravity force on a given mass of slope material ($mg \sin \alpha$) can be increased by increasing the slope, and this can be accomplished by deposition of sediment at the top of the slope, by undercutting at the base of the slope, or by tectonic activity (e.g., an earthquake). The cohesion force (τ_c) can be reduced by physical disintegration of the slope material by weathering, by increasing the water content of clays, and by degrading organic matter. The effective normal force controlling friction ($mg \cos \alpha - L$) can be reduced by increasing the slope and by increasing the pore-water pressure. Excess pore-water pressure (over hydrostatic pressure) can be generated by rapid subaqueous **255**

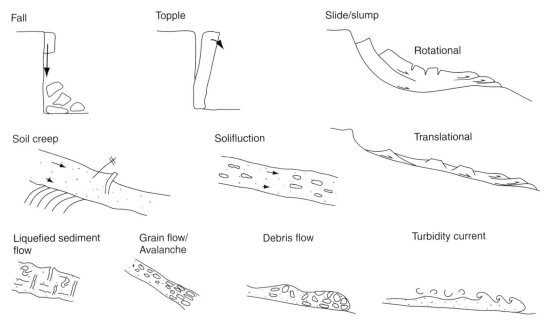

FIGURE 8.1. Main types of mass movements.

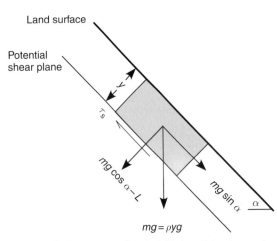

FIGURE 8.2. Forces controlling the initiation of shearing motion of sediment on a slope due to gravity.

deposition of sediment with low permeability (e.g., muds and fine-grained sands), by upward or outward movement of groundwater (such as through channel banks after floods or ebbing tides), by earthquake shaking, and by cyclic wave loading of the sea bed. The friction coefficient ($\tan \phi_c$) is decreased by wetting or shaking the sediment. In subaerial environments, heavy rains on saturated loose material devoid of binding vegetation are likely to initiate mass movements. Therefore, land areas that have suffered forest fires are prone to mass movements (Meyer *et al.*, 1992). Active

volcanic areas are particularly susceptible to mass movements (*lahars*) because of their high slopes, earthquake activity, extensive areas of poorly vegetated, fine-grained ash, and abundant water during eruptions from convective thunderstorms and melting of snow (e.g., Pierson and Scott, 1985; Best, 1992b; Pierson, 1995; Vallance and Scott, 1997). Climate and tectonic setting clearly play key roles in controlling mass movements. Eustatic sea-level changes also play a role, as explained in Chapter 20.

Rock falls, slides, and slumps

Initial failure of coherent rock or sediment masses can be by falling, toppling, or sliding (Figures 8.1 and 8.3). Shear planes in these different types of mass movement vary in their geometry and orientation, such that details of the force balance at the initiation of movement will be different in each case. Rock falls obviously require steep slopes and no support from below. Topples depend on tensile stresses rather that shear stresses. Slides vary greatly in the length and curvature of their basal sliding surfaces, and it is common to distinguish rotational slides (slumps) from translational slides. Rotational slides (slumps) commonly occur on relatively steep slopes such as stream banks, coastal cliffs, and artificially steepened slopes adjacent to buildings and roads. These slides are typically on the order of tens to hundreds of meters wide and long, and

FIGURE 8.3. (A) Rockfall in Washington State, USA. From Hyndman and Hyndman (2006). (B) Slump and slide from western Austria (photo courtesy of Henry Posamentier).

meters thick, and their length/thickness ratio is commonly on the order of 10^0; see the reviews by Prior and Coleman (1979), Allen (1985), and Ritter *et al.* (2002). Translational slides are typically longer (length/thickness ratios on the order of 10^1) and their basal surfaces are flatter. Translational slides may be hundreds of meters to tens of kilometers long, and meters to hundreds of meters thick. All types of coherent sediment slides have the following features in common (Figure 8.3): (1) a bowl-shaped erosional scarp at the upstream end, with backtilted, faulted blocks; (2) a central section, possibly channelized, with transported blocks, and tensional cracks and faults, possibly producing horst and graben topography; and (3) a lobate downstream end with a

series of slope-parallel ridges produced by folding and thrust faulting, and possibly sedimentary volcanoes (Chapter 12).

Once a coherent sediment mass starts to move downslope, it may, depending on its cohesion, remain coherent but become distorted by folding and faulting (as described above), or it may lose shear strength and disintegrate into a sediment gravity flow. Cohesionless sediment may disintegrate at the moment of failure. Apparently, the nature of disintegration and subsequent flow of fine sand depends on whether the mass failure was associated with liquefaction (excess pore-water pressure) or breaching (excess hydrostatic water pressure) (Van Den Berg *et al.*, 2002; Mastbergen and

Van Den Berg, 2003). A loosely packed fine sand may lose strength due to excess pore-water pressure (i.e., it becomes liquefied), and relatively low permeability inhibits upward escape of the pore fluid. In this case, a liquefied sediment flow may exist for a few minutes (until excess pore-water pressure is dissipated), unless it is transformed into a grain flow or a turbidity current (see below). Breaching failures occur as sand grains or thin slabs of sand start to move down oversteepened slopes. The initial oversteepening may be associated with channel erosion or with another kind of slope failure. Dilation of the shearing sand is inhibited by hydrostatic pressure, resulting in increased shear resistance. Thus sand must move down slopes that are steeper than the angle of repose at the top. A single breaching event may last for several hours, during which time the steep slope progressively retreats at a speed on the order of several $mm\,s^{-1}$. The rate of retreat increases linearly with the permeability of the sand. Breaching is important because subaqueous slopes can be eroded relatively rapidly, providing a long-lived, steady source of sand that can feed grain flows or turbidity currents.

The nature of motion of sediment gravity flows

Sediment is maintained in a dispersed state during downslope motion by grain collisions, cohesive strength associated with clays in suspension, buoyant support by the fluid–sediment mixture, turbulence, and in some cases upward-escaping pore fluid. Sediment gravity flows are commonly classified according to the *dominant* support mechanism(s), as follows:

- grain flows – grain collisions
- debris flows – cohesive strength and grain collisions
- turbidity currents – turbulence
- liquefied sediment flows – upward-escaping pore fluid

although this terminology varies slightly among workers in the field (e.g., Middleton and Hampton, 1973, 1976; Carter, 1975; Lowe, 1979, 1982; Nardin *et al.*, 1979; Middleton and Southard, 1984). On the basis of these definitions, grain flows, debris flows, and liquefied sediment flows are predominantly laminar (non-turbulent) flows. However, during the motion of a given sediment gravity flow, the different support mechanisms commonly vary in importance, and are

not mutually exclusive. Sediment gravity flows initially accelerate away from their point of origin and then decelerate to their final resting place. A grain flow or a debris flow that accelerates and becomes diluted with water or air may be transformed into a turbidity current. In the final stages of movement, turbidity currents (or the lower parts of them) may revert to grain flows or debris flows. If deposition is rapid, these flows may become liquefied sediment flows. More continuous movement of sediment gravity flows occurs if the sediment supply can be maintained, as with hyperpycnal (relatively high-density) plumes of sediment-laden water originating at river mouths, and with the breaching process described above.

Sediment gravity flows are resisted in their down slope motion by friction at their flow boundaries and internal shear resistance. This means that the flow velocity increases away from the flow boundaries, such that velocity gradients, $\partial u/\partial y$, exist (where u is the time-averaged velocity at a point, and y is the normal distance measured from the boundary, Figure 8.4A). If velocity gradients exist, shear stresses, τ, must also exist. The magnitude of the velocity gradient within a sediment gravity flow depends partly on the internal resistance of the flow to shearing, or its viscosity, μ_s. The relationship among the shear stresses, velocity gradient, and viscosity for laminar (non-turbulent) sediment gravity flows is expressed as

$$\tau = \tau_s + \mu_s\,\partial u/\partial y \tag{8.3}$$

where τ_s is the shear resistance (yield strength) due to cohesion and friction (Equation 8.1). Sediment gravity flows in which the viscosity is independent of the velocity gradient are called Bingham plastics (Figure 8.4B). However, the viscosity is dependent upon the velocity gradient in some sediment gravity flows (see Figure 8.4B and discussion below). The viscosity of sediment–fluid mixtures is a function of the fluid viscosity, μ (which is dependent on temperature), and the volume concentration of sediment, C (volume of grain/volume of grains plus fluid). A well-known empirical expression for the viscosity of sediment–fluid mixtures (Roscoe, 1953; Davidson *et al.*, 1977) is

$$\mu_s = \mu(1 - 1.35C)^{-2.5} \tag{8.4}$$

This equation predicts that the viscosity of a mixture is infinite when C reaches 0.74, which is the volume concentration of tightly packed grains that cannot be sheared. The fluid viscosity appears in this equation because the fluid is being sheared within the mixture.

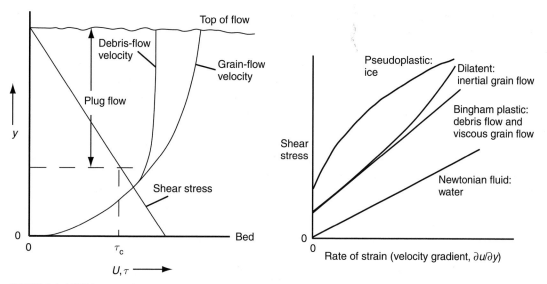

FIGURE 8.4. (A) Velocity gradients and shear stresses in sediment gravity flows. (B) A schematic representation of the relationship between shear stress and velocity gradient in sediment gravity flows, water, and ice.

However, a different approach to defining mixture viscosity was taken by Bagnold (1954, 1956), who maintained that mixture viscosity depends on grain concentration and fluid viscosity only at relatively low grain concentration and velocity gradient (Bagnold's viscous region). At high grain concentrations and velocity gradients (Bagnold's inertial region: Figure 8.4B), mixture viscosity depends on velocity gradient, and on grain concentration, diameter, and density, but is independent of fluid viscosity. This is apparently because the shear resistance is dominated by the relative motions of grains following grain collisions, and shearing of the interstitial fluid is relatively negligible.

Grain flows

Flow mechanics

The mechanics of grain flows are reviewed by Lowe (1976b), Savage (1979), Takahashi (1981, 2001), Haff (1983), Jenkins and Savage (1983), Campbell (1990, 2001), and Jaeger *et al.* (1996). Grain flows are laminar flows of cohesionless sediment. They are most common on the steep lee faces of bed forms such as ripples and dunes, but also include rock avalanches. Dynamic conditions similar to those of grain flows also occur within the bed-load layer of turbulent, sediment-transporting water and air flows, but the turbulent fluid rather than gravity is moving the grains. In order to demonstrate why grain flows are laminar

rather than turbulent, a flow Reynolds number can be defined as

$$Re = \rho_s U d / \mu_s \qquad (8.5)$$

where ρ_s is the mixture density, U is the mean flow velocity, and d is the grain-flow thickness. The viscosity of grain flows is commonly on the order of 10^2–10^3 Pa s, which is five to six orders of magnitude greater than that of water, and grain concentrations are commonly up to 0.5. Therefore, realistic values of $\rho_s = 2{,}000 \text{ kg m}^{-3}$, $U = 1 \text{ m s}^{-1}$, $d = 1 \text{ m}$, and $\mu_s = 100$ give a Reynolds number of 20, which is well below that needed for the transition to turbulence (500–2,000).

In order to maintain a steady, uniform grain flow, the downslope gravity force must equal the shear resistance at the solid lower boundary (ignoring the shear resistance at the upper boundary with air or water). Using Equation (8.1), and assuming that $L = 0$, α must equal ϕ for a steady, uniform grain flow to exist. Here $\tan \phi$ is called the dynamic (rather than static) friction coefficient. Experimental determination of $\tan \phi$ by Bagnold (1954) gave values between 0.37 and 0.75, with the value increasing from the inertial to the viscous region. There has been considerable debate over Bagnold's experimental technique and the veracity of his values of the dynamic friction coefficient, which has been reviewed by Bridge and Bennett (1992) and Straub (2001). Subsequent workers have increased the range of values from about 0.25 to 1. This means that grain flows require slopes of about 15° to 45° for steady, uniform motion.

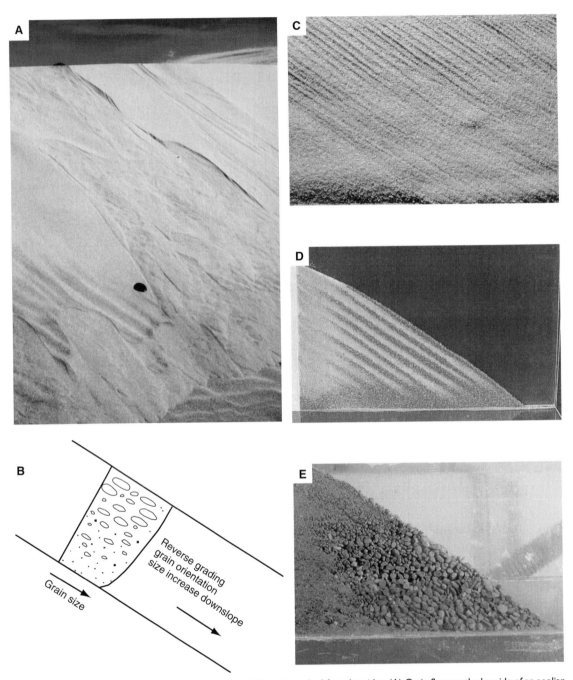

FIGURE 8.5. Grain-flow geometry, sorting processes, and deposits on bed-form lee sides. (A) Grain flow on the lee side of an aeolian dune. (B) Idealized sorting and fabric in grain flow. (C) An experimental grain-flow deposit, showing reverse-graded cross strata in medium sand. (D) An experimental grain-flow deposit, showing segregation of fine and coarse sand grains in cross strata, with coarse grains concentrated near the base. From Makse et al. (1997). (E) An experimental grain-flow deposit in sandy gravel, showing no cross strata but segregation of fine and coarse grains.

Grain flows down the lee sides of bed forms such as ripples and dunes travel at fractions of a meter per second. Such grain flows quickly stop moving at the base of the lee side, because of sudden reduction of the slope necessary for movement, such that the horizontal runout distance at the base of the slope is relatively small. These grain flows tend to have a lobate shape in plan, and are lenticular in cross section (Figure 8.5).

The length, width, and maximum thickness of these grain flows increase with the size of the wedge of sediment that accumulates on the lee side of the bed form prior to avalanching, which generally increases with the size of the bed form. For example, grain flows on ripples are millimeters thick and centimeters long, whereas grain flows on dunes are up to centimeters thick and meters long. As will be seen below, the time interval separating successive grain flows on bed-form lee sides is related to unsteady supply of sediment to the top of the lee slope. In contrast to grain flows on bed-form lee sides, rock avalanches in steep, tectonically active regions can reach speeds of tens of meters per second, and they tend to have long runout distances related to the initial speed of the avalanche, and relatively low frictional resistance due to intense grain collisions near the bed (Melosh, 1987). Excess air pressure at the base and upward escape of trapped air are apparently not causes for the mobility of these avalanches, because the density and viscosity of air are very small.

Sediment sorting and orientation in grain flows

The relative motion of grains of different sizes in grain flows results in preferential movement of the larger grains away from the flow boundary. This is most likely due to kinetic sieving (Middleton and Southard, 1984; Knight et al., 1993; Jullien et al., 2002), during which relatively small grains settle between the spaces created between the larger grains during grain shearing and upward dispersion. The dispersion of grains results in preferred downward movement of relatively small grains through the spaces between the larger grains. Kinetic sieving can be readily observed by shaking a can of mixed nuts sideways. The large nuts (brazils, walnuts, cashews, almonds) rise to the top and the small peanuts sink to the bottom. Some authors (Bagnold, 1954; Sallenger, 1979) have proposed that the upward movement of the largest and densest grains in grain flows is due to the existence of greater collisional forces (dispersive stress) between these grains. However, this mechanism has substantially been discredited (Legros, 2002), and kinetic sieving is undoubtedly more important. Nevertheless, it is the grain collisions that cause upward dispersion of the grains, allowing kinetic sieving to proceed.

The exact nature of kinetic sieving depends on the duration, speed, and thickness of the grain flow (which determine the effectiveness of grain-size sorting throughout the flow), and the relative sizes and shapes

of the grains (e.g., Makse et al., 1997; Kleinhans, 2004, 2005). Experimental data show that, in relatively thin grain flows, the larger grains move quickly towards the outer edge of the flow, and then move downslope more quickly than smaller grains. This results in a grain-flow deposit that becomes coarser-grained from the basal surface to the outer surface, but also coarsest at the base of the slope (Figure 8.5). If the largest grains are hindered in their movement to the base of the slope, due to frictional resistance, then the size sorting from basal surface to outer surface will be more marked. The frictional resistance depends on the relative sizes and shapes of the large and small grains. For example, if there is a big difference in size of the largest and smallest grains, the largest grains can more easily move down the outer edge of the grain flow. In addition, the laminar shear stresses can give rise to orientation of platy grains with their long axes more or less parallel to the flow boundary.

Grain-size sorting in grain flows on the lee sides of bed forms such as ripples and dunes is complicated by (1) temporal and spatial variation in the amount and grain size of sediment arriving at the top of the lee slope, related to flow unsteadiness and smaller bed forms superimposed on the host bed form; (2) intermittent grain flows at low sediment-transport rate becoming almost continuous at high sediment-transport rate; (3) grain-size sorting due to size-selective deposition of suspended sediment on the lee side in addition to sorting within grain flows; (4) limited travel distance of grain flows; (5) size-selective movement of sediment on the lee side by the water currents in the lee-side flow-separation zone; and (6) entrainment of grains from the top of a grain flow by a subsequent grain flow. These processes are discussed in detail in Chapter 5.

Grain-flow deposits

Grain flows stop moving when the driving force of gravity is reduced by a reduction in slope or flow thickness (due to lateral spreading). Deposition occurs by progressive reduction in relative motion of grains, starting at the top of the flow where the shearing stress is lowest, and working its way down to the base. The grain texture and fabric look essentially the same as during the final stages of movement. A single grain-flow deposit has a sharp, possibly erosional base, variably developed reverse grading, possibly an increase in grain size downslope, and platy grains oriented parallel to the basal shearing surface

(Figure 8.5). Grain-flow deposits are lenticular in shape, and range in thickness from millimeters to centimeters within ripples and dunes to over a hundred meters in the case of the megabreccias formed by rock avalanches (e.g., Selby, 1993; Yarnold, 1993; Friedman, 1998).

Debris flows

Flow mechanics

The geometry and mechanics of debris flows have been reviewed by Johnson (1970), Hampton (1972, 1975), Takahashi (1981, 2001), Costa and Wieczorek (1987), Nguyen and Boger (1992), Iverson (1997), and Whipple (1997). The videotape on debris flows by Costa and Williams (1984) is informative and entertaining. Debris flows are also predominantly laminar flows, but are characterized by containing cohesive mud, giving the consistency of wet concrete or cement. The cohesive, highly viscous, and dense muddy matrix is difficult to move away from the larger grains, and provides added buoyancy. The cohesive shear resistance, viscosity, and density of debris flows are typically 10^3–10^4 Pa, 10^2–10^3 Pa s, and 1,800–2,300 kg m^{-3}, respectively. Since shear stress decreases upwards in debris flows, shear stress in the upper parts may be less than the cohesive shear resistance (Figures 8.4 and 8.6). This part of the flow, where there is no velocity gradient, is called the *plug*. Maximum velocities are up to meters per second. Although debris flows are normally initiated on slopes exceeding 10°, they can run out onto slopes of 1° to 2°. A common feature of debris flows is wave-like surging of the flow (e.g. Major, 1997). Debris flows may be preceded and/or succeeded by hyperconcentrated flows or river flows. Indeed, some debris flows originate as water from a hyperconcentrated turbulent flow infiltrates into the ground.

Debris flows are commonly hundreds of meters to kilometers long, tens to hundreds of meters wide, and decimeters to meters thick. They commonly flow in channels that are bordered by levees or terraces constructed of debris flow deposits (Figure 8.6). The front of a debris flow has a lobate form (Figure 8.6A), and there may be more than one of these lobes at the end of a debris-flow channel. Also, each wave-like surge of a debris flow may have a lobate front (Major, 1997). The super-elevation of the tops of debris flows in curved channels, and the heights they attain as they run up sloping surfaces, have been used to estimate the velocity of the flow (e.g., Pierson, 1995). This calculation is

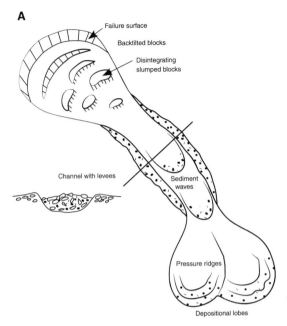

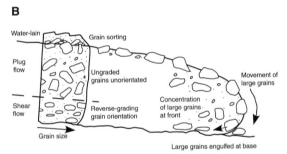

FIGURE 8.6. Debris-flow geometry, mechanics, and deposits. (A) Typical geometrical features of a debris flow, with distorted scale. (B) Typical sorting mechanisms and deposits of a debris flow. (C) A debris-flow channel with bouldery levees, near Portland, Oregon. Large boulders are concentrated at the top of the deposit. From Hyndman and Hyndman (2006). (D) A cross section of a debris-flow deposit (Saline Valley, California), showing ungraded, unoriented grains, and the largest sediment at the top of the deposit.

based on the assumption that all kinetic energy has been converted to potential energy at the highest point of the ascent.

The transition from laminar to turbulent flow in debris flows is not determined solely by the Reynolds number as in the case of grain flows and pure fluid flows, but also depends on the Bingham number (the ratio of cohesive shear strength to viscous shear stress), given by

$$Bi = \tau_c d/(U\mu_s) \qquad (8.6)$$

Figure 8.7 shows that, for Bingham numbers less than about 1, the Reynolds number is the sole determinant

C

FIGURE 8.6. Continued.

D

of the transition from laminar to turbulent flow. Thus, using realistic values of $\rho_s = 2{,}000 \, \text{kg m}^{-3}$, $d = 1 \, \text{m}$, and $\mu_s = 100 \, \text{Pa s}$, the transition to turbulence would require a mean flow velocity in excess of $25 \, \text{m s}^{-1}$, which is most unlikely. For Bingham numbers greater than about 10, the transition to turbulence occurs at progressively greater Reynolds numbers, because cohesive forces inhibit the development of turbulent eddies. With typical values of τ_c of $10^3 \, \text{Pa}$ for subaerial debris flows, the transition to turbulence at Bingham numbers in excess of 10 would require mean flow velocities greater than $25 \, \text{m s}^{-1}$. In subaqueous debris flows, grain concentration, viscosity, and cohesive strength

tend to be significantly lower than for subaerial flows. Such flows could more readily reach velocities sufficient to generate turbulence.

Sediment sorting and orientation in debris flows

The upper surfaces of gravelly debris flows commonly have boulders and cobbles that tend to become concentrated at the sides. If a channelized debris flow overtops its banks, these coarse grains are preferentially deposited as levees. Boulders and cobbles on the steep frontal surface of a debris flow may fall forwards and

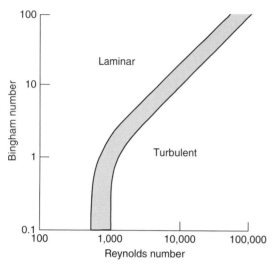

FIGURE 8.7. The laminar–turbulent transition zone for debris flows as a function of the Reynolds number and Bingham number. Modified from Hampton (1972).

down to the bed, and be overrun (and possibly oriented – imbricated) by the flow. Since there is little relative movement of grains in the plug, there is no mechanism for sediment sorting. Therefore, sediment within the plug is unsorted and has no fabric. However, in the basal shear zone, kinetic sieving is possible, and platy grains will be oriented parallel to the base of the flow by the laminar shear flow. Large grains in the base of the flow may be overrun by grains from the top of the flow, or may be eroded from the bed. Apparently, there is no obvious grain sorting associated with wave-like surges (Major, 1997). Sediment on the top surface of debris flows may be subsequently sorted by turbulent water flows, and gravel grains may be imbricated.

Debris-flow deposits

Deposition from debris flows occurs as they decelerate due to a decrease in bed slope or flow thickness, or loss of water to the bed. A decrease in flow thickness can occur as a channelized flow spreads out over its banks, or flows beyond its confining channel. This results in deposition adjacent to the channel as levees and deposition at the termination of channels as lobes (Figure 8.6). The geometry of the deposits will be the same as the final geometry of the flow: hundreds of meters to kilometers long, tens to hundreds of meters wide, and decimeters to meters thick. Debris-flow deposits change thickness laterally and in the downstream direction. As a flow decelerates and

deposition occurs, the plug extends towards the bed. Deposits of the plug would exhibit no grain-size sorting or grain orientation (Figure 8.6). However, it is expected that the deposits that were once part of the lower laminar shear zone will have grain orientation and inverse grading as in grain-flow deposits. Rapid deposition and upward escape of fluid may result in fluidization pipes (Best, 1992b). The base of the deposit will be sharp and erosional, and overlain by relatively fine-grained sediment with a few scattered coarser grains. Underlying sediments may be deposited by turbulent water flows. The tops of debris-flow deposits are commonly reworked by hyperconcentrated flows or streamflows, and may fine upwards if deposition occurred from these flows. Examples of subaerial and subaqueous debris-flow deposits (excluding volcaniclastic varieties called lahars) are given by Embley (1976, 1980), Damuth and Embley (1981), Pierson (1981, 1995), and Damuth et al. (1983). The seismic signature of large subaqueous debris-flow deposits is quite distinctive (see Chapter 18).

Turbidity currents

Introduction

Turbidity currents are turbulent density currents in which the excess density that drives their downslope motion is caused by suspended sediment (Figure 8.8). Turbulence created at the flow boundary helps maintain the suspension of sediment. The fluid in and above the turbidity current can be water or air. Turbidity currents under water commonly develop from grain flows, debris flows, or liquefied sediment flows as they accelerate and mix with more water. This may happen when a large volume of sediment is put into suspension by an earthquake, or when an oversteepened sediment slope slumps. These turbidity currents are relatively short-lived (lasting for hours) and surge-like. Relatively long-lived turbidity currents (lasting for days or weeks) may occur when dense, sediment-laden river water (hyperpycnal flow) flows into and beneath a standing body of water such as a lake or the sea (e.g., Prior et al., 1987; Mulder and Syvitski, 1995), and during the breaching process discussed above (Mastbergen and Van Den Berg, 2003). A subaerial, water-based grain flow or debris flow may flow into a standing body of water and become a turbidity current, and subaerial, channelized debris flows can also become turbidity currents (e.g., mudflows). Examples of subaerial turbidity currents involving air turbulence

FIGURE 8.8. The head region of an experimental turbulent density current, showing lobes and clefts in the front and billows along the upper surface. The density current is a milky saline solution flowing beneath freshwater. From Simpson (1997).

include avalanches of sediment and snow, and low-density pyroclastic surges. Turbidity currents commonly flow in self-formed or pre-existing channels, but also as unchannelized sheet flows.

Direct observations of major turbidity currents beneath the sea have not been made, which is not surprising in view of the fact that such turbidity currents may be hundreds of meters thick, and travel at speeds up to tens of meters per second. Inferences about the nature of submarine turbidity currents have been made on the basis of circumstantial evidence. For example, following the Grand Banks of Newfoundland earthquake in 1929, the sequential breaking of trans-atlantic telephone cables suggested a sediment gravity flow with a velocity of $7 \, \mathrm{m \, s^{-1}}$, and divers found fresh sediment deposited (Heezen *et al.*, 1954; Krause *et al.*, 1970; Piper *et al.*, 1999). However, cold, turbid river water flowing into clear lake water can be observed to flow a long way in subaqueous channels as turbidity currents.

As a result of difficulties in directly observing natural turbidity currents, many experimental studies have been undertaken; see reviews by Simpson (1982, 1997), Middleton (1993), Huppert (1998), Leeder (1999), and Kneller and Buckee (2000), and papers in McCaffrey *et al.* (2001). Experimental studies suffer from scaling problems, cohesive forces in the fine sediment used, and difficulties in measuring turbulent flow velocities in high-concentration, transient flows. Many experiments have been conducted using release of dense sediment from a reservoir at the upstream end of a flume (so-called lock–release experiments). These experiments suffer from excessive flow disturbance at the lock gate, reflection of the density flow from both ends of the flume, and overemphasis on the head of the turbidity current (Peakall *et al.*, 2001). More recent experiments with a steady supply of dense sediment upstream from the flume entrance are more realistic. New flow-measuring techniques for measuring turbulence in dense suspensions are also being developed (McCaffrey *et al.*, 2001).

Theoretical modeling of turbidity-current flow is also very difficult, because of the transient nature of the flow, significant spatial variations in turbulence and sediment concentration, and the intimate relationship between the sediment concentration and the nature of the turbulence. Simple models for the motion of turbidity currents assume steady, uniform flow and describe depth-averaged flow properties (e.g., Middleton, 1966a, b; Middleton and Southard, 1984). So-called box models (e.g., Dade and Huppert, 1994; Hogg and Huppert, 2001) are used to model the flow rate of finite-volume, low-concentration, surge-type turbidity currents moving into an ambient flow such as in a lock–release experiment, and also use depth-averaged flow properties. Box models have been used successfully to describe the length of a turbidity current as a function of time, and the spatially varying deposit thickness, under certain restricted conditions. The depth-averaged Navier–Stokes equations have also been used for modeling turbidity currents (e.g., Pantin, 1979; Parker *et al.*, 1986; Zeng and Lowe, 1997). However, depth-averaged flow models are not capable of describing many aspects of turbidity-current flow (e.g., Leeder, 1999; Kneller and Buckee, 2000). A complete description of the vertical and lateral variation in turbulence and sediment concentration in unsteady, non-uniform turbidity flows is most desirable; however, such numerical models have only recently been developed (e.g., Felix, 2001; Necker *et al.*, 2005), and are almost impossible to test by virtue of the lack of appropriate experimental data.

Flow mechanics

The density of turbidity currents in excess of that of the surrounding fluid, $\partial\rho$, is generally much less than that of other sediment gravity flows, because turbidity currents have much lower sediment concentrations. The density of a sediment–fluid mixture (from Equation (5.81)) is $\rho + (\sigma - \rho)\,C$, in which σ is the sediment density and C is the volumetric concentration of sediment. If ρ and σ are 1,000 and 2,650 kg m^{-3}, respectively, and C is of the order of 10^{-2}, then the dimensionless excess density, $\partial\rho/(\rho + \partial\rho)$, is also of the order of 10^{-2}. However, if C is on the order of 10^{-1}, the dimensionless excess density is on the order of 10^{-1}. Grain flows and bed-load-transport zones typically have grain concentrations of 0.3–0.5. Thus, turbidity currents with grain concentrations on the order of 10^{-1} may be referred to as *high-density*, whereas those with concentrations on the order of 10^{-2} or less may be referred to as *low-density*.

Turbidity currents normally flow with a recognizable head, body, and tail (or should we say front, middle, and back since turbidity currents are not animals?). The *head* region (Figures 8.8 and 8.9) is approximately twice the thickness of the body. The flow velocity in the head reaches a maximum at a distance above the bed of 0.1–0.2 times the head thickness (Simpson, 1997), as a result of friction at the upper and lower boundaries. The front of the head overhangs the bed and coincides with the position of maximum velocity. The flow diverges upwards and downwards from this position (Figures 8.8 and 8.9). Ambient fluid entrained at the base and front of the head gives rise to a pattern of flow-parallel lobes of relatively fast turbid fluid separated transverse to the flow by clefts of relatively slow sediment-free ambient fluid (Figure 8.9). The pattern of lobes and clefts is due to the inverse density gradient caused by turbid fluid overriding ambient fluid. The lobes contain counter-rotating Taylor–Gortler-type (streamwise) vortices, with near-bed flow diverging away from the center of the lobe towards the slower-moving fluid of the cleft. The lobe spacing is

proportional to the flow thickness. This flow pattern gives rise to erosion of flow-parallel furrows beneath the lobes, resulting in ridges beneath the clefts. As the flow-transverse spacing of the lobes and clefts is on the order of decimeters, the furrows may give rise to *gutter casts*. The longitudinal rill marks formed by viscous-sublayer streaks (Chapter 5) have a much smaller transverse spacing (about 10 mm). The upper boundary of the head has Kelvin–Helmholtz-type (transverse) vortices (eddies or billows) that become torn away in the turbulent wake behind the head (Figure 8.9). These turbulent eddies cause mixing of the turbidity current with the ambient fluid. Direct numerical simulation (DNS) of turbidity currents is capable of describing many of these features of fluid motion (Figure 8.9D); see Necker *et al.* (2005).

The mean velocity of the head of a turbidity current can be calculated theoretically in several ways (e.g., Benjamin, 1968; Middleton and Southard, 1984), and is given approximately as

$$U_h \approx k\{[\partial\rho/(\rho + \partial\rho)]\,gd_h\}^{0.5} \tag{8.7}$$

where k is approximately 0.7 for large Reynolds numbers, ρ is the density of the ambient fluid, $\partial\rho$ is the excess density of the turbidity current, g is the gravitational acceleration, and d_h is the thickness of the head. If $\partial\rho/(\rho + \partial\rho)$ is approximately 0.03 and d_h is 10 m, the velocity of the head is about 1.2 m s^{-1}. The velocity of the head is less than that of the body by 10%–20% because the head is continually losing turbid fluid in the turbulent wake, which reduces its excess density. This means that the body of the turbidity current is continually supplying turbid fluid to the head, and that the turbidity current progressively shortens once a steady body velocity is attained (i.e., the tail catches up with the head). This also ensures that a surge-type turbidity current has a finite "lifespan." However, the lifespan of a turbidity current is also limited by loss of excess density or thickness due to mixing with ambient fluid, lateral spreading, reduction in slope, and deposition. The lifespan of a surge-type turbidity current is typically on the order of hours, during which time it can travel a distance an order of magnitude greater than its starting length (Allen, 1985).

The vertical profile of time-averaged velocity in the *body* of a turbidity current is commonly divided into an inner region (beneath the maximum velocity, which occurs at 0.2–0.3 times the body thickness from the bed) and an outer region (above the maximum velocity) which includes an upper wake region (Figure 8.10). Kelvin–Helmholtz vortices occur along the top of the

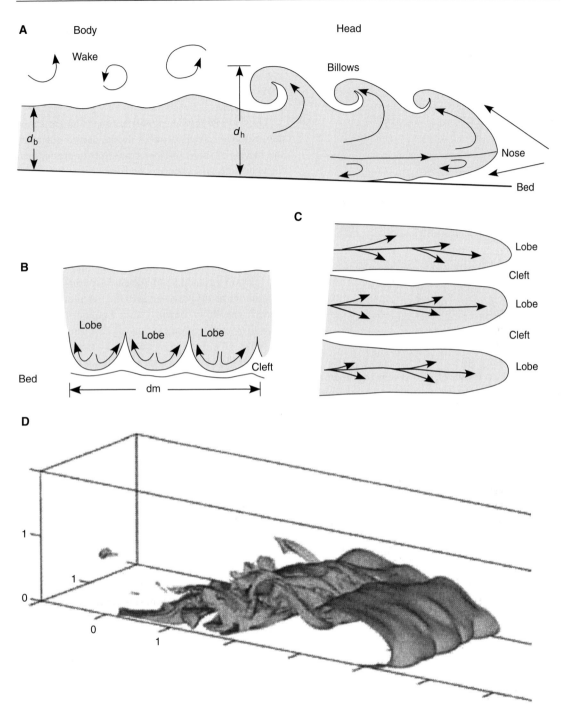

FIGURE 8.9. The mechanics of a turbidity-current head. (A) A vertical section parallel to the flow, showing the overhanging nose and billows (vortices). Arrows denote flow vectors relative to the mean flow velocity. (B) A vertical section normal to the front of the head, showing lobes of turbid fluid and clefts of entrained fluid near the bed. (C) A horizontal section near the bed at the front of the head, showing the pattern of flow in lobes and clefts. (D) A direct three-dimensional (3D) numerical simulation of a turbidity current following lock release. From Necker *et al.* (2005). The 3D shape of the 0.5 sediment concentration is shown at an instant in time. The flow direction is to the right.

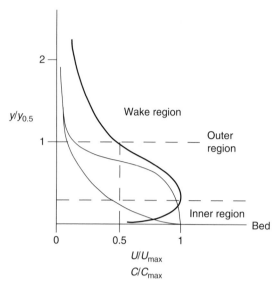

FIGURE 8.10. Typical vertical variations of the time-averaged flow velocity (thick line) and sediment concentration (thin lines) in the body of a turbidity current.

body as well as the head. If flow in the body of a turbidity current may be considered to be approximately steady and uniform, it is possible to obtain an expression for the mean velocity of the body of a turbidity current that is much wider than it is thick by balancing the downslope gravity force on the turbid flow with the frictional resistance at the upper and lower boundaries. The downslope gravity force per unit length and width is

$$[(\rho + \partial\rho) - \rho]gd_b S = \partial\rho \, gd_b S \qquad (8.8)$$

where d_b is the thickness of the body and S is the bed slope. The flow resistance per unit cross-sectional area is the sum of the shear stresses acting at the base, τ_o, and top, τ_i, of the body, which is a function of the mean flow velocity:

$$\tau_o + \tau_i = 0.125(f_o + f_i)(\rho + \partial\rho)U_b^2 \qquad (8.9)$$

where f_o and f_i are resistance coefficients for the bed and upper boundary, respectively. Equating the gravity and frictional forces gives

$$U_b^2 = 8gd_b S[\partial\rho/(\rho + \partial\rho)]/(f_o + f_i) \qquad (8.10)$$

This equation is analogous to the Darcy–Weisbach equation for steady, uniform river flow, except that for river flows there is no friction coefficient for the upper flow boundary (air resistance is negligible) and the relative density term is not necessary (air density is negligible compared with water density). In fact, the

lower part of a turbidity-current body is very similar to a turbulent boundary layer of a river flow (see Chapter 5), and a logarithmic velocity-profile law applies. The value of f_o is expected to be between about 0.01 and 0.04 for a planar to rippled bed of sand or mud.

The friction coefficient for the upper boundary of the turbidity current depends on the degree of turbulent mixing at this interface, which in turn depends on the excess density, velocity, and thickness of the current, as expressed using a densiometric Froude number:

$$Fr_b = U_b/\sqrt{gd_b[\partial\rho/(\rho + \partial\rho)]} \qquad (8.11)$$

If $Fr_b < 1$ (subcritical flow), there is relatively minor mixing and f_i is relatively small (on the order of 0.01). If $Fr_b \geq 1$ (supercritical flow), the degree of mixing and f_i increase markedly, and the turbidity current becomes diluted and slows down. Using Equation (8.10), if $(f_o + f_i) \approx 0.05$, $[\partial\rho/(\rho + \partial\rho)] \approx 0.05$, and slope $S = 0.017$ (1°), flow velocities of 1–10 m s^{-1} would occur for flows that are 1–100 m thick. Combining Equations (8.10) and (8.11), and assuming that $(f_o + f_i) \approx 0.05$, shows that supercritical flows would be obtained at slopes of 0.00625 (a fraction of a degree). Such slopes occur in submarine canyons and on deep-sea fans, so extensive mixing and dilution of turbidity currents are expected.

The turbulence intensity (or Reynolds stress) in the head and body of turbidity currents is greatest at the top and base, which is where the flow is decelerated, and where mixing between slow- and fast-moving fluid occurs (Kneller and Buckee, 2000; McCaffrey et al., 2001). The turbulence intensity increases at the top of the turbidity current if the densiometric Froude number exceeds unity. Density stratification within turbidity currents may decrease the turbulence intensity somewhat, but clearly does not suppress it entirely. However, as the velocity of a turbidity current decreases and deposition occurs, a high-density turbidity current may cease to be turbulent close to the bed.

Autosuspension is the term for the theoretical condition whereby the gravitational energy due to the excess weight of the turbidity current just balances the turbulent energy required to suspend the sediment (Bagnold, 1962; Pantin, 1979, 2001; Middleton and Southard, 1984; Parker et al., 1986). Under this theoretical condition, turbidity currents could flow steadily for large distances without dissipation. There has been much debate about the nature and existence of auto-suspension. According to Bagnold (1962), the rate of energy expenditure required to suspend sediment per unit volume of sediment is $(\sigma - \rho)gV_s$, where σ is the sediment density and V_s is its settling velocity. The energy input per unit volume of sediment by virtue of its downslope movement is $(\sigma - \rho)gSU$. Theoretically, when these two energy terms are equal, the moving suspended sediment creates enough turbulent kinetic energy to keep itself suspended. The criterion for auto-suspension is thus

$$SU > V_s \qquad (8.12)$$

According to this criterion, a turbidity-current body with a mean velocity of $1\,\text{m s}^{-1}$ on a slope of 0.0001 could approach autosuspension if the sediment settling velocity is less than $0.1\,\text{mm s}^{-1}$ (for a grain diameter of about 0.01 mm: silt). This analysis is seriously flawed for two main reasons. First, most of the energy associated with turbidity-current flow is required to overcome friction at the boundaries rather than for suspension of sediment. Also, increasing the concentration of suspended sediment causes suppression of the very turbulent kinetic energy required to suspend it. Defining the energy balance in turbidity currents properly is much more complicated than indicated above (Pantin, 1979; Parker et al., 1986). Second, the progressive mixing of the turbidity current with ambient fluid inevitably results in loss of excess density and thickness, progressively reducing the flow velocity. Nevertheless, autosuspension can apparently exist in the case of thick, fast-moving, fine-grained turbidity currents on relatively steep slopes.

Some turbidity currents can experience reversal of density gradient as they flow, as a result of changes in their temperature, salinity, or suspended-sediment concentration. For example, a freshwater, sediment-laden turbidity flow flowing into a marine basin may lose excess density by deposition, thereby becoming less dense than the saline marine water. Similarly, a hot pyroclastic surge may lose excess density by deposition, becoming less dense than ambient cooler air. Reversal of density gradient leads to *lofting* (rising of the current) and mixing with the ambient fluid (e.g., Stow and Wetzel, 1990; Sparks et al., 1993), which can ultimately lead to widespread deposition of relatively fine-grained sediment.

Turbidity currents commonly flow in submarine channels, and are capable of eroding them (see Chapter 18). Many submarine channels are sinuous, and include features that are familiar in rivers, such as point bars and bend cut-offs. Such channels may also be bordered by levees and crevasse splays, indicating that turbidity currents spill out of their channels. Overbank

flow of turbidity currents leads to stripping of the upper parts of the turbidity current from the main channelized flow (called *flow stripping*), resulting in changes in its character. Frictional resistance at the top of a sinuous, channelized turbidity current results in a pattern of secondary circulation that is different from that in a curved river channel (e.g. Peakall et al., 2000; Corney et al., 2006). However, the pattern of main flow and sediment transport in turbidity currents must be fundamentally similar to that in river channels because their geometries and processes are so similar (as discussed in Chapter 18).

Turbidity currents can run over and around topographic obstacles (such as basin margins, fault scarps, or submarine volcanoes), or be blocked by them and be reflected; see reviews by Leeder (1999) and Kneller and Buckee (2000), and also Van Andel and Komar (1969), Pickering and Hiscott (1985), Pantin and Leeder (1987), Kneller et al. (1991), Edwards et al. (1994), and Alexander and Morris (1994). The behavior of turbidity currents that encounter obstacles depends on the height of the obstacle relative to the thickness of the turbidity current, the velocity of the current, and the density stratification. A turbidity current can surmount an obstacle if its thickness is greater than the height of the obstacle. Some of the lower flow may be diverted around the obstacle or be blocked, whereas the upper part may flow over it (e.g., flow stripping). Flow in the lee of an obstacle may be separated, or may form standing waves and hydraulic jumps, depending on an internal Froude number (Figure 8.11). If a

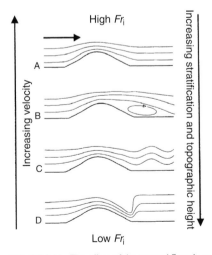

FIGURE 8.11. The effect of the internal Froude number on the behavior of turbidity currents downstream of topography. From Kneller and Buckee (2000). The internal Froude number is $U/(NH)$, where U is the mean current velocity, H the obstacle height, and N the frequency of unforced gravity waves on the density stratification.

turbidity current is blocked by an obstacle, the flow decelerates and thickens, forming a bore. The flow may also reverse direction and flow beneath and interact with the residual tail of the original flow. Internal, symmetrical, and translating waves (bores) may occur at the interface between the original and reflected flows. Deposition occurs where the flow is decelerated.

The nature of sediment transport

The typical vertical variation of concentration of suspended sediment in turbidity currents is shown in Figure 8.10. The concentration can either decrease exponentially upwards or decrease in a step-like pattern. The coarsest grains occur near the bed, but the concentration of the finest grains does not vary much vertically. The coarsest grains near the bed move as bed load, typically in the form of upper-stage plane beds and ripples. Given sufficient time, small dunes might form in relatively coarse-grained sediment. Wave-like bed forms may occur beneath sinusoidal waves to the lee of obstacles. In high-density turbidity currents undergoing rapid deposition from suspension, the high concentration of sediment near the bed, plus hindered settling and upward escape of interstitial fluid, may suppress turbulence to such a degree that upper-stage plane beds, ripples, and dunes cannot form.

Erosion and deposition by turbidity currents

The head is a region of erosion, and is apparently responsible for many erosional structures, ranging from flute marks, gutter casts, and tool marks up to channels and canyons that may be up to hundreds of meters deep and kilometers wide. Allen (1985) predicts that channels and flute marks are most common in the upstream (proximal) regions traversed by turbidity currents and that tool marks are more dominant in downflow locations. Furthermore, since the amplitude of flute marks decreases with the thickness and excess density of turbidity currents, flute marks should decrease in size in the down-flow direction.

Deposition occurs mainly from the body of a turbidity current. It occurs due to an initial sediment concentration in excess of what can be transported by the turbidity current, and subsequently by the sediment-transport rate decreasing in the flow direction. Such a downstream decrease in sediment-transport rate can be due to a reduction in flow velocity associated with spreading out (reducing thickness) or a lesser slope, or deceleration associated with obstacles. Deposition

typically occurs from channelized turbidity currents in submarine canyons and fan channels, and from sheet flows that spread out on the fan surface and adjacent abyssal plain. The deposit from a single turbidity current is typically centimeters to meters thick, and kilometers to tens of kilometers long.

It has been recognized since the early flume experiments of Kuenen and Migliorini (1950) and Middleton (1966a, b, c) that deposition from waning turbidity currents results in graded layers of sand and mud lying above sharp, erosional bases. Such deposits came to be called *turbidites*. This is bad terminological practice, because strata should really be named on the basis of what they look like, and many strata with the same sedimentological characteristics are not formed from turbidity currents. A well-known model for deposition from turbidity currents was proposed by Bouma (1962), on the basis of a study of the Annot Sandstone in France. This one-dimensional "Bouma sequence" (Figure 8.12A) now appears in most sedimentology textbooks, despite its serious shortcomings. Interpretations of the Bouma sequence have evolved somewhat over the years. The erosional base of the sequence is taken as due to the head of the turbidity current, whereas deposition is taken as due to the body. The basal, normally graded sand unit is associated with rapid deposition from suspension. The planar stratified sand unit is associated with deposition on an upper-stage plane bed, indicating flow deceleration, reduction of deposition rate, and establishment of a well-defined bed over which there was both bed load and suspended load movement. The small-scale cross-stratified sand unit (climbing-ripple type) indicates a further decrease in flow velocity but a still appreciable deposition rate. Convolute lamination is common in this unit, due to high pore-fluid pressure. The overlying planar laminated mud unit is associated with deposition from the tail of the turbidity current. As seen in Figure 8.12A, the units within the Bouma sequence have been assigned acronyms (T_a, T_b, etc.) and a unique interpretation. However, sequences of grain size and sedimentary structure similar to the Bouma sequence can be formed by unidirectional water flows (see Chapter 5), such that use of acronyms (and their attached interpretations) should be approached with circumspection.

Apart from the obvious terminological problems associated with the Bouma sequence, several other serious oversimplifications have been identified, including the following. (1) The sequence is an idealized composite of many different types of sequence. (2) A single sequence cannot represent all types of vertical

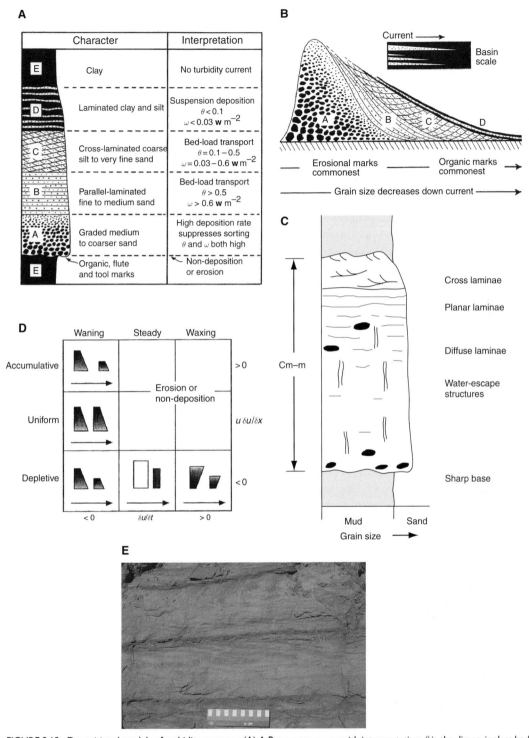

FIGURE 8.12. Depositional models of turbidity currents. (A) A Bouma sequence with interpretation; θ is the dimensionless bed shear stress and ω is the stream power per unit bed area. From Allen (1985). (B) The hypothetical lateral variation of a Bouma sequence. From Allen (1985). (C) A typical deposit of a high-concentration turbidity current. Based partly on Lowe (1982). (D) Hypothetical vertical and lateral variations in thickness and grain size of turbidity-current deposits as functions of the spatial and temporal variations in flow velocity and sediment-transport rate. After Kneller (1996). (E) Distal turbidites, showing a basal graded unit overlain by climbing ripple cross strata, with some convolute lamination (photo courtesy of Henry Posamentier and Roger Walker).

sequences, or lateral variations in the sequences. (3) The Bouma sequence applies only to a limited grain-size range, that is fine sand to mud. (4) The effects of high sediment concentrations and high deposition rate on bed forms and sedimentary structures are not recognized. (5) The effects of reversal of flow direction associated with obstacles are not recognized. These points are discussed in turn.

The Bouma sequence is idealized in that it is based on a compilation of many different types of sequential arrangement of the four basic units. Construction of this type of idealized sequence, which was popular in the 1960s, was an attempt to provide a simple summary model based on the limited knowledge of depositional mechanics at the time, rather than a comprehensive explanation. In reality, the idealized sequence is quite rare in the Annot sandstone and many other postulated deposits of turbidity currents.

A single, one-dimensional sedimentary sequence obviously cannot represent lateral variations in the deposits of turbidity currents. Subsequent workers have recognized such lateral variations (e.g., proximal to distal variations) in their models (Figure 8.12B); see Allen (1985) and Walker (1965). Lateral variations in the thickness of single turbidity-current deposits, and lateral and vertical variations in grain size and sedimentary structures, depend on spatial and temporal variations in flow velocity and sediment-transport rate (Kneller, 1996); see Figure 8.12D and Chapter 5. This is true for all unidirectional turbulent flows, and indeed it is very difficult to distinguish the deposits of turbidity currents from those of turbulent water flows.

The Bouma sequence applies to a limited range of grain size (fine sand to mud). Different types of depositional sequences occur with different grain sizes, because grain size influences possible bed forms and sediment concentrations (Stow and Shanmugan, 1980; Lowe, 1982). A persistent question has been that of why dunes and associated medium-scale cross strata apparently do not form beneath turbidity currents. The two main explanations have been that the sediment is too fine-grained for dunes to form (see Chapter 5) and that there is not enough time for dunes to form in a rapidly changing flow. Small dunes can easily form in the hours that a turbidity current can flow, and cross strata from dunes should occur in turbidity currents that transport sand and gravel.

Middleton (1966c) recognized that a gradual fining-upward sequence occurs only from deposition of low-concentration ($C < 0.3$) turbidity currents. For higher sediment concentrations and deposition rates, grain-size sorting and normal grading do not occur in the lower parts of the deposit (Middleton, 1966c; Lowe, 1982; Arnott and Hand, 1989; Allen, 1991; Kneller and Branney, 1995; Baas et al., 2004). The combination of high sediment concentration and rapid deposition rate gives rise to hindered settling, upward escape of pore fluid, and suppression of turbulence. This inhibits the formation of the characteristic bed forms associated with turbulent flow. The resulting deposit is unsorted and ungraded and may contain water-escape structures (vertical pipes and sheets, dish structures: Figure 8.12C and Chapter 12). Relatively large platy grains (e.g., mudstone clasts) that are oriented with their long axes parallel to the bed can occur within this unsorted sand bed (Postma et al., 1988). Such occurrences of oriented mudstone clasts might be considered to be typical of laminar grain flows and debris flows, because such clasts in turbulent flows are expected to be imbricated. However, if turbulence were suppressed in high-concentration turbidity currents during rapid deposition, grain imbrication would not be possible. Indistinct planar laminae and cross laminae commonly appear in the upper parts of these deposits, indicating their formation under lower-concentration, fully turbulent conditions (Figure 8.12C). The lack of sedimentary structures in these types of deposits led Shanmugan (1996, 1997) to suggest that they may have originally been sandy debris flows, especially if they contain a small amount of clay. Many other workers (e.g., Peakall et al., 2001) dismissed this suggestion.

Reversal of reflected turbidity currents results in multiple erosion surfaces and upstream-dipping cross strata within the deposit of a single flow (e.g., Pickering and Hiscott, 1985; Edwards et al., 1994). Some upstream-dipping cross strata in turbidity-current deposits had been interpreted as formed by antidunes, but have been re-interpreted as the deposits of reflected turbidity currents that reversed in direction. In fact, the most obvious feature of stratification formed by antidunes is superimposed trough fills (Chapter 5).

Liquefied sediment flows

Liquefied sediment flows occur when a sediment mass loses strength due to excess pore-water pressure (i.e., it becomes liquefied), and relatively low permeability inhibits upward escape of the pore fluid. In this case, a liquefied sediment flow may exist for a few minutes (until the excess pore-water pressure is dissipated), and can move down slopes as small as $3°$. During this dissipation period, the flow may be transformed into a

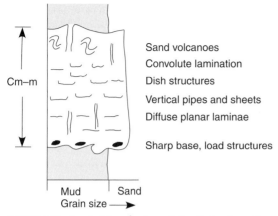

Cm–m

Sand volcanoes
Convolute lamination
Dish structures
Vertical pipes and sheets
Diffuse planar laminae

Sharp base, load structures

Mud   Sand
Grain size ⟶

FIGURE 8.13. A typical deposit of a sandy liquefied sediment flow.

grain flow or turbidity current. Rapid deposition from a grain flow, debris flow, or turbidity current may transform its basal part into a liquefied sediment flow. Thus, liquefied sediment-flow deposits may have the characteristics of the deposits of any of these other types of sediment gravity flows. Distinctive features are water-escape structures such as pipes, sheets, and dish structures (Lowe, 1976a); see Figure 8.13 and Chapter 12.

Solifluction

Solifluction is a type of sediment motion that occurs in permafrost regions where a layer of water-saturated surface sediment moves over the top of the permafrost. The surface layer becomes saturated because the permafrost is impermeable. The pore water acts as a lubricant, and the pore-water pressure may be elevated. The resulting sediment gravity flow is laminar, and may appear like a slow-moving debris flow. Flow rates range from 0.01 to 0.3 m per year (Saunders and Young, 1983), and maximum rates occur on slopes of 5° to 20°. Higher slopes facilitate drainage of water. Solifluction deposits are commonly decimeters to meters thick. Solifluction deposits will have platy grains oriented with their long axes parallel to the lower flow boundary in the zone where a laminar boundary layer exists. Otherwise there will be no preferred grain orientation or size sorting (Figure 8.14).

Soil creep

Soil creep is the imperceptible movement of soil particles downslope (Carson and Kirkby, 1972). Processes involved in downslope movement of soil include

breakage and reforming of clay-mineral bonds; expansion and contraction associated with freezing–thawing or wetting–drying; downslope filling of voids left by burrowing organisms, rotting vegetation, or mineral dissolution; washing of clay minerals downslope by groundwater; drag of snow cover; and contour ploughing (Figure 8.15). These mechanisms vary with the soil material and the climate. Soil creep is difficult to measure, but evidence of movement is obvious in the inclination of trees, fence posts, utility poles, and gravestones. The rate of surface movement ranges from 0.001 to 0.01 m per year (Saunders and Young, 1983). However, the rate of movement decreases downwards within the soil, and is negligible beyond a depth of about 1 m. A realistic velocity profile can be derived by assuming uniform, laminar flow and that the apparent viscosity of the soil increases downwards (Figure 8.15).

Earth surface slopes and sediment gravity flows

Earth surface slopes are closely related to the nature of the materials composing them and the modes of downslope movement of loose material, including sediment gravity flow, sheet flow, and streamflow. For reviews and compilations see Carson and Kirkby (1972), Young (1972), Brunsden and Prior (1984), Abrahams (1986), Selby (1993), and Ritter et al. (2002). It is common to distinguish transport-limited slopes (where sediment production by weathering exceeds the removal rate) and weathering-limited slopes (where the removal rate exceeds sediment production by weathering). Transport-limited slopes are common in humid climates and on low-relief landscapes, whereas weathering-limited slopes are common in arid climates and areas of high relief.

On transport-limited slopes, it is common to recognize an upper, convex part of the slope (ranging from 0° to 4°), a lower concave part (up to 4°), and a relatively straight section in between (Figure 8.16A). Sediment movement over the upper, convex part is attributed mainly to soil creep, and the convexity is related to the need for a downslope-increasing slope to allow creep movement of an increasing volume of weathered material. The volume of weathered material produced increases downslope as the catchment area increases. However, another explanation for the steep, upper parts of slopes may be the abundance of source areas for slides and slumps. The relatively straight section is an indication of steady, downslope motion by a dominant

FIGURE 8.14. A solifluction lobe (A) and solifluction deposits (B) from northern Sweden. Photos courtesy of Jan Boelhouwers.

type of sediment gravity flow (e.g., creep, debris flow, grain flow). The slope of this segment will depend on the flow mechanism, which is controlled by the climate, the overall slope, and the types of materials available for transport. The lower, concave slope section is a depositional region, and is related to a combination of sediment transport by sheet flow and streamflow, plus runout from debris flows or grain flows (Figure 8.16A).

Weathering-limited slopes tend to have relatively steep slopes of bare rock at the top (beneath a short convex slope) with a talus slope below, and then possibly a concave section at the base of the slope (Figure 8.16B). The bare rock section is dominated by rock falls, topples, and slides, and the slope is proportional to the rock strength (Selby, 1993). The slope of the talus is

typically 25° to 45°, depending on the materials supplied and the flow mechanism, which is mainly grain flow, with some creep and debris flow if clay is present.

A slope can be considered to be in a steady state (called dynamic equilibrium) if the input of sediment from above an area of slope is balanced by the output of sediment from that area. The output comprises the material transport through the area plus the new sediment produced by weathering. If the supply rate from above increases (for example, due to tectonism, climate change, or degradation of vegetation), the slope must steepen to maintain steady downslope motion, and this may involve a change in the dominant flow mechanism. The sediment that arrives at the base of a subaerial slope is normally transported away by river flows (less

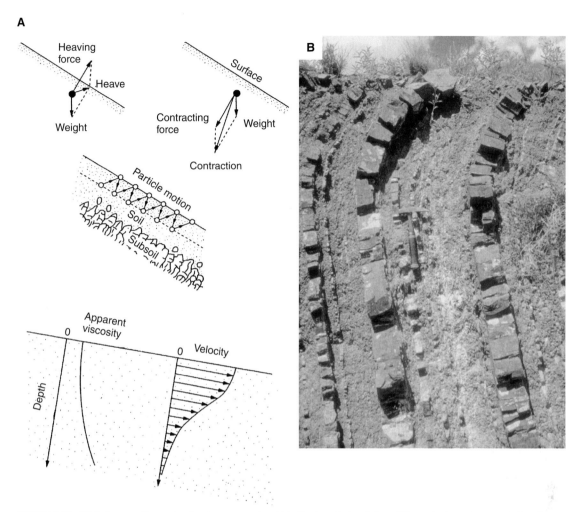

FIGURE 8.15. (A) Soil creep: the mode of grain movement and idealized vertical variation of flow velocity and apparent viscosity. After Allen (1985). (B) Bending of sedimentary layers to the right (downslope) due to soil creep. From Hyndman and Hyndman (2006).

commonly, by air flows). If the fluvial erosion rate exceeds the rate of supply by mass wasting and sheet flow, a steep erosional scarp will form and migrate upslope until the hillslope is steep enough to transport more sediment. If the fluvial erosion rate is less than the supply rate from the hillside, the lower part of the hillslope will decrease in slope. These principles are relevant to human activities on slopes, as will be seen below.

As loose material is transferred from the upper parts of slopes to the lower parts, and then possibly to river systems, the net loss of material results in slope retreat over the long term. There has been a long debate in the literature on exactly how slopes retreat, and one reason for this uncertainty is that it is not possible to observe directly how slopes change over long periods of time. Furthermore, it is not certain in many cases whether

the slopes observed today were formed by modern-day processes. Figure 8.17 shows the main types of slope retreat that have been proposed, but it is difficult to understand why these different types occur. Some light has been shed on this problem by quantitative theoretical modeling; see reviews by Bridge (2003, 2007). The basis of these models is the sediment continuity equation, which in one-dimensional form is

$$-C_b \, \partial h/\partial t = \partial i_x/\partial x + \partial i_x/(u_{sx} \, \partial t) \qquad (8.13)$$

where C_b is the volume concentration of sediment in the bed (1 − porosity), h is the bed height, t is time, i_x is the downstream sediment-transport rate by volume, x is the downstream distance, and u_{sx} is the downstream velocity of the sediment. The second term on the right-hand side is normally considered negligible. The next

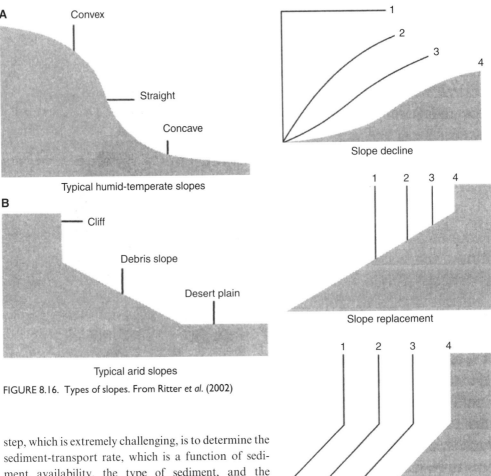

FIGURE 8.16. Types of slopes. From Ritter *et al.* (2002)

FIGURE 8.17. Types of slope retreat. From Ritter *et al.* (2002).

step, which is extremely challenging, is to determine the sediment-transport rate, which is a function of sediment availability, the type of sediment, and the mechanics of sediment transport (sediment gravity flow, sheet flow, streamflow). It is necessary to determine quantitatively what controls sediment production by weathering, where the various mass-movement processes occur, whether water flow is important, and whether it is in the form of sheet flow or streamflow!

Sediment transport on hillslopes associated with soil creep is commonly modeled using a linear diffusion approach, meaning that the sediment-transport rate is assumed to be proportional to slope, and the proportionality coefficient is called a diffusion coefficient. Diffusional transport may be limited by production of sediment by weathering. Landslides have been modeled by considering the shear strength of rock or soil and whether or not this material exceeds some critical hillslope angle. The rate of erosion of bedrock and cohesive sediment by moving surface water is commonly taken as proportional to $kA^{m}S^{n}$ in excess of some threshold value, where k is the bedrock erodibility, A is the area of the drainage basin, and S is the surface slope. Here A is a surrogate measure for stream water discharge, such

that the bedrock erosion rate becomes a function of stream power per unit channel length. The exponents m and n are not known well, although they are commonly assumed equal to unity. Sediment transport in channels and by overland flow is also commonly taken as proportional to a function of stream power in excess of some limiting value. Sediment transport by flowing water may also be limited by sediment production by weathering. Sheet flow and channel flow are not always explicitly distinguished in these models. In some models, the formation of channels is dependent on a channel-initiation function (which is proportional to functions of water discharge or drainage area and slope) exceeding some specified threshold.

Application of these modeling approaches to hillslope evolution shows that diffusional transport results in convex upward slopes and a gradual decline in slope over time (Figure 8.17). The concave-upward, lower parts of these slopes require sediment deposition or transport by water flows. To the best of our knowledge, slope replacement and parallel retreat (Figure 8.17) have not yet been modeled successfully. Parallel retreat requires that the various mass-wasting processes do not vary appreciably in time. By analogy with stream-bank profiles, parallel retreat requires active removal of sediment from the base of the slope (by river flow) and continued supply by slumping (or toppling or falling) followed by grain flow or debris flow. Continuing the analogy, slope replacement in stream banks occurs when the river is unable to remove the basal sediment but the steep upper slope continues to provide sediment to the debris/grain-flow apron.

Social and environmental impacts of sediment gravity flows

Numerous examples of serious loss of life and property resulting from mass movements have been reported in the scientific literature and the popular press, and they will not be repeated here. The most spectacular damage is caused by the physical impact of a large, rapid mass movement (damaging and burying people and buildings), the blockage of valleys (to create lakes), roads and railways, and displacement of water in lakes and the sea (causing water-surface waves, including tsunamis). Some of these examples are purely natural, whereas others are influenced by human activities. Humans help precipitate mass movement by undercutting and steepening slopes during construction of buildings, roads, and railways; creating potentially unstable slopes by piling up sediment to the angle of repose (e.g., during mining operations); and elevating groundwater levels (and thereby potentially raising pore-water pressures) and causing saturation of previously dry material during the building of dams and creation of reservoirs.

Mass movements are considered hazardous by humans only if they have significant impact on human life and property. An assessment of the mass-movement hazards in a region can be made (Selby, 1993) by researching previous surveys of mass-movement hazard; examination of time series of remote-sensing images (possibly with the assistance of GIS software) to determine active and past mass-movements; examination of geological, topographic, soil, and land-use maps to indicate potential sites for mass movement (on the basis of, for example, material types, orientation of potential failure planes, slopes, local groundwater tables); site surveys of specific mass movements, material properties of the surface and subsurface (requiring drilling), water-table levels, and potential sites for mass movement; and laboratory analysis of rock and soil samples.

A good way of avoiding natural mass-movement hazards is to move away from where they occur (e.g., next to active volcanoes and river banks). If this is not possible, various mitigation measures are available (Selby, 1993): removing or diverting the problem; reducing the ratio of moving/resisting forces; and implementation of warning systems. For example, water may be diverted away from a site of potential failure. Loose, potentially unstable blocks can be removed from a hillside. A road or railway line can be located away from a steep, unstable slope or put in a tunnel. Steep slopes may be stabilized by spraying with concrete, covering with steel-mesh curtains, use of rock anchors, or planting vegetation. Retaining walls can be built at the bases of slopes.

9 Generation and movement of volcaniclastic sediment

Introduction

Volcanoes cover only a small proportion of the Earth's land surface, and periods of activity of individual volcanoes are orders of magnitude less than inactive periods. The restricted spatial distribution and period of activity of volcanoes belie the importance of their effect on the Earth's surface, and specifically on (1) the chemical composition of the oceans and atmosphere; (2) the generation of airborne and waterborne volcanic sediment; (3) Earth's climate, through their effect on gaseous and particulate matter in the atmosphere; (4) generation of volcanic sediment gravity flows and tsunami; (5) the formation of distinctive topography that influences directions of fluid flow and sediment transport, including diversion of rivers and formation of lakes; and (6) destruction of many forms of life. As discussed in Chapter 2, the evolution of the Earth's lithosphere, atmosphere, hydrosphere, and biosphere, and major topographic features such as oceans and mountains, are intimately associated with volcanism and tectonism.

There is great variability in the topography associated with volcanoes, in the manner of eruption, in the composition and texture of the material erupted, and in the mode of transport of the erupted material. Furthermore, these features can vary with time, and even during a single eruption. Volcanic eruption may be in the form of relatively non-violent effusions of lava (as with Hawaiian volcanoes) or by violent explosions (as during the 1991 eruption of Mt. Pinatubo in the Philippines). Volcanic eruptions can take place into the atmosphere or into water, or both. Processes of volcanic eruption are not known very well because it is dangerous to be too close to an erupting subaerial volcano, and deep submarine volcanic eruptions have barely been observed. Furthermore, mega-eruptions (such as Tambora, Santorini, and the eruption of Mt. Mazama

that produced Crater Lake) have not occurred since detailed geological records have been taken. The solid products of explosive eruptions are referred to as *volcaniclastic* or *pyroclastic* sediments, or simply as *tephra*. These sediments are transported and deposited by the wind and by various types of sediment gravity flows. Erupted volcanic material is usually reworked and altered after deposition, and volcaniclastic sediments are commonly difficult to interpret because of this complicated history. Table 9.1 lists a number of historic volcanic eruptions and, for each, the volume of magma erupted, the volume and broad type of pyroclastic material produced, and the volume of material reworked contemporaneously with the eruption.

The long-term record of volcanic activity comes from ancient volcanic rocks such as lava flows, fills of volcanic vents, and volcaniclastic sediments. Volcaniclastic sedimentary rocks comprise about 27% of post-Archean sedimentary rocks (Fisher and Schminke, 1984). About $238,500\,km^3$ of Cenozoic pyroclastic deposits occur in the Great Basin of the southwestern USA, with deposits from individual eruptions of up to $2,500\,km^3$ (an order of magnitude more than from the Tambora eruption in 1815) and up to 70 meters thick. Volcanic rocks are particularly important for radiometric age dating, and have indicated that volcanic activity has been markedly episodic and localized throughout Earth history. Volcanic ash spread widely over the continents and oceans is very valuable for chronostratigraphic correlation of rocks over large distances. Table 9.2 lists some of the known Pleistocene eruptions in western North America, and Figure 9.1 shows the distribution of some of these ash beds.

Volcanic activity and its peripheral effects on Earth surface processes have had a fundamentally important effect on life throughout geological time. Degassing of the early Earth through volcanic activity led to the

TABLE 9.1. Examples of volcanic eruptions of the last 4,000 years (modified from Orton (1996) and Schminke (2004)). Note that, in most cases, much of the magma contributed significantly to the pyroclastic material.

Year	Volcano	Type of magma	Volume of magma (km^3)	Volume of pyroclastic fall (km^3)	Volume of pyroclastic density flows (km^3)	Volume of resedimented material (km^3)	Fatalities (approximate)
1991	Pinatubo (Philippines)	Dacite	~10	3	6	>4	800
1985	Nevado del Ruiz (Colombia)	Andesite	0.002		0.004	0.1	25,000
1980	Mt. St. Helens (USA)	Dacite	~0.2	1.2	~3	>0.1	60
1912	Katmai (Alaska, USA)	Rhyolite to andesite	13	17	11		?
1902	Pelée (Martinique)	Andesite	<0.2		~1		30,000
1883	Krakatoa (Indonesia)	Dacite	~10	1	~20		36,000
1815	Tambora (Indonesia)	Rhyolite	50	5	150		92,000
79	Vesuvius (Italy)	Phonolite	~4	6	4		3,300
3620 BC	Santorini (Greece)	Rhyolite	~30	25	29		?

TABLE 9.2. Large Pleistocene eruptions of western North America (from Sarna-Wojcicki and Davis, 1991)

Name	Age	Source	Volume (km^3)
Mazama	7 ka	Mt. Mazama ("Crater Lake"), Oregon	35
Rockland	400 ka	Lassen Peak, California	120
Lava Creek	620 ka	Yellowstone Park, Wyoming	1,000
Bishop	740 ka	Long Valley, California	500
Tshirege	1.15 Ma	Valles Caldera, New Mexico	300
Mesa Falls	1.27 Ma	Yellowstone Park, Wyoming	280
Otowi	1.47 Ma	Valles Caldera, New Mexico	300
Huckleberry ridge	1.97 Ma	Yellowstone Park, Wyoming	2,500

formation of the oceans and atmosphere. Periods of enhanced volcanic activity have led to global climate change (cooling), which may have played a role in major extinction events such as that at the end of the Cretaceous. Global cooling associated with single mega-eruptions such as that of proto-Krakatoa in the sixth century AD has been linked (not without controversy) to major migrations of land animals (including humans) and associated changes in human civilizations. There are many historic accounts of loss of life associated with volcanic eruptions (Table 9.1). However, volcanic activity also creates the conditions for renewal of life.

Explosive volcanicity and the formation of volcaniclastic sediments, and the effects of both lava flows and explosive eruptions on Earth surface processes and landforms, are the main topics of this chapter. General references on landforms, processes, and sediments associated with volcanic eruptions include Fisher and Schminke (1984), Heiken and Wohletz (1985), Cas and Wright (1987), Fisher and Smith (1991), Francis (1993), Orton (1996), Sparks and Gilbert (2002), Branney and Kokelaar (2003), and Schminke (2004).

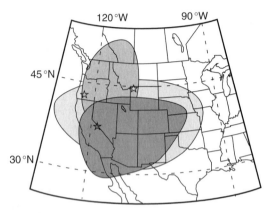

☆ Huckleberry ridge ash bed (Yellowstone caldera)

☆ Mazama ash bed (crater lake caldera)

☆ Bishop ash bed (long valley caldera)

FIGURE 9.1. The distribution of three Pleistocene ash beds, western USA. From Sarna-Wojcicki and Davis (1991).

The nature of volcanic eruptions and formation of pyroclastic sediment

Most (about 90%) active volcanoes today occur near (within a few hundred kilometers of) the constructional or destructional margins of tectonic plates (Figure 9.2), although some occur within plates above hot spots (e.g., Hawaii, Yellowstone). Volcanoes are found in four distinctive tectonic settings: divergent plate margins; convergent plate margins; oceanic plate interiors; and continental plate interiors. *Divergent plate boundaries* are associated mostly with submarine basaltic volcanicity on mid-oceanic ridges, although some ridges reach the water surface (e.g., Iceland). Basaltic volcanicity is not normally explosive because of the low viscosity of the magma and the small amount of volatiles (gases). *Convergent plate boundaries* contain most (80%) of the Earth's subaerial volcanoes (e.g., Vesuvius, Etna, Mt. St. Helens), and the volcanicity is typically rhyolitic and andesitic. Rhyolite volcanicity is typically explosive because of the high viscosity of lava and the abundance of volatiles. *Oceanic plate interiors* (e.g., Hawaii) are associated with basaltic volcanicity. *Continental plate interiors* (e.g., the East African Rift, Basin and Range, Yellowstone) are associated with basaltic and rhyolitic volcanicity. Examples of basaltic volcanicity are the Deccan Traps of India and the Columbia River basalts of the northwestern USA.

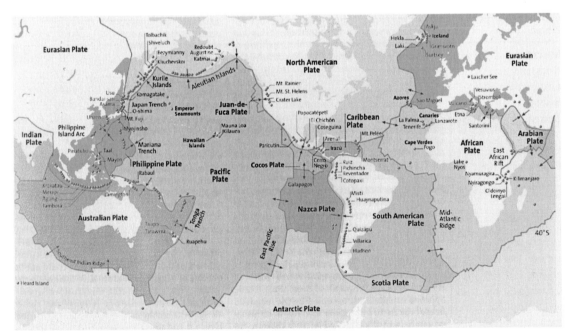

FIGURE 9.2. The distribution of volcanoes on Earth. From Schminke (2004).

FIGURE 9.3. Explosive eruptions. (A) Three regions in an explosive eruption column. After Orton (1996). (B) Eruption of Mount St. Helens. United States Geological Survey photograph taken on 18 May 1980, by Austin Post. (C) The 12 June 1991 eruption of Mt. Pinatubo, Philippines. United States Geological Survey photograph taken from Clark Air Base by Dave Harlow.

Compositional (density) zoning in magma chambers is common, and volatiles are concentrated near the top. Therefore, initial eruptions tend to be more explosive and acidic, whereas later eruptions may be more effusive and mafic. However, simultaneous eruption of mafic and felsic magmas is also possible.

Explosive, pyroclastic eruptions (Figure 9.3) are due to rapid expansion (increase in volume) of a gas phase (magmatic gas or water vapor). The volatiles in a rising magma become exsolved, resulting in an increase in gas pressure. If magma comes into contact with groundwater, surface water, or ice, the water or ice is superheated, and the transition to steam involves an increase in volume (and hence in vapor pressure). Explosive eruptions occur when the increased gas pressure exceeds the confining pressure of rocks, air, or water. Types of explosive eruptions are classified in terms of the degree of explosiveness (as measured by the degree of fragmentation of tephra) and the height of the eruption (as measured by the areal extent of the tephra).

The names given to the various categories of explosive eruptions are based on examples of volcanoes that have exhibited such eruptions (e.g., Hawaiian, Strombolian, Plinian). The term Plinian is used for the most explosive volcanoes and comes from Pliny the Younger, who described the eruption of Vesuvius in 79 AD. Other examples of devastating explosive eruptions are Tambora in 1815, Krakatoa in 1883, and Pelée (Martinique) in 1902.

Volcanic explosions result in ejections of molten drops of magma, and solid objects from the magma or blasted country rocks that become pyroclasts. Pyroclastic sediments are classified on the basis of their size and composition. Size terms for pyroclastic sediment are dust <0.063 mm; fine ash 0.063–0.25 mm; coarse ash 0.25–4 mm; lapilli 4–32 mm; blocks and bombs >32 mm. Blocks are fragments of pre-existing rock torn from the volcano by the explosion, whereas bombs are lumps of molten magma that solidified in flight. Compositionally, pyroclastic sediments are composed of three components: rock fragments, crystals, and broken bits of volcanic glass called *shards*. Rock fragments include pieces of pre-existing rock in the volcano (blocks: Figure 9.4A), or magma that solidified in flight (bombs). *Pumice* is rock composed of solidified glass froth, full of bubbles of gases dissolved in the magma, that exsolved in flight. Crystals commonly comprise broken euhedra of dipyramidal high-temperature quartz, sanidine (high-temperature K-feldspar) with common oscillatory zoning (Figure 9.4B), biotite flakes, amphiboles, pyroxene, and olivine. Glass shards (Figure 9.4C) are the most common and diagnostic component of pyroclastic sediment and form from blasted-apart froth of magma and exsolving gas. Glass shards are isotropic and commonly pale yellow. *Accretionary lapilli* are composed of one or more concentric layers of (originally wet) ash around a nucleus such as a crystal (Figure 9.4C). Volcanic glass weathers (or devitrifies) rapidly, and common alteration products include smectite (bentonite), chlorite, zeolite, opal, and chalcedony. Pyroclasts may be hot enough to be welded together, and welded particles are typically flattened (Figure 9.4D, E). The variability in composition and vesicularity of pyroclasts means that their density is very variable. Volcaniclastic sediments or sedimentary rocks are called tuff if composed mainly of ash, lapillistone if composed mainly of lapilli, and breccia or agglomerate if composed mainly of bombs (agglomerate) or blocks (breccia). The adjectives lithic, vitric, and crystal are used to connote predominance of rock fragments, glass shards, or crystals, respectively. Welded tuffs are sometimes called ignimbrites.

A typical pyroclastic eruption has three regions based on the relative importance of momentum and buoyancy of the erupted material (Figure 9.3A): (1) a lower gas-thrust region due to decompression and expansion of gas, with height of hundreds of meters to kilometers depending on initial velocity; (2) a convective region where the density of hot erupted material is lower than that of ambient air; and (3) an umbrella region where the density of the erupted material is the same as that of the surrounding air, but the eruption still has some upward momentum. The umbrella region can extend tens of kilometers above the Earth's surface, and a plume of wind-transported ash and dust starts there. Pyroclastic particles can be transported by the wind in the upper atmosphere (stratosphere) for long distances: many times around the Earth. Atmospheric dust following major eruptions results in spectacular sunsets. Figure 9.5 shows a famous painting by William

Ashcroft of the sunset on November 26, 1883 from the banks of the River Thames in Chelsea, London. The red colors are due to the dust in the atmosphere caused by the August 1883 eruption of Krakatoa. The eruption of Tambora in early 1815 is commonly cited as the cause of "the year without a summer" (1816) in northeastern North America, maritime Canada, and northern Europe. There were killing frosts into May and two June snowstorms in New England and New York that year.

The regions shown in Figure 9.3A are different depending on whether an eruption is subaqueous or subaerial, and the effects of water and steam on the eruption must be considered. With subaqueous eruption, expansion of magmatic volatiles or water in contact with magma is suppressed by the high confining pressure of ambient water. Also, there is greater resistance to upward movement than there is in air. Thus, the vertical extent of the gas-thrust region is reduced, and there is less mixing and heat exchange with the upper layers of the ambient water. Transport of ejecta through water into air is possible only in shallow water (<50 m deep). Pyroclasts erupted into either air or water either fall individually under the influence of gravity and ambient currents (sediment settling), or flow in sediment gravity flows, as discussed below.

Pyroclastic sediment settling (pyroclastic fall)

Relatively coarse-grained particles that settle locally from the plume margins can be broadly distinguished from the ash and dust that settle from the laterally extensive, high-level umbrella region (Figure 9.6). The largest particles (bombs and blocks) follow essentially ballistic trajectories and fall within a few kilometers of the volcanic vent. These particles may cause impact structures called *bomb sags* (e.g., Allen, 1982a), see Figure 9.4A. Coarse-grained ash and lapilli tend to fall out of the convective region, and their trajectories are influenced by upward-directed convection currents (Figure 9.6). The finest-grained ash and dust (usually constituting the greatest volume of pyroclasts) are transported large distances laterally by the wind before settling out, and their trajectories are influenced by air turbulence throughout their transport (Figure 9.1). Grains of differing size and composition can be transported in different directions in different layers of the atmosphere. The coarsest-grained and densest ash and dust settle out first, so that the grain size, grain-size range, and density of the cloud and deposited sediment

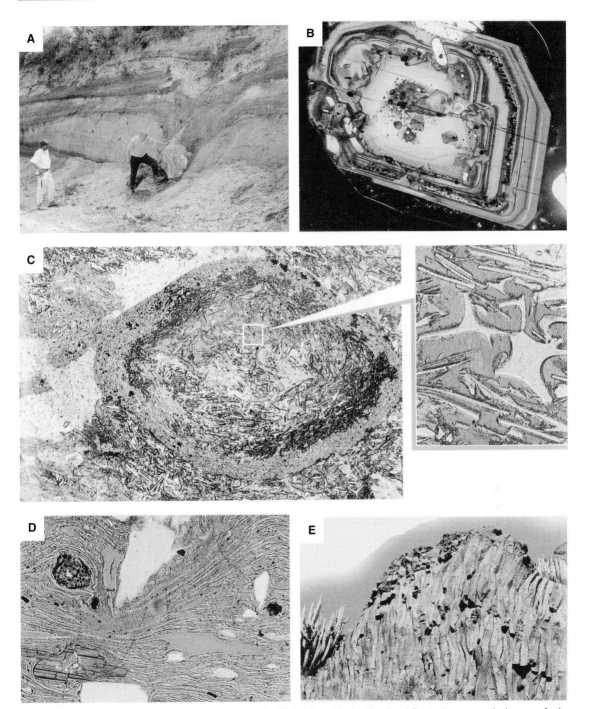

FIGURE 9.4. Volcaniclastic particles. From Schminke (2004). (A) A ballistic block of basalt $\sim$1.5 m in diameter at the bottom of a 4-m-deep bomb crater (sag) in pyroclastic material, Eifel, Germany. (B) A zoned sanidine crystal from Augustine Volcano, Alaska. The crystal also contains solid and liquid inclusions; it is 10 mm long. (C) A photomicrograph of tuff composed primarily of glass shards. Width of field $\sim$20 mm. Most of the field is taken up by an accretionary lapillus that has a core of shards and a rim of fine-grained ash. The inset shows detail of a shard in the lapillus. (D) A photomicrograph of welded glass shards. Light-colored crystals are feldspar; the blue crystal is amphibole. In places the shards have been fused into a homogeneous glass. Field of view $\sim$20 mm. (E) Columns in welded tuff resembling a lava flow.

FIGURE 9.5. William Ashcroft's sunset painting from the banks of the River Thames, Chelsea, England, in 1883. (See Plate 6 for a color version.)

FIGURE 9.6. Holocene pyroclastic fallout deposits blanketing the underlying topography, Japan. From Schminke (2004).

decrease downwind, as with dust deposited from a windstorm (Chapter 6). There is commonly an exponential decrease in thickness and grain size of the deposit with increasing distance from source, but with different rates of decrease for sedimentation from the convective region and from the umbrella region

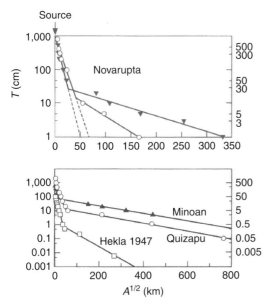

FIGURE 9.7. Thickness (T) versus distance from source (square root of isopach area, A) for pyroclastic fall deposits. From Orton (1996). Straight-line segments indicate exponential thinning, and change in slope represents transition from deposition from the convective region to deposition from the umbrella region.

(Figure 9.7). It is possible to calculate theoretically the spatial variation in thickness and grain size of the deposit given information on (1) the initial height, speed, and volume of grains of a given size and density; and (2) mean and turbulent variations in speed of wind and convection currents along the transport path. A phase of water transport prior to deposition would complicate such calculations. This information is commonly not available, and models for the thickness and grain-size distribution of ejecta deposits need further development (Allen, 1982a).

The thickness of deposits from a single eruption can range from centimeters to tens of meters. The deposits may fine or coarsen upwards or exhibit little vertical grading, depending on variations in the nature of eruption and subsequent wind and/or water transport, and may also have internal stratification due to fluctuations in these factors (Figure 9.6). However, ash and dust can subsequently be homogenized by bioturbation and pedogenesis, as with loess (Chapter 6). Subaqueous settling deposits can originate from subaerial or subaqueous eruptions.

Pyroclastic density currents

Pyroclastic density currents (Figure 9.8) are formed of mixtures of hot gas (from a variety of sources) and pyroclasts that are denser than the ambient fluid (air or water). They originate by increasing the density of an eruption column by overloading with pyroclasts and/or by decreasing the internal temperature as the hot gas and pyroclasts rise into the atmosphere. Such flows can be hundreds of meters thick, travel at up to hundreds of meters per second, and extend away from the source for up to hundreds of kilometers. Temperatures can exceed 700 °C. Deposit thicknesses can be from centimeters to many tens of meters (Figure 9.9). Pyroclastic density flows originate in a variety of ways. It used to be thought that most density flows originated from the periodic collapse of the eruptive column. However, during the Mt. St. Helens eruption, the explosive collapse of the north face of the mountain formed a lateral blast deposit and a debris avalanche followed by pyroclastic density flows (Figure 9.10). Freshly deposited volcanic ash on steep slopes is prone to remobilization as sediment gravity flows by heavy rainstorms associated with convective thunderstorms (a common companion of volcanic eruptions) and by earthquake shaking.

Pyroclastic density currents are traditionally subdivided into low-density, turbulent flows (pyroclastic surges) and high-density, laminar flows (pyroclastic flows). These are equivalent, respectively, to turbidity currents and debris-grain flows (discussed in Chapter 8), and their mechanics are similar. However, there are spatial and temporal transitions between these different types of sediment gravity flow, as discussed in Chapter 8. For example, some turbulent density currents may have such a high concentration of sediment near the bed that turbulence is suppressed there. This concentration gradient is not only caused by settling of suspended grains and mixing of the upper part of the flow with the ambient fluid: it is also enhanced by welding together of extremely hot particles, thereby increasing their size.

Pyroclastic surges are low-density, high-velocity pyroclastic flows that are relatively small in volume, and generally travel for only a few kilometers. The mechanics and deposits of these flows are essentially the same as those of turbidity currents (Chapter 8; Figure 9.11A, B). However, transport and depositional processes depend on whether the flows are wet or dry. Dry surges become less dense as they flow due to sediment settling and mixing with ambient air. When the hot surge becomes less dense than the ambient air, it rises into the atmosphere (lofting). In wet surges, the moisture facilitates adhesion of particles (thus enhancing sediment settling) and reduces the heating of entrained air.

FIGURE 9.8. Pyroclastic density currents. (A) From 7 August 1980, at Mt. St. Helens. United States Geological Survey photograph by Peter W. Lipman. (B) Down the north flank of Augustine Volcano, Alaska, 27 March 1986. United States Geological Survey AVO photograph by M. E. Young.

Planar strata and medium-scale cross strata are common in pyroclastic surge deposits (e.g., Allen, 1982a). The planar strata were presumably deposited on near-planar beds with low-relief bed waves. The medium-scale cross strata are commonly low-angle (less than 15°), although some reach the angle of repose. Cross strata dip upstream or downstream, or both, and may be of the climbing type. These types of cross strata are somewhat larger in set thickness than those formed by ripples in turbidity currents. The cross strata are associated with long, curved-crested bed forms that are decimeters high and meters long. The bed forms are variable in cross-sectional symmetry: some have the steepest side dipping upflow, others have it dipping downflow. Accretionary lapilli are abundant in the upflow-dipping cross strata. The bed forms responsible for these cross strata are probably dunes, antidunes, and chutes-and-pools, although Allen (1982a) has an alternative to the supercritical bed forms involving the moisture content and temperature of the surge.

Pyroclastic flows are high-concentration pyroclastic flows that are more voluminous and travel further than pyroclastic surges. Large flows can travel many tens of kilometers from their source. However, they are mainly laminar flows, and analogous to debris flows and grain flows. Pumice flows and ash flows consist predominantly of ash (glass shards and crystals) with some lapilli and small blocks. Deposits of these flows are called ignimbrites or ash-flow tuffs. Block and ash flows are coarser-grained and their deposits are called agglomerates or volcanic breccias. A hot ash flow (*nuée ardentes* or glowing cloud) was responsible for destroying Herculaneum during the eruption of Vesuvius in 79 AD.

Deposits of pyroclastic flows have a sharp base and may be up to tens of meters thick. The lower parts of ash flows may be planar- or cross-stratified, indicating an initially turbulent flow. However, most of the deposit is more or less unsorted and structureless with a lower region of reverse grading and grain orientation, as in debris and grain flows (Figure 9.11A and C;

FIGURE 9.9. Pyroclastic density-current deposits. An oblique aerial photograph of Mt. Pinatubo, Philippines (in the background) on 29 June 1991, with the eruptive column extending to the left. The Marella River Valley in the foreground is buried by white pyroclastic flow deposits up to 200 m thick. The flow deposits fill the valley, as opposed to pyroclastic fall deposits that blanket the topography.

Chapter 8). Fluidization structures may also be present, indicating hindered settling and upward escape of gas. The top of the deposit is commonly composed of relatively fine-grained ash that was deposited from a low-density, turbulent pyroclastic flow or from suspension fallout (Figure 9.11A).

Subaqueous pyroclastic flows occur due to either subaqueous eruption or flow of a subaerial flow into water. They are similar to subaerial pyroclastic surges and flows but are supported by water. Processes and deposits associated with lava flows, explosion craters, and calderas are not discussed here, but see Orton (1996) and Schminke (2004).

Related sediment gravity flows

Post-eruption sediment gravity flows are very common because loose pyroclastic material on steep slopes is prone to slumping, especially when water is added and when earthquakes occur. Volcanic eruptions are commonly accompanied by convective thunderstorms. Water may also come from melting snow and ice. These sediment gravity flows of pyroclastic material are commonly referred to as *lahars* (an Indonesian word) and range from debris flows to hyperconcentrated mudflows. Lahars caused much damage following the Mt. St. Helens eruption in 1980 and Pinatubo in 1991 (Figures 9.10 and 9.12). During the 1985 eruption of Nevado del Ruiz, Colombia, mudflows caused the majority of the 25,000 casualties.

Preservation of the stratigraphic record of volcanic activity

It should be possible to recognize syn-eruptive deposits and inter-eruptive periods in the stratigraphic record. During eruptive periods, pyroclastic deposits should be dominant, and they may reflect a change during the eruptive period from highly explosive to less explosive and effusive (lava flows), and more mafic. The supply of sediment during the eruptive period will generally be much greater than during the inter-eruptive period. The amount of erosion and

A

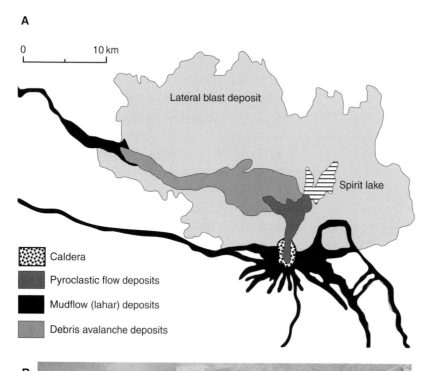

0 ___ 10 km

Lateral blast deposit

Spirit lake

Caldera

Pyroclastic flow deposits

Mudflow (lahar) deposits

Debris avalanche deposits

B

FIGURE 9.10. Effects of the 18 May 1980 eruption of Mt. St. Helens. (A) A map of the deposits. (B) Pyroclastic flows from Mt. St. Helens: upper flow with dual lobes deposited on 7 August 1980 (by flow shown in Figure 9.8A); lower flow (white) from 22 July 1980. United States Geological Survey photograph taken on 30 September 1980 by Lyn Topinka. The inset shows a close-up of the July flow front. United States Geological Survey photograph taken by Terry Leighley, Sandia Labs. (C) The valley of the river filled by a debris avalanche ($\sim$2.3 km^3) that comprised the pre-eruption north flank of the mountain. United States Geological Survey photograph by Lyn Topinka. (D) Pyroclastic (light) and avalanche (dark) debris in Spirit Lake. Holes are from melted ice blocks and steam explosions. From Schminke (2004). (E) Tree stumps blown off and covered by debris from lateral blast. From Schminke (2004).

deposition will be greatest immediately after the eruptive period because of the abundance of sediment and lack of vegetation.

Preservation of volcanic deposits requires abundant volcanic sediment, accommodation space, and minimal subsequent erosion and diagenetic alteration. However, erosion and diagenetic alteration are common, especially on subaerial volcanoes. Therefore, probably only the largest eruptions are recorded in the stratigraphic record.

Other effects of volcanic activity

Topography of the Earth

Volcanic activity results in topographic features such as volcanoes with craters and calderas, volcanic island arcs, mid-ocean ridges with rift valleys, and isolated volcanic islands, seamounts, atolls, and guyots. These topographic features affect flows of water and sediment on the Earth, and have characteristic associated depositional environments. These

FIGURE 9.10. Continued.

features are discussed in Chapters 18 and 20, and in Orton (1996).

Composition of surface rocks, air, and water

Greenhouse gases and dust from volcanic eruptions induce complicated patterns of global cooling and warming, as was evident with the particulates and gases (particularly sulfur) from the 1991 Mt. Pinatubo eruption (Figure 9.13, from Robock (2002)). The $\sim$20 megatons of SO_2 injected into the stratosphere also impacted the ozone layer. Hydrothermal activity can persist in volcanic areas for many years after eruptions have ceased. The hydrothermal circulation at the mid-ocean ridges, associated with the greenschist alteration of newly formed basalts, is thought to have a direct effect on the major-ion chemistry of the oceans (Demicco et al., 2005). Devitrification (weathering) of volcanic glass is the major source of zeolite minerals and also produces silica in solution (which can stimulate

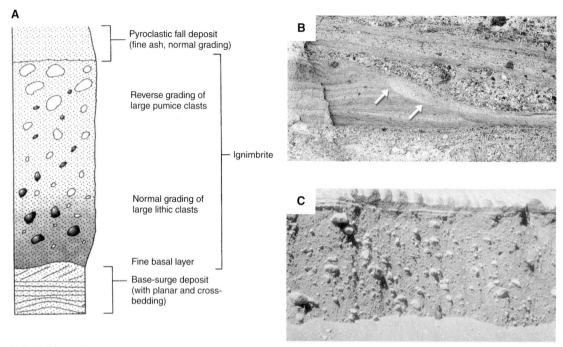

FIGURE 9.11. (A) An idealized sedimentary log of pyroclastic surge and flow deposits. After Sparks (1976). (B) Pyroclastic surge deposits from Eifel, Germany. From Schminke (2004). (C) Pyroclastic flow deposits from the 15 April 1990 eruption of Redoubt Volcano, Alaska. Photograph courtesy of US Geological Survey AVO, taken by C. Neal.

FIGURE 9.12. Volcanic mudflows (lahars – dark) issued from the Mt. St. Helens crater on 19 March 1982 over snow-covered terrain. See Figure 9.10 A. United States Geological Survey photograph courtesy of Tom Casadevall.

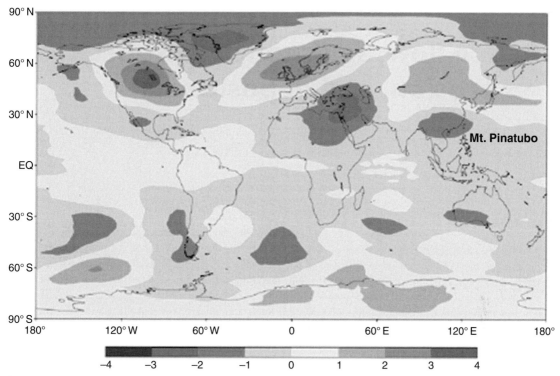

FIGURE 9.13. Lower-troposphere temperature anomalies in degrees centigrade (December 1991 to February 1992) attributed to the eruption of Mt. Pinatubo. From Robock (2002).

diatom blooms). Therefore, ancient volcaniclastic sediments are commonly associated with zeolites and chert.

Other Earth surface processes

Volcanic rocks and associated intrusions affect the nature of chemical weathering, and hence the topography of the landscape related to differential weathering.

Volcanic rocks are more susceptible to chemical weathering than country rocks in some cases and less susceptible in other cases. Unconsolidated pyroclastic sediments are prone to subsequent sediment gravity flows. Lava flows, pyroclastic flows, and subsequent sediment gravity flows affect drainage patterns, forming lakes and diverting rivers. Volcanic eruptions and associated gravity flows and earthquakes cause tsunami (e.g., Krakatoa in 1883).

10 Ice flow, sediment transport, erosion, and deposition

Introduction

Approximately 10% of the Earth's surface is covered with ice today, most of which is in the East and West Antarctic ice sheets. The present area of the Antarctic ice sheets is approximately 11.5 million square kilometers, its volume is about 26 million cubic kilometers, and its maximum thickness is about 4 km. These Antarctic ice sheets contain more than 90% of the Earth's ice and about 75% of the Earth's freshwater. Despite the coldness of the Antarctic climate, large lakes (e.g., Lake Vostok) have recently been discovered underneath the Antarctic ice. The other main ice accumulation is on Greenland, with an area of about 1.7 million square kilometers, a volume of about 2.7 million cubic kilometers, and a maximum thickness of about 3.4 km. The Greenland ice sheet contains about 7% of the Earth's freshwater. The remainder of the ice on the Earth, only 3% by area, is in small ice caps and valley glaciers in mountainous regions. In addition to the areas of permanent ice, about 20% of the Earth's surface is affected by *permafrost*, i.e., permanently frozen.

Permanent ice requires abundant precipitation as snow and temperatures low enough to permit the snow to accumulate. These conditions occur at high latitudes and/or high altitudes. The lower limit of permanent snow and ice (the snowline) is at sea level in polar regions, but is at an elevation of 3 km in the temperate Alps, and at 6 km in the equatorial Andes. Continental interiors at high latitudes, such as Siberia and Alaska, are very cold but do not receive enough snow for permanent ice to accumulate (because they are a long way from the ocean). This chapter is concerned with the formation of ice, the mechanics of ice movement, and the erosion, transport, and deposition of sediment from moving ice. Chapter 17 deals with the effects of these glacial processes (and other associated surface processes) on the formation of erosional and depositional topographic features, and the nature of their depositional sequences, in modern and ancient glacial environments. Glacial environments are treated separately from processes of ice flow because other processes and landforms, associated with flowing water and air, are also active in glacial environments. Useful reviews of glacial processes include Hambrey (1994), Paterson (1994), Menzies (1995, 2002), Benn and Evans (1998), Hooke (1998), Mickelson and Attig (1999), Martini *et al.* (2001), Ritter *et al.* (2002), Evans (2003), and Knight (2006).

Formation and physical properties of ice

Snow falls as ice crystals (hexagonal system) with intricate shapes, and freshly accumulated snow is very porous, with a bulk density of about $100 \, \mathrm{kg \, m^{-3}}$. Over a period of days to years, ice sublimates from crystal extremities and the vapor condenses and freezes near crystal centers, giving a granular crystalline texture. As this process continues, the porosity decreases and density increases. When the bulk density reaches $800 \, \mathrm{kg \, m^{-3}}$, the recrystallized snow is called ice. Most glacier ice has a density around $900 \, \mathrm{kg \, m^{-3}}$ and therefore can float on water. Glacier ice takes hundreds to thousands of years to form.

Ice also forms in supercooled turbulent water as flocs and aggregates of variously shaped crystals (called *frazil ice*), forming a kind of slush. Frazil ice forms in rivers, lakes, and oceans, including within and under the ice. *Anchor ice* is the term for ice that is precipitated from supercooled water onto the bed or ice-margin. *Sea ice* is formed from freezing of ocean water, which occurs at about −1.8 °C because seawater is saline. Sea ice develops from frazil ice. *Pack ice* is floating consolidated sea ice that is either detached from land and freely floating or jammed up against **292** land-attached ice. Pack ice differs from icebergs, which are calved from valley glaciers and ice sheets, and are thus composed of freshwater.

Mechanics of ice flow

Velocity, shear stress, and internal deformation of ice

Ice moves down pressure gradients under the influence of gravity. The pressure gradients may be associated with lateral variations in any combination of ice thickness, the elevation of the top of the ice, and the elevation of the land surface. Valley glaciers flow downhill through pre-existing river valleys; however, land-surface topography is not a prerequisite for the flow of continental ice sheets. Friction occurs at the boundaries of the flowing ice, as with water flow and air flow, and the ice furthest from the boundaries moves faster than the ice near the boundaries. Thus, a velocity gradient $\partial u/\partial y$ and shear stresses τ must exist in the flowing ice (Figure 10.1). The internal distortion of the ice is accomplished by slippage along lattice planes in the ice crystals and along the boundaries of ice crystals. The relative motion within the ice is slow enough that the flow is laminar. Velocity profiles through flowing ice (Figure 10.1) reveal large velocity gradients near the base and very small velocity gradients in the upper parts. Most of the shearing therefore occurs near the base, and the upper part acts almost like a rigid plug. The ice velocity need not be zero at the base of the ice because of the lubricating effect of meltwater or water-saturated sediment, as discussed below.

The relationship between the shear stresses and the velocity gradient in moving ice is given by

$$\tau = k(\partial u/\partial y)^n + \tau_c \qquad (10.1)$$

where k is an apparent viscosity, n is a variable exponent, and τ_c is a critical shear stress for movement (Figure 10.1B). Equation (10.1) is essentially Glen's law (Glen, 1952, 1955), but includes the additional critical-shear-stress term; see Middleton and Wilcock (1994). The ice does not flow unless the applied stress exceeds the critical value τ_c, which is on the order of 10^4–10^5 Pa. Ice is non-Newtonian and behaves like a pseudoplastic, and the parameters in Equation (10.1) depend on the absolute shear stress, ice temperature and pressure, and sediment, gas, and meltwater in the ice. The parameter n varies inversely with shear stress and with the mode of ice deformation, with a range of 0.2 to 1 and a mean

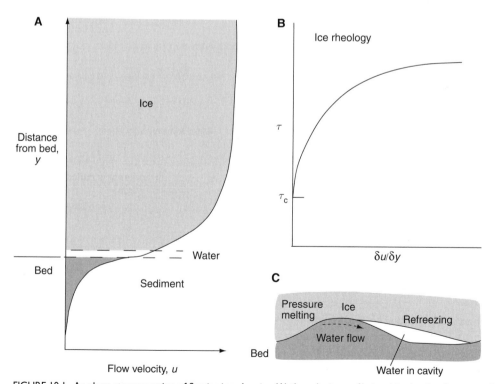

FIGURE 10.1. An along-stream section of flowing ice, showing (A) the velocity profile involving ice, basal water, and sediment; (B) the ice rheology; and (C) the mechanics of basal pressure melting, refreezing, and flow of meltwater into a separation-zone cavity.

value of 0.33. New experimental data and theory have led to re-evaluation of the ice-flow laws, and specifically the value of *n* in Glen's law; see various papers in Knight (2006). The apparent viscosity *k* increases rapidly with decreasing temperature, and is always large (on the order of 10^{15} kPa ^{3} s or more), which is why ice flow is always laminar (Paterson, 1994; Menzies, 2002). Temperature increases downwards in ice, except for the top few hundred meters, thus its viscosity decreases downwards. As pressure increases, the melting point of ice decreases. As both temperature and pressure increase downwards in ice, the melting point may be reached near the base, and meltwater is to be expected. Meltwater can also occur at the glacier base due to flow down crevasses or geothermal heat. Bedrock obstacles at the base of the ice may also cause local pressure-melting upstream of the obstacle and refreezing on the lee side where the dynamic pressure is reduced, which is referred to as *regelation* (Lliboutry, 1979; Weertman, 1979, 1986). Meltwater can remain in the cavities on the lee side of obstacles if the ice separates from the obstacle (Figure 10.1); see Kamb (1987). Meltwater can occur at the base of both temperate and polar glaciers and ice sheets. Meltwater and sediment within the ice influence motion in a complicated way, but both probably reduce viscosity locally (Benn and Evans, 1998; Knight, 2006).

If the ice does not slip over its bed (cold, dry ice), ice flow is by internal deformation (creep) only. If the ice slides over its bed on meltwater or water-rich sediment (warm, wet ice), much of the ice movement is accommodated by this basal sliding rather than by internal ice deformation (Figure 10.1). Waterlogged bed sediment can be sheared if the applied basal shear stress exceeds its shear strength (on the order of 10^4 Pa), as it commonly does (e.g., Boulton, 1979, 1996a, b, 2006; Boulton and Hindmarsh, 1987; Alley, 1991; Kamb, 1991). Waterlogged sediment is easier to shear than the ice because its critical shear resistance and viscosity are much less than those of ice. The shearing subglacial sediment behaves as either a Coulomb plastic or a Bingham plastic (Chapter 8), and the thickness of the shearing zone is on the order of decimeters to a meter or so. Whether the basal ice slides on a layer of water or is coupled to wet, shearing sediment is related to the amount of meltwater available and the permeability (hence grain texture and structure) of the sediment bed. Impermeable clays with high pore-water pressure are likely to promote ice sliding, whereas more permeable coarse-grained sediments or fractured sediments may allow drainage of meltwater, reduction in pore-water pressure, and direct coupling between the ice and

the shearing bed sediment (Boulton, 1996a, 2006). If the ice is strongly coupled to the sediment bed, entrainment of deformed sediment masses into the basal ice is more likely. It is becoming recognized that the amount of meltwater at the base of the ice and the composition of the sediment bed vary in space and time under any given glacier or ice sheet, such that the relative degree of basal sliding on meltwater and shearing of subglacial sediment will also vary in time and space (Evans *et al.*, 2006; Knight, 2006).

The average shear stress on or near the bed τ_0 for the case of steady, uniform flow of a valley glacier with width much greater than its thickness *d* can be calculated (as for water flow) by balancing the downslope weight component with the upslope frictional resistance created at the lower boundary (friction at lateral boundaries can be ignored):

$$\tau_0 = \rho g d \sin \beta \qquad (10.2)$$

where ρ is the ice density and β is the bed and ice-surface slope. Since *d* is commonly hundreds to thousands of meters, and slopes may range from 0.01 to 0.001, bed shear stress is commonly on the order of 10^2 kPa. The shear stress decreases linearly upwards from the bed to the surface of the ice in steady, uniform ice flow:

$$\tau = \tau_0(1 - y/d) \qquad (10.3)$$

Equations (10.1) and (10.3) can be used to describe the distribution of shear stress and velocity within a valley glacier with a parabolic cross section with no slip at the boundaries (Figure 10.2). However, general prediction

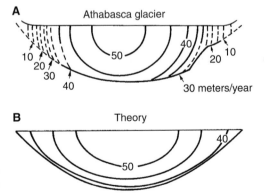

FIGURE 10.2. (A) The measured distribution of ice velocity within a cross section of Athabasca Glacier, and (B) comparison with the theoretical model of Nye (1957, 1965). From Paterson (1994). In the model, the basal sliding velocity is assumed to be uniform, but measurements indicate that the basal sliding velocity decreases from the thickest part to the edges. The friction at the sides is much greater than predicted.

of the ice velocity distribution within glaciers and ice sheets is much more complicated than this, as will be seen below.

The average velocity of ice measured in tunnels and boreholes ranges from on the order of 10^{-4} to 10^{-2} mm s^{-1} (10^0 to 10^3 m per year), which is many orders of magnitude slower than water flow. Valley glaciers move at 10–200 m per year, and ice streams in ice sheets move at 200–1,400 m per year (Paterson, 1994). Polar ice normally moves more slowly than does ice in temperate climates. The rate of movement of ice is commonly very variable in time and space, due to variations in the amount of basal meltwater and in the composition and structure of the bed sediment or rock. Periods of basal sliding on meltwater, shearing of subglacial sediment, and sticking on undeformable substrate alternate in time and space (Evans *et al.*, 2006; Knight, 2006). Glacier surges and ice streams in ice sheets are rapid ice movements, on the order of 10–100 times normal rates; see reviews by Kamb *et al.* (1985), Raymond (1987), Sharp (1988a, b), and Knight (2006). Glacier surges give rise to rapid glacial advance (up to meters per day) and/or thickening of the ice at the glacier snout. There are many theories for the origin of surges and ice streams (e.g., Clarke *et al.*, 1984; Kamb, 1987; Sharp *et al.*, 1994; Menzies, 2002), and most of them involve enhanced basal lubrication by meltwater. Surges are followed by periods of much slower motion, because time is required to re-build ice pressure at the base of the ice, which then causes basal melting and glacier surge. Ice surges are expected to be associated with increases in bed erosion and deformation, and large increases in meltwater discharge and sediment transport.

The up-valley parts of valley glaciers experience net accumulation of ice as snow fall and refreezing of melt water, whereas the down-valley parts experience loss of ice (ablation) by melting, sublimation, and calving as icebergs (Figure 10.3). The boundary between accumulation and loss of ice is called the *snowline* or *equilibrium line*. Upstream of the equilibrium line, ice velocity increases downstream and has a downward component (extending flow), whereas downstream of the equilibrium line ice velocity decreases downstream and has an upward component (compressive flow) (Nye, 1957). Seasonal accumulation of snow is normally recognizable by observation of distinctive layers within the glacier. As the ice moves down valley, these layers remain more or less planar because of the laminar flow pattern (Figure 10.3). If the rate of accumulation of ice exceeds the rate of loss, the long profile of

the top of the ice and the front of the glacier are steeper, and the flow velocity is faster, than if the rate of loss exceeds the rate of accumulation. However, glacier surges are short-term increases in velocity that are not necessarily related to the mass balance of net accumulation and loss of ice. Climate change can cause changes in rates of accumulation or ablation of ice, ice temperature distribution, and the amount of melt water. Such changes would be followed by changes in ice thickness, flow velocity, and surface slope, but with a time lag on the order of tens to hundreds of years for a valley glacier, and thousands of years for an ice sheet. Therefore, present-day advances and retreats of glaciers and ice sheets have little to do with present-day climate changes. It is actually very difficult to predict exactly how the geometry and dynamics of glaciers or ice sheets will respond to climate changes, as discussed further below.

Ice fractures, faults, and folds

Crevasses are cracks within the upper 50 m of the glacier that are produced by tension and brittle fracture (Figure 10.4). Crevasses may be *marginal*, due to the reduction in velocity and shearing at the sides of the glacier or ice stream; *longitudinal*, due to lateral spreading of the glacier in the downstream direction; and *transverse*, due to uplift of the glacier over an obstacle on the bed. *Transverse* crevasses are also associated with faults within the ice (Figure 10.4). A *Bergschrund* is a wide and deep transverse crevasse at the upstream end of a glacier, formed as the glacier is pulled away from the walls of a cirque in the summer (Figure 10.4). *Ogives* are alternations of light and dark (sediment-rich) bands that run transverse to the ice flow (see Figure 10.8B later). They are formed as the ice flows over the lee sides of bedrock obstacles (ice falls) and transverse crevasses form. During the summer, these crevasses are filled with water and sediment. During the winter, meltwater and sediment are not available to fill the crevasses, so the ice is light-colored. In addition to crevasses, faults are associated with brittle fracture of the ice, and folds are associated with plastic deformation. Normal faults are expected in the zone of extending flow, whereas folds and thrust faults are expected in the zone of compressive flow, especially near a glacier snout. Figure 10.4C shows folds in glacier ice (outlined by bands of sediment) that are related to differential movement of ice during surges. Ice streams typically have very crevassed margins.

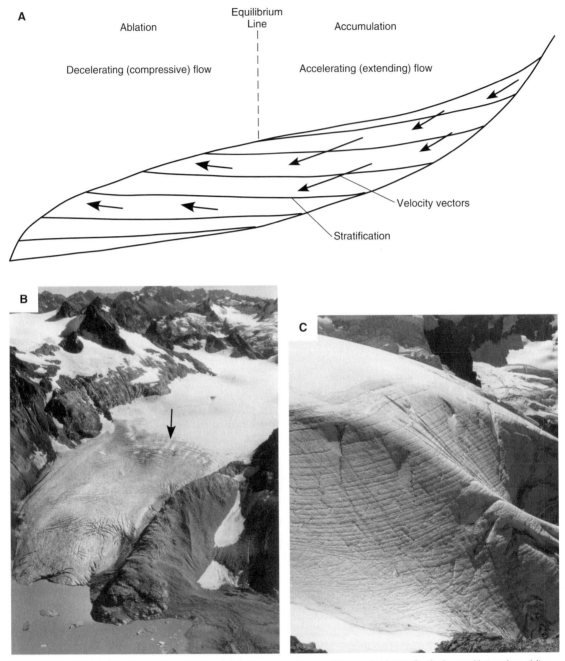

FIGURE 10.3. (A) A longitudinal section of a glacier showing zones of accumulation and ablation (loss), the equilibrium (snow) line, annual layers, and ice-velocity vectors. (B) The snowline (arrowed) on South Cascade Glacier. (C) Annual accumulation layers, Blue Glacier, Olympic Mountains. Images (B) and (C) are from Post and LaChapelle (2000).

Meltwater streams

Meltwater on, within, and under the ice comes from rain and from melting of ice and snow (Rothlisberger and Lang, 1987; Sharp *et al.*, 1998; Knight, 2006). The melting can be due to contact with relatively warm air or water, or due to frictional melting at moving ice boundaries, or geothermal heat. Meltwater is common in the ablation zone below the snowline, but is not restricted to this region. The water occurs and flows

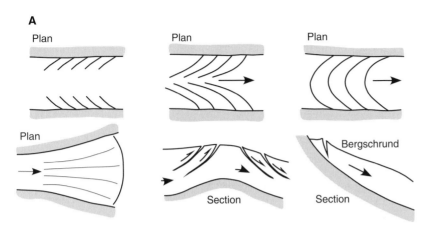

A

B

FIGURE 10.4. (A) Typical crevasse types. (B) A variety of crevasse patterns in the Roosevelt and Coleman Glaciers, Mount Baker, North Cascades. From Post and LaChapelle (2000). (C) Folds in ice (outlined by dark sediment bands) related to glacier surging. From Post and LaChapelle (2000).

within sheets, crevasses, pipes, cavities, and channels, and also within permeable subglacial sediment and rock (Figure 10.5). These hydraulic systems may be isolated or connected, depending on the temperature and velocity distribution of the ice. Spatial and temporal variations in the volume, connectivity, and velocity of water are to be expected.

Supraglacial meltwater occurs in dendritic channels, crevasses, and lakes that, in some cases, connect to meltwater systems within and under the ice. Meltwater within or under the ice that is under pressure is commonly discharged at the surface as vents and fountains (e.g., Lawson *et al.*, 1998). Anchor ice

grows in these vents and fountains from the supercooled meltwater, forming adjacent multi-tiered terraces that may be tens of meters wide and meters high (Figure 10.6). Sediment carried by meltwater is trapped in these ice terraces, and may be concentrated by seasonal melting and refreezing. Diurnal and seasonal variations in meltwater flow and freezing result in layering of the terrace ice with varying crystal sizes and sediment content. In contrast to meltwater vents and fountains, *moulins* are sinkholes analogous to those in karst terrains. Supraglacial meltwater flow may entrain and transport sediment, and abundance of meltwater may facilitate debris flows.

C

FIGURE 10.4. Continued.

Englacial and subglacial meltwater systems are not known as well as supraglacial systems because of observational difficulties; however, time-varying, interconnected sheets, pipes, cavities, and channels are expected (Holmlund, 1988). Subglacial meltwater may flow both immediately below the ice and within subglacial sediment and rock. Meltwater channels are formed in the basal ice if the bed is composed of inerodible bedrock (so-called R-channels), but can be carved into softer sediment beds below the ice (so-called N-channels). Lakes underneath the Antarctic ice sheet are now well documented (e.g., Siegert, 2000). Subglacial lakes are potential depositional sites, and may be a source of water for major floods (called *jokulhlaups*).

Subglacial meltwater at its pressure-melting point can become supercooled, resulting in formation of frazil ice and accretion of anchor ice to the base of the glacier (Alley *et al.*, 1998, 2003; Lawson *et al.*, 1998; Larson *et al.*, 2006). Such supercooling and freezing occur due to a reduction in fluid pressure, which occurs as the basal ice and meltwater flow upslope out of an area of low land topography. The upward land-surface slope must exceed the downward ice-surface slope by a factor of more than 1.2–1.7, depending on the air-saturation state of the meltwater. The formation of ice within the basal meltwater traps sediment and increases the amount of sediment in basal ice (see Figure 10.8E, F later). Temporal fluctuations in melt water discharge produce basal accretionary layers with varying proportions of ice and sediment. Such variations in the amount of sediment in basal ice affect the nature of sediment deposited at the ice margin and the depositional topography.

Icebergs

Icebergs are floating blocks of ice that have broken off (calved from) the downstream edges of ice sheets or glaciers that flowed into standing water (lakes and the sea). If the density of ice is about $900\,kg\,m^{-3}$ and the density of freshwater is $1,000\,kg\,m^{-3}$, approximately 90% of the mass of the iceberg is under water. As glaciers or ice sheets flow into standing bodies of water, the ice closest to the land remains grounded on the bed but the more distal part floats. The floating ice is in the form of a tapering wedge that thins into the water body to a cliff line, resulting from the effects of preferential melting into water. However, if the rate of iceberg calving exceeds that of melting, the cliff at the distal edge of the ice may retreat to the grounding line. Iceberg calving is due to tensile stresses exerted on the ice by gravity and water pressure associated with, for example, changes in sea level related to tides, storm waves and surges, and meltwater filling crevasses. As icebergs are created, some capsize. Icebergs vary greatly in size, from the size of a car or a large building

A

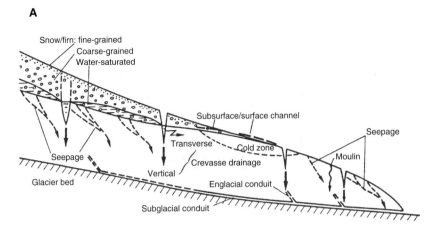

Snow/firn: fine-grained
Coarse-grained
Water-saturated
Subsurface/surface channel
Transverse
Cold zone
Seepage
Crevasse drainage
Moulin
Seepage
Vertical
Englacial conduit
Glacier bed
Subglacial conduit

FIGURE 10.5. (A) A model for meltwater drainage routes. From Rothlisberger and Lang (1987). (B) Dendritic meltwater streams and small hollows, Black Rapids Glacier, Alaska range. From Post and LaChapelle (2000).

B

to that of a small country. Arctic icebergs are typically tens to hundreds of meters thick and hundreds of meters in lateral extent. However, Antarctic icebergs tend to be larger and more tabular, with thicknesses of hundreds of meters and lateral extents of kilometers. Exceptionally large tabular icebergs from Antarctica can exceed 100 km in length. At present, about 280 km^3 of icebergs are formed per year in the Arctic. If the average volume of an Arctic iceberg is on the order of 10^{-3} km^3, this amounts to on the order of 10^5 icebergs per year. Antarctica provides about 1,800 km^3 of icebergs per year. Iceberg production is mainly during the spring and summer of the year, when meltwater is most abundant. Over a longer time scale, the rate of iceberg production is thought to increase during glacier surges, and at the end of periods of cooling climate and ice expansion.

Icebergs may, but do not necessarily, travel far from their source areas, depending on their size, the wind and water currents that drive them along, frictional resistance at their bases and sides, and the Coriolis force. In the northern hemisphere, the main iceberg source area is west Greenland, from which icebergs move to the north and then turn to the south. A small proportion of these icebergs can make it as far south as Newfoundland and the Grand Banks of the North Atlantic. A major source of icebergs from Antarctica is the ice shelves, and most of these icebergs are

FIGURE 10.6. A meltwater fountain flowing over an ice terrace (about 1.5 m high) into an outlet stream. From Lawson *et al.* (1998).

concentrated south of the Antarctic current convergence at about 60 °S. Icebergs can drift at up to tens of kilometers per day (1 knot is 44 km per day), and can last up to years (but generally less than 10 years) before melting. Therefore, they can travel thousands of kilometers from their source areas. In general, smaller icebergs travel further than large ones because they can be more easily moved by wind and water currents, and are less likely to become grounded.

It is well known that icebergs are a hazard to shipping at high latitudes, and the sinking of the *Titanic* is a testament to this. The towing of icebergs to population centers in lower latitudes in order to provide a source of freshwater has been considered, but has not yet been attempted because of problems such as iceberg breakage, wind drift, grounding, and lack of accessibility and control of the water supply.

Numerical models of glaciers and ice sheets

Numerical models of glaciers and ice sheets are now routinely constructed in order to understand and predict the spatial and temporal variations in ice thickness, topography, and flow rate, particularly in response to

changes in climate, sea level, and tectonism (see papers in Knight (2006)). For example, the behavior of glaciers and ice sheets during glacial–interglacial cycles and during global warming has been investigated using numerical models. The complexity of these models varies considerably depending on whether they are one-, two-, or three-dimensional, whether they are steady-state or dynamic, which parameters are modeled, and how.

A basic equation used in numerical models is the conservation of mass of ice:

$$\partial H / \partial t = b - \nabla(HU) \tag{10.4}$$

in which H is ice thickness, t is time, b is the net ice accumulation rate, ∇ is the 2D horizontal divergence operator, and U is the vertically averaged 2D horizontal velocity. The net ice accumulation rate is the accumulation rate (from snowfall and refreezing of meltwater) minus the ablation loss by melting, sublimation, and iceberg calving. Snow accumulation can be related to the local precipitation rate and temperature, but the amount of refreezing of meltwater is very difficult to calculate. Surface melting is a complicated function of factors such as incoming and outgoing (reflected) radiation, the latent heat of melting, conducted heat, heat advected by flowing meltwater, and snow and ice density. The rate of subglacial melting and iceberg calving is even more complicated, as discussed above. The vertically averaged horizontal velocity of ice must be calculated from a conservation-of-momentum equation (Equation (10.2) is a simplified version) plus an equation for the vertical distribution of ice-flow velocity (such as Equation (10.1)). However, the vertical distribution of ice-flow velocity depends critically on the form of Equation (10.1), the resistance to flow at the ice margins, the nature of the bed materials, the amounts of meltwater and sediment at the bed and within the ice, and the character of crevasses and faults.

In view of the difficulty in specifying many of these parameters in the equations of ice motion and mass balance, many of them are "parameterized" as empirical functions of what are considered to be the most important controlling variables. For example, the snow accumulation rate may be a function of the precipitation rate and air temperature averaged over specific time periods. The surface melting rate might be a function of air temperature. Marine melting and calving rates might be taken as dependent on water temperature, flow depth, ice thickness, etc. Such parameterization allows the effects of climate on ice geometry (thickness, areal extent) and dynamics to be modeled. However, there are several caveats. The "parameterization" demanded by the lack of detailed knowledge of the physics of glaciers and ice sheets means that the models are force-fitted to data by tuning empirical parameters that are not well understood. The feedbacks between changing ice geometry and dynamics and changing climate, sea level, ocean circulation, and tectonic loading of ice are difficult to model. For example, changing area and height of ice will affect in a complicated way the climate (air temperature, precipitation, and wind patterns), sea level and thermohaline circulation, and tectonic subsidence or uplift. There are also problems of relating model components and data of different spatial and temporal scales. Thus, claims to success of a numerical model should be treated with caution.

Sediment erosion, transport, and deposition

It is difficult to observe processes of erosion, sediment transport, and deposition underneath the ice, although some evidence is available from boreholes, tunnels, and geophysical surveys. As a result of this, much of our appreciation of these processes is based on theory, experiment, and inferences from erosional and depositional features left behind after the ice has melted. However, genetic interpretation of erosional and depositional features is far from straightforward because the features commonly result from a combination of various processes (involving flow of ice and water) that have acted at different times (e.g., Evans *et al.*, 2006a, b).

Ice erosion

Erosion by glacier ice occurs by ice wedging, plucking, and abrasion. *Ice wedging* occurs as ice forms in confined spaces in the surface of the bedrock or soil. The 9% increase in volume as water freezes results in tensile stresses on the rock or soil if the ice cannot expand towards the opening of the space. Angular fragments of rock or soil are then split away from the land surface to accumulate on top of the glacier or along its boundaries. *Plucking* occurs at the boundaries of moving ice, as blocks of fractured bedrock or soil are broken off and entrained in (frozen to the base of) the ice. This process is aided by freezing of meltwater within fractures and ice wedging. The shear strength of fractured or jointed rock is on the order of 200 kPa, which is commonly exceeded at the base of a glacier. Theoretical studies of plucking, reviewed by Iverson (1991) and Hallet (1996), have emphasized the combined effects of high shear stress applied by the ice and high pore-water pressures that occur locally near the crests of bedrock obstacles. If ice flow separates to the lee of the obstacle, plucked rock fragments may move relatively easily into the lee-side cavity. *Abrasion* occurs as the rock particles in the boundaries of the ice are dragged against the solid boundary of the ice. The larger grains make long grooves (*glacial striations*) or variously shaped, local gouge marks, whereas the finer particles tend to act like sandpaper and smooth off the boundary. Glacial striations and gouges (Figure 10.7A, B) give the direction of ice flow, and their length and straightness is evidence for the laminar flow of ice.

Glacial abrasion rates can be millimeters to centimeters per year. Abrasion rates increase as rock hardness decreases, as ice velocity increases, and as the effective normal stress (which is a function of ice thickness, downward movement of ice, and pore-water pressure) increases (Boulton, 1974; Hallet, 1996). If ice flows over bedrock obstacles, the erosion is concentrated on their upstream sides and crests, and

A

FIGURE 10.7. (A) Types of small-scale glacial erosion features. Modified from Benn and Evans (1998). (B) Glacial striations from Banff National park, Alberta, Canada (courtesy of Henk Berendsen). (C) Sediment rafts, separated by thrusts, sheared from the bed by moving ice (Evans et al., 2006a). (D) Bathymetric maps of high-latitude continental shelves, showing megascale glacial erosional marks and iceberg keel marks. From Dowdeswell in Knight (2006).

B

the lee sides are relatively protected from erosion. If ice is frozen to the bed, as in some polar regions, erosion by ice is minimal (fractions of a millimeter per year).

Ice moving over a wet shearing sediment bed can entrain rafts of this sediment if the water contained therein freezes to the base of the ice (Figure 10.7C). Such sediment rafts have basal thrusts and related

C

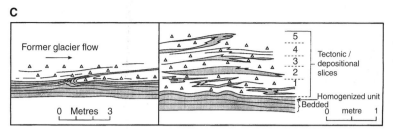

FIGURE 10.7. Continued.

D

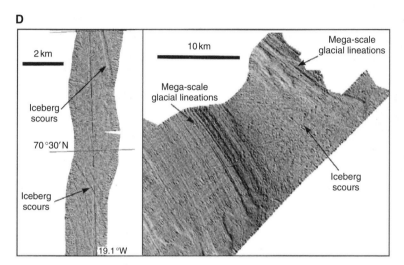

folds. Furthermore, protrusions from the basal ice (keels) can scour straight grooves in the subglacial sediment, and deform the adjacent sediment in the process. Ice-keel scours can be kilometers long, hundreds of meters wide, and meters deep. The bases of icebergs that collide with the sediment bed produce long and straight erosional scours, and turbate the bed sediment (Dowdeswell *et al.*, 1994). Iceberg scours can be hundreds of meters to kilometers long, meters to tens of meters wide, and meters deep (Figure 10.7D). On a smaller scale, sediment fragments protruding from the base of the ice can plough through bed sediment, rucking it up and compacting it.

Sediment transport in flowing ice

Sediment is transported mainly along the sides and bases of ice flows (Figure 10.8), and is concentrated here partly because of the proximity of abrasion and plucking, and partly because sediment carried by melt water is trapped as the meltwater becomes supercooled and freezes to the base of the ice (see above). The concentration of sediment around the perimeter of ice flows can be seen clearly as dark bands on the top

surface of the ice and in cross sections at ice margins (Figure 10.8). Multiple parallel bands of sediment seen in aerial views are produced by confluence of distinct glaciers. Local variations in the thickness and concentration of sediment around ice perimeters are due to local variations in the degree of ice erosion and in the amount of meltwater sediment caught up in the ice formed from supercooled meltwater. The sediment within basal ice formed by supercooling of meltwater occurs in a zone that is meters thick (3–15 m in the Matanuska Glacier: Larson *et al.* (2006); Figure 10.8E, F). The mean volumetric concentration of this sediment is 0.25–0.5, much higher than sediment concentrations resulting from ice entrainment (typically a few percent). Sediment-rich and sediment-poor ice layers are centimeters thick, but irregular lenses also occur. The sediment is mainly silt but contains a large range of grain sizes. Some of the sediment is rounded and sorted and contains primary sedimentary structures. This is in contrast to the unsorted and angular fragments produced by ice wedging and plucking.

Sediment also commonly occurs on the top of valley glaciers, and is derived from rock-falls, landslides, and sediment gravity flows (Figure 10.8). Seasonal variation

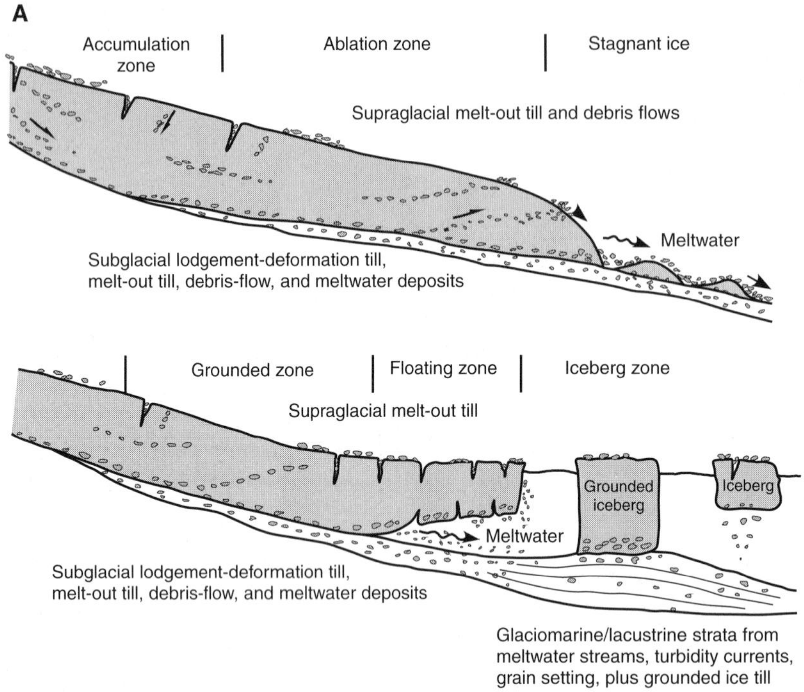

A

FIGURE 10.8. (A) Cross sections through terrestrial and coastal glaciers, showing locations and pathways of sediment transport, and the nature of deposits. (B) Dark bands of sediment in Yentna Glacier, Alaska range. From Post and LaChapelle (2000). Ogives are indicated by an arrow. Images (C) and (D) are of glacier termini, showing dark bands of sediment, Portage Glacier, Alaska. (E) Clear ice (white) overlying a meters-thick basal zone of debris-rich ice (dark), in turn overlying till, Matanuska Glacier, Alaska. From Lawson *et al.* (1998). (F) A 3.5-m layer of stratified pebbly sandy silt interbedded with ice from the basal debris-rich zone, Matanuska Glacier, Alaska. From Lawson *et al.* (1998).

in rock-falls and snow accumulation on temperate glaciers results in bands of sediment parallel to the upper surface of the ice. There can be no turbulent diffusion of sediment within the main body of a glacier, as with water flows and air flows. However, sediment finds its way into glaciers by being washed down crevasses, by being transported by meltwater streams within the glacier, and by being buried as snow accumulates above the snow line. Also, thrust planes at the glacier snout, produced by faster-moving upstream ice, carry subglacial and englacial sediment to high positions on the ice front, where it can melt out (Figure 10.8A).

The size of sediment carried by the ice varies widely from silt produced by crushing and grinding (rock flour) to huge boulders the size of a house. Much larger sediment grains can be carried by ice than can be carried by water or air. There is no mechanism for sorting the sediment within flowing ice because the flow is laminar, but platy grains moving in a laminar shear flow can be oriented with their long axes parallel to the flow lines. Jamming of large clasts together at the base of the ice may cause their imbrication. The larger grains may be smoothed and faceted or striated by abrasion. However, rounded, sorted, and stratified sediment can be incorporated into basal ice if it was transported by meltwater streams that became supercooled and froze, as discussed above.

For a basal layer of sediment-laden ice about 10 meters thick, a mean volumetric sediment concentration of 0.3, and an ice-flow rate of 100 m per year, the volumetric rate of sediment transport to the downstream terminus of the glacier would be about 300 cubic meters per unit width of glacier per year.

Sediment deposition from flowing ice

Deposition of a glacier's sediment load occurs due to frictional drag on sediment within basal ice and by melting of sediment-laden ice at the base and front of the glacier. General terms for sediment deposited directly from the ice are *till* and *diamicton* (Driemanis, 1988; Hambrey, 1994; Evans *et al.*, 2006a). The word till has genetic implications, but the word diamicton is a non-genetic term for a mixture of diverse grain sizes. Therefore, an unsorted mixture of mud, sand, and gravel with abundant fine-grained matrix (once commonly referred to as boulder clay) is a diamicton. Glacial diamictons are generally thought to be unsorted and unstratified, and to have angular to subrounded grains. However, results of recent studies of basal, sediment-laden ice indicate that grains that have been transported in meltwater and trapped in the basal ice as the meltwater froze can be rounded, sorted, and stratified.

FIGURE 10.8. Continued.

Different types of glacial till (diamicton) are commonly given different names based on their origin. For example, Evans *et al.* (2006a) recognize three distinctive types of subglacial till: *subglacial traction till*, *melt-out till*, and *glacitectonite* (Figures 10.9 and 10.10). Subglacial traction till is defined as sediment deposited from a glacier base by frictional drag on sediment within the basal ice and by pressure melting of sediment-laden basal ice. Such deposition will commonly be associated with shearing and homogenization of the sediment that is deposited. This definition includes so-called *lodgement till* and *deformation till* and is used because the processes of deposition and deformation of the sediment by moving ice are thought to be inextricable. This definition also includes sediment fills of cavities on the lee sides of obstacles on the bed. Subglacial melt-out till is defined as sediment that is released by melting or sublimation of stagnant or slowly moving, sediment-rich ice, and is not subsequently remobilized or deformed. Glacitectonite is defined as rock or sediment that has been deformed by subglacial shearing but retains some of the structure of the parent material. Glacitectonite is recognized by its folds, faults, and shear planes. Till containing abundant crushed and sheared sediment grains (*comminution till*) is encompassed by glacitectonite. Glacitectonite must be transitional with deformation till. A major problem in interpreting the origin of any given till is that it may well have resulted from a range of depositional and deformation processes at different times. Furthermore, tills formed at different times by similar or different processes may be superimposed or laterally adjacent. A particularly vexing problem is that of distinguishing among sediments deposited from melting ice, meltwater streams, and sediment gravity flows. Therefore, it is desirable to describe diamictons and associated sediments very carefully and to use descriptive terms prior to interpretation.

C

D

FIGURE 10.8. Continued.

Subglacial traction till (associated with lodgement and deformation) is deposited below the ice as a widespread, hummocky blanket that has been referred to as *ground moraine*. Such till blankets are typically meters to tens of meters thick, and deposition rates may be on the order of centimeters per year. The till tends to have high density (low porosity), is unsorted and structureless, but can contain sorted, stratified layers, and may, but need not, be fissile (Figure 10.10 A, B). Till characteristics apparently vary depending on the relative amount of ice sliding on meltwater and shearing of the wet subglacial sediment, which varies in time and space (Boulton *et al.*, 2001; Evans *et al.*, 2006a). Where meltwater was abundant, some of the deposited sediment may be size-sorted and stratified by the flowing water, whereas if basal sediment shearing was more important the sediment would contain shear planes (possibly polished and striated), oriented grains (with the long axes of platy grains parallel to the bed, and imbricated clusters in places), large grains with plough structures and fine-grained tails, and possibly thrusts and associated recumbent folds. Striated and faceted

FIGURE 10.8. Continued.

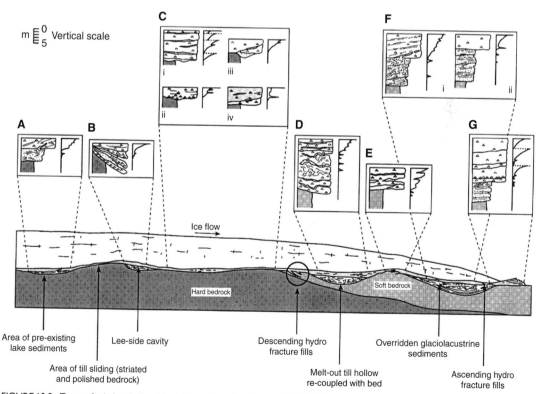

FIGURE 10.9. Types of subglacial till and their formation, after Evans *et al.* (2006). The shaded area immediately beneath the ice is subglacial traction till. Boxes show vertical sedimentary logs and the relative cumulative strain of subglacial deposits. These differ depending on the nature of the substrate, which may be composed of bedrock or pre-existing sediments. For details see Evans *et al.* (2006). The various sedimentary sequences are not necessarily coeval, and may be modified by subsequent glacial advances and retreats.

FIGURE 10.10. Types of subglacial till. (A) Compacted, fissile lodgement till with bullet-shaped and faceted clasts, from Skye, Scotland. (B) Deformation till with fold structures from Skipsea Till at Filey, England. (C) Melt-out till with an undeformed intra-till sand lens, from Alberta, Canada. Photos courtesy of Dave Evans.

above the cavity, and by sediment that is carried in ice that was liberated from the base of the ice (décollement). In the subglacial till studied by Boulton and Hindmarsh (1987), an upper A horizon was structureless, dilated (of high porosity), and had low shear strength, whereas the underlying B horizon was compact (of low porosity) and fissile, and had high shear strength. Apparently, sediment shearing is inhibited where the pore-water pressure is high and the grains are dilated. Temporal variations in the amount of melt water can give rise to alternations of structureless, unsorted diamicton and sorted, stratified sediments. Landforms and deposits associated with subglacial traction till are discussed in more detail in Chapter 17.

Melt-out (ablation) till is deposited at the margins of glaciers as they melt, including subglacially. Melt-out till is not as compacted as lodgement-deformation till, and it is not deformed (Figure 10.10C). Although melt-out till can have a random grain fabric, it can also have layers of silt, sand, and clay with scattered cobbles and boulders. Layered melt-out tills can originate from layered, sediment-laden basal ice, and the layers are preserved if the ice is confined as it melts, possibly by a covering of debris-flow deposits. Such layers would not be preserved if the sediment concentration in the ice was low and if meltwater could not be drained away easily. High pore-water pressures in melt-out tills are reflected in water- and sediment-escape structures, including clastic dykes (Chapter 12). Time variation of melting also produces layering in melt-out tills. Melt-out tills can be difficult to distinguish from lodgement tills that do not have deformation structures. Melt-out tills are commonly meters thick. It has been suggested that the thickness of subglacial melt-out tills must be limited by the amount of sediment carried by the ice; however, substantial thicknesses can accumulate by continued deposition from moving ice. Melt-out till is best known from mounds of sediment, called *end moraines*, deposited at the downstream margins of glaciers. The mound-like geometry of end moraines can be due to deposition from blocks of ice with a locally large concentration of sediment (dead-ice moraine), repeated deposition of sediment at the stationary terminus of a stagnant glacier (like a conveyor belt), and/or the bulldozing action at the front of a glacier (a push moraine). Push moraines are glacitectonites that also have deformation features like those of subglacial traction tills (see Chapter 17).

Gravity flow of pre-existing tills (especially water-logged melt-out tills), or gravity flow of the material carried on top of the glacier down the steep front of

grains are common in this type of till (Figure 10.10A). Tectonic slices (sediment rafts) emplaced from the basal ice may contain stratification as well as deformation structures (Figure 10.7C). Lee-side cavities can be filled with sediment transported by meltwater, with sediment released directly from the base of the ice the glacier, is very common. These debris-flow deposits have been called flow tills, and are similar to other types of till and deposits of solifluction (Chapter 8). Features that distinguish debris-flow deposits derived from tills from other types of debris-flow deposits are their striated and smoothed grains, and a distant origin of some grains. Indeed, *erratic blocks* are huge blocks of rock deposited from ice that have compositions very different from that of the bedrock beneath them. Details of debris-flow deposits are given in Chapter 8. Debris-flow deposits in ice-marginal environments are commonly lobate bodies that are decimeters to meters thick and occur interbedded with melt-out tills in sedimentary sequences that are meters thick. These deposits may be remobilized many times before final preservation.

FIGURE 10.11. Dropstone in laminated glacimarine silts, Ellesmere Island, Arctic Canada. Photo courtesy of Dave Evans.

Erosion, sediment transport, and deposition from meltwater

Meltwater streams under, within, and on top of glaciers are effective at entraining and transporting loose sediment or abrading bedrock, forming potholes, channels, and scour marks on obstacles (e.g., Sharpe and Shaw, 1989). Although it is not generally appreciated, chemical weathering can also contribute to erosion of boundary materials by virtue of the abundance of meltwater and freshly abraded and exposed mineral surfaces. Deposition from glacial meltwater streams occurs where they decelerate in space under the ice, and particularly as they emerge at the front of the ice. Deposits of meltwater streams are essentially the same as those of other unidirectional water flows (Chapter 13). However, if sediment is deposited in contact with the ice, various kinds of deformation structures form when the ice melts (Chapter 17). Deposits of meltwater streams are likely to be interbedded with tills in view of the variable amounts of meltwater beneath glaciers and ice sheets. Meltwater streams near ice margins are effective at reworking tills and debris-flow deposits.

Sediment transport and deposition from icebergs

Sediment that is carried by icebergs in lakes and the sea is deposited as the iceberg melts. Such deposits are generally poorly sorted and unstratified like melt-out till, and are associated with glaciolacustrine or marine deposits (Chapter 17). Individual rock fragments that fall from melting icebergs onto lacustrine or marine sediments are called *dropstones* or *lonestones* (Figure 10.11). Sediment deposited from icebergs close

to their source is likely to be very similar to the melt-out till deposited from grounded ice, and both of these types of till are prone to remobilization as sediment gravity flows. Iceberg-rafted and -deposited sediments tend to be coarser-grained than those on the deep floors of lakes and oceans, and are commonly referred to as Heinrich-event deposits. These deposits in deep-ocean sediments are recognized by observation of their relatively high proportion of lithic grains and magnetic susceptibility, and relatively low proportion of foraminifera (e.g., Bond and Lotti, 1995; Dowdeswell *et al.*, 1995). In the Quaternary oceanic deposits of the North Atlantic, these Heinrich layers range in thickness from centimeters to a meter (mean thickness 10–15 cm), extend laterally for thousands of kilometers, and decrease in thickness exponentially from their presumed source (the Laurentide ice sheet). Their mineral composition is clearly related to their source area. It has been calculated that the mean deposition rate of these layers is on the order of 10^{-1} mm per year, such that 1,000 years would be required to deposit a 10-cm-thick layer (Dowdeswell *et al.*, 1995). In some cases, such deposits have sharp bases, and are interpreted as representing the sudden arrival of an armada of icebergs. However, relatively coarse-grained layers in lacustrine and oceanic muds are not all deposited from icebergs: they can also originate from turbidity currents and sediment-rich plumes issuing from river mouths (Dowdeswell and O Cofaigh, 2002).

Interpretation of the origin and significance of iceberg deposits is difficult. The total volume of sediment deposited from an iceberg before it melts depends on the initial volumetric concentration of sediment in the ice and the initial volume of the iceberg. The areal distribution and local thickness of the deposited sediment depend on the rate of melting, the location of the sediment in the ice relative to where the melting occurs, the velocity of the iceberg, the distance traveled by the iceberg before it melts (the product of the average velocity and lifespan of the iceberg), transport of the sediment by basin currents as it falls to the bed, and subsequent reworking on the bed. The release of sediment from icebergs is quicker if the iceberg did not capsize and basal melting is dominant. If the iceberg capsized, sediment may be concentrated at the side or top of the iceberg, where melting is slower. A greater thickness of sediment can be deposited at a point if the number and volume of icebergs increases, all other things being equal.

The rate of sediment deposition from icebergs *averaged in time and space* can be calculated using simple assumptions about the volumetric rate of iceberg production, the concentration of sediment in the icebergs, and the area over which the sediment is to be spread (e.g., Dowdeswell *et al.*, 1995). If 1,000 km^3 of icebergs are produced per year from an ice sheet, and sediment-rich ice with a volumetric sediment concentration of 10% is restricted to the lowest 1% of the icebergs (such that the overall sediment concentration is 0.1% or 0.001), the total amount of sediment carried by the icebergs is 1 km^3 per year. If this sediment were distributed evenly over an area of ocean floor of 10^7 km^2, the mean deposition rate (not including porosity) would be 0.1 mm per year.

Temporal increases in the amounts of iceberg-deposited sediment at a point (Heinrich events) have been related to climate-controlled changes in glacier volume and flow rate, including break-up of ice shelves, and glaciologically controlled surges in glaciers and ice sheets (e.g., Bond and Lotti, 1995; Dowdeswell *et al.*, 1995, 1999; Kanfoush *et al.*, 2000; Hulbe *et al.*, 2004). Climatic changes have been linked both to Bond cycles and to Dansgaard–Oeschger cycles (Chapter 17). However, such interpretations should be undertaken cautiously. It is necessary to study the texture and structure of the deposits very carefully in order to recognize evidence for settling versus bed-load transport. Provenance studies can indicate the source region of the sediment, and thus the distance and direction of travel. Establishing the lateral distribution of iceberg-deposited sediment thickness is crucial to interpreting the direction of iceberg motion, and the number and volume of icebergs. Independent evidence of climate change (such as from isotope studies of foraminifera) is required, and dating of the deposits and climate changes is essential. If Heinrich layers from different source areas are not coeval in different parts of the ocean, a global climate change cannot be invoked to explain them (e.g., Dowdeswell *et al.*, 1999).

11 Biogenic and chemogenic depositional structures

Biogenic burrows, trails, and trackways: trace fossils

Introduction

Many organisms such as worms, molluscs, crustaceans, insects, sea urchins, and vertebrates move over surface sediment producing *trails* and *trackways*, or tunnel through sediment for food, reproduction, and shelter, producing *burrows* (Figure 11.1). Modern sediments and Phanerozoic sedimentary rocks deposited within the last 542 million years commonly contain well-preserved burrows and trackways, also called *trace fossils, ichnofossils*, or *Lebensspuren* (Osgood, 1987; Clarkson, 1998; Prothero, 2004). *Ichnology* is the study of trace fossils. In addition, certain clams and annelid worms can drill into such solid substrates as corals, rock, and wood for shelter, producing *borings*. The activities of organisms are also indicated by *fecal pellets* (Figure 11.1B). Ancient fecal pellets are commonly referred to as *coprolites* where they are conspicuous. The general biogenic disruption of primary stratification and the formation of randomly organized patches of sediment of varied texture and color is called *burrow mottling, bioturbation*, and *ichnofabric*. Unlike body fossils, ichnofossils are generally not transported, and are records of the behavior of animals that lived where their traces are found. This behavior is closely related to the sediment type and the Earth surface environment.

Laminated sediments associated with surface microbial biofilms known as *stromatolites* and *thrombolites* are generally not included in ichnology, nor are markings that result from the physical dragging of dead organisms or their body parts across the sediment surface. Some laminated sediments are produced by chemical precipitation alone, although it may be difficult to determine a chemogenic rather than biogenic origin. These other biogenic–chemogenic sedimentary structures are discussed after trace fossils.

Modern burrows, trails, and trackways

The only rational way to interpret ancient trace fossils is to understand their modern counterparts. Bromley (1996) gives a thorough discussion of modern traces and trace makers, and Howard and Reineck (1972), Howard and Frey (1975), Reineck and Singh (1980), and Reinharz *et al.* (1982) provide photographs and X-radiographs of various modern tracks, trails, and burrows, and their architects. Burrowing and track-making organisms exhibit a wide range of behavior. Many common bivalves (clams) are well-known burrowing organisms. Most clams live in semi-permanent burrows and extend siphons up into the overlying water for filter feeding and respiration (Figure 11.2A). Deposit-feeding bivalves can plough through the bottom sediment, ingesting it for its organic content as they go, but other clams (e.g., *Yoldia*) live in semi-permanent burrows and collect organic matter from the surface with their palps. Some clams are even predatory. Clams commonly produce distinctive sedimentary structures in response to erosion and deposition (Figure 11.2A). Some clams are very rapid burrowers and, when exhumed, can rapidly rebury themselves with their hydraulically powered foot. Clams can also produce spectacular escape structures when catastrophically buried by sediment (Figure 11.2A). Echinoderms such as sea cucumbers and sea urchins also burrow, and can produce a distinctive, backfilled burrow (Figure 11.2B, C). In marine settings, some whales, walruses, and fish scoop up mouthfuls of sediment for the invertebrates it contains. Fish also nest and rest on the bottom. Such traces are rarely recognized as trace fossils.

Every phylum has worm-like members and many of these are important infauna and epifauna, leaving a variety of tracks, trails, and burrows. Worms include filter feeders, subsurface and surface deposit feeders, scavengers, and predators. In addition to the annelid **311**

FIGURE 11.1 Modern trace fossils and trace makers. (A) Walking traces and burrows of fiddler crabs from tidal flats of the Colorado River delta. The blobs surrounding the burrow entrance center left are mastication pellets. (B) Fecal pellets accumulating in the troughs of wave ripples, Oosterschelde Estuary, the Netherlands. (C) Burrowing sea anemones in subtidal mud from Long Island Sound, eastern USA. Burrows are partially lined with parchment to which sediment is adhering. (D) A shallow excavation in the surface of a tidal flat exposing many parchment lined polycheate worm burrows.

worms, other burrowing groups include priapulid, sipunculid, and hemichordate worms. The tubes of many worms are lined with materials such as sand grains, fecal pellets, mucus, or tough organic chitinous material commonly referred to as parchment (Figure 2.1). The intertidal annelid known as the lugworm (Figure 11.3) constructs an unlined U-shaped burrow with the incurrent end marked by a small crater and the excurrent end marked with rope-like sediment casts. Some burrowing worms farm microbial communities in their burrows and the nearby sediment, some apparently have chemoheterotrophic microbes living in them that produce organic compounds without photosynthesis, and some produce burrows for reproduction. Figure 11.4 shows many different burrow networks, produced principally by polycheate worms, in the soft bottom muds of Chesapeake Bay, eastern USA.

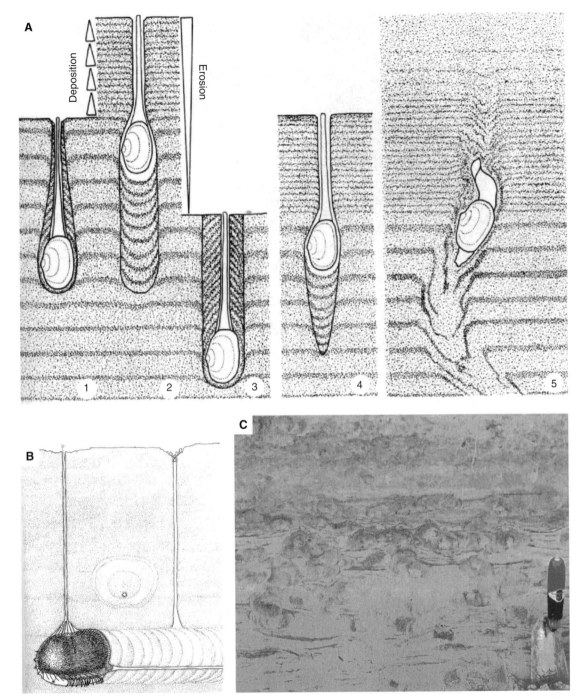

FIGURE 11.2. (A) Burrowing habits of the soft-shelled clam *Mya* in response to various erosional and depositional events. In (5) sudden deposition of a thick layer of sediment forces the animal to use its foot to pull it up and out, leaving a spiral escape trace. From Bromley (1996). (B) Sea urchin *Echnocardium* in its burrow, extending its siphon to the surface and backfilling its burrow as it proceeds. The animal's so-called "sanitary drain" leaves a small hole in the backfill. From Bromley (1996). (C) Recent sands disrupted by sea-urchin burrows exposed in a pit in recent deposits of the Oosterschelde Estuary, the Netherlands.

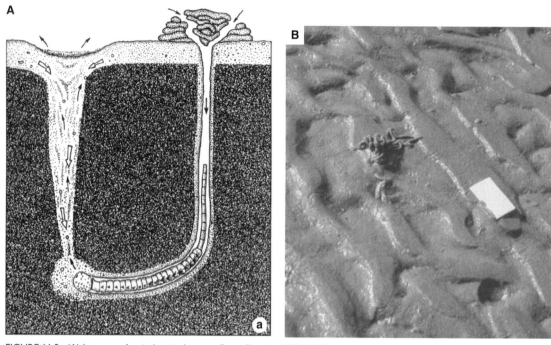

FIGURE 11.3. (A) Lugworm *Arenicola* in its burrow. From Bromley (1996). (B) A lugworm trace on a wave-rippled lower intertidal flat, Oosterschelde, Netherlands: incurrent pit lower right, excurrent pile of fecal castings center left; the scale bar is 5 cm.

An assemblage of burrows such as in Figure 11.4, together with any surface trails, is referred to as an *ichnocoenosis* (alternatively spelled *ichnocoenose*). An ichnocoenosis is analogous in some respects to the life assemblage (*biocoenosis*) and pre-transport death assemblage (*thanatocoenosis*) of body fossils. Bromley (1996) proposed a five-tiered ichnocoenosis for modern burrowing organisms under optimum conditions (Figure 11.5). Most of the organisms live in the uppermost 2–3 cm of the bed sediment, and this surface layer is thoroughly bioturbated by mobile burrowers, surface-dwelling benthic organisms such as fish and crabs, which disturb the sediment, and microscopic nematodes and other so-called meiofauna living between sediment grains. The second tier is typically 10–15 cm thick and is a *conveyor belt* comprising worms that ingest the sediment at this level and return it to the bed surface, where it is continually buried by new sediment and so eventually ends up back at the depth at which the conveyor begins. The third tier is ~20 cm below the bed surface and comprises open crustacean burrows such as that of the *Callianassid* shrimp. The fourth tier comprises echinoids and other deep, deposit-feeding polycheates, some of which may have symbiotic bacteria in their burrows and in themselves that derive energy and synthesize organic molecules by a variety of redox

reactions, such as sulfide reduction. Finally, perhaps as much as 1 m below the bed surface there are deep-burrowing, relatively immobile bivalves and other suspension feeders. The thicknesses of the various tiers can vary depending on the substrate conditions. However, there is a general separation of filter-feeding organisms from deposit-feeding organisms based on grain size of the sediment (see below), and episodes of erosion with net deposition may lead to superposition of tiers, making interpretation of tiering in ancient deposits difficult.

Controls on the nature of modern biogenic structures

Modern depositional environments contain variable numbers of epifaunal trace makers and infaunal burrowing organisms that behave in various different ways to produce a variety of types of burrows and tracks. The important controls on the type, behavior, diversity, and number of trace-making organisms on and in the sediment are (1) sediment grain-size and consistency, (2) water depth, (3) temperature, (4) salinity, (5) degree of oxygenation of the water, (6) food supply, (7) life habits and trace-making method of the organisms, and (8) sedimentation rate. These factors are interrelated in complicated ways.

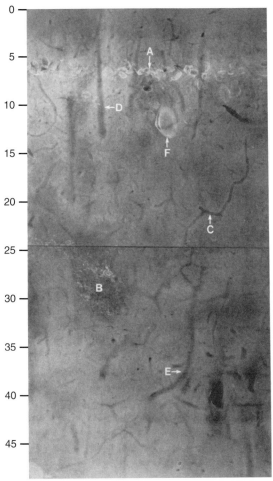

FIGURE 11.4. An X-radiograph of a box core taken from the mud bed of Chesapeake Bay. A denotes shell layer, B denotes a large burrow packed with shell debris, and C, D, and E are open burrow networks of various polycheate worms. F is an in-place clam. From Reinharz et al. (1982).

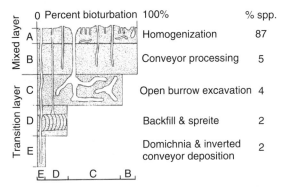

FIGURE 11.5. A model of tiering by burrowing organisms. From Bromley (1996). The width of column indicates the amount of bioturbation in a level. See the text for details and tier thicknesses.

Attached epibenthonic animals are most abundant on rock, and mobile epibenthic trace makers are most common on fine sand and muddy sand. Clay-rich mud with up to 70% porosity is quite soupy and cannot support the weight of large epibenthos. Only a few suspension-feeding organisms can bore into rock. Instead, inbenthic *suspension feeders* are most common and diverse in fine sand and coarse silt. *Deposit* feeders, in contrast, are most abundant in sandy mud and muddy sand and reach maximum numbers and diversity in mud with ~20% clay. Large deposit feeders are not common in soupy clays, but smaller forms can be abundant. Moving sands and gravels that comprise the bed load of river channels, tidal channels, tidal deltas, and beaches have the lowest diversity and numbers of benthic organisms, comprising a few highly mobile infaunal species that can quickly reburrow when exhumed by water currents. Organisms occur in sediment that ranges in consistency from soupy mud (such as the bottom of Chesapeake Bay in Figure 11.4) to compacted sand to rock. Taylor et al. (2003) applied the terms *soup ground, soft ground, loose ground, stiff ground,* and *firm ground* to sediment with varying consistency. Firm grounds generally develop in compacted and partly cemented mud, whereas hardgrounds (cemented bottom sediments: Chapter 19) are most common in sand, and typically contain borings made by rock-drilling organisms. Borings can commonly be identified because they cross-cut hardened sedimentary particles, such as shell fragments and cements. In addition, the walls of borings may rarely be encrusted by cave-dwelling organisms (*cryptobionts*). Firm grounds and hardgrounds are commonly interpreted as indicating sea-level change; see Chapter 20 and Pemberton et al. (1992).

The number and diversity of marine benthic trace makers increase from the nearshore into deeper water, reaching maxima at the upper continental slope. This is presumably because the temperatures there are more stable and there is less disturbance by storm currents in deeper water. The suspension-feeding benthic fauna declines rapidly down the continental slope, with only deposit feeders occurring in deep-ocean sediments. The distribution of benthos on the continental shelf and deep-sea bottom is notably patchy, with both the number of individuals and the number of species varying from place to place. This is most likely due to disturbance by currents and patchy settling of larvae of the various benthic animals (Chapter 2). The diversity of all marine organisms increases from polar regions to the tropics, and benthic

A

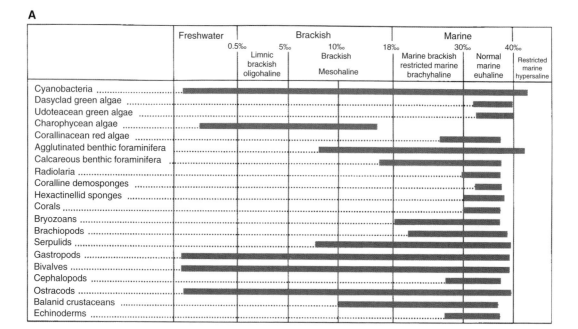

B

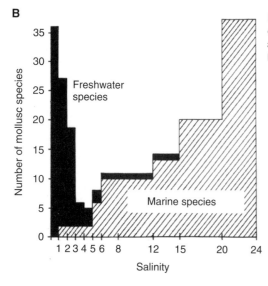

FIGURE 11.6. (A) Effects of salinity on major groups of marine organisms. From Schlager (2005). (B) The richness of mollusc species along the salinity gradient of the Randersfjord, Denmark. From Levinton (1982). By permission Pearson Education, Inc.

track makers and burrow makers are no exception. Salinity has a strong effect on the diversity of benthic organisms; most marine organisms cannot tolerate salinities much above 38‰–40‰, and there is a marked decrease in diversity through estuaries into freshwater (Figure 11.6).

The amount of dissolved oxygen in the water is another control on the number and diversity of benthic organisms. Ocean water under upwelling zones may locally have little or no dissolved oxygen and, where the sea bed is in these *dysaerobic* waters, only rare infaunal polycheate worms comprise the benthos. In stagnant water, such as in silled fjords or silled basins like the Black Sea, aerobic bacterial reduction of suspended organic matter may use up all the dissolved oxygen and anaerobic breakdown of organic matter may take place in the water column, generating toxic H_2S. Under these conditions, no benthic invertebrates can exist. The boundary between oxygenated waters and anoxic waters is usually found within the pore waters. Muddy sediments commonly contain a few percent organic matter, which comes from detrital organic matter adsorbed onto clay, particulate organic matter, mucus from burrowing organisms, and microbes living on and in the sediment. Aerobic microbes use up the dissolved oxygen in the pore waters in the upper few millimeters of muddy bottoms, and below this are zones where various anaerobic microbial communities continue to degrade the organic matter in the sediment using various redox reactions (Chapter 2). Burrowing organisms bring oxygenated water down into this otherwise anoxic setting. Thus, burrows commonly have oxygenated haloes around them that alter the sub-bottom microbial communities, and are important sites of early diagenesis (Chapter 19). Typical reactions include precipitation of iron minerals at the redox boundaries. Thus, burrowing commonly influences diagenesis, which, in turn, enhances and preserves the burrows. Diagenesis also affects burrows during compaction. Upon burial, compaction will collapse empty burrows, and even sediment-filled burrows may be significantly flattened.

The total number of individual benthic organisms (the biomass) depends on the food supply. Sediment surfaces in the photic zone are covered with macrophyte vegetation, algae, and cyanobacteria, providing food for grazing organisms. Suspended organic matter and plankton are available to filter feeders across the continental shelf and uppermost continental slopes, but below this the only source of food is the table scraps from the sunlit waters above.

The type and behavior (e.g., living, resting, moving, feeding, escaping) of an organism determines the geometry of a burrow or trackway, as shown above. The life habits of organisms also affect bottom sediment by (1) mixing, (2) increasing or decreasing the water content, (3) increasing or decreasing the grain size, (4) pelleting mud, and (5) adding variable amounts of organic mucus to the bottom. Polycheate worms of the shallow subtidal and intertidal ponds of Andros Island, in the Bahamas, are capable of reworking the upper 5 cm of sediment many times within the course of a year. Such reworking commonly cleans the sediment of mud as the deposit feeders pellet the mud and deposit it on the bottom (Garrett, 1977; Bromley, 1996). The pellets are reckoned to be more easily eroded and transported than mud–sand mixtures, and this transport has the potential to clog the feeding mechanisms of suspension feeders and generally inhibit the settling of larvae of other benthic organisms on the bottom. Organisms such as the *Callianassid* ghost shrimp filter the overlying water and *add* mud pellets to the sediment. *Callianassid* burrows (Figure 11.7) have a funnel-shaped incurrent tube that leads into a complicated

series of open gallery ways with dead ends. The tubes and gallery ways are lined with sand particles, pieces of plant debris, and other materials. The shrimp sets up currents through the burrow to feed on whatever comes in. A volcano-like pile of fecal pellets usually marks the excurrent end of the burrow. Pryor (1975) noted the distinctive shape and internal structure of *Callianassid* fecal pellets (Figure 11.7), and reckoned that the beach and shoreface population of *Callianassid* shrimp on Sapelo Island, Georgia (an area of about 18 km^2) removed and pelleted some 12.3 tonnes of suspended sediment per year.

Deposition and erosion can affect burrowing organisms, and the nature of trace fossils, by (1) introducing new sediment types that will influence the type of organism that can move through the sediment, and what it does; (2) burying or exposing organisms, forcing them to escape upwards or move downwards; and (3) affecting the amount of time and space available to bioturbate the sediment. High deposition rates leave deposits with a relatively low degree of bioturbation, whereas low deposition rates allow sediments to be completely bioturbated. As an example, deposition on continental shelves is commonly associated with storms that typically leave a graded layer of stratified sand up to a few tens of centimeters thick on top of muddy sediments that may have been eroded. Buried organisms that survived the storm may attempt to reach the sediment surface, generating escape burrows. In addition, opportunistic benthic species (*Skolithos* ichnofacies – see below) will be the first to recolonize the storm layer. Mud will gradually settle on the storm sand, and the early colonizers of the storm sand will eventually be replaced by ecological succession until a new stable community is established (*Cruziana* ichnofacies – see below). However, only the uppermost 10–15 cm of the new sediment will be bioturbated. In 1961, Hurricane Carla deposited an extensive graded sand bed off the Texas coast that was up to 15 cm thick (Hayes, 1967). The storm layer was thoroughly homogenized with the underlying sediment within a dozen years. Indeed, it is probable that most sandy mud and muddy sand began as separate layers of sand and mud that were mixed by burrowing organisms.

Modern and ancient marine sediments can range from having few burrows to complete bioturbation, depending on the factors described above. Numerical scales have been developed to attempt to quantify the degree of bioturbation in ancient sediments and sedimentary rocks (Droser and Bottjer, 1986, 1989; Bromley, 1996; Taylor *et al.*, 2003).

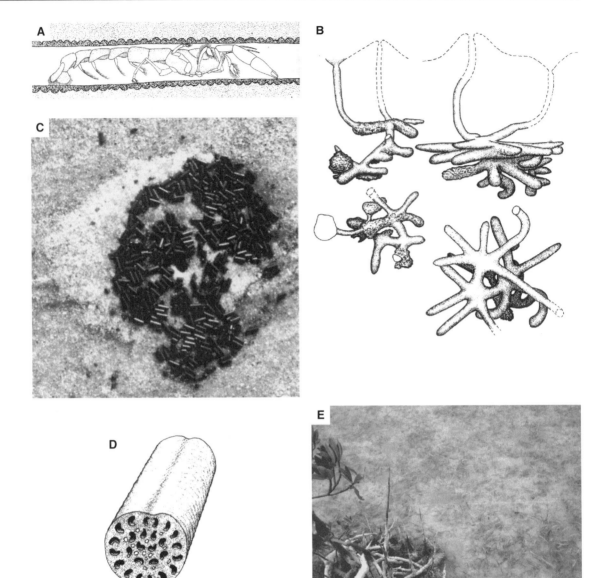

FIGURE 11.7. *Callianassid* shrimp and their burrows. (A) A shrimp in its lined burrow. From Bromley (1996). (B) *Callianassid* burrows in side and plan view. From Bromley (1996). (C) Fecal pellets of *Callianassid* shrimp around the excurrent end of a burrow. From Pryor (1975). (D) Detail of a *Callianassid* shrimp fecal pellet, showing its internal and external ornamentation. From Pryor (1975). (E) Shallow subtidal mud bottom from Florida Bay, USA, covered with turtle grass. White mounds are piles of debris (including fecal pellets) excavated from a *Callianassid* burrow, marking the excurrent end of the burrow.

Preservation modes of trace fossils

Burrows are three-dimensional tubes that infauna (animals that live below the sediment surface) make as they dig through unconsolidated sediment. Burrows are most obviously preserved if they are filled with sediment that is different than the host sediment. Sediment that fills burrows comes from (1) sediment ploughed up by the animals and stuffed back into the burrows, (2)

fecal pellets stuffed back into the burrow, and (3) sediment that passively filters into the burrow. The first two mechanisms usually leave *meniscus structure*, curved, meniscus-shaped laminae of sediment indicating episodic backfilling (Figure 11.2). Where an entire burrow structure is shifted vertically or laterally, preserved curving burrow traces are referred to as *Spreiten*. Post-depositional cements may also fill and preserve burrows. When filled by sediment or cement,

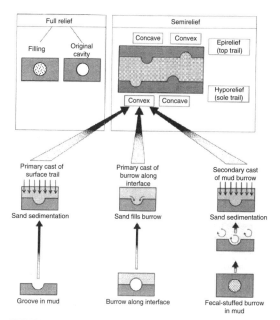

FIGURE 11.8. Preservation of trace fossils, with terminology. From Osgood (1987).

burrows are preserved in *full relief* (Figure 11.8). Tracks and trails are surface features commonly made by epifauna (organisms that live on the sediment surface). Thus, tracks and trails are preserved on bedding planes (ancient depositional surfaces) mostly in *semirelief*. However, it is common for burrowing organisms to work horizontally along buried bedding surfaces as well. Figure 11.8 includes further descriptive terminology that takes into account whether the trace is preserved on the base or top of a particular bed. Tracks and burrows made in firm mud are most commonly buried by sand beds that fill surface trails and partially to completely fill burrows.

In light of these considerations, the interpretation of ancient trace fossils is difficult and there is a large literature on the subject (Osgood, 1987; Ekdale and Pollard, 1991; Pemberton *et al.*, 1992; Clarkson, 1998; Hasiotis, 2002; Prothero, 2004). If analyzed systematically, ancient trace fossils can yield information about the anatomy and behavior of the trace makers, the nature of the environment, and the deposition rate. However, Bromley (1996) noted that (1) the same species can make various structures through different kinds of behavior; (2) the same burrows can be preserved in different ways; (3) some burrows are cooperative efforts between a main trace maker and symbiotic microscopic and macroscopic organisms that alter the burrow structures; and (4) in different Earth surface environments, different species can make identical structures. Furthermore, trace fossils do not indicate the age of rocks because a particular trace fossil can be made by various organisms, and behavior may evolve in different ways than the organisms themselves. However, major stages in the evolution of life are reflected in ichnofossils (Hasiotis, 2002).

Trace fossils are classified according to (1) interpretation of the trace maker's activity, (2) *ichnogenera*, (3) *ichnocoenosis*, and (4) *ichnofacies* – relatively well-defined recurring associations of traces. Ichnofacies are most commonly used in the interpretation of sedimentary rocks. The various classification schemes are described and discussed below.

Interpretation of behavior from trace fossils

Interpretation of what an organism was doing when it made the trace has led to definition of eight behavior types that are classified using Latinized terms as follows. (1) *Locomotion* (Latinized term "repichnia") traces are where an animal was moving on the sediment surface or within the sediment. (2) *Resting* or *hiding* (Latinized term "cubichnia") traces commonly preserve some of the morphology of the trace maker, such as trilobites and sea stars, and therefore grade into molds and casts of body fossils if they retain sufficient detail of anatomical features. (3) *Living* (Latinized term "domichnia") traces are the recognizable domicile of an infaunal animal, ranging from simple unbranched and unlined straight tubes, via U-shaped burrows, to large, complex structures with specific wall linings. (4) *Feeding* (Latinized term "fodichnia") traces are the burrows and trails of mostly soft-bodied, worm-like, deposit-feeding animals that were apparently grubbing through the bottom or along the surface, ingesting inorganic sediment, particulate organic matter, and the bacteria living on the organic matter, and expelling it as fecal pellets. (5) *Grazing* (Latinized term "pascichinia") traces are surface trails related to feeding traces, in which the organism (e.g., a snail) was specifically feeding on benthic microbial biofilms. (6) *Farming* (Latinized term "agrichnia") is a special type of trace where an organism apparently actively encouraged benthic biofilms or macroalgae to grow. (7) *Predation* traces (Latinized term "praedichnia") are produced by hunters and scavengers (e.g., polycheate worms and many arthropods). (8) *Escape* traces (Latinized term "fugichnia") are where mobile infaunal organisms such as clams or worms sought to escape predation, exhumation, or burial beyond their

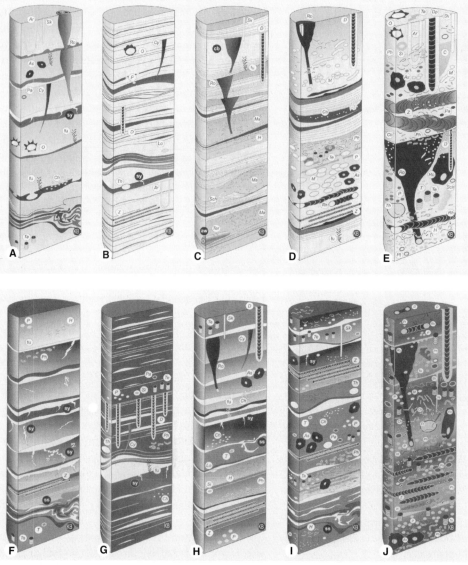

FIGURE 11.9. Trace fossils in cores through deltaic deposits. From MacEachern *et al.* (2005). Distal delta front environments include (A) river-dominated, (B) tide-dominated, (C) storm-dominated, (D) wave-dominated, and (E) non-deltaic lower shoreface. Prodelta environments include (F) river-dominated, (G) tide-dominated, (H) storm-dominated, (I) wave-dominated, and (J) non-deltaic offshore. *Ch* = Chondrites, *Cr* = Cosmorhaphe, *H* = Helminthopsis, *Ph* = Anconichnus/Phsycosiphon, *P* = Planolites, *Z* = Zoophycos, *Th* = Thallassinoides, *As* = Astersoma, *Ro* and *Rr* = Rosselia, *Pa* and *Pt* = Palaeophycus, *Rh* = Rhizocorallium, *Te* = Teichichnus, *T* = Terebellina, *S* = Siphonichnus, *Pb* = Phoebichnus, *Cy* = Cylindrichnus, *D* and *Dp* = Diplocraterion, *Sk* = Skolithos, *O* = Ophiomorpha, *Ar* = Arenicolites, *Ter* = Teredolites, *Ta* = Taenidium, *C* = Conichnus, *Lo* = Lockeia, *Ma* and *M* = Macaronichnus, *Sch* = Schaubcylindrichnus, fu = fugichnia, cb = cryoturbation, aw = allochthonous wood, sy = syneresis cracks, and ss = soft-sediment deformation.

normal depth range. Common examples are where clams burrow down into the mud to escape the person digging for them, or attempt to burrow up through a layer of sediment deposited on them. Escape burrows are common in shallow-marine storm deposits, tidal deposits, and river deposits. In environments where erosion and deposition alternate frequently, a quite detailed

record of that erosion and deposition can be reconstructed from the up and down motions of filter-feeding clams, anemones, or arthropods preserved in their dwelling burrows. The eight behavioral types listed above were originally coined to cover marine traces. Three additional terms used to describe terrestrial insects' nesting behavior are *aedificichnia* (Brown and

Ratcliffe, 1988), *calichnia* (Genise and Bown, 1994a, b), and *polychresichnia* (Hasiotis, 2003).

Description and interpretation of ichnogenera

Early stratigraphers, sedimentologists, and paleontologists considered trace fossils to be plants, so they were given Latinized (Linnaean) genus and species designations. By the time the true nature of trace fossils was realized it was too late, and trace fossils are now designated by Latinized *ichnogenus* names (i.e., they are capitalized and in italics, and most end in the suffixes *-ichnus, -ichnites, -craterion,* or *-opus*). Diagnostic descriptions of ichnogenera can be found in the *Treatise on Invertebrate Paleontology* (Häntschel, 1975). The diagnostic characteristics of an ichnogenus comprise a detailed description of the geometry of the trace including the size (width and length), the overall shape of the burrow (e.g., U-shaped, simple shaft), orientation of the trace with respect to bedding (which may change along the length of the trace), how the trace is preserved (e.g., full relief, convex hyporelief – Figure 11.8), whether the trace is unlined or lined and the nature of any wall-lining structure, whether the trace branches along its length and the geometry of the branching, the nature of the fill, whether spreite are present, and ornamentation away from the trace such as scrape marks or drag marks made by the trace-maker's appendages. Complete identification of trace-fossil ichnogenera in cores can be difficult (e.g., Figure 11.9). Ichnogenera are merely shorthand ways of referring to traces and, in most cases, the organism that made a specific trace is unknown. In one case where the organism that made a specific trace is known, the organism's genus is *Callianassa* and the burrow's ichnogenus is *Ophiomorpha*. Moreover, more than one organism can make the same trace and the organisms that made any particular trace undoubtedly changed through time. For these reasons, trace fossils have limited biostratigraphic value.

Definition and interpretation of ichnofacies or ichnocoenosis

Trace fossils are claimed to have value as environmental indicators, and the most common classification scheme for trace fossils is the organization of a limited number of ichnogenera into *ichnofacies* that have been related to depositional environment. Pemberton *et al.* (1992) distinguished nine ichnofacies designated by one

characteristic ichnogenus and several associated ichnogenera: (1) *Scoyenia*, (2) *Skolithos*, (3) *Psilonichnus*, (4) *Cruziana*, (5) *Zoophycus*, (6) *Nereites*, (7) *Glossifungites*, (8) *Trypanites*, and (9) *Teredolites*. The *Skolithos*, *Psilonichnus*, and *Cruziana* ichnofacies are characteristic of coastal and nearshore marine environments, whereas the *Zoophycus* and *Nereites* ichnofacies are characteristic of shallow-marine and deep-marine muddy environments (Figures 11.9–11.13). The *Glossifungites* ichnofacies is characteristic of marine firm grounds, whereas the *Trypanites* ichnofacies comprises borings made in cemented hardgrounds and other rock substrates. The *Teredolites* ichnofacies comprises wood-boring shipworms (*Teredos*) and related forms. The *Scoyenia* ichnofacies (Figure 11.13) is the only continental ichnofacies designated by Pemberton *et al.* (1992), although they discuss other continental trace fossils. Actually, many ichnogenera can occur in several different ichnofacies and the environmental implications of many of the ichnogenera have been called into question. Hasiotis (2004) recognized four major suites of continental ichnocoenoses: (1) alluvial, (2) lacustrine, (3)

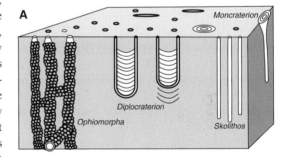

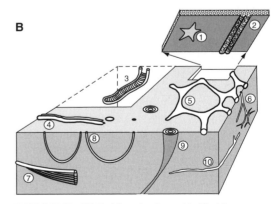

FIGURE 11.10. (A) *Skolithos* ichnofacies. Modified from Pemberton *et al.* (1992). (B) *Cruziana* ichnofacies: (1) *Asteriacites*, (2) *Cruziana*, (3) *Rhizocorallium*, (4) *Aulichnites*, (5) *Thalassinoides*, (6) *Chondrites*, (7) *Teichichnus*, (8) *Arenicolites*, (9) *Rossella*, and (10) *Planolites*. Modified from Pemberton *et al.* (1992).

eolian, and (4) transitional continental–marine. Such continental trace fossils are shown in Hasiotis (2002, 2003, 2004), Bridge (2003), and Figures 11.14 and 11.15.

Adolf Seilacher developed the concept of ichnofacies in the 1960s and 1970s and emphasized that water depth was the main control on the distribution of marine trace-making organisms. It is now known that marine ichnofacies do not have the depth implications as was originally thought, or even the main environmental implications. Many "marine" ichnogenera occur in fluvial and lacustrine deposits (see below). Instead, as detailed above, sediment type (texture, composition, and consistency), deposition rate, animal type and behavior, and physical parameters such as concentration of dissolved oxygen, salinity, and nutrient levels control trace-fossil distribution. The more common ichnofacies are briefly described here and more details can be found in Ekdale *et al.* (1984), Pemberton *et al.* (1992), Donovan (1994), and Hasiotis (2002, 2004).

The *Psilonichnus* ichnofacies is mostly known from modern beach backshore and dune environments, washover fans, and tidal flats, and is characterized by burrows excavated by crustaceans such as fiddler crabs and ghost crabs. Burrows of modern fiddler crabs and ghost crabs are curving tubes up to a meter or so deep that end in bulbous chambers located at the low-water level. The organisms can thereby keep their gills moistened. Similar ancient burrows would be called *Psilonichnus*. Entrances to fiddler-crab burrows are commonly surrounded by the mastication pellets of the crabs (Figure 11.1). Ants and other insects excavate galley ways beneath the sediment surface of modern tidal flats. Dune grasses may be found on beach sand dunes and salt-tolerant plants such as *Spartina*, *Salicornia*, and mangroves are common on modern tidal flats. Thus roots are to be expected in Cretaceous and younger examples of the *Psilonichnus* ichnofacies and may be mistaken for burrows.

The *Skolithos* ichnofacies (Figure 11.10A) is characterized by *Skolithos*, *Diplocraterion*, *Monocraterion*, and *Ophiomorpha* (although *Ophiomorpha* burrows are Pennsylvanian and younger). These ichnogenera are mostly interpreted as dwelling structures of filter feeders. Many Cambrian quartz sandstones are so riddled with *Skolithos* burrows they have been called *pipestone* or *piperock* (Figure 11.11A). *Ophiomorpha* burrows (Figure 11.11B) are made by Callianassid shrimps (Figure 11.7). *Diplocraterion* comprise a distinct U-shaped burrow where spreiten form similar U-shaped bridges between the main lateral burrow tubes. The spreiten record the vertical adjustments of the burrowing organism to erosion, where the burrow would deepen, and deposition, where the animal would have to reposition itself higher in the sediment (hence its one time appellation *Diplocraterion yoyo*). Finally, *Monocraterion* are simple vertical tubes with funnel-shaped entrances. The *Skolithos* ichnofacies is characteristic of sand beds with frequent sediment transport, erosion, and deposition, such as are found in beaches, offshore sand bars, tidal deltas, estuarine-channel point bars, and storm deposits. This ichnofacies is also found in submarine-fan deposits and on continental slopes swept by contour currents.

The *Cruziana* ichnofacies is characterized by about a dozen ichnogenera (Figure 11.10B) and is commonly interpreted as representing muddy shelf environments below fair-weather wave base and above storm-wave base. In this setting, *Skolithos* ichnofacies occurs in the sand portions of storm deposits whereas *Cruziana* ichnofacies occurs in the mud, so ichnofacies are dependent on sediment type in this case (Pemberton *et al.*, 1992). The *Cruziana* ichnofacies may extend below storm-wave base and is also found in subtidal portions of estuaries and tidal flats. Although this ichnofacies develops in interlayered sand and mud, vertical mixing of the sediments by the trace makers may render the sediment a poorly sorted mixture of sand and mud. The large variety of traces in this ichnofacies is formed by filter feeders, deposit feeders, carnivores, and scavengers. *Cruziana* is commonly interpreted to be the crawling trail of a trilobite. The two side-by-side furrows ornamented with little scratch marks are interpreted as being made by the legs of the trace maker. In a few rare instances, trilobite-body fossils have been found in *Cruziana* traces. *Thalassinoides* is a complex three-dimensional criss-cross network of centimeter-diameter tubes similar to burrow networks made by modern decapod crustaceans. Simple tubular burrows are also common in this ichnofacies, including simple horizontal burrows known as *Planolites* (Figure 11.11C), the U-shaped *Arenicolites* (Figure 11.11D), and the conical, curving *Rossillia*. Resting or dwelling traces of sea stars, known as *Asteriacites*, and a horizontal version of *Diplocraterion*, known as *Rhizocorallium*, are also members of this ichnofacies.

The *Zoophycus* ichnofacies also contains *Spirophyton* and *Phycosiphon* (Figure 11.12) and all three traces are interpreted as feeding structures where the organism was systematically probing and ingesting the bottom sediment. *Zoophycus* (Figure 11.11E) and *Spirophyton* have complicated, three-dimensional geometries, whereas *Phycosiphon* is a horizontal branched tube usually with

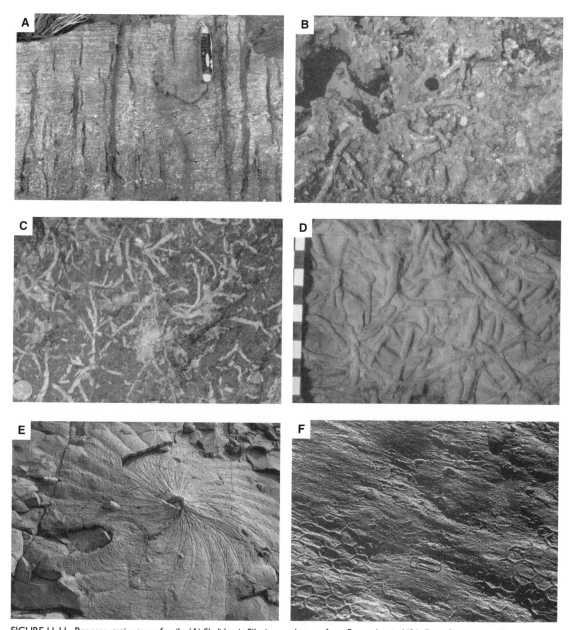

FIGURE 11.11. Representative trace fossils. (A) *Skolithos* in Silurian sandstones from Pennsylvania, USA. From Pettijohn and Potter (1964). By permission of Springer Science and Business Media. (B) *Ophiomorpha* (*Callianassid* shrimp burrow), Pleistocene of the Bahamas (courtesy of R. W. Mitchell). (C) *Planolites* in plan view, Cambrian of Alberta, Canada. (D) Plan view of *Arenicolites* from beneath showing U-shaped tubes, Cambrian of Pennsylvanian, USA. (E) *Zoophycus*. From Ricci Lucci (1995). (F) *Paleodictyon*. From Ricci Lucci (1995).

obvious spreiten. This ichnofacies is apparently diagnostic of settings with low dissolved-oxygen content and high levels of particulate organic matter. Presumably the low oxygen levels are due to stagnant conditions and the aerobic oxidation of particulate organic matter in the water column. The *Zoophycus* ichnofacies is not depth-dependent. It has been suggested that Paleozoic examples are from coastal lagoons and shallow shelf

settings, whereas Mesozoic and Cenozoic examples are from deeper, continental-slope and -rise settings (Pemberton *et al.*, 1992).

The *Nereites* ichnofacies is characterized by a variety of surface-feeding traces, and modern examples of this ichnofacies commonly occur on outer continental rises and across abyssal plains (Figure 11.12B). Ancient examples are almost invariably found in the E division

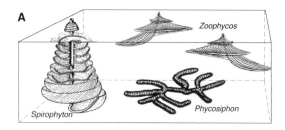

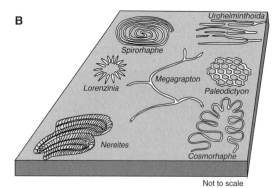

FIGURE 11.12. (A) *Zoophycus* ichnofacies. Modified from Häntschel (1975) and Pemberton *et al.* (1992). (B) *Nereites* ichnofacies. Modified from Pemberton *et al.* (1992).

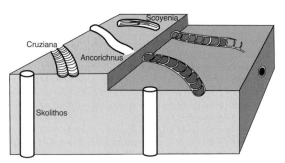

FIGURE 11.13. *Scoyenia* ichnofacies. Modified from Pemberton *et al.* (1992).

of turbidite deposits (i.e., quiet-water pelagic deposits – see Chapter 18). The *Nereites* ichnofacies apparently develops during the quiet intervals between turbidity-current depositional events. Although there is a variety of ichnogenera, they are generally not abundant. The signature ichnogenus, *Nereites*, is a curving, meandering trace ornamented by fecal pellets. *Spirorhaphe* and *Cosmorhaphe* are also curving, meandering traces but *Megagrapton* is a gently curved, branched trace. *Paleodictyon* (Figure 11.11F) comprises a stunning hexagonal array of traces, and is taken to represent a permanent microbe farm run by an unknown organism.

The most spectacular terrestrial trace fossils are track ways of dinosaurs and other vertebrates. Trackways have yielded valuable information about the locomotion of dinosaurs and large extinct mammals (Lockley, 1991; Lockley and Hunt, 1995; Prothero, 2004). Some trackways have been interpreted as showing evidence of herding behavior of sauropods, with large individuals on the outside and juveniles in the middle. These claims have been controversial (cf., Lockley, 1991; Prothero, 2004). Except for trackways, trace fossils in freshwater and terrestrial settings were reckoned to be sparse. However, trace fossils are now known to be very common in continental deposits from most sub-environments (Pemberton *et al.*, 1992; Retallack, 2001; Hasiotis, 2002,

2003, 2004; Bridge, 2003). They record the dwelling, burrowing, and surface movement of a wide variety of organisms, including worms, arthropods (insects and crustaceans), molluscs, and vertebrates. Plant-root traces occur in many terrestrial deposits, with the exception of deposits formed below the low-water level in channels and lakes. Kraus and Hasiotis (2006) describe using root traces to interpret soil moisture regimes. Until the 1980s, the only formal continental ichnofacies was the *Scoyenia* ichnofacies (Figure 11.13), comprising dwelling, feeding, and crawling traces produced mainly by arthropods in ephemeral lakes and floodplains. The full diversity and importance of continental trace fossils is now being realized and documented. A new *Mermia* ichnofacies has been proposed for lacustrine environments, and a *Coprinisphaera* ichnofacies (Genise *et al.*, 2000), named after dung-beetle nests, is an ichnofacies of insect traces (bees, wasps, ants, beetles, termites) in paleosols. In contrast to definition of these new ichnofacies, Hasiotis (2004) has constructed ichnocoenoses for alluvial, lacustrine, eolian, and transitional marine environments (e.g., Figure 11.14). Ichnocoenoses are assemblages of trace fossils that reflect biological communities (above and below ground). These biological communities, and hence ichnocoenoses, vary with sub-environment and climate and are more useful and finely subdivided than the broadly defined and poorly constrained ichnofacies. Examples of terrestrial trace fossils are given in Figure 11.15.

Continental trace fossils tend to be vertically zoned (tiered), reflecting the soil moisture content and the level of the water table (Hasiotis, 2002, 2004). For example, most insects and earthworms live above the water table, crabs and crayfish occupy a zone near the water table, and some organisms (oligochaete worms, molluscs, water-loving insects, shrimp) live below the water table. As the elevation of the water table varies seasonally, the traces made by these different categories

A

ALLUVIAL ICHNOCOENOSES

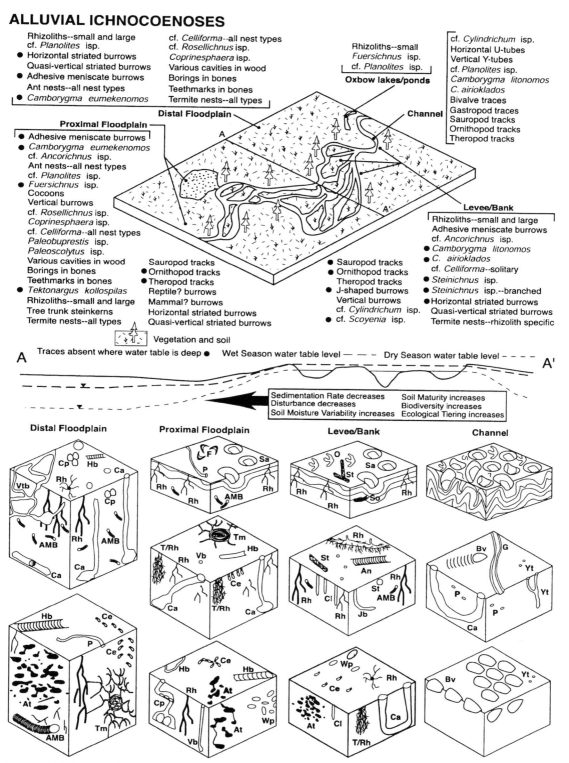

FIGURE 11.14. Continental ichnocoenoses from Hasiotis (2004): (A) fluvial, (B) lacustrine, (C) aeolian, and (D) continental–marine transition.

B

LACUSTRINE ICHNOCOENOSES

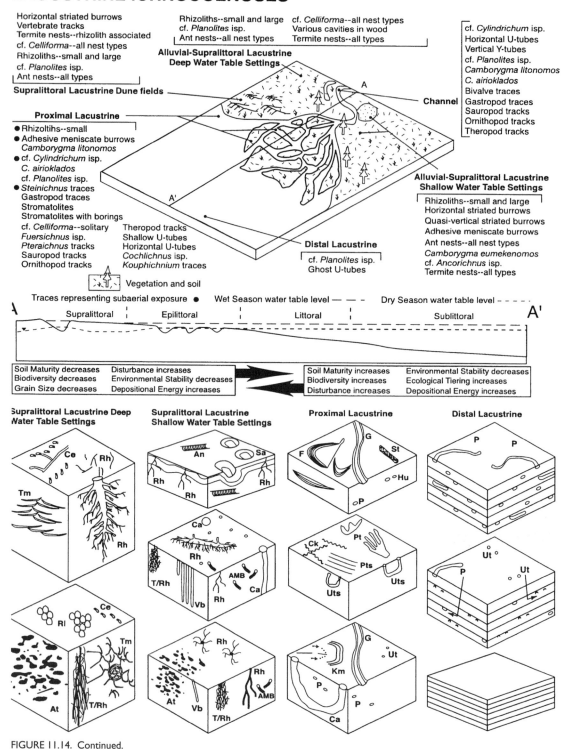

FIGURE 11.14. Continued.

C

EOLIAN ICHNOCOENOSES

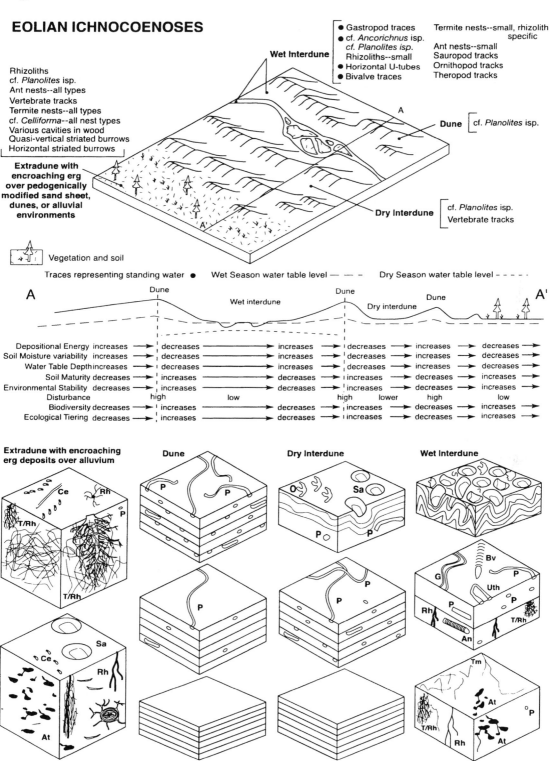

FIGURE 11.14. Continued.

D

TRANSITIONAL CONTINENTAL–MARINE ICHNOCOENOSES

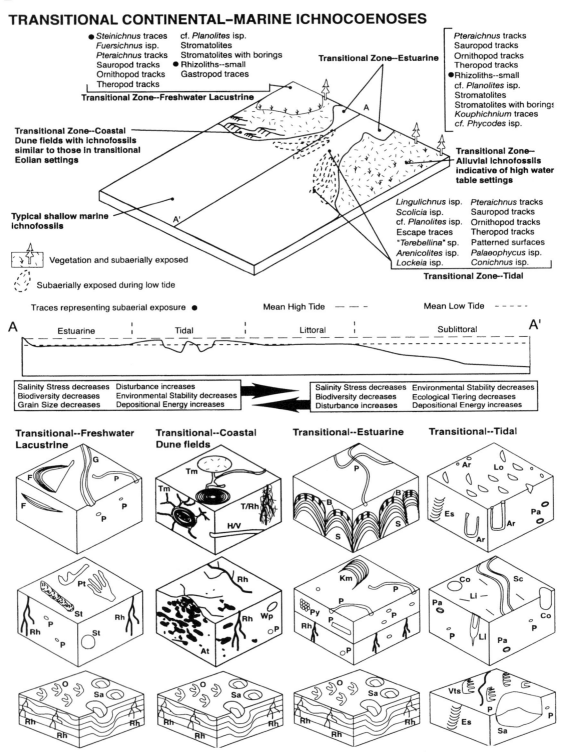

FIGURE 11.14. Continued.

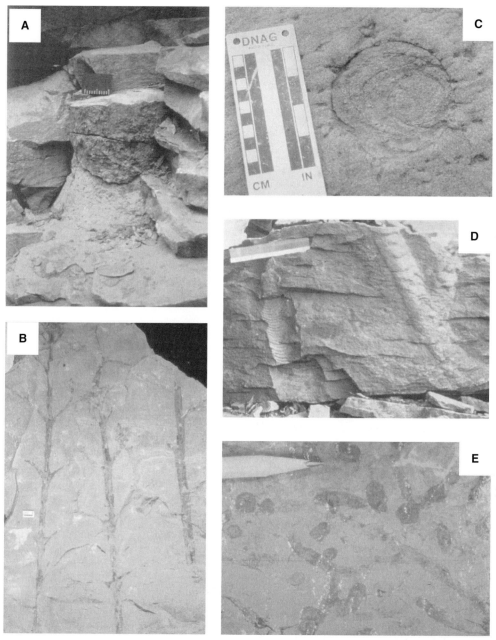

FIGURE 11.15. Paleozoic fluvial fossils from Devonian rocks of the northeastern USA or Canada: (A) a sandstone cast of a tree-trunk (0.1 m scale); (B) oriented plant axes (cm scale); (C) and (D) bivalve escape burrows; and (E) a common type of meniscate burrow. Mesozoic to Tertiary fluvial fossils: (F) mammal burrows; (G) a termite nest, about 1 m across; (H) a crayfish burrow; and (I) a dung-beetle nest with ball. Photos from Steve Hasiotis.

of organisms may become superimposed. The soil moisture content and the elevation of the water table are controlled by climate and by position on the floodplain relative to water bodies such as rivers and lakes. In arid climates, the amount of soil moisture is low, as is the average height of the water table. On moving from the edge of a lake or river onto a dry floodplain, the amount of soil moisture decreases, as does the mean elevation of the water table. This results in an increase in biodiversity, biotic exchange, burrowing depth, and

FIGURE 11.15. Continued.

degree of tiering from the water body to the floodplain (Figure 11.14A). For example, insect nests (of termites, soil bees, ants) tend to be deeper and larger where there are seasonal extremes of rainfall. Insect nests are smaller and shallower where the substrate is either dry or wet. Freshwater trace fossils reflect seasonal changes in water depth and salinity, and are thus controlled by climate also. Trace-fossil abundance and soil maturity increase as deposition rate decreases, because of the time available for bioturbation, which is an important part of pedogenesis (Hasiotis, 2002, 2004).

The stems and branches of higher plants make a number of distinctive tool marks. Flexible stems blown by the wind produce in sands circular grooves a constant radial distance from the stem. Coppice dunes and sand shadows are wind-blown accumulations of sand in the lee of shrubs. Water flows can produce scour pits around tree trunks that may be filled with cone-like inclined laminae. Finally, log jams in river channels produce an upstream scour and a downstream tail-like deposit.

Misidentification of trace fossils

Most burrows are tubular in shape, with diameters that range from less than a millimeter up to many centimeters, and are a few centimeters to several meters long. Some burrows can be confused with other millimeter-sized tubular sedimentary structures, such as root and stem casts and molds of higher plants, migrating gas-bubble trails, and molds or casts of cyanobacteria. Roots branch downwards, usually decrease in diameter at junctions as opposed to branching burrow networks, and end in microscopic root hairs. Roots may have carbonaceous linings and are mainly oriented vertically. As opposed to roots, burrows may be oriented in many different directions, may have complicated geometries, and are also commonly U-shaped. Another common mistake in interpreting continental and transitional continental–marine trace fossils is to misidentify cross sections of mudcracks as burrows or roots. Like burrow fills and root casts, mudcracks are commonly filled with a different material than that in which they occur. However, the latter are sheet-like and continuous, whereas the former are tubes. Differentiating gas-bubble trails from burrows is also difficult. Bubble trails can be nearly constant in diameter or can have bulges or variable diameter, and thus look remarkably similar to burrow networks as seen in resin casts from modern sediments. Also, although bubble traces tend to be vertical, other orientations have also been observed.

Evolutionary significance of trace fossils

Trace fossils are not useful for biostratigraphy, but do indicate major evolutionary events in Earth history. There is considerable debate as to whether animal phyla arose in a relatively short burst (a few tens of millions of years) of evolution just prior to the "Cambrian Explosion" or whether many hundreds of millions of years of evolution preceded the invention of shells. Geological considerations argue for the former whereas molecular clocks argue for the latter. Trace fossils hold great potential to settle the issue. Seilacher *et al.* (1998) claimed to have recovered trace fossils from 1.6-billion-year-old sandstones from India, and Rasmussen *et al.* (2002) describe purported 1.2-billion-year-old traces from Western Australia. These traces have been dismissed as mudcracks, but they are from the tops of sandstone beds. From the base of the Cambrian to the present, the depth and intensity of sediment reworking by organisms has increased. There has also apparently been a long-term increase in the depth to which animals can live beneath the sediment. The tiers are levels of organisms both above and below the sediment surface and their history is shown in Figure 11.16 (Ausich and Bottjer, 1982, 1985). Trace fossils are also commonly used to interpret sea-level changes, particularly for times when deposition in marine-shelf settings is curtailed due to rapid sea-level rise, and mineralized hardgrounds and firmgrounds develop with their attendant suites of trace fossils.

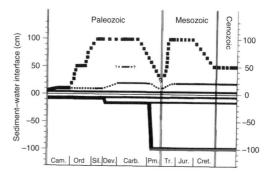

FIGURE 11.16. The history of tiering of suspension-feeding communities above and below soft substrates. Heaviest lines represent maximum levels, with other lines indicating tier subdivisions. Boundaries are dashed where inferred. Burrowing depths were limited until the Pennsylvanian and the development of meter-deep burrows of *Ophiomorpha*, polycheate worms, and other worm-like forms. Bivalves developed deeper burrows in the Early Mesozoic and gastropods began to burrow as well. Echinoids join the deep tier levels in the Mesozoic and, particularly, the Cenozoic. From Ausich and Bottjer (1985).

Trace fossils in continental deposits became abundant in the Silurian and Devonian periods as early land plants developed. Most of the early burrows were relatively small, lined and unlined dwelling and feeding burrows in various orientations, some of which had back filling (meniscate, spreiten) structures (e.g., *Beaconites, Meunsteria, Ancorichnus, Scoyenia, Planolites*). There were also vertical, cylindrical to irregular shafts formed by arthropods and aestivating lungfish, and the dwelling and escape burrows of bivalves. In addition, there was a variety of resting and crawling burrows (e.g., *Rusophycus, Cruziana*), tracks and trails of arthropods (e.g., *Diplichnites, Arthropodichnus*). The Carboniferous period saw continued development of the relationship between land plants and arthropods, and of trace-fossil assemblages. Mesozoic and Cenozoic trace fossils are dominated by the activities of worms, bivalves, spiders, insects (especially termites, ants, beetles, bees, and wasps), crustaceans, and mammals (Hasiotis, 2002, 2003). Continental trace fossils may indicate ancient soil moisture content, water-table levels, deposition rates, pedogenesis, and climatic changes (Hasiotis, 2004).

Biogenic and chemogenic laminated structures

Introduction

There is an array of biogenic–chemogenic laminated sedimentary structures, primarily composed of carbonate, that record growth by the addition of material to an external, generally irregularly shaped surface. These include *stromatolites, microbial laminites, thrombolites, tufas, travertines, cave deposits*, and *portions of soil carbonates*. Modern stromatolites, microbial laminites, tufas, and travertines form in many settings, including freshwater rivers, hot and cold springs, freshwater to saline lakes, carbonate tidal flats, carbonate shelves, around deep-sea hot springs, and around shelf and slope hydrocarbon seeps. Thrombolites are mainly crudely laminated structures similar in many respects to modern tufas and travertines that have been described mainly from ancient Proterozoic and Paleozoic carbonates. Cave deposits and soil carbonates are also laminated and form in diagenetic environments. Cave deposits (flowstones) in particular have well-developed chemogenic laminae composed of either radial crystalline carbonate or color-banded carbonate mud. They are easily recognized in modern deposits, but in ancient deposits they may be misinterpreted as a biogenic laminated structure such as a stromatolite.

The lamination in all of these structures (including the diagenetic cave and soil structures) is produced by three main mechanisms: (1) trapping and binding of sediment on mucilaginous microbial mats; (2) chemical precipitation due to the biological activity of microbial mats and colonies; and (3) chemical precipitation of minerals from supersaturated solutions. Unfortunately, laminae produced by these different mechanisms commonly resemble each other so much that it is not clear which laminae are made by which processes, even in modern examples. Laminae can also be produced in the various structures by different mechanisms at different times. In addition to lamination, many of these structures also commonly contain a number of enigmatic microstructures (microns to a millimeter in size) and macrostructures (millimeters to centimeters). Thus, modern stromatolites, microbial laminites, tufas, and travertines intergrade and resemble each other. In ancient deposits, thrombolites intergrade with stromatolites. Tufas and travertines from ancient deposits are increasingly being recognized and described, but they are still most commonly assigned a biogenic origin. This stems from the fact that all modern examples of these laminated biogenic–chemogenic structures invariably have a surface coating of bacteria (particularly cyanobacteria in sunlit environments), archaea, eukaryotic microbes (such as diatoms and macroalgae), and plants (such as mosses). The strong association of organisms with these structures has led to an almost *carte blanche* acceptance of their biogenic origin, although they may be chemogenic.

There has been renewed interest in the interactions between microbes and sediments since the late 1990s, fuelled by the recognition that the diverse biogenic–chemogenic laminated sedimentary structures of the Archean and Proterozoic might hold records of the hydrosphere, atmosphere, and biosphere of the premetazoan Earth. Microbes and, in particular, benthic microbial mats are important in the origin of many, but not all, modern biogenic–chemogenic laminated structures. In the Proterozoic Eon, benthic mats probably covered any sediment surface that was immobile for a few days. As grazing and deposit-feeding invertebrates arose and diversified, benthic microbial mats became restricted. Another impetus for studying microbe–sediment interactions comes from the search for extraterrestrial life. Imagine the uproar that would ensue if a Mars rover equipped with cameras beamed back a photograph of something that resembled an Earth-like biogenic–chemogenic laminated sedimentary structure.

Stromatolites

Description

Stromatolites are complex in-place structures with recognizable boundaries and are composed internally of millimeter-thick laminae, commonly of homogeneous mud (Figure 11.17). The laminae are usually convex-up, concordant, define the upper surface of the stromatolite, and usually curve down their sides. This definition of stromatolites excludes laterally-extensive, flat laminasets interpreted to have been deposited over benthic microbial mats. Such laminasets are commonly referred to as microbial laminites, planar stromatolites, or stratiform stromatolites, and are discussed separately. Stromatolites are common in ancient carbonate rocks, modern carbonate-shelf sediments, and a wide variety of chemical sedimentary deposits, including chert, deep-sea manganese crusts and nodules, phosphorites, and barites. Stromatolites in modern and ancient siliciclastic sediment, particularly tidal-flat mud, are becoming more fully appreciated. The term *protostromatolite* covers unlithified modern examples. For discussion of stromatolites and their origins see Walter (1976), Demicco and Hardie (1994), Grotzinger and Knoll (1999), Riding (2000), and many of the papers in Grotzinger and James (2000).

Individual stromatolites range in size from a few centimeters up to a few tens of meters in height and in width from a few millimeters to a few meters (Figures 11.17 and 11.18). Many stromatolites are higher than they are wide and such forms are either *club-shaped*, where the width of the stromatolite increases upwards, or *columnar*, where widths do not increase upwards. Laminae in club-shaped and columnar stromatolites are commonly curved and convex-up, but some Proterozoic columnar stromatolites have conical laminae (Figure 11.17D). Columnar and club-shaped stromatolites commonly branch upwards (Figure 11.17C), and branches may either be separated by sediment or grow side by side closely. Stromatolites that are wider than they are tall are *domal*, and slightly flattened, i.e., *ellipsoidal*. Ellipsoidal stromatolites tend to have laminae that wrap beneath overhangs, and domal and ellipsoidal stromatolites also commonly branch upwards. In plan view, stromatolites are commonly circular, but some are notably elongated, with their long axes oriented normal to the inferred paleoshoreline (Figure 11.18).

Stromatolites are generally arranged in larger-order structures referred to as *stromatolitic buildups* or *bioherms* (Figure 11.18C). *Buildup* is a useful descriptive term for an in-place accumulation of material of no particular origin, whereas a *bioherm* is a buildup inferred to be partly biogenic in origin. Stromatolitic buildups are common in Proterozoic carbonates and are usually a few meters to tens of meters thick, covering tens of square meters to hundreds of square kilometers. Smaller Proterozoic and most Paleozoic stromatolitic bioherms are relatively simple in structure. However, large stromatolitic bioherms can have quite complex internal structure where laminae are interstratified with less-distinctly laminated zones. Such structures are more akin to reefs.

Most laminae that comprise stromatolites are made of micron-scale mud that can be clay-rich. Other laminae are composed of pure carbonate mud (Figure 11.19B), well-sorted peloidal sands (Figure 11.19A), a clotted fabric of merged-peloids, and upward-growing radiating crystals of carbonate with laminae delineated by clay-rich material (Figure 11.19C). Most diagenetically altered stromatolites comprise a uniform interlocking mosaic of crystals, each a few tens of microns in diameter, arising from recrystallization and replacement (see Chapter 19). In some cases, micron-diameter tubes a few microns long, that apparently represent the molds of microbial filaments, are preserved in peloidal and mudstone laminae. It is common for these laminae types to be intermixed in a given stromatolite.

The origin of stromatolites

The laminae in stromatolites imply episodic addition of material to a mound with side slopes that can exceed the angle of repose of cohesionless sediment (Figure 11.19). This implies that the deposition of laminae is not a purely physical process. The best-known modern stromatolites are produced where sediment is trapped on mucilaginous organic slime produced by prokaryote-dominated biofilms in peritidal settings. This mechanism was first recognized by Black (1933) in a pioneering study of modern carbonate tidal flats on Andros Island. Ginsburg (1967) re-emphasized Black's contributions and, in the 1960s and 1970s, many other modern stromatolites produced by microbial biofilms were described from modern carbonate peritidal settings. Hardie (1977) described well-laminated, club-like stromatolites up to 60 mm high built between the entrances of fiddler-crab burrows along tidal creeks of northwest Andros Island (Figure 11.17B). Modern stromatolites are also known from a few subtidal shelves with carbonate mud, such as those of eastern Andros Island and Bermuda. These stromatolites are

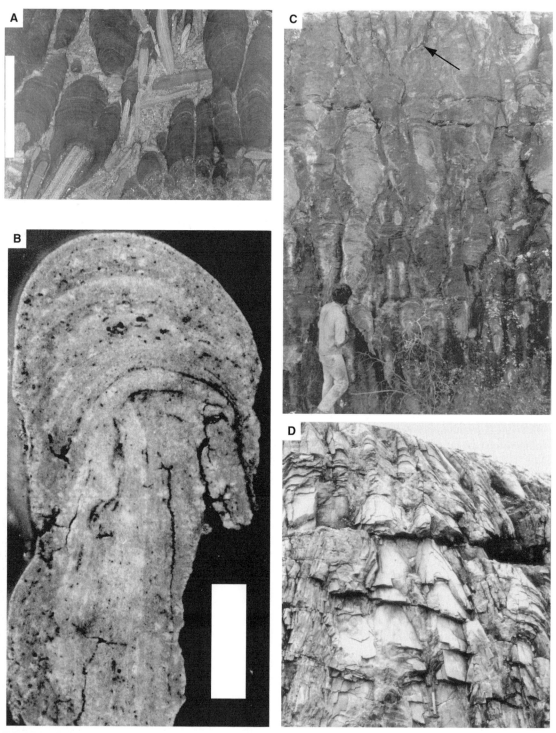

FIGURE 11.17. Stromatolite cross sections. (A) Stromatolites nucleated on flat intraclasts of planar and cross-stratified peloidal grainstone, Cambrian of western Maryland. The scale bar is 50 mm. (B) Epoxy-resin-embedded uncemented stromatolites from the banks of a tidal channel on Andros Island. Laminae have accreted over a vertically oriented intraclast. The scale bar is 10 mm. (C) Branching (arrow) columnar stromatolites from the Proterozoic Taltheilei Formation, Northwest Territories, Canada (photograph courtesy of Paul Hoffman). (D) Conical stromatolites (so-called *Conophyton*) from the Proterozoic Dismal Lake Group (photograph courtesy of John Grotzinger).

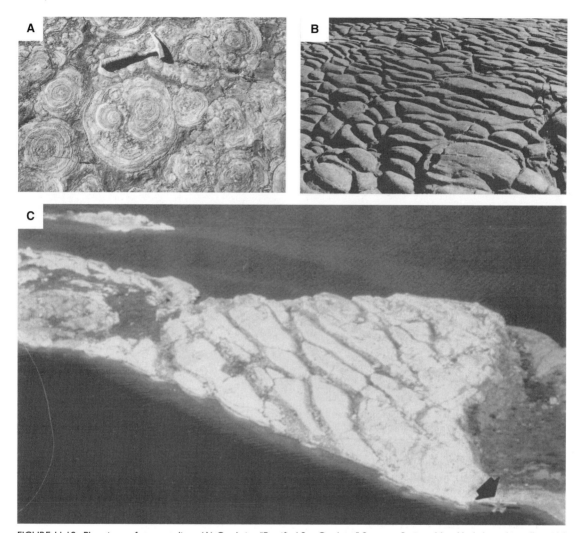

FIGURE 11.18. Plan views of stromatolites. (A) Cambrian "Petrified Sea Gardens," Saratoga Springs, New York, branching ellipsoidal forms. (B) Exhumed elongated, columnar stromatolites from tidal-flat deposits of the Proterozoic Taltheilei Formation, Northwest Territories, Canada. The long axis is perpendicular to the paleoshoreline (photograph courtesy of Paul Hoffman). (C) A large Proterozoic stromatolite reef composed of elongated *Conophyton* stromatolites (Figure 11.17D). Glaciers have scoured the inter-mound fill. The airplane on the beach at the arrow provides scale. Beechy Formation, Northwest Territories, Canada (photograph courtesy of John Grotzinger).

found in association with *oncoids*, the *algal biscuits* of Ginsburg *et al.* (1954). Solitary stromatolites (up to 2 m high and 1 m across) and clusters of coalesced stromatolitic mounds are found in actively migrating areas of Bahamian ooid shoals. Laminae are composed of very-fine to fine ooid and skeletal sand grains. Most laminae in these Bahamian stromatolites comprise centimeter-diameter, vertically oriented "fingers" that contain some concentric laminae, but are unlaminated for the most part. Less commonly, laminae span most of the way across a stromatolitic head, generally pinching out down the sides, but are also notably discontinuous.

Surface layers are unconsolidated, but within a few centimeters of the surface the stromatolites become cemented by fine-grained aragonite and magnesium calcite. Prior to Black (1933), modern stromatolites had been described mainly from freshwater settings (Hoffman, 1973), in which microbial biofilms (as well as higher plants such as mosses) were encrusted by calcium carbonate. This precipitated on or within the biofilm, either as a result of the local metabolic activities of certain components of the biofilm, or because the entire water mass was supersaturated with carbonate. As a result of these studies, stromatolites are most

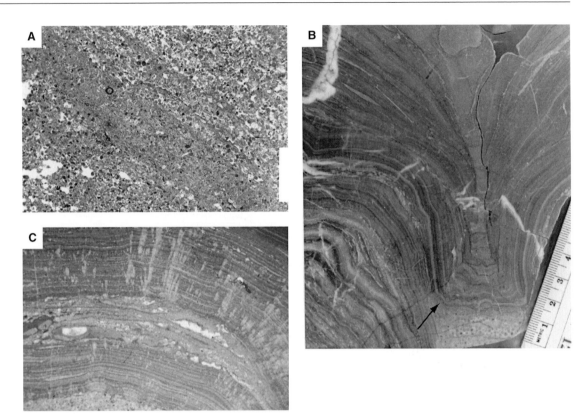

FIGURE 11.19. Various types of stromatolitic lamination. (A) A thin-section photomicrograph of a modern uncemented stromatolite (Figure 11.17B) composed of mud layers and layers of fine sand and silt-size pellets that adhered to a microbial mat. The scale bar is 1 mm. (B) Mudstone laminae in upper portions of this stromatolite pinch out down the sides of the stromatolite; however, mudstone laminae at the arrow maintain constant thickness (are isopachous), entomb pisoids, and bridge between adjacent stromatolites. Such isopachous laminae may be chemical precipitates. The scale bar is in cm. (C) A thin-section photomicrograph showing laminae composed of fine crystalline mosaics at top and bottom with pellets and mud laminae in the middle. Radiating crystal forms are suggestive of abiogenic carbonate precipitation, whereas mud and pellet layers suggest microbial trapping of grains. Field of view ~2 mm. Photograph courtesy of Brent Waters.

commonly interpreted as due to sediment trapping or precipitation of carbonate by a benthic microbial mat. In fact, some *definitions* of stromatolites specifically state that they are organo-sedimentary structures produced by the sediment trapping and precipitation of cyanobacterial mats. It is very difficult (if not impossible) to assign a specific origin to a stromatolite, especially in the field: "if a biological origin had to be demonstrated before a geological object could be called a stromatolite, the term would in most instances be inapplicable (or at least provisional) even after a careful search for associated microorganisms had been made" (Semikhatov *et al.*, 1979, p. 994).

Well-known modern peritidal "stromatolites" occur in Hamelin Pool, a hypersaline lagoon in Shark Bay, Western Australia (Figure 11.20) (Logan *et al.*, 1974; Hoffman, 1976). These are found from supratidal to subtidal areas with water depths of at least 3.5 m. They occur as isolated, club-shaped forms up to 0.75 m high, as coalesced clubs, and as elongated, isolated ridges oriented subparallel to the wave-approach direction, more or less perpendicular to the shoreline. These stromatolites range from laminated to unlaminated, and laminae are composed of mollusc shell fragments, peloids, and ooids (Figure 11.20B). The laminae are millimeters thick and lensoidal, and there is commonly angular discordance between adjacent sets of laminae. Laminae become indistinct towards stromatolite edges. The grains are cemented by fine-grained, acicular aragonite. Laminae are disrupted by numerous fenestrae, many of which have been enlarged by dissolution (Figure 11.20B). Playford and Cockbain (1976) reported that markers hammered into the structures show that there has been no addition of sediment for many years. These Shark Bay "stromatolites" and the giant Bahamian examples from subtidal channels are

A

FIGURE 11.20. (A) Stromatolites from Carbla Point, Hamelin Pool, Western Australia. (B) Cross sections of Shark Bay stromatolites, showing fenestrae and lamination. From Playford and Cockbain (1976).

B

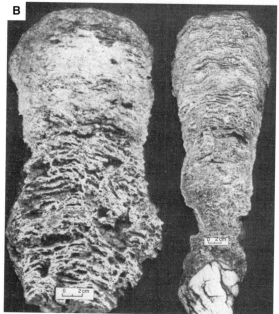

in the form of microbial populations (circa 3.5 Ga) and some Archean (and younger) chert stromatolites carry identifiable, well-preserved microfossils. Many Archean and some Paleoproterozoic stromatolites are most likely of abiogenic, chemically precipitated origin. Archean carbonates contain meter-diameter stromatolite-like mounds composed of large radiating crystals (Sumner and Grotzinger, 2000) and small, shrub-like masses of carbonate, commonly referred to as "tufa," also interpreted as abiogenic sea-floor precipitates (Demicco and Hardie, 1994; Pope and Grotzinger, 2000). It is not clear what the abundance of chemical precipitates in carbonate platforms of early Earth means in terms of the history of seawater chemistry (Grotzinger and James, 2000).

Stromatolites are most common and most spectacular in their size and complexity in Mesoproterozoic (1.6–1.0 Ga) and Neoproterozoic (1.0–0.54 Ga) carbonate rocks (Grotzinger and Knoll, 1999; Grotzinger and James, 2000). Mesoproterozoic and Neoproterozoic stromatolites are very common sedimentary structures in carbonates, occur in all environments of deposition including barrier reefs, and in some instances comprise entire formations hundreds of meters thick that cover thousands of square kilometers. Russian and Australian stratigraphers have given formal binomial genus and species names based on the external form of the stromatolite and its internal laminae to fifty or so stromatolites. They use stromatolites as biostratigraphic tools to subdivide the Upper Proterozoic stratigraphic column. European and American sedimentologists

not like typical muddy laminated stromatolites, and much more closely related to ancient thrombolites than to most ancient stromatolites. Thrombolites are described in the following section.

Evolution of stromatolites

Stromatolites occur in many types of sedimentary rocks of all ages. Archean stromatolites in cherts are generally regarded as direct evidence of the earliest life agree that there are several morphological motifs of Precambrian stromatolites but view stromatolites as the products of diverse biogenic and chemogenic processes dependent on light intensity, salinity, nutrient supply, current velocity, grain size, microbial-community diversity, carbonate-saturation state, and available carbonate-nucleation sites (Grotzinger and Knoll, 1999). Early cementation (a diagenetic process) is also implicated in the formation of many Proterozoic stromatolites. Moreover, Precambrian stromatolites probably grew in a variety of freshwater to marine settings (Hoffman, 1974). It is difficult to assess whether the various Precambrian stromatolite forms are characteristic of any particular carbonate shelf environment, although shelf-margin buildups appear to have the most diverse and largest forms. Stromatolites suffered an abrupt decline after the Early Ordovician, that corresponds to the expansion in number of bottom-grazing and burrowing invertebrates. Stromatolites apparently make brief comebacks in normal marine subtidal environments after major mass-extinction events (cf. Sheehan and Harris, 2004).

Thrombolites

Description

Thrombolites are in-place buildups of carbonate sediment that have a clotted fabric in addition to local laminae (Figures 11.21 and 11.22). The term is derived from the Greek *thrombos*, meaning a blood clot. Thrombolite was originally a field term used to describe Cambrian and Ordovician structures that lacked the prominent lamination of co-occurring stromatolites, that were similar in size and external shape. Thrombolites are most commonly described from Neoproterozoic, Cambrian, and Lower Ordovician rocks, where they range in size from mounds a decimeter high and a few centimeters across to masses tens of meters thick that are meters to tens of meters wide (Figure 11.21). Thrombolites also occur in later deposits, including oil reservoirs in the Mesozoic of the Gulf Coast of the USA (Mancini *et al.*, 2004). Commonly, individual thrombolites 1–2 meters high and up to 1 meter across are coalesced into a biohermal complex meters to hundreds of meters thick and tens to thousands of meters in diameter. Neoproterozoic thrombolitic bioherms comprise complexes of coalesced smaller buildups and sheets with thrombolitic fabric (Figure 11.23) (Grotzinger and James, 2000; Turner

et al., 2000a, b; Batten *et al.*, 2004). Stromatolites and thrombolites are commonly intimately associated in bioherms (Figure 11.23), mutually encrusting each other over a range of scales. Coalesced bioherms typically enclose irregularly shaped pockets of grainstone or flat-pebble conglomerate. References on Paleozoic thrombolites include Pratt and James (1982), Kennard and James (1986), Demicco and Hardie (1994), and Riding (2000). Mesozoic thrombolites are described by Mancini *et al.* (2004).

The most common internal structures of thrombolites are millimeter- to centimeter-diameter, amoeboid-shaped clots of carbonate mud that are dark gray to black, comprising 30%–40% of the rock, surrounded by lighter-colored mudstones or cement (Figure 11.22). Also common are finger-sized, discontinuous digitate columns with circular cross sections. These are also surrounded by mudstone or cement, with incomplete stromatolitic layers surrounding or interstratified with the mud clots (Figure 11.22A). A complete gradation apparently exists from carbonate mud columns with regular stromatolitic laminae through to unlaminated massive masses of carbonate mud. In a given thrombolite mound, laminated and partially laminated stromatolitic columns exist side by side with amoeboid mud clots. Unlaminated fingers and clots of carbonate mud are mottled due to irregular patches of carbonate mud a few tens of microns in diameter surrounded by a mosaic of microspar. Neoproterozoic and younger thrombolites include problematic microfossils such as *Renalcis* and unchambered *Renalcis*-like forms, *Epiphyton*, *Girvanella*, nondescript micron-diameter spheres and tubes up to a few hundred microns long (Figure 11.24). In Paleozoic thrombolites, the mudstones surrounding the various clots and stromatolitic masses commonly contain crinoid debris, brachiopod shell fragments, and trilobite shell fragments. Burrows that avoid skeletal fragments, and borings that cross-cut skeletal fragments and early cements, are also found in mudstones between the clots in Paleozoic examples. Geopetal cavities filled with laminated internal sediment and cements are also common and cross-cut the other elements (Figure 11.22B). The margins of the geopetally filled vugs are outlined by a carbonate mud rind a few microns thick that commonly truncates skeletal grains in Paleozoic examples. The internal sediment beneath the final void-filling cements can contain skeletal fragments, suggesting that the cavities were open to the surrounding sea floor. The cements commonly comprise a radially oriented, inclusion-rich fringe cement surrounding an internal blocky mosaic cement.

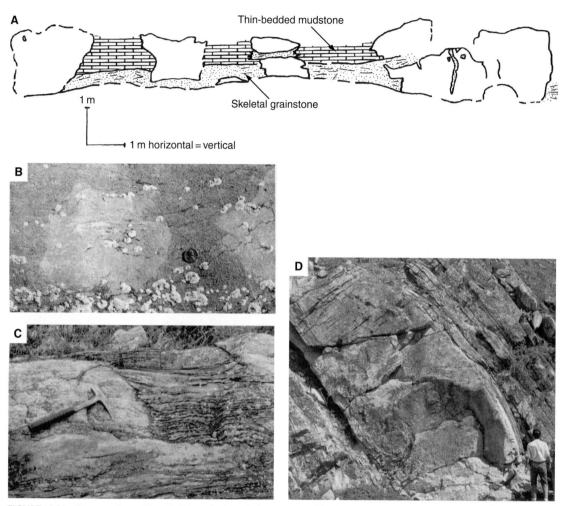

FIGURE 11.21. Cross sections of Lower Paleozoic thrombolites, western Maryland. (A) Meter-diameter thrombolites comprising irregularly shaped mounds separated by cross-stratified skeletal grainstone and thin-bedded mudstones, prepared from photomosaic. Images (B), (C), and (D) are field photographs of thrombolites of various sizes. The photograph of thrombolite buildup or a bioherm in (D) appears courtesy of Francis Pettijohn, from Pettijohn and Potter (1964). By permission of Springer Science and Business Media.

The origin of thrombolites

Thrombolites were hard, rigid bioherms with variable amounts of microbial framework when deposited. The attributes of biological reefs exhibited by thrombolites include depositional relief; biogenic framework, in this case built by microbes; synsedimentary cementation; and morphological changes related to water depth (see Chapter 15). Examples of thrombolitic barrier reefs, pinnacle reefs, and patch reefs are known (Turner *et al.*, 1993, 2000a, b; Demicco and Hardie, 1994; Grotzinger and James, 2000; Batten *et al.* 2004). The problematic microfossils making up the microbial framework are generally interpreted as calcified microbes, even in Proterozoic examples (Figure 11.23). In Paleozoic thrombolites, *Girvanella* and *Epiphyton* are interpreted as calcified filamentous microbes, whereas *Renalcis* is reckoned to be a calcified coccoid form. The enigmatic carbonate mud clots that give thrombolites their name have also been interpreted as being the result of the calcification of microbial mats and colonies (Grotzinger and James, 2000; Riding, 2000; Turner *et al.*, 2000a, b; Batten *et al.* 2004). Similar modern forms occur in freshwater tufas and travertines; see Chapter 16 and Laval *et al.* (2000). As in modern coral reefs, boring and perhaps dissolution of hard portions of the thrombolites apparently alternated with periods of internal (geopetal) sedimentation and cement growth, as

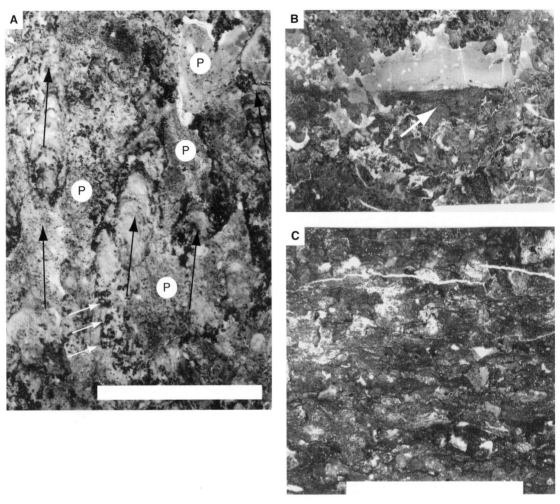

FIGURE 11.22. Cross sections of Paleozoic thrombolites. (A) Vertically oriented fingers with variably developed laminae (black arrows) separated by patches of skeletal packstone (p). White arrows point to clusters of *Renalcis* encrusting sides of fingers. The scale bar is 50 mm. (B) An irregular-shaped cavity in a thrombolite filled with peloidal grainstone (white arrow) that grades up to a burrowed mudstone. Microbial precipitates are truncated at the cavity margins. The scale bar is 50 mm. (C) Clotted mudstone fabric characteristic of thrombolites. The scale bar is 50 mm. Dark clots comprise *Renalcis* and *Renalcis*-like structures interpreted as calcified microbes.

judged from the vugs and their fillings. Indeed, the spar-filled voids commonly occur beneath overhangs of the carbonate-mud masses and beneath metazoan skeletons, indicating the presence of shelter voids within the thrombolites.

Evolution of thrombolites

Thrombolites apparently arise in the Early Neoproterozoic and become the dominant reef structures of the carbonate platforms and banks of the Lower Paleozoic (see Chapter 15). The large Neoproterozoic thrombolite reefs described by Turner *et al.* (2000a, b) and

Batten *et al.* (2004) are apparently pinnacle reefs. In the overall evolutionary trend of reefs, they represent a transition from the microbial stromatolite reefs of the Precambrian to the metazoan reefs of the Phanerozoic. Thus, the problematic microbial framework builders of the Cambro-Ordovician thrombolite reefs, such as *Epiphyton* and *Renalcis*, can perhaps be regarded as colonial calcareous organisms transitional in development between Precambrian non-skeletal microbes and Phanerozoic skeletal algae. In addition, in-place sponge or coelenterate-like skeletal metazoans begin to occur in thrombolites from the latest Neoproterozoic on (Wood *et al.*, 2002).

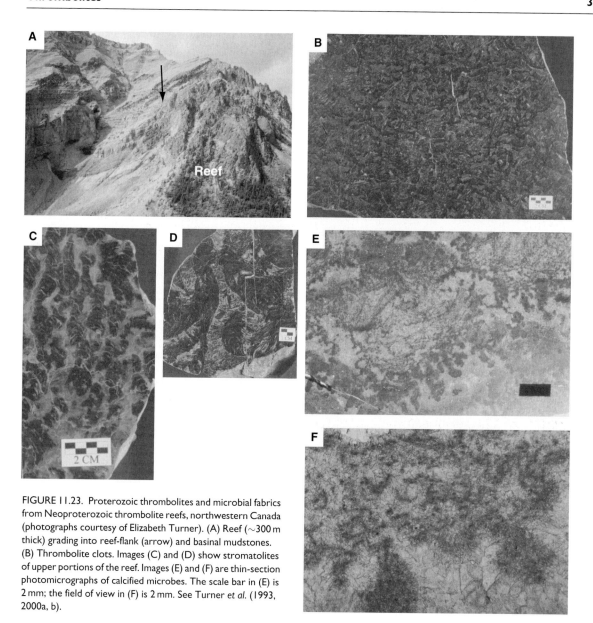

FIGURE 11.23. Proterozoic thrombolites and microbial fabrics from Neoproterozoic thrombolite reefs, northwestern Canada (photographs courtesy of Elizabeth Turner). (A) Reef (~300 m thick) grading into reef-flank (arrow) and basinal mudstones. (B) Thrombolite clots. Images (C) and (D) show stromatolites of upper portions of the reef. Images (E) and (F) are thin-section photomicrographs of calcified microbes. The scale bar in (E) is 2 mm; the field of view in (F) is 2 mm. See Turner *et al.* (1993, 2000a, b).

Shelf-edge bioherms preserved in place are rare around the North American shelf in the Lower Paleozoic; however, an important example composed of *Epiphyton–Girvanella* bioherms is found in the Lower Ordovician Grove Limestone exposed in the Frederick Valley of western Maryland. Such microbial reefs must have been quite widespread in Cambro-Ordovician times because blocks of *Epiphyton–Girvanella* boundstones are commonly found in breccias in off-shelf facies, in particular the famous Cowhead Breccias of Newfoundland (James, 1981)

and the Kicking Horse Rim of the Canadian Rockies (McIlreath, 1977). Thrombolites represent patch reefs on Paleozoic platforms. Thrombolites dominated by *Renalcis* and *Renalcis*-like forms are restricted to the interior of the shelves and their morphology changes systematically across the shelf. Thrombolites well away from the shelf edge typically do not contain *Renalcis* or *Renalcis*-like forms and instead are composed exclusively of carbonate-mud clots. In the Lower Cambrian, archeocyathids become an important element in thrombolites (James, 1983; James and Bourque,

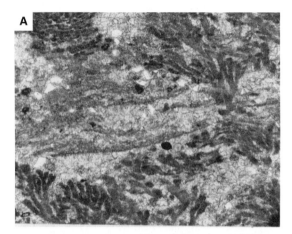

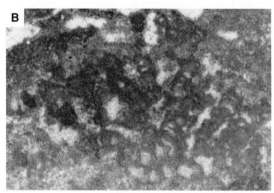

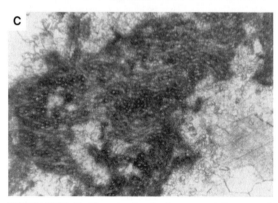

FIGURE 11.24. Problematic microfossils commonly interpreted as calcified microbes from thrombolites of western Maryland. (A) *Epiphyton*, width of field 2.5 mm. (B) *Renalcis*, width of field 2 mm. (C) *Girvanella*, width of field 2 mm. Images (B) and (C) appear courtesy of Chau Nguyen.

1992). In the Ordovician, corals, stromatoporoids, and bryozoans become important framework components of bioherms; however, thrombolite fabrics continue to be important components of reefs well into the Paleozoic. Much of what is referred to as "reef" between large, in-place, skeletal metazoans resembles

thrombolites. This is true, for example, of the so-called "Chazy" reefs of New York and the Devonian reefs of western Canada and the Canning Basin of northwestern Australia (Chapter 15).

Microbial laminites

Description

Microbial mats dominated by cyanobacteria are common on intertidal and supratidal mudflats (both carbonate and siliciclastic: see Chapter 16 for examples), principally because the harsh conditions of exposure preclude the presence of the various invertebrates that would otherwise forage on the mat. Laminated sediments are produced by episodic deposition on the mats, which help trap the sediment and then grow around or wriggle through the deposited laminae and re-establish themselves on the new sediment surface. Such structures are called *microbial laminites* and this term replaces the older terms algal laminite, cryptalgal laminite, stratiform stromatolites, planar stromatolites, and cryptmicrobial laminite (the prefix *crypt-* was meant to indicate that the microbial nature of the deposits was hidden or cryptic). In addition, the acronym MISS, from microbially influenced sedimentary structures, has been suggested (Noffke *et al.*, 2005). Modern and ancient microbial laminites comprise simple sets of laminae or composite sets of laminae (Figure 11.25). Modern microbial laminasets are up to a few tens of centimeters thick, whereas ancient microbial laminasets can be up to meters thick. Terms useful to describe the geometry of the laminae are *flat* (or *planar*), *wavy*, and *crinkled*, depending on the irregularity of the laminae.

There was a spate of studies of modern microbial laminates from carbonate tidal flats in the 1970s, and areas studied included the supratidal islands of Florida Bay (Ginsburg *et al.*, 1954), the intertidal and supratidal mudflats and coastal marshes of Andros Island (Black, 1933; Hardie, 1977), the intertidal and supratidal mudflats of the United Arab Emirates (the "Trucial Coast") of the Persian Gulf (Kendall and Skipwith, 1968; Kinsman and Park, 1976), and the intertidal sand and mudflats of Shark Bay in Western Australia (Davies, 1970; Logan *et al.*, 1974; Hoffman, 1976). Modern studies have concentrated on species makeup, particularly zoning of different microbes within mats, and the microbial enzymatic systems (e.g., Decho *et al.*, 2005; Visscher and Stolz, 2005).

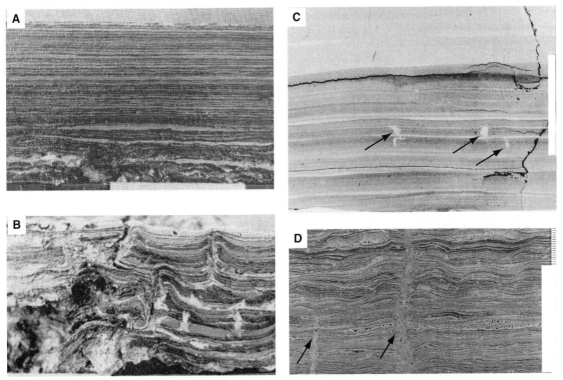

FIGURE 11.25. Modern and ancient microbial laminites. Images (A) and (B) show modern carbonate microbial laminites from upper intertidal flats of the United Arab Emirates, Arabian Gulf. From Kinsman and Park (1976). In (A), flat parallel laminae comprise white carbonate mud washed onto the flats during storms and dark cyanobacterial mats. The scale bar is 10 cm. Similar laminae in (B) are disrupted by desiccation cracks. Note also the irregular surfaces of the buried mats. The scale bar is 10 cm. (C) Planar laminae composed of alternations of very fine-grained peloidal grainstone (light) and dark mud interpreted as microbial laminae. Note the small mudcracks (arrows). The scale bar is 5 cm; photograph courtesy of R. W. Mitchell. (D) Wavy and crinkled microbial laminae composed of alternations of grainstones (dark) and mudstones (light). Note the deep desiccation cracks (arrows). The scale bar is 5 cm. Images (C) and (D) are from Lower Paleozoic rocks of western Maryland.

In modern microbial laminites, the type of mat is directly reflected in the resultant laminar structure. The type of mat is dependent on factors such as the amount of wetting and drying by seawater and rainwater (related in turn to elevation above sea level and the regional climate), frequency of deposition, amount and type of sediment, and salinity. Two general cyanobacterial mat types occur in modern settings: (1) motile filamentous *Oscillatoriaceae*-dominated mats that grow within and on the sediment surface (see Figure 15.41 later) and (2) immotile fleshy mats of non-oscillatoriacean cyanobacteria that grow only on the surface. The fleshy mats can be subdivided further into smooth fleshy mats or crinkled fleshy mats composed of various combinations of filamentous and coccoid forms. *Oscillatoriacean* mats from the supratidal flats of Andros Island produce flat, millimeter-thick laminae and alternating lenses (from low ripples and bed-load sheets) of fine

peloidal–skeletal sands and more continuous mud laminae (see Figure 15.41). The smooth fleshy mats of the Persian Gulf and Shark Bay are the most widely distributed type. Arabian Gulf laminites have laminae of fine sands to silts that vary from a millimeter to 10 mm thick, and may be lenticular. These alternate with thinner, muddy, organic-rich layers (the buried cyanobacterial mats) that are more laterally continuous (Figure 11.25A, B). Although laminae under smooth fleshy mats are flat to wavy, it is only local stromatolitic bumps over mat irregularities and roll-up structures (see below) that signal a cyanobacterial origin. In some areas of the Arabian Gulf and Shark Bay, subaerial desiccation of thick fleshy surface mats results in disruption into polygonal cracks (Figure 11.25B), and the upturned edges of these cracks become preferential sites for growth of the succeeding generation of cyanobacteria, producing oversteepened stromatolitic layering. Sediments below crinkled fleshy mats of the Arabian Gulf and Shark Bay are only vaguely laminated, or clotted, or massive. In lower intertidal ponds of Andros Island, fleshy surface mats result in wavy to crinkled, fenestral laminae and calcified cyanobacterial colonies (see Figures 15.40 and 15.41) (Hardie, 1977). The fenestrae are the result of desiccation and rotting of buried microbial colonies, desiccation, and insect burrows (see Chapter 12).

Interpretation

Laminated mudstones and sandstones are common in shallow-water carbonate shelf deposits, and such laminites are widely interpreted as microbial. Interpretation of crinkled laminae as microbial is commonly justified, especially where some of the laminae exceed the angle of repose of sediment. However, interpretation of flat and wavy laminae as microbial is not straightforward. Even wavy and crinkled laminae cannot always be ascribed to microbial mats, because non-organic processes can produce such laminae. For example, wavy laminae can be formed by wave ripples, and crinkled laminae can result from soft-sediment deformation. Therefore, independent evidence that a microbial mat covered the surface is required. Indicators of former microbial surface mats include roll-up structures (discussed below), calcified cyanobacterial colonies, and microbial filament molds and casts. Laminoid fenestrae (described below) are another common indicator of former microbial mats and desiccation cracks. Examples of ancient laminae interpreted to be microbial laminites are shown in Figure 11.25.

Roll-up structures

Roll-up structures (Simonson and Carney, 1999) have also been called jelly-roll structures, enrolling of mats, encapsulated and enrolled structures, and encapsulated roll-up structures (Demicco and Hardie, 1994). Roll-up structures resemble rolled-up carpets and are up to a meter long, a few tens of millimeters thick, and up to a few centimeters wide (Figure 11.26). They have been found on the modern tidal flats of Shark Bay, Western Australia (Davies, 1970), Virginia, and Baja California. Roll-up structures are distinguished from oncoids by their spiral geometry and three-dimensional, tube-like geometry.

Roll-up structures on modern tidal flats occur where smooth fleshy, non-oscillatoriacean cyanobacterial mats cover the sediment surface like a carpet, and crack and blister due to desiccation. Water currents

tear up the cyanobacterial mat and associated sediment (starting at mudcracks and other defects in the mat surface), undermine it, and roll it downcurrent. In addition to making roll-ups, water currents can slide sheets (up to a meter across) of cyanobacterially bound sediment along thrust faults. Displacement of the sheets is taken up by anticlines and brecciation of the carpet as well as roll-ups (see Figure 16.16 later). However, spiral folding and thrusting can also occur in slump fold sheets (see Chapter 12). Slump-generated structures occur in sheet-like disturbed zones on the order of 1 m thick and up to kilometers in lateral extent, whereas roll-up structures tend to be much smaller, isolated structures that pass laterally into undeformed sediment.

Although roll-up structures are rare in ancient shallow-marine deposits, they indicate that the sediment surface was covered by a cohesive but flexible

FIGURE 11.26. Microbial roll-up structures. (A) Modern roll-up from siliciclastic tidal flats of Colorado River delta, Baja California. The laminae are composed of interbedded mud and sand. (B) A thin sheet of alternating millimeter-thick peloidal laminae (dark) and several adjacent mudstone laminae (light) each less than a millimeter thick, doubled over into a roll about 12 cm long and 2 cm high. Overlying laminae drape and bury the roll-up structure. Cambrian from western Maryland. The scale bar is in mm.

mat that was most likely organic. Thus, roll-up structures are one of the most reliable criteria for inferring the influence of ancient microbial mats on sedimentation. The roll-up structures reported by Simonson and Carney (1999) occurred in Archean deep-shelf deposits.

Fenestrae

Introduction

A *fenestra* (plural: *fenestrae*; adjective: *fenestral*) is a void space in sediment that is larger than a simple grain-supported pore space. Thus, fenestrae must be bounded by sediment that has a degree of cohesion (due for example to surface tension, microbial mats, or cementation). The borders of fenestrae are the sediment grain boundaries, implying that the void was not the result of dissolution but a soft-sediment deformational feature. Fenestrae are most common in modern peloidal carbonate muds and in ancient carbonate mudstones that are either peloidal or have a clotted texture. However, fenestrae are also common in carbonate peloidal and intraclastic sandstones and siliciclastic beach sands, "bubble sands" (Chapter 15). Although fenestrae in modern sediments are open spaces, those in rocks are usually partially to completely filled by either cements that precipitated out of pore waters (see Chapter 19) or sediment that filtered into the hole after it was formed. *Birdseye* is an older American term for cement-filled fenestrae. Useful descriptions and reviews of ancient fenestrae can be found in Choquette and Pray (1970), Grover and Read (1978), and Demicco and Hardie (1994).

Types and origins of fenestrae

Fenestrae have four intergradational shapes: laminoid, tubular, spherical to subspherical, and irregular. Fenestrae are generally millimeter-scale features (Figure 11.27). Equant forms have diameters of a few millimeters whereas elongate forms are generally a few millimeters high and up to a few tens of millimeters long. Fenestrae commonly form by desiccation, as trapped bubbles of air or gases generated by bacterial degradation of organic matter, and by burial of microbial mats. The shapes of fenestrae are broadly correlated with these various origins as discussed below.

Laminoid fenestrae (Figure 11.27B) are usually formed either by desiccation or by the burial of microbial mats. Fischer (1964) referred to fenestrae as

shrinkage pores and interpreted them primarily as desiccation features associated with microbial mats. Shinn (1968b) produced subspherical, irregular, and laminoid fenestrae by experimentally wetting and drying homogeneous aragonite mud. These laminoid

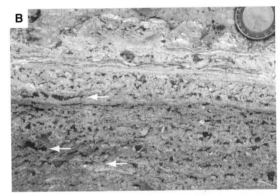

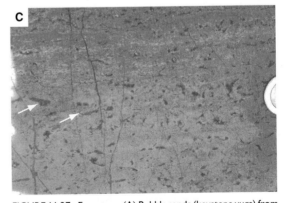

FIGURE 11.27. Fenestrae. (A) Bubble sands (keystone vugs) from Pleistocene beach grainstones, South Rock Sound Quarry, Eleuthera Island, Bahamas (photograph courtesy of R. W. Mitchell). (B) Laminoid fenestrae (white arrows) and irregular fenestrae disrupting microbial-laminated mudstone. (C) Tubular and irregular fenestrae, some of which are shelter pores (white arrows), beneath large, irregular grains. Images (B) and (C) are from the Triassic Esino Formation, Lombardic Alps, Italy.

fenestrae have both smooth walls that can be fitted back together and irregular walls that cannot. Those that have smooth matching sides are horizontal desiccation cracks also known as *sheet cracks*. Laminoid fenestrae also arise as molds of fleshy cyanobacterial biofilms buried by sediment. Drying out of surface cyanobacterial mats both before and after burial can wrinkle, lift, and separate them from the adjacent sediment. These laminoid fenestrae commonly have flat bottoms and irregular tops, especially where the mat has been dried out before being buried by a layer of sediment. Gases generated by organic decay can also lift subaqueous mats or buried mats to produce laminoid fenestrae.

Tubular fenestrae in modern sediments originate as burrows and as root holes (see above), but they can also be formed by the upward escape of gas bubbles produced by bacterial decomposition of organic matter within the sediment (see below and also under trace fossils).

Spherical and subspherical fenestrae are produced by air and gas bubbles trapped during deposition of the host sediment or generated by post-depositional decay of organic matter. Slight compaction of gas-bubble fenestrae leads to subspherical fenestrae with their long axes aligned parallel to bedding (common in modern siliciclastic muds). Gas bubbles in cores from the Bodensee are illustrated by Reineck and Singh (1980) and are common in modern muds from the bottom of Chesapeake Bay (Reinharz *et al.*, 1982). The small (1–2 mm) bubble fenestrae in the Bodensee are subspherical whereas the larger (2–3 mm) ones tend to be flattened and more irregular. Bubble fenestrae in the Chesapeake Bay mud are notably irregular, and many are subvertical irregular tubes apparently caused by upward migration of gas. Spherical and subspherical fenestrae known as *keystone vugs* in carbonates (Figure 11.27A) or *bubble sands* are commonly seen in modern beach sands and intertidal bars, where they can be associated with convolute bedding; see de Boer (1979) and Chapter 12. On the basis of experimental work, de Boer (1979) interpreted the bubbles as having formed during a rising-tide stage. After the sediment bed is submerged, trapped capillary air is compressed as the groundwater table rises, and is eventually converted into bubbles that separate sand grains.

Irregular fenestrae (Figure 11.27C) are formed in a variety of ways, but no good criteria exist to distinguish among them. They can form by desiccation, as has been demonstrated experimentally by Shinn (1968b), and by soft-sediment deformation of more regular-shaped fenestrae. However, many irregular fenestrae in carbonate sandstones are simply intergranular "shelter" pores beneath irregularly shaped intraclasts (Figure 11.27C). Shinn (1983b) reported laminoid to irregular *shelter pores* developed in subtidal hardgrounds on the Bahama Banks. These pores occur in peloidal sandstones beneath intraclasts and "grapestone" clasts that apparently were derived from reworking of hardground layers. These grain-supported interstices are not fenestrae.

Tufas, travertines, and cave deposits

Tufas and travertines

Carbonate mounds of variable shapes and sizes occur in modern lakes, marshes, hot and cold springs, rivers, waterfalls, deep-sea springs (Kelly *et al.*, 2005), and hydrocarbon seeps (Flügel, 2004). These deposits are generally called *travertine* after the locality in Italy near Rome originally known as Tivertino in Roman times but now called Tivoli. The quarries in the Campagna area between Rome and Tivoli provided travertine to build many of the edifices of ancient and modern Rome, and Tivoli travertine is found as facings on and within many buildings of the world. Porous, "spongy" deposits of travertine are usually called *tufa*, calcareous tufa, or plant tufa. *Flowstone* is a term used to describe smooth sheets of laminated carbonates that cover the walls and floors of caves and surface spring pools. Flowstones may carry the fabrics both of dense travertines and of porous tufas in adjoining laminae. Many of the huge carbonate mounds that formed in Holocene and pluvial Pleistocene lakes of the western USA are called tufas or bioherms and are further described in Chapter 16.

Travertines usually form sheets or discrete mounds up to meters high and wide around isolated geysers, cold springs, and hot springs (Figure 11.28, and see Figure 16.19 later). These springs occur subaerially and along the floors of lakes and the ocean. Large complex deposits of travertine develop around clusters of long-lived hot-spring vents. Such travertine deposits can cover tens of square kilometers and be tens of meters thick. Travertines are commonly laminated, and the laminae of travertines are usually continuous and isopachous with low-amplitude wavy forms (Figure 11.29A, C) commonly referred to as "microdigitate stromatolites" in ancient carbonates (Figure 11.29A). Internal fabrics of travertine laminae include homogeneous micrite with alternating dark and light-colored layers of equivocal origin (but most likely chemical precipitates), clotted

FIGURE 11.28. (A) Hot springs/geysers from Fly Ranch, Nevada, USA, showing typical travertine features, including mounds around spring orifices and rimmed pools. The largest mounds are ~3 m tall. Photograph courtesy of Tim Lowenstein. (B) Sheets of travertine/tufa coating bedrock (exposed at the white arrow) around Walker Lake, Nevada, USA. A person (indicated by the black arrow) is shown for scale. (C) Tufa composed of vertical calcite tubes that grew around moss filaments from Sitting Bull Falls, New Mexico, USA. The scale bar is 5 cm. (D) Deep-sea travertines from "Lost City Hydrothermal Field" located ~15 km from the main axial valley of the Mid-Atlantic Ridge on the Atlantis Massif. (1) A 10-m-tall chimney venting fluids. (2) A pinnacle vent ~4 m across. (3) A base of carbonate travertine built around a vent 30 m wide and 60 m tall. (4) Detail of carbonate growth at the end of a vent; the towers are ~1 m high. (5) The layer of carbonate overlying serpentine basement. From Kelly et al. (2005).

micrite and peloids aggregated into arborescent (shrub-like) shapes (Figure 11.29), and calcified filament molds of cyanobacteria. Some travertines are characterized by a primary crystalline fabric, which most typically takes the form of a radial array of needles or bladed crystals of calcite or aragonite with compromise boundaries

(see the description of "thinolite" tufas, Figure 16.19C later). Other features of many travertines, particularly larger hot-spring travertines, are their dam-and-pool configuration, and their common association with ooid-sand pisoids. Pisoids in spring deposits commonly exhibit reverse grading and isolated ooids and pisoids

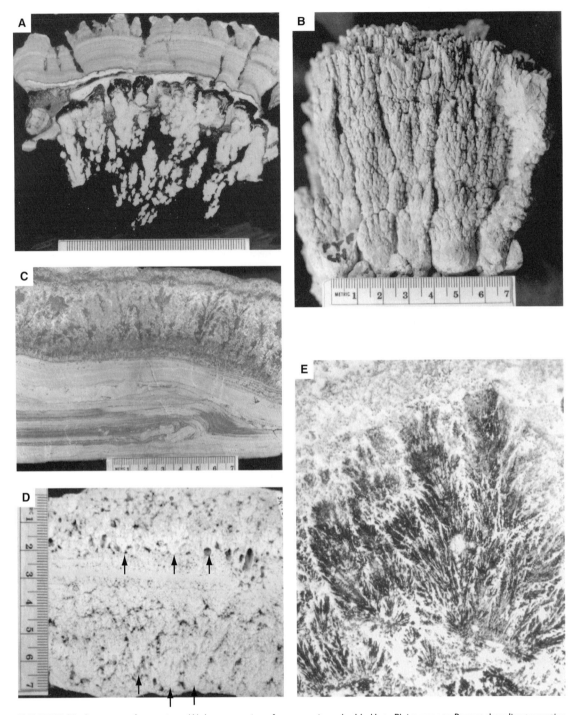

FIGURE 11.29. Structures of travertines. (A) A cross section of epoxy-resin-embedded Late Pleistocene to Recent shoreline travertine from Pyramid Lake, Nevada, USA, showing biogenic/chemogenic laminae overlying arborescent shrubs of so-called dendritic tufa. The scale is in mm. (B) Complete arborescent shrubs (dendritic tufa) from Pyramid Lake, Nevada, USA. (C) The Proterozoic counterpart of (A) with biogenic/chemogenic laminae encrusted by arborsercent shrubs, Pethei Group, Northwest Territories, Canada. The scale is in mm. (D) Tivoli travertine, showing arborescent shrubs (arrows) and porous nature. The scale is in mm. (E) A thin-section photomicrograph of Tivoli shrubs. Width of field 2 mm.

are commonly enclosed within laminated travertine (flowstones).

Tufas form discrete mounds up to tens of meters high and tens of meters thick that are primarily composed of a framework of calcareous tubules and pustules (Figures 11.28C and 11.29). Tufas also occur as sheets coating lake shores (Figure 11.28B). In modern tufas, the tubules are micron-thick calcareous sheaths that have precipitated around the filaments of cyanobacteria, bacteria, and fungi, and around the stems and fronds of green algae and mosses (Figure 11.28C). The pustules are sheaths around coccoid cyanobacteria and coccoid bacteria (Demicco and Hardie, 1994). The term tufa has also come to include layers of millimeter- to centimeter-high arborescent shrubs that consist of a framework of carbonate-mud clots and peloids of cryptocrystalline carbonate that are particularly common in Proterozoic carbonates (Figure 11.29). These arborescent shrubs are a common component of the Tivoli travertines. The shrubs were originally interpreted as calcified bacterial colonies (Chafetz, 1986; Chafetz and Folk, 1984), but at least some are now thought to originate by the process of diffusion-controlled carbonate precipitation (Pope and Grotzinger, 2000).

Travertines with primary crystalline textures are formed by direct chemical precipitation from supersaturated waters. The origin of non-crystalline travertine laminae and tufas is not straightforward, mainly because the mechanisms of calcification of microbial mats, non-calcareous eukaryote algae, and higher plants to make tufas are not certain. The carbonate coatings are not external skeletons, nor are the crystals precipitated by the organism within the outer rows of living cells in order to stiffen the thalli (as occurs with some modern soft calcareous green algae such as *Penicillus*). Instead, where living cyanobacteria are found in the early stages of calcification, the carbonate crystals are suspended in the mucilaginous sheaths around the filaments. The exact mechanism of carbonate precipitation is unknown, but where calcification is confined exclusively to the filament mucilage and there is no nucleation on associated inorganic particles, it seems most probable that the normal metabolic activities of the living organism itself (such as photosynthesis) have induced inorganic precipitation from the surrounding calcium bicarbonate-bearing waters.

Ancient travertines are known from both non-marine and marine settings. Lacustrine stromatolites with travertine and tufa fabrics occur in the Eocene Green River Formation of Wyoming, Colorado, and Utah (Smoot, 1983), in the Upper Miocene Reis Crater deposits of southern Germany (Riding, 1979), and in the Middle Miocene Barstow Formation of California (Cole *et al.*, 2005). Marine analogs are described by Immenhauser *et al.* (2005) and Portman *et al.* (2005). Many of the "microbial" features of the Proterozoic thrombolites and reef mounds described in Grotzinger and James (2000) are very similar to fabrics of modern non-marine tufas. Indeed, the distinction between thrombolites and microbial tufas is becoming blurred (e.g., Elrick and Snider, 2002). Most modern filamentous cyanobacterial tufas, such as the calcified *Scytonema* of Andros Island and the Everglades, look remarkably like the problematic filamentous fossils known as *Girvanella*, *Ottonosia*, and *Samphospongia*.

Cave deposits

Chemically precipitated carbonate deposits of caves are commonly termed flowstones. Flowstones can contain the fabrics both of dense travertines and of porous tufas in adjoining laminae. Laminated flowstones, stalagmites, stalactites, "cave popcorn," and cave ooids and pisoids (so-called "cave pearls") are usually referred to collectively as *speleothems* (Thrailkill, 1976). Stalagmites and stalactites are columnar-shaped deposits that can be up to tens of meters long and meters in diameter. Stalactites form around groundwater seeps from cave roofs whereas stalagmites form on cave floors beneath seeps. Springs along the sides and floors of caves commonly produce flowstones complete with dams, pools, and cave pearls. Thus, speleothems are underground travertines with common crystalline fabrics (Figure 11.30). Stalagmites and cave "popcorn" are good examples of non-isopachous, non-biogenic stromatolite-like structures with radial crystalline fabric.

Speleothems have a low preservation potential because caves form under conditions of net dissolution (Chapter 3). However, some remnants of speleothems may be preserved along old karst surfaces, where their presence will be crucial to the identification of the surface as one of karstic dissolution. Esteban and Klappa (1983) describe Cretaceous-age speleothems filling pale-okarst cavities. Traces of speleothems within a *terra rossa* soil separating shallow-marine deposits are evidence of a substantial relative sea-level fall and exposure of a carbonate shelf surface to karstic dissolution by a meteoric groundwater lens, as has occurred a number of times during the Pleistocene epoch. Further details can be found in James and Choquette (1988).

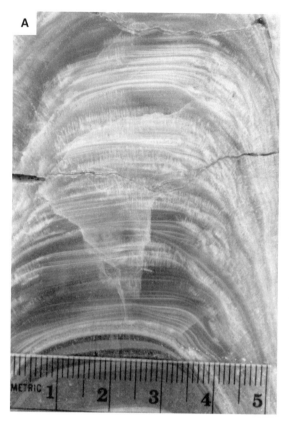

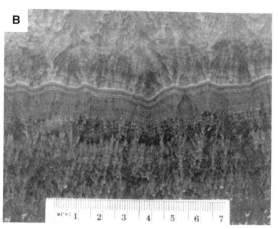

FIGURE 11.30. Flowstones. (A) A cross section of a stalagmite from a West Virginia, USA, cave, showing its well-laminated internal structure. (B) Laminated flowstone composed of upward-directed sprays of carbonate crystals. Laminae pass through the crystals. The scale is in mm. Sample courtesy of Paul Hoffman.

Calcareous soil crusts

Calcareous soil crusts are known as caliche, calcrete, kankar, nari, and duricrust. They are carbonate deposits precipitated within the B horizon of a soil, principally under semi-arid or arid climates (Chapters 3 and 16). Quaternary calcareous soils have been examined in many parts of the world (Goudie, 1973; Read, 1974; Reeves, 1977; Watts, 1977, 1980; Esteban and Klappa, 1983; Wright, 1986). Calcareous soil crusts range from isolated nodules of red and white microcrystalline calcite (that grew variously as cements, replacements, and by displacing sediment) to nodular layers up to many meters thick. Well-developed calcareous soils are generally organized as (1) a basal nodular chalky zone; (2) a coalesced nodule zone that may have a platy, semi-indurated structure; (3) a hard laminated crust; and (4) a red–brown soil with pisoids (coated grains). Calcareous soil crusts also commonly contain irregular fenestrae and dissolution vugs lined with clay cutans that indicate illuvial, unsaturated zone transport. *Microcodium* comprises sub-millimeter petal-shaped clusters of calcite microspar considered by Klappa (1978) to be a feature produced biogenically by soil

fungi and antiformal tepee structures (Chapter 12). Rhizoliths are carbonate casts of root holes. These carbonate sheaths or clusters of concretions are precipitated around roots as a result of the biochemical activities of the living plants or the postmortem decay of their roots. Esteban (1974) coined the term *alveolar texture* for a network of millimeter- to sub-millimeter-sized walls made of micrite that appears to be coalesced rhizoliths. Calcareous soil crusts are formed by the precipitation of calcite or magnesian calcite in the unsaturated zone from films of pore water drawn up to the surface by a combination of capillary draw and evaporative pumping.

Laminated carbonate soil crusts also develop on limestone bedrock underlying modern carbonate platforms when they were subaerially exposed by glacio-eustatic sea-level drops during the Pleistocene. These crusts are centimeter- to decimeter-thick and noteworthy for dense, color-banded mudstone laminae that commonly contain cyanobacterial or fungal filament molds, an assortment of micritic peloids, and coated grains (Figure 11.31). Laminated soil crusts are associated with karstic dissolution surfaces and *terra rossa* soils, red with the oxidized clayey residues of weathered

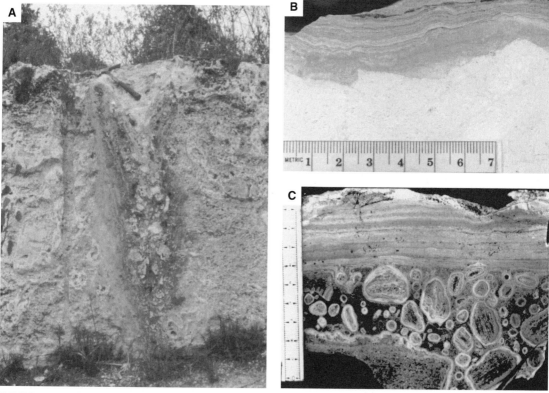

FIGURE 11.31. Soils and biogenic/chemogenic laminated soil crusts developed on carbonates. (A) A sinkhole developed on Pleistocene Key Largo Limestone, Windley Key, Florida, USA. The sinkhole is lined with laminated soil crust and filled with terra rossa soil and brecciated pieces of limestone. (B) A cross section of laminated soil crust developed on Key Largo Limestone. The scale is in mm. (C) Coated grains and laminated soil crust developed on Pleistocene limestones around Shark Bay, Western Australia. The scale is in cm. Photograph courtesy of J. Fred Read.

iron-bearing detrital components of the limestone or detrital components introduced as wind-blown dust. The well-known laminated crusts and the associated red–brown pisolitic breccias that fill solution pockets in the exposed Pleistocene coral reefs of the Key Largo Limestone of the Florida Keys are examples of such features in karstic soils (Figure 11.31).

Laminated soil crusts in an ancient sedimentary succession are indicators of subaerial exposure without detrital sedimentation for a significant period of geological time. This may be due to abandonment of a terrestrial sub-environment by river avulsion, abandonment of a supratidal flat by shoreline progradation, or a relative sea-level fall that was driven by either eustasy or tectonic movement. The time required to develop a laminated soil crust is on the order of 10^4 years. Laminated soil crusts caused by sea-level falls are particularly important in recognizing depositional sequences (Chapter 20).

12 Post-depositional deformation of soft sediment

Introduction

Some sedimentary structures indicate that sedimentary strata were deformed at the time of deposition or soon after, when the sediment was "soft" (Allen, 1982a, 1985). Biogenic deformation of soft sediment is discussed in Chapter 11, and purely physical deformation is considered here. The physical deformation of sedimentary strata requires low shear strength relative to the deforming force (commonly gravity or the surface tension of pore water in mud). Reduced shear strength may be associated with liquefaction, which occurs when the sediment grains become dispersed in a fluid and displace the fluid upwards as they settle downwards. However, deformed strata cannot be preserved if there is complete loss of shear strength and the sediment–fluid mixture flows. Low shear strength is commonly associated with high pore-water pressure and loose packing of grains (Chapter 8, Equation (8.1)) that can be produced by rapid deposition, low permeability, low cohesion, loading by landslides or water waves, and earthquake shaking. Rapid deposition of fine sand to silt is commonly conducive to soft-sediment deformation because of the low permeability and low cohesion of the sediment. The addition of water to sand confined between layers of impermeable mud is also conducive to an increase in pore-water pressure and a decrease of shear strength. The combination of high hydrostatic pressures on the bed under crests of large waves with low pressures under troughs results in cyclic shear stresses within bed sediment that can cause liquefaction. Cyclic wave loading is most effective where the wave height is comparable to the water depth. Earthquakes need to exceed Richter magnitude 5 to cause extensive liquefaction. The effects are felt further from the epicenter as earthquake magnitude increases, being felt hundreds of kilometers away for an earthquake of magnitude 8 (Allen, 1985).

Although ancient soft-sediment deformation structures are commonly taken as evidence of paleoseismicity, they commonly form in the absence of earthquakes.

Typical causes of deformation in "soft" sediment, producing characteristic sedimentary structures, are inverse density gradients, downslope slumping, upward escape of sediment–fluid mixtures, and fluid shear stress. These act together in many cases. In addition, sedimentary strata (particularly muds) may be deformed by contraction (shrinkage) associated with desiccation, freezing and thawing, and chemical dewatering. Finally, the early growth of carbonate or sulfate cements in sediments can lead to expansive deformation structures. These mechanisms and associated structures are discussed in turn below. Faulting of "soft" sediment is not considered here, but is mentioned in Chapters 8, 10, and 15.

Inverse density gradients

Inverse density gradients are due to upward-decreasing proportions of pore water in superimposed sedimentary strata. For example, fine sand with a porosity of 40% and bulk density of $2,000 \, \text{kg m}^{-3}$ may overlie mud with porosity of 80% and bulk density of $1,500 \, \text{kg m}^{-3}$. This results in what is referred to as a Rayleigh–Taylor instability. If the viscosity of the sediment is sufficiently low, the sand founders into the mud, forming bulbous protrusions of sand separated by relatively narrow ridges of mud (Figure 12.1). These bulbous protrusions of sand are called *load casts*. Load casts are commonly centimeters to meters long, and their size increases with original sand-bed thickness. The wave length of the load casts is larger for larger viscosity of the deforming sediment and for smaller difference in density between the sand and mud. In plan, load casts are equant to pillow-shaped, hence the common name ball-and-pillow structures. Stratification within the original

FIGURE 12.1. Load casts without, (A) and (B), and with (C) flame structures. Image (A) is from Pettijohn and Potter (1964). By permission of Springer Science and Business Media.

sand bed is deformed as the sand is loaded, and follows the lower surface of the load cast approximately.

The wedges of mud that are forced upwards between the load casts may divide into separate wedges, giving the appearance in cross section of flames in a fire (Figure 12.1C), thereby earning the name *flame structures*. The bulbous loads of sand may continue to move downwards into the mud and become detached from the parent sand layer above (Figure 12.2). These isolated load structures have been termed *pseudonodules* and *flow rolls*. Load structures may be preferentially initiated where there are locally thick parts of sand layers, such as those with ripples or flute casts. Figure 12.3 shows a series of ripples that were stacked on top of each other as they were loaded into underlying mud. The rotation associated with this loading caused deformation of the cross lamination within the ripples, producing a form of *convolute lamination*. Convolute lamination is particularly common in very fine sands containing climbing-ripple cross lamination associated with the deposits of river floods or turbidity currents, and has the form of box-like folds. This type of convolute lamination is probably mainly due to gravitational loading (Allen, 1982a).

Wrinkle marks on tidal flats appear to be formed at least partly by gravitational loading also (Allen, 1985). They form in laminated sands and muds. The sand is in the form of lenses with flat tops and bases, and narrow pinch-outs (Figure 12.4). Internal laminae follow the boundaries of the sand, as is the case with load casts. The sand may be removed subsequently, leaving impressions in the underlying mud. According to Allen (1985), wrinkle marks form following an ebbing tide, but in the absence of wave currents that would disrupt the structure.

Upward escape of fluid and sediment

Water and sediment under pressure from above can escape upwards in various ways. On a relatively small scale, water can move along local conduits with high permeability related to stratification. One example (Figure 12.5) is where water moves horizontally within permeable sand laminae beneath impermeable muddy layers. In places, the water can break through the mud layers, or the mud layers are discontinuous, allowing upward movement of pore water. The water may be moving fast enough to carry mud particles with it, leaving some conduits that are mud-free and others where mud is deposited. The resulting structures are called *dish* and *pillar structures* (Lowe and Lopiccolo, 1974; Lowe, 1975). Allen (1982a) discusses alternative origins for dish structures.

Beds of sand or mud with high pore-water pressure may lose strength and escape upwards vigorously through fissures or circular conduits (e.g., Nichols *et al.*, 1994). The water–sediment mixture may reach the surface as sedimentary volcanoes or fissure eruptions, which are underlain by sedimentary necks and dykes, respectively (Gill and Kuenen, 1958) (Figure 12.6). The volume lost with eruption of water and sediment is replaced by subsidence of the overlying sediment, producing downward-dipping strata. The pressure required to force the water and sediment upwards may be associated with downslope slumping of sediment (as described below) or an increasing weight of overburden.

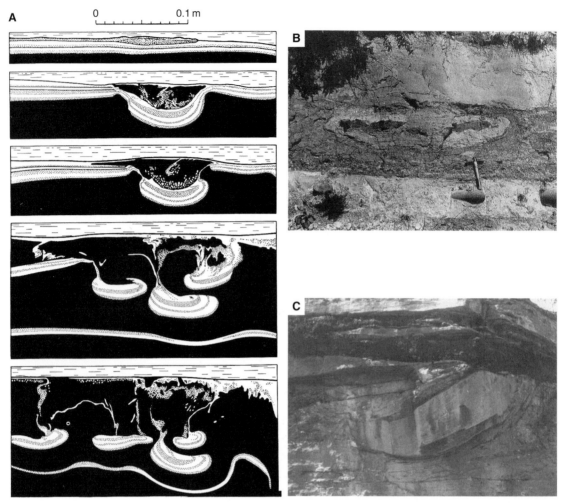

FIGURE 12.2. (A) Formation of detached load casts (pseudonodules). From Reineck and Singh (1973), based on experiments of Kuenen (1965). A shock was applied to initiate the loading. By permission of Springer Science and Business Media. (B) A pseudonodule. From Ricci Lucci (1995). (C) A sandstone pseudonodule in Devonian shale of New York State.

FIGURE 12.3. Piled, load-casted ripples. From Reineck and Singh (1980), after Dzulyński and Kotlarczyk (1962). By permission of Springer Science and Business Media.

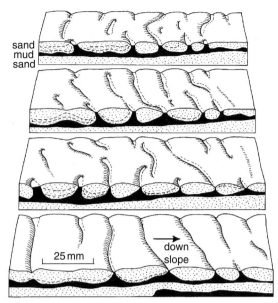

FIGURE 12.4. The geometry of wrinkle marks. From Allen (1985).

In the Mississippi delta region, large mud diapirs erupt at the surface as "mud lumps" and may form islands. These mud diapirs ascend in response to loading of mouth-bar sands over prodelta muds (Chapter 15).

Trapping of gas in porous sands, particularly in beaches and intertidal flats, leads to deformation of the sediment into *bubble sand* (Figure 12.7). De Boer (1979) published photographs of bubble-rich sands associated with convolute bedding from a modern, intertidal sand bar. On the basis of experimental work, he interpreted these bubbles as having formed during rising tide. After a bar is submerged, trapped capillary air is further compressed as the groundwater table rises. The trapped air is eventually formed into bubbles, which separate sand grains. Bubble sands also form in fluvial sand bars, especially when the river stage rises rapidly due to release of water from dams. Bubble sands are referred to as *keystone vugs* in carbonate grainstones. The release of gas generated by bacterial decomposition of organic matter can also deform soft sediments (Chapter 11).

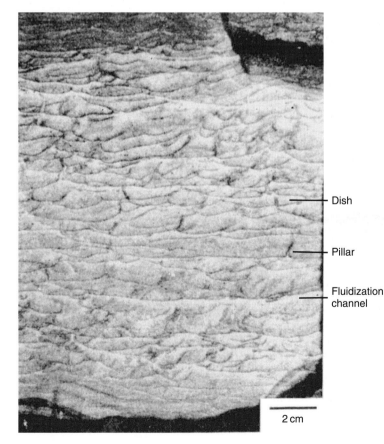

FIGURE 12.5. Fluid-escape structures: dishes, pillars, and horizontal fluidization channels. From Lowe and Lopiccolo (1974).

Dish

Pillar

Fluidization channel

2 cm

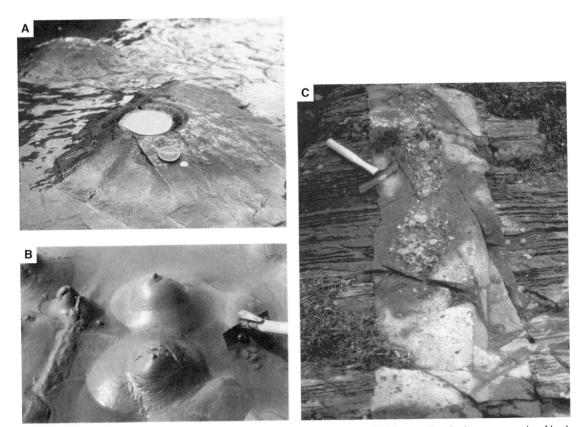

FIGURE 12.6. (A) A sand volcano from Carboniferous rock of County Clare, Ireland. (B) "Monroes," mud volcanoes on modern North Sea tidal flats. From Reineck and Singh (1980). By permission of Springer Science and Business Media. (C) A sedimentary dyke cutting laminated Proterozoic Espanola Formation, Sudbury, Ontario. The coarsest material has segregated into the center of the dyke.

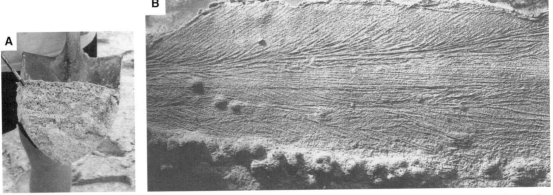

FIGURE 12.7. (A) Bubble sand from a beach adjacent to a tidal inlet. (B) Bubble sand overlain by herringbone cross-stratified sand, and an ebb-oriented dune, from a sandy tidal flat.

Downslope slumping

Downslope slumping of sediment under the influence of gravity was discussed in Chapter 8, and can occur over a large scale range (e.g., Horowitz, 1982; Allen, 1985; Owen, 1996). Sedimentary strata that remain coherent during slumping are normally deformed plastically (Figure 12.8). Characteristic structures associated with the downslope ends of coherent slumps are folds and thrust faults. The folds are normally overturned such that their axial planes dip upslope and

FIGURE 12.8. Folding in a downslope slump deposit (photo courtesy of Henry Posamentier and Roger Walker).

strike approximately normal to the slope (Figure 12.8). Two features distinguish this type of syn-depositional folding from regional tectonic folding: (1) the confinement of the deformed strata within undeformed strata and (2) the relationship between the orientation of faults and the paleoslope. The pressure on the sediment at the base of the slope created by the slumping commonly gives rise to elevated pore-water pressure and upward escape of fluid and sediment (to form sedimentary dykes and volcanoes).

Fluid shear stress

Overturned cross stratification (Figure 12.9) is cross strata overturned in the flow direction. The syn-depositional origin of this structure is deduced from the lack of deformation of cross strata above and below the deformed cross set and the fact that the direction of overturning is the same as the flow direction. The loosely packed cross-stratified sand within subaqueous dunes can have high enough pore-fluid pressure that the fluid shear stress on the crest of the dune can, according to Allen (1985), drag these layers in the flow direction (Figure 12.9). Allen and Banks (1972) proposed a theory for the origin of overturned cross

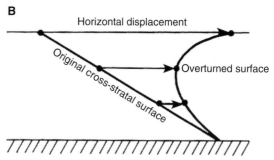

FIGURE 12.9. (A) Overturned cross strata. From Pettijohn and Potter (1964). By permission of Springer Science and Business Media. (B) Horizontal displacement required to form overturned cross strata.

strata that involves (1) a dune initially composed entirely of liquefied sediment; (2) fluid shear stress that decreases linearly from the top of the dune to zero at the base of the liquefied sediment zone; (3) laminar shearing within the very viscous liquefied sediment; and (4) thinning over time of the liquefied sediment zone from the bottom upwards as the liquefied sediment settles. This model can apparently explain the geometry of overturned cross strata. However, it is not clear why the liquefied sediment does not just flow down the steep lee slope of the dune, rather than waiting to be sheared by the flow above. Furthermore, the model predicts the geometry of the overturned cross strata as if the whole cross set had been preserved. It is well known, however, that perhaps only the lowest third of a cross set is preserved in a coset of cross strata, such that the overturned part of the cross set should not be preserved. Finally, if fluid shear stress acting on loosely packed sand is the explanation for overturned cross strata, they would be expected to be more common than is observed. These uncertainties warrant further research on this problem, and Allen (1982a) cites alternative explanations.

Contraction (shrinkage)

Shrinkage of sediment can be produced by desiccation of wet mud under the atmosphere, freezing of sediment in areas of permafrost, and (experimentally) by chemical dewatering of clays that have interlayer water molecules (*synaeresis*). Whatever the cause of the shrinkage, the result is a characteristic pattern of intersecting cracks (Figures 12.10 and 12.11). Cracks are initiated at "defects" of low tensile strength within the mud (e.g., large grains, burrows) and initially occur as isolated cracks of limited length (Lachenbruch, 1962). Such incomplete crack patterns, especially those with curving shapes, are commonly interpreted as due to synaeresis, or misinterpreted as animal trails. As crack growth continues, secondary cracks occur approximately normal to the initial crack, forming a trilete crack pattern (Figure 12.10). Eventually, the cracks become elongated and join up to form the characteristic intersecting pattern. Cracks normally decrease in width downwards, because the maximum contraction occurs at the land surface and decreases downwards until the weight of overlying sediment puts the sediment under compression. Details of shapes of cracks in plan and profile are given in Allen (1982a). The spacing and width of cracks increases with bed thickness. The mudcracks known as prism cracks are typically a few decimeters in diameter and the cracks are a few decimeters to a meter deep (Figure 12.11C). Prism cracks tend to be developed in laminated sediments deposited over benthic microbial mats and are common in carbonate muds. Mudcracks can also

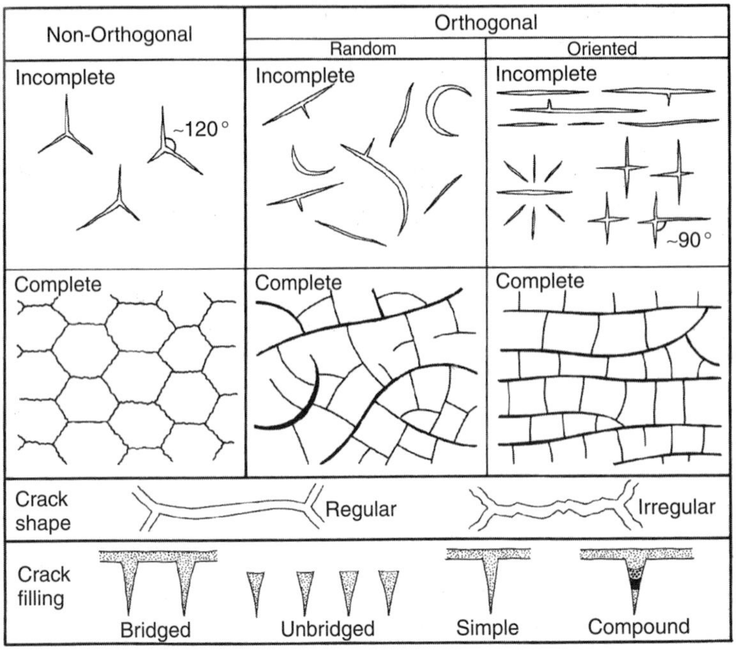

FIGURE 12.10. Classification of shrinkage cracks and fillings. From Allen (1982a).

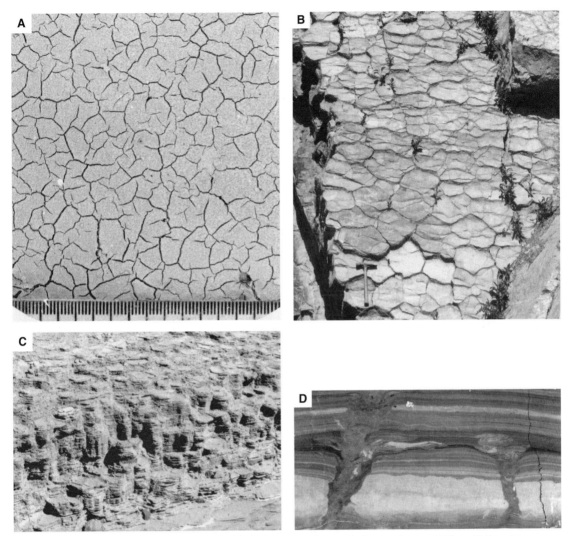

FIGURE 12.11. Shrinkage cracks due to desiccation. (A) Modern incomplete cracks from carbonate tidal flats. (B) Complete non-orthogonal shrinkage cracks from Silurian microbial laminites. From Pettijohn and Potter (1964). By permission of Springer Science and Business Media. (C) Modern prism shrinkage cracks from Colorado River delta tidal flats. (D) A cross section of shrinkage cracks with compound fills with complicated stratigraphies. Some layers can be traced or correlated to fill.

branch laterally along bedding surfaces, especially where the mudcracked sediments curl up into the familiar mud curls (Figure 12.12).

The commonest type of shrinkage crack in modern and ancient sediments is formed by atmospheric desiccation of mud. Multiple generations of cracks can be formed in one mud bed by repeated wetting and drying (Figure 12.11D). The upper surfaces of cracked mud may be ornamented with rain-drop imprints, which is further testament to their subaerial conditions of formation (Figure 12.12). In arid climates, desiccated cracks may be associated with salt pseudomorphs. Mudcracks are preserved by burial and filling of cracks with sediment. The sediment fills can be simple or record quite complicated filling history, in many cases directly relatable to cracked layers (Figure 12.11D). Crack fills in cross section are commonly slightly folded due to compaction. Desiccation of mud produces flakes of hard mud that can be transported and incorporated into other deposits as intraformational breccias. An abundance of such intraformational mud chips in sedimentary strata is normally an indication of desiccation of mud. Desiccated mud is one of the most important types of evidence for exposure to the atmosphere and for interpretation of terrestrial environments.

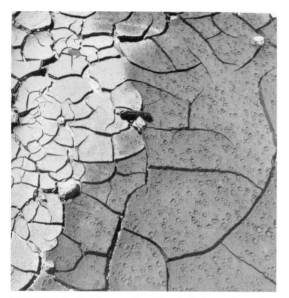

FIGURE 12.12. Shrinkage cracks with curled edges and rain-drop imprints. From Pettijohn and Potter (1964). By permission of Springer Science and Business Media.

Shrinkage cracks formed by thermal contraction are common in permafrost regions, and are termed ice-wedge polygons (see Figure 17.18 later). These features are similar in geometry to desiccation cracks in mud, but are markedly larger. The cracks are commonly centimeters to decimeters wide and meters deep, and the polygons are meters to tens of meters across.

Cracks formed by chemical dewatering (*synaeresis*) (e.g., White, 1961; Burst, 1965; Donovan and Foster, 1972) require unusual chemical conditions of varying salinity and specific types of swelling clay, such that they are expected to be a rarity in nature, and cannot form in carbonate muds. Synaeresis cracks have never been documented from a modern sedimentary environment. Those formed in the laboratory range from poorly developed cracks to networks that resemble familiar desiccation cracks. Demicco and Hardie (1994) concluded that most reports of ancient synaeresis cracks are misinterpreted subaerial desiccation cracks, which has serious implications for

FIGURE 12.13. (A) A carbonate tepee from Deep Lake (a coastal lake) in South Australia. From Tucker and Wright (1990). (B) A cross section of tepees from Triassic Latemar Limestone of the Dolomites in northern Italy. The figure center right is shown for scale. (C) The crest of a tepee antiform, dark-gray strata fractured and surrounded by light-colored brecciated fill. Field of view ∼1.5 m.

interpretation of subaerial versus subaqueous depositional environments. In particular, mudcracked sediments in modern playas experience repeated cycles of wetting and drying, leading to very complicated crack cross sections with associated breccias and pseudoanticlines. These features were once taken as common evidence of complicated synaeresis processes in saline lakes (Chapter 16).

Expansion due to early cement growth (tepee structures)

Tepee structures are formed by the expansive growth of carbonate or sulfate cements within surface sediments. Reviews and descriptions can be found in Dunham (1969a), Assereto and Folk (1980), Esteban and Pray (1983), Kendall and Warren (1987), and Demicco and Hardie (1994). Tepees are most common in modern gypsum and halite crusts but also occur in modern and ancient carbonates (Chapter 16). Modern tepee structures are known from calcareous soil profiles (where they are also known as pseudoanticlines), submarine hardgrounds in the Arabian Gulf, ephemeral lake deposits, and coastal salinas and tidal flats of Australia. In plan view, tepee structures comprise a polygonal network of ridges arched along polygonal cracks in cemented surface crusts (Figure 12.13A). The simplest cross-sectional forms of tepee ridges are trochoidal antiforms (Figure 12.13B, C). The limbs of each antiform consist of upward-turned strata, and opposing limbs meet along a fracture that marks the crest of the tepee (Figure 12.13B). Strata from one limb may be thrust over those from the other limb along this fracture, producing brecciated strata. Between the antiformal crests are broad, saucer-shaped synforms, and the distance between antiform crests ranges from a few tens of centimeters to a few tens of meters. Vertical relief between crests and troughs ranges from a few centimeters to a meter. The thickness of strata being disturbed is typically proportional to the width of the polygons.

Tepee structures usually occur in well-defined stratigraphic zones bounded at the top and bottom by undeformed strata (Figure 12.13). The tepee structures die out abruptly downwards against a more or less flat undisturbed layer. The synforms are commonly filled with strata that overlap the tepees. However, the arched crests of tepee structures are commonly truncated by a flat, overlying bedding surface (Figure 12.13). These stratal relationships demonstrate that the tepees are syn-depositional deformation features.

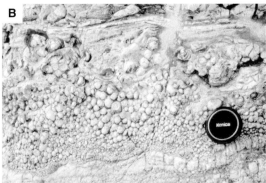

FIGURE 12.14. Fills associated with ancient carbonate tepees. (A) Pisoids (lower right) and laminated cements (the rest of the view). (B) Pisoids from the Capitan Formation (Permian). Note the fitted textures of pisoids on the left of the photograph.

The initial, minor buckling of tepee structures is the result of precipitation of cement within the intergranular voids of permeable beds to produce a compressionally disrupted hardground. The open fractures at the crests of the new antiforms become the principal conduits for water flow, resulting in preferential growth of cements within these fractures, which in turn leads to further disruption, in particular arching and pushing apart of polygon edges. Tepee structures normally have horizontal and vertical fractures and shelter voids that are filled with large cement crystals and internal sediments (Figure 12.14). As the tepee structures become more structurally complex and brecciated, the number of different generations of fractures and their fills increases. Bed-parallel fractures are widest and most abundant in crestal zones and die out towards the relatively undeformed synforms (Figure 12.13). Vertical and oblique fractures have matching walls that can be fitted back together, indicating simple tensional rupture without rotation. As the number of fractures increases, the host sediment becomes increasingly brecciated, with rotated blocks producing a complex network of shelter voids.

The voids formed by the fracturing are filled with a variety of internal sediments, pisoids, and cements that display geopetal fabrics (Figure 12.14). The internal sediment may contain red siliciclastic mud (*terra rossa*), volcaniclastic debris, carbonate mud (commonly peloidal), and fossils, including gastropods, nautiloids, and ostracods. The cements can indicate precipitation both under saturated-zone and under unsaturated-zone conditions (see Chapter 19). Unsaturated zone cements include micrite and microspar pendant (gravitational) cements and meniscus cements. Isopachous cements indicate saturated-zone conditions and include laminated micrite and/or radially oriented crystals. Another common cement type consists of small hemispheres of radially oriented crystals. Large, radially arranged calcite crystals (interpreted to be pseudomorphs of aragonite) can be up to decimeters in length (the "raggioni" of Assereto and Folk (1980)). Pisoids (Figure 12.14) occur not only as part of the fills of tepee fractures, but also in synforms, where they probably formed on the bottom of open pools. The outermost laminae of some pisoids exhibit gravitationally influenced downward growth and fitted polygonal fabrics (Dunham, 1969a, b), indicating in-place growth under unsaturated-zone conditions (Figure 12.14B). Pisoids also commonly undergo multiple episodes of fracturing and recoating.

In tepees formed in subtidal hardgrounds, the cements precipitated from seawater. In peritidal and coastal salina settings, the cementing pore water could be seawater, meteoric water, or both. In arid settings, the ambient groundwater is drawn up through the vadose zone by evaporative pumping, although tidal pumping may also contribute some flow. The large volumes of cement precipitated within tepee structures demand flow through the sediments of a large volume of pore water supersaturated with respect to alkaline-earth carbonates, and this in turn requires an extensive groundwater reservoir with a water table close to the surface.

Tepee structures formed in subaerial soils require upward movement of supersaturated groundwater in the vadose zone. Evaporation can be most effective at drawing groundwater up through a shallow vadose zone; so an arid or semi-arid climate along a coastal plain should be most favorable to subaerial tepee development, as for example is the case for the modern Australian coastal-plain environments (e.g., Handford *et al.*, 1984; Warren, 1982). Under intensely arid conditions, such as those of sabkhas and playas (see Chapter 16), evaporative pumping can be so effective that saline minerals such as gypsum and halite, which are far more soluble than alkaline-earth carbonates, are precipitated *en masse* in the vadose zone (see below), which can lead to the disruption of gypsum and halite surface crusts into tepee structures analogous to carbonate tepees (see Chapter 16).

Part 4
Environments of erosion and deposition

13 Rivers, alluvial plains, and fans

Introduction

Rivers and floodplains have always attracted humans because they provide a supply of water, power, and food. Rivers are used as sewers and for recreation (commonly simultaneously), for transportation, and as political boundaries. Rivers and floodplains are also sources of raw materials such as sand and gravel. These desirable aspects of rivers and floodplains have resulted in all manner of human construction within and adjacent to rivers, including buildings; irrigation canals; dams for water supply, flood control, and recreational lakes; and levees to control flooding. Channels have been straightened and banks stabilized to protect against bank erosion, to accelerate passage of floodwater, and to aid navigation. In addition, sediment in rivers and floodplains has been mined extensively. These human activities have commonly had a detrimental effect on rivers and floodplains, including overuse and pollution of water supply; disruption of freshwater ecosystems; lack of recharging of groundwater in floodplains and associated wetlands; reduction of deposition of fertile silt on floodplains; erosion downstream of dams and sedimentation in reservoirs; and erosion upstream of straightened, leveed channels, with enhanced flooding and deposition downstream. Catastrophic loss of life and damage to property can occur when engineering structures fail. These significant problems require effective management of rivers and floodplains, a degree of restoration of those that have been damaged, and recognition of the inherent fallibility of engineering structures. This in turn requires detailed understanding of their geometry and processes of water flow, sediment transport, erosion, and deposition.

Rivers have been eroding, transporting, and depositing sediment since the Earth's hydrosphere came into existence, and have had a major influence on changing the landscape. Deposition of fluvial sediment has occurred in river channels, floodplains, lakes, alluvial fans, and deltas as well as in the sea. These deposits are a record of the geometry, flow, and sedimentary processes of past fluvial environments, and these can be related to past climates, tectonic activity, and eustatic sea-level change. The only rational way of interpreting the origin of ancient fluvial deposits is by studying modern fluvial processes. These studies are conducted in the field and in the laboratory, and also include numerical modeling.

Fluvial deposits commonly contain economically important resources such as water, oil, gas, placer minerals, peat and coal, building stone, and sand and gravel. The extensive Pleistocene fluvioglacial outwash that occurs over much of the northern hemisphere forms important aquifers and rich sources of sand and gravel. Oil and gas are abundant in fluvial deposits in the USA (e.g., Alaska, Texas), Russia, China, Argentina, Venezuela, and the North Sea of northwest Europe. Fluvial deposits must be understood for effective exploration, development, and management of these resources. For example, it is commonly necessary to predict the geometry, proportion, and spatial distribution of subsurface river-channel deposits from limited borehole and seismic data.

There is a vast amount of published literature on rivers and floodplains. Some recent reviews and compilation volumes are Anderson *et al.* (1996), Bennett and Simon (2004), Bridge (2003), Carling and Dawson (1996), Chang (1988), Ikeda and Parker (1989), Julien (2002), Knighton (1998), Miall (1996), Petts and Calow (1996), Schumm *et al.* (2000), Tinkler and Wohl (1998), and Yalin (1992). Other useful sources of information are series of volumes that have arisen from conference series dealing, for example, with fluvial sedimentology, braided rivers, and gravel-bed rivers. Rivers and floodplains have been studied scientifically by a wide range **365** of geologists, geographers, and engineers, and are more easily studied than most other Earth surface environments. This is the reason for the vast literature, and why this chapter is so long.

Geometry of river systems

A river system is a series of connected river channels in a drainage basin (e.g., Figure 13.1). The drainage basin (or catchment area) is the area that contributes water and sediment to the river system, and is circumscribed by the line of maximum altitude called a drainage divide. The geometry of a river system is adjusted to maintain the continuity of water and sediment supplied from the drainage basin. For example, the surface area of a drainage basin is proportional to the volume of water that must be discharged. The plan geometry (or drainage pattern) of the typical river system shown in Figure 13.1A is composed of (1) a mountain belt with steep valley walls, narrow floodplains, and river channels bounded by both bedrock and coarse-grained alluvium; (2) an alluvial fan where the main channel emerges from the mountain belt, and where the channels may be both distributive and tributive (i.e., anastomosing); (3) broad alluvial valleys with floodplains and alluvial channels with a tributive drainage pattern; and (4) a delta where the main channel flows into a lake or the sea, and where the channels form a distributive pattern. Many other drainage patterns can occur.

A river system consists of channel segments of various sizes and each segment has its own drainage basin (Figure 13.2). In order to analyze the relative geometry of the stream segments and drainage basins of various sizes, each segment can be assigned a numerical *stream order* (e.g., Horton, 1945; Strahler, 1957); see Figure 13.2A. The number, length, slope, and drainage area of streams of successively lower order increase or decrease in geometric series, referred to as morphometric laws (Figure 13.2). The bifurcation ratio (Figure 13.2) depends on the degree of dissection of the drainage basin, and is commonly in the range 2–5 (Knighton, 1998). The length ratio and area ratio are commonly in the ranges 1.5–3 and 3–6, respectively, for most drainage systems (Knighton, 1998). An alternative stream-ordering scheme is to assign a *stream magnitude* (Shreve, 1966); see Figure 13.2B. In practice, objective ordering of stream segments is difficult because identification of stream segments of the smallest order or magnitude is dependent on the map scale and on the stream discharge when the map was constructed.

Drainage density is defined as the total channel length/drainage basin area, and is a measure of the degree of dissection of the drainage basin. Drainage density increases directly as the average distance between adjacent channels decreases, and the closer the stream sources are to the drainage divides. Drainage density increases with the precipitation rate (minus evaporation rate), but decreases with the permeability of the surface materials, the resistance to erosion of the surface material, and the degree of protection by vegetation. Drainage density tends to be high in semi-arid regions (on the order of tens to hundreds of kilometers per square kilometer) because of the temporally concentrated precipitation (and runoff) and lack of protective vegetation. Drainage density is lower in humid temperate regions (less than ten kilometers per square kilometer) because of the vegetation cover.

The *drainage-basin area*, A, increases with channel length, L, on moving down valley from the headwaters, according to

$$L = \text{coefficient} \times A^{\text{exponent}} \tag{13.1}$$

in which the coefficient and exponent vary from region to region (Knighton, 1998). The exponent is generally close to 0.6 (ranging from 0.5 to 0.7), meaning that the length of drainage basins increases with drainage-basin area more than their width does.

The position and orientation of river channels in drainage basins are controlled by geological structure, the areal distribution of different types of surface material, and antecedent drainage conditions (e.g., radial, dendritic, and rectilinear drainage patterns shown in Figure 13.3: see the review by Schumm *et al.* (2000)). Most of the world's largest rivers have valleys in long-established (over many millions of years) structural lows (Miall, 1996). Many rivers flow either parallel or transverse to the main structural grain of the landscape (e.g., Figure 13.3). However, drainage patterns change with time. River diversions are common and are caused by changes in local valley gradients associated, for example, with tectonism, deposition, erosion, and ice dams (e.g., Schumm *et al.*, 2000).

Long profiles of river valleys are graphs of the height of the valley floor plotted against the down-valley distance. Long profiles are commonly thought of as being concave upwards, or as a series of valley segments of constant slope between major tributaries, with slopes decreasing down valley. However, this is not always the case, especially in areas of active tectonism and in river systems where discharge does not increase down valley (Figure 13.4). It is also commonly stated that the

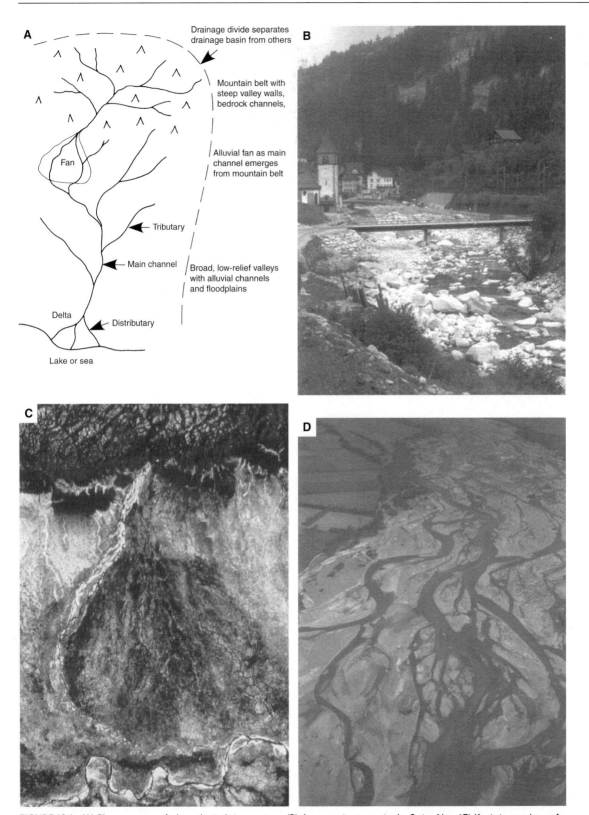

FIGURE 13.1. (A) Plan geometry of a hypothetical river system. (B) A mountain stream in the Swiss Alps. (C) Kosi river and megafan adjacent to the Himalayas (upper) and the Ganges River (lower), India. (D) A braided river and adjacent floodplain near the coast, Canterbury Plains, New Zealand.

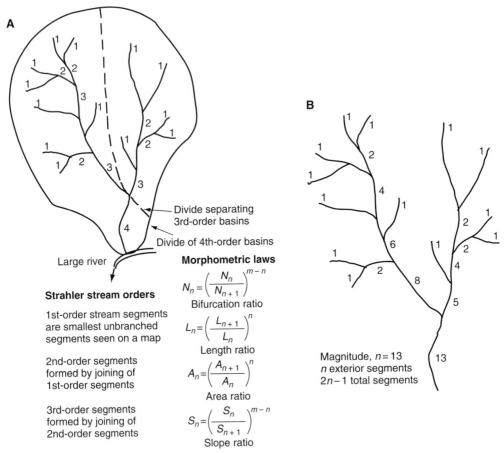

FIGURE 13.2. (A) Stream orders and morphometric laws of the drainage basin, according to Horton and Strahler. N_n, L_n, S_n, and A_n are the number, mean length, mean slope, and mean drainage area of stream segments of order n; m is the highest-order segment. (B) Definition of stream magnitude according to Shreve. Modified from Bridge (2003).

down-valley parts of river systems are areas of net deposition, whereas the up-valley parts are mainly erosional (e.g., Schumm, 1977). This view may have arisen from studies of mid-latitude rivers in which down-valley reaches have experienced deposition associated with postglacial sea-level rise whereas upland reaches have experienced uplift and erosion associated with unloading of glacial ice. This is much too simple a view. Near-coastal valleys commonly experience erosion (even during sea-level rise), and many areas within and immediately adjacent to upland areas have major amounts of deposition, associated with time-varying tectonic activity and climate (Blum and Tornqvist, 2000).

Origin and evolution of river systems

Studies of the origin and evolution of river systems have involved fieldwork, analysis of sequential maps or aerial photographs, laboratory experiments, and theoretical approaches. Initiation of a river channel requires a surface water flow of sufficient power (the product of water discharge and slope) to entrain and transport surface material, which in turn requires a critical drainage area and slope. In order for progressive erosion to occur, the transporting power and sediment-transport rate must increase downstream. Channel initiation has been associated with erosion by concentrated overland flow, by localized groundwater seepages, and by shallow landslides that produce local increases in slope and unvegetated surfaces that are relatively easily eroded (e.g., Montgomery and Dietrich, 1989, 1994; Dietrich and Dunne, 1993; Kirkby, 1994). Distinct modes of channel initiation are associated with different drainage area–slope combinations. For example, shallow landslides generally require relatively steep slopes. Erosion associated

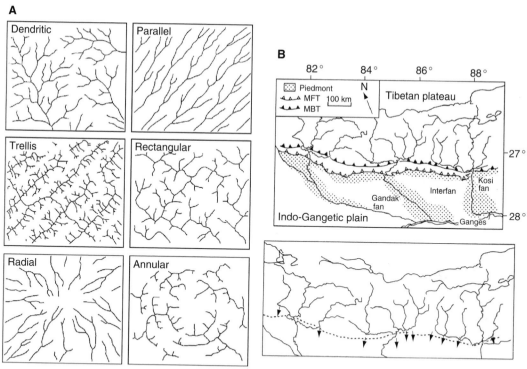

FIGURE 13.3. (A) Drainage patterns controlled by topography and geology. After Howard (1967). Dendritic patterns form on slopes underlain by homogeneous material. Trellis and rectangular patterns are controlled by fracture patterns or dipping sedimentary strata of unequal resistance to erosion. Radial dendritic patterns form on topographic domes formed of homogeneous material (e.g., ash cones). Annular patterns form on topographic domes with either concentric fracture patterns or rocks with unequal resistance to erosion. AAPG © 1967 reprinted by permission of the AAPG, whose permission is required for further use. (B) Modification of Himalayan rivers by progressive growth of anticlines associated with thrust fronts. From Leeder (1999), after Gupta (1997). The upper figure shows the present-day drainage system: MFT is the Main Frontal Thrust; MBT is the Main Boundary Thrust. The lower figure shows postulated courses of rivers (arrows) prior to formation of the thrust-related drainage divide.

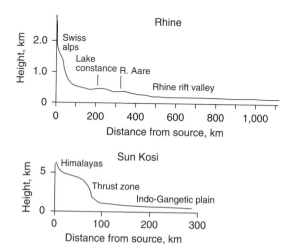

FIGURE 13.4. Long profiles of the Rhine and Sun Kosi Rivers. The Rhine profile has the typical concave-upward shape, with local undulations associated with tributary confluences. The Sun Kosi profile has a marked convex-upward and steep section associated with the thrust zone that separates the Himalayas from the Indo-Gangetic plain.

with groundwater seepages might not require as much drainage area as overland flow. However, it is unlikely that any of these processes of channel initiation acts independently, and the actual drainage area–slope combination when channel heads form is largely dependent on the erodibility of surface materials. Resistance of surface material to erosion in turn depends on its texture (e.g., grain size and shape) and on the degree of cohesion brought about by lithification, clay content, and vegetation. Spatial variation in the erodibility of surface material will contribute to the localization of erosion required for initiation of channels.

Experimental and field studies of the initiation and evolution of river systems on freshly exposed, regular surfaces composed of homogeneous material demonstrate that channel systems evolve by headward extension (elongation) of channels and development of branches (e.g., Figure 13.5A). The relative timing of headward extension and development of branches

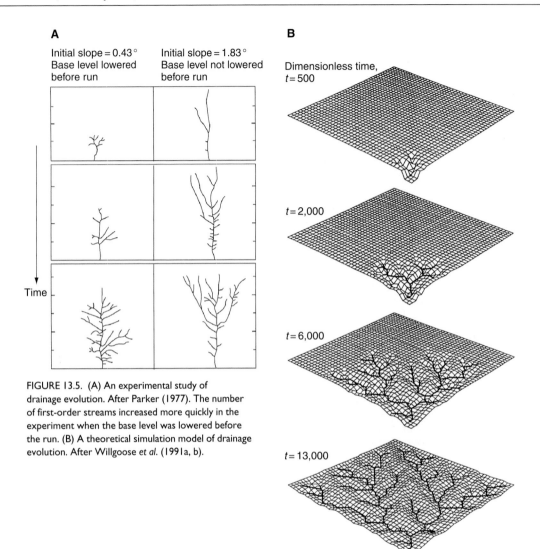

FIGURE 13.5. (A) An experimental study of drainage evolution. After Parker (1977). The number of first-order streams increased more quickly in the experiment when the base level was lowered before the run. (B) A theoretical simulation model of drainage evolution. After Willgoose *et al.* (1991a, b).

depends on particular experimental conditions, such as initial slope and spatial variations in slope (Knighton, 1998). During the early stages of river-system evolution, drainage density increases rapidly, but later it changes little with time.

An early qualitative theory for the evolution of drainage systems involved the amalgamation of numerous sub-parallel, closely spaced channels (called rills) flowing down a planar surface into a master channel by a process called *micropiracy* (Horton, 1945). Subsequently, channels developed on the valley slopes of the initial master channel, eventually forming tributaries, and so on. This theory does not agree with observations. Dunne (1980) developed a theory for the evolution of river systems from an initial stream head at a local seepage of groundwater. Headward

extension of the stream head increases the convergence of groundwater flow into the channel. Tributaries originate by seepage erosion of "susceptible" zones in the valley sides. However, many channel networks are not produced by seepage erosion.

Over the past 20 years, many process-based theoretical models for the evolution of drainage systems have been developed (Figure 13.5B); see the review by Bridge (2007). The erosion rate of bedrock and cohesive sediments by streams is commonly taken as proportional to kA^mS^n in excess of some threshold value, where k is the erodibility, A is the area of the drainage basin, and S is the surface slope. Here A is a surrogate measure for stream-water discharge, such that the erosion rate becomes a function of the stream power per unit channel length. The exponents m and n are not known well, although they are commonly assumed equal to unity. Sediment transport in channels and by overland flow is also commonly taken as proportional to a function of the stream power in excess of some limiting value, and may also be limited by sediment production by weathering. In some models, the formation of channels is dependent on a channel-initiation function (which is proportional to some function of slope and the discharge or drainage area) exceeding some specified threshold. Headward extension of channels and development of tributaries continues until the channel-initiation function decreases below the threshold because the drainage areas contributing water to the tributary channels progressively decrease. The influence of climate on stream erosion and sediment transport has been modeled by taking the water discharge to be proportional to the product of the effective precipitation rate per unit area and the drainage-basin area. River erosion and development of drainage systems are critically dependent on the temporal and spatial variation of rainfall events, the form of the erosion law, and the magnitude of the erosion thresholds relative to that of the erosive forces.

The development of drainage systems has also been approached from the point of view of the energy required to transport water and sediment from the land as efficiently as possible. Drainage systems are most efficient (require the least expenditure of energy) when flow resistance due to boundary friction is at a minimum. The best way of minimizing flow resistance is to have channel flow instead of overland (sheet) flow, and to have a relatively small number of large channels. However, channel initiation requires a certain amount of overland and/or shallow subsurface flow, both of which are associated with high flow resistance. Drainage networks are thought to develop such that the energy expenditure in any stream segment and in the whole system are at a minimum, and the expenditure of energy per unit bed area is constant (Rinaldo *et al.*, 1992, 1998; Rodriguez-Iturbe *et al.*, 1992; Rodriguez-Iturbe and Rinaldo, 1997). Although these assumptions regarding distribution of energy expenditure can be argued with, this theory agrees with the morphometric laws for drainage networks mentioned above.

Water supply to river systems

Water is supplied to rivers from precipitation in the drainage basin (catchment area). Some of the precipitation is returned to the atmosphere by evaporation and evapo-transpiration, but the remainder flows under the influence of gravity over the surface or through the ground towards rivers, floodplains, and lakes (Figure 13.6). *Overland flow* is commonly subdivided into *infiltration-excess overland flow* (due to the precipitation rate exceeding the infiltration capacity of the ground) and *saturation-excess overland flow* (due to the fact that the ground is saturated with water). The relative importance of these two types of overland flow depends on the precipitation rate relative to the permeability of the ground. Water may flow through the ground relatively rapidly near the surface (called *throughflow* or *subsurface storm flow*) or more slowly deeper down (*groundwater flow*).

Hydrographs

The water supply in a river system is normally expressed as the water discharge measured at various river cross sections within the drainage basin. Discharge data obtained from gauging stations on many rivers in the world are available from government agencies, and some are available in real time on the World Wide Web. Water discharge in rivers and floodplains varies in space and time. Discharge usually

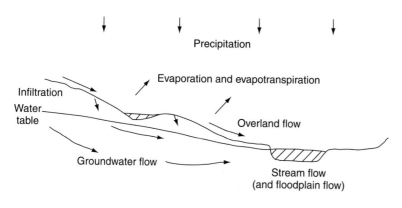

FIGURE 13.6. Movement of water on and through the ground. Modified from Bridge (2003).

Precipitation

Evaporation and evapotranspiration

Infiltration
Water table

Overland flow

Groundwater flow

Stream flow
(and floodplain flow)

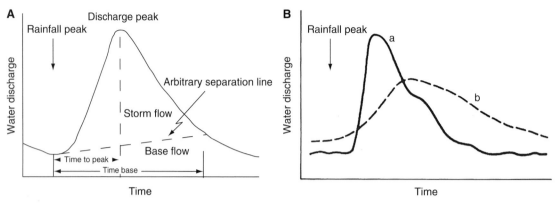

FIGURE 13.7. (A) Definition of a flood-discharge hydrograph. (B) Types of flood hydrographs. Modified from Bridge (2003).

increases in the down-valley direction as tributaries join, unless water is lost by evaporation and/or by infiltration. Temporal changes in water discharge may be associated with (in order of increasing time scale) daily and seasonal variations in snowmelt, individual storms, seasonal precipitation, sunspot cycles, drainage diversions, and Milankovitch cycles. A *hydrograph* is a graph of the time variation of water discharge at one cross section of a river (e.g., Figures 13.7 and 13.8). A single-flood hydrograph (Figure 13.7) is typically asymmetrical, with the peak discharge lagging behind peak precipitation because of the time it takes for water to move through the ground and over the land surface. Flood hydrographs are commonly separated into *quickflow* (or *stormflow*) and *delayed flow* (or *baseflow*) (Figure 13.7). Quickflow originates from overland flow and fast, shallow subsurface flow, whereas delayed flow is due to slow flow through the ground.

Differences in the shapes of flood hydrographs at a point depend on (1) the temporal and spatial distribution of precipitation; (2) the proportion of overland flow relative to groundwater flow reaching the river, as determined by geology, soil type, vegetation, land use, and antecedent conditions; and (3) the drainage-system geometry, which controls the amount of overland flow and groundwater flow relative to channel flow. The sharp-peaked hydrograph in Figure 13.7B may be due to a high rate of rainfall or snowmelt over a short period of time and/or a high proportion of overland flow relative to groundwater flow due to impermeability, a high water table, or lack of the retarding effects of vegetation. The broad hydrograph in Figure 13.7B may be due to precipitation distributed over time and/or high groundwater flow/overland flow due to high permeability, a low groundwater table, or vegetation cover. High drainage density and steep

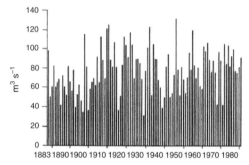

FIGURE 13.8. The long-term variation of the average annual discharge of the River Thames, London, UK. From Gustard (1996).

valley slopes tend to result in highly peaked hydrographs, because of the relative speed with which overland flow can reach the channels.

It is common for the time of peak discharge following a storm to occur progressively later in the down-valley direction, and for the shape of the flood hydrograph to change down valley. The translation and modification of a flood wave can be treated using the equations of motion for unsteady, non-uniform flow (Chapter 5), but with difficulty. Storage of the inflow to a reach of river, such as during overbank flow or the filling of a lake, leads to attenuation and delay of the flood peak in moving down valley. In order to predict the hydrograph shape as a function of time and space in a drainage basin, it is thus necessary to consider antecedent conditions, the temporal and spatial variations in effective precipitation, overland flow and groundwater flow, and the routing of river and overbank flow through the system. This is not a simple task; see the review by Fawthrop (1996).

Long-term streamflow hydrographs (e.g., Figure 13.8) are time series that are commonly

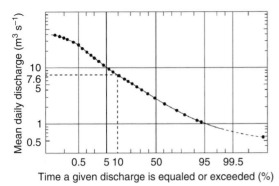

FIGURE 13.9. An example of a flow-duration curve. After Knighton (1998).

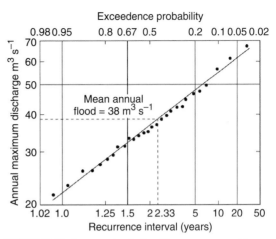

FIGURE 13.10. An example of a flood-frequency curve. After Knighton (1998).

analyzed in order to recognize trends, cycles, and random fluctuations in discharge (see the appendix). Mathematical models of hydrological time series are used for sequential generation of streamflow data for simulation purposes. Hydrograph shapes have been classified (e.g., Gustard, 1996) either by relating a particular hydrograph shape to climate (seasonal variations in precipitation and temperature) or by carrying out statistical analyses of hydrographs.

Flood frequency

The frequency of occurrence of water discharges of a given magnitude at a point on a river is commonly represented on a so-called *duration curve* (e.g., Figure 13.9), which shows the percentage of time that a particular discharge is exceeded or equaled. If the frequency distribution of discharge at a point is lognormal, the duration curve will be a straight line when plotted on normal probability paper (Figure 13.9). A flow-duration curve with a relatively low negative slope implies a large discharge range (i.e., flashy conditions), whereas a high negative slope implies less discharge variability.

The distribution of maximum annual discharges that occur over a period of years is called the *annual series*. The mean of the annual series is called the *mean annual flood*. If there are N years of record, and these maximum annual discharges are ranked, with the largest having rank $m = 1$ and the smallest having rank $m = N$, then the probability of an annual flood of magnitude x exceeding one with magnitude m is

$$p(x) = m/(N+1) \tag{13.2}$$

The mean return period (or recurrence interval) of this flood event is

$$T = 1/p(x) \tag{13.3}$$

and the cumulative probability is

$$f(x) = 1 - p(x) \tag{13.4}$$

A *partial duration series* is a series of independent flood discharges that exceed some specified threshold discharge. It is analyzed in the same way as an annual series, but the flood return period can be less than one year. Figure 13.10 is an example of an analysis of annual maximum discharges. An annual maximum discharge with a return period (recurrence interval) of 10 years is called the 10-year flood. A flood discharge of this magnitude is expected to occur *on average* once in a 10-year period *only if it can be assumed that past records of annual maximum discharges are representative of future floods*. Terms such as 100-year flood and 500-year flood are commonly used to refer to extreme maximum annual flood discharges from which we might need to protect ourselves (by building levees and flood walls for instance). But, if we have only 80 years of reliable flood records, how can we determine the discharge of a 100-year or 500-year flood? This is explained below.

Frequency distributions of extreme events such as maximum annual discharges are commonly positively skewed (i.e., the mode is displaced towards smaller values) and can be described by distribution functions such as exponential, gamma, lognormal, Gumbel EV1, and Pearson type III. In order to fit any of these functions to data, it is necessary to have a long period of records and no progressive changes in flood discharge, such as, for instance, those brought about by human interference with a river system. For the Gumbel EV1 function, the probability of exceeding the mean annual flood is 0.43, and its recurrence interval is 2.33 years. The modal (most probable) annual flood has a recurrence interval of 1.58 years. If annual maximum discharges follow the Gumbel EV1 function, a plot of maximum annual discharge versus cumulative probability or recurrence interval (with a special type of logarithmic scale) will be a straight line (e.g., Figure 13.10). The discharge of a 100-year or 500-year flood is estimated by extrapolating the straight line of the theoretical function beyond the discharge values that were used to define it (Figure 13.10). Such estimates can be greatly in error if the theoretical function is not appropriate or if past discharges cannot be considered to be representative of the future.

In river systems without comprehensive discharge records, future flood discharges of a given frequency may be estimated using multivariate statistical techniques. Such flood discharges are related empirically to catchment characteristics such as climate (rainfall), land use, soil type, basin geometry (area, slope), and drainage density (Fawthrop, 1996; Knighton, 1998). Problems with this approach include bias caused by the samples used to derive the statistical relationship, correlations among the independent variables, and failure to consider all relevant controlling variables. There are also physically based models for predicting flood discharges, but these are fraught with uncertainty because of the complicated nature of the problem.

Sediment supply to river systems

Sediment is supplied to river systems by weathering of exposed rocks and by downslope movement of the loose material under the influence of gravity (mass wasting) or with the assistance of overland water flow. The rate of weathering, and the texture and composition of weathered material, are controlled by at least (1) the composition, texture, and structure of exposed rocks; (2) the amount of precipitation; (3) temperature variations; and (4) vegetation (Chapter 3). These controls are in turn controlled by topography and climate. Relatively coarse-grained weathered material tends to be most common in cold climates and where there are steep slopes. Clays (produced by chemical weathering) are most voluminous in warm humid climates. Organic material makes its maximum contribution to the sediment supply in humid climates. The rate of mass wasting depends on the texture and composition of the loose material, the availability of water, vegetation, slope

angle, and ground motions associated with earthquakes (Chapter 8).

Dissolved material in river systems comes from rainfall, overland flow, and groundwater. Solute composition in groundwater reflects the geology of the drainage basin (Chapter 3). It is common in river systems to observe peaks in solute concentration under low-flow conditions, and changes in the composition of the dissolved species as groundwater flow starts to add to quickflow. The concentration of dissolved material decreases as water discharge increases, because of the decreasing importance of groundwater flow. However, the total dissolved load increases with discharge (Knighton, 1998). Humans can modify the nature of dissolved material in river systems adversely by adding pollutants to the atmosphere (e.g., producing acid rain), to the land surface, to groundwater, and directly to rivers.

The sediment supply to rivers and floodplains varies in space and time for the same reasons as for water supply. There is more variability in the case of sediment supply because of discrete mass movements such as debris flows and landslides. Peaks in sediment supply commonly lag behind peaks in water supply because sediment requires a threshold gravity and/or fluid force to initiate downslope movement, and sediment travels more slowly than the fluid (unless it is suspended load). Furthermore, sediment from mass wasting is commonly stored temporarily at the outer edges of floodplains.

Sediment yield from drainage basins has been estimated from measurements of suspended load and dissolved load in rivers, and rates of deposition in reservoirs (Leeder 1999; Bridge, 2003), and corresponds to denudation rates on the order of 0.01–1 mm per year. The average global denudation rate is 0.055 mm per year based on suspended solids and 0.01 mm per year based on dissolved material; however, these rates vary greatly across the globe (Knighton, 1998). Estimates of sediment yield are likely to be very inaccurate, partly because bed load is not considered. Since sediment loads are very variable in time and space, the accuracy of the estimates is greatly dependent on the sampling frequency and extent in time and space. It is also likely that humans have had a large impact on sediment yields in some regions, as a result of deforestation, agriculture, construction, and mining. Sediment yields may actually underestimate denudation rates of hillsides because much eroded sediment may be stored on floodplains. Therefore, great care must be exercised in extrapolating both sediment yields and derived denudation rates to the geological past.

TABLE 13.1. Independent and dependent variables in alluvial river systems as a function of time, modified from Schumm and Lichty (1963)

Variable	Decreasing time span ⟶		
Geology	Independent ⟶		
Climate	Independent ⟶		
Vegetation	Dependent ⟶ Independent ⟶		
Topography	Dependent ⟶ Independent ⟶		
Water and sediment supply	Dependent ⟶ Independent ⟶		
Channel geometry	Dependent ⟶ Independent		
Channel flow, sediment transport, erosion, and deposition	Dependent ⟶		

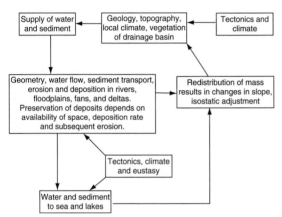

FIGURE 13.11. The relationships among materials, processes, and landforms in alluvial river systems. Modified from Bridge (2003).

It is suggested that sediment yields are greatest in semi-arid climates where vegetation is sparse, and the low annual-average precipitation commonly falls heavily in short-lived seasonal storms. Arid climates also have little protective vegetation, but precipitation and runoff are low. Very wet climates have dense forest vegetation to protect the surficial material from erosion. However, sediment yields depend on factors other than precipitation, runoff and vegetation cover, such as seasonality of precipitation and temperature, topographic relief, seismicity, rock and soil type, and land use by humans.

Controls on geometry, water flow, and sediment transport in river systems

Bedrock rivers have substantial amounts of hard rock forming their banks and beds, but also varying amounts of loose sediment. Erosion of rocks in river channels arises mainly from hydraulic pressures acting on loosened blocks, cavitation associated with extreme turbulence, and abrasion by moving sediment. Erosion of solid rock is a slow process, typically fractions of a millimeter per year. The geometry of bedrock channels is mostly determined by the erodibility of rocks, which is controlled by rock composition, texture, and structure; the nature of weathering; and the amount of mobile, abrasive sediment.

Alluvial rivers flow substantially within unconsolidated sediments that can be eroded, transported, and deposited by the river flow, even if only during floods. The ability of alluvial rivers to move their boundary sediments relatively easily means that they can readily change their geometry. The loose-sediment boundary both controls and is controlled by the water flow and sediment transport, such that there is an intimate and complicated relationship among channel geometry, water flow, and sediment transport. The geometry, water flow, sediment transport, erosion, and deposition in alluvial river systems are controlled by the supply of water and sediment, which are in turn controlled by the nature of the drainage basin and, ultimately, by regional climate and tectonics (Figure 13.11). Other controls on the geometry, flow, and sedimentary processes within river systems include changes in topography and "accommodation space" associated with tectonism and relative base-level changes. The interactions between the various independent and dependent variables occur over a range of time and space scales (Table 13.1). There are also time lags between changes in independent variables and responses of dependent variables, which are controlled by the rates and magnitudes of changes in independent variables, and by the ability of the dependent variables to change.

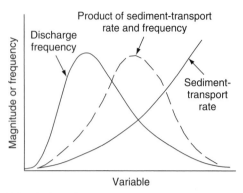

FIGURE 13.12. The dominant discharge defined by the magnitude and frequency of sediment transport associated with a range of water discharge. From Wolman and Miller (1960).

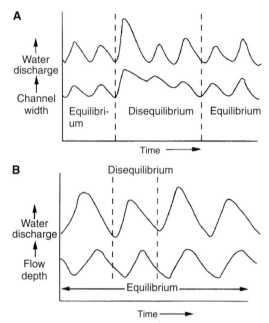

FIGURE 13.13. The definition of equilibrium and disequilibrium in stream geometry. (A) The channel width varies congruently with the water discharge (i.e., equilibrium) until an extreme discharge that causes excessive bank erosion and channel widening. The channel is in disequilibrium with subsequent water discharge and requires a finite lag time in order to regain equilibrium (by deposition). (B) The flow depth varies incongruently with the water discharge over seasonal floods (disequilibrium), but the flow depth and water discharge averaged over several floods are both constant (in equilibrium). There are correlations among width, depth, and discharge over various time intervals, whether or not equilibrium exists. Recognition of equilibrium and disequilibrium requires observations at sufficiently small time intervals. From Bridge (2003).

Adjustment of dependent variables such as channel width and depth to changes in independent variables requires erosion and deposition. Most erosional and depositional activity in alluvial rivers (and hence adjustment of channel geometry) is accomplished at relatively high flow stages. However, it is necessary to consider the frequency as well as the magnitude of depositional and erosional events. The *dominant discharge* or *channel-forming discharge* (see Figure 13.12) is near the *bankfull discharge*, the discharge that just fills the channel. Bankfull discharge commonly has a return period of between 1 and 2 years. The return period of bankfull discharge varies along the length of a given river and among rivers, because of differences in shape of flood hydrographs. Furthermore, the return period of bankfull stage for aggrading rivers is less than that for incising rivers. In reality, all discharges that are capable of transporting sediment will have some effect on channel geometry.

Adjustments of channel geometry to changes in water and sediment supply depend on the rates and magnitudes of the changes, and on the effectiveness of erosion and deposition in changing the channel form. If the flood (bankfull) discharges of water and sediment in a river did not vary appreciably over time spans on the order of decades, the channel geometry, flow, and sedimentary processes would be adjusted to these flood discharges, with relatively minor adjustments associated with changing flow stage. This adjusted, steady-state condition is referred to as *dynamic equilibrium* (also *graded* or *regime*). If there were a major change in the flood discharge of water and sediment in a river, caused for example by an exceptional (e.g., 1,000-year) flood, earthquake, channel diversion, or human activity (e.g., dams, mining, over-grazing), there would most likely

be a major, regional change in river geometry, flow, and sedimentary processes. Such exceptional (even catastrophic) short-term changes in the river system would cause it to be in *disequilibrium* with previous "normal" flood discharges, and a finite lag time would be required in order to re-establish dynamic equilibrium with subsequent "normal" conditions (see Figure 13.13). Whether the system is regarded as in equilibrium or in disequilibrium depends on the time interval of measurement relative to the lag time (Figure 13.13). Therefore, it is necessary to have measurements of the dependent and independent variables at short intervals but over long periods of time. Long-term (longer than 10,000 years), regional-scale changes in climate, tectonic activity, or relative sea level will probably result in changes in the supply of water and sediment to rivers and floodplains. Dynamic equilibrium may well be attained at any location, but the whole river system would be changing gradually and probably imperceptibly over time spans of decades to centuries. The controls on the geometry, flow, and sedimentary processes in alluvial river systems are discussed further below. Bedrock rivers are discussed subsequently.

Origin and evolution of alluvial river channels

Well-defined banks of rivers give rise to large-scale bed forms in river channels called *bars*, the dimensions of which are controlled by the flow width as well as the depth. The formation and evolution of bars result in a continuum of distinctive alluvial-channel patterns (plan geometry). This continuum has been demonstrated in experimental studies of natural and laboratory channels, and with theoretical models. Figure 13.14 (modified from Bridge, 2003) illustrates how common types of channel pattern can evolve from a straight, erodible channel with *constant water discharge and sediment supply*. Initially, single rows or multiple rows of lobate *alternate bars* form on the bed, depending on the supply of water and sediment (Figure 13.14). These alternate bars increase in length and height as they evolve until they are in equilibrium with water flow and sediment transport. The water flow over these bars is sinuous in plan, with wave length and width equivalent to those of the bars in a particular row. Thus, multiple rows of bars have multiple rows of sinuous flow paths. Bar lengths, L, are proportional to their width, w (and thus to the width of the sinuous flows), and their heights are comparable to the flow depth. Theoretical and experimental results show that L/w of alternate bars is commonly 3–7. Alternate bars are asymmetrical in along-stream cross section, may but need not have an avalanche face on the downstream side, and generally migrate in the downstream direction. The term *alternate bar* is used because, within a given row, the bars occur on alternating sides of the channel with progression downstream. The term *bar* is used because the bed-wave length is proportional to the flow width, and the bed-wave height is comparable to the flow depth. Bars are also referred to as *macroforms*. Alternate bars (or parts of them) have also been referred to as *unit bars, linguoid bars, side bars, transverse bars, cross-channel bars*, and *diagonal bars* and *riffles*. Smaller-scale bed forms such as dunes, ripples, and bed-load sheets are commonly superimposed on alternate bars.

The next stage in the evolution of channel patterns involves bank erosion and channel widening, resulting in a drop in water level and emergence of the highest parts of alternate bars. At this stage, three morphological units that grade into each other can be recognized: *bar tail, riffle* (or *tributary mouth bar* or *crossover*), and *bar head*. These units represent different positions on an ancestral alternate bar (Figure 13.14). If erosion of a concave bank is sufficiently rapid and intermittent, the eroded sediment may be transported and molded into a unit bar and deposited on a downstream convex bank as a distinct bar-head unit bar lobe or bar-tail unit bar (also known as a *scroll bar*). If deposition is less rapid and/or more continuous, sheets of sediment are deposited rather than discrete unit bars. The resulting accumulations of sediment in mid channel, or forming the convex banks on the inside of river bends, are referred to as *braid bars* and *point bars*, respectively. Braid bars and point bars are *compound bars* inasmuch as they are normally composed of parts of multiple unit bars (Figure 13.14). Migration rates of compound point bars and braid bars are normally considerably less than those of the unit bars that comprise them. The more mobile unit bars have been referred to as "free" bars and the less-mobile compound bars have been referred to as "forced" bars. Synonyms for braid bars include *medial, longitudinal, crescentic,* and *transverse bars*, as well as *sand flats*. Point bars are also referred to as *side bars* and *lateral bars*. As with the ancestral alternate bars, the overall geometry of point and braid bars is controlled by flow and sediment-transport conditions at channel-forming discharge.

Bank erosion and bar deposition create changing gradients of the bed and water surfaces, leading possibly to the formation of new channels that cut across braid bars and point bars (Figure 13.14). Some of these channels develop as a sheet of water accelerates down a steep bar surface and becomes progressively concentrated into channels that develop by headward erosion. Other (chute) channels develop as the flow takes advantage of the low areas (chutes) between adjacent unit bars. These *cross-bar channels* commonly develop their own bars, the geometry of which is controlled by the flow and sediment-transport conditions in these channels. Where these channels join another channel, solitary delta-like deposits with avalanche faces commonly form (e.g., chute bars, tributary mouth bars: Figure 13.14). A *cross-bar channel* may be enlarged progressively at the expense of an adjacent channel, thus becoming a new main channel. Chute cut-off of point bars is a familiar example of such behavior, and is promoted by the deposition of bar-head unit bars and the lack of filling of the low areas between these bars. In other cases, a channel on one side of a braid bar may become enlarged at the expense of the channel on

FIGURE 13.14. Evolution of channel patterns from growth of alternate bars, bank erosion, and channel widening. Modified from Bridge (2003). Image (A) shows the formation of a single row of alternate (unit) bars in a straight channel. The crestlines of unit bars (indicated by solid or dashed lines) may, but need not, be associated with angle-of-repose lee faces. Arrows represent the flow direction and stippled areas are topographic highs. With bank erosion and channel widening, the single row of alternate bars evolves into point (side) bars composed of bar-head lobes and bar-tail scrolls (i.e., compound bars). Continued growth of point bars is by episodic accretion of unit bars, although such discrete features are not always present. From Bridge (1993), and based on numerous laboratory experiments and river studies. (B) An experimental single row of alternate bars evolving into point bars. Flow is towards the observer. (C) A view of bar-tail scroll in the foreground looking downstream towards the riffle and bar head of the next side bar downstream. River Feshie, Scotland. (D) A point bar on Madison River, Montana, showing bar-head lobes, bar-tail scrolls, and cross-bar channels. Flow is to the right. (E) As for (A), but with cross-bar channels and associated channel-mouth bars. (F) A view looking upstream of a chute channel and chute bar (on the right). The vegetated island is the bar head. The cut bank is on the left. Calamus River, Nebraska. (G) A bar-tail scroll with a cross-bar channel and mouth bar (foreground). The main channel is in the background. The cross-bar channel has its own point

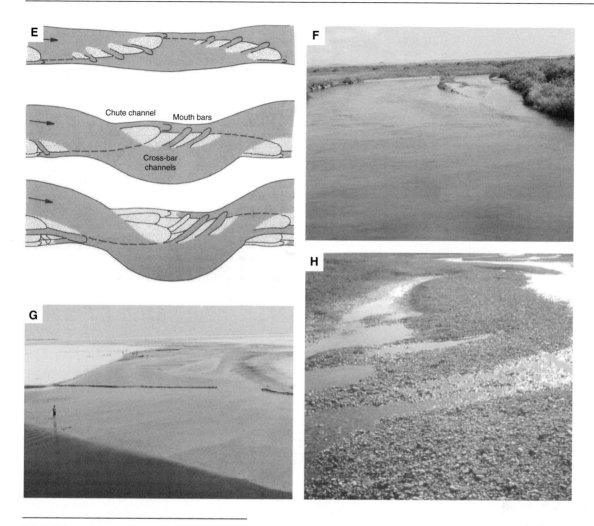

FIGURE 13.14. Continued. bars. Brahmaputra River. (H) A view looking upstream of a bar-tail scroll with cross-bar channels and mouth bars in inner-bank swale. Bar-head lobes can be seen in the background. Sagavanirktok River, Alaska. (I) Double rows of alternate (unit) bars in straight alluvial channels evolving into braid bars and point (side) bars. (J) A series of compound braid bars, each with several bar-head lobes and bar-tail scrolls. Flow is to the right. South Saskatchewan River, Canada. (K) As for (I), but with cross-bar channels and channel-mouth bars. (L) Compound braid bars and cross-bar channels in Sunwapta River, Canada (photo courtesy of Peter Ashmore). Flow is towards the observer. This type of braiding pattern would have evolved from multiple rows of alternate bars. See Bridge (1985). (M) Single lobate unit bars occur in the channel at (a). An incipient compound braid bar occurs at (b). The bar head is composed of a single, partly emergent lobate unit bar that is dissected by a series of cross-bar channels. These cross-bar channels have small solitary mouth bars at their downstream ends. The bar tail is composed of two emergent scroll bars and at least one incipient submerged scroll bar. A well-developed compound braid bar occurs at (c). The bar head is composed of at least six lobate unit bars, and the bar tail is composed mainly of a single scroll bar. Cross-bar channels are also evident. Rakaia River, New Zealand. Photo from Jim Best. Flow is from right to left, and the width of view is about 500 m.

the other side. Thus, channel segments are commonly abandoned and filled, resulting in accretion of bars and channel fills to the floodplain or to other bars. Previously abandoned channels are commonly re-occupied as the process of channel migration continues.

These patterns of bar growth and channel change can occur at constant discharge; however, they also occur in natural rivers where discharge varies in time. Some channel geometries arise specifically as a result of changing discharge, particularly the dissection and modification of emerging bars, which might give a braided appearance to a river at low-flow stage that does not exist at high-flow stage. Thus, it is important to study rivers over a large range of stage over an extended time span, because their appearance is controlled by the history of changing flow stage.

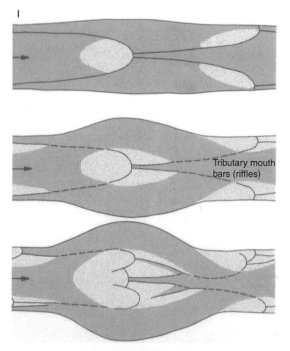

FIGURE 13.14. Continued.

Plan geometry (pattern) of alluvial river channels

Alluvial channel patterns have traditionally been classified as straight, meandering, and braided (Leopold
and Wolman, 1957), but this classification is unsatisfactory because the classes are not mutually exclusive
and different parameters are used to define the various
patterns. Another popular classification of channel
patterns divides channels into straight and meandering
(single channels of varying sinuosity), and braided and
anastomosing (multiple channels of varying sinuosity).
This classification is also flawed, as will be seen below.
Channel patterns should be defined by two quantifi-
able parameters: the *degree of channel splitting* around
braid bars or islands and the *sinuosity* of the channel
segments. However, the classification of channel patterns is not a simple task, as explained below.

Channels and bars normally experience a complex
history of erosional and depositional modification over
a range of flow stages, even though their overall form is
controlled by flood discharge. Thus channels and bars

K

L

M

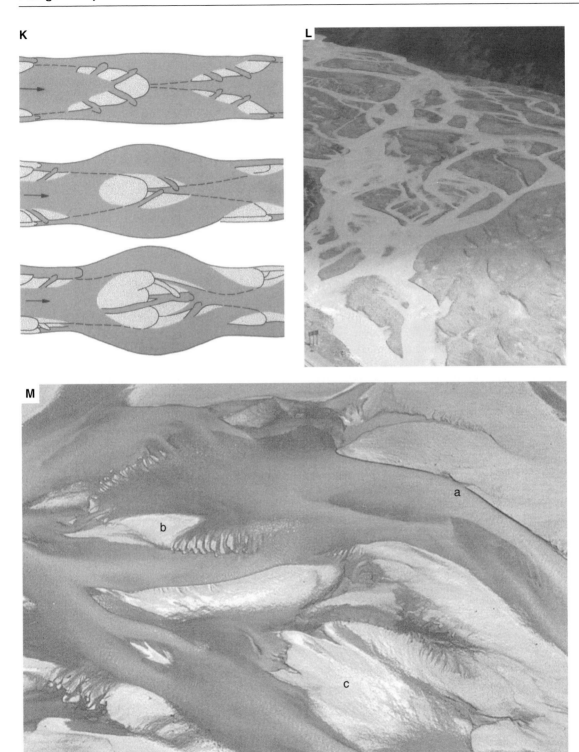

FIGURE 13.14. Continued.

will generally look different at different flow stages, and from year to year. However, channel patterns are normally defined using maps or aerial photographs that show the channel pattern at an instant in time. It is therefore necessary to decide at what flow stage the channel pattern will be described. Then, it is necessary to consider which of the channel segments and bars will be used to define the nature of channel splitting and the channel sinuosity. For example, should the relatively minor cross-bar channels be considered together with the main channels, even if all are active at the same flow stage? There is no simple answer to this question, since any cross-bar channel may evolve into a main channel. This dilemma has led to various attempts to assign different orders to channels and bars in rivers (Bridge, 2003). Then, channel pattern is described using channels of a specific order, normally main (order 1) channels.

A distinction is sometimes made between *mid-channel bars* and *islands*. Mid-channel bars are unvegetated, submerged at bankfull stage, and "transient," whereas islands are vegetated, emergent at bankfull stage, and more "stable." However, many bars have vegetated and unvegetated parts. The degree of development of vegetation on channel bars is related to the amount of time the bar surface has been exposed above the seasonal low-water mark, the nature of the sediment surface exposed, and the type of vegetation available for colonization. These are in turn controlled by the history of erosion and deposition, the sediment available to the river, and the climate. Depending upon these factors,

freshly emergent unvegetated bars may, or might not, become progressively vegetated as they grow. Thus, it is difficult to assess when a bar becomes an island. Furthermore, such a distinction between bars and islands artificially separates depositional forms that may have a common geometry and genesis.

The splitting of channels around bars or islands has been referred to as *braiding* and *anastomosing*. Although the terms braiding and anastomosing have been used synonymously (e.g., Leopold and Wolman, 1957), a distinction has more recently been made between rivers with channels that split around bars or islands (braiding) and those that split around floodplain areas (anastomosing or anabranching). Anastomosing main-channel segments are generally longer than a curved channel segment around a single braid bar or point bar, and the patterns of flow and sediment transport in adjacent anastomosing segments are essentially independent of each other. This means that each anastomosing segment contains bars appropriate to the imposed discharge and sediment load, enabling assignment of its own channel pattern on the basis of the degree of channel splitting around bars and sinuosity. This means also that the terms anastomosing and braiding are not mutually exclusive, and should not be used together in a single classification (see above). In fact, many braided rivers appear to be both braided and anastomosing (Figure 13.15).

The *degree of braiding* (*braiding index*) is best measured as the mean number of active channels or braid bars per

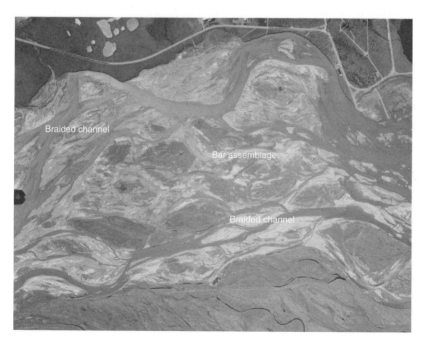

FIGURE 13.15. Sagavanirktok River, Alaska, showing braided channels that are also anastomosed. The anastomosed channels are separated by bar assemblages. Modified from Bridge (2003).

Braided channel

Bar assemblage

Braided channel

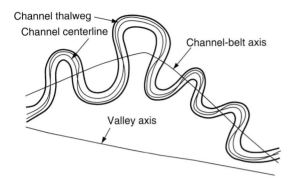

FIGURE 13.16. Lengths used to define sinuosity. From Bridge (2003).

transect across the channel belt. In order to use this approach, it is necessary to define what channel orders will be considered. This approach to defining the degree of braiding is desirable because it is useful to know how many channels and bars will be encountered when traversing the width of a channel belt. Other terms to describe the nature of braiding are discussed by Bridge (2003).

The *sinuosity* of single-channel rivers is defined as channel–thalweg length/valley length, channel length/valley length, or channel length/channel-belt axis length, the latter taking account of channel-belt sinuosity (Figure 13.16). An arbitrary sinuosity of 1.5 has been used to separate low-sinuosity rivers from high-sinuosity rivers, and rivers with sinuosity of less than 1.5 have been called straight (Leopold and Wolman, 1957; Rust, 1978). The perceptive reader will realize that straight rivers must have a sinuosity of unity. Definition of sinuosity for multiple-channel rivers requires definition of some average length of channel segments of a given order at a given flow stage (Rust, 1978). Multiple-channel rivers with sinuosity less than 1.5 were called braided, whereas those with sinuosity greater than 1.5 were called anastomosing. As demonstrated previously, this classification is untenable.

The sinuosity of a single-channel river can be related to the variance of the channel direction, and this relationship can be used to estimate paleochannel sinuosity from paleocurrent ranges measured in ancient channel deposits. The sinuosity of single-channel rivers has also been described using autocorrelation or spectral analysis (see the appendix) of series of changes in the direction or curvature of the channel. The plan geometry both of meandering and of braided rivers has been subjected to fractal analysis, but without producing a great deal of enlightenment.

Controls of alluvial channel pattern

Channel-forming discharge is a single discharge measure that can be assumed to represent the range in flood discharges that is responsible for the steady-state geometry of alluvial channels. Channel-forming discharge is commonly taken as bankfull discharge. However, bankfull discharge may be difficult to define in nature because the bankfull level cannot always be measured accurately. The bankfull level has variously been taken as the elevation of the adjacent floodplain surface, the edge of vegetation, and that corresponding to the minimum channel width/depth ratio. Therefore, a frequency-based measure of flood discharge may be desirable, such as the mean annual flood or the flood with a recurrence interval of, say, 1–2 years.

Channel patterns are controlled by channel-forming discharge and by the amount and type of sediment (defined by grain size, shape, and density) supplied during flood stages. The width-integrated sediment transport rate, i_w, depends on the stream power per unit channel length, $\rho g Q S$, where ρ is the water density, g is the gravitational acceleration, Q is the water discharge, and S is the bed- and water-surface slope. However, the sediment-transport rate also depends on the grain types available for transport and the flow resistance associated with channel roughness (including bed waves such as bars and dunes). The stream power, grain type, and channel roughness are not independent of each other. For example, the mean grain size of bed sediment is commonly proportional to the channel slope, and the channel roughness is closely related to grain size and stream power. In general, an increase in QS for a given type of sediment supply will result in an increase in sediment-transport rate, and a change in channel pattern (e.g., an increase in the degree of braiding). The sediment-transport rate must equal the sediment-supply rate for steady-state channel geometry (i.e., in dynamic equilibrium). If the supply of sediment to a steady-state channel is increased in time or space, aggradation will occur and the channel pattern will change (e.g., there might be an increase in braiding index) even if water discharge, valley slope, and grain size do not change. Conversely, if the sediment-supply rate is reduced, degradation will occur and the degree of braiding may decrease.

The continuum of steady-state channel patterns has been represented empirically on plots of channel-forming discharge, Q_f, versus valley slope, S_v (Figure 13.17). Valley slope is used as a surrogate for the rate of sediment transport, i_w, because the sediment-transport rate is very

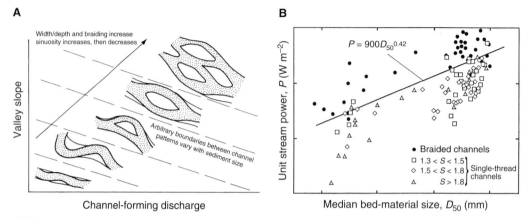

FIGURE 13.17. (A) Gradual variation of equilibrium channel patterns with channel-forming water discharge, valley slope, and sediment size. (B) Prediction of channel patterns as a function of stream power per unit bed area and bed material size; $S =$ sinuosity. After Van Den Berg (1995).

difficult to measure in rivers. Since $i_w \propto Q_f S_v / D_m$, an increase in slope for a given water discharge and mean grain size (D_m) corresponds to an increase in width-integrated sediment-transport rate. Thus, the transitions between the various channel patterns shown in Figure 13.17 must depend on the mean grain size of the sediment load as well as the channel-forming discharge and valley slope. The degree of braiding and the width/depth ratio of channels increase with $Q_f S_v$ for a given sediment size. Braiding occurs at lower $Q_f S_v$ as the sediment size decreases. The sinuosity of channels increases with their width/depth ratio and sediment size for single-channel streams, but decreases with width/depth ratio for multiple-channel rivers. Van Den Berg (1995) effectively defined the transition from meandering to braided rivers as a function of stream power per unit bed area and the median grain size of bed sediment (Figure 13.17). The discrimination between channel patterns using discharge, slope, and bed-material size is imprecise, partly because of inconsistent definitions of channel-forming discharge, slope, and channel pattern.

A common myth is that *discharge variability* is greater for braided rivers than for single-channel rivers. This myth probably originated from the early studies of proglacial braided rivers in mountainous regions of North America, where discharge varied tremendously during snowmelt. In contrast, many single-channel rivers were studied in temperate lowland regions where discharge variations were moderated by groundwater supply. In fact, discharge variability does not have a major influence on the existence of the various channel patterns, because they can all be formed in laboratory channels at constant discharge, and many rivers with a given discharge regime have along-stream variations in channel pattern.

Another common myth is that rivers that transport large amounts of bed load relative to suspended load have relatively low sinuosity and a high degree of braiding (Schumm, 1977; Miall, 1996). Such *bed-load streams* have been associated with relatively easily eroded banks of sand and gravel, large channel slope and stream power, such that they are *laterally unstable*. In contrast, rivers with relatively large suspended loads were postulated to be characteristic of undivided rivers of higher sinuosity. Such *suspended-load* streams were associated with cohesive muddy banks, low stream gradient and power, and *lateral stability*. This myth probably arose because early studies of braided rivers were carried out in mountainous areas of sandy–gravelly outwash and those of single-channel sinuous streams were performed in temperate lowlands (e.g., the US Great Plains). In fact, many braided rivers carry large amounts of sand and silt in suspension (e.g., Brahmaputra in Bangladesh, the Yellow River in China, Platte in Nebraska), and many single-channel, sinuous rivers carry sands and gravels as bed load (Madison in Montana, South Esk in Scotland, Yukon in Alaska).

Vegetation helps stabilize cut banks and bar surfaces, given adequate time and conditions for development. Partially lithified bank sediments (e.g., those with calcretes or silcretes) may have an effect on bank stability similar to that of vegetation. Such bank stabilization allows the existence of relatively steeply cut banks and may hinder lateral migration of channels. However, there is no conclusive evidence (contrary to popular belief) that vegetation or early cementation

Plan

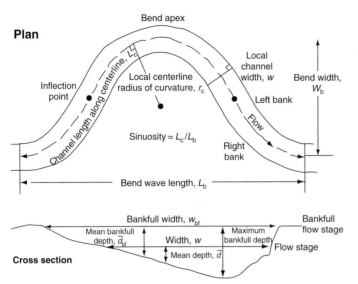

Bend apex

Inflection point

Channel length along centreline, L_c

Local centreline radius of curvature, r_c

Local channel width, w

Bend width, W_b

Left bank

Flow

Sinuosity = L_c/L_b

Right bank

Bend wave length, L_b

Bankfull width, w_{bf}

Mean bankfull depth, $\bar{d}_{bf}$

Width, w

Maximum bankfull depth

Bankfull flow stage

Flow stage

Mean depth, $\bar{d}$

Cross section

FIGURE 13.18. Definition of geometrical parameters of single curved channels. Most geometrical parameters vary with the flow stage, and parameters such as width, mean depth, and radius of curvature vary along the length of the channel. From Bridge (2003).

has a significant influence on the steady-state channel pattern, as long as the flood flow is capable of eroding banks and transporting sediment. Indeed, all channel patterns can be produced in the laboratory in the absence of vegetation. Low-powered streams might not be capable of eroding banks and transporting appreciable amounts of sediment, thus allowing vegetation to encroach into the channel (e.g., streams in lowland swamps). This is probably one reason for the existence of straight alluvial channels. However, straight channels are rare compared with sinuous and braided channels.

Theoretical *stability analyses* predict that channel pattern is controlled by channel width, depth, slope, and Froude number; see Bridge (2003) for a review. Theoretically, the major control of braiding is w/d (being >50 for braiding to occur). Unfortunately, width and depth are not independent variables, but depend on the supply of water and sediment. Several theoretical analyses of the controls of steady-state channel patterns involve the concept of *minimization of energy expenditure* (or maximum efficiency) in transporting sediment; see Bridge (2003) for a review. With these approaches, water discharge, sediment supply, and valley slope are considered to be independent (given) variables and the dependent variables (flow velocity, channel slope, width, depth) are calculated using the equations of motion of water and sediment. To predict the steady-state channel pattern, the assumption of minimization of stream power per unit channel length (or minimum S for a given Q) is introduced. The basis for this assumption, and the methods of discriminating among different channel patterns, cannot be justified in general; see the

discussion in Bridge (2003). However, these analyses do point to a fundamental requirement for the existence of sinuous channels: the steady-state channel slope must be less than the valley slope.

Geometry of alluvial channels at the compound-bar–bend scale

Alluvial channels are essentially a series of channel bends (curved channel segments), or bends joined by zones of channel convergence (*confluences*) or divergence (*diffluences*). Straight channel segments are rare in nature. Since channel geometry varies in time and space with variation in water flow and sediment transport, and evolves in time as individual channels develop and become abandoned, it is desirable to measure channel geometry over a range of spatial and temporal scales. Variability of the width and depth of individual channel segments will be discussed first, followed by variability of the geometry of channel bends, as defined by bend wave length, radius of curvature, sinuosity, and bend width (Figure 13.18). Then the geometry of confluences and diffluences will be discussed.

Cross-sectional geometry of curved channel segments

Bankfull channel geometry associated with the main-channel patterns is illustrated in Figure 13.19. Cross sections of curved channel segments are asymmetrical (triangular in shape), whereas cross sections of straight channel segments are symmetrical (trapezoidal in shape). The locus of maximum depth (*talweg* or

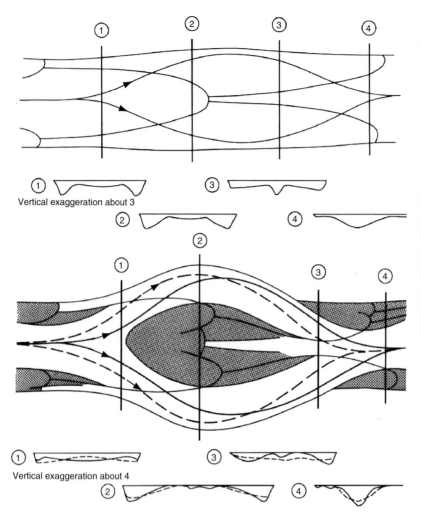

FIGURE 13.19. Idealized channel geometry for three simple channel patterns. Equivalent alternate bar patterns in straight channels are shown (upper diagrams) for comparison. Bankfull cross sections are smoothed (no mesoforms or cross-bar channels are shown) and have vertical exaggerations by factors of approximately 3 or 4. Broken lines in cross sections represent the low-flow-stage geometry. Solid and broken lines with arrows represent loci of maximum flow velocity for the high- and low-flow stage, respectively. The single arrow and E represent potential locations of flow diversion and erosion in response to encroachment of the tributary bar (riffle) of the major channel. Modified from Bridge (2003).

thalweg) occurs near the outer (cut) bank of a curved channel segment and near mid channel in the diffluence and confluence regions of a braided channel. In curved channel segments, the lower part of the cut bank is inclined at the angle of repose, but the upper part may be almost vertical. The mean inclination of the inner (depositional) bank may be up to 10°, and increases as the radius of curvature/width ratio decreases (i.e., as the sinuosity increases), as the bed-sediment size decreases, and as the flow stage increases. The smoothed depositional banks shown in Figure 13.19 would in reality have superimposed bed forms (e.g., dunes, unit bars) and perhaps slope discontinuities. Discontinuities may occur where permanently vegetated upper-bar surfaces pass into lower, unvegetated surfaces. Lower-bar platforms may be produced by rapid erosion of the cut bank and deposition on the lower bar instead of over the whole bar surface.

At the entrance and exit to channel bends, the radius of curvature is infinite, and the cross section is approximately trapezoidal in shape (the bed is almost flat except at the banks). However, the middle of the channel might be slightly elevated (i.e., a *riffle*). In single-channel rivers, this zone is called the *crossover*, a term used by riverboat operators. In order to avoid running aground, riverboat operators prefer the deepest water near the outer banks of bends. At the end of a bend, it is necessary to *cross over* the river towards the outer bank of the next bend along the river course. The mean bankfull depth at crossovers is close to the mean bankfull depth of the whole curved channel segment. This depth is on average about 60% of the bankfull thalweg depth in the bend.

As discharge decreases after floods, deposition near the upstream tips of point bars and braid bars tends to move the talweg towards the outer banks, resulting in reversal of the cross-sectional asymmetry of the

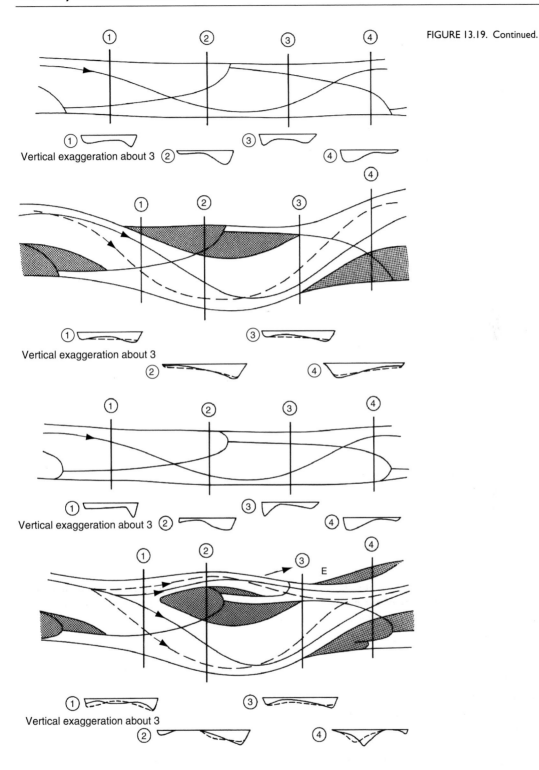

FIGURE 13.19. Continued.

channels in this region (Figure 13.20). In most parts of curved channels, the across-stream (transverse) slope of the depositional bank tends to be reduced with falling discharge by deposition in the talweg and erosion of the topographically high areas of the bar and riffle. Such falling-stage modification in channel geometry depends on the ability of the flow to erode and deposit. This might not be possible if the bed becomes armored or shallow

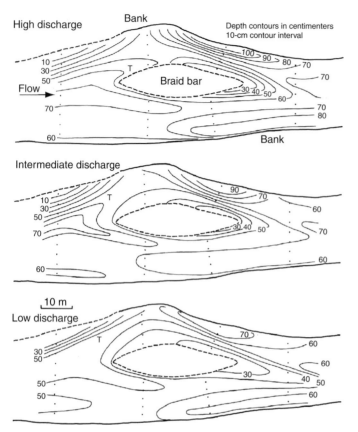

FIGURE 13.20. Channel geometry at high, intermediate, and low discharge associated with a braid bar, Calamus River, Nebraska. From Bridge and Gabel (1992). Depth contours are in centimeters, and contours are at 10-cm intervals. Dots are flow-measurement stations. The thalweg of the left-hand channel near the upstream end of the bar (T) moves towards the left bank as stage falls.

parts of channel segments become emergent. During the rising-flow stage, the opposite happens, with scouring of the talweg and deposition on bars and riffles.

At-a-station hydraulic geometry of channels

Channel width, mean depth, and radius of curvature may vary along the length of a curved channel segment, and among channel segments, at a given water discharge. For a given channel segment, some of these parameters will vary over time with discharge. There will also be temporal changes in channel geometry as individual segments form, evolve, and become abandoned. It is common practice to plot the logarithms of channel width w, mean depth d, and mean flow velocity U against the logarithm of water discharge Q for a given cross section or station (Figure 13.21), defining what is called *at-a-station hydraulic geometry*. Linear regression analysis of such data results in power functions of the form

$$w = aQ^l \tag{13.5}$$

$$d = bQ^m \tag{13.6}$$

$$U = cQ^n \tag{13.7}$$

in which the coefficients and exponents vary from section to section and from channel to channel. This variation arises because hydraulic geometry is not controlled solely by water discharge: it is controlled also by the grain size and transport rate of the sediment (that are related to flow resistance), and by the bank erodibility (that is related to cohesion of bank material and vegetation). For example, sandy and gravelly channels commonly have relatively large width/depth ratios, leading to small m and large l. In contrast, muddy channels commonly have relatively small width/depth ratios, leading to large m and small l. Hydraulic geometry also varies around a given curved channel segment. For example, as discharge increases and water spreads out onto the upper surfaces of point bars and braid bars, the channel width increases relatively rapidly, whereas the mean flow depth and flow velocity may stop increasing or even decline (Figure 13.21). These changes do not occur in crossover (riffle) sections, where the channel is more trapezoidal in section. In these sections, width increases relatively slowly with increasing discharge, whereas mean depth and mean

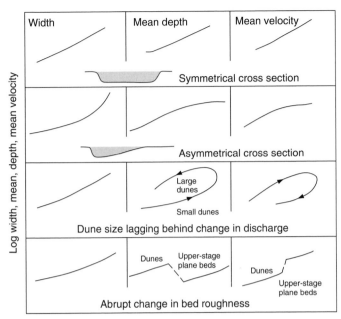

FIGURE 13.21. Variable shapes of at-a-station hydraulic-geometry curves.

velocity increase relatively steadily. In braided rivers, the hydraulic geometry of individual channel segments cannot be related to the whole-stream discharge because the discharges of the individual channels and the whole stream might not vary congruently.

The scatter of data points on linear hydraulic-geometry plots is normally large (Bridge, 2003), due to factors such as (1) measurement errors; (2) evolution of the channel geometry during the observation period; and (3) nonlinear relationships among logarithms of variables. Nonlinear depth–discharge curves and velocity–discharge curves arise because flow resistance varies with discharge (Figure 13.21). Discontinuities and loops in such curves are, respectively, due to abrupt changes in flow resistance with discharge and the fact that flow resistance lags behind changes in discharge (e.g., due to dune lag). Another problem with hydraulic-geometry equations is that they have dimensional coefficients, which limits their general application. Dimensionless parameters are much more desirable. The variability of linear hydraulic-geometry relationships among rivers, reaches of the same river, transects in a given reach, and channel segments is so great (for reasons listed above) as to be of little general predictive value.

At-a-station channel width, mean depth, mean flow velocity, and water-surface elevation (stage height) are routinely measured at gauging stations in order to calculate water discharge ($Q = wdU$), and to establish a statistical regression relationship between the water discharge and the stage height. Once the relationship between the discharge and the stage height has been established, the discharge is routinely calculated from the stage height rather than by repeatedly measuring channel width, depth, and flow velocity. However, the original discharge–stage relationship may be inaccurate if an inappropriate (e.g. linear) regression line was fitted to these data and the data scatter was large. Furthermore, the discharge–stage relationship can change through time if the geometry (including roughness) of the channel cross section changes. Such changes can be related to downstream migration of bars, erosion and deposition in the channel during exceptional floods, and human interference with the river. It is essential to have reliable discharge–stage relationships, because these are used to predict floodwater stages, and specifically whether artificial levees and flood walls will be overtopped during high discharges.

Spatial variations in hydraulic geometry of channels

The hydraulic geometry at a particular discharge can also be compared for different stations along one river, and for stations along different rivers. The channel-forming discharge (bankfull or flood discharge) is normally chosen for this purpose, although other discharge measures have been used. If the hydraulic geometries of rivers are to be compared, the data should come from similar types of cross sections, such as crossover (riffle) sections. As with at-a-station hydraulic geometry, spatial variation in hydraulic geometry is commonly represented by log–log plots of bankfull channel width, mean depth, and flow velocity against bankfull discharge. Linear regression of these data result in power equations such as

$$w_{bf} = xQ_{bf}^{0.5} \tag{13.8}$$

$$d_{bf} = yQ_{bf}^{0.4} \tag{13.9}$$

$$U_{bf} = zQ_{bf}^{0.1} \tag{13.10}$$

in which the subscripts refer to bankfull conditions. The exponents vary somewhat from the values given (but must always sum to 1, by virtue of continuity), and the proportionality coefficients vary with the other controlling variables, as discussed previously (Knighton, 1998). There have been many empirical and theoretical efforts to include these other controlling variables in the downstream hydraulic-geometry equations. There is still much scope for development of empirical hydraulic-geometry equations that contain all of the important controlling variables and that are based on a large sample of rivers worldwide.

The bankfull width/bankfull depth ratio generally increases with the discharge of water and sediment (see Equations (13.8) and (13.9)), and therefore as the channel pattern changes from straight single-channel via sinuous single-channel (meandering) to braided. For a given discharge of water and sediment, the bankfull width/bankfull depth ratio also increases as the grain size of the bed and banks increases. Therefore, empirical regression equations relating bankfull width and bankfull depth (reviewed by Bridge (2003)) depend upon the water and sediment discharge (and hence channel pattern) of the rivers used in the sample. This is one of the reasons for the large standard errors of the regression equations.

The so-called *regime approach* to predicting the geometry of alluvial channels was used by engineers who were designing irrigation canals in India and Pakistan. These canals had sand beds and cohesive muddy banks, and were therefore subject to erosion and deposition. The design problem was to determine the width, depth, and slope of a straight steady-state (regime) channel (no net erosion and deposition) with a trapezoidal cross section, given the water discharge and valley slope. If necessary, the channel slope could be reduced in stepwise fashion using drop structures (dams). The channels would remain straight as long as the Froude number was less than 0.3. The regime

equations express width, depth, and slope as power functions of the water discharge and the nature of bed and bank materials. Needless to say, such empirical equations were applicable only to the channels from which data were derived.

Empirical and theoretical studies of single, sinuous (meandering) channels indicate that the channel width, bend wave length, and mean radius of curvature are proportional to the square root of the channel-forming discharge. Thus the bend wave length/width ratio is commonly in the range 10–14. The riffle-to-riffle spacing in rivers is approximately half of a bend wave length, and is commonly in the range of 5–7 channel widths. The radius of curvature/bankfull width ratio generally lies in the range 2–3, although it can be considerably larger. Ratios between 2 and 3 are apparently associated with maximum flow resistance and bank-erosion rate. If the ratio falls below about 2, the flow tends to separate from the inner bank. These relationships among channel width, bend wave length, and mean radius of curvature seem to apply also to braided river segments. The curvature of channels on either side of braid bars is commonly such as to give braid-bar length/maximum bar width ratios of 3–4. In many rivers, the lengths of compound bars and bends do not change appreciably as discharge falls seasonally below the channel-forming discharge. However, if a channel segment experiences a long-term reduction in discharge, a series of bars and bends that have a shorter wave length than the original may develop. Many sinuous channel belts with wave lengths larger than their constituent bends are probably related to such long-term reduction in discharge.

Channel sinuosity is a complicated function of water and sediment supply. It appears to increase with water and sediment discharge from straight single channels to sinuous single channels, but then decreases towards the transition to braiding (Figure 13.17). The sinuosity of single-channel rivers with a given discharge increases with valley slope to some limiting value, and then decreases abruptly near the transition to braiding. In general, an increase in valley slope is associated with an increase in sediment-transport rate. Changes in valley slope may be related to changes in sediment supply and/or to tectonism. Rivers with the same sinuosity might have different radii of curvature, bend width, and length, and plan-form asymmetry (Figure 13.22). If the distance between upstream and downstream inflection points and the point of minimum radius of curvature is the same, the bend is called symmetrical; otherwise it is asymmetrical (Figure 13.22). A bend

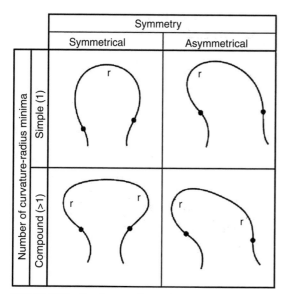

FIGURE 13.22. Classification of meander shapes; r marks minima in radius of curvature, and dots are points of inflection. Modified from Allen (1982a).

with only one minimum radius of curvature is called simple, whereas one with more than one minimum is called compound (Figure 13.22). Meander bends apparently tend to become more asymmetrical with increasing sinuosity. Compound meander bends may be associated with increasing channel length and sinuosity, or with superposition of reduced discharge (hence smaller bends) on a formerly larger one.

Experimental and field studies of single, sinuous (meandering) channels show that the (*meander*) *bend width* (incorrectly termed meander amplitude by some) tends to change with time as bends develop. For example, an initially straight channel will increase in sinuosity and bend width to a limiting value, after which the meander will either migrate downstream at constant width or experience cut-off. This behavior explains why bend width correlates poorly with discharge or other controlling variables. The *channel-belt width* is generally larger than the maximum width of meander bends if there has been a long history of bend development, migration, and cut-off in an area. The channel-belt width has been related empirically to the bankfull channel width and depth, but the empirical regression equations have a large standard error, related partly to variations in channel pattern (Bridge, 2003). The channel-belt width is linearly related to channel width, but the coefficient of proportionality depends on the maximum sinuosity and wave length of bends. Since the channel pattern is related to sediment supply and bank resistance, the relationships

among channel-belt width, channel width, and depth vary with the nature of bank materials.

Geometry of confluence zones

Figure 13.23 illustrates the geometry of confluence zones. Important features are (1) the confluence angle; (2) whether the incoming channels are oriented symmetrically or asymmetrically relative to the confluence direction; (3) the relative widths, depths, flow velocities, and discharges of the incoming channels; (4) complications arising from more than two joining channels; (5) the maximum depth, width, and length of the confluence scour zone; and (6) the nature of the side bars adjacent to the scour zone. These features are difficult to define because they vary with time and discharge. Confluence angles commonly range from 15° to 110°, and generally increase as the sinuosity of the joining channels increases. The relative discharge is the difference between the incoming discharges, and typically changes with time as the upstream channels change. Braided channels entering confluence zones tend to be riffle areas (Figure 13.19), so the depth does not vary much across the channel width. However, with the plan geometry shown in Figure 13.19, a talweg is commonly present near the outer bank during the high-flow stage, but may be near the inner bank during the low-flow stage. Avalanche faces may be present where the channels enter the confluence scour zone if the confluence angle exceeds 20° (i.e., tributary-mouth bars, chute bars: Figure 13.23). Generally, their crestlines are oblique to the channel direction, and this obliquity increases with cross-sectional asymmetry and flow stage, and if one channel becomes dominant over the other. The avalanche face of the dominant channel may be almost parallel to its inner bank, and may migrate into the minor channel, leading to an increase in erosion of the outer bank of the minor channel, or an upstream diversion. Where a cross-bar channel joins a larger main channel, only one avalanche face may be present, associated with the mouth bar of the cross-bar channel (Figure 13.23). Tributary-mouth bars may grow forwards, increase in height, and change in crest orientation, as discharge increases. Their crests tend to be eroded at falling flow stages, and may become dissected, resulting in a complex confluence zone (Figure 13.23).

The orientation and geometry of confluence scour zones are influenced by the confluence angle and the relative discharge of the joining channels. As the confluence angle increases, the scour zone changes from trough-shaped to more basin-like. If the discharges of

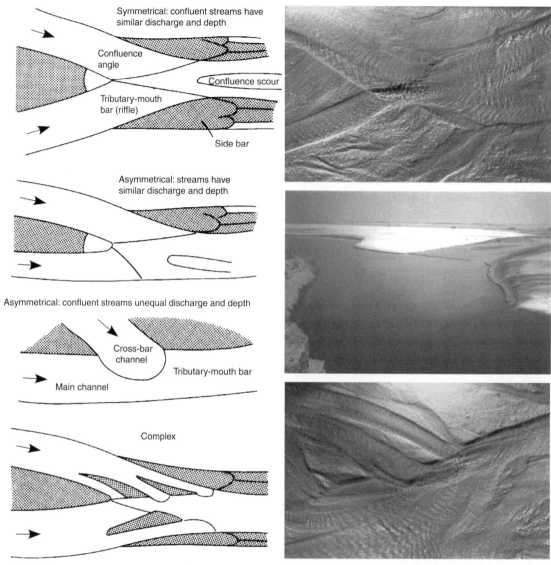

FIGURE 13.23. Idealized plan geometries of various types of confluences. From Bridge (1993a). The top photo is an example of a symmetrical confluence, and the bottom photo is an example of a complex confluence. Photos (courtesy of Peter Ashmore) are from an experimental study of braiding. The middle photo is an example of an asymmetrical confluence with a cross-bar channel terminating in a tributary mouth bar, Brahmaputra River, Bangladesh.

the entering channels are similar, the long axis of the scour tends to bisect the confluence angle. If one channel is dominant, the scour zone tends to be parallel to the direction of this channel. The maximum depth in the confluence scour zone is commonly up to four times the average depth of the incoming channels. This maximum scour depth increases nonlinearly with the confluence angle for a given relative discharge, and, for a given confluence angle, the maximum scour depth increases as the discharges in the entering channels tend to equality. If one channel is overwhelmingly dominant, the maximum scour depth is equivalent to that in a single-channel bend, and the dependence on the confluence angle is lost. The maximum scour depth may also be related to the cross-sectional asymmetry of the entering channels and to the presence of multiple entering channels.

As confluence angles and relative discharges change with flow stage, so will the orientation and geometry of the scour zone. The confluence angle may decrease as the flow stage increases, and the confluence scour zone may move downstream. Thus, at high-flow stages, deposition may occur in the upstream end of the confluence scour zone, whereas at low-flow stages these high-stage deposits are eroded. If one confluent channel becomes dominant, the maximum scour depth may be decreased and the scour zone moved towards the outer bank of the subordinate channel, possibly inducing bank erosion. The position of the scour zone will also shift in response to erosion of the outer banks of both entering channels and concomitant deposition near the downstream tip of the upstream braid bar. The confluence scour will also change in orientation and geometry if there is an avulsive shift of an upstream channel.

The confluence scour zone passes laterally into side bars that are analogous to the downstream parts of point bars in single curved channels. However, flow separation about vertical axes is more likely to occur over these side bars than over point bars. There is only one bar present if the confluence is asymmetrical (one straight-through channel; Figure 13.23). Longitudinal ridges of sediment on these bars, running parallel to the edges of the scour zone, result from episodic deposition, and are dynamically equivalent to bar-tail scrolls.

The widths and cross-sectional areas of the single channels downstream of confluences are generally less than those of the entering channels combined, and mean flow velocities are greater. Lyell (1830) explained this in the case of tributary junctions as due to there being less flow resistance in the single channel, since it has a smaller wetted perimeter than the two entering channels.

Geometry of diffluence zones

Downstream of confluence zones, the maximum depth decreases and, in braided rivers, the channel commonly widens and splits around a braid bar. In straight unbraided channels, a lobate unit bar would occur in this mid-channel position (Figure 13.19). Migratory unit bars also occur in this diffluence zone, possibly related to episodic bank erosion upstream. If the diffluence angle is large and/or one of the dividing channels is dominant, there is a tendency for the upstream side bar to build into the subordinate channel from the outer bank, thus tending to block its entrance.

Classification of geometry of channel cross sections

The geometry of channel cross sections has traditionally been related to the bed and bank materials. Such materials influence bank erodibility, flow resistance, and the ability of the flow to modify channel geometry. For example, Schumm (1977) classified channels on the basis of the percentage of silt and clay in the channel boundary, width/depth ratio, and sinuosity. Rosgen (1994, 1996) adopted a similar approach and classified channel cross sections on the basis of the size of sediment forming the perimeter of the channel, the entrenchment ratio (flood-prone width/bankfull width), sinuosity, width/depth ratio, and water-surface slope. Although this classification is widely used, it is unsatisfactory because (1) the different types of channel cross section are not mutually exclusive; (2) the categories are not defined objectively; and (3) the along-stream variation in channel geometry is not recognized. Others have classified channel geometry depending upon whether the boundary is bedrock, gravel, or sand (e.g., Howard, 1987: Parker, 1978a, b), on whether the channel width and depth are comparable to the size of the boundary material or much larger than it (Church, 1996), and on whether the boundary materials are cohesive or non-cohesive.

Theories for geometry of alluvial channels

Minimum-energy theories

Leopold and Langbein (1962) postulated the tendency towards minimum rate of energy expenditure per unit channel length, which is the same as uniform distribution and minimum variance of stream power per unit length ($\rho g Q S$). Since Q increases downstream in a river system with tributaries, this postulate implies that S must decrease downstream. Leopold and Langbein (1962) also postulated a tendency for uniform distribution of stream power per unit bed area ($\rho g Q S/w$), which would tend to reduce the degree of downstream decrease in S demanded by the previous postulate. The resulting channel form would therefore be a "quasi-steady" state between these two tendencies, while obeying the basic hydraulic laws (Langbein and Leopold, 1964). Langbein (1964) further showed that, in fulfilling these energy-distribution requirements and hydraulic laws, there is minimum variance of the components of stream power (w, d, U, and S), such that no single variable absorbs a disproportionate share of the variation. This minimum-variance theory was applied to the hydraulic geometry of channels, with claims to success.

The minimum-variance theory was also applied to the plan form of alluvial channels. Bagnold (1960) proposed that the common value of r_m/w of around 2.3 was that at which the flow resistance is *minimum* within the channel, suggesting some kind of energy-conservation principle. It has also been claimed that the mean depth, velocity, and slope are adjusted so as to decrease the along-stream variance of flow resistance in river bends compared with that in straight channels (Leopold and Langbein, 1966). This is apparently manifested in a more uniform water-surface slope in meandering channels than in straight channels. However, such claims are not generally supported by hydraulic data from rivers, and it has been suggested by others that the plan form of rivers is associated with *maximum* flow resistance.

The theory of minimum variance was applied to river meanders by Leopold and Langbein (1966) and Langbein and Leopold (1966), who maintained that "meanders are the result of erosion–deposition processes tending toward the most stable form in which the variability of certain essential properties is minimized." The most probable path taken by a meandering channel is taken as that where the sum of squares of changes in channel direction per unit length is minimized. This path is a *sine-generated curve* given by

$$\phi = \omega \sin(2\pi s/M) \qquad (13.11)$$

where ϕ is the deviation angle from the mean down-valley direction, ω is the maximum value of ϕ, s is the distance along the path, and M is the total path distance (Figure 13.24). This equation fits the plan shapes of symmetrical bends very well, but it is too regular to describe a whole series of river meanders. Furthermore, the sine-generated curve cannot describe the shape of asymmetrical and compound meanders.

Chang (1988) used the assumption of minimization of stream power per unit channel length (or minimum S for a given Q) to predict channel width and depth in addition to channel pattern. The discharge of water and sediment and the median sediment size are taken as independent (specified) variables. Then, equations for bed-load sediment transport and flow resistance are used to calculate channel width, depth, and slope. The steady-state channel width and depth are assumed to be those for which the channel slope (and flow resistance) is a minimum. The steady-state width and depth are those intermediate between small width/depth ratio (large bank resistance) and large width/depth ratio (large bed resistance). Chang's theory agrees well with a limited amount of empirical data. Chang (1988) also applied minimum-stream-power theory to channel geometry in meanders. However, he invoked some dubious assumptions, and the predicted widths, depths, and radii of curvature do not agree very well with observations. Other related approaches to

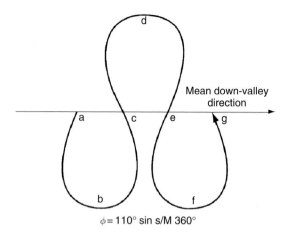

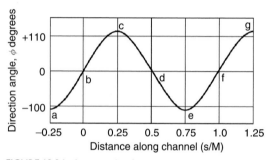

FIGURE 13.24. An example of a sine-generated curve in plan (upper) and variation of direction angle with distance along the channel (lower). After Langbein and Leopold (1966).

predicting channel geometry include minimizing the rate of expenditure of potential energy per unit mass of water, maximizing the sediment-transport rate, minimizing the Froude number, and a probabilistic entropy-maximization approach. Deficiencies of these "extremal hypotheses" for predicting channel geometry are reviewed by Knighton (1998) and Bridge (2003).

Channel-stability theories

Theoretical stability analysis has been successful at predicting the formation, geometry, and migration rate of alternate bars. Theoretical stability analyses have also predicted the stability conditions for single sinuous and braided rivers, and in some cases have yielded relationships for bend wave length. The bend wave length is proportional to the square root of the discharge, but agreement between theory and field data is only modest.

Hydraulic theories

Hydraulic theories for steady-state channel width and depth are based on the equations of motion for water

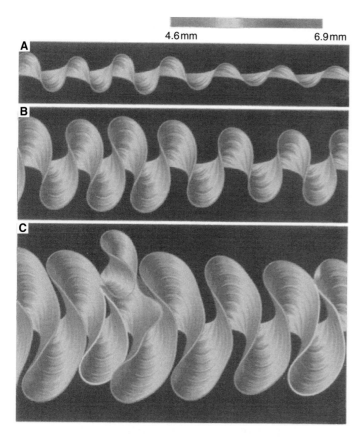

4.6mm 6.9mm

FIGURE 13.25. The meander-belt model of Sun *et al.* (2001c). The shade scale is for the grain size of surface sediment. The maps indicate meander-belt development after (a) 2,762 years, (b) 3,263 years, and (c) 4,000 years.

and sediment, and on the condition that the bed can be mobile while the banks are immobile (non-eroding). In the case of streams with sandy and silty boundaries, non-eroding banks were presumed by Parker (1978a) to be associated with diffusion of suspended sediment from the main channel to the banks. However, in real sand-bed rivers, sediment movement is intimately associated with secondary currents and spatial variations in flow, and stable banks are commonly associated with cohesive sediment. In the case of gravel-bed rivers, Parker (1978b, 1979) assumed that gravel banks will remain stable if the bed shear stress in the channel is only a little more than the threshold of sediment motion. In fact, bed shear stress commonly greatly exceeds this threshold for sediment movement, as is evidenced by common bed forms such as dunes and bed-load sheets. It is not surprising, therefore, that Parker's theories do not agree well with natural river data. However, more refined models of this type for channels with mobile beds but stable banks agree quite well with observations (e.g., Vigilar and Diplas, 1997, 1998; Cao and Knight, 1998).

The plan geometry of meander bends has also been treated analytically, by using hydraulic models for the variation of flow velocity and depth in bends, plus empirical equations for the bank-erosion rate; see the review by Bridge (2003). Despite deficiencies in the bank-erosion equations (see later), these models predict realistic plan forms (e.g., Figure 13.25). The modeled meander bends are symmetrical (close to sine-generated curves) for relatively low sinuosity, and become asymmetrical at high sinuosity. The lateral and downstream growth rate of meanders is initially rapid but slows with time as the bend width increases.

Flow in alluvial channels at the compound-bar–bend scale

Measurement of the spatial and temporal variation of flow velocity and direction in natural rivers is very difficult, requiring specialized equipment and deployment procedures (see the appendix). Such studies are particularly difficult in large rivers in flood. As a result of this, many studies of river flow have been performed in laboratory channels. Flow patterns in alluvial channels at the scale of compound bars and bends are revealed after averaging in time over turbulence, and averaging in space over small-scale bed forms such as

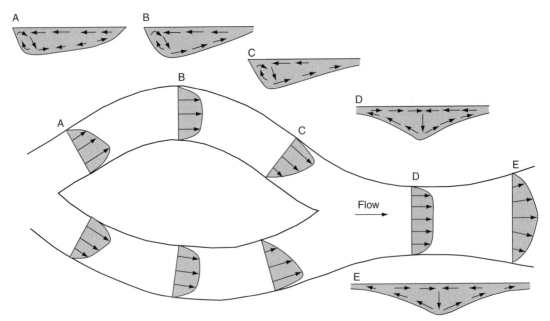

FIGURE 13.26. A simplified representation of the pattern of near-bankfull water flow for curved channels of similar geometry adjacent to a braid bar (sections A–C), and the downstream confluence (section D) and diffluence (section E) regions. Flow patterns associated with unit bars and smaller bed forms are not included for simplicity. Arrows on the map represent vectors of depth-averaged, downstream flow velocity. The cross sections show the channel geometry and flow-velocity vectors (near the surface and bed) for the cross-stream flow. Cross-stream flow velocities are typically an order of magnitude less than downstream flow velocities. These flow patterns and the channel geometry change with the flow stage. From Bridge (2006).

ripples and dunes. Flow patterns at channel-forming flow stage are discussed first, followed by those in changing flows.

Flow in straight channel segments

Straight alluvial channels are rare in nature, but studies of flow in straight laboratory channels have revealed important three-dimensional, depth-scale and width-scale turbulent flow structures that are probably responsible for initiation of dunes and sand ribbons, and the various modes of alternate bars (see Chapter 5).

Flow in curved channel segments

Most river channels are curved in plan. The flow in single curved channels (i.e., meander bends) is broadly equivalent to the flow in curved channels around braid bars. Curved flow around any type of channel bar results in (1) a cross-stream (transverse) component of water-surface slope towards the center of curvature and an associated cross-stream flow pattern (referred to as secondary, spiral, or helical); and (2) convective acceleration and deceleration of the depth-averaged downstream (primary) flow associated with bar topography (Figure 13.26). The secondary flow pattern arises primarily because of an imbalance through the flow depth of the curvature-induced centrifugal force and the pressure gradient associated with the cross-stream sloping water surface (Figure 5.16). Cross-stream flow velocities are generally an order of magnitude less than downstream flow velocities. In addition, spatial variation in the depth-averaged downstream flow velocity results in a net outward cross-stream flow in the upstream segments of bends (cross section A in Figure 13.26) and a net inward cross-stream flow in the downstream segments (cross section C in Figure 13.26), and these cross-stream flows modify the curvature-induced cross-stream (secondary) flows. These interacting flow patterns cause the maximum depth-averaged flow velocity to cross from the inner convex bank at the bend entrance towards the outer concave bank with progression around the bend (Figure 13.26). The bed shear stress tends to vary in a similar way to the depth-averaged flow velocity. However, because the flow resistance tends to be controlled mainly by the local bed configuration (e.g., ripples, dunes) rather than by bar topography, the bed shear stress does not always have a simple relationship with the depth-averaged flow velocity.

As the radius of curvature of a channel bend decreases (bed sinuosity or "sharpness" increases) the following flow parameters increase: (1) the magnitude of the local maximum depth-averaged velocity; (2) the deviation of the depth-averaged velocity vector from the mean channel direction; (3) the magnitude of the cross-stream flow components; and (4) the cross-stream slopes of the bed and water surface. In low-sinuosity channels, the cross-stream flow components are one or two orders of magnitude less than the downstream flow components, and the maximum angle of deviation of the local flow direction from the mean downstream direction is only a few degrees, making it difficult to measure. In high-sinuosity channels, the cross-stream flow components are much greater relative to the downstream flow, and deviation angles of local flow directions from the mean downstream direction may be up to tens of degrees. At bend entrances, the inward-directed near-bed cross-stream flow due to flow curvature tends to be counteracted by the outward-directed depth-averaged flow, such that there is an overall outward flow, increasing in magnitude from the bed upwards (Figure 13.26, cross section A). In the downstream parts of bends, the inward-directed depth-averaged flow adds to the inward-directed cross-stream flow at the bed. Similar changes in flow parameters also occur as the sediment-transport rate increases, or the bed roughness decreases.

Immediately adjacent to steep cut banks, weak secondary circulation commonly occurs in the opposite sense to the main secondary circulation (Figure 13.26), and is a corner-flow structure that results from turbulence anisotropy (discussed in Chapter 5). Also, flow-separation vortices with vertical axes occur near the inner bank of the downstream parts of channel bends with small radii of curvature ($r_c/w < 2$). Such flow separation is caused by adverse pressure gradients associated with curvature-induced cross-stream water-surface gradients in combination with a reduced downstream water-surface slope resulting from local flow expansion and deceleration over the downstream part of a bar. At the crossover sections between single-channel bends, secondary flow is weak but may comprise a double spiral with bottom flow converging towards the topographically elevated channel center.

All of these flow patterns change character with changing discharge. The depth-averaged flow velocity, bed shear stress, water depth, and width generally decrease as the discharge decreases. Water flows in a more sinuous course around emerging compound bars as the discharge decreases, resulting in relatively strong cross-stream components of depth-averaged flow near the bend entrance, an increase in magnitude of secondary flow, and movement of the maximum depth-averaged flow velocity from the inner to the outer bank over a relatively shorter distance. In braided rivers, the relative discharges of the various channels commonly change with flow stage, particularly as relatively shallow channels become inundated at high-flow stage or abandoned at low-flow stage. New flow patterns may arise at high-flow stages as new channels are cut, particularly where high flow velocities take advantage of local topographic depressions. The relatively high water level on the outside of a curved channel may lead to spilling of water over a low section of bank, and possibly the development of a new channel. Falling-stage flow may also be diverted into low areas of the emerging bar, leading in some cases to incision of existing or new cross-bar channels.

Flow in confluence zones

Confluences occur at the downstream ends of braid bars, as cross-bar channels join main channels, and as tributaries join trunk streams. Confluence flow has been compared to the flow in joining river bends and to wall jets (because the cross-sectional area of the joined channels is less than the sum of the two joining channels). The flow in confluences (Figures 13.26 and 13.27) comprises (1) the entrance zone; (2) the confluence mixing zone, where the streams join asymptotically and mix; and (3) zones of flow separation at the upstream tip of the confluence and associated with side bars adjacent to the confluence scour. Flow in these zones depends at least on the confluence angle and the relative discharges of the joining channels, both of which vary with the flow stage.

Entrance zones are equivalent to the riffle or cross-over zones of single curved channels, and tributary-mouth bars (including chute bars) may be present. At high-flow stage, the depth-averaged velocity over the mouth-bar crest is expected (depending on the confluence geometry) to be greatest near the outer bank, and there should be a cross-stream component of flow towards the center of the confluence throughout the flow depth, since curvature-induced secondary flows are negligible (Figure 13.27). However, because the crestline of the mouth bar is oblique to flow direction, the separated and re-attached flow downstream of the crestline will have a component of near-bed flow towards the outer bank, whereas the surface flow will continue to be directed towards the center of the

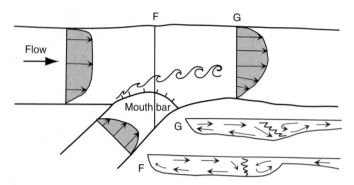

FIGURE 13.27. A simplified representation of the pattern of near-bankfull flow for a confluence in which a relatively small channel with a tributary mouth bar joins a larger channel. Symbols are as for Figure 13.26. Based on the work of Jim Best and Andre Roy. The photo (courtesy of Jim Best) shows Kelvin–Helmholtz vortices developed at the interface between confluent streams, one having a high concentration of suspended sediment compared with the other.

confluence (Figure 13.27). During low-flow stage, shifts in the maximum-velocity loci may result in changes in mouth-bar crest height and orientation, and dissection of the mouth-bar crest may produce a complex, multi-channel flow. The distance from the crest of the mouth bar to the reattachment zone is expected to be proportional to the height of the avalanche face, thus changes in the height and orientation of the mouth bar will influence the geometry of the scour zone associated with the re-attaching boundary layer.

The curvature and convergence of the joining channels in a confluence zone should inevitably give rise to super-elevation of the water surface in mid channel, and secondary flow with near-bed flow components towards the outer banks (Figure 13.26). The water-surface super-elevation and the magnitude of the secondary-flow components tend to increase as the confluence angle increases (or as the radius of curvature of joining channels decreases). By analogy with flow in single curved channels, the maximum high-stage velocity should occur near the center of the confluence scour, with lower velocities near the outer banks (Figure 13.26). The analogy with flow in back-to-back meanders is valid for symmetrical confluences; however, the flow pattern in confluences is also dependent upon (1) flow acceleration associated with a decrease in cross-sectional area; (2) flow separation downstream from mouth-bar crests; (3) inequality in the depths of confluent channels; and

(4) the enhanced turbulence of the mixing layer between the joining streams (Figure 13.27).

The mixing zone between the two joining streams results from their velocity differential, and can commonly be seen clearly if the suspended-sediment loads of the two streams differ (Figure 13.27). The relative velocity leads to Kelvin–Helmholtz-type vortices with near-vertical axes that become entrained in the faster flow. The turbulence intensity in the mixing zone is considerably greater than that in either adjacent stream. If one stream is slower near the bed than the other (for example, due to flow separation in the lee of a mouth bar, or inequality in the velocity or depth of the joining streams), the mixing layer is distorted near the bed towards the slower stream (Figure 13.27). This may result in upwelling of the fluid from one stream into the region occupied by the other, greatly affecting the mixing of sediment from the joining streams, and enhancing curvature-induced secondary flow.

Zones of flow separation with vertical axes of flow rotation may occur near the downstream tips of braid bars and over the side bars adjacent to confluence scour zones. These separation zones are most marked for large confluence angles (high channel sinuosity). In asymmetrical confluences, the length and width of the separation zone increases with the confluence angle and the discharge of the tributary relative to the main channel.

During rising-flow stage, a partially abandoned channel segment may receive water from its downstream junction with an active channel, i.e., water flows upstream into the channel. This situation also occurs in ephemeral rivers where water may start to flow in only part of the river system, resulting in flow of water from an active tributary up the inactive tributary from their confluence. Thus, paleocurrents that are oriented opposite to the down-valley direction in ancient river deposits may be associated with flow separation, flows up inactive channels, or tidal influence. Distinguishing among these alternatives requires careful analysis.

Flow in diffluence zones

Water flow in channel diffluences (upstream tips of braid bars or where distributary channels split) is poorly known. In the upstream parts of diffluence zones in braided rivers, the maximum-velocity locus is in mid channel, and splits downstream such that each thread of maximum velocity is close to the upstream tip of the braid bar downstream. Relatively complicated patterns of convergence and divergence of depth-averaged velocity associated with bar-scale bed

topography occur in this zone. Flow throughout the depth tends to diverge away from trough areas, and to converge towards bar crests. With large diffluence angles, zones of flow separation with vertical axes may occur near the outer banks of the channel entrances. At present, it is not possible to predict the steady-state geometry and hydraulics of the diffluent channels, or the flow conditions that lead to abandonment of one diffluent channel in favor of the other. It is possible that compound-bar migration into one diffluent channel may cause an obstruction that diverts more discharge into the other diffluent channel. Laboratory and field studies addressing these problems are in progress.

Sediment transport in alluvial rivers at the compound-bar–bend scale

Sediment transport in alluvial rivers varies in time and space due to turbulence, the movement of bed waves (e.g., ripples, dunes, unit bars, point bars, and braid bars), random changes in sediment supply (e.g., bank slumping, break-up of armored or cemented beds), local channel cutting and abandonment, weather-related changes in discharge and sediment supply, and tectonically induced changes in sediment supply. Such variation occurs over time intervals of seconds to many years, and over lengths ranging from millimeters to kilometers. Detection of progressive or periodic changes in sediment transport therefore depends on the sampling interval and extent in both time and space. Methods of measuring sediment transport are discussed in the appendix. The mean grain size and rate of sediment transport over compound bars, averaged in time over turbulence, and averaged in space over relatively small-scale bed forms such as ripples, dunes, and unit bars are discussed here. The bed forms that move over compound bars will also be discussed.

Rate and mean grain size of sediment transport

Both the bed-load transport rate and the mean grain size over sandy point bars and braid bars during the channel-forming flow stage increase with the depth-averaged flow velocity and bed shear stress (Figure 13.28). All of these parameters are largest near the inner bank at the upstream end of a bar and near the outer bank at the downstream end. However, for streams with a substantial gravel fraction, the locus of the maximum bed-load transport rate stays close to

A

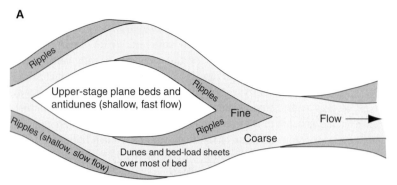

B

FIGURE 13.28. (A) A schematic diagram of the distribution of bed-load grain size and sub-bar-scale bed forms in sandy and gravelly rivers at the bankfull flow stage. Ripples occur only in sands with grain diameter less than about 0.7 mm. The boundary between coarse and fine sediment is actually gradational. From Bridge (2006). (B) Downstream fining of bed material on the exposed part of a point-bar surface. The plane gravel bed passes downstream into rippled sand.

the center of the channel even though the locus of maximum grain size follows that of the bed shear stress. This is because the largest gravel sizes (upstream near the inner bank, downstream near the outer bank) tend to be close to their threshold of motion, such that their transport rate is low. The spatial variation in mean grain size of bed load over compound bars increases as the cross-stream slope of the bar increases, which increases as the radius of curvature/ width ratio decreases (sinuosity increases) and the sediment-transport rate increases. The average-sized bed-load grains tend to travel approximately parallel to the channel margins, implying a balance between the cross-stream downslope gravity force and upslope near-bed fluid force on these grains. Bed material normally fines downstream on the tops of both braid bars and point bars in modern sandy and gravelly rivers, and is relatively coarse-grained in the talwegs of the downstream segments (Figure 13.28). Thus, the spatial distribution of the mean grain size of the bed material

reflects the distribution of bed-load grain size and bed shear stress at high (channel-forming) discharges, as would be expected.

As the discharge changes, the changing patterns of bed shear stress, bed form, and available bed material result in complicated patterns of transport rate and mean grain size of the bed load. The lower bed shear stress during the low-flow stage generally results in smaller grain sizes and transport rates of bed load, or bed-load transport may cease. Armor layers can develop in gravelly sediments. Bed-load grains tend to move preferentially towards the outer banks as the stage falls, because the cross-stream downslope component of the gravity force acting upon them is greater than the upslope near-bed fluid force. Such downslope movement results in deposition and reduction of the cross-stream bed slope. Limited deposition during the low-flow stage results in veneers and patches of relatively fine-grained sediments, and possibly filling of openwork gravels with a sandy matrix.

FIGURE 13.29. Dunes preserved on the upper part of a point bar (Congaree River, South Carolina, USA) following a bankfull flood. These dunes are approximately 0.2 m high and 4 m long. Medium-scale trough cross strata are exposed in the trench. The scale in the trench is 0.15 m long, and the trench is about 0.75 m deep.

In confluences, the largest bed-load transport rates also generally occur where the flow velocity and bed shear stress are largest, provided that the bed is not armored. Bed-load particles travel more or less parallel to the channel banks and bed contours as they pass through the confluence scour zone, implying that bed load from the two joining channels experiences little mixing as it passes through the confluence. There is a tendency for bed-load grains to move downslope into the scour zone during falling stages, upslope out of the scour zone during rising-flow stages, and more or less parallel to the scour zone during channel-forming flow stages. The largest mean grain sizes occur in the base of the scour zone, whereas the finest mean grain sizes occur immediately upstream of the scour zone and near the banks adjacent to the downstream end of the confluence.

Systematic, detailed observations of the temporal and spatial variations of suspended-sediment load in rivers are rare. In general, suspended-sediment concentrations are relatively large where the bed shear stress and turbulence intensity are large and where the bed-material size is small. Suspended-sediment concentrations are very difficult to predict in confluence zones in view of the zones of mixing, upwelling, and flow separation.

Bed forms on bars

The variation of sediment-transport rate and bed-sediment size in alluvial channels is reflected in the bed forms superimposed both upon unit bars and upon compound bars. Dunes with curved crestlines are the most common bed forms on sandy unit bars, point bars, and braid bars during the high-flow stage (Figure 13.29). Upper-stage plane beds occur locally in shallow areas of high flow velocity. Dunes are also common in gravelly streams during the high-flow stage, as are lower-stage plane beds (with bed-load sheets, pebble clusters, and sand ribbons). Dune length and height vary locally with flow depth, and dune crestlines are commonly oblique to the local channel direction, due to across-channel variation in bed shear stress, bed-load transport rate, and dune height. Ripples in sand are normally restricted to areas of slow-moving water near banks. Antidunes and transverse ribs occur only rarely in fast, shallow water.

At flow stages lower than bankfull, dunes are generally shorter and lower, and the proportion of curved-crested dunes decreases relative to that of ripples and lower-stage plane beds. Dune geometry is commonly not in equilibrium with rapidly changing flow stage, particularly in shallow water, where dunes can become exposed by small decreases in water level.

The emergence and dissection of bars during falling-flow stages, and the modification of flow patterns, result in a diverse orientation of the superimposed bed forms. Dendritic channel systems that develop from groundwater flowing down the local slopes of emerging bars may terminate in small deltas. Desiccation cracks occur in local areas of mud deposition (e.g., in bed-form troughs and chute channels), and the encroachment of animals and plants onto the emerging bar surfaces results in root casts, burrows, and trails.

Erosional bed forms in channels

Erosional marks occur in the talwegs of channels cut into cohesive mud: e.g., flutes, gutter casts, rill marks, and pot holes. Crescent-shaped erosional marks also occur on bars on the upstream side of obstacles such as ice blocks, log jams, and lumps of bank material.

Theoretical steady-state water flow, sediment transport, and bed topography in alluvial channels

There are many theoretical models for the water flow in river bends, but not as many consider the interaction among water flow, sediment transport, and bed topography; see the review by Bridge (2003). These models differ primarily in the degree of simplification of the equations of motion of fluid and sediment. The most complete (and physically enlightening) models are complex and it is difficult to solve their mathematical equations. Furthermore, most of these models have not been tested against extensive data sets from natural rivers or laboratory flumes, so it is very difficult to assess their general validity. Bridge (1992) developed a simple, approximate theoretical model for the inter-action among steady water flow, bed topography, and the transport rate and mean grain size of bed load in river bends. This model was compared with a large range of data from natural rivers and laboratory channels, and performed well (Bridge, 2003). The most sophisticated model for water flow, sediment transport, grain-size sorting, erosion, and deposition in river bends is that of Sun *et al.* (2001b, c), but it has not yet been tested extensively against real-world data. Theories for the interaction among water flow, sediment transport, and bed topography in braided rivers are not well developed, but notable attempts have been made by Bridge (1993a) and by Murray and Paola (1994, 1997).

Erosion and deposition at the compound-bar–bend scale

Processes of erosion and deposition

Erosion and deposition at the compound bar–bend scale are due to (1) adjustments of bed topography associated with varying discharge; (2) bank erosion and associated compound-bar deposition, occurring mainly during the high-flow stage; and (3) cutting of new channels, enlarging of existing channels, and filling of others. During the rising-flow stage, erosion tends to occur in bend talwegs, confluence scours and the upstream ends of compound bars, whereas these areas receive deposits during falling stages (Figure 13.30). In contrast, the highest parts of bar tails tend to be areas of deposition during the high-flow stage, with erosion as the stage falls. Such adjustments in bed topography are normally associated

with episodic bank erosion and deposition of the eroded sediment on downstream compound-bar margins. Such bar migration during the high-flow stage is normally by lateral and downstream accretion, and may involve episodic accretion of distinct unit bars. The upstream ends of compound bars are sites of erosion during the high-flow stage, but may be depositional as the stage falls. Channel migration is also associated with cutting of new channels, enlargement of existing channels, and abandonment and filling of others. This is because unit bars commonly migrate into and block channel entrances, and low areas adjacent to unit bars can accommodate diverted discharge.

Erosion and deposition within channels can be explained using the two-dimensional form of the sediment continuity equation:

$$-C_b\, \partial h/\partial t = \partial i_s/\partial s + \partial i_n/\partial n + \partial i_s/(u_{ss}\, \partial t) + \partial i_n/(u_{sn}\, \partial t)$$
$$\quad (1) \qquad\qquad (2) \qquad (3) \qquad (4) \qquad\qquad (5)$$
$$(13.12)$$

where i_s and i_n are channel-wise and across-channel components of the volumetric sediment-transport rate, and u_{ss} and u_{sn} are corresponding mean velocities of sediment grains. Commonly, terms (4) and (5) are much smaller than terms (2) and (3), and can be ignored. If the bed topography is in equilibrium with a steady (high) flow, there is no erosion and deposition, term (1) is zero, terms (2) and (3) are equal and opposite in sign, and terms (4) and (5) are zero (when averaged over turbulence and bed forms smaller than bars). However, as discharge in a curved channel decreases, the cross-channel bed slope becomes too steep for the reduced discharge, and the cross-channel downslope gravity force on the bed load exceeds the upslope-directed fluid drag. This force inequity results in an imbalance between terms (2) and (3), erosion of the top of the bar, and deposition in deeper areas (Figure 13.30). Imbalance between terms (2) and (3) can also occur in equilibrium, steady flows due to "random" events such as bank slumps and the passage of unit bars.

Erosion in talwegs near banks during floods leads to a steepening of the lower part of banks that may lead to slumping. The important factors governing bank erosion are the degree of basal erosion and steepening of the bank (increasing the downslope component of the gravity force on the bank material), the resistance of the banks to slumping, and the ability of the flow to remove the slumped bank material. The shear strength of bank material (see also Chapter 8) is given by

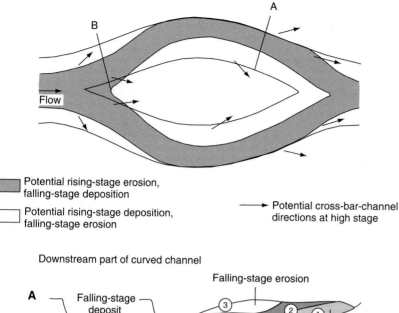

FIGURE 13.30. Theoretical locations of erosion and deposition in a simple braided-channel pattern during changing flow stage, and cross sections, (A) and (B), showing theoretical erosion and deposition during changing flow stage plus channel migration. (1) low-stage geometry; (2) flood-stage geometry; (3) flood-stage geometry following cut-bank erosion and bar deposition; and (4) final low-stage geometry. From Bridge (1993a).

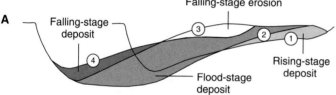

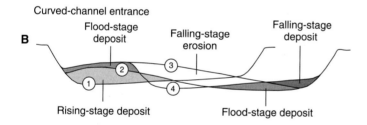

$$\tau_s = \tau_c + (mg\cos\alpha - L)\tan\phi_c \qquad (13.13)$$

where m is the sediment mass, g is the gravitational acceleration, τ_s is the shear strength, τ_c is the shear resistance due to cohesion (which is enhanced by soil cements, clays, and binding vegetation), the term in brackets is the effective force normal to the Earth's surface (equal to the normal weight component minus any upward-directed force such as that due to excess pore-water pressure), and $\tan\phi_c$ is the static friction coefficient. The excess pore-water pressure increases during falling-flow stages when water flows out of the banks. When the downslope component of the gravity force exceeds the shear resistance of the bank material, a slump occurs (Figure 13.31). Slumps are typically on the order of decimeters to meters wide and meters to tens of meters long, and occur intermittently

in time and space. Slumps may remain coherent or disintegrate and become grain flows. The slumped material may be entrained by the flow but, if not, it will contribute to armoring of the bed and protection of the bank against further erosion. Thus, the bank-erosion rate depends on the magnitudes of terms (2) and (3) of Equation (13.12) at the toe of the bank, that in turn must be controlled by spatial gradients of $\tau_o - \tau_c$ (the difference between the bed shear stress and that critical for entrainment of bed grains). The critical entrainment stress, τ_c, is large for large cohesive slumped blocks, thereby reducing the bank-erosion rate. In contrast, banks of cohesionless sand and gravel are relatively more easily eroded. Thus, the bank-erosion rate is high where the cohesiveness of the bank materials is low (lack of clay, vegetation, and cement), where the banks are wet and groundwater flows out of the banks during the

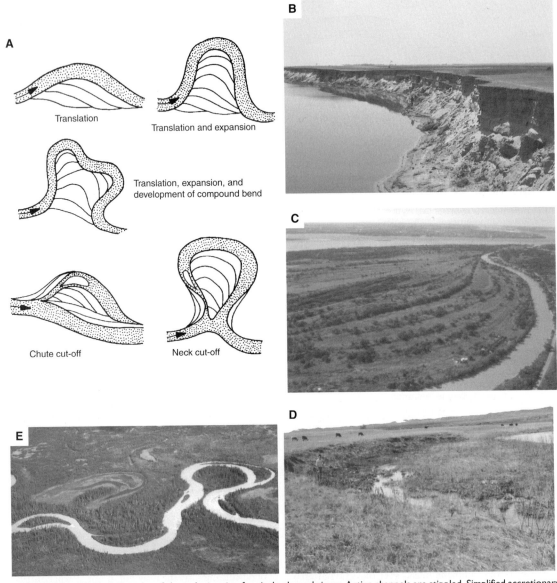

FIGURE 13.31. (A) Typical modes of channel migration for single-channel rivers. Active channels are stippled. Simplified accretionary units on point bars (separated by lines) are actually composed of unit bars (bar-head lobes and bar-tail scrolls). (B) Active bank erosion with slightly vegetated bank slumps disintegrating into grain flows. (C) Accretion topography on a point bar adjacent to a narrow, filling channel (photo courtesy of Jim Best). (D) A channel abandoned following chute cut-off, with a pond preserved in the downstream section (foreground). (E) Incipient neck cut-off (right) and examples of chute cut-off and abandoned meander loops.

falling-flow stage, where the sediment-transporting power of the river is large, and where there are large spatial gradients in near-bank fluid shear stress. In cold climates, freeze–thaw action can lead to an increase in bank-erosion rate, and undercutting of frozen banks by river erosion can enhance the degree of slumping. Water waves from strong winds and boat wakes also enhance bank erosion. Bank-erosion rates range up to the order of channel widths per year.

Cut-bank erosion and associated expansion (and deceleration) of flow into the eroded space will inevitably result in deposition on adjacent bar surfaces. Likewise, if deposition occurs on a bar due to a local increase in sediment supply from upstream, narrowing of the channel cross section may induce erosion of the adjacent bank. Such deposition may be associated with the accretion of distinct unit bars onto braid bars or point bars, but is not always. *Accretion topography* (ridges and swales) results from the episodic accretion of unit bars (Figure 13.31). A large amount of bank erosion during major floods may be associated with preferential deposition on the lower parts of bars (producing a platform) while the upper parts do not receive as much sediment.

Patterns of channel migration

In curved channel segments, the maximum flow velocity and bed shear stress impinge on the concave cut bank progressively further downstream as the discharge increases. Therefore, if most bank erosion is limited to the flood stage, it will occur mainly in the downstream parts of bends (Figure 13.31). If bank erosion also occurs during a relatively low-flow stage, it may occur along most of the length of the curved channel. In addition, as the radius of curvature of a curved channel decreases, the maximum velocity and bed shear stress increase and occur progressively further downstream. Therefore, the degree of talweg scouring (and hence bank erosion) increases as the radius of curvature decreases, and the zone of maximum erosion shifts downstream. This explains why, as a straight channel becomes more sinuous by bank erosion and bar deposition (called bend expansion), the rate of bank erosion normal to the valley direction decreases exponentially until some maximum steady-state sinuosity is attained. As the sinuosity increases, the downstream component of bank erosion increases relative to the component normal to the valley direction until the bend migrates only downstream (called bend translation). The steady-state width of a meander belt must therefore be related to the radius of curvature of the meander bends.

Theoretical models of bank erosion in single curved channels relate the bank-erosion rate to the near-bank flow velocity or depth. If the erosion rate is related to the near-bank flow velocity, the meander bends grow in width but most bend migration is downstream. If the bank-erosion rate is related to the near-bank flow depth, the width of bends increases rapidly and downstream migration is relatively small. The models predict that meanders are initially symmetrical (sine-generated curves) but become asymmetrical as the sinuosity increases (Figure 13.25). Most bank-erosion models do not consider flow unsteadiness or the unpredictable entrainment stresses of slumped blocks, and they normally contain empirical coefficients that must be calibrated with field data.

The curved channels adjacent to braid bars migrate in the same way as single channels adjacent to point bars.

Therefore, braid bars may migrate predominantly downstream, or may have a component of lateral growth in addition (Figure 13.32). Confluence zones commonly migrate downstream in response to downstream migration of entering channels, especially for symmetrical confluences of channels with similar geometry. The lack of room for across-stream migration of side bars in confluence zones may lead to the cutting of chute channels, and the downstream migration of side bars may lead to blocking of the entrance of a downstream braid channel. Lateral migration of confluences is possible in the case of asymmetrical confluences and those with entering channels differing in discharge and/or geometry.

Cutting and filling of channels

Cutting and filling of channels occur in single-channel and braided rivers, but the process is not well understood. The probability of *chute cut-off* increases as channel sinuosity increases. Channel diversion appears to be associated with high-stage scouring of upstream talwegs and deposition of the eroded sediment as obstacles in channel entrances downstream. Chute channels then take advantage of the low areas adjacent to inner, convex banks or between older unit bars on the surfaces of point bars and braid bars. These chute channels may then be enlarged at the expense of the original channel. Once the process of channel abandonment has been initiated, the channel is progressively filled with sediment from the upstream end. Blockage of the downstream end of the filling channel is caused by downstream migration of the downstream tip of the enlarging channel's bar (Figure 13.31). *Neck cut-off* occurs in single-channel rivers that are so sinuous that cut banks of successive bends can meet, leading to cut-off, abandonment of the meander loop and formation of an ox-bow lake (Figure 13.31). Mark Twain described in his book *Life on the Mississippi* how neck cut-offs isolate formerly valuable land (the cut-off point bar) and bring the river close to formerly inferior land, which increases drastically in value. Reduction in river length by neck cut-off was also a great advantage to riverboat captains. Thus, the temptation to lend a helping hand to the process of neck cut-off was irresistible to some ditch-digging scoundrels. Such scoundrels may have discovered a new meaning to the word neck cut-off if they were caught in the act! Cut-offs can be problematical if the river acts as a property boundary or political boundary (as is very common). Furthermore, the effects of the increased flow velocity and sediment-transport rate through the new channel can have detrimental effects downstream.

A

Translation with symmetrical channels

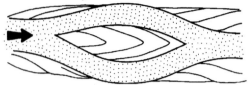

Translation and expansion with symmetrical channels

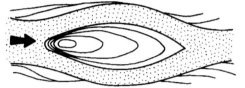

Translation and changes in relative discharge of channels

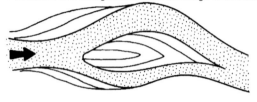

C

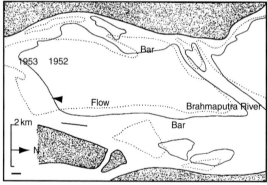

B

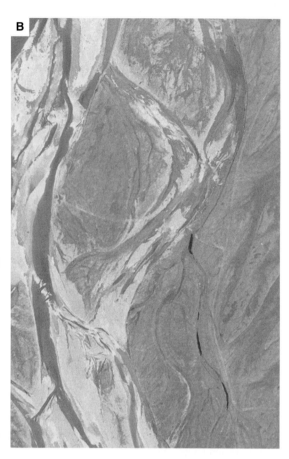

FIGURE 13.32. (A) Typical modes of channel migration for simple braided-river patterns. From Bridge (2003). Simplified accretionary units on braid bars (separated by lines) are actually composed of unit bars (bar-head lobes and bar-tail scrolls). The braid bar grows asymmetrically for cases in which discharge in one channel is increasing at the expense of the other channel. (B) The photo (Sagavanirktok River, northern Alaska) shows a braid bar with accretion topography, with the active channel to the left and the filling channel to the right. The filling channel (about 120 m wide) contains lobate unit bars. (C) Downstream migration of braid bars on the Brahmaputra River. From Coleman (1969).

Effect of riparian vegetation on flow and sedimentary processes in rivers

Riparian vegetation is that which borders stream channels. This vegetation, specifically the binding effects of plant roots, reduces bank-erosion rates and can stabilize point bars and braid bars such that it is difficult to erode their upstream ends; see reviews in Bridge (2003) and Bennett and Simon (2004). In addition, vegetation within the flow increases flow resistance (e.g., by inducing flow separation and enhancing turbulence intensity), reduces flow velocity and sediment-transport capacity, induces deposition, and inhibits erosion. The nature of these effects depends on the size, shape, flexibility, orientation, concentration (density), and degree of submergence of the vegetation. Vegetation has been used to restore meandering in artificially straightened channels. Fallen trees and log jams in rivers can have detrimental or beneficial effects. Log jams cause water flow to be diverted over, under, and around them, leading to bed erosion upstream, under, and adjacent to the log jam. Log jams can also cause diversion of the river. On the other hand, trees that have accumulated in the channel next to banks can reduce the flow velocity adjacent to the banks, encourage deposition, and protect the banks from erosion.

River engineering and restoration

River engineering

River engineering or training concerns the efforts made by humans to engineer river channels and floodplains in the interests of water-supply provision, prevention of bank erosion, maintenance of navigation channels, and flood control. River engineering involves bank protection; building of dykes (groins, spurs) into the channel; straightening, widening, and deepening of channels; grade-control structures; building of dams; and building of levees and floodways (Figure 13.33); see reviews by Chang (1988), Julien (2002), and Bridge (2003). Entertaining accounts of efforts to control the Mississippi are given by McPhee (1989) and Barry (1997).

Before undertaking any engineering work on a river, it is desirable to analyze the nature of water flow and sediment transport in the channel to be modified, and the effect that the engineering will have. This would normally involve use of a sediment-routing model, and possibly scale-model experiments. Then, it is necessary to decide whether the benefits of the proposed engineering work outweigh the (typically large) financial cost of planning, building, and maintaining the structure, the impact on freshwater ecosystems, and the damage to the natural beauty of the river. Engineering works commonly fail or have unanticipated detrimental effects because of insufficient prior analysis. For example, regulation of water flow and sediment transport downstream of dams can have a damaging effect on freshwater ecosystems. Breaching or overtopping of dams and artificial levees can have devastating effects on lives and property (a recent example being the effects of hurricane Katrina on New Orleans). The rapid routing of floodwater in modified channels bounded by artificial levees may result in an increase in the risk of flooding downstream. Inhibition of regular flooding of floodplains reduces recharging of groundwater and provision of soil nutrients carried by floodwater, and has a damaging effect on floodplain ecosystems. Further problems have been created by deforestation of floodplains and by implementation of questionable agricultural practices (e.g., excessive use of fertilizers and pesticides). Pollution of rivers, floodplains, and groundwater associated with urban development is also a pressing problem.

River restoration

River managers are turning increasingly to river and floodplain restoration, as a result of problems with previous attempts at river and floodplain engineering, and with agricultural and industrial development along rivers and floodplains. Restoration of rivers and floodplains normally enhances recreational possibilities and aesthetic beauty. River restoration ranges from complete to partial return to the pre-disturbance state. Techniques of river restoration include planning of use of land, surface water, and groundwater; replacement of specific freshwater biota (e.g., fish); restoration of natural flood events; introduction of fresh sediment; reinstatement of natural channel geometry; re-vegetation of channels and their banks; increasing communication between river and floodplain by changing bed levels and by removing obstacles to flow; livestock control; and creation of flood storage and wetland areas to control pollution and downstream flooding. Restoration projects have made extensive use of vegetation. As a result, there is renewed interest in understanding the interactions among vegetation, water flow, sediment transport, and the morphology of channels and floodplains. There is much work to do on this problem.

Depositional models for channel and bar deposits

The deposits of all Earth surface environments comprise various scales of strataset. In river deposits, the scales of the stratasets depend on the scale of the associated topographic feature (e.g., bar, dune), the time and spatial extent over which deposition occurred, and the degree of preservation. Four scales of strataset are shown in Figure 13.34: (1) a complete channel belt (e.g., a sandstone-conglomerate body); (2) deposits of individual compound channel bars and channel fills (sets of large-scale inclined strata, also known as storeys); (3) depositional increments (large-scale inclined strata) on compound channel bars and in channel fills formed during distinct floods; and (4) depositional increments associated with passage of bed forms such as dunes, ripples, and bed-load sheets (sets of medium-scale and small-scale cross strata and planar strata). An additional scale of strataset may also be recognized, because a large-scale inclined stratum deposited on a compound bar may be simple or compound if formed

FIGURE 13.33. Bank erosion and human activities. (A) Failed riprap and serious bank erosion along a creek in New York State. Remains of concrete riprap and bridge piers lie in the river. (B) Bank erosion undercutting a railway line in northwest Argentina. (C) River and roads moved during a flood in central America, but this bridge did not. (D) Bank protection on an Alpine river, Switzerland. Images (E) and (F) show riprap along the Brahmaputra River, Bangladesh (photos courtesy of Jim Best). Modified from Bridge (2003).

by migration of a unit bar (Figure 13.34). The geometry of the various scales of strataset is related to the geometry and migration of the associated bed form. In particular, the length/thickness ratio of stratasets is similar to the wavelength/height ratio of associated bed forms (Bridge and Lunt, 2006). Furthermore, the wave length and height of bed forms such as dunes and bars are related to channel depth and width. Therefore, the thickness of a particular scale of strataset (i.e.,

medium-scale cross sets and large-scale sets of inclined strata) will vary with river dimensions.

An understanding of how these different scales of strataset are distributed within channel belts requires detailed knowledge of the water flow, sediment transport, and bed geometry during erosion and deposition, and the pattern of channel migration. It also requires detailed observations of the three-dimensional distribution of channel deposits in modern rivers. Methods

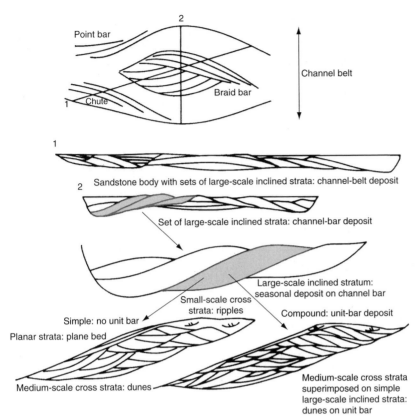

FIGURE 13.34. Different superimposed scales of river deposits, illustrated with cross sections through an idealized braided river. The cross sections (1) and (2) show sets of large-scale inclined strata (storeys) formed by deposition on the braid bar and the side bar. The shaded part of a storey (formed by incremental deposition on the side of the braid bar) is expanded below to show detail of large-scale inclined strata. A large-scale inclined stratum (shaded) can be simple (formed during a single flood) or compound (associated with deposition on a unit bar over one or more flood periods). Large-scale inclined strata contain smaller-scale stratasets associated with passage of dunes, ripples, and bed-load sheets over bars. From Bridge (2003).

of obtaining these data are given in the appendix. Sufficient data from modern rivers are now available to construct quantitative and qualitative, three-dimensional, depositional models. It is possible to construct generally applicable models of river deposits because there is remarkable similarity in the geometry, flow and sedimentary processes, and deposits of rivers of widely differing scale and grain size. For example, unit bars and dunes occur in nearly all river types during the high-flow stage, and the evolution and migration of different types of channels are similar. However, some differences between gravel-bed and sand-bed rivers are that there are more bed-load sheets and associated planar strata in gravel-bed rivers, increasing as the mean grain size increases; and more abundant ripples (small-scale cross strata) and upper-stage plane beds (planar strata) in sand-bed rivers, increasing as the mean grain size decreases. These differences are reflected in the depositional models below.

Quantitative depositional models

Quantitative, dynamic, three-dimensional depositional models of river-channel deposits have been developed

only recently, and such models are at a rudimentary stage (e.g., Willis, 1989; Bridge, 1993a; Sun *et al.*, 2001b, c). These types of models require prediction of the interaction among bed topography, water flow, sediment-transport rate, mean grain size of bed load, and bed forms within channels of prescribed geometry. The flow conditions are assumed to be steady and bankfull, with the bed topography, water flow, and sediment transport in equilibrium. The models apply to either single-channel bends with an associated compound point bar or two channel bends separated by a braid bar. The plan forms of the channels are sine-generated curves, and features such as unit bars and cross-bar channels are not considered. The channels migrate by bank erosion and bar deposition, and change geometry with time. Net vertical deposition is negligible over the time spans considered in the models. Despite being simplified, these models give important insights into the nature of channel deposits that did not come from earlier static one- and two-dimensional models. Examples of these models shown in Figures 13.35 and 13.36 illustrate a number of fundamentally important aspects of river-channel deposits.

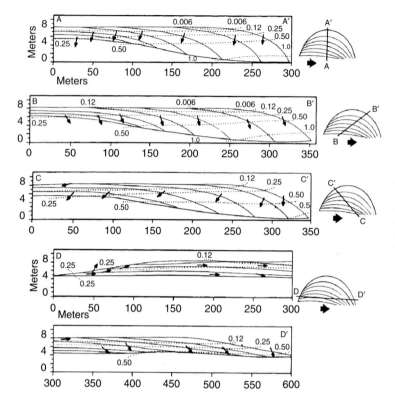

FIGURE 13.35. An example of a quantitative model of point-bar deposits (Willis, 1989). Meander plans to the right indicate downstream and lateral growth of a point bar in discrete increments, and positions of cross sections in various orientations. Cross sections indicate the basal erosion surface of point-bar deposits, large-scale inclined strata due to incremental deposition, contours of mean grain size (dotted lines annotated in mm), and current orientations relative to the cross section (arrows pointing down indicate flow out of the plane of cross section). Point-bar deposits thicken, and inclination of large-scale inclined strata increases, from left to right in sections (A)–(C).

(1) As channels migrate by lateral and downstream migration, the deposits from different parts of channel bars become vertically superimposed (e.g., bar-head deposits overlying bar-tail deposits, bar-tail deposits overlying confluence-scour deposits).

(2) Systematic spatial variations in the thickness of channel deposits, and the inclination and orientation of large-scale strata, are due to bed topography and the mode of channel migration. For example, it is common for channel-bar deposits to thicken (by up to a factor of two), and for large-scale strata to steepen, towards a cut bank (channel-belt margin) or confluence scour.

(3) Lateral and vertical variations in grain size and sedimentary structures are controlled by the bed topography, flow, sediment transport, and bed forms, and by the mode of channel migration. Channel-bar deposits normally fine upwards, but they also commonly exhibit little vertical variation in grain size. Some channel-bar deposits coarsen at the top if bar-head regions are preserved.

These models do not consider the somewhat complicated flow structures in channel diffluences and confluences, and in filling channels. They also do not describe erosion and deposition during unsteady flows.

There is therefore much scope for development of these quantitative models. Qualitative depositional models contain more details, but lack precision.

Qualitative depositional models

The qualitative depositional models shown in Figures 13.37–13.41 comprise (1) maps showing idealized active and abandoned channels, compound bars, and lobate unit bars; (2) cross sections showing large-scale inclined strata and their internal structures, associated with migration of compound bars, unit bars, and their superimposed bed forms; and (3) vertical logs of typical sedimentary sequences through various parts of compound-bar deposits and channel fills. The cross sections and vertical logs differ somewhat between gravelly–sandy rivers (Figures 13.38 and 13.39) and sandy rivers (Figure 13.40 and 13.41), and the cross sections differ between single channels and braided channels. The bar-head regions of the compound bars have formed by accretion of the fronts of lobate unit bars, and their bar-tail regions have formed by accretion of the sides of lobate unit bars (i.e., scroll bars) (Figure 13.37). Thus, compound-bar growth and migration has been mainly by lateral and downstream

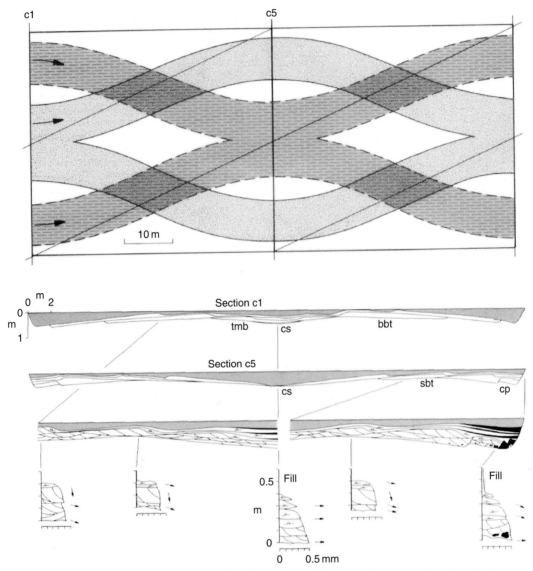

FIGURE 13.36. An example of a quantitative model of braid-bar deposits (Bridge, 1993a). The upper figure shows the plan geometry of initial braided channels (stippled) and migrated channels (dashed). The braid bar migrated downstream in four discrete increments. Cross sections show the basal erosion surface of bar deposits, large-scale inclined strata due to incremental deposition, and details of spatial variations in deposit thickness, grain size, sedimentary structure, and paleocurrents. The deposit thickness and inclination of large-scale inclined strata vary systematically. Bar sequences generally either fine upwards or have little vertical variation in grain size. The dominant internal structure in this example is medium-scale trough cross strata (formed by dunes), with subordinate small-scale cross strata (formed by ripples); cs = confluence scour, tmb = tributary-mouth-bar deposits, cp = counterpoint deposits, bbt = braid bar-tail deposits, and sbt = side-bar-tail deposits.

accretion. The abandoned channel in the braided-channel model is being filled with unit-bar deposits (Figure 13.37), and its upstream end was blocked by a compound point bar. The abandoned channel in the meandering-channel model (Figure 13.37) is filled with unit-bar deposits only at the upstream end, and the downstream end is a lake.

Channel-bar deposits

Figures 13.37, 13.38, and 13.40 show how the geometry and mode of migration of river channels and bars control the geometry and orientation of *large-scale inclined strata*. Large-scale strataset (storey) thickness in a channel belt can vary laterally by a factor of two or

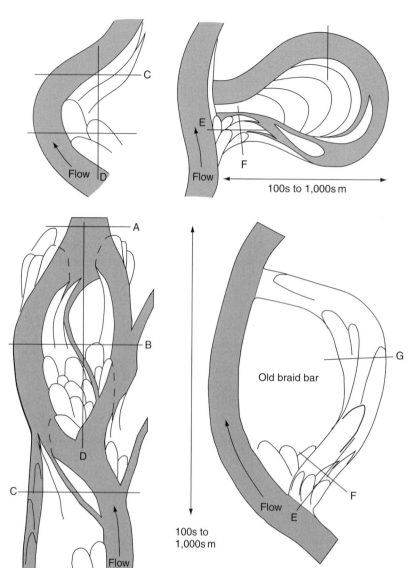

FIGURE 13.37. Qualitative models of channel deposits. Maps of meandering and braided rivers showing active and abandoned channels, compound bars, and some lobate unit bars. Stippled areas are filled with water during the low-flow stage, and unit bars within these areas are not shown. Bar heads of compound braid and point bars have formed by accretion of the fronts of lobate unit bars, and their bar tails have formed by accretion of the sides of lobate unit bars (i.e., scroll bars). Thus, compound bar growth has been mainly by lateral and downstream accretion. The upstream end of the abandoned braid channel was blocked by a compound point bar, and the channel is being filled with unit bar deposits. The upstream and downstream ends of the abandoned meandering channel were also blocked by bar deposits, but unit bars in the filling channel do not fill the entire abandoned channel. Cross sections and vertical sedimentary logs are shown in Figures 13.38–13.41. Cross sections with letters correspond to those in Figure 13.38. From Bridge (2006).

more. In places, large-scale stratasets thicken laterally as the large-scale strata increase in inclination. Some sets have large-scale strata inclined predominantly in one direction, whereas others have convex-upward or concave-upward stratal inclinations. The deposits of braided and unbraided rivers can be distinguished on the basis of these patterns of large-scale inclined strata in cross section. This distinction between channel patterns cannot be made from vertical lithofacies profiles, contrary to published opinions. The definitive depositional evidence for braiding in ancient deposits is cross sections through braid bars with adjacent, coeval channels and confluences. Examples of these patterns of

large-scale inclined strata from modern channel belts are shown in Figure 13.42.

The large-scale inclined strata shown in Figures 13.38 and 13.40 rarely have such systematic inclinations, and both discontinuities and discordances are common. Discontinuities in inclination may be associated with the occurrence of unit bars (discussed below) and lower-bar platforms (Figure 13.42). Discordances in large-scale inclined strata form through discharge fluctuations, shifts in channel position, and (as discussed below) in relation to the formation of cross-bar channels.

The number of large-scale inclined strata comprising most of the thickness of a set (storey) is commonly

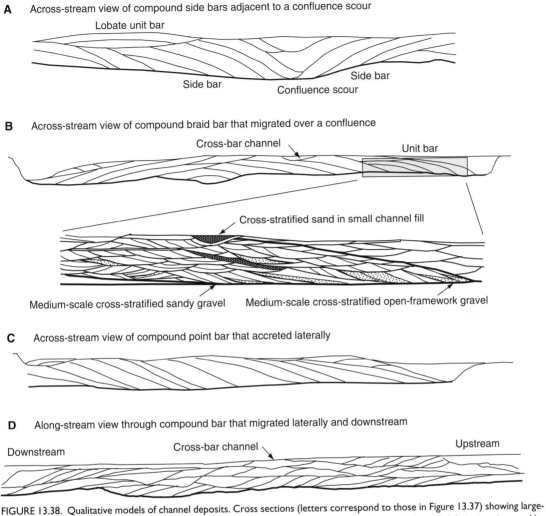

A Across-stream view of compound side bars adjacent to a confluence scour

Lobate unit bar

Side bar
Side bar
Confluence scour

B Across-stream view of compound braid bar that migrated over a confluence

Cross-bar channel
Unit bar

Cross-stratified sand in small channel fill

Medium-scale cross-stratified sandy gravel   Medium-scale cross-stratified open-framework gravel

C Across-stream view of compound point bar that accreted laterally

D Along-stream view through compound bar that migrated laterally and downstream

Downstream
Cross-bar channel
Upstream

FIGURE 13.38. Qualitative models of channel deposits. Cross sections (letters correspond to those in Figure 13.37) showing large-scale inclined strata, and their internal structures, associated with migration of compound bars, unit bars, and their superimposed bed forms for gravelly–sandy rivers. Cross sections are hundreds of meters to kilometers wide and meters thick. Vertical exaggerations are 5–10. From Bridge (2006).

between one and ten, depending on the rate of channel migration relative to channel-bar width. For example, if the channel migrates a distance equivalent to the apparent bar width during a single depositional event, the bar sequence will comprise a single large-scale stratum. If ten depositional episodes are required to migrate one bar width, ten large-scale strata will be formed. The amount of channel migration during a flood is commonly on the order of 10^{-1} times the channel width.

Downstream translation of bars results in preferential preservation of bar-tail deposits, and erosional truncation of bar-head deposits. Bar-tail deposits fine upwards (Figures 13.39 and 13.41), and the vertical range of mean grain size in such sequences increases with channel sinuosity. Bar sequences with little vertical variation in mean grain size occur where bend-apex deposits build over bar tails. Such sequences may coarsen at the top if the bar head migrates over bar-tail deposits. Bar-head deposits can only be preserved if the upstream part of the bar is not extensively eroded. Thus, different types of vertical sequence of lithofacies depend mainly on the position in the bar and on the mode of channel migration rather than on channel pattern. In some channel deposits, upper-bar deposits and lower-bar deposits can be distinguished by their differences in grain size and sedimentary structure, and by the more common presence of buried vegetation in upper-bar

E   Along-stream view through upstream end of large channel fill: lateral and downstream growth of compound bar

Upstream   Unit-bar deposit   Downstream

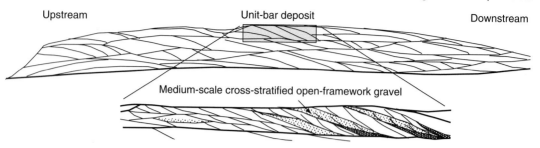

Medium-scale cross-stratified open-framework gravel

F   Across-stream view of upstream end of large channel fill: lateral accretion and channel filling

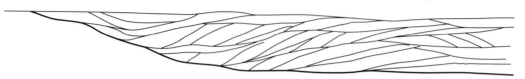

G   Across-stream view of downstream end of large channel fill: scroll-bar accretion and channel filling

Scroll bars

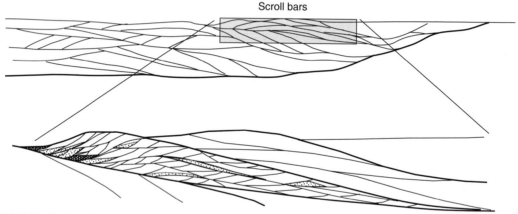

FIGURE 13.38. Continued.

deposits (Figures 13.41 and 13.42). Upper-bar deposits commonly increase in thickness in the down-bar direction, whereas lower-bar deposits decrease in thickness.

Individual large-scale inclined strata can be recognized by vertical changes in grain size and sedimentary structure: commonly they fine upwards at the top (Figures 13.39, 13.41, and 13.42C). The internal structure of large-scale inclined strata in sandy and gravelly braid bars and point bars is normally dominated by medium-scale trough cross strata, by virtue of the ubiquitous presence of curved-crested dunes on bar surfaces during the high-flow stage. Sets of medium-scale cross strata are commonly decimeters thick, that is an order of magnitude thinner than the channel-bar thickness. Gravelly deposits may have relatively more planar strata with imbricated clasts (formed by migration of

bed-load sheets on lower-stage plane beds). Planar strata formed by the migration of low-relief bed waves on upper-stage plane beds are common in the upper parts of sandy braid bars and point bars. Small-scale cross-stratified sand from ripple migration tends to be restricted to high areas near banks, but can occur in other positions in low-stage deposits. Climbing types of small-scale cross strata are common. Small-scale cross-stratified, bioturbated sand commonly occurs interbedded with vegetation-rich mud as centimeter-thick units in the upper-bar deposits near channel banks. Antidune cross stratification and transverse ribs occur rarely in the upper parts of sandy and gravelly channel bars, that is where flow is fast and shallow.

Large-scale inclined strata may be sheet-like, or mound-like if associated with deposition of unit bars

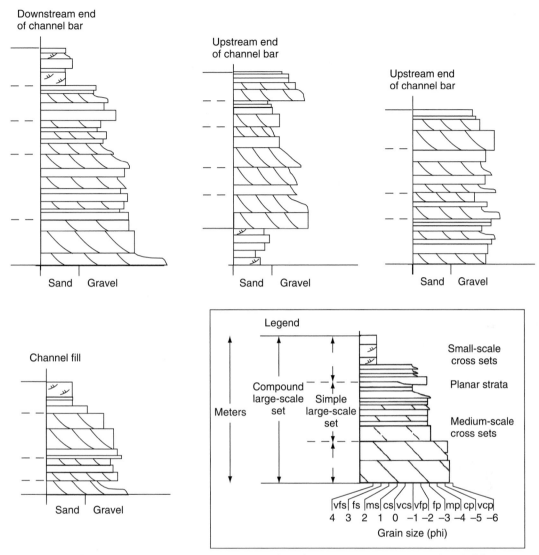

FIGURE 13.39. Qualitative models of channel deposits. Vertical logs of typical sedimentary sequences through different parts of compound bar deposits and channel fills for gravelly–sandy rivers. From Bridge (2006).

(Figures 13.37, 13.38, and 13.40). Unit bars commonly grow in height and migrate downstream by accretion of the deposits of the smaller-scale bed forms that migrate over them: bed-load sheets, dunes, or ripples. During their growth and migration, such unit bars are asymmetrical in along-stream profile with a relatively steep downstream side that is angled at less than the angle of repose. Therefore, cross-set bases and planar strata tend to be inclined at a relatively low angle (up to about 10°), reflecting the geometry of the unit bar over which they are migrating. The lee face of a unit bar may reach the angle of repose as it migrates, and superimposed bed forms would then halt at the crest of

the unit bar, from where their sediment avalanches. Therefore, the low-angle inclined set boundaries within unit bars pass laterally into angle-of-repose cross strata, defining a smaller scale of large-scale inclined strata than that associated with the compound bar on which the unit bar is superimposed (Figures 13.38, 13.40, and 13.43). The angle-of-repose cross strata associated with unit bars are commonly referred to as planar cross strata, and thought to be characteristic of braided rivers. In reality, unit bars and their deposits form in all river types, angle-of-repose cross strata are not normally the dominant internal sedimentary structure, and such cross strata look planar only in sections

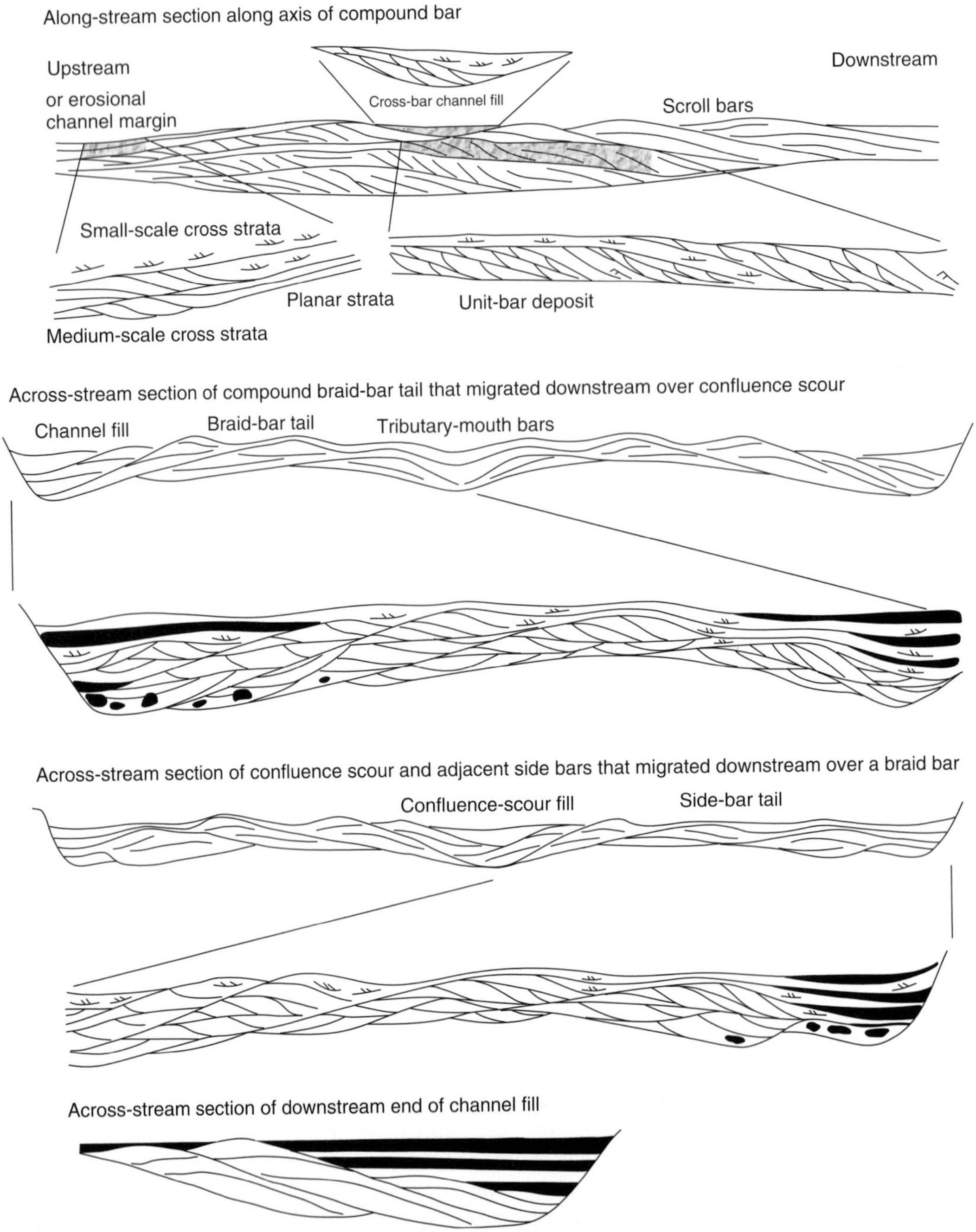

Along-stream section along axis of compound bar

Upstream
or erosional
channel margin

Cross-bar channel fill

Downstream

Scroll bars

Small-scale cross strata

Planar strata

Unit-bar deposit

Medium-scale cross strata

Across-stream section of compound braid-bar tail that migrated downstream over confluence scour

Channel fill   Braid-bar tail   Tributary-mouth bars

Across-stream section of confluence scour and adjacent side bars that migrated downstream over a braid bar

Confluence-scour fill   Side-bar tail

Across-stream section of downstream end of channel fill

FIGURE 13.40. Qualitative models of channel deposits. Cross sections showing large-scale inclined strata, and their internal structures, associated with migration of compound bars, unit bars, and their superimposed bed forms for sandy rivers. Cross sections are hundreds of meters to kilometers across and meters to tens of meters thick. Vertical exaggerations are approximately 5. From Bridge (2006).

Legend for sedimentological logs

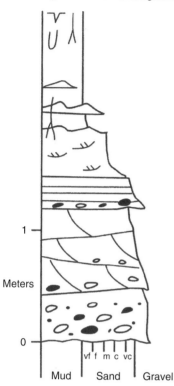

Mudstone with minor sandstone lenses, vegetation-rich layers, root casts, burrows, and desiccation cracks. Degree of bioturbation increases upwards.

Lenticular to wavy bedding with asymmetrical ripples

Small-scale cross strata (set thickness <0.02 m) with asymmetrical ripples

Planar strata

Medium-scale cross strata (Set thickness drawn to scale)

Sandy gravel-stone including intraformational mud clasts (black)

FIGURE 13.41. Qualitative models of channel deposits. Vertical logs of typical sedimentary sequences through different parts of compound bar deposits and channel fills for sandy rivers. From Bridge (2006).

that are small relative to the unit bar. Angle-of-repose cross strata formed by unit bars can easily be confused with those due to dune migration. Unit bar deposits associated with bar-head lobes and bar-tail scrolls tend to occur in the upper parts of compound bar deposits, whereas those associated with tributary-mouth bars occur nearer the base (Figure 13.40).

Relatively small channels cutting into the upper parts of channel bars, particularly between unit bars, are cross-bar channels. Cross-bar channels commonly develop their own bars, the geometry of which is controlled by the flow and sediment-transport conditions in these channels. Where a cross-bar channel joins a main channel, solitary delta-like deposits with avalanche faces commonly form (e.g., chute bars: Figure 13.43A). The maximum depth of cross-bar channels is commonly less than a third of the maximum depth of main channels, but any cross-bar channel could be enlarged to become a main channel (e.g., by chute cut-off).

Channel-fill deposits

The deposits of channel fills are dependent on the history of flow through the channel following the

beginning of abandonment. Abandonment is normally initiated by growth of a bar into the entrance to the channel. If the angle between the enlarging channel and the filling channel is relatively small, as in low-sinuosity rivers, flow is maintained in the filling channel so that bed load can be deposited (Figure 13.37). Such bed load is commonly moved as unit bars with superimposed ripples or dunes. Although bed load may extend a considerable way into such filling channels, the downstream ends will receive mainly fine-grained suspended sediment and organic matter from slowly moving water. With larger angles of divergence, both ends of the abandoned channel are quickly blocked (Figure 13.37) so that most of the channel fill is relatively fine-grained and organic-rich due to deposition from suspension in ponded water.

Channel fills generally fine upwards, reflecting progressively weaker flows during filling (Figures 13.39 and 13.41). They also generally fine down channel as water flow decelerates in that direction. The relatively coarse bed-load deposits at the upstream end of the channel fill tend to fine upwards, since they represent progradation of bar-tail deposits into the channel entrance (Figure 13.41). Bed-load deposits in channel

Downstream parts of channel bars

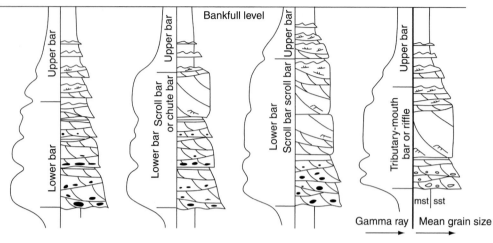

Upstream parts of channel bars

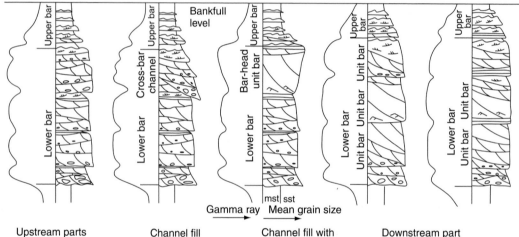

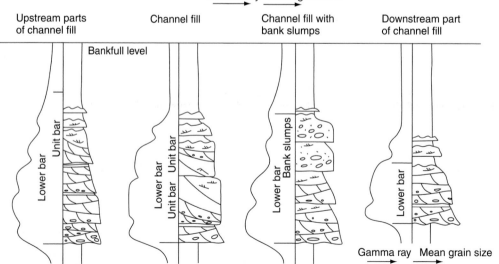

FIGURE 13.41. Continued.

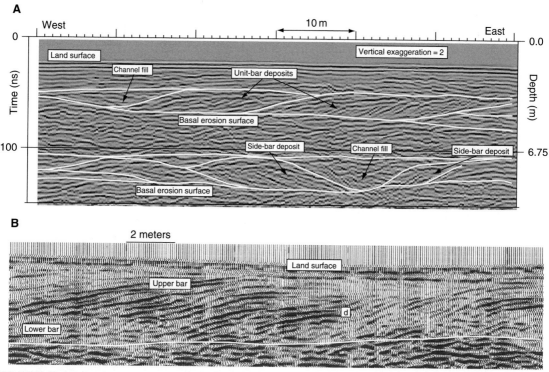

FIGURE 13.42. Large-scale inclined strata in ground-penetrating-radar (GPR) profiles through braided and meandering river deposits. Profiles are oriented across-channel. (A) Sagavanirktok River, Alaska, showing a compound-bar deposit with large-scale inclined strata dipping to the west and passing into a channel fill. Variable inclination of large-scale strata is associated with deposition on unit bars. Lower in the profile, a confluence fill is bounded on both sides by side (point) bars. The large-scale strata within the side bars increase in inclination, and their basal erosion surfaces become deeper, towards the confluence scour. (B) South Esk River, Scotland, showing point-bar deposits with large-scale inclined strata dipping to the left. The basal erosion surface of point-bar deposits is about 3 m below the land surface. Discordance in inclination of large-scale strata is marked by d. Upper-bar deposits have more laterally continuous radar reflections than do lower-bar deposits. (C) A trench showing lower-bar deposits (medium-scale trough cross-stratified sand) overlain by upper-bar deposits (small-scale cross-stratified and burrowed sand interbedded with dark layers of vegetation-rich silt). The dark layers are low-flow deposits, and define the upper parts of large-scale inclined strata. From Bridge (2006).

fills may show evidence of accretion on progressively smaller bars (e.g., rows of unit bars) as the discharge is reduced (Figures 13.41 and 13.44). Small deltas may prograde into entrances of abandoned channels containing ponded water, thereby producing coarsening-upward sequences. Sediment gravity flows from cut banks may accumulate in talwegs as poorly sorted, structureless deposits (Figure 13.41). The suspended-load deposits drape over existing bed topography. Horizontal suspended-load deposits lap onto inclined channel margins (Figure 13.40). In humid climates, peat may accumulate in the ponded water of channel fills. In arid climates, evaporitic tufas may form.

Channel-fill deposits grade laterally into channel-bar deposits, and it may be difficult to distinguish them in the subsurface. Channel-fill sequences can look very similar to channel-bar tail deposits. The deposits of the relatively small bars within channel fills may look similar to the deposits within cross-bar channels. The fine-grained parts of channel fills may look very similar to overbank deposits, including lacustrine deposits.

Evidence of the falling-flow stage

Evidence of the falling-flow stage in channel deposits includes fining of grain size and associated changes in sedimentary structures in the upper parts of large-scale inclined strata (Figures 13.39, 13.41, and 13.42C). Cross strata associated with dunes and unit bars may have current ripples and wave ripples superimposed, and possibly mud drapes with abundant plant debris. Rill marks oriented parallel to depositional slopes represent falling-stage drainage channels, and cross-stratified sand wedges represent the small deltas that form as

FIGURE 13.42. Continued.

these channels flow into standing water (Figure 13.45). Desiccation cracks occur in emergent mud drapes, and rooted plants can colonize areas exposed during the low-flow stage. The level of these features in channel sequences gives an indication of the low-stage level. Burrowing and surface-browsing animals are most active above and below the water table following floods, and escape burrows may occur within the flood deposits.

Paleocurrent orientations

Paleocurrent orientations recorded in channel deposits depend on (1) the orientation of the bed forms and associated sedimentary structures that vary with bed-form type, their position in the channel, and river stage; and (2) what part of the channel bar or channel fill is preserved (Allen, 1966). The mean orientation of structures such as pebble imbrication and various scales of cross strata generally correspond to local water-flow

FIGURE 13.43. Unit-bar deposits. (A) The view looking upstream of a Sagavanirktok bar-tail scroll bar with cross-bar channels that pass to the left into mouth bars within an inner-bank swale (about 2 m wide). Two bar-head unit bars can be seen in the background. (B) A unit bar with steep downstream face on a point bar, Congaree River, South Carolina. This unit bar is about 0.5 m high. (C) A trench through the front of the unit bar in (B), showing medium-scale cross strata (a) formed by dunes migrating over the unit bar. Cross strata associated with sand avalanching down the steep front of the unit bar are limited in extent (b). From Bridge (2006).

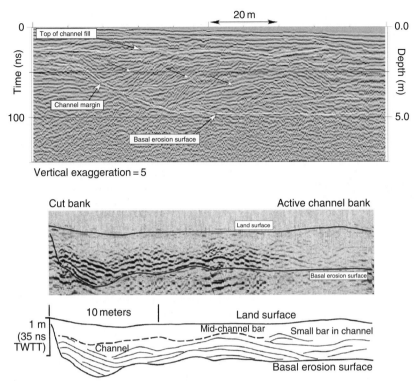

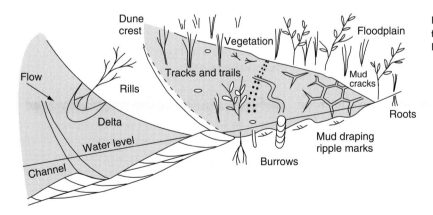

FIGURE 13.44. Channel-fill deposits in across-channel GPR profiles. The upper profile (Sagavanirktok River, Alaska) shows compound large-scale inclined strata (with their boundaries marked by small arrows) formed by individual gravelly unit bars that filled the channel. Within these compound large-scale strata are smaller-scale large-scale inclined strata formed by episodic migration of unit bars. The channel fill is approximately 4 m thick, and the vertical exaggeration of the profile is 5. The lower profile (Calamus River, Nebraska) has high-amplitude reflections (medium-scale cross stratified sand) overlain by low-amplitude reflections (bioturbated small-scale cross-stratified sand interbedded with vegetation-rich sand). The channel fill contains deposits of small bars. The vertical exaggeration of the profile is 3. From Bridge (2006).

FIGURE 13.45. Falling-stage features at channel margins. Modified from Bridge (2003).

directions. However, near banks, these paleocurrent indicators tend to be oblique to the local flow direction, as are the dips of angle-of-repose cross strata in scroll bars. Local paleoflow directions may be associated with deposition over a range of paleoflow stages and strengths, and will not necessarily be parallel to the orientation of the high-stage channels. Furthermore, it is expected that paleoflow directions from the downstream parts of ancient channel bars will be preserved preferentially. As a result, the mean paleocurrent azimuth for any particular structure in a channel deposit need not be parallel to the mean channel orientation, and the range of azimuths will probably differ from the range of local channel orientations. Thus, great care must be exercised in interpreting local channel orientations (and channel sinuosity) from paleocurrent data (e.g., Allen, 1966).

Porosity and permeability of channel deposits

Since river-channel deposits form important aquifers and hydrocarbon reservoirs, knowledge of their porosity and permeability is essential. The porosity and permeability of channel deposits vary spatially with variation in texture and structure, and such variation occurs over various scales associated with the different scales of strataset. Information on the three-dimensional variation in porosity and permeability over this range of scales is lacking in general. Porosity and permeability decrease upwards with grain size within some channel-bar deposits. The porosity and permeability of individual channel-bar deposits are expected to decrease downstream because bar-tail deposits are likely to be finer-grained than bar-head and mid-bar deposits. Furthermore, the finest-grained deposits will occur as low-flow drapes within large-scale inclined strata and in the uppermost large-scale inclined strata of bar tails. Low-permeability strata are also expected in relatively fine-grained channel fills that will be concentrated near the margins of channel belts.

Superimposed channel bars, channel fills, and channel belts

The spatial distribution of the deposits of individual channel bars and channel fills could not be included easily in the models above because it is very difficult to predict how individual channel segments and bars will migrate and become preserved within channel belts. Vertical superposition of channel-bar and channel-fill deposits in single channel belts can occur by superposition of a cross-bar channel on a main-channel bar and by migration of one main-channel over another (Figure 13.46). In the latter case, the degree of preservation of the overridden bar depends on the relative elevations of the two superposed basal erosion surfaces. The likelihood of preservation of the lower parts of the eroded bar increases with the vertical deposition rate relative to the lateral migration rate of the superposed bar, and the variability of

channel scour depth and bar thickness. The relative importance of deposition rate/lateral migration rate of bars and the variability of channel scour depths (bar heights) in controlling the amount of preservation of truncated bars can be assessed using Equation (5.112). In general, the variability of scour depths is the main control.

Object-based stochastic models have been used to distribute channel deposits within channel belts; e.g., see the reviews by Bridge (2003, 2006). The common approach in these models is (1) define shapes of channels; (2) position a series of channels randomly within an aggrading channel belt; and (3) define sedimentary facies, porosity, and permeability within the channels. None of these approaches (e.g., Figure 13.47A,B) correctly represents the nature of channel deposits in channel belts, which are in fact composed predominantly of parts of channel bars with relatively minor volumes of channel fills (see below). It is necessary to define shapes of objects properly (Figure 13.47C). Numerical simulation of channel deposits within channel belts is in its infancy.

Vertical superposition of channel-bar and channel-fill deposits can also result from superposition of distinct channel belts without intervening floodplain deposits (Figure 13.46). In cores and well logs, it may be difficult to distinguish between superimposed channel bars and channel fills in a single channel belt with superimposed channel belts (Figure 13.46). The ability to make this distinction hinges on the ability to interpret correctly the various superimposed scales of strataset, as explained below.

Qualitative interpretation of ancient channel deposits

In order to make the best use of the qualitative depositional models to interpret ancient channel deposits, it is necessary to have (1) detailed descriptions of large outcrops; (2) thorough understanding of the geometry, flow and sedimentary processes, and modes of migration of modern channels and bars; and (3) knowledge of how channel-bar and channel-fill deposits appear in variously oriented two-dimensional sections. Figure 13.48 shows descriptions of some well-exposed Miocene fluvial deposits from the Siwaliks of northern Pakistan, that have allowed interpretation of depositional environment in great detail (Willis, 1993a). Figure 13.49 shows other examples of relatively simple qualitative interpretations of ancient channel deposits.

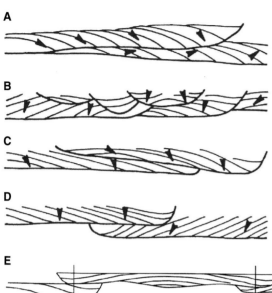

A

B

C

D

FIGURE 13.46. Superposition of channel bars and fills (storeys) within channel-belt sandstone bodies. Thick lines are basal erosion surfaces of compound channel bars, thin lines are large-scale inclined strata, and arrows are idealized paleocurrent directions relative to the outcrop plane (down is out of the outcrop). (A) Downstream migration and climbing of one bar over another bar. (B) Superposition of channel bars and fills of different depths and widths. (C) Superposition of channel bars of similar size but different sinuosities and orientations. (D) Superposition of different channel belts. (E) Superposition of channel bars in different channel belts and within channel belts. It may be difficult to distinguish these two different types of superposition using vertical sedimentary logs or gamma-ray logs (compare the two logs on the right-hand side of the figure). Images (A)–(D) from Willis (1993a); (E) from Bridge and Tye (2000). AAPG © 2000 reprinted by permission of the AAPG, whose permission is required for further use.

E

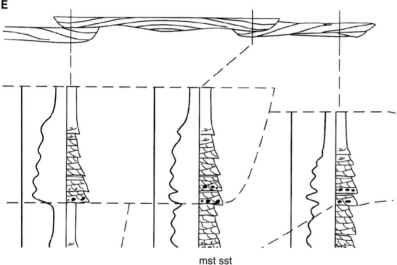

mst sst

Gamma ray   Mean grain size

Quantitative interpretation of ancient channel deposits from outcrops

The most sophisticated quantitative interpretation of channel geometry, flow, sediment transport, and channel migration requires the same kind of information as that mentioned above for qualitative interpretation, but it must be quantitative. The quantitative models described above have been used to interpret both meandering and braided river deposits (Bridge, 2003); see Figure 13.50. Willis (1993a) was able to reconstruct quantitatively the width, depth, mean velocity, slope, wave length, and sinuosity of individual channel segments in Miocene Siwalik braided-river deposits, and to estimate channel-belt widths and the degree of

braiding. Channel bars were interpreted as having migrated mainly by downstream translation and bend expansion, but also by channel switching within the channel belts.

Other, less sophisticated, mainly empirical methods of quantitative interpretation of paleochannel hydraulics and geometry are routinely applied; see reviews in Bridge (1978), Ethridge and Schumm (1978), Williams (1988), and North (1996). Grain-size data have yielded estimates of the threshold bed shear stress for bed-load or suspended-load motion, although this information is of limited value unless the grains were close to these thresholds at the time of deposition. Komar (1996) pointed out potential errors in some of the methods for determining the threshold bed shear stress for

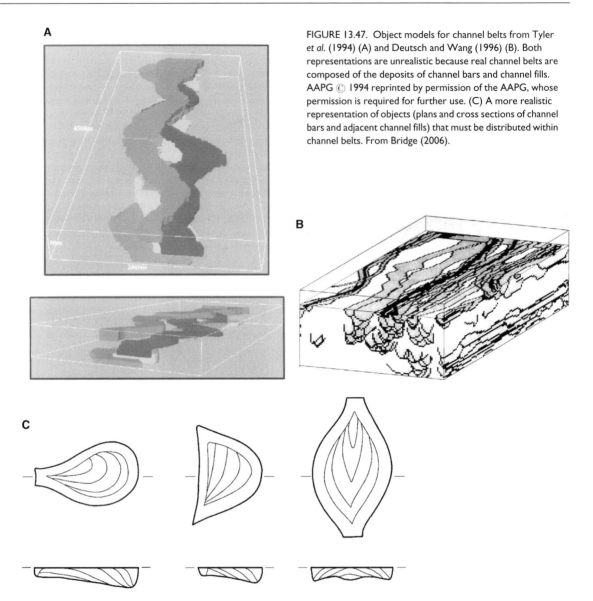

A

B

C

FIGURE 13.47. Object models for channel belts from Tyler *et al.* (1994) (A) and Deutsch and Wang (1996) (B). Both representations are unrealistic because real channel belts are composed of the deposits of channel bars and channel fills. AAPG © 1994 reprinted by permission of the AAPG, whose permission is required for further use. (C) A more realistic representation of objects (plans and cross sections of channel bars and adjacent channel fills) that must be distributed within channel belts. From Bridge (2006).

movement of gravel. Grain-size data have yet to be used seriously to estimate quantitatively flow parameters such as bed shear stress during transport and deposition. Sedimentary structures and preserved bed forms indicate the geometry of bed forms that existed at the time of deposition. If these bed forms are assumed to be in equilibrium with the flow, estimates of ranges of bed shear stress or flow velocity and depth can be made (see Figure 5.74). The mean thickness of various scales of cross strata can give estimates of the mean heights of the bed forms responsible (e.g., ripples, dunes, bars), and the heights of dunes and bars can in turn be related to the flow depth. Estimates of flow velocity, depth, and friction coefficients from grain-size

data and reconstructed bed forms have been used to estimate channel slope using uniform-flow formulae. Unfortunately, the flow equations used for these procedures are commonly misapplied or are inappropriate. It is also commonly difficult to understand what the reconstructed flow velocities and depths actually mean. Are they local or spatially averaged values, and what flow stages do they represent?

The average sinuosity of paleochannels can be estimated from the maximum range of paleocurrent directions observed in a single channel-belt deposit, provided that paleocurrent indicators are analyzed carefully. Paleocurrent indicators should represent the local paleochannel direction (e.g., medium-scale trough cross strata

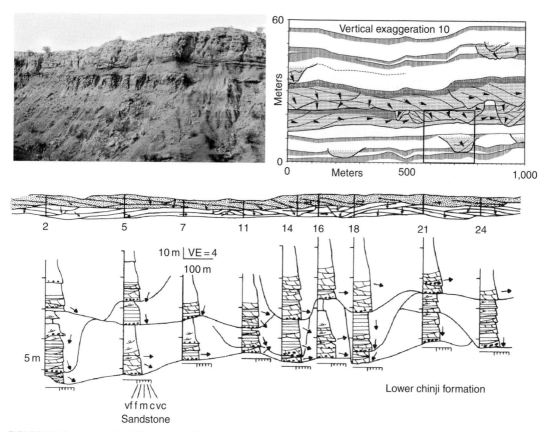

FIGURE 13.48. An example of Miocene Siwalik deposits of northern Pakistan. The position of the photo is shown as a box in the figure to the right, which shows channel sandstone bodies (stippled) with large-scale inclined strata, floodplain mudstones (unshaded), and well-developed paleosoils (vertical line ornament). The lower figure gives more sedimentological details of the extensive sandstone body shown in the upper-right figure (which includes logs 5–18). Line diagrams modified from Willis (1993a).

FIGURE 13.49. (A) Devonian river-channel deposits from southwest Ireland with large-scale inclined strata dipping to the left (at the top of the cliff) and fine-grained channel fills. A person is shown on the lower left for scale. (B) Carboniferous river-channel deposits from northwest Germany, showing large-scale inclined strata dipping to the left. The basal erosion surface is immediately above the head of the person.

from lower-bar deposits), but the preferential preservation of bar-tail deposits must be recognized. Calculation of sinuosity from paleocurrent ranges requires a functional relationship between these two parameters.

On Holocene to Pleistocene floodplains, it is commonly possible to observe the geometry and plan form of paleochannels. In some well-exposed fluvial deposits, the width and depth of channels and bars,

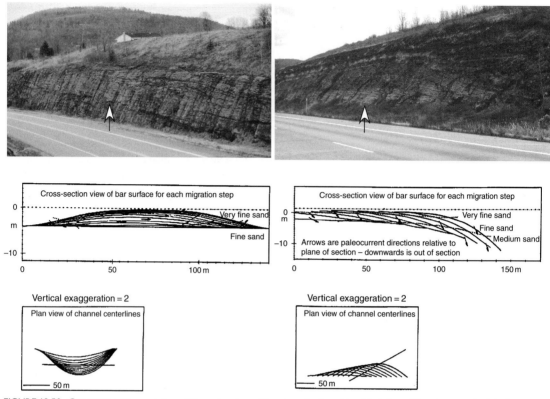

FIGURE 13.50. Quantitative interpretation of Devonian channel-bar deposits from New York State. The photo on the left shows a channel-bar deposit viewed parallel to the flow direction (the basal erosion surface is marked by an arrow). The reconstructed channel-bar geometry and migration pattern are show below. The photo and line diagrams on the right show a channel bar viewed approximately normal to the flow direction (an arrow indicates the basal erosion surface). Channel reconstructions from Willis (1993b).

and (exceptionally) the length and sinuosity of channel bends, can be observed directly (Bridge, 1978; North, 1996). In most cases, only one or two of these geometric parameters can be observed, and reliance has been placed on empirical equations derived from modern rivers to predict other geometric parameters. Observed or calculated geometric parameters of rivers are then used, in some cases together with sedimentary data, to calculate the channel-forming discharge using empirical regression equations (Bridge, 1978; Ethridge and Schumm, 1978; Dury, 1985; Williams, 1988; Bridge and Mackey, 1993a; North, 1996). This procedure is fraught with problems, including the inadequacy of empirical regression equations and their misapplication, as discussed in Bridge (2003). Unfortunately, such empirical equations that relate geometric and sedimentary characteristics of ancient rivers to their paleodischarge have largely formed the basis for the field of paleohydrology. The hydrology of major paleofloods has gained attention recently (e.g., Baker et al., 1988; Martini et al., 2002). One novel way of assessing paleoflood levels is to

determine the level of backwater deposits in canyons (Kochel and Baker, 1988). The depth and velocity of floods associated with catastrophic draining of ice-dammed lakes have been interpreted by examining large bed forms on floodplains, e.g., channelized scablands (Baker and Nummedal, 1978).

Estimation of paleochannel depth from subsurface data

It is desirable to estimate the thickness and width of subsurface channel-belt deposits in view of their importance as aquifers and hydrocarbon reservoirs (Bridge and Tye, 2000). The first step is estimation of the maximum paleochannel depth. In order to do this, channel-belt sands and gravels must be distinguished from floodplain sands, and the various scales of strataset must be distinguished, particularly those due to individual floods, single channel bars, and channel fills within one channel belt, and individual channel belts (the method is shown in Figure 13.51). This requires

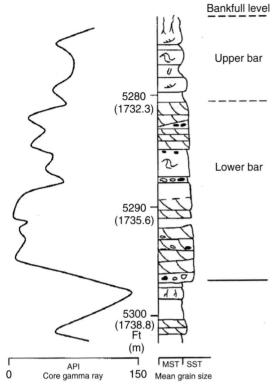

FIGURE 13.51. Estimation of flow depth from cores and gamma-ray logs through channel-bar deposits. From Bridge and Tye (2000). The maximum bankfull-flow depth can be estimated from the thickness of channel-bar deposits (7.5 m). This estimate can be checked using the mean thickness of medium-scale cross sets (formed by dunes) from the lower part of bar deposits (0.29 m), giving a mean dune height of about 0.85 m. This dune height is appropriate for estimated water depths during formation of 6–7 m. AAPG © 2000 reprinted by permission of the AAPG, whose permission is required for further use.

knowledge of spatial variations in grain size, sedimentary structures, paleocurrents, and the degree of stratal disruption (determined from cores and well logs). However, it is difficult to distinguish meters-thick fining-upward sequences associated with sandy to muddy fills of main channels from a relatively thin sequence of overbank sandstone (e.g., a levee, crevasse splay, or lacustrine delta) overlain by muddy floodbasin deposits. Also, upper-bar deposits look very similar to near-channel floodplain deposits. Superimposed channel bars or fills are difficult to distinguish from single channel bars or channel fills. The thickness of all single, untruncated channel bars or fills (from the tops of channel belts) must be measured to get an estimate of the range of maximum channel depths. An independent check on the estimates of bankfull flow depth is to use the relationships between the

thicknesses of medium-scale cross sets and dunes and the flow depth (Bridge, 2003).

Estimation of widths of single-channel-belts from subsurface deposits

Four commonly used methods for estimating the widths of isolated channel belts are (1) well-to-well correlation; (2) using empirical regression equations relating maximum channel depth, channel width, and channel-belt width; (3) measurement of outcrop analogs; and (4) amplitude analysis of three-dimensional seismic horizon slices. These are discussed in turn.

Well-to-well correlation

Well-to-well correlation of channel-belt sandstone bodies using wire-line logs has been the most common method for estimating channel-belt widths and orientations (Figure 13.52). The spatial resolution of this technique is no better than the average well spacing. The validity of this technique is very much dependent on the correlation rules utilized. Once a suitable horizontal datum has been chosen for the wells to be correlated, it is necessary to establish whether sandstone bodies at similar stratigraphic levels in different wells can be correlated. In order to make this assessment, it is essential to have a reasonable genetic interpretation of the sandstone body, and a model for its lateral extent and lateral variation in thickness and lithofacies. Well-to-well correlation is commonly compromised by lack of a realistic model for the possible lateral extent and lateral variation of sandstone bodies, and erroneous assumptions such as that (1) sandstone bodies positioned at the same stratigraphic level must be connected between adjacent wells; (2) vertical sequences through channel deposits indicate the paleochannel pattern and hence the geometry of channel-belt sandstone bodies; and (3) sandstone-body width/thickness ratios are closely related to paleochannel pattern. Modern and ancient channel belts have width/maximum channel depth ratios of between 700 and 20 (Bridge and Mackey, 1993a). It is commonly stated that this ratio will be larger for braided rivers than for meandering rivers. This is a moot point when utilizing core and wire-line log data, because such a distinction between paleochannel patterns cannot be made. Furthermore, this supposition is not generally correct. For example, the channel-belt width/maximum bankfull depth ratio for the meandering lower Mississippi

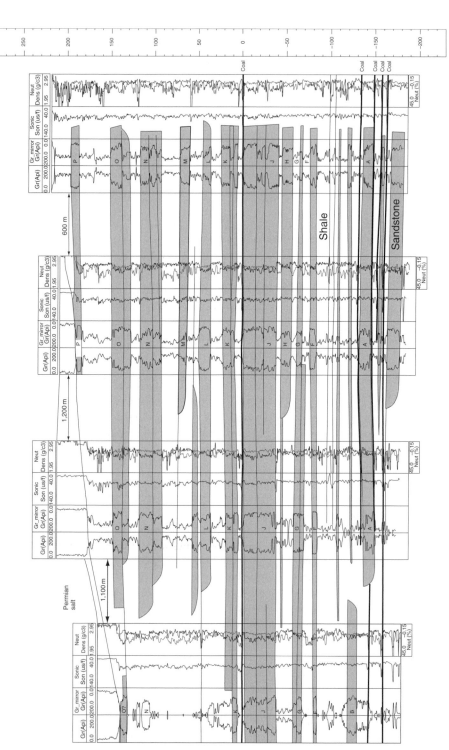

FIGURE 13.52. The channel-belt width is commonly estimated using well-to-well correlation of channel-belt deposits if channel belts are wider than the well spacing. This requires an estimate of the expected channel-belt width from the channel-bar thickness. However, it is difficult to distinguish relatively thick floodplain sands from channel-bar deposits and the sandy parts of channel fills. Fine-grained channel-fill deposits are difficult to distinguish from floodplain shale. In this example, the deposits are interpreted as channel-belt deposits (stippled), floodplain sands (stippled), or floodbasin shales (unstippled). The datum used to aid correlation is the coal seam in the middle of the section. Channel-belt deposits (letters A–P) were recognized on the basis of their (large) thickness and gamma-ray (GR) patterns. The correlation of channel deposits was based on estimated widths derived from channel-bar thickness, and hence maximum channel depth. It is not necessary to know whether the channel was meandering or braided to do this. Also, channel fills and floodplain sands help define channel-belt edges (e.g., L and H). Superimposed channel bars are distinguished using GR patterns (e.g., J). Floodplain sandstones are recognized and correlated on the basis of their (small) thickness, expected geometry, and relationship to channel-belt edges. From Bridge (2006).

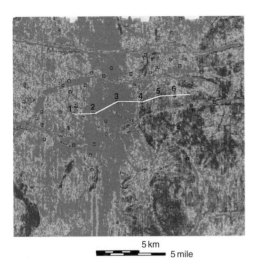
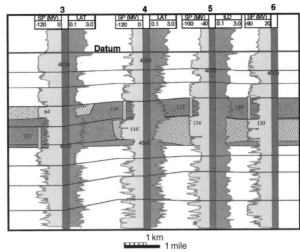

5 km
5 mile

1 km
1 mile

FIGURE 13.53. Amplitude analysis of a three-dimensional seismic horizon slice showing the width and channel pattern of channel-belt sandstone bodies. The cross section (right) shows correlated logs and the position of the horizon slice. Log 3 cuts the variable-width, straight channel belt. Logs 4–6 cut through a point bar and channel fill of a slightly older channel belt. From Bridge (2003).

River approximates that of the braided Brahmaputra River, being about 300.

Empirical regression equations

In order to make a critical assessment of the expected width of channel-belt sandstone bodies, it is first necessary to determine whether the sandstone body is a single channel belt, or a combination of channel belts. The widths of single channel belts can be estimated using empirical regression equations that relate maximum channel depth, channel width, and channel-belt width (Bridge and Mackey, 1993a; Bridge and Tye, 2000). This approach requires estimates of the maximum bankfull channel depth from cores or well logs. Since the empirical equations available have large standard errors, estimates of channel-belt width are imprecise.

Outcrop analogs

The use of outcrop analogs to interpret subsurface strata is very popular but has serious shortcomings. The interpreted depositional environments of the outcrop analog and the subsurface strata must obviously match. It is difficult to make detailed interpretations of depositional environments from typical subsurface data, and outcrop data can easily be misinterpreted. Outcrops are rarely extensive or numerous enough to allow unambiguous determination of the three-dimensional geometry and orientation of channels and channel belts. This is why it is desirable to use

analog data from Holocene depositional environments, where channel-belt dimensions can be determined easily, and the relationship between the nature of the deposits and the geometry, flow, and sedimentary processes of the environment can be established unambiguously.

Amplitude analysis of three-dimensional seismic horizon slices

Amplitude analysis of three-dimensional seismic horizon slices is the only method capable of yielding directly the width of channel belts, and imaging the channel pattern (sinuosity, channel splitting) of subsurface sandstone bodies (Figure 13.53). This is also the only method that can be used to predict the spatial distribution of channel-belt thickness and lithofacies. These are major advances. However, this method depends on the resolution of the seismic data relative to the thickness of the sandstone bodies imaged, and requires calibration by wire-line logs and cores. In general, the sandstone-body thickness must be greater than approximately 10 m.

Estimation of width of superimposed-channel-belts from subsurface deposits

Widths of superimposed channel belts can be estimated with the help of models of alluvial stratigraphy (Bridge and Mackey, 1993b; Mackey and Bridge, 1995), and depend on the proportion and degree of connectedness of channel-belt deposits in a cross section (Figure 13.54). For any channel-deposit

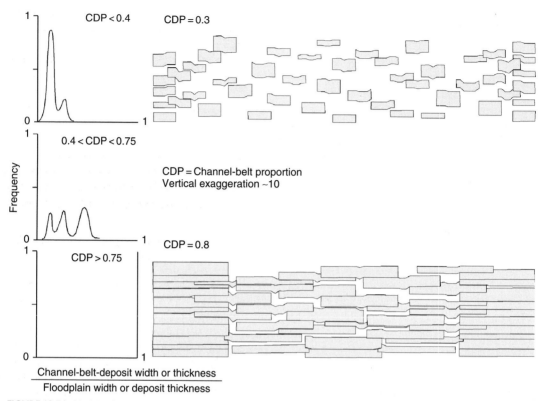

FIGURE 13.54. Models of superimposed channel belts. Channel-belt connectedness increases with the channel-deposit proportion (CDP). For CDP <0.4, most channel belts (shown as stippled boxes in the cross-floodplain section on the upper right) are unconnected, such that frequency distributions (upper left) of the channel-deposit width or thickness (relative to the floodplain width or thickness) are bimodal, with a large mode equivalent to unconnected channel belts. As the CDP increases, more channel belts are connected, channel deposits become larger, and the frequency distribution of channel-deposit width or thickness becomes polymodal. For CDP >0.75 (lower figures), all channel belts are connected and the single-channel deposit is as wide and thick as the floodplain. From Bridge (2006).

proportion less than about 0.4, channel belts are unconnected and the sandstone-body width equals the channel-belt width. As the channel-deposit proportion increases, some channel belts become connected, the mean and standard deviation of width increase, and the frequency distribution becomes polymodal. If the channel-deposit proportion exceeds about 0.75, all channel belts are connected, and the single sandstone body has a width equal to the floodplain width.

Floodplains: definition

A floodplain is the strip of land that borders a river channel, and that is normally inundated during seasonal floods. Floodplains develop in all alluvial valleys and on alluvial fans and deltas, irrespective of the channel pattern. The interaction among water flow, sediment transport, and bed topography of floodplains is not known as well as for channels (Bridge, 2003). Soil

formation is important on floodplains, and these fertile soils and adjacent wetlands support diverse ecosystems. Floodplains are underlain by deposits formed both in river channels and in floodplain environments.

The geometry of floodplains

Floodplain geometry has been reviewed by Allen (1965, 1970), Brierley *et al.* (1997), and Bridge (2003). Floodplain width can be up to tens of channel-belt widths, and the floodplain length is usually many times the floodplain width (except in the case of alluvial fans). Floodplains normally contain active and abandoned *alluvial ridges* that rise several decimeters to meters above adjacent lowlands, called *floodbasins*. Alluvial ridges contain active and abandoned channels and bars (the channel belt), levees, and crevasse channels and splays (Figure 13.55). *Levees* are discontinuous, wedge-shaped ridges around active and

FIGURE 13.55. Geometry of floodplains. (A) An alluvial ridge with active channel and levee (background) and abandoned channels. Parana River, Argentina. Photo from J. Best. (B) An active channel and levee (left) passing to the right into a wet floodbasin. Cumberland Marshes, Saskatchewan. Photo from N. Smith. (C) An active crevasse splay from Saskatchewan River. The floodbasin is bordered by an older channel belt (background). Cumberland Marshes. Photo from N. Smith. (D) A crevasse channel cut into crevasse splay passing into a dry floodbasin (background). Brahmaputra River. (E) Desiccated muddy floodbasin with a channel belt in the background. From Bridge (2006).

abandoned channels. Levee widths are up to about four channel widths. Levee surface slopes vary greatly spatially, and locally may be many times the slope of the adjacent main channel. Levee crests may be up to meters higher than adjacent floodbasins, increasing in height with river size, deposition rate, and stage of development. Newly formed levees are narrower and steeper in cross profile than older ones. This is due to initial deposition near the channel of relatively coarse-grained sediment. As the levees grow, the deposition rate of the coarse-grained material near the levee crest is reduced, but fine-grained material continues to be deposited in the floodbasin. Levees commonly have channels cut into their surfaces. The larger ones are called *crevasse channels* and split downslope into smaller distributaries surmounting fan- or lobe-shaped mounds of sediment called *crevasse splays* (Figure 13.55). In some cases, levees comprise a series of adjacent crevasse splays. Crevasse splays on major rivers are commonly hundreds of meters to kilometers long and wide, and can have a wide range of geometry and crevasse-channel development. Insofar as crevasse channels operate only during floods, they are ephemeral channels. Crevasse channels can have their own levees and terminal mouth bars. The distal margins of crevasse splays can either thin gradually, or end abruptly with a steep (angle-of-repose) slope. Crevasse splays that terminate in permanent lakes are similar to lacustrine deltas.

Floodbasins are much longer than they are wide, and are commonly segmented into subsidiary basins by crevasse splays, alluvial ridges of tributary channels, or abandoned alluvial ridges. Permanent lakes and marshes may be present in wet climates, whereas lakes are ephemeral in dry climates. Permanent lakes are particularly common on coastal plains and where there is local tectonic subsidence or sea-level rise. Floodplain drainage channels are common, and those that flow into lakes form deltas. Plant cover induces deposition by decelerating sediment-laden flows, and protects surface sediment from entrainment by wind or water. Plant cover is sparse in arid and semi-arid climates, and surface sediment may be moved by the wind.

The geometry of floodplains changes with time as a result of channel migration within the channel belt; migration, cutting, and filling of floodplain channels; large-scale movements of channel belts (avulsions); local tectonism; and progressive deposition or erosion. The geometry of floodplains also varies in the along-valley direction, as discussed below.

Water flow on floodplains

Comprehensive field studies of water flow and sediment transport over floodplains during overbank floods have not been performed because of obvious logistical problems and a lack of willing student assistants. Water flow on floodplains is complicated by its transient nature, variable floodplain width and surface topography (channels, depressions, mounds of sediment such as levees and crevasse splays), vegetation, and structures produced by humans and other animals. Therefore, flow and sediment-transport patterns on natural floodplains are not well known, and numerical models of water flow and sediment transport are poorly developed. However, many laboratory experimental studies of overbank flow adjacent to river channels have been undertaken, mostly with steady flows over simple channel–floodplain geometry, and with immobile boundaries without sediment movement. Despite these simplifications, the experiments have elucidated some important features of floodplain flows when water covers the floodplain and flow is steady and mainly down valley. The fastest flow in simple channel–floodplain models is in the main channels, and the flow velocity and bed shear stress on the floodplain diminish away from channel margins (Figure 13.56). The most abrupt lateral decrease in flow velocity occurs at the channel–floodplain margin, in association with vortices with horizontal axes that exchange water between the channel and the floodplain. As the depth of flow on the floodplain increases relative to that in the main channel, the flow velocity on the floodplain also

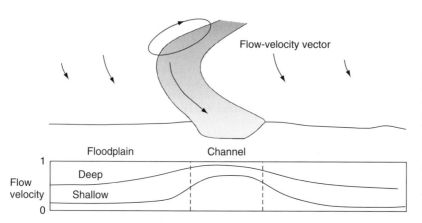

FIGURE 13.56. A model of flow over floodplains based on a laboratory study. The perspective diagram indicates a decrease in downvalley flow velocity with distance away from the channel, and large-scale vortices at the channel margin. The across-floodplain variation in flow velocity depends on the depth of overbank flow relative to the channel depth. The interaction between overbank flow and channel flow pattern is not shown. From Bridge (2006).

increases, and the reduction in velocity at the channel margin is not as great as with shallow floodplain flows (Figure 13.56). With wide floodplains (width/depth ratio >100, say), the velocity of the floodplain flow decreases with distance from the channel only in a zone close to the channel margin. On real floodplains, flow velocities will be relatively high in zones of flow convergence and relatively low in expanding flow zones, and will be greatly influenced by variations in bed roughness.

Changes in water flow and sediment transport in space and over flood periods control erosion and deposition on floodplains. However, there have been no systematic experimental studies of spatial and temporal variation of floodplain flows during flood waves, and only limited observations in natural environments. Therefore, the following description of the nature of an overbank flood is necessarily generalized and hypothetical. During the *rising-flow stage*, water gets to the floodplain initially through crevasse channels, over low parts of levees, and by overland flow from the valley margins. At this stage, the water level in the main channels may be much higher than that in the flood basin, leading to accelerating, erosive flows from the channel to the floodbasin. During *peak flood*, water completely covers the floodplain and flows more or less down valley. The water flows in a broad shallow channel (the floodplain) within which there are narrower, deeper channels (the main channels and crevasse channels). The flow patterns during this stage are described above and in Figure 13.56. During the *falling flow-stage*, water flows back into the main channels through floodplain drainage channels and via the groundwater. Floodplain lakes gradually diminish as the water level goes down.

Sediment transport and bed forms on floodplains

Sediment is transported over the floodplain as bed load and suspended load during overbank floods. The sources of sediment are the main channel and tributary channels, the valley sides, and the floodplain itself. A large range of sediment sizes is commonly available, from mud to gravel. The coarser-grained sediment available may be in the form of mud pellets (consolidated by desiccation or bioturbation), soil concretions, and organic debris (bones, shells, plant fragments). Indeed, much of the mud that is transported on floodplains may be in the form of pellets. Sediment is routed onto the floodplain from the main and tributary channels via smaller channels, sheet flows, and the large-scale vortices at channel margins. Bed-load sand is

transported mainly as ripples and upper-stage plane beds. However, dunes occur also, especially in floodplain channels, and antidunes occur in very rapid, shallow flows. Waves in lakes mold the bed into wave ripples. Realistic numerical models of sediment transport on floodplains are not available at present.

Deposition and erosion on floodplains

Marked flow deceleration and decreasing turbulence intensity near the margins of channels (on levees and crevasse splays) generally result in the greatest deposition rates of the coarsest-grained sediments there, thus explaining the origin of alluvial ridges. Relatively high rates of overbank deposition also occur in zones of flow deceleration such as abandoned channels and floodplain lakes, thus tending to smooth out floodplain relief. However, this is not always the case, particularly if the depression acts as a channel. The lowest deposition rates of the finest-grained sediments occur in flood basins distant from channels and areas that are relatively elevated. Although the grain size of sand on modern floodplain surfaces tends to decrease away from the channel belt, mud usually accumulates as a more or less continuous blanket. Deposition rates (averaged over hundreds of years) on modern floodplains decrease exponentially with distance from the active channel belt. This means that the cross-valley floodplain slope and elevation of alluvial ridges above floodbasins increase with time. The *average* thickness of sediment deposited on floodplains during seasonal floods is on the order of millimeters to centimeters. However, the (compacted) floodplain deposition rate averaged over millions of years is on the order of 0.1 mm per year, because long-term floodplain deposition is interrupted by lateral shifts of alluvial ridges (avulsions) and by periods of erosion. Erosion occurs where flow is accelerated in locally narrow or topographically high floodplain sections and where vegetation cover is poor. Freshly exposed sediment surfaces are modified by the wind, by plant growth, and by the activities of animals. Desiccation cracks appear in muddy sediments, and in arid climates salts may be precipitated in the soil as a result of evaporation of surface and groundwater.

Flood frequency and floodplain deposition

Flooding of active floodplains occurs once per year on average. A constant recurrence interval of flooding has been taken to indicate that deposition on floodplains does not lead to buildup of floodplains above channels, provided that the channels are not aggrading or degrading. This was explained by invoking a low deposition rate or erosion by overbank flows, and erosion of floodplain sediments by laterally migrating channels. The implication of this hypothesis is that floodplains should be underlain mainly by channel deposits with only thin overbank deposits at the surface. However, thick (relative to channel deposits) overbank deposits do accumulate beneath modern and ancient floodplains, particularly under conditions of net aggradation. Under aggradational conditions, deposition occurs also in the channel, such that alluvial ridge development does not imply elevation of the levee tops above the channel bed. Therefore, the frequency of overbank flow should not be decreased by development of alluvial ridges. In contrast, degradational conditions would lead to incision of the channel below the floodplain surface and a decrease in the frequency of flooding of the floodplain.

Floodplain deposits

The basic sedimentation units on floodplains are millimeter- to decimeter-thick stratasets deposited during overbank flooding events (Figures 13.57 and 13.58). Basal erosion surfaces are present if erosion preceded deposition. Upward-fining stratasets indicate deposition during temporally decelerating flows, whereas those that coarsen upwards and then fine upwards reflect deposition during accelerating and then decelerating

flows. Grain sizes and internal structures depend on local flow conditions and sediment availability. Many floodplain deposits are planar-stratified and small-scale cross-stratified, fine to very-fine sands interbedded with silt and clay. The planar laminae were formed on upper-stage plane beds, and the small-scale cross strata were formed by current ripples. Small-scale cross strata are commonly of the climbing-ripple type and associated with convolute lamination. Ripple forms at the top of the sandy part of a strataset may be draped with mud that has desiccation cracks. Wave-ripple marks can be superimposed on these ripples, marking the ponding stage of the waning flood. The upper parts of these stratasets are commonly bioturbated with root casts and animal burrows. Layers of drifted vegetation are common in the overbank deposits of humid climates. In general, overbank deposits closest to the main channel (on levees and crevasse splays) are similar to the upper-bar deposits described above, but decrease in grain size (and change internal structure) with distance from channels. The stratasets may be sheet-like, wedge-shaped, or lenticular, depending on the local environment of deposition.

Qualitative depositional models for floodplain deposits

Levee, crevasse-splay, and lacustrine-delta deposits

Depositional models for levees, crevasse splays, lacustrine deltas, and floodbasins are still rudimentary,

FIGURE 13.57. A typical flood-generated sedimentation unit on a floodplain near the main channel of the Brahmaputra River, Bangladesh. Planar laminae overlie an erosion surface (not shown), in turn overlain by small-scale cross laminae (of climbing-ripple type). Load structures are common in the small-scale cross sets. The sedimentation unit fines upwards and is capped by bioturbated mud. From Bridge (2003).

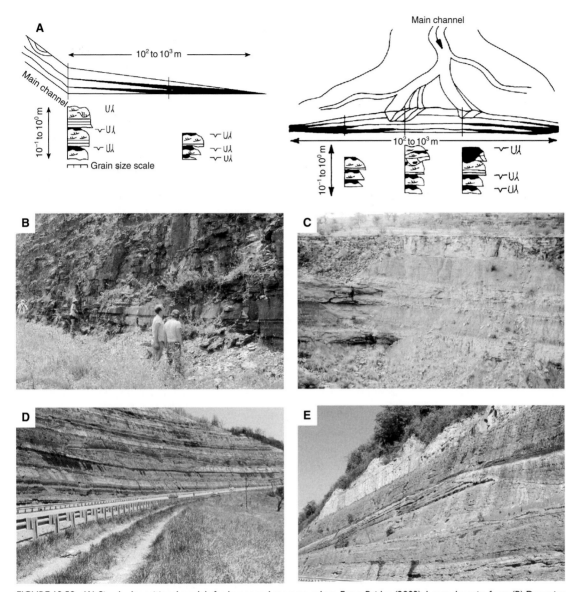

FIGURE 13.58. (A) Simple depositional models for levees and crevasse splays. From Bridge (2003). Levee deposits from (B) Devonian of New York and (C) Miocene of northern Pakistan. Levee deposits thin away from main channels. Images (D) and (E) show crevasse-splay deposits from Carboniferous rock of eastern Kentucky. Sections are normal to the flow direction, and show channels cutting through lobate sandstone bodies.

despite increased study over the past two decades as reviewed by Bridge (2003). The three-dimensional geometry of crevasse-splay deposits is different from that of levee deposits (Figure 13.58). Crevasse-splay deposits tend to be coarser-grained and thicker than levee deposits, although the edges of crevasse-splay deposits furthest from the main channel may be difficult to distinguish from levee deposits. Flood-generated stratasets of crevasse splays are similar to those of levees, but medium-scale cross strata (formed by dune migration) are more common in sandy crevasse-splay deposits. Channel-bar and channel-fill deposits are common in crevasse splays (Figure 13.58), and these may be difficult to distinguish from the main-channel deposits. The channels on crevasse splays are expected to be smaller on average than those in main channels, and exhibit evidence of periodic cessation of discharge (e.g., desiccation-cracked mud layers,

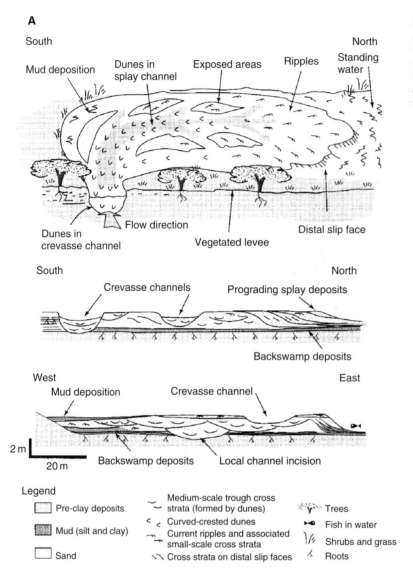

A

South

North

Mud deposition · Dunes in splay channel · Exposed areas · Ripples · Standing water

Dunes in crevasse channel · Flow direction · Vegetated levee · Distal slip face

South

North

Crevasse channels · Prograding splay deposits

Backswamp deposits

West

East

Mud deposition · Crevasse channel

2 m

20 m

Backswamp deposits · Local channel incision

Legend

	Pre-clay deposits
	Mud (silt and clay)
	Sand

Medium-scale trough cross strata (formed by dunes)
Curved-crested dunes
Current ripples and associated small-scale cross strata
Cross strata on distal slip faces

Trees
Fish in water
Shrubs and grass
Roots

FIGURE 13.59. Depositional models for crevasse splays. (A) Modified from Bristow *et al.* (1999), based on the Niobrara River, Nebraska. (B) From Pérez-Arlucea and Smith (1999), based on Cumberland Marshes, Canada.

root casts, burrows, tracks and trails throughout). However, some crevasse channels may be similar in size to main channels (especially immediately prior to an avulsion), and some main channels may be ephemeral just like crevasse channels. The margins of levees, crevasse splays, and lacustrine deltas can slope at up to the angle of repose. If they reach the angle of repose, the marginal deposits will resemble those of Gilbert-type deltas (Figures 13.59 and 13.60).

Within levees, crevasse splays, and lacustrine deltas, groups of flood-generated stratasets may occur in distinctive vertical sequences that are up to meters thick and perhaps hundreds of meters in lateral extent (Figures 13.58, 13.59, and 13.61). Coarsening-upward sequences are produced by progradation of the sediment bodies into floodbasins or lakes, and upward-fining sequences are produced by abandonment. Such progradation and abandonment may be associated with migration and abandonment of individual channels in a crevasse splay or lacustrine delta, migration and cut-off of channels within the active channel belt, or avulsion of the whole channel belt. However, these sequences might also be related to regional changes in sediment supply and deposition rate associated with, for example, climatic change, tectonism, or relative sea-level changes. Distinguishing among these various possibilities is no trivial task, requiring observation of floodplain deposits of a given age across the full extent of the floodplain.

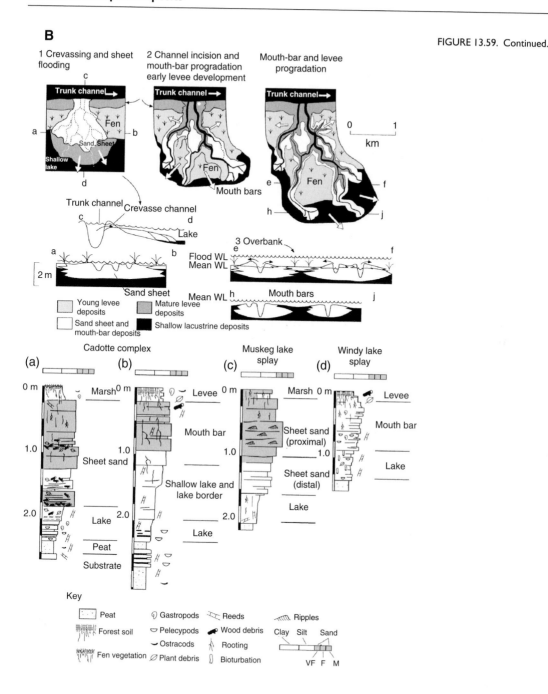

FIGURE 13.59. Continued.

Floodplain channel fills

Channel fills in floodplain deposits may be associated with abandoned main channels, crevasse channels (including floodplain drainage channels), and tributary channels. After their ends have become plugged with bed-load sediment, these abandoned channels become lakes, and receive mainly suspended-load sediment. Deposits are typically small-scale cross-stratified sands and muds. Lacustrine deltas may form at the entrances to these lakes. Channel fills normally contain abundant plant debris, shells of freshwater molluscs, vertebrate bones (given the appropriate climate and stage in Earth history), burrows, tracks, and trails (see below).

FIGURE 13.60. Floodbasin deposits. Sandstone-filled desiccation cracks in mud (now eroded) that was draping small wave-ripple marks (wave lengths on the order of a centimeter) and current-ripple marks. The wave-ripple marks are symmetrical at the top on the left, but become asymmetrical and smaller in wavelength towards the lower right. Their crestlines also change orientation as they become more asymmetrical. These changes in wave-ripple marks are associated with shoaling water at the edge of an ephemeral floodplain pond. The current ripples in the lower half of the photo indicate a unidirectional flow to the right. From Bridge (2003).

Floodbasin deposits

Floodbasin deposits are the finest sediments available for transport (e.g., silts and clays), with subordinate millimeter- to centimeter-thick sheets and lenses of sand containing small-scale cross strata and both wave- and current-ripple marks (Figure 13.60). Rippled pelleted muds are common. Evidence of subaerial exposure in floodbasin deposits is desiccation cracks and raindrop imprints in mud, and wind-blown sand. Burrows and root casts, and layers of drifted plant material, may be abundant, depending upon climate. However, exceptional sheet floods may deposit thick (up to a meter or so) sheets of sediment over large areas of a floodplain. Internal structures may be medium-scale cross strata and planar strata, formed from dune migration and upper-stage plane beds, respectively.

Floodplain lake deposits

Stratasets formed in floodplain lakes are sheet-like and millimeters to centimeters thick (see also Chapter 14). Evidence for waves in lakes includes cross-stratified sands and silts with relatively small wave-ripple marks, and planar laminae (Figure 13.61). Burrows, tracks, and trails are common as long as the lakes are oxygenated and not hypersaline, and root traces occur in shallow-water deposits. Common shelly fossils are pelecypods, gastropods, and ostracods (Figure 13.59). If the sediment supply is low, chemical or biochemical precipitation of deposits may be important. For example, carbonate mud may be formed by calcareous cyanobacteria. A common association of features in such deposits is centimeters-thick strata of calcite or dolomite mud with pellets, ostracods, burrows, and evidence of cyanobacterial filaments and mats. In swampy areas with low sediment supply in temperate and humid climates, peat may accumulate. In arid climates, evaporites typically form.

Upward-coarsening sequences (decimeters to meters thick, and up to kilometers across) are expected from progradation of lacustrine shorelines, whereas fining-upward sequences are expected from retrogradation (e.g., Figures 13.59 and 13.61). Lacustrine deposits also commonly exhibit evidence for temporal changes in the sediment supply as well as in depth and areal extent of the lake (e.g., vertical changes in grain size, composition, internal structure, and thickness of strata). As with other overbank deposits, such changes may be local and associated with migration of channels, levees, and crevasse splays, local tectonism; or regional and associated with changes in climate, relative sea level, and tectonism.

Floodplain soils

Weathering and soils were discussed in Chapter 3. Alluvial soil features have been reviewed by Retallack

FIGURE 13.61. Lacustrine deposits. (A) A coarsening-upward sequence (about 4 m thick) from laminated shale (a) via cross-laminated and ripple-marked sandstone and siltstone (b) to channel-fill sandstone (c). From Carboniferous rock of eastern Kentucky. Interpreted as progradation of crevasse splay into lake, analogous to Cumberland Marshes (Figure 13.59). (B) A coarsening-upward sequence (about 4 m thick) from laminated shale (a) to sandstone with angle-of-repose cross strata (b). From Carboniferous rock of northwestern Germany. Interpreted as progradation of the distal slipface of a crevasse splay into a lake, analogous to the Niobrara example in Figure 13.59. From Bridge (2006).

(1997, 2001), Kraus (1999), and Bridge (2003). Typical features of alluvial soils include (1) horizons (e.g., A, B, C); (2) textural evidence of leaching of soluble materials and elluviation of clays from A horizons and precipitation of secondary minerals and illuviation of clays in B horizons; (3) disruption of original structures by burrowing organisms, plant roots, changes in moisture content, and growth of secondary minerals; and (4) characteristic coloration and mottling associated with chemical alteration of parent material and formation of new minerals. The degree of development of these features in alluvial soils depends on time, deposition rate, climate, vegetation, topography relative to the water table, and source materials. Well-developed soils with horizons require on the order of 10^3 years to form, and a relatively low deposition rate, less than the order of millimeters per year (Leeder, 1975). Such soils are typically decimeters to meters thick.

Soils and paleosoils commonly vary laterally (the related assemblage of soil types is called a *catena*) and in time, and various models have been proposed for such variations. A common type of catena on floodplains is associated with decreases in elevation, deposition rate, and grain size with distance from the channel belt (Figure 13.62). Soils on sandy, well-drained levees and crevasse splays have a relatively thick oxidized and leached zone (the zone of aeration) underlain by a gleyed horizon (the saturated zone). The lower, muddy, poorly drained floodbasin deposits are more gleyed. If soils undergo extremes of wetting and drying, calcium carbonate is leached from the zone of aeration and accumulates as glaebules in the capillary fringe above the water table. Poorly drained floodbasin soils may have glaebules of both calcium carbonate and iron oxide. Local or widespread aggradation on floodplains may result in "drying-out" vertical sequences of soils (Figure 13.63). A decreasing deposition rate from channel belt to floodbasin may also result in an increasing degree of soil development further away from channel belts, and change in the degree of soil development in vertical sequences of paleosoils has been related to varying proximity of channel belts (Brown and Kraus, 1987; Kraus, 1987; Kraus and Aslan, 1999) (Figure 13.62). Furthermore, the relationship between increasing soil maturity and decreasing deposition rate

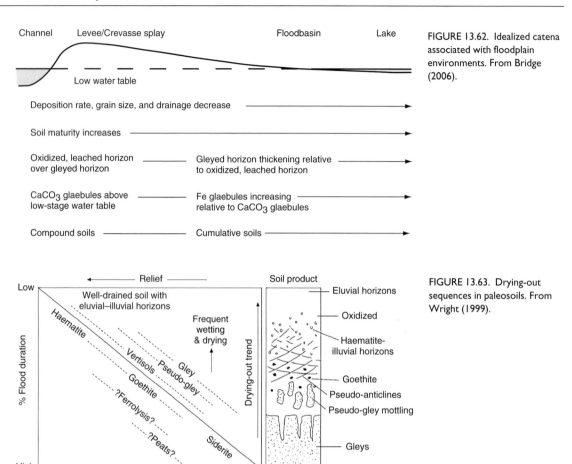

FIGURE 13.62. Idealized catena associated with floodplain environments. From Bridge (2006).

FIGURE 13.63. Drying-out sequences in paleosoils. From Wright (1999).

has been related to long-term, large-scale changes in deposition rate (Allen, 1974; Wright and Marriott, 1993; Willis and Behrensmeyer, 1994; McCarthy and Plint, 1998; Retallack, 2001). However, some caution is necessary, because soil maturity (pedofacies) is not controlled solely by the deposition rate. In sequences of floodplain deposits where paleosoils vary in their maturity, it is necessary to determine whether these variations are local or regional, and whether they are related to variations in deposition rate or other factors. This is no easy task.

Fossils in fluvial deposits

The hard parts of animals and plants are common in fluvial deposits, depending upon the stage in evolution of organisms on Earth. Most terrestrial vertebrates are known from river deposits. Vertebrates include fish, amphibians, reptiles, and mammals. Invertebrates include molluscs (bivalves, gastropods), arthropods (crustaceans and insects), and annelids (earthworms,

nematodes). Plant remains (e.g., plant stems (axes), roots, leaf impressions, pollen, and spores) are present in abundance in Devonian and younger sediments (Gensel and Edwards, 2001) (Figure 11.15). These body fossils can occur *in situ* or as transported remains. Some fossils are transported only a short distance, thus remaining within their local habitat. Transported plants and bones are normally sorted by size, shape, and density, oriented by the current, and abraded due to contact with other hard objects during transport. The abundance and diversity of fossils varies greatly among fluvial sub-environments; for details see Behrensmeyer and Hook (1992). Abundance and diversity are related to those of the original ecosystem, the nature of transport and burial, and post-depositional modification of the material. Good preservation is generally favored by rapid deposition and burial in topographically low sites, and by negligible chemical or biological degradation in the burial environment. Trace fossils (ichnofossils) are also very common in fluvial deposits from most sub-environments, and are reviewed in Chapter 11 (Hasiotis, 2002). They record the dwelling, burrowing, and surface movement of a wide variety of organisms, including worms, arthropods (insects, crustaceans), molluscs, and vertebrates. Plant-root casts occur in many fluvial deposits, with the exception of deposits formed below the low-water level in channels and lakes.

The types of organisms and their modes of preservation in fluvial deposits have changed over time because of (1) evolution of life on land and (2) change in depositional environments related to tectonism and climate change. It is commonly difficult to discern whether an evolutionary sequence of organisms is related to changes in the organisms or to changing depositional environments. There are many more fluvial fossils available in Quaternary and Tertiary sediments than in Mesozoic and Paleozoic sediments. This is partly due to the evolution of life on land, and partly due to the greater exposed volume of the youngest deposits. Some important stages in the evolution of life on land relating to fossil preservation are given in Behrensmeyer and Hook (1992), Buatois *et al.* (1998), Driese *et al.* (2000), Driese and Mora (2001), and Shear and Selden (2001).

River diversions (avulsions) across floodplains

Observations of the nature of avulsion

Avulsion is the relatively abrupt shift of an entire channel belt from one location to another on the floodplain (Bridge and Leeder, 1979; Mackey and Bridge, 1995; Jones and Schumm, 1999; Smith and Rogers, 1999; Stouthamer and Berendsen, 2000, 2001; Berendsen and Stouthamer, 2001; Bridge, 2003; Slingerland and Smith, 2004). Evidence for avulsions is abandoned channel belts on Holocene floodplains. Initiation of an avulsion is favored by energetic overbank water flow, a high slope from the channel to the floodplain relative to the slope down the channel (related to a high channel-belt aggradation rate and differential subsidence across the floodplain), and a pre-existing flood-plain channel and easily eroded sediment near the avulsion site. Avulsions normally occur during floods, although the high water levels required for avulsions may be created by downstream blockage of the channel associated, for example, with ice, vegetation, or sediment. An avulsion is typically initiated either by enlargement of a channel on a crevasse splay or by intersection of the main channel with a pre-existing

channel. The new channel follows the maximum flood-plain slope on its way towards the locally lowest part of the floodbasin. Abandoned channel belts may block the path of avulsing channels; however, an abandoned or active channel belt may be taken over by the avulsing channel. The transfer of water discharge from the old channel belt to the new channel, and the associated change in channel pattern, may occur over years to centuries. Inter-avulsion periods for a given channel belt (defined as the period of activity of the channel belt minus the avulsion duration) range from decades to thousands of years. Successive avulsions may be initiated from a specific section of a valley, particularly in the case of alluvial fans where the highest deposition rate is near the fan apex (giving rise to nodal avulsion: Figure 13.64). In other cases, successive points of avulsion may shift progressively upstream with decreasing avulsion period, until there is an abrupt down-valley shift in the location of the avulsion points. In yet other cases, avulsing channel belts appear to move progressively in one direction across floodplains or fans (e.g., Kosi River; Figure 13.64).

Avulsion and anastomosis

If the different channels in avulsive river systems co-exist for a finite length of time, the river system can be classified as distributive (with divergent channel belts) or anastomosing (if channel belts split and rejoin). Such river patterns are typical of alluvial fans, crevasse splays, and deltas. These are environments with high deposition rates and active growth of alluvial ridges, producing the gradient advantages conducive to avulsion (Smith and Smith, 1980; Smith *et al.*, 1989, 1998; Törnqvist, 1993, 1994: Makaske, 2001). The water and sediment discharge (and hence channel geometry) of individual channel-belt segments may change with time. The channel segments in anastomosing river systems can have any type of channel pattern (e.g., meandering, braided). Therefore, it makes no sense to have depositional models for anastomosing river systems that are distinguished from models for braided or meandering rivers (as occurs in much of the literature on fluvial deposits).

Effects of sedimentation rate on avulsion

The avulsion frequency increases (the inter-avulsion period decreases) with increasing channel-belt deposition rate (Mackey and Bridge, 1995; Heller and Paola, 1996; Ashworth *et al.*, 2004). This is because a high

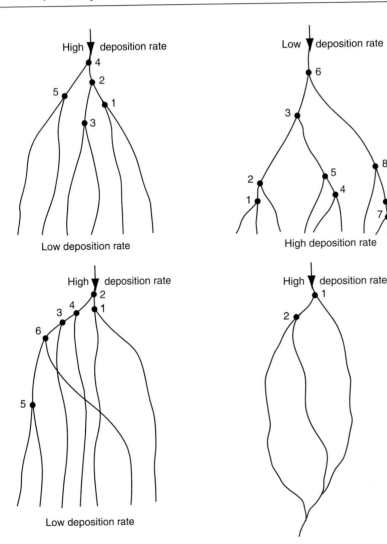

FIGURE 13.64. Typical patterns of channel-belt avulsion, depending upon the spatial variation in deposition rate and pre-existing floodplain topography. Channel belts are represented by lines. Numbers are locations of avulsion in chronological order. A full explanation is given in Bridge (2003). From Bridge (2006).

deposition rate in a channel-belt causes rapid growth of the alluvial ridge above the floodplain and an attendant rapid increase in cross-valley slope of the alluvial ridge relative to its down-valley slope. The channel-belt aggradation rate can be influenced by base-level change, climate change in the hinterland, or tectonic activity, as explained below.

Effects of base-level change on avulsion

Inter-avulsion periods decrease (avulsion frequency increases) as the relative base level rises, due to the down-valley decrease in river slope and sediment-transport rate that causes aggradation and growth of alluvial ridges (e.g., Törnqvist, 1994; Ethridge *et al.*, 1999). Stouthamer and Berendsen (2001) found an increase in avulsion frequency with increasing rate

of base-level rise for the Rhine–Meuse delta, but not an associated decrease in inter-avulsion period. This means that the number of coeval channels must increase when there is an increasing rate of base-level rise. Another effect of a rising base level is that points of avulsion shift up valley as the sea transgresses over the land. Avulsion is apparently more likely to be associated with crevasse channels and splays when the aggradation rate is high, and with re-occupation of pre-existing channels when the aggradation rate is relatively low (Aslan and Blum, 1999; Morozova and Smith, 1999; Stouthamer, 2005).

The response of rivers to base-level fall depends on factors such as the slope and sediment type of the exposed land areas, and the rate of fall (Schumm, 1993; Wescott, 1993). If the slope of the exposed surface is greater than the steady-state channel slope, there may be channel incision and/or the channel may become more sinuous. The avulsion frequency may be expected to decrease in this region, but not necessarily further up valley. An upstream avulsion may limit the amount of incision further down valley (Leeder and Stewart, 1996). If the slope of the exposed surface is less than the steady-state channel slope, there may be deposition and/or the channel may become straight. Under these conditions, the avulsion frequency may be increased.

Effects of climate change on avulsion

Changes in climate in the drainage basin can affect avulsion by changing the discharge regimes and sediment supply to the rivers. For example, during and immediately following glacial periods, the magnitude and variability of water and sediment supply may be increased, resulting in higher avulsion frequencies. This trend may be reversed during warmer periods. In addition, changes in the base levels of lakes and the sea during glaciations should affect avulsion frequencies (Morozova and Smith, 1999), and ice dams could cause river diversions.

Effects of tectonism on avulsion

Tectonically induced changes in river and floodplain gradients may result in avulsions (Mackey and Bridge, 1995; Peakall *et al.*, 2000; Schumm *et al.*, 2000; Bridge, 2003). Avulsion may occur as a direct response to an Earth movement or (more likely) in response to a gradual, tectonically induced change in floodplain topography. Avulsing channels concentrate in areas of tectonic subsidence, and avoid areas of tectonic uplift. However, if channels occupying subsided areas have a high aggradation rate, subsequent river diversions may be directed away from the zone of maximum subsidence. Tectonically induced changes in river and floodplain gradients may also result in aggradation and an increase in the probability of avulsion, or incision and a lower probability of avulsion.

Theoretical models of avulsion

Mackey and Bridge (1995) suggested that the probability of avulsion at a given location along a channel belt increases with a *discharge ratio* (maximum flood discharge for a given year/threshold flood discharge necessary for an avulsion) and a *slope ratio* (cross-valley slope at the edge of the channel belt/local down-valley slope of the channel belt). They assumed

that avulsions are initiated during extreme discharge events when the erosive power of the stream is greatest, and that a sufficient stream-power advantage exists for a new course to be established on the floodplain. The stream power and sediment-transport rate per unit channel length are proportional to the discharge–slope product. Therefore, if the discharge–slope product and sediment-transport rate of an overbank flow can exceed that of the channel flow, the sediment-transport rate increases from the channel to the floodplain, and erosion and enlargement of a crevasse channel will be possible. The discharge in a growing crevasse channel will initially be less than that in the main channel. Therefore, the overbank slope must be much greater than the channel slope in order for the sediment-transport rate to increase from the main channel to the floodplain. In the limiting case in which the discharges of the main channel and the developing crevasse channel are equal, the slope in the crevasse channel must exceed that of the main channel for the avulsion to proceed. The water level in a floodbasin must be lower than that in the main channel to allow water to flow away from the main channel through a crevasse channel. When the water-surface elevations of the main channel and the floodbasin are the same, there can be no such crevasse-channel flow. Therefore, crevasse-channel enlargement can operate only during certain stages of overbank flooding. Accordingly, it may take a number of periods of overbank flooding for a crevasse channel to enlarge to the point of avulsion.

Mackey and Bridge (1995) used their avulsion-probability model to simulate avulsions where floodplain slopes vary in space and time due to variations in deposition rate, tectonic tilting, and faulting within the floodplain (Figures 13.64 and 13.65). An *increase in channel-belt deposition rate down valley* produces a decrease in channel-belt slope down valley but an increase in cross-valley slope as the alluvial ridge grows, as is likely to happen during base-level rise. Under these circumstances, the avulsion probability increases through time, and is greatest in the down-valley part of the floodplain where channel-belt slopes are smallest but cross-valley slopes are largest. The model predicts that avulsion is initiated in the down-valley part of the floodplain and successive avulsion locations shift up valley with a progressive decrease in inter-avulsion period. This is due to continued, uninterrupted growth of alluvial ridges (and increase in avulsion probability) up valley of previous avulsion locations. New channel-belt segments down valley from avulsion locations do not have time to aggrade significantly and develop alluvial

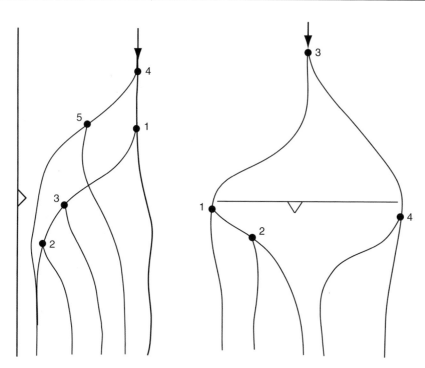

FIGURE 13.65. Typical effect of faults and tilting on channel-belt avulsion. Channel belts are indicated by lines. Numbers are locations of avulsion in chronological order. Downthrown sides of faults are indicated by triangles. A full explanation is given in Bridge (2003). From Bridge (2006).

ridges, and therefore have low avulsion probabilities. After a finite number of avulsions stepping up valley, the progressive decrease in channel-belt slopes in the down-valley part of the floodplain causes an abrupt shift in the locus of avulsion to this location. Although model results agree broadly with what is observed in nature (Mackey and Bridge, 1995), the model does not take into account the effects of the increase in slope of the new channel as it leaves the old channel. This steep slope would result in channel incision and upstream retreat of a knickpoint in the vicinity of the point of avulsion. Therefore, the probability of avulsion would be greatly reduced immediately upstream of the location of a recent avulsion.

A *decrease in channel-belt aggradation rate down valley* produces a decrease in channel-belt slope down valley, but an increase over time in channel-belt slope at any point on the floodplain, as is typical of alluvial fans and where the base level is falling. The avulsion probability decreases with time because the overall channel-belt slope increases. However, the avulsion probability is high in the up-valley parts of the floodplain where cross-valley slopes are increased by a high aggradation rate. The concentration of avulsions up valley where the deposition rate is high produces nodal avulsions, which are characteristic of alluvial fans.

The Mackey–Bridge model predicts that avulsion periods vary greatly, depending on the stage of growth of alluvial ridges. Sections of alluvial ridges that are well developed are associated with short avulsion periods (say, decades or centuries), whereas newly formed channel belts might not experience an avulsion for on the order of a thousand years. Also, the obstruction to flow caused by pre-existing alluvial ridges may cause subsequent channel belts to be clustered preferentially on one side of the floodplain with a distinctive *en échelon* pattern (Figure 13.64). This means that other parts of the floodplain distant from the active channels experience relatively low overbank deposition rates for extended periods of time, allowing soils to develop.

Tectonic tilting and faulting within the floodplain increase the avulsion probability locally, according to Mackey and Bridge (1995) (Figure 13.65). Channel belts shift away from zones of uplift and towards zones of maximum subsidence. However, if channel-belt aggradation keeps pace with fault displacement or tilting, alluvial-ridge topography causes channels to shift *away* from areas of maximum subsidence. Although these predictions agree broadly with data from modern rivers (e.g., Alexander and Leeder, 1987; Leeder, 1993; Peakall, 1998; Peakall *et al.*, 2000; Schumm *et al.*, 2000), data on the relationship between tectonism and avulsion are insufficient to test model predictions in detail.

Slingerland and Smith (1998) made the only analytical approach to the cause of avulsion. Their model is based on simplified equations of motion for fluid and sediment applied to simple channel geometry. The crux of this model is that the concentration of suspended sediment at the entrance to a crevasse channel leading from a deeper main channel is different from the equilibrium concentration that should exist in the crevasse channel. Then, depending on local hydraulic conditions, the crevasse channel would deepen or fill with sediment until the equilibrium sediment concentration is reached. A condition such that the channel progressively deepens is taken as a criterion for avulsion. Avulsions are predicted to occur wherever the slope ratio exceeds about 5. The importance of this first analytical approach notwithstanding, the geometry and physical processes are treated at a very simple level. Bed-load sediment transport is not treated, so the model cannot explain avulsions in the many rivers that transport mainly bed load during floods.

Sun *et al.* (2002) constructed a numerical model of channel flow and sediment-transport dynamics linked to channel avulsion for the case of fluvial fan deltas. They used semi-empirical, regime-type models for predicting channel geometry, flow, and sediment transport, assuming steady and uniform flow conditions. To simulate channel bifurcations, they assumed that the water discharge to each bifurcating channel is proportional to the square root of the channel slope for that channel. A channel avulsion occurs if the channel elevation at the bifurcation is greater than the elevation at some reference point on the fan. Therefore, a degree of channel super-elevation is required for an avulsion to occur. If an avulsion is predicted, the channel follows the line of steepest descent, but is allowed some statistical variation in its course.

Karssenberg *et al.* (2004) have improved the Mackey–Bridge model by treating channel-belt bifurcation and avulsion separately. Thus several channel belts can co-exist on the floodplain, forming distributive and anastomosing river systems. A channel-belt bifurcation can lead to an avulsion under specific circumstances, related to the discharge and slope ratios. The new model can simulate the occupation of a pre-existing floodplain channel by an avulsing channel. Following avulsion, the new channel belt increases gradually in width as channel migration proceeds within it.

Effects of avulsions on erosion and deposition

Diversion of a channel belt to a new area of floodbasin may be preceded by extensive development of crevasse splays (Smith *et al.*, 1989). It has even been suggested that most floodplain deposits may be formed during the period of crevasse-splay deposition preceding an avulsion (Smith *et al.*, 1989; Kraus, 1996; Kraus and Gwinn, 1997; Kraus and Wells, 1999). This is difficult to establish, however, and crevasse splays can be deposited without having an avulsion associated with them.

Avulsions may also be recognized in ancient floodplain deposits without actually observing the diverted channel deposits (e.g., Elliott, 1974; Bridge, 1984; Behrensmeyer, 1987; Farrell, 1987, 2001). For example, initiation of an avulsion may be recorded in floodplain deposits by an erosion surface overlain by relatively coarse-grained deposits (associated with a major overbank flood), and overlying deposits may be different from those that were deposited prior to the initiation of the avulsion. If the channel belt moved to a more distant location on the floodplain than hitherto, the new flood-generated stratasets on the floodplain may be thinner and finer grained and could be associated with a different flow direction. If the deposition rate is decreased, soils may become more mature (Leeder, 1975; Brown and Kraus, 1987; Kraus, 1987). If the channel belt moved closer to a given floodplain area, the new overbank flood deposits could be thicker and coarser-grained, and coarsening-upward sequences may occur due to development and progradation of levees and crevasse splays (e.g., Elliott, 1974; Farrell, 1987, 2001; Pérez-Arlucea and Smith, 1999). However, meter-scale overbank sequences that fine upwards or coarsen upwards can also be produced by progradation or abandonment of different levees/crevasse splays from a fixed channel belt, or by regional changes in sediment supply and deposition rate (see the previous section).

Avulsion by channel re-occupation is very difficult to recognize in the stratigraphic record. It has been claimed (Mohrig *et al.*, 2000; Stouthamer, 2005) that channel re-occupation will result in relatively thick, multistorey channel belts and multiple levees, but these stratigraphic features can easily be produced by episodic deposition in a single channel belt. The only conclusive evidence for channel re-occupation is the preservation of deposits indicating long time periods (such as mature soils) between the deposits of superimposed channel belts.

Non-avulsive shifts of channel belts across floodplains

It has been suggested that channel belts are capable of gradual migration across their floodplains by preferential bank erosion along one side of the channel belt and net deposition on the other side (Allen, 1965, 1974; Coleman, 1969; Thorne *et al.*, 1993; Peakall *et al.*, 2000). Such migration is possibly a response to across-valley tectonic tilting of the floodplain, and gives rise to so-called asymmetrical meander belts (Alexander and Leeder, 1987; Leeder and Alexander, 1987). However, evidence for such movement is rare or equivocal in modern rivers. Peakall *et al.* (2000) suggested that gradual shift of channel belts will occur instead of avulsion if tectonic tilt rates are relatively low. However, the only physical model for the effect of floodplain tilting on channel migration (Sun *et al.*, 2001a) predicts that a river could migrate towards or away from the down-tilted area, depending upon flow characteristics. Avulsions are the most important means of moving channel belts around most natural floodplains.

Along-valley variations in channels and floodplains

River and floodplain geometry, water flow, sediment transport, erosion, and deposition vary along valley as tributaries join, as water is lost by evaporation and/or infiltration, as geological features such as faults and lava flows change the valley slope, as the valley width changes, and as rivers enter bodies of water such as lakes and seas.

Along-valley variation of river slope and geometry

Long profiles of river channels and river valleys (along-stream variation in bed elevation) are commonly considered to be concave upwards and fitted by an exponential curve (Sinha and Parker, 1996; Morris and Williams, 1997, 1999a, b; Rice and Church, 2001). However, long profiles are not always concave upwards, especially in tectonically active areas (Figure 13.4). It has been suggested that down-valley decrease in alluvial channel slope may be related to the tendency for minimization of spatial changes in the rate of energy expenditure per unit channel length, which is proportional to the channel-forming discharge–slope product, QS (e.g., Langbein and Leopold, 1966). If the discharge of water increases in the down-valley direction, due to tributaries joining, the channel slope should decrease if QS is a conservative quantity. This is only a partial explanation, however, because the distribution of energy expenditure along alluvial channels is

not fully understood. The commonly observed down-valley decrease in mean grain size of channel sediment is due mainly to down-valley decrease in bed shear stress within the river system, such that the coarsest grains are progressively lost in the down-valley direction (Chapter 5). Down-valley reduction in bed-sediment size due to progressive abrasion is of minor importance, as described in Chapter 5.

Downstream increase in channel-forming discharge of water and sediment causes a downstream increase in bankfull channel width and depth. For a given sediment supply, width increases more than depth as the discharge increases (Equations (13.8) and (13.9)), such that the width/depth ratio might be expected to increase down valley together with the discharge. Related to this is that the channel pattern might be expected to change downstream from single-channel to braided. Actually, it is quite common to see the opposite, because, although the discharge increases downstream, the channel slope and mean grain size may decrease.

Temporal changes in river long profiles

Building of a dam across a river, and the associated formation of an upstream reservoir, result in decreases in the channel slope immediately upstream of the reservoir (the backwater effect) and in the rate and size of sediment being transported, leading to deposition and possibly a change in channel pattern and avulsion. The sediment supply (and possibly also water supply) immediately downstream of the dam would be reduced, resulting in erosion and coarsening of the sediment load. However, degradation below dams may be reduced as a result of reduction river flow or in bed erodibility associated with armoring, vegetation growth, or the appearance of bedrock in the channel. An increase in supply rate and grain size of sediment from upstream, due for example to tectonism or climate change, would necessitate an increase in slope if the channel-forming discharge remained constant. This increase in slope could be accomplished in the short term by a decrease in channel sinuosity and an increase in the degree of braiding, and in the long term by valley deposition or erosion.

Effects of tectonism on along-valley variation of rivers and floodplains

Tectonic activity affects the slopes of rivers and floodplains, and their water and sediment supply, over a range of spatial and temporal scales. For example,

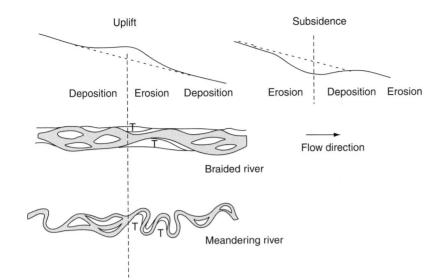

FIGURE 13.66. The effect of tectonism on along-stream changes in channel patterns. Based partly on Ouchi (1985). Erosion occurs where the slope is increased, and where the sediment supply is reduced due to upstream deposition, producing terraces (T), a decrease in degree of braiding, or an increase in sinuosity of meandering rivers. Deposition occurs where the slope is decreased, and downstream of erosion zones, producing an increase in braiding and local avulsion. From Bridge (2003).

periodic activity of a fault crossing a river valley may result in a local change in valley slope, river diversion, and a change in channel pattern over a period of several hundred or thousand years. On a much larger scale, the whole river system may be affected by tectonic activity for millions of years if it is located in a tectonically active mountain belt with adjacent sedimentary basin (e.g., Basin and Range Province, Himalayas, and Indo-Gangetic alluvial basins). Relatively short-term, local tectonic influences are discussed here; longer-term, regional tectonic influences are discussed in Chapter 20.

Local tectonic activity includes the effects of movement on relatively small folds and faults that directly influence land topography. Tectonic activity can also act indirectly by controlling the local base level and the slope of rivers entering standing water. Tectonic and volcanic activity can cause blockages of river valleys (creating lakes and local changes in slope) with lava, volcanic ash, and debris flows. Changing slopes within alluvial valleys can result in diversions of rivers and changing channel patterns. Changing topography of the floodplain relative to the groundwater table can influence floodplain flow and sediment transport, and the development of soils. These topics are discussed in recent reviews by Schumm et al. (2000) and Burbank and Anderson (2001).

Activity of a fault or fold may result in a depression or ridge with its axis approximately normal to the valley (Figure 13.66). Valley slope decreases upstream of an uplifting ridge axis, and increases downstream of the axis. The opposite occurs in the case of a subsiding depression. The short-term response to these slope changes might be a decrease in river sinuosity and degree of braiding where the valley slope is decreased, and an increase in river sinuosity and degree of braiding where the valley slope is increased. In regions of slope decrease, the floodplain may become permanently swampy, or the frequency and height of overbank flooding may increase. In other cases, a former floodplain lake may be reduced in level. Deposition is expected where the slope is decreased, and erosion (and possibly terrace formation) is expected where the slope is increased. If there is an abrupt transition from the zone of erosion and river incision to a zone of deposition downstream, the deposit may be in the form of an alluvial fan (see below).

The response of the river to such active tectonic deformation of the land surface depends on the rate of deformation relative to the rate of river erosion and deposition, in turn related to stream power and bank erodibility. If the rate of erosion and deposition is sufficiently large, along-stream changes in slope arising from tectonic activity will be reduced, and the river may tend to its former state. However, if the rate of erosion and deposition is not sufficient to remove tectonic topography, river diversion may occur. River diversion is most likely in areas of low valley slope and relatively large cross-valley slope. If the river cannot be diverted around the topographic obstruction, it may become anastomosing. When a river crosses an active strike–slip fault, it may be offset by continued fault movement. Alluvial fans on the downthrown sides of such faults may be separated laterally from the stream that provided the fan sediment.

FIGURE 13.67. An asymmetrical channel belt, Senguerr River, Argentina. The channel belt occupies one side of the floodplain, as a result of tectonic tilting. The floodplain is bordered by dissected uplands (background) or a terrace margin (foreground). From Bridge (2006).

Along-valley tectonic structures such as faults result in tilting of the floodplain in the across-valley direction. A common response of rivers to cross-valley tilting is periodic diversion towards the down-tilted area, producing asymmetrical channel belts where the active channel occupies the lowest part of the flood-plain, and abandoned channel belts occur on the up-tilt side (Figures 13.65 and 13.67). The low area of the floodplain will also be flooded preferentially and experience an increase in deposition rate. If the deposition rate temporarily exceeds the tectonic sub-sidence rate, rivers will not always occupy the zone of maximum subsidence. An increase in deposition rate in down-tilted sides of floodplains tends to reduce the development stage of soils, whereas the up-tilted sides experience a lower deposition rate and an increase in soil maturity.

Alluvial fans

Definition, occurrence, and geometry of alluvial fans

Alluvial fans and deltas are alluvial deposits that have distinctive plan shapes and distributive channels bordered by interchannel areas (Miall, 1996; Leeder, 1999). Deltas build into standing bodies of water. If fans build into standing bodies of water, they are referred to as *fan deltas*. Deltas range in size from small lobes where cross-bar channels enter slow-moving water in bar-tail areas, via lacustrine deltas on flood-plains, to the coastal deltas of major rivers. Major

coastal deltas are discussed in Chapter 15. Alluvial fans can be small crevasse splays on floodplains adjacent to channels, or the well-known fans that occur at the margins of small, fault-bounded valleys, or the mega-fans (e.g., the Kosi fan) that occur where major rivers flow from mountain ranges onto broad alluvial plains. The term *terminal fan* has been used for fans in arid areas where non-floodwater flow percolates into the ground before reaching beyond the fan margins. Alluvial fans are commonly classified according to their surface slopes and the relative importance of sediment gravity flows, sheet floods, and streamflows in forming them.

Alluvial fans occur in every climate where a river course passes sufficiently rapidly from an area of high slope to one of low slope. The abrupt change of slope results in a downstream decrease in bed shear stress and sediment-transport rate, which leads to local dep-osition. Alluvial fans commonly occur adjacent to fault scarps, and the preservation of fan deposits is enhanced by the subsidence of the hanging wall. Alluvial fans are best known in poorly vegetated, arid areas where infrequent violent rainstorms cause high sediment loads. However, alluvial fans also occur in humid and periglacial conditions. Alluvial fans may pass downstream into alluvial plains, tidal flats, beaches, perennial lakes, eolian dune fields, or playa lakes. The locally rapid deposition and fixed source of water and sediment give rise to the fan shape (a segment of a cone cut downwards from its apex; Figure 13.68). The fan radius varies from hundreds of meters to more than a hundred kilometers, increasing with the rate of supply of water and sediment from the catchment. The fan area increases with catchment area approximately linearly, although the exact relationship depends on climate, catchment geology, and the depo-sition rate relative to the basin subsidence rate. Fans tend to be more or less evenly spaced and laterally coalesced along faulted mountain fronts, and their areas may increase along the fault as the amount of throw increases.

Although the channel system on fans appears dis-tributive, not all channels are active at once. Diversion of flood flows among different channels (avulsion) is common. The trunk stream at the apex of the fan is commonly entrenched, as discussed below. In humid climates, it is normal for at least one of the channels to be perennial and to continue beyond the fan, but this is not the case in arid climates, where the channels are ephemeral and the fans are "terminal." In humid fans (e.g., Kosi), some of the smaller channels may originate

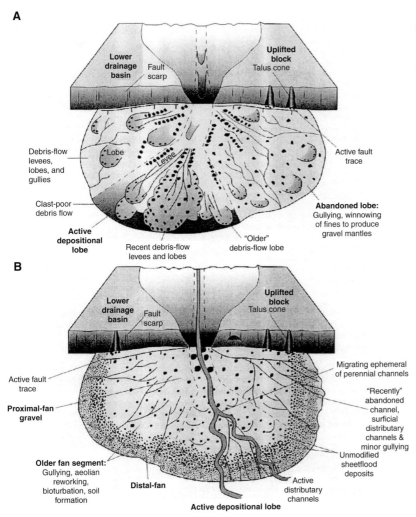

A

Lower drainage basin

Fault scarp

Uplifted block
Talus cone

Debris-flow levees, lobes, and gullies

Lobe

Levee

Active fault trace

Clast-poor debris flow

Active depositional lobe

Recent debris-flow levees and lobes

"Older" debris-flow lobe

Abandoned lobe: Gullying, winnowing of fines to produce gravel mantles

B

Lower drainage basin

Fault scarp

Uplifted block
Talus cone

Active fault trace

Proximal-fan gravel

Migrating ephemeral of perennial channels

"Recently" abandoned channel, surficial distributary channels & minor gullying

Unmodified sheetflood deposits

Older fan segment: Gullying, aeolian reworking, bioturbation, soil formation

Distal-fan

Active distributary channels

Active depositional lobe

FIGURE 13.68. Geometry and sedimentary processes of (a) debris-flow-dominated and (b) streamflow-dominated alluvial fans. From Leeder (1999).

on the fan surface from groundwater springs, instead of originating in the hinterland.

The long profile of fans is normally concave upwards, with slope decreasing down fan. The average slope generally decreases as water and sediment supply increase, and as the grain size of sediment supplied decreases. As slopes increase to the order of 10^{-2} or 10^{-1}, sediment gravity flows (debris flows and mudflows) become more common. Channelized debris flows are associated with distinct levees, and terminate in lobate deposits (Chapter 8). Cross-valley profiles of fans are convex-upwards. Fan geometry is complicated by incision of fan channels and formation of a new fan downstream of the incised fan, and by trimming of the fan toe by a river flowing along the fault-controlled valley.

Flow and sedimentary processes on alluvial fans

River channels on alluvial fans are braided if there is a large supply of water and sediment from the hinterland. Groundwater-fed fan channels are generally smaller and may have a different channel pattern (e.g., meandering). In arid fans, channels decrease in size down fan because water infiltrates or may become a sheet-flood. As the slope or depth of water decreases down fan, the grain size of surface sediment decreases. Overbank areas have both channelized flow and sheet-floods. Debris flows, grain flows, and mudflows are particularly common in channelized apex areas where slopes are steepest.

Channel diversions (avulsions) are common on alluvial fans during floods, and occur following a period of

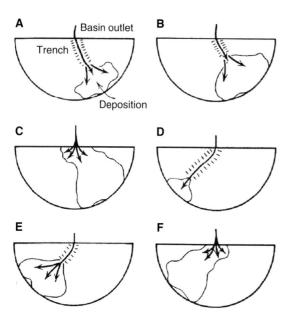

FIGURE 13.69. The evolution of experimental alluvial fans. From Schumm *et al.* (1987). (A) Fan-head channel entrenchment leads to deposition of a lobe downstream. (B) The deposition zone moves upstream and the channel is backfilled. (C) The channel is filled and deposition at the fan head increases the slope there. (D) The channel moves to a new area of fan and is entrenched. Images (E) and (F) as for (B) and (C). Reprinted with permission of John Wiley & Sons, Inc.

aggradation near the fan apex (Schumm *et al.*, 1987). This local aggradation causes the cross-valley slope of the fan apex to increase relative to the downstream slope, thereby facilitating avulsion. Once avulsion has initiated, the new channel is incised into the fan surface, with the maximum incision occurring near the fan apex where the new course is steepest (*fan-head entrenchment*). A sediment lobe is deposited downstream of the entrenched channel. As fan aggradation proceeds, the entrenched course of the new channel may start to aggrade by backfilling, starting from the lower-fan region (Figure 13.69). Eventually, the fan-head trench becomes filled with sediment, and the conditions become favorable for another channel avulsion. According to the fan model of Bridge and Karssenberg (2005), rapid deposition of a distinct lobe occurs in the region to which the channel is diverted, and this is associated with frequent channel bifurcations and anastomosis. It is possible that fan-head aggradation, avulsion, and entrenchment are associated with pulses of sediment supply from the hinterland, perhaps associated with episodic tectonism or climate change.

Large-scale, long-term deposition on alluvial fans, discussed further in Chapter 20, is related to (1) down-fan decreases in deposition rate and mean grain size as the slope decreases; (2) lateral changes in deposition rate and sedimentary facies from channel to interchannel areas; (3) the relative proportion of the deposits of channel flows, sheet-floods, and sediment gravity flows; (4) avulsion and related changes in deposition rate; (5) the fan-shaped paleocurrent pattern; and (6) spatial variations in sedimentary facies as a result of movements on basin–margin faults or climatic changes affecting the catchment. Spatial variations in sedimentary facies are therefore related to seasonal depositional events, the migration of channel bars and the filling of channels, the progradation of levees, crevasse splays or debris-flow lobes, the abrupt shifting of channels and associated loci of deposition, and changes in water and sediment supply to the fan associated with basin–margin tectonism or climate change in the catchment.

Alluvial architecture and its controls

Definition of alluvial architecture

Extensive accumulations of fluvial deposits in sedimentary basins formed over millions of years normally exhibit distinctive spatial variations in mean grain size, and in the geometry, proportion, and spatial distribution of channel-belt and floodplain deposits (referred to as *alluvial architecture* by Allen (1978)). These spatial variations occur over a range of scales (e.g., vertical sequences of strata may be tens to hundreds of meters thick: Figure 13.70 and Chapter 20), and regional unconformities (erosion surfaces that extend laterally for hundreds of kilometers) may underlie these sequences of strata. Definition of alluvial architecture requires extensive exposures, and/or high-resolution (preferably three-dimensional) seismic data, and/or many closely spaced cores or borehole logs, and accurate age dating. Since such data are commonly lacking or incomplete, it is necessary to "fill in" three-dimensional space in order to produce a complete (and hypothetical) representation of alluvial architecture. Furthermore, since it is not possible to observe directly the processes of development of alluvial architecture, it is necessary to use models to interpret and predict alluvial architecture. Most quantitative models of alluvial architecture (Bryant and Flint, 1993; Koltermann and Gorelick, 1996; North, 1996; Anderson, 1997; Bridge, 2003, 2006) are either process-based (process-imitating) or stochastic (structure-imitating), as will be seen below. These comments about the definition and interpretation of alluvial architecture actually

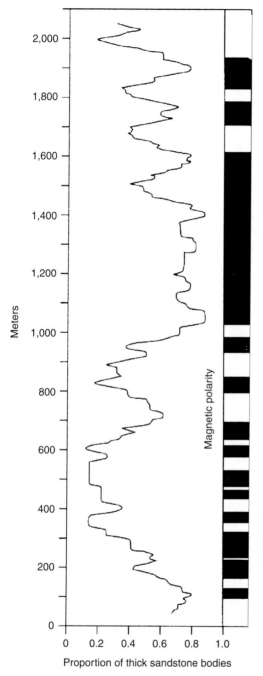

FIGURE 13.70. A large-scale sequence of fluvial deposits from the Miocene Siwaliks of northern Pakistan. The sandstone-body (channel-deposit) proportion decreases, increases, and decreases upwards over 2 km of strata, but also varies over 100–200 m and tens of meters. The photos show zones of high and low sandstone-body proportion. From Bridge (2006).

apply to the architecture of deposits in all depositional environments.

Controls of alluvial architecture

Alluvial architecture is primarily controlled by (1) the geometry and sediment type of channel belts and floodplains; (2) the rate of deposition or erosion in channel belts and on floodplains; (3) local tectonic deformation within the alluvial valley; and (4) the nature of channel-belt movements (avulsions) over floodplains. These *intrinsic* (intra-basinal) controlling factors are in turn controlled by *extrinsic* (extra-basinal) factors such as tectonism, climate, and eustatic sea-level

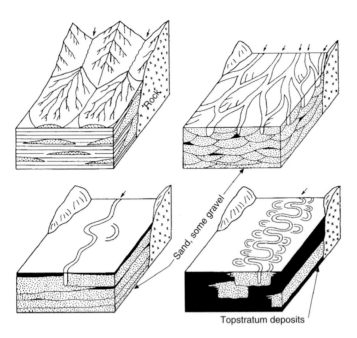

Topstratum deposits

FIGURE 13.71. Qualitative alluvial architecture models of Allen (1965). The braided-river model does not show an adjacent floodplain, and the depiction of braided-river deposits is unrealistic. The low-sinuosity-river model indicates erroneously that channels sweep gradually across their entire floodplains.

changes. For example, the geometry and sediment type of channel belts and floodplains are controlled by water and sediment supply (rate and type) that are in turn controlled by the source rocks, topographic relief, climate, and vegetation of the drainage basin. Long-term deposition in alluvial valleys is due to a long-term decrease in sediment-transport rate in the down-flow direction, which can be accomplished by increasing the sediment supply up valley (by tectonic uplift, climate change, or river diversions) and/or by decreasing the sediment-transport rate down valley (by flow expansion associated with tectonic subsidence or base-level rise). Subsequent erosion depends upon increasing the sediment transport in the down-flow direction, such as is caused by basin uplift or base-level fall. Local tectonic deformation in alluvial valleys can cause local changes in channel and floodplain geometry and location, deposition, and erosion. Channel-belt movements across floodplains (avulsions) are influenced by the severity of floods and the development of cross-floodplain slopes associated with alluvial-ridge deposition and local tectonic deformation. The intrinsic controls on alluvial architecture will be examined below, but the extrinsic controls are examined in Chapter 20.

Process-based (process-imitating) models of alluvial architecture

The earliest hypothetical models of alluvial architecture were qualitative and essentially two-dimensional

(Allen, 1965, 1974). Allen (1965) hypothesized that low-sinuosity, single-channel, and braided channels migrated rapidly (swept) across their floodplains, giving alluvium dominated by sheet-like to lenticular channel deposits (Figure 13.71). In contrast, single-channel, high-sinuosity streams were taken to migrate within well-defined meander belts that experienced periodic avulsion, leading to ribbon-like channel belts that were set in a relatively high proportion of fine-grained alluvium (Figure 13.71). These hypotheses are actually incorrect, but Allen's (1965) models were useful for stimulating more sophisticated quantitative approaches. Allen (1974) elaborated on his earlier models by considering the effects on alluvial architecture of climate and base-level change, degradation, and various modes of channel migration. More recent qualitative, two-dimensional (2D) models predict how alluvial architecture depends on changes in aggradation rate and valley width during a change in relative sea level (e.g., Shanley and McCabe, 1993, 1994; Wright and Marriott, 1993).

Leeder (1978) developed the first quantitative, process-imitating model of alluvial architecture. The channel-belt and floodplain deposits were modeled within a single cross-valley section. Channel-belt width, maximum channel depth, and floodplain width were specified. The channel belt was allowed to move by avulsion across the floodplain as aggradation (balancing subsidence) continued. The period of time separating avulsions and the location of avulsing channels in the

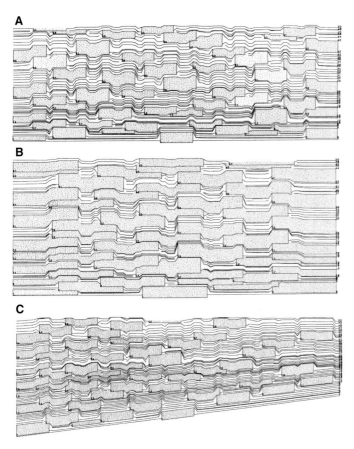

A

B

C

FIGURE 13.72. Examples of the quantitative 2D alluvial-architecture model of Bridge and Leeder (1979), showing the effects on alluvial architecture of varying channel-belt width, (A) and (B), and tectonic tilting (C). Channel belts are stippled blocks and lines are floodplain surfaces at the time of avulsion.

floodplain section were chosen randomly within defined limits. Leeder defined 2D measures of the proportion of channel-belt deposits and their degree of connectedness that depend upon the aggradation rate, avulsion frequency, channel-belt cross-sectional area, and floodplain width. Allen's (1978, 1979b) later approach followed Leeder's closely, but Allen added a function to allow diverted channel belts to avoid high floodplain areas underlain by pre-existing channel belts. The 2D approach was extended (Bridge and Leeder, 1979; Bridge and Mackey, 1993a, b) by considering also the effects on alluvial architecture of compaction, tectonic tilting of the floodplain, and variation of the aggradation rate with distance from channel belts (Figure 13.72), and predicting the width and thickness of channel sandstone/conglomerate bodies comprising single or connected channel belts (Figure 13.54). The channel-deposit proportion and the sandstone-body width and thickness increase as the bankfull channel depth and channel-belt width increase, and as the floodplain width, aggradation rate, and inter-avulsion period decrease. The channel-deposit

proportion and connectedness also increase in locally subsided areas of floodplain. These models show that the proportion of channel deposits (net-to-gross) has nothing to do with whether the river channel is meandering or braided, as suggested in Allen's early models. These quantitative 2D models have been tested against limited field data (e.g., Leeder et al., 1996; Mack and Leeder, 1998; Peakall, 1998; Törnqvist and Bridge, 2002), and have been used (and misused) widely to interpret and model alluvial architecture; see the references in Bridge (2003, p. 334). However, 2D models are unable to simulate realistically down-valley variations in the location and orientation of individual channel belts. This is possible only with three-dimensional (3D) models.

Mackey and Bridge (1995) developed the first 3D process-imitating model of alluvial architecture. The floodplain contains a single active channel belt (Figure 13.73). Changes in floodplain topography are produced by spatial and temporal variation of the deposition rate in channel belts and floodplains, by compaction, and by tectonism. The location and timing of

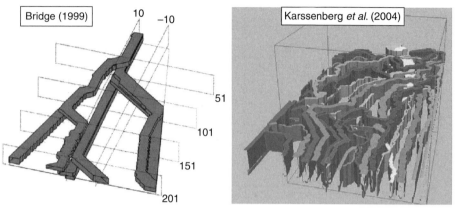

FIGURE 13.73. Quantitative 3D alluvial-architecture models from Bridge (1999) and Karssenberg *et al.* (2004). Shaded objects are channel belts.

avulsions are determined by local changes in cross-valley floodplain slope relative to the channel-belt slope, and by flood magnitude and frequency (discussed above). The diverted channel follows the locus of maximum flood-plain slope. Major differences between this model and the 2D models are the treatment of avulsion location and period as dependent variables, and constraints on the location of avulsing channels by the points of avul-sion and topographic highs on the floodplain. The behavior of the avulsion model was discussed above. Avulsions occur preferentially where there is a decrease in channel-belt slope and/or an increase in cross-valley slope that may be related to spatial variations in depo-sition rate, and/or tectonism, and/or base-level change. Evolution of alluvial ridges over time in various parts of the floodplain greatly influences the timing and location of avulsions. This may result in sedimentary sequences that increase upwards in channel-deposit proportion and connectedness, capped with overbank deposits with well-developed soils. Such sequences may take on the order of 10^3–10^5 years to form, durations comparable to those of cycles attributed to tectonism or climate change. Predictions of the Mackey–Bridge model agree with the somewhat limited data from modern rivers, and the model has been applied to interpreting and predicting the alluvial architecture of ancient deposits (Bridge, 2003, p. 336).

The Mackey–Bridge model has many shortcomings, and does not consider (1) changes in channel pattern and channel-belt width, formation of floodplain lakes, and deposition and erosion in the channel-belt and flood-plain associated with tectonism; (2) tributaries and downstream increases in channel-belt size; (3) wide-spread erosion (degradation) of channel–floodplain

systems; and (4) diversion of a channel belt into a pre-existing channel. However, model development is under way, and has been greatly aided by new data from Holocene fluvial and deltaic settings (Bridge, 2003). The component models for aggradation, avul-sion, and the development of channel belts following avulsion have been improved greatly (Karssenberg and Bridge, 2004). New channel belts are formed by chan-nel bifurcation, and multiple channel belts may develop and co-exist. In some cases, a pre-existing channel belt is abandoned in favor of another one, and an avulsion occurs. Degradation of the channel and flood-plain is treated using a diffusion–advection approach, such that upstream migration of knickpoints and for-mation of incised channels and terraces can be simu-lated. The effect of cyclic variation in degradation and aggradation on avulsion and alluvial architecture can be simulated (allowing a link to sequence-stratigraphic models).

Models of sediment deposition that are based on solution of the fundamental equations of motion of water and sediment are referred to as *sediment-routing models*. Engineers apply these models to relatively sim-ple flow and sedimentation problems in modern rivers and floodplains. Sedimentologists have also used them to explain phenomena such as downstream fining of bed material in rivers, and large-scale sorting of placer minerals such as gold (e.g., Van Niekerk *et al.*, 1992; Vogel *et al.*, 1992; Bennett and Bridge, 1995; Robinson and Slingerland, 1998a, b; Robinson *et al.*, 2001). Koltermann and Gorelick (1992) used a simple sediment-routing model to investigate how the large-scale patterns of deposition in alluvial fans are related to climatically controlled changes in water and sediment supply, tectonism, and base-level change. Sediment-routing models have not yet been used to simulate alluvial architecture, but they probably hold out the most hope for rational simulation of fluvial deposition and erosion.

Prediction of alluvial architecture of subsurface deposits: stochastic models

The most common approach to predicting the alluvial architecture of subsurface hydrocarbon reservoirs and aquifers is to (1) determine the geometry, proportion, and location of various types of sediment bodies (e.g., sandstones, shales) from well logs, cores, seismic data, or GPR data; (2) interpret the origin of the sediment bodies; (3) use outcrop analogs to predict more sediment-body characteristics; and (4) use stochastic (structure-imitating) models to simulate the sedimentary architecture between wells and the rock properties within sediment bodies such as channel-belt sandstones. Stochastic (structure-imitating) models are either object-based (also known as discrete or Boolean) or continuous, or both (Srivastava, 1994; Koltermann and Gorelick, 1996; North, 1996; Deutsch, 2002; Bridge, 2003). A common combined approach is to use object-based models to simulate the distribution of objects (e.g., channel-belt sandstone bodies and floodplain shales), and then use continuous models for simulating "continuous" variables such as porosity and permeability within the objects.

With object-based models, the geometry and orientation of specified objects (e.g., channel-belt sandstone bodies or discrete floodplain shales) are determined by Monte Carlo sampling from empirical distribution functions derived mainly from outcrop analogs. "Conditioned simulations" begin by placing objects such that their thickness and position correspond to the available well data. Then, objects are placed in the space between wells until the required volumetric proportion is reached. Objects are placed more or less randomly, although arbitrary overlap/repulsion rules may be employed to produce "realistic" spatial distributions of objects (Figure 13.74).

Continuous stochastic models have been used mainly to simulate the spatial distribution of continuous data such as permeability, porosity, and grain size. With these models, the prediction of a parameter value at any point in space depends on its value at a neighboring site. The conditional probabilities of occurrence are commonly based on an empirical semivariogram (see the appendix). These approaches have been modified to

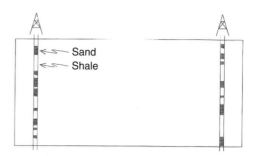

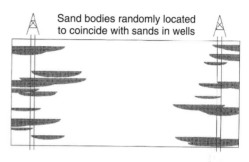

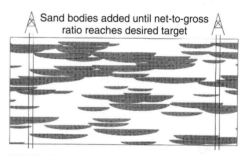

FIGURE 13.74. An example of 2D object-based stochastic modeling of alluvial architecture. Modified from Srivastava (1994). AAPG © 1994 reprinted by permission of the AAPG, whose permission is required for further use.

predict the distribution of discrete facies by using indicator semivariograms and simulated annealing (e.g., Johnson and Dreiss, 1989; Bierkens and Weerts, 1994; Deutsch and Cockerham, 1994; Seifert and Jensen, 1999, 2000) (Figure 13.75). A variant of the indicator-semivariogram approach is transition-probability (Markov) models in which the spatial change from one sediment type (e.g. channel sandstone) to another (e.g., floodplain mudstone) is based on the probability of the transition (see the appendix). The probability of spatial transition to a particular sediment type depends on the existing sediment type, and this dependence is called a Markov property. The matrix of probabilities of transition from one sediment type to another can be used to simulate sedimentary sequences in one, two, or three dimensions (e.g., Doveton, 1994; Tyler et al., 1994; Carle et al., 1998; Elfeki and Dekking, 2001) (Figure 13.75).

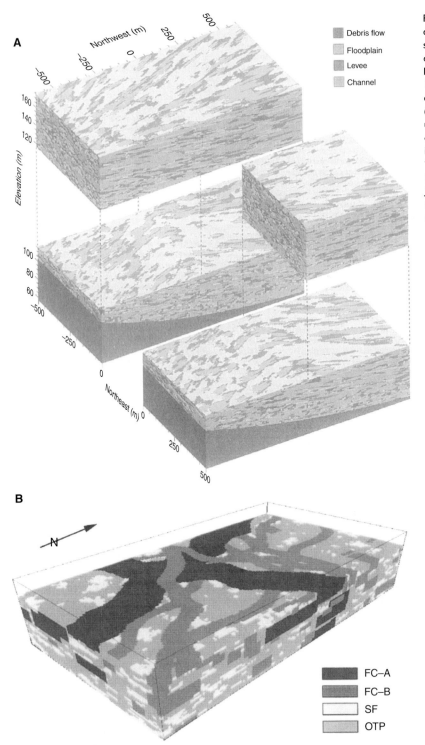

FIGURE 13.75. Three-dimensional stochastic model simulations. (A) Alluvial-fan deposits simulated using a Markov model (Carle *et al.*, 1998). Note the unrealistic depiction of channel deposits. (B) Fluvial deposits simulated using an object model for channel belts and a sequential indicator simulator for the "background" sheet-flood and lacustrine deposits (Seifert and Jensen, 2000). Note the unrealistic distribution of channel-belt orientations.

Debris flow
Floodplain
Levee
Channel

FC–A
FC–B
SF
OTP

It is commonly difficult to define the input parameters for stochastic models, especially the semivariograms and transition-probability matrices in lateral directions. The shapes, dimensions, and locations of objects in object-based models are difficult to define realistically (Figure 13.75). If definition of the dimensions of objects relies upon use of outcrop analogs and determination of paleochannel patterns from subsurface data, there may be serious problems (Bridge and Tye, 2000). Results obtained with process-based models demonstrate that the spatial distribution of channel-belt sandstones is not random. The generation of unrealistic shapes, dimensions, and spatial distributions of sediment types means that it is difficult to get the model to fit observed data and predict reservoir/aquifer behavior. Furthermore, since stochastic models do not simulate processes of deposition, they cannot give any insight into the origin of the alluvial architecture, and they have no predictive value outside the data region.

Despite all of the problems with stochastic models, they are widely used in preference to process-based models. This is mainly because software that is relatively easy to use is commercially available. Furthermore, process-based forward models are considered difficult to fit to subsurface data, and the models and software are not well developed. However, process-based models provide genetic interpretations of deposits, and can predict more realistic sedimentary architecture than can structure-imitating stochastic models. Karssenberg et al. (2001) have demonstrated that fitting of process-based models to well data using an essentially trial-and-error approach is possible in principle. Such an approach involves multiple runs of a process-based model under various input conditions, and optimization of the fitting of output data to observed data. Process-based models are being developed (e.g., Karssenberg et al., 2004), including development of software so that models can be fitted to subsurface data (the inversion approach). Another approach is to use output from process-based forward models to provide input for stochastic models that can be more easily conditioned with subsurface data.

Large-scale erosion in rivers and floodplains

Deposition in alluvial valleys is commonly punctuated by long periods of widespread erosion, resulting in the formation of incised valleys and river terraces. Long-term, large-scale erosion in alluvial valleys results from

increases in sediment-transport rate in the down-flow direction, such as are caused by basin uplift or base-level fall, or by a climatically influenced decrease in upstream sediment supply. Sediment-routing models have not been applied widely to long-term, large-scale erosion. However, erosion of fault scarps, valley slopes and river channels has been modeled using the diffusion approach (Bridge, 2003, p. 339; Bridge, 2007). The diffusion equation is derived from the sediment continuity equation for steady flows and a sediment-transport-rate equation. The one-dimensional sediment continuity equation is

$$-C_b \, \partial y/\partial t = \partial i_x/\partial x \qquad (13.14)$$

where C_b is the volume concentration of sediment in the bed (equals $1 - $ porosity), y is the bed elevation, t is time, i_x is the rate of sediment transport by volume in the downstream direction, and x is the downstream distance. The sediment-transport rate is assumed proportional to the bed slope:

$$i_x = k \, \partial y/\partial x \qquad (13.15)$$

where k is a coefficient. In reality, the sediment-transport rate is not proportional to the bed slope, so k also depends on a range of other parameters. Combining these two equations results in the diffusion equation:

$$-\partial y/\partial t = K \, \partial^2 y/\partial x^2 \qquad (13.16)$$

where $K = k/C_b$, which is the sediment-transport coefficient or diffusion coefficient, with units of area per unit time. The diffusion equation is thus a simple, slope-based method of describing erosion and deposition. However, the diffusion coefficient also depends on the surface slope, the scale of the system, and an array of other parameters such as sediment size, resistance to sediment motion, and the type of transporting medium. Diffusion coefficients are typically taken to be on the order of 10^3–10^8 m^2 per year for river-channel erosion and deposition, and on the order of 10^{-2}–10^{-4} m^2 per year for erosional retreat of fault scarps and sea cliffs.

The diffusion approach can be used to model the time variation of channel and floodplain erosion arising from an imposed increase in channel slope. This might occur as a result of base-level lowering or local tectonic movements, or as an avulsing channel flows over the steep edge of an alluvial ridge. The point in the river profile where the increased slope and erosion start is called the knickpoint (Figure 13.76). As river erosion proceeds, the knickpoint moves progressively up valley.

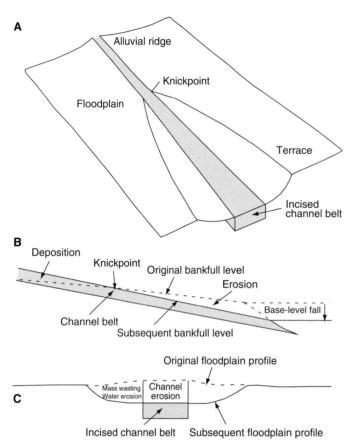

FIGURE 13.76. A conceptual model for channel-belt incision, knickpoint and riverbank retreat, and terrace formation resulting from a fall in base level. From Bridge (2003).

Incising channels can still migrate laterally and form floodplains, but these surfaces would lie beneath and within older, abandoned floodplains. River terraces are parts of floodplains that have become elevated above the bankfull level of the active channel as a result of widespread channel incision. Terrace risers would also experience erosional retreat associated with mass wasting (creep, debris flows), overland flow, and gullying. As a wave of channel incision moves upstream, the downstream parts of the valley experience channel erosion and degradation of terrace risers for the greatest period of time. Therefore, the floodplain of the incising channel should decrease in width in the up-valley direction (Figure 13.76).

Distinct episodes of degradation and aggradation can result in a series of terraces of different heights and valley fills with a complicated internal structure (Figure 13.77 and see Figure 20.13 later), and the sequence of degradation and aggradation may be very difficult to discern from the pattern of river terraces. In order to reconstruct the timing and location of degradation and aggradation in valley fills, it is necessary to establish the relative ages of the terrace surfaces and the

deposits beneath them. The relative age of terrace surfaces has been estimated using the degree of weathering of the surface deposits and using the degree of degradation of terrace scarps (using diffusion models of scarp retreat). Matching of heights of paired terraces on either side of valleys has also been used to establish age equivalence. However, terraces might not be paired on either side of valleys because of the variable age and elevation of a given abandoned floodplain, because a once-present floodplain was later eroded or obscured by colluvium from valley walls, or because of tectonic deformation of the terrace. Therefore, it should not be expected that a particular terrace should have an age and elevation that are constant and easily distinguishable from those of others.

Bedrock rivers

Bedrock rivers normally occur in the upstream parts of river systems where erosion is dominant (Tinkler and Wohl, 1998). Although substantial amounts of hard rock form their banks and beds, varying amounts of loose sediment also occur. Water flow and sediment

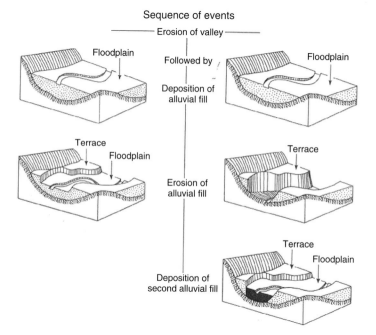

Sequence of events

Erosion of valley

Floodplain

Followed by

Deposition of
alluvial fill

Floodplain

Terrace

Floodplain

Terrace

Erosion of
alluvial fill

Terrace

Floodplain

Deposition of
second alluvial fill

FIGURE 13.77. Examples of river terraces. (A) Discontinuous terraces preferentially located on the inside of a bend of a river in New Zealand (photo courtesy of Jim Best). (B) Terraces from the Andean foothills, northwest Argentina. (C) Various ways of forming terraces, according to Leopold et al. (1964).

transport in bedrock rivers are controlled by water and sediment supply, and channel geometry. In contrast to alluvial rivers, the geometry of the channel is not readily modified by the flow during seasonal floods. The geometry of bedrock channels is greatly determined by its erodibility, as discussed below. Erosion of rocks in river channels arises mainly from hydraulic pressures, cavitation, and abrasion by moving sediment, and is brought about to a limited extent by dissolution. Hydraulic pressures entrain (or pluck) blocks of rock that have planes of weakness such as bedding, joints, and fractures. Cavitation is the process whereby vapor bubbles created by intense turbulence implode and cause shock waves that may weaken rock or dislodge small fragments. Erosion of solid rock is a slow process, typically removing fractions of a millimeter per year, and is therefore difficult to measure. The theoretical erosion rate of bedrock by streams is commonly taken as proportional to kA^mS^n in excess of some threshold value, where k is the erodibility, A is the drainage-basin area, and S is the surface slope. Here A is a surrogate measure for the stream-water discharge, such that the erosion rate becomes a function of the stream power per unit channel length. The exponents m and n are not known well, although they are commonly assumed equal to unity. Erodibility depends on factors such as rock composition, texture, and structure; the nature of weathering; and the presence of a moving sediment load (Whipple et al., 2000; Bridge, 2007).

FIGURE 13.78. Features of bedrock rivers: (A) and (B) potholes, gorges, rapids, and waterfalls in Watkins Glen, New York State; and (C) bedrock meander on San Juan River, Utah (photo courtesy of Henk Berendsen).

Erosional landforms associated with bedrock channels include potholes, flutes and grooves, rapids, waterfalls, and gorges (Figure 13.78) (Richardson and Carling, 2005). Potholes are cylindrical holes drilled into the bedrock bed by a combination of cavitation and sediment abrasion. Rapids and waterfalls occur at abrupt changes in slope, related to hard rock protrusions, fault scarps, or erosionally incised valleys (e.g., hanging valleys due to glacial erosion: Chapter 18). The extreme turbulence and dynamic pressure at the base of a waterfall result in erosion of a deep scour and undercutting of the waterfall cliff. This leads to slumping of the cliff and upstream migration of the waterfall, leaving a gorge in its wake downstream (Figure 13.78). The famous Niagara Falls on the Canadian side have retreated at more than a meter per year since 1850. Spectacular, well-known examples of rock gorges are the Grand Canyon in Arizona, and Canyonlands National Park in Utah (Figure 13.78).

14 Lakes

Introduction

Lakes cover approximately 1% of the Earth's continental areas. They range in size from small ponds and wetlands less than a few hundred square meters up to large lakes such as Lake Superior in North America ($82,410$ km^2). The brackish Caspian Sea ($374,000$ km^2) is generally regarded as the world's largest lake, but the Caspian Sea and Black Sea are both remnants of the Tethys ocean, and have been connected and disconnected from the Mediterranean Sea and the world ocean during the past 5 Myr (Meybeck, 1995). Lake depths vary greatly. Lake Baikal in Russia ($31,500$ km^2) is up to $1,640$ m deep, the maximum depth of Lake Superior is $\sim$400 m, and most large lakes are a few tens of meters to a few hundreds of meters deep. Some large lakes in semi-arid areas can be quite shallow. Lake Chad in central Africa ($25,000$ km^2) has a maximum depth of 7 m whereas Great Salt Lake, Utah, USA ($4,600$ km^2) has a maximum depth of $\sim$8 m. The depth of large ephemeral lakes may be as much as 5 m during floods.

Lakes are important because they are reservoirs of freshwater, sources of fish, shipping corridors, sites of recreational activity, and sites of chemical sedimentation. Large lakes influence regional climates and hydrological regimes. Many modern lakes respond to environmental changes in quite dramatic fashion. For example, the Aral Sea (the "poster child" for environmental degradation) on the border between Kazakhstan and Uzbekistan was the fourth largest lake in the world in the 1960s, with an area of $\sim$66,000 km^2 and a depth of $\sim$10 m or so, and was large and fresh enough to support large fishing communities. However, diversion of the main inflow rivers for irrigation led to a rapid decrease of the lake to approximately half its former area, with an increase in salinity. Ancient lake deposits are important sources of sand and gravel, terrestrial fossils, various evaporite minerals, and "oil shale" (kerogen – rich laminated rocks). Ancient lake deposits are also important because they are reckoned to preserve a detailed record of ancient environmental conditions in and around the lake, and especially the paleoclimate (discussed below). General references on lakes include Lerman (1978), Matter and Tucker (1978), Fouch and Dean (1982), Dean and Fouch (1983), Anadón *et al.* (1991), Imboden and Wüest (1995), Lerman *et al.* (1995), Talbot and Allen (1996), Leeder (1999), and Cohen (2003).

Classification of lakes

Lakes are commonly classified in terms of (1) their salinity; (2) whether they have outflows as well as inflows; and (3) their origin. A classification of lakes based on salinity recognizes fresh, brackish, and saline lakes. Drinking-water standards generally set an upper limit for potable "fresh" water at 500 ppm total dissolved solids (TDS). Brackish lakes range from 500 ppm to 5,000 ppm TDS, and saline lakes have > 5,000 ppm TDS. Brines in saline lakes may reach concentrations of 400,000 ppm TDS. The biota of lakes changes as salinity increases, with diatoms, freshwater aquatic plants, green algae, freshwater fish, snails, and clams giving way to benthic cyanobacterial microbial mats, archaea, bacterial plankton, salt-tolerant sedge plants, and unusual animals such as the salt-tolerant fish *Talapia* and the brine shrimp *Artemia* (a.k.a. "sea monkeys") (Hardie, Smoot, and Eugster, 1978). Evaporite deposition occurs predominantly in saline lakes.

Lakes that have an outlet of water are called hydrologically open, whereas those that do not have internal drainage and are called hydrologically closed. Some lakes change between open and closed during their history. The maximum lake level is controlled by the level of the outlet in open lakes. Lake-level fluctuations due to variations in water supply and evaporation tend to be much more marked in closed lakes.

A classification of lakes based on origin (Meybeck, 1995) recognizes (1) glacial lakes; (2) tectonic lakes; (3) fluvial lakes; (4) crater lakes; (5) coastal lakes; and (6) a miscellaneous category including solution lakes (sinkholes), wind-deflation lakes, landslide-dammed lakes, and volcanic-material-dammed lakes. Fluvial lakes on floodplains and so-called "ox-bow" lakes occupying abandoned river channels are described in Chapter 13. Coastal "lakes" receive seawater by storm washover or groundwater seepage and are discussed in Chapters 15 and 16. Lakes can fill both volcanic and meteor craters. Lakes in deflation depressions in aeolian sand-dune fields are described in Chapter 16. The two most important types of lakes in terms of number, area, and volume are glacial lakes (discussed in Chapter 17) and tectonic lakes.

Glacial lakes are found in and adjacent to areas that were glaciated during the Pleistocene. They are particularly numerous in areas deglaciated during the 18,000 years since the Last Glacial Maximum. Since recently deglaciated areas are most common in temperate areas, most glacial lakes are not saline. Lakes in glaciated areas originate either as parts of depositional complexes such as outwash plains, or as erosional features gouged deeply into bedrock (see Chapter 17). The Great Lakes of North America, and Great Slave Lake and Great Bear Lake in Canada, are due to glacial scour. The Great Lakes occur where Paleozoic sedimentary rocks were eroded off underlying topography in the Precambrian shield, whereas Great Slave Lake and Great Bear Lake are eroded into the Precambrian shield. Smaller erosional lakes gouged into bedrock such as New York's Finger Lakes are also common. Glacial lakes vary greatly in size, have quite irregular geometries, and can be hundreds of meters deep: floors of some of the Great Lakes and Finger Lakes are below sea level. Glacial-lake deposits are of course confined to periods of Earth history with extensive ice sheets (Chapter 17).

Many large lakes are located in tectonic basins that have existed for millions of years. Lakes are found in two kinds of tectonic basins: *rift basins* and *continental "sag" basins*. Active rift basins (such as the East Africa Rift) contain tremendous thicknesses (kilometers) of young sediments and commonly contain lakes (such as Lake Magadi, Lake Natron, and Lake Tanganyika). Lake Baikal in Siberia, purported to be the oldest and deepest (1,640 m) lake on Earth, also occupies a continental rift. The best known rift-basin

lakes (e.g. Owens Lake, Walker Lake, Mono Lake, and Great Salt Lake) are found in the Basin and Range Province of the southwestern USA. However, rift basins with lakes occur throughout the length of the North and South American Cordillera. Examples include the Basque Lakes of Canada, and Lake Titicaca and Lake Unuyi, which are located in fault-bound basins in the high Altiplano of the Andes. Some rift basins, such as the Dead Sea, may also have substantial strike–slip components. Many lakes in rift basins, such as Lake Baikal, Walker Lake in Nevada, Lake Tanganyika, and the Dead Sea, are elongated, and their shape is controlled by faults. Mono Lake, on the other hand, has a nearly circular geometry and the Great Salt Lake has a quite irregular geometry. The other category of tectonic basin where lakes are found is persistent mid-continent sags. Examples include Lake Chad in central Africa, Lake Victoria in east Africa, the Etosha Pan in Namibia, the Aral Sea in east Asia, and Lake Eyre in central Australia. Lake Eyre was first seen on aerial reconnaissance of central Australia. An expedition, complete with boats, was mounted to explore this lake, but it had dried up by the time the expedition arrived.

Tectonic basins that contain lakes are also commonly *closed basins* and have internal drainage. Although some of the lakes located in tectonic basins are fresh to brackish (e.g., Lake Baikal, Lake Victoria, Lake Titicaca, Lake Tahoe), many are saline, and either ephemeral or perennial. The aridity that causes saline lakes can be related to their location in subtropical high-pressure zones, or to rain-shadow effects (especially in the case of rift valleys surrounded by mountains).

Controls on lacustrine processes

Flow and sedimentary processes in lakes are affected by five main factors: (1) type of sediment and its rate of supply; (2) the composition of the lake waters; (3) water temperature; (4) the water-circulation pattern of the lake; and (5) water-level changes due to variable inflow, precipitation, and evaporation. These factors are discussed in turn below.

Sediment type and its rate of supply

Sediment supplied to and deposited in lakes includes (1) terrigenous grains; (2) biogenic grains (skeletal minerals and non-skeletal organic matter); and (3) chemogenic grains. The amount and type of biogenic and chemogenic sedimentary grains in lakes are controlled by the composition (chemical species and total salinity) of the lake waters (see below).

Most *terrigenous* sediment is delivered to lakes from rivers. Many glacial lakes have shores composed of unconsolidated glacial till or outwash that act as a local sediment source, especially if the deposits are thick and form cliffs around the lakeshore (e.g., the Great Lakes of North America). Sediment in lakes may also be wind-blown or ice-rafted. The input of such terrigenous sediment to lakes tends to vary with the seasons. However, lakes in temperate areas commonly freeze over in winter, so sediment input and bank erosion are minimal under those conditions. Volcanic ash also falls into lakes. In many cases, ash layers found in lakes can be dated and related to an eruption of a specific volcano (Chapter 9).

Biogenic sedimentary grains in freshwater and brackish-water lakes include snails (aragonite), clams (calcite and aragonite), ostracods (calcite), and diatoms (opaline silica). Diatoms are the major phytoplankton in freshwater and brackish-water lakes. With increasing salinity, molluscs become rare, whereas ostracods flourish and diatoms in the sunlit surface waters are replaced by cyanobacteria and archaea. Other important invertebrate organisms in freshwater and brackish-water lakes include chironomids (benthic insect larvae) and annelid worms. Brine shrimp in saline lakes produce organic skeletons and their egg cases are commonly entombed in lake sediments. Brine-shrimp fecal pellets are important sediment in saline lakes. Non-skeletal organic matter is an important component of lake sediments, ranging from a few percent of the sediment to as much as 60%. Such organic-rich sediments are sometimes referred to as *sapropels*.

For the most part, the only important *chemogenic sediment* in freshwater and brackish-water lakes is low-magnesium calcite. Sedimentation of inorganic carbonate in freshwater and brackish-water lakes occurs as (1) fine-grained precipitates directly from the surface lake water; (2) fine-grained precipitates around cyanobacteria and other aquatic vegetation in shallow water; and (3) thrombolite–like masses of lithified carbonate commonly referred to as "tufa." In saline lakes, carbonate minerals other than calcite precipitate, including aragonite, high-magnesium calcite, monohydrocalcite, and protodolomite. The mineral that will precipitate depends on the Mg^{2+}/Ca^{2+} ratio of the lake waters (Chapter 4). Inorganic precipitation of carbonate occurs in saline lakes in the same three ways as it does

in freshwater and brackish-water lakes. However, in saline lakes, chemically precipitated ooids are common in shoal areas, and carbonate cements are common in terrigenous and carbonate sediments in and around the lake. If the saline lake becomes concentrated enough, various evaporite minerals will form (Chapter 4).

A particular type of sediment (terrigenous, biogenic, or chemogenic) can be dominant in a given lake under certain conditions. Freshwater and brackish-water lakes with no major river input of terrigenous sediment contain marls composed of carbonate mud with scattered biogenic grains, non-skeletal organic matter, and limited terrigenous mud. Tufas are also common in these lakes and are more fully described in Chapters 11 and 16. In large freshwater lakes with major river input, the terrigenous input equals (rarely) or swamps (commonly) the biogenic and chemogenic sedimentation. In saline lakes, carbonates, terrigenous sediments, and evaporites all accumulate. This chapter focuses mainly on large freshwater and brackish-water lakes. Saline lakes are dealt with more thoroughly in Chapter 16.

The composition of the lake waters

The composition of the lake water includes both the chemical composition and the total concentration (salinity) of the water, which depend on (1) the chemical composition of the inflow water; (2) the ratio of inflow plus precipitation to evaporation; and (3) the biogenic and chemogenic sediments deposited from the lake water. The inflow water to lakes is mainly meteoric, chemical-weathering, river water and ground water. These are dilute Na^+, Ca^{2+}, Mg^{2+}, K^+, HCO_3^-, and SO_4^{2-} waters and, for any given inflow, the concentrations of the various ions depend on the composition of the rocks in the source area (see Chapter 3). Therefore, freshwater and brackish-water lakes have diverse chemical compositions. Moreover, many lakes in rift basins have substantial hydrothermal input of calcium chloride-type brines as well.

Small freshwater lakes may be fed entirely by groundwater. Large freshwater lakes in temperate and humid tropical climates are also fed by groundwater but are also commonly part of through-going drainage systems: that is, the lakes have streams entering through inlets and streams leaving through outlets. Lakes with outlets can maintain a near-constant water-surface elevation (the level of the outlet) for long periods of time. If there is sufficient inflow and outflow, lakes will remain fresh, but slight changes in inflow volume or the temperature on the basin floor can rapidly lead to increasing salinity or total desiccation. Lakes in closed basins (many in rift basins) in arid subtropical zones must have major rivers entering them if they are to remain fresh. Rift valleys with high, up-faulted mountains bordering them (such as in the Basin and Range Province of the southwest USA) are orographic deserts, with resultant arid conditions and saline lakes. Saline lakes occur all along the Rocky Mountains and the Andes, from Canada to Chile, with little regard to the climatic zones discussed in Chapter 2.

Equilibrium thermodynamics serves as a good guide to the minerals that will precipitate from lake waters. Just about all freshwater lakes are at or near saturation with calcium carbonate, and "whitings' (spontaneous precipitates of fine-grained calcium carbonate from surface lake water) are common in many lakes, including the Great Lakes of North America and the Finger Lakes of New York. Since the concentrations of dissolved calcium and bicarbonate in freshwater lakes are low, the small amounts of carbonate sediment made in this way can be swamped by the terrigenous mud entering the lake via rivers or eroded from the shores. As the ratio of evaporation to inflow of lake water increases, the salinity of the lake water increases, and the mass of chemogenic sediments that can be produced in the lake increases and can swamp the terrigenous input from rivers. In saline lakes that precipitate evaporite minerals, the sequence of minerals precipitated is dictated by the chemical composition of the inflow water, and in turn closely controls the composition of saline lake water in accordance with the principle of chemical divides (Eugster and Hardie, 1978) discussed in detail in Chapters 4 and 16. Finally, manganese nodules and crusts similar to those found in the deep ocean (Chapter 18) are found at the bottom of some of the deeper Great Lakes.

The biological productivity of lakes varies according to the dissolved nutrient content in the water. In freshwater and brackish-water lakes, algae and rooted aquatic angiosperms occur wherever sunlight reaches the bottom, and cyanobacteria and diatoms grow in sunlit surface water. Freshwater lakes with low salinity (a few tens to hundreds of ppm TDS) have low productivity, and visibility in these lakes can be tens of meters. At salinities greater than a few hundreds of ppm TDS, primary productivity in lakes increases, and can be quite high if nitrates and phosphates are present at elevated levels. Organic matter that settles out of the surface water into deep water is oxidized by bacteria, raising the concentration of dissolved CO_2 and reducing that of

dissolved O_2 of the deep lake water. The fate of organic matter in deeper portions of lakes is further described in the next section. Wooded swamps develop on the margins of some lakes. Little light penetrates into such swamps and surface photosynthesis is curtailed. Accumulation of falling trees and leaves and other vegetation may lead to acidic, reducing conditions and peat accumulation. This is also the case in bogs that develop in ice-block (kettle) holes in glaciated areas. In saline lakes, cyanobacterial and halophytic bacteria can bloom in vast numbers and tremendous amounts of organic material can accumulate in the sediments. This is a common origin of oil shale.

Water temperature

The density of lake water depends on its temperature and salinity. Density stratification in deep lakes has an important impact on the levels of dissolved O_2. Seasonal turnover in temperate freshwater lakes is a well-known phenomenon, and is due to the fact that the maximum density of freshwater is at a temperature of $\sim 4\,°C$, not at the freezing point. During the summer, solar heating produces a warm lens of low-density surface water sitting over a pool of densest water at $\sim 4\,°C$. Limnologists refer to the upper warm layer as the *epilimnion*, the transitional layer as the *metalimnion*, and the deep layer as the *hypolimnion*, and these layers are easily picked out on temperature profiles through the water column (Figure 14.1). The thermocline (the

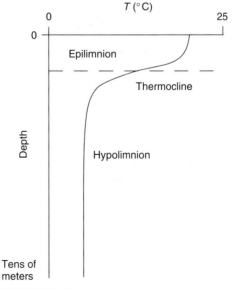

FIGURE 14.1. The temperature profile of a temperate-zone lake in summer, with terminology.

portion of the profile with the steepest temperature gradient) occurs in the metalimnion. Many people who swim in temperate lakes are well acquainted with this density stratification and the thermocline. Wind stress over the surface of the lake in summer mixes the surface water (the epilimnion) but does not affect the deeper, denser water. During the fall of the year, epilimnion temperatures drop, and at some point the lake becomes isothermal and there is no longer any density stratification (the water is isopycnal). At this point, wind stress over the surface of a lake can mix even deep lakes all the way to the bottom, renewing the dissolved oxygen in the deep lake waters. This mixing and renewal of dissolved oxygen content of the lake waters is referred to as *turnover*. As the surface water cools further into winter, the epilimnion again becomes less dense than the 4-°C bottom waters. If ice develops, it is much less dense than the water and floats on top of the lake. Upon spring thaw, the lake warms up and again reaches an isothermal, non-density-stratified condition. At this point, there is another turnover, and the deep O_2 is once again renewed. In most temperate lakes, this turnover is enough to keep the lake oxygenated all the way to the bottom.

Surface lake water is always well oxygenated due to turbulence caused by wind and waves, and by O_2 production by photosynthetic organisms. However, if the productivity of the surface water of the lake is high, enough particulate organic matter sinks into the deeper portions of the lake such that all of the dissolved O_2 there is used up by aerobic bacterial breakdown of the organic matter. Such a lake would be considered *eutrophic*, and anaerobic bacterial respiration might take over. The thermocline separates the oxygenated surface layer from the stagnant, anoxic water (hypolimnion) below. This is the situation in the density-stratified Black Sea, and in this case the water contains substantial sulfate that is reduced to H_2S by sulfate-reducing bacteria at depth. Hydrogen sulfide is a deadly poison to most eukaryotes. Many freshwater lakes do not contain much sulfate, so other anaerobic bacterial reactions take place in these eutrophic lakes.

In tropical areas, lakes remain density-stratified (*meromictic*) throughout the year and substantial amounts of methane, H_2S, and carbon dioxide can build up in the deeper waters below the oxygenated epilimnion. Rare occurrences of turnover in these lakes can lead to the mass asphyxiation of people living around the lakes, as was the case in Cameroon in 1984 when Nyos crater lake overturned and killed ~1,700 people, apparently mostly from carbon dioxide asphyxiation.

Thermal stratification may be reinforced by salinity stratification, and most saline lakes are permanently density-stratified due to salinity variations. Evaporation can take place only from the surface of a lake, so it is there that the densest water will be generated and then sink. Density gradients may be so large that even the shallowest saline lakes do not turn over with wind stress on the surface.

The density stratification due to spatial variation in temperature and salinity can influence the nature of mixing between lake water and incoming river water. Commonly, there is not much difference between incoming river water and receiving lake water, and mixing occurs on all fronts. Lake Geneva, on the other hand, is warmer than the glacial-derived river inflow that can carry large amounts of suspended sediment. In this situation, the river flows as a density current along the lake bed. In other cases, the incoming water may actually flow out along the thermocline instead of the bottom of the lake.

Water currents

Lakes have complex circulation patterns, mainly controlled by waves and wind-driven surface currents. Tides occur only in the largest lakes and they are never very strong. Detailed discussions of circulation in lakes can be found in Csanady (1978) and Imboden and Wüest (1995). The size of waves in lakes depends on the strength and duration of the wind, and the distance over which the wind blows (the fetch). In general, the largest waves can be generated on the largest lakes. However, waves in lakes are usually less important than waves on continental shelves, which receive long-period swell waves and hurricane-generated waves. Waves are most important along the shorelines of lakes. Wave-cut cliffs are common features of lake shorelines, but sand and gravel beaches and spits develop where sediments are available and the shorelines resemble "microtidal" marine shorelines. Wave currents transport progressively finer sediment with increasing distance (depth) offshore. Wave currents can mix lake surface layers or an entire isothermal water column with uniform density.

Sustained wind from one direction causes the surface water of lakes to flow in the direction of the wind, causing water-surface elevation (set-up) at the windward edge of the lake, as is the case in the ocean (Chapter 7). This also results in downwelling of water at the windward edge of the lake, water flow in the opposite direction to the wind deeper in the lake, and

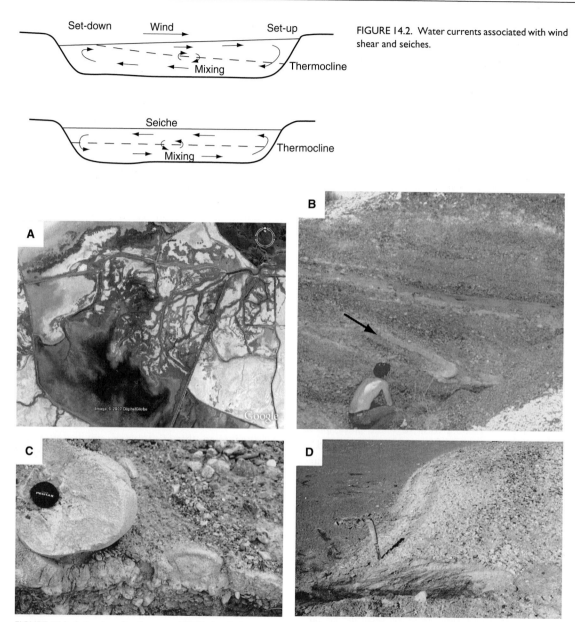

FIGURE 14.2. Water currents associated with wind shear and seiches.

FIGURE 14.3. Lake-shoreline features. (A) The bird's-foot delta where Bear River enters Great Salt Lake. North is to the top of the page. Width of view ~10 km. From Google Earth™ (B) Dipping bedsets of gravel and sand deposited in a Pleistocene "Gilbert-type" delta into Walker Lake, Nevada, USA. The bed at the arrow is a calcium carbonate "tufa." (C) Detail of the tufa in (B). (D) A trench in a small beach ridge marking the former position of the shoreline of Walker Lake. The beach ridge comprises lakeward-dipping sets of sand and gravel that truncate each other at low angles.

possibly upwelling of cold water at the opposite side of the lake (Figure 14.2). Such water circulation results in mixing. When the wind stress is removed, the water moves back across the lake, similarly to the storm-surge ebb that accompanies hurricane-generated waves on continental shelves. This surface wave may rebound from the opposite end of the lake and from the sides. This oscillation (sloshing) of the water surface is known as a *seiche*. Seiches can have amplitudes of greater than 1 meter and periods of a few hours. Seiche currents also have an important effect on mixing water layers and the suspended load in a lake (Figure 14.2). However, seiche-generated currents are generally not strong enough to move bed load. Seiches have been known to set up similar waves along the thermocline as well as the surface of the lake. These are called *internal waves* and would have much the same sediment-transport properties as surface seiche waves. In larger lakes, the Coriolis force can affect the circulation patterns, producing large-scale circulation gyres, Ekman spirals, and so on (see Chapter 7).

Lake water levels

Lakes commonly undergo short-term changes in water level due to seasonal variations in river discharge, precipitation, and evaporation. Water-level response to variations in inflow is less in deep lakes with steep shorelines, and in lakes with inlets and outlets. However, in shallow lakes with gently sloping beds, slight variations in inflow can lead to substantial changes in lake area. For example, much of the area around the Great Salt Lake is a flat plain that was the bottom of Pleistocene Lake Bonneville. Longer-term lake-level changes are due to climate change and tectonics. Changes in lake levels clearly affect the location and action of currents and the sites of chemical sedimentation. Lakes in closed basins are most susceptible to water-level changes.

Long-term changes in lake levels have been documented throughout the western USA. The Basin and Range Province is the site of numerous rift valleys, many of which contain fresh to saline lakes, playas (Chapter 16), and salt marshes. During the Last Glacial Maximum (LGM) around 18,000 years ago, lakes in this area (so-called *pluvial* lakes such as Lake Bonneville and Lake Lahonton) were much larger and deeper than they are today. Death Valley, now almost completely dry, was occupied by a lake 100 m deep. During the LGM, these large lakes filled their basins and emptied into adjacent basins over spill ways, and some developed into through-going drainage systems with outlets down the Colorado River. Over the last 18,000 years, these lakes responded rapidly to changes in climate and/or inflow patterns, resulting in large changes in their depths and sizes. Many, such as Owens Lake, dried up entirely when inflow streams from the Sierra Nevada were diverted to Los Angeles after the Second World War. In contrast, tropical lakes in Africa and Australia were low during glaciations and higher during interglacial periods.

Physical processes of deposition

Marginal and shallow-water deposition

Deposition along the margins of lakes is concentrated around river mouths. The nature of the deposits depends upon the suspended and bed load carried by the river, and the relative density of the river water and the receiving lake water. Overflow of rivers is common in saline lakes and this generates bird's-foot-type deltas (Figure 14.3A; Chapter 15). Where rivers enter freshwater lakes of nearly the same density there is rapid mixing, and bed load is deposited in a *Gilbert-type delta* (Figure 14.3B, C: details in Chapter 15). G. K. Gilbert described these types of deposits from around the shoreline of ancient Lake Bonneville, a large lake that occupied vast areas of the Utah desert, now reduced to the much smaller Great Salt Lake. Gilbert-type deltas are also common in glacial-lake deposits. If underflow occurs due to low river temperature or high suspended-sediment concentration, delta formation is inhibited and turbidity currents deposit sediment in the deeper portions of the lake. The depth of the lake must be large enough to allow delta deposition. Very shallow lakes can receive sheet floods that leave extensive sheets of sand containing climbing-ripple cross stratification (Smoot and Lowenstein, 1991). This was the case when the Colorado River avulsed into the Salton Sea basin, where it produced the current Salton Sea in 1910.

Wave activity will rework river sediment into beaches and spits (see Chapter 15). However, because of the generally low wave energy and frequent water-level fluctuation, these features are not as well developed as at marine shorelines (Figure 14.3D). Wave-cut platforms are common along the former shoreline traces of pluvial lakes in the American southwest. It is also common to find the beach ridges built by local reworking of the shoreline to be cemented by carbonate and to host an array of shoreline tufas. Wave activity in shallow water above storm-wave base typically results in bed-load transport and the formation of wave ripples and plane beds. Episodic deposition in these areas results in heterolithic (flaser, wavy, lenticular) bedding with wave-ripple-laminated and planar-laminated sands interbedded with mud (e.g., Figure 14.4).

Deep-water deposition

Deep-water sediment deposition in lakes is mainly from suspension (Figure 14.5). Suspended load of rivers is moved offshore by river currents, wave currents, or wind-shear currents. The suspended load in the surface water will settle in order of decreasing settling velocity. This will result in a fining of bottom sediment from the source(s) (proximal) out to the distant (distal) parts of the lake. The sediment distribution

FIGURE 14.4. (A) Wave-generated wavy bedding deposited in ~3–5 m depth in Owens Lake, western USA, prior to its desiccation. Sand lenses are composed of ooids. (B) A core from Late Pleistocene deposits of Pyramid Lake, showing disrupted wavy bedding. The light color is fine-grained, wave-rippled, terrigenous sand with ostracods, and the dark color is aragonite mud. The sediment is bioturbated, and a sand-filled desiccation crack is prominent in the center of the photo. The staple is 7 mm across. Photos courtesy of Joe Smoot.

will be affected in detail by the complex circulation due to wind-shear currents, seiche currents, and density stratification. In saline lakes, the flocculation of aluminosilicate muds will accelerate settling (see below). In addition, benthic filter-feeding organisms, such as clams, bryozoans, and sponges, and planktonic filter-feeding organisms, such as copepods, will capture inorganic as well as organic material and pelletize the suspended sediment, greatly increasing its settling velocity. The introduction of the prolific Eurasian zebra mussel into North American lakes has resulted in greater water clarity due to the prodigious filtering activities of these bivalves. Suspension settling of sediment commonly leads to very flat lake floors, with high-amplitude, bottom-parallel seismic reflections. However, in many lakes, the bacterial decomposition of organic matter in the basin-floor sediments leads to gas bubbles that deform the sediment layering, thus masking seismic reflections.

In addition to suspension deposition, underflow of river water may introduce coarser sediments into deep water. The main cause of density currents in most temperate lakes is temperature differences, not necessarily suspended-sediment concentration. Underflows of river water that do not contain sediment can still lead to reworking of the lake bottom. Many lakes, especially in tectonic basins and glacially scoured terrains, may have side slopes at and above the angle of repose. Deltas entering lakes also commonly build angle-of-repose slopes into deeper water. These slopes are subject to slumping, which gives rise to debris flows and turbidity currents that can further erode the slope and ultimately deposit sediment on the lake floor (Figure 14.5). Relatively coarse-grained, ungraded and graded strata in lake deposits could result from river-flood density currents and sediment gravity flows originating from slumping of the lake margins. Such density currents may lead to a system of bottom

B FIGURE 14.5. Portions of cores from (A) Holocene deposits on the floor of Walker Lake, western USA; and (B) Late Pleistocene deposits beneath the floor of Pyramid Lake, western USA. Arrows show turbidity-current deposits interstratified with suspension settle-out laminae. In (A), the dark laminae are mud, light laminae are silts and sands. In (B), light laminae are aragonite mud; dark laminae are clay and organic-rich. The staple is 7 mm. Photos courtesy of Joe Smoot.

channels and levees similar to those found on submarine fans (Chapter 18).

Chemical and biological processes of deposition

Calcium carbonate and silica

Whether or not biogenic and chemogenic sediments will be an important component of the lake sediment depends on the amount of such sediment produced and how much it is diluted by terrigenous sediment. In saline lakes, biogenic and chemogenic sediment can be dominant. However, in freshwater lakes with substantial river inflow, the biogenic and chemogenic component of sediments is usually swamped by the terrigenous input.

Carbonate sedimentation in freshwater and brackish-water lakes is due to a variety of processes. In shallow, benthic areas in the photic zone, cyanobacterial mats, green algae, and aquatic plants flourish. Cyanobacteria take in CO_2 for photosynthesis, increase the pH, and increase the saturation of calcium carbonate around their filaments, inducing precipitation of calcium carbonate. This so-called *periphyton* carbonate is released as carbonate mud upon death of the filament. Periphyton carbonate mud is a substantial contributor of carbonate

sediment to the Everglades in southern Florida. *Chara* is a freshwater green alga, commonly called stonewort. The fronds and stems of *Chara* and related genera (collectively known as charophytes) are usually encrusted by fine-grained carbonate sediment, which is also precipitated out of the water by the photosynthetic removal of CO_2 from the water adjacent to the fronds and stems. The spore cases of charophytes are calcified and are known as calcispheres. Locally, charophytes can produce freshwater marls that form benches around the lake shore. Periphyton and charophyte mud can also settle over benthic microbial mats to form stromatolites. In places, the mats may grow around intraclasts and other particles, and wave agitation of these particles produces *oncoids*. Sand-sized carbonate sediment in freshwater and brackish-water lakes includes shells and shell fragments of clams, snails, ostrocodes, and conchostracans.

Although there are some rare carbonate-secreting planktonic algae in lakes, it appears that carbonate precipitation directly from lake-surface water (whitings) is mainly due to chemical precipitation, perhaps triggered by uptake of carbon dioxide by cyanobacterial blooms. The most important photosynthesizing plankton in lakes are the diatoms, that produce an array of beautiful microscopic tests composed of opaline silica.

In saline lakes, the only common calcifying invertebrates are ostracods. Although charophytes are restricted to freshwater lakes, microbial mats are very common in saline lakes and can produce large amounts of calcium carbonate sediment. Ooids are common in shallow parts of saline lakes (Figure 14.4A) but fairly rare in freshwater lakes, where they are confined to springs.

Another important site of calcium carbonate precipitation in lakes is where springs enter either along a shoreline or at the lake bottom. In many cases, the mixing of spring water and lake water induces carbonate precipitation as massive tufas and travertines. The freshwater Green Lake near Syracuse, New York, has shoreline terraces composed of tufa. Tufas in saline lakes are more fully described and illustrated in Chapter 16. Saline lakes are important sites of evaporite deposition, and this also is described in Chapter 16.

Lamination of lake-floor sediments

Lake-floor sediments are deposited by settling of suspended sediment and by sediment gravity flows, and the episodic nature of these processes (described

FIGURE 14.6. Laminated "oil shales" from the Wilkens Peak Member of the Eocene Green River Formation, western USA. Dark laminae are organic (kerogen)-rich, and light laminae are dolomitic. Lamina irregularities in (A) are due to compaction and small faults. Irregularities in (B) are due to compaction and the organic layers, some of which may have originated as bottom-growth microbial mats. The scales are 2 cm long. Photos courtesy of Joe Smoot.

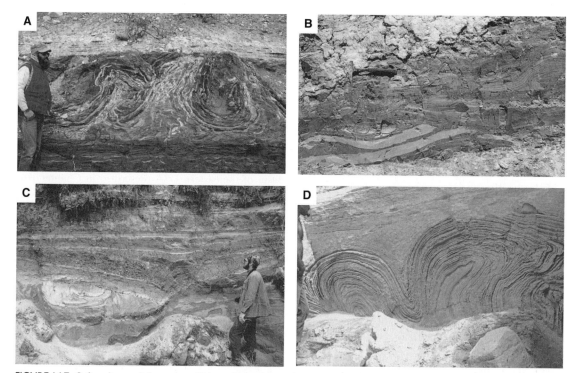

FIGURE 14.7. Soft-sediment deformation of lake and lake-delta sediments, western USA. (A) Brownish Late Pleistocene–Holocene laminated sediments from Mono Lake that were deformed and erosionally overlain by deltaic gravels. (B) Detail of deformed sediments from the base of the outcrop in (A). (C) Deformed Holocene deltaic sediments from Mono Lake. (D) Lake-floor sediments of Pleistocene Lake Bonneville contorted due to slumping. All deformation was probably triggered by seismic shocks.

above) gives rise to lamination (Figure 14.5). Laminated sediments are commonly composed of two components: mud laminae, formed by more or less continuous settling from suspension, and coarser sand and silt laminae that represent deposition during river floods and/or sediment gravity flows. In some lakes, short-lived events such as algal blooms or whitings may produce additional lamina types (organic-rich, diatom-skeleton rich, or carbonate-mud rich) that will occur with the terrigenous laminae (if present). In saline lakes, blooms of cyanobacterial and halophytic bacteria can produce tremendous amounts of organic material that can accumulate in the sediments if it is not oxidized. Organic-rich muds may eventually become oil shales (Figure 14.6). If the bottom water of the lake is oxygenated, lamination will be bioturbated and destroyed by benthic organisms, especially if the bioturbation rate/deposition rate ratio is high. In areas without benthic organisms, due to anoxia or a high deposition rate, the laminae are preserved. The laminae may be related to short-term variations in sediment supply (associated with individual floods, storms, or sediment gravity flows), seasonal changes in the weather, or longer-term changes in climate or tectonic activity. In ancient lake sequences, laminae are commonly interpreted to be annual, and are referred to as "varves." Varves were originally described from glacial lake deposits and are further discussed in Chapter 17. Although some variations in lake chemistry and temperature are annual, plankton blooms are notably sporadic, as are floods on rivers, and it is likely that many laminae in lake sequences are not annual.

Syn- and post-depositional deformation of lake sediments

Deformation of lake sediments can be due to bioturbation, generation of gas from decay of organic matter, and gravitational effects. Bioturbation, lacustrine trace fossils, and gas generation are described in Chapter 11 (Bridge, 2003; Hasiotis, 2004). Lake sediments accumulating in areas of active tectonics commonly exhibit abundant soft-sediment deformation features induced by earthquakes. Earthquakes can also trigger slumping of lake-margin sediments, and folded and otherwise deformed laminated sediments are common in ancient lake deposits (Figure 14.7).

Recognition of ancient lake deposits, with examples

Lake deposits might look superficially like normal marine deposits, but there are several distinguishing features, related to physical processes, mineralogy, and the biogenic component. Evidence of tidal currents would be lacking, and evidence for large waves and strong wave currents would be uncommon. Deltaic deposits would most likely contain evidence of fluvial dominance. There would also be evidence for frequent water-level changes, especially in the case of closed-basin lakes. Frequent and significant fluctuations in the chemical composition and salinity of lake water should give rise to significant differences in the mineralogy of lake laminae. The absence of fully marine fauna, and the possible presence of non-marine fauna such as certain gastropods, bivalves, ostracods, fish, and other vertebrates, should be distinctive. Since lakes are sensitive to climate changes, the fauna may be under stress, and therefore of low diversity.

Many ancient lake deposits accumulated in closed basins. Important examples include the Pleistocene to Holocene, so-called "pluvial" lakes of the southwestern USA, the Eocene Green River Formation of the western USA, and the Lockatong Formation of the Newark Basin of the eastern USA (described in Chapter 16). Pleistocene and older glacial-lake deposits are known for their "dropstones" and varved sediments (Chapter 17). Talbot and Allen (1996) provide other examples of ancient lake deposits.

15 Coasts and shallow seas

Introduction

Coasts and shallow seas are marine areas that lie adjacent to land. Coasts include a broad range of environments, including deltas, strandplains, barrier islands, beaches, estuaries, tidal inlets, tidal flats, lagoons, and rocky cliffs. Shallow seas (water depths up to 200 m) include continental shelves adjacent to the deep ocean and partly enclosed seas (epicontinental seas) such as the North Sea, Hudson Bay, and the Arabian Gulf. "Seas" that are completely enclosed by land, such as the Black Sea and the Caspian Sea, are discussed in Chapters 14 and 16 dealing with lakes. Continental shelves are commonly on the order of 100 km across (range 2–1,000 km) and nearly all have a distinct break in slope at their oceanward edge (known as the shelf–slope break). The oceanward edge of a shelf may be marked by sand shoals or by organic reefs. The depth of the shelf–slope break ranges from 0 m (at shelf margins with organic reefs) to 200 m, but averages 125 m. Thus, most continental shelves have slopes on the order of 1 in 1,000.

Sediments of coasts and shallow seas are mainly terrigenous in origin, with varying amounts of biogenic, chemogenic, and volcanogenic sediment. However, biogenic carbonate sediment can be dominant in areas where the input of terrigenous sediment is minimal, such as distant from the mouths of large rivers. Shallow seas with carbonate sediments are best known for their organic buildups (such as coral/algal reefs) and their ooid and skeletal sand shoals, that tend to occur at the shelf–slope break. Water flow and sediment transport at coasts (associated with wind waves, wind-shear currents, and tides: see Chapter 7) are intimately related to those in the adjacent shallow seas, which is why these environments are being treated together.

Coasts are prime sites for human settlements, with their associated buildings, harbors, and navigation channels. Problems arise as coasts become developed, some due to natural processes and some due to human activities, e.g., sedimentation in navigated waterways; beach erosion and damage to buildings due to seasonal storms and longer-term sea-level rise; exacerbation of beach erosion by degradation of vegetation associated with excessive transit of pedestrians and vehicle use; and loss of sediment supply to beaches by longshore transport from river mouths, due to damming of rivers upstream and building of jetties at the coast. Very expensive engineering solutions to these problems include dredging coastal waterways; building jetties, groins, and sea walls; and beach replenishment. Unfortunately, many engineering approaches fail due to insufficient knowledge of the flow and sedimentary processes at coasts.

In shallow marine environments, stormy seas, strong wave and tidal currents, and vigorous sediment movement on the sea bed can cause serious problems for ships in congested shipping lanes (e.g., the English Channel), and for oil and gas drilling platforms (e.g., the North Sea, Gulf of Mexico). Detailed knowledge of shallow-marine water flow and sediment transport is required. Human activities (e.g., specific types of fishing, tourist visits to coral reefs) have also caused considerable damage to shallow marine ecosystems that must be understood before protective measures can be implemented.

Ancient coastal and shallow-marine sandstone (and gravelstone) bodies, particularly those encased in mudstone, are important sources of fluids such as water, oil, and gas (e.g., Mesozoic to Cenozoic strata in North America and the Gulf of Mexico). Such sandstone bodies may be deposits of distributary channels and mouth bars on deltas, tidal inlets, and associated flood- and ebb-tidal deltas, beaches, or shallow-marine sand waves. Calcareous sandstone–gravelstone bodies may be associated with organic buildups (e.g., coral reefs) and shoal deposits, both of which are particularly prone to diagenetic modification of primary porosity and permeability. The geometry and sedimentary character (including porosity and permeability) of these sandstone bodies depend on the geometry, flow, and sedimentary processes in the specific depositional environments, and the nature of subsequent diagenesis. Successful evaluation, production, and management of these precious resources require detailed knowledge of the nature and origin of their sedimentary hosts.

Reviews and compendia of articles on coastal and shallow marine environments are numerous. Those primarily describing settings with terrigenous (siliciclastic) sediments include Alexander *et al.* (1998), Allen (1970), Allen and Pye (1992), Berendsen and Stouthamer (2001), Bergman and Snedden (1999), Black *et al.* (1998), Carter and Woodruffe (1994), Colella and Prior (1990), Dalrymple *et al.* (1994), Davis (1985), Davis and Fitzgerald (2004), de Boer *et al.* (1988), Flemming and Bartholoma (1995), Giosan and Bhattacharya (2005), Hunt and Gawthorpe (2000), Knight and McLean (1986), Komar (1998), Leatherman (1979), Leeder (1999), Morgan (1970), Nemec and Steel (1988), Nio *et al.* (1981), Nittrouer and Wright (1994), Nummedal *et al.* (1987), Perillo (1995), Rahmani and Flores (1984), Reineck and Singh (1980), Rhodes and Moslow (1993), Scholle and Spearing (1982), Sidi *et al.* (2003), Smith *et al.* (1991), Soulsby and Bettess (1991), Stride (1982), Swift *et al.* (1991), Walker and James (1992), Whateley and Pickering (1989), and Woodroffe (2003). Shallow marine carbonate environments are described in Bathurst (1975), Demicco and Hardie (1994), Enos and Perkins (1979), Flügel (2004), Ginsburg (2001), Hardie (1977), Hardie and Shinn (1986), James and Clark (1997), James and MacIntyre (1985), Logan *et al.* (1970, 1974), Purser (1973), Read (1985), Reading (1996), Schlager (2005), Scholle *et al.* (1983), Tucker and Wright (1990), Walker and James (1992), and Wilson (1975).

Controls on the geometry, water flow, and sediment transport of coasts and shallow seas

Terrigenous sediment supply

The type of terrigenous sediment and its rate of supply to coasts and shallow seas depend ultimately on that supplied by rivers, although erosion of the coast and sea bed supplies small amounts of sediment that can be redistributed locally. The sediment supply to rivers, in turn, depends on the land area and relief, rock types, vegetation, climate, and tectonic and igneous activity of the catchment area (Chapter 13). The sediment-transport rate of rivers is proportional to the product of water discharge and channel slope, and is inversely related to the mean grain size. Abundant mud and sand are supplied to coasts and shallow seas by large rivers flowing over gently sloping coastal plains (e.g., Mississippi, Niger, Amazon, Yellow). Abundant gravel is supplied to coasts by rivers flowing down the steep sides of fault-bounded basins (e.g., Copper). Major rivers of the Earth occupy regions of long-lived tectonic subsidence, such that major sources of sediment supply to coasts and shallow seas have remained fixed in position for many millions of years. Deltas occur at the mouths of major rivers, resulting in vast volumes of deposited sediment that accumulated over many millions of years. The supply of terrigenous sediment to coasts and shallow seas varies with time due, for example, to river floods and storms at sea, avulsions of coastal rivers, and tectonism and changes in climate in the hinterland. Shallow seas and coasts distant from major riverine sediment sources are where biogenic and chemogenic sediments can accumulate, and climate is an important control on the type of sediments deposited, as will be discussed next.

Climate

Climate affects the sediment supply to coasts and shallow seas, and the movement of water and sediment. The supply of terrigenous sediment from rivers is influenced by the climate of the catchment area (see above), and climate affects the production of biogenic and chemogenic sediment of coasts and shallow seas. Most of the calcium carbonate sediment of biogenic origin on modern tropical shelves comes from green algae and hermatypic corals, whereas calcium carbonate ooids and some of the muds are precipitated directly from supersaturated seawater. Climate influences the production of this biogenic and chemogenic sediment through its control of water temperature and salinity. Organic productivity (resulting in biogenic grains) is greatest in clear, warm–tropical, shallow seas of normal salinity with an appropriate nutrient supply. In arid regions with hypersaline seawater, organic activity and production of biogenic sediment are low, and evaporites are precipitated. Climate indirectly influences organic productivity by affecting the water currents that transport sediment and nutrients. Climate affects the movement of water, and hence sediment transport, by being responsible for wind waves and wind shear currents (including geostrophic flows), and by controlling the density of river water and seawater. Density currents can result from differences in salinity or temperature.

Organic activity

The abundance of marine animals and plants is controlled not only by water temperature and salinity (climate), but also by light intensity, turbidity, the concentration of dissolved oxygen, and nutrient supply (which is influenced by water depth and currents). Benthic animals live mainly in the upper 10–15 cm of the sediment bed (infauna) or on the surface (epifauna). Their morphology and life habits are adapted to their physical, chemical, and biological environment, which results in distinctive associations with other biota and with sediment type (Chapter 11). For example, sandy beds subjected to frequent bed-load movement, erosion, and deposition are occupied by relatively few species. Infauna (burrowers) that can easily move up and down in the sediment dominate such dynamic settings, and epifauna (surface dwellers) have large heavy shells that can withstand physical abrasion. The animals (e.g., bivalves) are mainly suspension feeders that filter food from water passing through their guts. Muddy beds with high concentrations of suspended sediment (such as near river mouths) also have few species. However, muddy beds with low sediment-transport rates support a larger number of epifaunal and infaunal species that are suspension feeders and deposit feeders. The biota on and in the bed changes with time as the conditions of water flow and sediment transport change.

Both filter feeders and deposit feeders aggregate the micron-sized clay particles into sand-sized pellets that can move as bed load (see Chapter 4). Furthermore, if the deposition rate is relatively low, deposit feeders can completely destroy primary sedimentary structures as they burrow for food and shelter. Benthic animals and plants can also stabilize and build up the bed by secreting sticky organic films (mucus), growing sediment-trapping frameworks, and inducing chemical precipitation (e.g., reefs, cyanobacterial mats, mangrove swamps: Chapter 11).

Cyanobacteria are ubiquitous on areas of the sea bed that sunlight reaches, and can develop a luxuriant microbial mat if benthic organisms that graze on them are excluded. Supratidal mudflats along coastlines in temperate climates are covered by distinctive salt-marsh grasses. These salt-tolerant grasses are replaced by mangroves in tropical climates. Angiosperms are also common, and a few genera of flowering plants (e.g., eel grass and turtle grass) have adapted to live subtidally. These plants have a stem known as a rhizome from which the roots descend and the leaves ascend. Such sea grasses form extensive meadows and can be important habitats for benthic invertebrates, fish, and mammals such as manatees.

Water chemistry

The salinity (about 35‰) and major-ion composition of surface oceanic water do not vary significantly today (see Chapter 4). However, important local variations in water chemistry can lead to characteristic changes in organic activity and chemogenic sediment formation. For example, low salinity and high mud concentration in water around the mouths of rivers result in low species diversity and population density. Periodically high concentrations of silica near river mouths may lead to diatom blooms. Elevated salinity occurs in restricted shallow seas in arid climates, leading to precipitation of evaporites (see Chapter 16). Most calcium carbonate sediments are biogenic, but they may be precipitated inorganically where temperature and salinity are locally elevated. The composition of seawater has changed throughout geological time (Chapter 2), and the effects on carbonate sediment production and evaporite deposition are discussed in Chapter 4.

Upwelling of phosphate-rich oceanic water at shelf edges increases organic productivity, and may lead to precipitation of phosphate-rich sediments. Upwelling water commonly has a low concentration of dissolved oxygen because of bacterial decomposition of the suspended organic matter. In areas of low sedimentation rate, iron-bearing minerals such as glauconite and chamosite may form at the sediment–water interface and become a substantial fraction of the sediment.

Physical properties of seawater

The density of seawater depends on its temperature, salinity, and suspended-sediment concentration. Density currents arise due to these temporal and spatial variations in fluid density. Relatively cold and/or sediment-laden river water may sink below seawater (hyperpycnal flow), and may become a turbidity current. However, river water may also flow over the surface of the sea (hypopycnal flow) by virtue of its relatively low salinity. Hypersaline seawater near an arid coast may sink and flow seawards, its place being taken by an onshore flow of normal seawater. Such density currents can occur at any depth in the sea, depending on where and how the density differences originate.

Water currents

Other types of water currents occur in shallow seas in addition to density currents. Rivers entering the sea are a type of jet, the flow of which is related to the river discharge, the geometry of the river and sea bed, and the relative density of the river water and seawater. Water currents associated with wind waves and surface wind shear are particularly important on shallow seas and coasts facing large oceans (Chapter 7). Cyclic changes in hydrostatic pressure on the bed associated with water waves can also induce slumps and sediment gravity flows (Chapter 8). Tidal currents are particularly important in semi-enclosed seas, elongated bays, and gulfs. Wind-related currents and tidal currents are responsible for redistributing sediment deposited at river mouths.

Sea-level change

Relative sea-level change can be due to any combination of tectonic uplift or subsidence, eustatic sea-level rise or fall, and erosion or deposition (Chapter 20). Most coastal areas of the Earth are currently experiencing a eustatic rise in sea level of approximately 1–2 mm per year. This is a continuation of a rise in sea level of approximately 125 m that has taken place since about 18,000 years ago at the Last Glacial Maximum (LGM). Apparently, there have been times since the LGM when the rate of sea-level rise has been as much as 10 mm per year! Some coastal areas of the Earth are currently rising fast enough to outpace the general global (eustatic) sea-level rise and are thereby experiencing falling sea level. Such areas are adjacent to active mountain belts (such as in Taiwan and New Zealand) and northerly continental areas (e.g., northeastern North America and Scandinavia) that are experiencing isostatic rebound following melting of large ice sheets since the last glacial maximum (Chapter 17).

Relative sea-level change affects the nature of water currents, sediment transport, erosion, and deposition. The relative importance of terrigenous versus biogenic and chemogenic deposition may also be linked to sea-level changes. *Rising sea level* results in marine transgression, increased deposition (sediment trapping) in drowned river valleys, reduced deposition in coastal areas, and possibly starving of deeper marine areas of sediment. River avulsions are expected to be more frequent as a result of the increased deposition rate in coastal rivers. River valleys flooded with seawater are estuaries, and the irregularity of estuarine coastlines is expected to increase the magnitude of tidal currents relative to that of wind-related currents (Chapter 7). New areas of shallow shelf surrounded by land masses (e.g., the North Sea) are also expected to lead to increased tidal range and currents. Wind-related currents would not be as effective at moving sediment on the deepening water of the shelf. These areas might not receive terrigenous sediment (or only minor amounts of mud) and this situation would be conducive to the formation of biogenic and chemogenic sediment. A low deposition rate in marine areas gives rise to so-called condensed sequences that are intensively burrowed and mineralized. If sediment is not being transported to the shelf edge, or not produced there, submarine canyons would become inactive.

Falling sea level results in marine regression, shallowing of the sea, and a decrease in shelf width. Wind-related currents are expected to become more important than tidal currents, resulting in a change in coastline configuration. Under some circumstances, former wave-cut platforms and beaches are elevated into flights of coastal terraces. In addition, rivers may become incised, and the eroded sediment would then be transported onto the marine shelf. Active transport of sediment on the shelf may lead to reactivation of submarine canyons and deep-sea fans.

Classification of coasts and shallow seas

A delta (so-named because its plan form was thought to resemble the capital Greek letter Δ) is a coastal protrusion formed where a river channel enters a basin (a lake or sea) and deposits more sediment than can be carried away by currents in the basin. All major rivers have some kind of delta. Coasts dominated by fluvial deposition may have fan deltas or finger-like distributaries with adjacent levees that extend into the sea, depending on the relative density of the river water and basin water. If the currents in the sea or lake can move away all of the sediment provided by the river, there will be no protrusion at the coast, such as the modern Amazon. Coasts dominated by wind-related currents and sediment transport have characteristic physiographic features such as beaches and barrier islands, and adjacent shallow seas have linear sand ridges that are either transverse to the direction of dominant wave approach or more or less parallel to the wind-shear direction. Coastal areas dominated by tidal currents and sediment transport have well-developed tidal channels and tidal flats, and shallow seas have areas of sand molded into linear ridges and sand waves. The linear sand-ridge crests are more or less parallel to the main tidal current directions. Coasts where both wave and tidal currents are important tend to have beaches cut by numerous tidal channels. Rocky coasts occur where the rate of sediment removal exceeds supply. Organic buildups in shallow seas occur where the supply of terrigenous sediment is minor and local organic activity produces or traps more sediment than can be carried away by water currents. This overview of the genetic classification of coasts and shallow seas is elaborated upon below.

The geometry and deposits of coasts and shallow seas have been classified in terms of the relative importance of river, wave, and tidal currents in transporting and depositing sediment (e.g., Fisher et al., 1969; Coleman and Wright, 1975; Galloway, 1975; Hayes, 1979; Homewood and Allen, 1981; Boyd et al., 1992; Dalrymple et al., 1992; Orton and Reading, 1993). The early versions of these classifications (e.g., Figure 15.1A) are qualitative and simplistic, and difficult to apply. Quantitative measures of the sediment-transporting power of river, wave, and tidal currents are needed (Figure 15.1C). The sediment-transporting power of river currents can be expressed as the (flood-stage) discharge–slope product. The sediment-transporting power of waves could be indicated by storm-wave height or by storm-wave power. Tidal-current strength can be expressed in terms of the maximum tidal range. Tidal range has been classified as microtidal (<2 m), mesotidal (2–4 m), and macrotidal (>4 m). Coasts with microtidal range are commonly thought of as wave-dominated, and those with macrotidal range are tide-dominated.

The simple classifications mentioned above do not generally take into account all of the factors controlling the morphology of coasts and shallow seas, or how the controlling factors vary in space and time. The mean grain size of the available sediment must be considered, because the sediment-transport rate for a given flow strength increases as the grain size decreases (Orton and Reading, 1993; Figure 15.1C). The effects of wind-shear (including geostrophic) currents and turbidity currents are not accounted for. The relative density of river water and seawater, the flow velocity and geometry of the river, and the water depth in the nearshore region have an important effect on the nature of deposition at river mouths. Thus, Postma (1990) classified coarse-grained, river-dominated deltas in terms of the entering river system (single- or multi-channel), the relative density of river and basin water, and the depth of the basin. Also, the frequency and duration of the sediment-transporting activity of the various currents is as important as their magnitude. It is common in deltas for one or more of the distributary channels to be carrying the bulk of the sediment load, and for fluvial outbuilding to be dominant over removal by basin currents. Such an area would be experiencing a relative fall in base level. Other areas of the same delta may be experiencing subsidence (due to compaction or tectonism) and reworking by basin currents. At some stage, the locus of main deposition may shift to another distributary, and the abandoned area becomes one of subsidence, marine transgression, and reworking by basin currents. Thus different parts of deltas may be simultaneously under conditions dominated by either fluvial outbuilding or reworking by basin currents (Figure 15.2). An entire delta can also be abandoned due to an avulsion up valley, resulting in subsidence, marine transgression, and reworking by basin currents. The relative importance of river, wave, and tidal currents can also vary in space due to the configuration of the sea and coastline (Figure 15.2), and can vary with time during marine transgressions and regressions related to tectonism and eustasy. Dalrymple et al. (1992) and Boyd et al. (1992) classified coasts as a function of wave and tidal power, the relative contributions of sediment from the river and the sea, and whether marine transgression or regression was proceeding (Figure 15.1B). This classification erroneously implies that deltas are unlikely to occur during marine transgressions, and that estuaries and barrier beaches are unlikely to occur during marine regressions. None of the classifications cited above considers the role of organic buildups.

The remainder of this chapter is concerned with details of the processes, landforms, and sedimentary deposits of various types of coasts and shallow seas. Following a discussion of flow and sedimentary processes at river mouths, coasts and shallow seas are considered under the following headings: deltas; wave-dominated coasts and shallow seas; tide-dominated coasts and shallow seas; coasts and shallow seas with calcium carbonate sediments; and rocky coasts. The chapter ends with engineering, environmental, and economic aspects. Large-scale and long-term aspects of coastal and shallow-marine environments and their deposits are given in Chapter 20.

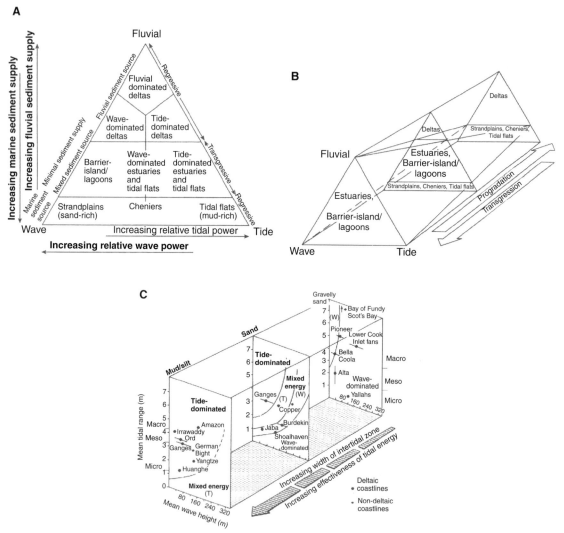

FIGURE 15.1. Classifications of shorelines and shallow seas on the basis of the relative importance of river, wave, and tidal currents. From Reading and Collinson (1996).

Flow and sedimentary processes at river mouths

When a turbulent jet of fluid (a river) enters a standing body of water (lake or sea), fluid at the margin of the jet mixes with the basin water, and the jet spreads out and decelerates (Figure 15.3). Deceleration of the flow results in deposition of the sediment load. Bed load is deposited close to the river mouth in the form of a mouth bar (also called a middle-ground bar). The finer sediment in suspension is carried further into the basin before deposition. Deposition of suspended sediment is concentrated along the sides of the outflow, because this is where deceleration is greatest. The

nature of this mixing, spreading, deceleration, and sediment deposition depends on the velocity and geometry (width, depth) of the river, the depth variation of the basin, the density difference between the river and basin water, and the grain size and density of the transported sediment (Bates, 1953; Wright, 1977; Edmonds and Slingerland, 2007). The density difference between river water and receiving water depends on contrasts in salinity, temperature, and suspended-sediment concentration.

Relative motion between two fluids results in wave-like instabilities along the boundary between them, and these may develop into turbulent (Kelvin–Helmholtz) vortices that move away from the boundary, thereby

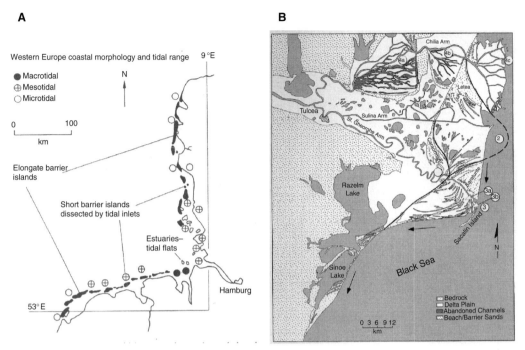

FIGURE 15.2. Examples of spatial variations in effects of river, wave, and tidal currents. (A) The decreasing length of coastal barrier islands in northwest Europe due to increasing tidal range. (B) Fluvial-dominated lobes (4) and wave-dominated lobes (3) of the Danube delta. From Bhattacharya and Giosan (2003).

mixing the two fluids. The degree of mixing along this boundary zone depends on the Reynolds number and (if there is a density contrast) Richardson number. As the river flow is always fully turbulent, the fluid mixing is dominated by dynamic pressure, and the fluid viscosity is irrelevant. The Richardson number is a measure of the stabilizing influence of the density gradient. In most cases, the density difference is not sufficient to remove turbulence, but may dampen it. The nature of mixing along the base of an outflow that is less dense than the basin fluid (or along the top of one that is more dense than the basin fluid) is also controlled by the nature of internal gravity waves, which is controlled by the densiometric Froude number (Equation (8.14)). If this Froude number exceeds 1, gravity waves can grow in amplitude and break, which enhances mixing of the two fluids. The densiometric Froude number increases as the outflow velocity increases, as the outflow thickness decreases, and as the density contrast decreases. Friction with the sediment bed also contributes to deceleration of the outflow, and causes widening of the spreading angle (Figure 15.3). Bed friction is particularly significant for shallow receiving basins.

If the densities of the river and basin water are similar (*homopycnal*), as in the case of a river with low suspended-sediment load flowing into a freshwater lake or a turbid river flowing into a brackish bay, the river and basin water mix in three dimensions. Edmonds and Slingerland (2007) numerically modeled the flow and sediment transport at the mouth of a river for the homopycnal case (Figure 15.3B), using a three-dimensional flow model that solves the Reynolds-averaged Navier–Stokes equations (conservation of mass and momentum: Chapter 5). This flow model was linked to models for transport of bed load and suspended load in order to predict the evolution of middle-ground (mouth) bars and associated submarine levees. The model demonstrated how the mouth bar (and associated levees) grew initially in height, then prograded downstream as it aggraded, and then stopped prograding downstream and increased in width, effectively causing bifurcation of the water stream (Figure 15.3B). Progradation and aggradation proceeded until the depth over the bar top became less than about 0.4 of the channel depth upstream. At this point, the bar started to cause more diversion of the flow around its sides rather than acceleration over its stoss side. The distance of the crest of the mouth bar from the river mouth increased with increasing channel width, mean depth, and mean flow velocity, and with

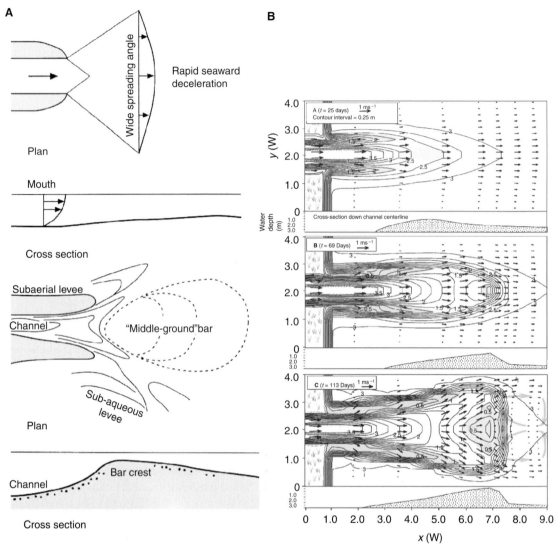

FIGURE 15.3. Effects of the relative density of river water and basin water on the nature of mixing and deposition at river mouths. In cases (A)–(C) the densities of river water and basin water are similar (*homopycnal*). In (A), the basin is shallow relative to the river, and river water is decelerated by bed friction in addition to turbulent mixing. The characteristic mouth bar bordered by bifurcated channels is common in crevasse splays prograding into lakes and interdistributary bays. (B) A numerical simulation of the evolution of a middle-ground bar. From Edmonds and Slingerland (2007). Bathymetric contours and velocity vectors are shown. In (C), the basin is deep relative to the river, and the river water is decelerated by turbulent mixing rather than bed friction (inertia-dominated). Rapid deposition of coarse sediment gives rise to mouth-bar fronts at the angle of repose, as in Gilbert-type fan deltas. (D) River water less dense than saline seawater (*hypopycnal*). The river water is buoyantly supported, allowing the river water to extend considerable distances into the basin as a narrow plume. Subaqueous levees result from rapid deceleration and deposition at the edges of the plume. This is typical of Mississippi River distributaries. Figures (A), (C), and (D) from Leeder (1999), based on Wright (1977).

decreasing grain size (for constant sediment density). The bar crest ranged from about three to seven channel widths from the river mouth as it evolved. The friction coefficient is an important parameter because it controls the rate of deceleration and spreading of the flow. The slope (and hence depth variation) of the basin floor is important because it controls the rate of deceleration

and spreading of the flow and the space available for deposition.

If a river with either a relatively low temperature or high suspended-sediment load flows into a freshwater lake (*hyperpycnal*), the river water flows beneath the basin water as a density current, and much of the sediment load bypasses the coast (Chapter 8). If a river

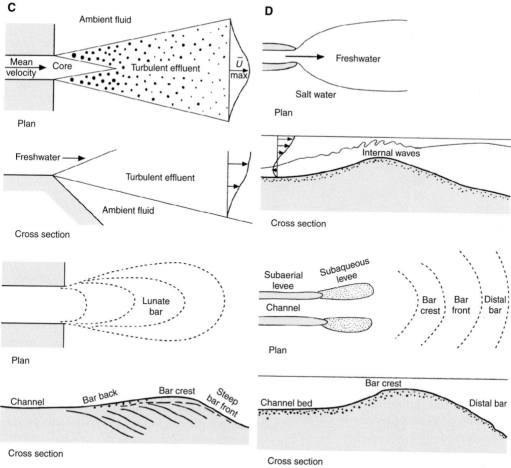

FIGURE 15.3. Continued.

enters a saline marine basin (*hypopycnal*), the fresh river water usually flows over the denser seawater as a *buoyant plume*. The degree of mixing is less than in the homopycnal case because of the density gradient and the relatively smaller cross-sectional area exposed to drag. In this case, the river outflow can extend for tens or hundreds of kilometers from the shoreline, allowing suspended sediment to travel well into the basin before deposition (Figure 15.3D). When clay suspended in river water mixes with seawater, it may flocculate, thereby increasing its settling velocity. The densiometric Froude number of the plume over the mouth bar may exceed unity, causing increased mixing. However, such mixing decreases the density contrast and slows the plume, thereby reducing the Froude number.

The nature of mixing of river outflow and basin fluid, and the associated deposition, are also affected by the currents (due to waves, wind shear, and tides) in the basin. Sediment gravity flows associated with rapid deposition on mouth bars or sediment-laden underflows serve to move and deposit sediment offshore. Furthermore, the relative importance of the river outflow and the basin currents in transporting sediment will vary seasonally. Therefore, patterns of sediment transport, erosion, and deposition at river mouths are extremely difficult to predict.

Hypopycnal conditions (i.e., a saline wedge) can also exist *within* a river channel (or estuary) before it enters the basin if tidal currents are insufficient to mix the fresh and saline water. Saline wedges can reach hundreds of kilometers inland at low river stage. If tidal currents are strong, the fresh and saline water in rivers and estuaries is well mixed. The degree of mixing of fresh and saline water in rivers and estuaries depends on temporal variations in tidal range and wave currents, and river discharge. During river floods, salt water may be completely evacuated from river channels (Wright and Coleman, 1973).

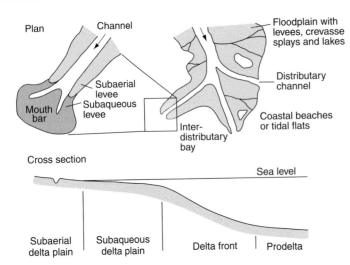

FIGURE 15.4. Definition and geometry of the delta plain and delta front.

Deltas

The major rivers of the world and their associated deltas have been the sites of development of great civilizations (e.g., Nile, Tigris–Euphrates) because of their abundance of natural resources. The deltas of the Mississippi, Amazon, Niger, and Ganges–Brahmaputra, for example, are home to millions of people. Deltas are associated with long-term tectonic subsidence and deposition. Thick and laterally extensive deltaic deposits commonly contain important reserves of oil, gas, and coal. Examples include the Carboniferous strata of North America and Europe, the Cenozoic deposits of the US Gulf Coastal Plain and Gulf of Mexico, and the Jurassic Brent Group of the North Sea region.

Geometry, flow, and sedimentary processes

Deltas can be divided into delta plain, delta front, and prodelta geomorphic regions (Figure 15.4). *Delta plains* are areas of low slope (on the order of 1 in 10,000) that may be subaerial or subaqueous. Subaerial parts are floodplains (with drainage channels, levees, and crevasse splays) traversed by distributary and anastomosing river channels. Subaqueous parts comprise marshes, lakes, lacustrine deltas, interdistributary bays, and tidal flats (e.g., Coleman *et al.*, 1964; Coleman and Gagliano, 1965; Tye and Coleman, 1989a, b). The type of vegetation on delta plains depends on the geographic and climatic setting. Swamps support woody vegetation, whereas marshes support non-woody plants such as grasses, reeds, and rushes. In tropical areas, the lower delta plain may be covered with mangroves, whereas salt marshes occur in temperate climates. In arid climates, delta plains may be almost devoid of vegetation, and can be sites of evaporite formation and aeolian dunes. Delta plains are affected mainly by river currents, and to a lesser extent by tidal currents and wave currents, so in this sense they are similar to normal alluvial floodplains. Edmonds and Slingerland (2007) reviewed the hydraulic geometry of distributary channel networks on delta plains and concluded that the width, depth, and length of channel segments decrease nonlinearly downstream as the channels became progressively bifurcated. The decrease in width and depth can be explained in terms of the decreasing discharge and by hydraulic geometry equations (Chapter 13). The decreasing length of channels can be explained in some cases by the process of bifurcation around middle-ground bars during delta progradation. As the delta progrades outwards, the distance of the points of bifurcation from the river mouths decreases together with the decreasing width and depth of successive bifurcating channels.

River flow in the lower reaches of rivers can be impeded by flood tides or storm-induced surges, enhancing the deposition rate in these areas. An increase in deposition rate may in turn lead to an increase in the frequency of river avulsion (Chapter 13). Tide-influenced channels on delta plains contain various kinds of sand bars, and are similar in form and process to estuaries. These channels are typically associated with highly asymmetrical tidal currents (Chapter 7), with flood-tidal currents tending to dominate the upper parts of channel bars, and tidal ebb currents enhancing fluvial currents on the lower parts of channel bars.

Areas of standing water (interdistributary bays, lagoons, and lakes) are greater on delta plains than on floodplains. Wave currents in these environments can transport sediment onshore, alongshore, and offshore. Strong tidal influence in interdistributary bays and lagoons is reflected in tidal flats and tidal drainage channels, the size and abundance of which increase with tidal range (Hayes, 1979). Since crevasse splays and levees have a high probability of prograding into the standing water of the delta plain, evidence of wave and tidal activity in their deposits (Chapter 8) might be more noticeable than in normal floodplains.

The generally high water tables on delta plains tend to result in more reducing conditions immediately below the sediment surface. Siderite nodules and pyrite replacements of plant material commonly form in sediments of interdistributary bays (e.g., Coleman and Prior, 1982). Considerable thicknesses of peat can accumulate in wet areas (marshes, swamps) of low sediment supply and high subsidence rate, provided that the appropriate vegetation is present and that plant litter is buried rapidly to prevent oxidation (e.g., McCabe, 1984; Kosters, 1989). Peat tends to occur in mires that can be raised meters above the delta plain. The peat-accumulation rate can be up to decimeters per year in some tropical deltas. Such raised mires form barriers to water and sediment movement and channel migration.

The *delta front* forms a rim around the delta plain and is the area where the sediment-laden river enters open water. Bed load is deposited immediately as the flow is decelerated, whereas suspended sediment is added to the delta front mostly beyond the zone of bed-load deposition. As the grain size or settling velocity of suspended sediment decreases, it settles progressively further from the river mouth, depending in detail on the effects of nearshore wave, wind-shear, and tidal currents. As discussed above, deposition of bed load and suspended load at the river mouth forms a mouth bar, which causes splitting of the river channel. A high deposition rate of suspended sediment along the sides of the outflow may give rise to submarine levees that pass downstream into the side of the mouth bar (Figures 15.3 and 15.4). The height of mouth bars and associated levees is typically on the order of 10^{-1} times the bankfull channel depth. The width and length of a mouth bar are typically several to many times the channel width, and its crestline is typically on the order of channel widths from the river mouth (e.g., Coleman and Gagliano, 1965; Wright and Coleman, 1974; Wright, 1977; Roberts *et al.*, 1980; Van Heerden and Roberts, 1988; Rodriguez *et al.*, 2000; Tye, 2004; Edmonds and Slingerland, 2007).

Deposition on the delta plain and delta front is episodic, being associated with distinct floods, migration of

bed forms such as ripples, dunes, and bars, and discrete sediment gravity flows. Mouth bars and submarine levees can grow and prograde seawards by many meters during major floods. During the progradation of deltas or crevasse splays into interdistributary bays and lakes (homopycnal flow), river-channel flow is split around mouth (middle-ground) bars into new distributaries, that in turn form new mouth bars further basinwards. These deltas (or subdeltas: Coleman and Gagliano (1964)) grow basinwards and laterally by repeated channel diversions, and growth and coalescence of mouth bars. Examples are the Atchafalaya delta of Louisiana (van Heerden and Roberts, 1988) and the Cumberland Lake delta in Saskatchewan (Edmonds and Slingerland, 2007) (Figure 15.5B). In contrast to these examples, extension of the main distributaries of the Mississippi delta by seaward progradation of mouth bars and associated levees under hypopycnal conditions, with little redistribution of sediment by marine currents, has resulted in a so-called bird's-foot plan form (Figure 15.5A). The distributaries extend into the Gulf of Mexico for tens of kilometers, but some distributaries have been maintained artificially by engineers.

The seaward slope of the mouth bar and delta front depends on the basin topography, the nature of deposition, and subsequent remobilization of sediment. It can range from less than a degree (as in the case of the Mississippi delta) to the angle of repose in the case of a Gilbert-type delta. A slump on a delta front may turn into a debris flow or mudflow, and then into a turbidity current. Slumping may be encouraged by earthquakes, wave loading during storms, and reduction in shear strength either by rapid deposition or by generation of methane gas as organic matter decays. Slump structures give a distinctive topography to the delta front: fault scarps, slump blocks, channels, and lobes (Prior and Coleman, 1979; Coleman *et al.*, 1983; Coleman, 1988); see Figures 8.7 earlier and 15.9 later. On the Mississippi delta front, slump and debris-flow channels can be tens to hundreds of meters wide and kilometers long. Turbidity currents also originate from hyperpycnal, sediment-laden river flows.

Deposition on the basin floor (the *prodelta* region) is also episodic, being associated with flood-related variation in suspended-sediment load and with turbidity currents. Episodic deposition on delta fronts and basin floors may also be related to diversions of river channels, possibly related to deposition of obstructing mouth bars at river mouths (Figure 15.5). Wave, wind-shear, and tidal currents are also responsible for sediment transport, erosion, and deposition on the delta front and in

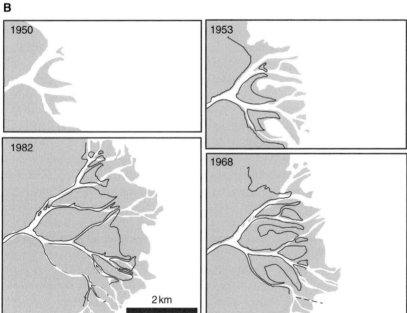

FIGURE 15.5. (A) The Mississippi delta showing the bird's-foot pattern of distributaries produced by seaward progradation of mouth bars without significant reworking by waves. Some of the straight distributaries are artificially maintained. The barrier-island coastline to the right is a wave-reworked front of an abandoned delta lobe. (B) Development of the Cumberland Lake delta in Saskatchewan by growth and coalescence of mouth bars, channel splitting around mouth bars, and avulsions. Modified from Edmonds and Slingerland (2007).

the prodelta region, and may have a dominating influence on their geometry, as will be seen below.

Processes and deposits of Gilbert-type deltas

Deltas that have fronts at the angle of repose, and that are not greatly influenced by basin currents, are commonly referred to as Gilbert-type deltas, named after the person who first described them at the margins of Pleistocene Lake Bonneville in Utah (Gilbert, 1885). These deltas are particularly common where meltwater rivers flow into proglacial lakes (Chapter 17) and sediment is composed of gravel, sand, and silt. As the delta progrades basinwards, the deposits of the river channel

FIGURE 15.6. (A) A gravel fan delta with a turbid plume at the mouth of a distributary channel, Bow Lake, Banff National Park, Canada. From Miall and Smith (1989). (B) A section through a Gilbert-type fan delta from the Canadian Rockies with a flat surface of the delta plain and an angle-of-repose delta front. Topsets, foresets, and bottomsets are marked. (C) A closer view of foresets, deposited by sediment gravity flows. (D) A close-up of bottomsets, showing turbidity-current deposits. Images (B)–(D) from Bridge (2003).

and delta plain are superimposed on those of the delta front. Thus, a three-fold division of deltaic deposits can be recognized (Gilbert, 1885) (Figure 15.6). The relatively flat-lying and coarsest deposits of the delta plain are referred to as *topsets*. The angle-of-repose deposits of the delta front are called *foresets*, and the finest, low-inclination deposits of the basin floor are called *bottomsets*. The terms foreset and bottomset have also (confusingly) been applied to cross strata formed by ripples and dunes.

A Gilbert-type delta is a form of fan delta, in that it is a streamflow-dominated alluvial fan that progrades directly into a standing body of water (Nemec and Steel, 1988; Colella and Prior, 1990; Miall, 1996; Reading, 1996). However, other types of fan deltas do not have delta-front slopes as steep as the angle of repose, and they may build into standing bodies of water that have more energetic basin currents. Those that build into the sea are discussed in the next section.

Processes and deposits of river-dominated marine deltas

Sediment deposits of river-dominated delta plains are very similar to those of alluvial plains (Chapter 13), but some differences may be discernible. Rivers flowing on very low slopes through muddy deltaic plains tend to have single channels with relatively low width/depth ratios, and to carry relatively fine-grained sediment. Although dunes cover most of the bed during high-flow stages, ripples and upper-stage plane beds also occur. Therefore, medium-scale trough cross-stratified sand should occur in the lower parts of channel-bar deposits, but small-scale cross strata and planar strata are expected in the upper parts. Since channel avulsion occurs relatively frequently on delta plains (at intervals on the order of hundreds of years), and delta plains are very wide compared with channel-belt widths, the proportion and connectedness of channel-belt deposits are expected to be low (Chapter 13). As a channel-belt becomes abandoned and river flow diminishes, it will subside, and the lower reaches may be influenced by marine erosion and deposition. Further information on the river avulsions of the Mississippi and Rhine–Meuse deltas is given below and in Chapter 13.

Levees, crevasse splays, and drainage channels on delta plains may well prograde into lakes or interdistributary bays (e.g., Coleman *et al.*, 1964; Coleman and Gagliano, 1964, 1965; Elliott, 1974; Tye and Coleman, 1989a, b; Perez-Arlucea and Smith, 1999). Crevasse splays and lacustrine deltas typically develop, prograde, and become abandoned over periods of 100–150 years. Crevasse channels that debouch into standing water have mouth bars and submarine levees (the deposits are described in Chapter 13). The upper parts of flood-related depositional units on levees and crevasse splays display evidence of standing water and shoreline features, such as wave-ripple marks and planar laminae. Interdistributary bays and lakes receive fine-grained sediment during floods, and these sediments may be reworked by waves during storms. Typical sedimentation units are centimeters-thick layers of fine to very-fine sand capped by mud. The sand contains ripple marks and associated small-scale cross strata formed both by river currents and by wave currents. Laminated clays occur in deeper water in locations distant from river channels, and these may

be organic-rich and anoxic. Peat may also accumulate in marshes and swamps distant from sediment sources. Delta-plain sediments are typically extensively bioturbated by plant roots, crustaceans, and worms (see Chapter 11), and organic remains (e.g., plant fragments, charophytes, ostracods, bivalves, gastropods) are common. Siderite nodules and pyrite replacements form post-depositionally. Meters-thick, coarsening-upward sequences result from progradation of delta-plain channels, crevasse splays, or levees into bays or lakes, and fining-upward sequences are formed due to subsequent abandonment (Chapter 13; Figure 15.7).

In most cases of rivers flowing into the sea, fresh river water flows over the denser seawater as a buoyant plume (hypopycnal flow). Deposition of the coarsest sediment on the delta front occurs in the form of distributary mouth bars and associated submarine levees (Figure 15.4). If basin currents are not powerful enough to rework these deposits, the distributary channel and associated mouth bar and levees extend seawards like a "finger" (Figure 15.8). In the Mississippi, these so-called *bar-finger sands* (Fisk *et al.*, 1954; Fisk, 1961) are tens of kilometers long, kilometers wide, and many tens of meters thick. As discussed below, their thickness varies greatly due to post-depositional deformation.

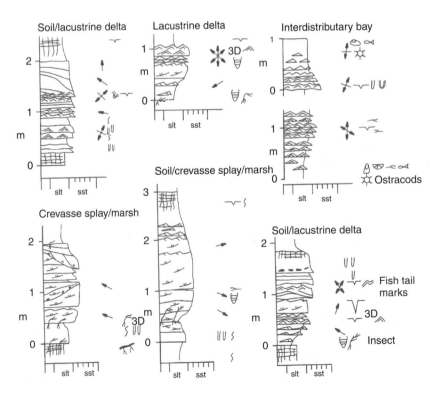

FIGURE 15.7. Examples of interpreted delta-plain deposits from Devonian rock of northeast North America. From Griffing *et al.* (2000). The legend is Figure 1.6.

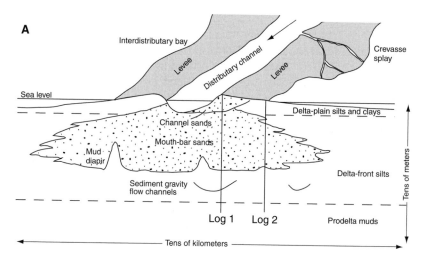

FIGURE 15.8. A model for deposits of a river-dominated, marine delta front. (A) A cross section through mouth-bar and delta-front deposits, showing locations of vertical logs (B). The legend for the logs is Figure 1.6.

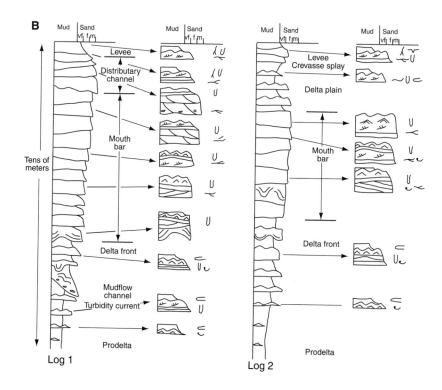

The sandy distributary channels and mouth bars are covered with dunes (forming medium-scale cross strata) with minor upper-stage plane beds (forming planar laminae) during river flood, although current ripples (forming small-scale cross strata) appear on the seaward side of mouth bars where the flow decelerates. Finer-grained sediment (with abundant disseminated plant remains) and bed forms indicative of low flow power (e.g., ripples) occur as floods wane. In the

Mississippi delta, stratasets deposited during floods in distributary channels and mouth bars can be up to meters thick, and they are inclined seawards at less than one degree. Submarine levees are mainly covered with upper-stage plane beds and current ripples in silty sand during floods, but are draped with silt and clay during waning-flow stages. Their internal structures are therefore planar laminae and climbing-ripple cross-laminae with muddy draping laminae. Depositional units in submarine levees associated with flooding events are typically centimeters to decimeters thick. Seawards of mouth bars, silts and clays are deposited, becoming finer-grained towards the prodelta region. Depositional units from flooding events are similar to those on submarine levees.

The deposits on mouth bars and submarine levees are likely to be reworked by wave and tidal currents during and after floods, possibly forming planar laminae, hummocky–swaley cross strata, and wave-ripple laminae (Figure 15.8). The high deposition rates and relatively low salinity in the vicinity of major river mouths are hostile to organisms. As a result of this, shelly fossils are not common, and bioturbation is not well developed. Plant remains are, however, common in delta-front sediments. Burrows and shelly fossils become more common in laminated prodelta silts and clays. The surface expression of slumps from the delta front is scarps, slumped blocks, channels, and depositional lobes (Figure 15.9 and Chapter 8). Deposits of debris flows and turbidity currents that originated in slumps occur in prodelta regions (Figure 15.8).

Syn-depositional soft-sediment deformation structures are important on delta fronts, being associated with inverse density gradients brought about by progradation of mouth-bar sands over prodelta muds with high porosity and pore-water pressure. High pore-water pressure in very-fine-grained sands and muds is produced by rapid deposition, and enhanced by generation of methane gas. Small-scale structures are load casts (Figure 15.8). Larger-scale structures associated with subsurface flow of sediment are mud diapirs, small rift valleys, and growth faults (e.g., Morgan et al., 1968) (Figure 15.9). Mud diapirs occur near the mouths of Mississippi distributaries after a period of rapid deposition, and form temporary islands (called mud lumps) (Figure 15.9A). Rift valleys can be meters to tens of meters deep, hundreds of meters wide, and kilometers long, and the downfaulted region is where mud underneath has flowed away (Figure 15.9C). Growth faults are syn-sedimentary faults that strike parallel to the shoreline and dip basinwards, the dip decreasing with depth (Figure 15.9). The downthrown side is basinwards. The fault motion is episodic, and mouth-bar sand accumulates in the downthrown region. The thicker sand deposits on the downthrown side are commonly deformed into rollover anticlines and antithetic faults (Figure 15.9D). These soft-sediment deformation structures cause the thickness of bar-finger sands to vary greatly.

River avulsion causes the water and sediment supply to part or the whole of a delta to be cut off, leading to subsidence, marine transgression, and reworking of the top of the delta on its seaward side. The Holocene avulsion history of the Mississippi delta is quite well known, although the exact dating of avulsions is still under revision (e.g., Scruton, 1960; Kolb and Van Lopik, 1966; Frazier, 1967; Boyd et al., 1989; Saucier, 1994; Törnqvist et al., 1996) (Figure 15.10). So-called regional avulsions originate in the lower Mississippi valley with a period on the order of 1,000–1,500 years, and have resulted in up to six Holocene delta lobes (each on the order of 10^4 km^2 in area) that are distributed across a delta plain that is about 400 km in the across-stream direction and 200 km in the along-stream direction (Figure 15.10) (Bridge, 1999). More frequent, local avulsions, associated with distributary switching, occur within delta lobes. Reworking of delta fronts by marine currents following abandonment involves (1) formation of sand shoals or barrier islands as waves move sediment alongshore from mouth bars and levees; (2) overwash of shoals or islands and landward migration; (3) formation of a bay containing fossiliferous muds and sands; and (4) peat formation further inland (e.g., Penland et al., 1988; Kosters and Suter, 1993; Bohacs and Suter, 1997) (Figure 15.11). The seaward margins of former delta lobes can be recognized by arcuate strings of islands and shoals (e.g., the Chandeleur islands; Figure 15.5). The shoals and islands may extend preferentially to one side of a distributary, depending on the angle of wave approach and direction of longshore currents. The greatest volume of deposits is due to delta growth and progradation, and the reworked part is only a surficial veneer. Figure 15.11B illustrates the lateral variability of the reworked deposits. The same types of reworked deposits can also occur during delta progradation if the marine currents are powerful enough to redistribute sediment as it is introduced by rivers (see the section on wave- and tide-dominated deltas). It is also necessary to consider the effects on delta deposition and erosion of marine transgressions and regressions due to tectonism or eustasy (e.g., Boyd et al., 1989).

The modern "bird's-foot" pattern of the Mississippi River Delta is in part due to the distributaries being in relatively deep water near the edge of the shelf, where water currents are weak. It is maintained in this position by the US Army Corps of Engineers. Earlier lobes of the Mississippi were in more shoreward positions of shallower water, and had many more distributary channels

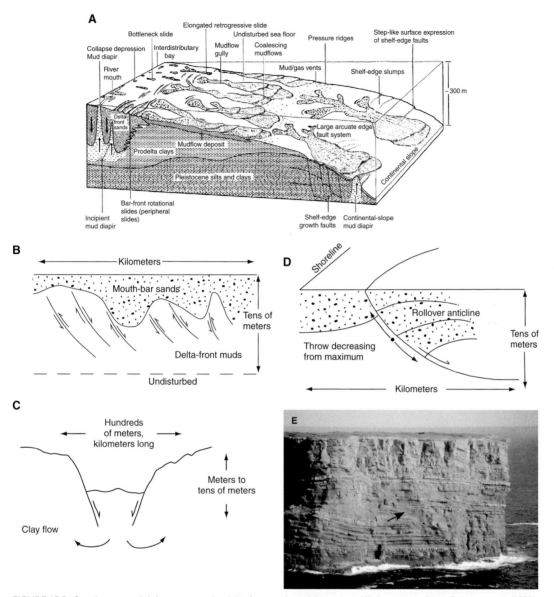

FIGURE 15.9. Syn-depositional deformation in the delta-front and prodelta region. (A) A summary from Coleman *et al.* (1983). Details of mud diapirs (B), delta-front gullies (C), and growth faults (D). (E) A growth fault (arrow) in Namurian delta-front deposits, Co. Clare, Ireland (photo courtesy of Jim Best).

(Gould, 1970). Where there are multiple distributaries and stronger wave currents, the bar-finger sands meld into a continuous sand sheet. Such a sand sheet was deposited by the Lafourche delta and was subsequently buried by younger delta-plain deposits.

A depositional model of a fluvial-dominated delta must reflect the episodic deposition during floods and storms in the various sub-environments; the different scales of progradation and abandonment related to regional and local avulsions; the effects on sedimentary

architecture of the channel-belt width/delta-plain width, deposition rate, subsidence rate, and avulsion period; syn-depositional soft-sediment deformation; and the effects on deposition and erosion of marine regression and transgression associated with eustasy or tectonism. A detailed, quantitative, three-dimensional model is not available because of the lack of data, although such models are under development (Karssenberg *et al.*, 2004). Figures 15.7, 15.8, 15.11, and 15.12 show components of a model based on

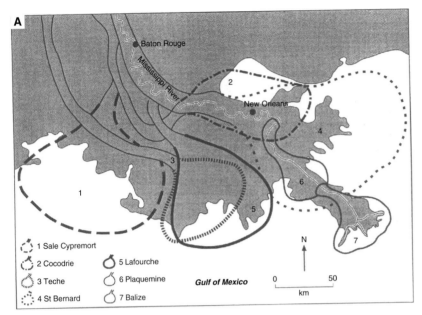

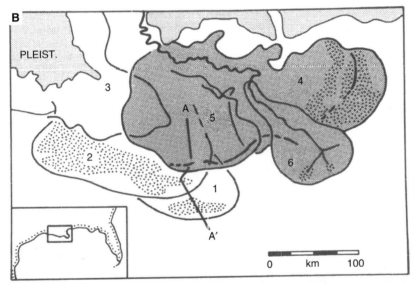

FIGURE 15.10. Holocene delta lobes of the Mississippi according to (A) Kolb and Van Lopik (1966), from Leeder (1999); and (B) Boyd et al. (1989), from Bhattacharya and Walker (1992): 1 is the oldest lobe, and successively higher-numbered lobes are younger.

progradation of a fluvial-dominated delta into a wave-dominated sea, and subsequent abandonment. The model shows a laterally variable coarsening-upward sequence associated with delta progradation comprising sharp-based distributary channel sands bordered by a complex series of coarsening-upward and fining-upward sequences formed by progradation and abandonment of levees and crevasse splays in lakes and interdistributary bays; beaded finger-like mouth-bar sands with sharp or gradational bases; and sediment gravity flows and soft-sediment deformation in the delta-front and prodelta regions. The relatively thin abandonment phase is represented by a landward progression of sandy shoals, islands, and washovers; lagoonal muds; and extensive accumulations of peat on former delta-plain deposits.

Effects of basin currents on delta-front geometry and deposits

On most delta fronts, basin currents transport much of the river sediment alongshore and offshore, creating characteristic coastline geometry and deposits. On *wave-dominated delta fronts*, sand is moved alongshore from mouth bars to form beach ridges (Figure 15.13A). Seaward protrusion of distributaries and mouth bars is

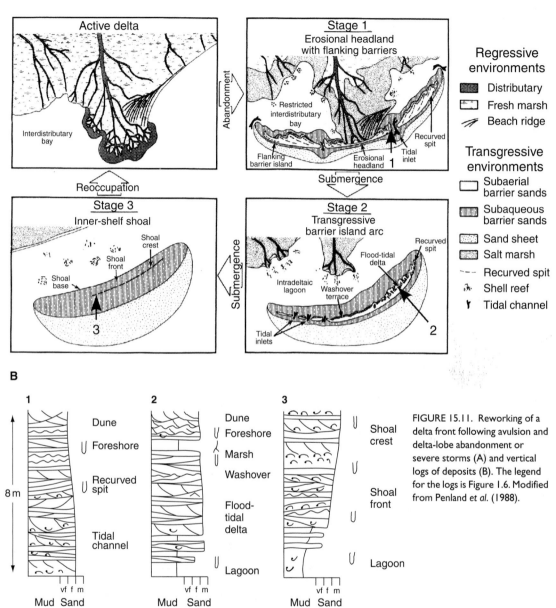

FIGURE 15.11. Reworking of a delta front following avulsion and delta-lobe abandonment or severe storms (A) and vertical logs of deposits (B). The legend for the logs is Figure 1.6. Modified from Penland et al. (1988).

therefore greatly subdued, and the coastline resembles the abandonment phase of a river-dominated delta front. Progradation involves mainly beaches, but also wave-modified mouth bars. Mud is moved offshore and along-shore. *Tide-dominated delta fronts* are characterized by tidal channels and bars, and by tidal flats (Figure 15.13B). Many marine delta fronts are influenced in different places and at different times by both wave and tidal currents, and contain barrier beaches cut by tidal chan-nels, with tidal flats and lagoons on their landward side.

Coastal physiography and deposits formed by waves, tides, and other currents are discussed below.

Wave-dominated coasts and shallow seas

Large-scale geometry and origin

The most obvious attribute of wave-dominated coast-lines is beaches, including those on spits, tombolos, and barrier islands (Figure 15.14). The beach is defined as

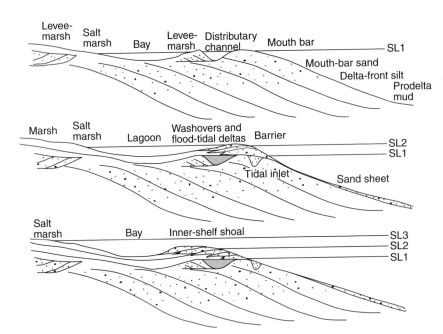

FIGURE 15.12. A fluvial-dominated-delta stratigraphy model, showing deposits associated with active delta progradation, subsequent delta abandonment, and associated changes in relative sea level (SL1, 2, and 3). Details of deposits are given in Figures 15.7, 15.8, and 15.11.

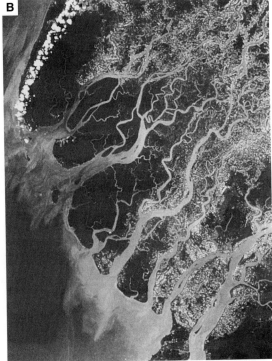

FIGURE 15.13. (A) The wave-dominated Paraibo-do-Sul delta. (B) The tide-dominated Irrawaddy delta.

the area between mean low tide and the highest level reached by the sea (Figure 15.15). The *shoreface* is seawards of the beach, defined as the area between mean low tide and the fair-weather wave base, and may contain linear sand ridges. The *foreshore* of the beach lies between mean low tide and mean high tide (i.e., it is the intertidal region). Two or three longshore bars commonly occur on the foreshore of the beach and further seawards below the breaker zone (Figure 15.15C, D). The *backshore* extends from mean high tide to the level of the highest tides (i.e., it is the supratidal region), which is the base of the aeolian dunes, if such are present (Figure 15.15A). The backshore may be separated from the foreshore by a low ridge with a steep

seaward slope, called a berm. The berm may be ornamented with beach cusps (Figure 15.15B). Aeolian dunes commonly form ridges on the landward edge of beaches, and these may be cut by washover channels that pass landwards into washover fans (Figure 15.14C). The region on the landward side of a beach can be an older

FIGURE 15.14. Wave-dominated coastlines. (A) A spit extending from a mainland beach, with tidal flats and a lagoon on the landward side. (B) A recurved spit with distinct curved beach ridges. (C) A barrier island with washovers on the landward side adjacent to tidal flats and a lagoon, North Carolina (from Google Earth™). (D) A beach with beach cusps, Pearl Beach, Sydney, New South Wales (from Rob Brander). (E) A tombolo at Port Sur State Historic Park, California (from Ann Dittmer). (F) North Carolina Capes Lookout and Fear (from Google Earth™).

beach ridge, a floodplain (with marshes and lakes), a lagoon, or a supratidal flat. The width of beaches varies from a few tens of meters to hundreds of meters, depending on beach slope and tidal range. The average sea-bed profile extending from the beach offshore is commonly concave upwards (i.e., the bed slope decreases offshore), as explained below. Empirical equations for the variation in water depth d with distance offshore x are (Komar, 1998):

$$d = Ax^b \tag{15.1}$$

where b ranges from $\frac{2}{3}$ to $\frac{1}{2}$, A depends on the grain size and wave characteristics, and

$$d = B(1 - e^{-kx}) \tag{15.2}$$

The average beach slope increases with the mean sediment size. Beaches have been classified depending on whether wave energy is dissipated by friction over large areas of undulatory sea bed, as occurs on low-slope

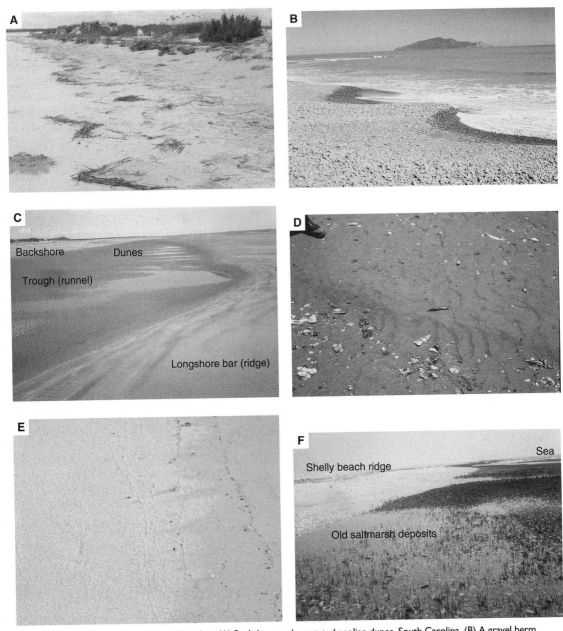

FIGURE 15.15. Geometrical features of beaches. (A) Backshore and vegetated aeolian dunes, South Carolina. (B) A gravel berm with beach cusps at Otaki, North Island, New Zealand (photo from Rob Brander). (C) A longshore bar (ridge) and trough (runnel) with dunes, South Carolina. (D) A dune with superimposed ripples, and transported shells, in the trough (runnel) in (C). (E) Swash marks on the foreshore of a beach, Mexico. (F) Erosion and landward migration of a beach, exposing older saltmarsh deposits, South Carolina.

sandy beaches, or whether wave energy is reflected, as occurs on steep gravelly beaches (Wright and Short, 1984). There is every gradation in space and time between these dissipative and reflective beaches. For example, fair-weather beach profiles tend to be more reflective and storm beach profiles tend to be more dissipative (see below).

Progradation of beaches results in ridge-and-swale topography associated with accretion of successive beach ridges (Figure 15.13A). Chenier plains are successions of beach ridges (cheniers) that are separated from each other by mudflats (Otvos and Price, 1979; Augustinius, 1989). Episodic deposition leading to ridge-and-swale topography is related to changes in sediment supply, such as those

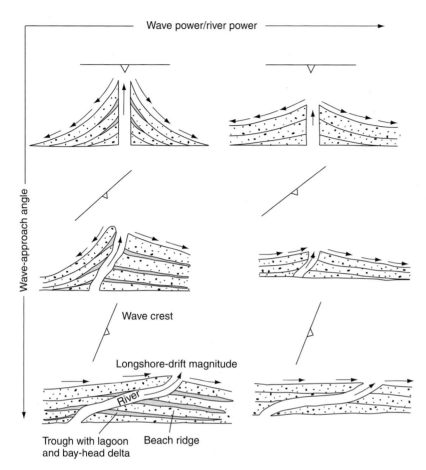

FIGURE 15.16. The plan geometry of wave-influenced shoreline as a function of the wave power, river power, and wave-approach angle. Based on Komar (1973, 1977) and Bhattacharya and Giosan (2003).

caused by major storms, delta switching, or climate change. Deposition of cheniers on the Suriname coast (northwest of the mouth of the Amazon) is related to alongshore migration of shore-connected mudbanks. Interbank regions are areas of erosion, leading to the formation of cheniers (Augustinius *et al.*, 1989).

Spits are beach ridges that extend into bays, estuaries, tidal inlets, and river mouths due to longshore transport of sand or gravel (Figure 15.14A, B). Wave refraction around the tips of spits results in so-called recurved spits (Figure 15.14B). Long-term but episodic deposition and progradation of spits results in ridge-and-swale topography. Tombolos are spits that connect the mainland with an island (Figure 15.14E). Mudbanks in Florida Bay comprise tombolos that generally run between mangrove-covered tidal-flat islands. These mudbanks are probably related to longshore transport of mud controlled by wave refraction around the islands. Spits are responsible for diverting the mouths of rivers and blocking harbors in coastal inlets, stimulating various engineering projects (see below).

Barrier islands are hundreds of meters to kilometers wide and kilometers to tens of kilometers long, and are shorter as the tidal range increases due to cutting by tidal channels. Barrier islands can be formed by longshore transport of sediment from a river mouth, like a spit (Figure 15.14), or can be formed by drowning of a mainland beach (e.g., Swift, 1975; Oertel, 1985; Reinson, 1992). An origin by upward growth of a submarine shoal is not now thought to be viable. The drumstick shape of some barrier islands in mesotidal settings (Figure 15.26E) is attributed to increased deposition near alongshore-migrating tidal inlets due to accretion of abandoned ebb-tidal delta deposits onto the downdrift barrier island (Hayes, 1979).

The development and maintenance of wave-dominated coasts with beaches and barrier islands require a source of sediment, wave currents capable of moving the sediment, and time. The plan form of wave-dominated coastlines is related to the balance between the episodic input of sediment at river mouths and the removal of this sediment by wind-wave and wind-shear currents (Komar, 1973, 1977; Bhattacharya and Giosan, 2003) (Figure 15.16). Sediment may also be supplied to beaches by wave erosion of the sea bed during periods of falling sea level (Dominguez et al., 1987; Dominguez, 1996). As sediment is deposited at river mouths, the protrusion of the coastline gives rise to longshore currents. In the case of wave crests parallel to the overall shoreline, the magnitude and direction of longshore currents are symmetrically distributed on either side of the river-mouth protrusion. The magnitude of longshore currents for a given wave power increases as the angle between the local shoreline and wave crests increases. The protrusion at the coastline grows outwards until the lateral sediment transport due to longshore currents is balanced by the sediment input, and the deposition rate along the coastline protrusion is constant. This steady-state plan configuration is maintained as sediment continues to be delivered by rivers and the coastline progrades seawards. The angle between the river-mouth protrusion and the mean shoreline orientation decreases asymptotically away from the river mouth, because the magnitude of longshore currents must decrease away from the river mouth in order for deposition to occur. The maximum angle between the river-mouth protrusion and the mean shoreline orientation increases as the sediment-input rate increases relative to the wave power. The shoreline protrusion is very small and the shoreline is almost linear in plan if the wave power greatly exceeds the river power. The rates of sediment transport and deposition by river currents and longshore currents vary seasonally and are not congruous. Therefore, the shoreline configuration will oscillate in and out of steady state. Episodic deposition on prograding beaches causes beach ridges and swales, as discussed above. The swales may be elongated lagoons with bayhead deltas at their riverward ends (Figure 15.16).

As the angle between wave crests and the mean shoreline orientation increases, the longshore current magnitude is asymmetrically distributed along the shoreline protrusion (Figure 15.16). This causes the deposition rate on the updrift side of the river to decrease relative to that on the downdrift side. The angle of the shoreline relative to the mean shoreline orientation is also greater on the updrift side. With highly oblique wave approach and low river input, the updrift side of the river may be a zone of erosion (Figure 15.16).

Large-scale shoreline protrusions with spacings of tens to hundreds of kilometers are called *capes* (Figure 15.14F). Good examples are found along the coasts of North and South Carolina (eastern USA):

Capes Hatteras, Lookout, Fear, and Romain. Many possible origins have been proposed for capes, including giant eddies associated with wind-shear currents (e.g., the Gulf Stream), longshore variations in wave energy, long-wavelength edge waves, and the effects of offshore shoals. The coincidence of the mouths of major rivers with the capes of the Carolinas suggests that they may be the remnants of deltas that were active during glacial times (Hoyt and Henry, 1971).

All of these physiographic features change with time due to seasonal erosion and deposition associated with river floods and storms at sea, due to diversions of rivers, and, in the long term, due to climate change, tectonism, and eustasy.

General patterns of water flow, sediment transport, erosion, and deposition

Water flow and sediment transport associated with waves and wind-shear currents in shallow seas and coasts were discussed in Chapter 7 and will be summarized only briefly here. Wave-induced currents near the sea bed increase in magnitude with wave height and as waves move into shallower water. The asymmetry of oscillatory wave currents also increases as waves move into shallow water, and the associated net fluid drift is shorewards near the bed and offshore higher above the bed (in the absence of wind-shear currents, tidal currents, and river currents). Cell-like circulation near the shore includes longshore drift in the breaker, surf and swash zones, and offshore-directed rip currents. The cell-like circulation pattern (especially longshore currents) is modified by oblique wave approach. Wind-shear currents (including geostrophic currents) are normally associated with wave currents, and can substantially modify their magnitude and direction. The magnitude and direction of all of these currents vary in space and time with the weather, and with longer-term changes in relative sea-level associated with tectonism, eustasy, erosion, and deposition.

The grain size of bed load and suspended load is expected to increase with the near-bed water velocity, given an appropriate sediment supply. Substantial settling of suspended mud to the bed occurs only where bed load transport is negligible. Therefore, the mean grain size of the bed material is generally expected to decrease offshore from the breaker zone. Also, alongshore and alongshelf variation in mean grain size of bed material is expected in response to spatial variation in water depth and velocity of the various currents. The near-bed water velocity (and hence the mean grain size

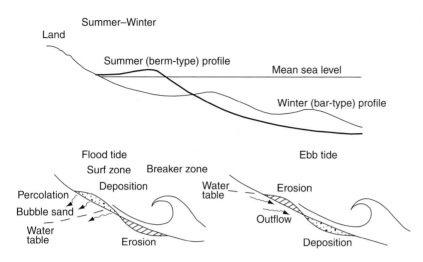

FIGURE 15.17. Idealized changes in beach-shoreface profiles over time due to winter (large-wave) and summer (small-wave) conditions, and tidal ebb and flood. After Komar (1998).

of bed material) varies with time due to the passage of storms. Most of the bed material will be related to these high-flow events, and the relatively finer-grained bed material moved during quiet-water periods forms a veneer on the surface. On modern shelves, a substantial proportion of the bed area is covered with immobile sand and gravel, particularly in deeper water (Shepard, 1932, 1973; Emery, 1952). These sediments accumulated on alluvial plains and coasts during the much lower (say 130 m) sea levels during the Pleistocene.

Since the water depth increases offshore according to a power or exponential function (Equations (15.1) and (15.2)), the bed slope also decreases offshore. If it is assumed that the bed slope is in steady state with water flow and sediment transport, the net onshore movement of sediment must be equal to the net offshore movement. Furthermore, if the near-bed orbital velocity of water is greater in the onshore direction than in the offshore direction, the near-bed sediment-transport rate in the onshore direction can only equal that in the offshore direction if an additional offshore force acts on the bed sediment. This additional force is the downslope gravity force. If the components of the near-bed water velocity are equal in magnitude and duration, there is no need for a bed slope to attain a steady-state bed profile. As the asymmetry of the near-bed water currents increases towards the shoreline, so must the bed slope in order to maintain a steady-state bed profile. This force balance is also related to the mean grain size, because the gravity force acting on bed grains increases with grain size (and density) as well as slope, and the asymmetry of the near-bed orbital velocity increases with velocity. In general, the mean grain size of bed sediment increases with the bed slope under steady-state conditions.

The swash zone differs from the sea bed in that the water in the swash zone may infiltrate the bed instead of returning in the surface backwash. The onshore flow velocity can therefore greatly exceed the offshore flow velocity, and a steep beach slope would be required in order to maintain a steady-state beach profile. In the extreme case of a beach that is so permeable that there is no surface return flow (backwash), the steady-state beach slope represents a balance between the onshore fluid drag force and offshore gravity force: that is, the beach face is at the angle of repose. This is why highly permeable gravel beach faces are commonly at the angle of repose.

The beach-shoreface profile is well known to vary with wave conditions (small waves during fair weather, large waves during storms). On some coasts, storm conditions are prevalent in winter, whereas fair-weather conditions prevail in summer. On other coasts (e.g., the US east coast), storm conditions do not necessarily occur in winter, as is evident to hurricane watchers. Figure 15.17 shows an example of a winter profile called a bar profile (because of the presence of longshore bars) and a summer profile called a berm profile (because of a relatively flat backshore and a berm with a steep seaward side). During the large-wave conditions of winter storms, the berm is eroded, and sediment is moved offshore onto longshore bars. The upper beach is built up during the summer. Although this appears to be a common seasonal variation on coasts, it is a simplistic picture, because other currents (tides, wind shear) are likely to be active (Chapter 7). During a flooding tide, the water table is relatively low, so some of the swash percolates into the beach, reducing the amount of backwash. This results in erosion in the surf zone and deposition in the swash zone. The reverse happens during the ebb tide because the water table is relatively high, and backwash is enhanced by groundwater flowing out of the beach. These variations occur over the various tidal periods (semidiurnal, neap–spring).

Wind-shear currents also affect beach-shoreface profiles. During a major storm, an onshore-directed bottom current created by offshore-directed wind-shear may add to storm-enhanced oscillatory currents, thereby increasing the overall magnitude and asymmetry of near-bed currents in the onshore direction. This should result in steepening of the sea-bed profile, coarsening of bed sediment, erosion of the offshore sea bed, and possibly deposition on longshore bars and the beach. The corresponding offshore-directed upper flow would transport suspended sediment offshore, where it would be deposited. If the wind-shear current was directed offshore at the bed (due to an onshore component of wind shear), the asymmetry of the oscillatory currents would be reduced or reversed, the bed profile would tend to become less steep, the beach would be eroded, bed sediment would be moved offshore and deposited, and some bed sediment might be transferred shorewards in overwash and deposited on washover fans.

There have been many attempts at predicting steady-state bed profiles on the basis of wave processes and cross-shore sediment transport, as reviewed by Komar (1998). Some of these models (e.g., SBEACH) work reasonably well, but they are mainly empirical. Process-based models are under development, but their construction and testing against natural data are very challenging. Such models need to be capable of predicting the temporal and three-dimensional spatial variation of water flow and sediment transport associated with wind waves, edge waves, wind shear, and tides. They should also be capable of accounting for sorting of bed sediment by grain size, shape, and density. For example, if deposition occurs due to a decrease in sediment transport in a given direction, the sediment grain size would decrease in the same direction. This level of sophistication has not yet been attained. Furthermore, in view of the temporal variation in currents of various kinds, it is questionable whether any sea-bed or beach profile is in steady state with water flow and sediment transport, because of the vast amounts of sediment that would need to be moved if flow conditions changed. Much of the present-day outer continental shelf is not in steady state because coarse-grained sediment on the bed cannot be moved.

Flow, sedimentary processes, and deposits for deep offshore zones

Storm wave base on continental shelves varies depending on the wave regime. On the Atlantic coast of the central and southern USA, storm wave base is about 20 m. On open Pacific shelves, storm wave base may exceed 100 m and wave ripples may be active in depths as great as 200 m. Exceptionally strong wind-shear currents can occur in deep water on the edge of the continental shelf, e.g., the Agulhas current on the southeast African shelf (Flemming, 1980), giving rise to large sand dunes on the bed. Areas of sea bed that are below storm wave base for most of the time, and that are not greatly affected by other currents, receive relatively small amounts of suspended mud during storms or river floods. The relatively immobile bed sediment provides a stable bed surface that can support an abundant and diverse infauna and epifauna. If the input of terrigenous sediment is relatively low, poorly sorted, muddy accumulations of mainly epifaunal shell material occur, including benthic foraminifera, gastropods, barnacles, bryozoans, red algae, and some infaunal and epifaunal bivalves. The amount of infauna may be low because food supply is lacking and the substrate is unsuitable for burrowing. If abundant muddy sediment accumulates on the bed but does not exceed the turbidity tolerance of the biota, both infauna and epifauna flourish because the mud provides a stable substrate for burrowing, and organic matter tends to be adsorbed onto the surfaces of mud particles, providing food for deposit feeders. In modern oceans, bivalves, sea urchins, and various polychaete worms dominate the infauna. This infauna comprises both suspension and deposit feeders, with burrows of the sea urchins and polycheate worms being mainly horizontal. Surface and subsurface trails of foragers, commonly gastropods and decapod crustaceans (e.g., crabs, lobsters), are common, and in some cases have bizarre geometric forms (see Chapter 11). Trace fossils belong to the *Cruziana* and *Zoophycos* ichnofacies (see Chapter 11). During exceptional storms, the epifauna and infauna may be eroded, moved around, and concentrated in patches, and mud may temporarily be suspended. Therefore, offshore-shelf deposits are expected to be sheets of laminated and bioturbated mud containing scattered infaunal fossils and local concentrations of epifaunal and infaunal shelly material (Figure 15.18). Deposits of turbidity currents that originated on delta fronts may be interbedded with these muds. The tops of such deposits would exhibit little evidence of reworking by waves and wind-shear currents.

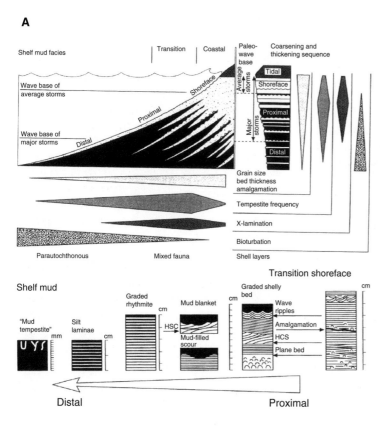

FIGURE 15.18. Summaries of coast-to-offshore changes in deposits associated with storm waves according to (A) Aigner and Reineck (1982), based on the modern North Sea, and (B) Allen (1985). Images (C)–(E) show examples of storm-wave deposits from Devonian rock of New York containing hummocky cross strata, planar strata, and wave-ripple cross strata. (C) Interbedded sandstones and shales typical of the zone between storm wave base and fair-weather wave base. (D) A sandstone-dominant section with amalgamation of storm beds (to the right), which is typical of immediately below fair-weather wave base. (E) Amalgamated sandstone storm beds typical of the shoreface. The upper parts of storm beds are bioturbated. Bases of storm beds are arrowed.

Flow, sedimentary processes, and deposits for shallow offshore zones

The zone between storm wave base and fair-weather wave base lies between depths of about 20 m and 100 m on most modern continental shelves. Sand is moved on and near the bed by wave currents and wind-shear currents during storms, but during fair weather there is no bed-load movement and mud may be deposited from suspension on the bed. Therefore, this zone contains interbedded sand and mud. Infauna and epifauna are abundant in this zone. Suspension feeders, typically bivalves in modern settings, live in vertically oriented burrows in the sand (*Skolithos* ichnofacies) and sediment feeders and grazers burrow through the mud above and below sand layers (*Cruziana* ichnofacies) (Pemberton *et al.*, 1992) – see Chapter 11. Modern deposit feeders include sea urchins, sea cucumbers, and a variety of polycheate worms.

During storms, wind-shear currents normally accompany wave currents, and these combined flows generally decrease in magnitude offshore and/or alongshore (e.g., Murray, 1970; Forristall *et al.*, 1977; Hequette and Hill, 1993); see Chapter 7. Turbidity

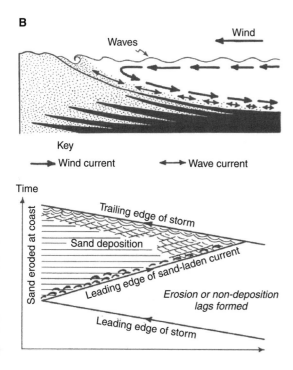

FIGURE 15.18. Continued.

currents may also occur. The bed is molded into hummocks, plane beds, or small-scale wave ripples, depending on the near-bed orbital velocity, orbital diameter, and mean grain size (Chapter 7). In the presence of an additional unidirectional current (due to wind shear or turbidity current), hummocks and wave ripples will be asymmetrical in profile. These bed forms change in space and time with changing water currents. The bed will be eroded if there is a spatially accelerating net near-bed flow, and deposition occurs under spatially decelerating net flows. The epifauna and infauna are substantially disrupted during such erosional and depositional activity. Shells commonly comprise the coarsest sediment available for transport. The largest shells may be concentrated in lags above erosion surfaces, and others may be transported and sorted into coarse-grained lenses. The shells are not transported very far, and so are representative of the local fauna but probably not of their relative abundance (because of hydraulic sorting). Burrowing organisms (suspension feeders) can normally move up and down during

deposition and erosion events. Epifauna disturbed during storms can recolonize the bed during fair weather.

Storm-generated sedimentation units are commonly underlain by an erosion surface (Figures 7.24 and 15.18). Erosion surfaces may contain channels up to meters wide and decimeters deep, and smaller-scale features such as gutter casts, flute marks, and tool marks (Chapter 5). These indicate the activity of a unidirectional-flow component, commonly directed offshore or alongshore. Sedimentation units are normally centimeter- to decimeter-thick sheets and lenses of sand (Figures 7.24 and 15.18): they commonly thin and fine offshore and/or alongshore (e.g., Hayes, 1967; Morton, 1981; Aigner and Reineck, 1982; Allen, 1982a, 1985; Figueredo *et al.*, 1982; Snedden and Nummedal, 1991; Swift and Thorne, 1991). These storm deposits tend to fine upwards and are capped by mud deposited during waning flows and fair weather. In some cases, mud is deposited on the basal erosion surface instead of sand. Typical vertical and lateral sequences of sedimentary structures and textures in sedimentation units are shown in Figures 7.24 and 15.18. The various sequences of hummocky cross stratification, planar lamination, and wave-ripple lamination reflect changing bed forms as current conditions change in time and space. If a unidirectional-flow component is present, cross strata may have a preferred orientation. If so, it may be very difficult to distinguish storm deposits from deposits of tidal currents and shallow-marine turbidity currents. Wave-ripples and their cross laminae are normally of the three-dimensional type, but they are two-dimensional at the top of sand units and further offshore. The upper parts of the sedimentation units are bioturbated. The proportion of mud in the sedimentation units tends to decrease shorewards, either because mud was not deposited during fair weather, or because it was eroded prior to deposition of the overlying sandy unit.

Sequences of storm-generated sedimentation units that are meters thick, extend laterally for tens of kilometers, and either fine upwards or coarsen upwards indicate systematic temporal variation in current strength and sediment-transport rate. This could be related to shifting positions of river mouths, lateral migration of shelf sand ridges (see below), or regional variations in climate or sea level (Craft and Bridge, 1987; Walker and Plint, 1992; Storms and Swift, 2003).

Flow, sedimentary processes, and deposits for shorefaces

In water depth less than about 20 m, fair-weather wave currents can move fine-grained sand on the bed, and finer-grained sediment is kept in suspension. Therefore, storm-related sedimentation units are devoid of mud. The so-called amalgamated storm beds deposited in this zone are distinguished by erosional bases overlain by coarse shelly layers, and bioturbated tops (Figure 15.18). Hummocky cross stratification (HCS), swaley cross stratification (SCS), and planar strata are the dominant internal structures. Wave ripples and associated cross strata (normally of three-dimensional type) occur only near tops of storm beds. Nearer to the coast, ripple crests tend to parallel the coast and more interfering patterns occur due to wave reflection.

Flow, sedimentary processes, and deposits for shorefaces: linear sand ridges

Linear sand ridges occur on the inner shelf where wind-shear (geostrophic) currents are important during storms. These ridges are 3–10 m high and 1–2 km wide, have side slopes of typically 1 or 2 degrees, and

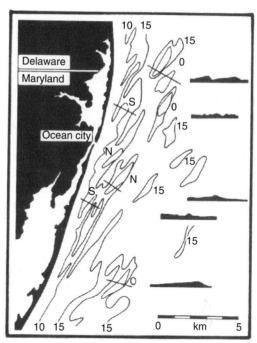

FIGURE 15.19. Linear shelf ridges offshore from Maryland, USA. Modified from Swift and Field (1981) by Walker and Plint (1992).

are up to tens of kilometers long. On the eastern shelf of the USA (Virginia, Maryland), crestlines of ridges are oriented at 15° to 35° from the coastline, making an angle of about 20° with the direction of wind-shear currents (Swift and Field, 1981) (Figure 15.19). These ridges are asymmetrical in profiles normal to crestlines, with the steeper slope facing obliquely offshore. They migrate in the direction of the steepest side (obliquely offshore) during storms. The mean grain size is greatest on ridge crests and decreases into troughs. The ridges are covered with wave-modified dunes and ripples during storms. Internally, these sand ridges are composed of low-angle (a few degrees) inclined sedimentation units formed during storms. The main internal sedimentary structure of storm beds is either HCS or (more likely) angle-of-repose trough cross strata. Storm units become thinner and finer-grained offshore. Sand ridges that migrate offshore into deeper water produce a coarsening-upward sequence up to 10 m thick (e.g., Figure 15.19).

The formation, geometry, and migration behavior of linear sand ridges are analogous to those of tidal sand ridges and linear aeolian dunes, and can be explained using the model of Huthnance (1982a, b). Some sand ridges on the continental shelf are moribund, and may have formed during a stage of lower sea level (Field, 1980). However, some moribund shelf ridges might be drowned beach ridges, also formed during periods of a lower sea level.

Some large-scale, shore-attached sediment ridges on the inner shelf are transverse to bottom currents. These sediment ridges can be composed of mud (Rine and Ginsburg, 1985; Augustinius, 1989) or sand (Komar, 1998) and are associated with wave currents of obliquely approaching waves or wind-shear currents.

Elongated ridges (tens of kilometers long, a few kilometers wide, and meters thick) of gravelstone or sandstone that are subparallel to the paleoshoreline and enclosed in marine mudstone in ancient deposits have been explained in various ways. One explanation is that they represent linear sediment waves formed either by wind-shear currents or by deposition from sediment that is transported alongshore from the mouths of rivers (e.g., Tillman and Martinsen, 1984, 1987; Gaynor and Swift, 1988). However, Walker and Plint (1992) dismiss this explanation in favor of one involving localized shoreface erosion during a period of unsteadily rising sea level, and subsequent deposition of lenses of sand or gravel. Much of their evidence is based on interpretation of well logs and cores rather than detailed sedimentological information, and therefore cannot be considered conclusive. More research, particularly numerical modeling, needs to be done on this problem.

Flow, sedimentary processes, and deposits for shorefaces and beaches: longshore bars

Two or three longshore bars commonly occur on the foreshore of the beach and the inner shoreface in water depths less than 5–10 m (Figures 15.20 and 15.21). These bars are asymmetrical in profile, with the steeper side (up to the angle of repose) facing onshore. The seaward-facing slope is up to a few degrees. The bars are up to 1 or 2 m high, and tens to hundreds of meters in wavelength, and the height and length of successive bars decrease shorewards. The alongshore length of the bars is generally several hundred meters, and the bars nearer shore tend to be shorter because they are cut by rip currents. In plan form, these bars may have a crescentic form, especially those further seawards (Figure 15.22). Waves break on the crests of longshore bars. Bar surfaces may be planar, or covered with hummocky wave ripples or dunes. When the tide goes out, longshore bars become partly to completely emergent, waves break on the crest of the next bar seawards and shallow water rushes up and down the back of the emergent bar. Bar troughs become runnels of ridge-and-runnel topography. Water flows along runnels as the tide comes in and out, forming current ripples and dunes (e.g., Greenwood and Sherman, 1993). Wave ripples are formed in ponded water in runnels.

The sediment needed to form longshore bars comes partly from erosion of the berm and beach face during storms, but could also come from offshore. Breaking waves have always figured prominently in explanations of longshore bars; see the review by Komar (1998). The positions and heights of longshore bars have been related to breaking of waves of various heights. The troughs of bars have been ascribed to erosion following wave breaking, whereas the backwash is responsible for transporting sediment back offshore to accumulate on the bar. The asymmetry of near-bed wave currents

Beach progradation with longshore bars and berms

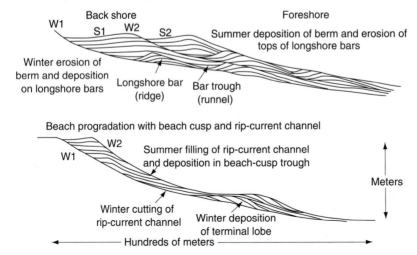

FIGURE 15.20. Cross sections showing idealized deposits associated with (A) progradation of a beach and upper shoreface with longshore bars and a berm that change geometry and position with wave size; and (B) progradation of a beach and upper shoreface where a rip current cuts through longshore bars. In these cases, wave size is taken as increasing from summer to winter (see also Figure 15.17), although this is not so along all coasts. Labels S1, S2, W1, and W2 refer to successive summer and winter bed profiles.

FIGURE 15.21. Beach deposits from South Carolina, USA. Images (A) and (B) show planar laminae in backshore-to-foreshore deposits. (C) Landward-dipping cross strata associated with a longshore bar (ridge) migrating into a trough (runnel) on the foreshore. The crest of the ridge is covered with antidunes and rhomboid ripples, both indicating supercritical (fast, shallow) flow of the wave swash and backwash. (D) Landward-dipping cross strata (formed by a longshore bar) overlain by planar laminae (formed on a plane bed) in foreshore deposits. Images (E) and (F) show an oyster-shell beach and eroded berm. The oyster shells were eroded from backbarrier marsh deposits exposed in the foreshore.

could produce the asymmetry of the bar, as also could asymmetrical near-bed wave currents combined with a strong onshore unidirectional current associated with wind shear. Longshore bars have also been related to the net drift under standing waves (including edge waves) associated with wave reflection from beaches (Chapter 7). The bars would be formed under the antinodes. The association of some longshore bars with edge waves is suggested by the relationship between crescentic longshore bars and rip currents (Figure 15.22). It may well be that several mechanisms act together to form longshore bars (e.g., Holman and

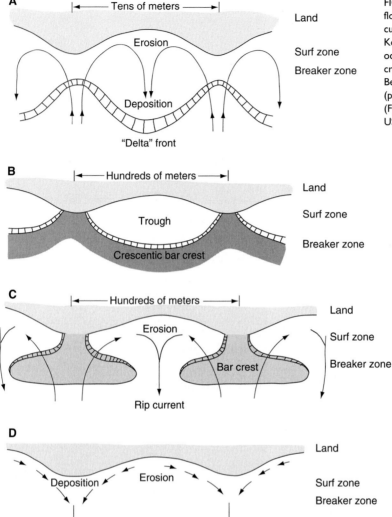

FIGURE 15.22. Variable geometry and flow patterns associated with beach cusps and longshore bars. Based on Komar (1998). Types (B) and (C) can occur together. Examples of rhythmic crescentic bars are from (E) Tairu Beach, North Island, New Zealand (photo from Rob Brander) and (F) Hatteras Island, North Carolina, USA (photo from Robert Tye).

Sallenger, 1993), and further studies are required to resolve this.

On the inner shoreface, longshore bars are cut by channels formed by rip currents or wind-shear currents (Chapter 7). Channels are decimeters deep and meters wide, decrease in depth offshore, and may have a depositional lobe at the downstream end (e.g., Gruszczynski et al., 1993) (Figure 15.20). The channels may contain seaward-facing dunes or ripples, giving rise to offshore-dipping, angle-of-repose cross strata. These channels may migrate laterally, forming decimeter-thick sequences of cross-stratified sand resting on an erosional base. Successive storm beds deposited on the channel sides will dip at a low angle in the direction of channel migration (i.e., large-scale inclined strata).

Longshore bars grow in height, become more asymmetrical, and migrate shorewards as the wave height increases during storms. This shoreward migration also commonly occurs over a neap–spring tidal cycle. As longshore bars migrate onshore they produce shoreward-dipping laminae, inclined at up to the angle of repose. Previous trough (runnel) deposits will be buried by deposits of the bar lee face (Figures 15.20 and 15.21). Longshore bars may be subdued during fair-weather periods, and the surf zone is a planar surface dipping seawards. However, the inner shoreface may contain dunes or hummocky wave ripples (e.g., Clifton et al., 1971; Davidson-Arnott and Greenwood, 1974; Greenwood and Mittler, 1985; Greenwood and Sherman, 1986; Hart and Plint, 1989).

E

FIGURE 15.22. Continued.

F

Flow, sedimentary processes, and deposits for beaches: surf and swash zones

The surf zone is landwards of the breaker zone, the rush of water up the beach following wave breaking is called the *swash*, and the return flow is called the *backwash*. The water velocity in the surf zone following wave breaking is directly proportional to the maximum near-bed orbital velocity at the point of breaking, which increases with breaker height. The maximum longshore currents occur in the breaker zone and surf zone. The velocity of these longshore currents increases

with the angle of wave approach and the maximum orbital velocity at breaking, and can reach meters per second. Dunes that migrate longshore are commonly developed under such conditions. Komar (1998) gives the mean velocity of longshore currents in the middle of the surf zone (averaged over periodic variations due to edge waves) as

$$U_L = 2.7 U_m \sin \alpha \ \cos \alpha = 1.17 (gH_b)^{0.5} \ \sin \alpha \ \cos \alpha \tag{15.3}$$

where U_m is the maximum orbital velocity at the time of wave breaking, H_b is the breaker height, and α is the angle of wave approach. Equation (15.3) fits experimental data very well. Actually, U_L increases from the break point into the surf zone and then decreases to the landward edge of the swash zone. In detail, the longshore current velocity varies with the bottom topography, with variable wave geometries and directions, tides, and wind-shear currents. Although quite good numerical models for the distribution of longshore currents exist (e.g., Longuet-Higgins, 1970a, b; Larson and Kraus, 1991), the effects of all of the factors listed above have not been evaluated. The longshore sediment-transport rate is proportional to $H_b^{5/2} \sin \alpha$ and $H_b^2 U_L \sin \alpha$ (Komar, 1998).

In general, the bed-sediment size decreases from the breaker zone to the landward edge of the swash zone, due to decreases in water velocity and turbulence. Alongshore variation in bed-sediment size is due to cell circulation related to edge waves, and larger-scale variations in wave characteristics and flow velocity. Broken and sorted shell material is common in the surf and swash zones. Burrows or trails are generally rare here, although some organisms (such as some bivalves and polycheate worms, and the common mole crab) can flourish in the swash zone of sandy beaches. Sandy beaches are commonly planar, with some antidunes and rhomboid ripples. Planar beach lamination is very common (Clifton, 1969) and comprises subparallel sets of laminae that dip seawards at several degrees. Set boundaries are erosional and themselves dip seawards at low angles (Figures 15.20 and 15.21). Cross cutting sets of planar laminae at different inclinations are formed by deposition on beach faces that change elevation and slope frequently depending on the wave regime. As explained above, beach slopes increase with the mean grain size, and some gravel beaches can attain the angle of repose. The gravel is relatively well sorted and rounded, with marked sorting by shape (spherical gravel below the breaker zone and disks more common further landwards). Disk- and tabular-shaped gravel can be inclined onshore or offshore, or can be closely packed with the long axes near vertical. Only the most robust shells survive on gravel beaches. Gravel beaches are planar in the surf and swash zone, so internal structures are planar strata.

Flow, sedimentary processes, and deposits for beaches: berms and beach cusps

The backshore may be separated from the foreshore by a nearly flat surface with a steep seaward slope, called a berm. The berm elevation is that of the highest level of swash run up, which is related to the wave height, beach slope, and sediment permeability (controlling percolation). Berms are deposited during low-wave conditions. Deposition occurs on the seaward side of the berm due to deceleration of the swash. Large waves may overtop the berm and deposit sediment, thereby giving a slight landward inclination to the top of the berm. Storm waves tend to erode berms, but, paradoxically, can deposit sediment to high levels on berms while eroding them.

The berm may be ornamented with beach cusps (also called giant cusps and rhythmic topography), which are bowl-shaped erosional features bounded by spurs (Figures 15.14D, 15.15B, and 15.22). Although beach cusps are formed by erosion, the lateral spurs may be associated with deposition. Their spacing along the beach is a few meters to hundreds of meters, and is proportional to the wave height and edge-wave length. The cusp height increases with the beach slope and wave period. Some beach cusps have a delta-like lobe of sediment immediately offshore. Others have rip-current channels offshore, and deposits near their edges that pass laterally into crescentic longshore bars. Rip currents may cut through these longshore bars (Figure 15.22).

The water flow and sediment-transport patterns associated with beach cusps, and explanations for their origin, are contradictory (Figure 15.22) (Komar, 1998). For example, offshore-directed water flow can converge towards either the center of the cusp or the lateral spurs. The two main theories for the origin of beach cusps involve edge waves and enlargement of chance depressions in the beach. The edge-wave origin is suggested by the proportionality between cusp spacing and edge-wave length, and the association of at least the larger cusps with rip currents. With this theory, the spurs correspond to positions of edge-wave nodes, the cusps correspond to antinodes, and water diverges away from spurs and converges into the center of the cusps. Erosion of the cusp is related to increasing longshore current velocity and sediment transport as the water converges into a rip current. This theory does not explain the observation of rip currents that occur seawards of spurs. The alternative theory is that a depression in the beach results in wave refraction, such that wave crests deviate away from the depression, resulting in longshore sediment transport away from the depression. Continuity of water mass requires that this alongshore water motion must be returned seawards further alongshore. The alongshore currents and sediment transport directed away from the depression are expected to cause erosion and enlargement of the hollow. The eroded sediment would be carried seawards, where it may be deposited at the edge of the hollow, as a spur. This theory does not explain beach cusps in which seaward flow occurs in the center of the cusp and deposits a delta immediately offshore. The big questions therefore seem to be where the water converges and flows offshore, and where most deposition occurs. Furthermore, steady-state beach cusps should have wave crests parallel to the beach contours, so that there is no alongshore transport of water and sediment, and therefore no net offshore movement of water and sediment. Numerical models for the formation of beach cusps, reviewed by Komar (1998), do not take account of all of the processes involved, and therefore have not resolved this problem.

Deposition on berms with beach cusps during summer should result in seaward-dipping strata in trough-shaped sets due to filling of beach cusps (Figure 15.20). Such troughs are expected to be meters to hundreds of meters across (parallel to the shoreline), tens of meters long, and up to meters deep. These trough fills may be capped by nearly flat planar strata deposited on top of the berm. Successive periods of winter erosion and summer deposition would produce a series of overlapping trough fills.

Flow, sedimentary processes, and deposits for beaches: backshore

The backshore of a beach is the area between the foreshore and the aeolian dunes (Figure 15.15A). The seaward edge of the backshore may be marked by a berm. The backshore is inundated only during highest water level, and is therefore dry most of the time. Wind can act on the surface for extended periods, winnowing fine to medium sand and transporting it into the aeolian dunes. Broken shells that cannot be moved by the wind are common on the surface of the backshore, as are driftwood and grass stems. Insects and crustaceans ("beach hoppers") live in seaweed, driftwood, and other storm debris left on the backshore. In addition, larger decapod crustaceans such as land crabs and ghost crabs are found here. Some crabs, such as ghost crabs, excavate burrows down to the water table so that they can keep their gills moist. Such burrows may be up to 2–3 m long. Backshore sediment is coarser-grained than that on the foreshore because of the wind deflation of the fine–medium sand fraction. Backshore

deposits are mainly planar-laminated, rooted, and burrowed sand, with lag concentrations (single layers) of convex-upward shells.

Flow, sedimentary processes, and deposits for aeolian dunes

Aeolian dunes on beaches and barrier islands are formed as onshore winds entrain sediment from the upper, dry parts of the beach and transport it as dunes. Aeolian dunes on dry, unvegetated coasts are mainly transverse dunes and barchans (Chapter 6). The dunes increase in height onshore, but rarely reach more than 5 m high. In wet, vegetated regions, deposition on dunes is promoted by plants, and dune shapes are irregular. Pyramidal and parabolic dunes are common. These dunes are meters high and up to hundreds of meters long and wide. Cross strata are variable in direction and occur in cross-cutting trough-shaped sets (Figure 15.23A). The sand is fine- to medium-grained and texturally distinct from the beach sand. Substantial dune fields can be found landwards of some beaches.

Flow, sedimentary processes, and deposits for washovers

Washovers form during storms and exceptionally high tides when the sea level is elevated relative to the top of the beach, see e.g., Leatherman et al. (1977), Schwartz (1982), and papers in Stone and Orford (2004). Seawater flows through low areas in the aeolian dunes and expands in the back-beach area. Sediment carried through the dunes is deposited as a fan-shaped body on the landward side of the dunes on an alluvial plain, old beach ridge, tidal flat, or marsh, or in a lagoon (Figures 15.24 here and 15.28 later). Washovers may occur several times in the same place, or they may change location along the beach. The washover channels through the aeolian dunes may be tens to hundreds of meters wide and up to meters deep. The washover fans are typically on the order of hundreds of meters to kilometers long and wide. A fan deposit may be up to 1 m thick, decreases in thickness inland, and the perimeter may, but need not, be at the angle of repose. The surface of washover fans is normally an upper-stage plane bed, but antidunes and ripples occur in places. The base of the fan may be planar or channelized, and is commonly an erosion surface with a shell lag. Internal structures are mainly planar laminae, although angle-of-repose cross strata may occur at

FIGURE 15.23. Aeolian dunes on North Carolina beaches: (A) Backshore and eroded aeolian dunes, showing trough cross strata; and (B) and (C) vegetated aeolian dune ridges and (more vegetated) swales.

the perimeter (Figure 15.24B). The sand is extensively burrowed by fiddler crabs and further disrupted by plant roots, and the top is likely to be reworked by the wind. Several washover deposits may be superimposed. Thin sand sheets and lenses occurring interbedded with tidal-flat or lagoonal muds are probably washover deposits. Old washovers commonly are colonized by

FIGURE 15.24. (A) Washover sand deposit extending into a back-barrier saltmarsh. (B) A trench through a washover deposit, showing bioturbated, planar-laminated sand. Flow is from right to left.

Spartina marshes or mangroves and incorporated into the marsh sediments behind the barriers.

Flow, sedimentary processes, and deposits for back-barrier lagoons

The nature of lagoons behind barrier islands depends on tidal range and climate (Hayes, 1979; Boothroyd *et al.*, 1985; Ashley, 1988; Nichols, 1989; Oertel *et al.*, 1989). In microtidal (wave-dominated) coasts, access from the lagoon to the sea is limited because tidal inlets are rare, thus lagoons are brackish in humid climates and hypersaline in arid climates. Lagoons along meso-tidal coasts are smaller than those along microtidal coasts, have normal marine salinity, and are influenced by tidal currents. Mud is deposited episodically from suspension in lagoons. The mud is laminated but also extensively bioturbated, mainly by polycheate worms and bivalves. Snails commonly rework surface sediments. Rippled sand layers may be formed by wave action. In humid climates, abundant plant material is washed in by rivers, and salt marshes or mangrove swamps are common in intertidal areas. Shelly fossils are common, especially bivalves and gastropods. In arid climates, lagoonal deposits have low organic content and include cyanobacterial mats, desiccation cracks, and evaporites. Bay-head deltas occur where a river flows into the lagoon, and these are expected to be similar in geometry and deposits to the lacustrine deltas and crevasse splays of delta plains described above (see Figure 15.7).

Depositional models for wave-dominated coasts and shallow seas during marine transgression and regression

Marine transgressions and regressions, and associated deposits, depend on the relative effects of deposition and erosion, tectonic subsidence and uplift, and eustatic sea-level change (see also Chapter 20). If sediment supplied from rivers is deposited on coasts and the sea bed, with no tectonic subsidence, uplift, or eustatic sea-level change, the coast would prograde seawards (i.e., marine regression). As deposits of the beach and shoreface prograde over those of the off-shore zone, a coarsening-upward sequence (mud to mud–sand to sand and/or gravel) is produced that could be tens of meters up to about 100 m thick (e.g., Bernard *et al.*, 1962; Curray *et al.*, 1969; Hunter *et al.*, 1979; Howard and Reineck, 1981; McCubbin, 1982; Davis and Clifton, 1987; Storms and Swift, 2003) (Figure 15.25). In sequences that coarsen up to gravel, the gravel tends to be restricted to the beach and upper-shoreface deposits. The lateral extent of such coarsening-upward sequences could be on the order of kilometers to hundreds of kilometers. Superimposed on this sequence might be smaller-scale coarsening-upward sequences (up to 10 m thick and extending laterally for up to tens of kilometers) due to progradation of individual linear sand ridges, river-mouth bars, or plumes of sand drifted alongshore from river mouths. Bases of the mouth-bar deposits might be sharp and erosional, due to either the proximity of channel flow or abrupt diversion of a river to the region.

Under these depositional conditions with no tectonic subsidence, uplift, or eustatic sea-level change, facies boundaries are horizontal (Figure 15.25). If deposition and marine regression occur with varying combinations of tectonic uplift and subsidence and eustatic sea-level change, facies boundaries as seen in sections normal to the coast will be inclined upwards or downwards (Figure 15.25). Marine regressions under conditions of falling sea level are referred to as forced regressions (Posamentier *et al.*, 1992; Walker and Plint, 1992; Posamentier and Allen, 1999; Hunt and Gawthorpe, 2000). This term seems pointless and potentially confusing, in that all marine regressions are forced by some combination of factors. It is commonly stated that shoreface deposits are sharp-based

A

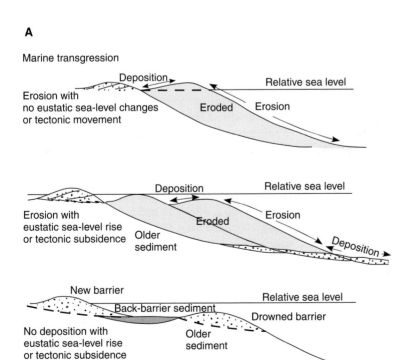

FIGURE 15.25. Wave-dominated coastal depositional models associated with (A) marine transgressions (including shoreface retreat and barrier-drowning models) and (B) marine regressions. (C) A vertical log through prograding beach to the shoreface. (D) A vertical log through the landward side of a landward-migrating barrier beach. The legend to the logs is Figure 1.6.

B

FIGURE 15.25. Continued.

Marine regression

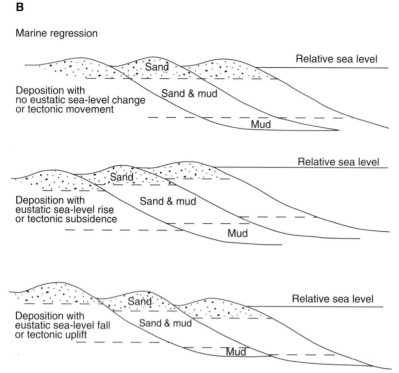

and erosional with forced regression. There is no physical reason why this should always be the case (see Chapter 20).

Marine transgression occurs where tectonic subsidence and/or sea-level rise exceeds the deposition rate. The natures of the deposits under various conditions of deposition or erosion, tectonic subsidence or uplift, and sea-level change are given in Figure 15.25. In general, fining-upward sequences are formed during marine transgression, as offshore muds are deposited over nearshore sands and/or gravels, and facies boundaries dip seawards. In the case of minor deposition or erosion compared with sea-level rise and/or subsidence, deposits formed behind the beach (i.e., on floodplains, tidal flats, lagoons, or washovers) are truncated by erosion and overlain by a shelly gravel lag and offshore marine deposits (Kraft, 1971; Demarest and Kraft, 1987; Kraft *et al.*, 1987; Nummedal and Swift, 1987; Ashley *et al.*, 1991) (Figure 15.25). This is because beach and shoreface deposits are eroded during marine transgression, and this eroded sediment is deposited either offshore or in washovers. In some cases, rising sea level and/or subsidence may cause the sea to drown a beach before it can be eroded, thereby forming a barrier island or

shelf sand ridge (Swift, 1975; Rampino and Sanders, 1980; Wellner *et al.*, 1993).

Tide-dominated coasts and shallow seas

Geometry and origin

Tidal currents are considered to be dominant in macrotidal settings (e.g., estuaries and tide-dominated deltas) where wave currents are normally subordinate; however, tidal currents are also locally important in mesotidal settings (e.g., barrier beach coasts) where wave currents are also important (Figure 15.26). Three depth zones can be defined whatever type of coast. The *subtidal* zone is below the mean low-tide level, where tidal currents and wave currents are active all the time, and includes the deeper parts of tidal channels and deltas, lagoons, and the open sea. The *intertidal* zone lies between mean low tide and mean high tide, and includes the upper parts of tidal channels and deltas, intertidal flats, and beaches. The *supratidal* zone is above the mean high-tide level, is inundated only during the highest tides and storms, and includes salt marshes, mangrove swamps, and the uppermost parts of beach ridges (backshore, aeolian dunes, and washovers).

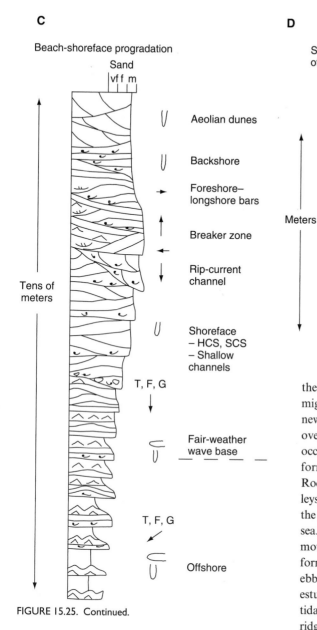

C

Beach-shoreface progradation

Sand
vf f m

U Aeolian dunes

U Backshore

→ Foreshore–
longshore bars

↑ Breaker zone

← Rip-current
channel

Tens of
meters

U Shoreface
– HCS, SCS
– Shallow
channels

T, F, G

Fair-weather
wave base

T, F, G

Offshore

FIGURE 15.25. Continued.

D

Shoreward progradation
of barrier island

Sand
vf f m

U Aeolian
dune

U Washovers

U Salt marsh

Meters

Lagoon or
tidal flats

The aerial extent of these zones depends on the tidal range and the type of coastline (Allen, 1970; Hayes, 1979; Dalrymple, 1992; Reinson, 1992; Davis and Fitzgerald, 2004). For example, subtidal lagoons are relatively large in microtidal coasts with long barrier islands, but are relatively small in mesotidal coasts with short barrier islands (Figures 15.26 and 15.27). In such mesotidal, barrier-island coasts, intertidal and supratidal flats with meandering drainage channels cover a large area. Although tidal inlets through the barrier islands are limited in area, large areas may be occupied by the flood-tidal and ebb-tidal deltas associated with these inlets (Figures 15.26 and 15.27). Tidal inlets migrate laterally, and can be abandoned in favor of a new location. Some tidal inlets develop from washovers (Moslow and Heron, 1978), and some inlets occupy the sites of former river valleys cut during a former sea-level lowstand (Tye, 1984; Fitzgerald, 1993; Rodriguez et al., 2000). Estuaries (drowned river valleys) are funnel-shaped in plan, with a river entering at the narrow end and the broad end passing into the open sea. A bay-head delta may occur at the river mouth. The mouth of the estuary may be partially barred by a wave-formed barrier in mesotidal settings and have associated ebb-tidal and flood-tidal deltas (Figure 15.26B). In estuaries, the subtidal area is large relative to the intertidal and supratidal areas. Channel bars and tidal sand ridges are important components both of estuaries and of tide-dominated delta distributaries. Tidal sand ridges and sand waves, and areas of bare sea bed with sand ribbons, occur in tide-dominated shallow seas.

Tidal channels are on the order of meters or tens of meters deep and hundreds of meters to kilometers wide at bankfull stage. Their deepest parts are well below the mean sea level. Tidal channels are commonly curved, with asymmetrical cross sections like river channels. The channels normally contain mid-channel bars, point bars, or linear tidal-current ridges. A tidal inlet may have an adjacent bar that resembles a spit. These channel-scale bed forms are generally covered with

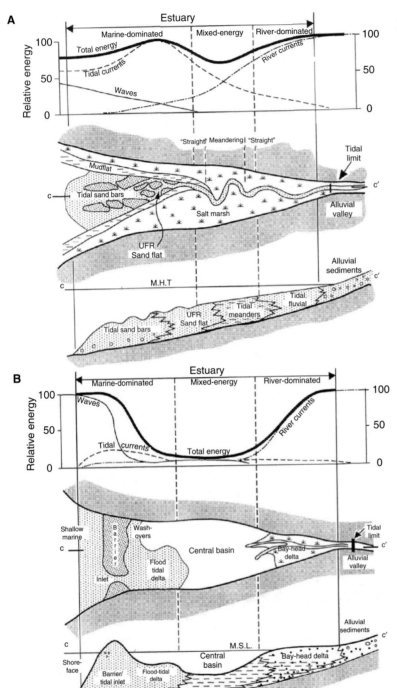

FIGURE 15.26. Coastlines (or parts of them) strongly influenced by tidal currents. (A) An idealized tide-dominated estuary according to Dalrymple et al. (1992). (B) An idealized wave- and tide-influenced estuary according to Dalrymple et al. (1992). (C) An estuary with tidal sand bars, Solway Firth, Scotland. (D) An idealized wave- and tide-influenced barrier-beach–tidal-inlet system with subtidal, intertidal, and supratidal zones marked (Allen, 1970). (E) Short barrier islands with beach ridges, tidal inlets with ebb-tidal deltas, and tidal flats with meandering tidal creeks landwards of the barriers, Capers Island and Price Inlet, South Carolina.

dunes (up to meters high and hundreds of meters long). Dunes change shape and orientation with the tide, but are normally dominated by either flood-tidal currents (higher up) or ebb-tidal currents (lower down). During slack-water periods, dunes may be covered with smaller dunes, current ripples, and wave ripples.

Flood-tidal deltas form on the landward side of tidal inlets through barrier beaches or barred estuary

C

D

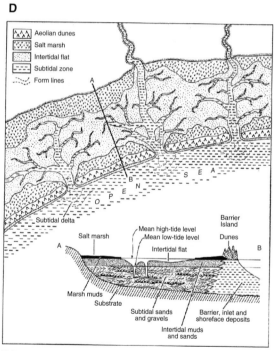

E

FIGURE 15.26. Continued.

mouths (Hayes, 1979; Boothroyd, 1985; Davis and Fitzgerald, 2004) (Figures 15.26–15.28). They result from flow expansion, deceleration, and sediment deposition. They may be hundreds of meters to kilometers long and wide, and meters thick. They decrease in thickness in the landward direction away from the tidal inlet. Typically, the tidal inlet channel splits into two flood-dominated channels around a type of mouth bar called a flood ramp (Figure 15.28). The downstream, highest part of this flood ramp is referred to as an ebb shield. This flood-dominated part of the delta is bordered by ebb-dominated channels that converge into the tidal inlet (Figure 15.28). Flood-tidal deltas are commonly covered with flood- or ebb-oriented dunes.

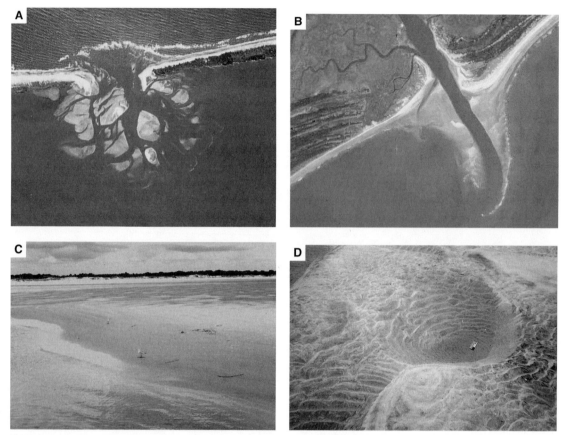

FIGURE 15.27. (A) A microtidal barrier island with tidal inlet and flood-tidal delta, Drum Inlet, North Carolina. Breakers show the outline of the small ebb-tidal delta. Images (B)–(D) show a mesotidal barrier island with ebb-tidal delta, Price Inlet, South Carolina. (C) The view from the main ebb channel towards the barrier beach with the swash platform in the foreground and the marginal flood channel nearer the beach. (D) Ebb-oriented dunes at the margin of the main ebb channel.

Flood-oriented dunes occur on the topographically high parts of flood tidal deltas (the flood ramp), whereas ebb-oriented dunes occur in the peripheral channels. In barrier-island coastlines, flood-tidal deltas pass landwards and laterally into lagoons and tidal flats. In barred estuaries, flood-tidal deltas pass landwards into broad channels and tidal flats. Since flood-tidal deltas build into areas with relatively weak wave and tidal currents, they can be preserved with little modification.

Ebb-tidal deltas form due to ebb-flow expansion on the seaward side of tidal inlets and barred-estuary mouths (Figures 15.27 and 15.28). They are similar in form and process to tide-influenced distributary-mouth bars, and differ mainly in the relative contribution of river currents to the ebb-tidal currents. Ebb-tidal deltas may be hundreds of meters to kilometers long and wide, and meters thick, and decrease in thickness seawards. Ebb-tidal deltas are reworked by wave currents, and are therefore generally less extensive than flood-tidal deltas. The part of the ebb-tidal delta directly seawards of the tidal inlet (the main ebb channel and terminal lobe) is normally covered with ebb-oriented dunes, whereas marginal areas are covered with wave-formed longshore bars (swash bars) that are tens of meters long and wide, and up to a meter or so high. The areas on the sides of ebb-tidal deltas adjacent to the barrier island also contain flood-tidal channels containing flood-oriented dunes. During slack-water periods, ebb-tidal deltas and superimposed dunes are covered with wave- and current-ripple marks.

Intertidal flats occur on tops of bars in tidal channels, on flood- and ebb-tidal deltas, at the margins of estuaries, and behind mesotidal barrier islands (Figures 15.26 here and 15.31 later). The width of

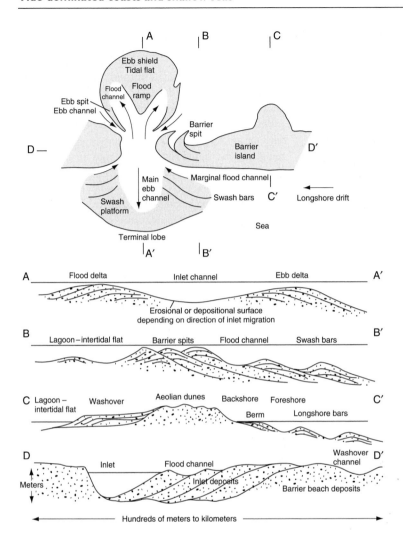

FIGURE 15.28. An idealized model for the geometry, migration, and large-scale depositional features of tidal inlets, flood- and ebb-tidal deltas, and washovers.

intertidal flats between mean high and low tide depends on the tidal range and the slope of the bed, and can range up to kilometers. Tidal flats have been subdivided into lower and upper parts, depending on the distribution of grain size, bed forms, and biogenic activity (Evans, 1965), as will be seen below. Intricate dendritic channel networks are characteristic of tidal flats (Figures 15.26 here and 15.32 later). Intertidal channels are sinuous and have point bars, although lateral channel migration may be very limited. Low levees on intertidal channels may be capped by cyanobacterial mats. The geometry of meandering tidal creeks was analyzed by Barwis (1978), Duc and Tye (1987), and Marani *et al.* (2002).

Salt marshes occur in the supratidal zone in temperate climates where wave and tidal currents are generally weak (see Figures 15.26 here and 15.33 later). Salt marshes comprise a dozen or so species of salt-tolerant plants (e.g., *Spartina*, *Salicornia*, *Juncus*, *Distichlis*, *Festuca*, *Puccinellia*) that tend to be roughly zoned according to elevation. The zonation is related to salt tolerance and competition among species (Allen and Pye, 1992). These plants are effective at trapping and binding sediment. Intertidal channels may extend into the salt marshes, and their widths decrease exponentially landwards. The edges of salt marshes next to intertidal flats commonly have low cliffs, isolated clumps of salt marsh, and furrows that bound upstanding ribs. These features are caused by differential erosion and deposition of the salt marsh and tidal flat, particularly during severe storms. Mangrove swamps in tropical climates comprise shrubs and trees that occupy intertidal as well as supratidal settings. The mangroves (normally only two species) have a dense prop-root mat that binds sediment, promotes deposition, and produces a reticulate drainage pattern.

Tidal sand waves in shallow seas are long-crested bed waves oriented normal to the tidal current direction (Allen, 1982a; Belderson et al., 1982; Hulscher, 1996; Dyer and Huntley, 1999) (see Figure 15.34 later). Their length parallel to the tidal current is 50–300 m (maximum 1 km), their length along the crest is hundreds of meters, and their height is normally 1–8 m (maximum 25 m). The sand-wave size increases approximately with depth. Side slopes are 1 to 10 degrees and sand waves may be symmetrical or asymmetrical in cross section. Superimposed dunes are common. These sand waves are the tidal analog of wave ripples according to Allen (1982a).

Tidal sand ridges occur in groups at mouths of tide-dominated estuaries and deltas and on tide-dominated shelves (see Figure 15.34 later). They are 10–40 m high, 1–2 km across, and 20–60 km long, have side slopes of less than 5 degrees, and are asymmetrical in cross section. Ridges are almost parallel to the net flow and sediment-transport direction (off by about 20°), and each side may be dominated by either flood- or ebb-tidal currents. Trough areas are erosional and covered with a gravel lag. The ridges are composed of shelly sand, and commonly have superimposed dunes migrating obliquely up towards the crest. Some tidal sand ridges on shelves are no longer active (moribund), having formed under some past current regime.

Linear stability analysis of tidal currents and sediment transport (e.g., Huthnance, 1982a, b; Hulscher, 1996; Dyer and Huntley, 1999; Walgreen et al., 2002) can predict the initial formation of tidal sand waves and tidal current ridges. Their occurrence and geometry depend on parameters such as the bed shear stress, water depth, tidal period, and turbulent eddy viscosity. Transverse sand waves can apparently grow much more quickly than the longitudinal tidal current ridges. However, as with all linear stability analyses, not all of the controlling variables are considered, and continued growth and migration of bed waves cannot be predicted.

Sand ribbons on erosional sea beds are parallel to the tidal current, up to a meter high, kilometers long, up to 200 m wide, and spaced at about four times the water depth (see Figure 15.34 later). Sand ribbons may have superimposed dunes. Sand ribbons may be considered to be incipient tidal current ridges with insufficient sediment supply. The origin of sand ribbons related to longitudinal vortices was discussed in Chapter 5.

General patterns of water flow, sediment transport, erosion, and deposition

In mesotidal barrier-island settings, tidal currents are concentrated in the tidal inlets and in the tidal flats landwards of the barriers, and wave currents are concentrated on the seaward side of the barriers. Tidal currents are active throughout estuaries, and wave currents may be significant in the large areas of open water. In all tide-influenced areas, the tidal-current velocity generally increases with tidal range and as depth decreases (Chapter 7). However, the maximum tidal-current velocity is normally greater in channels (up to the order of meters per second) than on tidal flats (up to the order of decimeters per second), because tidal currents are forced to converge into channels, and tidal currents are damped over broad shallow areas. Another common feature of tidal currents is their time–velocity asymmetry, whereby the maximum flood currents occur near high tide, and the maximum ebb currents occur near low tide. This leads to different preferred paths for the main flood and ebb currents. Maximum ebb currents are normally restricted to channels. In estuaries, the degree of asymmetry increases up estuary (landwards), so that flood currents become faster and more dominant up estuary (Lauff, 1967). However, river currents add to the ebb currents. Simple symmetrical, standing tidal waves associated with wave reflection are unlikely in most tide-influenced coasts and shallow seas, and some degree of time–velocity asymmetry is expected.

A saline wedge can extend into a tide-influenced delta distributary or estuary if tidal currents are insufficient to mix the fresh and saline water (i.e., microtidal). Saline wedges can reach hundreds of kilometers inland during the low-river stage. If tidal currents are strong, the fresh and saline water in rivers and estuaries is well mixed (Lauff, 1967). The degree of mixing of fresh and saline water in rivers and estuaries depends on temporal variations in tidal range and wave currents, and on the river discharge. During river floods, salt water may be completely evacuated from river channels.

Tidal currents are capable of moving sand and gravel on the bed and fine sand to mud in suspension. The largest grains occur on the bed where the combined tidal and wave currents are strongest. The asymmetry of tidal currents, and spatial segregation of maximum flood and ebb currents, result in distinct net sediment-transport paths and bed-sediment size trends (Chapter 7). The sediment-transport rate and mean bed-grain size may increase or decrease along these paths (e.g., Figure 7.28). In mesotidal barrier-island coasts, the maximum tidal-current velocities and grain sizes are in the inlets, but decrease both shorewards and seawards. The tidal-current velocity on the seaward side of the inlets (on ebb-tidal deltas) is enhanced by strong wave currents, so the bed sediment is relatively coarse there. In estuaries, the maximum bed-sediment size occurs at the mouth of the estuary, where tidal and wave currents are strong, and near the river mouth, where river and flood-tidal currents are strong. Maximum flood currents become dominant up estuary as the tidal wave becomes more asymmetrical, but river flood currents add to the ebb current. A zone of relatively slow flow occurs in the middle of a barred estuary, due to deceleration both of river currents and of tidal currents as they expand into this region, and as river flows are decelerated by being backed up by flood tides or storm tides (Figure 15.26). This is an area of extensive mud deposition, due to both flocculation of clay as freshwater and seawater mix, and deceleration of river and tidal flows. Near the bed, the concentration of mud in suspension increases markedly, below the so-called *lutocline* (Mehta, 1989). The mud just above the bed is maintained in a fluid state because of hindered settling. Mud is also deposited on upper intertidal flats and lagoons where currents are weak. In most tidal settings, mud is deposited on the upper tidal flats during high tides, but is not removed during falling tide because of tidal-current asymmetry and because mud is more difficult to erode than to deposit (Chapter 7). The sediment deposited in estuaries is supplied from the river and from offshore (Dyer, 1989).

Flow, sedimentary processes, and deposits in tidal channels and deltas

Examples of studies of flow, sedimentary processes, and deposits in tidal channels and deltas are Hoyt and Henry (1967), Klein (1970), Kumar and Sanders (1974), Hayes (1975), Hine (1975), Wright *et al.* (1975), Barwis (1978), Hubbard *et al.* (1979), Oertel (1979, 1988), Boersma and Terwindt (1981), Nummedal and Penland (1981), Van Den Berg (1982), Berelson and Heron (1985), Moslow and Tye (1985), Fitzgerald (1988, 1993), Imperato *et al.* (1988), Smith (1988), Woodroffe *et al.* (1989), Dalrymple *et al.* (1990), G. P. Allen (1991), Sha and de Boer (1991), Harris *et al.* (1992, 1996), Berné *et al.* (1993), Siringan and Anderson (1993, 1994), Tye and Moslow (1993), Dalrymple and Rhodes (1995), Davis and Flemming

(1995), Hori *et al.* (2002), Dalrymple *et al.* (2003), and Willis (2005). Flow velocity in tidal channels commonly reaches 1 or 2 meters per second in places, which is comparable to that of a river in flood. Flows of this magnitude in water depths on the order of meters can transport sand and gravel on the bed in the form of dunes. The coarsest sediments on the bed are shelly, gravelly sands and the finest bed sediments are fine sands. In view of the characteristic asymmetry and time variation in the tidal currents, the dunes vary in geometry and orientation over a tidal cycle, but are normally either flood- or ebb-dominated. Cross stratification formed by dunes is normally in the form of tidal bundles (Chapter 7), with medium-scale cross strata dominantly dipping in the flood or ebb direction, superimposed wave and current ripples, and possibly single or double mud drapes. The thickness of the cross sets is commonly on the order of tens of centimeters. Herringbone cross stratification is rarely produced by dunes. Waves are also active in tidal channels, particularly in the shallower parts. These can form wave-ripple marks (interfering types commonly), plane beds, and longshore bars.

Curved tidal channels migrate laterally by erosion of cut banks and deposition on adjacent bars. The rate of tidal-channel migration is limited if the banks are composed of mud rather than sand. Tidal inlets migrate laterally as a result of channel curvature, and due to longshore drift, which causes sediment to be deposited on the updrift side of a tidal inlet as a spit. Rates of tidal-inlet migration can be up to tens of meters per year. Longshore drift may cause a tidal inlet to be progressively deflected in a preferred direction (Figure 15.28). At some point, the inlet may cut a new path through the diverting spit. In other words, the tidal inlet switches position and migrates laterally within a well-defined zone (Moslow and Tye, 1985). New tidal inlets can be cut through barrier islands, and others can be abandoned and filled up with sediment. The ebb-tidal delta of an abandoned tidal inlet would be reworked and removed by wave currents. However, the flood-tidal delta would be preserved (Figure 15.27A). Fan-shaped deposits in a back-barrier setting may be abandoned flood-tidal deltas or washovers, or both.

Lateral migration of tidal channels, and cutting and abandonment of tidal channels, are episodic and occur during major river floods, high tides, and storms. Tidal-channel deposits rest on an erosional surface (Hoyt and Henry, 1967; Kumar and Sanders, 1974; Tye, 1984; Moslow and Tye, 1985; Tye and Moslow, 1993) (Figures 15.28–15.30). Material that has slumped from

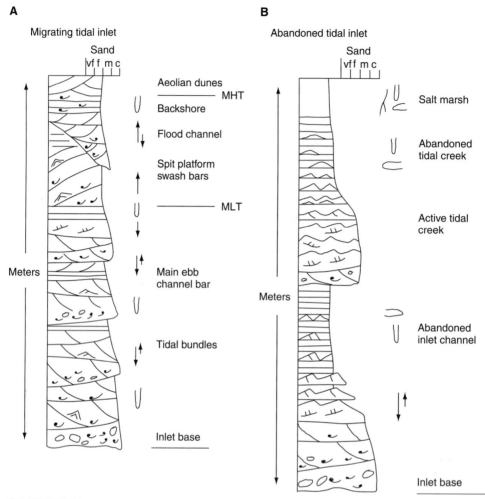

FIGURE 15.29. Idealized vertical logs through (A) a migrating tidal inlet, (B) an abandoned tidal inlet, (C) a flood-tidal delta, and (D) an ebb-tidal delta. The legend for the logs is Figure 1.6.

the cut bank (lumps of cohesive mud, large plant fragments, and large shells) may accumulate on this erosion surface if it cannot be transported away. The grain size will change through the thickness of the bar deposits, depending on the spatial distribution of bed-sediment size during deposition and the mode of channel migration: fining upwards, coarsening upwards or exhibiting little variation in grain size. Cross strata in the lower parts of bar sequences are expected to be ebb-oriented, whereas those higher up are flood-oriented (Figures 15.29 and 15.30). The coarsest-grained cross strata commonly contain shelly, gravelly layers, especially near their bases. The shells are large and robust, typical of nearshore communities. Episodic deposition on tidal-channel bars should result in large-scale inclined strata, as is the case with river-channel bars,

dipping at up to about 10°. The upper parts of individual large-scale inclined strata, deposited during the low-flow stage, are relatively fine-grained (Figure 15.29). The upper parts of tidal channel-bar deposits may be cut by flood-tidal channel fills containing flood-dominated cross strata (Figure 15.29). At the landward end of a flood-tidal channel, at the spit side of a tidal inlet, sediment may be deposited on an angle-of-repose slope that dips landwards and is on the order of meters high (Figures 15.29 and 15.30). Wave-ripple marks, current-ripple marks, and associated cross laminae occur commonly on dune surfaces. Planar laminae formed by waves are common in the upper, intertidal parts of channel-bar deposits. Longshore bars and flood-oriented dunes (sand waves) give rise to landward-dipping planar cross

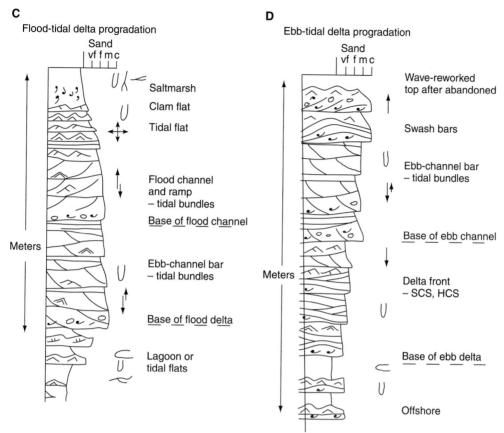

C
Flood-tidal delta progradation

Sand
vf f m c

Saltmarsh
Clam flat
Tidal flat

Flood channel
and ramp
– tidal bundles
Base of flood channel

Meters

Ebb-channel bar
– tidal bundles

Base of flood delta

Lagoon or
tidal flats

D
Ebb-tidal delta progradation

Sand
vf f m c

Wave-reworked
top after abandoned

Swash bars

Ebb-channel bar
– tidal bundles

Base of ebb channel

Meters

Delta front
– SCS, HCS

Base of ebb delta

Offshore

FIGURE 15.29. Continued.

strata in sets up to a meter thick in the upper parts of channel-bar deposits (Figure 15.29). A few burrows of the *Skolithos* ichnofacies may occur in tidal-inlet deposits (see Chapter 11). Abandonment of tidal channels results in filling with sediment, although this may involve several periods of filling and reopening (Moslow and Tye, 1985) (Figure 15.29).

Flood-tidal deltas also have laterally migrating channels and bars, such that their deposits are expected to be similar to adjacent tidal-inlet deposits. However, the distributary channels in flood-tidal deltas will be shallower and narrower than the tidal inlet, and the bars will be broader. As maximum tidal flow velocity decreases in the landward direction, bed-sediment size decreases, and bed forms change from dunes to ripples. Wave ripples and current ripples may be of comparable abundance in the distal parts of flood-tidal deltas. Therefore, flood-tidal deltas become thinner and finer-grained landwards (Berelson and Heron, 1985) (Figure 15.28). Channelized parts of flood-tidal deltas will have deposits resting on a basal erosion surface and

may fine up, coarsen up, or have constant grain size, depending on the distribution of bed-sediment size. Any vertical change in texture may be manifested in a variation in mud content. The main internal structure will be medium-scale cross strata in tidal bundles (Figure 15.29). The more distal, unchannelized parts will coarsen upwards as a result of progradation of the sandy delta over lagoonal muds. Lagoonal muds will contain lenses of wave- and current-rippled sand (lenticular bedding) with abundant burrows of polycheate worms. The overlying deltaic sands would have wave and current ripples and associated cross laminae, with layers and lenses of mud (i.e., flaser, wavy, and lenticular bedding). The sequence would be on the order of meters thick and change upwards from lenticular via wavy to flaser bedding (Figure 15.29). Lateral migration of tidal inlets may produce laterally extensive sheets (hundreds of meters to kilometers long and wide, and meters thick) of flood-tidal delta sands. Abandonment of tidal inlets would cause the top of the flood-tidal delta to be covered by relatively

A

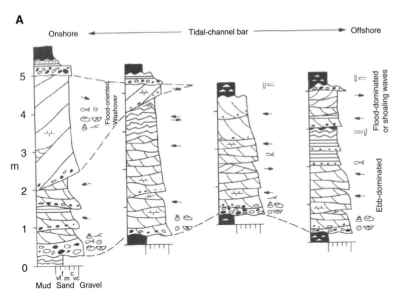

FIGURE 15.30. Devonian coastal deposits of New York State. From Bridge (2000). (A) The interpreted tidal-inlet or tidal-channel bar deposit. (B) The interpreted lateral transition from tide-influenced channel to channel-mouth bar or ebb-tidal delta. The legend for the logs is Figure 1.6.

B   Major tidal-channel bar transitional to mouth bar / nearshore sands

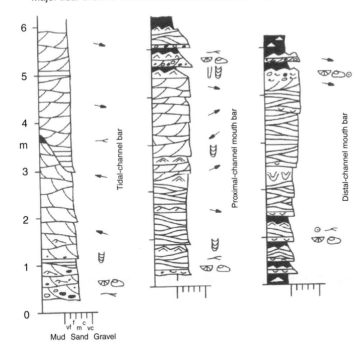

fine-grained deposits of intertidal flats, lagoons, and washovers (Figure 15.29).

Ebb-tidal deltas (and tide-influenced distributary-mouth bars) are also expected to have similar deposits to tidal channels, except at their distal, unchannelled edges, where wave currents may become more important than ebb-tidal currents and combined ebb-tidal and river currents. The base of these deposits may be

sharp and erosional close to the tidal channel, especially if the channel and delta or bar formed relatively abruptly (Imperato *et al.*, 1988) (Figure 15.29). In other cases, there may be a gradual coarsening-upward sequence due to gradual progradation of the delta or bar over deeper and finer-grained marine deposits. In wave-dominated shallow seas, the distal parts of ebb-tidal deltas or mouth bars will be covered with planar beds and hummocks during storms. Marginal parts are covered with longshore (swash) bars and flood-dominated tidal dunes or sand waves. Therefore, deposits of distal ebb-tidal deltas or mouth bars should be on the order of meters thick, and lie above storm beds (sands with HCS, planar strata, wave ripples, and cross laminae, interbedded with muds). The distal delta or mouth bar is composed mainly of SCS, HCS, and planar strata formed during storms, with wave ripples marking periods of fair weather (Figure 15.29). These storm beds are typically decimeters thick. Towards the margins and more landward parts of the delta or bar, the upper parts should contain landward-dipping (angle-of-repose and lower-angle) cross strata formed by longshore bars and wave-influenced flood-oriented dunes. Wave- and current-ripple marks and associated cross laminae are superimposed on these structures. A typical sedimentary sequence within an ebb-tidal delta would decrease in thickness from the tidal channel to the margins of the delta, and the ratio of low-angle cross strata (HCS, SCS) to angle-of-repose cross strata would increase (Figure 15.29). Abandonment of a tidal channel and reworking of the ebb-tidal delta by waves would produce a deposit very similar to that due to abandonment of a deltaic distributary mouth bar (Figure 15.11). The lateral migration of tidal inlets (and associated ebb- and flood-tidal deltas) through barrier islands, and the cutting and abandonment of tidal inlets, mean that a large proportion of barrier-island deposits (40% according to Tye and Moslow (1993)) may be composed of tidal-channel and delta deposits rather than wave-formed beach and shoreface deposits.

Flow, sedimentary processes, and deposits for intertidal flats and associated environments

Intertidal flats have been studied intensely (e.g., Evans, 1965; Reineck and Singh, 1980; Carling, 1981; Collins *et al.*, 1981; Weimer *et al.*, 1982; Terwindt, 1988; Frey *et al.*, 1989; Woodroffe *et al.*, 1989; Alexander *et al.*, 1991; G. P. Allen, 1991; Dalrymple *et al.*, 1991; Eisma, 1997). Tidal currents on intertidal flats have velocities on the order of tens of centimeters per second, and directions that are greatly influenced by the topography of bars and islands. Waves can come from many different directions because of wave refraction and reflection by various shapes of bars and islands. Wave sizes and currents are much less in back-barrier settings than on the seaward side of barriers and in open estuaries. The great variability of current direction and strength associated with waves and tides results in complicated bed-form patterns on intertidal flats. Current and wave ripples of different generations are commonly superimposed, ripple crests are commonly truncated or rounded by subsequent currents as water levels fall, and evidence of falling and rising water level is present in the form of bubble sand, water-level marks, and rill marks (Figure 15.31).

The lower parts of intertidal flats on sand bars adjacent to channels and on flood- and ebb-tidal deltas are composed predominantly of fine to very-fine sands molded by tidal currents into small dunes and current ripples, and by wave currents into plane beds and wave ripples. The dominant internal structure is cross lamination formed by wave or tidal currents, with subordinate medium-scale cross strata and planar strata (Figure 15.31). Mud is rarely deposited in the troughs of ripples during slack water. This mud is dried out at low tide, re-eroded during flood tide, and incorporated into the deposits as mud chips. The mobility of the bed sediment in this zone limits the number of infaunal species, but two or three species abound: the lugworm *Arenicola* and the shrimp *Callianassa*. *Arenicola* constructs parchment-lined U-shaped dwelling burrows (Figure 11.3). *Callianassa* constructs more complicated burrows with an incurrent funnel leading to a series of radiating chambers from 10 cm to meters deep (Figure 11.7). There is also an excurrent siphon. The excurrent ends of *Arenicola* and *Callianassa* burrows are marked by volcano-like mounds of fecal pellets and sediment excavated from the burrows.

The upper parts of intertidal flats are dominated by mud deposition, in the absence of strong and persistent wave currents (as on beaches). Sand is transported and deposited by maximum tidal currents (mainly flood currents) and during storm-wave action. However, mud is deposited during the weaker phases of the flood tide. The mud is transported both as pellets carried in bed load and as suspended floccules. Most mud is transported as aggregates in the silt to fine-sand size range. The resulting deposits are flaser, wavy, and lenticular bedding (with wave- and current-ripple marks), and finely laminated mud and sand (laminites), as discussed in Chapter 7. Sequences of thinning-upward or fining-upward strata are associated with neap–spring cycles or seasonal storms (Figure 15.31C). Mud is commonly desiccated and cracked at low tide, and the chips are later eroded and incorporated into the deposits. Upper intertidal flats contain abundant burrowing worms, bivalves, gastropods, crabs, and shrimps, which destroy stratification and produce

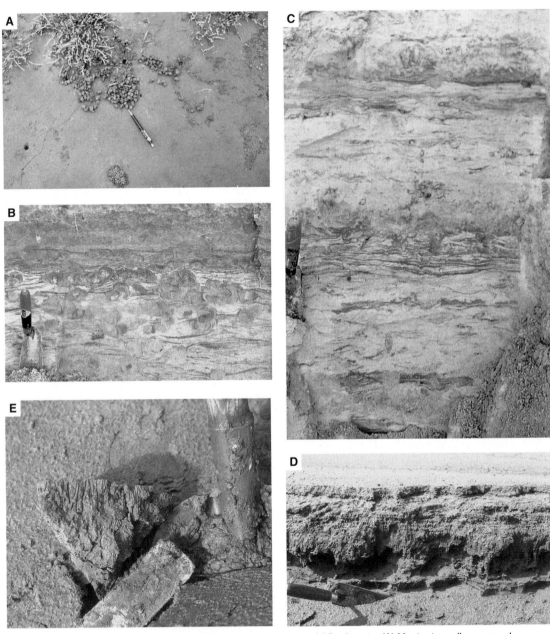

FIGURE 15.31. Images (A)–(E) show burrowing of temperate upper-intertidal flat deposits. (A) Mastication pellets next to burrows of the fiddler crab (*Uca*). (B) Flaser bedding with heart-urchin burrows. (C) Burrowed flaser to wavy bedding, with fining upwards due to either changing wave conditions or the neap–spring tidal cycle. Images (D) and (E) show vertical and U-shaped burrows of worms in laminated sand–mud. Images (F)–(I) show shells on temperate upper-intertidal flats. (F) Gastropods feasting on cyanobacteria. (G) Mussel beds. (H) Oysters adjacent to saltmarsh. (I) Transported (mainly) bivalve shells. Images (J)–(L) show bed forms and sedimentary structures on tidal flats (see also Chapter 7). (J) Tidal dunes and two generations of superimposed wave ripples on a lower-intertidal flat. (K) Medium-scale cross strata and flaser bedding from a lower-intertidal flat. (L) Two sets of superimposed wave ripples on an upper-intertidal sand flat.

FIGURE 15.31. Continued.

pellets (Figure 15.31). Cyanobacteria and cyanobacterially laminated muds are ubiquitous on the upper parts of intertidal flats. The normal grazing on the cyanobacteria by gastropods is curtailed by the prolonged exposure of these flats.

Meandering intertidal drainage channels have the same type of deposits as the flats themselves; however, the sedimentary strata are deposited on laterally migrating point bars (e.g., Bridges and Leeder, 1976; DeMowbray, 1983; Thomas *et al.*, 1987). These deposits have an erosional base overlain by a lag of shells and lumps of mud that came from the cut bank. Overlying bar deposits contain large-scale inclined strata, composed in many cases of heterolithic bedding (Figure 15.32). Much of the deposition on intertidal flats may be by lateral deposition in channels rather than vertical deposition on broad flats. The thickness of intertidal-flat deposits (and maximum depth of intertidal channels) can indicate the tidal range if analyzed properly.

Deposition in supratidal areas occurs on salt marshes and mangrove swamps in humid areas and on salt-encrusted *sabkhas* in arid areas. Sabkhas are described more fully in Chapter 16. Deposition in supratidal areas occurs during the highest tides, and the level to which sediment can accumulate is limited by this highest tide level. During severe storms, the salt-marsh surface and edge next to the tidal flat can be eroded, in some cases while deposition occurs on the adjacent tidal flat (Allen and Pye, 1992). This lost sediment is then gradually replaced by subsequent deposition. Supratidal deposits are mainly mud with subordinate thin and irregular layers of very-fine sand and silt (Figure 15.33). Oyster mounds occur at the edge of the salt marsh (Figure 15.31H), and shells may be transported and deposited onto the supratidal flats

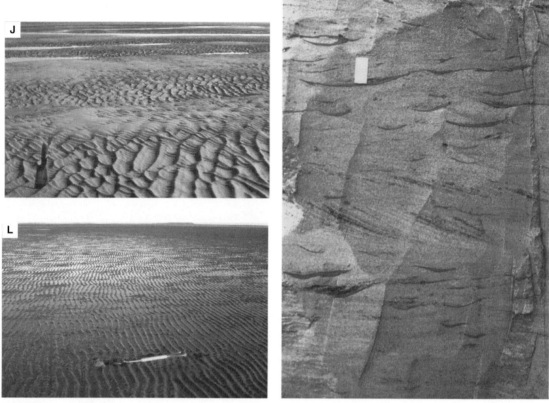

FIGURE 15.31. Continued.

FIGURE 15.32. Examples of intertidal channels and deposits (heterolithic bedding) exposed in cut banks at low tide. From northwest-European intertidal flats.

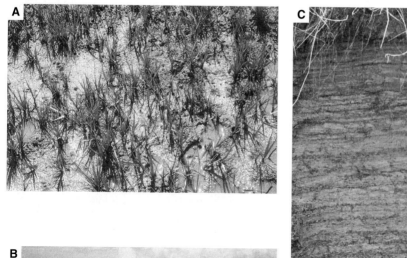

FIGURE 15.33. Temperate supratidal marshes, (A) and (B), and associated laminated mud deposits (C).

during storms. A major component of supratidal deposits is roots and decaying buried plant material (i.e., rice grass *Spartina*). Burrowing animals are abundant. In salt marshes along the Atlantic coast of the USA, ribbed mussels and fiddler-crab burrows riddle the exposed edges of salt-marsh peats along tidal creeks. Cyanobacterial mats and associated fauna are particularly noticeable in muddy pond deposits on the marsh surface. Some supratidal flats behind barrier islands are covered with relatively coarse sediments during washovers.

Flow, sedimentary processes, and deposits for tide-dominated seas

Reviews of flow and sedimentary processes in tide-dominated seas include Stride (1982), McCave (1985), Dalrymple (1992), Bergman and Snedden (1999), and Willis (2005). As discussed in Chapter 7, tidal currents in shallow, tide-dominated seas are strongly influenced by topographically driven flow convergence and wave reflection, and amphidromic points related to Coriolis deflection. These influences result in elliptical paths of water particles; time–velocity asymmetry of tidal currents; spatial variation in tidal range and degree of time–velocity asymmetry; and preferred net water-flow and sediment-transport paths in which the net flow velocity and sediment-transport rate either increase or decrease. Sediment-transport paths are parallel to coastlines and extend for hundreds of kilometers (Figure 7.28). Also, superimposed wave currents and other oceanic currents may enhance one tidal current path. Typical changes in sediment transport and deposition along a path are discussed below.

Erosional areas exhibit patches of bare sea bed with a gravel lag composed commonly of mollusc, bryozoan, and ophiuroid debris. The gravel lag may be covered in places by the sand ribbons described above, possibly with superimposed dunes. Downcurrent of this zone of erosion and limited sediment supply is a zone of sand formed into dunes and, if there is enough sand, into tidal-current ridges and sand waves with dunes superimposed (Figure 15.34). Tidal-current ridges migrate laterally, and may become

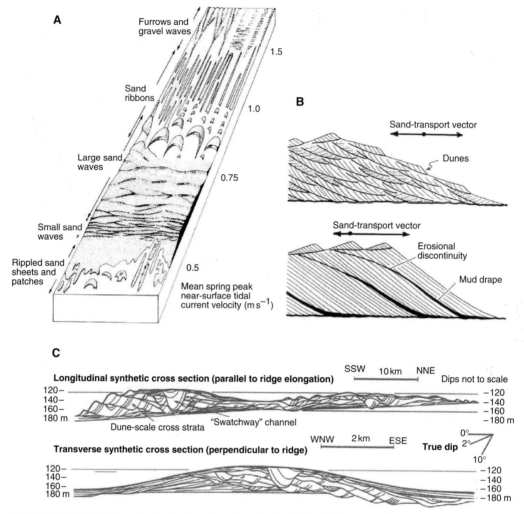

FIGURE 15.34. (A) Bed forms in tide-dominated seas. From Stride (1982). (B) Marine sand-wave depositional models (Allen, 1985). (C) Stratal geometry in tidal sand ridges in the Celtic Sea. From Willis (2005), after Reynaud et al. (1999).

sinuous and split into two ridges. Sand waves migrate in the direction of the dominant current. Internal structures of tidal-current ridges and sand waves are poorly known, since there is only limited evidence from shallow seismic profiles and cores (e.g., Berné et al., 1988, 1998; Reynaud et al., 1999; Willis, 2005). Seismic records indicate that the largest scale of bedding is associated with migration of the bed form itself: low-angle inclined strata (Figure 15.34). Cores have medium-scale cross strata formed by migration of superimposed dunes. Depositional models of Allen (1980, 1982a) are given in Figure 15.34. The inclination and character of large-scale inclined strata, and the degree of dominance of flood or ebb currents represented in medium-scale cross strata, depend on

the asymmetry of tidal currents. Further down the transport path is rippled fine sand that is well burrowed. Mud occurs at the end of the transport path, which is normally in deep water, where waves are ineffective, and near amphidromic points, where the tidal range and currents are negligible (see Chapter 7). Net deposition and downcurrent migration of these zones would produce a coarsening-upward sequence up to meters thick.

Depositional models for tide-dominated coasts and shallow seas during marine transgression and regression

Models for erosion and deposition in mesotidal and macrotidal, estuarine coastal, and shallow-marine

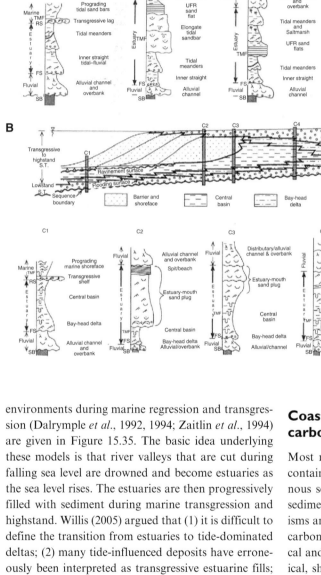

FIGURE 15.35. (A) A depositional model for a tide-dominated estuary that experienced sea-level fall, marine transgression, filling, and shoreface progradation (Dalrymple *et al.*, 1992). (B) A depositional model for a wave-dominated estuary that experienced sea-level fall, marine transgression, filling, and shoreface progradation (Dalrymple *et al.*, 1992). (C) A cross section through a tide-dominated estuary that experienced initial river incision during falling sea level, then marine transgression with estuarine deposition and filling, followed by drowning of interfluves and continued marine transgression and deposition. From Dalrymple (1992). (D) A depositional model for a macrotidal, wave-influenced estuary that experienced net deposition. From G. P. Allen (1991). (E) A model for progradation and retreat of a tide-dominated deltaic coastline. From Willis (2005). (F) A model of a tide-dominated deltaic sandstone layer, with sequence-stratigraphic terminology. From Willis (2005). Major erosion surfaces can have different origins.

environments during marine regression and transgression (Dalrymple *et al.*, 1992, 1994; Zaitlin *et al.*, 1994) are given in Figure 15.35. The basic idea underlying these models is that river valleys that are cut during falling sea level are drowned and become estuaries as the sea level rises. The estuaries are then progressively filled with sediment during marine transgression and highstand. Willis (2005) argued that (1) it is difficult to define the transition from estuaries to tide-dominated deltas; (2) many tide-influenced deposits have erroneously been interpreted as transgressive estuarine fills; and (3) the origin of major erosion surfaces in tide-dominated marine deposits is difficult to interpret unambiguously. Willis (2005) gives various ancient examples of tide-dominated delta deposits, and his depositional models for tide-dominated deltas are shown in Figure 15.35.

Coasts and shallow seas with calcium carbonate sediments

Most modern coastal and shallow-marine sediments contain shell fragments and, if the supply of terrigenous sediment is low, dominantly calcium carbonate sediments accumulate. However, the types of organisms and particles, and rates of production of calcium carbonate sediment, vary substantially between tropical and temperate shallow-marine seas. Modern tropical, shallow-water carbonates have been studied, for example, in the Florida–Bahama Banks area (Shinn *et al.*, 1969; Gebelein, 1974; Enos and Perkins, 1977; Hardie, 1977), the Arabian Gulf (Shearman, 1963; Kinsman, 1966; Butler, 1969; Kendall and Skipwith, 1969; Purser, 1973; Alsharhan and Kendall, 2003), and Shark Bay, Western Australia (Logan *et al.*, 1970,

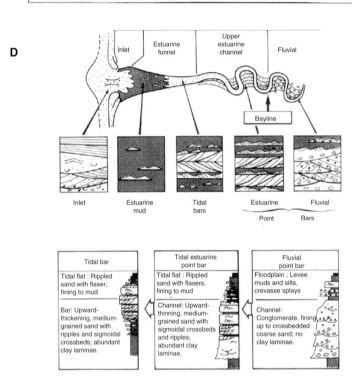

FIGURE 15.35. Continued.

1974). Temperate areas are less well known (James and Clarke, 1997), but the Great Australian Bight off South Australia has been well studied (James *et al.*, 2001). Other references for shallow-marine carbonate deposition include Enos (1983), Halley *et al.* (1983), Inden and Moore (1983), James (1983), Shinn (1983b), Wilson and Jordan (1983), Tucker and Wright (1990), James and Bourque (1992), James and Kendall (1992), Jones and Desrochers (1992), Pratt *et al.* (1992), Demicco and Hardie (1994), Wright and Burchette (1996), Flügel (2004), and Schlager (2005).

Tropical shallow-water environments with low terrigenous-sediment supply host a veritable "carbonate factory" of biogenic and non-biogenic carbonate-sediment production today and also did so in the past. Biogenic production of calcium carbonate is controlled mainly by water temperature, salinity, depth, and clarity. Much of the calcium carbonate mud and sand on modern shelves is produced by green algae that need warm, shallow, clear water. Hermatypic corals also typically occur in tropical shallow-water environments, and these have symbiotic photosynthetic unicellular microorganisms known as *Zooxanthellae* in their tissues. These are dinoflagellate protozoans and, like green algae, require warm, sunlit (i.e., shallow and clear) and siliciclastic-mud-free water to thrive. Lees (1975) referred to the biota of tropical shallow-water carbonate environments as the *chlorozoan* association, emphasizing the importance of the green algae (**Chloro**phyta) and corals (**Zoan**tharia). The chlorozoan association also contains a wide range of other benthic organisms, but mainly molluscs.

E

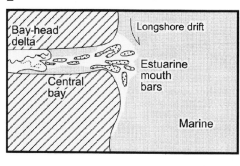

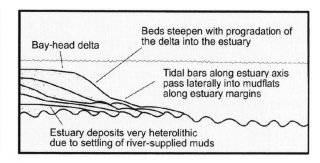

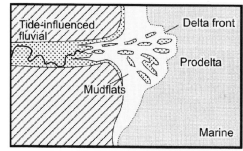

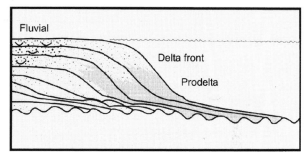

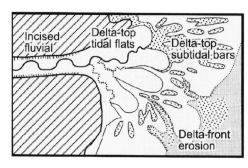

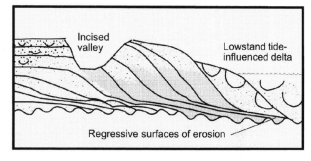

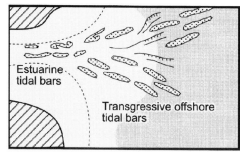

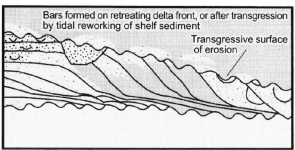

FIGURE 15.35. Continued.

F

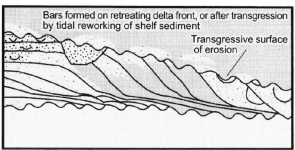

The warm, shallow conditions of the chlorozoan association also favor non-biogenic precipitation of calcium carbonate both as sedimentary particles (mud and sand-sized ooids) and as early diagenetic cements. Precipitation of calcium carbonate as mud takes place in *whitings*, clouds of suspended aragonite needles ~4 microns long that occur in surface water on shallow-marine carbonate shelves (and in lakes). Whitings occur where deep ocean water rises onto a shallow platform, warms, begins to evaporate, and degasses CO_2 (see Chapter 4). Whitings are perhaps triggered by blooms of unicellular photosynthetic organisms that also take CO_2 out of the water. Ooids are also considered to be chemical precipitates created as seawater warms, evaporates, and degasses as it washes onto a shallow platform. Ooids typically accumulate as tidal shoals at the shelf margin or beaches. Carbonate cementation on and just below the sea floor is also important in shallow, tropical-marine carbonate deposition. Reefs and carbonate-sand shoals become cemented and commonly build the shelf margin up to sea level. The cemented edges of carbonate shelves allow them to maintain steep slopes into the adjacent basin. Carbonate-mud pellets produced by deposit-feeding organisms become cemented on the sea floor and aggregated into lumps known as *grapestones*. Erosion of these cemented sediments results in various types of intraclasts.

The outer portions of many continental shelves in temperate areas (e.g., substantial areas of the continental shelf around South Africa and the entire shelf off South Australia, called the Great Australian Bight) have abundant sand- and gravel-sized shells of calcium carbonate. These so-called "cold-water carbonates" comprise the *foramol* (**fora**minifera, **mol**lusc) association of Lees (1975), and the principal sediment contributors are bivalves, gastropods, and benthic foraminifera, with lesser input by echinoderms, barnacles, bryozoans, and coralline red algae. The carbonate mud in these areas is mostly broken and worn shell fragments and is not common. Ooids are exceedingly rare, as are early cements.

Overall geometry, flow, and sedimentary processes

The geometry, flow, and sedimentary processes of coasts and shallow seas where carbonates accumulate differ slightly from those where terrigenous (siliciclastic) sediments accumulate. Wave-dominated coasts

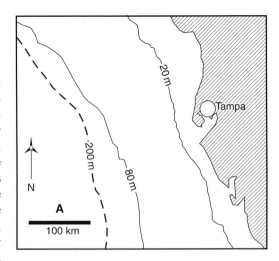

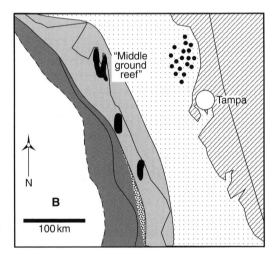

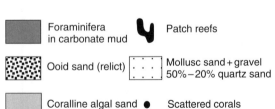

FIGURE 15.36. (A) Bathymetry and (B) surface sediments of the subtropical carbonate shelf deposits of western Florida – an example of a modern carbonate ramp. Modified from Ginsburg and James (1974). By permission of Springer Science and Business Media.

and continental shelves in the tropics to subtropics where carbonates are accumulating include western Florida (Figure 15.36), the Yucatan Peninsula, Northwestern Australia, and South Australia. These gently sloping (on the order of 1 in 1000) continental shelves are similar to most terrigenous-sediment shelves with a shelf–slope break at 125–200 m water depth (Figure 15.36A), and are commonly called "carbonate ramps." On the western Florida shelf, ∼1 m of Holocene sediment covers older deposits. The proportion of carbonate in the sediment increases from ∼50% near shore to >80% at the shelf–slope break where foraminiferal oozes are accumulating (Ginsburg and James, 1974) (Figure 15.36B). The percentage of carbonate in the sediment may either increase or decrease into the deeper portions of the basin. The grain size of the sediment also typically changes across shelves. On the western Florida shelf, there are sands and gravels on the inner shelf, sands on the middle shelf, and mud near the shelf–slope break. The dominant types of carbonate sediment grains on the western Florida shelf occur in shore-parallel belts (Figure 15.36B). Mollusc fragments are abundant on the inner shelf where they are mixed with quartz sand. At about 60 m water depth, sand-sized fragments of coralline red algae occur, whereas ooid sands are found at depths of 80–100 m. The ooid sands are reckoned to be relict sediments deposited in ∼5 m of water depth during the Holocene rise in sea level (Ginsburg and James, 1974). Middle Ground Patch Reef rises some 15 m off the sea floor at about 40 m depth and comprises branching and massive corals, bryozoans, and green algae surrounded by mollusc sand. Patch reefs to the south comprise coralline red algae. Scattered corals occur on the inner shelf north of Tampa.

In contrast to carbonate shelves (ramps), *rimmed carbonate shelves* are relatively shallow seas (depth less than about 10 m) rimmed by reefs or other shoals that abruptly pass seawards into kilometers-deep (abyssal) water (Figure 15.37). Modern examples of rimmed shelves include the Great Barrier Reef, the Bahama Banks off the southeastern USA, and southeast Florida. All of these areas are far from sources of terrigenous sediment, are located in the tropics, and are <30 m deep. The Bahama Banks comprise a series of high rise banks separated from the main North American continent by the kilometer-deep Straits of Florida. In the geological past, epicontinental seas underlain by continental crust have hosted smaller versions of modern-day rimmed shelves.

The fundamental difference in flow and sedimentary processes between open-marine shelves and rimmed shelves is the location of the maximum current velocity, sediment-transport rate, and sediment size. With rimmed shelves, the maximum wave and tidal currents occur at the rim, and the shallow interior shelf and coast experience relatively weak currents even during storms. Therefore, coarse-grained sediment is concentrated at the rim, and most of the shelf is covered with mud or sand composed of mud pellets. The Great Bahama Banks form a more or less flat shelf covering some 100,000 km², and nowhere over that area is the water depth greater than 20 m. Bathymetry and bed sediments of a portion of the Great Bahama Bank are shown in Figure 15.38, and most of the shelf interior west of Andros Island is covered by pelleted mud, with oolitic and skeletal sands restricted to shoals at the shallow rims. On open carbonate shelves (ramps), the velocities of wave and tidal currents, and hence the sediment-transport rate and grain size, increase progressively shorewards. The flow, sedimentary processes, and deposits of carbonate-shelf coastal environments are the same as those described previously for terrigenous coasts and shelves, but the sand- and gravel-sized sediment comprises skeletal debris, intraclasts, ooids, and varying amounts of siliciclastic sediment. In addition, patch reefs and other organic buildups (described below) can occur sporadically on carbonate shelves, but only rarely as large, continuous structures. Therefore, further discussion will concentrate on rimmed shelves.

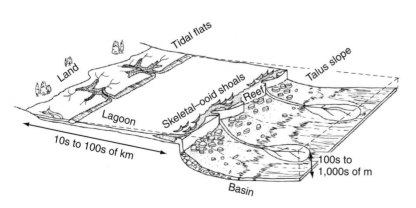

FIGURE 15.37. A cartoon showing the main features of a rimmed carbonate shelf.

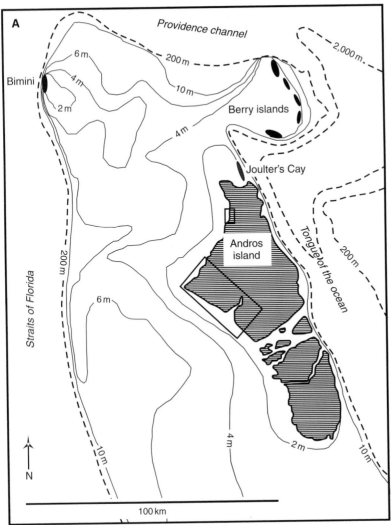

FIGURE 15.38. (A) Bathymetry of the northwestern Great Bahama Bank. Modified from Bathurst (1975). (B) Bed-surface sediments of a portion of the Great Bahama Bank. Modified from Enos (1974). Boxes outline areas of tidal-flat deposition of northwest Andros Island (Figure 15.39) and southwest Andros Island (Figure 15.43).

Geometry, flow, sedimentary processes, and deposits of coasts of rimmed shelves

Coasts on rimmed shelves occur either in the interior of the platform adjacent to the mainland or around islands on the platform. Shorelines in the interior of rimmed shelves tend to be muddy tidal flats, since the wave energy is dissipated at the shoals, reefs, and islands along the shelf margin. For example, most of the Florida Keys are fringed by dense mangrove-covered muddy tidal flats similar to those that comprise the tidal-flat islands of Florida Bay. On the Great Bahama Bank, islands such as Andros (Figure 15.38) and Eleuthera comprise mainly aeolian and shallow-marine carbonate sands deposited during various times in the Pleistocene that were

thoroughly cemented by exposure to freshwater during the last glacial maximum. On the other hand, Joulters Cay, off the northernmost tip of Andros Island, represents a series of Holocene ooid sand beaches that aggraded to sea level and prograded eastwards (Harris, 1979).

Muddy carbonate tidal flats accumulate west of Andros Island (in the lee of the trade winds) in a microtidal setting with a somewhat rainy climate (Figure 15.38). The tidal flats of northwestern Andros Island comprise a 3–4-km-wide "channeled belt" and a 4–8-km-wide "inland algal marsh" (Figure 15.39) (Hardie, 1977). The coast of the channeled belt has beach ridges, and the meandering tidal channels have levees, both on the order of a few decimeters high above mean high tide (Hardie, 1977). The beach ridges and

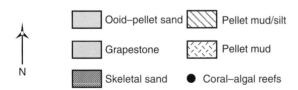

FIGURE 15.38. Continued.

Legend:
- Ooid–pellet sand
- Grapestone
- Skeletal sand
- Pellet mud/silt
- Pellet mud
- Coral–algal reefs

N

levees compartmentalize the tidal flats into what are referred to as "ponds" (Figure 15.39). The crests of the beach ridges and the tidal-channel levees are within 20 m of the adjacent offshore lagoon or tidal channel, and the beach ridges and levees slope back from these high points for a few hundred meters into the ponds. At their mouths, the tidal channels are up to 100 m across and up to 2.5 m deep, and commonly scour down to the underlying Pleistocene rock. The channels narrow and shallow back into the pond areas. The tidal range is ∼45 cm at the tidal-channel mouths and decreases inland to ∼25 cm in the ponds. The inland algal marsh is a flat plain ∼20 cm above the mean high-tide level.

A combination of physical, biological, and chemical processes is important on the tidal flats of northwestern Andros Island. The beach ridges and levees have a zonation of microbial-mat communities from their crests down into the ponds (Figure 15.40A). The highest portions of the beach ridges and levees have a

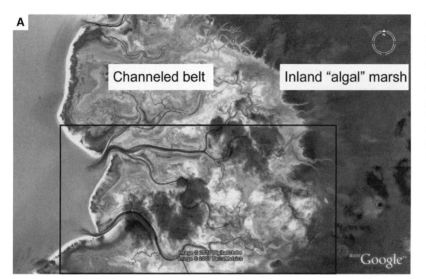

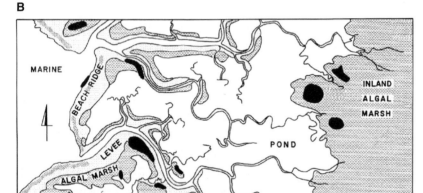

FIGURE 15.39. Northwest Andros Island tidal flats, location given in Figure 15.38A. (A) A satellite view showing the offshore, channeled belt, and freshwater "algal" (cyanobacterial) marsh (from Google Earth™). (B) A map of surface features of the area outlined in (A). From Hardie (1977).

smooth mat of filamentous mobile oscillatoriacean cyanobacteria that are intertwined around the fine-grained pellet sands that comprise the levee and beach-ridge crests (Figure 15.40B). Further down the slopes of the beach ridges and levees, the oscillatoriacean mat is gradually replaced by fleshy tufts of immobile cyanobacteria. Desiccation cracks and mud–chip conglomerates are common over this transition (Figure 15.40C). The fleshy mats become denser towards the pond and cover the surface entirely just above the high-tide level of the pond (Figure 15.40D). The cyanobacteria are abundant here because they are frequently wetted by high tides and storm surges. About half-way down the beach ridges and levees, the sediment surface becomes cemented by high-magnesium calcite (up to 45 mol% Mg, i.e.,

"protodolomite") that forms a black surface pavement. In the intertidal ponds, grazing of cerithid gastropods keeps the surface free of cyanobacteria, and polycheate worms infest the muddy sediment (Figure 15.40E). Red and black mangroves grow sporadically on the beach ridges, levees, and ponds.

During fair-weather conditions, the water flowing in and out of the tidal creeks is crystal clear because it is not transporting suspended sediment. Thus, deposition does not normally occur on these tidal flats during tidal cycles. Deposition of bed load and suspended load over the tidal flats occurs only during winter storms and tropical hurricanes. As storm floods overtop the beach ridges and levees, the smooth oscillatoriacean mat produces a sticky film of mucopolysaccharides that traps a continuous lamina of carbonate mud.

FIGURE 15.40. Surface features of Andros Island tidal flats. (A) An overview showing light-colored levee crests and the levee back slope (dark colored due to microbial mat). (B) Levee crest (∼10 m across) adjacent to the tidal channel. (C) Levee slope with desiccation cracks and rip-up conglomerates. The lens cap is 50 mm in diameter. (D) Fleshy cyanobacterial colonies on the levee back slope above intertidal pond. (E) An intertidal pond. Mangrove seedlings ∼10 cm high. (F) Inland microbial marsh flooded with freshwater.

Discontinuous laminae of fine-grained pellet sand a few millimeters thick are also commonly deposited on the highest portions of beach ridges and levees during storms. The mobile oscillatoriaceans eventually work their way back up through the sediment and recolonize the sediment surface. The beach-ridge and levee crests are thus underlain by flat laminated mud with lenses of fine-grained pellet sand (Figure 15.41A). The discontinuous, fleshy microbial mat on the pond-side slopes of beach ridges and levees is buried by a layer of mud that settles out of suspension from storm waters. As the water recedes, the buried fleshy mats die, desiccate, and rot. The mud covering the mats deforms as the mats deteriorate, and becomes further disrupted by desiccation cracks and fenestrae. The sediment beneath these portions of the beach ridges and levees therefore comprises crinkled laminated mud with desiccation cracks, and laminoid and irregular fenestrae (Figure 15.41B). The luxuriant fleshy mats of the lower portions of the beach ridges and levees are commonly calcified as they extract CO_2 from seawater. When storm floodwater inundates the ponds, suspended carbonate mud and silt-sized pelleted mud settles and buries these calcified microbes with strata up to a few centimeters thick. This sediment is quickly reworked by organisms.

The tidal channels have point bars and mid-channel bars composed of mud and pelleted mud (fine sand and silt-sized), and these channels migrate laterally. Bed forms in the channels are not commonly recorded, probably because of bioturbation. Intertidal banks of the tidal channels are riddled with fiddler-crab (*Uca*) burrows. Below mean high tide, algae, turtle grasses, and foraminifera grow on the bed surface and *Callianassid* shrimp burrows are common. During extremely low tides, exposed portions of channel bars are disrupted by desiccation cracks (Chapter 11). Slump blocks of muddy bank sediment, clasts of the protodolomite crust, and skeletal fragments are found in the channel thalwegs.

A model for deposition on the northwest Andros Island tidal flats (Figure 15.42) shows that most of the sediment is burrowed carbonate mud, commonly pelleted into fine-sand and silt-sized aggregates, that was deposited in the intertidal ponds. This mud also has cerithid gastropod shells scattered throughout, as well as roots of mangroves. Beach-ridge and levee deposits are also pelleted mud with millimeter-thick lenses of pellet sand; however, these deposits are distinctly layered, with the different microbial-mat communities being responsible for different lamination styles (Figure 15.42). The middle of the beach-ridge and levee

deposits is marked by a discontinuous cemented layer of protodolomite. Desiccation cracks, fenestrae, mangrove roots, grass roots, and insect burrows disrupt the sediment, except for the beach ridges and levee crests. The tidal-channel deposits are composed of muddy sediment similar to the pond deposits overlying a basal erosion surface marked by slump blocks, clasts of cemented protodolomite pavement, and coarse-grained skeletal debris. However, the pelleted muds of tidal-channel deposits have deep desiccation cracks in the upper subtidal deposits, *Callianassid* burrows, and lenses of muddy skeletal sand.

The inland "algal" marsh of northwest Andros Island is covered by a luxuriant growth of immobile fleshy filamentous cyanobacteria, and aragonite mud is deposited from suspension over these mats only rarely, during the most powerful tropical hurricanes. These muds are desiccated following the hurricane flood. During the rainy season (winter), this marsh is covered by ponded rainwater (Figure 15.40F). Since there is less Ca^{2+} in rainwater than there is in seawater, these cyanobacterial mats vary from lightly calcified periphyton to more thoroughly calcified tufas, depending on the amount of time between episodes of mud deposition. Also, because the Mg^{2+}/Ca^{2+} ratio of this freshwater is less than 2, the mineral that precipitates is low-magnesium calcite. The deposits of the inland freshwater marsh are therefore layers of *aragonite* mud disrupted by desiccation cracks alternating with *calcite* periphyton muds and calcified microbial colonies (Figure 15.41C).

The northeastern Andros Island tidal flats resemble the barrier-island–tidal-flat systems of the Atlantic coast of the USA in that they are being eroded on their seaward edge as the sea level rises. Laminated beach-ridge and levee deposits exist up to tens of meters offshore, where they are currently being eroded and buried by offshore lagoonal muds. This is analogous to the salt-marsh peats that are exposed in front of many Atlantic barrier islands (Figure 15.15F).

The southwestern portion of Andros Island comprises a ~50 km length of Holocene tidal flats up to 25 km across (Figure 15.43). Tidal channels are not common, but the tidal flats nearest the coast contain vegetated "V"-shaped ridges (called hammocks) up to 2 m high, about 200 m across, and up to 5 km long (Figure 15.43). Gebelein (1974) interpreted these ridges as marking the openings of now-filled channels. The tidal flats are underlain by a sheet of muddy sediments up to 4 m thick that has prograded some 25 km southwestwards in the past 7,000 years (Gebelein, 1974; Gebelein

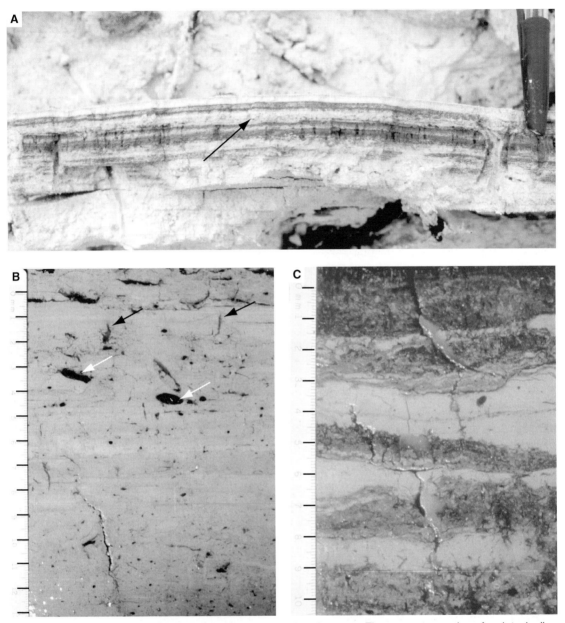

FIGURE 15.41. Tidal-flat deposits of Andros Island. (A) Levee-crest planar lamination. The arrow points to a lens of sand-sized pellets. The pen cap gives the scale. (B) "Crinkled" laminae from the levee backslope, disrupted by desiccation cracks (dark arrows) and mangrove roots (light arrows). The scale is in cm. (C) A freshwater marsh deposit. Light-colored aragonite storm layers are separated by organic-rich low-magnesium calcite muds. The scale is in cm. Images (B) and (C) courtesy of Bob Ginsburg.

et al., 1980) (Figure 15.44A). The intertidal and supratidal sediments resemble those of the northwestern Andros tidal flats and represent ponds, freshwater marshes, vegetated beach ridges, and extensive high-magnesium calcite-cemented layers (Figure 15.44B). Gebelein (1974) and Hardie and Shinn (1986) reckoned that the southwestern Andros Island tidal flats

accumulated by the progressive seaward establishment of beach ridges. In each case, the intertidal pond behind the new barrier fills with sediment while the older sediment-filled ponds become freshwater cyanobacterial marshes.

Other notable areas of carbonate tidal-flat deposition include the eastern shore of Shark Bay (Figure 15.45)

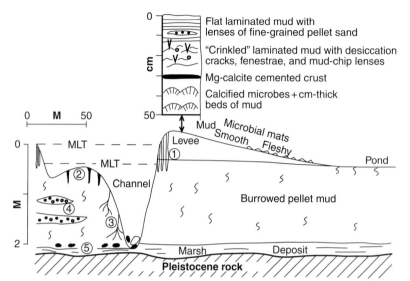

FIGURE 15.42. A model of deposits of northwestern Andros Island tidal flats. The inset gives details of levee deposits as they would appear in a core. Circled numbers refer to (1) *Uca sp.* (fiddler crab) burrows; (2) deep prism cracks in a tidal channel bar; (3) *Callianassa* burrows; (4) a packstone lens in tidal-channel deposits; and (5) the basal erosion surface of a tidal channel overlain by channel lag clasts of cemented protodolomite crust.

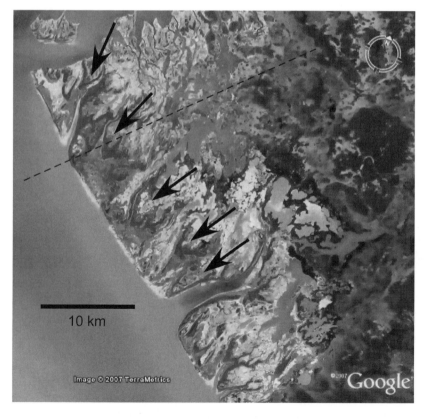

FIGURE 15.43. A satellite photograph of the prograding tidal flat of southwestern Andros Island, location given by the rectangle in Figure 15.38B. Arrows point to V-shaped hammocks that Gebelein (1974) reckoned to mark the mouths of former tidal channels. The dotted line shows the approximate location of sections in Figure 15.44. Photograph from Google Earth.[TM]

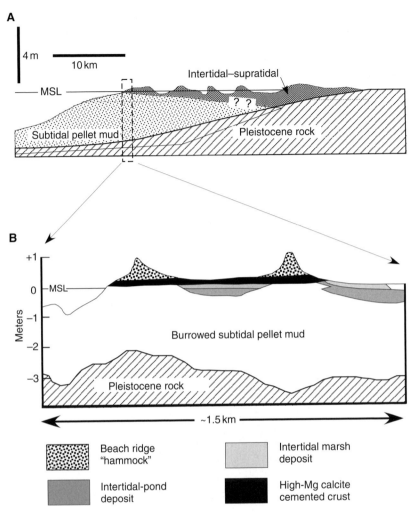

FIGURE 15.44. A cross section of southwestern Andros Island tidal flats. Modified from Gebelein (1974). Mean sea level is labeled MSL. (B) Details of the seaward edge of the tidal-flat deposit, showing the distribution of intertidal and supratidal deposits. From Gebelein et al. (1980).

and the United Arab Emirates. These tidal flats are accumulating under semi-arid to arid conditions, and are noted for their microbially laminated intertidal-flat deposits (Chapter 11) and for the intrasediment growth of evaporite minerals. The tidal flats of the United Arab Emirates are further described in Chapter 16.

Geometry, flow, sedimentary processes, and deposits of shallow shelves of rimmed shelves

Marine carbonate shelves with rims are up to 100 km wide and less than 10 m deep. They are protected from strong wave and tidal currents, and thereby suffer seasonal and daily variations in temperature and salinity. The shallow shelf is commonly thought of as a *carbonate factory* and this is certainly true of areas within about 10–20 km of the shelf edge, where open-ocean seawater constantly washes onto the shelf (Demicco and Hardie, 2002). In the shelf interior, the salinity may be up to double that of normal seawater, and the water is saturated with respect to both aragonite and calcite. *Codiacean* green algae are abundant on the shelf and include the prolific sand-maker *Halimeda*, the mud-making *Penicillus*, and their allies (Figure 15.46). Both of these algae produce aragonite, but coralline red algae (such as *Neogoniolithon*) produce high-magnesium calcite sand and gravel as well. In addition, foraminifera, molluscs, and echinoderms also produce abundant sand- to gravel-sized fragments, which can be further broken down by epibionts and abraded into mud. Sediments are extensively bioturbated by crustaceans, including the shrimps *Callianassa* and *Alpheus* and other invertebrates, particularly polycheate worms and echinoids (see Chapter 11). Large areas of mud bottom are stabilized by sea grass, with the rhizome and root

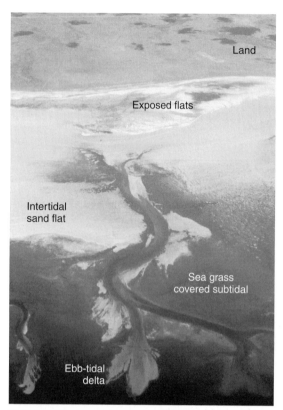

FIGURE 15.45. An aerial view of carbonate tidal flats accumulating along the eastern margin of Shark Bay, Western Australia, showing exposed supratidal, cyanobacterial-covered flats, light-colored intertidal sand flats, and a dark subtidal zone covered by a marine sea grass similar to turtle grass. A tidal channel cuts into the intertidal flat and subtidal sea grass bank, is flanked by subaqueous levees, and terminates in an ebb-tidal delta.

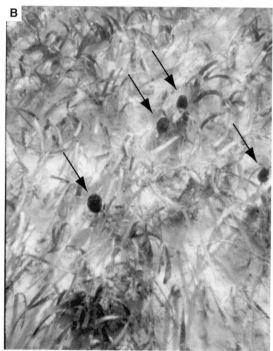

FIGURE 15.46. (A) A general view of a subtidal shelf bottom covered by green algae. Larger, segmented *Halimeda* (marked with an *H*) are present on the right in the foreground with smaller *Penicillus* (marked with a *P*). Photograph courtesy of Bob Ginsburg. (B) A sea-grass-covered mud bottom with *Penicillus* (at arrows).

network making a tough organic framework that resists erosion (Figure 15.46).

The mud and sand produced on the shelves are transported by wave and tidal currents, but the extensive bioturbation makes it unlikely that bed forms and sedimentary structures are preserved. In the inner portions of shelves, mud mounds and banks commonly develop. Mudbanks in the eastern parts of Florida Bay (Figure 15.47A) are linear features a few hundred meters across, 1–2 m high, and up to kilometers long, that run between tidal-flat islands. The tidal-flat islands are up to a few square kilometers in area and mostly covered in mangroves, although many have open intertidal to supratidal ponds covered with cyanobacterial mats (Figure 15.47C, D). In the western portions of Florida Bay, the mudbanks are larger and more irregular in shape, and the islands are scattered

on them. Florida Bay mudbanks were once thought to be a product of baffling and subsequent deposition of transported sediment by the sea grasses on the banks, with additional sediment produced in place by a host of small, calcareous organisms (known as epibionts) that lived on the grass. However, it appears likely that the mud mounds in Florida Bay are tombolos and spits that are influenced by wave refraction around the tidal-flat islands. Cores taken from the banks reveal bioturbated mud with sand- and gravel-sized skeletal fragments and the rhizomes of sea grasses (Figure 15.47B). Cores through the tidal-flat islands show that they are

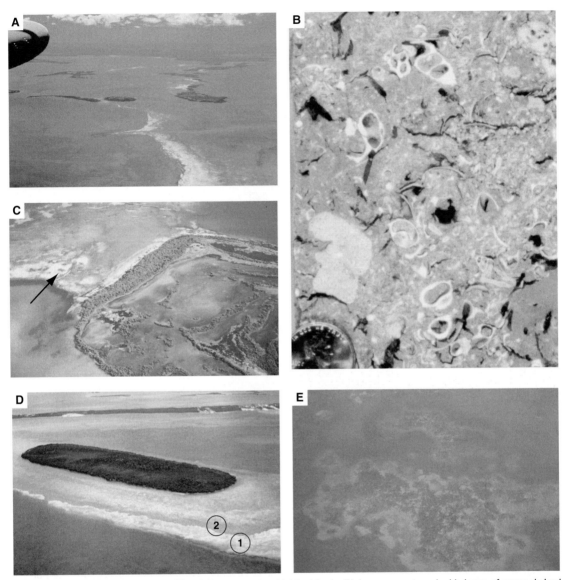

FIGURE 15.47. (A) Linear mudbanks of Florida Bay connecting tidal-flat islands. (B) An epoxy-resin-embedded core of coarse skeletal mudstone from the mudbank shown in (A). Dark tubes are mostly sea-grass roots. The coin (2 cm) gives the scale. (C) A tidal-flat island in Florida Bay, showing an open intertidal pond covered by cyanobacterial mats, with an adjacent mudbank at the arrow. (D) A tidal-flat island on the seaward side of Florida Keys (an elongated island can be seen in the middle distance). Circled numbers denote zones of red algae (1) and finger corals (2) rimming the mudbank. (E) A patch reef of dark domal corals surrounded by a light-colored halo of coarse sediment. Photographs (A) and (D) courtesy of Bob Ginsburg.

mostly underlain by pond deposits and cyanobacterially laminated muddy tidal-flat deposits, indicating that they have always been tidal flats that have migrated laterally perhaps a few hundred meters at most (Figure 15.48) (Enos and Perkins, 1979). Holocene tidal-flat islands on the ocean side of the Florida Keys are surrounded by seagrass-covered mudbanks that shoal up to sea level. These mud shoals have a strict zonation of biota facing the open shelf, particularly red algae and finger corals arrayed in depth zones (Figure 15.47D).

The mud bed of shallow shelves is locally cemented into hardgrounds (Chapter 11) that provide a substrate for corals, sponges, and other rock-encrusting organisms to colonize. In some areas of the shelves, the corals have established circular patch reefs (Figure 15.47E).

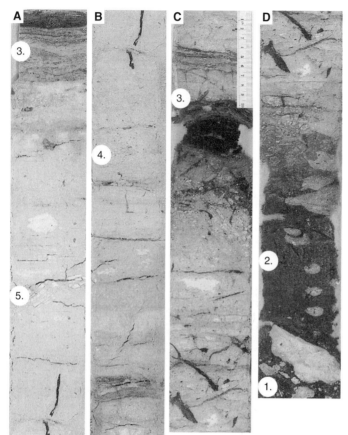

FIGURE 15.48. An epoxy-resin-embedded core 1.7 m thick from a tidal-flat island in Florida Bay. The base of the core is at D; the top of the core is at A. Circled numbers denote (1) pieces of Pleistocene rock, (2) freshwater peat, (3) two layers of cyanobacterial laminated muds of supratidal marsh, (4) burrowed mud of intertidal ponds, and (5) high-Mg-calcite-cemented crust.

Cementation is also common along the edges of the western Bahama Banks and across the open reaches of the bank to the southwest of Andros Island (Figure 15.38B), where sand-sized mud pellets are cemented individually or into aggregates known as grapestone. In these regions, some wave-ripple cross stratification or planar stratification formed during storm activity may survive the burrowing that is so intense in the muddier, quieter shelf.

Geometry, flow, sedimentary processes, and deposits for rim shoals

Strong wave and tidal currents at the edge of rimmed shelves produce sand shoals composed of skeletal sands and ooids. Frequent transport of ooids ensures that they are rounded and evenly coated with layers of carbonate mud. The water depth may be only a few meters over these shoals and, in many cases, ooid deposits have aggraded above mean sea level and are exposed at low tide. References on marginal sand shoals (principally of the Bahamas) include Ball (1967), Hine (1977), Hine and Neumann (1977), Dravis (1979, 1983), Harris (1979), and Halley *et al.* (1983).

The morphology of marginal shoals varies depending on the relative influence of waves and tidal currents. Where wave currents are dominant, barrier-like shoals form, and these are cut by channels that lead to subtidal spillover lobes on the shelf side of the shoal (Figure 15.49A). These lobes are about 1.5 m high, up to 1 km long, and up to 500 m across. Angle-of-repose slopes form the shelfward edges of the lobes. Sequential aerial photographs indicate that the barriers and lobes are storm-wave-built features, and they are covered with dunes, wave-built bars, and wave ripples. White Bank is an analogous skeletal-sand deposit along the margin of the Florida Reef tract.

Tidal sand ridges composed of ooids occur in groups along stretches of the shelf margin of the Great Bahama Banks, where tidal resonance in the adjacent embayments generates tidal ranges up to 2 m

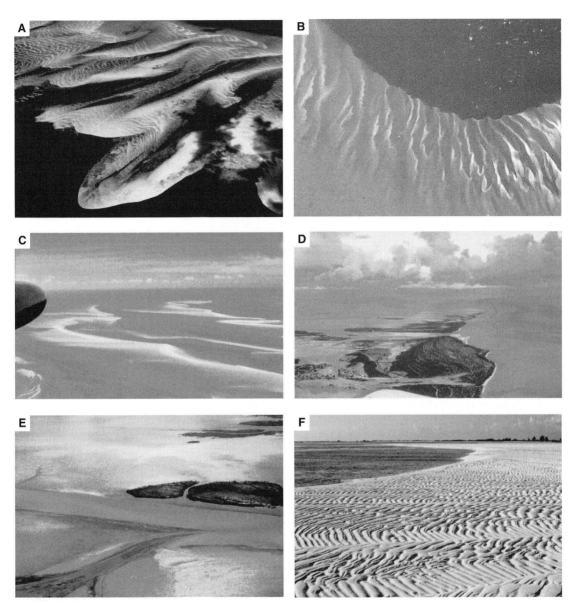

FIGURE 15.49. Bahamian ooid shoals. (A) Subtidal spill over lobes, with the shelf on the lower left of the photograph. From Pettijohn and Potter (1964). Width of belt ~1 km. By permission of Springer Science and Business Media. (B) A satellite view of tidal sand ridges along the southern Tongue of the Ocean. Ridges are ~10 km long. Courtesy of NASA Earth from Space. (C) An oblique aerial photograph of the tidal sand ridges shown in (B). The ridge in the center of the photograph is ~10 km long. Courtesy of Bob Ginsburg. (D) An oblique aerial view of Joulters Cay barrier island (~1 km wide), Tongue of the Ocean to the right of the photograph. (E) A tidal channel (~200 m wide) through Joulters Cay. Tidal flats can be seen in the background, with spill over the lobe in the foreground. (F) Rippled ooid sands from an ebb-tidal delta at the mouth of the tidal channel. Courtesy of Mitch Harris.

and strong tidal currents (e.g., the southern end of the Tongue of the Ocean; Figure 15.49B, C). The ridges are curved in plan, but their crests are approximately parallel to the direction of net water flow and sediment transport. The ridges are up to 5 m high, 400–800 m across, and up to 10 km long. The side slopes of the ridges are less than 5 degrees. The ridges are asymmetrical in cross section, and one side may be dominated either by flood- or by ebb-tidal currents. The ridges have superimposed dunes migrating obliquely up towards the crest. Many ridges are cut by spillover lobes similar to those found on other ooid shoals of the shelf edge. The areas between the ridges are covered with sea grasses.

Joulter's Cay (Figure 15.49D–F) (Harris, 1979), north of the northern tip of Andros Island, is an example of an ooid shoal comprising a barrier island, tidal channel, and tidal-flat complex similar to those of mesotidal siliciclastic coasts (although the spring tidal range is only ~80 cm). Cores through the shoal indicate that it began as a barrier spit prograding northwards from the tip of Andros Island (Harris, 1979). Today, most of the area is an intertidal to shallow-subtidal sand flat that covers 400 km^2 bankwards of the main barriers. The tidal flats are composed mainly of mixed ooid-pellet sands, are covered by sea grasses, green algae, and mangroves, and host a large number of *Callianassa* burrows. The deposits beneath these flats are up to 2–7 m of burrowed ooid-pellet sand. The barrier islands are vegetated, 2–3 km long, and 200–600 m across, and comprise a complex of beach ridges up to 3 m above sea level, spaced ~50 m apart with some freshwater ponds in the swales (Figure 15.49D). The barrier islands are some 5 km from the shelf edge. Active and abandoned tidal channels up to 200 wide cut through the barrier islands. The channels terminate in ebb-tidal deltas on the seaward side, and become narrower and shallower in the flood-tidal direction, although some terminate in spillover lobes on the sandflat (Figure 15.49E). Dunes in the tidal channels reverse orientation with the ebb and flood tide. Ooids are forming mostly on the front of the barrier islands, in the tidal channels, and on the ebb-tidal deltas (Figure 15.49F). The internal structures of all of these ooid-pellet sand deposits are expected to be similar to those of siliciclastic mesotidal coasts.

Aragonite cements are precipitated in addition to ooids, forming ooid-sandstone pavements that can be broken up into intraclasts. Large solitary stromatolites and clusters of stromatolites up to 2 m high occur in tidal channels in ooid shoals of the Bahamas (Dravis, 1983; Dill *et al.*, 1986, 1989). Although the outer layers of these stromatolites are unconsolidated, they become cemented with fine-grained aragonite and high-magnesium calcite within a few centimeters of the surface. It is generally thought that early cementation along the shelf edge coupled with the prolific sediment production of carbonate in shoals and reefs accounts for the rimmed shelf geometry (Wilson, 1975; Schlager, 2005).

Reefs and other organic buildups

The *Oxford English Dictionary* defines a reef as "a narrow ridge or chain of rocks, shingle, or sand, lying at or near the surface of the water." Sedimentologists generally reserve the term reef for a wave-resistant structure composed of limestone built to sea level by organisms. Implicit in this definition is that a substantial portion of a reef is rock, as opposed to loose sediment. About half of a modern reef is rock built mainly by corals and red algae that encrust each other, and the remainder is unconsolidated and cemented skeletal sand and gravel. Another type of organic buildup is the mudbanks of Florida Bay described above. These are generally composed of carbonate mud with lesser amounts of skeletal sand. However, some mud mounds on the Florida shelf are covered with thickets of finger corals and red algae that produce distinctive skeletal gravels. Some mud mounds are covered with sea grasses, and it has long been argued that sea-grass meadows (or crinoid meadows in the Paleozoic era) might have acted as baffles, trapping sediment and contributing to the carbonate buildup. Epibionts living on modern sea-grass banks produce large amounts of sediment that essentially accumulates in place upon the death of the organisms. There has been a long debate in the geological literature about what constitutes a reef in the geological record. Important references on geological aspects of modern and ancient reefs and other organic buildups, commonly referred to as mud mounds, bioherms, or reef mounds, include James (1983), Tucker and Wright (1990), James and Bourque (1992), Kiessling *et al.* (2002), and Flügel (2004).

General references to the biology and ecology of modern coral–algal reefs include Barnes and Hughes (1982), Levinton (1982), Birkeland (1997), and Mann (2000). Coral–algal reefs have high primary productivity and the greatest biodiversity of all marine ecosystems, which is astonishing in light of the fact that most reefs occur on the edges of carbonate shelves bordering open ocean areas, some of the least productive environments on Earth. The average net primary productivity of a coral reef is on the order of 2,500 grams of carbon fixed as organic material per square meter per year. By comparison, a tropical rainforest fixes approximately 2,200 g m^{-2} yr^{-1}, whereas open-ocean productivity is only 200 g m^{-2} yr^{-1}, comparable to that of desert areas. In addition to their high primary productivity, corals on reefs can fix large amounts of CaCO$_3$ as aragonite, building skeletons that can be up to many meters in diameter.

The most important factor governing reef productivity is the nature of the coral animal itself (Figure 15.50). Corals are colonial cnidarians that are composed of polyps and a sheet of tissue that connects the polyps. An aragonite skeleton ornamented with cups that the polyps sit in is secreted from the tissue of the coral. The

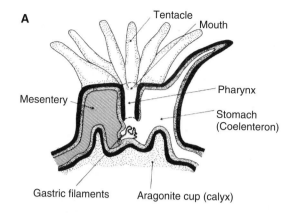

FIGURE 15.50. Coral animals. (A) A cutaway view of a coral polyp. From Barnes and Hughes (1982). (B) The surface of a small coral colony, showing polyps withdrawn in calyxes. (C) Another small coral colony at night with polyps extended for feeding.

polyps are rings of tentacles equipped with stinging cells (cnidoblasts) that encircle a one-way mouth that opens into a stomach partitioned by sheets of tissue called mesenteries. Reefs are built by hermatypic corals whose tissue layers are enfused with symbiotic

dinoflagellate protozoans known as *zooxanthellae*. The taxonomy of zooxanthellae is not certain. Most are assigned to a single genus and different corals have different species of zooxanthellae in them, as do hydrozoans, anemones, and even the large tridacnid clams of the Indo-Pacific area. During daylight hours, the carnivorous polyps withdraw (Figure 15.50B) and the zooxanthellae photosynthesize. Some of their photosynthetic products diffuse out into the corals, where they provide nutrition. Thus, during the day the coral polyps are not generally extended and the corals present various colors depending on the photosynthetic pigments of the species of zooxanthellae in the coral. At night, the tentacles of the polyps extend and the coral is a micropredator feeding on plankton caught by the tentacles (Figure 15.50C). The metabolic waste of polyp digestion diffuses into the coral tissue and fertilizes the zooxanthellae.

Modern coral–algal reefs have rather strict ecological requirements for maximum health and diversity that are based on the needs of the coral animal and its symbiotic algae. These are (1) optimal water temperatures of 26–28 °C (tolerance range 18–36 °C); (2) near-normal marine salinities of 33‰–35‰ (tolerance range 27‰–40‰); (3) sunlight; (4) current and wave energy sufficient to move nutrients and plankton onto the reef and remove sediment and waste; and (5) a stable substrate (i.e., rock) for initiation of reef growth. Modern hermatypic coral–algal reefs require sunlight in order for zooxanthellae to carry out photosynthesis. Sunlight is obscured by suspended sediment, and coral polyps cannot tolerate being buried by large amounts of suspended sediment. The various wave lengths of sunlight are absorbed by seawater to different degrees, and very little light of any wave length can penetrate deeper than 100 m. Coral growth rates closely follow this extinction of sunlight with depth (Schlager, 2005), see Chapter 4, so coral growth rates are highest in the shallowest water and fall off exponentially with depth. Hermatypic corals do not grow beneath ~100 m.

Two important types of modern coral–algal reefs are (1) shelf-edge reefs and (2) patch reefs. For the reasons given above, reefs are best developed on tropical shelf edges facing the prevailing wind and experiencing upwelling. However, many atolls have well-developed reefs completely encircling them. Well-known shelf-edge reefs are the Great Barrier Reef of Australia, the Belize Reef, the Florida Reef Tract, and the Red Sea reefs. Shelf-edge reefs are kilometer-wide features that, together with sand shoals, comprise the rim of carbonate shelves. Atoll reefs and fringing

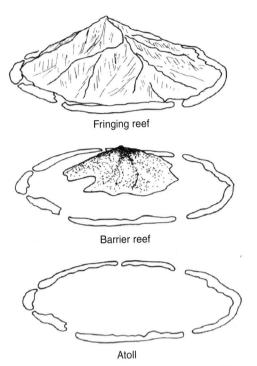

Fringing reef

Barrier reef

Atoll

FIGURE 15.51. Successive formation of fringing reef, barrier reef, and coral atoll around and on top of a subsiding volcanic edifice. From Barnes and Hughes (1982).

reefs of Indo-Pacific Islands are smaller versions of shelf-edge reefs built in the shallow areas of volcanic oceanic edifices. Darwin was the first to point out the evolutionary history of atolls as reefs that kept pace with subsidence of the edifices (Figure 15.51). Drilling on Pacific atolls in preparation for H-bomb tests in the 1950s confirmed this idea. Atolls with fringing reefs that could not grow up fast enough to keep pace with subsidence became submerged, flat-topped sea-mounts known as *guyots*. Patch reefs are generally found in shallow shelf interiors some distance from shelf edges (Figure 15.47E). Patch reefs tend to be more or less circular in plan, and are smaller and less biologically diverse than those on shelf edges, because conditions for coral growth in platform interiors are not optimal.

Sub-environments of modern shelf-edge reefs are (1) back reef, (2) reef flat, (3) reef crest, (4) reef front, and (5) fore reef (Figure 15.52A). Reef flats tend to be hundreds of meters across, reef crests are 10–20 m across, and the reef front and fore reef are hundreds of meters across. The Florida reef tract is discontinuous, with reef segments up to a few kilometers long separated by gaps up to a few tens of kilometers long. This is

because the Florida reef tract is at the lower limit of the temperature range for optimum reef development. The Belize Reef, the Great Barrier Reef, and many atoll reefs are much more continuous, with kilometer-long stretches of reef broken by passes with storm-built spill-over lobes of skeletal sands on the landward side. Sediments in and around reefs are up to boulder size, and organic processes such as encrusting and baffling produce a variety of rock and sediment types that are not easily classified. A somewhat interpretive classification for reef limestone types developed by Embry and Klovan (1971) (see the appendix) is used here.

The back reef is contiguous with the shelf, with water depths of a few meters to a few tens of meters. In fringing reefs, the back reef may be only a few hundred meters wide, but it is 60–70 km across on the Great Barrier Reef. Areas of sea grass in back-reef areas effectively trap moving sediment, resulting in bafflestone textures. Broken pieces of coral that are transported from the reef into the back reef during storms can either resume growth or die, producing floatstones. The reef flat is essentially a pile of reef debris built during storms that commonly has an avalanche face on its back-reef side. Much of the rubble is sand- and gravel-sized fragments of coral and green algae, resulting in grainstones and rudstones. In Indo-Pacific reefs, the reef flat is commonly a pavement-like encrustation of red algae (a bindstone). The reef crest is the shallowest portion of the reef and is commonly exposed at the lowest tides. The reef front comprises most of the living reef and it extends down to depths of ∼30 m. On modern reefs, the reef front has a distinctive "spur and groove" topography (Figure 15.53A). The spurs are coral-encrusted masses (Figure 15.53B–E) that are a few tens of meters wide and extend hundreds of meters perpendicular to the shelf edge. The spurs are thickest at their seaward edges and thin towards the reef crest into which they merge. The grooves are channels a few tens of meters wide between the spurs (Figure 15.53A). Grooves shoal and disappear towards the reef crest. Grooves are floored by coarse sand- and gravel-sized coral fragments and green-algae debris (Figure 15.53F). Sand in grooves may be covered with wave ripples with heights of a few centimeters and wave lengths of up to a meter. Sediments of the reef front are framestones, bafflestones, and bindstones. The fore reef is an apron of sand and gravel (grainstone to rudstone) that was transported from the reef by sediment gravity flows. On the Belize Reefs, the fore-reef sediment accumulates at the foot of a steep wall up to 1,000 m high. The area of the living part of a reef (the reef crest and

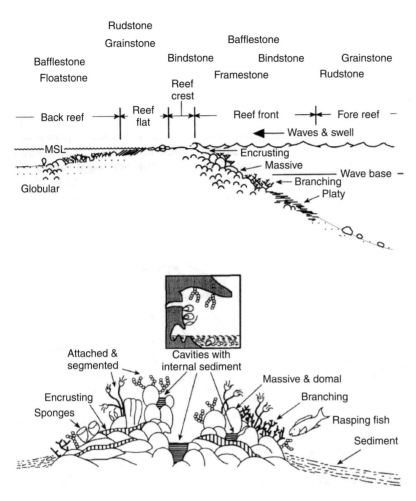

FIGURE 15.52. (A) A cross-section profile of a reef showing the various geomorphic subdivisions, the general shapes of the corals zoned according to general wave energy, and the general rock type deposited in each setting. From James and Bourque (1992). (B) A diagram illustrating the various processes that build reef rock. From James and Bourque (1992).

front) is small relative to that covered by the debris generated from a reef. Such debris may extend for kilometers behind and in front of the living reef.

Modern reefs exhibit a zonation of coral-growth forms across the reef front and reef crest (Figure 15.52A) that is related to the variable wave energy (which is maximum at the shallowest portion of the reef) and attenuation of sunlight with depth. The corals in the deepest portions of the reef (~10–30 m) are delicately branching or plate-like. In depths of 2–10 m, spurs are above wave base and encrusted by massive domal corals such as *Montastrea*, and the brain corals *Dipluria* and *Meandrema*. The reef crest and shallowest parts of the spurs are populated by massive encrusting red algae and two rapidly growing (up to 10 mm per day) branched corals, *Acrapora palmatta* and *Acrapora cervicornis* (Figure 15.53E). These corals get severely broken during storms, and therefore supply a good deal of the coral debris that

accumulates in the reef flat and fore reef. These *Acrapora* corals do not occur on patch reefs.

The processes contributing to reef deposition and growth are shown in Figure 15.52B. Up to 50% of a modern reef spur is composed of in-place bioprecipitates of corals, hydrozoans, and skeletons of red algae encrusted on each other as a limestone framework. This information was gained in the 1950s, when geologists dynamited reef spurs. The corals, hydrozoans, and red-algae encrusters that build the reef framework are irregular in shape, therefore the framework portion of a reef also has an irregular geometry. Where corals abut each other, the polyps of different species poison one another. Corals are also eaten by a variety of organisms (Figure 15.53C, D), broken by storms, and stepped on by tourists, so they have irregular growth forms. The parts of coral surfaces that are dead are rapidly colonized by other organisms (Figure 15.53C). Rapidly growing calcified algae and invertebrates with

FIGURE 15.53. Coral reefs. (A) An aerial photograph showing fore-reef rubble in the lower portion of the photograph, and reef front comprising spur and groove morphology in the central part of the photograph, with the reef crest and white rubble of the reef flat at the top. Spurs are ~100 m long. Courtesy of Bob Ginsburg. (B) A shallow portion of a spur near the reef crest. Moose horn coral (*Acrapora palmatta*) provides shelter for a school of fish. Various sea fans and calcified hydrozoans (so-called "fire coral") comprise most of the spur wall beneath the *Acrapora*. (C) A close-up of the spur. Most of the reef surface is the globose common star coral *Asterias*. However, note the sea anemone living in the cavity in the reef (arrow) and the bare patches where corals have been killed. Width of field ~1 m. (D) Clown parrotfish in a cave in the side of a spur. This fish uses its beak-like jaws to crop the coral rock, contributing coral sand to the reef-filling sediment. (E) Branching stag-horn coral (*Acrapora cervicornis*) just below the reef crest. Courtesy of Bob Ginsburg. (F) Large wave ripples and detached blocks of coral at the deep end of a groove. The barracuda is ~1.5 m long.

short life cycles cover many parts of the reef surface. Modern-day organisms of this type include *Halimeda* and so-called soft corals. As these species die and disintegrate, they add much sand-sized debris to the coral debris that fills the cavity network between and under the corals. This cavity network hosts a specialized cryptobiotic group of organisms, such as modern stromatoporoids, adapted to live in reef caves. The reef is also home to many bioeroders, organisms that drill into, eat, and otherwise damage the in-place corals and red algae. Bioeroders include filter-feeding polycheate worms, anemones, and bivalves (Figure 15.53C). In addition, boring sponges, polycheate worms, snails, sea urchins, and fish species such as parrot fish eat the coral (Figure 15.53D). The bioeroders contribute sediment grains that filter into the cavities among the skeletal elements. The cavity network of modern reefs is also cemented by large crystals of aragonite and smaller crystals of magnesian calcite. This cementation is induced by the warming and degassing of CO_2 of the relatively cool seawater that flows onto the shelves.

Sedimentary sequences in carbonate deposits

Carbonate shelves can have relatively high rates of production and deposition of sediment. For example, $\sim 4 \times 10^{10}$ kg yr^{-1} of carbonate mud and sand is produced in the area of the Bahamas shown in Figure 15.38 (Demicco and Hardie, 2002), which is roughly one tenth of the $(30-40) \times 10^{10}$ kg yr^{-1} of terrigenous mud and sand delivered to the Gulf of Mexico by the Mississippi River (Berner and Berner, 1996). Although some of this Bahamas sediment is transported off the platform, much of it is transported towards Andros Island, where it accumulates on tidal flats (Traverse and Ginsburg, 1966). The tidal flats of southwestern Andros Island have prograded seawards by up to 25 km over adjacent subtidal, inner-shelf muds during the past few thousand years (Figure 15.44). Modern carbonate tidal flats in other settings, such as the arid United Arab Emirates (Chapter 16), have likewise prograded many kilometers during the past few thousand years.

Models of carbonate tidal-flat deposits based on these observations depict them as sheets (generally 1–5 m thick but rarely up to 10 m, increasing with tidal range) covering hundreds to thousands of square kilometers. Within each sheet, the deposits comprise (from base to top): (1) a discontinuous, thin basal gravel representing initial flooding of the area during a marine transgression; (2) grainstone to packstone deposited in the subtidal shelf; (3) mudstones to wackestones deposited on intertidal flats, ponds, and levees; and (4) mudstones deposited on supratidal flats or marshes above sea level during storms. The basal transgressive deposits are underlain by freshwater peat and marsh deposits in many areas of the Florida–Bahamas platform, which are similar to sediments now accumulating in the Everglades. In humid areas such as Florida and the Bahamas, supratidal deposits are freshwater marls with periphyton or calcified microbial mats, whereas supratidal deposits contain a variety of evaporite minerals in arid and semi-arid settings (Chapter 16).

Many ancient tidal-flat deposits have a basal accumulation of intraformational conglomerate interpreted as being pieces of hardened bed sediment that were eroded and transported during marine transgressions. The subtidal deposits of most modern carbonate shelves are burrowed mud, as they are in some interpreted Phanerozoic subtidal deposits (Figure 15.54A). However, some Phanerozoic, and many Proterozoic, carbonate deposits attributed to subtidal shelves and adjacent tidal flats have wave and current ripples and associated cross lamination (Figure 15.54B). In intertidal sediments there is commonly a decrease in burrowing upwards, with a concomitant increase in microbially influenced laminated mudstones. Desiccation fenestrae and mudcracks become more common through the intertidal to supratidal transition. Proterozoic deposits also commonly exhibit microbial-mat-influenced sedimentary features in supposedly lower intertidal and subtidal settings (Chapter 11). Channel deposits from ancient carbonate tidal-flat deposits have only rarely been reported (Figure 15.55), which is puzzling in view of their importance in modern tidal-flat environments.

Carbonate tidal flats are reckoned to prograde under conditions of both rising and falling sea level, due to the high sediment-production rates on carbonate shelves. Many ancient carbonate-shelf deposits comprise repeated vertical sequences (or cycles) of tidal-flat deposits such as those shown in Figure 15.54. The cycles are asymmetrical, with little if any depositional record of initial sea-level rise, the first significant deposits being subtidal. In general, sediment grain size decreases upwards within cycles, faunal diversity and burrowing decrease upwards, desiccation features increase upwards, and microbial-mat-induced laminae increase in density upwards. The amount of dolomite also commonly increases upwards within cycles. The time represented by these cycles is on the order of 10^4-10^5 years, and this estimate comes from two sources. First, modern carbonate tidal flats prograde at 0.5–5 km per 1,000 years, so a tidal-flat deposit that was 10–100 km wide represents between 20,000 and 200,000 years of progradation. Second, Hardie and Shinn (1986) divided estimates for the ages of cyclic carbonate sequences by the number of cycles present, and got similar time estimates for deposition of a single cycle. The origin of these kinds of carbonate tidal-flat cycles is hotly debated (see Chapter 20).

Modern carbonate subtidal shelves (sometimes referred to as "shelf lagoons") and many ancient

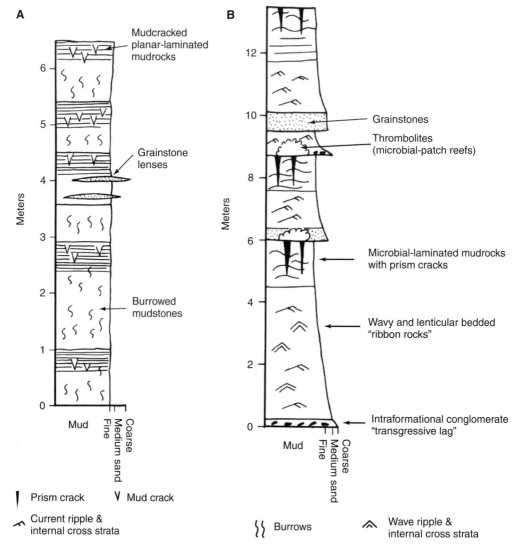

FIGURE 15.54. Measured logs of Paleozoic tidal-flat deposits from western Maryland, USA. (A) Middle Ordovician cycles mainly comprising burrowed subtidal mudstones grading up to intertidal planar laminites with microbial fabrics and desiccation cracks. (B) Cambrian tidal flat cycles. Further details are available in Demicco and Hardie (1994).

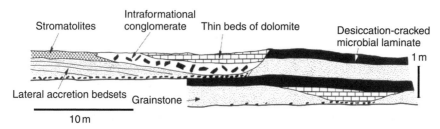

FIGURE 15.55. Tidal-channel deposits from the Cambrian Waterfowl Formation of the southern Canadian Rocky Mountains reconstructed from closely spaced measured sections and physically tracing beds along the outcrops between sections. From Waters et al. (1989).

examples comprise bioturbated muds with variable amounts of skeletal fragments, i.e., wackestone and packstones. However, some Phanerozoic and many Proterozoic subtidal shelf carbonates have decimeter- to meter-thick beds of medium-scale cross-stratified and hummocky cross-stratified grainstones (Figure 15.56), and resemble siliciclastic tide-dominated or wave-dominated shallow subtidal deposits. In addition, thrombolites (microbial patch reefs – Chapter 11), metazoan patch reefs, and mud mounds (see below) are common components of shelf-lagoon deposits.

The Dolomites region of the Southern Alps in northern Italy has well-exposed Middle Triassic carbonate shelf deposits surrounded by basinal sediments. The Latemar buildup is about 5 km in diameter and consists of a core of flat-lying shelf-lagoon deposits about 700 m thick surrounded by a rim of shelf-margin deposits that grade outwards into a wide apron of steeply dipping foreslope deposits (Figure 15.57A) (Bosellini, 1984; Goldhammer et al., 1987, 1990; Goldhammer and Harris, 1989). The shelf deposits are composed of laterally traceable, 0.1–5-m-thick cycles. Each cycle (Figure 15.57B) is composed, from base to top, of: (1) bioturbated, very-fine sand-sized, skeletal-peloidal wackestone to packstone with an open marine fauna of gastropods, crinoids, dasycladacean algae, and foraminifera; (2) medium to very-coarse sand-sized skeletal grainstone to packstone; (3) coarse-sand to gravel-sized pisolitic, lithoclast grainstone; and (4) 0.05–0.15-m-thick dolomite consisting of skeletal-intraclast grainstone to packstone overprinted by vadose calcrete fabrics (Chapter 11). There are about 300 laterally traceable cycles that in many places are organized into five-part "megacycles" with upsection decrease in cycle thickness (Figure 15.57C). Goldhammer et al. (1987, 1990) interpreted this shelf sequence as a record of repeated subaerial exposure and marine flooding of the Latemar shelf related to two superimposed time scales of sea-level oscillation (~20,000 and ~100,000 years). This is further discussed in Chapter 20.

Cross-stratified ooid and/or skeletal grainstones can be common near shelf edges and within tidal deltas. Parts of the city of Miami, Florida (USA), are built on Pleistocene ooid sand bars deposited during the last interglacial period (Halley and Evans, 1983; Schlager, 2005). The old sand bars comprise the Atlantic Coastal Ridge, with the intervening old tidal channels locally referred to as glades (Figure 15.58A) (Tucker and Wright, 1990). Barrier

islands of ooid sands that run northeast–southwest nearly at the modern coastline developed on old tidal deltas (Halley and Evans, 1983; Tucker and Wright, 1990). Further to the southwest, the lower Florida Keys are cored by Pleistocene ooid tidal bars, with the Holocene channels occupying the original Pleistocene inter-bar channels (Figure 15.58B). Much of the Miami oolite comprises 0.05–1-m-thick sets of medium-scale, cross-stratified ooid grainstone grouped into meters-thick cosets with shell lags at the base (Figure 15.58C). Reactivation surfaces and mud drapes occur in the cross-stratified ooid grainstones, and sea-anemone burrows are rare. Cosets of cross-stratified ooid grainstone are separated by finer-grained ooid–peloid grainstones, commonly disrupted by *Callianassa* shrimp burrows.

The deposits of ancient reef fronts and crests are commonly referred to as *reef cores*. Reef-core deposits tend to be limestones and dolostones up to many hundreds of meters thick. One of the most spectacular ancient reefs is a Devonian barrier reef exposed along the northern edge of the Canning Basin in the deserts of northwestern Australia (Figure 15.59) (Playford, 1984). The reef cores, well exposed in river gorges, are up to 2 km thick and 100 m wide, and are composed of framestones of stromatoporoids, corals, and calcified microbes such as *Renalcis* (Figures 15.59 and 15.60A). Cavities among the frame-builders are filled with laminated sediment and early marine cements (Ward, 1996). Cryptobionts, sediments, and early marine cements also surround talus blocks of the reef flat (Figure 15.60B). The shelf deposits comprise cyclic tidal-flat deposits composed of stromatoporoid-rich subtidal grainstones and packstones that grade upwards to intertidal mudstones with desiccation fenestrae (Read, 1973). The fore-reef talus near the reef is represented by angle-of-repose strata that grade basinwards into low-angle terrigenous shale. There is a well-developed zonation of fossils across the basin, reef, and back-reef areas (Figure 15.60C).

Patch reefs on the shelf are generally circular in plan, and are preserved in the rock record as buildups variously described as reef mounds or bioherms. Reef-crest deposits are commonly missing, and the reef is represented by a halo of reef-derived debris, commonly referred to as *reef-flank* deposits in ancient examples. The upper Florida Keys are a series of northeast-to-southwest trending islands that developed as a line of patch reefs during the last interglacial period (Stanley, 1966; Multer, 1975). The reef deposits are up to 60 m thick, and are known as the Key Largo Formation

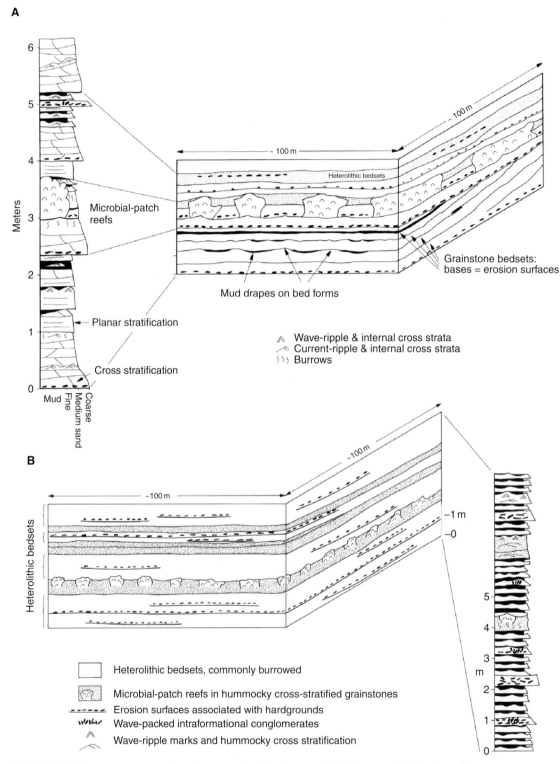

FIGURE 15.56. Measured sections and two-dimensional bedding diagrams of shelf deposits of (A) Cambrian Conococheague Formation, western Maryland, USA, interpreted as a tide-dominated shelf, and (B) Ordovician Beekmantown Group, western Maryland, USA, interpreted as a wave-dominated shelf. Further details are available in Demicco and Hardie (1994).

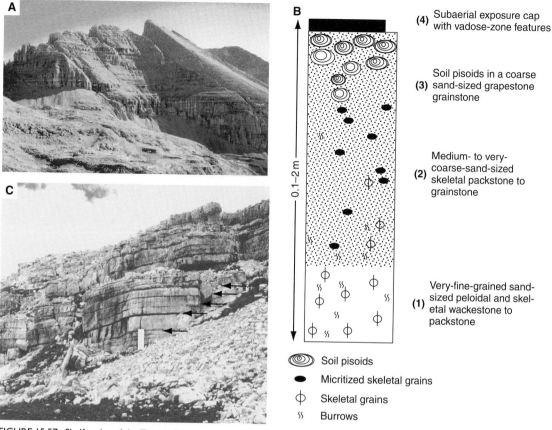

FIGURE 15.57. Shelf cycles of the Triassic Latemar Limestone, Dolomite Mountains, northern Italy. (A) A general view of the eroded interior of the Latemar shelf. Horizontal layers are defined by recessive-weathering dolomitic vadose soils. (B) Shelf cycle of the Latemar Limestone, showing coarsening-upward layers capped by a soil interpreted as recording exposure of the shelf due to falling sea level. From Demicco and Hardie (1994). (C) A group of five Latemar cycles that become progressively thinner up section. Arrows point to successive soil caps. The white scale bar is ~2 m tall. See Goldhammer *et al.* (1987, 1990) for further details.

(it is exposed at Fossil Reef State Park in an old quarry on Windley Key). The reef core is composed of ~40%–50% in-place coral skeletons, mainly domal forms in growth position that abut and overgrow each other in places (Figure 15.61). The corals are bored by bivalves and surrounded by a stratified coarse skeletal grainstone, including *Halimeda* fragments (Figure 15.61B, C). The reef-crest branching corals *Acrapora* are conspicuously absent. Ancient patch reefs range from framestones with coarse inter-framework debris (such as the Key Largo Limestone) to mud-rich bafflestones. Mud-rich patch reefs are commonly referred to as mud mounds, and may also accumulate on fore-reef slopes or on deep shelves (Tucker and Wright, 1990; James and Bourque; 1992; Flügel, 2004; Schlager, 2005). Well-known examples of mud mounds are the Carboniferous "Walsortian" mounds composed of crinoid- and bryozoan-bearing muds and

spar-filled cavities referred to as *stromatactis*. Schlager (2005) emphasized the importance of microbes in constructing many of these mounds, particularly those in deeper settings. Carbonate buildups also occur around hydrocarbon seeps, and hot and cold springs on the sea floor (Flügel, 2004).

An ecological succession in the development of ancient patch reefs and reef mounds is commonly recognized. Patch reefs are initiated on skeletal lime-sand shoals as they are colonized and stabilized by calcareous green algae, sea grasses, or crinoids that bind the sediment. Following stabilization, scattered branching metazoans, such as algae, bryozoans, corals, rudist bivalves, and sponges, become established and baffle sediment. Subsequently, reef-building metazoans (mainly branching forms) become established and provide sheltered environments for other attached organisms. Rapid diversification of species leads to a major phase

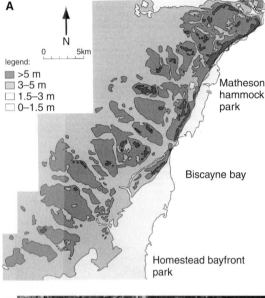

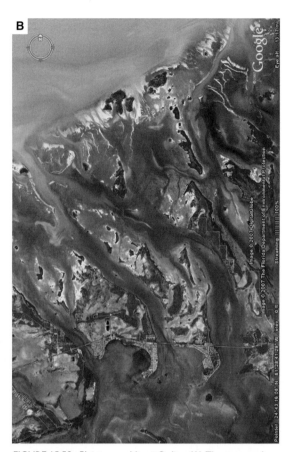

FIGURE 15.58. Pleistocene Miami Oolite. (A) The topography of the Miami Oolite in southeast Florida, USA, showing the morphology of ooid sand bars and intervening channels. From Schlager (2005). (B) The shapes of the lower Florida Keys reflect the underlying tidal-bar topography of the Miami Oolite. Holocene tidal channels follow Pleistocene channels through the tidal sand ridges. Field of view ~10 km × 7.5 km. From Google Earth™. (C) An outcrop photograph of Miami Oolite. Bioturbated bed at the arrow is overlain by a bedset of medium-scale cross stratification. Cross-strata sets show reactivation surfaces and rare mud drapes. The bedset boundary at the top (thick line) is an erosional surface at the base of the next bedset.

of reef-frame building, and eventually a few encrusting species come to dominate the patch reef.

Reef-building organisms have changed throughout geological time. As Bob Ginsburg put it, "... reefs are like Shakespeare's plays. The roles are constant but the actors change." James and Bourque (1992) summarized the abundance of reefs and changing reef biota through time (Figure 15.62). Stromatolites and thrombolites (Chapter 11) built by microbes comprise shelf-edge reefs, patch reefs, and mounds throughout the Proterozoic. In the latest Neoproterozoic, microbial mats are accompanied by mineralized metazoans (Wood et al., 2002). In the Phanerozoic, shelf-margin reefs were especially well developed in the Devonian,

FIGURE 15.59. Devonian barrier reef fringing the Canning Basin in northwest Australia. (A) An aerial view looking southwest over exhumed reef tract to basin beyond. Reef is dissected and exposed in river gorges. (B) A photograph and (C) a line drawing of reef exposed in Windjana Gorge. Images (A) and (B) from James (1983); (C) from Playford (1984). AAPG © 1984, 1984 reprinted by permission of the AAPG, whose permission is required for further use.

Triassic, Jurassic, and Neogene. A number of major extinctions altered reef communities (Figure 15.62). Calcified microbes and stromatolites enjoyed brief resurgences as important reef-builders for a few millions of years after these events until the next group of reef-building metazoans evolved. Corals, stromatoporoids, other sponges, bryozoans, green algae, and calcified microbes have been common reef builders. These organisms have been joined from time to time by more bizarre forms, including the Permian productid brachiopods and the Cretaceous rudist bivalves. These groups developed elongated, cylindrical shells with an encrusting growth habit. The rudists, in particular, were very important Cretaceous frame-building reef organisms.

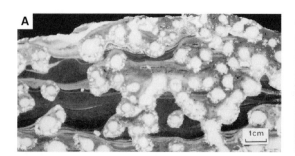

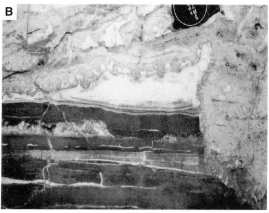

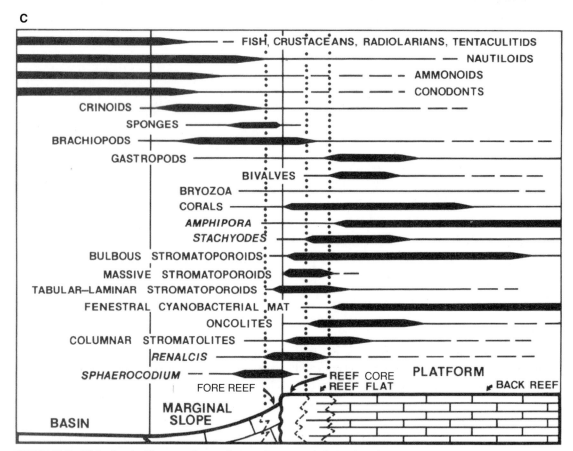

FIGURE 15.60. (A) Reef rock of Devonian Canning Basin reefs composed of white branches of one stromatoporoid colony encrusted by the calcimicrobe *Renalcis* (small grape-like white clusters). Dark laminated internal sediment fills holes around the framework elements. From James (1983). (B) Reef-flat rubble blocks forming a cavity. The cavity roof is encrusted by *Renalcis*, the cavity is partially filled with laminated internal sediment, and the top of the cavity is filled with white early marine cement. The lens cap is ~50 mm in diameter. From James (1983). (C) A diagrammatic representation of the abundance and zonation of fossil remains around the Canning Basin reefs. From Playford (1984). AAPG © 1983, 1984 reprinted by permission of the AAPG, whose permission is required for further use.

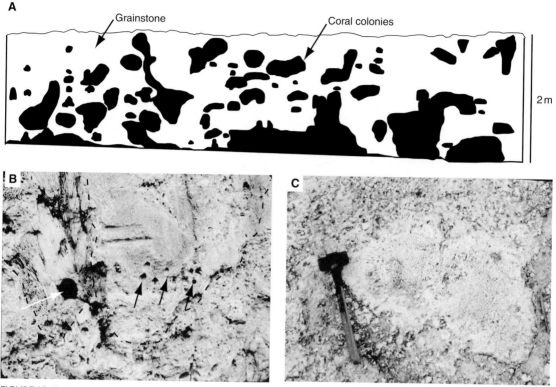

FIGURE 15.61. Pleistocene Key Largo Limestone exposed in Fossil Reef State Park, Windley Key, Florida, USA. (A) A trace of quarry wall showing the location and shape of in-place coral heads. Modified from Multer and Hoffmeister (1975). (B) Two abutting corals in the wall of Windley Key Quarry. The lens cap (white arrow) is 50 mm in diameter. Black arrows point to bivalve borings in the coral. Corals are surrounded by coarse, poorly bedded skeletal grainstone. (C) A brain-coral colony exposed on the quarry floor. The mottled nature of the surrounding grainstone is suggestive of burrow networks.

Rocky coasts

Rocky coasts occur where the rate of sediment removal exceeds the supply. Erosion of rock by waves is probably the most important process on rocky coasts. This erosion is due to the hydraulic pressure of breaking waves (including air pressure as air is compressed within spaces in rocks) and abrasion of rock by loose debris carried by the water, especially in the surf zone. The biggest waves have the greatest erosional effect. Wave erosion is concentrated within the intertidal zone, and causes undercutting of cliffs, leading to mass wasting. Mass-wasting deposits that accumulate at the base of cliffs inhibit further undercutting until they are removed by wave currents, whence undercutting of the base of the cliff can resume. Therefore, cliff erosion is episodic. Time-averaged rates of cliff erosion vary widely: e.g., on the order of 0.1 m per year for sedimentary rocks, and meters to tens of meters per year for unconsolidated sediments. Weakness in the rocks is exploited by waves, and the greatest erosion

rates are associated with unconsolidated sediments, mudstones, and highly fractured rocks. Refraction and diffraction of waves by shoreface features (rocks, bars) can concentrate waves on specific areas, thereby increasing local erosion rates. Such areas of enhanced coastal erosion can migrate alongshore if shoreface bars migrate alongshore. The sediment produced and eroded from the cliff is moved offshore or alongshore.

Erosion of rocky coasts may result in a notch in the cliff near the high-tide level (Figure 15.63). Continued undercutting of the cliff, mass wasting, and removal of the loose material produce a wave-cut platform that extends from the low-tide to the high-tide levels. These intertidal platforms can be hundreds of meters wide (depending on the tidal range and on the inclination of the platform). If sea level rises, the wave-cut platform can increase in width. Wave-cut platforms cause waves to break further from cliffs, which regulates the amount of erosion. Differential erosion of the cliffs, which is related to variations in rock erodibility, results

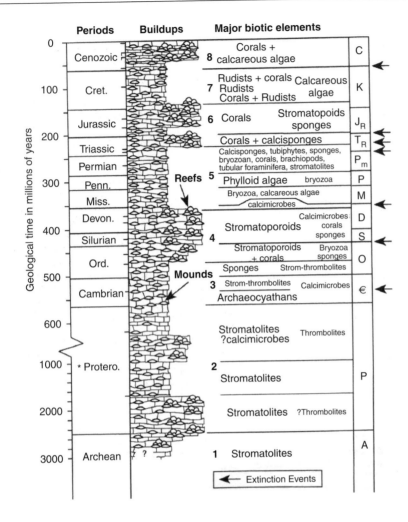

FIGURE 15.62. A summary of reef development through time. The horizontal scale of the diagrammatic column describes the relative abundance and types of reefs and reef mounds. Eight major reef-building fauna are recognized, many separated by extinction events. From James and Bourque (1992).

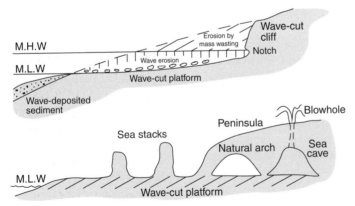

FIGURE 15.63. Formation of wave-cut platforms, sea caves, natural arches, and sea stacks.

in characteristic features such as sea caves, natural arches, stacks, blowholes, and geos (Figures 15.63 and 15.64). Blowholes result from collapse of the roof of a sea cave. Natural arches are formed in headlands by meeting of the back walls of sea caves eroded into either side of the headland. A sea stack is produced by collapse of the top of a natural arch or erosion around a particularly resistant block of rock. A geo is a linear coastal inlet produced by exploitation of a fault zone. Lulworth Cove (Figure 15.64E) is a famous cove in

FIGURE 15.64. Features of rocky coasts. (A) A wave-cut platform and sea caves, Co. Kerry, Ireland. (B) A geo (wave erosion along a fracture zone), Co. Kerry, Ireland. (C) Sea stacks from Victoria, Australia. (D) A natural arch, Durdle Door, Dorset, England. (E) Lulworth Cove, Dorset, England. The coastline and back of the cove are composed of resistant limestone, but the widest part of the cove is eroded into less resistant clay.

Dorset, England, that formed by preferential erosion of weak rocks.

The sediment eroded from rocky cliffs may be moved offshore to build a terrace beyond the wave-cut platform, or may be moved alongshore to form pocket beaches (crenulate-bay beaches) between rocky headlands (Figure 15.65). Wave refraction is associated with rocky headlands (Figure 15.65). If wave crests approach parallel to the shoreline, their refraction by headlands results in longshore currents directed away from the

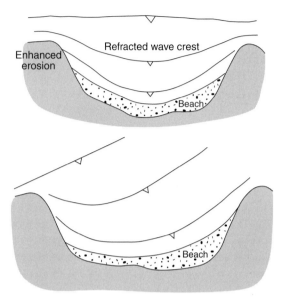

FIGURE 15.65. Pocket (crenulate-bay) beaches between headlands with steady-state shorelines parallel to refracted wave crests.

center of the beach towards each headland, resulting in a curved beach in plan (Figure 15.65). Oblique wave approach will cause the beach to be displaced towards the downdrift headland (Figure 15.65). Plan forms of beaches in the lee of rocky headlands (or artificial breakwaters) are approximated empirically by logarithmic spirals (the radius of curvature is an exponential function: see Komar (1998). In all of these cases, the beach shoreline is parallel to refracted or reflected wave crests.

The erosion of rocky headlands and the movement of sediment alongshore onto beaches is a coastline-smoothing process. The rate of smoothing depends on the original coastal morphology, the wave energy, and the resistance of the coastal materials: it will therefore vary in space and time. A generalized evolutionary scheme for smoothing of an embayed coast (e.g., drowned river valleys) composed of homogeneous materials is illustrated in Figure 15.66.

Coastal–marine engineering and environmental concerns

Coasts are sites for human settlements, resulting in intense human activity and many buildings on and behind beach ridges and along coastal channels and bays. Natural processes affect these settlements in various ways: sedimentation in navigated waterways (coastal rivers and river mouths); beach erosion due to

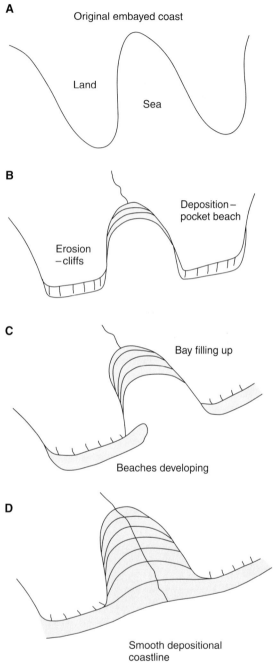

FIGURE 15.66. A model for the evolution of rocky coasts by erosion and deposition.

seasonal storms and longer-term sea-level rise; damage to beach buildings during storms due to high water and washovers. However, problems are also associated with human activities: erosion of beach ridges due to loss of vegetation associated with excessive human-related

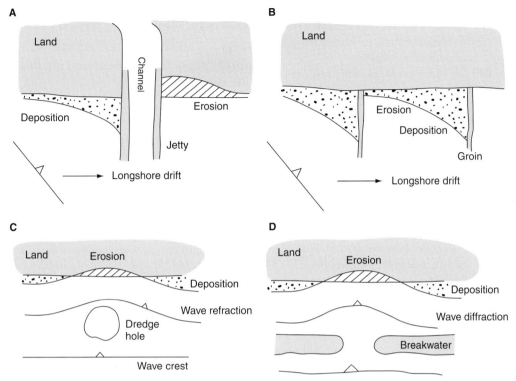

FIGURE 15.67. Coastal engineering problems: (A) jetties for coastal channels, (B) groins on beaches, (C) effects of an offshore dredge quarry, and (D) offshore breakwaters.

traffic; damage (including erosion) to beaches by vehicle use; loss of sediment supply to beaches from rivers due to damming of rivers and building jetties. Engineering solutions to these problems include dredging coastal waterways; building jetties, groins on beaches, sea walls, and offshore breakwaters; beach replenishment; putting beach buildings on stilts; limiting access to beaches; and replanting of vegetation. Many of these solutions cost millions of dollars.

A very common scenario is reduction of fluvial sediment supply to beaches due to dams on rivers, and building of jetties at river mouths to keep a navigation channel open. The supply of sediment from longshore drift is prevented from passing the river mouth by the jetties. As a result of this, beach sediment is eroded on the downdrift side of the river (Figure 15.67). One solution to this is building groins to stop beach erosion; another is beach replenishment. Beach replenishment involves finding an appropriate source of sediment and moving it to the beach. In some cases, sediment is dredged from an offshore sediment accumulation and pumped onshore (Figure 15.67). A problem with this may be that the offshore sediment is different from the beach sediment, containing too many shells and more mud. Also, the hole

left by the excavation will inevitably be filled in over time because of the tendency towards a steady-state bed profile. Depending on the water-current conditions, the source of the sediment to fill the hole may be the beach itself. Figure 15.67 shows how an offshore sand quarry (for beach replenishment) or a gap in an offshore breakwater can result in wave refraction and/or diffraction, localized longshore drift, and a characteristic pattern of enhanced erosion and deposition. Intricate and expensive patterns of breakwaters and groins are implemented to avoid such problems. Various numerical models using sediment-transport equations are available to simulate the effects of jetties, offshore breakwaters and dredge holes, and piles of sediment dumped on beaches (Komar, 1998).

Realistic measures for dealing with various problems at coastlines require appreciation of natural forms and processes. A long-term rise in sea level in the absence of tectonic uplift or a significant increase in sediment supply will inevitably cause beaches to move landwards. There is evidence for this in older (Pleistocene) beach ridges preserved inland. Storms will inevitably cause flooding, wind damage, and local beach erosion and deposition. Therefore, buildings on beach ridges will inevitably be damaged periodically. Excessive vehicle traffic on beaches will inevitably cause erosion. If the sediment supply to beaches from rivers is cut off, beach erosion will result.

The solutions to these problems are therefore obvious. If avoidance of expensive engineering works is desirable, here are some suggestions. Allow rivers to flow and transport sediment in a more natural way. Encourage river mouths to point in the downdrift direction rather than offshore, so that sediment from updrift does not accumulate near the mouth. If river-mouth dredging is necessary, move the dredged sediment to the downdrift beach. Ban vehicle use on beaches. Educate people about the folly of building houses on dynamic, unstable islands and beaches.

Humans are also active in shallow marine environments. Their activity includes shipping, and oil and gas drilling platforms. Strong currents and stormy seas can cause problems in congested shipping lanes in narrow waterways (e.g., the English Channel). Erosion of sediment around drilling platforms can cause them to topple over in heavy seas.

Economic aspects of coastal and shallow-marine deposits

Coastal and shallow-marine sands and gravels are important reservoirs for oil, gas, and water. Examples are deposits of the Cretaceous North American Western Interior Seaway; the Triassic of Alaska's North Slope; the Cenozoic of the Gulf coastal plain and Gulf of Mexico, USA; and the Paleozoic and Mesozoic of the North Sea and surrounding areas. The oil fields of the Middle East, are, for the most part, hosted in carbonate-platform sands of Cretaceous age. Carboniferous coals are also closely associated with coastal deposits. Coastal sands are also used for building, glass-making, and children's sandboxes, and quarried for heavy (placer) minerals.

16   Arid environments

Introduction

Arid and semi-arid environments occur over about a third of Earth's surface today. They are defined by a lack of precipitation relative to evaporation and evapotranspiration, and the resulting sparseness of vegetation. Arid environments are commonly referred to as deserts, although they are not actually deserted. They are a characteristic ecosystem in which organisms are adapted to their harsh environment. However, human activities are greatly curtailed in arid environments. Distinctive processes of weathering, erosion, and sediment transport involving water and wind (Chapter 6) result in distinctive landscapes, such as areas of bare rock molded into strange shapes by weathering and wind erosion, aeolian sand seas, and ephemeral rivers and lakes (playa lakes). Aeolian sand tends to have high porosity and permeability (depending on diagenesis), providing important aquifers and hydrocarbon reservoirs. Aeolian sand also tends to be quartz-rich and clay-free, which makes it useful for glass-making and sand pits for children. Aeolian silt (called *loess*) can be spread widely, and forms the basis of very fertile soil. Saline minerals are commonly deposited in closed continental lake basins and along coastlines in arid areas as water becomes progressively concentrated by evaporation (thus evaporite minerals: see Chapter 4). Cenozoic playa-lake deposits may contain valuable evaporite minerals such as trona and borate. Gypsum and halite comprise about 5% of the stratigraphic record and many of these salts accumulated in marginal-marine basins along desert coastlines throughout the Phanerozoic. Thick Miocene salt deposits beneath the Mediterranean Sea suggest that entire small ocean basins in arid settings can reach halite saturation.

Past arid climates and climate change are indicated by ancient desert deposits and wind-blown silt in oceanic mud and ice cores. Since arid zones are related to latitude, rain-shadow effects of mountain ranges, and distance from the sea, ancient desert deposits can indicate the past relative positions of the poles, major continental land masses, and mountain ranges, having implications for continental drift. General references on deserts and their deposits include Glennie (1970), Cooke and Warren (1973), Brookfield and Ahlbrandt (1983), Frostick and Reid (1987), Pye and Tsoar (1990), Smoot and Lowenstein (1991), Brookfield (1992), Cooke *et al.* (1993), North and Prosser (1993), Pye (1993), Pye and Lancaster (1993), Abrahams and Parsons (1994), Lancaster (1995, 2005, 2007), Kocurek (1996), Livingstone and Warren (1996), Thomas (1997), Goudie *et al.* (1999), and Warren (2006).

Climate, location, and types of arid environment

Deserts are generally defined as regions where the mean annual rainfall is less than 300 mm. A United Nations study of desertification defined aridity as the ratio of precipitation to potential evaporation plus evapotranspiration. Hyperarid areas have ratios less than 0.03 whereas arid areas have ratios between 0.03 and 0.2. Hyperarid and arid regions comprise some 20% of the Earth's surface. Semi-arid areas have ratios between 0.2 and 0.5. Actually, there is every gradation between hyperarid and semi-arid. The precipitation that occurs is seasonal and commonly sporadic, frequently associated with convective thunderstorms. Infiltration of rainfall and overland water flow is considerable, such that rivers are ephemeral. Since evaporation exceeds rainfall, standing water does not remain very long. In general, these climatic conditions exist where air is descending and being heated by compression, and has lost most of its moisture by precipitation elsewhere. Hot deserts (e.g., the Sahara, Arabia, Atacama, Namib, Simpson) are concentrated around latitudes **563**

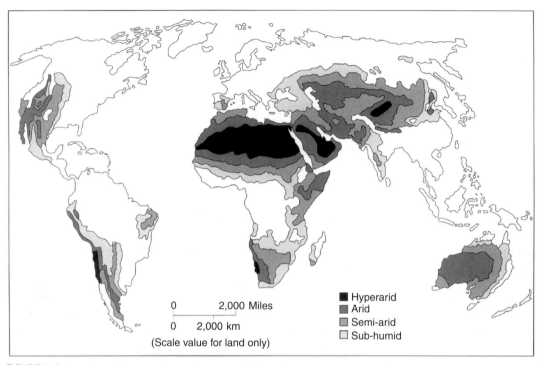

FIGURE 16.1. Locations of deserts and semi-deserts on the Earth. From Leeder (1999). Hyperarid zones have ratios of precipitation/ (evaporation + evapotranspiration) less than 0.03. For arid zones, this ratio is between 0.03 and 0.2, and for semi-arid zones it is between 0.2 and 0.5.

20° to 30° on either side of the equator, where the atmospheric pressure is high and air is descending (Figure 16.1; also see Chapter 2). An exception is southeast Asia, because of the monsoonal air circulation. Lee-side (or rain-shadow) deserts occur on the downwind side of major mountain ranges such as the Andes, Sierra Nevada, and Himalayas. Air is cooled as it rises over the mountains, resulting in precipitation. On the lee side of the mountain range, this dry air is heated by compression as it descends, reducing the possibility of further precipitation. Lee-side deserts occur in Argentina, throughout the western USA and southwestern Canada, and in Tibet. Deserts also occur in continental interiors that are distant from the oceans and hence from sources of water, e.g., the Gobi Desert in Asia. These deserts are sometimes referred to as cold deserts. The global distribution of deserts has varied considerably through geological time, particularly over glacial–interglacial cycles, as discussed below.

Sub-environments of deserts are (1) eroded bare rock with thin, discontinuous soils; (2) ephemeral (mainly) and perennial (rarely) streams on alluvial fans and floodplains; (3) aeolian sand seas (ergs); (4) ephemeral (mainly) and perennial (rarely) saline lakes;

and (5) coastlines. Some of these sub-environments are commonly associated in areas of internal drainage known as *closed basins*. Closed basins comprise about 30% of modern desert areas, and range in area from hundreds to hundreds of thousands of square kilometers. Thick (thousands of meters) accumulations of desert sediments occur today in fault-controlled closed basins. Such basins are characteristic of the Basin and Range province of the western USA and southern Canada, the Altiplano region of the Andes, the East African rift valleys, and the Tibetan Plateau. Examples include Death Valley and Saline Valley in California, Solar de Uyuni in Bolivia, the Dead Sea in Israel and Jordan, Lake Magadi in the East African Rift, and the Qaidam Basin of China. In addition, tectonically stable, cratonic closed basins (so-called *sag basins*) are found in the arid and semi-arid regions of the Earth, including the Lake Chad Basin in central Africa and the Lake Eyre Basin of central Australia. The great aeolian sand seas of the Sahara region, Arabian Peninsula, central Asia, and central Australia occur in these types of basins. Somewhat thinner (hundreds of meters thick) sediment accumulations occur in sag basins. Important ancient examples of closed basins include the Mesozoic rift basins of eastern North America and the Eocene Green River Formation of Wyoming.

In all closed basins, there is a characteristic transition of sub-environments from bare rock surrounding the basin to ephemeral or perennial lakes in the hydrographic center of the basins. In smaller fault-controlled basins such as Saline Valley the characteristic sequence of major sub-environments is bare rock → alluvial fans with ephemeral streams → mudflats → ephemeral (common) or perennial (rare) lakes (Hardie *et al.*, 1978) (Figure 16.2A, B). Sand dunes tend to accumulate at one end of the basin, depending on the prevailing wind. In larger fault-controlled basins and sag basins such as Lake Chad Basin (Figure 16.2C, D) the major sub-environments are bare rock → floodplains with mainly ephemeral but rarely perennial rivers → ephemeral or perennial lakes (in basins with a perennial river). Sand seas are common in larger basins. Minor sub-environments of closed basins include lake shorelines (deltas, beaches, and carbonate tufas) and spring deposits (including carbonate tufas). The locations of springs in many closed basins are controlled mainly by faults and may occur along the bottoms of lakes in the basins.

Arid coastlines in the tropics are commonly characterized by carbonates and saline minerals that accumulate in supratidal flats, and ephemeral and perennial marginal lagoons. Salt-encrusted surfaces (saline pans) in inland closed basins or in marginal-marine settings are commonly referred to by the Arabic term *sabkha* or the Spanish term *playa* (beach). Other terms for such areas include *salars*, *salinas*, and *chotts*. In Earth's past, entire continental shelves and ocean basins located in arid and hyperarid regions have become restricted enough to form substantial saline deposits. For example, the Mediterranean Sea became restricted enough to deposit gypsum and halite during the Miocene. Whether the Mediterranean became desiccated enough to become one large salt pan and undergo repeated cycles of desiccation (Hsü *et al.*, 1972a, b) is a matter of some debate, but it appears unlikely that this extreme state was ever actually reached (Hardie and Lowenstein, 2004).

Desert environments are difficult to study because of the inhospitable climate and logistical challenges, and because some are used as terrorist training areas. Not many scientists would be prepared to measure the wind speed and sediment-transport rate in a desert during a sand storm! It may be difficult to relate large depositional landforms such as megadunes (Chapter 7) to prevailing wind regimes, because they are most likely to be disequilibrium forms. It is also difficult to dig trenches in dry sediment even if there were enough

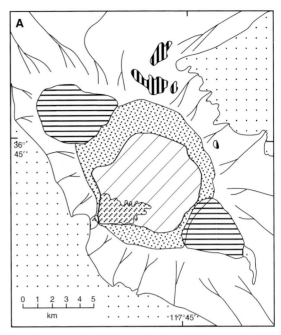

Saline Valley, California

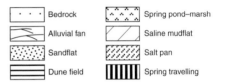

`· · ·`	Bedrock	`^ ^ ^`	Spring pond–marsh
	Alluvial fan		Saline mudflat
`· · ·`	Sandflat		Salt pan
	Dune field	‖‖‖‖	Spring travelling

FIGURE 16.2. (A) Sub-environments of Saline Valley, California, USA. From Smoot and Lowenstein (1991). (B) A photograph of Saline Valley with a view to the northwest. Photograph courtesy of T. K. Lowenstein. (C) Sub-environments of Lake Chad Basin, central Africa. From Smoot and Lowenstein (1991). (D) Lake Chad, central Africa. There is a flooded aeolian dune field on the top side of the photograph. Width of field ∼300 km. From NASA Earth from Space.

C

FIGURE 16.2. Continued.

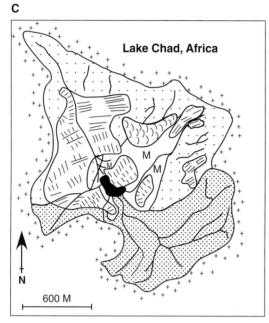

Lake Chad, Africa

 Perennial river floodplains

 Inactive or ephemeral braided and meandering river floodpains and deflation plains

 Aeolian dunefields with crest orientations

 Old-lake-bottom saline mudflat and dry mudflat

 Perennial saline lake

 Mountainous areas defining the basin

N

600 M

D

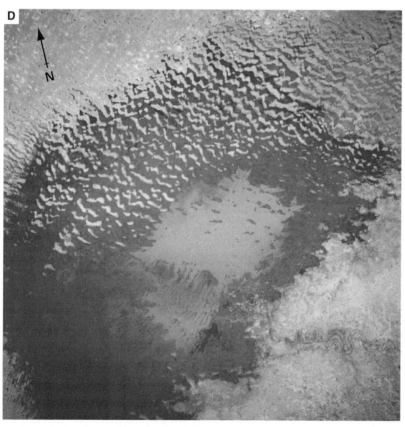

N

students to do the digging. The various sub-environments listed above are described more fully below.

Eroding bare rock with thin soils

Areas of eroding bare rock with thin soils constitute about 40% of desert areas. Soils are relatively thin and coarse-grained in deserts because of the lack of moisture and vegetation, and hence the lack of chemical weathering. However, both chemical and mechanical weathering operate in arid regions, albeit more slowly than in humid climates. Wind-blown dust that infiltrates deflated, armored gravelly surfaces can accumulate just below the surface, thereby decreasing permeability and enhancing chemical weathering. The dust can also serve as a source for solutes that accumulate in desert soils. Desert soils (aridisols) are typically immature, with poorly developed horizons. Microscopic soil organisms, particularly cyanobacteria and lichens, may form a scant cover to the soil, and may protect the soil from erosion and aid in the trapping and binding of wind-blown silt. This biota is also thought to contribute to the development of *desert varnish*, the smooth surficial coating on bare rock surfaces common in deserts. The A horizon in desert soils is relatively thin and lacking in organic remains, and the near-surface B horizon may well contain illuviated clay and precipitates of calcium carbonate, silica, or gypsum, forming duricrusts (see Chapter 3). The common red coloration of aridisols is evidence of chemical weathering and oxidation of iron-bearing minerals. When examining soils in deserts today, it should be realized that their character may be partly inherited from earlier, more humid climates.

The lack of soil in upland regions results in a characteristic topography such that the rocks at higher elevation are free of sediment, and soil accumulates as patches only in lower regions (Figure 16.3). This contrasts with the smooth topography of soil-covered slopes in humid areas. The exposed rocks exhibit a range of erosional features (e.g., yardangs; Chapter 6). Differential susceptibility of exposed rocks to weathering and erosion by wind or water yields many interesting landscape features (Figure 16.3).

Pediments are common in some arid regions, and are areas of (mainly) bare rock that occur at the

FIGURE 16.3. Erosional desert landscapes from the western USA.

boundary between upland areas and sediment-covered lowlands. They slope away from the mountain front at up to 11° (on average between 2° and 3°), and are mildly concave-upwards. Their surfaces contain channels and protrusions of rock that are resistant to weathering and erosion. Some low areas contain alluvium, and pediments generally pass downslope into alluvium-covered bedrock. Although their origin is controversial, they most likely form as a result of erosion of loose bedrock by channelized water flows and sheet-floods (e.g., Cooke and Warren, 1973; Cooke *et al.*, 1993). Their slopes are adjusted to transport the available loose material. Pediments exist because the sediment-transport capacity exceeds the sediment supply, and because of a lack of space for sediment accumulation. If sediment supply exceeded transport capacity, alluvial fans would appear in the place of pediments. It is most likely that pediments are not adjusted to present-day conditions and could have formed under different climatic conditions than those existing today.

Ephemeral streams, floodplains, and fans: geometry, processes, and deposits

Much of the lowland area of deserts is composed of the floodplains or alluvial fans of ephemeral rivers, possibly 20% of desert areas worldwide. Details of their geometry, flow and sedimentary processes, and deposits are in Chapter 13, and will not be repeated here. Water flows episodically in rivers in arid regions. Since the water table is generally below the surface, a large proportion of water infiltrates into the ground, especially if the surface is underlain by sand and gravel. Therefore, stream-discharge events do not tend to last very long, perhaps a few hours to days. Flood events can be generated by snowmelt or convective thunderstorms. Perennial large rivers with a sustained water supply can withstand infiltration losses as they cross the desert (e.g., the Nile, Colorado). Rivers that cross the desert but do not reach the sea have inland drainage systems that may end in perennial lakes such as Lake Chad (Figure 16.2C, D). In desert areas, the width and depth of rivers generally decrease downstream, because the discharge decreases downstream. Since the water table is relatively high beneath ephemeral rivers, riparian vegetation can be well developed, whereas vegetation is sparser on the floodplain (Figure 16.4).

An interesting feature of ephemeral river systems is that a localized convective thunderstorm may cause river flow in only one part of a drainage basin. If one flowing river is confluent with a dry river channel (its

FIGURE 16.4. Arid rivers, floodplains, and fans. (A) An alluvial-fan toe passing into a sparsely vegetated floodplain. The channel in the background is where vegetation becomes concentrated. (B) A close-up of the channel in the fan toe in (A). (C) An ephemeral channel (wadi). Images (A)–(C) are from Niger, Africa. (D) An alluvial fan from Death Valley, USA (courtesy of J. P. Smoot). (E) A channel bank from the fan-apex region, southwestern USA. (F) A low gravel bar surrounded by sandy, shallow ephemeral channels from the mid-fan area. (G) An unchanneled toe of fan gravels grading out into a sandy flat.

headwaters having received no precipitation), water starts to flow upstream into the dry river channel. As water starts to flow in a dry river channel following a thunderstorm, it can surprise travelers using the dry river bed as a convenient road, and the water could come from either the downstream or the upstream direction!

FIGURE 16.4. Continued.

Important features of flow and sediment transport in ephemeral rivers that have a bearing on the nature of erosion and deposition are (1) a flashy discharge regime whereby flowing water rapidly fills a dry channel to bankfull discharge and then gradually declines in discharge until the channel is dry again, all occurring over the order of hours or days; and (2) long periods of no water flow, allowing vegetation, desiccated mud, and wind-blown sand to accumulate in the bottom of the channel. This is what happens also in floodplain channels (e.g., crevasse channels) associated with perennial rivers. The deposits of ephemeral rivers are commonly considered to have certain distinctive features related to flashy discharge and large channel width/depth ratio (North and Taylor, 1996): predominance of planar-laminated and low-angle cross-stratified sand; and lack of well-defined channels or channels with large width/depth ratios. These features are actually not restricted to ephemeral rivers. The key sedimentary features that serve to distinguish the deposits of ephemeral

and perennial rivers are those that indicate drying out: plant roots, burrows, and mudcracks in the bottom of the channel. However, the main characteristics of channel deposits are controlled by flood conditions, and these occur in perennial and ephemeral channels alike.

Water flow over floodplains and alluvial fans in arid areas is particularly rare and short-lived. Infiltration of water results in common debris flows and mudflows (see Chapters 8 and 13). Alluvial fans in arid climates tend to be relatively steep because of the relative importance of sediment gravity flows (Figure 16.4; Chapter 13). Terminal fans are alluvial fans in which rivers do not flow beyond their margins. Floodplains and alluvial fans are typically covered with phreatophytes, plants with deep tap roots that reach the water table. The groundwater table commonly reaches saturation with alkaline-earth carbonates due to plant transpiration and evaporation. Carbonate-encrusted roots, nodules, and cements are thus common in floodplain and alluvial-fan deposits from arid regions. As the groundwater table becomes closer to the surface in low areas, the groundwater is usually so concentrated by evaporation that it is a brine, hence the paucity of vegetation in these areas (Figure 16.4). Soils on floodplains and alluvial fans in desert environments are aridisols as discussed in Chapter 3.

Bed surfaces of river channels, floodplains, and fans are subject to winds that deflate sand and silt and sand blast the remaining coarser sediment, resulting in an armored bed composed of pebbles (rarely faceted). Aeolian bed forms may be scattered over these surfaces. These areas are referred to as stony deserts or *reg*. The sediments were deposited by water flows or sediment gravity flows but are modified by the wind. Thin layers of wind-transported, planar-laminated sand may be difficult to distinguish from waterlain deposits.

Aeolian sand seas (ergs): geometry, processes, and deposits

Sand seas comprise only about 20% of desert areas worldwide (Figure 16.5); see the reviews by Kocurek (1991) and Lancaster (2007). Smaller aeolian dune fields are also common in closed basins. The sand comes mainly from entrainment from peripheral or internal streams, floodplains, and alluvial fans, with only a minor amount from sand blasting and weathering of adjacent bare rocks. Although siliciclastic sand is the most common, some dunes may be composed of gypsum (White Sands, New Mexico) or sand-sized intraclasts of clay (as in Australia). The principles of entrainment, transport, and deposition of sediment by wind are discussed in Chapter 6. Wind velocity and direction vary regionally in association with the subtropical high-pressure zones, but also with the topography of the land surface. Patterns of wind flow, sediment transport, erosion, and deposition can be ascertained from satellite images that show the orientation of aeolian dunes and erosional landforms such as yardangs. In general, the sand is deposited (eroded) where the wind velocity and sediment-transport rate decrease (increase) in space. However, the availability and movement of sediment on the bed are also influenced by its cohesion, which is related to its wetness (the level of the groundwater table), the percentage of clay, vegetation, and the degree of early cementation. Sand seas occur in regions of regional wind deceleration. The grain size of the deposited sand decreases in the direction of deceleration. Much of the silt and clay transported in suspension by the wind is transported away from the zone of sand deposition, and is deposited as the

wind further decelerates. This silt and clay can be transported over thousands of kilometers. Aeolian silt and clay can be recognized in oceanic deposits, and deposition rates on the order of 10^{-3}–10^{-1} mm per year have been calculated. Notable areas of Pleistocene to Holocene aeolian silt deposits known as *loess* include the mid-continent of the USA, eastern Europe including the adjacent Ukrainian steppes, nations around the Baltic, and east central China (Pye, 1987).

A typical distribution of bed forms in sand seas, from upwind to downwind and from middle to edges, is (1) sand sheets, zibars, and low barchan dunes with extensive deflated interdune areas; (2) a central area of large complex and compound dunes (draas); and (3) small transverse dunes and sand sheets on the leading edge (Figure 16.6). The character and internal structures of these bed forms are discussed in Chapter 6. It takes on the order of 10^3–10^5 years to form sand seas with bed forms the size of draas (megadunes). Although sand sheets without dunes occur due to limited sand supply, they may also be related to coarse sediment supply, the presence of vegetation, or a high water table (Fryberger *et al.*, 1979; Kocurek and Nielsen, 1986; Lancaster, 1995). In areas of actively migrating dunes or draas and abundant sand supply, interdune areas are not extensive (Kocurek, 1981; Ahlbrandt and Fryberger, 1981; Crabaugh and Kocurek, 1993). However, if sediment supply and dune activity are limited, interdune areas may constitute tens of percent of sand seas. Interdune areas are commonly more vegetated than the dunes, and may be armored surfaces with faceted pebbles, or sand sheets covered with wind ripples, zibar, and shadow dunes, or ephemeral ponds with evaporites. If the water table is relatively close to the surface, wet interdune areas are relatively more extensive.

Preservation of sand-sea deposits requires net deposition plus tectonic subsidence or a rising base level. Modern sand-sea deposits rarely exceed 100 m in thickness. It is commonly assumed that net deposition during bed-form migration will result in bed-form climbing and bases of cross sets that are inclined relative to the surface over which they are migrating. However, as discussed in Chapters 5 and 6, it is likely that the preservation of cross sets is more influenced by the variability of bed-form scour depths than by bed-form climbing. Figure 16.7 (from Crabaugh and Kocurek, 1993) shows examples of both bed-form climbing and variability of bed-form scour depths. The bed-form scour depth may in some cases be limited to the level of cohesive sediment beneath the surface (e.g., the level of the groundwater table: see Stokes

A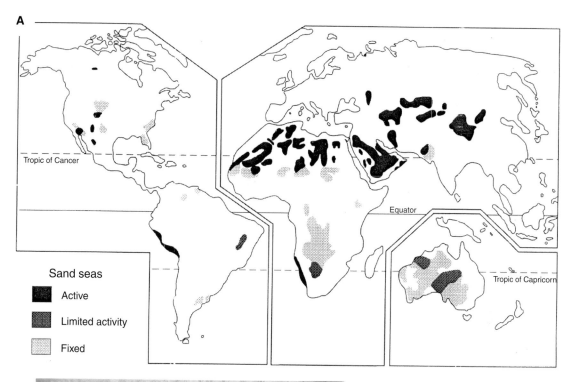

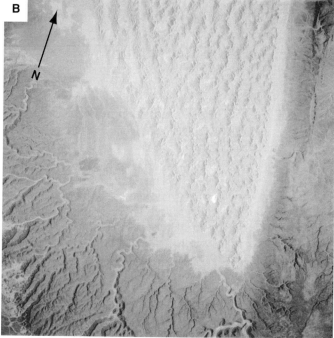

FIGURE 16.5. (A) The distribution of sand seas. From Lancaster (1995). (B) The Tifernine sand sea of southeastern Algeria surrounded by weathered sandstone. Width of field ∼150 km. From NASA Earth from Space.

(1968) and Kocurek *et al.* (1992)). If net deposition occurs during a period of rising base level, wet inter-dune deposits tend to be relatively thick and laterally extensive (Kocurek, 1996). Deposits formed at the edges of sand seas are likely to contain alternations of aeolian and waterlain deposits (e.g., Figure 16.8).

If a sand sea migrates laterally or downstream without appreciable net deposition, a coarsening-upward sequence would be expected in view of the downstream

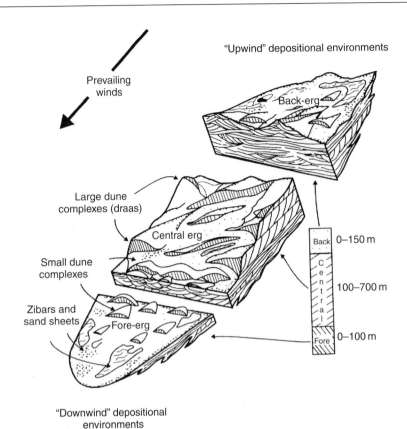

Prevailing winds

"Upwind" depositional environments

Back-erg

Large dune complexes (draas)

Central erg

Small dune complexes

Zibars and sand sheets

Fore-erg

Back   0–150 m

Central   100–700 m

Fore   0–100 m

FIGURE 16.6. The distribution of bed forms in sand seas. From Lancaster (1995).

"Downwind" depositional environments

decrease in bed-sediment size. If the bed-form height increases into the center of the sand sea, where sand is also thickest, and the cross-set thickness is related to the bed-form height, then the cross-set thickness will also increase upwards. However, if net deposition also occurs, some of the deposits from the upstream end of the sand sea may be preserved at the top of the sequence. These deposits would be representative of small dunes and sand sheets.

Bounding surfaces of aeolian cross strata and cross sets have been organized into a numbered hierarchy (e.g., Brookfield, 1977; Kocurek, 1981, 1988; Fryberger, 1993). However, bounding-surface-hierarchical schemes are very difficult to apply in nature, and do not adequately account for the complexity of aeolian cross strata; see Chapter 6 and Kocurek (1996). Recognition of different scales of stratasets and how they are superimposed in three-dimensional space is a much more useful exercise.

Perennial saline lakes: geometry, processes, and deposits

Perennial saline lakes in arid areas are maintained by the discharge of substantial rivers. Modern examples of

large perennial saline lakes include Lake Chad in North Africa (Figure 16.2C), Lake Balkhash in Kirgizstan, the Great Salt Lake in Utah, and the Dead Sea. These lakes typically have areas on the order of 10^3–10^4 km^2 and depths less than 10 m, but the Dead Sea has a maximum depth of 300 m. The Caspian Sea (area ~371,000 km^2, maximum depth ~980 m) is a land-locked mass of ancient seawater, presumably a remnant of the former Tethys Ocean that is now a brackish lake due to river inflow. Perhaps the most infamous perennial saline lake in the world is the Aral Sea on the border between Kazakhstan and Uzbekistan. In the 1960s this was the fourth largest lake in the world, with an area of ~66,000 km^2 and a depth of ~10 m. It was fresh enough to support fishing communities. Diversion of the main inflow rivers for irrigation led to a rapid decrease of the lake to approximately half its former area, with an increase in salinity.

Chapter 14 provides information about fresh- and brackish-water lakes, including sediment supply, thermal stratification patterns, and water circulation associated with waves, wind shear, and seiches. Perennial and ephemeral saline lakes are dealt with here. The same flow and sedimentary processes take place in

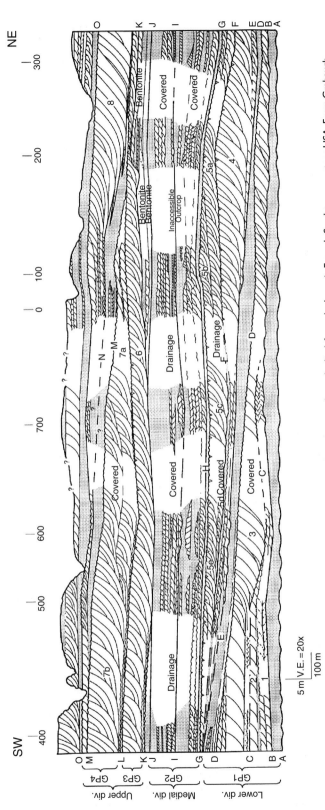

FIGURE 16.7. The large-scale distribution of aeolian cross strata and wet interdune deposits (gray shading) from the Jurassic Entrada Sandstone, western USA. From Crabaugh and Kocurek (1993). The bases of sets of aeolian cross strata show examples of upward climbing, downward climbing, and variation in scour depth. The scalloped pattern of cross strata is related to dune superimposition, and seasonal changes in bed-form type and direction. The restricted lateral extent of some of the interdune deposits, plus their having the same angle of climb as the cross-set boundaries, suggests migration of dunes over their own local interdune deposits. However, the horizontal, thick interdune deposits suggest a regional hiatus in deposition in the sand sea.

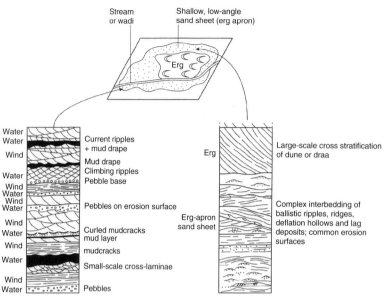

Stream or wadi

Shallow, low-angle sand sheet (erg apron)

Erg

FIGURE 16.8. Vertical sedimentary sequences typical of erg margins. From Leeder (1999), after Fryberger et al. (1979).

Water
Water
Wind — Current ripples + mud drape
Mud drape
Water — Climbing ripples
Wind — Pebble base
Water
Wind
Water — Pebbles on erosion surface
Wind
Water — Curled mudcracks mud layer
Wind — mudcracks
Water
Wind — Small-scale cross-laminae
Water
Wind
Water — Pebbles

Erg — Large-scale cross stratification of dune or draa

Erg-apron sand sheet — Complex interbedding of ballistic ripples, ridges, deflation hollows and lag deposits; common erosion surfaces

saline perennial lakes as in their fresher counterparts. The main differences are the degree of evaporation, the higher salinity and resulting enhanced density stratification, and the precipitation of evaporite minerals.

The composition of saline lake water is determined by the composition of the inflows, as with all lakes. The solute concentration of a saline lake depends on the ratio of inflow and precipitation to evaporation. Saline lakes have >5,000 ppm total dissolved solids (TDS). Brines in saline lakes may reach concentrations of 400,000 ppm TDS. The biota of lakes changes as the salinity increases, with diatoms, freshwater aquatic plants, green algae, freshwater fish, snails, and clams giving way to benthic cyanobacterial microbial mats, archaea, bacterial plankton, salt-tolerant sedge plants, and unusual animals such as the salt-tolerant fish *Talapia* and the brine shrimp *Artemia* (a.k.a. sea monkeys). The composition and solute concentration of saline lakes vary in space and time, as do the evaporite minerals precipitated. The spatial variation in solute concentration in saline lakes is due to local variations in the degree of evaporation relative to inflow, and gives rise to spatial zonation of the various evaporite minerals being precipitated. The fact that a particular evaporate mineral continues to be precipitated in a certain region of a saline lake despite continued evaporation is related to the continued inflow of low-salinity water, a process referred to as *reflux* (see Chapter 4). Time variations in the composition and solute concentration of saline lakes are discussed below.

Sediments deposited on the floor of a lake reflect the composition of the lake water, how concentrated it is, and whether the lake bottom is above or below wave base. Most saline lakes more than a few meters deep are density-stratified due to salinity gradients. Evaporation is limited to the surface of the lake in contact with the atmosphere. As evaporation proceeds, the surface brine is concentrated and saline minerals are precipitated, beginning with the alkaline-earth carbonates. Both the brine and the crystals sink into the deeper portions of the lake and less-saline water flowing into the lake will float out over the concentrated brines. Seasonal repetition of inflow, evaporative concentration, precipitation, and sinking of both the brine and chemical sediment are ubiquitous in perennial saline lakes. Extremely shallow areas of perennial saline lakes and isolated lagoons along the shoreline may reach saturation with the various saline minerals before the entire water mass becomes saturated. Sand-sized gypsum and halite crystals precipitated in the shallows of saline lakes above wave base are commonly reworked into ripples by wave- and wind-shear currents. Transported halite crystals in shallow areas can develop oolitic overcoats of halite.

The saline minerals that are precipitated depend on the chemistry of the inflow, according to the principle of chemical divides (Chapter 4). The first saline minerals to be precipitated are always the alkaline-earth carbonates, and which carbonate mineral will be precipitated depends on the Mg^{2+}/Ca^{2+} ratio of the lake water. Low-magnesium calcite is precipitated where

$Mg^{2+}/Ca^{2+} < 2$. Both high-magnesium calcite and aragonite are precipitated where $2 < Mg^{2+}/Ca^{2+} < 5$. Aragonite and perhaps magnesite form where $Mg^{2+}/Ca^{2+} > 5$. If the lake is at carbonate saturation, there will be a more or less steady rain of carbonate mud that may be pelleted in the surface waters by salt-tolerant filter-feeding organisms such as brine shrimp. The rain of carbonate mud is punctuated by settling of organic-carbon-rich material representing sporadic blooms of eukaryotic or prokaryotic microbes in the surface waters. In addition, river floods will introduce terrigenous mud in suspension. The laminae produced by these different types of sediment will extend over most of the bottom of the lake below wave base, and drape irregularities on the bottom, especially in deep lakes. The laminated oil shales of the Eocene Green River Formation were probably deposited in such a setting, but with only minor terrigenous mud input. The "oil" is actually kerogen, a complex assemblage of various organic compounds that needs to be raised to higher temperatures to become oil; and the inorganic laminae are carbonate, not siliciclastic. Thus, the Green River "oil shales" are actually kerogenous, laminated carbonates.

As saline lake water evaporates further, its composition remains dictated by the law of chemical divides (Chapter 4). As carbonate is precipitated, either the Ca^{2+} or the HCO_3^- is used up. If Ca^{2+} is used up, an alkaline brine (pH > 10, where the dominant carbonate species is CO_3^{2-}) forms with Na^{2+}, K^+, SO_4^{2-}, and Cl^- as other dissolved species. Upon further evaporation, such a brine cannot produce gypsum (since all the Ca^{2+} has been used up), but halite and a diagnostic suite of sodium carbonate minerals (natron, trona, or nahcolite) are precipitated instead, depending upon the temperature and pCO$_2$. Modern examples of this type of brine are found in Mono Lake, Alkali Valley, and Lake Magadi. The most famous ancient example of this type of saline deposit is the Eocene Green River Formation of the western USA. If, on the other hand, the HCO_3^- is used up during the initial formation of carbonates, the brine will become enriched in Ca^{2+}, and form neutral brine depleted in HCO_3^- (Table 4.4).

As evaporation continues, the lake can reach either gypsum saturation or sodium carbonate saturation (depending on which side of the chemical divide it was on), and enters a phase of co-precipitation of calcium carbonate and gypsum or co-precipitation of calcium carbonate and sodium carbonate. The Dead Sea is at a stage of co-precipitation of carbonate and gypsum, and micron-sized aragonite needles are precipitated together with gypsum crystals that are commonly

polygonal plates up to a few microns in diameter (Neev and Emery, 1967) or needles. However, the gypsum in the Dead Sea is apparently removed from the bottom sediments by the bacterial reduction of sulfate. Lakes at this stage of brine evolution produce alternating laminae (couplets) of calcium carbonate and gypsum, or calcium carbonate and sodium carbonate. Each couplet represents the evaporative concentration of a pulse of low-density (and -salinity) water that enters the lake and floats over the denser lake brine. As the surface water evaporates and becomes more saline, precipitation of calcium carbonate mud is followed by precipitation of gypsum mud (or deposition of sodium carbonate mud). By the time the water has become concentrated enough to precipitate gypsum and the sodium carbonate minerals, carbonate precipitation is minor, and a relatively pure layer of the more soluble mineral is deposited. Such couplets are commonly interpreted as annual layers (*varves*) but more likely represent sporadic storm-generated floods into the basin. Laminated chemogenic muds are the dominant sedimentary structures of stratified perennial saline lakes.

If the volume/surface area ratio of the lake is relatively small (i.e., the average depth is less than a few tens of meters), the entire water column can become permanently supersaturated with gypsum or sodium carbonate. In this case, gypsum or sodium carbonate crystals that fall to the bottom commonly serve as the nucleation points for additional syntaxial growth of gypsum (Figure 16.9) or sodium carbonate. The syntaxial growth layers are separated by drapes of carbonate mud in Figure 16.9A, and in Figure 16.9B the carbonate mud is incorporated as thin drapes within the large gypsum "selenite" crystals. The rounding of the selenite crystals from the Holocene Marion Lake (Figure 16.9B) indicates episodic dissolution due to temporary slight undersaturation of bottom water caused by undersaturated inflows.

In lake water that does not contain sodium carbonate, the precipitation of gypsum ($CaSO_4 \cdot 2H_2O$) marks another chemical divide. As gypsum is precipitated, either the Ca^{2+} is used up or the SO_4^{2-} is used up. If Ca^{2+} is used up, a calcium-free brine with Na^+, K^+, Mg^{2+}, Cl^-, and SO_4^{2-} will be the final product, and the diagnostic minerals formed will include sodium sulfate and magnesium sulfate. If the sulfate is used up, the final brine will be a Na^+, K^+, Mg^{2+}, Ca^{2+}, Cl^- water. Such calcium chloride water will produce diagnostic potassium chloride and calcium chloride minerals such as sylvite, tachyhydrite, and antarcticite.

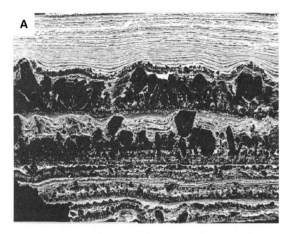

FIGURE 16.9. Carbonate and gypsum layers. (A) Dark layers of upward-directed gypsum crystals that nucleated on surface-formed crystals that sank to the bottom in a shallow brine body. Fine layers are alternations of carbonate and gypsum that precipitated at the surface and sank. (B) Large gypsum "selenite" crystals similar to (A); the layering is defined by light-colored layers of aragonite crystal mud. Some of the aragonite layers continue through the large crystals. In the lower center of the photograph the sharp terminations of the crystal were rounded off by slight dissolution during successive floods.

Evaporative concentration of tens to hundreds of times the original inflow composition are required in order to reach supersaturation with the next saline mineral, usually halite. The sodium carbonate brines will not precipitate gypsum and instead proceed to precipitation of halite, a sodium carbonate phase, or both. At halite saturation, halophytic archaea are usually the only organisms that can survive, and they dye the waters vivid shades of ochre, orange, and lavender. Halite crystals form at the surface of these lakes as square or rectangular plates, as stepped skeletal *hopper crystals*, or as cubes with slightly indented sides. Halite crystals can also coalesce into wafer-thin rafts. Initially, the halite crystals are buoyed by surface tension, but when they become too dense, or surfactants lower the surface tension, the hoppers and rafts sink to the bottom. If the volume/surface area ratio of the lake is large, the deeper waters do not become supersaturated, and the halite cubes and rafts simply accumulate in layers with so-called *cumulate textures*, piles of cubes that have settled through the water column (Figure 16.10). If the volume/surface area ratio of the lake is small, the bottom brine can become supersaturated with halite, and syntaxial growth continues on the halite rafts and cubes on the bottom, producing layers of halite crystals with distinctive *chevron* or *coronet* structures (Figure 16.11). The chevrons and coronets are growth faces of the crystals outlined by bands that are either rich or poor in fluid inclusions (Figure 16.11B, C). Finally, at the most advanced stages of drying, the so-called bittern salts (Chapter 4) may precipitate. Which salts precipitate at these very advanced stages of evaporation depends on the evaporation path the lake waters have taken. The salts precipitated on the bottom may be partially to completely dissolved if the bottom water is undersaturated, possibly due to lack of sufficient evaporation or due to periodic inflows of water with low salinity.

Individual floods bring in suspended terrigenous clays that settle and drape over the previous bottom growth layers (Figure 16.11A). Organic-rich layers representing cyanobacterial and archaea blooms may also drape the bottom. Laminae are thus defined by variations in the size of the halite cubes, by variations in the density of fluid inclusions in the crystals, by organic layers representing cyanobacterial and archaea blooms, and by terrigenous muds delivered to the lake by streams. Debris flow and turbidity-current deposits of terrigenous sediment can also be interbedded with perennial lake sediments. These arise from hyperpycnal river currents or slumps off the front of a delta or faulted sides of the basin (see Chapter 8).

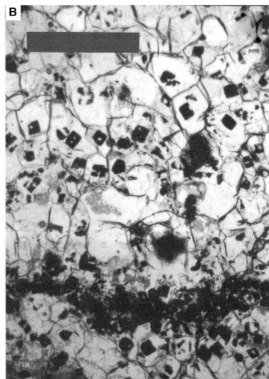

FIGURE 16.10. Thin-section photomicrographs of cumulate halite crystals from Miocene salts beneath the Mediterranean Sea. (A) A low-power view of layers of cumulate halite. The scale bar is 10 mm. (B) A medium-power view showing halite fabrics in two light laminae separated by a dark lamina at the bottom of the photograph. The dark centers of the cubes in the light layer are inclusion-rich crystals (interpreted as the original halite hoppers formed at the surface) surrounded by overgrowths that grew during crystal sinking and later welded into the interlocking mosaic now seen. The dark layers are formed by inclusion-rich crystals similar to the cores of the overlying halite without the overgrowths. The scale bar is 2 mm. Photographs courtesy of T. K. Lowenstein.

Ephemeral saline lakes: geometry, processes, and deposits

Ephemeral lakes include salt pans, mudflats, and a variety of shoreline features. *Salt pans* are shallow depressions floored with evaporite minerals, and are normally dry (Figure 16.12) except when rarely flooded (every few years). Salt pans may contain teepee structures (Figure 16.12; Chapter 12). Salt pans vary in area from less than 1 km² to thousands of square kilometers. The salt pan (*salar*) at Lake Uyuni in the Altiplano of Bolivia covers approximately 10,000 km². The ephemeral lake that forms during flooding may be much larger than the salt pan, and the mudflat left behind as the lake dries out and becomes more concentrated may be referred to as a *playa*. Mudflats are the sites of accumulation of terrigenous mud, although they may be covered by a thin efflorescent crust of evaporites. Lake Eyre in central Australia covered up to 8,000 km² at the height of flooding in 1949–1952. The mudflat at

Lake Eyre covers approximately two thirds of the area of the lake, whereas separate saline pans cover the other third. The shoreline features are described below.

Lowenstein and Hardie (1985) documented what they referred to as the saline-pan cycle. During a flood event (Stage I in Figure 16.13), surface water floods the old saline-pan surface and usually spreads beyond it. This water is usually undersaturated, and dissolves any thin efflorescent crusts on mudflats as well as the old salt-pan halite crust. The salt-pan halite is dissolved at the upper surface and along crystal boundaries to produce vertical pipes in the salt crust below the surface (Figure 16.14). Suspended terrigenous mud in the floodwater settles out and drapes the dissolution surface (Figure 16.14A), and some may filter down into the dissolution pipes. Mud layers are commonly a few millimeters to ten millimeters thick.

Flooding is followed by evaporative concentration of the water that eventually becomes saturated with halite (Stage II in Figure 16.13). Then, halite hopper

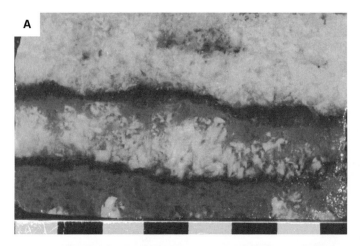

FIGURE 16.11. Bottom-growth fabrics in halite. (A) The dark gray layers are terrigenous mud introduced by flood waters. This is followed by halite cumulates that were overgrown after they had settled to the bottom. Light gray is polyhalite crystal mud that drapes the halite crystals. Polyhalite precipitation took place after saturation with halite. The scale-bar divisions are 10 mm. (B) A low-power thin-section photomicrograph of cumulate halite with syntaxial bottom overgrowths. The growth banding forms a distinctive "chevron" structure and is outlined by bands of dense fluid inclusions. Width of field 100 mm. (C) A close-up of fluid-inclusion banding in halite. The scale bar is 3 mm. All photographs courtesy of T. K. Lowenstein.

FIGURE 16.12. A saline pan in Saline Valley, California with teepee structures (Chapter 12). Note the alluvial fan in the background. Photograph courtesy of T. K. Lowenstein.

crystals nucleate on the surface and join into floating rafts (Figure 16.15A). The halite cubes and rafts eventually become too heavy to be supported by surface tension and sink to the bottom, where they rest on the mud (Figure 16.15A). This is the origin of many cubic halite impressions on the bases of sedimentary layers (Figure 16.15B). Syntaxial growth of halite continues on the bottom, producing chevron and coronet structures similar to those of the perennial saline lakes (Figure 16.14C). Growth bands in the halite are outlined by varying concentrations of fluid inclusions that were trapped as the crystals grew.

The lake finally becomes completely desiccated (Stage III in Figure 16.13). Groundwater that usually

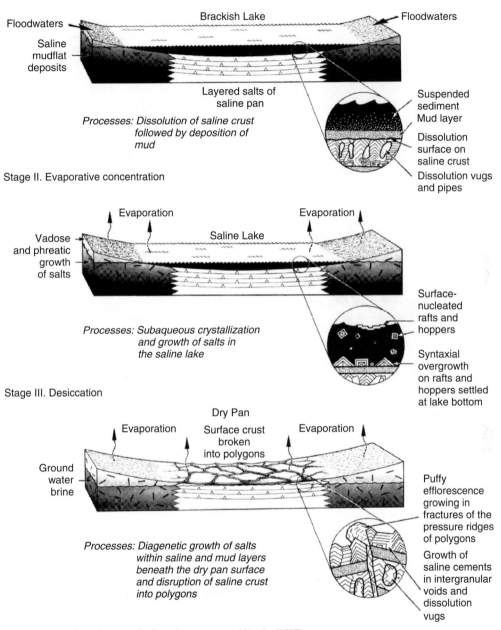

Stage I. Flooding

Floodwaters

Brackish Lake · Floodwaters

Saline
mudflat
deposits

Layered salts of
saline pan

Processes: Dissolution of saline crust
followed by deposition of
mud

Suspended
sediment
Mud layer

Dissolution
surface on
saline crust

Dissolution vugs
and pipes

Stage II. Evaporative concentration

Evaporation   Evaporation

Vadose
and phreatic
growth
of salts

Saline Lake

Processes: Subaqueous crystallization
and growth of salts in
the saline lake

Surface-
nucleated
rafts and
hoppers

Syntaxial
overgrowth
on rafts and
hoppers settled
at lake bottom

Stage III. Desiccation

Dry Pan

Evaporation   Surface crust   Evaporation
broken
into polygons

Ground
water
brine

Puffy
efflorescence
growing in
fractures of the
pressure ridges
of polygons

Growth of
saline cements
in intergranular
voids and
dissolution
vugs

Processes: Diagenetic growth of salts
within saline and mud layers
beneath the dry pan surface
and disruption of saline crust
into polygons

FIGURE 16.13. The salt-pan cycle. From Lowenstein and Hardie (1985).

flows when there is no overland flow into the ephemeral lake is concentrated by evaporation. This results in precipitation of calcium carbonate cements, calcareous concretions, and so-called *desert roses* (lenticular gypsum crystals that incorporate the surrounding sediment) in the alluvial fans and stream deposits around the margins of the lake. By the time the groundwater reaches the mudflats and saline pans at the center of the basin, all the Ca^{2+} has been removed by precipitation of calcium carbonate and gypsum. The groundwater is a concentrated Na^+, Cl^-, SO_4^{2-} brine, and halite is precipitated in the voids of the old halite crust. This addition of salt causes lateral growth of the salt crust, which then buckles into a polygonal *tepee* pattern (Figure 16.12). The

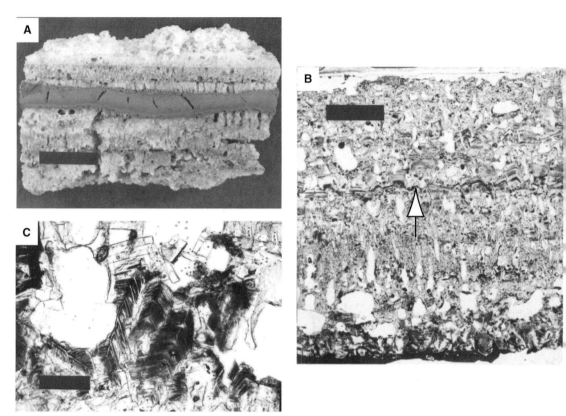

FIGURE 16.14. Features of halite surface crusts from modern salt pans. (A) Two subaerially exposed surface crusts separated by a mud layer deposited by a storm flood over the old surface. The scale bar is 50 mm. (B) A thin-section photomicrograph showing dissolution tubes and pipes cutting across the primary chevron fabrics of the halite in a surface crust. Note the dissolution surface (arrowed) that truncates the underlying vertical features. (C) Details of dissolution truncation of chevron halite. The scale bar is 2 mm. Photographs courtesy of T. K. Lowenstein.

pressure ridges at the boundaries of polygons fracture, producing pathways for additional evaporation. Commonly, a puffy efflorescent crust of halite grows in and over the cracks between the pressure ridges. The halite crusts of salt pans are typically tens of centimeters thick. Thus, the deposits of an ephemeral lake saline pan will comprise alternations of siliciclastic mud and halite crusts riddled by dissolution fenestrae (Figure 16.14).

Mudflats (playas) are those portions of ephemeral lake beds not covered by evaporites, except for a thin efflorescent crust of salt. Some mudflats have groundwater brine under them (saline mudflats, Figure 16.16) and others do not (dry mudflats, Figure 16.17) (Hardie et al., 1978). Mud is deposited from suspension by floodwater. As the lake dries out, its muddy bed surface is modified by footprints, burrowing animals, rainprints, and scour marks from wind-transported grains. With saline mudflats, groundwater brine is concentrated by evaporation, and saline minerals, usually gypsum or halite, grow in the mud. These tend to be single crystals that incorporate very little of the surrounding sediment. Gypsum crystals that remain bathed in the brine are commonly dehydrated into *nodular anhydrite* (Figure 16.16C). Nodular anhydrite is composed of micron-long laths that deform in the dewatering conversion, thus losing the original discoidal shape of the gypsum crystals. The surface of a saline mudflat is commonly covered by a cyanobacterial mat or a thin fluffy efflorescent crust of halite, depending on the amount of time the mudflat had standing water on it (Figure 16.16A). Water flooding over the cyanobacterial mat can find defects in the mat surface, and tear it up, producing roll-up structures (Figure 16.16B). Saline mudflats usually do not desiccate and contract to produce desiccation mudcracks. Where the saline mudflat is covered by efflorescence, wind-blown sand can accumulate in depressions in the crust surface. Floodwater commonly dissolves the halite crust, increasing the concentrations of Na^+ and Cl^- in the floodwater (significantly changing the

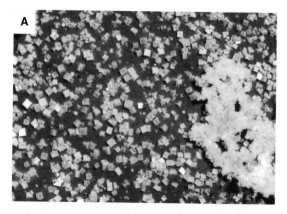

FIGURE 16.15. Nucleation of halite at the beginning of the salt-pan cycle. (A) Halite cubes, plates, and a raft of coalesced crystals (right). The out-of-focus cubes are floating at the brine–air interface. The in-focus cubes have sunk to the bottom and become embedded into the mud deposited by the last flood into the basin. Note the various angles at which the cubes have impacted with the bottom. The cubes are up to 8 mm in diameter. (B) Impressions of halite cubes on the bottom of a dolostone layer. A flood dissolved halite cubes similar to those shown in (A) and the crystal molds were filled and preserved by the dolomite.

chemistry of the inflow), and leaving behind the wind-blown sand as sand patches in otherwise completely chemoturbated muds. It is this change in chemistry of inflow waters due to recycling of halite surface deposits that accounts for the fact that most saline pans have deposits of halite as opposed to gypsum.

On dry mudflats, the groundwater brine table is so deep that these mudflats experience extreme desiccation (Figure 16.17). The cracks are vertical and horizontal and completely brecciate the mud, commonly to a depth of a few tens of centimeters (Figure 16.17B, C). Wind and rainfall erode the crack walls. The mud chips produced by desiccation may subsequently be eroded and incorporated into mud-chip breccias. The cracks become filled with complicated layers of wind-blown sand and mud chips. Clay minerals are commonly

smectites; hence they are capable of swelling and shrinking as they are repeatedly wetted and dried. This repeated wetting and drying, wind erosion, and crack-filling produces a completely brecciated soil riddled with complicated mudcrack patterns. These types of soils are commonly mistaken for lake deposits that had experienced synaeresis shrinkage of gel-like muds in the brines of saline lakes. However, cutans (elluviated clay linings – see Chapter 3) are common features of these soils, and are especially useful for dismissing previous interpretations of such deposits as being formed in deep lakes (Smoot and Lowenstein, 1991).

The "Racetrack Playa" in Death Valley National Park, California, is noted for large boulders that sit at the ends of furrows that the boulders have apparently carved as they were moved along the surface of the playa. The furrows are up to hundreds of meters long, and can curve and cross each other in complicated patterns. No one has ever caught the boulders in the act of moving, so the exact mechanism of motion is unknown. Suspicions include wind propulsion of the rocks while thin, lubricating sheets of water are around. However, ice rafts propelled by the wind are also a possibility.

Saline lacustrine shorelines and tufas: geometry, processes, and deposits

Shorelines of lakes are described in Chapter 14, and include wave-cut benches, beaches, spits, longshore bars, and deltas (Figure 16.18). These shoreline features are common in perennial saline lakes and are analogous to their marine counterparts (See Chapter 15), although there are some differences. Waves on most lakes are expected to be smaller than oceanic waves (due to the limited fetch), and wind-shear currents should also be less important. This results in relatively weaker wave and wind-shear currents at the shoreline, and lower sediment-transport rates. Ephemeral lakes are subject to rapid oscillations of lake level and many are extremely shallow. Beaches, spits, and bars are poorly developed in ephemeral lakes. Tidal currents do not exist in lakes. Deltas are most common in perennial lakes with perennial river inflow and may be of the bird's-foot type if freshwater flows into a saline lake, or of the Gilbert type if the lake is deep and differences in density between river water and lake water are minor (see Chapter 15). So-called sheet deltas in ephemeral lakes (Smoot and Lowenstein, 1991) are essentially crevasse splays that flow into standing water on floodplains (Chapter 13).

Carbonate mounds, variously referred to as tufas, algal tufas, and travertines, are common in and around

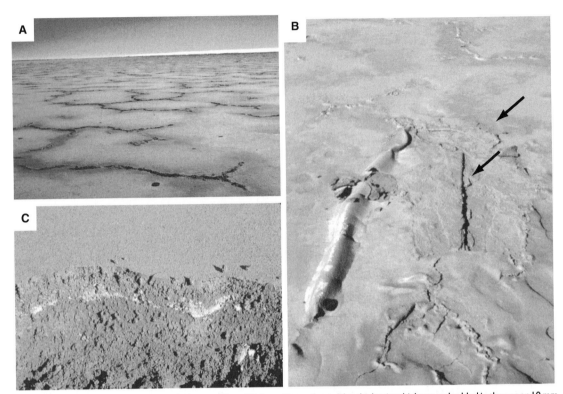

FIGURE 16.16. Sedimentary features of saline mudflats. (A) A mudflat surface with a thick microbial mat embedded in the upper 10 mm. Pressure ridges are due to mat growth. The discoloration on the surface is a halite efflorescent crust. The lens cap in the foreground is 50 mm in diameter. (B) A surface as in (A) where floodwater has deformed the cohesive surface into anticlines and microbial "roll-up structures" (arrow). The lens cap against the anticline on the left is 50 mm in diameter. (C) A trench in saline mudflat deposits, showing anhydrite nodules in thoroughly "chemoturbated" mud. The trench is ~0.5 m deep.

saline lakes (Figure 16.19), and range in size from centimeter-diameter spheres to complex mounds up to 30 m high. Carbonate tufas occur (1) as continuous sheets coating lake shorelines; (2) as isolated mounds widely scattered over the floors of saline lakes; and 3) at spring orifices within and outside lakes. Shoreline sheets of tufa and isolated mounds occur in saline lakes that are saturated with carbonates but not yet at gypsum saturation, and are common markers of Pleistocene lake levels in many of the closed basins of the American southwest. The spring-related mounds tend to be the most spectacular, and many of these that are not currently in the lake (such as the tufa towers around Mono Lake – Figure 16.19A) were formed during Pleistocene high lake levels. At subaerial spring orifices, the groundwater commonly degasses carbon dioxide, inducing carbonate precipitation. Within the lakes, the mixing of spring water with lake water commonly produces locally supersaturated water that rapidly precipitates carbonate and coats anything in the area. Many tufa mounds have internal

spring orifices preserved (Figure 16.19E). Similar tufa towers of carbonate minerals around spring orifices also occur in the ocean.

The so-called *thinolite* tufas around Walker Lake are composed of radiating crystals up to a meter in length (Figure 16.19C). These crystals are now composed of low-magnesium calcite but, as was noted by United States Geological Survey Geologists over 150 years ago, the crystals were originally some unknown mineral they dubbed *thinolite*. Shearman *et al.* (1989) correctly interpreted these crystals as pseudomorphs after the low-temperature mineral ikaite ($CaCO_3 \cdot 6H_2O$) originally described from the Ikai Fiord of Greenland. The occurrence of this mineral in Walker Lake attests to the low temperature of the lake waters at the Last Glacial Maximum while the tufas were being deposited.

The fabrics of other tufas are complicated amalgamations of small bush-like shrubs (commonly referred to as *arborescent tufa*), other less-distinct clotted areas, and stromatolitic laminae (e.g., Figure 16.19D), and very reminiscent of the thrombolites of the

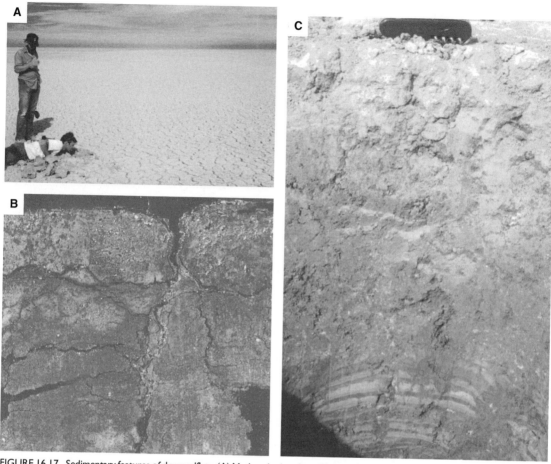

FIGURE 16.17. Sedimentary features of dry mudflats. (A) Mudcracked surface. (B) A box core showing brecciation of dry mudflat deposit due to desiccation. Width of field ∼200 mm. Courtesy of J. P. Smoot. (C) A laminated lake deposit composed of alternations of sand (light) and mud (dark) at the base of a trench has become progressively disrupted by mudcracks up to the modern dry mudflat surface.

FIGURE 16.18. Old shorelines (arrows) of Pleistocene Lake Bonneville, Utah, USA.

Proterozoic and early Paleozoic (see Chapter 11). Arborescent tufas are common in the building stone excavated from Tivoli in Italy, which is used for facing many buildings in the USA and UK. The Tivoli arborescent tufas nucleate with a distinct spacing and grow up and out until they begin competitive growth, after which they grow up together along compromise boundaries. These are not the features of bacterial

FIGURE 16.19. Carbonate lacustrine tufas. (A) Mono Lake. (B) Pyramid Lake. (C) "Thinolite" crystals (pseudomorphs after ikaite) from Pyramid Lake. (D) A sawn slab of "microbial" tufa from Great Salt Lake. (E) An eroded column from Pyramid Lake, showing the internal spring orifice (arrow) (courtesy of T. K. Lowenstein).

colonies that tend to overgrow each other; therefore, Tivoli arborescent tufas are most likely chemical precipitates. Subaqueous carbonate tufa mounds are usually covered with macrovegetation such as algae and mosses, as well as cyanobacterial colonies. Most workers would call upon the vegetation, particularly the cyanobacteria, to induce precipitation of the carbonate, but it is clear where old cans and other inert materials are being coated with carbonate that it is being precipitated inorganically. The arborescent tufas have also been claimed to be the result of the growth of bacterial colonies. Tufas from Great Salt Lake have calcified bacterial colonies very reminiscent

of the problematic forms *Renalcis* and *Givanella* so common in Proterozoic and Paleozoic thrombolites (Demicco and Hardie, 1994).

Arid coastlines

Shorelines in arid and hyperarid areas have essentially the same geometry, processes, and deposits as shorelines in humid climates (Chapter 15); however, there are some differences, as discussed below. The sediments of arid coasts will be terrigenous close to river input (e.g., Colorado River delta), but carbonate sediments accumulate distant from river mouths. In humid temperate and tropical regions, the upper supratidal flats are covered with either salt-marsh grasses or mangroves. In arid areas, vegetation does not grow on the supratidal flats, and saline minerals grow on and in the sediment instead. Along arid shorelines, back-barrier lagoons and drowned water courses of ephemeral streams are also commonly the site of deposition of evaporite minerals. Modern, well-studied arid shorelines rim the Persian Gulf, with carbonates on the Arabian Peninsula, and terrigenous shorelines around the Tigris–Euphrates delta and along the Iranian coast. The western coast of Australia is also the site of both terrigenous and carbonate shorelines, with Shark Bay being a well-studied site of carbonate deposition. The best-described terrigenous sabkha is the Colorado River delta at the northern end of the Gulf of California. General references on modern arid shorelines include Logan *et al.* (1970, 1974), Purser (1973), Shinn (1983b), James and Kendall (1992), Kendall (1992), Kendall and Harwood (1996), Alsharhan and Kendall (2003), and Warren (2006). At certain times in the Earth's past, saline deposits are postulated to have accumulated over entire continental shelves and in small ocean basins located in arid and hyperarid regions, although there are no modern examples. Supratidal sabkhas, saline lagoons, and desiccated shelves and ocean basins are discussed below.

Supratidal sabkhas: geometry, processes, and deposits

Sabkha (a transliteration of the Arabic for salt flat) is an arid, subaerially exposed salt-encrusted mudflat (Hardie, 2003b). Coastal sabkhas are high-intertidal to supratidal flats that can be up to 10 km across and extend for up to 100 km parallel to shore. The best-known ancient coastal sabkhas occur in carbonates, as do the most famous modern sabkhas, which form part of a complex of islands, lagoons, and tidal flats comprising the United Arab Emirates of the southern margin of the Arabian Gulf, an area once known as the Trucial Coast (Figure 16.20). The area has been urbanized and industrialized, so these descriptions are based on information from the 1960s and early 1970s. The barrier islands have beaches backed by aeolian dune ridges facing the Arabian Gulf and perpendicular to the dominant northeast winds. Intertidal and supratidal tombolos connect the barrier islands to the mainland. The Gulf in front of the islands has salinities of 40‰–50‰ and is a gently shoaling shelf composed of bioclastic carbonate sand. Ebb-tidal deltas composed

of aragonite ooids molded into dunes occur on the Gulf sides of the inlets between barriers. Low-relief coralgal patch reefs are situated in front of the islands away from the active ebb-tidal deltas, and the coral fauna is impoverished due to the high salinity. Channels up to several kilometers wide and up to 10 m deep feed from the tidal deltas into shallow lagoons. The channels are composed of mixed ooid–pellet–skeletal sand molded into dunes. The lagoons are floored by pelleted aragonite-needle mud and, together with the lower intertidal mudflat, are thoroughly burrowed by a variety of organisms, including mudskippers. The tidal range decreases from ~2 m to ~1 m from the barrier islands into the lagoons, the temperature of the water varies from 15 to 40 °C, and salinities in the lagoons can be around 70‰ (roughly twice normal seawater concentrations). Thickets of mangroves are scattered around the lagoons. The high-intertidal flats that fringe the lagoons are famous for the microbial mats that blanket the surface and the associated laminated sediments (Chapter 11).

The sabkha inland from the coastal lagoons is a nearly planar surface up to 10 km across that rises up 1–2 m landwards to where it overlaps Tertiary bedrock. The sabkha formed by seaward progradation of the tidal flats (Figure 16.21) that have been covered with up to 1 m of wind-blown quartz sand derived mainly from the desert inland. The progradational history of the tidal flat is revealed in cores and trenches through it (Figure 16.22, and see below). A brine table exists in the carbonate sediments beneath the quartz sand. The hydrology beneath the surface of the sabkha is complicated and includes areas of unconfined and semi-confined flow under artesian conditions (McKenzie *et al.*, 1980). The brine table parallels the sabkha surface, indicating groundwater flow towards the coastal lagoons. Initially, it was reckoned that the brine was fed primarily from the seaward side by sporadic storm washover and extreme high tides, and this type of groundwater system, with floodwater evaporating, sinking into the sediment, and flowing seawards again as groundwater, came to be known as *subsurface reflux*. It is now known that the hydrologic system beneath the sabkhas is complicated by significant input of continental water with elevated salinity from the landward side, and by rainfall. Evaporative concentration of subsabkha brines causes discoidal gypsum crystals to grow displacively in the aeolian quartz sand, and in places they are so numerous that they comprise a gypsum crystal mush (Figure 16.22B). Old gypsum crystals that sit in the brine-logged sediments are converted into nodules of anhydrite, and areas of gypsum crystal mush dehydrate

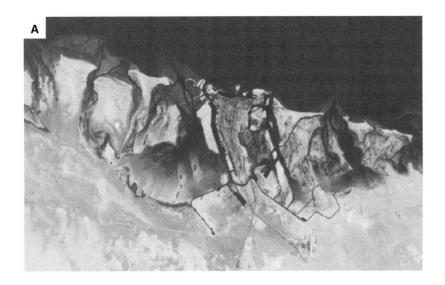

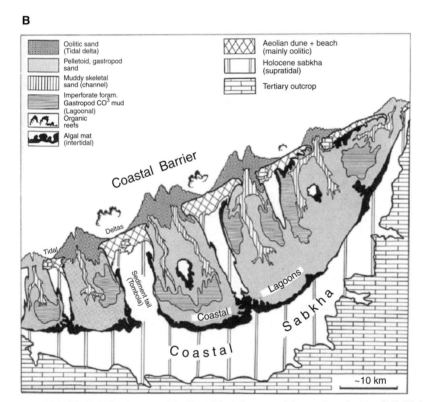

FIGURE 16.20. Tidal-flat deposits of the United Arab Emirates of the southern Arabian Gulf. (A) A photograph from space of the area (Earth from Space NASA archives). (B) A diagrammatic representation of the surface sedimentary environments of the United Arab Emirates. From Purser and Evans (1973). By permission of Springer Science and Business Media.

to form so-called "chicken-wire" anhydrite (see Figure 16.22C). In some of the sabkha sediments, continuous layers of nodular anhydrite are deformed and folded, and in many cases are truncated by an erosion surface that has been buried by the wind-blown sand (Figure 16.22D). The origin of the folding is unknown, but is commonly related to the volume change as gypsum is converted into anhydrite. Demicco and Hardie (1994) suggested that the folded and truncated layers of nodular anhydrite may be replacements of old subaerial gypsum

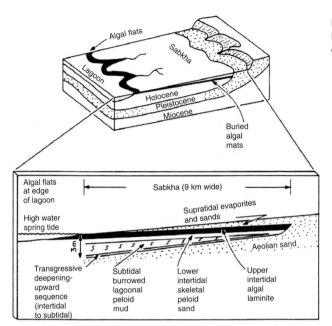

FIGURE 16.21. Holocene deposits of the United Arab Emirates tidal flats as an example of active progradation of tidal-flat deposits. From Demicco and Hardie (1994).

crusts deposited in ephemeral saline pans on the sabkha surface. These gypsum salt-pan crusts were deformed into tepee structures analogous to halite crusts (Figure 16.12) prior to burial. Isolated nodules of anhydrite, layers of deformed anhydrite, and "chicken-wire" anhydrite are commonly considered diagnostic of sabkhas, but similar features can be produced under quite different circumstances, including syn-depositionally around hydrothermal vents, and during burial diagenesis.

The intertidal mudflat sediments can be traced up to 9 km back from the current strandline beneath the present sabkha surface of the United Arab Emirates. Lower intertidal and subtidal burrowed pellet mud occurs beneath the laminated upper intertidal sediments (Figures 16.21 and 16.22). The subtidal-lagoon deposits rest on a veneer of intertidal deposits a few centimeters thick that cover subaerially deposited wind-blown sands. These sedimentary layers beneath the sabkha of the United Arab Emirates are up to 3 m thick and are interpreted as having been produced by the Holocene transgression of a gently dipping aeolian sand surface followed by progradation of laterally adjacent environments over each other during shoreline regression.

Modern saline lagoons: geometry, processes, and deposits

Coastal bodies of water (many of which are described in Chapter 15) include elongated lagoons behind microtidal and mesotidal barrier islands, and estuaries along mesotidal coasts. In addition to these lagoons, tectonic basins such as San Francisco Bay can also become flooded. In humid temperate and tropical areas, these coastal bodies of water usually stay at normal marine salinities (35‰), or become brackish where major rivers enter them. In semi-arid and arid settings, where freshwater input is low, lagoons behind barriers can become hypersaline. Examples include the Laguna Madre, behind Padre Island in Texas, and the lagoons in the United Arab Emirates described above. Hamelin Pool at the extreme landward end of Shark Bay in Western Australia is a hypersaline lagoon (salinity 60‰–65‰) that is isolated by a series of carbonate banks. The inner reaches of estuaries and bays in semi-arid and arid settings can also become hypersaline, and have commonly been dammed by humans and turned into salt-evaporator pans (e.g., parts of San Francisco Bay, and Scammon's Lagoon along the west coast of Baja California). In hyperarid settings, all coastal bodies of water may become hypersaline. If the inflow of seawater and freshwater becomes restricted enough, the waters of these saline lagoons can reach saturation with gypsum and halite. In hypersaline estuaries, there is commonly a lateral salinity gradient, with the more concentrated brines, and more soluble minerals, furthest inland.

The Gulf of Kara Bogas is a large ($\sim$3,360 km^2), circular depression up to 20 m deep off the Caspian Sea in Turkmenistan. The Gulf was originally separated from the Caspian by barrier islands cut by inlets and was a perennial saline lagoon fed by water flowing into

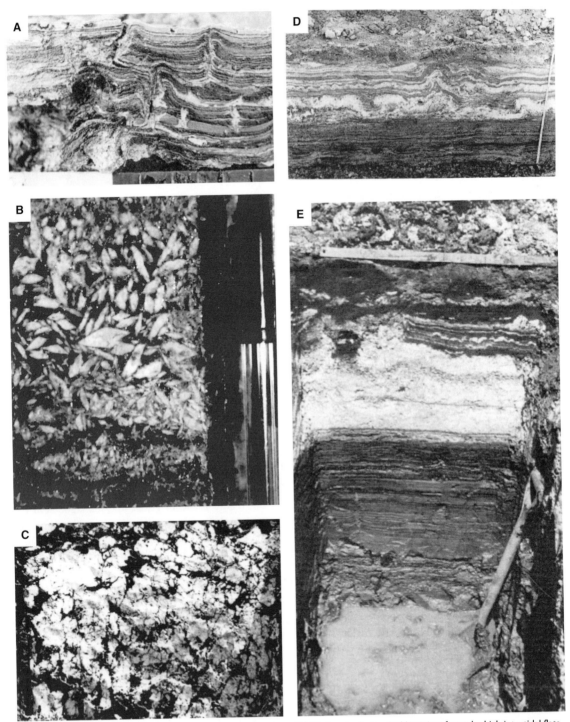

FIGURE 16.22. Deposits of the Arabian Gulf tidal flats and sabkha. (A) Mudcracked microbial laminites from the high intertidal flats surrounding the lagoons. White ticks on the ruler, lower right, are 10 mm. (B) A gypsum-crystal mush shown in an epoxy-resin-embedded core. These displacively grown crystals have the characteristic discoidal or lenticular morphology flattened normal to the c-axis. Some of the crystals engulf bits of the surrounding sediment. The pen cap gives the scale. (C) "Chicken-wire" nodular anhydrite, reckoned to be the dehydrated equivalent of a gypsum-crystal mush similar to that in (B). The machete (left) provides the scale. (D) Complexly deformed layers of nodular anhydrite buckled into tepee structures commonly referred to as enterolithic folding. Note the erosion of the buckled sediment. The bottom of the trench comprises high-intertidal microbial laminites. (E) The trench through the entire shallowing-upward tidal-flat cycle of the Arabian Gulf. The base of the trench filled with brine is in subtidal burrowed pellet mud. The shovel shaft and handle rest against high-intertidal microbial laminite. The white patch is "chicken-wire" anhydrite. The top of the trench below the ruler is layers of folded nodular anhydrite in wind-blown quartz sand.

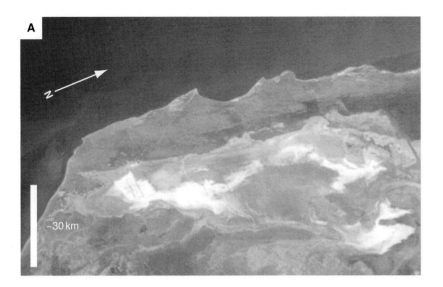

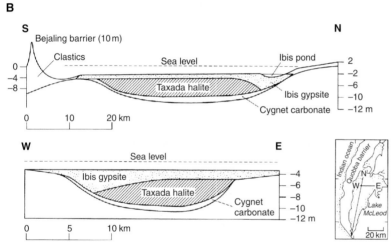

FIGURE 16.23. (A) Lake MacLeod, northwestern Western Australia. Photograph from NASA Earth from Space. (B) Schematic cross sections through the Holocene evaporites filling this former coastal estuary. From Kendall and Harwood (1996).

the Gulf. This water evaporated and deposited a 2-m-thick layer of magnesium sulfate salts. The lagoon was isolated from the Caspian in 1979 by a dam across the inlets, and the Gulf has been turned into an ephemeral saline lagoon.

Lake MacLeod (Figure 16.23) is an elongated, ephemeral, saline coastal lagoon on the northwest coast of Western Australia (Logan, 1987). The surface of Lake Macleod is 2–4 m below sea level, and is underlain by up to 12 m of Holocene gypsum and halite. Lake MacLeod occupies a tectonic depression that was also the drainage course of ephemeral rivers. The western boundary of Lake MacLeod is an anticline ridge that is covered by Pleistocene dunes and is up to 80 m high. The lake is now fed by seepage of seawater as groundwater through

the porous dunes along its western margin, and by continental groundwater and ephemeral overland flow.

Physical depositional processes in coastal saline lagoons are like their counterparts in temperate areas (Chapter 15). Lagoons behind microtidal barriers receive storm overwash of seawater and sediment and tidal exchange through their widely spaced inlets. Waves and wave-induced currents are important. Washover is also important in microtidal barriers but, as the tidal range increases, the importance of migrating tidal channels increases. Supratidal flats in saline lagoons are not covered by salt-marsh grasses or mangroves, and intertidal areas are normally covered by cyanobacterial mats. Evaporite minerals grow in the sediments. Thus, tidal flats around saline lagoons are

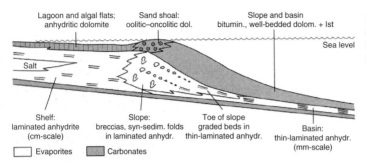

FIGURE 16.24. The morphology of the depositional basin during deposition of the latest Permian Zechstein evaporites. From Schlager (2005).

capped by sabkhas. If the water in the lagoon has a salinity <40‰, a limited number of species of benthic invertebrates can exist, and the plankton is still dominated by dinoflagellates. At salinities higher than this, cyanobaterial mats cover the lagoon floor wherever it is not frequently agitated by currents, and archaea and cyanobacteria are the main pelagic phytoplankton.

Evaporite minerals will be deposited if the lagoon water reaches saturation, and the chemical processes and sediments of saline lagoons depend on whether the lagoons are ephemeral or perennial. Deposits of perennial saline lagoons comprise storm layers of detrital material, commonly carbonate sands or pellet muds derived from the adjacent marine subtidal area, overlain by beds of bottom-growth gypsum (which may be converted into anhydrite) that grew as the fresh flood-waters evaporated and were concentrated. If evaporation in the lagoon proceeds for long enough, halite will nucleate on the surface, sink, and form a layer of chevron halite over the gypsum. Ephemeral marine lagoons undergo a cycle analogous to the saline-pan cycle, and their sedimentary deposits are similar.

Saline coastal lagoons receive the majority of their inflow from the adjacent sea through tidal exchange and storm washover, or as groundwater seepage. However, nearly all saline coastal lagoons also receive a component of meteoric groundwater or surface water. In some cases, this input of freshwater can alter the chemistry of lagoon water enough that the predicted sequence of evaporite minerals for seawater (Chapter 4) is not obtained. In addition, since the composition of seawater has changed throughout the Phanerozoic, the sequence of evaporite minerals precipitated from seawater has also likely changed significantly (Hardie, 1996).

Desiccated shelves and ocean basins: geometry, processes, and deposits

No modern continental shelves and ocean basins are restricted enough to allow salinity to increase to the point at which precipitation of gypsum or halite would occur. However, there are ancient saline deposits in which this seems to have been the case. One of the best known is the Permian Zechstein evaporites found in the subsurface throughout north central Europe (Stewart, 1963; Borchert and Muir, 1964; Braitsch, 1971; Schlager and Bolz, 1977; Schlager, 2005). Schlager and Bolz's (1977) model of Zechstein evaporites includes shelf, slope, and basin facies composed of both sulfates and carbonates (Figure 16.24). The sulfate shelf facies become halite and potash salts in the interior of the platform. Slope facies include breccias and syn-sedimentary folds developed in laminated anhydrite and graded turbidites of anhydrite. The basin facies comprises millimeter-thick laminae of anhydrite. Overlying the sulfates are carbonates with shelf facies of microbially laminated dolomite rich in anhydrite, a shelf-margin ooid shoal, and slope and basin deposits of organic-rich, laminated carbonate.

Marine basins hundreds to thousands of meters deep and saline lagoons hundreds of meters deep, formed in tectonic basins, can become concentrated enough to precipitate saline minerals. The evaporite sediments that accumulate in these settings have many parallels with evaporites that accumulate in perennial saline lakes >100 m deep. Alternating laminae of carbonate and anhydrite from the Permian Castile Formation (Figure 16.25) likely were deposited in a deep, stratified marine basin. Each carbonate–anhydrite couplet represents successive evaporative concentration of pulses of seawater entering the basin and floating over denser brine filling the basin. As the surface water evaporates, and becomes more saline, precipitation of carbonate mud is followed by precipitation of gypsum mud (now converted to anhydrite).

Supposedly "deep-basin" evaporites were recovered from two Deep Sea Drilling Project drill cores from beneath the floor of the Mediterranean Sea in the early 1970s. The gypsum, halite, and potash salts recovered

FIGURE 16.25. Laminated carbonate (dark) and anhydrite (light) from the Castile Formation (Permian) of west Texas and New Mexico. Each layer precipitated at the surface of a stratified brine body as chemogenic mud. The volume/surface area ratio was large enough to keep brine below supersaturation.

in these cores were of Messinian (latest Miocene) age and were interpreted by their discoverers as having been deposited in sabkhas similar to those in the Arabian Gulf (Hsü, 1972a, b; Ryan and Hsü, 1973). Since the evaporites were interbedded with foraminiferal oozes, it was postulated that the Mediterranean had been completely desiccated many times in the late Miocene and repeatedly refilled by gargantuan waterfalls over the Straits of Gibraltar. Hardie and Lowenstein (2004) re-examined the evidence for sabkha deposition of these evaporites. The diagnostic criteria for saline pans, namely bottom-growth crystals riddled with dissolution voids, are nowhere to be found in the Messinian evaporites. Indeed, the halite layers were cumulate textures, simple piles of cubes that have settled through the water column (Figure 16.10). Moreover, the nodular and chicken-wire anhydrite fabrics that were critical in the original interpretations are now known to be equivocal criteria for exposure. Furthermore, the cycles originally described in the Miocene evaporites do not closely resemble the modern deposits of the Arabian Gulf, since they are composed of laminated dolomite alternating with nodular anhydrite. Whereas the tidal-flat microbial laminites of the Arabian Gulf are intensely disrupted by mudcracks, the laminated dolomites are not. The *single* desiccation crack cited as proof of periodic subaerial exposure is a halite-filled fracture cutting halite and detrital silt. The significance of the Miocene evaporites recovered from beneath the Mediterranean is far from settled.

Time variations in arid environments

The areal extent of arid environments varies over time with climate change, and this has been demonstrated clearly with the Quaternary glacial–interglacial cycles. Evidence for such changes is the distribution of inactive (fixed) sand seas (Figure 16.5), the widespread distribution of aeolian silt (loess) across the northern hemisphere, changes in the deposition rate of aeolian silt and clay recorded in deep-sea sediments and ice cores, changes in the water levels, salinity and sediment supply to lakes, and changes in the nature of rivers and floodplains. The timing of such changes during the Quaternary has been established using a host of paleoclimate proxies and dating of sediments using radiocarbon and luminescence methods.

Quaternary evidence suggests that deserts were most widespread, and aeolian activity was at its maximum, during the Last Glacial Maximum (LGM) (about 20,000 years ago). During this period, sea-level and water tables were low, and large areas of freshly deposited, poorly vegetated fluvial sediment were available to be transported by relatively strong winds. Sand seas are expected to be actively growing under these conditions (Kocurek and Lancaster, 1999; Lancaster, 2007). More humid conditions came with the subsequent warming trend. However, there have been shorter-term cycles of aridity and humidity in the Holocene, and desertification is proceeding today in some regions. The change from arid to more humid conditions would be associated with higher groundwater tables and sea level, more vegetation growth, and smaller areas of sediment available for transport. Such conditions might cause sand seas, or parts of them, to become destroyed or inactive (stabilized), with sediment gravity flow and sheet wash of sediment down steep dune slopes, growth of vegetation, and soil development. As the groundwater table and sea level rise with increasing warmth and humidity, interdune areas may become perennial lacustrine environments, and perhaps be flooded by the sea, as was the case in Shark Bay in Western Australia. Rivers may change from ephemeral to perennial and extend across sand seas. Previously ephemeral saline lakes may become perennial, the water level would rise, and there would be an increase in the amount of suspended mud deposited in the lake (Lowenstein *et al.*, 1999). The details of such changes in Earth surface processes from glacials to interglacials would vary from region to region.

The cessation of activity of sand seas, ephemeral lakes, and rivers, and a change to more humid conditions, may be marked by erosional or non-depositional surfaces of regional extent (e.g., Stokes, 1968; Glennie,

1970; Loope, 1984; Talbot, 1985; Kocurek, 1988, 1996; Fryberger, 1993; Blakey *et al.*, 1996), and such surfaces have been called supersurfaces by some workers. These surfaces may result from climate change as discussed above, but could also be brought about by eustatic sea-level change in the absence of climate change, or by regional tectonic activity. They could also be caused by a reduction in sediment supply, leading to widespread wind deflation to the water table. These are the bounding surfaces of so-called sequences in sequence-stratigraphy jargon (Kocurek and Havholm, 1993). Supersurfaces are actually difficult to distinguish from less extensive interdune surfaces that bound cross sets, especially in areas of limited exposure. The kind of evidence that suggests a regional extent related to climate, eustasy or tectonism includes the existence of significantly different deposits above and below; distinct differences in character relative to other bounding surfaces; and correlation with other basin-wide surfaces such as marine flooding surfaces. If wet interdune deposits above supersurfaces are composed of muds and evaporites, they would form permeability barriers.

Large-scale depositional sequences

The two main types of arid depositional regions, as stated above, are (1) tectonically active fault basins, where rocky uplands typically pass laterally into alluvial fans and then into aeolian dunes and/or a playa lake; and (2) tectonically stable cratons, in which alluvial or erosional environments pass laterally into central sand seas (in which dune height increases inwards) and possibly into an inland playa, and large-scale atmospheric circulation patterns may be reflected in paleocurrents. The large-scale depositional sequence depends on these spatial changes in depositional setting, but also on temporal changes in sediment supply, sedimentary facies, and deposition rate related to changes in climate and base level, and on tectonic activity. For example, increasing aridity may cause progradation of sand seas over playa lakes and floodplains, producing an overall coarsening-upward sequence possibly tens to hundreds of meters thick. Decreasing aridity may produce the opposite effect. An increase in sediment supply upstream (or upwind) due to tectonic uplift or a climate change in the hinterland may result in progradation of alluvial fans over sand seas or playa lakes, and of sand seas over playa lakes.

Recognition of deposits of ancient aeolian sand seas is not particularly easy, because aeolian sedimentary structures and textures can look very similar to those of water-lain sediments. However, the wind-ripple laminae within sets of aeolian cross strata are distinctive (Chapter 7). Probably the best way of distinguishing desert aeolian deposits is by their association with other indicators of aridity: evaporites, aridisols, and lack of organic remains (bearing in mind evolution of life). Nevertheless, ancient desert deposits from, for example, the Permian Rotliegendes of northwest Europe (e.g., Glennie, 1986; Clemmensen, 1989; George and Berry, 1993), where ancient sand seas pass laterally into lacustrine clays and evaporites in the center of the basin, have been interpreted convincingly. Jurassic and Permian strata in the southwestern USA constitute other well-studied examples of ancient desert deposits (e.g., Kocurek, 1981; Loope, 1985; Langford and Chan, 1988, 1993; Clemmensen *et al.*, 1989; Crabaugh and Kocurek, 1993; Havholm *et al.*, 1993; Blakey *et al.*, 1996; Mountney and Jagger, 2004; Mountney, 2006). Drying-upward and wetting-upward sequences bounded by regionally extensive bounding surfaces have been interpreted from these examples.

Saline lakes in closed basins are sensitive to changes in the local hydrologic regime and to variable temperatures; therefore, these deposits are an excellent source of paleoclimatic data. For example, during the "climatic optimum" from about 10,000 to ~4,000 years B.P., lakes existed in many of the closed basins of the Sahara and Middle East (Hartmann, 1994). These were home to hippopotami and crocodiles, and served as water sources for hoofed animals. Now, oases in these areas are few and far between. The American Southwest has many well-known closed lake basins, including Death Valley, Owens Valley, Mono Lake, and Great Salt Lake. It was recognized in the 1800s that much larger lakes had filled these basins during the LGM and at other times prior to that (e.g., Lake Bonneville). Old lake levels in these basins can be located by identifying old shorelines and shoreline tufas.

A 185-m-long sediment core from Death Valley, California, provides a record of changes in lake levels, water chemistry, biota, and climate over the last 190,000 years (Figures 16.26 and 16.27) (Lowenstein, 2002). The salts were dated using uranium-series methods, and the sediments were interpreted by comparison with modern analogs described above. The temperature and chemistry of lake water were studied using fluid inclusions in halite crystals. Alternations of terrigenous mud and various types of halite-rich zones were interpreted as due to changes from a perennial lake, via a perennial saline lake, to an ephemeral saline lake (Figure 16.26).

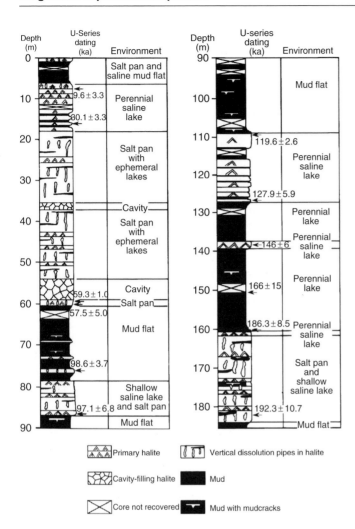

FIGURE 16.26. A stratigraphic column of core drilled in Bad Water Basin, central Death Valley, showing locations of U-series dates and interpreted depositional environments. From Lowenstein (2002).

Tufas some 100 m above the floor of Death Valley suggest that the shorelines were much higher more than 150,000 years ago. Remains of certain ostracods that are adapted to specific ranges of temperature and salinity also indicate climate changes (Figure 16.27). Specifically, those species deposited in lacustrine muds during the LGM are currently found in Alaskan lakes. Although climate clearly affected the nature of the lacustrine deposition, it is unknown how climate change and tectonic activity influenced the supply of sediment and solute from the hinterland, and how tectonic activity affected the depth of the lake.

Ancient lake deposits are also potential sources of ancient climatic data (Cohen, 2003). The Mesozoic rift basins of eastern North America are famous for their fossil fish. Franklin Van Houten of Princeton University was one of the first to describe cycles from the Triassic Lockatong Formation of the Newark Basin in New Jersey (Van Houten, 1964; Olsen, 1986). The laminated, organic-rich, pyrite-bearing shales with the fossil fish were interpreted as deposits of a freshwater to brackish-water lake. These dark shales are encased in siliciclastic and dolomitic mudstones with mudcracks. The mudcracks form complicated, branching networks and are filled with up to three generations of different sediments. Fenestrae filled with analcime or dolomite cements riddle the disrupted mudstones, and some may be pseudomorphs after gypsum or glauberite. Van Houten (1964) interpreted the disrupted mudstones as deep lake deposits that had been subjected to *synaeresis*: chemically induced shrinkage of the colloidal muds due to strong chemical gradients in concentrated lake and pore waters. However, Joseph P. Smoot of the US Geological Survey was the first to identify the disrupted mudstones correctly as the deposits of dry mudflats by analogy with the

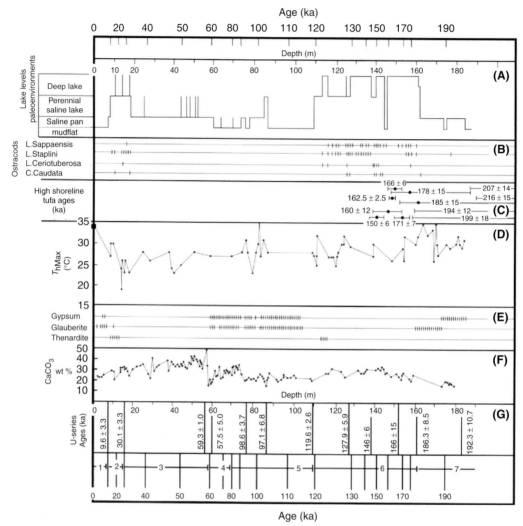

FIGURE 16.27. Various paleoclimate records interpreted from the Death Valley core logged in Figure 16.26. (A) Lake level and paleoenvironments. (B) Ostracods. (C) Uranium-series dates of tufa demarkating shorelines some 100 m above the floor of Death Valley. (D) Maximum paleotemperatures derived from homogenization temperatures of fluid inclusions in halite deposited in perennial and ephemeral paleolakes in the basin. (E) Sulfate minerals. (F) The weight percentage of the > 1-micron-size insoluble fraction. (G) Uranium-series dates of salts, interpolated ages and marine oxygen-isotope stages. From Lowenstein (2002).

desiccation-disrupted soils of dry playa mudflats in the southwest USA. This re-interpretation is generally, but not universally, accepted.

Economic aspects of arid-region deposits

Evaporite deposits have been mined for a variety of commodities, including gypsum (for wall board), halite, borate minerals, and potassium-bearing minerals (important components of fertilizers). Brines from coastal lagoons (such as Scammon's Bay) and from saline lakes (such as the Great Salt Lake) are processed

in evaporating pans to produce halite. The brines feeding modern saline lakes have been pumped for economic elements such as lithium. Salt diapirs create important oil traps in many areas, including the US Gulf Coast and the Middle East. Oil occurs in lacustrine sequences, particularly in China. The oil shales of the Eocene lake deposits of Wyoming constitute vast reserves of oil. Aeolian sands form important oil and gas reservoirs such as the Permian Rotleigendes Sandstone, an important hydrocarbon reservoir beneath the North Sea. Finally, many carbonate-sabkha tidal-flat deposits are hydrocarbon reservoirs.

17 Glacial and periglacial environments

Introduction

Approximately 10% of the Earth's surface is covered with ice at present. Most of this ice is in the Antarctic ice sheets (85% of ice by area), another major ice accumulation is on Greenland (12% of ice by area), and the remainder (3% by area) is in glaciers in mountainous regions. Conditions that allow permanent ice to exist (abundant precipitation as snow, and temperatures low enough to permit the snow to accumulate) occur at high latitudes, high altitudes, and near coasts. In addition to the areas of permanent ice, about 20% of the Earth's surface is permanently frozen (e.g., dry continental interiors at high latitudes such as Siberia, Canada, and Alaska).

In the past, large areas of the Earth's surface were covered in ice, and such *ice ages* have occurred at various times throughout Earth history (Figure 17.1). The last few millions of years of the Neogene ice age have been composed of cold periods (*glacials*) with intervening warmer periods (*interglacials*). During the Last Glacial Maximum (LGM) about 18,000 years ago, up to 30% of the Earth's surface was covered in ice, but the Earth is now in an interglacial period. The glacial–interglacial alternations of the past several million years, particularly the latest one, have left a remarkable record on the Earth's surface, and have affected human activities in many ways, perhaps even influencing our evolution. Present-day glacial processes are under intense scrutiny in order to understand them more fully (especially in the context of global warming), and to allow more confident interpretation of ancient glacial landforms and deposits. Ancient glacial deposits in particular are important because they record climate change, and are important sources of sand and gravel, groundwater, and even hydrocarbons.

Modern and Pleistocene glacial and periglacial environments are considered in this chapter. The chapter ends with a review of the most recent (Neogene) ice age

and the causes of ice ages. Relatively recent reviews of glacial environments and deposits, modern and ancient, include Anderson and Ashley (1991), Ashley *et al.* (1985), Benn and Evans (1998), Brodzikowski and Van Loon (1991), Dawson (1992), Dowdeswell and O Cofaigh (2002), Dowdeswell and Scourse (1990), Evans (2003), Eyles and Eyles (1992), Hambrey (1994), Knight (2006), Martini *et al.* (2001), Menzies (1995, 2002), and Miller (1996). Reviews of pre-Pleistocene glacial environments are Hambrey and Harland (1981) and Menzies (1996). Glaciotectonism is reviewed by Aber *et al.* (1989) and Van Der Wateren (2002).

Glacial environments

The main types of glaciers are *ice sheets*, *valley glaciers*, and *piedmont glaciers* (Figure 17.2). Ice sheets such as those that cover Antarctica and Greenland are kilometers thick, blanket large land areas, and include relatively fast-moving *ice streams* and coastal *ice shelves*. Valley glaciers flow through valleys, but may be restricted to small upland basins called *cirques*. Where valley glaciers meet the sea, they are called tidewater glaciers. Piedmont glaciers occur at the base of mountain ranges, where lobes of ice fed by valley glaciers extend up to tens of kilometers from the mountain front. Areas of land and sea standing adjacent to these various types of glaciers are strongly affected by glacial processes. Glacial environments under, next to, and beyond the ice are referred to as *subglacial*, *ice-contact* (or *ice-marginal*), and *proglacial*, respectively. Glacial environments also include meltwater streams and lakes within and under the ice.

Moving ice modifies the existing topography rather than creating a new one. Upland glaciated areas are dominated by erosion, whereas glacial deposits are concentrated in lowland areas, lakes, and the sea, as is **595**

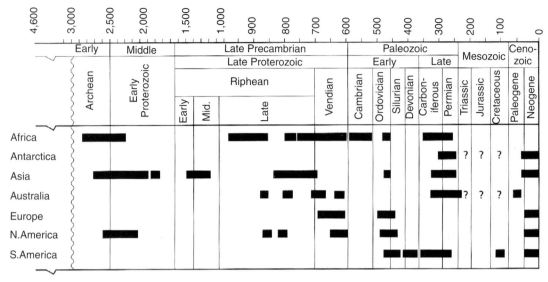

FIGURE 17.1. Ice ages on the Earth through time. From Eyles and Eyles (1992).

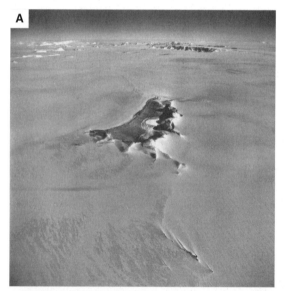

FIGURE 17.2. (A) An ice sheet with protruding mountain peaks (nunataks), Antarctica. (B) A Piedmont glacier with distorted sediment bands, Malaspina Glacier, St. Elias Mountains. (C) A valley glacier with terminal and lateral moraines, moraine-dammed lake, and outwash fan, Pattullo Glacier, British Columbia. All photos from Post and LaChapelle (2000).

the case in unglaciated regions. Glacial depositional environments can be subdivided into those under the ice, those next to the ice, and those beyond the ice, as above. Sediments are deposited directly from the ice, from sediment gravity flows, and from meltwater flows, and are commonly reworked by the wind. Also, these different types of deposits vary depending on whether the environment is on land, lacustrine, or marine. The preservation potential of glacial deposits is greatest in topographically low areas that are actively subsiding, particularly in lakes and the sea.

Glacial environments therefore involve the processes, landforms, and sediments associated not only with moving ice (as discussed in Chapter 10), but also with moving water, air, and sediment–fluid mixtures (as discussed in Chapters 5, 6, 7, and 8). Furthermore,

glacial environments are intimately related to fluvial, eolian, lacustrine, and marine environments (as discussed in Chapters 13, 14, 15, 16, and 18).

Environments dominated by glacial erosion

Small-scale erosional features (e.g., glacial striations) were discussed in Chapter 10. The larger-scale erosional forms described here were eroded by ice or meltwater, or both.

Roches moutonnées

Roche moutonnée is the name given to an asymmetrical ridge of rock produced by glacial erosion (Figure 17.3A).

FIGURE 17.3. Glacial erosional features. (A) Roche moutonnée from Serrada Estrela, Portugal. Photo from Goncalo Vieira. Ice flow is to the left. (B) Cirques in the Scottish Highlands. Photo from Richard Kesel. (C) The Matterhorn, Switzerland. (D) Cirques and arêtes.

These landforms are a few meters high and tens of meters long, and their size has been related to the spacing of planes of weakness (e.g., joints) in bedrock. They have a relatively low-angle, smooth, striated side that faces up ice flow, which is ascribed to glacial abrasion. A relatively steep, jagged side faces down ice flow, and is related to glacial plucking. Roches moutonnées are therefore used as indicators of the direction of ice flow.

Crag-and-tail features and rock drumlins

Knobs of rock (crags) that are relatively resistant to erosion may protrude above the general land surface, whereas softer surrounding rocks are eroded. In some cases, sediment (till) is deposited on the lee side of the crag (the tail). A famous example of a crag is Edinburgh Castle rock. A crag may become completely covered with till, forming a rock drumlin (a drumlin with a core of solid rock). Some glacial flutes (see below) are like very elongated rock drumlins or small-scale crag-and-tail features (Boulton, 1976; Benn, 1994). These features are also used to indicate the direction of ice flow.

Cirques, arêtes, and horns

A *cirque* (also known as a corrie or cwm in the UK) is a steep-walled hollow, shaped like a half bowl, in a mountain side (Figure 17.3). Cirques form beneath a patch of snow and ice, just above the snowline. Rock beneath the patch is preferentially eroded by ice wedging and flowing meltwater, leading to deepening of the cirque. A lip of rock occurs at the down-valley edge of the cirque. Downslope movement of the cirque glacier results in further erosion of the base of the cirque by plucking and abrasion. The lip of the cirque may also be eroded, producing striations and roches moutonnées. Cirques formed in glacial periods commonly contain lakes (called *tarns*) today. Juxtaposition of several adjacent cirques in glaciated mountains produces characteristic jagged peaks and ridges, called *horns* and *arêtes*, respectively (Figure 17.3). Well-known examples of horns are the Matterhorn in the Alps, and Mount Everest in the Himalayas.

Valleys modified by glacial erosion

Valleys that are modified by glacial erosion are typically U-shaped in cross section (Figure 17.4A). This is because a glacier that fills a valley erodes along its sides as well as its base. Numerical modeling of the formation of U-shaped valleys predicts that their characteristic parabolic shape can be formed within a few thousand years (Harbor, 1992). The erosion rate increases as the discharge and slope of the ice increase. A characteristic feature of such valleys is truncation of the interfluves of tributaries by the main valley glacier, forming truncated spurs with smoothed and striated surfaces (Figure 17.4A). The floors of the tributary valleys are commonly significantly higher than those of the main valley because the main-valley glacier could erode deeper. These tributary valleys are called hanging valleys, and commonly have waterfalls where they join the main valley (Figure 17.4B). Headward migration of such waterfalls produces gorges that are occupied by the tributary streams.

The long profiles of glacially eroded valleys commonly undulate, with relatively deep and shallow regions. This is mainly related to variability in erosive power of the glacier, and differential rock hardness is probably a secondary factor. Long, deep basins may well be related to surging glaciers or ice streams within ice sheets. Such basins may be occupied subsequently by lakes or fiords (if at sea level). The Finger Lakes in New York State are glacially deepened parts of valleys that are tens of kilometers long and hundreds of meters deep (Figure 17.4C). The Finger Lakes are associated with classic examples of hanging tributary valleys, waterfalls and gorges (Figure 17.4D). The floors of fiords can also be hundreds of meters below sea level. Glaciers can easily erode below sea level because they will not start to float until 90% of their thickness is under water. However, the present-day depth of fiords is also related to more than 100 m of sea-level rise and isostatic uplift due to glacial melting and unloading, in addition to the original depth of glacial erosion (see below).

Subglacial erosion by glacial meltwater

Subglacial meltwater flow, in some cases associated with catastrophic floods, is capable of eroding bedrock and basal till, and producing characteristic features that occur in the same regions as roches moutonnées, crag-and-tail features, and rock drumlins. Indeed, it is not always easy to distinguish the relative roles of ice and meltwater erosion. Typical features of meltwater erosion are smooth, rounded bedrock surfaces, potholes, crescentic scours around bedrock obstacles, and channels (or tunnel valleys) of various sizes (e.g., Kor *et al.*, 1991; Hooke and Jennings, 2006). Subglacial channels range in width from tens to hundreds of meters, in depth from meters to tens of meters, and in length up to tens of kilometers. They may be sinuous, have bases that vary greatly in elevation along their

FIGURE 17.4. Glacial erosional features: (A) a U-shaped valley with truncated spurs, Denali, Alaska; (B) a hanging valley with waterfall, Resurrection Bay, Alaska; (C) the Finger Lakes of New York State (photo from NASA); and (D) Watkins Glen, a gorge with waterfalls formed by headward erosion of a tributary valley, Finger Lakes region of New York State.

length, and be oriented in directions unrelated to the regional topographic slope, and commonly have poorly developed drainage systems (Kor *et al.*, 1991). Some of these channels cut across only the high parts of bedrock protrusions such as rock drumlins or valley-margin spurs. The channels are partly or completely filled with glaciofluvial sediment, and may even be completely buried. The characteristics of subglacial channels suggest that flow in the channels is not merely due to downslope gravity forces, so additional hydro-static pressures must be involved. The channel size and discharge vary in time and space with temperature, ice

movement, and filling with sediment. These channels are most likely to occur, and be preserved, under con-ditions of stagnant and retreating ice.

Some proglacial lakes are formed by damming of meltwater in tributary valleys by a main-valley glacier or by a glacial deposit (moraine). Lake levels may rise high enough to overtop the interfluves between adja-cent tributaries, leading to cutting of drainage channels through the lowest parts of interfluves (V-shaped notches). Levels of ancient proglacial lakes are indi-cated by these overflow channels plus shoreline terraces and tops of lake-head deltas.

Depositional environments under and next to ice

Ground moraines

Ground moraine is the widespread subglacial traction till (also known as lodgement-deformation till) that is deposited at the base of a glacier. Ground moraine from the latest Pleistocene ice sheets covers large swaths of northern North America and northern Eurasia, particularly Scandinavia and the UK. Ground moraine has a smooth upper surface consisting of gently sloping mounds and depressions, and is typically meters to tens of meters thick. Thickness variations may be related to variable distribution of sediment in the base of the glacier, differential deposition rate, or post-depositional remolding by ice and water flow. The upper surface of ground moraine contains distinctive land-forms such as drumlins, flutes, and Rogen. Ridges of sediment deposited on the lee side of obstacles, such as bedrock knobs and boulders (e.g., crag-and-tail features, and some fluted moraine and drumlins), are also common features of ground moraine.

Drumlins are streamlined hills of ground moraine with the shape of an inverted spoon. The blunt end of the drumlin faces upstream, whereas the pointed end faces downstream (Figure 17.5). Drumlins are typically tens of meters high and wide, and hundreds of meters to kilometers long. Groups of drumlins produce "basket-of-eggs" topography. Some drumlins are made entirely of till whereas others have bedrock cores. The origin of drumlins is controversial (e.g., Menzies and Rose, 1989; Eyles and Eyles, 1992). They have been ascribed to glacial molding of pre-existing till, plastering of till over bedrock obstacles, spatial variations in ice flow and sediment transport, filling of hollows in the base of the ice that were produced by erosion by meltwater, and erosion of pre-existing sediment by meltwater (e.g., Boulton, 1987; Shaw and Sharpe, 1987; Shaw *et al.*, 1989; Boyce and Eyles, 1991; Hindmarsh, 1998; Knight, 2006). *Fluted moraine* resembles very elongated drumlins. The flutes are typically up to 2 m high, 1–3 m wide, and hundreds of meters long, and have a trans-verse spacing of 0.5–1.5 km. They are composed of a large range of sediment types and bedrock, as with drumlins, and similar origins to those of drumlins have been proposed.

Rogen moraines (also called *ribbed moraines*) are repetitive ridges, elongated transverse to the direction of ice flow, that occur over areas formerly occupied by ice ranging from a few km² to a few thousand km²

(Figure 17.5C). Individual ridges range in height from meters to tens of meters, and have wave lengths of tens to hundreds of meters and lateral extents of hundreds to thousands of meters (Dunlop and Clarke, 2006). In plan, they tend to be arcuate and concave down ice (downstream pointing horns), and anastomosing. However, they are not always oriented this way, and actually have a large range of plan form. Although they normally have steeper lee sides than upstream sides, this is not always the case. Crestlines are commonly undulating and discontinuous. Rogen may pass upstream, downstream, and laterally into drumlins and flutes, but such transitions are not universal. Drumlins can actually occur on the Rogen moraines. Many origins have been proposed for Rogen moraines (e.g., Lundqvist, 1989, 1997; Hättestrand and Kleman, 1999; Hindmarsh *et al.*, 2003; Moller, 2006): (1) thrust stacking of sediment-rich basal ice followed by stagnation and melt-out; (2) subglacial megaflood deposits; (3) fracturing and extension of a frozen sediment bed into jigsaw-like segments separated by spaces; and (4) remolding of existing sediment ridges. Moller (2006) recently dismissed the view of Hättestrand and Kleman (1999) that Rogen are the remnants of horizontally extended, fractured, frozen subglacial sediments because of the lack of morphological, sedimentary and structural evidence, and doubts about the theoretical mechanism. Instead, Moller (2006) attributed their origin to glacial reshaping of transverse moraines that were originally ice-cored marginal moraines. Inter-ridge troughs were filled with deposits of debris flows and meltwater streams. As the ice cores melted, topography was inverted, producing linear ridges and hummocks. Hindmarsh *et al.* (2003) view drumlins and Rogen moraines as part of a continuum of ice-molded forms, and performed a stability analysis of the interface between basal ice and sheared wet sediment. Deformation of this surface into wavy forms depends on the viscosities of ice and sediment, shear stresses and fluid pressures, and the thickness of the ice and deforming sediment. This theory is considered to have promise and can predict approximately correct wave lengths of the various subglacial depositional bed forms described above.

Linear subglacial moraines are ridges ornamenting ground moraine that have been associated with the margins of ice streams (Stokes and Clark, 2001, 2002, 2003). The moraines are tens of kilometers long, hundreds of meters wide, and tens of meters high. The sediment is thought to come from erosion

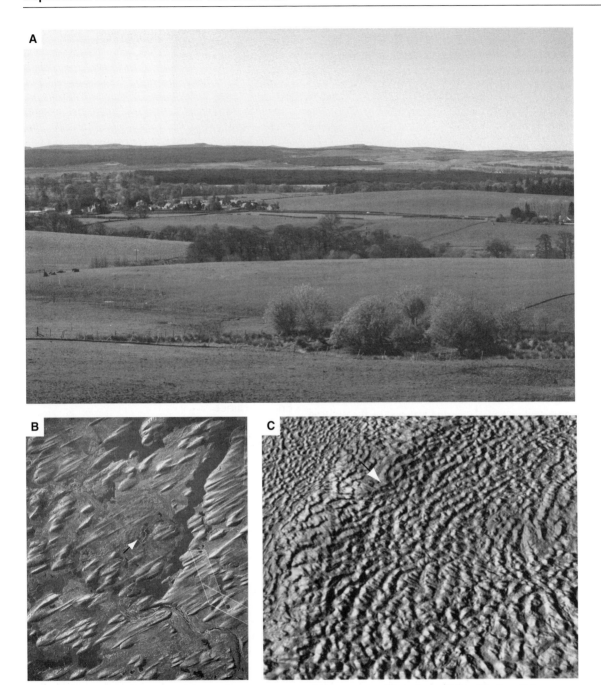

FIGURE 17.5. Drumlins at Croftamie, near Loch Lomond, Scotland. Photo courtesy of Dave Evans. (B) Drumlins from Livingstone Lake, Saskatchewan, Canada. From Benn and Evans (1998). The width of view is about 13 km. (C) A digital elevation map of Rogen moraine from Co. Monaghan, Ireland. From Clark and Meehan (2001). The field of view is approximately 20 km across in the foreground. Arrows indicate the ice-flow direction.

at the ice-stream margin that is smeared out by the shearing process. Erosional troughs are formed underneath the ice streams, but these are covered with deformation till and streamlined bed forms. Such erosional–depositional features also formed on the continental shelf when it was covered in ice during the LGM (Dowdeswell and O Cofaigh, 2002). It is likely that such linear moraines could be confused with the "medial moraines" formed at the boundaries of adjacent valley glaciers (see below).

FIGURE 17.6. Subglacial till in cross section through a drumlin, New York State.

Subglacial traction till typically rests on a regional erosion surface and is sheet-like. Internally, this till is composed of overlapping, stacked lenticular bodies with erosional bases (Eyles and Eyles, 1992; Menzies, 2002; Evans *et al.*, 2006); see Chapter 10. This is probably an indication of localized erosion and deposition during a single glacial period rather than the result of multiple glaciations. The erosional bases may contain groove marks and gouges. The till tends to be a high-density diamicton, generally unsorted and structureless but with sorted, stratified layers, and may, but need not, be fissile (Figure 17.6). Till characteristics apparently vary depending on the relative amount of ice sliding on meltwater and shearing of the wet subglacial sediment, which varies in time and space (Evans *et al.*, 2006). If meltwater was abundant, some of the deposited sediment may be size-sorted and stratified by the flowing water, whereas if basal sediment shearing was more important the sediment would contain shear planes (possibly polished and striated), oriented grains (with the long axes of platy grains parallel to the bed, and imbricated clusters in places), large grains with plough structures and fine-grained tails, and possibly thrusts and associated recumbent folds (Chapter 10). Striated and faceted clasts are common. Tectonic slices (sediment rafts) emplaced from the basal ice may contain stratification as well as deformation structures. Temporal variations in the amount of meltwater can give rise to alternations of structureless, unsorted diamicton and sorted, stratified sediments (Chapter 10; Figure 17.6). Deposits of subglacial meltwater streams are likely to be interbedded with, and cut into, subglacial traction till (Eyles *et al.*, 1982). Subglacial traction

till may be overlain by melt-out till and debris-flow deposits (end moraines – see below), or by deposits of outwash plains, lakes, or the sea (see Figures 17.8 and 17.20 later).

End moraines

End moraines are interconnected mounds of till, debris-flow deposits, and meltwater deposits forming irregular ridges along the margin of a glacier. End moraines of valley and piedmont glaciers range in height from a few meters to tens of meters, are up to kilometers long and wide, and may be responsible for damming lakes or diverting river courses. End moraines of continental ice sheets are up to hundreds of meters high and kilometers across, and form arcuate landforms hundreds to thousands of kilometers long. These comprise some of the most distinctive topography of the northern USA. End moraines are composed partly of melt-out (ablation) till that is deposited at the margins of glaciers as they melt. There are terminal, lateral, and medial parts of valley and piedmont glacial end moraines (Figure 17.7). Lateral and medial moraines owe their origin to the high concentration of sediment at the margins of glaciers, and to continued (conveyor-belt-like) deposition of this sediment during ice movement, and can easily be confused with shear-margin moraines. The mound-like geometry of terminal moraines can be due to deposition of a large concentration of sediment in stagnant ice (dead-ice moraine), repeated deposition of sediment at the stationary terminus of a stagnant glacier (like a conveyor belt), and/or the bulldozing action at the front of a

A

B

FIGURE 17.7. End moraines: (A) lateral and terminal moraine of Gornergrat, Switzerland (photo courtesy of Henk Berendsen); and (B) terminal moraine and outwash plain cut by a modern river, Denali, Alaska.

glacier (a push moraine: see Van der Wateren (2002)). A number of terminal moraines may occur in a valley as a glacier retreats episodically (recessional moraines). However, distinct terminal moraines in a valley may also be caused by different glacial periods. For continental ice sheets, the terminal and recessional moraines are used to document the deglaciation history. Gravity flow of waterlogged ablation tills or of the sediment on the steep front of the glacier produces debris-flow deposits (e.g., Boulton, 1972; Lawson, 1982; Paul and Eyles, 1990) that tend to overlie melt-out till or subglacial traction till (Figure 17.8). Kettle lakes associated with end moraines are formed by deposition of till or outwash around a stagnant ice mass, leaving a water-filled hollow as the ice melts.

Melt-out tills are commonly meters thick and are mainly diamictons that are neither as compacted as lodgement-deformation till, nor as deformed (Chapter 10). Although melt-out till can have a random grain fabric, it can also have layers of silt, sand, and clay with scattered cobbles and boulders. Layered melt-out tills can originate from layered, sediment-laden basal ice, and the layers are preserved if the ice is confined as it melts, possibly by a covering of debris-flow deposits. Time variation of melting also produces layering in melt-out tills. Melt-out tills can be difficult to distinguish from lodgement tills that do not have deformation structures. Push moraines have shearing and deformation (Figure 17.8B). Debris-flow deposits originating from till look like any other sediment-gravity-flow deposit, and establishment of a glacial origin requires careful examination (e.g., regarding striated rock fragments, association with lodgement till, or ice-contact stratified deposits; see below). In view of the abundance of meltwater associated with end moraines, interbedding of stratified, water-lain deposits is

A

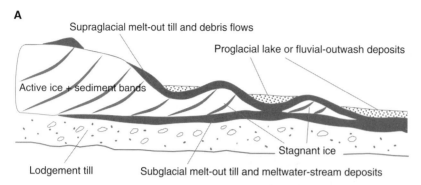

Supraglacial melt-out till and debris flows

Proglacial lake or fluvial-outwash deposits

Active ice + sediment bands

Lodgement till

Subglacial melt-out till and meltwater-stream deposits

Stagnant ice

FIGURE 17.8. (A) End-moraine deposits. Modified from Boulton (1972). The deposits become deformed as the stagnant ice melts. (B) Push-moraine deposits (deformed fluvial deposits) near Hattem, Veluwe, The Netherlands. Photo courtesy of Henk Berendsen.

B

common, as is soft-sediment deformation. Deposition of end moraines over or next to stagnant ice can also give rise to soft-sediment deformation structures (e.g., normal faults) as the ice melts. End-moraine deposits commonly overlie ground-moraine deposits (Figure 17.8).

Deposits of meltwater streams

Meltwater streams flow under, within, and on top of the ice. Deposition from meltwater streams occurs where they decelerate in space, and particularly as subglacial streams emerge at the front of the ice. Deposits of subglacial streams cut into the base of the ice form sinuous ridges called *eskers* (Figure 17.9). Roads and railway lines are commonly constructed along the crests of eskers. Eskers are typically tens of meters high, hundreds of meters across, and kilometers to tens of kilometers long, and are oriented parallel to the direction of ice flow. Eskers are best developed within a few kilometers of

an ice front, in the ablation zone. They are considered to occur mainly during stagnation and retreat of ice. *Beaded eskers* are produced by deposition of a succession of overlapping alluvial fans at an ice front that is retreating. The variety of esker patterns is reviewed by Warren and Ashley (1994) and Menzies (2002). In order to distinguish eskers from medial and shear-margin moraines, it is necessary to examine the nature of their deposits (Banerjee and McDonald, 1975); see below.

Meltwater stream deposits along the lateral margins of valley glaciers form so-called *kame terraces*. The term *kame* is old and well known, but is not explicit, so its continued use is not recommended. These ice-contact terraces are flat-topped with a steep deformed edge where the ice used to be (Figure 17.9). They can be confused easily with ordinary river terraces if internal deposits are not examined carefully. If a meltwater stream flows from the ice into a lake or the sea, an ice-contact delta occurs at the glacier margin. Such

A

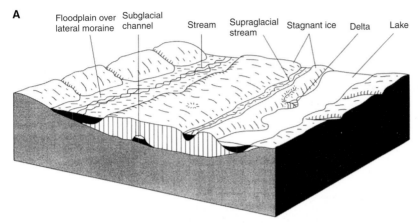

B

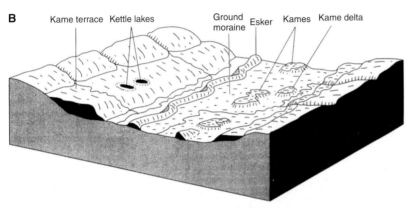

FIGURE 17.9. Ice-contact and subglacial landforms and deposits. (A) Stagnant-ice conditions. (B) After the glacier has retreated. Images (A) and (B) from Allen (1997), after Flint (1971). (C) Kame terraces at the margin of Tungnaarjokull, Iceland (photo courtesy of Dave Evans). (D) Esker ridges in Manitoba (photo courtesy of Henk Berendsen).

C

deposits used to be called *delta kames* (Rust and Romanelli, 1975; McCabe and Eyles, 1988). The upstream margins of ice-contact deltas are deformed as the ice melts. These deltas have flat tops and can be tens of meters high and hundreds of meters long and wide. Proglacial lakes that are formed by a main-valley glacier damming a tributary valley may have a delta at the mouth of the tributary river rather than next to the ice. The lacustrine deposits (see below) may be immediately adjacent to ice-contact terrace deposits associated with the main-valley glacier.

Deposits of meltwater streams are essentially the same as those of other river-channel deposits: sands and gravels with various scales and types of stratification

D

FIGURE 17.9 Continued.

(Chapter 13; Figure 17.10). However, if sediment is deposited next to and on top of the ice and then the ice melts, soft-sediment deformation structures form. Typically, the edge of the sediment next to the ice is let down by a series of normal faults (Figure 17.10). Meltwater-stream deposits are also likely to be associated with till and lacustrine deposits.

Many subglacial landforms and deposits, such as drumlins, ribbed moraines, fluted moraine, and channels have been interpreted as formed by subglacial meltwater megafloods rather than by ice movement (Shaw, 2006). Benn and Evans (2006) and Evans et al. (2006) discuss these interpretations at length, especially with respect to examples in Alberta, Canada, and propose revised interpretations that are more in line with what is known about processes beneath modern glaciers. This remains a controversial issue.

Time variations in the rate and type of deposition from ice and meltwater might be related to seasonal melting and freezing or to longer-term changes associated with surging glaciers and ice streams. For example, ice surges are expected to be associated with increased bed erosion and deformation, increased marginal deposition and deformation of end moraines, and large increases in meltwater discharge, sediment transport, and deposition. Repetition of these events produces stacked sequences of diamicton interbedded with stratified sediments.

FIGURE 17.10. A cross section through an esker (Lisburn, Northern Ireland), showing stratified sand and gravel, and marginal down-faulted blocks (arrow).

Depositional environments beyond ice

Outwash plains and deposits

Specific articles dealing with glaciofluvial environments are Jopling and McDonald (1975), Ashley *et al.* (1985), and Maizels (1989, 2002). Outwash plains or fans are the river-dominated areas downstream of ice margins (Figure 17.11). They are called *sandurs* (or *sandar*) in Iceland. Water discharge on outwash plains and fans varies with seasonal and diurnal melting and refreezing of ice, with summer rainstorms, with river diversions, and with bursting of ice-dammed lakes (*jokulhlaups*, which is Icelandic for "glacier bursts": see Bjornsson (1992)). During the initial stages of seasonal break-up of river ice, water flows on and around the ice. This stage is followed by a rapid increase (over several days) in water discharge: the snowmelt flood. Then the discharge gradually wanes during the summer, punctuated with minor discharge peaks associated with rainstorms (Figure 17.12A). Diurnal variation in water discharge may occur, especially in the summer (Figure 17.12B). Jokulhlaup flood hydrographs have a characteristic gradual rise to a sharp peak followed by a relatively steep decline (Figure 17.12C).

The annual sediment yield of rivers fed by glaciers is orders of magnitude greater than that of non-glacial rivers. Sediment discharge may well vary in time incongruently with meltwater discharge, because it is controlled by time variations in sediment supply within the glacier. By virtue of the high peak discharges of water and sediment, rivers on outwash plains and fans are typically braided and anastomosing, and river avulsion is frequent. The bed slope and mean grain size of bed sediment decrease, and channels tend to become less braided, in the downstream direction. Deposition of outwash sediment around blocks of stagnant ice produces hollows that might later become kettle lakes.

Outwash-plain deposits (Figure 17.13) are typically stratified sands and gravels with stratasets typical of braided rivers (see Chapter 13). The evidence in these deposits for the cold environment is kettle-lake deposits, scours around (since-melted) ice blocks, evidence for diurnal variations in water discharge, and deposits associated with catastrophic floods (jokulhlaups). Diurnal variations in discharge might be associated with repetitive fine-grained drapes on sandy–gravelly cross-stratal surfaces (Ashley *et al.*, 1985). Catastrophic floods may be associated with exceptionally coarse deposits with thick sets of strata associated with very deep channels and very large dunes. However, Smith *et al.* (2006) showed, using a series of high-resolution, airborne LIDAR surveys prior to and following a jokulhlaup, that, although a large amount of erosion and deposition occurred during the flood, subsequent

FIGURE 17.11. The outwash plain associated with a Greenland glacier (photo from Henk Berendsen).

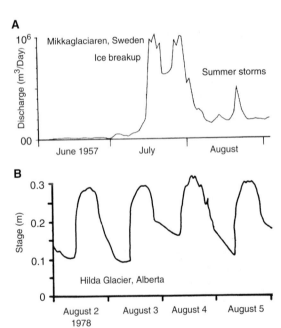

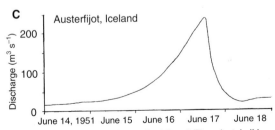

FIGURE 17.12. Ice-melt hydrographs, (A) and (B), and a jokulhlaup hydrograph (C). Modified from Ashley *et al.* (1985).

minor floods removed much of the depositional evidence. A well-known example of the results of a mega-megaflood is the Channelized Scablands of the Columbia Plateau (east Washington State) resulting from the bursting of an ice dam and catastrophic draining of glacial lake Missoula in western Montana (Baker, 1973, 1990) (Figure 17.14). Peak discharges during such catastrophic floods may have been on the order of tens of millions of cumecs, which is two orders of magnitude greater than the annual discharge of the world's largest river (the Amazon). Outwash-plain deposits typically occur downstream of terminal moraines and become finer-grained downstream. Glacial advance and retreat may result in outwash-plain deposits on top of till (see below). Apparently, outwash-plain deposits are likely to be eroded by the ice during glacial re-advance.

Loess blankets and sand seas

Wind blowing across unvegetated moraines and outwash plains is very effective at entraining and transporting sand and silt. Wind directions are typically away from the glaciers because the air pressure over cold regions is normally high. Such *katabatic* winds occur before, during, and after glaciations. Sand is moved as sand dunes, and the Nebraska Sandhills is one of the biggest ancient sand seas in the world associated with the deglacial period of the LGM (Ahlbrandt and Fryberger, 1980). Silt suspended by the wind is deposited as well-sorted, yellow (oxidized) silt called *loess* (Figure 6.3). Loess associated with the last ice age blankets much of the northern parts of North America, Europe, and Asia. Its thickness ranges

FIGURE 17.13. Outwash-plain deposits from western Canada.

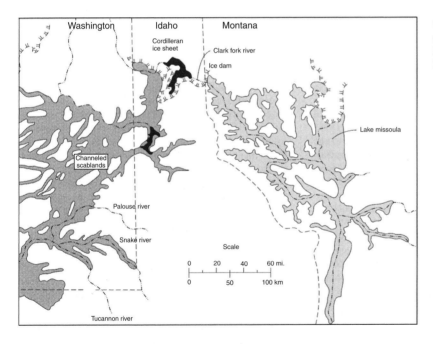

FIGURE 17.14. A map of channeled scablands and glacial Lake Missoula. From Easterbrook (1999).

from 1–2 m to tens or hundreds of meters. Loess is very fertile, and has had a major impact on agricultural success.

Glaciolacustrine and glaciomarine environments and deposits

Reviews and compendia of glaciolacustrine and glaciomarine environments have been published by Anderson and Ashley (1991), Ashley (2002), Dowdeswell and

O Cofaigh (2002), Dowdeswell and Scourse (1990), Eyles *et al.* (1985), and Molnia (1983). Fiords are discussed by Domack and Ishman (1993), King (1993), Powell and Molnia (1989), and Syvitski *et al.* (1987). Many glaciers flow directly into lakes or the sea. Proglacial lakes also form as ice and/or moraines dam peripheral drainages. Water bodies immediately adjacent to continental ice sheets may be due to depression of the crust by the weight of the ice (see below). As the ice sheets melt, the crust rebounds while sea level rises,

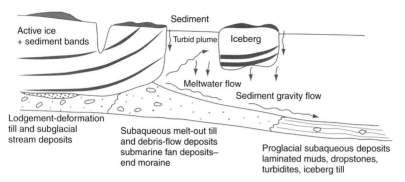

FIGURE 17.15. The ice margin at a sea or lake, showing flow and sedimentary processes and deposits. Modified from Powell (1990).

leading to complicated alternations of marine, lacustrine, and terrestrial conditions next to retreating glaciers. As glaciers or ice sheets flow into standing bodies of water, the ice closest to the land remains grounded on the bed but the more distal part floats. The floating ice is in the form of a tapering wedge (thinning towards the water body), due to the effects of melting (Figure 10.8). However, if the rate of iceberg calving exceeds that of melting, the cliff at the distal edge of the ice may retreat to the grounding line (Figure 17.15). The nature of incoming water flow and sediment transport in the lake or sea depends on the sediment supply, the difference in density between the incoming water and the basin water (which is dependent on differences in temperature, salinity, and concentration of suspended sediment), and subsequent reworking by basin currents. Lakes deeper than about 10 m have thermal layering of water, with a relatively warm *epilimnion* above a colder *hypolimnion*, separated by a relatively sharp boundary (*thermocline*). This layering varies seasonally and is most stable in summer, but is disrupted by wind-induced currents that mix the water body. Wind-induced currents affect the bed (reworking deposits) only in very shallow lakes and in shallow water near the shore of deeper lakes. In marine-shelf environments, thermal layering does not occur, and shallow-marine deposits are reworked by tidal currents, wave currents, and wind-shear currents.

Very high deposition rates from ice occur at the ice margin because water quickly melts ice. More sediment is deposited from temperate glaciers (up to meters per year) than from polar glaciers (millimeters to centimeters per year) because there is more melting. Furthermore, higher deposition rates occur when ice streams flow into standing water. Ground moraines and end moraines are formed beneath grounded ice (including grounded icebergs) in these subaqueous environments, with end moraines occurring near the grounding line (Figure 17.15). The moraines are composed of subglacial traction till and melt-out till. Melt-out till is also deposited into the water from all sides of floating ice and icebergs (Chapter 10; Figures 10.8 and 17.15). Subglacial traction till and melt-out till in glaciolacustrine and glaciomarine environments are similar in character to those in terrestrial environments. However, sediment-gravity deposits are expected to be more common because of the abundance of water. Fossils are well preserved in melt-out till and sediment-gravity-flow deposits. Individual rock fragments that fall from melting icebergs (*dropstones or lonestones*; Chapter 10) are normally associated with localized melt-out till deposited from icebergs, and such till will be interbedded with other types of lacustrine or marine deposits. These deposits may contain long and straight erosional scours and evidence of turbation caused by bases of icebergs being dragged along the bed (Dowdeswell *et al.*, 1994).

Meltwater streams flowing into lakes and the sea may become overflows, interflows, or underflows, depending on their point of entry from the ice and the relative density of the meltwater and the basin water. In glaciolacustrine environments, seasonal variations in water temperature, thermal layering, and suspended-sediment concentration are important controls on sediment transport and deposition. Salinity differences are also important in glaciomarine environments.

Subglacial meltwater streams deposit their bed load immediately upon appearing from the ice, giving rise to submarine fans or deltas. Such fans may have avalanche faces (i.e., Gilbert-type deltas: Figure 15.6). Debris flows and turbidity currents transport sediment episodically down the front of deltas. Such sediment gravity flows are commonly associated with slump scars, channels, and compressional hummocks and depositional lobes at the base of the slope (Chapter 8). Episodes of sediment-gravity-flow activity at any point on the delta front may be associated with seasonal

FIGURE 17.16. Varves from southeast Iceland (courtesy of Dave Evans).

variations in meltwater flow and sediment transport, or with channel switching on the delta top. More continuous flow of sediment gravity flows occurs if breaching occurs (Chapter 8). Gilbert-type deltas have *topsets* associated with channel migration; *foresets* associated with slump scars, sediment gravity flows down the delta front, and compressional hummocks; and *bottomsets* associated with distal sediment gravity flows and sediment deposited by settling from plumes (Figure 15.6). Submarine fans or deltas need not necessarily have angle-of-repose foresets, but otherwise similar types of facies occur.

Suspended sediment from meltwater streams is transported in plumes of muddy water away from the mouth of the meltwater stream. Such plumes may extend into the lake or marine basin for kilometers to tens of kilometers, and flow at speeds of centimeters to meters per second for days. Sediment-charged plumes may be diverted by surface winds or by the Coriolis force. Sediment plumes spread out and decelerate with distance into the basin, progressively depositing suspended sediment. The amount and grain size of deposited suspended sediment decreases in the direction of plume motion. The amount and grain size of suspended sediment vary seasonally (with larger amounts of coarser sediment being delivered during summer melting), but also with shorter-term variations in meltwater discharge.

In glaciolacustrine environments, if river water is colder than lake water (as it will be in spring) and/or contains a high concentration of suspended sediment, a sediment-charged plume moves as an underflow (hyperpycnal flow) along the bed of the lake. It is quite common for plumes to flow along the thermocline as an interflow, especially during the summer when thermal stratification is best developed. It is rare for sediment-charged meltwater to be less dense than lake water, so overflows (hypopycnal flows) are uncommon. There are also occasions when the densities of the lake and river water are similar, and the resulting homopycnal inflow is like an axial jet (see Chapter 15). Lake-floor sediments are rhythmic deposits of light-colored silts and darker organic-rich clays in millimeter- to centimeter-thick bands called *varves* (Figure 17.16). Varves have traditionally been interpreted as recording seasonal variations in the supply of suspended sediment, and as such have been used for age dating. In this traditional interpretation, dark organic clays are deposited during the winter when the lake is frozen and there is minimal river input. The coarser layers are associated with summer conditions when meltwater input from rivers is significant and various currents are active in the lake. However, this is much too simplistic a view. Variations in sediment type can be due to episodic turbidity flows and variations in suspended-sediment supply that are unrelated to seasonal melting of ice. Criteria for distinguishing annual rhythmites from other types are given by Ashley *et al.* (1985). Reworking of lake deposits by basin currents (wind- and wave-induced) is likely to occur only in shallow water, such as near shorelines. Lacustrine deposits affected by water currents (e.g., above the thermocline) tend to be homogenized by bed transport and bioturbation. Lacustrine shoreline deposits occur in terraces along some valleys, and such shoreline terraces have been used to establish lake levels during glacial periods (see below).

In glaciomarine environments, sediment gravity flows occur not only on the slopes of moraines, fans,

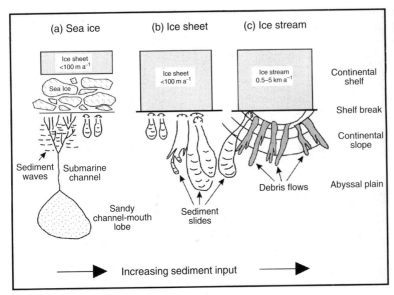

FIGURE 17.17. A conceptual model of landforms and sedimentary processes on ice-influenced continental margins in relation to the rate of sediment supply from sea ice, icesheets, and ice streams. From Dowdeswell *et al.* (2002).

and deltas, but also on the continental slope if deposits reach the shelf edge. During the last ice age when ice reached the edge of the continental shelf, high deposition rates from ice streams resulted in many sediment gravity flows down the continental slope, to be deposited on large deep-sea fans. Such deposits record ice-sheet expansion and recession. However, large fans were not present at the termini of slower-moving parts of ice sheets (Figure 17.17) (Dowdeswell *et al.*, 2002). Suspended-sediment plumes in the sea may be overflows and interflows because of the cold saline seawater. In glaciomarine shelf environments, bed sediment is reworked by waves, wind-shear currents, and tides. Thus, silt–clay rhythmites and clay may be nothing to do with seasonal variation in sediment supply. Glaciolacustrine and glaciomarine deposits produced by glacial advance and retreat would coarsen upwards (basin floor–delta–till), then fine upwards (see below).

Periglacial settings

Periglacial environments have no glaciers, but are still very cold, with average annual temperatures between about 2 °C and −15 °C (Martini *et al.*, 2001; Ritter *et al.*, 2002). The ground is permanently frozen down to depths of hundreds of meters (i.e., *permafrost*), although some thawing of the upper 1–2 m of the ground (the active layer) occurs in summer. Also, lenses of water occur in the frozen ground. Permafrost regions have tundra vegetation, which is mainly low

shrubs and grass. Weathering is by freezing and thawing, and downslope movement of sediment on hillslopes is by solifluction (see Chapter 8). Rivers have a typical frozen–melting hydrograph (Figure 17.12A), and the large water discharges following melting of ice are capable of extensive bank erosion, even though parts of the bed and banks may be frozen.

Typical landforms in permafrost regions include ice-wedge polygons, pingos, thermokarst, and sorted stones (patterned ground) (Figure 17.18). Ice-wedge fissures are shrinkage cracks formed by thermal contraction (Figure 17.18A, B). The fissures are centimeters to decimeters wide and decimeters to meters deep, and taper downwards. They become filled with alternations of water, snow, ice, and sediment over time, and gradually become enlarged. Such enlargement may involve bending of adjacent sediment layers. The fissures are preserved by being filled with sediment (Figure 17.18A). The polygonal arrangement of ice-wedge fissures is analogous to desiccation cracks in mud (Chapter 12). The polygons bounded by the fissures are meters to tens of meters across.

Pingos are isolated dome-shaped hills, meters to tens of meters high and up to hundreds of meters across (Figure 17.18C). They are ice-cored, and are formed as local subsurface accumulations of water freeze and expand. The local accumulation of water may have underlain a dried-up lake, or could be supplied from a local groundwater source. Tension cracks are common on the crests of pingos, and lakes of melted ice may occur in these crestal areas.

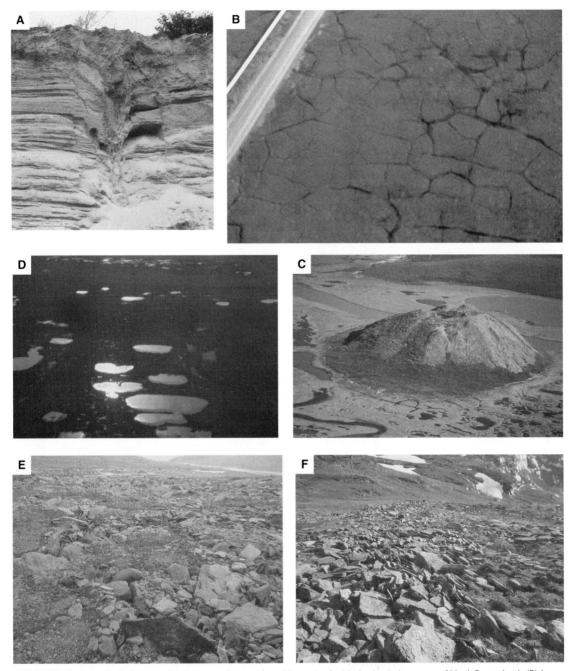

FIGURE 17.18. Permafrost features. (A) An ice wedge cast from Vorstwig, the Netherlands (courtesy of Henk Berendsen). (B) Ice-wedge polygons from the north slope of Alaska. (C) Pingo from the MacKenzie delta, North West Territory, Canada (from Henk Berendsen). (D) Thermokarst lakes from the north slope of Alaska. (E) Sorted stones in a polygonal pattern, northern Sweden. (F) A blockfield from northern Sweden. Images (E) and (F) courtesy of Jan Boelhouwers.

Thermokarst is formed as local melting of the permafrost below the surface results in lakes on the surface (Figure 17.18D). The lakes are commonly elongated parallel to the prevailing wind direction, and this is probably related to preferred wave erosion of the bank of the lake that faces the wind.

Relatively large stones can become arranged on the land surface into distinctive spatial patterns, including circles, polygons, stripes, steps, and nets (Figure 17.18). In general, the stone size increases as the size of a feature (polygon, circle, stripe) increases, and stone size decreases with depth. Stone stripes become prevalent as the ground slope exceeds about 3°. Several different origins for stone-size sorting have been proposed. In general, stone-size sorting might be considered analogous to the kinetic sieving observed in grain flows and for artificially vibrated grains within solid containers. The grains become dispersed upwards by the lateral shearing motion, and the large grains move upwards because the fine grains can move downwards more readily into the spaces created by the dispersion. In solid containers, the large grains move upwards in the center of the container while small grains move downwards near the walls. Under periglacial conditions, the upward dispersion is associated with periodic freezing and thawing within the surface active layer. Freezing causes elevation of sediment grains, but the smaller ones can move back downwards more easily during thawing. The relative movement of stones of different sizes has also been related to inverse density gradients created during thawing of ice.

Solifluction deposits have platy grains oriented with their long axes parallel to the lower flow boundary (Chapter 8), and may also have interstratified sheetwash deposits. Solifluction deposits may also occur as tongue-shaped lobes (tens of meters wide and meters thick), indicating episodic movement similar to a debris flow. Such lobes are also somewhat similar in shape to rock glaciers. *Rock glaciers* are tens of meters thick, hundreds of meters wide, and kilometers long, and range from true glaciers overlain by abundant debris to mixtures of rock, ice, and water that are more like debris flows. *Blockfields* are accumulations of boulders up to 20 m thick, hundreds of meters long, and tens of meters wide, that occur on slopes of up to about 12° (Figure 17.18). They have been ascribed to solifluction, rock glaciers, or avalanches.

There has been much interest in periglacial processes and deposits over the past half century, partly because of the realization that large areas of the Earth experienced periglacial conditions during the LGM. Interpretation of ancient glaciations requires detailed knowledge of periglacial landforms and depositional features. In addition, human activities related to military installations, tourism, and development of economic resources (particularly oil and gas) have expanded more and more into cold regions. Periglacial conditions present major challenges to human existence, and to the building of infrastructure such as oil fields, roads, and pipelines. Such human activities, if uncontrolled, can easily damage fragile periglacial ecosystems.

Evolution of glacial environments over glacial and interglacial periods

In order to predict the nature of erosion and deposition over glacial–interglacial cycles, it is necessary to consider the isostatic effects of the changing distributions of ice and water, and eustatic sea-level changes, which can be quite complicated (Figure 17.19). A 3,000-m-thick ice sheet could depress the lithosphere by about 900 m if isostatic balance could be attained. However, such deformation is resisted by the flexural rigidity and viscosity of the lithosphere, such that the depression is more like 600 m. Furthermore, since the lithosphere is viscoelastic rather than elastic, there is a time lapse between imposition of the ice load and depression of the lithosphere. Lithospheric deformation under the weight of the ice is accompanied by a peripheral bulge which is on the order of tens of meters high (about 20 m for the Laurentide ice sheet), with a crest that is 200–300 km from the edge of the ice (Figure 17.19). As an ice sheet thickens and expands in area, the peripheral bulge increases in height and moves outwards from the center of the ice. As an ice sheet wanes, the peripheral bulge is reduced in height and migrates towards the center of the ice, and the lithosphere is uplifted. Rates of uplift (and depression) depend on rates of ice growth and decay, and, because the lithosphere is viscoelastic, decrease with time. Maximum rates of uplift were many centimeters per year following the LGM, and Scandinavia and Hudson Bay are still rising at rates up to 10 mm per year (see below). Measurements of glacial uplift have been important in determining the rheological properties of the crust and upper mantle. During a glacial maximum, the eustatic sea level may be lowered by about 150 m, but the oceanic lithosphere must rise isostatically by about 50 m to account for the reduction in volume of seawater. Furthermore, the total volume of unfrozen seawater decreases (thermal contraction) as the water temperature decreases. As ice sheets wane, there is a eustatic sea-level rise due to melting of ice and thermal expansion of the seawater, but the addition of more seawater will cause some isostatic subsidence of the sea floor. Sea-level data derived from coral-reef terraces (assumed to have formed at near the mean sea level) distant from ice sheets indicate that the absolute eustatic rise in sea level over the past 18,000 years is

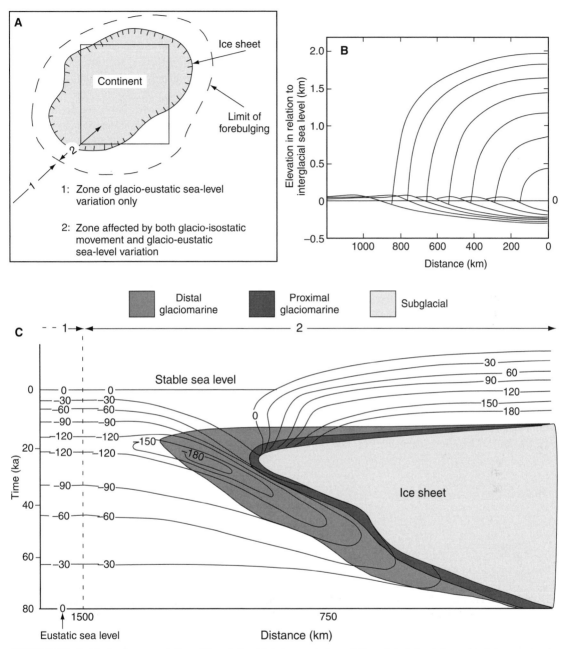

FIGURE 17.19. An idealized representation of the relationship between glacio-eustatic and glacio-isostatic sea-level change around a glaciated continental margin. From Eyles and Eyles (1992). (A) A map showing the zone of subsidence and the limit of the forebulge surrounding a continental ice sheet. (B) A cross section showing upward and outward growth of an ice sheet and migration of the peripheral zone of subsidence and forebulge. (C) The spatial and temporal variation of the relative sea-level and associated depositional environments at the ice-sheet margin.

about 120 m (7 mm per year), with a maximum rate of rise of about 15 mm per year (see below). However, such records must be looked at cautiously, because the rate of tectonic uplift or subsidence beneath the coral reefs might not be known well.

The subsided lithosphere adjacent to an ice sheet is potentially a site for deposition of hundreds of meters of sediment. Since the amount of subsidence adjacent to the ice is much greater than the eustatic sea-level fall, this will be a region of base-level rise, and is likely to be occupied by lakes or the sea (i.e., marine or lacustrine transgression occurs as the ice expands). Regions further from the ice will experience either little change in base level or a base-level fall. A base-level fall results in marine or lacustrine regression and possibly incision of the continental shelf, lake floors, or rivers and floodplains. If the continental shelf is exposed or partially covered with ice, sediment can reach the edge of the shelf and travel down the continental slope into ocean deeps. As the ice melts, the region close to the shrinking ice will experience greater isostatic uplift than eustatic rise in sea level, and would become an area of net erosion rather than net deposition. Catastrophic drainage of enormous proglacial lakes may occur due to a combination of uplift and base-level rise. In contrast, regions distant from the retreating ice would experience rising base level and deposition.

Hypothetical depositional sequences produced by glacial advance and then retreat based on Boulton (1990) and Eyles and Eyles (1992) are shown in Figure 17.20. Ice-marginal deposits do not generally survive glacial advance, and most sediment is deposited during glacial retreat. In terrestrial environments near the ice, the hypothetical sequence from base upwards is (1) subglacial traction (lodgement-deformation) till (ground moraine) on deformed substrate interbedded with subglacial meltwater-channel deposits; (2) melt-out till and debris-flow deposits (ice margin–end

moraine) interbedded with ice-contact-stratified deposits; (3) outwash and/or lacustrine deposits; (4) eolian deposits. In terrestrial environments further from the ice, the hypothetical sequence is (1) outwash and/or lacustrine deposits; (2) eolian deposits. In marine environments near the ice, the hypothetical sequence from base upwards is (1) coarse ice-contact deposits – melt-out till and submarine outwash; (2) offshore laminites and ice-rafted diamicton; (3) erosion due to crustal rebound. In marine-shelf environments further from the ice, the hypothetical sequence is: (1) nearshore ice-rafted diamicton and sediment-gravity-flow deposits; (2) offshore laminite and ice-rafted diamicton. On the continental slope and rise, deposits formed during glacial advances are associated with mass movements (slumps and slides, debris flows, and turbidity currents) and growth and progradation of submarine fans (see Chapters 8 and 18). These deposits are channelized or sheet-like, and range from unsorted and ungraded sands and gravels to stratified sands. Diamictons deposited from icebergs are interbedded. More mud is deposited during glacial retreat. Glacial–interglacial cycles are represented by coarsening-upward sequences (mud-dominant to sand-dominant) followed by fining-upward sequences that may be on the order of hundreds of meters thick (Chapter 18).

A more recent model for erosion and till deposition during advance and retreat of a terrestrial ice sheet

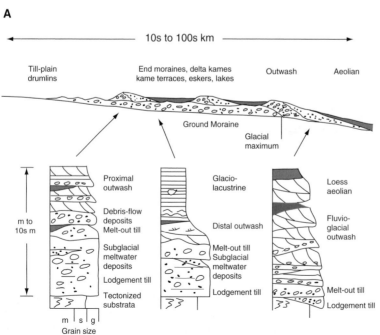

FIGURE 17.20. Sedimentary sequences produced by glacial advance and retreat. Modified from Miller (1996). (A) Terrestrial environments. (B) Shallow-marine environments. (C) The temporal and spatial variation of deposition on a continental margin over a glacial–interglacial cycle. From Boulton (1990).

B

FIGURE 17.20. Continued.

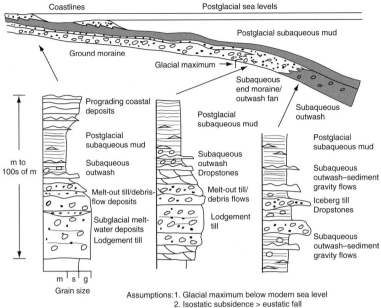

Assumptions: 1. Glacial maximum below modern sea level
2. Isostatic subsidence > eustatic fall
3. Eustatic rise preceded isostatic rebound

C

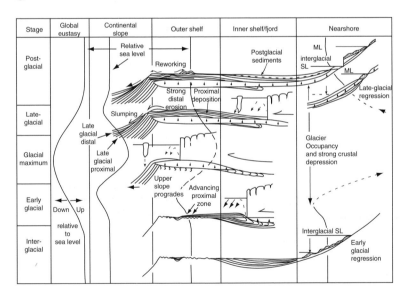

with a deforming bed (Boulton, 2006) is shown in Figure 17.21. In the zone up glacier from the equilibrium line, the ice is accelerating and erosive. Down glacier from the equilibrium line, the ice is decelerating and depositional, and a downstream thickening till is formed. However, as the ice sheet grows and advances, the wave of downstream till deposition followed by

upstream erosion also advances. The resulting deposit (Figure 17.21) has a downstream zone of thick, continuous till deposition on an uneroded substrate, a middle zone where till deposited during the glacial advance is partially eroded prior to deposition of a retreat-phase till, and an upstream zone where till deposited during the advance phase plus some pre-till material is eroded,

A

B

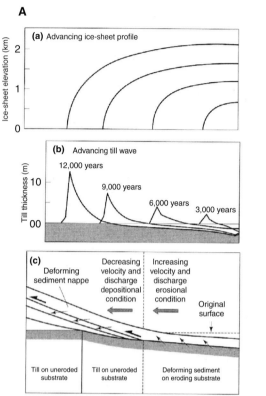

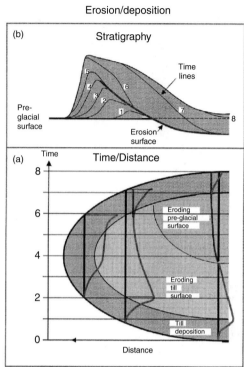

FIGURE 17.21. (A) A model of an advancing ice-sheet profile (a), advancing till wave (b), and along-stream variations in ice-flow velocity, bed erosion, and till deposition (c). (B) Idealized erosion and deposition through a simple glacial cycle of advance and retreat. The along-stream variation of erosion and deposition at a point in time is shown in (a). The structure of the resultant till, including time lines and erosion surfaces, is given in (b). From Boulton (2006).

and the erosion surface is overlain by till deposited during a late stage of glacial retreat.

The Neogene ice age

Historically, the Cenozoic era was divided into the Tertiary period (comprising, from oldest to youngest, the Paleocene, Eocene, Oligocene, Miocene, and Pliocene epochs) and the much shorter Quaternary period (comprising the Pleistocene and Holocene epochs). Nowadays, the Paleogene period comprises the Paleocene, Eocene, and Oligocene epochs, and the Neogene period comprises the Miocene, Pliocene, and Pleistocene epochs and the Holocene (recent) interval. Although the Pleistocene epoch is well known for its pronounced glacials and interglacials, global cooling was initiated in the Paleogene and developed through the Neogene. Reviews of the Neogene glaciation can be found in Dawson (1992), Benn and Evans (1998), Easterbrook (1999), Martini *et al.* (2001), Zachos *et al.* (2001), and the papers in Knight (2006).

In the early Eocene, mid-latitude regions had a tropical climate, whereas polar regions had a temperate climate, and were probably mainly ice-free. This warm period is commonly called the Early Eocene Climatic Optimum (EECO). Global cooling was initiated in the Eocene around 50 million years ago, and by the Oligocene Australia had separated from Antarctica enough that the Antarctic Circumpolar Current developed and prevented warm currents from the north reaching the Antarctic coast. The increasing coldness of Antarctica increased the temperature difference with the equatorial region, thus increasing the energy of atmospheric circulation. Another influence on global cooling is considered to be the collision between India and Asia (starting around 50 million years ago), the uplift of the Himalayas and Tibetan plateau, initiation of the Asian monsoon, and increased chemical weathering of newly uplifted rocks (which decreases the atmospheric concentration of carbon dioxide). A marked increase in ice volume and decrease in ocean temperature around Antarctica occurred about 36 million years ago.

TABLE 17.1. *Classical (and outdated) Pleistocene glacial and interglacial periods used in various regions*

Glacial/interglacial	European Alps	North Europe	North America
Glacial	*Wurm*	*Weichsel*	*Wisconsin*
Interglacial	R-W	Eem	Sangamon
Glacial	*Riss*	*Saale*	*Illinoian*
Interglacial	M-R	Holstein	Yarmouth
Glacial	*Mindel*	*Elster*	*Kansan*
Interglacial	G-M	Cromer	Aftonian
Glacial	*Gunz*	*Menap*	*Nebraskan*

TABLE 17.2. *Climate changes and approximate ages since the last glacial*

Years before present	Climate change
20,000–18,000	Glacial maximum of Laurentide Ice Sheet
16,000–14,000	Warming
12,000–11,000	Cold – younger Dryas
11,000–8,500	Warming and drying
8,200	Cooling
2,500	Cooling – little ice age

Large glaciers existed in Antarctica during the Oligocene, and ice sheets had covered Antarctica by 25 million years ago. Climatic amelioration occurred between about 20 and 14 million years ago, during the early Miocene, but then global cooling resumed. Greenland had been glaciated by 7 million years ago, and glaciers were forming in North America and Eurasia by 6 million years ago. Another period of minor warming occurred around 5 million years ago, but major cooling resumed about 3 million years ago.

Since about 2.4 million years ago, there have been ice sheets over the land at high northern latitudes and elevations in response to global climate change (Figure 17.22). This glaciation had an important effect on the landscapes and surface sediments of North America, northern Europe, northern Asia, and mountainous regions such as the Alps, Himalayas, and Andes. With the exception of Antarctica, the southern hemisphere was not as widely affected because the proportion of ocean is much greater. The Pliocene–Pleistocene ice age comprised up to 21 major cold periods (glacials when the mean annual temperature was ~6 °C colder than it is today, and a third of the land was covered with ice) with intervening warmer periods (interglacials like today when only a tenth of the land area was covered with ice). The main glacial–interglacial period was on the order of 41,000 years for the first 900,000 years of the Pliocene–Pleistocene ice age, but was approximately 100,000 years thereafter. The glacial–interglacial periods were originally recognized on continents by classic geological methods such as superposition of ground moraines and relative dating by development of buried soils on the tills. Only some of the glacial–interglacial periods were recognized and given different names in different parts of the world, and the classical terminology shown in Table 17.1 is now considered inadequate in view of new dating and oxygen-isotope data (see below). Many more glacials and interglacials have been recognized in the oxygen-isotope data. Also, the most recent glacial advance has removed much of the landform and sedimentary evidence for even the three other glacials listed in Table 17.1. Furthermore, a range of shorter-period climate changes (tens of thousands of years to decades) is becoming increasingly recognized (Table 17.2).

Evidence for glaciations

The history of glacial advances and retreats is recorded directly in landforms and deposits as described above. However, there is more information on the decay of ice sheets and glaciers than on their growth, and most of the evidence is for the last glacial period because much evidence of earlier glaciations has been eroded or covered. The position of glacier margins is recorded in end moraines. Directions of ice motion are recorded in drumlins, Rogen moraine, deposits to the lee of rock obstacles, and erosional striations. The upper surface (and hence thickness) of ice is recorded by the elevation of erosional and depositional features. The absolute or relative age of glacial deposits is established using radiometric dating, luminescence dating, fission-track dating of associated volcanic ash, paleomagnetic

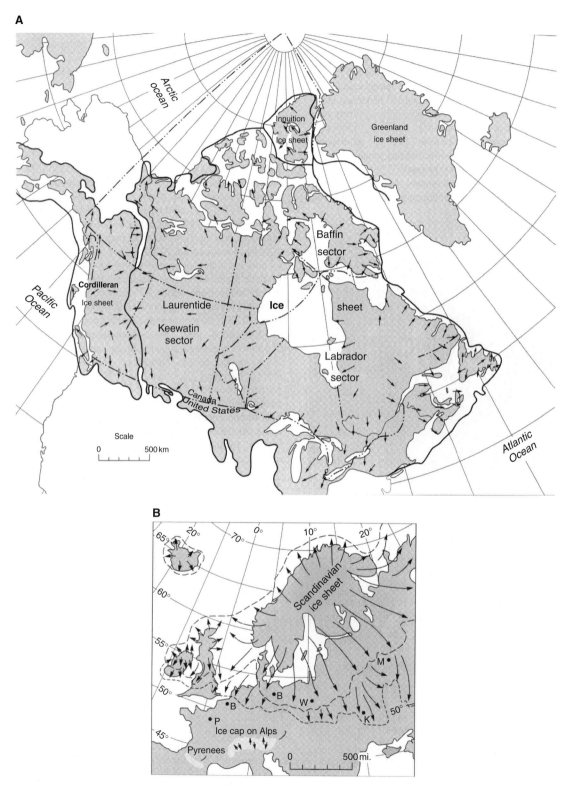

FIGURE 17.22. The approximate extent of Laurentide (A) and Scandinavian (B) ice sheets during the last Pleistocene glaciation. From Easterbrook (1999). Arrows show the direction of ice flow.

dating, dendrochronology, varve chronology, lichenometry, measurement of the degree of weathering and soil development, cosmogenic-nuclide dating of surface exposure, and fossils. Climate changes are best recorded in ratios of oxygen isotopes obtained from ice cores and microfossils in deep-sea cores, and also from paleoecological examination of microfossils such as foraminifera and pollen. Other, less direct forms of evidence for glaciations (discussed below) include isostatic adjustments to loading and unloading of ice, eustatic sea-level changes, changes in the supply of water and sediment to rivers and lakes, catastrophic flooding associated with bursting of ice-dammed lakes, and indications of changes in transport and deposition of sediment by wind. Reconstruction of the history of glacial advances and retreats is also aided by numerical models of ice geometry and dynamics that are linked to changes in the atmosphere, the oceans, and the lithosphere (see Chapter 10 and papers in Knight (2006)).

Evidence from deep-sea cores

Biogenic oozes from the deep ocean (Chapter 18) contain microfossils such as foraminifera (heterotrophic protozoans with calcareous shells), radiolarians (heterotrophic protozoans with siliceous shells), coccoliths (autotrophic protozoans with calcareous shells), and diatoms (autotrophic protozoans with siliceous shells). These planktonic and benthic organisms indicate past oceanic temperature and circulation, and hence climate. Imbrie and Imbrie (1979) provided a comprehensive review of the use of microfossils in climatic reconstructions. Warm-water species and cold-water species can be distinguished and, as long as the species are extant, ocean-temperature changes can be quantified, commonly down to decades. In addition, the oxygen-isotopic composition of their shells indicates the oxygen-isotopic composition, water temperature, and (to a lesser extent) salinity of the seawater they grew in. Glacial-stage ocean water is relatively enriched in ^{18}O. This is because water molecules with ^{16}O are easier to evaporate than those with ^{18}O, so that water vapor, precipitation, and hence ice sheets are enriched in ^{16}O relative to ^{18}O. The degree of enrichment increases as the temperature difference between air and water increases. The relative proportion of ^{18}O and ^{16}O is measured and expressed in dimensionless form as

$$\delta^{18}O = \left(\frac{\left(^{18}O/^{16}O \right)_{sample} - \left(^{18}O/^{16}O \right)_{standard}}{\left(^{18}O/^{16}O \right)_{standard}} \right) \times 1,000$$

The standard is referred to as SMOW (standard mean ocean water). If $\delta^{18}O$ is positive, the sample is enriched in ^{18}O (isotopically heavy), indicating that it is glacial-stage ocean water.

The basis for using the oxygen-isotopic composition of calcareous microfossils to indicate ocean water temperature of the past is that, for a given $\delta^{18}O$ of seawater, the $\delta^{18}O$ of the shell is a function of the ambient water temperature, but only for some species. Both benthic and planktonic foraminifera were collected from some of the same cores in the CLIMAP project. It was assumed that oceanic bottom water has a constant temperature of about 4 °C, so that the $\delta^{18}O$ of selected benthic foraminifera indicates the $\delta^{18}O$ of the bottom water, which varies with the waxing and waning of ice sheets. With this knowledge of the temporal variation of the $\delta^{18}O$ of oceanic bottom water, the surface water temperature of the oceans can be calculated using data from planktonic foraminifera. It turns out that most of the variation in the isotopic composition of planktonic foraminifera is due to changes in the isotopic composition of the oceans as the ice sheets wax and wane. However, there are difficulties in the use of deep-sea organisms for indication of either ambient water temperature or isotopic composition: (1) the $\delta^{18}O$ of some shells does not indicate ambient water temperature because certain microfossils exhibit a so-called "vital effect" and do not follow equilibrium fractionation; (2) shells may be transported and reworked; (3) there is preferential preservation of shells that are resistant to dissolution; (4) there are depth variations in water temperature and salinity and lack of knowledge of the depth where a shell was precipitated; and (5) there are uncertainties in age of shells. Nevertheless, deep-sea records can be combined with ice-core records and records from continental environments, making it possible to recognize glacials and interglacials (Figure 17.23). In addition, evidence of variable surface temperature in deep-sea cores is recorded in variable amounts of iceberg deposits (so-called Heinrich events). Other evidence of cold periods is an increase in wind-blown dust.

Evidence from ice cores

Thicker portions of the Antarctic ice sheets contain a record of snow and ice accumulation over about 420,000 years (Figure 17.23), whereas the Greenland ice-sheet record is considerable shorter. The $\delta^{18}O$ in ice indicates air temperature, becoming less negative (isotopically heavier) as the air temperature increases because less ^{16}O is trapped in the ice. Gases trapped in the ice can

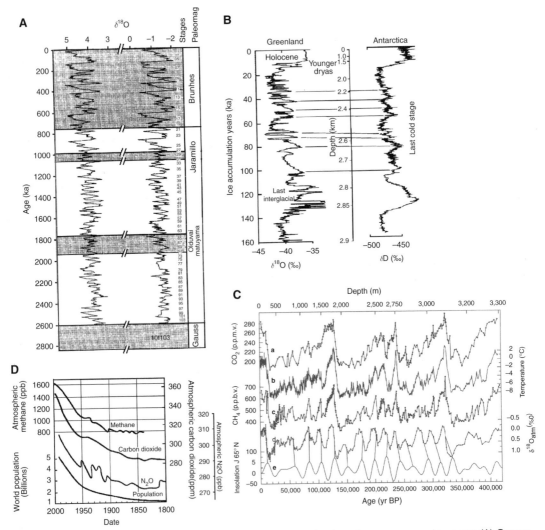

FIGURE 17.23. Glacials and interglacials indicated using oxygen-isotope data from deep-sea cores and ice cores. (A) Oxygen isotopes from benthic (left) and planktonic (right) foraminifera, correlated with paleomagnetic stages. (B) Oxygen isotopes from Greenland and Antarctic ice cores. (C) Variations in temperature, and in levels of carbon dioxide and methane, during the last 420,000 years from the Vostock (Antarctica) ice core. (D) Variation in level of greenhouse gases and human population over the past 200 years. Images (A), (B), and (D) from Martini *et al.* (2001); (C) from Petit *et al.* (1999). Reprinted by permission from Macmillan Publishers Ltd: *Nature*.

indicate the concentration of carbon dioxide in the atmosphere, which decreases during glacial periods. The amount of wind-blown dust in the ice is taken to increase during glacials. However, dating of the ice is very uncertain because it involves assumptions about temporal variations in the rate of ice accumulation, ice thickness, flow pattern, and environment of ice accumulation.

The oxygen-isotope curves from ice cores clearly show the main glacials and interglacials over the past 420,000 years, but also numerous higher-frequency climatic oscillations (temperature changes of 5–8 °C), as

mentioned above. For example, Bond cycles are 10,000–15,000-year cooling trends followed by abrupt warming. Dansgaard–Oeschler events consist of abrupt warming over decades to centuries followed by gradual cooling over 1,000–2,000 years. The oxygen-isotope curves for the Antarctic and Greenland ice cores (Figure 17.23) have broadly similar shapes, and are similar to the deep-sea core data, indicating that the main glacials and interglacials were simultaneous worldwide, suggesting that they were caused by a mainly external forcing mechanism (see below).

However, apparently not all of the smaller-scale climate cycles are global events. The Younger Dryas is a thousand-year cold interval (stade) bounded by periods of rapid (decadal) climate change, and apparently can be widely recognized. The Younger Dryas has been related to sudden release of a large volume of meltwater into the North Atlantic (perhaps during a phase of draining of glacial Lake Agassiz), which interrupted the circulation of North Atlantic Deep Water. However, there is disagreement over whether the Younger Dryas is a global event, and whether changes to the circulation of North Atlantic Deep Water would result in more or less simultaneous global cooling. The Vostock Antarctic ice core (Figure 17.23) reveals a close correlation between temperature and atmospheric levels of carbon dioxide and methane. The correlation between human population and concentration of carbon dioxide is obvious in Figure 17.23. The ice cores also show that volcanic eruptions have had a frequent and important effect on global climate over the past 110,000 years. The huge emissions of ash from mega-eruptions can obscure the Sun and cause cold conditions for years or decades.

Isostatic adjustments to loading and unloading of ice

The lithosphere is depressed under the weight of ice sheets, and the zone of depression decreases away from the ice to a peripheral bulge (Figure 17.19). Isostatic uplift occurs as the ice melts. In Scandinavia, up to 830 m of uplift has occurred in the past 13,000 years (Figure 17.24). The maximum rate of uplift was on the order of centimeters per year and is now up to 10 mm per year. Around Hudson Bay, the maximum amount of uplift is 250 m (millimeters per year at present), and another 100–150 m of uplift will be required in order to achieve isostatic balance (Figure 17.24). An assessment of future isostatic uplift, a delayed response to glacial unloading, is made using gravity surveys. A negative free-air gravity anomaly indicates that isostatic balance has not yet been achieved. Isostatic uplift is recognized in raised and tilted beaches and river terraces (Figure 17.25). Spatial variation in the elevation of shorelines and river terraces of a given age is measured to produce isobase maps. These maps can be used to show the amount of isostatic uplift as ice retreats (e.g., Figure 17.25). In the case of a shoreline, the actual isostatic uplift is a combination of the observed uplift and base-level change. In order to reconstruct uplift history, it

is necessary to be able to date the geomorphic surface and make assumptions about spatial variations in its age.

Eustatic sea-level changes

Sea level was about 120 m lower than today during the LGM (Figure 17.26). The eustatic sea level during a glacial period is controlled by the amount of water in the ice and by the temperature of the ocean water. Evidence for a lower sea level during the last glacial is provided by shoreline deposits at the edge of the continental shelf; the observed increase in sediment transport down the continental slope and deep-sea fan deposition; and submerged beaches and coral-reef flats. However, it is difficult to measure eustatic sea-level changes at the coastline because of the effects of tectonic subsidence and uplift associated with glacial loading and unloading, and because of effects of erosion and deposition of sediment. The glacioeustatic sea-level rise during the last 18,000 years, independently of the effects of isostatic uplift following ice removal, is obtained from coral reefs (in Barbados and New Guinea) using the assumption that the reef builds to sea level (Figure 17.26). However, these reefs have been uplifted, so the tectonic uplift rate must be estimated. The Florida Keys is made of coral reefs that now lie above sea level (Figure 17.26), but these reefs were formed during the last Sangamon interglacial when sea level was higher than it is today. This higher Sangamon sea level is also indicated by old beach ridges along the coast of Georgia and the Carolinas (Figure 17.26).

Changes in supply of water and sediment to rivers and lakes

Increases in precipitation and sediment production during glacial periods are thought to result in greater discharge of water and sediment into rivers, such that rivers are likely to be relatively large and braided, with frequent river avulsions. Lake levels are higher during glacials, and lower during interglacials. More transport and deposition of wind-blown sediments occurs during glaciation and especially deglaciation, when vegetation cover is poorly developed over large areas of sediment. Evidence for these changes is found in the elevation of lake shorelines; alternations between muddy (cold and wet) and saline (warm and dry) lake deposits; and alternations between deposits of wind-blown sand and those of rivers and floodplains. For example, the salt lakes in the southwestern USA (Death Valley, California; Lake Bonneville, Utah) were much bigger

A

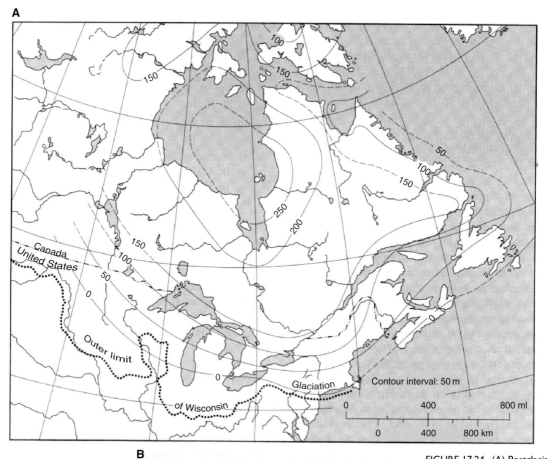

B

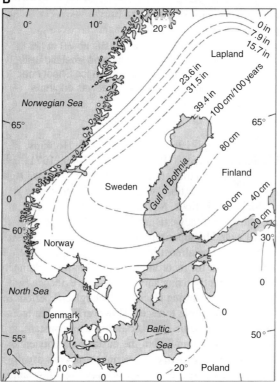

FIGURE 17.24. (A) Postglacial isostatic uplift following retreat of the Laurentide ice sheet. (B) The present rate of uplift of the Scandinavian region. From Easterbrook (1999).

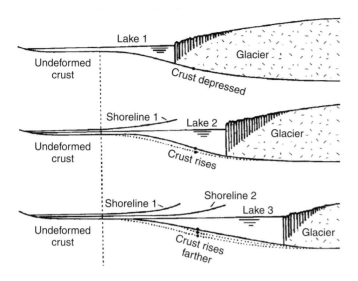

FIGURE 17.25. Uplift of successively younger shorelines in response to glacial retreat and spatially varying isostatic uplift. From Martini *et al.* (2001).

and received more mud deposits during the wetter glacial periods. Ancient river systems are preserved beneath present-day Sahara sand seas. The Great Lakes region of the northern USA contained much larger lakes, including glacial Lakes Agassiz and Missoula (Figure 17.27). Catastrophic floods due to bursting of ice dams of these lakes occurred during deglaciation, as discussed above. Volcanic eruptions under glaciers (e.g., Iceland, Mount St. Helens) produced a lot of meltwater and water vapor, leading to explosive eruptions, and the formation of jokulhlaups, debris flows, and mudflows (lahars).

Changes in land biota and soils

Extinctions and migrations of land and marine animals resulted from changes in temperature and the creation of land bridges (Martini *et al.*, 2001). Geographic redistribution of vegetation zones and soil types also occurred in response to episodes of global cooling and warming. The tundra, temperate, and tropical biozones were shifted southwards and compressed during glacial periods. The tundra biozone includes mosses, lichen, sedges, and dwarf willows. The cool temperate (boreal) zone includes spruce, fir, aspen, poplar and birch. The warm temperate zone includes mixed deciduous hardwood and fir forests, and grasslands. Soil types are related to climate and vegetation. Thus poorly developed soils (inceptisols and entisols) occur in tundra zones, Spodosols occur beneath spruce-fir forests, and a diverse assortment of alfisols, mollisols (grassland soils), and vertisols occurs in temperate climates. Ultisols and oxisols occur mainly in tropical climates.

Peat bogs are common on emergent lake and sea beds at high latitudes.

Pre-Neogene glaciations

All of the forms of evidence used for reconstructing the Quaternary glaciations are not available when trying to interpret older glaciations. In order to reconstruct pre-Neogene glaciations, reliance must be placed on the depositional record, on reconstructions of the positions of lithospheric plates using paleomagnetic methods, on interpreted records of eustatic sea-level change and the CO_2 content of the atmosphere, and on interpreted fossil evidence. Since glacial till can look very similar to gravity flow deposits, use of deposits to indicate glacial environments is no easy task.

It has been suggested (Kirschvink, 1992; Hoffman *et al.*, 1998; Hoffman and Schrag, 2002) that the entire Earth was glaciated (so-called snowball Earth) at least two separate times during the Cryogenian system of the Neoproterozoic. The main evidence for this is estimated paleolatitudes of interpreted diamictons within 10° of the paleoequator, and an apparent global negative shift in $\delta^{13}C$ of more than 10 parts per thousand in marine carbonates immediately prior to the glaciations. In addition, diamictons from suspected snowball-Earth episodes are overlain by distinctive so-called *cap carbonates*. Snowball-Earth episodes are thought to be caused by a so-called "runaway" increase in the albedo of the Earth. During global glaciations, oceanic and terrestrial photosynthesis ceases, so the oceans become enriched in heavy ^{13}C, which is normally preferentially removed by photosynthesis. Normal ocean–atmosphere

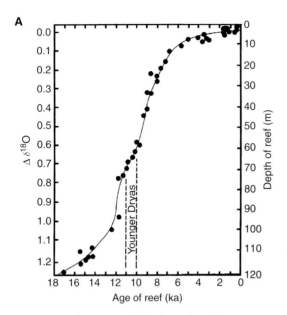

FIGURE 17.26. (A) A glacio-eustatic sea-level curve constructed from Barbados coral reefs. From Martini et al. (2001). (B) An aerial photo of part of the South Carolina coast, showing the Sangamon shoreline (arrow).

interactions are curtailed due to pervasive oceanic ice cover, so atmospheric CO_2 no longer diffuses into the oceans and builds up. Likewise, with limited HCO_3^- diffusing into the oceans, oceanic Ca^{2+} builds up. Following meltdown of the ice sheets, there would have been a period of rapid precipitation of carbonate from the highly supersaturated oceans. This scenario has been challenged vigorously because the diamictons have substantial meltwater deposits prior to the cap carbonates that are supposed to signal the catastrophic meltdown. Thick diamictons and meltwater deposits suggest a vigorous hydrologic system that would hardly be possible under the conditions of a snowball Earth (Christie-Blick et al., 1999). Olcott et al. (2005) document extensive black shales deposited on the South American craton during the glaciations. Biomarkers from the carbon that enriches the shales indicate that there was a surface ocean microbial eukaryotic ecosystem.

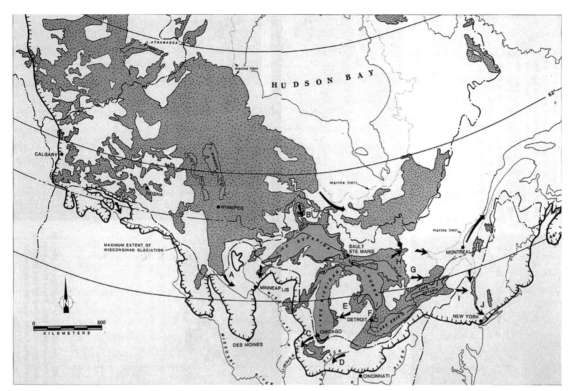

FIGURE 17.27. The area covered by proglacial lake sediment (shaded) during retreat of the Laurentide Ice Sheet and the major drainage routes into the Mississippi River and Atlantic Ocean. From Benn and Evans (1998).

Causes of ice ages

Any theory for the origin of glaciations must explain (1) the occurrence of ice ages (icehouse conditions) for millions of years, separated by warmer climates (greenhouse conditions) that last for hundreds of millions of years; (2) the cyclicity of global climate (a mid-latitude temperature range of up to 6 °C) within ice ages associated with glacials and interglacials, and stadials and interstadials within the glacial periods; and (3) the synchroneity of glacials and interglacials worldwide. Theories for the origin of ice ages include (1) changes in solar radiation reaching the Earth; (2) the effects of lithospheric plate movements and associated volcanic activity on the formation or break-up of supercontinents, positioning continents in polar regions, changing the topographic relief of the Earth's surface, the carbon dioxide content of the atmosphere, and oceanic circulation; (3) the interaction among the distribution of land and sea, the biosphere, the atmosphere, and the oceanic water circulation; and (4) autocyclic factors related to surface and atmospheric albedo. As discussed below, no single mechanism can explain all aspects of glaciations. Furthermore, it is very difficult to

distinguish cause and effect. Data-constrained quantitative calculations of the effects of the various mechanisms suggested for global climate change are needed in order to assess them. Indeed, numerical models are being used widely to evaluate causes of ice ages and alternating warm and cold periods within ice ages (Knight, 2006). These models need to be developed, however, because at present they rely excessively on poorly constrained empirical parameters (see Chapter 10).

Changes in solar radiation reaching the Earth

The Sun's radiation has increased by about 30% over the age of the Earth, but there is no evidence for periodic changes in solar radiation to produce ice ages or the shorter-term climatic variations. However, sunspot cycles have an 11-year period. During low sunspot activity, weather zones move towards the equator, increasing temperature gradients, intensifying atmospheric circulations, and producing colder and wetter climate in general. The solar system rotates around the center of the Milky Way about every 300 million years, passing through stationary nebulae of hydrogen-rich particles that could reduce the radiation reaching the Earth. However, this is very uncertain, and there is not a distinct 300-million-year cycle of ice ages throughout Earth history.

Milankovitch cycles are variations in solar insolation associated with the changing position of the Earth's axis of rotation and its orbit around the Sun (Figures 2.15 and 2.16). The obliquity cycle with a mean period of 41,000 years is due to variation in the angle of inclination of the Earth's axis from $22°$ to $24.5°$. The precession cycle, with period between 19,000 and 23,000 years, is due to the wobble of the Earth's rotational axis. The eccentricity cycle is associated with changes in the Earth's elliptical orbit around the Sun, with periods of approximately 100,000, 400,000, 1.3 million, and 2 million years. Milankovitch cycles interact with each other to cause changes in the areal distribution of insolation (seasonality) rather than in the total amount of insolation. The timing of Pliocene–Pleistocene glacials and interglacials is apparently related to the 41,000- and 100,000-year Milankovitch cycles; however, it is unknown exactly how variation in insolation affects climate. Cyclic variations in seasonality (due to the precession and obliquity cycles) cannot be the main cause of ice ages, because ice ages have not occurred throughout geological time. Also, Milankovitch cycles alone cannot explain sudden climatic changes (e.g., Dansgaard–Oeschler cycles) over short time periods, apparently globally synchronous glacials and interglacials, and sudden terminations of glacial periods.

Lithospheric plate movements and associated volcanic activity

The movement of lithospheric plates and associated volcanic activity are responsible for the formation or break-up of supercontinents; positioning continents in polar regions; and changing the topographic relief of the Earth's surface, the carbon dioxide content of the atmosphere, and the oceanic circulation. Major landmasses or enclosed seas near the poles are essential for glaciation. In the case of the Neogene glaciation, Antarctica was located over the South Pole, and Greenland and the Arctic Ocean were located over the North Pole. In the Permo-Carboniferous, the supercontinent Gondwanaland was located over the South Pole. Periods of greater movement of lithospheric plates and break-up of continents are associated with greater volcanic activity at the edges of plates. The volcanic activity is associated with increases in the

concentration of CO_2 in the atmosphere, and potential global warming. However, increases in volcanic ash in the atmosphere could block solar radiation, causing global cooling. Periods of reduced plate movements and existence of supercontinents apparently precede icehouse conditions, e.g., the Permo-Carboniferous glaciation. Another effect of lithospheric plate movements is uplift of mountains such as the Himalayas and the Tibetan Plateau. Such large high landmasses cause global cooling and have a major effect on atmospheric circulation (monsoons). Such uplift has been associated with greater chemical weathering, which reduces the amount of carbon dioxide in the atmosphere. However, cooling associated with uplift may actually reduce the amount of chemical weathering. Chemical weathering may be expected to be at a maximum in warm climates and where there are large continental masses. Although the uplift and growth of the Asian continent may have contributed to the Quaternary glaciation, major orogenic uplifts in themselves do not always lead to ice ages (e.g., the Caledonian and Acadian orogenies). The areal distribution of continental landmasses and oceans has a major effect on atmospheric and oceanic circulation, as discussed below.

Interactions among the biosphere, atmosphere, and oceans

The interaction between the atmosphere and the oceans has a major impact on global climate. For example, the Gulf Stream transports warm, saline water from tropical latitudes to high latitudes, and causes much warmer weather at high latitudes (northwestern Europe) than would otherwise occur. Cold, saline currents (e.g., North Atlantic Deep Water (NADW)) move from polar regions along the deep parts of oceans towards lower latitudes (see Figure 18.18 later). These types of currents enhance the exchange of CO_2 between the oceans and the atmosphere (involving dissolution and precipitation of $CaCO_3$). Furthermore, it is considered desirable to have free circulation of water within oceans between continents in order to allow redistribution of heat around the Earth. If such currents did not exist, and CO_2 exchange were reduced, the concentration of CO_2 in the atmosphere would be reduced, and the temperature would decrease. In glacial times, the currents shown in Figure 18.18 probably did not exist. Continental shelves were exposed or covered with ice. Large quantities of freshwater were added to the oceans in polar regions, reducing deep oceanic circulation (e.g., NADW). These changes in oceanic circulation are indicated by microfossils. The effects of large ice sheets on ocean circulation can be seen today in Antarctica, where there is a circum-continental deep-water current that does not interact strongly with warm water from lower latitudes. The Younger Dryas has been explained by some workers as due to a reduction in oceanic circulation associated with large increases in meltwater discharge to the North Atlantic, but this theory is controversial.

The carbon dioxide content in the atmosphere is not just controlled by atmospheric and oceanic circulation. Long-term changes in CO_2 content are also influenced by (1) volcanic eruptions; (2) photosynthesis and formation of calcareous skeletons; (3) burial of organic deposits such as peat, coal, and limestone; and (4) weathering of siliciclastic rocks. As mentioned above, increases in volcanic activity increase the CO_2 content of the atmosphere. Photosynthesis removes CO_2 from the atmosphere, and the formation of calcareous skeletons releases CO_2 into the atmosphere. The rates of these processes are influenced by temperature, oceanic circulation, and organic productivity. It has been suggested that the evolution of algae at the end of the Proterozoic, and the evolution and expansion of land plants in the late Paleozoic, were responsible for a decrease in atmospheric CO_2 content, and hence global cooling. However, there was no similar increase in plants prior to the Quaternary glaciation. Burial of peat and coal is required in order to remove CO_2 from circulation, so deposition in actively subsiding basins (i.e., during periods of active plate movement such as the Devonian, Jurassic, Cretaceous, and early Tertiary) is also needed. Similarly, extensive shelf seas with abundant calcareous sediments existed in the late Proterozoic, late Paleozoic, and late Mesozoic. An increase in removal

of CO_2 by chemical weathering requires an increase in land area (as during assembly of supercontinents and tectonic uplift) and increases in temperature, moisture, and vegetation on land.

Perhaps the influence of the CO_2 content of the atmosphere on global climate has been overemphasized. Carbon dioxide is not the most important greenhouse gas in the atmosphere, water vapor being much more important. The largest repository of CO_2 on the Earth is the oceans, and these regulate the concentration of CO_2 in the atmosphere, via dissolution and precipitation of calcium carbonate, and controlling organic productivity. Thus, it is possible that other factors control global warming and cooling, and the CO_2 content of the atmosphere is merely a reflection of such climate change.

Autocyclic factors

Once an ice sheet has formed, it would expand rapidly due to (1) reflection of solar radiation away from the Earth; (2) the increased temperature gradient between the cold ice and warmer sea, leading to more storms, cloudiness, and precipitation; (3) lowering of sea level and exposure of land, thus decreasing the air temperature (especially in winter); and (4) cooling of the ocean by the lower air temperature. However, as the ocean is cooled and reduced in area, the evaporation and precipitation is decreased, thus ice retreats. Such an increase in ice sheets and reduction in exposed land area may also increase the CO_2 content of the atmosphere due to decreases in the rate of chemical weathering and photosynthesis. The increase in CO_2 content would thus initiate global warming. The effectiveness of these autocyclic, self-regulation mechanisms at creating cyclic changes in climate is not known. However, they are certainly not responsible for causing ice ages or glacial–interglacial cycles.

18　Deep seas and oceans

Introduction

The deep sea is defined as water depth greater than 200 m, thereby excluding the continental shelf (Chapter 15). Deep-sea environments are present over 65% of the Earth's surface, and include ocean basins, oceanic trenches, island-arc basins, marginal seas, and downfaulted continental borderlands (Figure 18.1). Although these environments are in different physiographic and tectonic situations, they have certain features in common. There is a continuous supply to the upper parts of the sea of fine-grained organic and inorganic sediment, which settles slowly to the bed (to form pelagic and hemipelagic sediment). Sediment also commonly travels down basin slopes and submarine canyons as sediment gravity flows (turbidity currents, debris flows, and grain flows: Chapter 8) and is spread over submarine fans and basin plains. Water currents due to tides and waves are relatively unimportant, but currents due to wind shear and density contrasts (related to variations in temperature and salinity) are important. These currents are especially significant in moving sediment within the water column and on the bed of deep-sea basins.

Major advances in technology during the latter half of the twentieth century led to revolutionary developments in ocean science, including development of the theories of sea-floor spreading and plate tectonics; acquisition of much more detailed knowledge of ocean-floor geometry and sediment deposits; the realization that sediment gravity flows and deep-sea water currents are responsible for transporting sand and gravel to and around the deep sea, and that these sediment deposits may contain vast quantities of hydrocarbons; great increases in knowledge of oceanic water chemistry, physics, and biology, and how these are linked to the atmosphere and global climate change; and the discovery of new ecosystems associated with

mid-ocean-ridge volcanicity and within sediment pore waters.

Useful general references dealing with oceanography are Weaver and Thompson (1987), Open University (1989), Pernetta (1994), Summerhayes and Thorpe (1996), Thurman and Trujilo (2004), and Pinet (2006). Many references review sediment transport and deposition, particularly associated with deep-sea fans: Hsü and Jenkyns (1974), Bouma *et al.* (1985), Pickering *et al.* (1989, 1995), Weimer and Link (1991), Stow (1994), Weimer *et al.* (1994), Hartley and Prosser (1995), Clark and Pickering (1996), Lizitsin (1996), Stow *et al.* (1996, 2002), Posamentier and Kolla (2003), and Posamentier and Walker (2006). Basic references on deep-sea environments dominated by carbonate sediments include James and Ginsburg (1979), Crevello and Schlager (1980), Cook *et al.* (1983), Cook and Mullins (1983), Enos and Moore (1983), Scholle *et al.* (1983), McIlreath and James (1984), Tucker and Wright (1990), Coniglio and Dix (1992), Ginsburg (2001), and Schlager (2005).

Problems in studying deep-sea environments and their deposits

Deep-sea environments and their deposits are largely inaccessible, and study of the deep sea poses many problems related to its great area, volume, and depth of water. Ships are relatively slow-moving, and deployment of equipment and people to oceanic depths and pressures is very challenging, as is their accurate positioning. The general morphology and sediment cover of ocean basins was established in part during the voyage of HMS Challenger (1872–1876). At that time it was concluded that the ocean floors were still and covered with mud. In the twentieth century, it was realized that the oceans held the key to many questions about the evolution of the Earth and its atmosphere. In addition, knowledge of

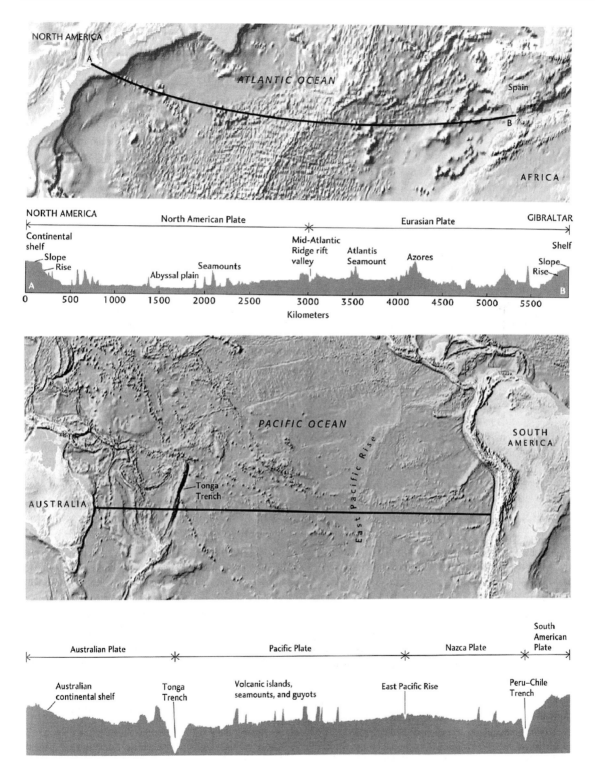

FIGURE 18.1. Maps and cross sections showing the physiography of the Atlantic and Pacific Oceans. From Press *et al.* (2004).

deep-sea environments was needed for the laying of communications cables, submarine operations, and plans to dispose of human waste. Enormous advances in our knowledge of the deep sea, related to major development of technology, came in the latter half of the twentieth century. Satellites were used for altimetry and other types of remote sensing, and highly accurate positioning (GPS). Greatly improved drilling of the ocean floor came with the Deep Sea Drilling Project (DSDP) and the Ocean Drilling Project (ODP), and now with the Integrated Ocean Drilling Project (IODP). Side-scan sonar (e.g., GLORIA, TOBI) was used to image the ocean floor and shallow subsurface in great detail. Deep-sea submarines (e.g., Alvin) were developed. Automated, moored submarine recorders and samplers have yielded an enormous amount of information on water temperature, salinity, turbidity, flow velocity, water-surface elevation, ocean biology, and sediment type and movement. Oxygen-isotopic studies of calcareous microfossils have been used to deduce past oceanic water temperatures. Three-dimensional seismic reflection studies have revealed a gold mine of information on subsurface sedimentary strata and geomorphology (e.g., Posamentier and Kolla, 2003; Posamentier and Walker, 2006).

Use of these new measurement techniques resulted in development of the revolutionary theories of sea-floor spreading and plate tectonics starting in the 1960s. Knowledge of the spatial and temporal variation of the physiography of the ocean floor, and the physical, chemical, and biological properties of water and sediment, and their movement, has greatly improved. It is now realized that the ocean floor is far from still, and that vast quantities of sand and gravel are transported down oceanic margins to be deposited on deep-sea fans. The use of submarines has revealed much new information on deep-sea volcanicity and biology, including the discovery of a new ecosystem associated with mid-oceanic-ridge volcanicity (black smokers; e.g., giant tube worms).

Geometry of deep-sea environments

The main deep-sea environments are basin margins, basin plains, and mid-oceanic ridges and volcanic hills. As will be seen below, the geometry and nature of development of these environments are related to controlling factors such as sediment supply, tectonic activity, volcanism, and sea-level change (e.g., Mutti,

1985; Shanmugan and Moiola, 1988; Reading and Richards, 1994).

Basin margins

Reading and Richards (1994) distinguished basin-margin environments according to the mud/sand ratio of the sediment supply, and to whether the sediment is supplied at a point (submarine canyon and fan), at multiple points (ramp), or as a linear source (slope apron). The geometry of basin margins also varies depending on whether or not they are tectonically active (rather than passive), fed by major rivers, or glaciated. Furthermore, ocean margins bordering carbonate shelves with marginal shoals or reefs differ somewhat from those with a smoother transition from shelf to deep sea (Chapter 15). In the case of "passive" continental margins adjacent to spreading oceans (Chapter 2), the oceanic edge of the continental shelf passes into a relatively steep *continental slope*, with a gradient ranging from about 1/50 to 1/5 (1 to 10 degrees) (Figure 18.1). Continental slopes are a few tens of kilometers across, and contain submarine canyons, particularly opposite the mouths of major rivers. Submarine canyons have widths of kilometers and depths of hundreds of meters. Their locations are controlled partly by the mouths of major rivers that supply sediment to canyon heads, and partly by pre-existing topography along their courses created by erosion, deposition, and tectonism. Continental slopes also contain gullies (channels), depressions, and mounds related to the movement of slides and sediment gravity flows, collectively called slope aprons. Sediment drifts deposited by bottom currents (called contourite drifts and contour currents, respectively) also occur on continental slopes, although they are more common on continental rises (Stow *et al.* 2002).

The base of the continental slope of passive continental margins is marked by a relatively abrupt decrease in slope (gradients 1/100 to 1/700: a fraction of a degree) comprising the *continental rise*. The continental rise is hundreds of kilometers across, and contains *submarine fans* adjacent to slope canyons. Fans can be tens to hundreds of kilometers across and along stream, and up to kilometers thick, and have concave-upward along-stream profiles. Average slopes are 10^{-3}–10^{-2}, with the larger, mud-rich fans having lower slopes than the smaller, sand-rich fans. The terms inner, middle, and outer fan are used commonly, and defined in Figure 18.2. Geometrical features of slopes and submarine fans are summarized in Figure 18.3.

Fan channels are up to kilometers wide and tens to hundreds of meters deep. Fan channels have a range of channel patterns, including meandering and braided, and contain features such as point bars, braid bars, and cut-offs (Figure 18.4). Width/depth ratios of channels are shown in Figure 18.5. Some channels become distributive and decrease in size downfan, and pass into terminal lobes (Figure 18.6). Quantitative studies of submarine-fan geometry include Flood and Damuth (1987), Clark *et al.* (1992), and Clark and Pickering (1996). Levees generally occur adjacent to fan channels (Figure 18.7). Levees on the inner fan can be hundreds of meters high and tens of

kilometers wide. However, they tend to decrease in height and width downfan. Levees tend to be higher on the right (or left) side of the channel in the northern (or southern) hemisphere because of the effect of the Coriolis force on overbank flows. Crevasse channels and splays occur adjacent to some channels (Figure 18.8). Other geometric features on submarine fans and channels include giant sediment waves (possibly dunes or antidunes) that are meters to tens of meters high and tens of meters to kilometers long, and arcuate scours that are meters deep and hundreds of meters across (there may be slump scarps or flute-like current scours) (Figure 18.9).

The continental rise is also where contourite drifts of sediment (deposited by contour currents) are most likely to occur, although they occur also on continental slopes and abyssal plains (Stow *et al.*, 2002). Contourite drifts are typically up to a few hundred meters thick, tens to hundreds of kilometers long, and cover areas on the order of 10^3–10^5 km^2. Their form can be sheet-like, elongated mounds, and fan-shaped (at the mouths of submarine channels). Their form may also correspond to the shape of a space that confined the contour current. These drifts are composed of the sediment that was available for transport, typically muds to fine-grained sands. Their upper surfaces may be covered with bed forms such as giant waves, dunes, ripples, and erosional lineations.

Sediment beneath the continental slope and rise of passive continental margins is several kilometers thick, borders continental-shelf deposits upslope and deep-oceanic sediment downslope, and rests upon oceanic–continental crust. With a tectonically active

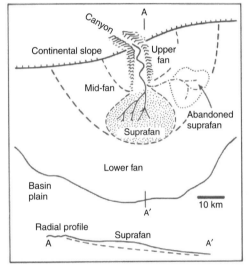

FIGURE 18.2. Definition of upper (inner), middle, and lower (outer) fan. From Walker (1992), after Normark (1978).

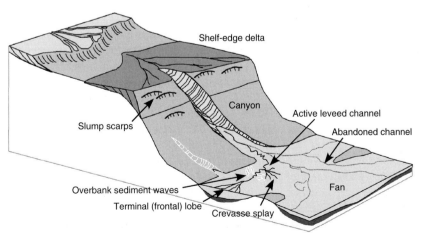

FIGURE 18.3. A summary of geometric features of slopes and submarine fans. Modified from Posamentier and Walker (2006).

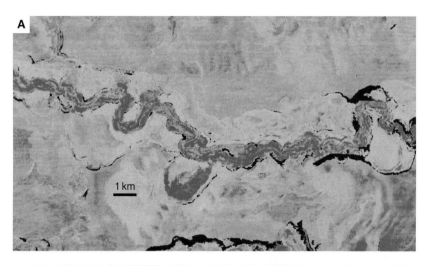

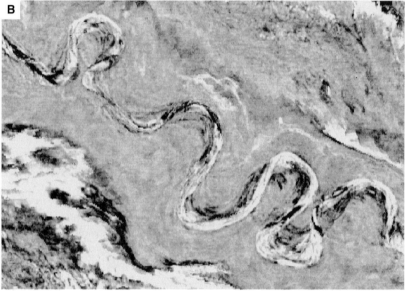

FIGURE 18.4 (A) A meandering channel belt with abandoned meander loops, as seen on an amplitude map from a three-dimensional seismic time slice (sea floor plus 80 ms). (B) A seismic-amplitude-extraction horizon slice showing meandering channels with lateral accretion topography and incipient cut-offs. Flow is to the southeast. The meander loops have migrated by bend expansion and downvalley translation, as is common for river meanders. The field of view is about 10 km across. Joshua Channel, Gulf of Mexico. From Posamentier and Walker (2006).

(convergent) continental margin (Chapter 2), the continental rise is replaced by an *oceanic trench* that may be up to 11 km deep (Figure 18.1). The deepest oceanic trench (11,033 m) is the Marianas trench, east of the Mariana Islands near Japan. Sediment deposits within oceanic trenches are deformed during subduction into an "accretionary wedge" (see Chapter 20). In some cases, the oceanic trench is adjacent to a volcanic island arc, with a forearc basin on the trench side, and a back-arc basin on the continental side. In other cases, the oceanic trench is adjacent to a continental shelf.

Continental slopes adjacent to rimmed carbonate shelves (e.g., Florida, the Bahama Banks, Belize, Jamaica, the Great Barrier Reef of Australia) commonly differ from those adjacent to siliciclastic shelves, primarily in their larger slopes and the absence of *major* submarine canyons and fans. The upper portions of nearly all siliciclastic continental slopes rarely reach slopes of 10° and most are about 4°, regardless of the water depth they lead to (Figure 18.10A). Carbonate slopes can exceed 30°, and deeper basins have steeper marginal slopes (Figure 18.10A). Many of the

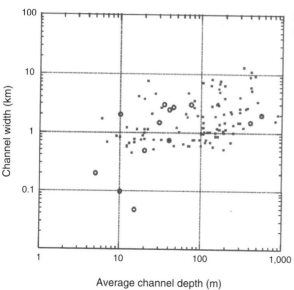

FIGURE 18.5. The width and depth of modern submarine channels. From Clark and Pickering (1996).

- Modern submarine channel data
- Data from Eocene Hecho Group submarine channels

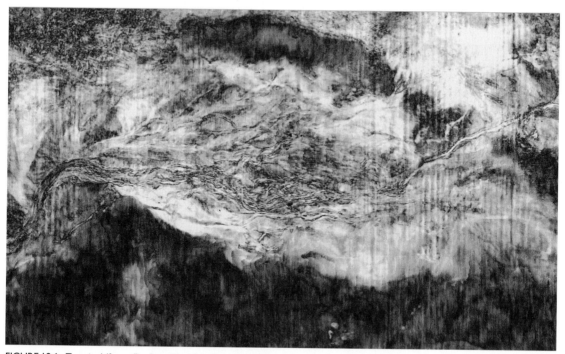

FIGURE 18.6. Terminal (frontal) splay with a distributary channel network (flow to the right) as seen in a seismic-amplitude-extraction horizon slice. The field of view is about 25 km across. Gulf of Mexico. From Posamentier and Walker (2006).

steepest carbonate slopes are made of rock and can have stretches of rocky cliffs that are many hundreds of meters high. The steepest carbonate slopes composed of unconsolidated sediment are talus piles of coarse sand and gravel (Figure 18.10B).

The Florida–Bahama Banks have a complicated geomorphology and variable marginal slopes (Figure 18.11). The margins around the Bahama Banks are commonly referred to as erosional, bypass, or accretionary, depending on their height and

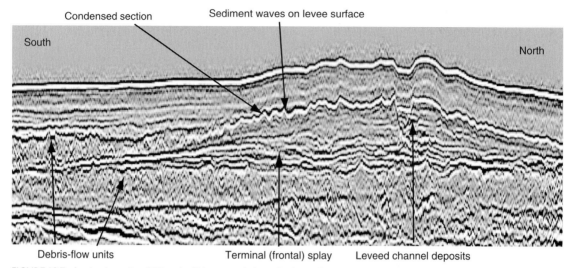

Condensed section

Sediment waves on levee surface

South

North

Debris-flow units   Terminal (frontal) splay   Leveed channel deposits

FIGURE 18.7. A seismic section (300 ms by 10 km across) through a buried leveed channel, showing a vertical sequence of debris-flow deposits overlain by frontal splay (outer-fan) deposits, then levee–channel (middle-fan) deposits. The levee surface contains sediment waves. More debris-flow deposits occur on the side of the levee–channel complex. Makassar Strait, Indonesia. Courtesy of Henry Posamentier.

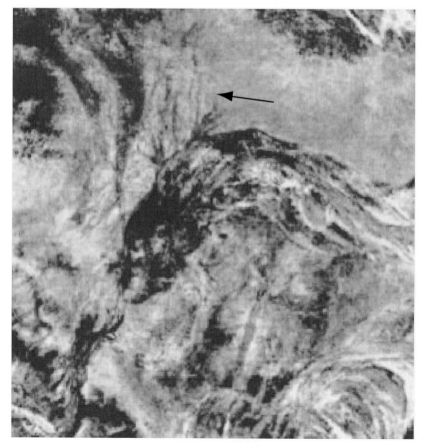

FIGURE 18.8. Crevasse splay (arrowed) associated with a channel bend, as seen on an amplitude extraction of a seismic horizon slice. The field of view is about 10 km across. Gulf of Mexico. Modified from Posamentier and Walker (2006).

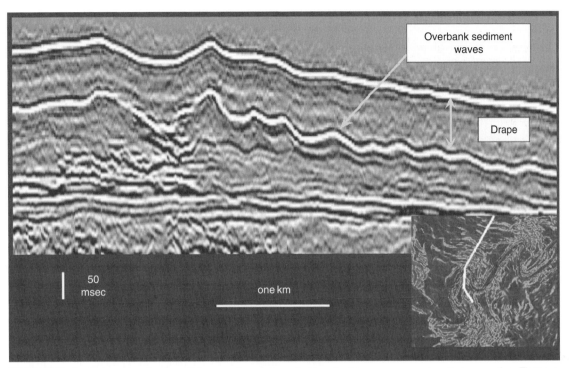

FIGURE 18.9. A seismic section showing a Pleistocene leveed channel with sediment waves preserved on the levee surface. The inset shows a reflection-dip-magnitude map for the levee surface with the location of the seismic section. The wave amplitude is greatest adjacent to the outside of the downstream parts of channel bends. Makassar Strait, Indonesia. Courtesy of Henry Posamentier.

steepness (Figure 18.11B) (Schlager and Ginsburg 1981). The steep *erosional* slopes (vertical cliffs in some places) have basal talus cones or sediment-gravity-flow deposits composed of material derived from the modern reefs and shoals at the platform margin and from the older rocks exposed in the cliff face (Figure 18.12). These slopes are maintained by catastrophic failure along faults, erosion by deep-sea currents, dissolution of aragonite sediment from groundwater seeps, and bioerosion. The *accretionary* slopes have deposition primarily along their upper reaches, whereas the *bypass* slopes have deposition primarily along their lower reaches. Mullins (1983) pointed out that the margin types defined by Schlager and Ginsburg (1981) change character along their lengths, and that a steep (commonly >45°) marginal escarpment rings the entire Bahamas Bank from bank-edge depths of 20–40 m to depths of 150–180 m (Figure 18.11).

Gullied slopes are common in the Bahamas (profile 2 in Figure 18.11, and Figure 18.13), as they also are on siliciclastic continental slopes (slope aprons). These slopes are divided into a 3°–5° upper slope some 10–20 km across and a gentler, broader lower slope

that leads to the basin floor (Figure 18.13A). The upper slope is cut by evenly spaced submarine gullies (channels) that are 0.25–0.5 km across and 20–150 m deep. The lower slope has a hummocky topography in some areas. These slopes are mostly composed of carbonate mud. The marginal escarpment and upper portion of the gullied slope are thoroughly cemented hardgrounds that pass downwards to a zone where the cementation is confined to discrete nodules (Figure 18.13B). Seismic profiles of the gullied slope (Figure 18.13C) indicate that rotational and translational slumps are common. Crevello and Schlager (1980) documented that slumping of another gullied slope generated debris flows and associated turbidity currents that left deposits up to 3 m thick that cover 2,100 km^2 of the floor of Exuma Sound (Figure 18.14). The debris-flow deposits are composed of clasts of cemented slope deposits and platform-derived mud and sand. Two cone-shaped sand "drifts" off the northwestern corners of the Great Bahama Bank and Little Bahama Bank (labeled 3 in Figure 18.11) were presumably shaped by the Gulf Stream that flows northwards through the Straits of Florida.

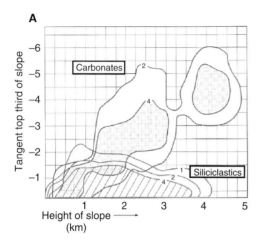

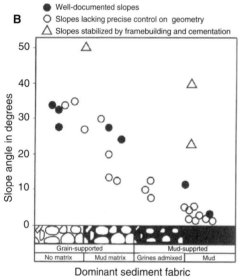

FIGURE 18.10. (A) Modern slope angles of upper portions of 413 carbonate and 72 siliciclastic slope deposits. Contours are percentages of total deposits that occur in each block of the diagram (0.25 km × 0.05 tangent of slope). The 4% contour is shaded. (B) The influence of sediment composition on the angles of carbonate slopes. Both images from Schlager (2005).

Basin plains

Basin plains are the relatively flat, extensive plains that form the floors of deep seas. The deep-ocean floor (depth >4 km) occupies ~29% of the Earth's surface. Basin plains are primarily depositional features and the sediment thickness typically ranges from a few hundred meters to kilometers, depending on the amount of tectonic subsidence. Abyssal plains occupy large areas of the deep-ocean floor seawards of continental rises, particularly in the Atlantic and Indian Oceans, and typically lie at depths of 4–6 km (Figure 18.1).

Gradients are generally less than 1/1000, although deep-sea channels and contourite drifts occur, as do volcanic hills with relief commonly of hundreds of meters. Sediment is typically a few hundred meters thick beneath abyssal plains and buries small abyssal hills and other irregularities of the basaltic oceanic crust. Basin plains also occur in oceanic trenches and in rift basins in areas of crustal extension (e.g., mid-ocean ridges and back-arc basins).

Volcanic hills and mid-oceanic ridges

Most of the floor of the Pacific Ocean, and about half the floor of the Atlantic and Indian Ocean, are covered with volcanic hills or mountains. Abyssal hills are less than 1 km high and 10 km in diameter, whereas sea-mounts are greater than 1 km high and tens of kilometers in diameter. Their diameter/height ratio is typically about 20. Some volcanic hills or mountains have craters or calderas up to kilometers across and hundreds of meters deep. Most of these volcanoes are extinct (and aseismic), but some are active. Active volcanoes are found at the ends of linear chains of shield volcanoes, particularly in the Pacific Ocean. The Hawaiian Islands are thought to be the result of the drift of the Pacific Plate over a more or less fixed hotspot in the mantle beneath. These volcanic islands rise well above sea level from abyssal depths (total height of up to 9 km), and their basal diameters are on the order of hundreds of kilometers. The largest, youngest Hawaiian island currently sits over the hotspot and has active basaltic volcanism. The other islands in the chain become older, more eroded, and lower in elevation above sea level (due to increasing thermal subsidence) with increasing distance from the hotspot.

If the oceanic volcanic islands are in the tropics, they develop fringing reefs as they become older, lower, and more eroded. Coral atolls are edifices of coral built atop subsided volcanoes where the corals and other carbonate-secreting organisms have been able to grow upwards to keep pace with the thermal subsidence of the volcano. Perhaps the most famous atoll is Bikini Atoll, one of the Marshall Islands in the central Pacific, which was the site of nuclear-bomb tests by the USA in the 1940s and 1950s. As a volcano subsides, there is a progression from fringing reefs to barrier reefs (when the volcano is still above water) to an atoll (when the volcano is under water). Guyots are flat-topped seamounts that are sunken atolls. In these cases, the corals of the atoll could not grow upwards quickly enough to keep up with the subsidence of the

A

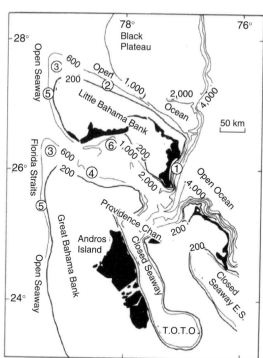

FIGURE 18.11. Slopes around the margins of the Bahama Banks. (A) Bathymetry and types of platform margins based on Mullins (1983). Contours are in meters. Numbers refer to profiles in (C); north is to the top of the page; TOTO, Tongue of the Ocean; and ES, Exuma Sound. (B) Slope profiles of the Bahamas classified according to Schlager and Ginsburg (1981). The vertical exaggeration is 4. From Tucker and Wright (1990). (C) Profiles of various slopes labeled in (A). Modified from Tucker and Wright (1990). Note that the vertical and horizontal scales are different.

B

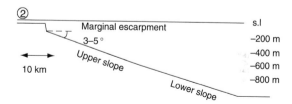

C

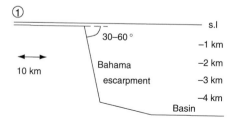

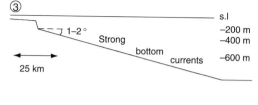

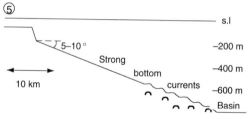

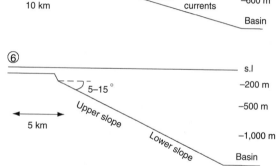

FIGURE 18.12. Talus slope at the foot of the submarine escarpment off the Belize reefs. From James and Ginsburg (1979). (A) Slope composed of plates of coral broken from the reef and large blocks of limestone surrounded by *Halimeda*-rich sand. The block at the arrow is ~1 m in diameter. (B) A close-up of another limestone block some 5 m high resting on talus slope.

seamount. Subsided volcanic edifices may be 2–3 km below sea level, and rise the same distance above abyssal plains.

The sides of active submarine volcanoes are typically covered with sheet lavas and pillow lavas, and large blocks of (chilled) lava that have been broken up by thermal contraction. Pyroclastic debris may also come from subaerial and shallow submarine eruptions. The loose material on subaerial and submarine volcanoes can be transported downslope in sediment gravity flows (Chapter 8), and also moved around by rivers and a variety of marine currents. However, as the volcanoes become inactive and subside further below sea level, they become covered with non-volcanic sediment, as will be seen below.

Mid-oceanic ridges (~23% of the Earth's surface) comprise an Earth-circling, submarine mountain belt that is 1,500–2,000 km wide, stands 1–3 km above the general level of the sea floor, and reaches the surface of the ocean in places (e.g., Iceland, Figure 18.1). Mid-oceanic ridges are regions of volcanicity, earthquakes, and ocean-floor spreading (Chapter 2). Mid-oceanic ridges are cut by transverse fracture zones containing

deep canyons and rugged ridges, and these fracture zones continue out into the adjacent ocean floor. The geometry of mid-ocean ridges is related to the rate of sea-floor spreading. Mid-oceanic ridges in the Atlantic and Indian Oceans (where the spreading rate is relatively slow) have considerable local relief, with a central rift valley 25–50 km wide and 1–2 km deep. The East Pacific Rise (where the spreading rate is fast) is broader and not as high (1–2 km) as the Atlantic and Indian Ocean ridges, and does not have as pronounced a central axial rift. The central axial rifts of the mid-oceanic ridges are sites of submarine volcanic vents and hydrothermal vents (black and white smokers: Chapter 4), and ridge flanks are sites of less intense hydrothermal circulation. The flanks of the ridges are covered with sheet and pillow lavas, and broken chunks of lava accumulate as talus. The basins adjacent to the volcanic ridges are filled with sediments, as described below. The progressive decrease in height of the mid-oceanic ridges away from their central axes is interpreted as being the result of cooling, contraction, and increasing density of the ocean crust (thermal subsidence). The basin-plain sediments that lie over the basaltic oceanic crust become progressively thicker down the flanks of the ridges away from the central rifts. Both the basalts and the sediments become older with increasing distance from the mid-oceanic ridge, due to sea-floor spreading.

General properties and movement of water and sediment in deep seas

Seawater composition, temperature, and movement

Ocean water is part of a coupled oceanic–atmospheric circulation system that distributes mass and energy around the Earth: it is driven by the unequal distribution over the Earth of heat energy from the Sun. The oceanic circulation system involves plumes of water and sediment emanating from the mouths of large rivers; surface currents related to wind drag on the water surface; pressure-gradient currents associated with variations in the elevation of the ocean surface; and thermohaline currents due to density differences arising from variations in ocean temperature and salinity (Chapter 2). These water currents are modified by the Coriolis force, bed friction, and friction between water bodies moving at different speeds. Interfaces between oceanic water bodies with different speeds and/or densities are commonly wavy, and they may

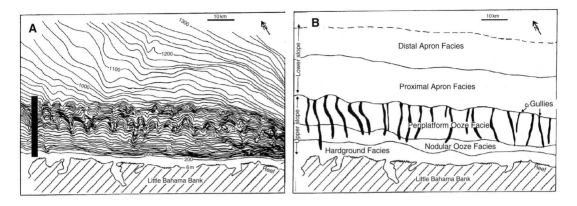

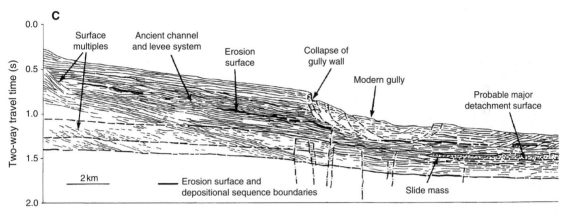

FIGURE 18.13. Topography and deposits of the northeastern slope of Little Bahama Bank (located approximately at the location denoted by 2 in Figure 18.11A. From Coniglio and Dix (1992). (A) A contour map, with contour intervals of 20 m below the 200-m contour. (B) A sediment surface map, with gullies indicated by thick black lines. (C) An interpreted seismic line through the upper slope, at the approximate location indicated by a bar in (A).

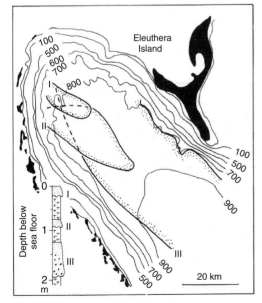

FIGURE 18.14. Extents of three debris-flow and turbidity-current deposits (labelled I, II, and III) on and beneath the surface of Exuma Sound. From Tucker and Wright (1990).

be associated with vortical motions (eddies) that mix the water bodies. Water-surface waves also cause mixing of water bodies down to depths of a couple of hundred meters.

Surface-water circulation

The upper 100 m or so of the ocean experiences direct interaction with the Sun's radiation and with the Earth's atmosphere. In this near-surface region, the water temperature, salinity, turbidity, light penetration, and oxygen content vary greatly with depth and laterally. Wind-shear currents are very important in this surface layer, and tend to move in circulatory patterns (gyres) related to global wind patterns, Coriolis deflection, and continental configuration (Figure 18.15). The major subtropical gyres in the Pacific and Atlantic Oceans result from the action of the tropical trade winds and the higher-latitude westerly winds. The gyres in the northern hemisphere rotate clockwise, whereas those in the southern hemisphere

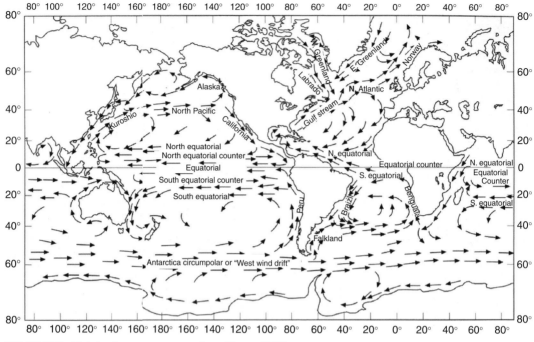

FIGURE 18.15. Global surface ocean currents. From Chester (2003).

rotate counter-clockwise. The currents associated with these gyres commonly have specific names (Figure 18.15). The water surface is relatively high (super-elevation in excess of 1 m) over the subtropical gyres because the Coriolis force causes Ekman transport of surface water towards their centers (Figure 18.16). The super-elevation of the water surface and the center of circulation of these gyres are displaced to the western sides of the oceans. This results in relatively narrowly confined and fast surface currents on the western sides of the oceans (so-called western intensification), but broad and slow currents on the eastern sides. Surface ocean currents normally have velocities on the order of 10^{-1} m s^{-1}. However, intensified surface ocean currents on the western sides of oceans may have flow velocities in excess of 1 m s^{-1}. These currents and the eddies along their edges (Figure 18.17) can extend to the ocean floor, affecting the deep thermohaline currents, and causing deep-water "storms" (Hollister and McCave, 1984).

The well-known Gulf Stream flows through the Straits of Florida, northwards towards Cape Hatteras, and then across the North Atlantic towards northwest Europe. The western edge of the Gulf Stream is marked by a large temperature gradient

(Figure 18.17) known as the "Cold Wall," and a zone of intense shear marked by concentrations of floating junk and rapid change from turbid, cold continental-slope water to the distinctive warm and deep blue of the Gulf Stream. The Gulf Stream meanders and generates eddies with vertical axes that extend to depths of kilometers, causing "storms" in the ocean that have lifetimes of a few years (Figure 18.17). Incidentally, the warm eddies are responsible for periodic invasions of exotic warm-water species (such as the Portuguese Man-O'-War or the killer shark of *Jaws*) into coastal areas of eastern North America.

Convergence of the trade winds in the equatorial region gives rise to westward-flowing equatorial currents, Ekman transport towards the poles, and eastward-flowing counter currents (Figure 18.15). There are some major differences between the equatorial currents in the Atlantic and Pacific Oceans. The Atlantic Equatorial Counter Current is restricted to the east side of the tropical Atlantic. The south Equatorial Current in the Atlantic divides where it impinges on Brazil and a portion flows north into the Caribbean. There is also a remarkable subsurface current in the equatorial Pacific called the *Equatorial Undercurrent* that has no counterpart in the Atlantic (Figure 18.15). This current is an

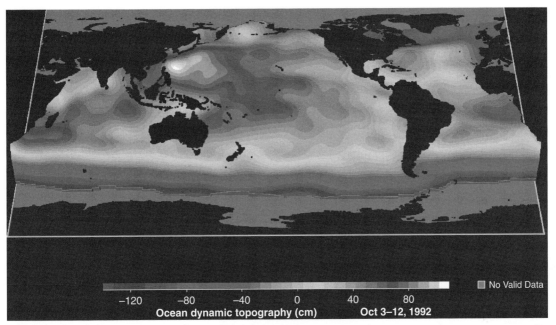

FIGURE 18.16. Ocean dynamic topography derived from satellite altimetry, showing high water levels where water is warm and in the center of subtropical gyres. From NASA Jet Propulsion Lab.

eastward-flowing current embedded in the westerly-flowing South Equatorial Current in the Pacific.

The Atlantic, Pacific, and Indian Oceans communicate in the so-called *Southern Ocean* around Antarctica. Wind-driven circulation in the Southern Ocean is the westerly flowing *Antarctic Circumpolar Current* (a.k.a. the West Wind Drift), a deep, relatively slow $(4–15\,\text{cm}\,\text{s}^{-1})$ current that encircles Antarctica.

Langmuir circulation comprises longitudinal vortices oriented parallel to the wind direction that occur in the upper part (meters to tens of meters) of oceans and lakes. The vortices are comparable to the Taylor–Gortler vortices present in shear flows in rivers and in the atmosphere (see Chapters 5 and 6), and occur as pairs of vortices with counter-rotating helical flow that affect mixing in the upper layer of the oceans or lakes. Pairs of vortices typically have transverse spacing of tens of meters but this can reach 300 m, and the vortices extend downwards from the water surface for meters to tens of meters. Langmuir circulation is visible on the water surface because of parallel streaks of flotsam (windrows) that occur where surface water converges between adjacent vortices. If the wind changes direction, the streaks will be realigned in a matter of minutes. Langmuir circulation appears to be due to interactions between the wind-shear current

in the surface layer and Stokes drift from wind waves (Chapter 7) (Pickard and Emery, 1990).

An important aspect of surface ocean currents is their effect on vertical water motions (upwelling and downwelling). Upwelling occurs at the boundaries of subtropical and subpolar gyres, because of surface-water flow away from these boundaries (e.g., due to Ekman transport). Upwelling also occurs where the surface flow is away from continental margins (e.g., off the western coasts of Africa and the Americas). Upwelling brings nutrients such as phosphates and nitrates to the surface, increasing the productivity of plankton. This increased productivity works its way up through the food chain. The increase in biomass results in an increase in deposits of the hard parts of plankton (e.g., diatoms, radiolarians) and fish, and possibly also phosphatic sediment grains.

The intensity of the surface ocean–atmospheric circulation varies with time over a range of time scales. For example, temporal reduction in the strength of the trade winds and the velocity of westward-directed surface currents over periods of decades causes relatively warm surface water to move from the west Pacific to the east Pacific (ENSO, or El Niño Southern Oscillation). An increase in current velocity causes the surface water of the eastern Pacific to cool (called La Niña). This affects the distribution of precipitation on

A

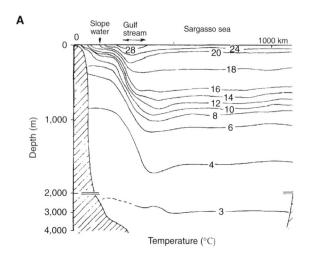

FIGURE 18.17. The Gulf Stream: (A) a west–east cross section showing the large temperature gradient on the western margin of the Gulf Stream; and (B) an infrared satellite image showing the meanders and eddies of the Gulf Stream in the western North Atlantic. From NASA Jet Propulsion Lab.

B

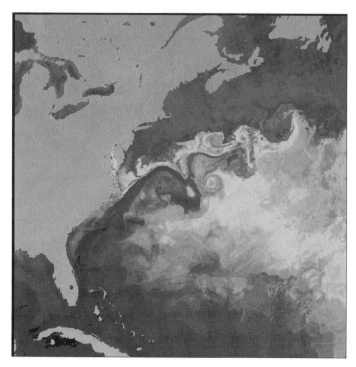

the continents and oceanic upwelling around the Pacific rim. Seasonal or decadal variations in upwelling and biological productivity result in temporal and spatial variation in biogenic sediment production.

Deep-water (thermohaline) circulation

Water circulation below the surface layer is due mainly to spatial variations in water density, which are related to the water temperature and salinity (thermohaline circulation). Temperature variations originate from the unequal distribution over the Earth's surface of the Sun's radiant energy. Salinity variations arise from variation of the evaporation rate, melting and freezing of ice (related to the Sun's radiant energy), and input of freshwater from large rivers. Thermohaline circulation in the Atlantic Ocean can be seen by considering a cross section of temperature

A

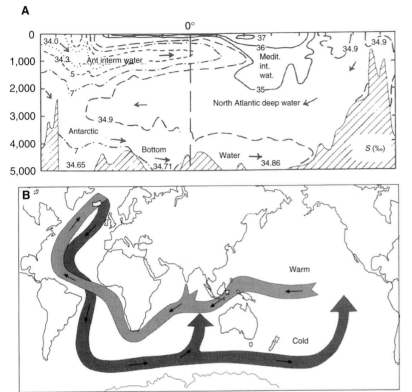

FIGURE 18.18. Thermohaline circulation. (A) A cross section of the western Atlantic Ocean, showing salinity. Major water masses are labeled. From Pickard and Emery (1990). (B) The "global ocean conveyor" according to Broecker and Dent (1989). Deep current is dark gray; return intermediate-depth currents are light gray.

and salinity (Figure 18.18A). The lens of warm, saline water at the surface (known as *Atlantic Surface Water*) is warmed by solar radiation, and is made more saline by the preferential evaporation of water from the subtropical zones of standing high pressure. Nestled below the Atlantic Surface Water in the northern Atlantic is a lens of *Mediterranean Intermediate Water* that comprises warm salty water that flows over the sill at the Straits of Gibraltar (Figure 18.18A). Most of the Atlantic is filled with *North Atlantic Deep Water* (NADW). The source of NADW is cooling of seawater in the North Atlantic between Greenland and Europe. The slow flow of NADW is restricted in places by submarine topography, and is influenced by the Coriolis force. Bottom water moving from polar regions towards the equator is deflected towards the western sides of ocean basins by the Coriolis force, where it flows parallel to topographic contours, mainly along continental rises (hence the name contour currents). The NADW is detectable all the way to the southern Atlantic Ocean. Atlantic Surface Water mixes with a thin layer of *Antarctic Surface Water* at the *Antarctic Ocean Convergence* at latitude about 50° S. This mixture forms *Antarctic Intermediate*

Water that sinks and flows northward to latitude about 15° N, insinuating itself beneath Atlantic Surface Water and on top of North Atlantic Deep Water. On the Weddell Sea continental shelf of Antarctica, the freezing of ice and extremely cold temperatures generate a cold, salty, and therefore extremely dense water mass that flows down the continental slope and northwards beneath the NADW. This *Antarctic Bottom Water* is also principally found in the western deep portions of the Atlantic and can be detected as far north as New Jersey. The southern Atlantic Ocean off the Weddell Sea is one of the areas where NADW preferentially returns to the surface, where it mixes with the *Antarctic Circumpolar Water*. Antarctic Circumpolar Water is found beneath the Antarctic Circumpolar Current and is flowing into the plane of the cross section (Figure 18.18A) from the Pacific Ocean.

The NADW receives a continual rain of particulate organic matter from the photosynthetic surface waters above it. The particulate organic matter comprises copepod and other fecal pellets (fancifully referred to as "marine snow"), and vertebrate and invertebrate carcasses of various sizes. As the particulate organic matter settles through the water column into the deeper water masses, it is subject to bacterial decomposition. As a result, the dissolved-oxygen content of NADW is continually lowered during its southward journey, and the dissolved organic content (i.e., nutrients) of NADW increases concomitantly. The continual transfer of particulate organic matter and nutrients from the surface waters into the aphotic deep ocean is known as the *biological pump*. Where deep water returns to the surface, as NADW does off the Weddell Sea, it returns the dissolved nutrients back to the surface and effectively fertilizes the surface waters. The southern Atlantic Circumpolar Water has very high productivity and lots of animals. In fact, filter-feeding whales that consume planktonic crustaceans (commonly known as krill) were abundant in these waters and attracted nineteenth- and twentieth-century whaling fleets.

In the Pacific Ocean, *Pacific Surface Water* mixes with Antarctic Surface Water at the Antarctic Ocean Convergence and forms *Pacific Antarctic Intermediate Water* that flows northwards, but not to the extent that Atlantic Antarctic Intermediate Water does. Because the Pacific communicates with the Arctic Ocean only across the very shallow Bering Sea, very little deep water forms in the north Pacific. Instead, the Pacific Ocean and Indian Ocean are both filled with cold water (known as *Common Water*) that originates off Antarctica. Common Water is entrained by the steady, deep Antarctic Circumpolar Current and is dragged around into the Atlantic where it is known as Antarctic Circumpolar Water. Common Water is very uniform in composition and has a lower dissolved-oxygen content and a higher content of nutrients and dissolved inorganic carbon than does NADW. The continental margin of western North and South America is one of the places where Common Water returns to the surface, and the dissolved-oxygen level below these fertilized areas approaches zero below the thermocline (this is further discussed below). The implication is that Common Water has a longer residence time (i.e., it is older) than that of NADW.

Broecker and Denton (1989) postulated that, to achieve salt balance, NADW forms part of a much larger, global system of deep flow now popularly referred to as the "global conveyor belt" (Figure 18.18B). The NADW is entrained by the Antarctic Circumpolar Current, and this deep conveyor then distributes NADW throughout the oceans. Broecker and Denton (1989) postulated a return conveyor that operates at intermediate depths (Figure 18.18B). It takes a water molecule 10^3–10^4 years from sinking at the North Pole to return to the surface, and maximum upward flows are calculated to be on the order of millimeters per day. As discussed in Chapter 17, the occurrence of the last ice age has been attributed to a shutting down of this global conveyor belt. Briefly, if the flow of NADW ceases, the carbon dioxide content of the deep oceans increases at the expense of that in the atmosphere, and the Earth cools. When the Earth is cold enough, the global conveyor belt can restart.

Thermohaline-current velocities are typically on the order of $10^{-2}\,\mathrm{m\,s^{-1}}$, but locally can be on the order of $10^{-1}\,\mathrm{m\,s^{-1}}$ where flow is forced to converge by the Coriolis force and by submarine topography. Although contour (bottom) currents flow approximately parallel to slope contours, their direction and velocity are quite variable (Stow *et al.*, 2002). Temporal variations in current velocity and direction can be seasonal, and fluctuations in velocity have been related to the influence of wind-driven flows (the deep-sea "storms" mentioned above). Longer-term periodicity of thermohaline currents is expected to result from climatic periodicity. Results of isotopic studies of benthic foraminifera suggest that generation of NADW has varied over glacial–interglacial periods due to changes in the extent of the northern-hemisphere ice sheet.

Sediment supply

The sources of sediment supply to the deep sea include (1) continental-shelf sediments transported to the edge of the shelf by fluvial, wind-wave, wind-shear, and tidal currents; (2) sediment derived from cemented shelf-edge deposits such as organic buildups and carbonate sand shoals; (3) sediment released by melting of icebergs; (4) dust from the atmosphere, associated with eolian transport, volcanic ash, and meteorites; (5) hard parts of planktonic (unicellular) organisms living in the upper 100 m of the ocean; (6) chemical precipitates on and just below the sea bed; (7) pellets produced by burrowing organisms; and (8) skeletons of deep-water organisms such as ahermatypic corals, sponges, and echinoderms.

Sediment derived from the bed of the continental shelf and shelf edge is mainly terrigenous or biogenic (especially carbonate), and can include various proportions of mud, sand, and gravel depending on what is available and the nature of the shelf water currents. Fragments of cemented shelf-edge deposits are mainly carbonates. Iceberg melt-out sediment is terrigenous and ranges in size from mud to gravel. Sediment from the atmosphere and from unicellular organisms is fine-grained (clay size), and is commonly referred to as *pelagic* sediment (the dictionary definition of *pelagic* is "of the open sea"). The term *hemipelagic* (literally, half pelagic) is used to refer to mud that is partly derived from the continental shelf and that is transported in suspension onto the continental slope and rise (Stow and Tabrez, 1998). Hemipelagic mud deposited adjacent to carbonate shelves (platforms) is commonly referred to as *periplatform* sediment. Abundant volcaniclastic sediment may come from active volcanoes, such as those associated with island arcs, parts of mid-ocean ridges, and other hotspots.

Downslope movement of sediment by gravity

Falls and slides of consolidated material, slumps, and sediment gravity flows (debris flows, grain flows, turbidity currents) move down basin margins and down submarine channels and the sides of volcanic hills. They originate by (1) oversteepening of slopes as a result of sediment deposition (e.g., by shelf currents, deep-sea currents, or sediment gravity flows); (2) oversteepening and/or a decrease in shear strength associated with earthquakes; (3) cyclic wave loading, especially associated with tsunamis and major storms; (4) hyperpycnal flows from turbid rivers; and (5) submarine lava flows and pyroclastic flows. During the last glacial period, sea level was in some cases near the edge of the continental shelf, such that fluvial and shallow-marine water currents transported sediment directly onto the top of the continental slope. Oversteepening of the top of the continental slope led to frequent sediment gravity flows and active deposition on continental slopes (slope aprons) and fans on continental rises. However, not all coastlines around the world were near the edge of the continental shelf during the last glacial lowstand of sea level. The association of volcanic eruptions and earthquakes is well known to generate submarine sediment gravity flows of loose volcanic material (Chapter 9), in some cases causing tsunami (Chapter 7).

The geometry and mechanics of downslope movement of debris under the influence of gravity are discussed in Chapters 8 and 12. Surface mapping and seismic data indicate that slumps and slides are widespread in the deep sea, especially on continental slopes and in submarine channels (Figure 18.3). They can move on slopes as low as one or two degrees. Debris flows, grain flows, and high-density turbidity currents are common within submarine canyons and fan channels, but lower-density turbidity currents are more common in interchannel regions and on the distal, unchannelized parts of fans and basin plains. Sediment gravity flows occur just about every year on some modern submarine fans, although evidence for the occurrence or otherwise of sediment gravity flows is necessarily difficult to obtain. Sediment gravity flows were probably much more frequent during sea-level lowstands (glacial maxima) than they are today, due to there being much more wind-wave influence and a greater deposition rate at the top of the continental slope.

Deposits of sediment gravity flows (details in Chapter 8) depend on the sediment source, and range from mud to sand and gravel. The sediment may be terrigenous (siliciclastic), biogenic, volcanic, or a combination. Sediment gravity flows off partly cemented carbonate shelf edges comprise a wide range of limestone types, including grainstones and packstones (calcareous sandstones with varying amounts of mud) and poorly sorted muddy gravelstones with clasts up to 10 m in diameter. Volcanic gravity flows can contain pyroclastic material and chunks of lava flows that were broken as they were chilled. Fossils in sediment-gravity-flow deposits are commonly transported shallow-marine shells, but deep-water benthonic or nektonic fauna occur in associated pelagic and hemipelagic muds. Trace fossils can be common in the pelagic and hemipelagic sediment between sediment-gravity-flow deposits, and were formed post-depositionally by a variety of infauna and epifauna (see Chapter 11).

Iceberg transport and deposition

Iceberg melt-out deposits are locally developed, unsorted diamicts with dropstones (lonestones). Such deposits will likely be difficult to distinguish from debris-flow deposits. However, they should stand out from deposits of turbidity currents and pelagic deposits. An increase in the amount of iceberg deposits interbedded with pelagic deposits (Heinrich events) is related to an increase in the amount and size of melting icebergs, which could have a variety of causes (Chapter 17).

Vertical settling of fine-grained sediment

Fine-grained sediment that accumulates by continuous slow settling from the upper ocean is commonly termed pelagic or hemipelagic sediment. Although this settling process occurs across the sea, deposits are clearly discernible only where deposition is not dominated by sediment gravity flows or contour currents. Indeed, deposits of mud from the final flow stages of turbidity currents are not considered to be pelagic sediments. Pelagic sediments are commonly classified according to the proportions of non-biogenic clay, calcareous biogenic clay, and siliceous biogenic clay (Stow *et al.*, 1996). Biogenic ooze contains more than 70% biogenic clay, is called siliceous ooze if more than 50% of the biogenic component is siliceous, and is called calcareous ooze if more than 50% of the biogenic component is calcareous. Brown clay contains more than 70% non-biogenic siliciclastic clay. Other names are given to intermediate pelagic compositions. Hemipelagic sediments are muds that are somewhat coarser than pelagic clays and contain a substantial quantity of sediment derived in suspension from the adjacent shelf.

Biogenic clay is composed of the hard parts of phytoplankton (calcareous coccolithophores, opaline siliceous diatoms, organic-walled dinoflagellates) and zooplankton (calcareous foraminifera and pteropods, opaline siliceous radiolarians). Foraminifera and coccolith plates are composed of low-Mg calcite, whereas pteropod tests (pelagic snails) are composed of aragonite. Although foraminifera and coccoliths are now responsible for about 75% of deep-sea carbonate sediment, these organisms arose only in the Mesozoic. Prior to this, biogenic pelagic sediments would have been mainly siliceous. There is a steady rain (or snow) of this calcareous, siliceous, and organic matter towards the sea bed. Some of the material is in the form of fecal pellets. The rate of production of this sediment is greatest in the upper 100 m of the ocean, and depends on the water chemistry, temperature, sunlight, water currents, and nutrient supply. Regions of highest biological productivity, and hence of greatest production of biogenic sediment, are where the nutrient supply is greatest; that is, where rivers enter the ocean and where there is upwelling of water rich in phosphate and nitrate. However, accumulation of biogenic sediment on the ocean floor also depends on what acts to concentrate it (e.g., water currents) and what tends to decompose it.

Surface ocean water (the top 100 m) is well oxygenated, as is most deep, moving ocean water. Therefore, dead organic matter is readily oxidized (decomposed). However, mid-depth ocean water is normally deficient in oxygen, especially at depths of 0.3–1.5 km. This is due to a maximum in the bacterial oxidation of organic matter. Furthermore, anoxic conditions can occur in

semi-enclosed seas that do not have vigorous water circulation. Density stratification related to salinity or temperature in these environments further reduces mixing and oxygenation of deep water. Organic matter can accumulate on the sea bed due to very high biological productivity or lack of oxidizing conditions, or both. However, slow reduction of some of this organic matter by anaerobic bacteria is possible, and results in the association between organic matter and iron sulfide and other metallic minerals (Wignall, 1994). Such organic accumulations may become source rocks for hydrocarbons if buried deeply and heated up over extended time periods.

Except for the surface layer (the top 100 m), ocean water is undersaturated with respect to calcium carbonate. Therefore, calcium carbonate should dissolve as it falls through the water. However, the dissolution rate is very slow until the depth exceeds about 3 km (the *lysocline*), below which there is selective dissolution of planktonic foraminifera species. The *carbonate compensation depth* (CCD) is the depth at which the rate of settling of carbonate through the water equals the rate of its dissolution. Little or no calcium carbonate sediment occurs below the CCD. The CCD varies in time and space with the temperature of deep ocean water. The CCD is ~4.5–5 km in the Atlantic Ocean, ~5 km in the Indian Ocean, and 3–5 km in the Pacific Ocean. The aragonite compensation depth is much shallower (usually less than 3 km), and in the tropical Pacific Ocean it is only a few hundred meters down (Chester, 2003). Calcareous oozes are therefore found in areas of high biogenic productivity above the CCD, particularly on the flanks of mid-ocean ridges (Figure 18.19). Siliceous oozes do not have such a depth restriction, and occur on the bed below regions of high biological productivity (Figure 18.19). Today, diatoms are most common at high latitudes, whereas radiolarians are more common at low latitudes. Biogenic oozes are not common near the centers of surface-current gyres because the biological productivity is low in these areas (Figure 18.19). Time-averaged deposition rates of biogenic oozes are typically 10^{-2}–10^{-1} mm per year. Temporal variations in supply rate or deposition rate of biogenic sediment should give rise to laminae or beds, but these are commonly destroyed by bioturbation and/or diagenesis.

The opaline silica of siliceous oozes is slowly transformed into microcrystalline silica (chert, flint) upon burial, such that the resulting rock is a nodular or bedded chert (Chapter 4). Lamination plus or minus organic matter has the same implications as for the

A

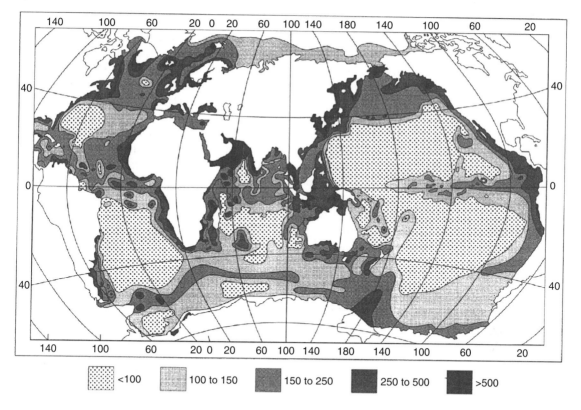

▦ <100	▦ 100 to 150	▦ 150 to 250	▦ 250 to 500	▦ >500

B

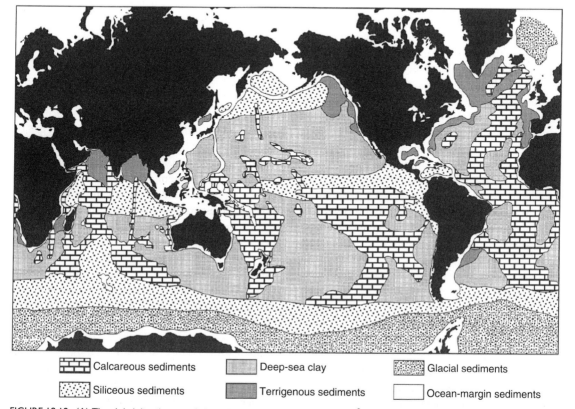

▦ Calcareous sediments	▦ Deep-sea clay	▦ Glacial sediments
▦ Siliceous sediments	▦ Terrigenous sediments	☐ Ocean-margin sediments

FIGURE 18.19. (A) The global distribution of phytoplankton productivity in mg cm^{-2} per day. From Stow *et al.* (1996). (B) The global distribution of types of pelagic sediment and other sediments on the ocean floors. From Stow *et al.* (1996).

calcareous oozes. Biogenic oozes with mixtures of calcareous and siliceous material may, following diagenesis, become fine-grained limestones containing chert nodules, as is the case with the Cretaceous Chalk of northwest Europe.

Brown siliciclastic clays are very fine-grained (80% of particles are smaller than 30 microns), are non-biogenic in origin, and contain less than 30% calcium carbonate. They are common in depths greater than 4 km where biogenic clay is not abundant (Figure 18.19). Deposition rates are typically on the order of 10^{-3} mm per year. Some of the clay forms from diagenetic alteration of volcanic material.

Cyclic variation in the relative proportions of calcium carbonate, opaline silica, and inorganic clay in pelagic deposits has been related to climate changes. For example, colder oceanic bottom water during glacial periods might raise the carbonate compensation depth, hence causing the proportion of calcium carbonate to decrease at a point. However, this is a simplistic view. The relative proportions of various constituents depend also on at least the surface biological productivity (related to upwelling) and transport by thermohaline currents.

A regular, decimeter-scale alternation between aragonite-rich and calcite-rich muds in slopes around the Bahama Banks is also interpreted as being related to glacial eustatic changes. Aragonite-rich zones record times when the shelves were flooded, whereas calcite-rich zones record times when the shelf was exposed and the slope received calcite mud from foraminifera and coccoliths. However, these alternations have also been interpreted as recording preferential dissolution of aragonite or recrystallization as calcite.

On continental slopes and rises where settling of terrigenous mud is significant, hemipelagic sediments are green to dark-gray muds with less than 15% biogenic grains and more than 30% terrigenous silt. It is difficult to determine whether these muds were transported basinwards by wave currents, tidal currents, hyperpycnal flows, or low-density turbidity currents, or whether they represent mud reworked by thermohaline currents (e.g., contour currents) (Stow and Tabrez, 1998). Whatever the origin, this hemipelagic mud is commonly deposited from a *nepheloid layer* of low sediment concentration (up to the order of 10^{-1} mg per liter or ppm) in the lowest few hundred meters of the sea. Deposition rates are typically on the order of 10^{-1} mm per year. On slopes adjacent to modern carbonate shelves (platforms), hemipelagic (periplatform) sediments are composed mainly of aragonite needle

muds. Close to explosive volcanoes, hemipelagic sediments may be mainly volcanic dust. In cold regions, hemipelagic sediment may be derived partly from melting icebergs. Episodic deposition of hemipelagic sediments should result in layering, but, unless bioturbation is inhibited by anoxic bottom conditions, such layering is likely to be destroyed by bioturbation. As with pelagic sediments, anoxic bottom conditions can result in laminated, organic-rich hemipelagic deposits.

The common complete bioturbation of pelagic and hemipelagic sediments occurs mainly because of their slow rate of accumulation under oxygenated bottom conditions. Burrows (e.g., *Zoophycos*, *Chondrites*, *Planolites*) and fecal pellets are typically produced by epibenthic and inbenthic deposit feeders (see Chapter 11).

Sediment transport by deep-water (thermohaline) water currents

Thermohaline currents and some deep wind-driven currents transport sediment laterally both within the deep sea and along the deep-sea bed, and therefore affect the distribution of pelagic and hemipelagic sediment, and sediment transported and deposited by sediment gravity flows and hyperpycnal flows. Contour currents (flowing approximately parallel to slope contours) transport bed sediments around the edges of major oceans, particularly along the continental rise. The direction of contour currents is actually quite variable because they are influenced by submarine topography and the presence of large eddies with vertical axes (Stow *et al.*, 2002). Contour-current velocities are typically on the order of 10^{-2} m s^{-1}, but locally can be on the order of 10^{-1} m s^{-1} where flow is forced to converge by the Coriolis force and by submarine topography. Current velocity and direction also vary with time on various scales, such as seasonally. Furthermore, fluctuations in velocity may be related to the influence of deep wind-driven currents. Contour-current velocities of up to tens of centimeters per second can transport fine sand to mud-size grains, but exceptionally vigorous currents can transport coarse sand and gravel. The grain size transported is strongly correlated with the current strength, as expected (see Chapter 5). Bed forms in cohesionless sediment include current lineations, ripples, and dunes. However, persistent bioturbation tends to destroy the smaller bed forms. Erosion and deposition associated with contour currents depend on whether the flows converge and accelerate in space (causing erosion) or diverge and decelerate in space (causing deposition). Surfaces of erosion may contain longitudinal scours on the order of meters deep, tens of meters wide, and hundreds of meters to kilometers long. Time-averaged deposition rates can range from 10^{-2} to 10^{0} mm per year, are partly dependent on the time scale of averaging, and obviously vary in time and space. Deposits of contour currents are commonly called *contourites*. Use of such an interpretive name is unwise unless the origin of the deposit is certain (which is not normally the case with ancient deposits).

Larger-scale deposits interpreted as being due to contour currents are called *contourite drifts* (Faugeres et al., 1993; Stow, 1994; Stow et al., 2002). Contourite drifts are typically up to a few hundred meters thick and tens to hundreds of kilometers long and wide, and cover areas on the order of 10^{3}–10^{5} km^{2}. In form they can be sheet-like, elongated mounds, and fan-shaped (at the mouths of submarine channels). Their shape may also conform to the space that confined the contour current during deposition. Their upper surfaces may be covered with bed forms such as current lineations, ripples, dunes, and giant waves. Giant sediment (mud) waves, tens of meters high and hundreds of meters to kilometers long, have near-parallel crestlines, and are symmetrical to mildly asymmetrical in cross section. High-resolution seismic profiles indicate that they tend to have internal layering that is either concordant with the surface form (indicating bed-form stationarity plus deposition) or slightly discordant (indicating migration, commonly upstream, in addition to deposition). However, these sediment waves can be oriented variously relative to the current direction, and can migrate upstream or downstream. The origin of these waves is unknown, but they could possibly arise from internal waves in stratified thermohaline currents or from standing lee waves to the lee of an obstacle such as a channel margin and levee. Such giant sediment waves also occur on submarine levees, oriented transverse to the inferred overbank turbidity current flow, suggesting that they are also related to symmetrical waves formed from low-concentration turbidity currents.

Contourite deposits have a wide range of grain size and composition, due to the variability in sediment supply and velocity of the depositing currents (Stow and Lovell, 1979; Stow and Piper, 1984; Hollister, 1993; Shanmugan, 1993; Stow and Faugeres, 1993, 1998; Carter and McCave, 1997; Stow et al., 2002). Periodicity of depositional contour currents results in discrete cross-laminated and planar-laminated sediment layers up to the order of centimeters thick. The

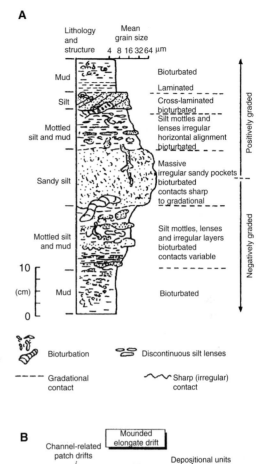

FIGURE 18.20. (A) A facies model for a sequence of contour-current deposits. From Stow et al. (1996). (B) Idealized principal seismic characteristics of contourite drifts. From Stow et al. (2002).

layers may be inverse-graded, normal-graded, or ungraded, and may have sharp or gradational upper and lower boundaries. However, contourites (especially muddy ones) tend to be very bioturbated, such that internal laminae and bed boundaries are commonly not discernible. Contourites commonly occur grouped in decimeter-scale coarsening-upward and fining-upward units (Stow, 1994) (Figure 18.20A) that have been related to changes in flow velocity over the order of thousands to tens of thousands of years. The boundaries and larger-scale internal structures of contourite drifts can be discerned using high-frequency seismic measurements (Stow *et al.*, 2002) (Figure 18.20B). The boundaries are commonly continuous, high-amplitude reflectors. The internal reflectors tend to be discontinuous, subparallel, low-amplitude reflectors that indicate upward and lateral accretion of the drift. Contourites will normally be interbedded with turbidity-current deposits and hemipelagic muds, and these different types of deposits may be difficult to distinguish in cores, seismic sections, or outcrops.

Chemical precipitation and early diagenesis of deep-sea sediment

Chemical precipitates (e.g., Mn and Fe nodules and crusts) occur in areas where the physical sedimentation rate is negligible (millimeters per 1,000 years), particularly associated with brown clay. Ferromanganese crusts occur on basaltic rock and boulders, and ferromanganese nodules have a variety of discoidal to spherical shapes and range in diameter up to a few centimeters. Most have nuclei composed of volcanic fragments, shark's teeth, or foraminiferal skeletons, surrounded by concentric layers of oxides and hydroxides of Mn and Fe (Chester, 2003). Seawater, pore water, and, in some cases, hydrothermal solutions provide the dissolved Mn and Fe. Precipitation is thought to be due to oxidation–reduction reactions at and below the sediment–water interface. Phosphate crusts and nodules form in anoxic bottom waters, as explained in Chapter 4. Borings are made in these crusts and nodules by benthic organisms.

As discussed in Chapter 4, hydrothermal fluids circulating around mid-ocean-ridge systems and venting at the sea bed (black and white smokers) result in precipitation of tufas ("chimneys") composed of mixtures of pyrite, chalcopyrite, sphalerite, pyrrhotite, and anhydrite (black smokers) or pyrite, amorphous silica, barite, and anhydrite (white smokers). These deposits are associated with a prolific fauna of crabs, giant clams, and tube worms that are dependent upon sulfur-reducing bacteria. Collapse of chimneys produces loose sediment. Alteration of the iron sulfides results in yellow, hydrated iron oxides.

Aragonite and calcite dissolve as the water depth increases, and aragonite reorganizes into calcite in some areas of subsurface water flow. However, Mg calcite commonly cements the sands or muds accumulating around carbonate shelves. The Mg content of the cements is related to the temperature of the water from which they precipitate (Chapter 19). The degree of *in* *situ* cementation ranges from aerially extensive hardgrounds to scattered nodules. These sediments may well be transported downslope in sediment gravity flows. Ahermatypic corals and other deep-water organisms such as sponges are commonly found in extensive buildups initiated on hardgrounds.

Anoxic and hypersaline deep-sea deposits

Organic-rich muds (which become black shales following diagenesis) in deep-sea deposits have a total organic content of 1%–15%, and contain iron sulfide and traces of various other metallic minerals. They are generally taken to indicate anoxic bottom water and relatively rapid burial (allowing preservation of organic matter) and/or a high supply rate of organic matter (related to high surface productivity or terrigenous supply of organic matter). So-called oceanic anoxic events have been recognized in the sedimentary record (e.g., Cretaceous, late Devonian to early Carboniferous) and have been related to burial of enhanced amounts of organic matter. Suggested causes for such events include greenhouse climate conditions, sluggish oceanic circulation, and increased biological productivity (Arthur *et al.*, 1984; Arthur and Hageman, 1994; Wignall, 1994).

Hypersaline deposits in the deep sea are discussed in Chapter 16. Miocene deposits beneath portions of the Mediterranean Sea containing evaporites have been attributed to the repeated complete desiccation of the Mediterranean (the Miocene Mediterranean Salinity Crisis) followed by gigantic floods over the Straits of Gibraltar (see Chapter 16 for discussion of this somewhat outlandish idea).

Classification of deep-sea sedimentary facies

Deep-sea sedimentary facies have been classified on the basis of texture, fabric, and internal structure (Mutti and Ricci Lucci, 1972; Pickering *et al.*, 1986, 1989; Ghibaudo, 1992; Clark and Pickering, 1996; Stow, 1996) in a similar way to fluvial deposits. Each facies is given a number-and-letter code. Also as for fluvial deposits, architectural elements and bounding-surface hierarchies have been set up for deep-sea deposits (Mutti and Normark, 1987, 1991; Ghosh and Lowe, 1993; Clark and Pickering, 1996). Architectural elements are defined on the basis of their facies and shape as seen in profile or plan, and are independent of scale. A given shape could be given more than one name (e.g., channel, scour, lens, wave form). The various two-dimensional shapes could be different views of the same three-dimensional form. Thus, the various elements are not mutually exclusive as they must be in a rational classification. A bed form or a scour is an interpreted topographic feature, not a two-dimensional sediment shape. None of the architectural element classifications address the key issue that the shape and dimensions of a deposit are not the same as those of the original topographic form. The assignment of different orders of bounding surfaces is not easily done on outcrops, and is impossible using cores (see the discussion in Bridge (1993b)). To describe and understand sedimentary strata, it is necessary to recognize how distinctive strataset types of different scales are superimposed in three-dimensional space.

Flow and sediment transport in various deep-sea environments

Continental slopes and associated submarine channels and fans

The flow and sedimentary processes of continental slopes and associated submarine channels depend on whether they are tectonically active or passive; the proximity of coastlines, particularly major river deltas; and the spatial distribution of sediment supply at the top of the slope. Faulted continental margins may have a steep, fault-controlled continental slope with very active slumps, slides, and sediment gravity flows. If a delta margin or some other source of abundant sediment is close to the top of the continental slope, deposited sediment may oversteepen the slope, resulting in slumps and sediment gravity flows. In the case of a nearby delta front, growth faults, and mud and salt diapirs, may also affect the slope. The location of these sediment-transport processes along the slope will depend on the spatial distribution of tectonism and sediment supply. The position of sediment supply may shift as a result of river avulsion, and this may cause a shift in position of submarine canyons and fans. Depending on the spatial distribution of sediment transport to and on the slope, the slope may be erosional or depositional, or both, in different places. Depositional slopes will contain deposits of debris flows, grain flows, and high-density turbidity currents. These so-called mass-transport complexes can be tens of meters thick and tens of kilometers across. They may have concave-upward basal-erosion surfaces. Their seismic signature is structureless or contorted, chaotic, low-amplitude reflections (Figure 18.21).

Submarine canyons (and smaller channels) eroded in continental slopes may have originated by: (1) shelfward retreat of a slump scar at the top of the slope, associated with shelf water currents; (2) river erosion during sea-level lowstand and exposure of the shelf edge; or (3) erosion by sediment gravity flows, especially during sea-level lowstand. Water flow and sediment transport in the upper parts of canyons and channels are influenced by ocean tides, large wind waves, geostrophic (wind-shear) currents, and possibly river-related currents. However, sediment transport, erosion, and deposition in submarine channels are dominated by debris flows, grain flows, and turbidity currents. Huge slumped blocks and coherent slides are also common in canyons. If slumping, sliding, and sediment-gravity-flow activity decline, canyons may become filled with hemipelagic mud. Therefore, submarine-canyon deposits formed during an active period followed by inactivity might comprise elongated lenses of sand and gravel (up to kilometers wide and long, and hundreds of meters thick) plus displaced, deformed blocks of consolidated sediment encased in hemipelagic mud. The seismic expression of channel-filling deposits depends on variation in grain size. Reflections in gravels, sands, and muds associated with sediment gravity flows may be of relatively high amplitude and continuous to discontinuous. Reflections in hemipelagic muds are of low to moderate amplitude and discontinuous or contorted. Slumps and slides may be chaotic, contorted, low-amplitude reflections or acoustically transparent (Figure 18.22). Compressional thrusts and folds may also be visible (Posamentier and Kolla, 2003; Posamentier and Walker, 2006).

Modern submarine fans have been studied extensively since the 1970s, particularly those off California (Piper and Normark, 1983; Normark and Piper, 1985), and in the Amazon (Damuth et al., 1983b, c, 1988, 1995; Flood et al., 1991), Indus (Droz and Bellaiche, 1985, 1991; Kolla and Coumes, 1985, 1987; McHargue, 1991; Kenyon et al., 1995), and Mississippi (Bouma et al., 1985; Weimer, 1990, 1995; Twichell et al., 1991). For reviews see Bouma et al. (1985), Pickering et al. (1989, 1995), Weimer and Link (1991), Weimer et al. (1994), Clark and Pickering (1996), and Posamentier and Walker (2006). These studies have indicated that there is a large variety of fan dimensions and shapes, controlled by the nature of sediment supply and pre-existing sea-bed topography. Fan activity is sensitive to sea-level change (being most active during lowstands: Posamentier et al. (1991)) and also to

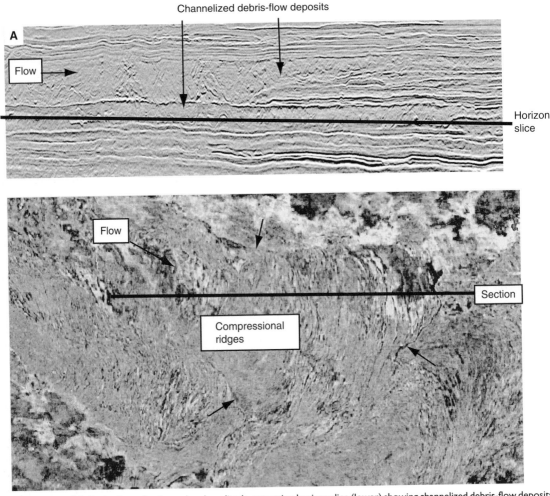

FIGURE 18.21. (A) A seismic section (upper) and amplitude-extraction horizon slice (lower) showing channelized debris-flow deposits. The thickness of the seismic section is about 1s, and its length is 20 km. The section shows two debris-flow units with erosional bases. The lower unit is shown in the horizon slice and presents evidence of compressional ridges. (B) A seismic section and amplitude-extraction horizon slice through a debris-flow deposit. The horizon slice shows the lobate nose of the debris flow. Eastern Gulf of Mexico. Modified from Posamentier and Walker (2006).

climatically influenced changes in river discharge and sediment supply to submarine canyons. Fan activity is also related to variations in sediment supply related to river avulsions.

The flow and sedimentary processes on submarine fans are remarkably similar to those on subaerial alluvial fans (Chapter 13), although there are also differences (Peakall *et al.*, 2000). The upstream feeder channel contains turbulent and laminar sediment gravity flows, and fan deposition results from a spatial decrease in sediment-transport rate associated with flow expansion and decrease in slope. Flow, sediment transport, and deposition are episodic. The channels on fans can be meandering or braided, and they migrate

laterally by gradual lateral accretion on bars and by channel cutting and filling. Channel belts periodically move over the fan by (mainly nodal) avulsion. Some channels are mainly erosional, some are aggradational, but most are both. Interchannel areas are analogous to alluvial floodplains, and include levees and crevasse splays adjacent to the channel. Depositional (terminal) lobes occur at the mouths of some fan channels, as is the case with alluvial fans. However, the various features of submarine channels discussed below (channel bars, levees, crevasse splays, terminal lobes) are not unique to submarine fans. They are found also in submarine channels that occur on continental-slope aprons (without fans) and on basin plains.

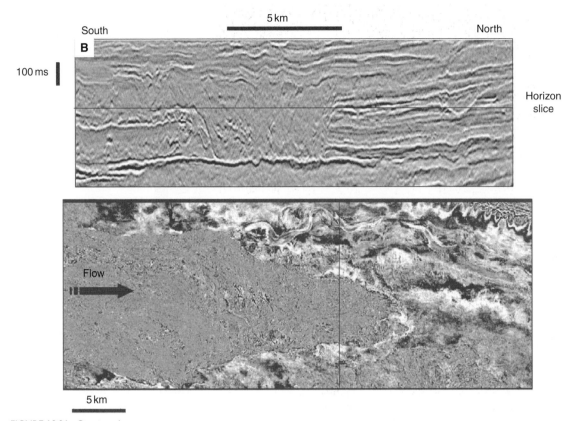

South   5 km   North

100 ms

Horizon
slice

Flow

5 km

FIGURE 18.21. Continued.

The inner fan commonly contains a single meandering channel bordered by levees (Figure 18.4). The meandering channel has a plan geometry that is remarkably similar to that of alluvial channels (Flood and Damuth, 1987; Clark and Pickering, 1996). Evidence for episodic bank erosion, point-bar migration, and channel cutoff is provided by slumped blocks along channel margins, accretion topography, and abandoned channels (Figure 18.23). The channel may also present evidence for incision or deposition, or both (Figure 18.23). The detailed pattern of flow and sediment transport in these submarine channels may be expected to differ somewhat from subaerial flows, mainly because of the presence of an upper boundary layer in addition to one at the bed (e.g., Peakall *et al.*, 2000). This means that the patterns of secondary circulation will differ. However, the geometry and sediment transport in curved channels are controlled mainly by the primary depth-averaged flow under bankfull flow conditions, and these are expected to be similar for subaerial and submarine channels.

Deposits of submarine channels are expected to be arranged into different scales of stratasets, as for river-channel deposits (Chapter 13): single-event deposits associated with discrete debris flows or turbidity currents; deposits of individual braid bars, point bars, and channel fills; and laterally and vertically stacked channel-bar and channel-fill deposits within channel belts (Figures 18.23 and 18.24). The nature of these submarine-channel deposits is not known as well as that of river-channel deposits. The single-event deposits are mainly deposits of turbidity currents and debris flows (see Chapter 8), which can range from centimeters to meters thick and up to kilometers in lateral extent. Deposits of channel bars should be low-angle inclined turbidites and debris-flow deposits in erosionally based sets that may be tens to hundreds of meters thick and extend laterally for many kilometers (width/thickness ratios on the order of 10^2). These bar deposits may fine and thicken upwards. Channel-fill deposits should have erosional banks, fine and thin upwards, and contain a high proportion of fine-grained turbidites and mud at the top. Slumps and slides down channel margins (mass-transport complexes) can be up to tens of meters thick and tens of kilometers across, and are composed of disturbed and contorted bedding. Such

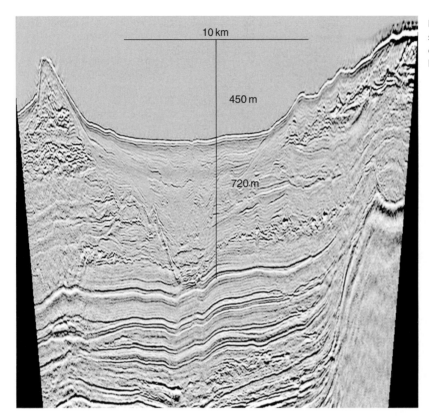

FIGURE 18.22. A seismic section through a Mississippi canyon and fill. Modified from Posamentier and Walker (2006).

material is expected near the base of many channel-bar or channel-fill deposits. Vertical sequences of channel-bar and channel-fill deposits within channel-belt-levee units can be up to hundreds of meters thick, indicating vertical aggradation of hundreds of meters. Average deposition rates in channels may be tens of millimeters per year. The largest channels and greatest amounts of vertical aggradation are in inner-fan channels. It is important not to make the common mistake of confusing individual channels with channel belts that contain the deposits of many channel bars and channel fills.

Only the larger scales of strataset (generally tens of meters thick, rarely meters thick) can be discerned on seismic sections. The seismic signature of channel deposits composed of interbedded gravel, sand, and mud is high-amplitude, discontinuous reflections. Channel margins may be recognized, and lateral-accretion surfaces can be recognized exceptionally (Figure 18.23). Single lithologies and slumps and slides are seismically transparent. All of these scales are potentially interpretable from wireline logs, cores, and outcrops, although many outcrops are too small to allow one to discern very-large-channel deposits. The

geometry of ancient submarine-channel deposits is discussed at length by Clark and Pickering (1996).

Overbank flow of channelized turbidity currents is common, as evidenced by levees, crevasse channels, and splays (Figures 18.6–18.8). Overbank flow has two main effects: (1) it exerts drag on the underlying channel flow, influencing the pattern of secondary circulation; and (2) diversion of turbid water away from the channel, a process referred to as flow stripping (Piper and Normark, 1983; Bowen et al., 1984), thereby reducing flow thickness and excess density. Flow stripping reduces the lifespan of the turbidity current. As the overbank flow moves away from the channel and mixes with ambient fluid, it is decelerated and deposition occurs. This is why levees and crevasse splays exist. Average deposition rates on the near-channel parts of levees can be millimeters to centimeters per year. Levees are commonly lower on the left side of the main channel in the northern hemisphere, because of the influence of the Coriolis force on the flow direction. In the southern hemisphere, levees would be lower on the right side of the main channel, but near the equator (where the Coriolis force is negligible) there would be no preferred levee height. Another feature that appears

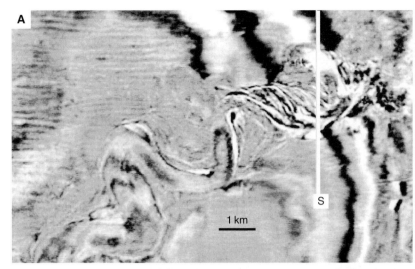

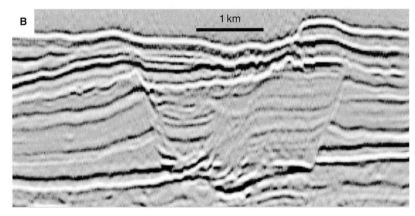

FIGURE 18.23. (A) A seismic-amplitude-extraction horizon slice showing a meandering channel belt (flow to the southwest), with the corresponding section showing channel-bar and-fill deposits, offshore Nigeria. Modified from Posamentier and Walker (2006). (B) A seismic section showing channel point-bar and fill deposits with adjacent levee deposits, Makassar Strait, Indonesia. The base and top of the lenticular channel-levee sequence are marked by high-amplitude continuous reflections. The channel deposits are high-amplitude discontinuous reflections. The levee deposits are low-amplitude continuous reflections. The channel deposits indicate that the channel has migrated to the right by more than 1 km and aggraded vertically by about 50 m. Modified from Posamentier and Walker (2006).

to be associated with overbank flow and sediment transport by turbidity currents is transverse sediment waves (meters to tens of meters high and hundreds of meters to kilometers long) that look like antidunes. The geometry, migration character, and internal structures of these bed forms are discussed above. Another feature of overbank environments (and probably also channels) is arcuate scours that are up to meters high and hundreds of meters across. These may be large erosional flute marks or slump scars.

Typical deposits of levees and crevasse splays are "distal turbidites" (Figure 18.25 and Chapter 8). Individual centimeter- to decimeter-thick turbidites become thinner and finer-grained away from the channel and downfan. Relatively small crevasse channel fills are expected. Furthermore, by analogy with fluvial deposits, coarsening-upward and fining-upward sequences on the order of meters to tens of meters thick might be expected from progradation and

abandonment of crevasse splays or parts of levees. The large sediment waves may be associated with meters-thick sets of low-angle inclined strata. Slides, slumps, and sediment gravity flows are initiated on levees, and some of these deposits may be contorted or structureless. Meters-deep scour fills are expected. The overall levee sequence is wedge-shaped, and can be up to hundreds of meters thick and tens of kilometers wide. Levee deposits in seismic sections give low-amplitude, continuous to discontinuous reflections (Figures 18.7, 18.9, and 18.23). Slumps and slides on levees are seismically transparent.

The middle and outer fan is characterized by distributary channels with levees that lead to terminal lobes. The channels decrease in width and depth downfan, and the levees decrease in height downfan. Overall, the channel and levee grain size decreases downfan, and the grain size of levees decreases away from channels. Channels migrate laterally by bank erosion and

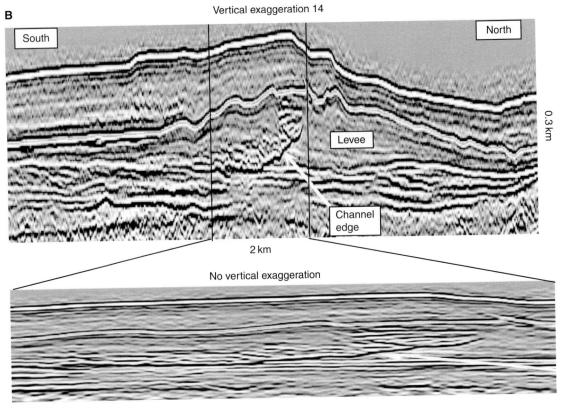

FIGURE 18.23. Continued.

bar deposition, and by channel cutting and filling, as for inner-fan channels. A given channel termination may be associated with a number of terminal lobes that are active at different times. Nodal avulsion of channel belts is prevalent on the middle-fan region, although it also occurs on the inner fan. Normally only one channel belt is active at a time. The fact that fan channels were most active during sea-level low-stands is suggested by hemipelagic mud drapes that are forming today. Channel avulsions on the Amazon fan occurred at intervals of 5,000–10,000 years during the Pleistocene.

According to Posamentier and Kolla (2003) and Posamentier and Walker (2006), progressive overbank flow stripping and mud deposition leave relatively less mud available for deposition on levees further down-channel. This would result in a downstream decrease in levee height but a downstream increase in sand/mud ratio in levee deposits. The maximum sand/mud ratio is considered to occur at the apex of the terminal lobes. Furthermore, a decrease in the sand/mud ratio of the sediment supply is thought to cause a downstream shift

in the position of the terminal lobe. These ideas need to be tested.

Progradation and abandonment of channels and terminal lobes result in coarsening-upward and fining-upward units that are meters to tens of meters thick and kilometers in lateral extent. These may well be similar to sequences associated with crevasse splays and channels. Avulsion of channel-levee complexes leads to stacked, laterally offset channel-belt-levee sequences, as is exemplified by the Amazon and Indus fans (e.g., Damuth et al., 1988; Weimer and Link, 1991; Pickering et al., 1995) (Figure 18.26). These sequences may be up to hundreds of meters in maximum thickness and tens to hundreds of kilometers long and wide. These sequences look remarkably like those produced using alluvial-stratigraphy models (e.g. Bridge and Mackey, 1993b).

Channel deposits are finer-grained and thinner with distance downfan. Channels decrease in size on the outer fan, and the surface relief is much less. The turbidites and hemipelagic mud are more laterally continuous and sheet-like. Individual beds are centimeters to

FIGURE 18.24. Interpreted submarine channel deposits, showing large-scale inclined strata resulting from lateral channel migration (from right to left). (A) Upper Miocene, San Clemente, California. (B) Gull Island Formation, western Ireland. From Posamentier and Walker (2006).

meters thick and extend laterally for tens of kilometers. The lower fan might be sandier than the middle fan (1) due to lack of confinement of the lower, coarse part of turbidity currents, or (2) because the lower fan is composed of older, coarser sediment. The seismic signature of outer-fan deposits is high- to low-amplitude, continuous and parallel reflections.

Basin plains

References on basin plains and their deposits include Hsü and Jenkyns (1974), Scholle *et al.* (1983), and Weaver and Thomson (1987). Basin plains experience low-density turbidity currents and bottom (contour) currents from time to time, plus hemipelagic and pelagic sedimentation. Tectonically controlled channels and volcanic hills in basin plains can cause deflection, convergence, and divergence of turbidity currents and bottom currents. Turbidity currents can be reflected and ponded by topography (Chapter 8). The nature of erosion and deposition on basin plains thus depends on the amount, composition, and grain size of the sediment supply, the nature of the basin margins (number, size, and locations of sediment-entry points), the topography (including accommodation space) of the basin plain, the flow characteristics of turbidity currents and

FIGURE 18.25 (A) Interpreted overbank deposits from Marnoso-arencea, Italy, showing a thickening-upward sequence. (B) Interpreted muddy levee deposits in a 10-cm-wide core from the Lange Formation, offshore Norway. The light-colored sandstone beds contain climbing-ripple cross strata and convolute lamination. (C) Tabular, sheet-like turbidites with thickening- and thinning-upward sequences, Lower Ross Formation, western Ireland. Pictures Courtesy of Henry Posamentier.

FIGURE 18.25 Continued.

bottom currents, and the relative importance of the various erosional and depositional mechanisms. These factors are likely to be related to sea level and tectonic activity.

Basin-plain deposits are generally sheet-like, with rare channel fills. The deposits of abyssal plains are expected to be interbedded fine-grained (distal) turbidites, contour-current deposits, and hemipelagic muds near basin margins, with more laterally extensive and finer-grained pelagic deposits towards basin centers. Basin plains within mid-ocean-rift systems are expected to contain abundant pelagic deposits and sediment-gravity-flow deposits composed of volcanic and pelagic sediments. There are good opportunities for reflection and ponding of turbidity currents in rift basins (van Andel and Komar, 1969) (Figure 18.27). Basin plains over oceanic trenches are expected to contain relatively thin pelagic and hemipelagic sediments and abundant sediment-gravity-flow deposits comprising terrigenous and volcanic grains. Oceanic-trench sediments are typically preserved in deformed sequences that are thrusted up onto continental margins (accretionary prisms; Chapter 20). Basin-plain deposits in modern settings range from hundreds of meters thick in gradually subsiding abyssal plains to kilometers thick in actively subsiding basins (rift basins, oceanic trenches). Single large volcanoes have flexural moats around their bases due to the downward flexure of the crust by their enormous weight. Such moats will have relatively thick sediment accumulations.

Volcanic hills and mid-oceanic ridges

The sides of active submarine volcanoes are typically covered with sheet lavas and pillow lavas, and large blocks of (chilled) lava that have been broken up by thermal contraction. Pyroclastic debris may also come from subaerial and shallow-submarine eruptions. Volcanic debris is commonly altered to montmorillonite and smectite clays. In mid-ocean ridges, volcanic material is also associated with blocks of serpentinized peridotite, debris from metalliferous deposits formed by hydrothermal fluids (black and white smokers), and various other chemogenic deposits. Pelagic deposits also occur on volcanic hills and mid-oceanic ridges, and may become the dominant sediment on inactive volcanoes (Figure 18.27). The type of pelagic sediment depends partly on the depth relative to the carbonate

A

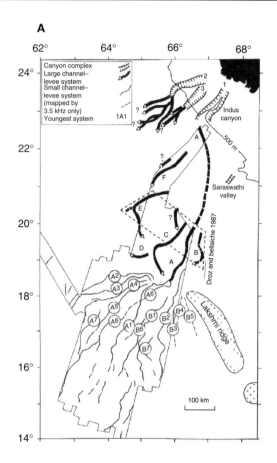

FIGURE 18.26. (A) Canyons and channel–levee systems of the Indus Fan; A to F is youngest to oldest, and I to 7 is youngest to oldest. (B) A seismic profile through channel–levee complex F with interpretation. (C) A model for Indus fan deposition. From Kenyon et al. (1995). (D) A map of an Amazon deep-sea fan, showing the active channel, many abandoned channels, and marginal debris flows. Numbers are core sites. (E) A cross section of the Amazon fan based on seismic and core data, showing stacked, offset levee–channel units arranged into complexes separated by debris-flow deposits and pelagic deposits. Images (D) and (E) are from Leeder (1999).

C

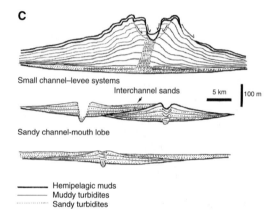

B

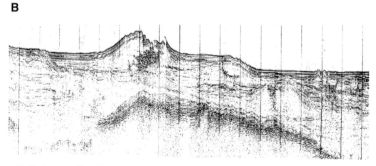

Vertical exaggeration x 22

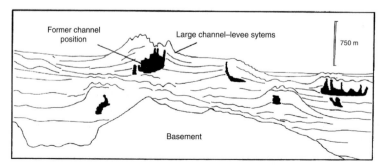

D

FIGURE 18.26. Continued.

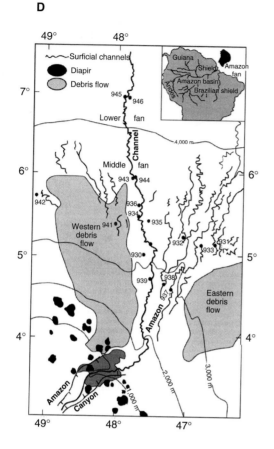

E

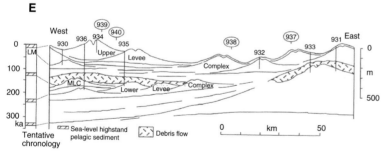

compensation depth. The loose material on submarine volcanoes is commonly transported downslope into adjacent basins by sediment gravity flows (Chapter 8). The sediment gravity flows will be a mixture of volcanic material, metalliferous and other chemogenic sediment, and pelagic sediment. Ancient mid-ocean-ridge deposits (called *ophiolites*) can be found in upthrusted accretionary prisms. These deposits are a mixture of basalt sheets (lava flows and dykes), serpentinized peridotite, pillow lavas, metalliferous deposits, fine-grained pelagic limestones and cherts, and sediment-gravity-flow deposits comprising fragments of all of the above.

Submarine volcanoes that build above sea level and then become inactive and subside have certain distinctive features. As volcanoes approach sea level, the volcanicity becomes more explosive, generating more volcaniclastic material. The volcanic debris and pelagic sediment on the volcano can be moved by wave and tidal currents in addition to sediment gravity flows. Upon subaerial exposure, the volcano is subjected to weathering, and erosion by sediment gravity flows (e.g., lahars) and river flows. Deposition of sediment at the coast may be in the form of flat-topped wedges with seaward slopes at the angle of repose. In tropical

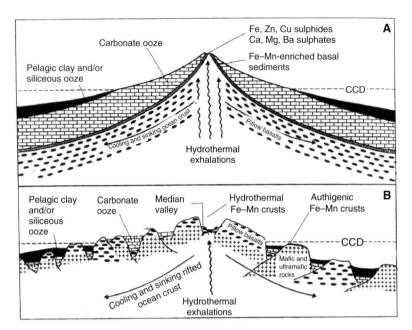

FIGURE 18.27. Sediment distributions on (A) a fast-spreading (Pacific-type) mid-ocean ridge and (B) a slow-spreading (Atlantic-type) mid-ocean ridge. After Orton (1996).

climates, fringing coral reefs will form (Chapter 15). As the volcano becomes inactive and subsides below sea level, the fringing reef becomes an atoll, and loose material again is subjected to wave and tidal currents. Such deposits will gradually be replaced by pelagic deposits as the depth of water increases. Thus, a characteristic sequence of basaltic rocks will be overlain by hundreds of meters of shallow-water marine deposits overlain by pelagic limestones and cherts.

Large-scale stratigraphic models of deep-sea fans and associated environments

Early submarine-fan models (e.g., Normark, 1970, 1978; Mutti and Ricci Lucci, 1972; Walker and Mutti 1973; Walker, 1979, 1978; Normark et al., 1979) were based on a fan divided into inner (upper), middle, and outer (lower) parts (Figure 18.28). The depositional elements were channels, levees, interchannel areas, and middle-fan depositional lobes. Avulsion of channels, plus growth of some depositional lobes and abandonment of others, resulted in coarsening-upward and fining-upward sequences superimposed upon an overall coarsening-upward fan progradation sequence (Figure 18.28). Autocyclic depositional mechanisms were emphasized. This model was not based on extensive knowledge of modern fans, but was heavily biased towards interpreted rocks, and is

clearly too simplistic. As more studies of modern environments were undertaken using greatly improved technology, the variety and complexity of these environments was realized. Also their morphology and deposits depend on sediment supply, sea-level change, and tectonic setting (e.g., Mutti, 1985; Posamentier et al., 1988, 1991; Reading and Richards, 1994). Modern facies models are much more complicated and firmly grounded in studies of modern settings (Figure 18.28). For example, a coarsening- and thickening-upward sequence that is tens of meters thick and kilometers to tens of kilometers across might be formed by progradation of a terminal lobe, progradation of a crevasse splay, or some externally controlled change in turbidity-current activity. The evolution of deep-sea depositional models is a testament to the value of studies of modern environments, and the folly of deriving facies models from interpreted rocks.

Some modern deep-sea fans (e.g., Amazon, Mississippi) are underlain by a mass-transport complex with an erosional base, which is in turn overlain by sheet-like sandy strata (possibly outer-fan deposits), and then overlain by overlapping, stacked channel-levee complexes (middle-fan deposits?). However, mass-transport complexes are not always present, and they are not restricted to basal positions. These large-scale fan sequences are bounded by hemipelagic muds. Individual fan sequences may in turn be stacked next to

A

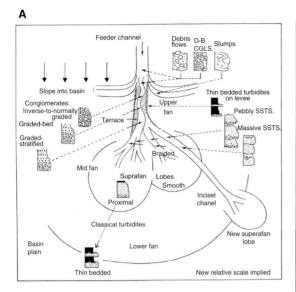

B

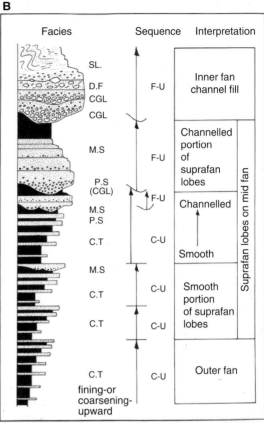

FIGURE 18.28. In (A) and (B) are shown early facies models for submarine fans (Walker, 1979). In (C) and (D). are shown submarine-fan models that differ in terms of the type of basal deposit. From Walker (1992).

and overlapping each other (Figure 18.29). The age of these sequences is actually Pliocene to Recent.

Individual fan sequences bounded by hemipelagic mud have been interpreted to be related to Pliocene–Pleistocene glacioeustatic sea-level cycles (e.g., Bouma *et al.*, 1989; Weimer, 1990; Posamentier *et al.*, 1991; Twichell *et al.*, 1991; Kolla and Perlmutter, 1993; Posamentier and Kolla, 2003). Falling sea level has been associated with erosion of continental slopes by slumping, sliding, and canyon cutting, giving rise to mass-transport complexes. Falling sea level is also associated with an increasing supply of sediment to the top of the slope as coastlines and rivers become closer to the slope edge. This results in an increase in activity of turbidity currents and other sediment gravity flows. During sea-level lowstand, fans grow and prograde. Adjacent and overlapping fans are due to changing positions of the submarine-canyon feeder system (Weimer, 1990, 1995) (Figure 18.29). With the subsequent sea-level rise, coastlines move inland, the sediment supply rate and grain size

decrease substantially, and turbidity-current activity wanes. Sea-level highstand is associated with deposition of hemipelagic mud, although there are exceptions, such as where canyon heads are close to river mouths.

As with most models, this sequence-stratigraphic model is oversimplified because it is not based on an extensive set of data and is entirely qualitative. It is strongly biased towards the allocyclic controls of sea-level and climate change at the expense of autocyclic controls, and tectonic setting is not considered. Age dating is not accurate enough to demonstrate unequivocally the temporal relationship between sea-level change and stages of fan development (Walker, 1992). Temporal and spatial changes in side slopes of basins (including intra-slope basins) can be due to tectonic activity or progressive filling with sediment. Such slope changes can result in changes in the downstream extent, thickness, and degree of sorting in turbidity-current deposits, and the relative positions of channelized and unchannelized turbidity flows (Posamentier and Kolla, 2003).

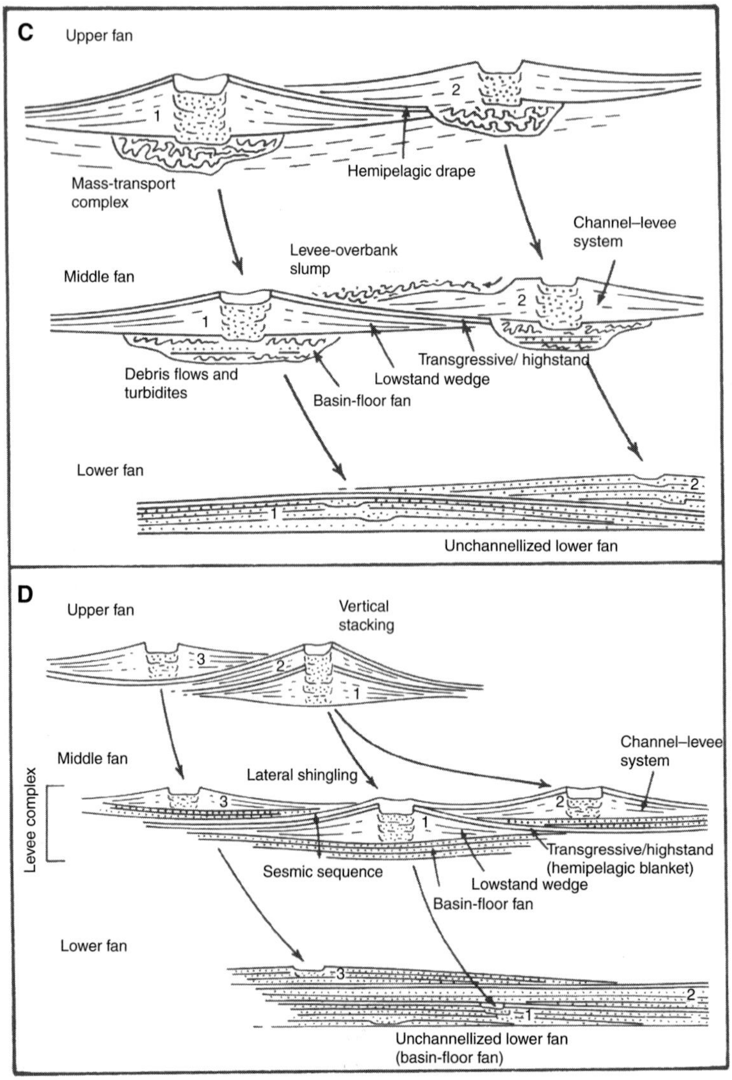

FIGURE 18.28. Continued.

Large-scale depositional sequences are expected to vary among different types of deep-sea basin. For example, large ocean basins should contain widespread sandstone sheets deposited by sediment gravity flows, large channels and paleocurrent patterns associated with large fans, and large amounts of pelagic sediment. Oceanic trenches above subduction zones would have relatively narrower and more elongated depositional sequences, the different facies would not be widely separated, and deformation is expected. Accretionary prisms of thrusted and obducted sediment may be separated by individual sedimentary basins within the trench slope (Chapter 20). Furthermore, rift basins may occur in the peripheral bulge on the oceanward side of the trench. In marginal-sea basins, fans would be smaller, sediment may be coarser, and there would be less pelagic sediment than in oceanic basins. Rift basins associated with mid-oceanic ridges and peripheral bulges would contain relatively small fans with sediment-gravity-flow deposits composed of pelagic sediment, volcanic material, and possibly coral-reef debris (if volcanic uplands reached the water surface).

Large-scale stratigraphic models for carbonate slopes

Carbonate-slope deposits are famous for their sediment-gravity-flow deposits, as is the case for siliciclastic slopes (Figure 18.30). However, submarine fans are notably absent from modern carbonate slopes, although they

A

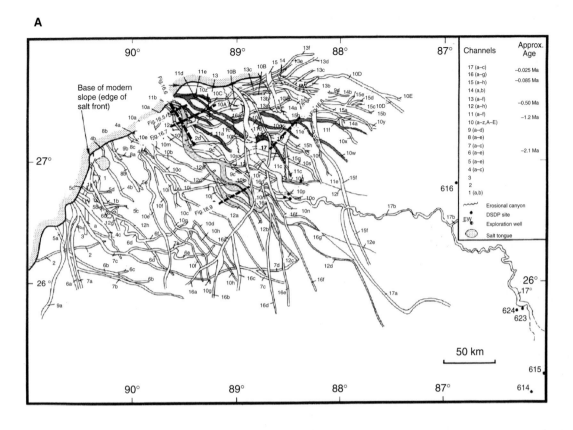

B

WSW   ENE

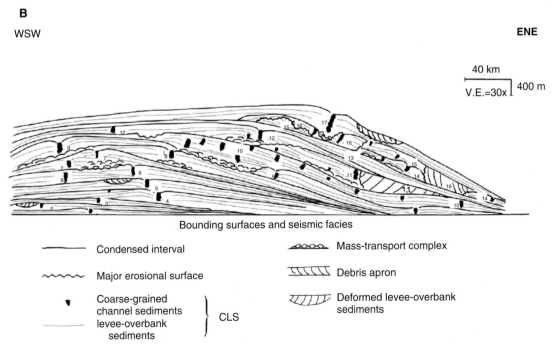

Bounding surfaces and seismic facies

FIGURE 18.29. (A) A map showing the locations of 17 channel–levee units of the Mississippi deep-sea fan. Number 1 is the oldest and number 17 is the youngest. (B) A cross section through the Mississippi fan, showing the offset stacking pattern of channel-levee units. From Weimer (1995).

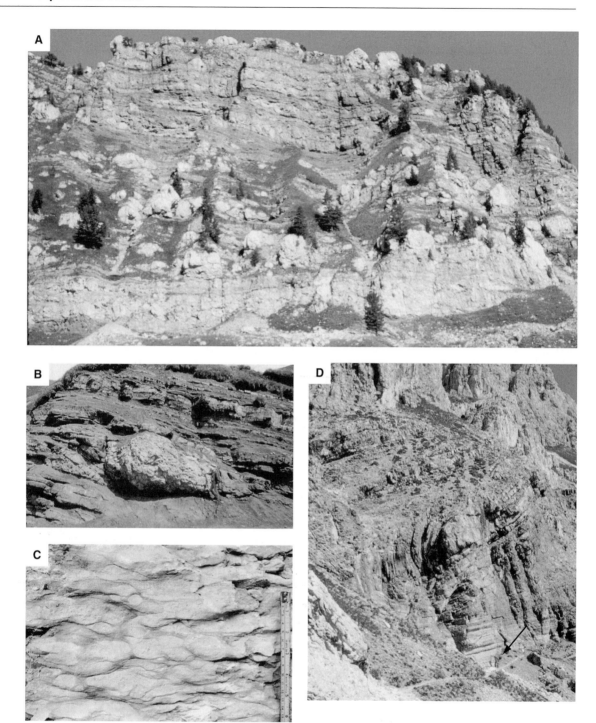

FIGURE 18.30. Ancient carbonate slope deposits around Triassic platforms that comprise the Dolomite Mountains of northern Italy. (A) A thick debris-flow deposit at the base of a hill overlain by stratified mixed periplatform and pelagic siliciclastic mudstones. Many loose rockfall blocks (locally referred to as *Cipiti*) up to 5 m in diameter lie in the mudstones above the debris-flow deposit. Trees provide the scale. (B) A large boulder (locally referred to as "The Locomotive") in siliciclastic slope deposits. (C) Nodular bedded periplatform ooze due to early cementation. The ruler on the right is ∼1 cm wide. (D) A soft-sediment fold in slope mudstones interpreted as the result of a rotational slump. The arrow points to a figure on the footpath along the base of the cliff.

have been interpreted from ancient deposits. This is because the collapse of the slope and parts of the rimmed shelf produces a "line source" of sediment rather than the point source of a submarine canyon. In contrast to the case of siliciclastic slopes, sediment gravity flows on carbonate slopes can contain huge blocks of cemented sediment from the platform-edge shoals and the slopes (Figure 18.30). Isolated boulders (*cipiti* in Italian) are common in the talus-slope deposits in the Dolomites of northern Italy (Figure 18.30). Talus piles are commonly difficult to separate from debris-flow deposits because carbonate mud can be preferentially removed or added to the deposit during diagenesis (Chapter 19). Truncation surfaces and soft-sediment deformational features in muddy slope deposits are associated with both translational and rotational slumps comprising debris-flow deposits (Figure 18.30).

Ancient carbonate-slope deposits are composed of periplatform (hemipelagic) muds interbedded with pelagic sediments and a variety of sediment gravity deposits, like their modern counterparts (Figure 18.30). Pre-Mesozoic pelagic deposits are mostly siliciclastic muds and cherts composed of siliceous plankton or sponge spicules. Periplatform muds tend to occur in centimeters-thick beds separated by siliciclastic or cherty pelagic layers that may be of equal or lesser thickness than the periplatform mud. In many cases, these layers are delicately laminated, implying that there was no bioturbation, presumably due to anoxic or low-oxygen conditions in the basins. In other cases, the limestones are thoroughly bioturbated, and there is some dispute as to whether the more siliciclastic-rich layers reflect greater pelagic siliciclastic mud input or are where pervasive dissolution of more soluble carbonate components has produced a diagenetic layering (Chapter 19). This is particularly important because these marl-limestone layers are commonly analyzed for Milankovitch cycles by time-series analysis (Chapter 20). Following

the Jurassic, pelagic carbonate skeletal debris becomes increasingly important in muddy slope facies. Partial early cementation of the muds results in nodular fabrics (Figure 18.30).

Large-scale depositional models of carbonate slopes are based mainly on interpretations of Quaternary sedimentation around the Bahama Banks and a theoretical "sequence-stratigraphic" model similar to the siliciclastic models (Sarg, 1988; Schlager, 2005) (Chapter 20). Sarg (1988) concluded that most carbonate sediment is shed from shelf edges and basin slopes during lowstands of sea level. However, this model is not supported by data from modern carbonate slopes such as the Bahama Banks. During periods of low sea level, carbonate production around the Bahamas Banks was limited to a narrow rim around the bank ~100 m below the top of the bank. Few turbidites formed during times of low sea level, and these comprised material from skeletal shoals and reefs that had migrated down the sides of the shelf. In addition, the exposed portions of the Bahama Banks underwent cementation and dissolution during times of low sea level, further reducing the sediment supply from the bank top. During periods of high sea level, when the Bahamas Banks were flooded, the basins around the banks received more turbidites than they did when the platform was exposed, and many of these turbidites contain shelf-derived ooids, peloids, and aragonite mud. This "highstand shedding" has been documented for other carbonate platforms in the Caribbean, Indian, and Pacific Oceans, and contradicts Sarg's (1988) model. Thus, the timing of appreciable deposition on carbonate platforms and basins is opposite to that of their Pliocene–Pleistocene siliciclastic counterparts discussed above. The generation of carbonate megabreccias due to collapse of the slope and shelf margin has still not been tied conclusively to sea level.

Part 5
Sediment into rock: diagenesis

19 Diagenesis

Introduction

Diagenesis is the physical, biochemical, and chemical changes that occur within sediments after deposition, and involves processes such as compaction, cementation, dissolution, and recrystallization (Table 19.1). These processes change the texture, structure, and mineralogy of sediments, and commonly increase the bulk density and reduce the porosity and permeability, thereby transforming the sediments into sedimentary rocks (i.e., lithification). The generation, migration, and entrapment of hydrocarbons happen during diagenesis, and many economic metalliferous deposits (e.g., iron, copper, lead, zinc) that occur in sedimentary rocks owe their origin to diagenesis. Diagenesis occurs as sediments become progressively buried beneath the Earth's surface, and lasts until they are moved into the temperature and pressure range of greenschist-facies metamorphism, or are exposed to another cycle of weathering and erosion. Syn-depositional soft-sediment deformation features such as desiccation cracks, convoluted laminae, and fenestrae are not normally considered to be diagenetic features (Chapter 12), nor are biogenic structures such as burrows, tracks, and trails (Chapter 11).

It is important to understand diagenetic features in sedimentary rocks because (1) they must be distinguished from the primary sedimentary features that are used to interpret the environment of deposition; (2) diagenetic changes in mineralogy must be taken into account in provenance studies (Chapter 3); (3) they provide a record of the burial history of sediments that can be related to factors such as the original depositional environment, the deposition rate, sea-level changes, and tectonic activity; and (4) their influence on porosity and permeability is a major concern to hydrogeologists and petroleum geologists.

This chapter starts with definition of the temperature and pressure conditions of diagenesis, and is followed by an overview of diagenetic processes and products, and the factors that control diagenesis. The remainder of the chapter is concerned with details of how diagenetic processes (compaction, cementation, dissolution, and recrystallization) affect different kinds of sediments in various sedimentary environments. General references on diagenesis include Pettijohn *et al.* (1972), Bathurst (1975), Pettijohn (1975), Wilson (1975), Scholle (1979), Berner (1980), Longman (1980), Adams *et al.* (1984), Burley *et al.* (1985), McIlreath and Morrow (1990), Horbury and Robinson (1993), Giles (1997), Adams and Mackenzie (1998), Morad *et al.* (2000), Tucker (2001), Burley and Worden (2003), Scholle and Ulmer-Scholle (2003), and Flügel (2004). In addition, many of the articles in Middleton (2003) deal directly with diagenesis.

Conditions of diagenesis

Diagenesis occurs in a temperature (T) range of roughly 0–300 °C at pressures (P) of up to 1 kilobar (Figure 19.1). Normal geothermal gradients near the Earth's surface are on the order of 20–30 °C km^{-1}, such that 300 °C is reached at a depth of about 10 km. However, geothermal gradients may reach 100 °C km^{-1} in areas of high heat flow, and may be as low as 10 °C km^{-1} in areas of low heat flow. In areas of high heat flow, heat transported (advected) by flowing groundwater can result in local near-surface temperatures on the order of 100 °C. A hydrostatic gradient for pure water reaches a pressure of 1 kilobar at a depth of approximately 10 km, whereas a pure lithostatic gradient for Earth material with a density of 2,750 kg m^{-3} reaches a pressure of 1 kilobar at a depth of approximately 3.5 km. Granular frameworks with water-filled pore

TABLE 19.1. An outline of diagenetic processes

(1) Mechanical compaction – rearrangement of grains due to increased pressure.
(2) Cementation – precipitation of material into a void from groundwater.
(3) Dissolution – solid component dissolves into the groundwater.
    (A) Includes "pressure solution," whereby strain gradients in crystals promote dissolution.
(4) Isochemical reorganization of material *in situ*.
    (A) Recrystallization (e.g., small calcite crystals → bigger calcite crystals).
    (B) Polymorphic change (e.g., small aragonite crystals → bigger calcite crystals).
    (C) "Neomorphism" – where it is impossible to tell A from B.
(5) Non-isochemical reorganization *in situ* – commonly referred to as *replacement*.
    (A) Weathering (see Chapter 3).
    (B) "Albitization" – Ca feldspars (common) and K feldspars replaced by Na feldspars.
    (C) Clay-mineral reactions – changes in clay-mineral assemblages with depth.
    (D) "Dolomitization" – replacement of aragonite or calcite by dolomite.
    (E) Magnesium calcite replacements (high-Mg calcite → low-Mg calcite and vice versa).
    (F) Oxidation–reduction reactions – common with iron and manganese minerals.
    (G) Dehydration – e.g., of gypsum ($CaSO_4 \cdot 2H_2O \rightarrow CaSO_4 + 2H_2O$).
    (H) "Chertification" – typically carbonates replaced by chert.
    (I) Zeolite replacements – particularly common in volcaniclastic glasses.
    (J) Calcrete growth – replacement of soil material with $CaCO_3$.

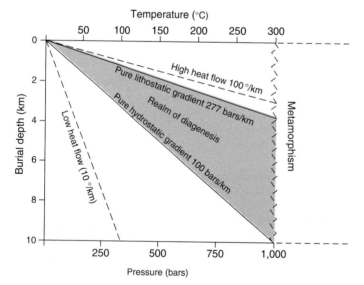

FIGURE 19.1. Conditions of diagenesis.

spaces reach a pressure of 1 kilobar between these depths. Therefore, diagenesis takes place from the surface to at most 10 km burial depth. At the *T* and *P* conditions of diagenesis, solid diffusion reactions within and among minerals are orders of magnitude too slow to effect significant changes. Instead, aqueous chemical and biochemical processes involving pore fluids of variable composition dominate diagenetic reactions.

It is common to distinguish *early diagenesis*, *burial diagenesis*, and *exhumation diagenesis*. Exhumation diagenesis occurs during uplift and erosion, and is essentially weathering (see Chapter 3). Early diagenesis is considered to occur at or just below the sediment surface, down to 2 km depth, at temperatures of about 70 °C, and burial diagenesis is considered to occur at greater depths and temperatures (Morad *et al.*, 2000). These distinctions are based on specific diagenetic

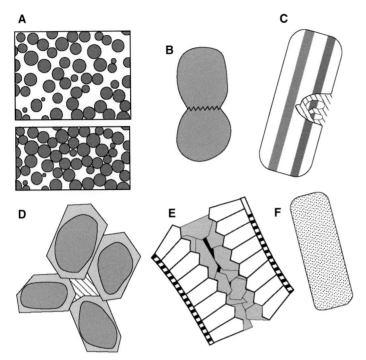

FIGURE 19.2. Diagrammatic representations of common diagenetic processes and products. (A) Compaction. (B) Pressure solution resulting in the interpenetration of two quartz grains. (C) Partial dissolution of a twinned feldspar grain with remnants of the original grain surrounded by a later cement (diagonal ruled pattern). (D) Overgrowths of quartz cements (light gray) over rounded detrital grains (dark gray). The overgrowths have euhedral terminations into a former void that was later filled by a second generation of cement (diagonal ruled pattern). (E) Shell fragments (ruled pattern) serve as the substrate for the growth of cement crystals normal to the shell surfaces out into a large pore where they had euhedral terminations. These first-formed cements were later buried by a second-generation anhedral mosaic of cement (gray) that almost completely filled the pore (remaining pore space black). (F) A feldspar crystal pseudomorphed by clay minerals that replaced the original grain.

processes and products in various types of sediments (discussed below). However, there is no clear-cut boundary between early diagenesis and burial diagenesis that applies to all sediment types.

Overview of diagenetic processes and products

Diagenesis begins at or near the sediment surface. Weathering takes place in any sediment deposited under subaerial conditions but is usually described separately from diagenesis (Chapter 3). As sediments are progressively buried, the temperature (T) and lithostatic pressure (P) increase, and the grains are continually bathed in pore fluids of variable composition that may be stagnant or flowing at millimeters to meters per day. As the lithostatic pressure increases, the volume of the sediment is reduced as the sediments are compacted and pore water is expelled. Compaction results in deformation and reorientation of sediment grains, a reduction in porosity, and an increase in bulk density. Under conditions of increasing T and P, some sedimentary minerals and organic matter formed at the low temperature and pressure of surface weathering, chemical precipitation, and organic growth (e.g., clays, iron

oxides, iron sulfides, iron silicates, carbonates, evaporites, and organic matter; Chapters 3 and 4) are taken out of their fields of chemical stability. Thus, new minerals form and organic matter is transformed. The new minerals can grow into void spaces as cements or one mineral may grow at the expense of another. These diagenetic processes can be organized into six broad categories, but may act simultaneously in a given sediment, and multiple episodes of diagenetic change commonly affect sediments.

(1) *Physical compaction* is the reorganization of grain packing by reorientation, deformation, and breakage of grains due to increasing overburden pressure during sediment burial (Figure 19.2A). The volume of the sediment is decreased and the bulk density is increased as pore space is reduced and water is expelled. Physical compaction is most important in sediments that have the highest initial porosity, and those that do not undergo early cementation (that reduces porosity). Larger-scale physical deformation also occurs due to overburden pressure. For example, buried evaporite deposits composed of minerals such as anhydrite and halite are commonly deformed into salt domes and diapirs. Mud diapirs are common in deltas,

and also form due to differential gravitational loading (Chapter 15). On a smaller scale, anhydrite layers and nodules are also commonly deformed, apparently as a result of the change in volume accompanying the dewatering of gypsum to form anhydrite.

(2) *Chemical compaction* (also known as *pressure solution*) is associated with dissolution at grain-to-grain contacts where focused lithostatic (or tectonic) pressure and elevated temperature lower the solubility of minerals (Figure 19.2B). This process leads to sutured (interpenetrating) grain-to-grain contacts and the jagged *stylolites* common in carbonate rocks and many sandstones.

(3) *Chemical dissolution* is distinguished from pressure solution in that it is not associated with localized pressure (Figure 19.2C). Chemically unstable grains are preferentially removed by chemical dissolution, leaving holes that may comprise parts of grains, whole grains, or large volumes (caves) as in carbonate and evaporite karst (Chapter 3). Relatively soluble minerals such as carbonates and evaporites are most susceptible to chemical dissolution, but feldspars and other aluminosilicate minerals in sandstones and shales are also commonly dissolved. Compaction may reduce porosity formed by dissolution, especially in sandstones and shales.

(4) *Cementation* is the chemical or biochemical precipitation of a new mineral in the pore space of sediment (Figure 19.2D, E). The pore space may be primary or produced partly by dissolution. Cementation may be concentrated in local areas up to a few decimeters in size (known as *nodules* or *concretions*), concentrated in certain layers, or spread more homogeneously throughout the sediment. Some cementation involves displacing (or pushing aside) the surrounding sediment grains. Evaporite and carbonate minerals commonly grow in this way as nodules (see Chapters 11 and 16). Cementation is most significant in relatively coarse-grained sediments with high initial porosity and permeability.

(5) *Mineral replacement* is the replacement of an existing mineral with a new one (Figure 19.2F). Mineral replacement generally occurs by piecemeal dissolution of one mineral and precipitation of the new one. If this occurs on a small scale, the structure of the original grain can be preserved in detail. However, the original structure will be lost if there is large-scale dissolution prior to precipitation of the new mineral (as a cement).

Recrystallization is where a pre-existing mineral is reorganized into a new geometric arrangement of the same mineral, with or without preservation of the structure of the original mineral grain. Mineral replacement and recrystallization are most common in the chemogenic sediments such as carbonates, evaporites, cherts, and iron-rich sediments (details are given below). Common replacement reactions in terrigenous sediments include the transformation of smectite to illite in shales, and the replacement of potassium feldspar and calcium feldspar by sodium feldspar (*albitization*) in sandstones.

The term *authigenic* (meaning "formed in place") is commonly used indiscriminately to describe sedimentary minerals such as (1) chemogenic and biogenic grains formed in the depositional area, (2) diagenetic cements and nodules that grew within the sediment, and (3) replacement and recrystallization products. Explicit terms should be used for these features, and the term authigenic should be abandoned. *Neomorphism* is an old term introduced by Folk (1965b) to describe situations where one polymorph of a mineral replaces a pre-existing polymorph, such as the replacement of aragonite by calcite. The term neomorphism was intended to cover situations where it was not clear whether calcite had recrystallized or whether calcite had replaced aragonite. *Metasomatism*, a term most commonly used by metamorphic petrologists, refers to wholesale mineralogical alteration, both iso-chemical and non iso-chemical, interpreted as due to the large-scale introduction of aqueous solutions.

(6) *Organic-matter transformations* in sediment take place initially as bacteria feast upon buried organic matter, and later at depth as a series of complicated *maturation* processes that result in the production of gas and oil. Terrestrial plant material is also progressively altered during burial to give lignite, bituminous coal, and finally anthracite coal.

Factors that control diagenesis

Most diagenetic processes involve dissolution and/or precipitation associated with aqueous solutions moving through the pore spaces in sediments, and reduction of these pore spaces by compaction. Therefore, it is easy to understand why the main factors that control diagenesis are (1) the chemical composition and texture of the starting sediments; (2) the chemical composition of the pore fluids; (3) the temperature, pressure, pH, and pe, which control the chemical stability of the mineral phases; (4) the porosity and permeability; and (5) the rate and direction of movement of pore fluids. These factors, discussed below, are intimately interrelated.

Chemical composition and texture of the starting materials

Traditionally, the diageneses of coarse-grained terrigenous sediments (sands, gravels), siliciclastic mud, carbonate sand and mud, and chemogenic sediments are treated separately, and this approach is followed here. The various sedimentary minerals have different stability fields in terms of T and P and the pH and pe of the pore fluid. Also, the shapes and sizes of the grains in these different sediment types have important implications for compaction. Most terrigenous sands and gravels are composed of spherical to ellipsoidal grains of quartz, feldspars, and rock fragments, although mica grains are plates. Siliciclastic clay minerals in muds are micron-diameter plates (see Chapter 3). Carbonate sands include spherical to ellipsoidal peloids and ooids, but also oddly shaped skeletal fragments (Chapter 4). Modern aragonite mud comprises 4-micron-long aragonite needles, and blockier, micron-diameter calcite crystals. Ooids and peloids are aragonite and/or calcite, as are shell fragments. The chemical compositions and shapes of evaporite minerals are described in Chapters 4 and 16.

Chemical composition of the pore fluids

Important near-surface (early diagenetic) pore fluids include dilute, meteoric (chemical weathering) water (Ca^{2+}, Na^+, HCO_3^-) described in Chapter 3, seawater (Na^+, Mg^{2+}, Cl^-, SO_4^{2-}), evaporated meteoric water (see Chapter 16), and evaporated seawater (see Chapter 4). In terrestrial settings, the most common near-surface pore water is meteoric, although evaporative brines are common in deserts (see Chapter 16). Such pore water in the unsaturated (vadose) zone may be oxidizing, but is usually reducing in the saturated (phreatic) zone (Chapter 3). In near-surface transitional environments (such as deltas, beaches, and tidal flats) the pore water on the landward side is usually meteoric, whereas it is either seawater or evaporated seawater on the seaward side. The seawater is transported onto the supratidal areas during storms

and exceptionally high tides, resulting in a complex transition zone between the landward meteoric groundwater and basinward seawater or brine. Sea-level changes can drastically affect the pore-water composition of coastal sediments (Morad et al., 2000; Stonecipher, 2000). Conditions of early (shallow) diagenesis vary significantly between continental-shelf areas (<200 m deep), where the water temperature can range up to 30 °C and pressures are less than ~8 bars, and deep marine settings (>200 m deep), where temperatures are ~4 °C and pressures are ~400–1,000 bars. Furthermore, in some cases of early diagenesis the pore water remains in contact with the atmosphere or the overlying water mass and active groundwater flow is taking place, whereas in other cases the pore water is cut off from the atmosphere or overlying water mass and stagnant within a few meters of the sediment surface.

During deeper burial diagenesis, the pore waters are no longer in direct contact with the atmosphere or a surface water body. In sedimentary basins on and adjacent to continents, these pore waters are usually anoxic brines with a wide range of salinities depending on the starting composition of the initial pore waters, the diagenetic reactions taking place, and the addition of formation water from other parts of the basin. In some basins, there may be substantial flows of meteoric groundwater at depths of 2–3 km. Beneath the deep ocean floor, hydrothermally altered seawater from deeper metamorphic alteration of basaltic ocean crust can affect diagenesis in sediments overlying the crests and flanks of mid-ocean ridges (Chapter 4). However, the pore water in most sediments accumulating on the deep-sea floor is seawater. Below a few kilometers depth, many sedimentary basins (e.g., the US Gulf Coast, European North Sea basin, and western Canada basin) contain so-called *oilfield brines* with salinities that range from 10,000 to >300,000 mg l^{-1} (Hanor, 1994), see Table 2.6. The most important ions in the majority of these brines are Na^+, Ca^{2+}, and Cl^-, which distinguishes them chemically from the common near-surface pore fluids listed above. The origin of these basinal brines is controversial: they may result from dissolution of evaporites in the subsurface, or they may be residual evaporite fluids that have interacted with the sediments during diagenesis (e.g., Carpenter, 1978). Lowenstein et al. (2003) suggested that the difficulty in interpreting the origin of these brines stems mainly from assuming that modern seawater was the starting fluid. It is now known that the composition of seawater has varied throughout the Phanerozoic (Chapters 2 and 4). According to Lowenstein *et al.* (2003), the origin of basinal brines is easier to interpret if an age-appropriate composition of seawater is used. This view on the origin of basinal brines has been challenged by Hanor and McIntosh (2006) and further work is needed to sort out the origin and histories of these highly saline brines.

T, P, pH, and pe

The equilibrium constants for various chemical reactions in sediments vary with *T* and *P* (Chapters 3 and 4). Moreover, the pH and pe (oxidation–reduction potential) also control mineral reactions in the subsurface, although oxidation–reduction reactions are most common in early diagenesis and exhumation diagenesis. Quartz and calcite are two of the most common cements in sedimentary rocks of all compositions. The solubility of quartz increases with increasing temperature, whereas the solubility of calcite decreases with increasing temperature. The solubility of CO_2 increases with increasing pressure, and this also affects the solubility of calcite.

Porosity and permeability

Porosity is defined as the volume of void space divided by the total volume of grains plus void space, and *permeability* is a measure of the ability of fluids to flow through unconsolidated sediments or rocks. Measurement techniques for porosity and permeability are given in the appendix, and depositional factors that determine porosity and permeability are discussed in Chapters 4 and 5 and in the appendix. Porosity and permeability tend to be correlated in sedimentary rocks, although this is not necessarily the case for unconsolidated sediments. Cementation and compaction reduce the porosity and permeability in a sediment or sedimentary rock whereas dissolution increases them.

Depending on the grain size and sorting, the original porosity of modern terrigenous sands ranges from 0.25 to 0.55 (average ~0.45). For comparison, the porosity of uniform spheres of equal size with face-centered cubic packing is 0.48, whereas rhombohedrally packed spheres have a porosity of 0.26 (Allen, 1982a, 1985). Experimental random packs of spheres such as ball bearings, and computer simulations of closest-sphere packing, give a porosity of approximately 0.36. The porosity of oblate spheroids (ellipsoids) depends on the aspect ratio of the three principal diameters of the ellipsoid, and ranges from 0.32 to 0.26. The porosity of

various packing arrangements of solid spheres is independent of the absolute diameter of the spheres. The porosity of mixtures of different-sized spheres is much harder to predict theoretically. Computer-generated packing of spheres of two sizes gives porosity of 0.36–0.12. Well-sorted natural sands have porosities of 0.25–0.55, whereas poorly sorted muddy sands have porosities of 0.45–0.7. The porosity of modern skeletal carbonate sands ranges from 0.40–0.70. Experiments show that the porosity of randomly packed non-spherical grains of more or less the same size range ranges from 0.31 for common dried beans to as much as 0.77 for flattened shells (Vinopal and Coogan, 1978; Enos and Sawatsky, 1981).

Permeability is a parameter from Darcy's law that describes the resistance to fluid flow through a material. Darcy's law can be expressed as

$$Q = \frac{-k\rho g}{\mu}\frac{\partial h}{\partial x} = \frac{k\partial p}{\mu \partial x} \tag{19.1}$$

where Q is the fluid discharge through unit cross section of sediment or rock, k is the permeability (in units of length squared), ρ is the fluid density, g is the gravitational acceleration, μ is the fluid viscosity, and $\partial h/\partial x$ and $\partial p/\partial x$ are, respectively, the hydraulic gradient and pressure gradient in the direction of flow. Permeability defined in this manner depends only on the properties of the material, not on the fluid, which could be meteoric water, dense brine, or hydrocarbons. Permeability varies approximately over the range 10^{-20}–$10^{-7}\,\mathrm{m}^2$ in sediments and sedimentary rocks. A curious older unit used to describe permeability is a *darcy*, which is defined as the permeability that allows a velocity of $1\,\mathrm{cm\,s}^{-1}$ of a fluid with viscosity 1 centipoise under a pressure gradient of 1 atmosphere cm^{-1}. A permeability of 1 darcy is approximately $10^{-12}\,\mathrm{m}^2$. The variation of permeability with grain texture, fabric, and internal structure is discussed in Chapters 4 and 5 and in the appendix. In general, permeability increases with mean grain size in unconsolidated sediments, but is greatly influenced by diagenetic changes, particularly compaction and cementation.

Flow in the basin

One of the key issues in diagenesis is whether the pore fluid is stationary or flowing. If it is stationary, chemical diffusion through the fluid is the main mode of chemical transport. If the fluid is moving, advection (the transport of dissolved chemical constituents together with the flowing pore water) becomes

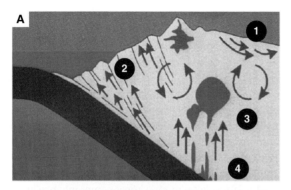

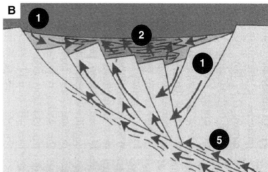

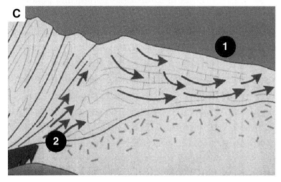

FIGURE 19.3. Diagrammatic representations of mechanisms that can drive fluid flow in (A) subduction zones, (B) extensional basins, and (C) compressional (foreland) basins. From Person *et al.* (1996). Fluid-flow mechanisms include (1) topography-driven; (2) compaction/tectonically driven; (3) convection-driven; (4) fluid production during mineral dehydration or juvenile magmatic water derived from intrusions; and (5) seismogenic pumping, whereby pore spaces are abruptly extended and compressed during earthquakes.

important. The pore-water flow is dependent on permeability, so measurable flows in sedimentary basins are usually confined to aquifers (porous and permeable geological formations). However, fluid flows even in relatively impermeable clay-rich units can be significant over geological time spans. Such flows might be driven by compaction or dehydration of clay, and may be particularly significant in migration of oil and gas from clay-rich rocks that originally contained the organic matter (*source rocks*). Heat can also be transported by flowing groundwater.

The driving force responsible for moving pore fluid is the pressure gradient ($\partial p/\partial x$ in Equation (19.1)). Pressure gradients in sedimentary basins may result from: (1) a hydraulic gradient caused by spatial variations in hydraulic head (topography-driven flow), as for example occurs between hills and valleys (Chapter 3); (2) lateral variations in lithostatic pressure due to differential gravitational loading and directed tectonic stresses such as with converging tectonic plates (compaction/tectonically driven flow); (3) spatial variations in fluid density related to temperature or salinity (convection-driven flow); (4) production of fluid during diagenesis by dewatering of minerals (fluid-production-driven flow); and (5) earthquake-driven flow (seismogenic pumping) (Figure 19.3) (Person *et al.*, 1996). Different flow-driving mechanisms occur

in different portions of sedimentary basins of different tectonic origins and at different times during the evolution of a sedimentary basin.

The important feature of pore-fluid-flow systems in sedimentary basins is that the hydrological, mechanical, thermal, and chemical mass-transfer processes that occur are closely coupled (Person *et al.*, 1996, p. 62), meaning that they all take place simultaneously and affect the flow, and are in turn affected by the flow. Coupled models of fluid flow, heat transport, and diagenetic chemical mass-transfer interactions on a basin scale are an active area of research (Horbury and Robinson, 1993; Montañez *et al.*, 1997). Specific examples of such models include a model of the Rhine rift valley (Person and Garven, 1992) and the paleohydrologic reconstruction of the Gulf of Mexico of Harrison and Summa (1991). Garven *et al.* (1993) modeled hydraulic-head-driven flows in order to explain a variety of diagenetic deposits, including Mississippi Valley lead–zinc ores and tar sands. Yao and Demicco (1995, 1997) modeled kilometer-scale dolomitization of shelf carbonates using a similar flow system. Models of pore-fluid motion due to spatial variations in heat or salinity (Phillips, 1991; Wilson *et al.*, 1990) have lagged behind head-driven models. Fluids forced under pressure out of clays by compaction and clay-mineral structural changes flow into surrounding permeable strata.

In foreland basins, pressure gradients may be due to combined differential loading and laterally directed stresses (Oliver, 1986; Cathles, 1993). Hydraulic gradients in pore water will also be produced by differential uplift in some basins. In view of the episodic nature of tectonism, it is probable that tectonically induced fluid flow in sedimentary basins is also episodic.

Details of diagenetic processes and products and their controls

Details of the effects of compaction, dissolution, cementation, replacement, and recrystallization on the composition, texture, and structure of sediments and sedimentary rocks are discussed below. In order to understand the controls on these diagenetic features it is necessary to establish their relative timing (e.g., different generations of cement) from field studies and petrographic study of hand samples and thin sections. The absolute timing of the various diagenetic features can be obtained by radiometric dating of cement minerals, particularly clays (Lee *et al.*, 1989). Paleomagnetic dating is also useful. The temperature conditions of diagenesis can potentially be established using measurements of stable-isotopic composition, fluid-inclusion paleo-thermometry, and temperature-dependent mineral assemblages. Although the pressure conditions of diagenesis are difficult to ascertain, they can be constrained by burial and tectonic history and, if temperatures are known, also by assumed geothermal gradients. The chemistry and origin of the fluids that filled the pores, and how the pore-water composition changes with time, can be ascertained using techniques such as trace-element geochemistry, stable-isotope geochemistry, and fluid-inclusion studies. Establishment of the fluid-flow paths for the delivery and removal of ions (the "plumbing system") relies on knowledge of the vertical and lateral distribution of the diagenetic features, calculations of the masses of minerals involved, and simulations of the flow systems that are consistent with the observations.

Compaction

Introduction

Physical and chemical compaction are caused by the increasing overburden pressure as sediments are progressively buried, although tectonic compression can have the same effects. Physical compaction results in the breakage, deformation, and reorientation of grains

(see Figure 19.4A–D and the appendix). Chemical compaction (also known as *pressure solution*) is associated with dissolution at grain-to-grain contacts where focused lithostatic (or tectonic) pressure increases the solubility of minerals (Figure 19.4E). This process leads to interpenetration of adjacent grains with sutured grain-to-grain contacts and *stylolites*, which are irregularly shaped surfaces of dissolution (especially in carbonates) where insoluble residues have been concentrated (Figure 19.4F). Pressure solution may be enhanced by minerals between compacting quartz grains. Mica flakes in sandstones commonly penetrate into quartz grains with little deformation of the mica flakes (Bjørkum, 1996). Detailed textural and chemical-microprobe observations of sutured quartz grains in sandstones indicate that the suture is commonly a film of potassium- and aluminum-rich material. This has led to the speculation that pressure solution in quartz is promoted by micas or clays between the quartz, and that it may take place at pressures of ~10 bars or less (Bjørkum, 1996).

As the overburden pressure increases during burial, the bulk volume of the sediment decreases as the pore space is reduced and water is expelled. Thus, the porosity, bulk density, and permeability are progressively reduced as compaction increases with increasing depth of burial. Terrigenous sand, carbonate sand, terrigenous mud, and carbonate mud react somewhat differently to compaction due to differences in initial shapes and packing of the mineral grains, and susceptibility to pressure solution. Physical compaction is most important in sediments with high initial porosity, such as terrigenous muds. Chemical compaction is most important in sediments that are highly soluble under pressure. References on physical and chemical compaction include Chilingarian and Wolf (1988), Giles (1997), Bjørlykke (2003), and Renard and Dysthe (2003).

Empirical curves of porosity versus depth (e.g., Figure 19.5) are called *compaction curves* (Baldwin and Butler, 1985; Giles, 1997; Ehrenberg and Nadeau, 2005). Many older compaction curves may include the effects of porosity loss by cementation as well as compaction, but more recent ones (e.g., Paxton *et al.*, 2002) do not. As a result, it is common to see a number of distinctly different porosity–depth curves for a given sediment type.

Terrigenous sandstones and carbonate grainstones

Porosity–depth curves for terrigenous sands (Figure 19.5A) exhibit great variability. Porosity–depth

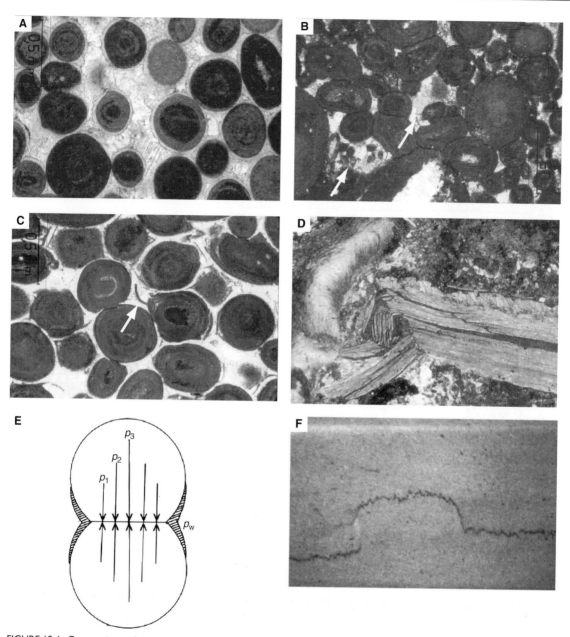

FIGURE 19.4. Compaction and pressure solution. (A) Uncompacted ooid–peloid grainstone. Width of field 2 mm. Courtesy of Tim Lowenstein. (B) Compacted ooid–peloid grainstone with sutured polygonal contacts and long contacts between many of the grains. Note the more open areas (white arrows) protected from compaction by pre-compaction early cementation. Width of field 2 mm. Courtesy of Tim Lowenstein. (C) Compacted ooid–peloid grainstone where early isopachous cements have been crushed and spalled (arrows). Width of field 2 mm. Courtesy of Tim Lowenstein. (D) A skeletal fragment (bivalve shell) fractured due to loading. Width of field 1.6 mm. From Scholle and Ulmer-Scholle (2003). AAPG © 2003 reprinted by permission of the AAPG, whose permission is required for further use. (E) A model of the differential lithostatic pressure at the contact of two spherical quartz grains. Pressures p_1 to p_3 are different from each other and from the pore-fluid pressure, p_w. The pressure differences lead to a chemical potential gradient down which silica diffuses to precipitate as overgrowth cements (ruled pattern). From Hutcheon (1990). (F) A stylolite in limestone. The height of the photograph is ~15 cm.

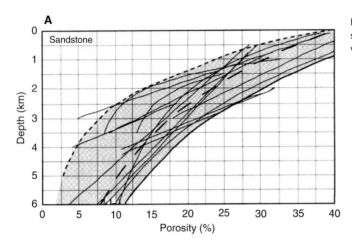

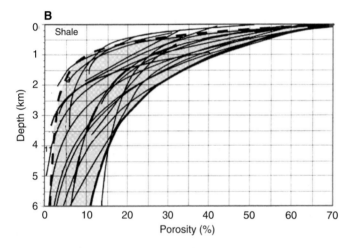

FIGURE 19.5. Porosity versus depth for sands and sandstones (A) and mud and mudstones (B) from various sedimentary basins. From Giles (1997).

curves for carbonates typically do not distinguish whether the rocks are grainstones or mudstones (Ehrenberg and Nadeau, 2005). Paxton *et al.* (2002) compiled a data set to show that typical sandstones with ~40% pre-cement porosity at the surface are reduced in porosity to ~30% at a burial depth of 1.5 km. Between 1.5 km and 2.5 km burial depth, the pre-cement porosity is further reduced to ~25% and stays near this value to burial depths of up to 6.7 km. Volume loss in sandstones due to compaction amounts to ~15% of the total rock volume, which is why primary sedimentary structures of sandstones are not greatly modified by burial compaction. Cementation during burial is important for decreasing the porosity of sandstone with depth. The degree of compaction in sandstones can be assessed by measuring the *pre-cement porosity*, which is the sum of the primary porosity and the volume of cement that filled the primary pore space. Comparison of this value with a true compaction curve can give a rough idea of the depth at which the cements

were introduced. The number of grain-to-grain contacts per grain increases with increasing compaction and the types of contacts between grains also change systematically with burial depth (see the appendix). As compaction proceeds, point contacts become long contacts and ultimately sutured contacts. In addition, grains such as peloids and intraclasts that are composed of mud are commonly squashed between harder grains, deforming into a fine-grained, matrix-like material. In such cases, the squashed soft grains may be hard to recognize and differentiate from fine-grained pore fills. Plate-like grains (such as mica flakes and skeletal fragments) that are compressed between harder grains can bend, fracture, splinter, and break.

Chemical compaction (pressure solution) enhances the reduction of porosity in terrigenous sandstones, but particularly so for carbonates. Pressure solution in terrigenous quartz sands begins at a burial depth of ~3 km and temperatures of >100 °C, whereas it can begin in carbonates after only 200–300 m of burial.

Stylolites (Figure 19.4F) are particularly common in carbonate rocks (Pettijohn, 1975; Scholle and Halley, 1985; Railsback, 2003) but are also found in sandstones and evaporites. They are irregular, wavy surfaces caused by interpenetration of the two sides. The wave height of the interpenetration ranges from a fraction of a millimeter to decimeters, and the wave length ranges from a few centimeters to meters. The three-dimensional geometry of stylolites ranges from somewhat conical to regular rectilinear columns and interdigitating sockets. The surface of the stylolite is marked by clays, iron minerals, organic matter, and other insoluble residues that have been left along the seam during dissolution. Stylolites are usually parallel to bedding planes, and in some places accentuate them, but tectonic pressures can give rise to stylolites that cross-cut bedding at high angles. Some geologists have postulated the existence of pervasive pressure solution that generates seams of clays and perhaps dolomite without recognizable suturing (Logan and Seminiuk, 1976; Wanless, 1979; Bathurst, 1987); see the discussion by Demicco and Hardie (1994). Stylolites in carbonates may reduce the thickness of the original sedimentary sequence by 20%–35%, and it has long been argued that pressure solution of carbonates provides a source for the carbonate cement that is typically found in carbonate rocks. This case has also been made for terrigenous clastics, with pressure solution providing the material for syntaxial quartz overgrowths (Figure 19.4E, and see below). Quartz overgrowth cements are common in older sandstones and less common in younger ones.

Terrigenous muds

The porosity of freshly deposited terrigenous mud ranges from 0.6 to 0.8. Terrigenous mud is composed of platy clay minerals and variable amounts of quartz and feldspar silt (Chapter 3). In addition, many terrigenous and carbonate muds also contain silt- and sand-sized mud peloids (Bennett et al., 1991). Peloids are commonly fecal pellets of marine invertebrates, but may also be eroded remnants of pre-deposited mud (known as intraclasts). Another source of peloids in aluminosilicate muds is floccules, which owe their origin to the electro-chemical properties of phyllosilicates. The important point here is that the relatively large initial porosity of mud is associated with porosity within and between the peloids. Terrigenous muds and mud peloids are composed of clay sheets that have a "house of cards" structure, whereas modern carbonate muds and mud peloids have a structure like

a pile of toothpicks with all possible orientations. The porosity between peloids is the same as would be found from random packing of ellipsoids.

Siliciclastic muds rich in clay minerals are most profoundly affected by compaction because of their large initial porosity. Most compaction takes place in the upper 1 km before mineralogical changes occur in the clay minerals (Figure 19.5B). Significant mineralogical changes occur only when temperatures reach 100 °C at a depth of about 3 km, where the porosity has already been reduced to about 10%. Re-orientation of platy clay minerals upon compaction results in fissile shale with the fissility normal to the vertical stress due to overburden (Bennett et al., 1991). By virtue of its orientation, this diagenetic fissility can be confused with primary lamination. X-radiographs of box cores from modern muds reveal primary sedimentary features and complicated burrow networks. These features are commonly obliterated by the compaction.

Carbonate muds

Carbonate muds do not suffer the profound physical reorganization upon compaction that siliciclastic muds do: therefore, primary sedimentary features may still be recognizable (Demicco and Mitchell, 1982; Shinn and Robbin, 1983; Mitchell, 1985; Demicco and Hardie, 1994; Goldhammer, 1997). Preservation of some types of primary sedimentary structures in some carbonate muds suggests that these carbonate muds were cemented early, although this is still an open question. Two main factors allow preservation of sedimentary structures in shallow-marine carbonate muds: (1) aragonite-needle muds apparently compact far less than clay-rich muds; and (2) early cementation (pre-compaction lithification) is widespread in carbonate environments (see below). Carbonate muds in modern shallow-marine environments have porosities in the range of 70%–80%, and consist mainly of a loose aggregate of ~4-μm-long aragonite needles. Shinn and Robbin (1983) showed in squeezing experiments that the porosity of such muds cannot be reduced below about 35% by mechanical compaction, presumably because the packed needles behave as a grain-supported framework. A fully compacted carbonate mud will therefore have suffered a maximum reduction in volume of 50%–55% and will have the porosity of compacted sand, but with a lower permeability because of the small pore throats. Carbonate mud deposited during times of *calcite seas* (Chapter 2) probably comprised equant calcite crystals such as comprise

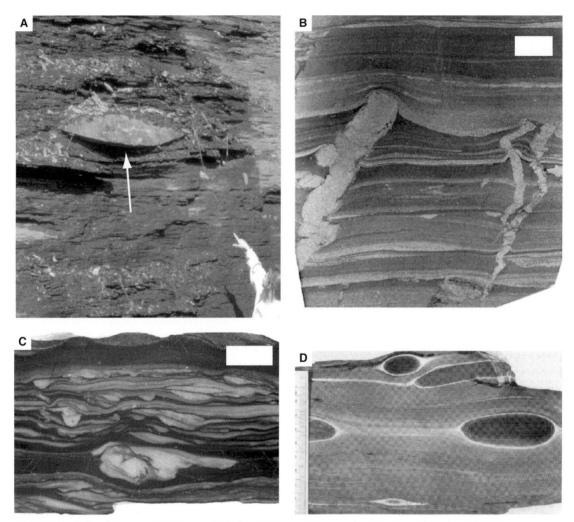

FIGURE 19.6. Differential compaction features. (A) Differential compaction around a small channel fill (arrow), Carboniferous rock of Kentucky. The original channel morphology has been lost. (B) Sand-filled mudcracks folded due to compaction of the adjacent mud. The scale bar is 10 mm long. From Pettijohn and Potter (1964). (C) Lenticular bedding where shapes of sand ripples have been distorted both by differential compaction and by soft-sediment deformation. The scale bar is 5 mm long. (D) Limestone concretions in a laminated dolomitic mudstone, Silurian rock of western New York. Laminae bend over and under the concretions, demonstrating the occurrence of volume reduction in the uncemented mudstone. The scale is in mm. From Demicco and Hardie (1994).

carbonate mud in non-marine settings where the water has Mg^{2+}/Ca^{2+} ratios of <2 (Chapter 4). Such a calcite mud would be less affected by compaction than an aragonite-needle mud. Primary sedimentary structures such as laminae, mudcracks, and sediment-filled burrows are somewhat distorted but survive compaction (Shinn and Robbin, 1983). However, open burrows, vugs, and fenestrae will be eliminated by this degree of compaction. Organic matter will be reduced to discontinuous wispy seams that resemble micro-stylolites, and soft peloids will be merged into a clotted muddy paste, producing a texture like that of the *structure grumeleuse* of Cayeaux: see Bathurst (1975).

Differential compaction

The various types of sedimentary rock that occur together in a sedimentary succession will normally compact differently, giving rise to differential compaction. A common example is interbedded terrigenous muds and sands, for which the muds are reduced in porosity much more than the sands at a given depth. Resulting features of differential compaction include originally flat-topped, sandy channel fills that become lenticular in shape (Figure 19.6A), and originally straight sand-filled mudcracks that become folded (Figure 19.6B). In addition, the shapes of sandstone beds in wavy and lenticular bedding can be affected by differential compaction of mudstone layers (Figure 19.6C). Early-cemented concretions may experience minimal compaction and commonly preserve the burrowing and peloidal nature of the original sediments, whereas the surrounding terrigenous muds are compressed around them (Figure 19.6D).

Over-pressuring

When the porosity in sediments is reduced by compaction, the expelled pore water might not be able to escape, especially if surrounding rocks are impermeable. In this case, some of the overburden pressure is transmitted into the pore waters and the sediment is said to be *over-pressured*. Over-pressuring also prevents continued mechanical compaction of the over-pressured sediment, and therefore contributes to the variability of compaction curves for various sediments. Over-pressuring can occur quite close to the surface and may result in sand- and mud-volcanoes at the surface (Chapter 12). The subsurface release of over-pressure is also thought to generate so-called "hydro-fracturing" in some instances. Over-pressured sediments are sought-after targets because they may be valuable petroleum and gas reservoirs. However, over-pressured zones are a danger to oil-well drilling.

Dissolution and secondary porosity

Dissolution of chemically unstable grains may remove parts of grains, whole grains, or large volumes (caves) as in carbonate and evaporite karst (Chapter 3). The dissolution holes are commonly referred to as secondary pores to distinguish them from the primary intergranular and intragranular pores (see the appendix for further terminology on porosity). Secondary porosity commonly hosts hydrocarbon accumulations and increases the permeability of carbonate aquifers. In terrigenous sandstones in the Gulf Coast and North Sea regions, geochemical and textural evidence suggests that feldspars (particularly potassium feldspar) and detrital kaolinite are dissolved at burial depths greater than ~3.5 km and illite is precipitated as a cement. Since it is thought that there is no net import or export of materials dissolved in the pore fluids, there is little change in porosity due to grain dissolution and cementation (Giles and de Boer, 1990; Chuhan *et al.*, 2000, 2001). In other cases, textural evidence suggests that large amounts of feldspar (perhaps up to 25% of the initial rock volume) can be dissolved in deeply buried sandstones, the dissolved material is removed from the system, and the overburden pressure collapses the pore network, leaving only 3% identifiable secondary pores (Harris, 1989). Therefore, the mineralogy of sandstones can be altered substantially by dissolution.

Cementation

Introduction

Cementation is the chemical or biochemical precipitation of a new mineral into the pore space of a rock, and acts with physical and chemical compaction to reduce porosity and permeability. Cements are best known from limestones and terrigenous sandstones and gravelstones because of their high porosity and permeability. Cementation occurs in the pore spaces between (intergranular) and within (intragranular) sediment grains, and within larger openings such as fenestrae, vugs, caverns, and fractures. Cementation may be concentrated in certain layers, local areas known as concretions (also known as nodules and glaebules), or spread more homogeneously throughout the rock. Concretions commonly form early in the diagenetic history of a rock and thus can preserve primary sedimentary structures and textures (see above). Some cements displace (or push aside) the surrounding sediment as they develop, and such displacive cements commonly occur as nodules (e.g., evaporite minerals in sabkhas and carbonate nodules in calcareous soils). Sources of the dissolved materials that are precipitated as cements include the original chemical components of the pore water (meteoric, seawater, or oil-field brine) and new dissolved components added by dissolution of unstable grains, replacement reactions, and pressure solution. Sequences of cements of different mineralogy or morphology are common and may be punctuated by dissolution. Growth history in cements can be revealed by cathodoluminescence (Burley, 2003) or by detecting fluid inclusions parallel to crystal growth faces. Useful reviews of cementation are provided by Bricker (1971), Bathurst (1975), Scholle (1979), Adams *et al.* (1984), Schneidermann and Harris (1985), McIlreath and Morrow (1990), Morad *et al.* (2000), Tucker (2001), Scholle and Ulmer-Scholle (2003), and Flügel (2004).

Cement mineralogy

The most common cements in modern carbonates are aragonite, high-Mg calcite, and low-Mg calcite. The

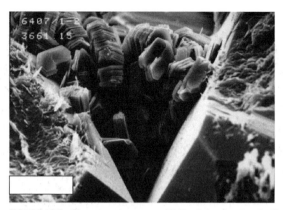

FIGURE 19.7. Clay-mineral cements. (A) Kaolinite showing the "book" morphology typical of cements. The scale bar is 10 microns long. The large blocky surfaces are quartz overgrowths. (B) Illite, showing fine filamentous structures. The scale bar is 10 microns long. Photomicrographs courtesy of Tim Lowenstein.

mineral that forms is apparently dependent on the Mg^{2+}/Ca^{2+} ratio of the fluid and the temperature (Chapter 4). Where the Mg^{2+}/Ca^{2+} ratio is <2, low-Mg calcite typically forms, and this is the case in most dilute meteoric groundwater. Where the Mg^{2+}/Ca^{2+} ratio ranges from ~ 2 to ~ 5, high-Mg calcite and/or aragonite form. This is the case for seawater on shallow tropical shelves where aragonite and high-Mg calcite cements typically form in beachrocks and reefs. High-Mg calcite cements are generally found in temperate and deep-water settings. Füchtbauer and Hardie (1976, 1980) showed that the Mg content of modern Mg-calcite cement is a function of the temperature of formation, and modern calcite cement formed in cold deep waters and in cool temperate areas is low-Mg calcite. Aragonite forms where the Mg^{2+}/Ca^{2+} ratio is >5 (as is found in many saline lakes). The most common cements in ancient limestones are low-Mg calcite with lesser amounts of dolomite. This implies that the aragonite and high-Mg calcite cements that are so common in modern environments are unstable during burial (see below). Other cements in carbonates include ankerite, siderite, gypsum, anhydrite, halite, and quartz.

In coarse-grained terrigenous sediments, the most common cements by far are quartz (including hydrated amorphous opal and a fibrous variety of quartz known as chalcedony), calcite, and clay minerals. Other, less common cements in terrigenous sediments include hematite, iron oxyhydroxides ("limonite"), gypsum, anhydrite, barite, feldspars, and zeolites. In addition, mineralogical changes in detrital clay minerals in sandstones may cement matrix-rich sandstones together.

Cement morphology

Recognition of pore-filling cements on the basis of their morphology is not always easy, because detrital carbonate and terrigenous mud may also occur in pore spaces. Upon burial, these fine-grained detrital components (or fine-grained diagenetic components including opal) may recrystallize into an interlocking mosaic of larger crystals that superficially resembles cement. This is not usually a problem with terrigenous mud because clay-mineral cements have distinctive crystalline structures (Figure 19.7), but it is with carbonate detrital (or diagenetic) mud. Bathurst (1975) lists criteria to separate carbonate cement crystals from masses of recrystallized carbonate that replaced previous, finer-grained carbonate (Table 19.2). Many of these criteria are also useful for identifying cements in terrigenous sedimentary rocks. Figure 19.8 (from Flügel, 2004) describes some of the more common types of carbonate cement and synonyms. Carbonate cements are commonly referred to as *spars*, apparently named after the variety of calcite known as Iceland spar. This is confusing (see below) and this term should be avoided. The terms in Figure 19.8 are useful also for cements in terrigenous sediments. Chalcedony (fibrous opal, Chapter 4) commonly comprises acicular, fibrous, or botryoidal cements, and granular, blocky, and syntaxial quartz cements are common.

Cements can be divided into *syntaxial* and *epitaxial*. Syntaxial cements have the same mineralogy and crystallographic orientation as the host mineral grain, which is normally a single crystal. This means that, as a microscope stage is rotated under cross-polarized light, the grain and the overgrowth become extinct

TABLE 19.2. Criteria to differentiate recrystallized material from cement crystals (modified from Bathurst, 1975)

(1) The cement crystals are between well-sorted and abraded framework grains, which are in depositional contact with one another.
(2) There are two or more generations of crystals.
(3) There are no relict structures (ghosts) of former grains in the crystals.
(4) Peloids and other particles composed of fine-grained material are not altered.
(5) Micrite coats and envelopes are not altered.
(6) Fine-grained detrital material is present and not altered.
(7) Contacts between crystals and detrital grains are sharp.
(8) The margin of the crystal mosaic coincides with surfaces that were once free, such as the surfaces of skeletal particles, ooids, other grains, or molds of leached shell fragments.
(9) The crystals line cavities that they do not completely fill.
(10) The crystals occupy the upper part of a cavity whose lower portion is filled by an internal geopetal sediment.
(11) The crystals have the form of a pore filling or encrustation such as a tufa or stalactite.
(12) The boundaries between the crystals are planar.
(13) The sizes of the crystals increase away from the substrate.
(14) The crystals have a preferred orientation of optic axes normal to the substrate.
(15) The crystals have a preferred shape orientation with long axes normal to the substrate.
(16) The crystals have 30%–70% "enfacial" junctions.

simultaneously. Syntaxial overgrowths are common on single-crystal (monocrystalline) grains of quartz, feldspar, zircon, and tourmaline in terrigenous sandstones, and on crinoid columnals (single crystals of high-Mg calcite) in limestones. Syntaxial overgrowths on quartz grains (Figure 19.9A) may be identified by *dust rims*, which pick out the original rounded perimeter of the grains, and euhedral quartz terminations in pore spaces. The dust rims usually comprise clay minerals, iron oxides, iron oxyhydroxides, and fluid inclusions. However, in many cases, the dust rims around sediment grains never formed or have been removed, and cathodoluminescence is required to identify the original grains (see the appendix for techniques). Less commonly, small crystals of cement can grow syntaxially on crystallites that make up larger (polycrystalline) grains such as invertebrate shells and ooids.

Epitaxial cements have a crystallographic orientation that is independent of the host grain and mineralogy that is either the same as the host grain or different. Common terms used to describe epitaxial cements include *fibrous*, *drusy*, and *equigranular mosaics* (illustrated in Figure 19.8). *Fibrous* cements (also called acicular or needle-like: Figure 19.9B) are commonly aragonite and the fibrous variety of quartz known as chalcedony. If cement nucleation sites are scattered, bundles of cement fibers occur, with curved outer edges that are mutually interfering, and are referred to as *spherical fibrous* or *botryoidal*. If there are many

nucleation sites, the needle-like crystals end up growing perpendicular to the edge of the host grain, and they all point into the void cavity. These are commonly called *radial fibrous*. *Drusy* (also called dog-tooth) cements, which are common in low-Mg calcite and quartz, display increasing crystal sizes into the pore space, where they can have crystal-face terminations (Figure 19.9C, D). At the base of a fibrous or drusy cement layer, it is common to see many small crystals that initially nucleated in multiple orientations. As the crystals grew, those with their fast-growing directions oriented nearly perpendicular to the edge of the substrate buried the crystals growing at less-favorable orientations.

Many radial-fibrous and drusy cements maintain the same thickness all the way around the void (they are called *isopachous* or *void-lining*: i.e., isopachous drusy cement). Isopachous cements are commonly interpreted as implying that the cementation is taking place in an area where all the pores are filled with pore water. Non-isopachous fibrous and drusy cements are rare, but fine-grained carbonate cements (crystal size <4 microns) are commonly non-isopachous. These are typically high-Mg calcite, but rarely low-Mg calcite or aragonite, and are commonly precipitated in the unsaturated (vadose) zone. They manifest the effects of the strong surface tension of water in contact with air. For example, *meniscus cements* (Figure 19.9E) between grains have curved walls of bubbles that bounded

Acicular: Needle-like crystals, growing normal to the substrate. Crystals elongated parallel to the c-axis, exhibiting straight extinction. Terminations are pointed or chisel-shaped, twinning is common. Width <10 μm, length to about 100 μm and more. Commonly forming isopachous crusts. Predominantly aragonite, but also Mg calcite. Marine phreatic.

Fibrous: Fibrous crystals, growing normal to the substrate. Crystals show a significant length elongation, usually parallel to the c-axis. Crystal shape is needle-like or columnar (Length to Width ratio >6:1, Width >10 μm). Size commonly fine to medium crystalline. Commonly forming isopachous crusts; common in inter-and intra-particle pores. Aragonite or high Mg calcite. Mostly marine phreatic, but also meteoric vadose and marine vadose (columnar crystal shape). Syn.: Radial fibrous.

Botryoidal: Pore-filling cement made of individual and coalescent spherules that range in size from tens of microns to several centimeters. Spherules consist of individual and compound fans which are themselves composed of elongated euhedral fibers with a characteristic sweeping extinction in cross-polarized light. Spherules exhibit discontinuous, concentric dust bands. Aragonite (calcite in caves). Usually marine and common in cavities of reefs and steep seaward slopes, but also known from caves and burial environments, Syn.: Spherulitic.

Radiaxial fibrous: Large, commonly cloudy and turbid, inclusion-rich crystals with undulose extinction. Size medium to coarse crystalline, rarely extending several millimeters in length, commonly about 30–300 μm. Crystal length/width ratio 1:3 to 1:10. Crystals show a pattern of subcrystal units. Within each subcrystal that diverges away from the substrate an opposing pattern of distally convergent optic axes occurs, caused by a curvature of cleavage and twin lamellae. Undulose extinction of subcrystals or subcrystal units is used in distinguishing subtypes. Commonly forming isopachous crusts. Calcite. Marine phreatic and burial.

Dog Tooth: Sharply pointed acute or rarely blunted calcite crystals of elongated scalenohedral or rhombohedral form, growing normal and subnormal to the substrate (grain surface or atop earlier cements). Crystals are a few tens to a few hundred microns long. Calcite. Commonly meteoric and shallow-burial but also marine-phreatic and hydrothermal. Syn.: Bladed scalenohedral cement, bladed prismatic cement, dentate cement, scalenohedral cement.

Bladed: Crystals that are not equidimensional and not fibrous. They comprise elongated crystals somewhat wider than fibrous crystals (length/width ratio between 1.5:1 and 6:1) and exhibiting broad flattened and pyramid-like terminations. Crystal size up to 10 μm in width and between less than 20 and more than 100 μm in length. Crystals increase in width along their length. Commonly forming thin isopachous fringes on grains. Commonly high-Mg calcite but also aragonite. Marine phreatic (abundant in shallow-marine settings) and marine vadose.

Dripstone: Pendant cement characterized by distinct thickening of cement crusts beneath grains or under the roofs of intergranular and solution voids. The cement forms on droplets beneath grains after the bulk of the mobile water has drained out of the pores, leaving a thicker water film at the lower surface of the grains. Forms typically gravitational, beard-like patterns. Predominantly calcite. Formed above the water table with the meteoric vadose zone (where it is commonly associated with meniscus cement) but also found in marine vadose zone (beachrocks and intertidal and supratidal crusts) where it is aragonite. Syn.: Gravitational cement, microstalactitic cement, microstalactitic drusy cement, stalactitic cement, aragonite dripstone cement.

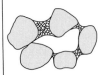

Meniscus: Cements precipitated near grain-to-grain contacts in pores containing both air and water. Exhibits curved surfaces due to meniscus formed by surface tension. Calcite. Characteristically zoned in the meteoric-vadose zone, but may also occur in vadose-marine beachrocks.

Drusy: Void-filling and pore-lining cement in intergranular and intraskeletal pores, molds, and fractures. Characterized by equant to elongated, anhedral to subhedral crystals. Size usually > 10 μm with crystal size increasing towards the center of the void. Non-ferroan calcite. Near surface meteoric as well as burial environments. Syn.: Drusy calcite spar mosaic, drusy equant calcite mosaic.

FIGURE 19.8. Types of carbonate cement. Modified from Flügel (2004). Springer-Verlag © used by permission.

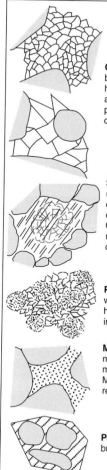

Granular (above) and blocky (below): Comprise equidimensional mosaics and are distinguished on the basis of crystal size, with granular cements generally tens of microns in diameter and blocky cements hundreds of microns. Crystals are generally subhedral, and, since all pore space is filled, crystal terminations are rare. High percentages of enfacial junctions. High-Mg calcite, low Mg calcite, and dolomite. Meteoric phreatic and vadose environments and burial. Similar textures can arise from the neomorphic growth of carbonate mud or earlier cements (see Table 19.2).

Syntaxial overgrowth cement: Substrate-controlled overgrowth of a single crystal around a host grain (commonly a high-Mg calcite echinoderm fragment). Overgrowth is in crystallographic (hence optical) continuity with the host grain. Overgrowths on echinoderm fragments are commonly zoned. Color differences between the skeletal grain and the overgrowth cement can be conspicuous. Calcite. Overgrowths from near-surface marine, vadose-marine, and meteoric-phreatic environments are inclusion-rich and cloudy, deep burial overgrowths clear. Syn.: Grain-overgrowth cement, syntaxial echinoderm cement, syntaxial rim cement, syntaxial overgrowth rim cement.

Peloidal microcrystalline cement: Peloidal fabric composed of well-sorted peloids generally <100 μm within a microcrystaline (<40 μm) granular mosaic. Peloids consist of mud-sized crystals with a radiating halo that grades out to the cement. Calcite and aragonite. Lakes and shallow-marine environments, common in reefs where it is probably microbially induced.

Microcrystalline or micrite cement: Micron-sized crystals, in SEM seen to comprise curved, rhombic mosaic. Forms thin coatings around grains, lines intraskeletal pores, fills pores completely or makes up meniscus and dripstone cements. Mg calcite or aragonite. Marine and meteoric vadose and phreatic zones. Microcrystalline cement coatings can resemble micrite envelopes and microcrystalline pore-filling cement resembles detrital mud. Commonly associated with peloidal cements.

Poikilitic cement: Single crystal surrounds a number of grains. Calcite. Marine phreatic, meteoric phreatic, burial.

FIGURE 19.8. Continued.

the cement-precipitating water droplet. Also fairly common in the vadose zone are *gravitational* or *pendant cements*, where droplets were suspended at the tops of cavities and deposited "micro-dripstones." Macintyre (1985) reported a unique peloidal morphology for carbonate-mud cements precipitated in reef cavities by bacteria. Clay, iron oxide, and iron oxyhydroxide cements in terrigenous sedimentary rocks are analogous to carbonate-mud cements. Clay cements in soils also commonly exhibit meniscus and pendant structures. Fine-grained cements in carbonates and clay cements in terrigenous sediments are perhaps the most difficult cement types to identify correctly.

Epitaxial, equigranular mosaic cements (also known as blocky cements) of calcite and quartz are very common in ancient limestones and terrigenous sandstones. Here the crystals comprise an anhedral interlocking aggregate (Figure 19.9C). Equigranular mosaics probably arise where the growth rate of crystals does not vary greatly, and the crystal habit is equant. Poikilitic cements are a variety of blocky mosaics in which several grains are overgrown by a single crystal (Figure 19.9F). Gypsum and barite *desert roses* are examples of poikilitic cements. Terrigenous sandstones are commonly cemented by poikilitic calcite. In hand specimens, the large cement crystals shine when sunlight is reflected off their crystal faces, giving rise to shiny patches in the rock, referred to as luster mottling.

It is common to find more than one generation of cement in a coarse-grained sedimentary rock. Meyers (1974, 1991) used cathodoluminescence and cross-cutting relationships to establish a *cement stratigraphy*. Figure 19.9C gives an example of a sequence of different cements.

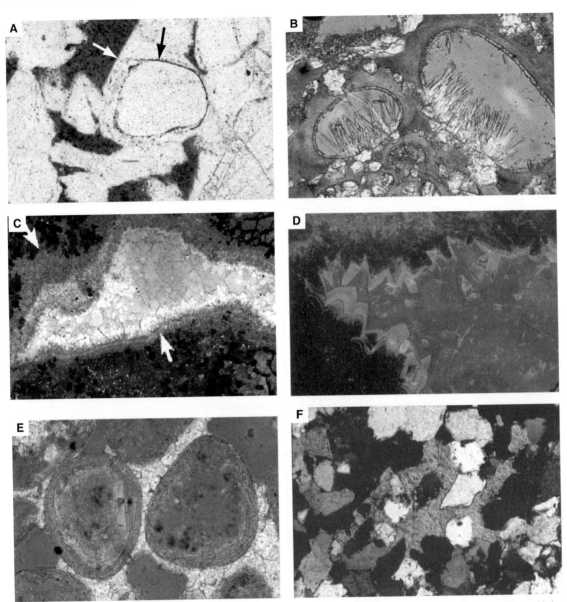

FIGURE 19.9. Thin-section photomicrographs of various cement types. (A) Syntaxial quartz overgrowths under cross-polarized light. Dark arrows point to "dust rims" outlining the detrital grain boundary; light arrows point to euhedral termination of quartz overgrowth in pores (filled with blue epoxy resin). Width of field 2 mm. Courtesy of Tim Lowenstein. (B) Fibrous aragonite cement in an intraparticle pore. Width of field 0.42 mm. From Scholle and Ulmer-Scholle (2003). (C) Two generations of carbonate cement in a reef cavity. The arrows point to a cloudy, fibrous first-generation cement lining the cavity that was buried by a later blocky mosaic. Field of view ~10 mm. From Scholle and Ulmer-Scholle (2003). (D) A cathodoluminescence view of drusy cement expanding into a large void. The luminescence is excited by manganese but repressed by iron and reveals slight geochemical variability in the cements. Field of view 11 mm. From Scholle and Ulmer-Scholle (2003). (E) Meniscus cements comprised of fine blocky carbonate. Note the rounded pore at the lower center of the photograph. Field of view 0.6 mm. From Scholle and Ulmer-Scholle (2003). (F) Large patches of poikilitic calcite cement crystals. Field of view 2 mm. Courtesy of Tim Lowenstein. Figures from Scholle and Ulmer-Scholle AAPG © 2003 reprinted by permission of the AAPG, whose permission is required for further use.

Mineral replacement and recrystallization

Introduction

Mineral replacement refers to a new mineral forming in the place of an existing one. Replacement is usually not isochemical, so the new mineral has a chemical composition different from that of its predecessor, and therefore new ions need to be imported and/or exported from the sediment (Figure 19.10B). However, replacement can involve a polymorphic transition whereby the chemical composition of the two minerals may be nearly the same but the crystal structure is changed. Mineral replacement generally occurs by piecemeal dissolution of one mineral and precipitation of the new one. If this occurs on a fine scale, the structure of the original grain can be preserved in detail (e.g., silicification or pyrite replacement of the cell structure of plants: Figure 19.10C), implying that the replacement is taking place through a thin film or replacement front (Putnis, 2002). However, the original structure will be lost if there is large-scale dissolution prior to precipitation of the new mineral (as a cement), as is commonly encountered in carbonates, where the outline (or *ghost*) of a former grain picked out by solid inclusions or a *micritic envelope* marking the former shell is preserved in a coarse, interlocking mosaic of carbonate crystals (Figure 19.9C). Displaced or dropped nuclei preserved in some "replaced" ooids clearly indicate substantial dissolution and the generation of voids at least tens of microns in diameter (Figure 19.10D).

Some of the more common replacement reactions in diagenesis include (1) oxidation–reduction reactions, typically involving iron minerals; (2) replacement of aragonite by calcite; (3) dolomitization, namely replacement of aragonite or low-Mg calcite by dolomite; (4) dehydration, such as of gypsum to anhydrite, and opal to chert; (5) replacement of high-Mg calcite by low-Mg calcite; (6) albitization – the replacement of calcium feldspars or potassium feldspars by nearly pure sodium feldspar (albite); (7) clay-mineral reactions, typically whereby smectites are transformed into illite; (8) zeolite replacement of volcanic glass; and (8) silicification – replacement of carbonate, plant remains, bone, or other fossils by chert. Some examples of replacement and recrystallization fabrics are shown in Figure 19.10.

Mineral recrystallization refers to reorganization of mineral crystals into a new geometric arrangement of the same mineral, and reduction of the surface free energy of the minerals is recognized as the driving mechanism. Recrystallization most commonly involves re-growth from one or more neighboring crystals, resulting in the change from a fine mosaic of crystals into a coarser one. Examples are the growth of micron-diameter calcite or quartz crystals (chert) into sub-millimeter-diameter mosaics of calcite and quartz. Carbonate mud, in particular, commonly recrystallizes and increases in particle size, ultimately forming an interlocking mosaic of crystals each a few tens of microns in diameter. Interlocking mosaics of calcite crystals, each a few microns in diameter, are generally known as *microspar* mosaics. Such mosaics in ancient rocks are almost always calcite, but detrital carbonate mud can be either aragonite or calcite. Folk (1965b) introduced the term *neomorphism* to describe those situations where one polymorph of a mineral replaces a pre-existing polymorph, such as the replacement of aragonite by calcite. However, it is commonly not clear whether calcite has recrystallized or whether calcite has replaced aragonite. It is also difficult to distinguish microspar mosaics that represent recrystallized muds from those that originated as cements (Figure 19.10F) (Bathurst, 1975, Table 2.2).

Examples of various types of compaction, cementation, replacement, and recrystallization are discussed below for the main rock types: terrigenous sandstones and muds; carbonate sands and muds; evaporites; and carbonaceous sediments. Early diagenesis of siliceous, iron-bearing, and phosphatic sediments was discussed in Chapter 4.

Diagenesis of terrigenous sands and gravels

Diagenesis of coarse terrigenous sediments is described in Pettijohn *et al.* (1972), Hutcheon (1990), Morad *et al.* (2000), Stonecipher (2000), and Chuhan *et al.* (2001). Early subaerial diagenesis of terrigenous sediments includes continued incongruent weathering of feldspars, amphiboles, and micas, especially under humid, moist conditions, typically in the unsaturated soil zone (Chapter 3). Clay pseudomorphs of silicate minerals provide evidence of continued weathering after deposition. Under semi-arid conditions, calcretes, silcretes, and ferricretes form in the B soil horizon (Chapters 3 and 4), and, under more arid, evaporitic conditions, saline evaporite minerals are likely to grow in the sediment as crystals and nodules (Chapter 16). Concretionary growth of minerals in soil zones involves cementation and replacement. Groundwater flow is

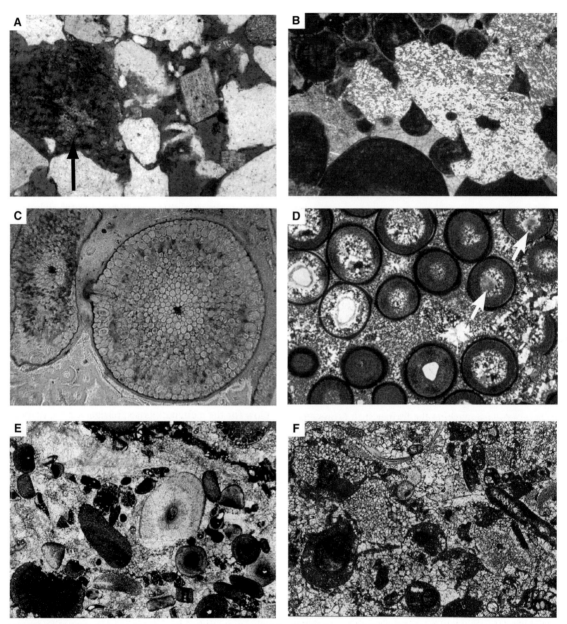

FIGURE 19.10. Dissolution and replacement fabrics. (A) A partially dissolved potassium feldspar grain (arrow). Width of field 2 mm. Courtesy of Tim Lowenstein. (B) Ooids and calcite cement replaced by anhydrite. Width of field 2 mm. Courtesy of Tim Lowenstein. (C) Silicified stems of the Lower Devonian vascular plant *Rhynia*, one of the earliest vascular plants known. Considerable detail of the plant structure was preserved during silicification. Width of stem ~3 mm. From www.uni-muenster.de/GeoPalaeontologie/Palaeo/ Palbot/rhynneu3.htm. (D) Silica-replaced ooids. Some of the ooids nucleated on quartz grains. Inclusions of carbonate preserved in the chert preserve a "ghost" of the ooid structure, but some dissolution is implied by the displaced nuclei of some ooids (white arrows). Width of field 6 mm. From Scholle and Ulmer-Scholle (2003) AAPG © 2003 reprinted by permission of the AAPG, whose permission is required for further use. (E) Crinoid stems, showing various degrees of replacement by dark-colored iron oxide. Width of field 5.5 mm. From Adams *et al.* (1984). (F) A neospar mosaic that probably represents aggrading neomorphism of an original carbonate mud. Note the embayment of many peloid boundaries. Compare this with Table 19.2. Width of field ~3 mm. From Adams and MacKenzie (1998).

likely in the saturated (phreatic) zone in subaerial settings (Bear, 1972; Freeze and Cherry, 1979). Under oxidizing, evaporative conditions of groundwater flow, concretions cemented by calcite are likely to form in sandstones. However, if the groundwater is anoxic, siderite may form concretionary cement instead of calcite.

Transitional subaerial to marine environments with sands include beaches and sandy tidal flats. In tropical and subtropical settings, evaporation of seawater from the pores of beach sediments leads to precipitation of needles of aragonite, and blocky cements of magnesium calcite forming beach rocks. Meteoric groundwater flow in transitional settings can mix with seawater, leading to the dissolution of calcite (see below under carbonate diagenesis for more on mixing zones).

In shallow-marine shelves, carbonate cements can precipitate and form hardgrounds in both carbonate and siliciclastic sediments. The mineralogy of the carbonate is controlled by the temperature, with aragonite and high-magnesium calcite in tropical and subtropical settings, and low-magnesium calcite in polar and tropical areas. Dissolution of marine shells is a possible source for carbonate cement in hardgrounds in terrigenous sediments. Calcite cements are common in shallow-marine sandstones, where they typically either have a patchy concretionary distribution or occur as continuous cemented layers. The concretions are usually meters thick and tens to hundreds of meters wide, whereas the continuous layers may be traceable for hundreds of kilometers (Taylor *et al.*, 2000; Dutton *et al.*, 2002). These concretionary cements appear to have formed within a few hundred meters of the surface prior to appreciable compaction and are interpreted as being due to either infiltration of meteoric fluid or precipitation from seawater pore fluids. In shallow-marine environments, where marine pore water is stagnant and rapidly depleted in oxygen, common early-diagenetic processes include the precipitation of iron silicates such as glauconite, phosphates, and iron oxyhydroxides (see Chapter 4). These minerals are commonly precipitated as replacements of clay fecal pellets, as oolitic grains at or near the surface, and as concretionary cements at shallow depths. These redox reactions typically occur at the boundaries between reduced and oxidized pore fluids. In completely reducing pore fluids, iron and manganese sulfides form as fine-grained concretionary cements and framboids (Chapter 4). Where volcanogenic hot springs discharge on shelves in the vicinity of volcanic islands in Indonesia, iron oxide ooids occur in layers up to 1 meter thick (Heikoop *et al.*, 1996; Morad *et al.*, 2000).

In deep-sea sands, patchy cementation by siderite and manganese oxide occurs. In addition, zeolites, clays, opal, and chalcedony occur in turbidity-current sands that are rich in volcanogenic glass. Alteration of glass shards into clays is apparently responsible for a great deal of the clay that is found in muddy sandstones (graywackes), and compaction deforms these weathered grains into the matrix-like groundmass common in many ancient turbidites. Early calcite cements, apparently precipitated from seawater pore fluids, occur in submarine fan-channel deposits (Dutton *et al.*, 2003).

Major compaction of coarse-grained terrigenous sediments occurs in the upper few kilometers of burial, and further porosity reduction in terrigenous sandstones is apparently due to cementation. Burial diagenesis in continental basins, where the pore fluids below 2–3 km are typically NaCl-rich brines with total dissolved solid concentrations of >30,000 ppm, differs from that in the deep-sea floor, where the pore fluids are mostly trapped seawater. In continental basins, at 2–3 km burial depth where the temperatures are typically 70–100 °C, the most important diagenetic changes are precipitation of carbonate cements (typically calcite), and the beginning of albitization of feldspars. Although substantial amounts of potassium feldspars and detrital or early-diagenetic kaolinite may dissolve during burial diagenesis, collapse of the sediment framework due to the compaction also occurs, so secondary porosity in terrigenous sandstones typically accounts for only ~5% of the total porosity. Other important burial-diagenetic processes of terrigenous sandstones include continued conversion of plagioclase feldspars and undissolved potassium feldspars into albite, conversion of smectite clays into mixed illite–smectite and then into illite (see below), precipitation of clays as cements, and pressure solution of quartz grains (apparently providing much of the quartz for cementation). Clay layers intercalated with the sandstones are sources of materials for cementation in the sandstones and also act as aquitards that affect pore-water circulation.

Diagenesis of terrigenous muds and associated organic matter

Organic matter is commonly co-deposited with terrigenous muds, and most muds contain a percent or so of organic matter (Potter *et al.*, 2005). Early-diagenetic chemical reactions in muds are driven mostly by the bacterial decomposition of organic matter, with little change in the clay minerals. The initial reduction of organic matter takes place by aerobic respiration, which can be represented simply as

$$CH_2O + O_2 \rightarrow CO_2 + H_2O \qquad (19.2)$$

In terrigenous muds, the oxygen is used up quickly, usually within a few millimeters of the surface. Diffusion of oxygen through pore water is limited. In the deep ocean, where there is little organic input to the sediments, the oxidized zone can extend down to 10 m into the sediment (Chester, 2003). However, on shelves and especially near deltas, the sediments can contain up to 10% organic matter, and the oxidizing zone may only be a few millimeters thick (although burrowing organisms can take oxygenated water to a depth of up to a meter in such sediments). Bacteria that use electron acceptors other than oxygen to break down organic matter are found beneath the initial oxidized zone. There is a general depth zonation of the bacteria in muddy sediments (Jorgenson, 2000). The uppermost are the denitrifying bacteria:

$$5CH_2O + 4NO_3^- \rightarrow 2N_2 + 4HCO_3^- + CO_2 + 3H_2O \qquad (19.3)$$

Beneath these are the manganese reducers:

$$CH_2O + 3CO_2 + H_2O + 2MnO_2 \rightarrow 2Mn^{2+} + 4HCO_3^- \qquad (19.4)$$

below which are the iron reducers:

$$CH_2O + 7CO_2 + 4Fe(OH)_3 \rightarrow 4Fe^{2+} + 8HCO_3^- + 3H_2O \qquad (19.5)$$

The sulfate reducers are particularly important because modern seawater contains abundant sulfate:

$$CH_2O + SO_4^{2+} \rightarrow H_2S + 3HCO_3^- \qquad (19.6)$$

If iron is available from iron reducers, iron sulfides such as greigite and mackinawite (precursors to pyrite) form. This is the source of the common pyrite found in mudstones (Berner, 1980). Beneath the sulfate reducers, methanogenic archaea (Chapter 2) derive their energy by using hydrogen to reduce CO_2 according to the reaction

$$CO_2 + 4H_2 \rightarrow CH_4 + O_2 \qquad (19.7)$$

The prokaryotes in the fine-grained muds of the sea floor are estimated to comprise as much as one third of the biomass of Earth and can be found in sediments up to 800 m below the sea floor (D'Hondt et al., 2002, 2004).

Mud (unlike sand) undergoes substantial compaction within the upper few tens of meters of burial

(Figure 19.5B). This is because muds are generally composed of floccules and pellets that internally have open house-of-cards structure. Upon deeper burial, the clay minerals begin to recrystallize, the most important change being the transformation of smectite into illite. Many of the descriptions of this transition came from oil wells in shales centered on depths of ~3–4 km in the Gulf Coast sediments of the USA, where temperatures were 50–120 °C (Aronson and Hower, 1976; Hower et al., 1976; Boles and Franks, 1979; Roberson and Lahann, 1981). The transformation of smectite into illite is not a direct conversion, but instead takes place through an intermediary *mixed-layer* illite–smectite. This is accompanied by dissolution of potassium feldspar, which is interpreted as providing the potassium needed for the reorganization of the clay minerals. The interlayer waters of the smectite clays are released, as are silica, Ca^{2+}, and Na^+. The sodium goes into albitization of plagioclase and any undissolved potassium feldspar. The water so produced may go towards developing overpressure in the muds, and is suspected of being involved in the migration of hydrocarbons that are maturing in the muds at the same time. The silica released by the reaction is postulated to be the source of quartz cementation in the mudstones or in adjacent sandstones. Given slightly different starting materials, the final clay is chlorite and the intermediary is an illite–chlorite mixed clay.

The maturation of organic matter in the mud generates hydrocarbons and organic acids, and perhaps CO_2, which is postulated to enhance porosity in rocks by dissolving carbonate components. The organic matter produces oil, oil mixed with gas, and finally dry gas (inorganically produced methane). This transformation takes place through the so-called oil window, generally at temperatures of 60–160 °C. The best sources of kerogen for this transformation are marine or lacustrine plankton.

Diagenesis of carbonates

Early subaerial diagenesis

Primary sources of information about the products and processes of early subaerial diagenesis came from Pleistocene reef, shoal, lagoon, and aeolian limestones in Bermuda and the Caribbean that were subaerially exposed during the glacioeustatic sea-level lowstands and subjected to early subaerial diagenesis by meteoric waters (Bathurst, 1975; James and Choquette, 1990a, b, c). The carbonate sediments were cemented by low-Mg calcite precipitated from meteoric groundwater both in

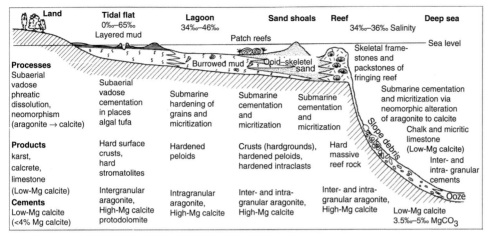

FIGURE 19.11. Diagenetic features across the Bahama Banks. From Demicco and Hardie (1994).

the unsaturated (vadose) zone and in the saturated (phreatic) zone. In the unsaturated zone, fine-grained meniscus cements, pendant cements, and vadose silt are common. Blocky calcite cements and isopachous fringe cements of low-Mg calcite occur in the saturated zone. Cementation was accompanied by partial to complete alteration of metastable aragonite and high-Mg calcite particles into stable low-Mg calcite (carbonate *stabilization*). At the same time, considerable dissolution of the original sediments occurred as the undersaturated meteoric groundwater percolated downwards (karstification; see Chapter 3).

Contemporary cementation of tidal-flat sediments on Andros Island (Bahamas) occurs in those supratidal settings where sedimentation takes place only during hurricane-driven onshore surges, spaced many years apart (Hardie, 1977). In these areas, centimeter-thick, calcrete-like surface crusts form by precipitation in the vadose zone of aragonite, high-Mg calcite, and calcian dolomite cements (Shinn *et al.*, 1969; Hardie, 1977; Lasemi *et al.*, 1989; Demicco and Hardie, 1994); see the section on supratidal crusts in Chapter 11. The driving force is evaporative pumping of seawater up from the marine phreatic zone (Hardie, 1977) that also raises its salinity and seems to aid nucleation of calcian dolomite at surface temperatures (Hardie, 1987); but see Carballo *et al.* (1987) for a different view of the hydrologic processes involved in the formation of supratidal dolomite crusts.

Early submarine diagenesis

Holocene subtidal carbonate sediments that were never exposed to meteoric water have also been cemented on the sea floor to varying degrees by aragonite and high-Mg calcite crystals precipitated from seawater (Bricker, 1971; Bathurst, 1975, 1980; James and Choquette, 1990a, b). Re-examination of ancient shallow-marine carbonate rocks has confirmed that early marine cements are abundant throughout the rock record (James and Choquette, 1990a, b). The marine cements in the sediments of the present-day Bahama Banks (Figure 19.11) include intergranular and intragranular cements in the periplatform oozes of the deep basins, the foreslope debris, the bank-margin reefs, the bank-margin sands, the shelf-lagoon peloidal muds, and the tidal-flat peloidal muds. Cementation in these settings occurs where the sedimentation and erosion rates are very low, the host sediments have the same mineralogy as the cements and thus can act as seeds, and large volumes of supersaturated seawater are pumped through the sediments.

Reefs produce an open framework with a substantial network of connected voids through which large volumes of seawater are flushed every day by tides and waves, providing ideal sites for extensive growth of void-lining cements unhindered by sedimentation or erosion. In many barrier reefs and fringing reefs around the world, the intra-framework voids, the internal sediments of these voids, and the skeletal gravels and sands between framework masses are extensively cemented by radial fibrous (botryoidal) aragonite and magnesian calcite (James and Choquette, 1990b); see Chapter 15. It is possible that at least some of the carbonate precipitation within these reef voids is mediated by microbial activity (Macintyre, 1985).

Hardgrounds are submarine sediment surfaces that became lithified before a subsequent sediment layer

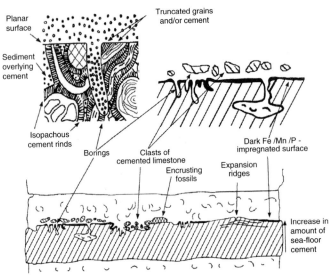

Planar surface

Truncated grains and/or cement

Sediment overlying cement

Isopachous cement rinds

Borings

Clasts of cemented limestone

Encrusting fossils

Expansion ridges

Dark Fe /Mn /P - impregnated surface

Increase in amount of sea-floor cement

FIGURE 19.12. Diagnostic features of submarine hardgrounds. From Flügel (2004), after James and Choquette (1983).

was deposited and occur both on shallow-water shelves and on deeper, off-shelf slopes and basins. Bathurst (1975), James and Choquette (1990b), and Flügel (2004) give criteria for recognizing ancient submarine hardgrounds (Figure 19.12). Sands of beaches, shelf-lagoons, and marginal shoals are cemented by aragonite or high-Mg calcite (12–14 mol% $MgCO_3$). Hardgrounds on beaches are referred to as *beachrock*. The degree of cementation decreases gradually from the marginal sand shoals to the peloidal muds of the restricted inner shelf-lagoon, as does the intragranular cementation of peloids (Hardie and Shinn, 1986) and in lithoclasts commonly referred to as *grapestones*. Such cementation occurs in layers up to a decimeter thick that occur at the sediment–water interface as surface crusts (submarine hardgrounds *sensu stricto*) or within the near-surface sediment as layers parallel to the sediment–water interface. Cementation is most intense at the upper surface of the layer and decreases with depth. Multiple layers may be found at depth, with up to four layers found beneath the Persian Gulf (Shinn, 1969). The hardgrounds of the peloidal sands of the Persian Gulf subtidal areas are buckled into tepees (Shinn, 1969); see Chapter 12. In the Bahamas, hardgrounds within shelf-marginal sand shoals are broken into intraclasts by storms (Harris, 1979; Halley *et al.*, 1983). In addition, cements occur in pelagic oozes, in mixtures of pelagic and periplatform-derived oozes, in turbidites on slopes and shallow basins, and in "lithoherms." The nodules of limestone produced by differential cementation of shallow (<500 m) slope deposits off the Bahama Banks may serve as modern analogs for differential cementation of ancient shelf deposits; see the reviews by Tucker and Wright (1990) and Demicco and Hardie (1994).

Firmgrounds are sediments in the initial stages of hardground formation that can be broken up easily and transported as cohesive intraclasts. The partly cemented walls of large tubular burrows are especially easily disinterred from firmgrounds by storm erosion, and transported as tubular intraclasts recognizable in cross sections as grainstone "doughnuts" with a smooth sharp inner boundary and a rougher outer boundary. Such tubular fragments of *Callianassa* shrimp burrows are commonly found together with flat-pebble lithoclasts at the base of lee slopes of the modern ooid shoals on the Bahama Banks. Analogous tubular intraclasts have been identified in ancient shallow-shelf carbonates of the Lower Ordovician in the central Appalachians and the Middle Triassic in northern Italy.

The basic requirement for development of a hardground or firmground is that there is no erosion or net deposition for a period of time long enough to allow precipitation of sufficient cement to hold the sandy sediment firmly together. In modern hardgrounds and firmgrounds, cements occupy as little as 5% of the intergranular void space, but it is not known how long it takes to reach this degree of cementation. However, Shinn (1969) reports that early Grecian pottery is incorporated into hardgrounds in the Persian Gulf, indicating that they can form in a matter of a few thousand years. Beach rocks have been found with incorporated fragments of middle- to late-twentieth-century soda bottles. Thus, it is probable that

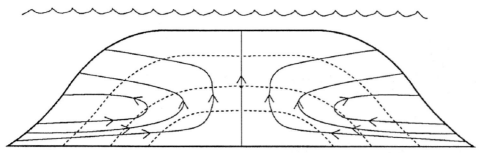

FIGURE 19.13. A diagrammatic representation of slope convection (also called Kohout convection) developed in a carbonate shelf surrounded by isothermal seawater. Dashed lines represent isotherms within the shelf due to geothermal heating; solid lines represent flow lines of seawater circulating through the shelf deposits. From Wilson *et al.* (1990). Reprinted by permission of the *American Journal of Science*.

submarine hardgrounds can develop in tens to thousands of years. The hiatuses in submarine deposition reflected by hardgrounds are reflected in "omission surfaces" or "condensation surfaces" (see Chapter 20). An irregular upper surface of a hardground is sometimes referred to as an "erosion surface," but it is most likely that rasping snails and limpets have molded such surfaces (Demicco and Hardie, 1994).

Deep-burial carbonate diagenesis

Burial cementation on carbonate shelves can be from seawater, meteoric water, or oilfield brine. Seawater can penetrate downwards for tens to hundreds of meters into the underlying sediment if the density of the warm surface seawater was raised by evaporative concentration until it exceeded that of the cooler seawater occupying the pores of the subsurface sediments. Such conditions exist on the Great Bahama Bank and have the potential to generate a significant downward flux of seawater (Simms, 1984; Wilson *et al.*, 2001) that could result in pervasive cementation of those permeable subsurface sediments that acted as flow conduits (shallow-burial marine cementation). Another process for driving large volumes of seawater supersaturated with respect to aragonite and magnesian calcite through large volumes of subsurface shelf sediments is slope convection (Wilson *et al.*, 1990; Phillips, 1991; Wilson *et al.*, 2001). This kind of lateral and upward flow of the deep ocean water surrounding the shelf has been documented for the present-day Florida shelf (Kohout *et al.*, 1977; Simms, 1984; Phillips, 1991). Slope convection is driven by the temperature contrast between cold seawater and geothermally heated pore fluids in the center of the shelf (Figure 19.13). In addition, this warming may increase the $CaCO_3$ supersaturation sufficiently to induce precipitation of pore cements. Like the early marine cementation of the

surface sediments of modern buildups, any shallow-burial marine cementation resulting from pervasive subsurface flow of seawater should occur preferentially in the more permeable sediments. Ooid–peloid–skeletal sands and intraclast gravels should be favored over carbonate muds. These ideas on shallow-burial marine cementation are speculative and it remains to be demonstrated to what extent, if any, such mechanisms are responsible for the pervasive cementation by aragonite and magnesian calcite found in ancient shallow-marine carbonates.

Burial compaction of carbonate sediments, particularly carbonate muds, is controversial (Demicco and Hardie, 1994), revolving around two issues: (1) a view that carbonate muds are cemented very early and do not compact; and (2) the role of stylolites in carbonate rocks. There is ample evidence for compaction of carbonate sands and gravels, including broken grains, porosity reduction, and the number and type of grain contacts (see the appendix). However, unlike in terrigenous muds, in which compaction eliminates primary sedimentary structures, those in carbonate mudstones are usually well preserved. Demicco and Hardie (1994) show many features of compaction in carbonate mudstones. Ehrenberg and Nadeau (2005) show depth–porosity data for carbonate reservoir rocks, and Goldhammer (1997) shows depth–porosity curves for carbonate mudstones that, like many sandstones, seem to reach a maximum porosity value of about 20% at depth. Stylolites in carbonates were discussed previously.

Dolomitization

Modern carbonate sediments generally comprise three minerals: aragonite, low-Mg calcite, and high-Mg calcite with up to 20–30 mol% magnesium (Chapter 4). Mesozoic and older limestones generally comprise low-Mg calcite and ordered, near-ideal dolomite

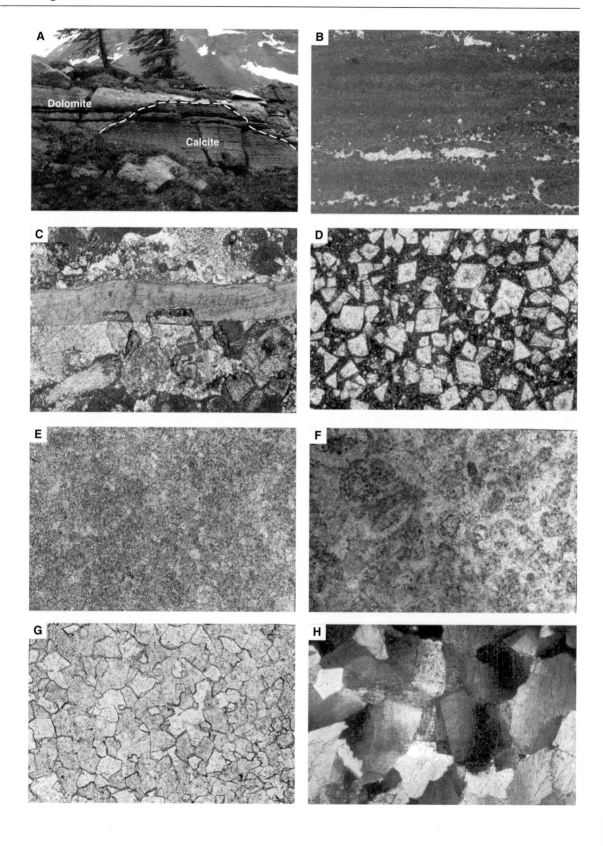

(CaCO$_3$·MgCO$_3$). Many Proterozoic carbonates are composed of the latter. Ordered, ideal dolomite cannot be synthesized at temperatures below 100 °C, and no organism makes a shell from ordered ideal dolomite. These observations, the abundance of dolomite in the geological record, and the inability to synthesize ordered, ideal dolomite at low temperatures suggest that the dolomite in the rock record is diagenetic in origin. The hypothesis that dolomite is a replacement of pre-existing calcite is supported by several lines of evidence (Figure 19.14). Ancient dolomite may faithfully preserve sedimentary structures and calcite grain types or can occur as an interlocking mosaic of dolomite crystals with no preserved sedimentary features. Dolomite occurs commonly as rhombohedra with diameters of a few tens of microns (Figure 19.14C, D). These euhedra commonly penetrate calcite shell fragments (Figure 19.14C), peloids, and ooids, and ghosts of the replaced grains are preserved in the dolomite (Figure 19.14E, F). The ghosts comprise solid and fluid inclusions engulfed in the dolomite. It is also common to see dolomite–calcite boundaries cross-cut primary bedding in carbonates (Figure 19.14A). Interlocking mosaics of dolomite are perhaps the most common replacement texture (Figure 19.14G). Dolomite with curved crystal faces, curved twin planes, and sweeping extinction is known as *saddle* or *baroque* dolomite and also comprises replacement mosaics (Figure 19.14H). However, it is common to find saddle-dolomite crystals that protrude into open pores and comprise cements, at least in part. The most enigmatic dolomites are fine-grained dolomite muds that preserve the sedimentary textures of the sediment (Figure 12.12B). Such dolomites may have originated as high-Mg calcite (protodolomite) mud (so-called "primary" dolomite), or they may be very-fine-grained replacements.

No uniform terminology is used to describe dolomite. Friedman and Sanders (1967) used terms based on the crystal texture and fabric. The shapes of individual dolomite crystals can be euhedral, subhedral, or anhedral. A dolostone composed of euhedral crystals is called idiotopic, one composed of subhedral crystals is hypidiotopic, and, if the rock is composed of anhedral crystals, it is xenotopic. A particularly common texture of ancient dolomites is a porous, euhedral to subhedral mosaic (so-called "sucrosic" texture) in which some of the dolomite crystals terminate in voids as crystal faces. Gregg and Sibley (1984) and Sibley and Gregg (1987) proposed classification schemes for dolomite rock textures.

High-Mg calcite with Ca–Mg compositions that overlap the ideal composition of dolomite can be synthesized in the laboratory and is correctly referred to as protodolomite (Graf and Goldsmith, 1956) – see Chapter 4. Natural protodolomite can be found as a very minor cement component in some modern carbonates (particularly in evaporitic settings) (Illing et al., 1965; Shinn, 1968a; Patterson and Kinsman, 1982). Well-crystallized, high-magnesium calcites (commonly referred to as dolomite, but without X-ray-diffraction ordering peaks) can be quite abundant in Tertiary carbonates (Jones and Luth, 2002; Melim and Swart, 2001).

Finding the diagenetic conditions under which dolomitization takes place is the crux of the so-called "dolomite problem." The dolomite problem has been re-solved in a series of successive bandwagons (Hardie, 1987). Following the discovery of high-Mg calcite cements in the sabkhas of the Arabian Gulf in the 1960s, dolomite was considered to be the result of the conversion of aragonite into dolomite by saline waters that had precipitated gypsum in the upper sabkha sediments and were thus enriched in Mg^{2+}. The dense, evaporated seawater brines flowed downwards through the sabkha sediments and towards the sea, and apparently dolomitized the sediments that they flowed through. This idea actually originated from the observation that dolomite in the Permian reef deposits of New Mexico occurred underneath and

FIGURE 19.14. Dolomite textures. (A) A field photograph of an outcrop approximately 5 m high where the calcite–dolomite boundary cross-cuts the bedding. (B) A dolomite mudstone with complete preservation of original sedimentary textures. Field of view 5 mm. From Adams and MacKenzie (1998). (C) A brachiopod shell and adjacent cement partially replaced by two dolomite crystals in the center of the photomicrograph. Field of view ~2.75 mm. From Adams and MacKenzie (1998). (D) Euhedral rhombs of dolomite a few tens of microns in diameter partially replacing carbonate mudstone. Field of view 2.4 mm. From Scholle and Ulmer-Scholle (2003). In (E) and (F) are shown two views of completely dolomitized carbonate. The original fabric is nearly completely obscured beneath the dolomite mosaic in (E), but, by placing a white sheet of paper between the thin section and the stage, original intraclastic and peloidal textural "ghosts" are revealed in (F). The field of view for (E) and (F) is 4 mm. From Adams and MacKenzie (1998). (G) An interlocking anhedral mosaic of coarsely crystalline dolomite. Field of view 2 mm. From Adams and MacKenzie (1998). (H) Dolomite with curved crystal faces (and twin planes) and sweeping extinction is known as "saddle" dolomite and occurs both as replacement mosaics and as cements. Field of view 4.4 mm. From Scholle and Ulmer-Scholle (2003). AAPG © 2003 reprinted by permission of the AAPG, whose permission is required for further use.

seawards of evaporite deposits (Adams and Rhodes, 1960). However, the high-Mg calcites in many areas where so-called modern dolomites are found were cements, not replacements, and in many cases the postulated hydrology was later shown to be incorrect (Hardie, 1987). In the 1970s, the so-called dorag or *mixing-zone model* (Badiozamani, 1973) came into vogue. In this model, the mixing of seawater and freshwater in coastal carbonate aquifers was postulated to create groundwater undersaturated with calcite (and aragonite) yet oversaturated with dolomite. After years of searching for extensive dolomites in such modern mixing zones, only minor high-Mg calcite cements have been located (see Hardie (1987) for a thorough review). Recently, high-Mg calcite has been found in mud associated with bacteria, and cultures of the bacteria have produced dolomite (van Lith *et al.*, 2003; Roberts *et al.*, 2004). A number of recent papers on the microbial origin of dolomite has shown that this new bandwagon is under way.

In addition to the solutions to the dolomite problem suggested above, there have been studies on individual dolomite occurrences that have been constrained by fluid-inclusion data (Wilson *et al.*, 1990; Yao and Demicco, 1995, 1997). These studies have gone back to the original idea that dolomite can be produced by high-temperature burial and postulated analytical and computer models of fluid flow responsible for the dolomitization that more or less reproduced the geometry of the dolomite alteration. Finally, computer models of circulation through the Great Bahama Bank, where Cenozoic protodolomite can be abundant (Wilson *et al.*, 2001), have shown that seawater can also be an effective dolomitizing fluid. In summary, virtually every aqueous fluid on Earth could be involved, and every diagenetic setting from subaerial to deep burial has been postulated as a site of dolomite replacement.

Evaporites

Evaporite diagenesis is discussed by Demicco and Hardie (1994), Hardie *et al.* (1985), and Spencer and Lowenstein (1990), is touched upon in Chapter 16. Most sulfate forming in modern environments is gypsum, which grows as euhedra to subhedra with variable amounts of included sediment. Large poikilitic gypsum cements are known as sand roses or desert roses. Gypsum may be replaced by anhydrite in near-surface sediments and upon deeper burial, usually in <1 km depth. When gypsum dewaters to anhydrite, there is a 38% loss of solid volume, but the total volume of

anhydrite plus water is greater than that of the original gypsum. At or near the surface, the gypsum crystals are converted into nodules of anhydrite, and the volume loss was suspected of causing deformation of anhydrite layers in the sabkhas of the Persian Gulf (Chapter 16). At greater burial depths, the transformation of gypsum into anhydrite, and concomitant release of water, may lead to local over-pressuring. It is also speculated that the water liberated by this reaction from large thicknesses of sulfate may dissolve the evaporites, flow into adjacent rocks, and precipitate sulfate or halite cements in originally evaporite-free rocks. Upon uplift and exposure to meteoric water, the anhydrite may be replaced by gypsum, usually within a few hundred meters of the surface.

Sulfate-reducing bacteria such as *Desulfovibrio* are important in the global sulfur cycle (Chapter 2). Much of the gypsum precipitated from the surface water of the Dead Sea does not build up on the bottom. In the deeper water, oxygen is scarce, and the sulfate-reducing bacteria break down the gypsum. Bacteria are also active in the diagenesis of sulfate deposits. Microbial degradation of petroleum is apparently responsible for much of the world's sulfur reserves (Barnes *et al.*, 1990). Sulfur is commonly encountered in the cap rocks of salt domes. The sulfate from anhydrite in the salt domes is apparently used by anaerobes such as *Desulfovibrio* to metabolize petroleum associated with the salt domes, resulting in sulfides, portions of which are subsequently oxidized to form sulfur. This takes place in shallow exhumation environments.

Peat-coal

In freshwater swamps and bogs, much of the organic matter that accumulates (such as wood, spores, or marsh plants) escapes aerobic bacterial oxidation and accumulates to form peat. On burial, the increases in overburden pressure and temperature systematically change the physical and chemical properties of the peat according to the coal series:

Peat → Lignite (Brown coal) → Bituminous coal → Anthracite

As the "rank" of the coal increases, (1) the carbon content increases, (2) the oxygen and hydrogen contents decrease, (3) methane and carbon dioxide are released, and (4) the optical reflectance of the coal increases. The last of these properties is commonly used to determine the diagenetic conditions to which organic matter has been exposed (see the appendix).

Part 6
Long-term, large-scale processes: mountains and sedimentary basins

20 Tectonic, climatic, and eustatic controls on long-term, large-scale erosion and deposition

Introduction

Earth surface processes that operate over relatively short time spans (less than a million years) and small areas (such as floodplains, beaches, and deep-sea fans) have been described up to this point. Interpretation of the landforms and sediment deposits that originate from such small-scale, short-term processes is based on field observations, laboratory experiments, and theoretical models. Mountain ranges and sedimentary basins are large-scale features of the Earth that formed over much longer time periods (millions of years). Sedimentary basins commonly contain stratigraphic sequences of diverse types and scales that were deposited in the various sedimentary environments described in Part 4 of this book (Leeder, 1993, 1999; Miall, 1996, 1997). However, these sequences are much more extensive than those described in previous chapters (hundreds of meters to kilometers thick and tens to hundreds of kilometers in lateral extent) and may contain regional unconformities that indicate major interruptions to deposition.

Definition of the three-dimensional (3D) geometry and spatial distribution of different stratal types in sedimentary basins requires some combination of extensive exposures, high-resolution (preferably 3D) seismic data, many closely spaced cores or borehole logs, and accurate age dating (the methodology is given in the appendix). Such data are commonly lacking or incomplete, and it is necessary to "fill in" 3D space in order to produce a complete (and hypothetical) model of sedimentary architecture. Furthermore, since it is not possible to observe directly the processes of development of sedimentary architecture over such large spatial and temporal scales, it is necessary to use models to interpret and predict subsurface sedimentary architecture (see the appendix). Such models typically describe the effects of tectonic uplift and subsidence,

climate change, and eustatic sea-level change on the nature of erosion, sediment transport, and deposition. It is actually very difficult to distinguish the influences of these different factors on the nature of stratigraphic sequences. Another very difficult question to answer is whether a vertical change in sedimentary facies in a sequence is due to lateral progradation of a sloping surface across which the facies change (i.e., Walther's law of facies), whether the conditions of deposition at a point changed through time, or both. These difficult issues are discussed in this chapter. Reviews of the definition and interpretation of large-scale, long-term Earth surface processes, involving mountains and sedimentary basins, include Allen and Allen (2005), Busby and Ingersoll (1995), Einsele (1992), Leeder (1999), and Miall (1997).

Description of strata in sedimentary basins

The 3D geometry and spatial distribution of different stratal types in sedimentary basins (so-called stratigraphic architecture) must be pieced together from available exposures, seismic profiles, cores, and borehole logs. These data sources provide an extremely small sample of the volume of rock of interest (North, 1996). The "filling in" of strata between the places where data are available involves age dating, lateral correlation of stratigraphic units, and commonly numerical simulation (see the appendix for details of methods). Therefore, building up a picture of the 3D stratigraphic architecture in a sedimentary basin is dominantly interpretive rather than descriptive, and the resulting picture is a model rather than reality. The term *stratigraphy* literally means the description of strata. It is commonly used today to mean the description, dating, and correlation of relatively thick **703**

TABLE 20.1. Some of the larger-scale sedimentary sequences recognized in the stratigraphic record

Order	Thickness	Duration	Examples	Possible causes
1	Many km	200–400 Myr	Cretaceous, Cambro-Ordovician marine transgressions	Tectono-eustasy associated with break-up of supercontinents
2	Few km	10–100 Myr	Sloss system cycles, Catskill clastic wedge	Tectono-eustasy: development of mid-ocean ridges, continental collisions, orogeny
3	100s of m	1–10 Myr	Transgressive–regressive sequences, carbonate grand cycles	Tectonic, continental glaciations?
4, 5	m to 10s of m	0.01–0.5 Myr	Carboniferous deltaic cycles, parasequences	Tectonic, glacioeustatic, autocyclic

FIGURE 20.1. A 1-km-thick (fluvial) stratigraphic sequence from the Miocene Siwaliks, Potwar Plateau, northern Pakistan. Grey sandstones and red mudstones vary in thickness and relative proportion throughout the sequence.

stratigraphic units in sedimentary basins. It is therefore part of *sedimentology*.

All sedimentary basins contain stratigraphic sequences in which sedimentary properties such as bed thickness, grain size, sedimentary structures, and composition vary spatially repeatedly, and commonly cyclically. In any sedimentary basin, there are many different superimposed types and scales of sedimentary sequence (Leeder, 1993, 1999; Miall, 1996, 1997) (Figure 20.1; Tables 20.1 and 20.2). The repetitive or cyclic nature of such superimposed sequences is commonly defined statistically, as described in the

TABLE 20.2. Fluvial depositional sequences

Nature of strataset	Thickness (m)	Lateral extent (m)	Duration (years)	Cause
Small- and medium-scale cross-set	10^{-3}–10^0	10^{-1}–10^3	10^{-6}–10^0	Migration of ripples, dunes
Cosets, composite sets (large-scale inclined stratum)	10^{-3}–10^0	10^{-1}–10^3	10^{-1}–10^0	Single flood deposit
Large-scale inclined strataset	10^0–10^1	10^1–10^3	10^0–10^3	Channel bar or fill; crevasse splay; levee
Group of large-scale inclined stratasets	10^0–10^1	10^1–10^4	10^1–10^4	Channel belt; crevasse-splay–channel–levee complex
Channel-belt/overbank cycles; coarsening-upward or fining-upward overbank sequences; paleosol-bounded sequences	10^0–10^2	10^1–10^4	10^3–10^5	Avulsions (with or without tectonism); progradation or abandonment of levees, crevasse splays, and lake deltas; regional change in sediment supply
Changes in mean grain size, geometry, and proportion of channel-belt deposits	10^2	10^4–10^5	10^5	Changes in deposition rate, avulsion frequency, channel-belt dimensions associated with tectonism, glacioeustasy, or climate change
Changes in mean grain size, geometry, and proportion of channel-belt deposits	10^2–10^3	10^5–10^6	10^6–10^7	Changes in deposition rate, avulsion frequency, and channel-belt dimensions associated with regional tectonism and/or climate change
Basin fill	10^3	10^6	10^7–10^8	Sea-floor spreading, orogeny, break-up of continents

appendix. Another important feature of sedimentary sequences in sedimentary basins is that they may be partially bounded by regional unconformities or erosion surfaces. Cursory examination of some of the larger-scale sequences indicates that they record repeated changes in depositional environment (e.g., marine to terrestrial, arid to humid, cold to warm) and deposition rate, and times when widespread erosion occurred. The reasons for such changes are very difficult to interpret. A major goal of many stratigraphers is to interpret the origin of the different scales of sedimentary sequence and the bounding unconformities. This requires detailed sedimentological training.

Techniques of description and preliminary interpretation of seismic profiles developed considerably in the latter half of the twentieth century, due to expansion of the field of *seismic stratigraphy*, particularly by the Exxon research group (e.g., Vail *et al.*, 1977). Seismic stratigraphy involves detailed description of the attributes of seismic reflections, including their amplitude, lateral continuity, and geometric arrangement (Figure 20.2). An important development was the recognition on seismic profiles of sets of inclined, conformable strata (on the order of hundreds of meters thick and kilometers across) that are bounded by unconformities. It is commonly assumed that the unconformities are more or less isochronous, and that individual seismic reflectors separate isochronous strata, such that seismic stratigraphy is supposed to be a way of putting large-scale sets of strata into a chronostratigraphic framework.

Extension of these ideas to outcrops, cores, and well logs led to a sub-discipline called *sequence stratigraphy*. Sequence stratigraphy is defined as the subdivision of sedimentary basin fills into so-called genetic packages bounded by unconformities and their correlative conformities. These genetic packages are divided into isochronous stratasets (called parasequences in some cases) analogous to stratasets separated by seismic reflectors. Seismic stratigraphic *sequences* and sequence stratigraphic *sequences* are defined as "third order" (Table 20.1), whereas parasequences are defined as either fourth or fifth order, depending on the investigator. This particular geological bandwagon has produced a large explosion of literature; reviews include

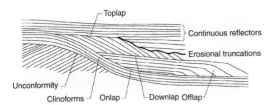

FIGURE 20.2. Seismic-stratigraphy terminology. From Nichols (1999). Toplap appears to be the same as an unconformity.

Vail *et al.* (1977), Wilgus *et al.* (1988), Van Wagoner *et al.* (1990), Posamentier *et al.* (1993), Weimer and Posamentier (1993), Van Wagoner and Bertram (1995), Emery and Myers (1996), Miall (1997), Posamentier and Allen (1999), and Schlager (2005). Sequence stratigraphy is more about interpretation than description, and a large terminology has developed in order to describe interpreted sedimentary features (Figure 20.3). Terms such as *systems tract* are not explicit, and they do not fit in any other classification of sedimentary strata. Furthermore, the non-genetic, scale-independent term *sequence* has apparently been highjacked and used in a sense that has genetic and scale implications. It will be noticed that sequence-stratigraphy terms are related to marine transgressions and regressions, which were originally interpreted as due to eustatic sea-level changes. As discussed below, marine transgressions and regressions are not solely controlled by eustasy. Indeed, the sequence-stratigraphy

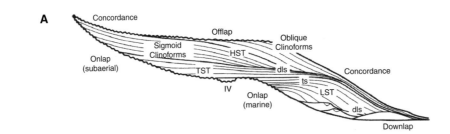

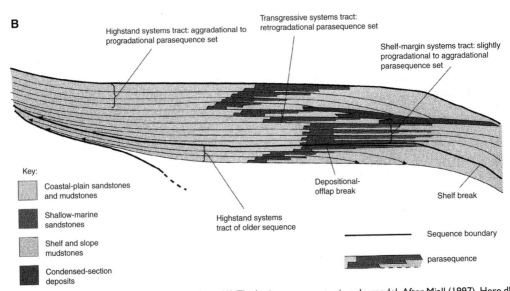

FIGURE 20.3. Sequence-stratigraphy terminology. (A) The basic sequence-stratigraphy model. After Miall (1997). Here dls is downlap surface, ts is transgressive surface, IV is incised valley, LST is lowstand systems tract, TST is transgressive systems tract, and HST is highstand systems tract. This example is one of a range of models. (B) A sequence-stratigraphy model of fluvial to shallow-marine deposits, showing the definition of parasequences and parasequence sets within a sequence. After Emery and Myers (1996).

movement has led to many misinterpretations of sedimentary sequences, as explained below.

A review of controls on erosion and deposition in various depositional environments

Deposition at any scale ultimately depends upon (1) a supply of sediment, (2) a place to put it, (3) the manner in which the sediment is put in that place, and (4) the degree of subsequent erosion. Changes (in space and time) in the nature of the deposits are controlled by changes in these factors. Regarding sediment supply, the rate and type of terrigenous sediment produced by weathering and delivered to depositional environments by mass wasting are controlled by source-rock type, topographic relief, climate, and vegetation, which are in turn controlled by the tectonics, igneous activity, and climate of the drainage basin. Abundant supply of terrigenous sediment is associated with large mountain ranges and large rivers to deliver sediment to floodplains, lakes, and the sea. Biogenic and chemical sediments are produced and deposited in abundance where the terrigenous sediment supply is small. This occurs in extensive subaqueous areas of low relief far away from mountain belts (hence it is dependent on tectonics and relative sea level), and where organisms that secrete calcium carbonate or silica are abundant (hence it is dependent on water and air temperature, salinity, water depth, and water currents).

Space to put sediment (commonly called *accommodation*) depends on tectonic subsidence (including compaction) and base level (e.g., lake level or sea level) in the basin. Base level, in turn, is intimately related to plate tectonics, climate change (particularly the amount of continental ice), and the amount of sediment filling the space. Deposition of physically transported sediment must always result from a decrease in sediment-transport rate in the down-flow direction. Subsequent erosion depends upon an increase in sediment transport in the down-flow direction, such as would be caused by basin uplift or base-level fall. The manner in which sediment is deposited in the space, or is subsequently eroded, depends upon the geometry, flow, and sedimentary processes in the various depositional environments.

The geometry, flow, and sedimentary processes of river channels and floodplains (e.g., channel patterns, the nature of avulsions) are controlled by water and sediment supply, intra-basinal tectonism, and base level. Long-term aggradation can be accomplished on a large scale by either increasing sediment supply up a valley (by tectonic uplift, climate change, or river diversions) and/or by decreasing the sediment-transport rate down valley (by flow expansion associated with tectonic subsidence or base-level rise). The existence and processes of arid and glacial environments are obviously controlled by climate and the distribution of land and sea. The geometry, flow, and sedimentary processes of coasts and marine shelves depend on the type of sediment available; the relative influence of rivers, tides, wind waves, and wind-shear currents on sediment transport; and the type and abundance of organisms. Tides are dependent on the distribution of the land and sea (e.g., shelf width, and spatial variations in the width and depth of seaways), whereas wind waves and associated wind-shear currents are dependent on climate, water depth, and the distribution of land and sea (controlling fetch). Thus, sea level relative to land level (controlled by tectonism and climate) is expected to influence coasts and marine shelves strongly. Deep-sea flow and sedimentary processes are associated with oceanic currents driven by the wind and variations in water density (related to water temperature and salinity), sediment gravity flows, variations in type of pelagic sediment, and the topography of the ocean floor (which is mainly controlled by tectonism and volcanicity). Climate exerts a strong control on oceanic (thermohaline) currents, the distribution of pelagic sediment, and eustatic sea level. Sea level influences the interaction between shelf and deep-sea processes, and the activity of sediment gravity flows down ocean-basin margins. Tectonism influences deep-sea currents and sediment by virtue of its control of sea-bed topography and the activity of sediment gravity flows. Tectono-volcanicity influences sea-bed topography and seawater chemistry.

It is clear from the discussion above that the main controls on Earth surface processes, landforms, and sediment deposits are tectonism, climate, and eustasy. These controls are commonly referred to as *extra-basinal* (or *extrinsic* or *allocyclic*), implying that they are independent of processes within the sedimentary basin. *Intra-basinal* (or *intrinsic* or *autocyclic*) controls are those related to factors such as seasonal floods, river avulsions, sediment compaction, and storms at sea. However, these are ambiguous terms in that tectonism can be both intra-basinal and extra-basinal, and so-called intra-basinal controls can be influenced by extra-basinal factors. Furthermore, as demonstrated below, there are complicated interactions among tectonism, climate, and eustatic sea-level change, such that it is commonly difficult to isolate the cause for a particular sedimentological change in a sedimentary basin.

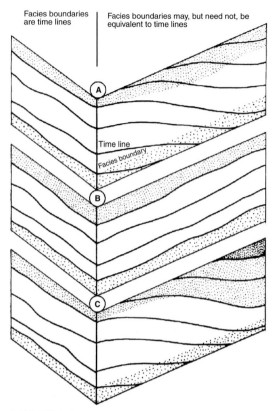

Facies boundaries are time lines

Facies boundaries may, but need not, be equivalent to time lines

A

Time line

Facies boundary

B

C

FIGURE 20.4. Problems with interpretation of vertical changes in lithofacies. (A) Progradation of a sloping surface across which lithofacies change produces a vertical sequence of lithofacies, according to Walther's law. Facies boundaries are diachronous in the view parallel to the dip, but are isochronous in the view parallel to the strike. (B) A vertical sequence of lithofacies produced by an abrupt temporal change in depositional conditions on a flat surface. Facies boundaries are isochronous everywhere. (C) A vertical sequence of facies produced by both spatial and temporal changes in depositional conditions (the most likely case).

Another difficult question to answer is whether a vertical change in sedimentary facies (of any scale) is due to lateral progradation of a sloping surface across which facies change (i.e., Walther's law of facies), or whether the conditions of deposition at a point changed through time (Figure 20.4). In fact, the two mechanisms occur simultaneously, and distinguishing their relative effects requires extensive exposures and accurate age dating (see the appendix).

Early attempts to interpret large-scale sedimentary sequences in sedimentary basins

Reviews of early studies of sedimentary basins are given in Reading (1986), Dott (1992), and Emery and Myers

(1996), and details will not be repeated here. Early (pre-nineteenth-century) stratigraphers recognized that sedimentary strata could be subdivided into major *systems* separated by unconformities, even if the relative ages could not initially be determined (see the appendix). Some workers attributed the systems and unconformities to major marine transgressions and regressions associated with eustatic (global) sea-level changes. The erosional unconformities were associated with marine transgression. Others doubted that the unconformities and marine transgressions were isochronous, and considered that the marine transgressions and regressions were more related to regional tectonism than to eustasy. More recently, Sloss (1963) defined six unconformity-bounded sequences in the Phanerozoic of the USA and considered them to be recognizable worldwide, but could be locally modified by tectonism. This was the precursor of the seismic and sequence-stratigraphy movements that started in earnest in the 1970s and 1980s.

Later (nineteenth- and early-twentieth-century) stratigraphers studied large thicknesses of sedimentary strata preserved in orogenic belts (mountain ranges) such as the Alps, Appalachians, and Himalayas. These orogenic belts presented evidence of tectonic deformation, metamorphism, and igneous activity in addition to large amounts of deposition. The sedimentary strata were considered to have been deposited in large depressions called *geosynclines*. Certain patterns of deposition, tectonism, metamorphism, and igneous activity were associated with various types of geosyncline, and the evolution of a given type of geosyncline. The relationship between the types of deposits and geosynclinal evolution was first proposed for the Alps, and the terms defined there were applied to other orogenic belts. For example, associations of thin fine-grained limestones, shales, radiolarian cherts, and ophiolites (serpentinized basalts and gabbros) were considered to be deep-sea deposits, and were called *pre-flysch*. These deposits were then followed in time by thick sequences of marine sandstones and shales called *flysch*. Flysch deposits were associated with rapid basin filling, uplift, and deformation. The latest deposits were thick sequences of sandstones, conglomerates, and shales (called *molasse*) interpreted as being fluvial–deltaic in origin. The largely undeformed molasse was deposited at the margins of the geosyncline after the main orogenic movements, and the sediment came from the adjacent rising mountains. With such studies of orogenic belts, the association of depositional sequences with tectonism and igneous activity is difficult to escape. These ideas about geosynclines were put into a plate-tectonic

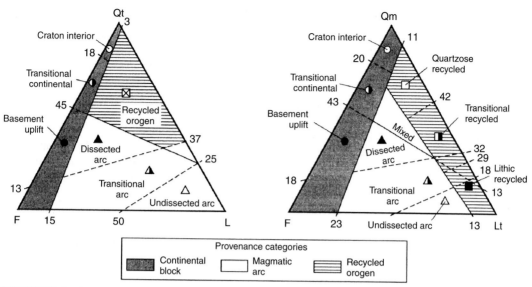

FIGURE 20.5. Sandstone compositions associated with various source (provenance) terrains. From Tucker (2001), after Dickinson (1985). Here Qt is all quartz grains (monocrystalline and polycrystalline), Qm is monocrystalline quartz grains, F is feldspar grains, L is unstable (volcanic, sedimentary) lithic (rock) fragments, and Lt is all lithic fragments, including polycrystalline quartz.

context starting in the 1960s, and the term geosyncline is no longer used. However, the terms molasse and flysch are still used today, although not necessarily with the same meanings as originally defined.

Krynine (1942) is well known for relating sediment composition to tectonism, by virtue of the control of the composition of terrigenous sediment by source-rock types, weathering, uplift, and subsidence. For example, mineralogically mature sandstones (quartz arenites: see the appendix) and limestones were associated with tectonic quiescence, recycling of sediments, and extensive chemical weathering. Periods of tectonic uplift, minimal chemical weathering, and rapid erosion, transport, and deposition in rapidly subsiding basins result in mineralogically immature sandstones (litharenites and arkoses: see the appendix). These concepts were put into a plate-tectonic framework by Dickinson and co-workers (e.g., Dickinson, 1985) (Figure 20.5). However, there are many exceptions to these idealized schemes, depending on the specifics of the tectonic and geographic setting.

General quantitative models for the influence of tectonism, climate, and eustasy on erosion and deposition

Paola (2000) classified general quantitative models for the influence of tectonism, climate, and eustasy on long-term, large-scale erosion and deposition as *geometric* and *dynamic*. The earlier, geometric models assumed a prescribed geometry for the land surface, and concerned the influence of sediment supply rate, subsidence, and eustatic sea-level change on the geometry of sediment deposits, normally in two-dimensional (2D) cross sections. Dynamic models consider the link between changing topography of the land surface and the sediment-transport rate, which must be linked to the nature of fluid flow. Paola also makes a distinction between models that use the basic equations of fluid flow and sediment transport rather than those that use simpler rules to imitate flow and sedimentary processes, and also between coupled models that examine the links between different processes (i.e., tectonic subsidence and deposition rate), and uncoupled models that do not. These general models are discussed first, followed by those designed to consider specific effects of tectonism, climate change, and eustatic sea-level change.

Terrestrial environments

Numerical models for erosion of the subaerial landscape and the evolution of drainage systems in upland regions have been developed recently. They are discussed in Chapter 13, and reviewed by Paola (2000) and Bridge (2007). The various models differ in the way erosion and deposition are modeled, but there are many similarities. Sediment transport on hillslopes associated with mass-wasting processes such as soil creep is commonly modeled using a linear diffusion approach (i.e., the sediment-transport rate is taken to be dependent on slope), and such transport may be limited in the model by the rate of production of sediment by weathering. Landslides have been modeled by considering the shear strength of rock or soil and whether or not this material exceeds some critical hillslope angle. The rate of erosion of bedrock and cohesive sediments by streams is commonly taken as proportional to kA^mS^n in excess of some threshold value, where k is the erodibility of the bedrock, A is the area of the drainage basin, and S is the surface slope. Here A is a surrogate measure for stream water discharge, such that the bedrock-erosion rate becomes a function of the stream power per unit channel length. Sediment transport in channels and by overland flow is also commonly taken as proportional to a function of the stream power (which is proportional to the product of the water discharge and the slope) in excess of some limiting value. Sediment transport by flowing water may also be limited by sediment production by weathering. Grain-size sorting during sediment transport has been treated at a rudimentary level in some models. In view of the fact that erosion, development of drainage systems, and deposition are, according to these models, controlled by slope, drainage area, and effective precipitation per unit area, the influences of tectonism, climate, and eustasy on these variables can be assessed. For example, the influence of climate on stream erosion and sediment transport can be modeled by taking the water discharge to be proportional to the product of the effective precipitation rate per unit area and the drainage-basin area (e.g., Bridge, 2007).

Paola *et al.* (1992) developed a theoretical model of large-scale variations in mean grain size in alluvial basins, as viewed in a cross section in the direction of water flow and sediment transport (see also developments by Paola *et al.* (1999) and Marr *et al.* (2000)). Sediment transport was modeled using a linear diffusion approach (i.e., with the sediment-transport rate proportional to the slope), with the rate of sediment transport for a given slope (and diffusivity of sediment) being controlled mainly by the water discharge and the channel pattern (braided or single-channel). Grain-size sorting in the model was based on the first-order assumption that gravel will dominate a deposit until all gravel in transport has been exhausted, at which point deposition of sand will begin. Paola *et al.* (1992)

examined the response of an alluvial basin to sinusoidal variation of four extrinsic controlling variables: the rate of sediment supply, diffusivity of sediment, tectonic subsidence rate, and proportion of gravel in the sediment supply (Figure 20.6). The basin response depends strongly on the time scale over which variation in the controlling variables occurs. "Slow" and "rapid" variations are defined as those that vary with periods that are respectively longer or shorter than a so-called *basin equilibrium time*, defined as the square of the basin length divided by the sediment diffusivity. Changes in the rates of uplift of source areas, erosion, sediment supply, and subsidence are not linked (coupled) in this model, although they are in nature. Despite the simplifications in this model, Heller and Paola (1992) applied it to three alluvial basins to help determine whether conglomerate progradation was coincident with tectonic uplift and increases in erosion and sediment supply (i.e., *syntectonic*) or not (*antitectonic*).

Slingerland *et al.* (1994) applied a simple sediment-routing model for river channels to predict changes in the depth and longitudinal profiles of rivers as a result of downstream changes in water discharge, sediment supply, and channel width. This model was also coupled with models for tectonic subsidence and uplift, varying sea level, and variable sediment supply related to climate change, in order to assess the effects of these variables on long profiles of rivers (and hence on erosion and deposition). These simple sediment-routing models (Figure 20.7) are potentially more realistic than diffusion-based fluvial models, but they still require development before they are capable of simulating the nature of sediment deposits in detail. Robinson and Slingerland (1998a, b) were the first to combine a sophisticated sediment-routing model for river channels (MIDAS) with basin subsidence and aggradation.

The sediment-routing model of fluvial deposition called **SEDSIM** (Tetzlaff and Harbaugh, 1989) is based on solution of the simplified equations of motion of water and sediment. It is claimed to be able to simulate flow, sediment transport, erosion and deposition in river-channel bends, braided rivers, alluvial fans, and deltas. However, there are fundamental problems with the basic assumptions and construction of this model, particularly in the treatment of sediment transport, erosion, and deposition (North, 1996). It has not been tested by detailed comparison with real-world data (Paola, 2000). Tetzlaff and Priddy (2001) and Griffiths *et al.* (2001) describe the development of

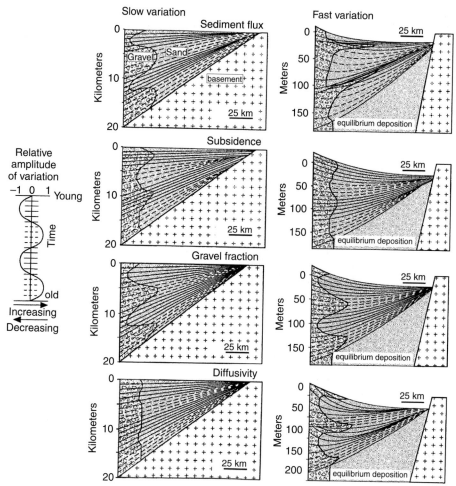

FIGURE 20.6. Hypothetical variations in distribution of sand and gravel in a sedimentary basin subjected to periodic variation in sediment flux, subsidence, gravel fraction, and sediment diffusivity. Modified from Paola *et al.* (1992). Figures on the left are for slow variation, and lines are isochrons drawn at 1-million-year intervals. Figures on the right are for fast variation and isochrons are drawn for every 10,000 years.

SEDSIM into STRATSIM, in order to simulate a range of fluvial and marine sedimentary processes acting over time intervals of hundreds of thousands of years. Unfortunately, few details of the workings of most of these models are given, or of how or whether they were tested against data from natural sedimentary environments (Paola, 2000).

Coastal and marine environments

Many general models for coastal and marine environments (including sequence-stratigraphic models) are 2D geometric models (Paola, 2000). These consider the effects of changes in sediment supply, tectonic subsidence, and eustatic sea-level change on marine transgressions and transgressions, and on the geometry of various types of sediment (Figure 20.8, from Paola, 2000). Models of marine carbonate deposition normally take the sediment-production rate as dependent on the water depth, and some models allow lateral transport of sediment (Demicco, 1998). However, carbonate production over shelves is not a simple depth-dependent variable (Demicco and Hardie, 2002). In general, the same stratigraphy can be produced by various combinations of the input variables.

Dynamic models for marine environments are not as well developed as those for terrestrial environments. The linear diffusion approach to sediment transport has been applied to shallow marine environments, using lower diffusion coefficients than for rivers (Figure 20.9). However, it is not really known whether

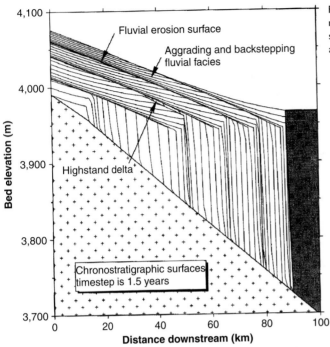

FIGURE 20.7. The two-dimensional fluvial–deltaic model of Slingerland *et al.* (1994) that links a simple sediment-routing model to an elastic-flexure model and sinusoidal variation in sea level.

this coefficient should be related to the water depth or the various marine sedimentary processes operating. Storms (2003) developed a 2D process-based model (BARSIM) for episodic erosion and deposition of multiple grain sizes in wave-dominated shoreface–shelf environments. Episodic erosion and deposition were determined using the mass-conservation equation plus stochastic prediction of storm and fair-weather "events." However, the relationship among bed topography, water flow, and sediment transport is empirical, not based on basic physical equations. Storms and Swift (2003) used BARSIM to explore the effects of changes in sea level, sediment supply, and storm-wave regime on sedimentary sequences in wave-dominated shallow-marine environments. More sophisticated dynamic models make use of the basic equations for water flow and sediment transport. For example, patterns of water flow in the Devonian Catskill Sea and the Cretaceous Western Interior Seaway (both in the USA) were calculated in order to examine the relative influence of tidal currents and wind-driven currents on sediment transport (Slingerland, 1986; Erickson *et al.*, 1990). The INSTAAR models (e.g., Overeem *et al.*, 2005) calculate patterns of deposition and sediment sorting on continental margins over time periods of up to tens of thousands of years. These models include prediction of floods, river-mouth dynamics (hyperpycnal and hypopycnal plumes), ocean storm currents, slope instabilities (turbidity currents and debris flows), compaction, tectonic subsidence, and eustatic sea-level changes.

Tectonic effects in various types of sedimentary basins

Tectonic activity controls the rate, amount, and location of uplift and subsidence, the gradient of the land surface, the position of rivers, and the position of the land and sea. Thus tectonic activity is also intimately related to climate, vegetation, and eustasy. These factors in turn control rates, amounts, and locations of sediment production, erosion, sediment transport, deposition, and facies types. Thick accumulations of sediments occur in a variety of different tectonic settings (e.g., foreland, rift, passive margin, and strike–slip basins), and stratigraphic architecture varies with tectonic setting (Busby and Ingersoll, 1995; Miall, 1996; Leeder, 1999). Some general principles relating to the influence of tectonics on sedimentation are discussed first. Then, general models for the variation in deposition rate and facies with sediment supply and subsidence will be discussed, followed by models for selected types of basin. More comprehensive details on the interaction among tectonics, volcanicity, and sediment accumulation in the complete range of types of sedimentary basins can be found in Busby and Ingersoll (1995).

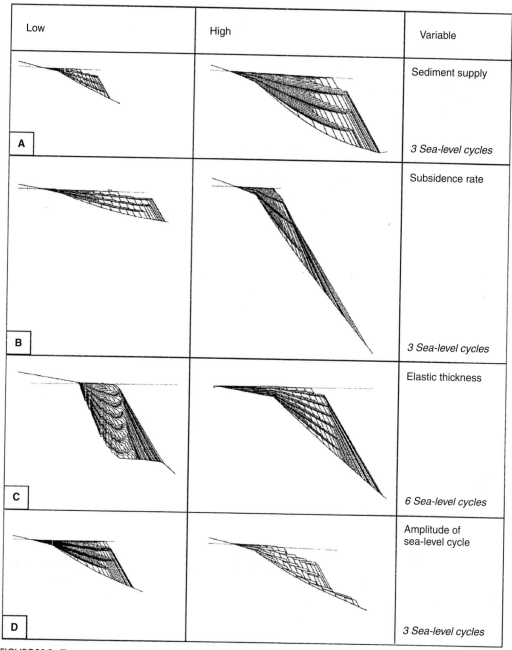

FIGURE 20.8. Two-dimensional models for the geometry of coastal–marine strata under varying sediment supply, tectonic subsidence, and sea-level change. From Paola (2000).

Tectonic uplift and subsidence can be gradual or episodic (Schumm *et al.*, 2000). Rates of uplift and subsidence averaged over many thousands of years are commonly on the order of 10^{-1} mm per year (ranging from 10^{-2} to 10^{1} mm per year: Leeder (1999)). Instantaneous translation on single faults of on the order of meters typically occurs at intervals on the order of 10^{3} years, and would be associated with major earthquakes. Prolonged movement on faults, or growth of folds over periods of 10^{4}–10^{5} years, could therefore result in vertical or horizontal motions on the order of tens to hundreds of meters. Such rates

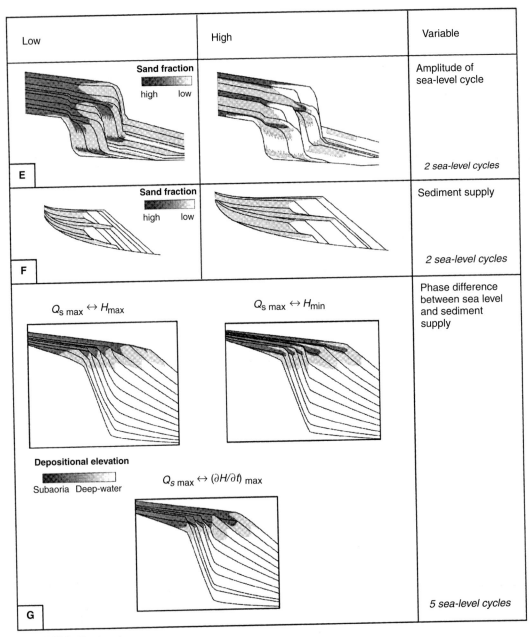

FIGURE 20.8. Continued.

and amounts of motion can be produced within tectonic plates (by intraplate stresses) as well as at plate margins (Miall, 1996).

Tectonic uplift and subsidence also vary spatially. For example, the fault zone at the margin of a basin comprises many distinct fault segments that may be active at different times, and each fault segment has spatially variable throw along its length (e.g., Gawthorpe and Leeder, 2000). Thus uplift of a mountain belt or subsidence of a basin comprises the sum of motions along all of the various active structures.

Episodic vertical ground motions of up to a meter along a single fault that ruptures along tens of kilometers may locally influence the channel pattern and course of a river, but are unlikely to have a big impact on an entire sedimentary basin. However, a succession of movements on many related faults, over say 10^5

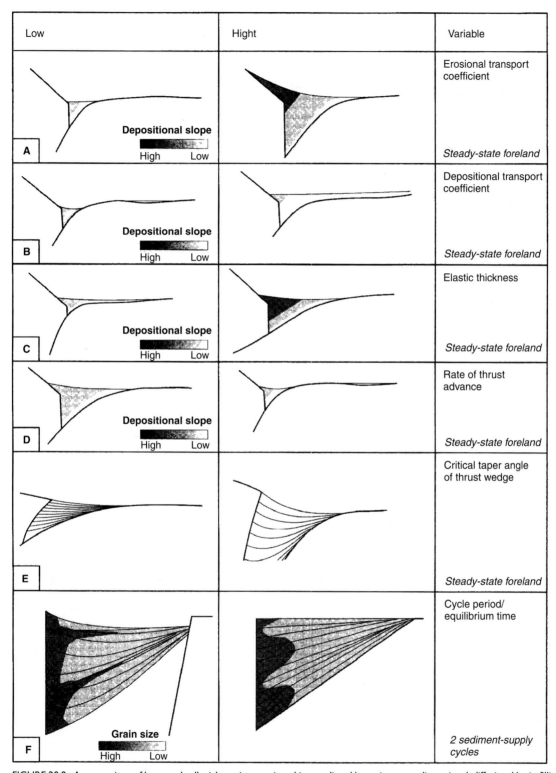

FIGURE 20.9. A comparison of large-scale alluvial–marine stratigraphies predicted by various two-dimensional, diffusional basin-filling models. From Paola (2000).

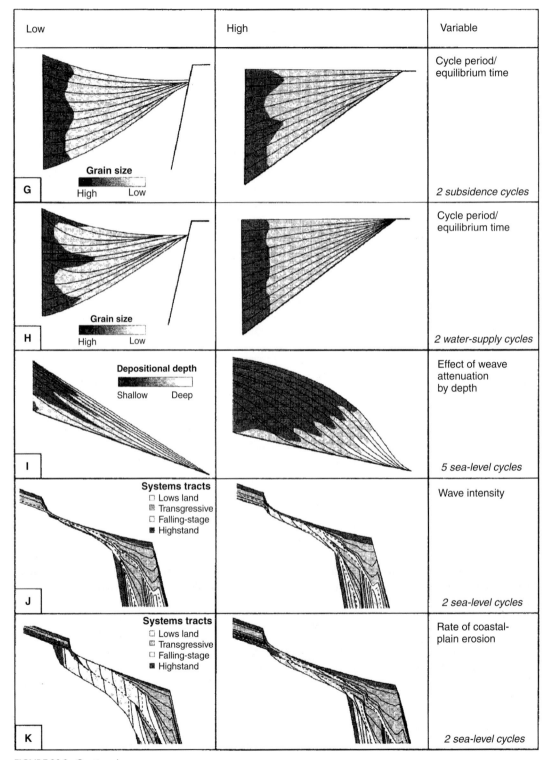

FIGURE 20.9. Continued.

years, may produce tens to hundreds of meters of ground motion, affecting basin lengths of hundreds of kilometers. Such movements are responsible for diversion of major rivers, and growth or dissipation of alluvial fans (Gawthorpe and Leeder, 2000). If such tectonic motions were cyclic in nature, they could explain cyclicity in sedimentary facies over vertical thicknesses of tens to hundreds of meters and lateral extents of up to hundreds of kilometers.

Evidence for tectonic activity in sedimentary basins

Evidence for tectonic activity within and adjacent to sedimentary basins includes variation in (1) sediment thickness and lithofacies in relation to syn-sedimentary folds and faults, (2) paleocurrent direction, and (3) sediment provenance. The sediment fills of extensional and compressional basins are closely related to uplift of adjacent uplands due to translation along basin-marginal faults. However, variations in sediment thickness and lithofacies in sedimentary basins may also be related to changes in base level or climate, although such changes may occur over different spatial and temporal scales. For example, diffusion models of erosion in drainage basins (e.g., Tucker and Slingerland, 1997; Allen and Densmore, 2000) indicate that the increase in sediment supply to rivers resulting from tectonically induced headward erosion may occur at a much slower rate than the increase in sediment supply due to an increase in precipitation in the drainage basin. This is because upstream migration of a wave of river erosion (a knickpoint) is a local process that does not affect all of the drainage basin simultaneously. In contrast, an increase in precipitation may cause an increase in erosion of the whole drainage basin simultaneously. Such a widespread, simultaneous increase in sediment supply could also be produced by a general increase in topographic slope due to uplift and tilting of the drainage basin.

A change in mean paleocurrent direction in river deposits may indicate tectonically induced changes in slope direction or river diversions. However, a change in paleocurrent direction can also be caused by deposition-induced changes in slope direction, and tectonic activity does not always cause river directions to change dramatically. Changes in sediment provenance may also have a tectonic cause, possibly related to unroofing of new rock types as a result of uplift, or a river diversion in the hinterland. But erosional exposure of new rock types does not require tectonic uplift, and river diversions in the hinterland do not need to result in a change in provenance of the sediment supplied to the river. Tectonic activity is also commonly associated with volcanic eruptions, and all of the associated features such as catastrophic floods and mudflows, and the damming of rivers with debris flows, ash, and lava.

Compressional basins

Examples of compressional basins are the Paleozoic basins of the Catskill Clastic wedge, developed along the western side of the Appalachians in eastern USA and Canada, the Mesozoic to Tertiary basins on the eastern side of the Rockies in North America and the Andes in South America, the Tertiary Siwaliks of the Indo-Gangetic basins on the south edge of the Himalayas, and the molasse of the European Alps. There are various kinds of compressional basins (e.g., foreland, foredeep, retroarc: Allen and Allen (2005) and Miall (1996)). They are formed by crustal thickening arising from compression, thrusting, and folding. The area adjacent to the thickened crust (the basin) subsides due to gravitational loading (Figure 20.10). An important aspect of the crustal flexure is a flexural bulge at the periphery of the basin. If the crust has high flexural rigidity and viscosity, the basin is relatively shallow and wide; otherwise it is deep and narrow. Thus, temporal changes in the rheological properties of the crust result in changes in the shape of the basin.

An episode of lithospheric thickening and loading results in uplift of mountains and increases in valley slopes, erosion rate, and sediment supply in the vicinity of the uplift. It also results in subsidence and deposition in the basin, and growth of the peripheral bulge and its migration towards the basin. Erosion of uplifted mountains and deposition in the basin causes further isostatic uplift of the highlands and subsidence in the basin. The erosional and depositional responses to uplift and subsidence depend upon the relative timing, positions, and rates of these events, and these are very difficult to ascertain. For example, the nature of subsidence in response to loading depends on lithospheric rheology (specifically whether it is elastic or viscoelastic) and how rheological properties change in time with temperature and pressure. The responses of weathering, erosion, and sediment supply to changes in source-rock type, elevation, slope, and vegetation are very difficult to predict.

The early quantitative models for compressional basins considered flexural isostatic response to loads

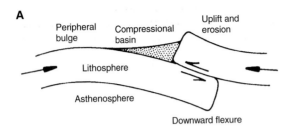

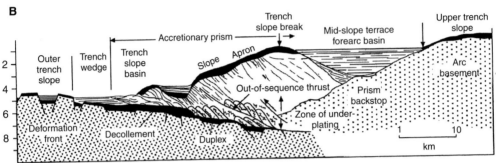

FIGURE 20.10. (A) Main features of compressional basins. (B) Compressional basins associated with a subduction zone and deep-sea trench. From Leeder (1999).

created by tectonic thrusting and sediment deposition, as represented in 2D sections parallel to the direction of thrusting (e.g., Beaumont, 1981; Jordan, 1981; Quinlan and Beaumont, 1984; Beaumont *et al.*, 1988). Erosion and deposition were not modeled explicitly. Erosion and deposition were modeled explicity by Flemings and Jordan (1989) (see also Flemings and Jordan, 1990; Jordan and Flemings, 1990, 1991) using a linear diffusion approach. The model of Paola *et al.* (1992) discussed above is also essentially a model for stratigraphy in compressional basins, but it does not consider the mechanics of uplift and subsidence. Paola (2000) compared the stratigraphy predicted by these various 2D, diffusion-based models (Figure 20.9). The treatment of surface processes in these models is simplistic, and the 2D (rather than 3D) treatment of tectonics and sedimentation can be misleading.

A major step forwards in the modeling of deposition at the river system and basin scale was the treatment of surface processes and drainage basin evolution in 3D, and linking them to climate, tectonic activity, and base-level change (e.g., Beaumont *et al.*, 1992; Kooi and Beaumont, 1994, 1996; Johnson and Beaumont, 1995; Allen and Densmore, 2000; Coulthard *et al.*, 2002; Garcia-Castellanos, 2002; Tucker *et al.*, 2002; Clevis *et al.*, 2003, 2004a, b) (Figure 20.11). These 3D basin models indicate that the availability of sediment, which is related to the weathering rate and bedrock

erodibility, exerts a strong control on the sediment-transport rate and basin deposition. Changes in the rate of transport of sediment to basins lag behind episodes of tectonic uplift, because of limits to sediment availability (weathering and bedrock erodibility) and the time it takes for sediment to move downslope through the drainage network. Sediment may be stored temporarily in an orogen because of the development of intermontane basins related to local thrusting and folding (e.g., Tucker and Slingerland, 1996). The lag time may be on the order of 10^4–10^6 years.

Episodic uplift and subsidence lead to episodic progradation and retrogradation of fluvial gravels and coastlines, but the relative timing of these events is equivocal (Bridge, 2003). In models that assume an elastic lithosphere, tectonic subsidence in the basin is an immediate response to thrusting, lithospheric thickening, and loading. Thus, the subsidence rate may exceed the sediment supply and basin deposition rate during lithospheric loading. Relatively coarse sediment produced as a result of uplift is deposited close to the source, and marine transgression may occur. Such basins have been called *under-filled*. Subsequent to uplift (with a time lag on the order of 10^4–10^6 years), the rate of sediment supply may begin to exceed the subsidence rate, and as deposition proceeds there is progradation of relatively coarse sediment across the basin and possibly marine regression. Such basins have been called *over-filled.* Prograding coarse material is called *antitectonic* in this case because it is not coincident with the tectonic uplift. In models that assume a viscoelastic lithosphere, the sediment supply and deposition rate may exceed the subsidence rate during thrusting and uplift, such that relatively coarse sediment fills the basin and prograades basinwards (i.e., *syntectonic*), possibly resulting in marine regression. Subsequently, the subsidence rate exceeds the deposition rate, the coarsest sediment is limited to areas near the uplift, and marine transgression may occur. The predictions of these models can be changed dramatically by making different assumptions about the responses of both erosion and transport of sediment to uplift.

Uplift is likely to be associated with climate change in the mountain belt and surrounding basins (Bridge, 2003, p. 352). Climate changes are strongly linked to basin stratigraphy, mainly because of the strong link among rainfall, water discharge, bedrock-erosion rate, and sediment-transport rate in rivers. Climate also indirectly affects basin stratigraphy through its effect on vegetation and weathering rate, which influence both effective precipitation and sediment production. Changes in rate of transport of sediment to basins may lag behind changes in rainfall by on the order of hundreds to thousands of years. Burbank (1992) suggested that periods of accelerated isostatic uplift associated with climatically induced increases in erosion rate should result in deposits that do not vary greatly in thickness across the basin, because uplift is not associated with downwarping of the basin. Eustatic sea-level fluctuations will add more complexity to basin stratigraphy, especially in near-coastal fluvial deposits (see below).

A

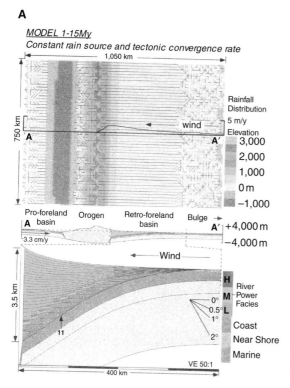

B

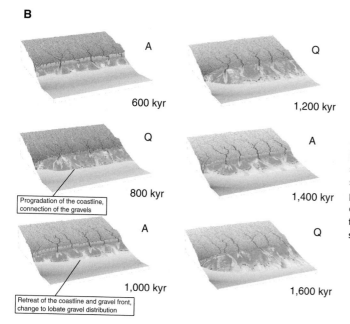

FIGURE 20.11. (A) A three-dimensional (3D) compressional basin model with surface processes (Johnson and Beaumont, 1995). (B) A 3D compressional basin model with surface processes and grain-size sorting (Clevis et al., 2003). Perspective views of successive stages of landscape subjected to tectonic pulsations. Phases of tectonic activity (A) reflected by retreat of the coastline and gravel front. Quiescent periods (Q) associated with progradation of the coastline and gravel front. (C) A 3D compressional basin model with surface processes and grain-size sorting (Clevis et al., 2003). Cross sections (c)–(d) show the distribution of gravel during tectonic pulsation (of 200,000-year period). Gravel progrades during tectonic quiescence. Sinusoidal sea-level fluctuation (period 100,000 years, amplitude 20 m) superimposed on tectonic pulsation is shown in (e).

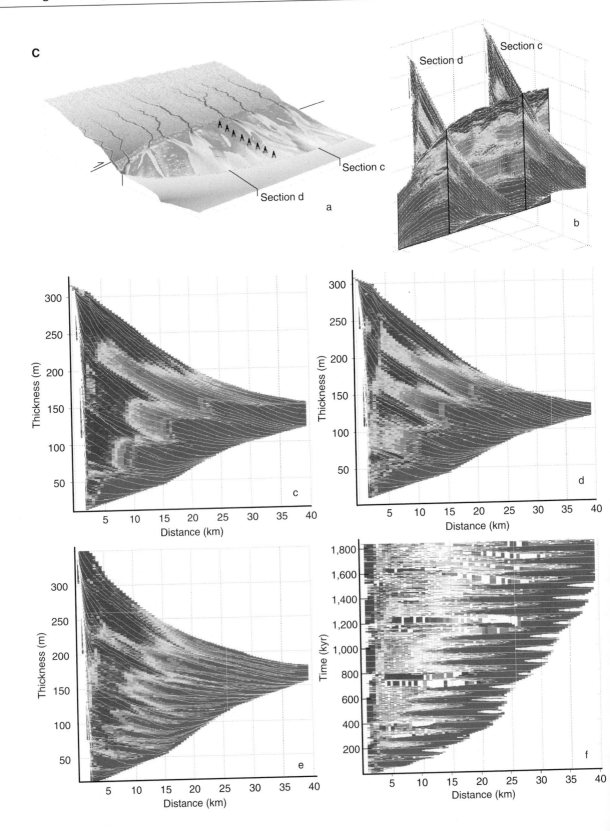

FIGURE 20.11. Continued.

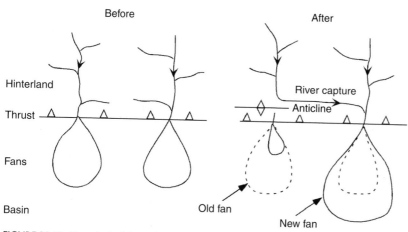

FIGURE 20.12. Hypothetical changes in alluvial-fan size in a foreland (compressional) basin due to hinterland tectonics and river capture. From Bridge (2003). An increase in water and sediment supply to the growing fan should result in an increase in channel size and frequent avulsion, hence increasing the channel-deposit proportion.

The large spatial and temporal scales considered in these basin models preclude consideration of individual depositional environments such as channels and floodplains, and important processes such as channel-belt avulsion and local tectonic movements. Therefore, these models do not predict detailed depositional architecture. However, it is possible to link changes in depositional architecture qualitatively with changes in subsidence rate, deposition rate, and grain size. A high deposition rate of relatively coarse sediment in basins that are backtilting should result in a relatively high avulsion frequency close to the uplands, producing fans with nodal avulsion, as observed in the Himalayan foreland (e.g., the Kosi fan). Overlapping channel belts on such fans result in sandstone-conglomerate bodies with large width/thickness ratios, with the channel-deposit proportion and connectedness increasing towards the mountain belt (Mackey and Bridge, 1995). Paleoslopes and river courses are approximately normal to the edge of the mountain belt (Figure 20.12). Periods of relatively low deposition rate of relatively finer sediment should be associated with a relatively low avulsion frequency, and the possibility of a relative rise in base level may lead to the highest avulsion frequencies occurring in regions distant from the edge of the mountain belt. In this case, rivers may be flowing parallel or oblique to the axis of the foreland basin. Garcia-Castellanos (2002) discusses the tectonic influences on the orientation of rivers flowing across compressional basins.

Episodes of thrusting, loading, and the development of peripheral bulges will also have an effect on the sediment supply and water depth (and hence sedimentary facies) in marine environments, although it is uncertain exactly how. Condensed, wave-reworked carbonate sequences have been interpreted as occurring on the crests of developing peripheral bulges (Brett and Baird, 1985). Thrusting effects in marine forearc basins (deep-sea trenches) are very complicated (Figure 20.10B). A peripheral bulge may occur on the ocean side of the deep-sea trench. A combination of the flexure within the bulge and the gravitational force on the subducting lithosphere may result in rift basins in the outer trench slope. Small sedimentary basins may also form between the individual thrust slices of the accretionary prism.

Tectonic uplift occurs at different rates at different times in different parts of a mountain belt, which may result in diversions of rivers within mountain belts (e.g., Tucker and Slingerland, 1996; Gupta, 1997). Thus, the supply of sediment and water to rivers entering the basin, and the positions of the entry points of rivers, may vary in space as well as time. Variations in water and sediment supply in different rivers in the same basin may, but need not, be coeval. Furthermore, changes in climate in different parts of the mountain belt (especially during glaciations) could also result in both congruent and incongruent changes in the discharges and sizes of rivers flowing from the mountain belt. Figure 20.12 illustrates diversion of a river by a thrust-related anticline near the edge of a compressional basin, resulting in a reduction in the supply of water and sediment to a basin-marginal fan. The size and slope of this fan may then become more influenced by tectonic subsidence than by sediment progradation. The river that receives the diverted flow will experience an increased discharge of water and sediment. Its basin-marginal fan would experience an increase in deposition rate, and the size and slope of the fan would become dominated by this sediment progradation. According to the sediment-routing model used by Robinson and Slingerland (1998a, b), downstream fining of river-bed material would not be as effective on the growing fan as it would on the shrinking fan. Increases in the discharges and sizes of rivers, and in the deposition rate and avulsion frequency, on the growing fan would probably result in increasing channel-deposit proportion and connectedness. Such increases in deposition rate in a cross section oriented normal to the thrust belt and basin axis might not be related to a change in uplift rate or subsidence rate in the same cross section. Thus, although tectonism (with or without climate change) may occur over a broad region over a long period of time, the depositional responses in different parts of the basin might not be the same. This illustrates the need for modeling tectonics and sedimentation in three dimensions rather than two.

Compressional basins normally undergo increases in deposition rate and mean grain size with time. If it is assumed that the deposition rate is approximately equal to the subsidence rate, this trend can be interpreted as an increase in subsidence rate as the thrust front encroaches, and/or as the rate of uplift increases (see Figure 20.14 later). The only way to test the various basin-filling models that relate uplift, subsidence, and deposition is to have independent estimates of the timing and magnitude of uplift and deposition: such information is not normally available. If the thrust front moves relative to the compressional basin at the same speed as tectonic plates (on the order of 10 mm per year), and the subsidence rate is on the order of 1 mm per year, then 100 km of convergence would produce 10 km of sediment thickness in 10 million years. Shorter-term episodes of tectonic uplift and subsidence in compressional basins possibly range from on the order of a million years for major regional thrusting episodes to on the order of a thousand years for meter-scale throws on single faults.

Extensional basins

Examples of extensional basins are the Triassic–Jurassic rift basins developed during the opening of the Atlantic Ocean, and now underlying parts of northwest Europe, northeast North America, Africa, and South America (Fitzgerald *et al.*, 1990; Olsen, 1990; Steel and Ryseth, 1990; Miall, 1996; Gupta and Cowie, 2000). Other well-known extensional basins occur in the Basin and Range province of the western USA. Passive continental margins that formed on the Atlantic margins following the rifting stage (Figure 20.13B) have extensive thicknesses of Cretaceous and Tertiary deposits, much of which is alluvial and deltaic (e.g., Galloway, 1975, 1981, 1989a, b). Failed rift basins include the Viking Graben of the North Sea, the East Shetland Basin, and the Benue Trough. Back-arc basins also tend to be extensional.

Extensional basins are caused by lithospheric stretching and thermal subsidence (Leeder, 1999; Allen and Allen, 2005). Lithospheric stretching causes brittle fracture and normal faulting in relatively shallow parts of the lithosphere, but thinning by plastic deformation in deeper parts (Figure 20.13A). Upwelling of hot asthenosphere beneath the area of thinned lithosphere increases the thermal gradient and causes decreasing density and thermal expansion of the lithosphere. This results in both isostatic and expansional uplift at the margins of the thinned lithosphere. Sediment is eroded from these peripheral uplifts (and other local uplifts) and deposited in the extensional basin (Figure 20.13A). As the lithosphere cools, density increases and subsidence occurs as a result of isostasy and thermal contraction. The rate of thermal subsidence decreases with the square root of time. Loading of sediment deposited in the basin eventually causes downward flexure of the lithosphere (depending on its rheological properties) and onlap of sediment at the basin margins (Figure 20.13A). Since the flexural rigidity increases as the lithosphere cools, the zone of onlap increases in width with time, and the basin becomes wider and shallower. Thus lapping of marine sediments onto fluvial sediments is not necessarily due to eustatic sea-level rise, as is commonly thought.

A feature of extensional basins that distinguishes them from compressional basins is that the subsidence rate (related to thermal contraction) decreases with time (Figure 20.14). If one side of an extensional basin becomes a "passive" rifted continental margin (e.g., the Atlantic), and moves away from the spreading center for on the order of 100 million years (1000 km at 10 mm per year), and the average subsidence rate is on the order of 0.1 mm per year, kilometers of sediment could accumulate in the basin. During this time, the sea level may be rising eustatically because of mid-oceanic-ridge growth.

Gawthorpe and Leeder (2000) modeled continental and marine erosion and deposition in extensional basins in relation to the initiation, growth,

A

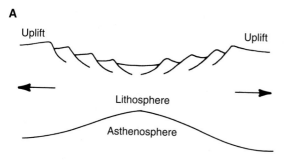

Uplift   Uplift

Lithosphere

Asthenosphere

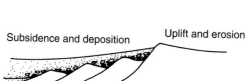

Subsidence and deposition   Uplift and erosion

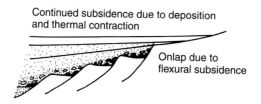

Continued subsidence due to deposition
and thermal contraction

Onlap due to
flexural subsidence

FIGURE 20.13. (A) Main features of extensional basins. (B) A cross section of the Atlantic continental margin off New Jersey, USA. A Triassic–Jurassic rift basin with evaporites overlain by Lower Cretaceous carbonate-rimmed shelf. Modified from Sheridan (1974). By permission of Springer Science and Business Media.

B

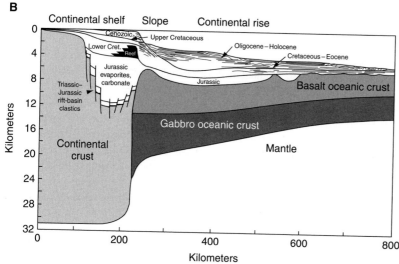

propagation, and death of arrays of normal faults (Figures 20.15 and 20.16). During the fault-initiation stage, isolated fluvio-lacustrine basins occur in the hanging walls of propagating normal-fault segments. Antecedent river courses start to become influenced by the emerging fault-related topography, and some rivers may be diverted in and along developing rift basins. Incision of new drainage systems in the uplifted footwalls leads to the development of small, regularly spaced alluvial fans. The size of the drainage basins and fans decreases towards the fault tips. The centers of these developing rift basins may be occupied by aeolian sands, ephemeral or perennial lakes, or axial rivers with floodplains (Figure 20.15A).

Lateral propagation and joining of fault segments leads to enlargement and coalescence of rift basins,

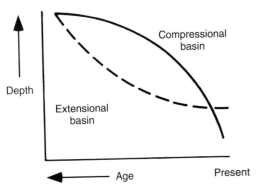

FIGURE 20.14. Subsidence rate versus time in extensional and compressional basins.

and further development of drainage systems (Figure 20.15B). The largest fans are associated with the largest antecedent rivers, and with areas distant from fault tips. Lakes occur in the areas of maximum subsidence, distant from fault tips. Figure 20.15B illustrates the effects of climate change superimposed on tectonic activity. Increases in precipitation and runoff combined with low sediment supply lead to incision of drainage basins, floodplains, and fans and deposition downstream, possibly in expanded lakes. Such incision does not occur if the increase in precipitation is accompanied by an increase in sediment supply. Major changes in climate affecting catchments tend to be reflected relatively rapidly (within on the order of thousands of years) in downstream alluvial deposition. Episodic uplift of footwalls causes a wave of incision to move up the catchment, providing an increased sediment supply to alluvial fans. However, it may take tens of thousands of years for this to be reflected in an increase in fan growth, if at all (Allen and Densmore, 2000). Incision of fan channels and local growth of fans can also be associated with more or less random increases in sediment supply and river avulsions. Episodes of fan progradation into the basin result in upward-coarsening sequences on the order of tens of meters thick, whereas fan recession results in fining-upward sequences.

Continued linkage of adjacent fault segments results in definition of elongated half-graben basins (Figure 20.15C), allowing development of major axial rivers and floodplains between the basin-edge fans. Such growth of axial rivers may be associated with an increased rate of deposition in rift-basin lakes. Axial channel belts tend to move towards the downthrown side of the basin, resulting in relatively thick channel deposits with high channel-deposit proportion and

connectedness, interbedded with alluvial-fan deposits. Up-tilted sides are likely to have low channel-deposit proportion and connectedness, and well-developed soils. Active tectonism and surface faulting ensure frequent avulsions and episodes of high erosion and deposition rates. Deposition in the basin may also be affected by periodic damming of rivers by earthquake-induced landslides or volcanic eruptions (Alexander *et al.*, 1994).

These active stages in the evolution of extensional basins are associated with relatively high rates of erosion of uplifted areas, and high rates of deposition in local fault-bounded rift basins. In contrast, the "fault-death" stage is associated with relatively low deposition rates and increasing basin areas, and possibly marine transgression. Figure 20.15D shows progradation of axial rivers into a lake or the sea, and relatively low deposition rates of axial rivers and floodplains that are being shifted away from the footwall by deposits of reworked fans. Low channel-deposit proportion and connectedness are expected from the low avulsion frequency and larger floodplain widths, but high values are expected if the deposition rate is very low, allowing extensive reworking of the deposits. A high avulsion frequency is expected in local areas affected by sea-level rise.

In the case of coastal-marine environments during the fault-initiation stage (Figure 20.16A), there are numerous isolated depocenters that become partially linked at sea-level highstand to form elongated marine gulfs. River reorganization results in initiation of deltas and fan deltas at the upstream ends and sides of the emerging marine gulfs. The sediment delivered by rivers is reworked by tidal and wave currents. Tidal currents are expected to be particularly important in shallow marine gulfs (Chapter 15). Details of deposits are controlled by spatial and temporal variations in tectonic subsidence and sediment sources, and by eustatic sea-level changes.

The stage of fault interaction and linkage is illustrated for sea-level highstand (Figure 20.16B) and lowstand (Figure 20.16C). Lateral propagation and interaction between fault segments lead to coalescence and enlargement of some depocenters. Continued growth of river drainages along uplifted regions results in growth of fan deltas along the sides of elongated marine basins, and increasing activity of sediment gravity flows down basin sides, especially near the base of footwalls. Depositional lobes formed by sediment gravity flows in deep marine basins may be preferentially stacked close to the base of footwalls. The

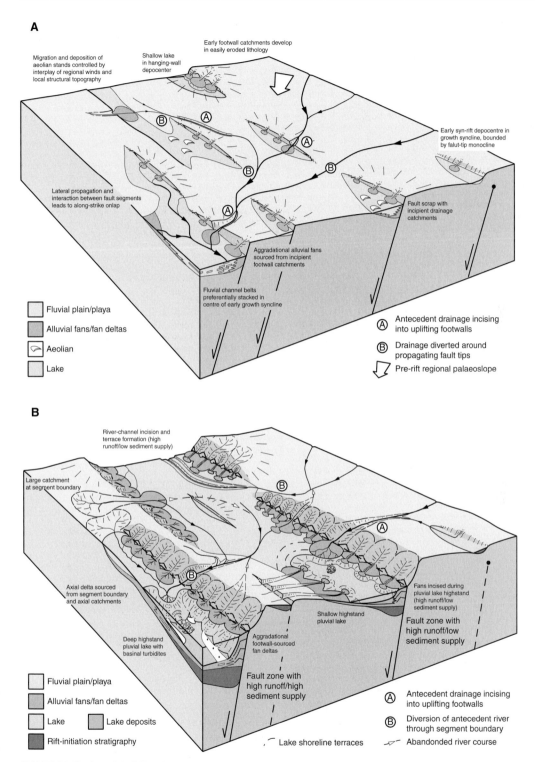

A

Migration and deposition of
aeolian stands controlled by
interplay of regional winds and
local structural topography

Shallow lake
in hanging-wall
depocenter

Early footwall catchments develop
in easily eroded lithology

Early syn-rift depocentre in
growth syncline, bounded
by falut-tip monocline

Lateral propagation and
interaction between fault segments
leads to along-strike onlap

Fault scrap with
incipient drainage
catchments

Aggradational alluvial fans
sourced from incipient
footwall catchments

Fluvial channel belts
preferentially stacked in
centre of early growth syncline

Fluvial plain/playa

Alluvial fans/fan deltas

Aeolian

Lake

(A) Antecedent drainage incising
into uplifting footwalls

(B) Drainage diverted around
propagating fault tips

Pre-rift regional palaeoslope

B

River-channel incision and
terrace formation (high
runoff/low sediment supply)

Large catchment
at segment boundary

Axial delta sourced
from segment boundary
and axial catchments

Deep highstand
pluvial lake with
basinal turbidites

Aggradational
footwall-sourced
fan deltas

Shallow highstand
pluvial lake

Fans incised during
pluvial lake highstand
(high runoff/low
sediment supply)

Fault zone with
high runoff/low
sediment supply

Fault zone with
high runoff/high
sediment supply

Fluvial plain/playa

Alluvial fans/fan deltas

Lake   Lake deposits

Rift-initiation stratigraphy

Lake shoreline terraces

(A) Antecedent drainage incising
into uplifting footwalls

(B) Diversion of antecedent river
through segment boundary

Abandonded river course

FIGURE 20.15. A model of alluvial architecture in evolving extensional basins. From Gawthorpe and Leeder (2000). (A) The initiation
stage. (B) The interaction and linkage stage. (C) The through-going fault stage. (D) The "fault-death" stage.

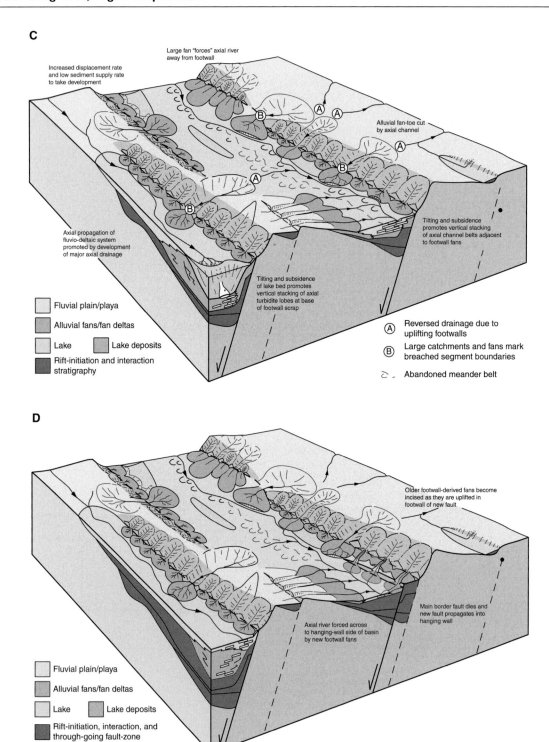

FIGURE 20.15. Continued.

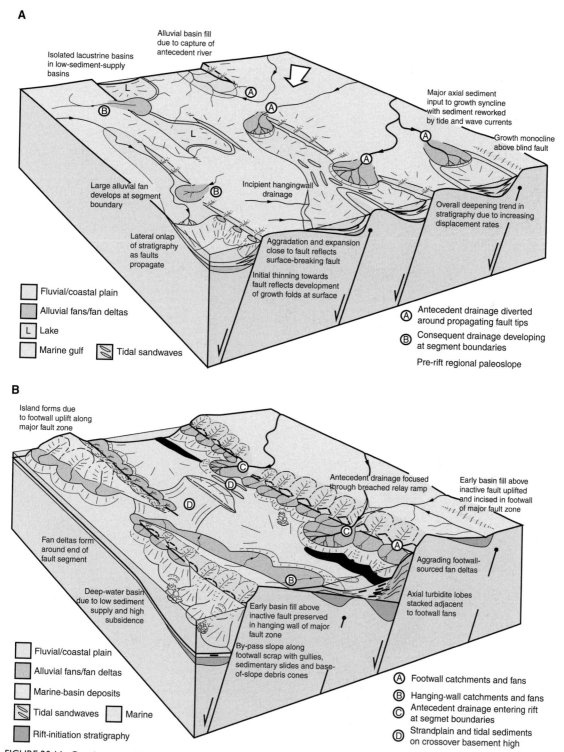

FIGURE 20.16. Gawthorpe and Leeder (2000) models for sedimentary architecture of coastal/marine environments in extensional basins. (A) The initiation stage. (B) The interaction and linkage stage during sea-level highstand. (C) The interaction and linkage stage during sea-level lowstand. (D) The through-going fault-zone stage.

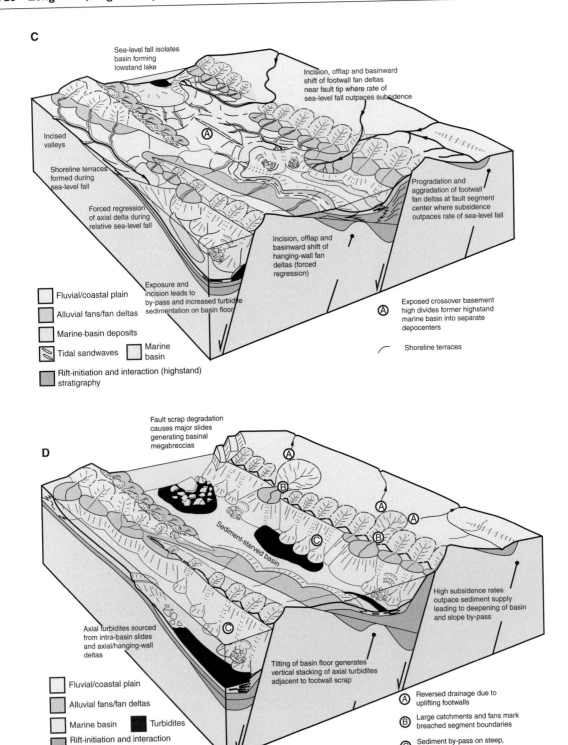

FIGURE 20.16. Continued.

rate of sediment supply to marine basins depends on the area of the adjacent land that is being eroded. Local tectonic highs (islands) result in local coastlines and enhanced tidal currents where a marine basin narrows. Eustatic sea-level fall might result in subaerial exposure and incision of shallow marine regions if the rate of sea-level fall is greater than the subsidence rate. This increases the land area available for supplying sediment, produces more isolated marine basins and lakes, causes a seaward shift of depositional environments (so-called forced regression), and increases sediment gravity flows onto basin floors.

At the through-going fault-zone stage (Figure 20.16D), there is a limited number of major fault zones with high displacement rates and structural relief. Here, tectonic subsidence outpaces sediment supply, leading to marine transgression of marginal fan deltas and coastlines, and marine basins starved of sediment. Sediment gravity flows and rock slides originate from sediment-starved, degrading fault scarps.

Effects of climate on Earth surface processes, landforms, and sediments

Variation of climate over the Earth's surface

Climate varies over the Earth's surface as a result of latitudinal variation in solar heating and associated patterns of air flow (Chapter 2). It is well known that there are broad decreases in average temperature from equatorial to tropical to temperate to polar regions. Precipitation tends to be relatively high in equatorial and temperate regions, where air is commonly ascending, but relatively low in tropical and polar regions, where there is descending air. This is why the hot deserts are concentrated in the latitudes of 15–30 degrees north and south of the equator. However, temperature and precipitation are greatly influenced by land topography and the distribution of the land and sea. Air temperature decreases with increasing elevation, such that the snowline is at sea level at the poles but is at an elevation of 6 km at the equator. Topographic relief gives rise to wet climates on the windward side of mountain ranges, but dry and warm climates on the leeward side (the rain shadow). Thus orographic deserts can occur well outside the subtropical high-pressure belts. Continental interiors distant from major water bodies tend to be relatively dry, and cold deserts occur in such settings (e.g., the Gobi). In some cases, large continental masses give rise to such extreme summer heating and low pressure that air

masses cross normal latitudinal boundaries (e.g., the Asian monsoon). There are many different classifications of climatic conditions on the Earth's surface, in terms of variations in mean annual temperature and precipitation, and in seasonality (Miall, 1996; Leeder, 1999). The terms used above are sufficient for the following discussion.

Evidence for paleoclimatic conditions in sedimentary deposits

Evidence of climate variations preserved in sediments and sedimentary rocks is discussed here, followed by paleoclimatic reconstructions based on this evidence. Reconstruction of Phanerozoic paleoclimates is discussed in many publications (e.g., Frakes, 1979; Parrish and Barron, 1986; Crowley and North, 1991; Frakes et al., 1992; Parrish, 1993, 1998; Wright et al., 1993; Miall, 1996; Bradley, 1999; Blum and Törnqvist, 2000).

Rivers of all sizes and channel patterns occur in all climatic zones today. The suggestion that vegetation (hence climate, or stage in geological history) has an important influence on channel patterns (i.e., dense vegetation causes low discharge variability, enhances bank cohesion, and promotes the occurrence of single-channel sinuous rivers) was dismissed in Chapter 13. Nevertheless, regional changes through time in channel dimensions and pattern (arising from changes in water and sediment supply) may be related to climatic changes. There are many examples of a change from large, braided rivers to progressively smaller meandering rivers associated with the change from glacial to interglacial conditions in the Quaternary, as reviewed in Blum and Törnqvist (2000).

Seasonal variability of river discharge may have an influence on some aspects of channel deposition. In high-latitude/altitude rivers, there is commonly a pronounced low-flow (or even frozen) period, and a relatively short spring-to-summer snowmelt flood period comprising a number of flood peaks. Furthermore, there may be marked diurnal variations in flow resulting from day-time thawing of ice but night-time freezing. Rivers in seasonal temperate and tropical climates may also have relatively long low-flow periods and single or multi-peak flood periods in the spring and summer. Seasonal variation in discharge is likely to be reflected in variation in grain size and sedimentary structures. However, it is unlikely that there would be discernible differences in the deposits of rivers in these different climates, insofar as they all have low-flow periods and multi-peak flood periods when bankfull level would be approached. However, diurnal variations in flow in cold climates might be discernible in the character of the cross stratification in dunes and unit bars (Smith, 1974). The level of base flow in perennial rivers can be recognized using features of subaerial exposure, such as desiccation cracks, footprints, and plant roots. In fact, ephemeral rivers (defined as having a completely dry course for some of the time) in dry climates can be recognized by observing features of subaerial exposure at the bottom of the channel. Many published sedimentological criteria (e.g., North and Taylor, 1996) for the recognition of ephemeral channels (e.g., predominance of planar laminated and low-angle cross-stratified sandstone, a lack of well-defined channels or channels with high width/depth ratios, downstream decrease in channel size) are invalid because they are not distinctive. When assessing the paleoclimatic significance of ephemeral river channels, it is important to make sure that such channels are not overbank channels, which are expected to be ephemeral in any climate. Furthermore, although a reduction in channel size in a downstream direction could be due to loss of discharge as the river flowed across an arid land, it could also be due to a distributive channel system.

Deposits of subaerial sediment gravity flows are commonly taken as being characteristic of ephemeral rivers in arid climates, particularly on alluvial fans. Actually, the only prerequisites for a sediment gravity flow are an appropriate sediment supply, water, and a relatively large slope. Sediment gravity flows occur on the steep slopes of alluvial fans in both arid and wet climates. Furthermore, sediment gravity flows commonly occur down the cut banks of channels, and can be preserved if the channel is being abandoned and filled.

The climatic significance of paleosols was discussed in Chapter 3. Some types of floodplain soil are indicative of paleoclimate in the depositional area, but soil features are controlled by a combination of parent materials, stage of formation, deposition rate, and groundwater level, as well as by climate and vegetation. The stage in development of paleosols has been related indirectly to climate, by associating well-developed paleosols with climatically controlled periods of low deposition rate or erosion. The composition of clay minerals is commonly related to the degree of chemical weathering, and hence to climate and vegetation. For example, kaolinite and gibbsite are related to the deep weathering in humid low latitudes. However, the composition of clays depends on parent materials, the nature of weathering during their journey from their site of origin to their burial site, and diagenetic changes. Therefore, there are many uncertainties in the paleoclimatic meaning of clay minerals.

Coal in swampy environments is not a climate indicator but rather an indicator of non-saline wetlands where precipitation exceeds evaporation (McCabe, 1984; McCabe and Parrish, 1992). Lacustrine deposits interbedded with overbank deposits may yield information about climate and changes in climate. For example, evaporite minerals indicate saline lakes in arid climates, and climatically induced changes in lake level may be reflected in changes in the relative abundance of evaporites and terrigenous mud and sand. Aeolian deposits on floodplains indicate wind action upon unvegetated sediment surfaces. Although such deposits may merely reflect wind acting on freshly deposited flood sediments, an association with evaporites may indicate more prolonged aridity.

Glacial deposits such as till and ice-contact-stratified drift that are interbedded with non-glacial deposits are a clear indication of paleoclimate (Chapter 17). However, continental glaciations also bring changes to Earth surface environments where there is no evidence of deposition directly from or next to the ice. These changes associated with glacial and interglacial periods are discussed below.

Lakes are sensitive to climatically induced changes in water supply, and associated rises and falls in lake level can leave obvious beaches and tufa-encrusted shorelines that can be dated radiometrically (Chapters 14 and 16). During arid periods and falls in lake level, rivers may be incised and coarse sediment may be deposited on the lake floor as lowstand deltas or fans. Carbonates, evaporites, or other chemical sediments may also be deposited on the floor of a low-level lake or in salt pans if the lake dries out completely. During wet periods and high lake level, abundant sediment is delivered to the lake, resulting in progradation of deltas and formation of depositional lobes from sediment gravity flows. Terrigenous mud is deposited on the lake floor instead of carbonates and evaporites. The carbonate versus terrigenous content of lake sediments is commonly used as a proxy for dry versus wet climates. The fossils in lakes commonly depend on the salinity of the lake, with brine shrimp and ostracods characteristic of saline lakes, and fish, diatoms, and molluscs in freshwater. Modern ostracod species, in particular, are sensitive to salinity and temperature, whereas diatom species are only temperature-dependent.

Deposits of sand seas (ergs) and associated ephemeral streams and inland playas are good indicators of climate. Wind reworking of alluvial and lake shoreline sands is also common in arid and semi-arid climates where vegetation is lacking. Transport and deposition of aeolian silt and clay in the sea and on ice are greater during glacial periods.

The supratidal portions of carbonate tidal-flat deposits record climatic conditions, with dry climates indicated by evaporites in supratidal sabkhas (Chapter 16), and calcified microbes, microbial fresh-water-marsh deposits, and soil crusts common under wetter conditions (Chapter 15). Evaporites are also common in siliciclastic tidal-flat deposits in arid climates. Carbonate rocks containing abundant mudstones, green algae, ooid grainstones, shelf patch reefs, and shelf-margin barrier reefs are interpreted as indicating shallow seas in subtropical to tropical climates. Marine evaporites are commonly interstratified with carbonates, and these have long been used to indicate climates in ancient subtropical high-pressure belts that became arid enough to allow sulfate and halite precipitation. Both shallow-marine and deep-marine environments can be taken over by evaporite deposition. The most famous example of this is the Miocene evaporites beneath the floor of the Mediterranean Sea that were thought to indicate that complete desiccation of the sea had occurred many times (Chapter 16).

The climate controls wind-wave regimes and surface ocean currents. The wind-wave regime controls the strength of wave currents and wind-shear currents, the depth of wave base, and the morphology and deposits of a shoreline and shallow-marine shelf (Chapters 7 and 15). However, these features are also controlled by sea-level changes. Climate also has an important effect on deep thermohaline ocean currents (e.g., contour currents, upwelling) and pelagic sediments in the deep ocean (Chapter 18). The carbonate compensation depth (CCD) varies with climate, particularly if climate is influenced by the atmospheric concentration of CO_2. At the Paleocene–Eocene boundary, there was apparently a rapid rise in the CCD throughout the oceans of Earth, and a warming of perhaps 4–5 °C in the deep ocean that changed deep-ocean circulation patterns (Tripati and Elderfield, 2005; Zachos *et al.*, 2005). Prior to this time, the Mesozoic era is noted for a number of so-called oceanic anoxic events that resulted in widespread deposition of organic-rich black shales on continental shelves. These shales serve as the source beds for some of the world's largest oil reserves. These anoxic events in the Cretaceous were probably related to a decrease in ocean ventilation due to warm polar oceans (Jenkyns *et al.*, 2004).

Organisms (in particular planktonic foraminifera shells and plant pollen) and tree rings have long been used to study climate change, and use of other fossil organisms for this purpose is growing (Chapter 17). Fossil plants and animals may suggest the paleoclimate, especially if the species or close relatives are still extant. In the case of extinct groups, paleoclimates may be deduced by comparison with the habitats of comparable extant species (Bosence and Allison, 1995; Francis and Smith, 2002; Flügel, 2004; Stanley, 2005).

Large-scale stratigraphic features

Regional paleogeographic reconstructions can lead to delineation of the areal extent and relative locations of mountain ranges, lowlands, and oceans. The elevations of mountains can be estimated from some types of tectonic model (e.g., Beaumont *et al.*, 1988). Paleoslopes and paleodischarges of rivers can be estimated from sedimentary data. Latitudes can be reconstructed from paleomagnetic data, and the paleoclimate can be estimated using sedimentological and paleontological data. Wind-circulation patterns may be deduced from aeolian deposits. As a result of this, rain-shadow effects can be estimated, as can the likelihood of monsoonal climates. Global climate models have been applied to paleogeographies of the past in order to predict the paleoclimate and to assess the paleoclimatic evidence that comes from the rocks themselves. For example, global climate models have been used to analyze the onset of the Asian monsoon in response to the collision of the Indian subcontinent with Asia, and the uplift of the Himalayas and the Tibetan Plateau. Such models also attempt to explain the onset of global cooling in the Tertiary and the advent of continental glaciation.

Climate change over time

Long-term climate change recorded in sediment deposits has been related to changes in solar output, planetary orbital geometries, geographic distribution of continents, oceanic circulation patterns, atmospheric composition, or any combination of these. The last 600 Myr of Earth's history has featured 10^8-year swings from "icehouse" climates (the latest Precambrian, late Ordovician–early Silurian, Permo-Carboniferous, and late Neogene–Holocene) to "greenhouse" climates. Icehouse to greenhouse swings are coincident with long-term sea-level changes, changes in the mineralogy of oceanic carbonates ("aragonite seas" versus "calcite seas" (Sandberg, 1983)), changes in evaporite mineralogy (Hardie, 1996), and changes in the chemistry of seawater (Lowenstein *et al.*, 2001). Modeling and proxy data for pCO_2 have suggested that most (but not all) icehouse to greenhouse fluctuations were mirrored by variations in the carbon dioxide content of the Earth's atmosphere (Crowley and Berner, 2001), but see Boucot and Gray (2001) and Veizer *et al.* (2000) for dissenting views. A long-term carbon cycle controls the carbon dioxide content of the atmosphere. Berner (1991, 1994) and Berner and Kothavala (2001) modeled this carbon cycle as controlled by factors such as the degassing of carbon dioxide from the mantle, recycling of subducted organic and inorganic carbon at magmatic arcs, burial of carbon as organic carbon and as carbonates, and consumption of atmospheric carbon dioxide by carbonate and silicate weathering. Silicate and carbonate weathering are modeled as being almost exclusively controlled by land vegetation. Coincident with the notable climatic swing at the end of the Paleozoic was the development and spread of land plants and their root systems (Algeo *et al.*, 2001; Driese and Mora, 2001). Larger and deeper roots led to thicker and better-developed paleosols after the Devonian, and to the appearance of specific types of paleosols such as histosols, alfisols, ultisols, and spodosols (Chapter 3). An increase in the rate of weathering due to the spread of land plants, and enhanced preservation of buried organic matter because of its resistance to microbes, are thought to have led to a major drop in the atmospheric level of CO_2 from the Devonian to the early Carboniferous, resulting in global cooling (Algeo *et al.*, 2001; Berner, 2001; Driese and Mora, 2001). Similarly, Berner's model suggests that the spread of angiosperms after the Cretaceous may have at least partly influenced the global cooling of the later Tertiary.

Cyclicity in climate is related to variations in the Earth's orbit around the Sun and in the Earth's own rotation, resulting in cyclic changes in solar radiation (Chapter 2). Such Milankovitch cycles have discrete periods (on the order of 10^4–10^5 years) and amplitudes that interact in a complex way. These cycles had a major influence on the volume of continental ice (and hence on sea level) during the Pleistocene. There has been a high-amplitude 100,000-year period to glacial–interglacial cycles during the middle-to-late Pleistocene, but lower-amplitude, 40,000-year cycles dominated prior to that. Evidence for these climatic cycles is found in studies of depositional sequences, oxygen isotopes of marine microfossils, and comparative studies of pollen, spores, and marine microfossils. More recent work on ice cores and marine microfossils indicates the existence of cycles of abrupt warming followed by longer-term cooling over thousands to tens of thousands of years. These global climatic cycles are apparently related to interactions among the atmosphere, ice masses, and thermohaline circulation of the oceans (Blum and Törnqvist, 2000). Regional climate changes are more complex than global changes, because of the way that changes in the global atmospheric circulation are manifested in different regions with differing proportions and elevations of land. Some short-period (decades and less) fluctuations in regional climate are related to sunspot cycles or ocean–atmosphere interactions such as El Niño.

The direct effects of Quaternary glacial periods on fluvial, lacustrine, and marine environments and their deposits are discussed in Chapter 17. The effects of Quaternary climate change on rivers and their deposits have been studied widely (Gregory, 1983; Knox, 1983, 1996; Gregory *et al.*, 1987, 1996; Bull, 1991; Starkel *et al.*, 1991; Benito *et al.*, 1998; Blum and Törnqvist, 2000). Some of the changes in rivers and their deposits that are expected over glacial–interglacial cycles are summarized below and in Table 20.3. This summary is based on much data, but especially data from the Mississippi valley and delta plain, the Gulf of Mexico coastal plain, and the Rhine–Meuse delta. Glacial and early deglacial periods are associated with ice-related crustal loading and drainage diversions in upland areas; increases in flood discharge, mean grain size, and rate of sediment supply; and low sea level. This results in progradation from uplands of relatively coarse sediment and steepening of valley slopes, increases in channel-belt size, and changes in channel pattern. High avulsion frequency and high channel-belt/floodplain width in up-valley regions should result in high channel-deposit proportion and connectedness. Falling sea level causes incision or deposition in coastal areas (and changing channel patterns), depending on factors such as the slope of the exposed surface, rate of sea-level fall, and up-valley avulsions. The channel-deposit proportion and connectedness in coastal areas are increased by larger channel-belt size and smaller deposition rate, but decreased by low avulsion frequency and high floodplain width.

TABLE 20.3. Responses of Mississippi Valley and Delta Plain to glacial–interglacial cycles (following Autin et al., 1991; Saucier, 1994)

Parameter	Uplands	Alluvial valley	Coastal plain
Deglaciation–interglacial (Holocene)			
Tectonism	Isostatic uplift	Subsidence increasing towards coast	
Sea level			Rising: transgression
Water supply		Decreasing	Decreasing
Sediment supply		Decreasing	Decreasing
Mean sediment size		Intermediate	Fine
Channel-belt width and depth		High	Low
Channel pattern		Meandering	Low sinuosity
Floodplain width		Low	Very high
Floodplain slope		Intermediate	Low
Deposition/erosion rate	Erosion	Terrace formation up valley	Deposition rate increasing to coast
Avulsion frequency		Intermediate	High
Channel-deposit proportion		High	Low
Glacial maximum–deglaciation (Pleistocene)			
Tectonism	Crustal loading	Subsidence decreasing towards coast	
Sea level			Low to rising
Water supply		High	High
Sediment supply		High	High
Mean sediment size		High	Intermediate
Channel-belt width and depth		High	High
Channel pattern		Braided	Braided–meandering
Floodplain width		High	Low
Floodplain slope		High	Intermediate
Deposition/erosion rate	Erosion	Maximum deposition to minor erosion	Valley incision to deposition
Avulsion frequency		High, decreasing	Low, increasing
Channel-deposit proportion		High	High, decreasing

Interglacial periods are associated with isostatic uplift and re-vegetation of uplands; reduction in flood discharge, grain size, and rate of sediment supply; and rising sea level. This results in erosion and terrace formation in up-valley regions. Reduced deposition rates, avulsion rates, and channel sizes in mid-valley regions lead to decreased channel-deposit proportion and connectedness. Rising sea level leads to increasing deposition rate and avulsion frequency in coastal areas, as well as increasing channel-deposit proportion and connectedness. However, up-valley shift in avulsion points and abandonment of delta lobes leads to high overbank deposition rates and small channel belts there, decreasing the channel-deposit proportion and connectedness.

These predictions are complicated by local changes in sediment supply, deposition rate, channel pattern, and avulsion frequency associated with local tectonism or climate change, which is not necessarily congruent with regional changes. Indeed, Blum and Törnqvist (2000) describe examples of allostratigraphic units within valley fills that record widespread episodes of river incision and aggradation that are interpreted as due to a combination of interglacial–glacial cycles and shorter-term climate changes. Each allostratigraphic unit has an erosional base, is on the order of meters to tens of meters thick, extends across valley for kilometers to tens of kilometers, and is capped by a well-developed paleosol (e.g., Figure 20.17). Finally, predictions for present-day temperate regions cannot necessarily be expected to apply to regions with different climate and vegetation, because changes in water and sediment supply may be different.

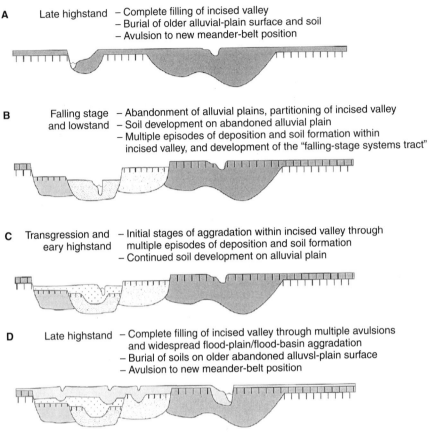

A Late highstand – Complete filling of incised valley
– Burial of older alluvial-plain surface and soil
– Avulsion to new meander-belt position

B Falling stage – Abandonment of alluvial plains, partitioning of incised valley
and lowstand – Soil development on abandoned alluvial plain
– Multiple episodes of deposition and soil formation within
incised valley, and development of the "falling-stage systems tract"

C Transgression and – Initial stages of aggradation within incised valley through
eary highstand   multiple episodes of deposition and soil formation
– Continued soil development on alluvial plain

D Late highstand – Complete filling of incised valley through multiple avulsions
and widespread flood-plain/flood-basin aggradation
– Burial of soils on older abandoned alluvsl-plain surface
– Avulsion to new meander-belt position

FIGURE 20.17. Evolution of river terraces and valley fills during periods of falling and rising base level, for the Gulf of Mexico coastal plain. From Blum and Price (1998). Multiple episodes of erosion and deposition are caused by climate-related variation in water and sediment supply, and by channel avulsion.

Koltermann and Gorelick (1992) explained changes in alluvial-fan deposits due to climate change using a simple sediment-routing model. Floods were generated using a stochastic simulator, and variations in water discharge and sediment transport during large floods were linked to Quaternary paleoclimate change. The effects of a basin-bounding fault, compaction, and base-level change were also considered in this model. Periods of high rate of deposition of relatively coarse sediment (fan progradation) were associated with the periods of wet, cool climate when flood discharges of water and sediment were high. Fine-grained sediment was deposited during warmer, drier climate. Transcurrent movement on the basin-margin fault caused the fan to move horizontally relative to the feeder stream, such that successive progradations of the fan were offset relative to each other, producing a 100-m-thick sequence where the thickness of successive fan deposits either increased or decreased upwards.

Arid environments became more arid during Quaternary glacial periods and less arid during interglacials (Chapter 16). Transport and deposition of aeolian silt and clay (in the sea and on ice) were greater during glacial periods. Arid environments were also affected indirectly by glacioeustasy, such that coastal arid environments had increasing interdune and lacustrine areas relative to sand seas as the sea level rose.

Marine areas directly affected by glaciers are discussed in Chapter 17. However, many of the effects of Quaternary glacials and interglacials on marine environments were related to the effects of glacioeustatic sea-level changes on terrigenous sediment supply, sediment transport by wave and tidal currents, and sediment gravity flows down continental slopes (as discussed below). In addition, changes in wind patterns, air and water temperature, and water salinity over glacials and interglacials caused changes in surface ocean currents, thermohaline currents, and the distribution of contourites and pelagic deposits (Chapter 18).

Many fluvial depositional sequences from meters to tens of meters thick (representing 10^4–10^5 years) in rocks older than the Pleistocene have been interpreted in terms of Milankovitch glacioeustasy (e.g., Olsen, 1990, 1994; Read, 1994; Van Tassell, 1994). Milankovitch climatic cycles have been claimed to be recorded in pre-Pleistocene deposits of other environments also; for reviews and compilations see Einsele *et al.* (1991), de Boer and Smith (1994), House and Gale (1995), and Hinnov (2000). In particular, there has been much speculation about Milankovitch climate cycles driving meter-scale sea-level oscillations that are responsible for the cyclicity of ancient carbonate platform deposits (Hardie and Shinn, 1986; Demicco and Hardie, 1994; Hinnov, 2000; Schlager, 2005). An example from the Triassic carbonates of the Italian Dolomite Mountains is given in Chapter 15. Many carbonate sedimentologists hold the extreme view that all carbonate parasequences (meter-scale cycles) are the result of sea-level oscillations in tune with Milankovitch orbital rhythms. Also, 100-m-thick, 1–10-million-year cycles have been interpreted as due to the growth and decay of continental ice.

There are many potential problems with such interpretations. It is not possible to establish the period of these older depositional sequences accurately if they are less than 10^5 years old (because of dating limitations); therefore, it is not possible to establish the age equivalence of depositional sequences with periods less than 10^5 years in different parts of the world. The links among solar radiation, climate, sediment production, river and sediment discharges, and sea level are tenuous and poorly known in the absence of major continental ice sheets. Milankovitch-cycle periods are similar to those associated with other mechanisms (e.g., tectonism) that can produce sedimentary cycles (Steel *et al.*, 1977; Algeo and Wilkinson, 1988; Fraser and Decelles, 1992). However, some workers have appealed to the 5:1 ratio of the periods of the different Milankovitch cycles (e.g., 20,000-year and 100,000-year cycles) to justify orbital forcing of sedimentary cycles.

Eustatic effects on erosion and deposition

Eustatic sea-level changes depend greatly upon tectonics and climate, and are due to (1) changes in the amount of continental ice (sea-level changes of more than 100 m over time scales of 10^4–10^6 years); (2) changes in the volume of ocean basins, occurring over periods of 10–100 million years; and (3) changes in the amount of water on the surface of the Earth due to outgassing from the interior of the Earth throughout geological time. Relative sea-level changes depend also on local tectonism and deposition and erosion. In most cases, the effects of relative and eustatic sea-level changes are combined; therefore, it is generally difficult to recognize a particular eustatic sea-level change in deposits worldwide, even if the deposits could be dated accurately enough. Exceptions are where major eustatic sea-level changes occur over long periods of time.

The Exxon research group (e.g., Vail *et al.*, 1977) interpreted sequences and associated bounding surfaces (defined within the seismic and sequence-stratigraphy paradigm) as due to eustatic sea-level changes. These interpretations have been criticized by many individuals, notably Miall (1986, 1991, 1996) on the grounds that (1) the method for determining relative sea-level change from seismic records and cores is flawed; (2) other interpretations of the sequences are not considered; and (3) the data used to correlate the sequences and interpreted sea-level changes worldwide are not generally available. More recent approaches to interpreting the influence of eustasy on stratigraphic sequences are discussed below.

Continental environments

There are now many different (sequence-stratigraphic) models for the effects of relative sea-level change on deposition rate and alluvial architecture (Bridge, 2003). Most of them are qualitative and only 2D, and do not adequately represent all of the controls on alluvial deposition. Miall (1991, 1996) has criticized some of the earlier models of the effects of sea-level change on near-coastal alluvial deposition. His main point is that a relative fall in sea level is not normally associated with alluvial aggradation, except for the newly exposed part of the sea bed, and even then only under special circumstances. Whether or not a river valley is incised or aggraded during sea-level fall depends, among other things, on the slope of the exposed shelf relative to that of the river valley (Figure 20.18). In general, effects of sea-level change are expected to decrease up valley.

The effect of eustatic sea-level change on fluvial processes near shorelines depends on co-variation in tectonic subsidence or uplift, and deposition or erosion. However, certain generalizations can be made about the nature of near-coastal alluvial deposition associated with *eustatic sea-level changes that dominate*

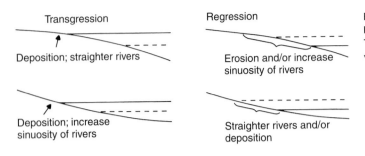

FIGURE 20.18. Effects of rising and falling sea level on a fixed land surface that varies in slope. The dashed line is the original position of the water level.

over changes in tectonic uplift/subsidence and deposition/ erosion. During *sea-level rise* (and marine transgression), slopes of rivers and floodplains are reduced near shore due to backwater effects, and the width of valleys increases due to drowning. These changes result in a decrease in the grain size of transported sediment, deposition, changes in channel pattern, and an increase in the frequency of avulsion. Increases in avulsion frequency increase the channel-deposit proportion and connectedness, but increases in deposition rate and width of flood plains decrease them. During *falling sea level* (and marine regression), the exposed land surface may experience erosion or deposition, depending on the slope of the exposed surfaces (Schumm, 1993; Wescott, 1993; Wood *et al.*, 1993; Posamentier and Allen, 1999) (Figure 20.18). Erosion will occur on relatively steep slopes, given enough time and flow power. However, erosion of channels and floodplains is a long-term process, and the short-term response may be to increase the sinuosity of rivers. Also, upstream avulsion may cause channel abandonment before incision is complete (Leeder and Stewart, 1996). Incision of channels and floodplains is associated with terrace formation, reduction of valley width, and up-valley migration of knickpoints. The avulsion frequency is expected to be low in areas of erosion. The channel-deposit proportion and connectedness may increase during sea-level fall because of a reduced deposition rate or erosion and reduced floodplain width, but may decrease as a result of a reduced avulsion frequency.

Most alluvial sequence-stratigraphy models have an erosional base to the "sequence" (a so-called Type 1 unconformity) due to valley incision arising from a relative fall in base level (Shanley and McCabe, 1993, 1994; Wright and Marriott, 1993; Gibling and Bird, 1994; Miall, 1996) (Figure 20.19). The erosional base in these models is overlain by superimposed channel-belt deposits, which supposedly accumulated under conditions of low deposition rate and restricted floodplain width (due to channel incision). Such deposits comprise the so-called *lowstand systems tract*. Many workers assume that zones of high channel-deposit proportion in alluvial deposits represent the basal parts of "sequences" and that the basal erosion surface of the lowest sandstone body represents an incised floodplain (e.g., Aitken and Flint, 1995; Flint *et al.*, 1995; Hampson *et al.*, 1997, 1999; Davies *et al.*, 1999). In many cases, evidence for an incised floodplain is lacking. Criteria for incised floodplains (valleys) include (Dalrymple *et al.*, 1994; Posamentier and Allen, 1999) the following: (1) erosional relief that is greater than the thickness of a single-channel fill; (2) multiple, vertically stacked channel bars within the incised-valley fill; (3) evidence for extended periods of non-deposition (mature paleosols) on interfluves (incised floodplain surfaces); and (4) alluvial channel deposits resting erosively upon shallow marine sands and muds. Commonly, a large amount of erosional relief on the base of a single-channel deposit can be misinterpreted as an incised-valley margin (Salter, 1993; Best and Ashworth, 1997). Thick and laterally extensive amalgamated channel deposits are commonly important oil and gas reservoirs. Therefore, such deposits should be interpreted accurately when predicting their thickness, lateral extent, and bounding facies. If zones of high channel-deposit proportion are incorrectly interpreted as incised-valley fills, their extent normal to the valley direction will be underestimated, and their extent parallel to the valley will be overestimated.

Fluvial deposition may be occurring up valley while incision is occurring near the coast due to falling sea level (Blum and Price, 1998; Blum and Törnqvist, 2000; Törnqvist *et al.*, 2000; Van Heijst and Postma, 2001), so that the basal erosion surface of the sequence will not be coeval or correlatable inland. An ancient example of this type of lowstand valley fill is given by Willis (1997), who explained it in terms of a low rate of sea-level change relative to the fluvial sediment supply. Interestingly, major erosional surfaces may be associated with climate-related fluvial incision or estuarine channels during periods of rising sea level.

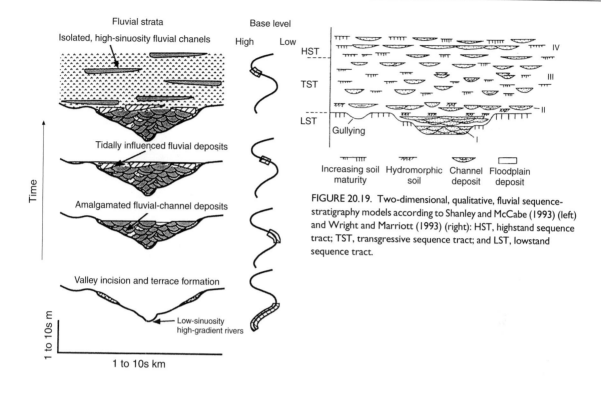

FIGURE 20.19. Two-dimensional, qualitative, fluvial sequence-stratigraphy models according to Shanley and McCabe (1993) (left) and Wright and Marriott (1993) (right): HST, highstand sequence tract; TST, transgressive sequence tract; and LST, lowstand sequence tract.

According to the sequence-stratigraphy models, the deposits above the lowstand systems tract were deposited under conditions of relatively high deposition rate on a broad alluvial plain, in response to a rising relative base level (the so-called *transgressive systems tract*). The channel-deposit proportion and connectedness are relatively low as a result. The association of a relatively high proportion of overbank mud with rising sea-level and marine transgression was recognized early on by McCave (1968, 1969). Törnqvist (1993, 1994) associated high avulsion frequency and anastomosing rivers with periods of rapid base-level rise and deposition. According to Gibling and Bird (1994) and Heckel *et al.* (1998), extensive coals occur just below the *maximum-flooding surface*, when sea level is near its highstand. However, others predict coal development immediately above the incised valley fill, associated with the *initial-flooding surface* (Aitken and Flint, 1995; Flint *et al.*, 1995; Hampson *et al.*, 1997, 1999; Davies *et al.*, 1999). Paleosols in the transgressive systems tract are likely to reflect a high groundwater table (Wright and Marriott, 1993). Deposits associated with the maximum-flooding surface may contain evidence of marine influence. The *highstand systems tract* is also associated with a relatively low channel-deposit proportion according to Shanley and McCabe (1993). However, Wright and Marriott (1993) predict an increase in channel-deposit proportion here, and an increase in soil maturity, both related to a reduced deposition rate. Retallack (2001) also predicts increasing soil maturity as sea level rises.

The alluvial sequence-stratigraphic models discussed above all include a basal incised-valley fill related to sea-level fall, and predict variations in channel-deposit proportion and connectedness as a function of the deposition rate and floodplain width. However, it is known that sea-level fall is not always accompanied by valley incision, and the nature of erosion and deposition in alluvial systems is controlled also by climate and tectonism. Furthermore, the channel-deposit proportion and connectedness are controlled also by factors such as the avulsion frequency and channel-belt size. Therefore, it is unlikely that extant 2D alluvial sequence-stratigraphy models are generally applicable. A quantitative, 3D fluvial sequence-stratigraphy model is long overdue (Karssenberg *et al.*, 2004).

Marine environments

Falling sea level results in marine regression, rivers flowing on pre-existing sea bed, a decrease of water depth on the shelf, and a decrease in shelf width. Rivers may become incised, thus increasing sediment supply to the marine shelf. Wave and tidal currents acting on a shelf

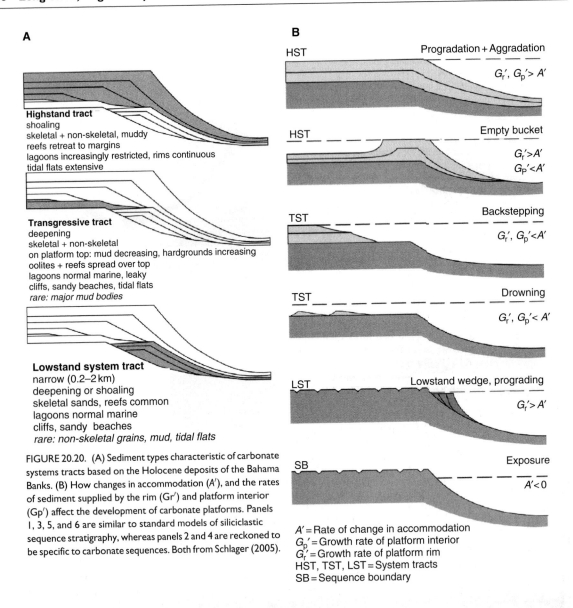

A

Highstand tract
shoaling
skeletal + non-skeletal, muddy
reefs retreat to margins
lagoons increasingly restricted, rims continuous
tidal flats extensive

Transgressive tract
deepening
skeletal + non-skeletal
on platform top: mud decreasing, hardgrounds increasing
oolites + reefs spread over top
lagoons normal marine, leaky
cliffs, sandy beaches, tidal flats
rare: major mud bodies

Lowstand system tract
narrow (0.2–2 km)
deepening or shoaling
skeletal sands, reefs common
lagoons normal marine
cliffs, sandy beaches
rare: non-skeletal grains, mud, tidal flats

B

HST   Progradation + Aggradation
G_r', $G_p' > A'$

HST   Empty bucket
$G_r' > A'$
$G_p' < A'$

TST   Backstepping
G_r', $G_p' < A'$

TST   Drowning
G_r', $G_p' < A'$

LST   Lowstand wedge, prograding
$G_r' > A'$

SB   Exposure
$A' < 0$

FIGURE 20.20. (A) Sediment types characteristic of carbonate systems tracts based on the Holocene deposits of the Bahama Banks. (B) How changes in accommodation (A'), and the rates of sediment supplied by the rim (Gr') and platform interior (Gp') affect the development of carbonate platforms. Panels 1, 3, 5, and 6 are similar to standard models of siliciclastic sequence stratigraphy, whereas panels 2 and 4 are reckoned to be specific to carbonate sequences. Both from Schlager (2005).

A' = Rate of change in accommodation
G_p' = Growth rate of platform interior
G_r' = Growth rate of platform rim
HST, TST, LST = System tracts
SB = Sequence boundary

with reduced depth and width are expected to be more effective at transporting sediment along and across the shelf. The outer shelf may be incised, starting at the edge of the continental slope. Increases in erosion and deposition at the edge of the shelf will increase the activity of sediment gravity flows down the continental slope, leading to growth of submarine fans (Chapters 15 and 18).

Rising sea level results in marine transgression, drowning river valleys (to become estuaries) and other coastal areas, and increasing the depth and width of marine shelves. Increasing coastline irregularity is expected to result in an increase in the magnitude of tidal currents relative to that of wind-related currents (Chapter 7). Wind-related currents would not be as effective at moving sediment on the deepening water of the shelf. The space created at the coast may result in an increase in deposition rate of terrigenous sediment, possibly starving the deeper marine areas of terrigenous sediment, which is conducive to formation of biogenic and chemogenic sediment. Calcium carbonate sediments are particularly important in shallow water where the supply of terrigenous sediment is negligible, due to distance from a major continental sediment source. A low deposition rate in marine areas gives rise to so-called condensed

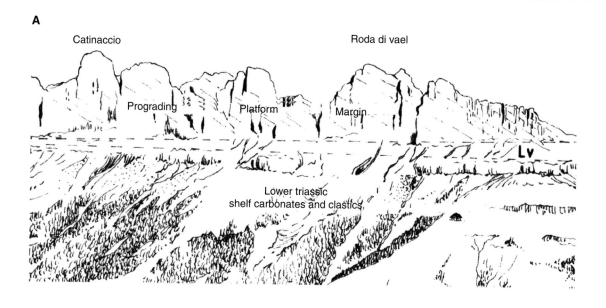

FIGURE 20.21. The Middle Triassic Catinaccio rimmed platform from the Dolomite Alps of northern Italy with a photograph (B) and interpretive diagram (A) from Bossellini (1984). The large-scale bedding surfaces nearly at the angle of repose mark a prograding talus slope of reef rubble that has been built out into the basin for approximately 15 km.

sequences that are intensively burrowed and mineralized. If sediment is not being transported to the shelf edge, or not produced there, submarine canyons and fans would become inactive and covered with suspended mud.

Sequence-stratigraphic models for coastal and shallow-marine environments (Chapter 15) typically have a basal coarsening-upward sequence (tens of meters to a hundred meters or so thick) that is formed during marine regression. This sequence may, but need not,

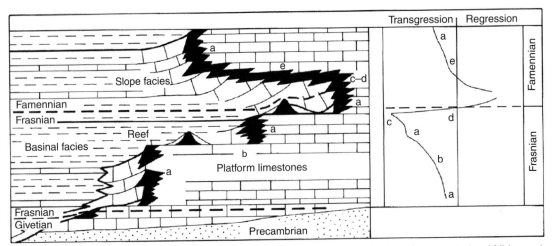

FIGURE 20.22. The response of Devonian Canning Basin Reefs to variable rates of sea-level rise (transgression) and fall (regression). From Tucker and Wright (1990), after Playford (1984).

have an erosional base. In general, a somewhat thinner fining-upward sequence is produced during marine transgression. In some cases, the transgressive sequence is underlain by an erosion surface, produced by beach and shoreface erosion during marine transgression. Superimposed on this overall coarsening–fining-upward sequence are smaller-scale sequences related to intra-basinal processes such as channel mouth-bar migration and abandonment.

Sequence-stratigraphic models for deep-sea fans (Chapter 18) normally have an erosion surface overlain by a mass-transport complex formed during falling sea level. This is overlain by sheet-like deposits of the outer fan and then channel–levee complexes of the inner and middle fan, recording fan growth and progradation during lowstand and rising sea level. Hemipelagic muds deposited during highstands caps the sequence. However, this simplistic view must be modified to consider the following complications: (1) the effects of tectonic activity; (2) mass-transport complexes are not always restricted to the base of the fan sequence; and (3) feeder-channel avulsions on the shelf edge may shift the locus of deposition.

A standard model for the sequence stratigraphy of carbonate-rimmed shelves is summarized in Figure 20.20. Carbonate sequences are thought to be generated by relative sea-level oscillations of a few tens of meters on time scales of 10^5–10^6 years (Loucks and Sarg, 1993; Schlager, 2005). During periods of low sea level, the exposed loose sediments on the rimmed shelves are chemically weathered into soils, and dissolved or cemented by freshwater (forming karst

features). Cementation halts physical erosion, so only chemical weathering takes place. During periods of low sea level, deposition is restricted to narrow zones along the shelf slopes (the lowstand systems tract). As sea level rises, the shelf becomes flooded and carbonate deposition resumes (the so-called transgressive systems tract), but reefs are not present because they cannot grow up at the same rate as a rapid rise in sea level. If reefs grow during this stage, they grow on the shelf rather than at the shelf edges. As the rate of sea-level rise slows, and the highstand of sea level is reached, maximum rates of carbonate production can occur. Under these conditions, reefs and shoals become established at the shelf edge, carbonate mud accumulates in tidal flats in the shelf interior, and carbonate sand and mud are washed off the platform into the adjacent basin (forming the so-called highstand systems tract). The large-scale strata shown in Figure 20.20 have distinctive geometries that can be recognized in seismic lines. Furthermore, it is claimed that the proportions of the different environments of carbonate shelves (i.e., reefs, shoals, tidal flats) change on passing between the three distinct systems tracts, and that changes among the rate of change of accommodation potential and the growth potentials of the shelf and rim (sediment supply) influence shelf-margin geometry (Figure 20.20) (Schlager, 2005).

Since most modern reefs are bordered by kilometer-deep ocean basins, the carbonate debris transported off the shelf during sea-level highstands is deposited in water that is thousands of meters deep. However, many ancient rimmed shelves with shelf-edge reefs developed in shallower epicontinental seaways (such as the modern Hudson Bay, North Sea, or Arabian Gulf), and the shelf slope dropped off into a relatively shallow (on the order of 100 m) marine basin. In this case, the shelf-margin sands and reefs can prograde out into the basin over the slope talus during highstands of sea level.

This occurred for Triassic platforms in the Dolomite Alps region of Italy (Figure 20.21) (Bossellini, 1984), and in the Permian reef complexes of west Texas. The Devonian reef margin of the Canning Basin manifests a complicated history of shelf-margin reef growth in response to sea-level changes (Figure 20.22).

References

Aber, J. S., Croot, D. G., & Fenton, M. M. 1989. *Glaciotectonic Landforms and Structures.* Dordrecht: Kluwer Academic Publishers.

Abrahams, A. D. 1986. *Hillslope Processes.* Boston: Allen and Unwin.

Abrahams, A. D., & Parsons, A. J. 1994. *Geomorphology of Desert Environments.* London: Chapman and Hall.

Adams, A. E., & MacKenzie, W. S. 1998. *A Colour Atlas of Carbonate Sediments and Rocks under the Microscope.* London: Manson Publishing.

Adams, A. E., MacKenzie, W. S., & Guilford, C. 1984. *Atlas of Sedimentary Rocks under the Microscope.* New York: John Wiley and Sons.

Adams, J. E., & Rhodes, M. L. 1960. Dolomitization by seepage refluxion. *American Association of Petroleum Geologists Bulletin* **44**, 1912–1921.

Ahlbrandt, T. S., & Fryberger, S. G. 1980. Eolian deposits in the Nebraska Sand Hills. *United States Geological Survey Professional Paper* **1120A**.

1981. Sedimentary features and significance of interdune deposits. In *Recent and Ancient Nonmarine Depositional Environments: Models for Exploration* (ed. F. G. Ethridge and R. M. Flores), pp. 293–314. Tulsa, OK: SEPM (Society for Sedimentary Geology).

Aigner, T., & Reineck, H. -E. 1982. Proximality trends in modern storm sands from the Helgoland Bight (North Sea) and their implications for basin analysis. *Senckenbergiana Maritima* **14**, 183–215.

Aitken, J. F., & Flint, S. S. 1995. The application of high-resolution sequence stratigraphy to fluvial systems: a case study from the Upper Carboniferous Breathitt Group, eastern Kentucky, USA. *Sedimentology* **42**, 3–30.

Alexander, C. R, Davis, R. A., & Henry, V. J. 1998. *Tidalites: Processes and Products.* Tulsa, OK: SEPM (Society for Sedimentary Geology).

Alexander, C. R., Nittrouer, C. A., DeMaster, D. J., Park, Y. A., & Park, S. C. 1991. Macrotidal mudflats of the southwestern Korean coast: a model for interpretation of intertidal deposits. *Journal of Sedimentary Petrology* **61**, 805–824.

Alexander, J., Bridge, J. S., Cheel, R. J., & Leclair, S. F. 2001. Bed forms and associated sedimentary structures formed under supercritical water flows over aggrading sand beds. *Sedimentology* **48**, 133–152.

Alexander, J., Bridge, J. S., Leeder, M. R., Collier, R. E. L., & Gawthorpe, R. L. 1994. Holocene meander belt evolution in an active extensional basin, southwestern Montana. *Journal of Sedimentary Research* **B64**, 542–559.

Alexander, J., & Leeder, M. R. 1987. Active tectonic control on alluvial architecture. In *Recent Developments in Fluvial Sedimentology* (ed. F. G. Ethridge, R. M. Flores, & M. D. Harvey), pp. 243–252. Tulsa, OK: SEPM (Society for Sedimentary Geology).

Alexander, J., & Morris, S. 1994. Observations on the experimental, nonchannelised, high-concentration turbidity currents and variations in deposits around obstacles. *Journal of Sedimentary Research* **64A**, 899–909.

Algeo, T. J., Scheckler, S. E., & Maynard, J. B. 2001. Effects of the Middle to Late Devonian spread of vascular plants on weathering regimes, marine biotas, and global climate. In *Plants Invade the Land* (ed. P. G. Gensel & D. Edwards), pp. 216–236. New York: Columbia University Press.

Algeo, T. J., & Wilkinson, B. H. 1988. Periodicity of mesoscale Phanerozoic sedimentary cycles and the role of Milankovitch orbital modulation. *Journal of Geology* **96**, 313–322.

Allen, G. P. 1991. Sedimentary processes and facies in the Gironde estuary: a recent model for macrotidal estuarine systems. In *Clastic Tidal Sedimentology* (ed. D. G. Smith, G. E. Reinson, B. A. Zaitlin, & R. A. Rahmani), pp. 29–40. Calgary, Alberta: Canadian Society of Petroleum Geologists.

Allen, J. R. L. 1965. A review of the origin and characteristics of recent alluvial sediments. *Sedimentology* **5**, 89–191.

1966. On bedforms and paleocurrents. *Sedimentology* **6**, 153–190.

1970. *Physical Processes of Sedimentation.* London: George Allen and Unwin.

1973. Features of cross-stratified units due to random and other changes in bed forms. *Sedimentology* **20**, 189–202.

1974. Studies in fluviatile sedimentation: implications of pedogenic carbonate units, Lower Old Red Sandstone, Anglo-Welsh outcrop. *Geological Journal* 9, 181–208.

1978. Studies in fluviatile sedimentation: an exploratory quantitative model for architecture of avulsion-controlled alluvial suites. *Sedimentary Geology* 21, 129–147.

1979a. A model for the interpretation of wave-ripple marks using their wavelength, textural composition and shape. *Journal of the Geological Society of London* 136, 673–682.

1979b. Studies in fluviatile sedimentation: an elementary geometrical model for the connectedness of avulsion-related channel sand bodies. *Sedimentary Geology* 24, 253–267.

1980. Sand waves: a model of origin and internal structure. *Marine Geology* 26, 281–328.

1982a. *Sedimentary Structures: Their Character and Physical Basis.* Amsterdam: Elsevier Science Publishers.

1982b. Mud drapes in sand-wave deposits: a physical model with application to the Folkestone beds (early Cretaceous, southeast England). *Philosophical Transactions of the Royal Society of London* 306A, 291–345.

1985. *Principles of Physical Sedimentology.* London: George Allen and Unwin.

1991. The Bouma division A and the possible duration of turbidity currents. *Journal of Sedimentary Petrology* 61, 291–295.

Allen, J. R. L., & Banks, N. L. 1972. An interpretation and analysis of recumbent-folded deformed cross bedding. *Sedimentology* 19, 257–283.

Allen, J. R. L., & Pye, K. 1992. *Saltmarshes.* Cambridge: Cambridge University Press.

Allen, P. A. 1981. Some guidelines in reconstructing ancient sea conditions from wave ripple marks. *Marine Geology* 43, M59–67.

1985. Hummocky cross-stratification is not produced purely under progressive gravity waves. *Nature* 313, 562–564.

1997. *Earth Surface Processes.* Oxford: Blackwells.

Allen, P. A., & Allen, J. R. 2005. *Basin analysis,* 2nd edn. Oxford: Blackwells.

Allen, P. A., & Densmore, A. L. 2000. Sediment flux from an uplifting fault block. *Basin Research* 12, 367–380.

Allen, P. A., & Homewood, P. 1984. Evolution and mechanics of a Miocene tidal sand wave. *Sedimentology* 31, 63–81.

Alley, R. B. 1991. Deforming-bed origin for southern Laurentide till sheets. *Journal of Glaciology* 37, 67–76.

Alley, R. B., Lawson, D. E., Evenson, E. B., Larson, G. J., & Baker, G. S. 2003. Stabilizing feedbacks in glacier bed erosion. *Nature* 424, 758–760.

Alley, R. B., Lawson, D. E., Evenson, E. B., Strasser, J. C., & Larson, G. J. 1998. Glaciohydraulic supercooling: a freeze-on mechanism to create stratified, debris-rich basal ice. 2. Theory. *Journal of Glaciology* 44, 563–569.

Alsharhan, A. S., & Kendall, C. G. St. C. 2003. Holocene coastal carbonates and evaporites of the southern Arabian Gulf and their ancient analogues. *Earth-Science Reviews* 61, 191–243.

American Society of Civil Engineers Task Force on Bed Forms in Alluvial Channels 1966. Nomenclature for bedforms in alluvial channels. *Journal of the Hydraulics Division ASCE,* 92, 51–64.

Amos, C. L., Li, M. Z., & Choung, K.-S. 1996. Storm-generated, hummocky stratification on the outer-Scotian shelf. *Geomarine Letters* 16, 85–94.

Anadón, P., Cabrera, L., & Kelts, K. 1991. *Lacustrine Facies Analysis.* Oxford: Blackwells.

Anbar, A. D., & Knoll, A. H. 2002. Proterozoic ocean chemistry and evolution: a bioinorganic bridge? *Science* 297, 1137–1192.

Anderson, J. B., & Ashley, G. M. 1991. *Glacial Marine Sedimentation: Palaeoclimatic Significance.* Boulder, CO: Geological Society of America.

Anderson, M. G., Walling, D. E., & Bates, P. D. 1996. *Floodplain Processes.* Chichester: Wiley.

Anderson, M. P. 1997. Characterization of geological heterogeneity. In *Subsurface Flow and Transport: A Stochastic Approach* (ed. G. Dagan & S. P. Neuman), pp. 23–43. Cambridge: Cambridge University Press.

Anderson, R. S. 1987. A theoretical model of aeolian impact ripples. *Sedimentology* 34, 943–956.

Anderson, R. S., & Bunas, K. L. 1993. Grain size segregation and stratigraphy in aeolian ripples modeled with a cellular automaton. *Nature* 365, 740–743.

Archer, A. W. 1998. Hierarchy of controls on cyclic rhythmite deposition: Carboniferous basins of eastern and mid-continental USA. In *Tidalites: Processes and Products* (ed. C. R. Alexander, R. A. Davis, & V. J. Henry), pp. 59–68. Tulsa, OK: SEPM (Society for Sedimentary Geology).

Arnott, R. W. C., & Hand, B. M. 1989. Bedforms, primary structures and grain fabric in the presence of sediment rain. *Journal of Sedimentary Petrology* 59, 1062–1069.

Arnott, R. W., & Southard, J. B. 1990. Experimental study of combined-flow bed configurations in fine sands, and some implications for stratification. *Journal of Sedimentary Petrology* 60, 211–219.

Aronson, J. L., & Hower, J. 1976. Mechanism of burial metamorphism of argillaceous sediment: 2, radiogenic argon evidence. *Geological Society of America Bulletin* 87, 738–744.

Arthur, M. A., Dean, W. E., & Stow, D. A. V. 1984. Models for the deposition of Mesozoic–Cenozoic fine-grained organic-car-bon-rich sediment in the deep sea. In *Fine-Grained Sediments: Deep-Water Processes and Facies* (ed. D. A. V. Stow & D. J. W. Piper), pp. 527–560. London: Geological Society of London.

Arthur, M. A., & Sageman, B. B. 1994. Marine black shales: depositional mechanisms and environments of ancient deposits. *Annual Review of Earth and Planetary Sciences* 22, 499–551.

Ashley, G. M. 1988. The hydrodynamics and sedimentology of a back-barrier lagoon–salt marsh system, Great Sound, New Jersey. *Marine Geology* **82**, 1–132.

1990. Classification of large-scale subaqueous bed-forms: a new look at an old problem. *Journal of Sedimentary Petrology* **60**, 160–172.

2002. Glaciolacustrine environments. In *Modern and Past Glacial Environments* (ed. J. Menzies), pp. 335–359. Oxford: Butterworth-Heinemann.

Ashley, G. M., Shaw, J., & Smith, N. D. 1985. *Glacial Sedimentary Environments*. Tulsa, OK: SEPM (Society for Sedimentary Geology).

Ashley, G. M., Wellner, R. W., Esker, D., & Sheridan, R. E. 1991. Clastic sequences developed during late Quaternary glacio-eustatic sea-level fluctuations on a passive margin: example from the inner continental shelf near Barnegat Inlet, New Jersey. *Geological Society of America Bulletin* **103**, 1607–1621.

Ashworth, P. J., Best, J. L., & Jones, M. 2004. Relationship between sediment supply and avulsion frequency in braided rivers. *Geology* **32**, 21–24.

Aslan, A., & Blum, M. D. 1999. Contrasting styles of Holocene avulsion, Texas Gulf Coastal Plain, USA. In *Fluvial Sedimentology VI* (ed. N. D. Smith & J. Rogers), pp. 193–209 J. Oxford: Blackwells.

Assereto, R., & Folk, R. L. 1980. Diagenetic fabrics of aragonite, calcite, and dolomite in an ancient peritidal-spelean environment: Triassic Calcare Rosso, Lombardia, Italy. *Journal of Sedimentary Petrology* **50**, 371–394.

Atwater, B. F., & Hemphill-Haley, E. 1997. Recurrence intervals for great earthquakes of the past 3,500 years at northeastern Willapa Bay, Washington. United States Geological Survey Professional Paper 1576.

Atwater, B. F., Musumi-Rokaku, S., Satake, K. *et al.* 2005. The Orphan Tsunami of 1700: Japanese clues to a parent earthquake in North America. United States Geological Survey Professional Paper 1707.

Augustinius, P. G. E. F. 1989. Cheniers and chenier plains: a general introduction. *Marine Geology* **90**, 219–229.

Augustinius, P. G. E. F., Hazelhoff, L., & Kroon, A. 1989. The chenier coast of Suriname; modern and geological development. *Marine Geology* **90**, 269–281.

Ausich, W. I., & Bottjer, D. J. 1982. Tiering in suspension-feeding communities on soft substrata throughout the Phanerozoic. *Science* **216**, 173–174.

1985. Phanerozoic tiering in suspension-feeding communities on soft substrata: implications for diversity. In *Phanerozoic Diversity Patterns* (ed. J. W. Valentine), pp. 255–274. New Jersey: Princeton University Press.

Autin, W. J., Burns, S. F., Miller, R. T., Saucier, R. T., & Snead, J. I. 1991. Quaternary geology of the lower Mississippi valley. In *Quaternary Nonglacial Geology: Conterminous U.S. The Geology of North America*, pp. 547–582. Boulder, CO: Geological Society of America.

Baas, J. H., van Kesteren, W., & Postma, G. 2004. Deposits of depletive high-density turbidity currents: a flume analogue of bed geometry, structure and texture. *Sedimentology* **51**, 1053–1088.

Badiozamani, K. 1973. The dorag dolomitization model – application to the Middle Ordovician of Wisconsin. *Journal of Sedimentary Petrology* **43**, 965–984.

Baeuerlein, E. 2004. *Biomineralization: Progress in Biology, Molecular Biology and Application*. Weinheim: Wiley-VCH.

Bagnold, R. A. 1941. *The Physics of Blown Sand and Desert Dunes*. London: Methuen.

1946. Motion of waves in shallow water. Interaction between waves and sand bottoms. *Proceedings of the Royal Society of London* **187A**, 1–16.

1954. Experiments on the gravity-free dispersion of large solid spheres in a Newtonian fluid under shear. *Proceedings of the Royal Society of London* **187A**, 49–63.

1956. The flow of cohesionless grains in fluids. *Philosophical Transactions of the Royal Society* **249A**, 235–297.

1960. Some aspects of the shape of river meanders. United States Geological Survey Professional Paper 282E, pp. 135–144.

1962. Auto-suspension of transported sediment: turbidity currents. *Philosophical Transactions of the Royal Society* **265A**, 315–319.

1963. Mechanics of marine sedimentation. In *The Sea: Volume 3* (ed. M. N. Hill), pp. 507–528. New York: Wiley.

1966. An approach to the sediment transport problem from general physics. United States Geological Survey Professional Paper 442-I.

1973. The nature of saltation and of "bed-load" transport in water. *Philosophical Transactions of the Royal Society* **332A**, 473–504.

Baker, V. R. 1973. Palaeohydrology and sedimentology of Lake Missoula flooding in eastern Washington. Geological Society of America Special Paper 144.

1990. Geological fluvial geomorphology. *Geological Society of America Bulletin* **100**, 1157–1167.

Baker, V. R., Kochel, R. C., & Patton, P. C. 1988. *Flood Geomorphology*. New York: Wiley.

Baker, V. R., & Nummedal, D. 1978. *The Channeled Scabland*. Washington, D.C.: NASA Planetary Geology Program.

Baldwin, B., & Butler, C. O. 1985. Compaction curves. *American Association of Petroleum Geologists Bulletin* **69**, 622–626.

Ball, M. M. 1967. Carbonate sand bodies of Florida and the Bahamas. *Journal of Sedimentary Petrology* **37**, 556–591.

Banerjee, I., & McDonald, B. C. 1975. Nature of esker sedimentation. In *Glaciofluvial and Glaciolacustrine Sediments* (ed. A. V. Jopling & B. C. McDonald), pp. 132–154. Tulsa, OK: SEPM (Society for Sedimentary Geology).

Barnes, M. A., Barnes, W. C., & Bustin, R. M. 1990. Chemistry and diagenesis of organic matter in sediments and fossil fuels. In *Diagenesis* (ed. I. A. McIlreath & D. W. Morrow), pp. 189–204. St. John's, Newfoundland: Geological Association of Canada.

Barnes, R. S. K., & Hughes, R. N. 1982. *An Introduction to Marine Ecology*. Oxford: Blackwells.

Barry, J. M. 1997. *Rising Tide: The Great Mississippi Flood of 1927 and How It Changed America*. New York: Simon and Schuster.

Barwis, J. H. 1978. Sedimentology of some South Carolina tidal creek point bars, and a comparison with their fluvial counterparts. In *Fluvial Sedimentology* (ed. A. D. Miall), pp. 129–160. Calgary, Alberta: Canadian Society of Petroleum Geologists.

Bates, C. C. 1953. Rational theory of delta formation. *American Association of Petroleum Geologists Bulletin* **37**, 2119–2162.

Bathurst, R. G. C. 1975. *Carbonate Sediments and Their Diagenesis*, 2nd edn. Amsterdam: Elsevier.

1980. Lithification of carbonate sediments. *Science Progress Oxford* **66**, 451–471.

1987. Diagenetically enhanced bedding in argillaceous platform limestones: stratified cementation and selective compaction. *Sedimentology* **34**, 749–778.

Batten, K. L., Narbonne, G. M., & James, N. P. 2004. Paleoenvironments and growth of early Neoproterozoic calcimicrobial reefs: platformal Little Dal Group, northwestern Canada. *Precambrian Research* **133**, 249–269.

Bear, J. 1972. *Dynamics of Fluids in Porous Media*. New York: Dover Publications.

Beard, B. L., Johnson, C. M., Cox, L. *et al.* 1999. Iron isotope biosignatures. *Science* **285**, 1889–1892.

Beaumont, C. 1981. Foreland basins. *Geophysical Journal of the Royal Astronomical Society* **65**, 291–329.

Beaumont, C., Fullsack, P., & Hamilton, J. 1992. Erosional control of active compressional orogens. In *Thrust Tectonics* (ed. K. R. McClay), pp. 1–18. London: Chapman and Hall.

Beaumont, C., Quinlan, G., & Hamilton, J. 1988. Orogeny and stratigraphy: numerical models of the Paleozoic in the eastern interior of North America. *Tectonics* **7**, 389–416.

Behrensmeyer, A. K. 1987. Miocene fluvial facies and vertebrate taphonomy in northern Pakistan. In *Recent Developments in Fluvial Sedimentology* (ed. F. G. Ethridge, R. M. Flores, & M. D. Harvey), pp. 169–176. Tulsa, OK: SEPM (Society for Sedimentary Geology).

Behrensmeyer, A. K., Damuth, J. D., DiMichele, W. A., Potts, R., Sues, H. -D., & Wing, S. L. 1992. *Terrestrial Ecosystems Through Time*. Chicago: University of Chicago Press.

Behrensmeyer, A. K., & Hook, R. W. 1992. Paleoenvironmental contexts and taphonomic modes. In *Terrestrial Ecosystems Through Time* (ed. A. K. Behrensmeyer, J. D. Damuth, W. A. DiMichele *et al.*), pp. 15–136. Chicago: University of Chicago Press.

Bekker, A., Holland, H. D., Wang. P. -L., Rumble, D. III. *et al.* 2004. Dating the rise of atmospheric oxygen. *Nature* **427**, 117–120.

Belderson, R. H., Johnson, M. A., & Kenyon, N. H. 1982. Bedforms. In *Offshore Tidal Sands* (ed. A. H. Stride), pp. 25–57. London: Chapman and Hall.

Bengtson, S. 1999. *Early Life on Earth: Nobel Symposium No. 84*. New York: Columbia University Press.

Benito, G., Baker, V. R., & Gregory, K. J. 1998. *Palaeohydrology and Environmental Change*. Chichester: Wiley.

Benjamin, T. B. 1968. Gravity currents and related phenomena. *Journal of Fluid Mechanics* **31**, 209–248.

Benn, D. I. 1994. Fluted moraine and till genesis below a temperate valley glacier: Slettmarkbreen, Jotunheimen, southern Norway. *Sedimentology* **41**, 279–292.

Benn, D. I., & Evans, D. J. A. 1998. *Glaciers and Glaciation*. London: Arnold.

2006. Subglacial megafoods: outrageous hypothesis or just outrageous? In *Glacier Science and Environmental Change* (ed. P. G. Knight), pp. 42–50. Oxford: Blackwells.

Bennett, R. H., Bryant, W. R., & Hulbert, M. H. 1991. *Microstructure of Fine-Grained Sediments: From Mud to Shale*. New York: Springer-Verlag.

Bennett, S. J., & Best, J. L. 1995. Mean flow and turbulence structure over fixed, two-dimensional dunes: implications for sediment transport and dune stability. *Sedimentology* **42**, 491–514.

1996. Mean flow and turbulence structure over fixed ripples and the ripple–dune transition. In *Coherent Flow Structures in Open Channels* (ed. P. J. Ashworth, S. J. Bennett, J. L. Best, & S. J. McLelland), pp. 281–304. Chichester: Wiley.

Bennett, S. J., & Bridge J. S. 1995. An experimental study of flow, bedload transport and bed topography under conditions of erosion and deposition and comparison with theoretical models. *Sedimentology* **42**, 117–146.

Bennett, S. J., Bridge, J. S., & Best, J. L. 1998. The fluid and sediment dynamics of upper-stage plane beds. *Journal of Geophysical Research* **103**, 1239–1274.

Bennett, S. J., & Simon, A. 2004. *Riparian Vegetation and Fluvial Geomorphology*. Washington, D.C.: American Geophysical Union.

Berelson, W. M., & Heron, S. D. 1985. Correlations between Holocene flood tidal delta and barrier island inlet sequences: Back Sound–Shackleford Banks, North Carolina. *Sedimentology* **32**, 215–222.

Berendsen, H. J. A., & Stouthamer, E. 2001. *Paleogeographic Development of the Rhine–Meuse Delta, the Netherlands.* Assen: Koninklijke Van Gorcum.

Berger, A., & Loutre, M.-F. 1991. Insolation values for the climate of the last 10 million years. *Quaternary Science Reviews* **10**, 297–317.

Bergman, K. M., & Snedden, J. W. 1999. *Isolated Shallow Marine Sand Bodies: Sequence Stratigraphic Analysis and Sedimentological Interpretation.* Tulsa, OK: SEPM (Society for Sedimentary Geology).

Bernard, H. A., LeBlanc, R. J., & Major, C. F. Jr. 1962. Recent and Pleistocene geology of southwest Texas, field excursion 3. *Geology of the Gulf Coast and Central Texas and Guidebook of Excursion*, pp. 175–225. Houston: Houston Geological Society.

Berné, S., Auffret, J. P., & Walker, P. 1988. Internal structures of subtidal sandwaves revealed by high-resolution seismic reflection. *Sedimentology* **35**, 5–20.

Berné, S., Castaing, P., LeDrezen, E., & Lericolais, G. 1993. Morphology, internal structure and reversal of asymmetry of large subtidal dunes in the entrance to Gironde estuary (France). *Journal of Sedimentary Petrology* **63**, 780–793.

Berné, S., Lericolais, G., Marsset, T., Bourillet, J., & De Batist, M. 1998. Erosional offshore sand ridges and low-stand shorelines: examples from tide- and wave-dominated environments of France. *Journal of Sedimentary Research* **68**, 540–555.

Berner, R. A. 1980. *Early Diagenesis: A Theoretical Approach.* Princeton, New Jersey: Princeton University Press.

1991. A model for atmospheric CO_2 over Phanerozoic time. *American Journal of Science* **291**, 339–376.

1994. GEOCARB II: A revised model of atmospheric CO_2 over Phanerozoic time. *American Journal of Science* **294**, 56–91.

2001. The effect of the rise of land plants on atmospheric CO_2 during the Paleozoic. In *Plants Invade the Land* (ed. P. G. Gensel & D. Edwards), pp. 173–178. New York: Columbia University Press.

Berner, E. K., & Berner, R. A. 1996. *Global Environment: Water, Air, and Geochemical Cycles.* Upper Saddle River, NJ: Prentice Hall.

Berner, R. A., & Holdren, G. R., Jr. 1977. Mechanisms of feldspar weathering: some observational evidence. *Geology* **5**, 369–372.

1979. Mechanism of feldspar weathering: II. Observations of feldspars from soils. *Geochimica et Cosmochimica Acta* **43**, 1173–1186.

Berner, R. A., Holdern, G. R. Jr., & Schott, J. 1985. Surface layers on dissolving silicates. *Geochimica et Cosmochimica Acta* **49**, 1657–1658.

Berner, R. A., & Kothavala, Z. 2001. GEOCARB III: a revised model of atmospheric CO_2 over Phanerozoic time: *American Journal of Science* **301**, 182–204.

Best, J. L. 1992a. On the entrainment of sediment and initiation of bed defects: insights from recent developments within turbulent boundary layer research. *Sedimentology* **39**, 797–811.

1992b. Sedimentology and event timing of a catastrophic volcaniclastic mass flow, Volcan Hudson, Southern Chile. *Bulletin of Volcanology* **54**, 299–318.

1993. On the interactions between turbulent flow structure, sediment transport and bedform development: some considerations from recent experimental research. In *Turbulence: Perspectives on Flow and Sediment Transport* (ed. N. J. Clifford, J. R. French, & J. Hardisty), pp. 61–92. Chichester: Wiley.

1996. The fluid dynamics of small-scale alluvial bedforms. In *Advances in Fluvial Dynamics and Stratigraphy* (ed. P. A. Carling & M. R. Dawson), pp. 67–125. Chichester: Wiley.

2005. The fluid dynamics of river dunes: a review and some future research directions, *Journal of Geophysical Research*, **110**, F04S02, doi:10.1029/2004JF000218.

Best, J. L., & Ashworth, P. J. 1997. Scour in large braided rivers and the recognition of sequence stratigraphic boundaries. *Nature* **387**, 275–277.

Bhattacharya, J. P., & Giosan, L. 2003. Wave-influenced deltas: geomorphological implications for facies reconstruction. *Sedimentology* **50**, 187–210.

Bhattacharya, J. P., & Walker, R. G. 1992. *Deltas. In Facies Models: Response to Sea Level Change* (ed. R. G. Walker & N. P. James), pp. 157–177. Ottawa: Geological Association of Canada.

Birkeland, C. 1997. *Life and Death of Coral Reefs.* New York: Chapman and Hall.

Birkeland, P. W. 1999. *Soils and Geomorphology*, 3rd edn. New York: Oxford University Press.

Bierkens, M. F. P., & Weerts, H. J. T. 1994. Application of indicator simulation to modeling the lithological properties of a complex confining layer. *Geodema* **62**, 265–284.

Bjørlykke, K. 2003. Compaction (consolidation) of sediments. In *Encyclopedia of Sediments and Sedimentary Rocks* (ed. G. V. Middleton), pp. 161–168. Dordrecht: Kluwer Academic Publishers.

Bjørkum, P. A. 1996. How important is pressure in causing dissolution of quartz in sandstones? *Journal of Sedimentary Petrology* **66**, 147–154.

Bjornsson, H. 1992. Jokulhlaups in Iceland: prediction, characteristics, and simulation. *Annals of Glaciology* **16**, 95–106.

Black, K. S., Paterson, D. M., & Cramp, A. 1998. *Sedimentary Processes in the Intertidal Zone*. London: Geological Society of London.

Black, K. S., Tolhurst, T. J., Paterson, D. M., & Hagerthey, S. E. 2002. Working with natural cohesive sediments. *Journal of Hydraulic Engineering ASCE* **128**, 2–8.

Black, M. 1933. The algal sediments of Andros Island, Bahamas. *Philosophical Transactions of the Royal Society* **122B**, 165–191.

Blakey, R. C., Havholm, K. G., & Jones, L. S. 1996. Stratigraphic analysis of eolian interactions with marine and fluvial deposits, Middle Jurassic Page Sandstone and Carmel Formation, Colorado Plateau, USA. *Journal of Sedimentary Research* **66**, 324–342.

Blatt, H. 1982. *Sedimentary Petrology*. San Francisco: W. H. Freeman.

Blatt, H., Middleton, G., & Murray, R. 1980. *Origin of Sedimentary Rocks*, 2nd edn. Englewood Cliffs, NJ: Prentice-Hall.

Blum, M. D., & Price, D. M. 1998. Quaternary alluvial plain construction in response to glacio-eustatic and climatic controls, Texas Gulf Coastal Plain. In *Relative Role of Eustasy, Climate and Tectonism on Continental Rocks* (ed. K. J. Shanley & P. J. McCabe), pp. 31–48. Tulsa, OK: SEPM (Society for Sedimentary Geology).

Blum, M. D., & Törnqvist, T. E. 2000. Fluvial responses to climate and sea-level change: a review and look forward. *Sedimentology* **47** (Supplement 1), 2–48.

Boardman, R. S., Cheetham, A. H., & Rowell, A. J. 1987. *Fossil Invertebrates*. Oxford: Blackwell Scientific.

Boersma, J. R. 1970. Distinguishing features of wave-ripple cross-stratification and morphology. Unpublished Ph.D. thesis, Utrecht University, the Netherlands.

Boersma, J. R., & Terwindt, J. H. J. 1981. Neap–spring tide sequences of intertidal shoal deposits in a mesotidal estuary. *Sedimentology* **28**, 151–170.

Bohacs, K., & Suter, J. 1997. Sequence stratigraphic distribution of coaly rocks: fundamental controls and paralic examples. *American Association of Petroleum Geologists Bulletin* **81**, 1612–1639.

Boles, J. R., & Franks, S. G. 1979. Clay diagenesis in Wilcox sandstones of southwest Texas: implications of smectite diagenesis on sandstone cementation. *Journal of Sedimentary Petrology* **49**, 55–70.

Bond, G. C., & Lotti, R. 1995. Iceberg discharges into the North Atlantic on millennial time scales during the last glaciation. *Science* **267**, 1005–1010.

Bondevik, S., Mangerud, J., Dawson, S., Dawson, A., & Lohne, O. 2005. Evidence for three North Sea tsunamis at the Shetland Islands between 8000 and 1500 years ago. *Quaternary Science Reviews* **24**, 1757–1775.

Boothroyd, J. C. 1985. Tidal inlets and tidal deltas. In *Coastal Sedimentary Environments* (ed. R. A. Davis), pp. 445–533. New York: Springer-Verlag.

Boothroyd, J. C., Friedrich, N. E., & McGinn, S. R. 1985. Geology of microtidal coastal lagoons: Rhode Island. *Marine Geology* **63**, 35–76.

Borchert, H., & Muir, R. O. 1964. *Salt Deposits; The Origin, Metamorphism and Deformation of Evaporites*. Princeton, NJ: Van Nostrand.

Bosence, D. W. J. 1989. Biogenic carbonate production in Florida Bay. *Bulletin of Marine Science* **44**, 419–433.

Bosence, D. W. J., & Allison, P. A. 1995. *Marine Paleoenvironmental Analysis from Fossils*. London: Geological Society of London.

Bosence, D. W. J., Rowlands, R. J., & Quine, M. L. 1985. Sedimentology and budget of a Recent carbonate mound, Florida Keys. *Sedimentology* **32**, 3137–343.

Bosscher, H., & Schlager, W. 1992. Computer simulation of reef growth. *Sedimentology* **39**, 503–512.

Bossellini, A. 1984. Progradation geometries of carbonate platforms: examples from the Triassic of the Dolomites, northern Italy. *Sedimentology* **31**, 1–24.

Boucot, A. J., & Gray, J., 2001. A critique of Phanerozoic climatic models involving changes in the CO_2 content of the atmosphere. *Earth-Science Reviews* **56**, 1–159.

Boulton, G. S. 1972. Modern Arctic glaciers as depositional models for former ice sheets. *Quarterly Journal of the Geological Society of London* **128**, 361–393.

1974. Processes and patterns of glacial erosion. In *Glacial Geomorphology* (ed. D. R. Coates), pp. 41–87. Binghamton, NY: Publications in Geomorphology.

1976. The origin of glacially fluted surfaces: observations and theory. *Journal of Glaciology* **17**, 287–309.

1979. Processes of glacier erosion on different substrata. *Journal of Glaciology* **23**, 15–37.

1987. A theory of drumlin formation by subglacial sediment deformation. In *Drumlin Symposium* (ed. J. Menzies & J. Rose), pp. 25–80. Rotterdam: Balkema.

1990. Sedimentary and sea level changes during glacial cycles and their control on glacimarine facies architecture. In *Glacimarine Environments* (ed. J. A. Dowdeswell & J. D. Scourse), pp. 15–52. London: Geological Society of London.

1996a. Theory of glacial erosion, transport and deposition as a consequence of subglacial sediment deformation. *Journal of Glaciology* **42**, 43–62.

1996b. The origin of till sequences by subglacial sediment deformation beneath mid-latitude ice sheets. *Annals of Glaciology* **22**, 75–84.

2006. Glaciers and their coupling with hydraulic and sedimentary processes. In *Glacier Science and*

Environmental Change (ed. P. G. Knight), pp. 3–22. Oxford: Blackwells.

Boulton, G. S., & Hindmarsh, R. C. A. 1987. Sediment deformation beneath glaciers: rheology and geological consequences. *Journal of Geophysical Research* **92B**, 9059–9082.

Boulton, G. S., Dobbie, K. E., & Zatsepin, S. 2001. Sediment deformation beneath glaciers and its coupling to the subglacial hydraulic system. *Quaternary International* **86**, 3–28.

Bouma, A. H. 1962. *Sedimentology of Some Flysch Deposits*. Amsterdam: Elsevier.

Bouma, A. H., Coleman, J. R., Stetling, C. E., & Kohl, B. 1989. Influence of relative sea level on the construction of the Mississippi Fan. *Geo-Marine Letters* **9**, 161–170.

Bouma, A. H., Normark, W. R., & Barnes, N. E. 1985a. *Submarine Fans and Related Turbidite Systems*. New York: Springer-Verlag.

Bouma, A. H., Stetling, C. E., & Coleman, J. M. 1985b. Mississippi Fan, Gulf of Mexico. In *Submarine Fans and Related Turbidite Systems* (ed. A. H. Bouma, W. R. Normark, & N. E. Barnes), pp. 143–150. New York: Springer-Verlag.

Bourgeois, J., Hansen, T. A., Wiberg, P. L., & Kauffman, E. G. 1988. A tsunami deposit at the Cretaceous–Tertiary boundary in Texas. *Science* **241**, 567–570.

Bowden, K. F. 1983. *Physical Oceanography of Coastal Waters*. Chichester: Ellis Horwood.

Bowen, A. J. 1969. Rip currents, 1: theoretical investigations. *Journal of Geophysical Research* **74**, 5467–5478.

Bowen, A. J., & Inman, D. L. 1969. Rip currents, 2: laboratory and field observations. *Journal of Geophysical Research* **74**, 5479–5490.

Bowen, A. J., Normark, W. R., & Piper, D. J. W. 1984. Modelling of turbidity currents of Navy Submarine Fan, California Continental Borderland. *Sedimentology* **31**, 169–185.

Boyce, J. I., & Eyles, N. 1991. Drumlins carved by deforming till streams below the Laurentide Ice Sheet. *Geology* **19**, 787–790.

Boyd, R., Dalrymple, R. W., & Zaitlin, B. A. 1992. Classification of clastic coastal depositional environments. *Sedimentary Geology* **80**, 139–150.

Boyd, R., Suter, J., & Penland, S. 1989. Sequence stratigraphy of the Mississippi delta. *Transactions of the Gulf Coast Association of Geological Societies* **39**, 331–340.

Bradley, R. S. 1999. *Paleoclimatology*. San Diego, CA: Academic Press.

Brady, N. C., & Weil, R. R. 2002. *The Nature and Properties of Soils*, 13th edn. Upper Saddle River, NJ: Prentice-Hall.

Braitsch, O. 1971. *Salt Deposits; Their Origin and Composition*. New York: Springer-Verlag.

Branney, M., & Kokelaar, P. 2003. *Pyroclastic Density Currents and the Sedimentation of Ignimbrites*. London: Geological Society of London.

Brayshaw, A. C., Davies, G. W., & Corbett, P. W. M. 1996. Depositional controls on primary permeability and porosity at the bedform scale in fluvial reservoir sandstones. In *Advances in Fluvial Dynamics and Stratigraphy* (ed. P. A. Carling & M. R. Dawson), pp. 374–394. Chichester: Wiley.

Brett, C. E., & Baird, G. C. 1985. Carbonate-shale cycles in the Middle Devonian of New York: an evaluation of models for the origin of limestones in terrigenous shelf sequences. *Geology* **13**, 324–327.

Brewer, R. 1976. *Fabric and Mineral Analysis of Soils*, 2nd edn. New York: Krieger.

Bricker, O. P. 1971. *Carbonate Cements*. Baltimore, MA: The Johns Hopkins University Press.

Bridge, J. S. 1978. Paleohydraulic interpretation using mathematical models of contemporary flow and sedimentation in meandering channels. In *Fluvial Sedimentology* (ed. A. D. Miall), pp. 723–742. Calgary, Alberta: Canadian Society of Petroleum Geologists.

1984. Large-scale facies sequences in alluvial overbank environments. *Journal of Sedimentary Petrology* **54**, 583–588.

1985. Paleochannel patterns inferred from alluvial deposits: a critical evaluation. *Journal of Sedimentary Petrology* **55**, 579–589.

1992. A revised model for water flow, sediment transport, bed topography and grain size sorting in natural river bends. *Water Resources Research* **28**, 999–1023.

1993a. The interaction between channel geometry, water flow, sediment transport and deposition in braided rivers. In *Braided Rivers* (ed. J. L. Best & C. S. Bristow), pp. 13–72. London: Geological Society of London.

1993b. Description and interpretation of fluvial deposits: a critical perspective. *Sedimentology* **40**, 801–810.

1997. Thickness of sets of cross strata and planar strata as a function of formative bed-wave geometry and migration, and aggradation rate. *Geology* **25**, 971–974.

1999. Alluvial architecture of the Mississippi Valley: predictions using a 3D simulation model. In *Floodplains: Interdisciplinary Approaches* (ed. S. B. Marriott & J. Alexander), pp. 269–278. London: Geological Society of London.

2000. The geometry, flow and sedimentary processes of Catskill rivers and coasts. In *New Perspectives on the Old Red Sandstone* (ed. P. F. Friend & B. P. J. Williams), pp. 85–108. London: Geological Society of London.

2003. *Rivers and Floodplains: Forms, Processes, and Sedimentary Record*. Oxford: Blackwells.

2006. *Fluvial Facies Models*. In *Facies Models Revisited* (ed. H. W. Posamentier and R. G. Walker), pp. 85–170. Tulsa, OK: SEPM (Society for Sedimentary Geology).

2007. Numerical modelling of alluvial deposits: recent developments. In *Analogue and Numerical Forward Modelling of Sedimentary Systems; from Understanding to Prediction* (ed. P. L. de Boer, G. Postima, P. Kukla, K. van des Zwan, & P. Burgess). Oxford: Blackwells.

Bridge, J. S., & Bennett, S. J. 1992. A model for the entrainment and transport of sediment grains of mixed sizes, shapes and densities. *Water Resource Research* **28**, 337–363.

Bridge, J. S., & Best, J. L. 1988. Flow, sediment transport and bedform dynamics over the transition from upper-stage plane beds: implications for the formation of planar laminae. *Sedimentology* **35**, 753–763.

Bridge, J. S., & Gabel, S. L. 1992. Flow and sediment dynamics in a low sinuosity, braided river: Calamus River, Nebraska Sandhills. *Sedimentology*, **39**, 125–142.

Bridge, J. S. & Karssenberg, D. 2005. Simulation of flow and sedimentary processes, including channel bifurcation and avulsion, on alluvial fans. *8th International Conference on Fluvial Sedimentology*, Delft, The Netherlands, August 7–12, 2005.

Bridge, J. S., & Leeder, M. R. 1979. A simulation model of alluvial stratigraphy. *Sedimentology* **26**, 617–644.

Bridge, J. S., & Lunt, I. A. 2006. Depositional models of braided rivers. In *Braided Rivers II* (ed. G. H. Sambrook Smith, J. L. Best, C. S. Bristow, & G. Petts), pp. 11–50. Oxford: Blackwells.

Bridge, J. S., & Mackey, S. D. 1993a. A theoretical study of fluvial sandstone body dimensions. In *Geological Modeling of Hydrocarbon Reservoirs* (ed. S. S. Flint & I. D. Bryant), pp. 213–236. Oxford: Blackwells.

1993b. A revised alluvial stratigraphy model. In *Alluvial Sedimentation* (ed. M. Marzo & C. Puidefabregas), pp. 319–337. Oxford: Blackwells.

Bridge, J. S., & Tye, R. S. 2000. Interpreting the dimensions of ancient fluvial channel bars, channels, and channel belts from wireline-logs and cores. *American Association of Petroleum Geologists Bulletin* **84**, 1205–1228.

Bridge, J. S., & Willis, B. J. 1994. Marine transgressions and regressions recorded in Middle Devonian shore-zone deposits of the Catskill clastic wedge. *Geological Society of America Bulletin* **106**, 1440–1458.

Bridges, P. H., & Leeder, M. R. 1976. Sedimentary model for intertidal mudflat channels with examples from the Solway Firth, Scotland. *Sedimentology* **23**, 533–552.

Brierley, G. J., Ferguson, R. J., & Woolfe, K. J. 1997. What is a fluvial levee? *Sedimentary Geology* **114**, 1–9.

Bristow, C. S., Bailey, S. D., & Lancaster, N. 2000. The sedimentary structures of linear sand dunes. *Nature* **406**, 56–59.

Bristow, C. S., Lancaster, N., & Duller, G. A. T. 2005. Combining ground penetrating radar surveys and optical dating to determine dune migration in Namibia. *Journal of the Geological Society of London* **162**, 315–321.

Bristow, C. S., Pugh, J., & Goodall, T. 1996. Internal structure of aeolian dunes in Abu Dhabi determined using ground-penetrating radar. *Sedimentology* **43**, 995–1003.

Bristow, C. S., Skelly, R. L., & Ethridge, F. G. 1999. Crevasse splays from the rapidly aggrading, sand-bed, braided Niobrara River, Nebraska: effect of base-level rise. *Sedimentology*, **46**, 1029–1047.

Brodzikowski, K., & Van Loon, A. J. 1991. *Glacigenic Sediments*. Amsterdam: Elsevier.

Broecker, W. A., & Denton, G. H. 1989. The role of ocean–atmosphere reorganizations in glacial cycles. *Geochimica et Cosmochimica Acta* **53**, 2465–2501.

Broecker, W. A., & Takahashi, T. 1966. Calcium carbonate production on the Bahama Bank. *Journal of Geophysical Research* **71**, 1575–1602.

Bromley, R. G. 1996. *Trace Fossils: Biology, Taphonomy and Applications*, 2nd edn. London: Chapman and Hall.

Brookfield, M. E. 1977. The origin of bounding surfaces in ancient aeolian sandstones. *Sedimentology* **24**, 303–332.

1992. Eolian Systems. In *Facies Models: Response to Sea Level Change* (ed. R. G. Walker & N. P. James), pp. 143–156. St. John's Newfoundland: Geological Association of Canada.

Brookfield, M. E., & Ahlbrandt, T. S. 1983. *Eolian Sediments and Processes*. Amsterdam: Elsevier.

Brown, T. M., & Kraus, M. J. 1987. Integration of channel floodplain suite, I. Developmental sequence and lateral relations of alluvial paleosols. *Journal of Sedimentary Petrology* **57**, 587–601.

Brown, T. M., & Ratcliffe, B. C. 1988. The origin of *Chubutolithes* lhering, ichnofossils for the Eocene and Oligocene of Chubut Province, Argentine. *Journal of Paleontology* **62**, 163–167.

Brunsden, D., & Prior, D. B. 1984. *Slope Instability*. New York: Wiley.

Bryant, E. 2001. *Tsunami: The Underrated Hazard*. Cambridge: Cambridge University Press.

Bryant, I. D., & Flint, S. 1993. Quantitative clastic reservoir geological modeling: problems and perspectives. In *Geological Modeling of Hydrocarbon Reservoirs* (ed. S. Flint & I. D. Bryant), pp. 3–20. Oxford: Blackwells.

Buatois, L. A., Mangano, M. G., Genise, J. F., & Taylor, T. N. 1998. The ichnologic record of the continental invertebrate invasion: evolutionary trends in environmental expansion, ecospace utilization, and behavioral complexity. *Palaios* **13**, 217–240.

Bull, W. B. 1991. *Geomorphic Responses to Climate Change*. New York: Oxford University Press.

Bullock, P., Fedoroff, N., Jungerius, A., Stoops, G., & Tursina, T. 1988. *Handbook of Soil Thin Section Description*. Albrighton: Waine Research.

Burbank, D. W. 1992. Causes of recent Himalayan uplift deduced from deposited patterns in the Ganges basin. *Nature* **357**, 680–683.

Burbank, D. W., & Anderson, R. S. 2001. *Tectonic Geomorphology*. Oxford: Blackwells.

Burley, S. D. 2003. Cathodoluminescence (applied to the study of sedimentary rocks). In *Encyclopedia of Sediments and Sedimentary Rocks* (ed. G. V. Middleton), pp. 102–106. Dordrecht: Kluwer Academic Publishers.

Burley, S. D., Kantorowicz, J. D., & Waugh, B. 1985. Clastic diagenesis. In *Sedimentology: Recent Developments and Applied Aspects* (ed. P. J. Brenchley & B. P. J. Williams), pp. 189–228. London, Blackwell Scientific Publications.

Burley, S. D., & Worden, R. H. 2003. *Sandstone Diagenesis: Recent and Ancient*. Malden, MA: Blackwells.

Burnett, W., & Riggs, S. R. 1990. *Phosphate Deposits of the World, Volume 3, Neogene to Modern Phosphorites*. Cambridge: Cambridge University Press.

Burst, J. F. 1965. Subaqueously formed shrinkage cracks in clay. *Journal of Sedimentary Petrology* **35**, 348–353.

Busby, C., & Ingersoll, R. V. 1995. *Tectonics of Sedimentary Basins*. Oxford: Blackwells.

Bustillo, M. A. 2003. Silcrete. In *Encyclopedia of Sediments and Sedimentary Rocks* (ed. G. V. Middleton), pp. 659–660. Dordrecht: Kluwer Academic Publishers.

Butler, G. P. 1969. Modern evaporite deposition and geochemistry of coexisting brines, the sabkha, Trucial Coast, Arabian Gulf. *Journal of Sedimentary Petrology* **39**, 70–89.

Campbell, C. S. 1990. Rapid granular flows. *Annual Review of Fluid Mechanics* **22**, 57–92.

Campbell, C. S. 2001. Granular flows in the elastic limit. In *Particulate Gravity Currents* (ed. W. D. McCaffrey, B. C. Kneller, & J. Peakall), pp. 83–89. Oxford: Blackwells.

Campbell, N. A., Reece, J. B., & Mitchell, L. G. 1999. *Biology*, 5th edn. Menlo Park, CA: Addison Wesley Longman.

Cao, S., & Knight, D. W. 1998. Design of hydraulic geometry of alluvial channels. *Journal of Hydraulic Engineering, ASCE* **124**, 484–492.

Carballo, J. D., Land, L. S., & Miser, D. E. 1987. Holocene dolomitization of supratidal sediments by active tidal pumping, Sugarloaf Key, Florida. *Journal of Sedimentary Petrology* **57**, 153–165.

Carle, S. F., Labolle, E. M., Weissmann, G. S. *et al.* 1998. Conditional simulation of hydrofacies architecture: a transition probability/Markov approach. In *Hydrogeologic Models of Sedimentary Aquifers* (ed. G. S. Fraser & J. M. Davis), pp. 147–170. Tulsa, OK: SEPM (Society for Sedimentary Geology).

Carling, P. A. 1981. Sediment transport by tidal currents and waves: observations from a sandy intertidal zone (Burry Inlet, South Wales). In *Holocene Marine Sedimentation in the North*

Sea Basin (ed. S. D. Nio, R. T. E. Schuttenhelm, & T. C. E. Van Weering), pp. 65–80. Oxford: Blackwells.

1999. Subaqueous gravel dunes. *Journal of Sedimentary Research* **69**, 534–545.

Carling, P. A., & Dawson, M. R. 1996. *Advances in Fluvial Dynamics and Stratigraphy*. Chichester: Wiley.

Carling, P. A. & Shvidchenko, A. B. 2002. A consideration of the dune : antidune transition in fine gravel. *Sedimentology*, **49**, 1,269–1,282.

Carpenter, A. B. 1978. Origin and chemical evolution of brines in sedimentary basins. In *Thirteenth Annual Forum on the Geology of Industrial Minerals* (ed. K. S. Johnson & J. A. Russell), pp. 60–77. Tulsa, OK: Oklahoma Geological Survey.

Carson, M. A., & Kirkby, M. J. 1972. *Hillslope Form and Process*. Cambridge: Cambridge University Press.

Carter, D. J. T. 1982. Prediction of wave height and period for a constant wind velocity using the JONSWAP results. *Ocean Engineering* **9**, 17–33.

Carter, L., & McCave, I. N. 1997. The sedimentary regime beneath the deep Western Boundary Current inflow to the southwest Pacific Ocean. *Journal of Sedimentary Research* **67**, 1,005–1,017.

Carter, R. M. 1975. A discussion and classification of subaqueous mass-transport with particular application to grain flow, slurry-flow, and fluxoturbidites. *Earth-Science Reviews* **11**, 145–177.

Carter, R. W. G., & Woodroffe, C. D. 1994. *Coastal Evolution: Late Quaternary Shoreline Morphodynamics*. Cambridge: Cambridge University Press.

Cas, R. A. F., & Wright, J. V. 1987. *Volcanic Successions: Modern and Ancient*. London: Allen & Unwin.

Cathles, L. M. 1993. A discussion of flow mechanisms responsible for alteration and mineralization in the Cambrian aquifers of the Ouachita–Arkoma Basin–Ozark system. In *Diagenesis and Basin Development* (ed. A. D. Horbury & A. G. Robinson), pp. 99–112. Tulsa, OK: American Association of Petroleum Geologists.

Cayeux, L. 1935. *Les roches sédimentaires de France: roches carbonatées*. Paris: Masson. [English translation by Carozzi, A. V. 1970. *Sedimentary Rocks of France: Carbonate Rocks*. Darien, CN: Hafner Publishing.]

Cerling, T. E. 1999. Stable carbon isotopes in paleosol carbonates. In *Palaeoweathering, Palaeosurfaces and Related Continental Deposits* (ed. M. Thiry & R. Simon-Coicon), pp. 43–60. Oxford: Blackwells.

Cerling, T. E., & Craig, H. 1994. Geomorphology and in-situ cosmogenic isotopes. *Annual Review of Earth and Planetary Sciences* **22**, 273–317.

Chafetz, H. S. 1986. Marine pisoids: a product of bacterially induced precipitation. *Journal of Sedimentary Petrology* **56**, 812–817.

Chafetz, H. S., & Folk, R. L. 1984. Travertines, depositional morphology and the bacterially constructed constituents. *Journal of Sedimentary Petrology* **54**, 289–316.

Chang, H. H. 1988. *Fluvial Processes in River Engineering*. New York: Wiley.

Chave, K. E. 1954. Aspects of the biogeochemistry of magnesium 1. Calcareous marine organisms. *Journal of Geology* **62**, 266–283.

Chave, K. E., Smith, S. V., & Roy, R. J. 1972. Carbonate production by coral reefs. *Marine Geology* **12**, 123–140.

Cheel, R. J., & Leckie, D. A. 1992. Coarse-grained storm beds of the Upper cretaceous Chungo Member (Wapiabi Formation), southern Alberta, Canada. *Journal of Sedimentary Petrology* **62**, 933–945.

Chester, R. 2003. *Marine Geochemistry*, 2nd edn. Oxford: Blackwells.

Chilingarian, G. V., & Wolf, K. H. 1988. *Diagenesis*. Amsterdam: Elsevier.

Choquette, P. W., & Pray, L. C. 1970. Geological nomenclature and classification of porosity in sedimentary carbonates. *American Association of Petroleum Geologists Bulletin* **54**, 207–250.

Christie-Blick, N., Sohl, L. E., & Kennedy, M. J. 1999. Considering a Neoproterozoic snowball Earth. *Science* **284**, 1087.

Chuhan, F. A., Bjørlykke, K., & Lowrey, C. J. 2000. The role of provenance in illitization of deeply buried reservoir sandstones from Haltenbanken and north Viking Graben, offshore Norway. *Marine and Petroleum Geology* **17**, 673–689.

2001. Closed-system burial diagenesis in reservoir sandstones: examples from the Garn Formation at Haltenbanken Area, offshore mid-Norway. *Journal of Sedimentary Petrology* **71**, 15–26.

Church, M. 1996. Channel morphology and typology. In *River Flows and Channel Forms* (ed. G. Petts & P. Calow), pp. 185–202. Oxford: Blackwell Science Limited.

Clark, C. D., & Meehan, R. T. 2001. Subglacial bedform morphology of the Irish Ice Sheet reveals major configuration changes during growth and decay. *Journal of Quaternary Science* **16**, 483–496.

Clark, J. D., & Pickering, K. T. 1996. *Submarine Channels: Processes and Architecture*. London: Vallis Press.

Clark, J. D., Kenyon, N. H., & Pickering, K. T. 1992. Quantitative analysis of the geometry of submarine channels: implications for the classification of submarine fans. *Geology* **20**, 633–636.

Clarke, G. K. C., Collins, S. G., & Thompson, D. E. 1984. Flow, thermal structure, and subglacial conditions of a surge-type glacier. *Canadian Journal of Earth Sciences* **21**, 232–240.

Clarkson, E. N. K. 1998. *Invertebrate Palaeontology and Evolution*, 4th edn. Oxford: Blackwell Scientific.

Clemmensen, L. B. 1989. Preservation of interdraa and draa plinth deposits by the lateral migration of large linear draas (Lower Permian Yellow Sands, northeast England). *Sedimentary Geology* **65**, 139–151.

Clemmensen, L. B., Olsen, H., & Blakey, R. C. 1989. Erg-margin deposits in the Lower Jurassic Moenave Formation and Wingate Sandstone, southern Utah. *Geological Society of America Bulletin* **101**, 759–773.

Clevis, Q., de Boer, P., & Nijman, W. 2004a. Differentiating the effect of episodic tectonism and eustatic sea-level fluctuations in foreland basins filled by alluvial fans and axial deltaic systems: insights from a three-dimensional stratigraphic forward model. *Sedimentology* **51**, 809–835.

Clevis, Q., de Boer, P., & Wachter, M. 2003. Numerical modeling of drainage basin evolution and three-dimensional alluvial fan stratigraphy. *Sedimentary Geology* **163**, 85–110.

Clevis, Q., de Jager, G., Nijman, W., & de Boer, P. L. 2004b. Stratigraphic signatures of translation of thrust-sheet top basins over low-angle detachment faults. *Basin Research* **16**, 145–163.

Clifton, H. E. 1969. Beach lamination: nature and origin. *Marine Geology* **7**, 553–559.

Clifton, H. E., & Dingler, J. R. 1984. Wave-formed structures and palaeoenvironmental reconstruction. *Marine Geology* **60**, 165–198.

Clifton, H. E., Hunter, R. E., & Phillips, R. L. 1971. Depositional structures and processes in the non-barred, high energy nearshore. *Journal of Sedimentary Petrology* **41**, 651–670.

Clout, J. M. F., & Simonson, B. M. 2005. Precambrian iron formations and iron formation-hosted iron ore deposits. *Economic Geology 100th Anniversary Issue*, 643–679.

Cohen, A. S., 2003. *Paleolimnology: The History and Evolution of Lake Systems*. Oxford: Oxford University Press.

Cole, J. M., Rasbury, E. T., Montañez, I. P. *et al.* 2005. Petrographic and trace element analysis of uranium-rich tufa calcite, middle Miocene Barstow Formation, California, USA. *Sedimentology* **51**, 433–453.

Coles, D. F. 1956. The law of the wake in turbulent boundary layer. *Journal of Fluid Mechanics* **1**, 191–226.

Colella, A., & Prior, D. B. 1990. *Coarse-Grained Deltas*. Oxford: Blackwells.

Coleman, J. M. 1969. Brahmaputra River: channel processes and sedimentation. *Sedimentary Geology* **3**, 129–239.

1988. Dynamic changes and processes in the Mississippi River delta. *Geological Society of America Bulletin* **100**, 999–1015.

Coleman, J. M., & Gagliano, S. M. 1964. Cyclic sedimentation in the Mississippi River deltaic plain. *Transactions of the Gulf Coast Association of Geological Societies* **14**, 67–80.

1965. Sedimentary structures: Mississippi River deltaic plain. In *Primary Sedimentary Structures and Their Hydrodynamic Interpretation* (ed. G. V. Middleton), pp. 133–148. Tulsa, OK: SEPM (Society for Sedimentary Geology).

Coleman, J. M., Gagliano, S. M., & Webb, J. E. 1964. Minor sedimentary structures in a prograding distributary. *Marine Geology* **1**, 240–258.

Coleman, J. M., & Prior, D. B. 1982. Deltaic environments. In *Sandstone Depositional Environments* (ed. P. A. Scholle & D. R. Spearing), pp. 139–178. Tulsa, OK: American Association of Petroleum Geologists.

Coleman, J. M., Prior, D. B., & Lindsay, J. F. 1983. Deltaic influences on shelfedge instability processes. In *The Shelfbreak: Critical Interface on Continental Margins* (ed. D. J. Stanley & G. T. Moore), pp. 121–137. Tulsa, OK: SEPM (Society for Sedimentary Geology).

Coleman, J. M., & Wright, L. D. 1975. Modern river deltas: variability of processes and sand bodies. In *Deltas, Models for Exploration* (ed. M. L. Broussard), pp. 99–149. Houston, TX: Houston Geological Society.

Collins, M. B., Amos, C. L., & Evans, G. 1981. Observations of some sediment transport processes over intertidal flats, The Wash, UK. In *Holocene Marine Sedimentation in the North Sea Basin* (ed. S. D. Nio, R. T. E. Schuttenhelm, & T. C. E. Van Weering), pp. 81–98. Oxford: Blackwells.

Coniglio, M., & Dix, G. R. 1992. Carbonate slopes. In *Facies Models, Response to Sea Level Changes* (ed. R. G. Walker & N. P. James), pp. 349–373. St. John's, Newfoundland: Geological Association of Canada.

Cook, H. E., Hine, A. C., & Mullins, H. T. 1983. *Platform Margin and Deep Water Carbonates*. Tulsa, OK: SEPM (Society for Sedimentary Geology).

Cook, H. E., & Mullins, H. T. 1983. Basin margin environments. In *Carbonate Depositional Environments* (ed. P. A. Scholle, D. G. Bebout, & C. H. Moore), pp. 540–617. Tulsa, OK: American Association of Petroleum Geologists.

Cook, P. J., & Shergold, J. H. 1986. *Phosphate Deposits of the World, Volume 1, Proterozoic and Cambrian Phosphorites*. Cambridge: Cambridge University Press.

Cooke, R. U., & Warren, A. 1973. *Geomorphology in Deserts*. London: Batsford.

Cooke, R. U., Warren, A., & Goudie, A. 1993. *Desert Geomorphology*. London: UCL Press.

Corney, R. K. T., Peakall, J., Parsons, D. R. *et al.* 2006. The orientation of helical flow in curved channels. *Sedimentology* **53**, 249–257.

Costa, J. E., & Wieczorek, G. F. 1987. *Debris Flows/ Avalanches: Process, Recognition, and Mitigation*. Boulder, CO: Geological Society of America.

Costa, J. E., & Williams, G. P. 1984. *Debris Flow Dynamics*. United States Geological Survey Open File Report 84–606

(VHS Videotape). Reston, VA: United States Geological Survey.

Cotter, E., & Link, J. E. 1993. Deposition and diagenesis of Clinton ironstones (Silurian) in the Appalachian Foreland basin of Pennsylvania. *Geological Society of America Bulletin* **105**, 911–922.

Coulthard, T. J., Macklin, M. G., & Kirkby, M. J. 2002. A cellular model of Holocene upland river basin and alluvial fan evolution. *Earth Surface Processes and Landforms* **27**, 269–288.

Crabaugh, M., & Kocurek, G. 1993. Entrada sandstone: an example of a wet aeolian system. In *The Dynamics and Environmental Context of Aeolian Sedimentary Systems* (ed. K. Pye), pp. 103–126. London: Geological Society of London.

Craft, J. H., & Bridge, J. S. 1987. Shallow-marine sedimentary processes in the Late Devonian Catskill Sea, New York State. *Geological Society of America Bulletin* **98**, 338–355.

Crawford, D. A., & Mader, C. L. 1998. Modeling asteroid impact and tsunami. *Science of Tsunami Hazards* **16**, 21–30.

Crevello, P. D., & Schlager, W. 1980. Carbonate debris sheets and turbidites, Exuma Sound, Bahamas. *Journal of Sedimentary Petrology* **50**, 1121–1148.

Crick, R. E. 1986. *Origin, Evolution, and Modern Aspects of Biomineralization in Plants and Animals*. New York: Plenum Press.

Cross, T. A. 1990. *Quantitative Dynamic Stratigraphy*. Englewood Cliffs, NJ: Prentice-Hall.

Crowley, T. J., & Berner, R. A. 2001. CO_2 and climate change. *Science* **289**, 270–277.

Crowley, T. J., & North, G. R. 1991. *Paleoclimatology*. New York: Oxford University Press.

Csanady, G. T. 1978. Water circulation and dispersal mechanisms. In *Lakes: Chemistry Geology and Physics* (ed. A. Lerman), pp. 21–64. New York: Springer-Verlag.

Curray, J. R., Emmel, F. J., & Crampton, P. J. S. 1969. Holocene history of a strand plain, lagoonal coast, Narayit, Mexico. In *Coastal Lagoons – A Symposium* (ed. A. A. Castanares & F. B. Phleger), pp. 63–100. Mexico City: Universidad Nacional Autónoma.

Dade, W. B., & Huppert, H. E. 1994. Predicting the geometry of channelized deep-sea turbidites. *Geology* **22**, 645–648.

Dade, W. B., Nowell, A. R. M., & Jumars, P. A. 1992. Predicting the erosion resistance of muds. *Marine Geology* **105**, 285–297.

Dalrymple, R. W. 1984. Morphology and internal structure of sand waves in the Bay of Fundy. *Sedimentology* **31**, 365–382.

1992. Tidal depositional systems. In *Facies Models: Response to Sea Level Change* (ed. R. G. Walker & N. P. James), pp. 195–218. St. John's, Newfoundland: Geological Association of Canada.

Dalrymple, R. W., Baker, E. K., Hughes, M., & Harris, P. T. 2003. Geomorphology and sedimentology of the muddy, tide-domonated, Fly River delta, Papua New Guinea. In *Tropical Deltas of South-East Asia – Sedimentology, Stratigraphy, and Petroleum Geology* (ed. F. H. Sidi, D. Nummedal, P. Imbert, H. Darman, & H. Posamentier), pp. 147–174. Tulsa, OK: SEPM (Society for Sedimentary Geology).

Dalrymple, R. W., Boyd, R., & Zaitlin, B. A. 1994. *Incised Valley Systems: Origin and Sedimentary Sequences.* Tulsa, OK: SEPM (Society for Sedimentary Geology).

Dalrymple, R. W., Knight, R. J., Zaitlin, B. A., & Middleton, G. V. 1990. Dynamics and facies model of a macrotidal sand-bar complex, Cobequid Bay–Salmon River Estuary (Bay of Fundy). *Sedimentology* **37**, 577–612.

Dalrymple, R. W., Makino, Y., & Zaitlin, B. A. 1991. Temporal and spatial patterns of rhythmite deposition on mudflats in the macrotidal Cobequid Bay–Salmon River Estuary, Bay of Fundy. In *Clastic Tidal Sedimentology* (ed. D. G. Smith, G. E. Reinson, B. A. Zaitlin, & R. A. Rahmani), pp. 137–160. Calgary, Alberta: Canadian Society of Petroleum Geologists.

Dalrymple, R. W., & Rhodes, R. N. 1995. Estuarine dunes and barforms. In *Geomorphology and Sedimentology of Estuaries* (ed. G. M. Perillo), pp. 359–422. Amsterdam: Elsevier.

Dalrymple, R. W., Zaitlin, B. A., & Boyd, R. 1992. Estuarine facies models: Conceptual basis and stratigraphic implications. *Journal of Sedimentary Petrology* **62**, 1130–1146.

Damuth, J. E., & Embley, R. W. 1981. Mass-transport processes on Amazon cone: western equatorial Atlantic. *American Association of Petroleum Geologists Bulletin* **65**, 629–643.

Damuth, J. E., Flood, R. D., Kowsmann, R. O., Belderson, R. H., & Gorini, M. A. 1988. Anatomy and growth pattern of Amazon deep-sea fans as revealed by long-range side-scan sonar (GLORIA) and high-resolution seismic studies. *American Association of Petroleum Geologists Bulletin* **72**, 885–911.

Damuth, J. E., Flood, R. D., Pirmez, C., & Manley, P. L. 1995. Architectural elements and depositional processes of Amazon deep-sea fan imaged by sidescal sonar (GLORIA), bathymetric swath mapping (SeaBeam), high-resolution seismic, and piston-core data. In *Atlas of Deep Water Environments: Architectural Style in Turbidite Systems* (ed. K. T. Pickering, R. N. Hiscott, N. H. Kenyon, F. Ricci Lucci, & R. D. A. Smith), pp. 105–121. London: Chapman and Hall.

Damuth, J. E., Jacobi, R. D., & Hayes, D. E. 1983a. Sedimentary processes in Northwest Pacific Basin revealed by echo-character mapping studies. *Geological Society of America Bulletin* **94**, 381–395.

Damuth, J. E., Kolla, V., Flood, R. D. *et al.* 1983b. Distributary channel meandering and bifurcation patterns on the Amazon deep-sea fan as revealed by long-range side-scan sonar (GLORIA). *Geology* **11**, 94–98.

Damuth, J. E., Kowsmann, R. O., Flood, R. D., Belderson, R. H., & Gorini, M. A. 1983c. Age relationships of distributary channels on Amazon Deep-Sea fan: implications for fan growth pattern. *Geology* **11**, 470–473.

Davidson, J. F., Harrison, D., & Guedes de Carvalho, J. R. F. 1977. On the liquidlike behaviour of fluidized beds. *Annual Review of Fluid Mechanics* **9**, 55–86.

Davidson-Arnott, R. G. D., & Greenwood, B. 1974. Bedforms and structures associated with bar topography in the shallow water wave environment, Kouchibouguac Bay, New Brunswick, Canada. *Journal of Sedimentary Petrology* **44**, 698–704.

Davies, G. R. 1970. Algal-laminated sediments, Gladstone Embayment, Shark Bay, Western Australia. In *Carbonate Sedimentation and Environments, Shark Bay, Western Australia* (ed. B. W. Logan, G. R. Davies, J. F. Read, & D. E. Cebulski), pp. 169–205. Tulsa, OK: American Association of Petroleum Geologists.

Davies, P. J., & Till, R. 1968. Stained dry cellulose peels of ancient and recent impregnated carbonate sediments. *Journal of Sedimentary Petrology* **38**, 234–237.

Davies, S., Hampson, G., Flint, S., & Elliott, T. 1999. Continent-scale sequence stratigraphy of the Namurian, Upper Carboniferous and its applications to reservoir prediction. In *Petroleum Geology of Northwest Europe: Proceedings 5th Conference* (ed. A. J. Fleet & A. A. R. Boldy), pp. 757–770. London: The Geological Society.

Davis, R. A. 1985. *Coastal Sedimentary Environments.* New York: Springer-Verlag.

Davis, R. A., & Clifton, H. E. 1987. Sea-level change and the preservation of wave-dominated and tide-dominated coastal sequences. In *Sea-level Fluctuations and Coastal Evolution* (ed. D. Nummedal, O. H. Pilkey, & J. D. Howard), pp. 167–178. Tulsa, OK: SEPM (Society for Sedimentary Geology).

Davis, R. A., & Fitzgerald, D. M. 2004. *Beaches and Coasts.* Oxford: Blackwells.

Davis, R. A., & Flemming, B. W. 1995. Stratigraphy of a combined wave- and tide-dominated intertidal sand body: Martens Plate, East Frisian Wadden Sea, Germany. In *Tidal Signatures in Modern and Ancient Sediments* (ed. B. W. Flemming & A. Bartholoma), pp. 121–132. Oxford: Blackwells Publication.

Dawson, A. G. 1992. *Ice Age Earth: Late Quaternary Geology and Climate.* London: Routledge.

1994. Geomorphological effects of tsunami run-up and backwash. *Geomorphology* **10**, 83–94.

1999. Linking tsunami deposits, submarine slides and offshore earthquakes. *Quaternary International* **60**, 119–126.

Dawson, A. G., Foster, I. K. L., Shi, S., Smith, D. E., & Long, D. 1991. The identification of tsunami deposits in coastal sediment sequences. *Science of Tsunami Hazards* **9**, 73–82.

Dean, W. E., & Fouch, T. D. 1983. Lacustrine. In *Carbonate Depositional Environments* (ed. P. A. Scholle, D. G. Bebout, &

C. H. Moore), pp. 97–130. Tulsa, OK: American Association of Petroleum Geologists.

de Boer, P. L. 1979. Convolute lamination in modern sands of the estuary of the Oosterschelde, the Netherlands, formed as the result of entrapped air. *Sedimentology* **26**, 283–294.

de Boer, P. L., & Smith, D. G. 1994. *Orbital Forcing and Cyclic Sequences*. Oxford: Blackwells.

de Boer, P. L., Van Gelder, A., & Nio, S. D. 1988. *Tide-influenced Sedimentary Environments and Facies*. Boston: D. Reidel Publishing Company.

Decho, A. W., Visscher, P. T., & Reid, R. P. 2005. Production and cycling of natural microbial exopolymers (EPS) within a marine stromatolite. *Palaeogeography, Palaeoclimatology, Palaeoecology*, **219** 71–86.

Deer, W. A., Howie, R. A., & Zussman, J. 1962. *Rock Forming Minerals, Volume 3: Sheet Silicates*. New York: Wiley.

Defant, A. 1958. *Ebb and Flow*. Ann Arbor, MI: University of Michigan Press.

Deffeyes, K. S., Lucia, F. J., & Weyl, P. K. 1965. Dolomitization of Recent and Plio-Pleistocene sediments by marine evaporite waters on Bonaire, Netherlands Antilles. In *Dolomitization and Limestone Diagenesis, A Symposium* (ed. L. C. Pray & R. C. Murray), pp. 71–80. Tulsa, OK: SEPM (Society for Sedimentary Geology).

Degens, E. T., & Ross, D. A. 1969. *Hot Brines and Recent Heavy Metal Deposits in the Red Sea: A Geochemical and Geophysical Account*. New York: Springer-Verlag.

Demarest, J. M. II, & Kraft, J. C. 1987. Stratigraphic record of Quaternary sea levels: implications for more ancient strata. In *Sea-level Fluctuations and Coastal Evolution* (ed. D. Nummedal, O. H. Pilkey, & J. D. Howard), pp. 223–239. Tulsa, OK: SEPM (Society for Sedimentary Geology).

Demicco, R. V. 1983. Wavy and lenticular-bedded carbonate ribbon rocks of the Upper Cambrian Conococheague Limestone, western Maryland. *Journal of Sedimentary Petrology* **53**, 1121–1132.

1998. CYCOPATH 2D – a two-dimensional model of cyclic sedimentation on carbonate platforms. *Computers & Geosciences* **24**, 405–423.

Demicco, R. V., & Gierlowski-Kordesch, E. 1986. Facies sequences of a semi-arid closed basin: the Lower Jurassic East Berlin Formation of the Hartford Basin, New England, U.S.A. *Sedimentology*, **33**, 107–118.

Demicco, R. V., & Hardie, L. A. 1994. *Sedimentary Structures and Early Diagentic Features of Shallow Marine Carbonate Deposits*. Tulsa, OK: SEPM (Society for Sedimentary Geology).

2002. The "carbonate factory" revisited: a reexamination of sediment production functions used to model deposition on carbonate platforms. *Journal of Sedimentary Research* **72**, 849–857.

Demicco, R. V., Lowenstein, T. K., Hardie, L. A., & Spencer, R. J. 2005. Model of seawater composition for the Phanerozoic. *Geology* **33**, 877–880.

Demicco, R. V., & Mitchell, R. W. III 1982. Facies of the Great American Bank in the central Appalachians. In *Central Appalachian Geology, Field Trip Guidebook* (ed. P. T. Lyttle), pp. 171–266. Falls Church, VA: American Geological Institute.

De Mowbray, T. 1983. The genesis of lateral accretion deposits in recent intertidal mudflat channels, Solway Firth, Scotland. *Sedimentology* **30**, 425–435.

De Mowbray, T., & Visser, M. J. 1984. Reactivation surfaces in subtidal channel deposits, Oosterschelde, southwest Netherlands. *Journal of Sedimentary Petrology* **54**, 811–824.

De Raaf, J. F. M., Boersma, J. R., & Van Gelder, A. 1977. Wave-generated structures and sequences from a shallow marine succession, Lower Carboniferous, County Cork, Ireland. *Sedimentology* **24**, 451–483.

Deutsch, C. V. 2002. *Geostatistical Reservoir Modeling*. New York: Oxford University Press.

Deutsch, C., & Cockerham, P. 1994. Practical considerations in the application of simulated annealing to stochastic simulation. *Mathematical Geology* **26**, 67–82.

Deutsch, C. V., & Wang, L. 1996. Hierarchical object-based stochastic modeling of fluvial reservoirs. *Mathematical Geology* **28**, 857–880.

D'Hondt, S., Rutherford, S., & Spivak, A. J. 2002. Metabolic activity of subsurface life in deep-sea sediments. *Science* **295**, 2067–2070.

D'Hondt, S., Jørgensen, B. B., Miller, D. J. *et al.* 2004. Distributions of microbial activities in deep subseafloor sediments. *Science* **306**, 2216–2221.

Dickinson, W. R. 1976. *Plate Tectonic Evolution of Sedimentary Basins*. Tulsa, OK: American Association of Petroleum Geologists.

1985. Interpreting provenance relations from detrital modes of sandstones. In *Provenance of Arenites* (ed. G. G. Zuffa), pp. 333–361. Dordrecht: Reidel.

Dickinson, W. R. 1995. Forearc basins. In *Tectonics of Sedimentary Basins* (ed. C. Busby & R. Ingersoll), pp. 221–262. Oxford: Blackwell Science.

Dickson, J. A. D. 1966. Carbonate identification and genesis as revealed by staining. *Journal of Sedimentary Petrology* **36**, 491–505.

2002. Fossil echinoderms as monitor of the Mg/Ca ratio of Phanerozoic oceans. *Science* **298**, 1222–1224.

Diem, B. 1984. Analytical method for estimating paleowave climate and water depth from wave ripple marks. *Sedimentology* **32**, 705–720.

Dietrich, W. E., & Dunne, T. 1993. The channel head. In *Channel Network Hydrology* (ed. K. Beven & M. J. Kirkby), pp. 175–219. Chichester: Wiley.

Dill, R. F., Kendall, C. G. St. C., & Shinn, E. A. 1989. *Giant Subtidal Stromatolites and Related Sedimentary Features, Lee Stocking Island, Exumas, Bahamas. IGC Field Trip* **T 373**, Washington, D.C.: American Geophysical Union.

Dill, R. F., Shinn, E. A., Jones, A. T., Kelly, K., & Steinen, R. P. 1986. Giant subtidal stromatolites forming in normal salinity water. *Nature* **324**, 55–58.

Dingler, J. R. 1979. The threshold of grain motion under oscillatory flow in a laboratory wave channel. *Journal of Sedimentary Petrology* **49**, 287–294.

Domack, E. W., & Ishman, S. 1993. Oceanographic and physiographic controls on modern sedimentation within Antarctic fjords. *Geological Society of America Bulletin* **105**, 1175–1189.

Dominguez, J. M. L. 1996. The São Francisco strandplain: a paradigm for wave-dominated deltas? In *Geology of Siliciclastic Shelf Seas* (ed. M. De Baptist & P. Jacobs), pp. 217–231. London: Geological Society of London.

Dominguez, J. M. L., Martin, L., & Bittencourt, A. C. S. P. 1987. Sea-level history and Quaternary evolution of river mouth-associated beach-ridge plains along the east-southeast Brazilian coast: a summary. In *Sea-level Fluctuations and Coastal Evolution* (ed. D. Nummedal, O. H. Pilkey, & J. D. Howard), pp. 115–127. Tulsa, OK: SEPM (Society for Sedimentary Geology).

Donovan, R. N., & Foster, R. J. 1972. Subaqueous shrinkage cracks from the Caithness Flagstone series (Middle Devonian) of NE Scotland. *Journal of Sedimentary Petrology* **42**, 309–317.

Donovan, S. K. 1994. Insects and other arthropods as trace-makers in non-marine environments and paleoenvironments. In *The Paleobiology of Trace Fossils* (ed. S. K. Donovan), pp. 200–220. Baltimore, MA: The Johns Hopkins University Press.

Dott, R. H. Jr. 1992. Eustasy: the ups and downs of a major geological concept. Boulder, CO: Geological Society of America.

Dott, R. H., & Bourgeois, J. 1982. Hummocky stratification: significance of its variable bedding sequences. *Geological Society of America Bulletin* **93**, 663–680.

Dove, P. M., DeYoreo, J. J., & Weiner, S. 2003. *Biomineralization.* Chantilly, VA: The Mineralogical Society of America.

Doveton, J. H. 1994. Theory and applications of vertical variability measures from Markov chain analysis. In *Stochastic Modeling and Geostatistics* (ed. J. M. Yarus & R. L. Chambers), pp. 55–64. Tulsa, OK: American Association of Petroleum Geologists.

Dowdeswell, J. A., Elverhoi, A., Andrews, J. T., & Hebbeln, D. 1999. Asynchronous deposition of ice-rafted layers in the Nordic seas and North Atlantic Ocean. *Nature* **400**, 348–351.

Dowdeswell, J. A., Maslin, M. A., Andrews, J. T., & McCave, I. N. 1995. Iceberg production, debris rafting, and the extent and thickness of Heinrich layers (H-1, H-2) in North Atlantic sediments. *Geology* **23**, 301–304.

Dowdeswell, J. A., & O Cofaigh, C. 2002. *Glacier-influenced Sedimentation on High-latitude Continental Margins.* London: Geological Society of London.

Dowdeswell, J. A., O Cofaigh, C., Taylor, J. *et al.* 2002. On the architecture of high-latitude continental margins: the influence of ice-sheet and sea-ice processes in the Polar North Atlantic. In *Glacier-influenced Sedimentation on High-latitude Continental Margins* (ed. J. A. Dowdeswell & C. O Cofaigh), pp. 33–54. London: Geological Society of London.

Dowdeswell, J. A., & Scourse, J. D. 1990. *Glacimarine Environments.* London: Geological Society of London.

Dowdeswell, J. A., Whittington, R. J., & Marienfeld, P. 1994. The origin of massive diamicton facies by iceberg rafting and scouring, Scoresby Sund, East Greenland. *Sedimentology* **41**, 21–35.

Dravis, J. J. 1979. Rapid and widespread generation of Recent oolitic hardgrounds on a high energy Bahamian Platform, Eleuthera Bank, Bahamas. *Journal of Sedimentary Petrology* **49**, 195–208.

1983. Hardened subtidal stromatolites. *Science* **219**, 385–386.

Drever, J. J. 1994. The effect of land plants on weathering rates of silicate minerals. *Geochimica et Cosmochimica Acta* **58**, 2325–2332.

1997. *The Geochemistry of Natural Waters*, 3rd edn. Englewood Cliffs, NJ: Prentice-Hall.

Drever, J. J., & Stillings, L. L. 1997. The role of organic acids in mineral weathering. *Colloids and Surfaces A: Physicochemical and Engineering Aspects* **120**, 167–181.

Driemanis, A. 1988. Tills: their genetic terminology and classification. In *Genetic Classification of Glacigenic Deposits* (ed. R. P. Goldthwait & C. L. Matsch), pp. 17–83. Rotterdam: Balkema.

Driese, S. G., & Mora, C. I. 2001. Diversification of Siluro-Devonian plant traces in paleosols and influence on estimates of paleoatmospheric CO_2 levels. In *Plants Invade the Land* (ed. P. G. Gensel & D. Edwards), pp. 237–253. New York: Columbia University Press.

Driese, S. G., Mora, C. I., & Elrick, J. M. 2000. The paleosol record of increasing plant diversity and depth of rooting and changes in atmospheric pCO_2 in the Siluro-Devonian. In *Phanerozoic Terrestrial Ecosystems: A Short Course* (ed. R. A. Gastaldo & W. A. DiMichele), pp. 47–61. Lawrence, KA: The Paleontological Society.

Droser, M. L., & Bottjer, D. L. 1986. A semiquantitative field classification of ichnofabric. *Journal of Sedimentary Petrology* **56**, 558–559.

1989. Ichnofabric of sandstones deposited in high energy nearshore environments: measurement and utilization. *Palaios* **4**, 598–604.

Droz, L., & Bellaiche, G. 1985, Rhone deep-sea fan: morphostructure and growth pattern. *American Association of Petroleum Geologists Bulletin* **69**, 460–479.

1991. Seismic facies and geological evolution of the central portion of the Indus fan. In *Seismic Facies and Sedimentary Processes of Submarine Fans and Turbidite Systems* (ed. P. Weimer & M. H. Link), pp. 383–402. New York: Springer-Verlag.

Duc, A. W., & Tye, R. S. 1987. Evolution and stratigraphy of a regressive/backbarrier complex: Kaiwah Island, South Carolina. *Sedimentology* **34**, 237–251.

Duke, W. L. 1985. Hummocky cross-stratification, tropical hurricanes, and intense winter storms. *Sedimentology* **32**, 167–194.

1990. Geostrophic circulation or shallow marine turbidity currents? The dilemma of paleoflow patterns in storm-influenced prograding shoreline systems. *Journal of Sedimentary Petrology* **60**, 870–883.

Duke, W. L., Arnott, R. W., & Cheel, R. J. 1991. Shelf sandstones and hummocky cross-stratification: new insights on a stormy debate. *Geology* **19**, 625–628.

Dumas, S., Arnott, R. W. C., & Southard, J. B. 2005. Experiments on oscillatory-flow and combined-flow bed forms: implications for interpreting parts of the shallow-marine sedimentary record. *Journal of Sedimentary Research* **75**, 501–513.

Dunham, R. J. 1969a. Early vadose silt in Townsend mound (reef), New Mexico. In *Depositional Environments in Carbonate Rocky, a Symposium* (ed. G. M. Friedman), pp. 139–182. Tulsa, OK: SEPM (Society for Sedimentary Geology).

1969b. Vadose pisolite in the Capitan reef (Permian), New Mexico and Texas. In *Depositional Environments in Carbonate Rocky, a Symposium* (ed. G. M. Friedman), pp. 182–190. Tulsa, OK: SEPM (Society for Sedimentary Geology).

Dunlop, P., & Clarke, C. D. 2006. The morphological characteristics of ribbed moraine. *Quaternary Science Reviews* **25**, 1668–1691.

Dunne, T. 1980. Formation and controls of channel networks. *Progress in Physical Geography* **4**, 211–239.

Dury, G. H. 1985. Attainable standards of accuracy in the retrodiction of palaeodischarge from channel dimensions. *Earth Surface Processes and Landforms* **10**, 205–213.

Dutton, S. P., Flanders, W. A., & Barton, M. D. 2003. Reservoir characterization of a Permian deep-water sandstone, East Ford field, Delaware basin, Texas. *American Association of Petroleum Geologists Bulletin* **87**, 609–627.

Dutton, S. P., White, C. D., Willis, B. J., & Novakovic, D. 2002. Calcite cement distribution and its effect on fluid flow in a deltaic sandstone, Frontier Formation, Wyoming. *American Association of Petroleum Geologists Bulletin* **86**, 2007–2021.

Dyer, K. R. 1989. Sediment processes in estuaries: future research requirements. *Journal of Geophysical Research* **94C**, 14,327–14,339.

Dyer, K. R., & Huntley, D. A. 1999. The origin, classification and modeling of sand banks and ridges. *Continental Shelf Research* **19**, 1285–1330.

Dzulyński, S., & Kotlarczyk, J. 1962. On load-casted ripples. *Annales de la Société Géologique de Pologne* **32**, 148–159.

Dzulyński, S., & Walton, E. K. 1965. *Sedimentary Features of Flysch and Graywackes*. Amsterdam: Elsevier.

Easterbrook, D. J. 1999. *Surface Processes and Landforms*, 2nd edn. New York: Macmillan Publishing Company.

Edmonds, D. A., & Slingerland, R. 2007. Mechanics of middle ground bar formation: implications for the formation of delta distributary channel networks. *Journal of Geophysical Research – Earth Surface* **112**, F02034, doi:10.1029/2006 JF000574.

Edwards, D. E., Leeder, M. R., Best, J. L., & Pantin, H. M. 1994. An experimental study of reflected density currents and the interpretation of certain turbidites. *Sedimentology* **41**, 437–461.

Ehrenberg, S. N., & Nadeau, P. H. 2005. Sandstone vs. carbonate petroleum reservoirs: a global perspective on porosity–depth and porosity–permeability relationships. *American Association of Petroleum Geologists Bulletin* **89**, 435–445.

Einsele, G. 1992. *Sedimentary Basins, Evolution, Facies and Sediment Budget*. Berlin: Springer-Verlag.

Einsele, G., Ricken, W., & Seilacher, A. 1991. *Cycles and Events in Stratigraphy*. Berlin: Springer-Verlag.

Eisma, D. 1997. *Intertidal Deposits*. Boca Raton, FL: CRC Press.

Ekdale, A. A., Bromley, R. G., & Pemberton, S. G. 1984. *Ichnology: The Use of Trace Fossils in Sedimentology and Stratigraphy*. Tulsa, OK: SEPM (Society for Sedimentary Geology).

Ekdale, A. A., & Pollard, J. 1991. Ichnofabric and ichnofacies. *Palios* **6**, 197–343.

Elfeki, A., & Dekking, M. 2001. A Markov chain model for subsurface characterization: theory and application. *Mathematical Geology* **33**, 569–589.

Elliott, T. 1974. Interdistributary bay sequences and their genesis. *Sedimentology* **21**, 611–622.

Elrick, M., & Snider, A. C. 2002. Deep-water stratigraphic cyclicity and carbonate mud mound development in the Middle Cambrian Marjum Formation, House Range, Utah. *Sedimentology* **49**, 1021–1047.

Embley, R. W. 1976. New evidence for occurrence of debris flow deposits in the deep sea. *Geology* **4**, 371–374.

1980. The role of mass transport in the distribution and character of deep-ocean sediments with special reference to the North Atlantic. *Marine Geology* **38**, 23–50.

Embry, A. F., & Klovan, J. E. 1971. A Late Devonian reef tract on north-eastern Banks Island, N.W.T. *Bulletin of Canadian Petroleum Geology* **19**, 730–781.

Emery, D., & Myers, K. J. (eds.) 1996. *Sequence Stratigraphy.* Oxford: Blackwells.

Emery, K. O. 1952. Continental shelf sediments off southern California. *Geological Society of America Bulletin* **63**, 1105–1108.

Enos, P. 1974. *Surface Sediment Facies Map of the Florida–Bahamas Plateau.* Boulder, CO: Geological Society of America.

1983. Shelf. In *Carbonate Depositional Environments* (ed. P. A. Scholle, D. G. Bebout, & C. H. Moore), pp. 267–295. Tulsa, OK: American Association of Petroleum Geologists.

1991. Sedimentary parameters for computer modeling. *Kansas Geological Survey Bulletin* **233**, 63–99.

Enos, P., & Moore, C. H. 1983. Fore-reef slope. In *Carbonate Depositional Environments* (ed. P. A. Scholle, D. G. Bebout, & C. H. Moore), pp. 507–538. Tulsa, OK: American Association of Petroleum Geologists.

Enos, P., & Perkins, R. D. 1977. *Quaternary Sedimentation in South Florida.* Boulder, CO: Geological Society of America.

1979. Evolution of Florida Bay from island stratigraphy. *Geological Society of America Bulletin* **90**, 59–83.

Enos, P., & Sawatsky, L. H. 1981. Pore networks in Holocene carbonate sediments. *Journal of Sedimentary Petrology* **51**, 961–985.

Erickson, M. C., Masson, D. S., Slingerland, R., & Swetland, D. W. 1990. Numerical simulation of circulation and sediment transport in the Late Devonian Catskill Sea. In *Quantitative Dynamic Stratigraphy* (ed. T. A. Cross), pp. 293–305. Englewood Cliffs, NJ: Prentice-Hall.

Ernst, W. G. 2000. *Earth Systems: Processes and Issues.* Cambridge: Cambridge University Press.

Esteban, M. 1974. Caliche textures and "Microcodium." *Bollettino della Società Geologica Italiana* **92**, 105–125.

Esteban, M., & Klappa, C. F. 1983. Subaerial exposure. In *Carbonate Depositional Environments* (ed. P. A. Scholle, D. G. Bebout, & C. H. Moore), pp. 1–54. Tulsa, OK: American Association of Petroleum Geologists.

Esteban, M., & Pray, L. C. 1983. Pisoids and pisolitic facies (Permian), Guadalupe Mountains, New Mexico and West Texas. In *Coated Grains* (ed. T. M. Peryt), pp. 503–537. New York: Springer-Verlag.

Ethridge, F. G., & Schumm, S. A. 1978. Reconstructing paleochannel morphologic and flow characteristics: methodology, limitations and assessment. In *Fluvial Sedimentology* (ed. A. D. Miall), pp. 703–721. Calgary, Alberta: Canadian Society of Petroleum Geologists.

Ethridge, F. G., Skelly, R. L., & Bristow, C. S. 1999. Avulsion and crevassing in the sandy, braided Niobrara River: complex response to base-level rise and aggradation. In *Fluvial Sedimentology VI* (ed. N. D. Smith & J. Rogers), pp. 179–191. Oxford: Blackwells.

Eugster, H. P. 1969. Inorganic bedded cherts from the Magadi area, Kenya. *Contributions to Mineralogy and Petrology* **22**, 1–31.

Eugster, H. P., & Hardie, L. A. 1978. Saline lakes. In *Lakes: Chemistry Geology Physics* (ed. A. Lerman), pp. 237–293. New York: Springer-Verlag.

Evamy, B. D. 1962. The application of a chemical staining technique to a study of dedolomitization. *Sedimentology* **2**, 164–170.

Evans, D. J. A. 2003. *Glacial Landsystems.* London: Arnold.

Evans, D. J. A., Phillips, E. R., Hiemstra, J. F., & Auton, C. A. 2006a. Subglacial till: formation, sedimentary characteristics and classification. *Earth-Science Reviews* **78**, 115–176.

Evans, D. J. A., Rea, B. R., Hiemstra, J. F., & O Cofaigh, C. 2006b. A critical assessment of subglacial mega-floods: a case study of glacial sediments and landforms in south-central Alberta, Canada. *Quaternary Science Reviews* **25**, 1638–1667.

Evans, G. 1965. Intertidal flat sediments and their environments of deposition in the Wash. *Quarterly Journal of the Geology Society of London* **121**, 209–245.

Ewers, W. E., & Morris, R. C. 1981. Studies of the Dales Gorge Member of the Brockman Iron Formation, Western Australia. *Economic Geology* **76**, 1929–1953.

Eyles, C. H., Eyles, N., & Miall, A. D. 1985. Models of glaciomarine sedimentation and their application to the ancient glacial record. *Palaeogeography, Palaeoclimatology, Palaeoecology* **51**, 15–84.

Eyles, N., & Eyles, C. H. 1992. Glacial depositional systems. In *Facies Models: Response to Sea Level Change* (ed. R. G. Walker & N. P. James), pp. 73–100. St. John's, Newfoundland: Geological Association of Canada.

Eyles, N., Sladen, J., & Gilroy, S. 1982. A depositional model for stratigraphic complexes and facies superimposition in lodgement tills. *Boreas* **11**, 317–333.

Faugeres, J.-C., Mezerais, M.-L., & Stow, D. A. V. 1993. Contourite drift types and their distribution in the North and South Atlantic ocean basins. *Sedimentary Geology* **82**, 189–206.

Farquhar, J., Bao, H., & Thiemens, M. 2000. Atmospheric influence of Earth's earliest sulfur cycle. *Science* **289**, 756–758.

Farrell, K. M. 1987. Sedimentology and facies architecture of overbank deposits of the Mississippi River, False River Region, Louisiana. In *Recent Developments in Fluvial Sedimentology* (ed. F. G. Ethridge, R. M. Flores, & M. D. Harvey), pp. 111–120. Tulsa, OK: SEPM (Society for Sedimentary Geology).

2001. Geomorphology, facies architecture, and high resolution, non-marine sequence stratigraphy in avulsion deposits, Cumberland marshes, Saskatchewan. *Sedimentary Geology* **139**, 93–150.

Faure, G. 1991. *Principles and Applications of Inorganic Geochemistry*, 2nd edn. New York: MacMillan Publishing Company.

Fawthrop, N. P. 1996. Modelling hydrological processes for river management. In *River Flows and Channel Forms* (ed. G. Petts & P. Calow), pp. 51–76. Oxford: Blackwell Science Ltd.

Felix, M. 2001. A two-dimensional numerical model for a turbidity current. In *Particulate Gravity Currents* (ed. W. D. McCaffrey, B. C. Kneller, & J. Peakall), pp. 71–81. Oxford: Blackwells.

Field, M. E. 1980. Sand bodies on coastal plain shelves: Holocene record of the U.S. Atlantic shelf off Maryland. *Journal of Sedimentary Petrology* **50**, 505–528.

Figueredo, A. G. J., Sanders, J. E., & Swift, D. J. P. 1982. Storm graded layers on inner continental shelves: examples from south Brazil and the Atlantic coast of the central United States. *Sedimentary Geology* **31**, 171–190.

Fischer, A. G. 1964. The Lofer cyclothems of the Alpine Triassic. *Kansas State Geological Survey Bulletin* **169**, 7–149.

1982. Long-term climatic oscillations recorded in stratigraphy. In *Climate in Earth History* (ed. Geophysics Study Committee), pp. 97–104. Washington, D.C.: National Academy Press.

Fisher, R. V., & Schminke, H. U. 1984. *Pyroclastic Rocks*. Berlin: Springer-Verlag.

Fisher, R. V., & Smith, G. A. 1991. *Sedimentation in Volcanic Settings*. Tulsa, OK: SEPM (Society for Sedimentary Geology).

Fisher, W. L., Brown, L. F., Scott, A. J., & McGowen, J. H. 1969. *Delta Systems in the Exploration for Oil and Gas*. Austin, TX: Bureau of Economic Geology, University of Texas.

Fisk, H. N. 1961. Bar finger sands of the Mississippi delta. In *Geometry of Sandstone Bodies – A Symposium* (ed. J. A. Peterson & J. C. Osmond), pp. 29–52. Tulsa, OK: American Association of Petroleum Geologists.

Fisk, H. N., McFarlan, E. Jr., Kolb, C. R., & Wilbert, L. J. Jr. 1954. Sedimentary framework of the modern Mississippi delta. *Journal of Sedimentary Petrology* **24**, 76–99.

Fitzgerald, D. M. 1988. Shoreline erosional–depositional processes associated with tidal inlets. In *Hydrodynamics and Sediment Dynamics of Tidal Inlets* (ed. D. G. Aubrey & L. Weishar), pp. 186–225. Berlin: Springer-Verlag.

1993. Origin and stability of tidal inlets in Massachusetts. In *Formation and Evolution of Multiple Tidal Inlets* (ed. D. G. Aubrey & F. J. Geise), pp. 1–61. Washington, D.C.: American Geophysical Union.

Fitzgerald, M. G., Mitchum, R. M. Jr., Uliana, M. A., & Biddle, K. T. 1990. Evolution of San Jorge Basin. *American Association of Petroleum Geologists Bulletin* **74**, 879–920.

Fitzpatrick, E. A. 1984. *Micromorphology of Soils*. London: Chapman and Hall.

Flemming, B. W. 1980. Sand transport and bedform patterns on the continental shelf between Durban and Port Elizabeth, southeast African continental margin. *Sedimentary Geology* **26**, 179–205.

Flemming, B. W., & Bartholoma, A. 1995. *Tidal Signatures in Modern and Ancient Sediments*. Oxford: Blackwells.

Flemings, P. B., & Jordan, T. E. 1989. A synthetic stratigraphic model of foreland basin development. *Journal of Geophysical Research* **94**, 3851–3866.

Flemings, P. B., & Jordan, T. E. 1990. Stratigraphic modeling of foreland basins: interpreting thrust deformation and lithospheric rheology. *Geology* **18**, 430–434.

Flint, R. F. 1971. *Glacial and Quaternary Geology*. New York: Wiley.

Flint, S. S., Aitken, J., & Hampson, G. 1995. Application of sequence stratigraphy to coal-bearing coastal plain successions: implications for the UK coal measures. In *European Coal Geology* (ed. M. K. G. Whateley & D. A. Spears), pp. 1–16. London: Geological Society of London.

Flood, R. D., & Damuth, J. E. 1987. Quantitative characteristics of sinuous distributary channels on the Amazon deep-sea fan. *Geological Society of America Bulletin* **98**, 728–738.

Flood, R. D., Manley, P. C., Kowsman, R. O., Appi, C. J., & Pirmez, C. 1991. Seismic facies and Late Quaternary growth of Amazon submarine fan. In *Seismic Facies and Sedimentary Processes of Submarine Fans and Turbidite Systems* (ed. P. Weimer & M. H. Link), pp. 415–433. New York: Springer-Verlag.

Flügel, E. 2004. *Microfacies of Carbonate Rocks: Analysis, Interpretation and Application*. Berlin: Springer-Verlag.

Folk, R. L. 1959. Practical petrographic classification of limestones. *American Association of Petroleum Geologists Bulletin* **43**, 1–38.

1965a. *Petrology of Sedimentary Rocks*. Austin, TX: Hemphil's.

1965b. Some aspects of recrystalization in ancient limestones. In *Dolomitization and Limestone Diagenesis; A Symposium* (ed. L. C. Pray & R. C. Murray), pp. 14–48. Tulsa, OK: SEPM (Society for Sedimentary Geology).

Folk, R. L., & McBride, E. F. 1976. The Caballos Novaculite revisited. Part I: origin of the novaculite members. *Journal of Sedimentary Petrology* **46**, 659–669.

Forristall, G. S., Hamilton, R. C., & Cardone, V. J. 1977. Continental shelf currents in Tropical Storm Delia: observation and theory. *Journal of Physical Oceanography* **7**, 532–546.

Fortin, D., & Langley, S. 2005. Formation and occurrence of biogenic iron-rich minerals. *Earth-Science Reviews* **72**, 1–19.

Fouch, T. D., & Dean, W. E., 1982. Lacustrine environments. In *Sandstone Depositional Environments* (ed. P. A. Scholle & D. Spearing), pp. 87–114. Tulsa, OK: American Association of Petroleum Geologists.

Frakes, L. A. 1979. *Climates Throughout Geologic History*. Amsterdam: Elsevier.

Frakes, L. A., Francis, J. E., & Syktus, J. I. 1992. *Climate Modes of the Phanerozoic*. Cambridge: Cambridge University Press.

Francis, J. E., & Smith, M. P. 2002. Paleoclimate reconstruction using fossils and lithologic indicators. *Palaeogeography, Palaeoclimatology, Palaeoecology* **182**, 1–143.

Francis, P. W. 1993. *Volcanoes: A Planetary Perspective*. Oxford: Oxford University Press.

Franseen, E. K., Watney, W. L., Kendall, C. G. St. C., & Ross, W. 1991. *Sedimentary Modeling: Computer Simulations and Methods for Improved Parameter Definition*. Lawrence, KA: Kansas Geological Survey.

Fraser, G. S., & DeCelles, P. G. 1992. Geomorphic controls on sediment accumulation at margins of foreland basins. *Basin Research* **4**, 233–252.

Frazier, D. E. 1967. Recent deltaic deposits of the Mississippi delta: their development and chronology. *Transactions of the Gulf Coast Association of Geological Societies* **17**, 287–315.

Fredsoe, J., & Deigaard, R. 1992. *Mechanics of Coastal Sediment Transport*. Singapore: World Scientific.

Freeze, R. A., & Cherry, J. A. 1979. *Groundwater*. Englewood Cliffs, NJ: Prentice-Hall.

Frey, R. W., Howard, J. D., Han, S. J., & Park, B. K. 1989. Sediments and sedimentary sequences on a modern macrotidal flat, Inchon, Korea. *Journal of Sedimentary Petrology* **59**, 28–44.

Friedman, G. M. 1959. Identification of carbonate minerals by staining methods. *Journal of Sedimentary Petrology* **29**, 87–97.

Friedman, G. M., & Sanders, J. E. 1967. Origin and occurrence of dolostones. In *Carbonate Rocks, Part A: Origin, Occurrence, and Classification* (ed. G. V. Chilingar, H. J. Bissell, & R. W. Fairbridge), pp. 167–348. Amsterdam: Elsevier.

Friedman, S. J. 1998. Rock-avalanche elements of the Shadow Valley Basin, Eastern Mojave Desert, California: processes and problems. *Journal of Sedimentary Research* **67**, 792–804.

Frostick, L. E., & Reid, I. 1987. *Desert Sediment: Ancient and Modern*. London: Geological Society of London.

Fryberger, S. G. 1993. A review of aeolian bounding surfaces, with examples from the Permian Minnelusa Formation, USA. In *Characterisation of Fluvial and Aeolian Reservoirs* (ed. C. P. North & D. J. Prosser), pp. 167–197. London: Geological Society of London.

Fryberger, S. G., Ahlbrandt, T. S., & Andrews, S. 1979. Origin, sedimentary features and significance of low-angle eolian "sand sheet" deposits, Great Sand Dunes National Monument and vicinity, Colorado. *Journal of Sedimentary Petrology* **49**, 733–746.

Füchtbauer, H., & Hardie, L. A. 1976. Experimentally determined homogeneous distribution coefficients for precipitated magnesium calcites: application to marine carbonate cements. *Geological Society of America Abstracts with Program* **8**, 877.

1980. Comparison of experimental and natural magnesium calcites. *IAS International Meeting, Bochum*. Bochum: International Association of Sedimentologists. pp. 167–169.

Galloway, W. E. 1975. Process framework for describing the morphologic and stratigraphic evolution of deltaic depositional systems. In *Deltas, Models for Exploration* (ed. M. L. Broussard), pp. 87–98. Houston, TX: Houston Geological Society.

1981. Depositional architecture of Cenozoic Gulf coastal plain fluvial systems. In *Recent and Ancient Nonmarine Depositional Environments: Models for Exploration* (ed. F. G. Ethridge & R. M. Flores), pp. 127–155. Tulsa, OK: SEPM (Society for Sedimentary Geology).

1989a. Genetic stratigraphic sequences in basin analysis I: architecture and genesis of flooding-suface bounded depositional units. *American Association of Petroleum Geologists Bulletin* **73**, 125–142.

1989b. Genetic stratigraphic sequences in basin analysis II: application to northwest Gulf of Mexico Cenozoic basin. *American Association of Petroleum Geologists Bulletin* **73**, 143–154.

Garcia-Castellanos, D. 2002. Interplay between lithospheric flexure and river transport in foreland basins. *Basin Research* **14**, 89–104.

Garrels, R. M., & MacKenzie, F. T. 1971. *Evolution of Sedimentary Rocks*. New York: W. W. Norton.

Garret, P. 1977. Biological communities and their sedimentary record. In *Sedimentation on the Modern Carbonate Tidal Flats of Northwest Andros Island* (ed. L. A. Hardie), pp. 124–158. Baltimore, MA: The Johns Hopkins University Press.

Garven, G., Ge, S., Person, M. A., & Sverjensky, D. A. 1993. Genesis of stratabound ore deposits in the mid-continent of North America. 1. The role of regional groundwater flow. *American Journal of Science* **293**, 497–568.

Gawthorpe, R. L., & Leeder, M. R. 2000. Tectono-sedimentary evolution of active extensional basins. *Basin Research* **12**, 195–218.

Gaynor, G. C., & Swift, D. J. P. 1988. Shannon Sandstone depositional model; sand ridge formation on the Campanian western interior shelf. *Journal of Sedimentary Petrology* **58**, 868–880.

Gebelein, C. D. 1974. *Guidebook for Modern Bahamian Platform Environments*. St. George, Bermuda: Geological Society of America.

Gebelein, C. D., Steinen, R. P., Garrett, P. *et al.* 1980. Subsurface dolomitization beneath the tidal flats of Central West Andros Island, Bahamas. In *Concepts and Models of Dolomitization* (ed. D. H. Zenger, J. B. Dunham, & R. L. Ethington), pp. 31–49. Tulsa, OK: SEPM (Society for Sedimentary Geology).

Genise, J. F., & Brown, T. M. 1994a. New Miocene scarabeid and hymenopterous nests and early Miocene (Santacrucian) paleoenvironments. *Ichnos* **3**, 107–117.

1994b. New trace fossils of termites (insecta: isoptera) from the late Eocene–early Miocene of Egypt, and the reconstruction of ancient isopteran behavior. *Ichnos* **3**, 155–183.

Genise, J. F., Mangano, M. G., Buatois, L. A., Laza, J. H., & Verde, M. 2000. Insect trace fossil associations in paleosols: the Coprinisphaera ichnofacies. *Palaios* **15**, 49–64.

Gensel, P. G., & Edwards, D. 2001. *Plants Invade the Land.* New York: Columbia University Press.

George, G., & Berry, J. K. 1993. A new lithostratigraphy and depositional model for the Upper Rotliegend of the UK sector of the southern North Sea. In *Characterisation of Fluvial and Aeolian Reservoirs* (ed. C. P. North & D. J. Prosser), pp. 291–320. London: Geological Society of London.

Ghibaudo, G. 1992. Subaqueous sediment gravity flow deposits: practical criteria for their field description and classification. *Sedimentology* **39**, 423–454.

Ghosh, B., & Lowe, D. R. 1993. The architecture of deep-water channel complexes, Cretaceous Venado Sandstone Member, Sacramento Valley, California. In *Advances in the Sedimentary Geology of the Great Valley Group, Sacramento Valley, California* (ed. S. A. Graham & D. R. Lowe), pp. 51–65. Los Angeles, CA: Pacific Section, Society of Economic Paleontologists and Mineralogists.

Gibling, M. R., & Bird, D. J. 1994. Late Carboniferous cyclothems and alluvial paleovalleys in the Sydney Basin, Nova Scotia. *Geological Society of America Bulletin* **106**, 105–117.

Gilbert, G. K. 1885. *The Topographic Features of Lake Shores.* Washington, D.C.: United States Geological Survey.

Giles, M. R. 1997. *Diagenesis: A Quantitative Perspective, Implications for Basin Modelling and Rock Property Prediction.* Dordrecht: Kluwer Academic Publishers.

Giles, M. R., & de Boer, R. B. 1990. Origin and significance of redistributional secondary porosity. *Marine and Petroleum Geology* **7**, 378–397.

Gill, W. D., & Kuenen, P. H. 1958. Sand volcanoes on slumps in the Carboniferous of County Clare, Ireland. *Quarterly Journal of the Geology Society of London* **113**, 441–460.

Ginsburg, R. N. 1967. Stromatolites. *Science* **157**, 339–340.

2001. *Subsurface Geology of a Prograding Carbonate Platform Margin, Great Bahama Bank: Results of the Bahamas Drilling Project.* Tulsa, OK: SEPM (Society for Sedimentary Geology).

Ginsburg, R. N., Isham, L. B., Bein, S. J., & Kuperburg, J. 1954. *Laminated Algal Sediments of South Florida and Their Recognition in the Fossil Record.* Coral Gables, FL: University of Miami Marine Laboratory.

Ginsburg, R. N., & James, N. P. 1974. Holocene carbonate sediments of continental margins. In *The Geology of Continental Margins* (ed. C. A. Burke & C. L. Drake), pp. 137–155. New York: Springer-Verlag.

Giosan, L., & Bhattacharya, J. P. 2005. *River Deltas – Concepts, Models, and Examples.* Tulsa, OK: SEPM (Society for Sedimentary Geology).

Glen, J. W. 1952. Experiments on the deformation of ice. *Journal of Glaciology* **2**, 111–114.

1955. The creep of polycrystalline ice. *Proceedings of the Royal Society of London* **228A**, 519–538.

Glenn, C. R., & Garrison, R. E. 2003. Phosphorites. In *Encyclopedia of Sediments and Sedimentary Rocks* (ed. G. V. Middleton), pp. 257–263. Dordrecht: Kluwer Academic Publisher.

Glennie, K. W. 1970. *Desert Sedimentary Environments.* Amsterdam: Elsevier.

1986. Early Permian Rotliegend. In *Introduction to the Petroleum Geology of the North Sea* (ed. K. W. Glennie), pp. 63–86. Oxford: Blackwells.

Goldhammer, R. K. 1997. Compaction and decompaction algorithms for sedimentary carbonates. *Journal of Sedimentary Petrology* **67**, 26–35.

Goldhammer, R. K., Dunn, P. A., & Hardie, L. A. 1987. High frequency glacio-eustatic sea level oscillations with Milankovitch characteristics recorded in Middle Triassic platform carbonates in northern Italy. *American Journal of Science* **287**, 853–892.

1990. Depositional cycles, composite sea-level changes, cycle stacking patterns, and the hierarchy of stratigraphic forcing: examples from Alpine Triassic platform carbonates. *Geological Society of America Bulletin* **102**, 535–562.

Goldhammer, R. K., & Harris, M. T. 1989. Eustatic controls on the stratigraphy and geometry of the Latemar buildup (Middle Triassic): the Dolomites of northern Italy. In *Controls on Carbonate Platform and Basin Development* (ed. P. D. Crevello, J. L. Wilson, J. F. Sarg, & J. F. Read), pp. 323–338. Tulsa, OK: SEPM (Society for Sedimentary Geology).

Goldrich, S. S. 1938. A study in rock weathering. *Journal of Geology* **46**, 17–58.

Goldsmith, J. R., & Graf, D. L. 1958. Relation between lattice constants and composition of the Ca–Mg carbonates. *American Mineralogist* **43**, 84–101.

Goldsmith, J. R., Graf, D. L., & Joensuu, O. I. 1955. The occurrence of magnesium calcites in nature. *Geochimica et Cosmochimica Acta* **7**, 212–230.

Gosse, J. C., & Phillips, F. M. 2001. Terrestrial *in situ* cosmogenic nuclides: theory and applications. *Quaternary Sciences Reviews* **20**, 1475–1560.

Goudie, A. S. 1973. *Duricrusts in Tropical and Subtropical Landscapes.* Oxford: Clarendon Press.

1999. Wind erosional landforms: yardangs and pans. In *Aeolian Environments, Sediments and Landforms* (ed.

A. S. Goudie, I. Livingstone, & S. Stokes), pp. 167–180. Chichester: Wiley.

Goudie, A. S., Livingstone, I., & Stokes, S. 1999. *Aeolian Environments, Sediments and Landforms*. Chichester: Wiley.

Gould, H. R. 1970. The Mississippi delta complex. In *Deltaic Sedimentation Modern and Ancient* (ed. J. P. Morgan & R. H. Shaver), pp. 3–30. Tulsa, OK: SEPM (Society for Sedimentary Geology).

Graf, D. L., & Goldsmith, J. R. 1956. Some hydrothermal syntheses of dolomite and protodolomite. *Journal of Geology* **64**, 173–186.

Grant, W. D., & Madsen, O. S. 1979. Combined wave and current interaction with a rough bottom. *Journal of Geophysical Research* **84**, 1797–1808.

Grass, A. J. 1970. Initial instability of fine sand beds. *Journal of the Hydraulic Division of the ASCE* **96**, 619–632.

Greeley, R., & Iverson, J. D. 1985. *Wind as a Geological Process: On Earth, Mars, Venus and Titan*. Cambridge: Cambridge University Press.

Green, M. O., Rees, J. M., & Pearson, N. D. 1990. Evidence for the influence of wave–current interactions in a tidal boundary layer. *Journal of Geophysical Research* **95C**, 9629–9644.

Greenwood, B., & Mittler, P. R. 1985. Vertical sequence and lateral transitions in the facies of a barred nearshore environment. *Journal of Sedimentary Petrology* **55**, 366–375.

Greenwood, B., & Sherman, D. J. 1986. Hummocky cross-stratification in the surf zone: flow parameters and bedding genesis. *Sedimentology* **33**, 33–45.

1993. Waves, currents, sediment flux and morphological response in a barred nearshore system. *Marine Geology* **60**, 31–61.

Gregg, J. M., & Sibley, D. F. 1984. Epigenetic dolomitization and the origin of xenotopic dolomite textures. *Journal of Sedimentary Petrology* **54**, 908–931.

Gregory, K. J. 1983. *Background to Palaeohydrology*. Chichester: Wiley.

Gregory, K. J., Lewin, J., & Thornes, J. B. 1987. *Palaeohydrology in Practice*. Chichester: Wiley.

Gregory, K. J., Starkel, L., & Baker, V. R. 1996. *Continental Palaeohydrology*. Chichester: Wiley.

Griffing, D. H., Bridge, J. S. & Hotton, C. L. 2000. Coastal–fluvial paleoenvironments and plant ecology of the early Devonian (Emsian), Gaspe Bay, Canada. In *New Perspectives on the Old Red Sandstone* (ed. P. F. Friend, & B. P. J. Williams), pp. 61–84. London: Geological Society of London.

Griffiths, C. M., Dyt, C., Paraschivoiu, E., & Liu, K. 2001. SEDSIM in hydrocarbon exploration. In *Geologic Modeling and Simulation: Sedimentary Systems* (ed. D. F. Merriam &

J. C. Davis), pp. 71–97. New York: Kluwer Academic/Plenum Publishers.

Grotzinger, J. P., & James, N. P. 2000. *Carbonate Sedimentation and Diagenesis in the Evolving Precambrian World*. Tulsa, OK: SEPM (Society for Sedimentary Geology).

Grotzinger, J. P., Jordan, T. H., Press, F., & Siever, R. 2007. *Understanding Earth*, 5th edn. New York: W. H. Freeman and Company.

Grotzinger, J. P., & Knoll, A. H. 1999. Stromatolites in Precambrian Carbonates: evolutionary mileposts or environmental dipsticks. *Annual Review of Earth and Planetary Sciences* **27**, 313–358.

Grover, G. Jr., & Read, J. F. 1978. Fenestral and associated vadose diagenetic fabrics of tidal flat carbonates, Middle Ordovician New Market Limestone, southwestern Virginia. *Journal of Sedimentary Petrology* **48**, 453–473.

Grunau, H. R. 1965. Radiolarian cherts and associated rocks in space and time. *Eclogae Geologicae Helvetiae* **58**, 157–208.

Gruszczynski, M., Rudowski, S., Semil, J., Slominski, J., & Zrobek, J. 1993. Rip currents as a geological tool. *Sedimentology* **40**, 217–236.

Gupta, S. 1997. Himalayan drainage patterns and the origin of fluvial megafans in the Ganges foreland basin. *Geology* **25**, 11–14.

Gupta, S., & Cowie, P. 2000. Processes and controls in the stratigraphic development of extensional basins. *Basin Research* **12**, 185–241.

Gustard, A. 1996. Analysis of river regimes. In *River Flows and Channel Forms* (ed. G. Petts & P. Calow), pp. 32–50. Oxford: Blackwell Science Ltd.

Guza, R. T., & Bowen, A. J. 1975. The resonant instabilities of long waves obliquely incident on a beach. *Journal of Geophysical Research* **80**, 4529–4534.

Guza, R. T., & Davis, R. E. 1974. Excitation of edge waves by waves incident on a beach. *Journal of Geophysical Research* **79**, 1285–1291.

Haff, P. K. 1983. Grain flow as a fluid-mechanical phenomenon. *Journal of Fluid Mechanics* **134**, 401–430.

Hallet, B. 1996. Glacial quarrying: a simple theoretical model. *Annals of Glaciology* **22**, 1–8.

Halley, R. B., & Evans, C. C. 1983. *The Miami Limestone. A Guide to Selected Outcrops and Their Interpretation (with a Discussion of Diagenesis in the Formation)*. Miami, FL: Miami Geological Society.

Halley, R. B., Harris, P. M., & Hine, A. C., 1983. Bank margin. In *Carbonate Depositional Environments* (ed. P. A. Scholle, D. G. Bebout, & C. H. Moore), pp. 463–506. Tulsa, OK: American Association of Petroleum Geologists.

Hambrey, M. 1994. *Glacial Environments*. London: UCL Press.

Hambrey, M., & Harland, W. B. 1981. *Earth's Pre-Pleistocene Glacial Record*. Cambridge: Cambridge University Press.

Hampson, G. J., Elliott, T., & Davies, S. J. 1997. The application of sequence stratigraphy to Upper Carboniferous fluvio-deltaic strata of the onshore UK and Ireland: implications for the southern North Sea. *Journal of the Geological Society of London* **154**, 719–733.

Hampson, G. J., Davies, S. J., Elliott, T., Flint, S. S., & Stollhofen, H. 1999. Incised valley fill sandstone bodies in Upper Carboniferous fluvio-deltaic strata: recognition and reservoir characterization of Southern North Sea analogues. In *Petroleum Geology of Northwest Europe* (ed. A. J. Fleet & S. A. R. Boldy), pp. 771–788. London: The Geological Society.

Hampton, M. A. 1972. The role of subaqueous debris flow in generating turbidity currents. *Journal of Sedimentary Petrology* **42**, 775–793.

1975. Competence of fine-grained debris flows. *Journal of Sedimentary Petrology* **45**, 834–844.

Handford, C. F., Kendall, A. C., Prezbindowski, D. R., Dunham, J. B., & Logan, B. W. 1984. Salina-margin tepees, pisolites, and aragonite cements, Lake MacLeod, Western Australia: their significance in interpreting ancient analogs. *Geology* **12**, 523–527.

Handford, C. R., Loucks, R. G., & Davies, G. R. 1985. *Depositional and Diagenetic Spectra of Evaporites – A Core Workshop*. Tulsa, OK: SEPM (Society for Sedimentary Geology).

Hanna, S. R. 1969. The formation of longitudinal sand dunes by large helical eddies in the atmosphere. *Journal of Applied Meteorology* **8**, 874–883.

Hannington, M. D., Jonasson, I. R., Herzig, P. M., & Petersen, S. 1995. Physical and chemical processes of seafloor mineralization at mid-ocean ridges. In *Seafloor Hydrothermal Systems: Physical, Chemical, Biological, and Geological Interactions* (ed. S. E. Humphris, R. A. Zierenberg, L. S. Mullineaux, & R. E. Thomson), pp. 115–157. Washington: D.C.: American Geophysical Union.

Hanor, J. S. 1994. Origin of saline fluids in sedimentary basins. In *Geofluids: Origin, Migration and Evolution of Fluids in Sedimentary Basins* (ed. J. Parnell), pp. 151–174. London: Geological Society of London.

Hanor, J. S., & McIntosh, J. C. 2006. Are secular variations in seawater chemistry reflected in the compositions of basinal brines? *Journal of Geochemical Exploration* **89**, 153–156.

Häntschel, W. 1975. *Trace Fossils and Problematica. Treatise on Invertebrate Paleontology, Part W Supplement 1*. Boulder, CO: Geological Society of America; and Lawrence, KA: The University of Kansas.

Harbaugh, J. W., Watney, W. L., Rankey, E. C. *et al.* 1999. *Numerical Experiments in Stratigraphy: Recent Advances in Stratigraphic and Sedimentologic Computer Simulations*. Tulsa, OK: SEPM (Society for Sedimentary Geology).

Harbor, J. M. 1992. Numerical modeling of the development of U-shaped valleys by glacial erosion. *Geological Society of America Bulletin* **104**, 1364–1375.

Hardie, L. A. 1977. *Sedimentation on the Modern Carbonate Tidal Flats of Northwest Andros Island, Bahamas*. Baltimore, MA: The Johns Hopkins University Press.

1984. Evaporites: marine or nonmarine? *American Journal of Science* **284**, 193–240.

1987. Dolomitization: a critical view of some current views. *Journal of Sedimentary Petrology* **57**, 166–183.

1990. The roles of rifting and hydrothermal $CaCl_2$ brines in the origin of potash evaporites: an hypothesis. *American Journal of Science* **290**, 43–106.

1991. On the significance of evaporites. *Annual Review of Earth and Planetary Sciences* **19**, 131–168.

1996. Secular variation in seawater chemistry: an explanation for the coupled secular variation in the mineralogies of marine limestones and potash evaporites over the past 600 m.y. *Geology* **34**, 279–283.

2003a Evaporites. In *Encyclopedia of Sediments and Sedimentary Rocks* (ed. G. V. Middleton), pp. 257–263. Dordrecht: Kluwer Academic Publishing.

2003b Sabkha, salt flat, salina. In *Encyclopedia of Sediments and Sedimentary Rocks* (ed. G. V. Middleton), pp. 584–585. Dordrecht: Kluwer Academic Publishing.

Hardie, L. A., & Eugster, H. P. 1970. The evolution of closed basin brines. *Mineralogical Society of America Special Paper* **3**, 273–290.

Hardie, L. A., & Lowenstein, T. K. 2004. Did the Mediterranean Sea dry out during the Miocene? A reassessment of the evaporite evidence from DSDP Legs 13 and 42 A cores. *Journal of Sedimentary Research* **74**, 453–461.

Hardie, L. A., Lowenstein, T. K., & Spencer, R. J. 1985. The problem of distinguishing primary from secondary evaporites. In *Proceedings of the Sixth International Symposium on Salt* (ed. B. C. Schreiber & H. L. Harner), pp. 11–39. Alexandria, VA: The Salt Institute.

Hardie, L. A., & Shinn, E. A. 1986. Carbonate depositional environments modern and ancient, part 3: tidal flats. *Colorado School of Mines Quarterly* **81**, 1–74.

Hardie, L. A., Smoot, J. P., & Eugster, H. P. 1978. Saline lakes and their deposits, a sedimentological approach. In *Modern and Ancient Lake Sediments* (ed. A. Matter & M. E. Tucker), pp. 7–41. Oxford: Blackwells.

Harms, J. C., Southard, J. B., & Walker, R. G. 1982. *Structures and Sequences in Clastic Rocks*. Tulsa, OK: SEPM (Society for Sedimentary Geology).

Harris, N. B. 1989. Diagenetic quartzarenite and destruction of secondary porosity: an example from the Middle

Jurassic Brent sandstone of northwest Europe. *Geology* **17**, 361–364.

Harris, P. M. 1979. *Facies Anatomy and Diagenesis of a Bahamian Ooid Shoal. Sedimenta 7*. Miami, FL: Comparative Sedimentology Laboratory, University of Miami.

Harris, P. T., Pattiaratchi, C. B., Cole, A. R., & Keene, J. B. 1992. Evolution of subtidal sandbanks in Moreton bay, eastern Australia. *Marine Geology* **103**, 225–247.

Harris, P. T., Pattiaratchi, C. B., Collins, M. B., & Dalrymple, R. W. 1995. What is a bedload parting? In *Tidal Signatures in Modern and Ancient Sediments* (ed. B. W. Flemming & A. Bartholoma), pp. 3–18. Oxford: Blackwells.

Harris, P. T., Pattiaratchi, C. B., Keene, J. B. *et al.* 1996. Late Quaternary deltaic and carbonate sedimentation in the Gulf of Papua foreland basin: response to sea-level change. *Journal of Sedimentary Research* **B66**, 801–819.

Harrison, W. J., & Summa, L. L. 1991. Paleohydrology of the Gulf of Mexico Basin. *American Journal of Science* **291**, 109–176.

Hart, B. S., & Plint, A. G. 1989. Gravelly shoreface deposits: a comparison of modern and ancient facies sequences. *Sedimentology* **36**, 551–557.

Hartley, A. J., & Prosser, D. J. 1995. *Characterization of Deep Marine Clastic Systems*. London: Geological Society of London.

Hartmann, D. L. 1994. *Global Physical Climatology*. San Diego, CA: Academic Press.

Harvie, C. E., Möller, N., & Weare, J. H. 1984. The prediction of mineral solubilities in natural waters: the Na–K–Mg–Ca–H–Cl–SO$_4$–OH–HCO$_3$–CO$_3$–CO$_2$–H$_2$O system to high ionic strengths at 25 °C. *Geochimica et Cosmochimica Acta* **48**, 723–752.

Harvie, C. E., Weare, J. H., Hardie, L. A., & Eugster, H. P. 1980. Evaporation of sea water: calculated mineral sequences. *Science* **208**, 498–500.

Hasiotis, S. T. 2002. *Continental Trace Fossils*. Tulsa, OK: SEPM (Society for Sedimentary Geology).

2003. Complex ichnofossils of solitary and social soil organisms: understanding their evolution and roles in terrestrial paleoecosystems. *Palaeogeography, Palaeoclimatology, Palaeoecology* **192**, 259–320.

2004. Reconnaissance of Upper Jurassic Morrison Formation ichnofossils, Rocky Mountain Region, USA: paleoenvironmental, stratigraphic, and paleoclimatic significance of terrestrial and freshwater ichnocoenoses. *Sedimentary Geology* **167**, 177–268.

Hättestrand, C., & Kleman, J. 1999. Ribbed moraine formation. *Quaternary Science Reviews* **18**, 43–61.

Havholm, K. G., Blakey, R. C., Capps, M. *et al.* 1993. Aeolian genetic stratigraphy: an example from the Middle Jurassic Page Sandstone, Colorado Plateau. In *Aeolian Sediments: Ancient and Modern* (ed. K. Pye & N. Lancaster), pp. 87–107. Oxford: Blackwells.

Hawley, N. 1981. Flume experiments on the origin of flaser bedding. *Sedimentology* **28**, 699–712.

Hayes, M. O. 1967. *Hurricanes as Geological Agents: Case Studies of Hurricanes Carla, 1961, and Cindy, 1963*. Austin, TX: Bureau of Economic Geology.

1975. Morphology of sand accumulation in estuaries: an introduction to the symposium. In *Estuarine Research: Volume II Geology and Engineering* (ed. L E. Cronin), pp. 3–22. London: Academic Press.

1979. Barrier island morphology as a function of tidal and wave regime. In *Barrier Islands – From the Gulf of St Lawrence to the Gulf of Mexico* (ed. S. P. Leatherman), pp. 1–27. London: Academic Press.

Heckel, P. H., Gibling, M. R., & King, N. R. 1998. Stratigraphic model for glacial–eustatic Pennsylvanian cyclothems in highstand nearshore detrital regimes. *Journal of Geology* **106**, 373–383.

Heezen, B. C., Ericson, D. B., & Ewing, M. 1954. Further evidence for a turbidity current following the 1929 Grand banks earthquake. *Deep-Sea Research* **1**, 193–202.

Heiken, G., & Wohletz, K. 1985. *Volcanic Ash*. Berkeley: University of California Press.

Heikoop, J. M., Tsujita, C. J., Risk, M. J., Tomascik, T., & Mah, A. J. 1996. Modern iron ooids from a shallow-marine volcanic setting: Mahengetang, Indonesia. *Geology* **24**, 759–762.

Heller, P. L., & Paola, C. 1992. The large scale dynamics of grain-size variation in alluvial basins, 2. Application to syntectonic conglomerate. *Basin Research* **4**, 91–102.

1996. Downstream changes in alluvial architecture: an exploration of controls on channel-stacking patterns. *Journal of Sedimentary Research* **B66**, 297–306.

Hequette, A., & Hill, P. R. 1993. Storm-generated currents and offshore sediment transport on a sandy shoreface, Tibjak Beach, Canadian Beaufort Sea. *Marine Geology* **113**, 283–304.

Hesp, P., Hyde, R., Hesp, V., & Zengyu, Q. 1989. Longitudinal dunes can move sideways. *Earth Surface Processes and Landforms* **14**, 447–451.

Hesse, R. 1990a. Origin of chert: diagenesis of biogenic siliceous sediments. In *Diagenesis* (ed. I. A. McIlreath & D. W. Morrow), pp. 227–251. St. John's, Newfoundland: Geological Association of Canada.

1990b. Silica diagenesis: origin of inorganic and replacement cherts. In *Diagenesis* (ed. I. A. McIlreath & D. W. Morrow), pp. 253–275. St. John's, Newfoundland: Geological Association of Canada.

Hills, J. G., & Mader, C. L. 1997. Tsunami produced by the impacts of small asteroids. *Annals of New York Academy of Sciences* **822**, 381–394.

Hindmarsh, R. C. A. 1998. Drumlinization and drumlin-forming instabilities: viscous till mechanisms. *Journal of Glaciology* **44**, 293–314.

Hindmarsh, R. C. A, Dunlop, P., Clark, C. D. 2003. Modeling the geomorphological effects of till redistribution; assessing a dynamic theory for Rogen moraine formation and drumlin formation. In *16th Congress of the International Union for Quaternary Research*. Reno, NV: INQUA.

Hine, A. C. 1975. Bedform distribution and migration patterns on tidal deltas in the Chatham Harbor estuary, Cape Cod, Massachusetts. In *Estuarine Research: Volume II, Geology and Engineering* (ed. L. E. Cronin), pp. 235–252. London: Academic Press.

　1977. Lily Bank, Bahamas: history of an active oolite sand shoal. *Journal of Sedimentary Petrology* **47**, 1554–1581.

Hine, A. C., & Neumann, A. C. 1977. Shallow carbonate bank margin growth and structure, Little Bahama Bank, Bahamas. *American Association of Petroleum Geologists Bulletin* **61**, 376–406.

Hinnov, L. 2000. New perspectives on orbitally forced stratigraphy. *Annual Review of Earth and Planetary Sciences* **28**, 419–475.

Hoffman, P. 1973. Recent and ancient algal stromatolites: seventy years of pedagogic cross-pollination. In *Evolving Concepts in Sedimentology* (ed. R. N. Ginsburg), pp. 178–191. Baltimore, MA: The Johns Hopkins University Press.

　1974. Shallow and deepwater stromatolites in Lower Proterozoic platform-to-basin facies change, Great Slave Lake, Canada. *American Association of Petroleum Geologists Bulletin* **58**, 856–867.

　1976. Stromatolite morphogenesis in Shark Bay, Western Australia. In *Stromatolites* (ed. M. R. Walter), pp. 261–271. Amsterdam: Elsevier.

Hoffman, P. F., Kaufman, A. J., Halverson, G. P., & Schrag, D. P. 1998. A Neoproterozoic snowball Earth. *Science* **281**, 1342–1346.

Hoffman, P. F., & Schrag, D. P. 2002. The snowball Earth hypothesis: testing the limits of global change. *Terra Nova* **14**, 129–155.

Hogg, A. J., & Huppert, H. E. 2001. Two-dimensional and axisymmetric models for compositional and particle-driven gravity currents in uniform ambient flows. In *Particulate Gravity Currents* (ed. W. D. McCaffrey, B. C. Kneller, & J. Peakall), pp. 121–134. Oxford: Blackwells.

Holland, H. D. 1972. The geologic history of seawater – an attempt to solve the problem. *Geochimica et Cosmochimica Acta* **36**, 637–651.

　1994. Early Proterozoic atmospheric change. In *Early Life on Earth, Nobel Symposium No. 84* (ed. S. Bengston), pp. 237–244. New York: Columbia University Press.

　2005. Sea level, sediments and the composition of sea water. *American Journal of Science* **305**, 220–239.

Hollister, C. D. 1993. The concept of deep-sea contourites. *Sedimentary Geology* **82**, 5–15.

Hollister, C. D., & McCave, I. N. 1984. Sedimentation under deep-sea storms. *Nature* **309**, 220–225.

Holman, R. A. 1983. Edge waves and the configuration of the shoreline. In *CRC Handbook of Coastal Processes and Erosion* (ed. P. D. Komar), pp. 21–34. Boca Raton, FL: CRC Press.

Holman, R. A., & Sallenger, A. H. 1993. Sand bar generation: a discussion of the Duck experimental series. *Journal of Coastal Research, Special Issue* **15**, 76–92.

Holmlund, P. 1988. Internal geometry and evolution of moulins, Storglaciaren, Sweden. *Journal of Glaciology* **34**, 242–248.

Homewood, P., & Allen, P. A. 1981. Wave-, tide-, and current-controlled sandbodies of Miocene molasse, western Switzerland. *American Association of Petroleum Geologists Bulletin* **65**, 2534–2545.

Honji, H., Kaneko, A., & Matsunaga, N. 1980. Flows above oscillatory ripples. *Sedimentology* **27**, 225–229.

Hooke, R. LeB. 1998. *Principles of Glacier Mechanics*. Englewood Cliffs, NJ: Prentice-Hall.

Hooke, R. L., & Jennings, C. E. 2006. On the formation of the tunnel valleys of the southern Laurentide ice sheet. *Quaternary Science Reviews* **25**, 1364–1372.

Horbury, A. D., & Robinson, A. G. 1993. *Diagenesis and Basin Development*. Tulsa, OK: American Association of Petroleum Geologists.

Hori, K., Saito, Y., Zhang, Q., & Wang, P. 2002. Architecture and evolution of the tide-dominated Changjiang (Yangtze) River delta, China. *Sedimentary Geology* **146**, 249–264.

Horita, J., Zimmermann, H., & Holland, H. D. 2002. Chemical evolution of seawater during the Phanerozoic: implications from the record of marine evaporites. *Geochimica et Cosmochimica Acta* **66**, 3733–3756.

Horowitz, D. H. 1982. Geometry and origin of large-scale deformation structures in some ancient windblown sand deposits. *Sedimentology* **29**, 155–180.

Horton, R. E. 1945. Erosional development of streams and their drainage basins: hydrophysical approach to quantitative morphology. *Geological Society of America Bulletin* **56**, 275–370.

House, M. R., & Gale, A. S. 1995. *Orbital Forcing Timescales and Cyclostratigraphy*. London: Geological Society of London.

Howard, A. D. 1967. Drainage analysis in geologic interpretation: a summary. *American Association of Petroleum Geologists Bulletin* **51**, 2246–2259.

Howard, A. D. 1987. Modeling fluvial systems: rock-, gravel- and sand-bed channels. In *River Channels: Environment and Process* (ed. K. S. Richards), pp. 69–94. Oxford: Blackwells.

Howard, J. D., & Frey, R. W. 1975. Estuaries of the Georgia coast, U.S.A.: sedimentology and biology, II. Regional animal-sediment characteristics of Georgia Estuaries. *Senckenbergiana Maritima* **7**, 33–103.

Howard, J. D., & Reineck, H.-E. 1972. Georgia coastal region, Sapelo Island, U.S.A.: sedimentology and biology, IV. Physical and biogenic sedimentary structures of the nearshore shelf. *Senckenbergiana Maritima* **4**, 47–79.

1981. Depositional facies of high-energy beach-to-offshore sequence, comparison with low energy sequence. *American Association of Petroleum Geologists Bulletin* **65**, 807–830.

Hower, J., Eslinger, E. V., Hower, M. E., & Perry, E. A. 1976. Mechanism of burial metamorphism of argillaceous sediment: 1. Mineralogical and chemical evidence. *Geological Society of America Bulletin* **87**, 725–737.

Hoyt, J. H., & Henry, V. J. 1967. Influence of island migration on barrier island sedimentation. *Geological Society of America Bulletin* **78**, 77–86.

1971. Origin of capes and shoals along the southeastern coast of the United States. *Geological Society of America Bulletin* **82**, 59–66.

Hsü, K. J. 1972a. Origin of saline giants: a critical review after the discovery of the Mediterranean evaporite. *Earth-Science Reviews* **8**, 371–396.

1972b. When the Mediterranean dried up. *Scientific American* **227**, 26–36.

Hsü, K. J., & Jenkyns, H. C. 1974. *Pelagic Sediments: On Land and Under the Sea*. Oxford: Blackwells.

Hubbard, D. K., Oertel, G., & Nummedal, D. 1979. The role of waves and tidal currents in the development of tidal-inlet sedimentary structures and sand body geometry: examples from North Carolina, South Carolina and Georgia. *Journal of Sedimentary Petrology* **49**, 1073–1092.

Hulbe, C. L., MacAyeal, D. R., Denton, G. H., Kleman, J., & Lowell, T. V. 2004. Catastrophic ice shelf breakup as the source of Heinrich event icebergs. *Paleoceanography* **19**, PA1004, doi: 10.1029/2003PA000890.

Hulscher, S. 1996. Tide-induced large-scale regular bedform patterns in three-dimensional shallow water model. *Journal of Geophysical Research* **101**, 20,727–20,744.

Hunt, D., & Gawthorpe, R. L. 2000. *Sedimentary Responses to Forced Regressions*. London: Geological Society of London.

Hunter, R. E. 1977. Basic types of stratification in small eolian dunes. *Sedimentology* **24**, 361–387.

1985. Subaqueous sand flow cross strata. *Journal of Sedimentary Petrology* **55**, 886–894.

Hunter, R. E., & Clifton, H. E. 1982. Cyclic deposits and hummocky cross-stratification of probable storm origin in the Upper Cretaceous rocks of the Cape Sebastian area, south-western Oregon. *Journal of Sedimentary Petrology* **52**, 127–143.

Hunter, R. E., Clifton, H. E., & Phillips, R. L. 1979. Depositional processes, sedimentary structures and predicted vertical sequences in barred nearshore systems, southern Oregon coast. *Journal of Sedimentary Petrology* **49**, 711–726.

Hunter, R. E., & Kocurek, G. 1986. An experimental study of subaqueous slipface deposition. *Journal of Sedimentary Petrology* **56**, 387–394.

Huntley, D. A., & Bowen, A. J. 1973. Field observations of edge waves. *Nature* **243**, 160–161.

Huntley, D. A., Guza, R. T., & Thornton, D. B. 1981. Field observations of surf beat, 1: progressive edge waves. *Journal of Geophysical Research* **83**, 1913–1920.

Huppert, H. E. 1998. Quantitative modeling of granular suspension flows. *Philosophical Transactions of the Royal Society of London* **356A**, 2471–2496.

Hutcheon, I. E. 1990. Aspects of diagenesis in coarse-grained siliciclastic rocks. In *Diagenesis* (ed. I. A. McIlreath & D. W. Morrow), pp. 165–176. St. John's, Newfoundland: Geological Association of Canada.

Huthnance, J. M. 1982a. On one mechanism forming linear sand banks. *Estuarine, Coastal and Shelf Science* **14**, 79–99.

1982b. On the formation of sand banks of finite extent. *Estuarine, Coastal and Shelf Science* **15**, 277–299.

Hyndman, D., & Hyndman, D. 2006. *Natural Hazards and Disasters*. Belmont, CA: Thomson Brooks/Cole.

Iida, K., & Iwasaki, T. 1983. *Tsunamis: Their Science and Engineering*. Dordrecht: Reidel.

Ikeda, S., & Parker, G. 1989. *River Meandering*. Washington, D.C.: American Geophysical Union.

Illing, L. V., Wells, A. J., & Taylor, J. C. M. 1965. Penecontemporary dolomite in the Persian Gulf. In *Dolomitization and Limestone Diagenesis, A Symposium* (ed. L. C. Pray & R. C. Murray), pp. 89–111. Tulsa, OK: SEPM (Society for Sedimentary Geology).

Imboden, D. M., & Wüest, A. 1995. Mixing mechanisms in lakes. In *Physics and Chemistry of Lakes*, 2nd edn. (ed. A. Lerman, D. Imboden, & J. Gat), pp. 83–138. New York: Springer-Verlag.

Imbrie, J., Hays, J. D., Martinson, D. G. *et al.* 1984. The orbital theory of Pleistocene climate: support from a revised chronology of the marine δ^{18}O record. In *Milankovitch and Climate: Understanding the Response to Astronomical Forcing* (ed. A. Berger, J. Imbrie, J. Hays, & G. Kukla), pp. 269–305. Dordrecht: Reidel.

Imbrie, J., & Imbrie, K. P. 1979. *Ice Ages: Solving the Mystery*. Cambridge, MA: Harvard University Press.

Imbrie, J., Mix, A. C., & Martinson, D. G. 1993. Milankovitch theory viewed from Devils Hole. *Nature* **363**, 531–533.

Imperato, D. P., Sexton, W. J., & Hayes, M. O. 1988. Stratigraphy and sediment characteristics of a mesotidal ebb-tidal delta, North Edisto Inlet, South Carolina. *Journal of Sedimentary Petrology* **58**, 950–958.

Immenhauser, A., Hillgartner, H., & van Benthum, E. 2005. Microbial–foraminiferal episodes in the Early Aptian of the southern Tethyan margin: ecological significance and possible relation to oceanic anoxia. *Sedimentology* **52**, 77–99.

Inden, R. F., & Moore, C. H. 1983. Beach. In *Carbonate Depositional Environments* (ed. P. A. Scholle, D. G. Bebout, & C. H. Moore), pp. 211–265. Tulsa, OK: American Association of Petroleum Geologists.

Iverson, N. R. 1991. Potential effects of subglacial water-pressure fluctuations on quarrying. *Journal of Glaciology* **37**, 27–36.

Iverson, R. M. 1997. The physics of debris flows. *Reviews of Geophysics* **35**, 245–296.

Jackson, R. G. 1975. Hierarchical attributes and a unifying model of bedforms composed of cohesionless sediment and produced by shearing flow. *Geological Society of America Bulletin* **86**, 1,523–1,533.

1976. Sedimentological and fluid-dynamic implications of the turbulent bursting phenomenon in geophysical flows. *Journal of Fluid Mechanics* **77**, 531–560.

Jaeger, H. M., Nagel, S. R., & Behringer, R. P. 1996. The physics of granular materials. *Physics Today* **49**, 32–36.

Jahren, A. H. 2004. Factors of soil formation: biota. In *Encyclopedia of Soils in the Environment* (ed. D. Hillel, C. Rosenzweig, D. Powlson *et al.*), pp. 507–512. New York: Academic Press.

James, N. P., & Choquette, P. W. 1983. Diagenesis 6. Limestones – the sea floor diagenetic environment. *Geoscience Canada* **11**, 161–194.

James, N. P., & MacIntyre, I. G. 1985. Carbonate depositional environments, modern and ancient. Part 1 – reefs, zonation, depositional facies and diagenesis. *Quarterly Journal of the Colorado School of Mines* **80**, 70pp.

James, H. L. 1954. Sedimentary facies of iron formations. *Economic Geology* **49**, 235–293.

James, N. P. 1981. Megablocks of calcified algae in the Cow Head Breccia, western Newfoundland: vestiges of a Cambro-Ordovician margin. *Geological Society of America Bulletin* **42**, 799–811.

1983. Reefs. In *Carbonate Depositional Environments* (ed. P. A. Scholle, D. G. Bebout, & C. H. Moore), pp. 345–462. Tulsa, OK: American Association of Petroleum.

James, N. P., Bone, Y., Collins, L. B., & Kyser, T. K. 2001. Surficial sediments of the Great Australian Bight: facies dynamics and oceanography on a vast cool-water carbonate shelf. *Journal of Sedimentary Research* **71**, 5549–567.

James, N. P., & Bourque, P. A. 1992. Reefs and mounds. In *Facies Models: Response to Sea Level Change* (ed. R. G. Walker & N. P. James), pp. 323–345. St. John's, Newfoundland: Geological Association of Canada.

James, N. P., & Choquette, P. W. 1988. *Paleokarst*. New York: Springer-Verlag.

1990a. Limestones – introduction. In *Diagenesis* (ed. I. A. McIlreath & D. W. Morrow), pp. 9–12. St. John's, Newfoundland: Geological Association of Canada.

1990b. Limestones – the sea floor diagenetic environment. In *Diagenesis* (ed. I. A. McIlreath & D. W. Morrow), pp. 13–34. St. John's, Newfoundland: Geological Association of Canada.

1990c. Limestones – the meteoric diagenetic environment. In *Diagenesis* (ed. I. A. McIlreath & D. W. Morrow), pp. 35–74. St. John's, Newfoundland: Geological Association of Canada.

James, N. P., & Clarke, A. D. 1997. *Cool Water Carbonates*. Tulsa, OK: SEPM (Society for Sedimentary Geology).

James, N. P., & Ginsburg, R. N. 1979. *The Seaward Margin of the Belize Reef*. Oxford: Blackwells.

James, N. P., & Kendall, A. C. 1992. Introduction to carbonate and evaporite facies models. In *Facies Models: Response to Sea Level Change* (ed. R. G. Walker & N. P. James), pp. 265–275. St. John's, Newfoundland: Geological Association of Canada.

Jenkins, J. G., & Savage, S. B. 1983. A theory for the rapid flow of identical, smooth, nearly elastic spherical particles. *Journal of Fluid Mechanics* **130**, 187–202.

Jenkyns, H. C., Forster, A., Schouten, S., & Damste, J. S. S. 2004. High temperatures in the Late Cretaceous Arctic Ocean. *Nature* **432**, 888–892.

Johnson, A. M. 1970. *Physical Processes in Geology*. San Francisco: W. H. Freeman.

Johnson, D. D., & Beaumont, C. 1995. Preliminary results from a planform kinematic model of orogen evolution, surface processes and the development of clastic foreland basin stratigraphy. In *Stratigraphic Evolution of Foreland Basins* (ed. S. L. Dorobek & G. M. Ross), pp. 3–24. Tulsa, OK: SEPM (Society for Sedimentary Geology).

Johnson, M. A., Kenyon, N. H., Belderson, R. H., & Stride, A. H. 1982. Sand transport. In *Offshore Tidal Sands* (ed. A. H. Stride), pp. 58–94. London: Chapman and Hall.

Johnson, N. M., & Driess, S. J. 1989. Hydrostratigraphic interpretation using indicator geostatistics. *Water Resources Research* **25**, 2501–2510.

Jones, B., & Desrochers, A. 1992. Shallow platform carbonates. In *Facies Models: Response to Sea Level Change* (ed. R. G. Walker & N. P. James), pp. 277–301. St. John's, Newfoundland: Geological Association of Canada.

Jones, B., & Luth, R. W. 2002. Dolostones from Grand Cayman, British West Indes. *Journal of Sedimentary Research* **72**, 559–569.

Jones, L. S., & Schumm, S. A. 1999. Causes of avulsion: an overview. In *Fluvial Sedimentology VI* (ed. N. D. Smith & J. Rogers), pp. 171–178. Oxford: Blackwells.

Jopling, A. V., & McDonald, B. C. 1975. *Glaciofluvial and Glaciolacustrine Sediments*. Tulsa, OK: SEPM (Society for Sedimentary Geology).

Jordan, T. E. 1981. Thrust loads and foreland basin evolution, Cretaceous western United States. *American Association of Petroleum Geologists Bulletin* **65**, 2506–2620.

Jordan, T. E., & Flemings, P. B. 1990. From geodynamical models to basin fill – a stratigraphic perspective. In *Quantitative Dynamic Stratigraphy* (ed. T. A. Cross), pp. 149–163. Englewood Cliffs, NJ: Prentice-Hall.

1991. Large-scale stratigraphic architecture, eustatic variation, and unsteady tectonism: a theoretical evaluation. *Journal of Geophysical Research* **96**, 6681–6699.

Jorgenson, B. B. 2000. Bacteria and marine biogeochemistry. In *Marine Geochemistry* (ed. H. D. Schulz & M. Zabel), pp. 173–207. Berlin: Springer-Verlag.

Julien, P. Y. 2002. *River Mechanics*. Cambridge: Cambridge University Press.

Jullien, R., Meakin, P., & Pavlovitch, A. 2002. Three-dimensional model for particle-size segregation by shaking. *Physical Review Letters* **69**, 640–643.

Kaiser, J. 2004. Wounding Earth's fragile skin. *Science* **304**, 1616–1618.

Kamb, B. 1987. Glacier surge mechanism based on linked cavity configuration of the basal water conduit system. *Journal of Geophysical Research* **92**, 9083–9100.

1991. Rheological nonlinearity and flow instability in the deforming bed mechanism of ice stream motion. *Journal of Geophysical Research* **96B**, 16,585–16,595.

Kamb, B., Raymond, C. F., Harrison, W. D. *et al.* 1985. Glacier surge mechanism: 1982–1983 surge of Variegated Glacier, Alaska. *Science* **227**, 469–479.

Kanfoush, S. L., Hodell, D. A., Charles, C. D. *et al.* 2000. Millennial-scale instability of the Antarctic Ice sheet during the last glaciation. *Science* **288**, 1815–1819.

Karssenberg, D., Törnqvist, T. E., & Bridge, J. S. 2001. Conditioning a process-based model of sedimentary architecture to well data. *Journal of Sedimentary Research* **71**, 868–879.

Karssenberg, D., Dalman, R., Weltje, G. J., Postma, G., & Bridge, J. S. 2004. Numerical modelling of delta evolution by nesting high and low resolution process-based models of sedimentary basin filling. In *Joint EURODELTA/ EUROSTRATAFORM Meeting*, Venice (Italy), pp. 1–37.

Katz, M. E., Finkel, Z. V., Grzebyk, D., Knoll, A. H., & Falkowski, P. G. 2004. Evolutionary trajectories and biogeochemical impacts of marine eukaryotic phytoplankton. *Annual Review of Ecology, Evolution, and Systematics* **35**, 523–556.

Keen, T. R., & Glenn, S. M. 1994. A coupled hydrodynamic–bottom boundary layer model of Ekman flow on stratified continental shelves. *Journal of Physical Oceanography* **24**, 1732–1749.

Kelly, D. S., Karson, J. A., Früh-Green, G. L. *et al.* 2005. A serpentine-hosted ecosystem: the Lost City hydrothermal field. *Science* **307**, 1428–1434.

Kemp, P. H., & Simons, R. R. 1982. The interaction between waves and a turbulent current: waves propagating with the current. *Journal of Fluid Mechanics* **116**, 227–250.

Kendall, A. C. 1992. Evaporites. In *Facies Models: Response to Sea Level Change* (ed. R. G. Walker & N. P. James), pp. 375–409. St. John's, Newfoundland: Geological Association of Canada.

Kendall, A. C., & Harwood, G. M. 1996. Marine evaporites: arid shorelines and basins. In *Sedimentary Environments: Processes, Facies and Stratigraphy*, 3rd edn. (ed. H. G. Reading), pp. 281–324. Oxford: Blackwell Science.

Kendall, C. G. St. C., & Skipwith, P. A. D'E. 1968. Recent algal mats of a Persian Gulf lagoon. *Journal of Sedimentary Petrology* **38**, 1040–1058.

1969. Geomorphology of a Recent shallow-water carbonate province: Khor al Bazam, Trucial Coast, southwestern Persian Gulf. *Geological Society of America Bulletin* **80**, 865–892.

Kendall, C. G. St. C., & Warren, J. 1987. A review of the origin and setting of tepees and their associated fabrics. *Sedimentology* **34**, 1007–1027.

Kennard, J. M., & James, N. P. 1986. Thrombolites and stromatolites: two distinct types of microbial structures. *Palios* **1**, 492–503.

Kennedy, J. F. 1963. The mechanics of dunes and antidunes in erodible-bed channels. *Journal of Fluid Mechanics* **16**, 521–544.

Kenyon, N. H., Amir, A., & Cramp, A. 1995. Geometry of the younger sediment bodies on the Indus Fan. In *Atlas of Deep Water Environments: Architectural Style in Turbidite Systems* (ed. K. T. Pickering, R. N. Hiscott, N. H. Kenyon, F. Ricci Lucci, & R. D. A. Smith), pp. 89–93. London: Chapman and Hall.

Kiessling, W., Flügel, E., & Golonka, J. 2002. *Phanerozoic Reef Patterns*. Tulsa, OK: SEPM (Society for Sedimentary Geology).

Kim, J., Dong, H., Seabaugh, J., Newell, S. W., & Eberl, D. D. 2004. Role of microbes in the smectite-to-illite reaction. *Science* **303**, 830–832.

King, L. H. 1993. Till in the marine environment. *Journal of Quaternary Science* **8**, 347–358.

Kinsman, D. J. 1966. Gypsum and anhydrite of Recent age, Trucial Coast, Persian Gulf. In *Proceedings of the Second Salt Symposium* (ed. J. L. Rau), Vol. 1, pp. 302–326. Cleveland, OH: Northern Ohio Geological Society.

Kinsman, D. J. J., & Park, R. K. 1976. Algal belt and coastal sabkha evolution, Trucial Coast, Persian Gulf. In *Stromatolites* (ed. M. R. Walter), pp. 421–433. Amsterdam: Elsevier.

Kirkby, M. J. 1994. Thresholds and instability in stream head hollows: a model of magnitude and frequency for wash processes. In *Process Models and Theoretical Geomorphology* (ed. M. J. Kirkby), pp. 294–314. Chichester: Wiley.

Kirschvink, J. L. 1992. Late Proterozoic low latitude glaciation: the snowball Earth. In *The Proterozoic Biosphere* (ed. J. W. Schopf & C. Klein), pp. 51–52. Cambridge: Cambridge University Press.

Klappa, C. F. 1978. Biolithogenesis of Microcodium; elucidation. *Sedimentology* **25**, 489–522.

Klein, C., & Beukes, N. J. 1992. Proterozoic iron formations. In *Proterozoic Crustal Evolution* (ed. K. C. Condie), pp. 383–418. Amsterdam: Elsevier.

Klein, G. deV. 1970. Depositional and dispersal dynamics of intertidal sand bars. *Journal of Sedimentary Petrology* **40**, 1095–1127.

Kleinhans, M. G. 2004. Sorting in grains flows at the lee side of dunes. *Earth-Science Reviews* **65**, 75–102.

2005. Grain-size sorting in grainflows at the lee side of deltas. *Sedimentology* **52**, 291–311.

Kleypas, J. A., Buddemeier, R. W., Archer, D. *et al.* 1999. Geochemical consequences of increased atmospheric carbon dioxide on coral reefs. *Science* **284**, 118–120.

Knauth, L. P. 2003. Siliceous sediments. In *Encyclopedia of Sediments and Sedimentary Rocks* (ed. G. V. Middleton), pp. 660–666. Dordrecht: Kluwer Academic Publishers.

Kneller, B. 1996. Beyond the turbidite paradigm: physical models for deposition of turbidites and their implication for reservoir prediction. In *Characterisation of Deep Marine Clastic Systems* (ed. A. Hartley & D. J. Prosser), pp. 29–46. London: Geological Society of London.

Kneller, B. C., & Branney, M. J. 1995. Sustained high density turbidity currents and the deposition of massive sands. *Sedimentology* **42**, 607–616.

Kneller, B. C., & Buckee, C. 2000. The structure and fluid mechanics of turbidity currents: a review of some recent studies and their geological implications. *Sedimentology* **47** (Supplement 1), 62–94.

Kneller, B. C., Edwards, D., McCaffrey, W., & Moore, R. 1991. Oblique reflection of turbidity currents. *Geology* **19**, 250–252.

Knight, J. B., Jaeger, H. M., & Nagel, S. R. 1993. Vibration-induced size separation in granular media: the convection connection. *Physical Review Letters* **70**, 3728–3731.

Knight P. G. 2006. *Glacier Science and Environmental Change.* Oxford: Blackwells.

Knight, R. J., & McLean, J. R. 1986. *Shelf Sands and Sandstones.* Calgary, Alberta: Canadian Society of Petroleum Geologists.

Knighton, A. D. 1998. *Fluvial Forms and Processes: A New Perspective.* London: Arnold.

Knoll, A. H., & Carroll, S. B. 1999. Early animal evolution: emerging views from comparative biology and geology. *Science* **284**, 2129–2137.

Knox, J. C. 1983. Responses of river systems to Holocene climates. In *Late Quaternary Environments of the United States, Volume 2, The Holocene* (ed. H. E. Wright & S. C. Porter), pp. 26–41. Minneapolis, MN: University of Minnesota Press.

Knox, J. C. 1996. Fluvial systems since 20,000 years BP. In *Continental Palaeohydrology* (ed. K. J. Gregory, L. Starkel, & V. R. Baker), pp. 87–108. Chichester: Wiley.

Kochel, R. C., & Baker, V. R. 1988. Palaeoflood analysis using slack water deposits. In *Flood Geomorphology* (ed. V. R. Baker, R. C. Kochel, & P. P. Patton), pp. 357–376. New York: Wiley.

Kocurek, G. 1981. Significance of interdune deposits and bounding surfaces in eolian dune sands. *Sedimentology* **28**, 753–780.

1988. First order and super bounding surfaces in eolian sequences – bounding surfaces revisited. *Sedimentary Geology* **56**, 193–206.

1991. Interpretation of ancient eolian sand dunes. *Annual Review of Earth and Planetary Sciences* **19**, 43–75.

1996. Desert aeolian systems. In *Sedimentary Environments: Processes, Facies and Stratigraphy*, 3rd edn. (ed. H. G. Reading), pp. 125–155. Oxford: Blackwells.

Kocurek, G., & Havholm, K. G. 1993. Eolian sequence stratigraphy – a conceptual framework. In *Recent Advances in and Application of Siliciclastic Sequence Stratigraphy* (ed. P. Weimer & H. Posamentier), pp. 393–409. Tulsa, OK: American Association of Petroleum Geologists.

Kocurek, G., & Lancaster, N. 1999. Aeolian sediment states: theory and Mojave Kelso Dunefield example. *Sedimentology* **46**, 505–516.

Kocurek, G., & Nielsen, J. 1986. Conditions favourable for the formation of warm-climate aeolian sand sheets. *Sedimentology* **33**, 795–816.

Kocurek, G., Townsley, M., Yeh, E., Havholm, K., & Sweet, M. L. 1992. Dune and dune-field development on Padre Island, Texas, with implications for interdune deposition and water-table controlled accumulation. *Journal of Sedimentary Petrology* **62**, 622–635.

Kohout, F. A., Henry, H. R., & Banks, J. E. 1977. Hydrogeology related to geothermal conditions of the Florida Plateau. In *The Geothermal Nature of the Florida Plateau* (ed. D. L. Smith & G. E. Griffin), pp. 1–41. Tallahassee, FL: Florida Bureau of Geology.

Kolb, C. R., & Van Lopik, J. R. 1966. Depositional environments of the Mississippi River deltaic plain – southeastern Louisiana. In *Deltas in Their Geologic Framework* (ed. M. L. Shirley & J. A. Ragsdale), pp. 17–61. Houston, TX: Houston Geological Society.

Kolla, V., & Coumes, F. 1985. Indus fan, Indian Ocean. In *Submarine Fans and Related Turbidite Systems* (ed. A. H. Bouma, W. R. Normark, & N. E. Barnes), pp. 129–136. New York: Springer-Verlag.

1987. Morphology, internal structure, seismic stratigraphy, and sedimentation of Indus fan. *American Association of Petroleum Geologists Bulletin* **71**, 650–677.

Kolla, V., & Perlmutter, M. A. 1993. Timing of turbidite sedimentation on the Mississippi Fan. *American Association of Petroleum Geologists Bulletin* **77**, 1129–1141.

Koltermann, C. E., & Gorelick, S. M. 1992. Paleoclimatic signature in terrestrial flood deposits. *Science* **256**, 1775–1782.

1996. Heterogeneity in sedimentary deposits: a review of structure-imitating, process-imitating and descriptive approaches. *Water Resources Research* **32**, 2617–2658.

Komar, P. D. 1973. Computer models of delta growth due to sediment input from rivers and longshore transport. *Geological Society of America Bulletin* **84**, 2,217–2,226.

1974. Oscillatory ripple marks and the evaluation of ancient wave conditions and environments. *Journal of Sedimentary Petrology* **44**, 169–180.

1976. *Beach Processes and Sedimentation*. Englewood Cliffs, NJ: Prentice-Hall.

1977. Modeling of sand transport on beaches and the resulting shoreline evolution. In *The Sea* (ed. E. Goldberg *et al.*) Vol. 6, pp. 499–513. New York: Wiley.

1996. Entrainment of sediments from deposits of mixed grain sizes and densities. In *Advances in Fluvial Dynamics and Stratigraphy* (ed. P. A. Carling & M. R. Dawson), pp. 127–181. Chichester: Wiley.

1998. *Beach Processes and Sedimentation*, 2nd edn. Englewood Cliffs, NJ: Prentice-Hall.

Komar, P. D., & Miller, M. C. 1973. The threshold of sediment motion under oscillatory water waves. *Journal of Sedimentary Petrology* **43**, 1101–1110.

1975a. On the comparison between the threshold of sediment motion under waves and unidirectional currents with a discussion of the practical evaluation of the threshold. *Journal of Sedimentary Petrology* **45**, 362–367.

1975b. The initiation of oscillatory ripple marks and the development of plane-bed at high shear stresses under waves. *Journal of Sedimentary Petrology* **45**, 697–703.

Kooi, H., & Beaumont, C. 1994. Escarpment evolution on high-elevation rifted margins: insights derived from a surface processes model that combines diffusion, advection, and reaction. *Journal of Geophysical Research* **99**, 12191–12209.

1996. Large-scale geomorphology: classical concepts reconciled and integrated with contemporary ideas using a surface processes model. *Journal of Geophysical Research* **101**, 3361–3386.

Kor, P. S. G., Shaw, J., & Sharpe, D. R. 1991. Erosion of bedrock by subglacial meltwater, Georgian Bay, Ontario: a regional view. *Canadian Journal of Earth Sciences* **28**, 623–642.

Kosters, E. C. 1989. Organic–clastic facies relationships and chronostratigraphy of the Barataria interlobe basin, Mississippi delta plain. *Journal of Sedimentary Petrology* **59**, 98–113.

Kosters, E. C., & Suter, J. R. 1993. Facies relationships and systems tracts in the late Holocene Mississippi delta plain. *Journal of Sedimentary Petrology* **63**, 727–733.

Kraft, J. C. 1971. Sedimentary facies patterns and geologic history of a Holocene marine transgression. *Geological Society of America Bulletin* **82**, 2131–2158.

Kraft, J. C., Chrzastowski, M. J., Belknap, D. F., Toscano, M. A., & Fletcher, C. H. 1987. The transgressive barrier–lagoon coast of Delaware: morphostratigraphy, sedimentary sequences and responses to relative rises in sea level. In *Sea-level Fluctuations and Coastal Evolution* (ed. D. Nummedal, O. H. Pilkey, & J. D. Howard), pp. 129–144. Tulsa, OK: SEPM (Society for Sedimentary Geology).

Kraus, M. J. 1987. Integration of channel and floodplain suites, II. Vertical relations of alluvial paleosols. *Journal of Sedimentary Petrology* **56**, 602–612.

1996. Avulsion deposits in lower Eocene alluvial rocks, Bighorn Basin, Wyoming. *Journal of Sedimentary Research* **66B**, 354–363.

1999. Paleosols in clastic sedimentary rocks: their geologic applications. *Earth-Science Reviews* **47**, 41–70.

Kraus, M. J., & Aslan, A. 1999. Palaeosol sequences in floodplain environments: a hierarchical approach. In *Palaeoweathering, Palaeosurfaces and Related Continental Deposits* (ed. M. Thiry & R. Simon-Coicon), pp. 303–321. Oxford: Blackwells.

Kraus, M. J., & Bown, T. M. 1986. Palaeosols and time resolution in alluvial stratigraphy. In *Paleosols, Their Recognition and Interpretation* (ed. V. P. Wright), pp. 180–207. Princeton, NJ: Princeton University Press.

Kraus, M. J., & Gwinn, B. M. 1997. Controls on the development of early Eocene avulsion deposits and floodplain paleosols, Willwood Formation, Bighorn Basin. *Sedimentary Geology* **114**, 33–54.

Kraus, M. J., & Hasiotis, S. T. 2006. Significance of different modes of rhizolith preservation to interpreting paleoenvironmental and paleohydologic settings: examples from Paleogene paleosols, Bighorn Basin, Wyoming, U.S.A. *Journal of Sedimentary Research* **76**, 633–646.

Kraus, M. J., & Wells, T. M. 1999. Recognizing avulsion deposits in the ancient stratigraphic record. In *Fluvial Sedimentology VI* (ed. N. D. Smith & J. Rogers), pp. 251–268. Oxford: Blackwells.

Krause, D. C., White, W. C., Piper, D. J. W., & Heezen, B. C. 1970. Turbidity currents and cable breaks in the western New Britain Trench. *Geological Society of America Bulletin* **81**, 2153–2160.

Krynine, P. D. 1942. Differential sedimentation and its products during one complete geosynclinal cycle. *Proceedings of the 1st Pan American Congress on Mining Engineering Geology, Part 1*, Vol. 2, pp. 537–560. Santiago: Chilean Institute of Mining Engineering.

1950. Petrology, stratigraphy and origin of the Triassic sedimentary rocks of Connecticut. *Connecticut State Geological and Natural History Survey Bulletin* **73**, 1–247.

Kuenen, P. H. 1965. Value of experiments in geology. *Geologie en Mijnbouw* **44**, 22–36.

Kuenen, P. H., & Migliorini, C. I. 1950. Turbidity currents as a cause of graded bedding. *Journal of Geology* **58**, 91–127.

Kuijper, C., Cornelisse, J. M., & Winterwerp, J. C. 1989. Research on erosive properties of cohesive sediments. *Journal of Geophysical Research* **95**, 14,341–14,350.

Kumar, N., & Sanders, J. E. 1974. Inlet sequences: a vertical succession of sedimentary structures and textures created by the lateral migration of tidal inlets. *Sedimentology* **21**, 491–532.

Kummel, B., & Raup, D. 1965. *Handbook of Paleontological Techniques*. San Francisco, CA: W. H. Freeman.

Kump, L. R., Brantley, S. L., & Arthur, M. A. 2000. Chemical weathering, atmospheric CO_2, and climate. *Annual Review of Earth and Planetary Sciences* **28**, 611–667.

Kump, L. R., Kasting, J. F., & Crane, R. G. 2004. *The Earth System*, 2nd edn. Upper Saddle River, NJ: Prentice-Hall.

Lachenbruch, A. H. 1962. Mechanics of thermal contraction cracks and ice-wedge polygons. Geological Society of America Special Paper **70**.

Lal, R. 2004. Soil carbon sequestration impacts on global climate change and food security. *Science* **304**, 1623–1627.

Lancaster, N. 1982. Linear dunes. *Progress in Physical Geography* **6**, 475–504.

1989. Star dunes. *Progress in Physical Geography* **13**, 67–91.

1995. *Geomorphology of Desert Dunes*. London: Routledge.

2005. Aeolian processes. In *Encyclopedia of Geology 4* (ed. R. C. Selley, R. L. M. Cocks, & I. R. Plimer), pp. 612–627. Oxford: Elsevier.

2007. Low-latitude dune fields. In *Encyclopedia of Quaternary Sciences* (ed. S. A. Elias). Amsterdam: Elsevier.

Lancaster, N., Kocurek, G., Singhvi, A. *et al.* 2002. Late Pleistocene and Holocene dune activity and wind regimes in the western Sahara Desert of Mauritania. *Geology* **30**, 991–994.

Land, L. S., Behrens, E. W., & Frishman, S. A. 1979. The ooids of Baffin Bay, Texas. *Journal of Sedimentary Petrology* **49**, 1269–1278.

Langbein, W. B. 1964. Geometry of river channels. *Journal of the Hydraulics Division, ASCE* **90**, 301–312.

Langbein, W. B., & Leopold, L. B. 1964. Quasi-equilibrium states in channel morphology. *American Journal of Science* **262**, 782–794.

1966. River meanders – theory of minimum variance. United States Geological Survey Professional Paper **422H**.

Langford, R. P., & Chan, M. A. 1988. Flood surfaces and deflation surfaces within the Cutler Formation and Cedar Mesa Sandstone (Permian), southeastern Utah. *Geological Society of America Bulletin* **100**, 1541–1549.

1993. Downwind changes within an ancient dune sea, Permian Cedar Mesa Sandstone, southeast Utah. In *Aeolian Sediments: Ancient and Modern* (ed. K. Pye & N. Lancaster), pp. 109–126. Oxford: Blackwells.

Larson, G. J., Lawson, D. E., Evenson, E. B. *et al.* 2006. Glaciohydraulic supercooling in former ice sheets? *Geomorphology* **75**, 20–32.

Larson, M., & Kraus, N. C. 1991. Numerical model of longshore current for bar and trough beaches. *Journal of the Waterways, Harbors and Coastal Engineering Division, ASCE* **117**, 326–347.

Lasemi, Z., Boardman, M. R., & Sandberg, P. A. 1989. Cement origin of supratidal dolomite, Andros Island, Bahamas, *Journal of Sedimentary Petrology* **59**, 249–257.

Lauff, G. H. 1967. *Estuaries*. Washington, D.C.: American Association for the Advancement of Science.

Laval, B., Cady, S. L, Pollack, J. C. *et al.* 2000. Modern freshwater microbialite analogues for ancient dendritic reef structures. *Nature* **407**, 626–629.

Lawson, D. E. 1982. Mobilisation, movement and deposition of active subaerial sediment flows, Matanuska Glacier, Alaska. *Journal of Geology* **90**, 279–300.

Lawson, D. E., Strasser, J. C., Evenson, E. B. *et al.* 1998. Glaciohydraulic supercooling: a mechanism to create stratified, debris-rich basal ice.1. Field evidence and conceptual model. *Journal of Glaciology* **44**, 547–562.

Lay, T., Kanamori, H., Ammon, C. J. *et al.* 2005. The Great Sumatra–Andaman earthquake of 26 December 2004. *Science* **308**, 1127–1133.

Leatherman, S. P. 1979. *Barrier Islands*. New York: Academic Press.

Leatherman, S. P., Williams, A. T., & Fisher, J. S. 1977. Overwash sedimentation associated with a large-scale northeaster. *Marine Geology* **24**, 109–121.

Leckie, D. A., & Krystinik, L. F. 1989. Is there evidence for geostrophic currents preserved in the sedimentary record of inner to middle-shelf deposits? *Journal of Sedimentary Petrology* **59**, 862–870.

Leclair, S. F. 2002. Preservation of cross-strata due to the migration of subaqueous dunes: an experimental investigation. *Sedimentology* **49**, 1157–1180.

Leclair, S. F., & Bridge, J. S. 2001. Quantitative interpretation of sedimentary structures formed by river dunes. *Journal of Sedimentary Research* **71**, 713–716.

Leclair, S. F., Bridge, J. S., & Wang, F. 1997. Preservation of cross-strata due to migration of subaqueous dunes over aggrading and non-aggrading beds: comparison of experimental data with theory. *Geoscience Canada* **24**, 55–66.

Lee, M., Aronson, J. L, & Savin, S. M. 1989. Timing and conditions of Permian Rotliegende sandstone diagenesis, southern North Sea; K/Ar and oxygen isotope data. *American Association of Petroleum Geologists Bulletin* **73**, 195–215.

Leeder, M. R. 1975. Pedogenic carbonates and flood sediment accretion rates: a quantitative method for alluvial arid-zone lithofacies. *Geological Magazine* **112**, 257–270.

1978. A quantitative stratigraphic model for alluvium with special reference to channel deposit density and interconnectedness. In *Fluvial Sedimentology* (ed. A. D. Miall), pp. 587–596. Calgary, Alberta: Canadian Society of Petroleum Geologists.

1980. On the stability of lower stage plane beds and the absence of current ripples in coarse sands. *Journal of the Geological Society of London* **137**, 423–430.

1982. *Sedimentology: Process and Product*. London: George Allen and Unwin.

1993. Tectonic controls upon drainage basin development, river channel migration and alluvial architecture: implications for hydrocarbon reservoir development and characterization. In *Characterization of Fluvial and Aeolian Reservoirs* (ed. C. P. North & D. J. Prosser), pp. 7–22. London: Geological Society of London.

1999. *Sedimentology and Sedimentary Basins: From Turbulence to Tectonics*. Oxford: Blackwells.

Leeder, M. R., & Alexander, J. A. 1987. The origin and tectonic significance of asymmetrical meander belts. *Sedimentology* **34**, 217–226.

Leeder, M. R., Mack, G. H., Peakall, J., & Salyards, S. L. 1996. First quantitative test of alluvial stratigraphy models: southern Rio Grande rift, New Mexico. *Geology* **24**, 87–90.

Leeder, M. R., & Stewart, M. 1996. Fluvial incision and sequence stratigraphy: alluvial responses to relative base level fall and their detection in the geological record. In *Sequence Stratigraphy in British Geology* (ed. S. P. Hesselbo & D. N. Parkinson), pp. 47–61. London: Geological Society of London.

Lees, A. 1975. Possible influences of salinity and temperature on modern shelf carbonate sedimentation. *Marine Geology* **19**, 159–198.

Legros, F. 2002. Can dispersive pressure cause inverse grading in grain flows? *Journal of Sedimentary Research* **72**, 166–170.

Leopold, L. B., & Langbein, W. B. 1962. The concept of entropy in landscape evolution. United States Geological Survey Professional Paper **500A**.

1966. River meanders. *Scientific American* **214**, 60–70.

Leopold, L. B., & Wolman, M. G. 1957. River channel patterns: braided, meandering and straight. United States Geological Survey Professional Paper **282-B**.

Leopold, L. B., Wolman, M. G., & Miller, J. P. 1964. *Fluvial Processes in Geomorphology*. San Francisco, CA: W. H. Freeman.

Lerman, A. 1978. *Lakes: Chemistry Geology and Physics*. New York: Springer-Verlag.

Lerman, A., Imboden, D., & Gat, J. 1995. *Physics and Chemistry of Lakes*, 2nd edn. New York: Springer-Verlag.

Levey, R. A. 1978. Bedform distribution and internal stratification of coarse-grained point bars, Upper Congaree River, South Carolina. In *Fluvial Sedimentology* (ed. A. D. Miall), pp. 105–127. Calgary, Alberta: Canadian Society of Petroleum Geologists.

Levinton, J. S. 1982. *Marine Ecology*. Englewood Cliffs, NJ: Prentice-Hall.

Li, M. Z., & Amos, C. L. 1999. Sheet flow and large wave ripples under combined waves and currents: field observations, model predictions and effects on boundary layer dynamics. *Continental Shelf Research* **19**, 637–663.

Lippman, F. 1973. *Sedimentary Carbonate Minerals*. New York: Springer-Verlag.

Livingston, D. A. 1963. Chemical composition of rivers and lakes. United States Geological Survey Professional Paper **440-G**.

Livingstone, I., & Warren, A. 1996. *Aeolian Geomorphology: An Introduction*. Harlow: Longman.

Lizitsin, A. P. 1996. *Oceanic Sedimentation*. Washington, D.C.: American Geophysical Union.

Lliboutry, L. 1979. Local friction laws for glaciers: a critical review and new openings. *Journal of Glaciology* **23**, 67–95.

Lockley, M. G. 1991. *Tracking Dinosaurs: A New Look at an Ancient World*. Cambridge: Cambridge University Press.

Lockley, M. G., & Hunt, A. P. 1995. *Dinosaur Tracks and Other Fossil Footprints of the Western United States*. New York: Columbia University Press.

Logan, B. W. 1987. *The MacLeod Evaporite Basin Western Australia*. Tulsa, OK: American Association of Petroleum Geologists.

Logan, B. W., Davies, G. R., Read, J. F., & Cebulski, D. E. 1970. *Carbonate Sedimentation and Environments, Shark Bay, Western Australia*. Tulsa, OK: American Association of Petroleum Geologists.

Logan, B. W., Hoffman, P., & Gebelein, C. D. 1974. Algal mats, crypt-algal fabrics and structures, Hamelin Pool, Western Australia. In *Evolution and Diagenesis of Quaternary Carbonate Sequences, Shark Bay, Western Australia* (ed. B. W. Logan, J. F. Read, G. M. Hagan *et al.*), pp. 140–193. Tulsa, OK: American Association of Petroleum Geologists.

Logan, B. W., Read, J. F., Hagan, G. M. *et al.* 1974. *Evolution and Diagenesis of Quaternary Carbonate Sequences, Shark Bay, Western Australia.* Tulsa, OK: American Association of Petroleum Geologists.

Logan, B. W., & Semeniuk, V. 1976. *Dynamic Metamorphism: Processes and Products in Devonian Carbonate Rocks, Canning Basin, Western Australia.* Sydney: Geological Society of Australia.

Longman, M. W. 1980. Carbonate diagenetic textures from near surface diagenetic environments. *American Association of Petroleum Geologists Bulletin* **64**, 461–487.

Longuet-Higgins, M. S. 1953. Mass transport in water waves. *Philosophical Transactions of the Royal Society* **245A**.

1970a. Longshore currents generated by obliquely incident waves, 1. *Journal of Geophysical Research* **75**, 6778–6789.

1970b. Longshore currents generated by obliquely incident waves, 2. *Journal of Geophysical Research* **75**, 6790–6801.

1981. Oscillating flow over steep sand ripples. *Journal of Fluid Mechanics* **107**, 1–35.

Loope, D. B. 1984. Origin of extensive bedding planes in aeolian sandstones: a defence of Stokes' hypothesis. *Sedimentology* **31**, 123–132.

1985. Episodic deposition and preservation of eolian sands – a Late Paleozoic example from southeastern Utah. *Geology* **13**, 73–76.

Loucks, R. G., & Sarg, J. F. 1993. *Carbonate Sequence Stratigraphy: Recent Developments and Applications.* Tulsa, OK: American Association of Petroleum Geologists.

Lowe, D. R. 1975. Water escape structures in coarse-grained sediments. *Sedimentology* **22**, 157–204.

1976a. Subaqueous liquefied and fluidized sediment flows and their deposits. *Sedimentology* **23**, 285–308.

1976b. Grain flow and grain flow deposits. *Journal of Sedimentary Petrology* **46**, 188–199.

1979. Sediment gravity flows: their classification and some problems of application to natural flows and deposits. In *Geology of Continental Slopes* (ed. L. J. Doyle & O. H. Pilkey), pp. 75–82. Tulsa, OK: SEPM (Society for Sedimentary Geology).

1982. Sediment gravity flows: 2. Depositional models with special reference to the deposits of high-density turbidity currents. *Journal of Sedimentary Petrology* **52**, 279–297.

Lowe, D. R., Anderson, K. S., & Braunstein, D. 2001. The zonation and structuring of siliceous sinter around hot springs, Yellowstone National Park, and the role of thermophilic bacteria in its deposition. In *Thermophiles: Biodiversity, Ecology, and Evolution* (ed. A.-L. Reysenbach, M. Voytek, & R. Mancinelli), pp. 143–166. New York: Kluwer Academic.

Lowe, D. R., & Braunstein, D. 2003. Microstructure of high-temperature (>73 °C) siliceous sinter deposited around hot springs and geysers, Yellowstone National Park: the role of biological and abiological processes in sedimentation. *Canadian Journal of Earth Sciences* **40**, 1611–1642.

Lowe, D. R., & Lopiccolo, R. D. 1974. The characteristics and origin of dish and pillar structures. *Journal of Sedimentary Petrology* **44**, 484–501.

Lowenstam, H. A. 1954. Factors affecting the aragonite : calcite ratios in carbonate-secreting marine organisms. *Journal of Geology* **62**, 284–322.

Lowenstam, H. A., & Weiner, S. 1989. *On Biomineralization.* Oxford: Oxford University Press.

Lowenstein, T. K. 2002. Pleistocene lakes and paleoclimates (0 to 200 ka) in Death Valley, California. *Smithsonian Contributions to the Earth Sciences* **33**, 109–120.

Lowenstein, T. K., & Hardie, L. A. 1985. Criteria for the recognition of salt-pan evaporites. *Sedimentology* **32**, 627–644.

Lowenstein, T. K., Hardie, L. A., Timofeeff, M. N., & Demicco, R. V. 2003. Secular variation in seawater chemistry and the origin of calcium chloride basinal brines. *Geology* **31**, 857–860.

Lowenstein, T. K., Li, J., Brown, C. *et al.* 1999. 200 k.y. paleoclimate record from Death Valley salt core. *Geology* **27**, 3–6.

Lowenstein, T. K., Timofeeff, M. N., Brennan, S. T., Hardie, L. A., & Demicco, R. V. 2001. Oscillations in Phanerozoic seawater chemistry: Evidence from fluid inclusions. *Science* **294**, 1086–1088.

Lundqvist, J. 1989. Rogen (ribbed) moraine – identification and possible origin. *Sedimentary Geology* **62**, 281–292.

1997. Rogen moraine – an example of two-step formation of glacial landscapes. *Sedimentary Geology* **111**, 27–40.

Lunt, I. A., & Bridge, J. S. 2006. Formation and preservation of open-framework gravel strata in unidirectional water flows. *Sedimentology* **53**, 1–17.

Lyell, C. 1830. *The Principles of Geology.* London: John Murray.

MacEachern, J. A., Bann, K. L., Bhattacharya, J. P., & Howell, C. D. Jr. 2005. Ichnology of deltas: organism responses to the dynamic interplay of rivers, waves, storms, and tides. In *River Deltas – Concepts, Models, and Examples* (ed. L. Giosan & J. P. Bhattacharya), pp. 49–85. Tulsa, OK: SEPM (Society for Sedimentary Geology).

Macintyre, I. G. 1985. Submarine cements – the peloidal question. In *Carbonate Cements* (ed. N. Schneidermann & P. M. Harris), pp. 109–116. Tulsa, OK: SEPM (Society for Sedimentary Geology).

Mack, G. H., & James, W. C. 1994. Paleoclimate and global distribution of paleosols. *Journal of Geology* **102**, 360–362.

Mack, G. H., James, W. C., & Monger, H. C. 1993. Classification of paleosols. *Geological Society of America Bulletin* **105**, 129–136.

Mack, G. H., & Leeder, M. R. 1998. Channel shifting of the Rio Grande, southern Rio Grande rift: implications for alluvial stratigraphy models. *Sedimentary Geology* **177**, 207–219.

Mackenzie, F. T., & Pigott, J. D. 1981. Tectonic controls of Phanerozoic sedimentary rock cycling. *Journal of the Geological Society of London* **138**, 183–196.

Mackey, S. D., & Bridge, J. S. 1995. Three dimensional model of alluvial stratigraphy: theory and application. *Journal of Sedimentary Research* **B65**, 7–31.

Mader, C. L. 1974. Numerical simulation of tsunamis. *Journal of Physical Oceanography* **4**, 74–82.

1988. *Numerical Modeling of Water Waves*. Berkeley, CA: University of California Press.

Madsen, O. S. 1977. A realistic model of the wind-induced Ekman boundary layer. *Journal of Physical Oceanagraphy* **7**, 248–255.

Maizels, J. K. 1989. Sedimentology, paleoflow dynamics and flood history of jokulhlaup deposits: paleohydrology of Holocene sediment sequences in southern Iceland sadur deposits. *Journal of Sedimentary Petrology* **59**, 204–223.

2002. Sediments and landforms of modern proglacial terrestrial environments. In *Modern and Past Glacial Environments* (ed. J. Menzies), pp. 279–333. Oxford: Butterworth-Heineman.

Major, J. J. 1997. Depositional processes in large-scale debris-flow experiments. *Journal of Geology* **105**, 345–366.

Makaske, B. 2001. Anastomosing rivers: a review of their classification, origin and sedimentary products. *Earth-Science Reviews* **53**, 149–196.

Makse, H. A., Havlin, S., King, P. R., & Stanley, H. E. 1997. Spontaneous stratification in granular mixtures. *Nature* **386**, 379–382.

Maliva, R. G., Knoll, A. H., & Simonson, B. M. 2005. Secular change in the Precambrian silica cycle: insights from chert petrology. *Geological Society of America Bulletin* **117**, 835–845.

Maliva, R. G., & Siever, R. 1989. Nodular chert formation in carbonate rocks. *Journal of Geology* **97**, 421–433.

Mancini, E. A, Llinas, J. C., Parcell, W. C. *et al.* 2004. Upper Jurassic thrombolite reservoir play, northeastern Gulf of Mexico. *American Association of Petroleum Geologists Bulletin* **88**, 1573–1602.

Mann, K. H. 2000. *Ecology of Coastal Waters: With Implications for Management*. Massachusetts, PA: Blackwell Science.

Mann, S. 2001. *Biomineralization: Principles and Concepts in Bioinorganic Materials Chemistry*. Oxford: Oxford University Press.

Marani, M., Lanzoni, S., Zandolin, D, Seminara, G., & Rinaldo, A. 2002. Tidal meanders. *Water Resources Research* **38**, 7–1 to 7–11.

Marr, J. G., Swenson, J. B., Paola, C., & Voller, V. R. 2000. A two-diffusion model of fluvial stratigraphy in closed depositional basins. *Basin Research* **12**, 381–398.

Marshall, J. F., & Davies, P. J. 1975. High magnesium calcite ooids from the Great Barrier Reef. *Journal of Sedimentary Petrology* **45**, 285–291.

Martini, I. P., Baker, V. R., & Garzon, G. 2002. *Flood and Megaflood Processes and Deposits*. Oxford: Blackwells.

Martini, I. P., Brookfield, M. E., & Sadura, S. 2001. *Glacial Geomorphology and Geology*. Englewood Cliffs, NJ: Prentice Hall.

Mastbergen, D. R., & Van Den Berg, J. H. 2003. Breaching in fine sands and the generation of sustained turbidity currents in submarine canyons. *Sedimentology* **50**, 625–637.

Matter, A., & Tucker, M. E. 1978. *Modern and Ancient Lake Sediments*. Oxford: Blackwells.

McCabe, M., & Eyles, N. 1988. Sedimentology of an ice-contact glaciomarine delta, Carey Valley, Northern Ireland. *Sedimentary Geology* **59**, 1–14.

McCabe, P. J. 1984. Depositional environments of coal and coal-bearing strata. In *Sedimentology of Coal and Coal-bearing Sequences* (ed. R. A. Rahmani & R. M. Flores), pp. 13–42. Oxford: Blackwells.

McCabe, P. J., & Parrish, J. T. 1992. Tectonic and climatic controls on the distribution and quality of Cretaceous coals. In *Controls on the Distribution and Quality of Cretaceous Coals* (ed. P. J. McCabe & J. T. Parrish), pp. 1–15. Boulder, CO: Geological Society of America.

McCaffrey, W. D., Kneller, B. C., & Peakall, J. 2001. *Particulate Gravity Currents*. Oxford: Blackwells.

McCarthy, P. J., & Plint, A. G. 1998. Recognition of interfluve sequence boundaries: integrating paleopedology and sequence stratigraphy. *Geology* **26**, 387–390.

McCave, I. N. 1968. Shallow and marginal marine sediments associated with the Catskill complex in the Middle Devonian of New York. In *Late Paleozoic and Mesozoic Continental Sedimentation, Northeastern North America* (ed. G. deV. Klein), pp. 75–108. Boulder, CO: Geological Society of America.

1969. Correlation of marine and nonmarine strata with example from Devonian of New York State. *American Association of Petroleum Geologists Bulletin* **53**, 155–162.

1970. Deposition of fine-grained suspended sediment from tidal currents. *Journal of Geophysical Research* **75**, 4151–4159.

1985. Recent shelf clastic sediment. In *Sedimentology: Recent Developments and Applied Aspects* (ed. P. J. Brenchley & B. P. J. Williams), pp. 49–65. London: Geological Society of London.

McCubbin, D. G. 1982. Barrier island and strand plain facies. In *Sandstone Depositional Environments* (ed. P. A. Scholle & D. R. Spearing), pp. 247–279. Tulsa, OK: American Association of Petroleum Geologists.

McHargue, T. R. 1991. Seismic facies, processes and evolution of Miocene inner fan channels, Indus submarine fan. In *Seismic Facies and Sedimentary Processes of Submarine Fans and Turbidite Systems* (ed. P. Weimer & M. H. Link), pp. 403–414. New York: Springer-Verlag.

McIlreath, I. A. 1977. Accumulation of a Middle Cambrian deep water, basinal limestone adjacent to a vertical, submarine carbonate escarpment, southern Rocky Mountains. In *Deep Water Carbonate Environments* (ed. H. E. Cook & P. Enos), pp. 113–124. Tulsa, OK: SEPM (Society for Sedimentary Geology).

McIlreath, I. A., & James, N. P. 1984. Carbonate slopes. In *Facies Models*, 2nd edn. (ed. R. G. Walker), pp. 245–257. St. John's, Newfoundland: Geological Association of Canada.

McIlreath, I. A., & Morrow, D. W. 1990. *Diagenesis*. St. John's, Newfoundland: Geological Association of Canada.

McKee, E. D. 1966. Structure of dunes at White Sands National Monument, New Mexico. *Sedimentology* 7, 3–69.

1979. A study of global sand seas. United States Geological Survey Professional Paper **1052**.

1982. Sedimentary structures in dunes of the Namib Desert, South West Africa. Geological Society of America Special Paper **188**.

McKee, E. D., & Gutschick, R. C. 1969. *History of the Redwall Limestone of Northern Arizona*. Boulder, CO: Geological Society of America.

McKee, E. D., & Tibbitts, G. C. Jr. 1964. Primary structures of a seif dune and associated deposits in Libya. *Journal of Sedimentary Petrology* 34, 5–17.

McKee, E. D., & Weir, G. W. 1953. Terminology for stratification and cross-stratification in sedimentary rocks. *Geological Society of America Bulletin* 64, 381–390.

McKenzie, J. A., Hsü, K. J., & Schneider, J. F. 1980. Movement of subsurface waters under the sabkha, Abu Dhabi, UAE, and its relation to evaporative dolomite genesis. In *Concepts and Models of Dolomitization* (ed. D. H. Zenger, J. B. Dunham, & R. L. Ethington), pp. 11–30. Tulsa, OK: SEPM (Society for Sedimentary Geology).

McLean, S. R. 1990. The stability of ripples and dunes. *Earth-Science Reviews* 29, 131–144.

McNeill, J. R., & Winiwarter, V. 2004. Breaking the sod: humankind, history, and soil. *Science* 304, 1627–1629.

McPhee, J. 1989. *Control of Nature*. New York: Farrar, Straus and Giroux.

Mehta, A. J. 1989. On estuarine cohesive sediment suspension behaviour. *Journal of Geophysical Research* 94, 14,303–14,314.

Melim, L. A., & Swart, P. K. 2001. Meteoric and marine-burial diagenesis in the subsurface of Great Bahama Bank. In *Subsurface Geology of a Prograding Carbonate Platform*

Margin, Great Bahama Bank: Results of the Bahamas Drilling Project (ed. R. N. Ginsburg), pp. 137–161. Tulsa, OK: SEPM (Society for Sedimentary Geology).

Melosh, H. J. 1987. The mechanics of large rock avalanches. In *Debris Flows/Avalanches: Processes, Recognition, and Mitigation* (ed. J. E. Costa & G. F. Wieczorek), pp. 41–49. Boulder, CO: Geological Society of America.

Menzies, J. 1995. *Modern Glacial Environments – Processes, Dynamics and Sediments*. Oxford: Butterworth-Heinemann.

1996. *Past Glacial Environments – Sediments, Forms and Techniques*. Oxford: Butterworth-Heinemann.

2002. *Modern and Past Glacial Environments*. Oxford: Butterworth-Heinemann.

Menzies, J., & Rose, J. 1989. Subglacial bedforms – drumlins, Rogen moraine and associated subglacial bedforms. *Sedimentary Geology* 62, 117–407.

Meybeck, M. 1995. Global distribution of lakes. In *Physics and Chemistry of Lakes*, 2nd edn. (ed. A. Lerman, D. Imboden, & J. Gat), pp. 1–35. New York: Springer-Verlag.

Meyer, G. A., Wells, S. G., Balling, R. C., & Jull, A. J. T. 1992. Response of alluvial systems to fire and climate change in Yellowstone National Park. *Nature* 357, 147–150.

Meyers, W. J. 1974. Carbonate cement stratigraphy of the Lake Valley Formation (Mississippian), Sacramento Mountains, New Mexico. *Journal of Sedimentary Petrology* 44, 837–861.

1991. Calcite cement stratigraphy: an overview. In *Luminescence Microscopy: Quantitative and Qualitative Analysis* (ed. C. E. Barker & O. C. Kipp), pp. 133–148. Tulsa, OK: SEPM (Society for Sedimentary Geology).

Miall, A. D. 1986. Eustatic sea level changes interpreted from seismic stratigraphy: a critique of the methodology with particular reference to the North Sea Jurassic record. *American Association of Petroleum Geologists Bulletin* 70, 131–137.

1991. Stratigraphic sequences and their chronostratigraphic correlation. *Journal of Sedimentary Petrology* 61, 497–505.

1996. *The Geology of Fluvial Deposits*. New York: Springer-Verlag.

1997. *The Geology of Stratigraphic Sequences*. Berlin: Springer-Verlag.

Miall, A. D., & Smith, N. D. 1989. *Rivers and Their Deposits. Slide Set No. 4*. Tulsa, OK: SEPM (Society for Sedimentary Geology).

Mickelson, D. M., & Attig, J. W. 1999. Glacial processes past and present. Geological Society of America Special Paper **337**.

Middleton, G. V. 1966a. Experiments on density and turbidity currents. 1: motion of the head. Canadian Journal of Earth Sciences **3**, 523–546.

1966b. Experiments on density and turbidity currents. 2: uniform flow of turbidity currents. *Canadian Journal of Earth Sciences* **3**, 627–637.

1966c. Experiments on density and turbidity currents. 3: deposition of sediment. *Canadian Journal of Earth Sciences* **4**, 475–505.

1993. Sediment deposition from turbidity currents. *Annual Review of Earth and Planetary Sciences* **21**, 89–114.

2003. *Encyclopedia of Sediments and Sedimentary Rocks.* Dordrecht: Kluwer Academic Publishers.

Middleton, G. V., & Hampton, M. A. 1973. Sediment gravity flows: mechanics of flow and deposition. In *Turbidites and Deep-Water Sedimentation* (ed. G. V. Middleton & A. H. Bouma), pp. 1–38. Los Angeles, CA: SEPM (Society for Sedimentary Geology) Pacific Section.

1976. Subaqueous sediment transport and deposition by sediment gravity flows. In *Marine Sediment Transport and Environmental Management* (ed. D. J. Stanley & D. J. P. Swift), pp. 197–218. New York: Wiley.

Middleton, G. V., & Southard, J. B. 1984. *Mechanics of Sediment Movement.* Tulsa, OK: SEPM.

Middleton, G. V., & Wilcock, P. R. 1994. *Mechanics in the Earth and Environmental Sciences.* Cambridge: Cambridge University Press.

Miller, J. M. G. 1996. Glacial sediments. In *Sedimentary Environments: Processes, Facies and Stratigraphy*, 3rd edn. (ed. H. G. Reading), pp. 454–484. Oxford: Blackwells.

Miller, M. C., & Komar, P. D. 1980. Oscillation ripples generated by laboratory experiments. *Journal of Sedimentary Petrology* **50**, 173–182.

Milliman, J. D., & Barretto, H. T. 1975. Relic magnesian calcite oolite and subsidence of the Amazon shelf. *Sedimentology* **22**, 137–145.

Mitchell, R. W. III. 1985. Comparative sedimentology of shelf carbonates of the Middle Ordovician St. Paul Group of the central Appalachians. *Sedimentary Geology* **43**, 1–41.

Mohrig, D., Heller, P. L., Paola, C., & Lyons, W. J. 2000. Interpreting avulsion process from ancient alluvial sequences: Guadalope–Matarranya system (northern Spain) and Wasatch Formation (western Colorado). *Geological Society of America Bulletin* **112**, 1787–1803.

Mojzsis, S. J., Arrhenius, G., McKeegan, K. D. *et al.* 1996. Evidence for life on Earth before 3,800 million years ago. *Nature* **384**, 55–59.

Moller, P. 2006. Rogen moraine: an example of glacial reshaping of pre-existing landforms. *Quaternary Science Reviews* **25**, 362–389.

Molnia, B. F. 1983. *Glacial–Marine Sedimentation.* New York: Plenum.

Montañez. I. P., Gregg, J. M., & Shelton, K. L. 1997. *Basin-wide Diagenetic Patterns: Integrated Petrologic, Geochemical, and Hydrologic Considerations.* Tulsa, OK: SEPM (Society for Sedimentary Geology).

Montgomery, D. R., & Dietrich, W. E. 1989. Source areas, drainage density, and channel initiation. *Water Resources Research* **25**, 1907–1918.

1994. Landscape dissection and drainage area-slope thresholds. In *Process Models and Theoretical Geomorphology* (ed. M. J. Kirkby), pp. 221–246. Chichester: Wiley.

Mora, C. I., & Driese, S. G. 1993. A steep, mid- to late Paleozoic decline in atmospheric CO_2: evidence from soil carbonate CO_2 barometer. *Chemical Geology* **107**, 217–219.

1999. Palaeoenvironment, palaeoclimate and stable carbon isotopes of Palaeozoic red-bed palaeosols, Appalachian Basin, USA and Canada. In *Palaeoweathering, Palaeosurfaces and Related Continental Deposits* (ed. M. Thiry & R. Simon-Coicon), pp. 61–84. Oxford: Blackwells.

Mora, C. I., Driese, S. G., & Colarusso, L. A. 1996. Middle to late Paleozoic atmospheric CO_2 levels from soil carbonate and organic matter. *Science* **271**, 1105–1107.

Morad, S., Ketzer, J. M., & De Ros, L. F. 2000. Spatial and temporal distribution of diagenetic alterations in siliciclastic rocks: implications for mass transfer in sedimentary basins. *Sedimentology* **47**, 95–120.

Morgan, J. P. 1970. *Deltaic Sedimentation: Modern and Ancient.* Tulsa, OK: SEPM (Society for Sedimentary Geology).

Morgan, J. P., Coleman, J. M., & Gagliano, S. M. 1968. Mudlumps: diapiric structures in Mississippi delta sediments. In *Diapirism and Diapirs* (ed. J. Braunstein & G. D. O'Brien), pp. 145–161. Tulsa, OK: American Association of Petroleum Geologists.

Moore, A., Nishimura, Y., Gelfenbaum, G., Kamataki, T., & Triyono, R. 2006. Sedimentary deposits of the 26 December 2004 tsunami on the northwest coast of Aceh, Indonesia. *Earth Planets Space* **58**, 253–258.

Morozova, G. S., & Smith, N. D. 1999. Holocene avulsion history of the lower Saskatchewan fluvial system, Cumberland Marshes, Saskatchewan–Manitoba, Canada. In *Fluvial Sedimentology VI* (ed. N. D. Smith & J. Rogers), pp. 231–249. Oxford: Blackwells.

Morris, P. E., & Williams, D. J. 1997. Exponential longitudinal profiles of streams. *Earth Surface Processes and Landforms* **22**, 143–163.

1999a. A worldwide correlation for exponential bed particle size variations in subaerial aqueous flows. *Earth Surface Processes and Landforms* **24**, 835–847.

1999b. Worldwide correlations for subaerial aqueous flows with exponential longitudinal profiles. *Earth Surface Processes and Landforms* **24**, 867–879.

Morse, J. W., Millero, F. J., Thurmond, V., Brown, E., & Ostlund, H. G. 1984. The carbonate chemistry of Grand Bahama Bank waters: after 18 years another look. *Journal of Geophysical Research* **89**, 3604–3614.

Morse, J. W., Wang, Q., & Tzio, M. Y. 1997. Influences of temperature and Mg : Ca ratio on $CaCO_3$ precipitates from seawater. *Geology* **25**, 85–87.

Morton, R. A. 1981. Formation of storm deposits by wind-forced currents in the Gulf of Mexico and the North Sea. In *Holocene Sedimentation in the North Sea Basin* (ed. S. D. Nio, R. T. E. Schuttenhelm, & T. C. E. Van Weering), pp. 385–396. Oxford: Blackwells.

Moslow, T. F., & Heron, S. D. 1978. Relict inlets: preservation and occurrence in the Holocene stratigraphy of southern Core Banks, North Carolina. *Journal of Sedimentary Petrology* **48**, 1275–1286.

Moslow, T. F., & Tye, R. S. 1985. Recognition and characterization of Holocene tidal inlet sequences. *Marine Geology* **63**, 129–151.

Mountney, N. P. 2006. Periodic accumulation and destruction of aeolian erg sequences in the Permian Cedar Mesa Sandstone, White Canyon, southern Utah, USA. *Sedimentology* **53**, 789–823.

Mountney, N. P., & Jagger, A. 2004. Stratigraphic evolution of an aeolian erg margin system: the Permian Cedar Mesa sandstone, SE Utah, USA. *Sedimentology* **51**, 713–743.

Mulder, T., & Syvitski, J. P. M. 1995. Turbidity currents generated at river mouths during exceptional discharges to the world's oceans. *Journal of Geology* **103**, 285–299.

Mullins, H. T. 1983. Modern carbonate slopes and basins of the Bahamas. In *Platform Margin and Deep Water Carbonates* (ed. H. E. Cook, A. C. Hine, & H. T. Mullins), pp. 4.1–4.138. Tulsa, OK: SEPM (Society for Sedimentary Geology).

Multer, H. G. 1975. *Field Guide to Some Carbonate Rock Environments, Florida Keys and Western Bahamas.* Madison, NJ: Farleigh Dickinson University.

Multer, H. G., & Hoffmeister, J. E. 1975. Petrology and significance of the Key Largo (Pleistocene) Limestone, Florida Keys. In *Field Guide to some Carbonate Rock Environments; Florida Keys and Western Bahamas* (ed. H. G. Multer) pp. 111–112. Madison, NJ: Farleigh Dickinson University.

Murray, A. B., & Paola, C. 1994. A cellular model of braided rivers. *Nature* **371**, 54–57.

1997. Properties of a cellular braided stream model. *Earth Surface Processes and Landforms* **22**, 1001–1025.

Murray, P. B., Davies, A. G., & Soulsby, R. L. 1991. Sediment pick-up in wave and current flows. In *Sand Transport in Rivers, Estuaries and the Sea* (ed. R. L. Soulsby & R. Bettess), pp. 37–44. Rotterdam: Balkema.

Murray, S. P. 1970. Bottom currents near the coast during hurricane Camille. *Journal of Geophysical Research* **75**, 4579–4582.

Murton, J. B., Peterson, R., & Ozouf, J.-C. 2006. Bedrock fracture by ice segregation in cold regions. *Science* **314**, 1127–1129.

Mutti, E. 1985. Turbidite systems and their relations to depositional sequences. In *Provenance of Arenites* (ed. G. G. Zuffa), pp. 65–93. Amsterdam: Reidel.

Mutti, E., & Normark, W. R. 1987. Comparing examples of modern and ancient turbidite systems: problems and concepts. In *Marine Clastic Sedimentology: Concepts and Case Studies* (ed. J. K. Legget & G. G. Zuffa), pp. 1–38. London: Graham and Trotman.

1991. An integrated approach to the study of turbidite systems. In *Seismic Facies and Sedimentary Processes of Submarine Fans and Turbidite Systems* (ed. P. Weimer & M. H. Link), pp. 75–106. New York: Springer-Verlag.

Mutti, E., & Ricci Lucci, F. 1972. Le torbiditi dell'Appennino Settentrionale: introduzione all'analisi di facies. *Memorie della Società Geologica Italiana* **11**, 161–199.

Myrow, P. M., & Southard, J. B. 1991. Combined-flow model for vertical stratification sequences in shallow marine storm-deposited beds. *Journal of Sedimentary Petrology* **61**, 202–210.

Nardin, T. R., Hein, F. J., Gorsline, D. S., & Edwards, B. D. 1979. A review of mass movement processes, sediment and acoustic characteristics, and contrasts in slope and base-of-slope versus canyon–fan–basin floor systems. In *Geology of Continental Slopes* (ed. L. J. Doyle & O. H. Pilkey), pp. 61–73. Tulsa, OK: SEPM (Society for Sedimentary Geology).

Nealson, K. H. 1997. Sedimentary bacteria: who's there, what are they doing, and what's new? *Annual Review of Earth and Planetary Sciences* **25**, 403–434.

Necker, F., Hartel, C., Kleiser, L., & Meiburg, E. 2005. Mixing and dissipation in particle-driven gravity currents. *Journal of Fluid Mechanics* **545**, 339–372.

Neev, D., & Emery, K. O. 1967. The Dead Sea: depositional processes and environments of evaporites. *Israel Geological Survey Bulletin* **41**.

Nemec, W., & Steel, R. J. 1988. *Fan Deltas: Sedimentology and Tectonic Settings.* London: Blackie.

Nezu, I., & Nakagawa, H. 1993. *Turbulence in Open-Channel Flows.* Rotterdam: Balkema.

Nguyen, Q. D., & Boger, D. V. 1992. Measuring the flow properties of yield stress fluids. *Annual Review of Fluid Mechanics* **24**, 47–88.

Nichols, G. 1999. *Sedimentology and Stratigraphy.* Oxford: Blackwells.

Nichols, M. J. 1989. Sediment accumulation rates and relative sea level rise in lagoons. *Marine Geology* **88**, 201–220.

Nichols, R. J., Sparks, R. S. J., & Wilson, C. J. N. 1994. Experimental studies of the fluidization of layered sediments and the formation of fluid escape structures. *Sedimentology* **41**, 233–253.

Nickling, W. G. 1994. Aeolian sediment transport and deposition. In *Sediment Transport and Depositional Processes* (ed. K. Pye), pp. 293–350. Oxford: Blackwells.

Nielsen, P. 1992. *Coastal Bottom Boundary Layers and Sediment Transport*. Singapore: World Scientific.

Nielson, J., & Kocurek, G. 1986. Climbing zibars in the Algodones. *Sedimentary Geology* **48**, 1–15.

Nio, S. D., Schuttenhelm, R. T. E., & Van Weering, T. C. E. 1981. *Holocene Sedimentation in the North Sea Basin*. Oxford: Blackwells.

Nio, S. D., Seigenthaler, C., & Yang, C. S. 1983. Megaripple cross-bedding as a tool for the reconstruction of the palaeohydraulics in a Holocene subtidal environment, S.W. Netherlands. *Geologie en Mijnbouw* **62**, 499–510.

Nio, S. D., & Yang, C. S. 1991. Diagnostic attributes of clastic tidal deposits: a review. In *Clastic Tidal Sedimentology* (ed. D. G. Smith, G. E. Reinson, B. A. Zaitlin, & R. A. Rahmani), pp. 3–28. Calgary, Alberta: Canadian Society of Petroleum Geologists.

Nittrouer, C. A., & Wright, L. D. 1994. Transport of particles across continental shelves. *Reviews of Geophysics* **32**, 85–113.

Noffke, N., Gerdes, G., & Klenke, T. 2005. Benthic cyanobacteria and their influence on the sedimentary dynamics of peritidal depositional systems (siliciclastic, evaporitic salty, and evaporitic carbonatic). *Earth-Science Reviews* **62**, 163–176.

Normark, W. R. 1970. Growth patterns of deep-sea fans. *American Association of Petroleum Geologists Bulletin* **54**, 2170–2195.

1978. Fan valleys, channels, and depositional lobes on modern submarine fans: characters for recognition of sandy turbidite environments. *American Association of Petroleum Geologists Bulletin* **62**, 912–931.

Normark, W. R., & Piper, D. J. W. 1985. Navy Fan, Pacific Ocean. In *Submarine Fans and Related Turbidite Systems* (ed. A. H. Bouma, W. R. Normark, & N. E. Barnes), pp. 87–94. New York: Springer-Verlag.

Normark, W. R., Piper, D. J. W., & Hess, G. R. 1979. Distributary channels, sand lobes, and mesotopography of Navy submarine fan, California Borderland, with applications to ancient fan sediments. *Sedimentology* **26**, 749–774.

North, C. P. 1996. The prediction and modelling of subsurface fluvial stratigraphy. In *Advances in Fluvial Dynamics and Stratigraphy* (ed. P. A. Carling & M. R. Dawson), pp. 395–508. Chichester: Wiley.

North, C. P., & Prosser, D. J. 1993. *Characterisation of Fluvial and Aeolian Reservoirs*. London: Geological Society of London.

North, C. P., & Taylor, K. S. 1996. Ephemeral–fluvial deposits: integrated outcrop and simulation studies reveal complexity. *American Association of Petroleum Geologists Bulletin* **80**, 811–830.

Notholt, A. J. G., Sheldon, R. P., & Davidson, D. F. 1989. *Phosphate Deposits of the World, Volume 2, Phosphate Rock Resources*. Cambridge: Cambridge University Press.

Nottvedt, A., & Kreisa, R. D. 1987. Model for the combined flow origin of hummocky cross-stratification. *Geology* **15**, 357–361.

Nummedal, D., & Penland, S. 1981. Sediment dispersal in Norderneyer Seegate, West Germany. In *Holocene Sedimentation in the North Sea Basin* (ed. S. D. Nio, R. T. E. Schuttenhelm, & T. C. E. Van Weering), pp. 187–210. Oxford: Blackwells.

Nummedal, D., Pilkey, O. H., & Howard, J. D. 1987. *Sea-Level Fluctuations and Coastal Evolution*. Tulsa, OK: SEPM (Society for Sedimentary Geology).

Nummedal, D., & Swift, D. J. P. 1987. Transgressive stratigraphy at sequence-bounding unconformities: some principles derived from Holocene and Cretaceous examples. In *Sea-level Fluctuations and Coastal Evolution* (ed. D. Nummedal, O. H. Pilkey, & J. D. Howard), pp. 241–260. Tulsa, OK: SEPM (Society for Sedimentary Geology).

Nye, J. F. 1957. The distribution of stress and velocity in glaciers and ice sheets. *Proceedings of the Royal Society of London* **239A**, 113–133.

1965. The flow of a glacier in a channel of rectangular, elliptic or parabolic cross-section. *Journal of Glaciology* **5**, 661–690.

Oertel, G. F. 1979. Barrier island development during the Holocene recession, SE United States. In *Barrier Islands* (ed. S. P. Leatherman), pp. 273–290. New York: Academic Press.

1985. The barrier island system. *Marine Geology* **63**, 1–18.

1988. Processes of sediment exchange between tidal inlets, ebb deltas and barrier islands. In *Hydrodynamics and Sediment Dynamics of Tidal Inlets* (ed. D. G. Aubrey & L. Weishar), pp. 297–318. Berlin: Springer-Verlag.

Oertel, G. F., Kearney, M. S., Leatherman, S. P., & Woo, J. 1989. Anatomy of a barrier platform: outer barrier lagoon, southern Delmarva Peninsula, Virginia. *Marine Geology* **88**, 303–318.

Ohfuji, H., & Rickard, D. 2005. Experimental synthesis of framboids – a review. *Earth-Science Reviews* **71**, 147–170.

Ohmoto, H., & Skinner, B. J. 1983. *The Kuroko and Related Volcanogenic Massive Sulfide Deposits*. New Haven, CT: The Economic Geology Publishing Company.

Olcott, A. N., Sessions, A. L., Corsetti, F. A., Kaufman, A. J., & de Oliviera, T. F. 2005. Biomarker evidence for photosynthesis during Neoproterozoic glaciation. *Science* **310**, 471–473.

Oliver, J. 1986. Fluids expelled tectonically from orogenic belts: their role in hydrocarbon migration and other geological phenomena. *Geology* **14**, 99–102.

Olsen, H. 1990. Astronomical forcing of meandering river behaviour: Milankovitch cycles in Devonian of East Greenland. *Palaeogeography, Palaeoclimatology, Palaeoecology* **79**, 99–115.

1994. Orbital forcing on continental depositional systems – lacustrine and fluvial cyclicity in the Devonian of East Greenland. In *Orbital Forcing and Cyclic Sequences* (ed. P. L. de Boer & D. G. Smith), pp. 429–438. Oxford: Blackwells.

Olsen, P. E. 1986. A 40-million-year lake record of Early Mesozoic orbital climatic forcing. *Science* **234**, 842–848.

Oltman-Shay, J., & Guza, R. T. 1987. Infragravity edge wave observations on two California beaches. *Journal of Physical Oceanography* **17**, 644–663.

Open University 1989. *Ocean Circulation*. Oxford: Pergamon Press.

Orton, G. J. 1996. Volcanic environments. In *Sedimentary Environments: Processes, Facies and Stratigraphy*, 3rd edn. (ed. H. G. Reading), pp. 485–567. Oxford: Blackwells.

Orton, G. J., & Reading, H. G. 1993. Variability of deltaic processes in terms of sediment supply, with particular emphasis on grain size. *Sedimentology* **40**, 475–512.

Osborne, P. D., & Greenwood, B. 1993. Sediment suspension under waves and currents: time scales and vertical structure. *Sedimentology* **40**, 599–622.

Osgood, R. G. Jr. 1987. Trace fossils. In *Fossil Invertebrates* (ed. R. S. Boardman, A. H. Cheetham, & A. J. Rowell), pp. 663–674. Oxford: Blackwell Scientific.

Otvos, E. G., & Price, W. A. 1979. Problems of chenier genesis and terminology – an overview. *Marine Geology* **31**, 251–263.

Ouchi, S. 1985. Response of alluvial rivers to slow active tectonic movement. *Geological Society of America Bulletin*, **96**, 504–515.

Overeem, I., Syvitski, J. P. M., & Hutton, E. W. H. 2005. Three-dimensional numerical modeling of deltas. In *River Deltas – Concepts, Models, and Examples* (ed. L. Giosan & J. P. Bhattacharya), pp. 13–30. Tulsa, OK: SEPM (Society for Sedimentary Geology).

Owen, G. 1996. Experimental soft-sediment deformation: structures formed by the liquefaction of unconsolidated sands and some ancient examples. *Sedimentology* **43**, 279–293.

Paola, C. 2000. Quantitative models of sedimentary basin filling. *Sedimentology* **47** (Supplement 1), 121–178.

Paola, C., & Borgman, L. 1991. Reconstructing random topography from preserved stratification. *Sedimentology* **38**, 553–565.

Paola, C., Heller, P. L., & Angevine, C. L. 1992. The large-scale dynamics of grain-size variation in alluvial basins. 1 – theory. *Basin Research* **4**, 73–90.

Paola, C., Parker, G., Mohrig, D. C., & Whipple, K. X. 1999. The influence of transport fluctuations on spatially averaged topography on a sandy, braided fluvial plain. In *Numerical Experiments in Stratigraphy: Recent Advances in Stratigraphic and Sedimentologic Computer Simulations* (ed. J. W. Harbaugh, W. L. Watney, E. C. Rankey *et al.*), pp. 211–218. Tulsa, OK: SEPM (Society for Sedimentary Geology).

Pantin, H. M. 1979. Interaction between velocity and effective density in turbidity flow: phase plane analysis, with criteria for autosuspension. *Marine Geology* **31**, 59–99.

2001. Experimental evidence for autosuspension. In *Particulate Gravity Currents* (ed. W. D. McCaffrey, B. C. Kneller, & J. Peakall), pp. 189–205. Oxford: Blackwells.

Pantin, H. M., & Leeder, M. R. 1987. Reverse flow in turbidity currents: the role of internal solitons. *Sedimentology* **34**, 1143–1155.

Parker, G. 1978a. Self-formed straight rivers with equilibrium banks and mobile bed. 1. The sand–silt river. *Journal of Fluid Mechanics* **89**, 109–126.

1978b. Self-formed straight rivers with equilibrium banks and mobile bed. 2. The gravel river. *Journal of Fluid Mechanics* **89**, 127–146.

1979. Hydraulic geometry of active gravel rivers. *Journal of the Hydraulic Division, ASCE* **105**, 1185–1201.

Parker, G., Fukushima, Y., & Pantin, H. M. 1986. Self-accelerating turbidity currents. *Journal of Fluid Mechanics* **171**, 145–181.

Parker, R. S. 1977. Experimental study of drainage basin evolution and its hydrologic implications. Colorado State University, Fort Collins, Colorado, Hydrology Paper **90**.

Parrish, J. T. 1993. Climate of the supercontinent Pangea. *Journal of Geology* **101**, 215–233.

1998. *Interpreting Pre-Quaternary Climate from the Geological Record*. New York: Columbia University Press.

Parrish, J. T., & Barron, E. J. 1986. *Paleoclimates and Economic Geology*. Tulsa, OK: SEPM (Society for Sedimentary Geology).

Paterson, W. S. B. 1994. *The Physics of Glaciers*, 3rd edn. Oxford: Pergamon.

Patterson, R. J., & Kinsman, D. J. J. 1982. Formation of diagenetic dolomite in coastal sabkha along Arabian (Persian) Gulf. *American Association of Petroleum Geologists Bulletin* **66**, 28–43.

Paul, E. A., & Clark, F. E. 1989. *Soil Microbiology and Biochemistry*. San Diego, CA: Academic Press.

Paul, M. A., & Eyles, N. 1990. Constraints on the preservation of diamict facies (melt-out tills) at the margins of stagnant glaciers. *Quaternary Science Reviews* **9**, 51–69.

Paxton, S. T., Szabo, J. O., Ajdukiewicz, J. M., & Klimentidis, R. E. 2002. Construction of an intergranular volume compaction curve for evaluating and predicting compaction and porosity loss in rigid-grain sandstone reservoirs. *American Association of Petroleum Geologists Bulletin* **86**, 2047–2067.

Peakall, J. 1998. Axial river evolution in response to half-graben faulting: Carson River, Nevada. *Journal of Sedimentary Research* **68**, 788–799.

Peakall, J., Felix, M., McCaffrey, B., & Kneller, B. 2001. Particulate gravity currents: perspectives. In *Particulate Gravity Currents* (ed. W. D. McCaffrey, B. C. Kneller, & J. Peakall), pp. 1–8. Oxford: Blackwells.

Peakall, J., Leeder, M., Best, J., & Ashworth, P. 2000a. River response to lateral ground tiltng: a synthesis and some implications for the modelling of alluvial architecture in extensional basins. *Basin Research* **12**, 413–424.

Peakall, J., McCaffrey, B., & Kneller, B. 2000b. A process model for the evolution, morphology, and architecture of sinuous submarine channels. *Journal of Sedimentary Research* **70**, 434–448.

Pemberton, S. G., MacEachern, J. A., & Frey, R. W. 1992. Trace fossil facies models: environmental and allostratigraphic significance. In *Facies Models: Response to Sea Level Change* (ed. R. G. Walker & N. P. James), pp. 47–72. St. John's, Newfoundland: Geological Association of Canada.

Penland, S., Boyd, R., & Suter, J. R. 1988. Transgressive depositional systems of the Mississippi delta plain: a model for barrier shoreline and shelf sand development. *Journal of Sedimentary Petrology* **58**, 932–949.

Pennisi, E. 2004. The secret life of fungi. *Science* **304**, 1620–1622.

Pérez-Arlucea, M., & Smith, N. D. 1999. Depositional patterns following the 1870's avulsion of the Saskatchewan River (Cumberland Marshes, Saskatchewan). *Journal of Sedimentary Research* **69**, 62–73.

Perillo, G. M. E. 1995. *Geomorphology and Sedimentology of Estuaries*. Amsterdam: Elsevier.

Pernetta, J. 1994. *Atlas of the Oceans*. London: Mitchell Beazly.

Person, M., & Garven, G. 1992. Hydrologic constraints on petroleum generation within continental rift basins: theory and application to the Rhine Graben. *American Association of Petroleum Geologists Bulletin* **76**, 468–488.

Person, M., Raffensperger, J. P., Ge, S., & Garven, G. 1996. Basin-scale hydrogeologic modeling. *Reviews of Geophysics* **34**, 61–87.

Peryt, T. 1983. *Coated Grains*. New York: Springer-Verlag.

Peterson, M. N. A., & von der Borch, C. C. 1965. Chert: modern inorganic deposition in a carbonate-precipitating locality. *Science* **149**, 1501–1503.

Petit, J. R., Jouzel, J., Raynaud, D. *et al.* 1999. Climate and atmospheric history of the past 420,000 years from the Vostock ice core, Antarctica. *Nature* **399**, 429–436.

Pettijohn, F. J. 1975. *Sedimentary Rocks*, 3rd edn. New York: Harper and Row.

Pettijohn, F. J., & Potter, P. E. 1964. *Atlas and Glossary of Primary Sedimentary Structures*. New York: Springer-Verlag.

Pettijohn, F. J., Potter, P. E., & Siever, R. 1972. *Sand and Sandstone*. New York: Springer-Verlag.

Petts, G., & Calow, P. 1996. *River Flows and Channel Forms*. Oxford: Blackwell Science Limited.

Phillips, O. M. 1991. *Flow and Reactions in Permeable Rocks*. Cambridge: Cambridge University Press.

Pickard, G. L., & Emery, W. J. 1990. *Descriptive Physical Oceanography*, 5th edn. Oxford: Pergamon Press.

Pickering, K. T., & Hiscott, R. N. 1985. Contained (reflected) turbidity currents from the Middle Ordovician Cloridorme Formation, Quebec, Canada: an alternative to the antidune hypothesis. *Sedimentology* **32**, 373–394.

Pickering, K. T., Hiscott, R. N., & Hein, F. J. 1989. *Deep Marine Environments: Clastic Sedimentation and Tectonics*. London: Unwin Hyman.

Pickering, K. T., Hiscott, R. N., Kenyon, N. H., Ricci Lucci, F., & Smith, R. D. A. 1995. *Atlas of Deep Water Environments: Architectural Style in Turbidite Systems*. London: Chapman and Hall.

Pickering, K. T., Soh, W., & Tiara, A. 1991. Scale of tsunami-generated sedimentary structures in deep water. *Journal of the Geological Society of London* **148**, 211–214.

Pickering, K. T., Stow, D. A. V., Watson, M. P., & Hiscott, R. N. 1986. Deep-water facies, processes and models: a review and classification scheme for modern and ancient sediments. *Earth-Science Reviews* **23**, 75–174.

Pierson, T. C. 1981. Dominant particle support mechanisms in debris flows at Mt Thomas, New Zealand, and implications for flow mobility. *Sedimentology* **28**, 49–60.

1995. Flow characteristics of large eruption-triggered debris flows at snow-clad volcanoes: constraints for debris-flow models. *Journal of Volcanology and Geothermal Research* **66**, 283–294.

Pierson, T. C., & Scott, K. M. 1985. Downstream dilution of a lahar: transition from debris flow to hyperconcentrated streamflow. *Water Resources Research* **21**, 1511–1524.

Pinet, P. R. 2006. *Invitation to Oceanography*, 4th edn. Boston: Jones and Bartlett.

Piper, D. J. W., Cochonat, P., & Morrison, M. L. 1999. The sequence of events around the epicenter of the 1929 Grand Banks earthquake: initiation of debris flow and turbidity current inferred from sidescan sonar. *Sedimentology* **46**, 79–97.

Piper, D. J. W., & Normark, W. R. 1983. Turbidite depositional patterns and flow characteristics, Navy submarine fan, California Borderland. *Sedimentology* **30**, 681–694.

Pitzer, K. 1973. Thermodynamics of electrolytes. I. Theoretical basis and general equations. *Journal of Physical Chemistry* **77**, 268–277.

Playford, P. E. 1984. Platform–margin and marginal–slope relationships in the Devonian reef complexes of the Canning Basin. In *The Canning Basin, Western Australia* (ed. P. G. Purcell), pp. 189–214. Perth: Geological Society of Australia and Petroleum Exploration Society of Australia.

Playford, P. E., & Cockbain, A. E. 1976. Modern algal stromatolites at Hamelin Pool, a hypersaline barred basin, Western Australia. In *Stromatolites* (ed. M. R. Walter), pp. 389–411. Amsterdam: Elsevier.

Plummer, L. N., Parkhurst, D. L., Fleming, G. W., & Dunkle, S. A. 1988. A computer program incorporating Pitzer's equations for calculation of geochemical reactions in brines. United States Geological Survey Water-resources Investigations Report **88–4153**.

Pond, S., & Pickard, G. L. 1983. *Introductory Dynamical Oceanography*, 2nd edn. London: Pergamon.

Pope, M. C., & Grotzinger, J. P. 2000. Controls on fabric development and morphology of tufas and stromatolites, Uppermost Pethei Group (1.8 Ga), Great Slave Lake, Northwest Canada. In *Carbonate Sedimentation and Diagenesis in the Evolving Precambrian World* (ed. J. P. Grotzinger & N. P. James), pp. 103–121. Tulsa, OK: SEPM (Society for Sedimentary Geology).

Portman, C. P., Andrews, J. E., Rowe, P. J., Leeder, M. R., & Hoodewerff, J. 2005. Submarine-spring controlled calcification and growth of large Rivularia bioherms, Late Pleistocene (MIS 5e), Gulf of Corinth, Greece. *Sedimentology* **52**, 441–465.

Posamentier, H. W., & Allen, G. P. 1999. *Siliciclastic Sequence Stratigraphy – Concepts and Applications*. Tulsa, OK: SEPM (Society for Sedimentary Geology).

Posamentier, H. W., Allen, G. P., James, D. P., & Tesson, M. 1992. Forced regressions in a sequence stratigraphic framework: concepts, examples, and exploration significance. *American Association of Petroleum Geologists Bulletin* **76**, 1687–1709.

Posamentier, H. W., Erskine, R. D., & Mitchum, R. M. 1991. Models for submarine fan deposition within a sequence stratigraphic framework. In *Seismic Facies and Sedimentary Processes of Submarine Fans and Turbidite Systems* (ed. P. Weimer & M. H. Link), pp. 127–136. New York: Springer-Verlag.

Posamentier, H. W., Jervey, M. T., & Vail, P. R. 1988. Eustatic controls on clastic deposition I – conceptual framework. In *Sea Level Changes; An Integrated Approach* (ed. C. K. Wilgus,

B. S. Hastings, C. G. St. C. Kendall *et al.*), pp. 109–124. Tulsa, OK: SEPM (Society for Sedimentary Geology).

Posamentier, H. W., & Kolla, V. 2003. Seismic geomorphology and stratigraphy of depositional elements in deep-water settings. *Journal of Sedimentary Research* **73**, 367–388.

Posamentier, H. W., Summerhayes, C. P., Haq, B. U., & Allen, G. P. 1993. *Sequence Stratigraphy and Facies Associations*. Oxford: Blackwells.

Posamentier, H. W., & Walker, R. G. 2006. Deep-water turbidites and submarine fans. In *Facies Models Revisited* (ed. H. W. Posamentier & R. G. Walker), pp. 397–520. Tulsa, OK: SEPM (Society for Sedimentary Geology).

Post, A., & LaChapelle, E. R. 2000. *Glacier Ice*. Seattle, WA: University of Washington Press.

Postma, G. 1990. Depositional architecture and facies of river and fan deltas: a synthesis. In *Coarse-Grained Deltas* (ed. A. Colella & D. B. Prior), pp. 13–27. Oxford: Blackwells.

Postma, G., Nemec, W., & Keinspehn, K. L. 1988. Large floating clasts in turbidites: a mechanism for their emplacement. *Sedimentary Geology* **58**, 47–61.

Potter, P. E., Maynard, J. B., & Depetris, P. J. 2005. *Mud and Mudstone*. Berlin: Springer-Verlag.

Powell, R. D. 1990. Glacimarine processes at grounding-line fans and their growth to ice-contact deltas. In *Glacimarine Environments* (ed. J. A. Dowdeswell & J. D. Scourse), pp. 53–73. London: Geological Society of London.

Powell, R. D., & Molnia, B. F. 1989. Glacimarine sedimentary processes, facies and morphology of the south-southeast Alaska shelf and fiords. *Marine Geology* **85**, 359–390.

Pratt, B. R., & James, N. P. 1982. Crypt-algal–metazoan bioherms of early Ordovician age in the St. George Group, western Newfoundland. *Sedimentology* **29**, 313–343.

Pratt, B. R., James, N. P., & Cowan, C. A. 1992. Peritidal carbonates. In *Facies Models, Response to Sea Level Change* (ed. R. G. Walker & N. P. James), pp. 303–323. St. John's, Newfoundland: Geological Association of Canada.

Press, F., & Siever, R. 1978. *Earth*, 2nd edn. New York: W. H. Freeman and Company.

Press, F., Siever, R., Grotzinger, J., & Jordan, T. H. 2004. *Understanding Earth*, 4th edn. New York: W. H. Freeman and Company.

Prior, D. B., Bornhold, B. D., Wiseman, W. J., & Lowe, D. R. 1987. Turbidity current activity in a British Columbia fjord. *Science* **237**, 1330–1333.

Prior, D. B., & Coleman, J. M. 1979. Submarine landslides – geometry and nomenclature. *Zeitschrift für Geomorphologie* **23**, 415–426.

Prothero, D. R. 2004. *Bringing Fossils to Life: An Introduction to Paleobiology*, 2nd edn. Boston, MA: McGraw-Hill.

Pryor, W. A. 1975. Biogenic sedimentation and alteration of argillaceous sediments in shallow marine environments. *Geological Society of America Bulletin* **86**, 1244–1254.

Purser, B. H. 1973. *The Persian Gulf: Holocene Carbonate Sedimentation and Diagenesis in a Shallow Epicontinental Sea.* New York: Springer-Verlag.

Purser, B. H., & Evans, G. 1973. Regional sedimentation along the Trucial Coast, SE Persian Gulf. In *The Persian Gulf: Holocene Carbonate Sedimentation and Diagenesis in a Shallow Epicontinental Sea* (ed. B. H. Purser), pp. 211–231. New York: Springer-Verlag.

Putnis, A. 2002. Mineral replacement reactions: from macroscopic observations to microscopic mechanisms. *Mineralogical Magazine* **66**, 689–708.

Pye, K. 1987. *Aeolian Dust and Dust Deposits.* London: Academic Press.

1993a. Late Quaternary development of coastal parabolic megadune complexes in northeastern Australia. In *Aeolian Sediments: Ancient and Modern* (ed. K. Pye & N. Lancaster), pp. 23–44. Oxford: Blackwells.

1993b. *The Dynamics and Environmental Context of Aeolian Sedimentary Systems.* London: Geological Society of London.

Pye, K., & Lancaster, N. 1993. *Aeolian Sediments: Ancient and Modern.* Oxford: Blackwells.

Pye, K., & Tsoar, H. 1990. *Aeolian Sand and Sand Dunes.* London: Chapman and Hall.

Quinlan, G. M., & Beaumont, C. 1984. Appalachian thrusting, lithospheric flexure, and the Paleozoic stratigraphy of the eastern interior of North America. *Canadian Journal of Earth Sciences* **21**, 973–996.

Rahmani, R. A., & Flores, R. M. 1984. *Sedimentology of Coal and Coal-bearing Sequences.* Oxford: Blackwells.

Railsback, L. B. 2003. Stylolites. In *Encyclopedia of Sediments and Sedimentary Rocks* (ed. G. V. Middleton), pp. 690–692. Dordrecht: Kluwer Academic Publishers.

Rampino, M. R., & Sanders, J. E. 1980. Holocene transgression in south-central Long Island, New York. *Journal of Sedimentary Petrology* **50**, 1063–1080.

Rasmussen, B., Bengston, S., Fletcher, I. R., & McNaughton, N. J. 2002. Discoidal impressions and trace-like fossils more than 1200 million years old. *Science* **296**, 1112–1115.

Raymo, M. E. & Ruddiman, W. F. 1992. Tectonic forcing of late Cenozoic climate. *Nature* **359**, 117–122.

Raymond, C. F. 1987. How do glaciers surge? A review. *Journal of Geophysical Research* **92**, 9121–9134.

Read, J. F. 1985. Carbonate platform facies models. *Bulletin of the American Association of Petroleum Geologists* **66**, 860–878.

1973. Carbonate cycles, Pillara Formation (Devonian), Canning Basin, Western Australia. *Bulletin of Canadian Petroleum Geology* **16**, 649–653.

1974. Calcrete deposits and Quaternary sediments, Edel Province, Shark Bay, Western Australia. In *Evolution and Diagenesis of Quaternary Carbonate Sequences, Shark Bay, Western Australia* (ed. B. W. Logan, J. F. Read, G. M. Hagan et al.), pp. 250–282. Tulsa, OK: American Association of Petroleum Geologists.

Read, W. A. 1994. High-frequency, glacial–eustatic sequences in early Namurian coal-bearing fluviodeltaic deposits, central Scotland. In *Orbital Forcing and Cyclic Sequences* (ed. P. L. de Boer & D. G. Smith), pp. 413–428. Oxford: Blackwells.

Reading, H. G. (ed.) 1986. *Sedimentary Environments and Facies,* 2nd edn. Oxford: Blackwells.

1996. *Sedimentary Environments: Processes, Facies and Stratigraphy,* 3rd edn. Oxford: Blackwells.

Reading, H. G., & Collinson, J. D. 1996. Clastic coasts. In *Sedimentary Environments: Processes, Facies and Stratigraphy* (ed. H. G. Reading), pp. 154–231. Oxford: Blackwells.

Reading, H. G., & Richards, M. 1994. Turbidite systems in deep-water basin margins classified by grain size and feeder system. *American Association of Petroleum Geologists Bulletin* **78**, 792–822.

Reeder, R. J. 1983. *Carbonates: Mineralogy and Chemistry.* Chelsea, MI: Mineralogical Society of America.

Reesink, A. J. H., & Bridge, J. S. 2007. Influence of superimposed bedforms and flow unsteadiness on formation of cross strata in dunes and unit bars. *Sedimentary Geology* doi:10.1016/j.sedgeo.2007.02.005.

Reeves, C. C. 1977. *Caliche: Origin, Classification, Morphology and Uses.* Lubbock, TX: Escado Books.

Reid, R. P., MacIntyre, I. G., & James, N. P. 1990. Internal precipitation of microcrystalline carbonate: a fundamental problem for sedimentologists. *Sedimentary Geology* **68**, 163–170.

Reineck, H.-E., & Singh, I. B. 1980. *Depositional Sedimentary Environments: With Special Reference to Terrigenous Clastics.* Berlin: Springer-Verlag.

Reineck, H.-E., & Wunderlich, F. 1968a. Zur Unterscheidung von asymmetrischen Oszillationrippeln und Stromungsrippeln. *Senckenbergiana Lethaia* **49**, 321–345.

1968b. Classification and origin of flaser and lenticular bedding. *Sedimentology* **11**, 99–104.

Reinharz, E., Nilsen, K. J., Boesch, D. F., Bertelsen, R., & O'Connell, A. E. 1982. *A Radiographic Examination of Physical and Biogenic Sedimentary Structures in the Chesapeake Bay.* Baltimore, MD: Maryland Geological Survey.

Reinson, G. E. 1992. Transgressive barrier island and estuarine systems. In *Facies Models: Response to Sea Level Change* (ed. R. G. Walker & N. P. James), pp. 179–194. St. John's, Newfoundland: Geological Association of Canada.

Reitner, J. 1993. Modern cryptic microbialite/metazoan facies from Lizard Island (Great Barrier Reef, Australia): formation and concepts. *Facies* **29**, 3–39.

Reitner, J., & Neuweiler, F. 1995. Mud mounds: a polygenetic spectrum of fine-grained carbonate buildups. *Facies* **32**, 1–70.

Renard, F., & Dysthe, D. 2003. Pressure solution. In *Encyclopedia of Sediments and Sedimentary Rocks* (ed. G. V. Middleton), pp. 542–544. Dordrecht: Kluwer Academic Publishers.

Retallack, G. J. 1997. *A Color Guide to Paleosols*. Chichester: Wiley.

2001. *Soils of the Past*, 2nd edn. Oxford: Blackwells.

2003. Weathering, soils, and paleosols. In *Encyclopedia of Sediments and Sedimentary Rocks* (ed. G. V. Middleton), pp. 770–776. Dordrecht: Kluwer Academic Publishers.

2005. Pedogenic carbonate proxies for amount and sea-sonality of precipitation in paleosols. *Geology* **33**, 333–336.

Reynaud, J. Y., Tessier, B., Proust, J. N. *et al.* 1999. Eustatic and hydrodynamic controls on the architecture of a deep shelf sand bank (Celtic Sea). *Sedimentology* **46**, 703–721.

Rhodes, B., Tuttle, M., Horton, B. *et al.* 2006. Paleotsunami research. *EOS* **87**, 205.

Rhodes, E. G., & Moslow, T. F. 1993. *Marine Clastic Reservoirs: Examples and Analogues*. New York: Springer-Verlag.

Rice, S. P., & Church, M. 2001. Longitudinal profiles in sim-ple alluvial systems. *Water Resources Research* **37**, 417–426.

Ricci Lucci, F. 1995. *Sedimentographica: Photographic Atlas of Sedimentary Structures*, 2nd edn. New York: Columbia University Press.

Richardson, K., & Carling, P. A. 2005. A typology of sculpted forms in open bedrock channels. Geological Society of America Special Paper **392**.

Ries, J. R. 2004. Effect of ambient Mg/Ca ratio on Mg frac-tionation in calcareous marine invertebrates: a record of the oceanic Mg/Ca ratio over the Phanerozoic. *Geology* **32**, 981–984.

Riding, R. 1979. Origin and diagenesis of lacustrine algal bioherms at the margin of the Ries Crater, Upper Miocene, southern Germany. *Sedimentology* **26**, 645–680.

2000. Microbial carbonates: the geologic record of calcified bacterial–algal mats and biofilms. *Sedimentology* **47**, 179–214.

Rinaldo, A., Rodriguez-Iturbe, I., & Rigon, R. 1998. Channel networks. *Annual Review of Earth and Planetary Sciences* **26**, 289–327.

Rinaldo, A., Rodriguez-Iturbe, I., Rigon, R. *et al.* 1992. Minimum energy and fractal structures of drainage networks. *Water Resources Research* **28**, 2183–2195.

Riggs, S. D. 1984. Paleoceanographic model of Neogene phosphorite deposition, U.S. Atlantic continental margin. *Science* **223**, 123–131.

Rine, J. M., & Ginsburg, R. N. 1985. Depositional facies of a mud shoreface in Suriname, South America. A mud analogue to sandy, shallow-marine deposits. *Journal of Sedimentary Petrology* **55**, 633–652.

Ritter, D. F., Kochel, R. C., & Miller, J. R. 2002. *Process Geomorphology*, 4th edn. New York: McGraw-Hill.

Roberson, H. E., & Lahann, R. W. 1981. Smectite to illite conversion rates: effects of solution chemistry. *Clay and Clay Minerals* **29**, 129–135.

Roberts, H. H., Adams, R. D., & Cunningham, R. H. W. 1980. Evolution of the sand-dominant subaerial phase, Atchafalaya Delta, Louisiana. *American Association of Petroleum Geologists Bulletin* **64**, 264–279.

Roberts, J. A., Bennett, P. C., González, L. A., Macpherson, G. L., & Milliken, K. L. 2004. Microbial precipitation of dolomite in methanogenic groundwater. *Geology* **32**, 277–280.

Robie, R. A., & Hemingway, B. S. 1995. *Thermodynamic Properties of Minerals and Related Substances at 298.15 K and 1 bar (10^5 Pascals) Pressure and Higher Temperatures*. Washington, D.C.: United States Geological Survey.

Robinson, R. L., & Slingerland, R. L. 1998a. Origin of fluvial grain-size trends in a foreland basin: the Pocono Formation of the central Appalachian basin. *Journal of Sedimentary Research* **A68**, 473–486.

1998b. Grain-size trends and basin subsidence in the Campanian Castlegate Sandstone and equivalent conglomer-ates of central Utah. *Basin Research* **10**, 109–127.

Robinson, R. A. J., Slingerland, R. L., & Walsh, J. M. 2001. Predicting fluvial–deltaic aggradation in Lake Roxburgh, New Zealand: test of a water and sediment routing model. In *Geologic Modeling and Simulation: Sedimentary Systems* (ed. D. F. Merriam & J. C. Davis), pp. 119–132. New York: Kluwer Academic/Plenum.

Robock, A. 2002. The climatic aftermath. *Science* **295**, 1242–1244.

Rodriguez, A. B., Hamilton, M. D., & Anderson, J. B. 2000. Facies and evolution of the modern Brazos delta, Texas; wave versus flood influence. *Journal of Sedimentary Research* **70**, 283–295.

Rodriguez-Iturbe, I., & Rinaldo, A. 1997. *Fractal River Basins: Chance and Self-organization*. Cambridge: Cambridge University Press.

Rodriguez-Iturbe, I., Rinaldo, A., Rigon, R. *et al.* 1992. Energy dissipation, runoff production, and the three-dimensional structure of river basins. *Water Resources Research* **28**, 1095–1103.

Roscoe, R. 1953. Suspensions. In *Flow Properties of Disperse Systems* (ed. J. J. Hermans), pp. 1–38. New York: Wiley Interscience.

Rosgen, D. L. 1994. A classification of natural rivers. *Catena* **22**, 169–199.

1996. *Applied River Morphology*. Fort Collins, CO: Wildland Hydrology.

Ross, D. A., & Degens, E. T. 1969. Shipboard collection and preservation of sediment samples collected during CHAIN 61 from the Red Sea. In *Hot Brines and Recent Heavy Metal Deposits in the Red Sea: A Geochemical and Geophysical Account* (ed. E. T. Degens & D. A. Ross), pp. 363–367. New York: Springer-Verlag.

Rothlisberger, H., & Lang, H. 1987. Glacial hydrology. In *Glacio-Fluvial Sediment Transfer – An Alpine Perspective* (ed. A. M. Gurnell & M. J. Clark), pp. 207–284. Chichester: Wiley.

Rowley, D. B. 2002. Rate of plate creation and destruction: 180 Ma to present. *Geological Society of America Bulletin* **114**, 927–933.

Rubin, D. M. 1987. *Cross-Bedding, Bedforms and Paleocurrents*. Tulsa, OK: SEPM (Society for Sedimentary Geology).

1990. Lateral migration of linear dunes in the Strzelecki Desert, Australia. *Earth Surface Processes and Landforms* **15**, 1–14.

Rubin, D. M., & Hunter, R. E. 1982. Bedform climbing in theory and nature. *Sedimentology* **29**, 121–138.

1985. Why deposits of longitudinal dunes are rarely recognized in the rock record. *Sedimentology* **32**, 147–157.

1987. Bedform alignment in directionally varying flows. *Science* **237**, 276–278.

Rust, B. R. 1978. A classification of alluvial channel systems. In *Fluvial Sedimentology* (ed. A. D. Miall), pp. 187–198. Calgary, Alberta: Canadian Society of Petroleum Geologists.

Rust, B. R., & Romanelli, R. 1975. Late Quaternary subaqueous outwash deposits near Ottawa, Canada. In *Glaciofluvial and Glaciolacustrine Sediments* (ed. A. V. Jopling & B. C. McDonald), pp. 177–192. Tulsa, OK: SEPM (Society for Sedimentary Geology).

Ryan, W. B. F., & Hsü, K. J. 1973. *Initial Reports of the Deep Sea Drilling Project*, Vol. 13. Washington, D.C.: U.S. Government Printing Office.

Sallenger, A. 1979. Inverse grading and hydraulic equivalence in grain-flow deposits. *Journal of Sedimentary Petrology* **49**, 553–562.

Salter, T. 1993. Fluvial scour and incision: models for their influence on the development of realistic reservoir geometries. In *Characterization of Fluvial and Aeolian Reservoirs* (ed. C. P. North & D. J. Prosser), pp. 33–51. London: Geological Society of London.

Sandberg, P. A. 1975. New interpretations of Great Salt Lake ooids and of ancient non-skeletal carbonate mineralogy. *Sedimentology* **22**, 497–538.

1983. An oscillating trend in Phanerozoic non-skeletal carbonate mineralogy. *Nature* **305**, 19–22.

Sarg, J. F. 1988. Carbonate sequence stratigraphy. In *Sea Level Changes: An Integrated Approach* (ed. C. K. Wilgus, B. S. Hastings, C. G. St. C. Kendall *et al.*), pp. 155–182. Tulsa, OK: SEPM (Society for Sedimentary Geology).

Sarna-Wojcicki, A. M., & Davis, J. O. 1991. Quaternary tephrochronology. In *Quaternary Nonglacial Geology, Conterminous U.S.* (ed. R. B. Morrison), pp. 93–116. Boulder, CO: Geological Society of America.

Satake, K. 2005. *Tsunamis: Case Studies and Recent Developments*. Dordrecht: Springer-Verlag.

Saucier, R. T. 1994. *Geomorphology and Quaternary Geologic History of the Lower Mississippi Valley*. Vicksburg, VA: Mississippi River Commission.

Saunders, I., & Young, A. 1983. Rates of surface processes on slopes, slope retreat and denudation. *Earth Surface Processes and Landforms* **8**, 473–501.

Saunderson, H. C., & Lockett, F. P. 1983. Flume experiments on bedforms and structures at the dune-plane bed transition. In *Modern and Ancient Fluvial Systems* (ed. J. D. Collinson & J. Lewin), pp. 49–58. Oxford: Blackwells.

Savage, S. B. 1979. Gravity flow of cohesionless granular materials in chutes and channels. *Journal of Fluid Mechanics* **92**, 53–96.

Schenk, C. J., Gautier, D. L., Olhoeft, G. R., & Lucius, J. E. 1993. Internal structure of an aeolian dune using ground-penetrating radar. In *Aeolian Sediments: Ancient and Modern* (ed. K. Pye & N. Lancaster), pp. 61–70. Oxford: Blackwells.

Schlager, W. 2005. *Carbonate Sedimentology and Sequence Stratigraphy*. Tulsa, OK: SEPM (Society for Sedimentary Geology).

Schlager, W., & Bolz, H. 1977. Clastic accumulation of sulphate evaporites in deep water. *Journal of Sedimentary Petrology* **47**, 600–609.

Schlager, W., & Ginsburg, R. N. 1981. Bahamian carbonate platforms – the deep and the past. *Marine Geology* **44**, 1–24.

Schlichting, H. 1979. *Boundary Layer Theory*. New York: McGraw-Hill.

Schminke, H.-U. 2004. *Volcanism*. Berlin: Springer-Verlag.

Schneidermann, N., & Harris, P. M. 1985. *Carbonate Cements*. Tulsa, OK: SEPM (Society for Sedimentary Geology).

Scholle, P. A. 1978. *A Color Illustrated Guide to Carbonate Rock Constituents, Textures, Cements, and Porosities*. Tulsa, OK: American Association of Petroleum Geologists.

1979. *A Color Guide to Constituents, Textures, Cements, and Porosities of Sandstones and Associated Rocks*. Tulsa, OK: American Association of Petroleum Geologists.

Scholle, P. A., Arthur, M. A, & Ekdale, A. A. 1983. Pelagic environments. In *Carbonate Depositional Environments* (ed. P. A. Scholle, D. G. Bebout, & C. H. Moore), pp. 620–691. Tulsa, OK: American Association of Petroleum Geologists.

Scholle, P. A., & Halley, R. B. 1985. Burial diagenesis: out of sight, out of mind! In *Carbonate Cements* (ed. N. Schneidermann & P. M. Harris), pp. 309–334. Tulsa, OK: SEPM (Society for Sedimentary Geology).

Scholle, P. A., & Spearing, D. R. 1982. *Sandstone Depositional Environments*. Tulsa, OK: American Association of Petroleum Geologists.

Scholle, P. A., & Ulmer-Scholle, D. S. 2003. *A Color Guide to the Petrography of Carbonate Rocks: Grains, Textures, Porosity, Diagenesis*. Tulsa, OK: American Association of Petroleum Geologists.

Schopf, J. W. 1983. *Earth's Earliest Biosphere*. Princeton, NJ: Princeton University Press.

Schubel, K. A., & Simonson, B. M. 1990. Petrography and diagenesis of cherts from Lake Magadi, Kenya. *Journal of Sedimentary Petrology* **60**, 761–776.

Schulz, H. D., & Zabel, M. 2000. *Marine Geochemistry*. Berlin: Springer-Verlag.

Schwartz, R. K. 1982. Bedform and stratification characteristics of some modern small-scale washover sand bodies. *Sedimentology* **29**, 835–849.

Schumm, S. A. 1977. *The Fluvial System*. New York: Wiley.

1993. River response to baselevel change: implications for sequence stratigraphy. *Journal of Geology* **101**, 279–294.

Schumm, S. A., Dumont, J. F., & Holbrook, J. M. 2000. *Active Tectonics and Alluvial Rivers*. Cambridge: Cambridge University Press.

Schumm, S. A., & Lichty, R. W. 1963. Channel widening and floodplain construction, Cimarron River, Kansas. *U.S. Geological Survey Professional Paper* **352-D**, pp. 71–88.

Schumm, S. A., Mosley, M. P., & Weaver, W. E. 1987. *Experimental Fluvial Geomorphology*. New York: Wiley.

Scruton, P. C. 1960. Delta building and the deltaic sequence. In *Recent Sediments, Northwest Gulf of Mexico* (ed. F. P. Shepard, F. B. Phleger, & T. H. Van Andel), pp. 82–102. Tulsa, OK: American Association of Petroleum Geologists.

Seifert, D., & Jensen, J. L. 1999. Using sequential indicator simulation as a tool in reservoir description: issues and uncertainties. *Mathematical Geology* **31**, 527–550.

2000. Object and pixel-based reservoir modeling of a braided fluvial reservoir. *Mathematical Geology* **32**, 581–603.

Seilacher, A, Bose, P. K., & Pflüger, F. 1998. Triploblastic animals more than 1 billion years ago: trace fossil evidence from India. *Science* **282**, 80–83.

Selby, M. J. 1993. *Hillslope Materials and Processes*, 2nd edn. Oxford: Oxford University Press.

Semikhatov, M. A., Gebelein, C. D., Cloud, P., Awramik, S. M., & Benmore, W. C. 1979. Stromatolite morphogenesis –

progress and problems. *Canadian Journal of Earth Sciences* **16**, 992–1015.

Sepkoski, J. J. Jr. 1981. A factor analytic description of the Phanerozoic marine record. *Paleobiology* **7**, 36–53.

Sha, L. P., & de Boer, P. L. 1991. Ebb-tidal delta deposits along the west Friesian islands (The Netherlands): processes, facies architecture and preservation. In *Clastic Tidal Sedimentology* (ed. D. G. Smith, G. E. Reinson, B. A. Zaitlin, & R. A. Rahmani), pp. 199–218. Calgary, Alberta: Canadian Society of Petroleum Geologists.

Shanley, K. W., & McCabe, P. J. 1993. Alluvial architecture in a sequence stratigraphic framework: a case history from the Upper Cretaceous of southern Utah, USA. In *The Geological Modeling of Hydrocarbon Reservoirs and Outcrop Analogues* (ed. S. Flint & I. D. Bryant), pp. 21–56. Oxford: Blackwells.

1994. Perspectives on the sequence stratigraphy of continental strata. *American Association of Petroleum Geologists Bulletin* **78**, 544–568.

Shanmugan, G. 1996. High-density turbidity currents: are they sandy debris flows? *Journal of Sedimentary Research* **66**, 2–10.

1997. The Bouma sequence and the turbidite mind set. *Earth-Science Reviews* **42**, 201–229.

Shanmugan, G., & Moiola, R. J. 1988. Submarine fans: characteristics, models, classification and reservoir potential. *Earth-Science Reviews* **24**, 383–428.

Sharp, M. 1988a. Surging glaciers: behaviour and mechanisms. *Progress in Physical Geography* **12**, 349–370.

1988b. Surging glaciers: geomorphic effects. *Progress in Physical Geography* **12**, 533–559.

Sharp, M., Jouzel, J., Hubbard, B., & Lawson, W. 1994. The character, structure and origin of the basal ice layer of a surge-type glacier. *Journal of Glaciology* **40**, 327–340.

Sharp, M., Richards, K. S., & Tranter, M. 1998. *Glacier Hydrology and Hydrochemistry*. Chichester: Wiley.

Sharp, R. P. 1963. Wind ripples. *Journal of Geology* **71**, 617–636.

Sharpe, D. R., & Shaw, J. 1989. Erosion of bedrock by subglacial meltwater, Cantley, Quebec. *Geological Society of America Bulletin* **101**, 1011–1020.

Shaw, J. 2006. A glimpse at meltwater effects associated with continental ice sheets. In *Glacier Science and Environmental Change* (ed. P. G. Knight), pp. 25–32. Oxford: Blackwells.

Shaw, J., Kvill, D., & Rains, B. 1989. Drumlins and catastrophic subglacial floods. *Sedimentary Geology* **62**, 177–202.

Shaw, J., & Sharpe, D. R. 1987. Drumlin formation by subglacial meltwater erosion. *Canadian Journal of Earth Sciences* **24**, 2316–2322.

Shear, W. A., & Selden, P. A. 2001. Rustling in the undergrowth: animals in early terrestrial ecosystems. In *Plants Invade the Land* (ed. P. G. Gensel & D. Edwards), pp. 29–51. New York: Columbia University Press.

Shearman, D. J. 1963. Recent anhydrite, gypsum, dolomite and halite from the coastal flats of Arabian shore of the Persian Gulf. *Proceedings of the Geological Society of London* **1607**, 63–65.

Shearman, D. J., McCugan, A., Stein, C., & Smith, A. J. 1989. Ikaite, $CaCO_3 \cdot 6H_2O$, precursor of the thinolites in the Quaternary tufas and tufa mounds of the Lahontin and Mono Lake basins, western United States. *Geological Society of America Bulletin* **101**, 913–917.

Sheehan, P. M., & Harris, M. T. 2004. Microbialite resurgence after the Late Ordovician extinction. *Nature* **430**, 75–78.

Shepard, F. P. 1932. Sediments on continental shelves. *Geological Society of America Bulletin* **43**, 1017–1034.

1973. *Submarine Geology*. New York: Harper and Row.

Shepard, F. P., & Inman, D. L. 1950. Nearshore circulation related to bottom topography and wave refraction. *Transactions of the American Geophysical Union* **31**, 555–565.

Sheridan, R. E. 1974. Altantic continental margin of North America. In *The Geology of Continental Margins* (ed. C. A. Burke & C. L. Drake), pp. 391–407. New York: Springer-Verlag.

Shields, A. 1936. Anwendung der Ähnlichkeitsmechanik und der Turbulenzforschung auf die Geschiebebewegung. *Mitteilungen der Preußischen Versuchsanstalt für Wasserbau und Schiffbau* **26**, 26 pp.

Shiki, T., Cita, M. B., & Gorsline, D. S. 2000. Seismoturbidites, seismites and tsunamiites. *Sedimentary Geology* (special issue) **135**, 1–322.

Shinn, E. A. 1968a. Selective dolomitization of recent sedimentary structures. *Journal of Sedimentary Petrology* **38**, 612–616.

1968b. Practical significance of birdseye structures in carbonate rocks. *Journal of Sedimentary Petrology* **38**, 611–616.

1969. Submarine lithification of Holocene carbonate sediments in the Persian Gulf. *Sedimentology* **12**, 109–144.

1983a. Birdseyes, fenestrae, shrinkage pores, and Loferites: a reevaluation. *Journal of Sedimentary Petrology* **53**, 619–628.

1983b. Tidal flats. In *Carbonate Depositional Environments* (ed. P. A. Scholle, D. G. Bebout, & C. H. Moore), pp. 171–210. Tulsa, OK: American Association of Petroleum Geologists.

Shinn, E. A., & Lidz, B. H. 1988. Blackened limestone pebbles: fire at subaerial unconformities. In *Paleokarst* (ed. N. P. James & P. W. Choquette), pp. 117–131. New York: Springer-Verlag.

Shinn, E. A., Lloyd, R. M., & Ginsburg, R. N. 1969. Anatomy of a modern carbonate tidal flat, Andros Island, Bahamas. *Journal of Sedimentary Petrology* **39**, 1202–1228.

Shinn, E. A., & Robbin, D. M. 1983. Mechanical and chemical compaction in fine-grained shallow-water limestones. *Journal of Sedimentary Petrology* **53**, 596–618.

Shinn, E. A., Steinen, R. P., Lidz, B. H., & Swart, R. K. 1989. Whitings, a sedimentological dilemma. *Journal of Sedimentary Petrology* **59**, 147–161.

Shreve, R. L. 1966. Statistical law of stream numbers. *Journal of Geology* **74**, 17–37.

Shvidchenko, A. B., & Pender, G. 2001. Macroturbulent structure of open-channel flow over gravel beds. *Water Resources Research* **37**, 709–719.

Sibley, D. F., & Gregg, J. M. 1987. Classification of dolomite rock texture. *Journal of Sedimentary Petrology* **57**, 967–975.

Sidi, F. H., Nummedal, D., Imbert, P., Darman, H., & Posamentier, H. 2003. *Tropical Deltas of Southeast Asia – Sedimentology Stratigraphy, and Petroleum Geology*. Tulsa, OK: SEPM (Society for Sedimentary Geology).

Siegert, M. J. 2000. Antarctic subglacial lakes. *Earth-Science Reviews* **50**, 29–50.

Simms, M. A. 1984. Dolomitization by groundwater-flow systems in carbonate platforms. *Transactions of the Gulf Coast Association of Geological Societies* **34**, 411–420.

Simone, L. 1980. Ooids: a review. *Earth-Science Reviews* **16**, 319–335.

Simonson, B. M. 2003a. Ironstones and iron formations. In *Encyclopedia of Sediments and Sedimentary Rocks* (ed. G. V. Middleton), pp. 379–385. Dordrecht: Kluwer Academic Publishers.

2003b. Origin and evolution of large Precambrian iron formations. Geological Society of America Special Paper **370**, pp. 231–244.

Simonson, B. M., & Carney, K. E. 1999. Roll-up structures: evidence of *in situ* microbial mats in Late Archean deep shelf environments. *Palaios* **14**, 13–24.

Simpson, J. E. 1982. Gravity currents in the laboratory, atmosphere and ocean. *Annual Review of Fluid Mechanics* **14**, 213–234.

1997. *Gravity Currents in the Environment and the Laboratory*, 2nd edn. New York: Cambridge University Press.

Singh, U. 1987. Ooids and cements from the Late Precambrian of the Flinders Range, South Australia. *Journal of Sedimentary Petrology* **57**, 117–127.

Sinha, S. K., & Parker, G. 1996. Causes of concavity in longitudinal profiles of rivers. *Water Resources Research* **32**, 1417–1428.

Siringan, F. P., & Anderson, J. B. 1993. Seismic facies architecture and evolution of the Bolivar Roads tidal inlet/delta complex, east Texas Gulf Coast. *Journal of Sedimentary Petrology* **63**, 794–808.

1994. Modern shoreface and inner-shelf storm deposits off the East Texas coast, Gulf of Mexico. *Journal of Sedimentary Petrology* **64**, 99–110.

Sleath, J. F. A. 1984. *Sea Bed Mechanics*. New York: Wiley.

Slingerland, R. L. 1986. Numerical computation of co-oscillating paleotides in the Catskill epeiric sea of eastern North America. *Sedimentology* **33**, 817–829.

Slingerland, R., Harbaugh, J. W., & Furlong, K. P. 1994. *Simulating Clastic Sedimentary Basins.* Englewood Cliffs, NJ: Prentice-Hall.

Slingerland, R. L., & Smith, N. D. 1998. Necessary conditions for a meandering-river avulsion. *Geology* **26**, 435–438.

2004. River avulsions and their deposits. *Annual Review of Earth and Planetary Sciences* **32**, 255–283.

Sloss, L. L. 1963. Sequences in the cratonic interior of North America. *Geological Society of America Bulletin* **74**, 93–114.

Smith, C. R. 1996. Coherent flow structures in smooth-wall turbulent boundary layers: facts, mechanisms and speculation. In *Coherent Flow Structures in Open Channels* (ed. P. J. Ashworth, S. J. Bennett, J. L. Best, & S. J. McLelland), pp. 1–39. Chichester: Wiley.

Smith, D. G. 1988. Modern point bar deposits analogous to the Athabasca Oil Sands, Alberta, Canada. In *Tide-Influenced Sedimentary Environments and Facies* (ed. P. L. DeBoer, A. Van Gelder, & S. D. Nio), pp. 417–432. Boston, MA: D. Reidel Publishing Company.

Smith, D. G., Reinson, G. E., Zaitlin, B. A., & Rahmani, R. A. 1991. *Clastic Tidal Sedimentology.* Calgary, Alberta: Canadian Society of Petroleum Geologists.

Smith, D. G., & Smith, N. D. 1980. Sedimentation in anastomosed river systems: examples from alluvial valleys near Banff, Alberta. *Journal of Sedimentary Petrology* **50**, 157–164.

Smith, L. C., Sheng, Y., Magilligan, F. J. *et al.* 2006. Geomorphic impact and rapid subsequent recovery from the 1996 Skeiðarásandur jokulhlaup, Iceland, measured with multi-year airborne lidar. *Geomorphology* **75**, 65–75.

Smith, N. D. 1974. Sedimentology and bar formation in the upper Kicking Horse River, a braided outwash stream. *Journal of Geology* **81**, 205–223.

Smith, N. D., Cross, T. A., Dufficy, J. P., & Clough, S. R. 1989. Anatomy of an avulsion. *Sedimentology* **36**, 1–23.

Smith, N. D., & Rogers, J. 1999. *Fluvial Sedimentology VI.* Oxford: Blackwells.

Smith, N. D., Slingerland, R. L., Pérez-Arlucea, M., & Morozova, G. S. 1998. The 1870s avulsion of the Saskatchewan River. *Canadian Journal of Earth Sciences* **35**, 453–466.

Smith, S. V., & Kinsey, D. W. 1976. Calcium carbonate production, coral reef growth, and sea level change. *Science* **194**, 937–939.

Smoot, J. P. 1983. Depositional subenvironments in an arid closed basin: the Wilkens Peak Member of the Green River Formation (Eocene), Wyoming. *Sedimentology* **30**, 801–827.

Smoot, J. P., & Lowenstein, T. K. 1991. Depositional environments of non-marine evaporites. In *Evaporites, Petroleum and Mineral Resources* (ed. J. L. Melvin), pp. 189–347. Amsterdam: Elsevier.

Snedden, J. W., & Nummedal, D. 1991. Origin and geometry of storm-deposited sand beds in modern sediments of the Texas continental shelf. In *Shelf Sand and Sandstone Bodies: Geometry, Facies and Sequence Stratigraphy* (ed. D. J. P. Swift, G. F. Oertel, R. W. Tillman, & J. A. Thorne), pp. 283–308. Oxford: Blackwells.

Sonnenfeld, P. 1984. *Brines and Evaporites.* Orlando, FL: Academic Press.

Sorby, H. C. 1879. The structure and origin of limestones. *Proceedings of the Geological Society of London* **35**, 56–95.

Soulsby, R. L., & Bettess, R. 1991. *Sand Transport in Rivers, Estuaries and the Sea.* Rotterdam: Balkema.

Southard, J. B., & Boguchwal, L. A. 1990. Bed configurations in steady unidirectional flows. Part 2. Synthesis of flume data. *Journal of Sedimentary Petrology* **60**, 658–679.

Southard, J. B., Lambie, J. M., Federico, D. C., Pile, H. T., & Weidman, C. R. 1990. Experiments on bed configurations in fine sands under bidirectional purely oscillatory flow, and the origin of hummocky cross-stratification. *Journal of Sedimentary Petrology* **60**, 1–17.

Sparks, R. S. J. 1976. Grain-size variations in ignimbrites and implications for the transport of pyroclastic flows. *Sedimentology* **23**, 147–188.

Sparks, R. S. J., Bonnecaze, R. T., Huppert, H. E. *et al.* 1993. Sediment-laden gravity currents with reversing buoyancy. *Earth and Planetary Science Letters* **114**, 249–288.

Sparks, R. S. J., & Gilbert, J. S. 2002. *Physics of Explosive Volcanic Eruptions.* London: Geological Society of London.

Spencer, R. J., & Lowenstein, T. K. 1990. *Evaporites.* In *Diagenesis* (ed. I. A. McIlreath & D. W. Morrow), pp. 141–163. St. John's, Newfoundland: Geological Association of Canada.

Spencer, R. J., Möller, N., & Weare, J. H. 1990. The prediction of mineral solubilities in natural waters: a chemical equilibrium model for the Na–K–Ca–Mg–Cl–SO$_4$–H$_2$O system at temperatures below 25 °C. *Geochimica et Cosmochimica Acta* **54**, 575–590.

Sposito, G. 1989. *The Chemistry of Soils.* New York: Oxford University Press.

Srivastava, R. M. 1994. An overview of stochastic methods for reservoir characterization. In *Stochastic Modeling and Geostatistics* (ed. J. M. Yarus & R. L. Chambers), pp. 3–16. Tulsa, OK: American Association of Petroleum Geologists.

Stallard, R. F., & Edmond, J. M. 1981. Geochemistry of the Amazon 1. Precipitation chemistry and the marine contribution to the dissolved load at the time of peak discharge. *Journal of Geophysical Research* **86**, 9844–9858.

1983. Geochemistry of the Amazon 2. The influence of geology and weathering environment on the dissolved load. *Journal of Geophysical Research* **88**, 9671–9688.

1987. Geochemistry of the Amazon 3. Weathering chemistry and the limits to dissolved inputs. *Journal of Geophysical Research* **92**, 8293–8302.

Stanley, S. M. 1966. Paleoecology and diagenesis of Key Largo Limestone, Florida. *American Association of Petroleum Geologists Bulletin* **50**, 1927–1947.

2005. *Earth System History*, 2nd edn. New York: W. H. Freeman and Company.

Stanley, S. M., & Hardie, L. A. 1998. Secular oscillations in the carbonate mineralogy of reef-building and sediment-producing organisms driven by tectonically forced shifts in seawater chemistry. *Palaeogeography, Palaeoclimatology, Palaeoecology* **144**, 3–19.

Stanley, S. M., Ries, J. B., & Hardie, L. A. 2002. Low-magnesium calcite produced by coralline algae in seawater of Late Cretaceous composition. *Proceedings of the National Academy of Sciences of the USA* **99**, 15323–15326.

Starkel, L., Gregory, K. J., & Thornes, J. B. 1991. *Temperate Palaeohydrology*. Chichester: Wiley.

Steel, R. J., Maehle, S., Nilsen, H., Roe, S. L., & Spinnanger, A. 1977. Coarsening-upward cycles in the alluvium of the Hornelen basin (Devonian, Norway): sedimentary response to tectonic events. *Geological Society of America Bulletin* **88**, 1124–1134.

Steel, R. J., & Ryseth, A. 1990. The Triassic–Early Jurassic succession in the northern North Sea: megasequence stratigraphy and intra-Triassica tectonics. In *Tectonic Events Responsible for Britain's Oil and Gas Reserves* (ed. R. F. P. Hardman & J. Brooks), pp. 139–168. London: Geological Society of London.

Stern, C. W., Scoffin, T. P., & Martindale, W. 1977. Calcium carbonate budget of a fringing reef on the west coast of Barbados, pt. 1, zonation and productivity. *Bulletin of Marine Science* **27**, 779–810.

Stewart, F. H. 1963. Marine evaporites. United States Geological Survey Professional Paper **440-Y**.

Stockman, K. W., Ginsburg, R. N., & Shinn, E. A. 1967. The production of lime mud by algae in South Florida. *Journal of Sedimentary Petrology* **37**, 633–648.

Stokes, C. R., & Clark, C. D. 2001. Palaeo-ice streams. *Quaternary Science Reviews* **20**, 1437–1457.

2002. Ice stream shear margin moraines. *Earth Surface Processes and Landforms* **27**, 547–558.

2003. Laurentide ice streaming on the Canadian Shield: a conflict with the soft-bedded ice stream paradigm. *Geology* **31**, 347–350.

Stokes, W. L. 1968. Multiple parallel-truncation bedding planes – a feature of wind deposited sandstone formations. *Journal of Sedimentary Petrology* **38**, 510–515.

Stokstad, E. 2004. Defrosting the carbon freezer of the north. *Science* **304**, 1618–1620.

Stone, G. W., & Orford, J. D. 2004. Storms and their significance in coastal morpho-sedimentary dynamics. *Marine Geology* **210**, 1–368.

Stonecipher, S. A. 2000. *Applied Sandstone Diagenesis – Practical Petrographic Solutions for a Variety of Common Exploration, Development, and Production Problems*. Tulsa, OK: SEPM (Society for Sedimentary Geology).

Storms, J. E. A. 2003. Event-based stratigraphic simulation of wave-dominated shallow-marine environments. *Marine Geology* **199**, 83–100.

Storms, J. E. A., & Swift, D. J. P. 2003. Shallow-marine sequences as the building blocks of stratigraphy: insights from numerical modeling. *Basin Research* **15**, 287–303.

Stouthamer, E., & Berendsen, H. J. A. 2000. Factors controlling the Holocene avulsion history of the Rhine–Meuse delta (The Netherlands). *Journal of Sedimentary Research* **70**, 1051–1064.

2001. Avulsion frequency, avulsion duration and interavulsion period of Holocene channel belts in the Rhine–Meuse delta, The Netherlands. *Journal of Sedimentary Research* **71**, 589–598.

Stouthamer, E. 2005. Reoccupation of channel belts and its influence on alluvial architecture in the Holocene Rhine-Meuse delta, the Netherlands. In *River Deltas – Concepts, Models, and Examples* (ed. L. Giosan & J. Bhattacharya), pp. 319–339. Tulsa, OK: SEPM (Society for Sedimentary Geology).

Stow, D. A. V. 1986. Deep clastic seas. In *Sedimentary Environments and Facies*, 2nd edn. (ed. H. G. Reading), pp. 399–444. Oxford: Blackwells.

1994. Deep sea processes of sediment transport and deposition. In *Sediment Transport and Depositional Processes* (ed. K. Pye), pp. 257–291. Oxford: Blackwells.

Stow, D. A. V., & Faugeres, J. C. (eds.) 1993. *Contourites and Bottom Currents. Sedimentary Geology* (special issue) **82**.

1998. *Contourites, Turbidites and Process Interaction. Sedimentary Geology* (special issue) **115**.

Stow, D. A. V., & Lovell, J. P. B. 1979. Contourites: their recognition in modern and ancient sediments. *Earth-Science Reviews* **14**, 251–291.

Stow, D. A. V., & Piper, D. J. W. 1984. *Fine-Grained Sediments: Deep-Water Processes and Facies*. London: Geological Society of London.

Stow, D. A. V., Pudsey, C. J., Howe, J. A., Faugeres, J. -C., & Vianna, A. R. 2002. *Deep-water Contourite Systems: Modern Drifts and Ancient Series, Seismic and Sedimentary Characteristics*. London: Geological Society of London.

Stow, D. A. V., Reading, H. G., & Collinson, J. D. 1996. Deep seas. In *Sedimentary Environments: Processes, Facies and*

Stratigraphy, 3rd edn. (ed. H. G. Reading), pp. 395–453. Oxford: Blackwells.

Stow, D. A. V., & Shanmugan, G. 1980. Sequences of structures in fine-grained turbidites: comparison of recent deep sea and ancient flysch sediments. *Sedimentary Geology* **25**, 23–42.

Stow, D. A. V., & Tabrez, A. 1998. *Hemipelagites: Facies, Processes, and Models*, pp. 317–338. London: Geological Society of London.

Stow, D. A. V., & Wetzel, A. 1990. Hemiturbidite: a new type of deep-water sediment. *Proceedings of the Ocean Drilling Program, Scientific Results* **116**, 25–34.

Strahler, A. N. 1957. Quantitative analysis of watershed geomorphology. *Transactions of the American Geophysical Union* **38**, 913–920.

Straub, S. 2001. Bagnold revisited: implications for the rapid motion of high-concentration sediment flows. In *Particulate Gravity Currents* (ed. W. D. McCaffrey, B. C. Kneller, & J. Peakall), pp. 91–109. Oxford: Blackwells.

Stride, A. H. 1982. *Offshore Tidal Sands*. London: Chapman and Hall.

Stumm, W., & Morgan, J. J. 1981. *Aquatic Chemistry: An Introduction Emphazing Chemical Equilibria in Natural Waters*. New York: John Wiley and Sons.

Sugden, D. E., & John, B. S. 1976. *Glaciers and Landscape*. London: Edward Arnold.

Summerhayes, C. P., & Thorpe, S. A. 1996. *Oceanography: An Illustrated Guide*. London: Manson.

Summerfield, M. A. 1983. Petrography and diagenesis of silcrete from the Kalahari Basin and Cape coastal zone. *Journal of Sedimentary Petrology* **53**, 895–909.

Sumner, D. Y., & Grotzinger, J. P. 2000. Late Archean aragonite precipitation: petrography, facies associations, and environmental significance. In *Carbonate Sedimentation and Diagenesis in the Evolving Precambrian World* (ed. J. P. Grotzinger & N. P. James), pp. 123–144. Tulsa, OK: SEPM (Society for Sedimentary Geology).

Sun, T., Meakin, P., & Jossang, T. 2001a. Meander migration and the lateral tilting of floodplains. *Water Resources Research* **37**, 1485–1502.

2001b. A computer model for meandering rivers with multiple bedload sediment sizes 1. Theory. *Water Resources Research* **37**, 2227–2241.

2001c. A computer model for meandering rivers with multiple bedload sediment sizes 2. Computer simulations. *Water Resources Research* **37**, 2243–2258.

Sun, T., Paola, C., Parker, G., & Meakin, P. 2002. Fluvial fan deltas: linking channel processes with large-scale morphodynamics. *Water Resources Research* **38**, 1151.

Swift, D. J. P. 1975. Barrier island genesis: evidence from the Middle Atlantic Shelf of North America. *Sedimentary Geology* **14**, 1–43.

Swift, D. J. P., & Field, M. E. 1981. Evolution of a classic sand ridge field; Maryland sector, North American inner shelf. *Sedimentology* **28**, 461–482.

Swift, D. J. P., Figueiredo, A. R., Freeland, G. L., & Oertal, G. F. 1983. Hummocky cross stratification and megaripples: a geological double standard? *Journal of Sedimentary Petrology* **53**, 1295–1317.

Swift, D. J. P., Han, G., & Vincent, C. E. 1986. Fluid processes and sea-floor response on a modern storm-dominated shelf: middle Atlantic shelf of North America. Part I: the storm-current regime. In *Shelf Sands and Sandstones* (ed. R. J. Knight & J. R. McLean), pp. 99–119. Calgary, Alberta: Canadian Society of Petroleum Geologists.

Swift, D. J. P., Oertel, G. F., Tillman, R. W., & Thorne, J. A. 1991. *Shelf Sand and Sandstone Bodies: Geometry, Facies and Sequence Stratigraphy*. Oxford: Blackwells.

Swift, D. J. P., & Thorne, J. A. 1991. Sedimentation on continental margins, I: a general model for shelf sedimentation. In *Shelf Sand and Sandstone Bodies: Geometry, Facies and Sequence Stratigraphy* (ed. D. J. P. Swift, G. F. Oertel, R. W. Tillman, & J. A. Thorne), pp. 3–31. Oxford: Blackwells.

Syvitski, J. P. M., Burrell, D. C., & Skei, J. M. 1987. *Fjords: Processes and Products*. New York: Springer-Verlag.

Takahashi, T. 1981. Debris flow. *Annual Review of Fluid Mechanics* **13**, 57–77.

2001. Mechanics and simulation of snow avalanches, pyroclastic flows and debris flows. In *Particulate Gravity Currents* (ed. W. D. McCaffrey, B. C. Kneller, & J. Peakall), pp. 11–43. Oxford: Blackwells.

Talbot, M. R. 1985. Major bounding surfaces in aeolian sandstones – a climatic model. *Sedimentology* **32**, 257–265.

Talbot, M. R., & Allen, P. A. 1996. Lakes. In *Sedimentary Environments: Processes, Facies and Stratigraphy*, 3rd edn. (ed. H. G. Reading), pp. 83–124. Oxford: Blackwell Science.

Taylor, A., Goldring, R., & Gowland, S. 2003. Analysis and application of ichnofabrics. *Earth-Science Reviews* **60**, 227–259.

Taylor, K. G., Gawthorpe, R. L., Curtis, C. D., Marshall, J. D., & Anwiller, D. A. 2000. Carbonate cementation in a sequence-stratigraphic framework: Upper Cretaceous sandstones, Book Cliffs, Utah–Colorado. *Journal of Sedimentary Research* **70**, 360–372.

Tennant, C. B., & Berger, R. W. 1957. X-ray determination of the dolomite–calcite ratios of a carbonate rock. *American Mineralogist* **42**, 23–29.

Terwindt, J. H. J. 1981. Origin and sequences of sedimentary structures in inshore mesotidal deposits of the North Sea. In

Holocene Marine Sedimentation in the North Sea Basin (ed. S. D. Nio, R. T. E. Schuttenhelm, & T. C. E. Van Weering), pp. 51–64. Oxford: Blackwells.

1988. Palaeo-tidal reconstructions of inshore tidal depositional environments. In *Tide-influenced Sedimentary Environments and Facies* (ed. P. L. DeBoer, A. Van Gelder, & S. D. Nio), pp. 233–263. Boston, MA: D. Reidel Publishing Company.

Tetzlaff, D. M., & Harbaugh, J. W. 1989. *Simulating Clastic Sedimentation.* New York: Van Nostrand Reinhold.

Tetzlaff, D. M., & Priddy, G. 2001. Sedimentary process modeling: from academia to industry. In *Geologic Modeling and Simulation: Sedimentary Systems* (ed. D. F. Merriam & J. C. Davis), pp. 45–69. New York: Kluwer Academic/Plenum Publishers.

Thomas, D. 1997. *Arid Zone Geomorphology.* London: Bellhaven/Hallsted Press.

Thomas, R. G., Smith D. G., Wood, J. M. *et al.* 1987. Inclined heterolithic stratification – terminology, description, interpretation and significance. *Sedimentary Geology* **53**, 123–179.

Thrailkill, J. 1976. Speleothems. In *Stromatolites* (ed. M. R. Walter), pp. 73–86. Amsterdam: Elsevier.

Thurman, H. V., & Trujilo, A. P. 2004. *Introductory oceanography,* 10th edn. Englewood Cliffs, NJ, Prentice-Hall.

Tillman, R. W., & Martinsen, R. S. 1984. The Shannon shelf-ridge sandstone complex, Salt Creek anticline area, Powder River Basin, Wyoming. In *Siliciclastic Shelf Sediments* (ed. R. W. Tillman & C. T. Seimers), pp. 85–142. Tulsa, OK: SEPM (Society for Sedimentary Geology).

Tillman, R. W., & Martinsen, R. S. 1987. Sedimentological model and production characteristics of Hartzog Draw Field, Wyoming. In *Reservoir Sedimentology* (ed. R. W. Tillman & K. J. Weber), pp. 15–112. Tulsa, OK: SEPM (Society for Sedimentary Geology).

Tinkler, K. J., & Wohl, E. E. 1998. *Rivers Over Rock: Fluvial Processes in Bedrock Channels.* Washington, D.C.: American Geophysical Union.

Thorne, C. R., Russell, A. P. G., & Alam, M. K. 1993. Planform pattern and channel evolution of Brahmaputra River, Bangladesh. In *Braided Rivers* (ed. J. L. Best & C. S. Bristow), pp. 257–276. London: Geological Society of London.

Törnqvist, T. E. 1993. Holocene alternation of meandering and anastomosing fluvial systems in the Rhine–Meuse delta (central Netherlands) controlled by sea-level rise and subsoil erodibility. *Journal of Sedimentary Petrology* **63**, 683–693.

1994. Middle and late Holocene avulsion history of the River Rhine (Rhine–Meuse delta, Netherlands). *Geology* **22**, 711–714.

Törnqvist, T. E., & Bridge, J. S. 2002. Spatial variation of overbank aggradation rate and its influence on avulsion frequency. *Sedimentology* **49**, 891–905.

Törnqvist, T. E., Kidder, T. R., Autin, W. J. *et al.* 1996. A revised chronology for Mississippi River subdeltas. *Science* **273**, 1693–1696.

Törnqvist, T. E., Wallinga, J., Murray, A. S. *et al.* 2000. Response of the Rhine–Meuse system (west-central Netherlands) to the last Quaternary glacio-eustatic cycles. *Global and Planetary Change* **27**, 89–111.

Traverse, A., & Ginsburg, R. N. 1966. Palynology of the surface sediments of Great Bahama Bank, as related to water movements and sedimentation. *Marine Geology* **4**, 417–459.

Trendall, A. F., & Morris, R. C. 1983. *Iron-Formation: Facts and Problems.* Amsterdam: Elsevier.

Trendall, A. F., & Blockley, J. G. 1970. The iron formations of the Precambrian Hamersley Group, Western Australia. Geological Survey of Western Australia Bulletin **119**.

Trewin, N. H. 1994. Depositional environment and preservation of biota in the Lower Devonian hot-springs of Rhynie, Aberdeenshire, Scotland. *Transactions of the Royal Society of Edinburgh: Earth Sciences* **84**, 433–442.

Trewin, N. H., & Rice, C. M. 1992. Stratigraphy and sedimentology of the Devonian Rhynie Chert locality. *Scottish Journal of Geology* **28**, 37–47.

Tripati, A., & Elderfield, H. 2005. Deep-sea temperature and circulation changes at the Paleocene–Eocene Thermal Maximum. *Science* **308**, 1894–1898.

Tsoar, H. 1982. Internal structure and surface geometry of longitudinal (seif) dunes. *Journal of Sedimentary Petrology* **52**, 823–831.

1983. Dynamic processes acting on a longitudinal (seif) dune. *Sedimentology* **30**, 567–578.

1984. The formation of seif dunes from barchans – a discussion. *Zeitschrift für Geomorphologie* **28**, 99–103.

1989. Linear dunes – forms and formation. *Progress in Physical Geography* **13**, 507–528.

Tsoar, H., & Pye, K. 1987. Dust transport and the question of desert loess formation. *Sedimentology* **34**, 139–153.

Tucker, G. E., & Slingerland, R. L. 1996. Predicting sediment flux from fold and thrust belts. *Basin Research* **8**, 329–349.

1997. Drainage basin responses to climate change. *Water Resources Research* **33**, 2031–2047.

Tucker, G. E., Lancaster, S. T., Gasparini, N. M., & Bras, R. L. 2002. The Channel-Hillslope Integrated Landscape Development Model (CHILD). In *Landscape Erosion and Evolution Modeling* (ed. R. S. Harmon & W. W. Doe III), pp. 349–388. New York: Kluwer Academic Publishing.

Tucker, M. E. 2001. *Sedimentary Petrology: An Introduction to the Origin of Sedimentary Rocks.* Oxford: Blackwell Scientific.

Tucker, M. E., & Wright, V. P. 1990. *Carbonate Sedimentology.* Oxford: Blackwells.

Turcotte, D. L., & Schubert, G. 2002. *Geodynamics*, 2nd edn. Cambridge: Cambridge University Press.

Turner, E. C., Narbonne, G. M., & James, N. P. 1993. Neoproterozoic reef microstructures from the Little Dal Group, northwestern Canada. *Geology* 21, 259–262.

2000a. Framework composition of Early Neoproterozoic calcimicrobial reefs and associated microbialites, Mackenzie Mountains, N.W.T., Canada. In *Carbonate Sedimentation and Diagenesis in the Evolving Precambrian World* (ed. J. P. Grotzinger & N. P. James), pp. 179–205. Tulsa, OK: SEPM (Society for Sedimentary Geology).

2000b. Taphonomic control on microstructures in Early Neoproterozoic reefal stromatolites and thrombolites. *Palaios* 15, 87–111.

Tuttle, M. P., Ruffman, A., Anderson, T., & Jeter, H. 2004. Distinguishing tsunami deposits from storm deposits along the coast of northeastern North America: lessons learned from the 1929 Grand Banks tsunami and the 1991 Halloween storm. *Seismological Research Letters* 75, 117–131.

Twichell, D. C., Kenyon, N. H., Parson, L. M., & McGregor, B. A. 1991. Depositional patterns of the Mississippi Fan surface: evidence from GLORIA II and high resolution seismic profiles. In *Seismic Facies and Sedimentary Processes of Submarine Fans and Turbidite Systems* (ed. P. Weimer & M. H. Link), pp. 349–364. New York: Springer-Verlag.

Tye, R. S. 1984. Geomorphic evolution and stratigraphy of Price and Capers Inlets, South Carolina. *Sedimentology* 31, 655–674.

2004. Geomorphology: an approach to determining subsurface reservoir dimensions. *American Association of Petroleum Geologists Bulletin* 88, 1123–1147.

Tye, R. S., & Coleman, J. M. 1989a. Depositional processes and stratigraphy of fluvially dominated lacustrine deltas: Mississippi Delta Plain. *Journal of Sedimentary Petrology* 59, 973–996.

1989b. Evolution of Atchafalaya lacustrine deltas, south-central Louisiana. *Sedimentary Geology* 65, 95–112.

Tye, R. S., & Moslow, T. F. 1993. Tidal inlet reservoirs: insights from modern examples. In *Marine Clastic Reservoirs: Examples and Analogues* (ed. E. G. Rhodes & T. F. Moslow), pp. 77–99. New York: Springer-Verlag.

Tyler, K., Henriquez, A., & Svanes, T. 1994. Modeling heterogeneities in fluvial domains: a review of the influence on production profiles. In *Stochastic Modeling and Geostatistics* (ed. J. M. Yarus & R. L. Chambers), pp. 77–89. Tulsa, OK: American Association of Petroleum Geologists.

US Department of Agriculture Soil Survey Staff 1993. *Soil Survey Manual*. Washington, D.C.: United States Department of Agriculture.

2003. *Keys to Soil Taxonomy*. Washington, D.C.: United States Department of Agriculture.

Usiglio, J. 1894. Analyse de l'eau de la Mediterranée sur la côte de France. *Annales de Chimie et de Physique* 27, 172–191.

Vail, P. R., Mitchum, R. M. Jr, Todd, R. G. *et al.* 1977. Seismic stratigraphy and global changes in sea-level. In *Seismic Stratigraphy – Applications to Hydrocarbon Exploration* (ed. C. E. Payton), pp. 49–212. Tulsa, OK: American Association of Petroleum Geologists.

Vallance, J. W., & Scott, K. M. 1997. The Osceola Mudflow from Mount Rainier: sedimentology and hazard implications of a huge clay-rich debris flow. *Geological Society of America Bulletin* 109, 143–163.

Van Andel, T. H., & Komar, P. D. 1969. Ponded sediments of the Mid-Atlantic Ridge between 22° and 23° north latitude. *Geological Society of America Bulletin* 80, 1163–1190.

Van Den Berg, J. H. 1982. Migration of large-scale bedforms and preservation of crossbedded sets in highly accretional parts of tidal channels in the Oosterschelde, S.W. Netherlands. *Geologie en Mijnbouw* 61, 253–263.

1995. Prediction of alluvial channel pattern of perennial rivers. *Geomorphology* 12, 259–279.

Van Den Berg, J. H., & Van Gelder, A. 1993. A new bedform stability diagram, with emphasis on the transition of ripples to plane bed in flows over fine sand and silt. In *Alluvial Sedimentation* (ed. M. Marzo & C. Puidefabregas), pp. 11–21. Oxford: Blackwells.

Van Den Berg, J. H., Van Gelder, A., & Mastbergen, D. R. 2002. The importance of breaching as a mechanism of sub-aqueous slope failure in fine sand. *Sedimentology* 49, 81–95.

Van der Wateren, F. M. 2002. Processes of glaciotectonism. In *Modern and Past Glacial Environments* (ed. J. Menzies), pp. 417–443. Oxford: Butterworth-Heinemann.

Van Dyke, M. 1982. *An Album of Fluid Motion*. Stanford, CA: Parabolic Press.

Van Heerden, I. L., & Roberts, H. H. 1988. Facies development of Atchafalaya Delta, Louisiana: a modern bayhead delta. *American Association of Petroleum Geologists Bulletin* 72, 439–453.

Van Heijst, M. W. I. M., & Postma, G. 2001. Fluvial response to sea-level changes: a quantitative analogue, experimental approach. *Basin Research* 13, 269–292.

Van Houten, F. B. 1964. Cyclic lacustrine sedimentation, Upper Triassic Lockatong Formation, central New Jersey and adjacent Pennsylvania. *Kansas Geological Survey Bulletin* 169, 497–532.

1985. Oolitic ironstones and contrasting Ordovician and Jurassic paleogeography. *Geology* 13, 722–724.

Van Houten, F. B. & Bhattacharyya, P. D. 1982. Phanerozoic oolitic ironstones – geologic record and facies model. *Annual Review of Earth and Planetary Sciences* 10, 441–457.

van Lith, Y., Warthmann, R., Vasconcelos, C., & McKenzie, J. A. 2003. Microbial fossilization in carbonate sediments: a result of the bacterial surface involvement in dolomite precipitation. *Sedimentology* **50**, 237–245.

2004. Sulfphate-reducing bacteria induce low-temperature Ca-dolomite and high Mg-calcite formation. *Geobiology* **1**, 71–79.

Van Niekerk, A. Vogel, A. K., Slingerland, R., & Bridge, J. S. 1992. Routing heterogeneous size–density sediments over a moveable bed: model development. *Journal of Hydraulic Engineering, ASCE* **118**, 246–262.

Van Rijn, L. C. 1990. *Principles of Fluid Flow and Surface Waves in Rivers, Estuaries, Seas, and Oceans.* Amsterdam: Aqua Publications.

1993. *Principles of Sediment Transport in Rivers, Estuaries and Coastal Seas.* Amsterdam: Aqua Publications.

Van Tassell, J. 1994. Cyclic deposition of the Devonian Catskill Delta of the Appalachian, U.S.A. In *Orbital Forcing and Cyclic Sequences* (ed. P. L. de Boer & D. G. Smith), pp. 395–411. Oxford: Blackwells.

Van Wagoner, J. C., & Bertram, G. T. 1995. *Sequence Stratigraphy of Foreland Basin Deposits.* Tulsa, OK: American Association of Petroleum Geologists.

Van Wagoner, J. C., Mitchum, R. M. Jr., Campion, K. M., & Rahmanian, V. D. 1990. *Siliciclastic Sequence Stratigraphy in Well Logs, Cores and Outcrop: Concepts for High Resolution Correlation of Time and Facies.* Tulsa, OK: American Association of Petroleum Geologists.

Veizer, J., Godderis, Y., & Francois, L. M. 2000. Evidence for decoupling of atmospheric CO_2 and global climate during the Phanerozoic eon. *Nature* **408**, 698–701.

Verway, J. 1952. On the ecology of distribution of cockle and mussel in the Dutch Waddensee, their role in sedimentation and the source of their food supply. *Archives Néerlandaises de Zoologie* **10**, 171–240.

Vigilar, G. G. Jr., & Diplas, P. 1997. Stable channels with mobile bed: formulation and numerical solution. *Journal of Hydraulic Engineering, ASCE* **123**, 189–199.

1998. Stable channels with mobile bed: model verification and graphical solution. *Journal of Hydraulic Engineering, ASCE* **124**, 1097–1108.

Vincent, C. E., Young, R. A., & Swift, D. J. P. 1982. On the relationship between bedload and suspended sand transport on the inner shelf, Long Island, New York. *Journal of Geophysical Research* **87**, 4163–4170.

Vinopal, R. J., & Coogan, A. H. 1978. Effect of particle shape on the packing of carbonate sands and gravels. *Journal of Sedimentary Petrology* **48**, 7–24.

Visscher, P. T., & Stolz, J. F. 2005. Microbial mats as bioreactors: populations, processes, and products. *Palaeogeography, Palaeoclimatology, Palaeoecology* **219**, 87–100.

Visser, M. J. 1980. Neap–spring cycles reflected in Holocene subtidal large-scale bedform deposits: a preliminary note. *Geology* **8**, 543–546.

Vogel K., Van Niekerk, A., Slingerland, R., & Bridge, J. S. 1992. Routing of heterogeneous size–density sediments over a moveable bed: model verification and testing. *Journal of Hydraulic Engineering, ASCE* **118**, 263–279.

Vose, R. S., Schmoyer, R. L., Steurer, P. M., Peterson, T. C., Heim, R., Karl, T. R., & Eischeid, J. K. 1992. *The Global Historical Climatology Network: Long-term Monthly Temperature, Precipitation, Sea Level Pressure, and Station Pressure Data.* Oak Ridge, TN: Oak Ridge National Laboratory Environmental Sciences Division.

Walgreen, M., Calvete, D., & de Swart, H. E. 2002. Growth of large-scale bedforms due to storm-driven and tidal currents: a model approach. *Continental Shelf Research* **22**, 2777–2793.

Walker, R. G. 1965. The origin and significance of the internal structures of turbidites. *Proceedings of the Yorkshire Geological Society* **35**, 1–32.

1978. Deep water sandstone facies and ancient submarine fans: models for exploration for stratigraphic traps. *American Association of Petroleum Geologists Bulletin* **62**, 932–966.

1979. Turbidites and associated coarse clastic deposits. In *Facies Models* (ed. R. G. Walker), pp. 91–103. St. John's, Newfoundland: Geological Association of Canada.

1984. Shelf and shallow marine sands. In *Facies Models*, 2nd edn. (ed. R. G. Walker), pp. 141–170. St. John's, Newfoundland: Geological Association of Canada.

1992. Turbidites and submarine fans. In *Facies Models: Response to Sea Level Change* (ed. R. G. Walker & N. P. James), pp. 239–263. St. John's, Newfoundland: Geological Association of Canada.

Walker, R. G., & James, N. P. 1992. *Facies Models: Response to Sea Level Change.* St. John's, Newfoundland: Geological Association of Canada.

Walker, R. G., & Mutti, E. 1973. Turbidite facies and facies associations. In *Turbidites and Deep-Water Sedimentation* (ed. G. V. Middleton & A. H. Bouma), pp. 19–157. Tulsa, OK: SEPM (Society for Sedimentary Geology).

Walker, R. G., & Plint, A. G. 1992. Wave- and storm-dominated shallow marine systems. In *Facies Models: Response to Sea Level Change* (ed. R. G. Walker & N. P. James), pp. 219–238. St. John's, Newfoundland: Geological Association of Canada.

Walter, M. R. 1976. *Stromatolites.* Amsterdam: Elsevier.

Walther, J. V. 2005. *Essentials of Geochemistry.* Sudbury, MA: A. Jones and Bartlett Publishers.

Wanless, H. R. 1979. Limestone response to stress: pressure solution and dolomitization. *Journal of Sedimentary Petrology* **49**, 437–462.

Ward, A. W., & Greeley, R. 1984. The yardangs at Rogers Lake, California. *Geological Society of America Bulletin* **95**, 829–837.

Ward, W. B. 1996. Evolution and diagenesis of Frasnian carbonate platforms, Devonian Reef Complexes, Napier Range, Canning Basin, Western Australia. Unpublished Ph.D. dissertation, State University of New York at Stony Brook, New York.

Wardle, D. A., Bardgett, R. D., Klironomos, J. N., Setata, H., van der Putten, W. H., & Wall, D. H. 2004. Ecological linkages between aboveground and belowground biota. *Science* **304**, 1629–1633.

Warren, J. K. 1982. The hydrological significance of Holocene tepees, stromatolites, and boxwork limestones in coastal Salinas in south Australia. *Journal of Sedimentary Petrology* **52**, 1171–1201.

2006. *Evaporites: Sediments, Resources and Hydrocarbons.* Berlin: Springer-Verlag.

Warren, W. P., & Ashley, G. M. 1994. Origins of the ice-contact ridges (eskers) of Ireland. *Journal of Sedimentary Research* **64**, 433–449.

Watabe, N., & Wilbur, K. M. 1976. *The Mechanisms of Mineralization in the Invertebrates and Plants.* Columbia, SC: University of South Carolina Press.

Waters, B. B., Spencer, R. J., & Demicco, R. V. 1989. Three-dimensional architecture of shallowing-upward carbonate cycles: Middle and Upper Cambrian Waterfowl Formation, southern Canadian Rocky Mountains. *Bulletin of Canadian Petroleum Geology* **37**, 198–209.

Watts, N. L. 1977. Pseudo-anticlines and other structures in some calcretes of Botswana and South Africa. *Earth Surface Processes and Landforms* **2**, 63–74.

1980. Quaternary pedogenic calcretes from the Kalahari (southern Africa); mineralogy, genesis and diagenesis. *Sedimentology* **27**, 661–686.

Weaver, P. P. E., & Thomson, J. 1987. *Geology and Geochemistry of Abyssal Plains.* London: Geological Society of London.

Weertman, J. 1979. The unsolved general glacier sliding problem. *Journal of Glaciology* **23**, 97–115.

1986. Basal water and high-pressure basal ice. *Journal of Glaciology* **32**, 455–463.

Weimer, P. 1990. Sequence stratigraphy, facies geometries, and depositional history of the Mississippi Fan, Gulf of Mexico. *American Association of Petroleum Geologists Bulletin* **74**, 425–453.

1995. Sequence stratigraphy of the Mississippi Fan (late Miocene–Pleistocene), northern deep Gulf of Mexico. In *Atlas of Deep Water Environments; Architectural Style in Turbidite Systems* (ed. K. T. Pickering, R. N. Hiscott, N. H. Kenyon, F. Ricci Lucci, & R. D. A. Smith), pp. 94–99. London: Chapman and Hall.

Weimer, P., Bouma, A. H., & Perkins, B. F. 1994. *Submarine Fans and Turbidite Systems.* Austin, TX: Society of Economic Paleontologists and Mineralogists Gulf Coast Section.

Weimer, P., & Link, M. H. 1991. *Seismic Facies and Sedimentary Processes of Submarine Fans and Turbidite Systems.* New York: Springer-Verlag.

Weimer, P., & Posamentier, H. 1993. *Siliciclastic Sequence Stratigraphy.* Tulsa, OK: American Association of Petroleum Geologists.

Weimer, R. J., Howard, J. D., & Lindsay, D. R. 1982. Tidal flats. In *Sandstone Depositional Environments* (ed. P. A. Scholle & D. A. Spearing), pp. 191–245. Tulsa, OK: American Association of Petroleum Geologists.

Wellner, R. W., Ashley, G. M., & Sheridan, R. E. 1993. Seismic stratigraphic evidence for a submerged middle Wisconsin barrier: implications for sea-level history. *Geology* **21**, 109–112.

Werner, B. T. 1995. Eolian dunes: computer simulations and attractor interpretation. *Geology* **23**, 1107–1110.

Wescott, W. A. 1993. Geomorphic thresholds and complex response of fluvial systems – some implications for sequence stratigraphy. *American Association of Petroleum Geologists Bulletin* **77**, 1208–1218.

Westbroek, P., & De Jong, E. W. 1983. *Biomineralization and Biological Metal Accumulation.* Dordrecht: D. Reidel Publishing Company.

Whateley, M. K. G., & Pickering, K. T. 1989. *Deltas: Sites and Traps for Fossil Fuels.* London: Geological Society of London.

Whipple, K. L. 1997. Open-channel flow of Bingham fluids: applications in debris-flow research. *Journal of Geology* **105**, 243–262.

Whipple, K. X., Hancock, G. S., & Anderson, R. S. 2000. River incision into bedrock: mechanics and relative efficacy of plucking, abrasion, and cavitation. *Geological Society of America Bulletin* **112**, 490–503.

White, W. A. 1961. Colloid phenomena in sedimentation of argillaceous rocks. *Journal of Sedimentary Petrology* **31**, 560–570.

Wignall, P. B. 1994. *Black Shales.* Oxford: Oxford University Press.

Wilgus, C. K., Hastings, B. S., Kendall, C. G. St. C. *et al.* 1988. *Sea-Level Changes: An Integrated Approach.* Tulsa, OK: SEPM (Society for Sedimentary Geology).

Wilken, R. T. 2003. Sulfide minerals in sediments. In *Encyclopedia of Sediments and Sedimentary Rocks* (ed. G. V. Middleton), pp. 701–703. Dordrecht: Kluwer Academic Publishers.

Wilkinson, B. H. 1979. Biomineralization, paleoceanography and the evolution of calcareous marine organisms. *Geology* **7**, 524–527.

Willgoose, G. R., Bras, R. L., & Rodriguez-Iturbe, I. 1991a. A physically-based coupled network growth and hillslope evolution model: 1. Theory. *Water Resource Research* **27**, 1,671–1,684.

1991b. A physically-based coupled network growth and hillslope evolution model: 2. Applications. *Water Resource Research* **27**, 1,685–1,696.

Williams, G. P. 1988. Paleofluvial estimates from dimensions of former channels and meanders. In *Encyclopedia of Sediments and Sedimentary Rocks* (ed. V. R. Baker, R. C. Kochel, & P. P. Patton), pp. 321–334. Chichester: Wiley.

Willis, B. J. 1989. Paleochannel reconstructions from point bar deposits: a three-dimensional perspective. *Sedimentology* **36**, 757–766.

1993a. Ancient river systems in the Himalayan foredeep, Chinji village area, northern Pakistan. *Sedimentary Geology* **88**, 1–76.

1993b. Interpretation of bedding geometry within ancient point-bar deposits In *Alluvial Sedimentation* (ed. M. Marzo & C. Puigdefabregas), pp. 101–114. Oxford: Blackwells.

1997. Architecture of fluvial-dominated valley-fill deposits in the Cretaceous Fall River Formation. *Sedimentology* **44**, 735–757.

2005. Deposits of tide-influenced river deltas. In *River Deltas – Concepts, Models, and Examples* (ed. L. Giosan & J. P. Bhattacharya), pp. 87–129. Tulsa, OK: SEPM (Society for Sedimentary Geology).

Willis, B. J., & Behrensmeyer, A. K. 1994. Architecture of Miocene overbank deposits in Northern Pakistan. *Journal of Sedimentary Research* **B64**, 60–67.

Wilson, A. M., Sanford, W., Whitaker, F., & Smart, P. 2001. Spatial patterns of diagenesis during geothermal circulation in carbonate platforms. *American Journal of Science* **301**, 727–752.

Wilson, E. N., Hardie, L. A., & Phillips, O. M. 1990. Dolomitization front geometry, fluid flow patterns, and the origin of massive dolomite: the Triassic Latemar Buildup, northern Italy. *American Journal of Science* **290**, 741–796.

Wilson, I. G. 1972a. Aeolian bedforms – their development and origins. *Sedimentology* **19**, 173–210.

1972b. Universal discontinuities in bedforms produced by wind. *Journal of Sedimentary Petrology* **42**, 667–669.

Wilson, J. L. 1975. *Carbonate Facies in Geologic History*. New York: Springer-Verlag.

Wilson, J. L., & Jordan, C. 1983. Middle shelf. In *Carbonate Depositional Environments* (ed. P. A. Scholle, D. G. Bebout, & C. H. Moore), pp. 297–343. Tulsa, OK: American Association of Petroleum Geologists.

Wolf, K. H., Easton, A. J., & Warme, S. 1967. Techniques of examining and analyzing carbonate skeletons, minerals and rocks. In *Carbonate Rocks* (ed. G. V. Chilingar, H. J. Bissell, & R. W. Fairbridge), pp. 253–341. Amsterdam: Elsevier.

Wollast, R. 1976. Kinetics of the alteration of K-feldspar in buffered solutions at low temperature. *Geochimica et Cosmochimica Acta* **31**, 635–648.

Wolman, M. G., & Miller, J. P. 1960. Magnitude and frequency of forces in geomorphic processes. *Journal of Geology* **68**, 54–74.

Wood, L. J., Ethridge, F. G., & Schumm, S. A. 1993. An experimental study of the influence of subaqueous shelf angles on coastal plain and shelf deposits. In *Siliciclastic Sequence Stratigraphy* (ed. P. Weimer & H. W. Posamentier), pp. 381–391. Tulsa, OK: American Association of Petroleum Geologists.

Wood, R. A., Grotzinger, J. P., & Dickson, J. A. D. 2002. Proterozoic modular biomineralized metazoan from the Nama Group, Namibia. *Science* **296**, 2383–2386.

Woodroffe, C. D. 2003. *Coasts; Form, Process and Evolution*. Cambridge: Cambridge University Press.

Woodroffe, C. D., Chappell, J., Thom, B. G., & Wallensky, E. 1989. Depositional model of a macrotidal estuary and floodplain, South Alligator River, Northern Australia. *Sedimentology* **36**, 737–756.

Wright, H. E. Jr., Kutzbach, J. E., Webb, T. III *et al.* 1993. *Global Climates since the Last Glacial Maximum*. Minneapolis, MN: University of Minnesota Press.

Wright, L. D. 1977. Sediment transport and deposition at river mouths: a synthesis. *Geological Society of America Bulletin* **88**, 857–868.

Wright, L. D., & Coleman, J. M. 1973. Variations in morphology of major river deltas as functions of ocean wave and river discharge regimes. *American Association of Petroleum Geologists Bulletin* **57**, 370–398.

1974. Mississippi River mouth processes: effluent dynamics and morphological development. *Journal of Geology* **82**, 751–778.

Wright, L. D., Coleman, J. M., & Thom, B. G. 1975. Sediment transport and deposition in a macrotidal river channel, Ord River, Western Australia. In *Estuarine Research II* (ed. L. E. Cronin), pp. 309–322. New York: Academic Press.

Wright, L. D., & Short, A. D. 1984. Morphodynamic variability of surf zones and beaches: a synthesis. *Marine Geology* **56**, 93–118.

Wright, V. P. 1986. *Paleosols: Their Recognition and Interpretation*. Oxford: Blackwell Scientific.

1999. Assessing flood duration gradients and fine-scale environmental change on ancient floodplains. In *Floodplains: Interdisciplinary Approaches* (ed. S. B. Marriott & J. Alexander), pp. 279–287. London: Geological Society of London.

Wright, V. P., & Burchette, T. P. 1996. Shallow-water carbonate environments. In *Sedimentary Environments: Processes,*

Facies and Stratigraphy, 3rd edn. (ed. H. G. Reading), pp. 325–394. Oxford: Blackwells.

Wright, V. P., & Marriott, S. B. 1993. The sequence stratigraphy of fluvial depositional systems: the role of floodplain sediment storage. *Sedimentary Geology* **86**, 203–210.

Yalin, M. S. 1977. *Mechanics of Sediment Transport*, 2nd edn. Oxford: Pergamon Press.

1992. *River Mechanics*. Oxford: Pergamon Press.

Yao, Q., & Demicco, R. V. 1995. Paleoflow patterns of dolomitizing fluids and paleohydrogeology of southern Canadian Rocky Mountains: evidence from dolomite geometry and numerical modeling. *Geology* **23**, 791–794.

1997. Dolomitization of the Cambrian carbonate platform, southern Canadian Rocky Mountains: dolomite front geometry, fluid inclusion geochemistry, isotopic signature, and hydrogeologic modeling studies. *American Journal of Science* **297**, 892–938.

Yarnold, Y. C. 1993. Rock-avalanche characteristics in dry climates and the effects of flow into lakes: insights from mid-Tertiary sedimentary breccias near Artillery Peak, Arizona. *Geological Society of America Bulletin* **105**, 345–360.

Young, A. 1972. *Slopes*. Edinburgh: Oliver and Boyd.

Young, I. M., & Crawford, J. W. 2004. Interactions and self-organization in the soil-microbe complex. *Science* **304**, 1634–1637.

Zachos, J., Pagani, M., Sloan, L., Thomas, E., & Billups, K. 2001. Trends, rhythms, and aberrations in global climate 65 Ma to present. *Science* **292**, 686–693.

Zachos, J. C., Röhl, U., Schellenberg, S. A. *et al.* 2005. Rapid acidification of the ocean during the Paleocene–Eocene Thermal Maximum. *Science* **308**, 1611–1615.

Zajac, I. S. 1974. *The Stratigraphy and Mineralogy of the Sokoman Formation in the Knob Lake Area, Quebec and Newfoundland*. Ottawa: Geological Survey of Canada.

Zeng, J., & Lowe, D. R. 1997. Numerical simulations of turbidity current flow and sedimentation: I. Theory. *Sedimentology* **44**, 67–84.

Zharkov, M. A. 1984. *Paleozoic Salt Bearing Formations of the World*. Berlin: Springer-Verlag.

Zaitlin, B. A., Dalrymple, R. W., & Boyd, R. 1994. The stratigraphic organization of incised-valley systems associated with relative sea-level change. In *Incised Valley Systems: Origin and Sedimentary Sequences* (ed. R. W. Dalrymple, R. Boyd, & B. A. Zaitlin), pp. 45–60. Tulsa, OK: SEPM (Society for Sedimentary Geology).

Index

Page numbers in italics refer to figures.

3D seismic data analysis 429, *429*
A soil horizon *67*, 68, *69*, *70*
abandoned-delta-lobe reworking 488, *491*
abandoned delta lobe *484*, 488, *490*
abandoned river channel 410, *412*, 417–419, 430, *431*
abyssal plain *631*, 638
accommodation of sediment 707
accretionary lapilli 282, *283*
accretionary wedge *13*, 14, *718*
acid–base reactions 49–50
acid mine runoff 55
acid rain 54
acid 49–50, *also see* pH
acidification of shallow ocean water 89
active (convergent) continental margin *631*, 633
activity 48
activity coefficient 48
aeolian-dune classification 200
aeolian-dune migration 202
aeolian-ripple laminae 205, *206*, *207*
aeolian sand seas 570–572, *571*, 608
aeolian sand seas: bed forms 570, *572*
aeolian sand seas: climate change 591–592
aeolian sand seas: deposits 570–572, *574*, 592
aeolian sand seas: interdune areas 570
aeolian sand seas: wind parameters 570
aeolian sand-patch fabric 581
aeolian silt and clay (loess) 198, *198*, 563, 570, 608, *616*
agglomerate 282
airborne pollutants 53, 54
Airy-wave theory 215–217, *216*
albitization 676
algae *see* green algae and red algae
algal biscuit *see* oncoid
Algoma-type iron formations 115
alkaline systems 54
alkalinity titration 90
allocyclic controls: large-scale sedimentary sequences 707
alluvial architecture *430*, 450–457, *451*, *452–456*

alluvial architecture: channel-belt connectedness 453, 485
alluvial architecture: channel-deposit proportion 453, 485
alluvial architecture: controls 451
alluvial architecture: definition 450, *451*
alluvial architecture: extrinsic (extra-basinal) controls 451
alluvial architecture: intrinsic (intra-basinal) controls 451
alluvial architecture: process-based models *430*, *452*, 452–455, *453*, *454*
alluvial architecture: stochastic models 455, *455–457*, *456*
alluvial-channel geometry *385*, 385–395, *386*, *387*, *388*, *389*, *391*, *392*
alluvial-channel geometry: theory 393–395
alluvial-channel patterns 380–385, 446
alluvial-channel patterns: controls 383–385, *384*, 446
alluvial-fan deposition 450
alluvial-fan lobes 450, *450*
alluvial fan 441, 444, 447, 448–450, *449*, *450*, 568, 569
alluvial fan: climate 734
alluvial fan: flow and sedimentary processes 449–450, *450*
alluvial fan: occurrence and geometry 448–449, *449*
alluvial ridge (channel belt) 430, *431*
alluvial river channels: bed forms on bars *400*, 401, *401*
alluvial river channels: erosion and deposition *400*, 402–405, *403*, *404*, *406*
alluvial river channels: sediment transport 399–401, *400*, *401*
alluvial river channels: water flow 395–399, *396*, *398*
alluvial-river erosion–deposition due to discharge variation 402, *403*
alluvial rivers 375, *375*
alternate bars in rivers 377, *378–381*
alveolar texture 350
amalgamated storm-generated stratasets *499–500*, 501
amphidromic point 240, *241*
anastomosing river 380, 382, *382*, 607
anchor ice 292, 297
ancient river-channel deposits in outcrop 422–426, *425–426*
ancient subsurface river-channel deposits 426–430, *427–429*, 430
Andros Island: tidal-channel deposits 536, *538*

Andros Island: tidal-channels 533
Andros Island: tidal-flat deposits 536, *538*
Andros Island: tidal flats 532–539, *534–535, 536, 537, 538–539*
angle of initial yield 172, 255
angle of repose 170, 255, 497
angular and tangential cross strata *172*
anions 47
ankerite 87
anoxic conditions 51, 317, 462, 466
anoxic conditions: prokaryotic metabolism 31–33
antidune cross strata 184–185, *185, 186*
antidunes 158, *181*, 181–182, *182, 237, 400*, 401, *503*
antitectonic 710, 719
aquifers 679
aragonite 87
aragonite-needle mud 96, *97*
aragonite seas 40, 87, *88,* 92, 101, *101, 109, 110*, 732
Aral Sea 462, 463, 572
arborescent shrubs 349, 582 *see also* tufa
archaea *see* prokaryotes
Archean *15*, 37
arête *597*, 598
arid environments 563–567, *564, 565, 566*
arid environments: aeolian sand seas 570–572, *571, 572, 573, 574*
arid environments: bare rock and soil *567*, 567–568
arid environments: climatic-change indicators 731, 734
arid environments: economic aspects 594
arid environments: ephemeral saline lakes 577–581, *578, 579, 580*
arid environments: ephemeral streams *568*, 568–570
arid environments: lake tufas and travertines *see* tufas
arid environments: large-scale depositional sequences 592–594
arid environments: marine basins 590–591, *591*
arid environments: marine coasts 565, 584–590, *586, 587, 588*
arid environments: marine lagoons 587–590, *589*
arid environments: marine shelves 590, *590*
arid environments: perennial saline lakes 572–576
arid environments: saline-lake coasts 581, *583*
arid environments: supersurfaces 592
arid environments: time variations 591–592
aridisols 567, 581
aridity 563
armored bed 153, 161
Artemia see brine shrimp
ash distribution from volcanic eruptions 278
asymmetrical and symmetrical tidal waves 239, *240*
asymmetrical and symmetrical wave ripples 230, *231–232*
asymmetrical channel belts *444*, 448, *448*
atmosphere: circulation 18–23, *20, 21, 21, 22*
atmosphere: composition 14
atmosphere: evolution 38–39
atmosphere: physical properties 14–16
atmosphere: temperature at Earth's surface 16, *16–17*

atmosphere: thermal structure *16*
atmospheric pressure 16, *18, 22*
atoll 545, *546*, 638, 664
authigenic 676
autocyclic controls: large-scale sedimentary sequences 707
automicrite 98
autosuspension 268

B soil horizon 67, 69, *69, 70*
back reef 546
backset laminae 185
backshore deposits 507
backshore 492, *494*, 507
backwash 219, 505
bacteria *see* prokaryotes
ball-and-pillow structures 352
banded iron formations 38, 104, 114, *116*
bankfull discharge 376, 383
bar-finger sands 486, *487*
barchan dunes: aeolian *201*, 202, *202, 203*
barrier island 488, 491, *491, 493*, 495, 510, *513–515*
barrier reef 638
BARSIM 712
basal sliding of ice 294
base 49–50, *see also* pH
base level: controls 707
baseflow 372, *372*
basin equilibrium time 710
Basin and Range 722
bauxite 60, 71
bay-head delta *495*, 509, 511, *512*
beach aeolian-dune deposits 507, *508, 511*
beach aeolian dunes 492, *494*, 507, *508, 513*
beach berm 492, *494*, 497, 502, *502*, 506
beach cusp 221, 492, *493, 494, 504*, 506
beach erosion 561, *561*
beach laminae 236, *237, 502, 503*
beach replenishment 561, *561*
beach ridge 494, *495*
beach-sediment grain size 506
beach slope 493, 497, *497*
beach *482*, 490, 491, *493, 494, 502*, 502–507, *503*
beachrock 693, 696
beaded esker 604
bed configuration 154
bed-sediment grain size under tidal currents 517
bed shear stress 130, 148, 151
bed shear stress of ice 294
bed states: unidirectional water flows 157, 158, 186–187, 188–191
bed form and sedimentary sequences of unsteady, non-uniform, unidirectional water flows 187, *191*
bed-form definitions 157, *157*
bed-form excursion 157, 165
bed-form lifespan 157, 165, 200

bed forms and sedimentary structures in muds 192–194
bed-load sediment in air flow 196–197
bed-load sediment in water flow *143*, 146–147, *152*
bed-load sheets 158, 160, *161*, *162*, 179, *180*, *400*, 401
bed-load transport rate in water flow 148–150
bedrock rivers 375, 458–461, *460*
bedrock rivers: erosion mechanics 459
bedrock rivers: gorge *460*, 461, 598, *599*
bedrock rivers: pothole *460*, 461, 598
bedrock rivers: rapids *460*, 461
bedrock rivers: waterfall *460*, 461, 598
benthic ocean ecosystems 35
Bergschrund 295, *297*
Bernoulli equation 123, *123*, 127, 137, 142, 167, 214
BIF *see* banded iron formation
Bingham number 262, *264*
Bingham plastic 258, *259*, 294
biofilm *see* microbial mat
biogenic sediments 85–86
bioherm 333
biosphere 30–36
biosphere: evolution 41–42
bioturbation 311, 486
birdseye *see* fenestrae
bird's-foot delta 483, *484*
bivalve burrows 311, *313*, 329
bivalve-mollusc skeletal grains 93, *94*
Bjerrum plot 54, *54*
black organic shales 652
Black Sea 462, 466
black smoker *112*, *113*, *114*, 652
blockfields *613*, 614
blocks: volcanic 281, *283*
blowholes 558, *559*
blowout dunes *202*, 202
bog iron ores 111
bomb sags: volcanic 282, *283*
bombs: volcanic 281, *283*
borings 311
botryoidal cements 687
bottomsets 485, 611
Bouma sequence 270, *271*
boundary Reynolds number *143*, 144
brachiopod skeletal grains 93, *94*
braid-bar expansion 405, *406*
braid-bar translation 405, *406*
braid bar 377, *379–381*
braided river *379–382*, 380, 382, 607
breaching 257
breaker zone *218*, 219, *504–505*, 505
brine shrimp 462, 464, 574
brucite 65
bryozoan skeletal grains *95*, 95
bubble sand *345*, 346, 355, *356*, *497*, 521
buffer layer 128, *128*

buildup 333
buoyancy force on sediment 140
buoyant plume *481*, 481, 486
burial diagenesis 674
burrow mottling 311, *312*, *313*
burrows *see* trace fossils
bursting 129, *129*, 134, 147

calcite seas 40, 87, *88*, 92, 101, *101*, *109*, *110*, 732
calcite *see* low-magnesium calcite
calcium carbonate coasts and shallow seas 527–549
calcium carbonate coasts: flow and sediment transport
  532–539, *532–535*, *538*, *540*
calcium carbonate coasts: tidal-flat deposits 536, 537, *537*,
  538, *539*, *540*, *542*, 549, *550*
calcium carbonate mud 96–98, *97*, 438, 540, *540*, *541*
calcium carbonate shallow seas: flow and sediment transport
  on rimmed shelves 539–542
calcium carbonate shallow seas: ramps *530*, 531
calcium carbonate shallow seas: rimmed shelves 531, *531*, *532*,
  549–551, *552*, *553*
calcium carbonate shallow seas: shelf-edge shoals 542–544,
  543, 551, *553*
calcium carbonate shallow seas: temperate (foramol) 530
calcium carbonate shallow seas: tropical (chlorozoan) 528
calcium carbonate: cement mineralogy 685
calcium carbonate: cement morphology 686, *688–690*
calcium carbonate: deep-burial diagenesis 697, *697*
calcium carbonate: early subaerial diagenesis 694–695
calcium carbonate: early submarine diagenesis 695,
  695–697, *696*
calcium carbonate: geochemistry 87–90, *90*
calcium carbonate: minerals 86–87, *88*
calcium carbonate: modern oceanic production 87, 90–91, 530
calcium carbonate: neomorphism 691, *692*
calcium carbonate: recrystallization 691
calcium carbonate: sediment-grain types 91–101
calcium carbonate: skeletal grains 91–96, *94–95*
calcium carbonate: soil crusts 350–351, *351 see also* calcrete
calcium carbonate: stabilization 695
calcium carbonate: supersaturation 88–89
calcium carbonate: supratidal crusts 534, 695
calcium carbonate: tidal-flat climatic indicators 731
calcium chloride brines 109
calcrete *70*, 71, 101
caliche *see* calcrete
Callianassa 317, *318*, 539
Callianassa pellets 99
Cambrian faunal explosion *41*, 42
Canning Basin "Great Barrier Reef" 551, *555–556*
cape *493*, 496
capillary fringe 54
carbon dioxide dissolved in rainwater 53–54
carbon dioxide in Earth's atmosphere 38, 732
carbon dioxide soil gas 55

carbon isotopes in paleosols 76
carbonate alkalinity 54, *54*
carbonate compensation depth 89, 648
carbonate compensation depth and climate change 731
carbonate continental slope sequences and glacio-eustasy 669
carbonate continental slopes: large-scale depositional models 666–669, *668, 738–740*
carbonate factory 539
carbonate fluorapatite 118
Caspian Sea 462, 572
Castile Formation 590, *591*
cathodoluminesence 689, *690*
cations 47
Catskill clastic wedge 717
causes of ice ages 627–629
causes of ice ages: autocyclicity 629
causes of ice ages: biosphere-ocean-atmosphere interaction 628
causes of ice ages: changes in solar radiation 627
causes of ice ages: plate tectonics and volcanism 628
cave deposits 349, *350*
cement stratigraphy 689, *690*
cement: mineralogy 685–686
cement: morphology 686–689, *688–690*
cement: recognition criteria 686
cementation *675, 676,* 685–689, *686, 688–690*
cenote *see* sinkhole
centrifugal force 122, 136, *136*, 238
CFA *see* carbonate fluorapatite
chalcedony 102, *102*
chamosite peloids 111
channel-bar accretion topography 404, *404*
channel-bar assemblage *382*
channel-bar head 377, *378–381*
channel-bar tail 377, *378–381*
channel bend length/channel width *385*, 390
channel cross-section geometry *385, 385–389, 386, 388, 389*
channel cross-section geometry: classification 393
channel migration patterns *395, 404,* 405, *406*
channel radius of curvature/channel width *385*, 390
channel riffle 377, *378–381*
channel sandstone body 407, *409*, 453
channel sinuosity 380, 383, *383, 384, 385*, 390, *391*, 447
channel width/depth *384, 385*, 390, 446
channel-belt width *385*, 391
channel-forming discharge 376, *376*, 383–385
Channelized Scablands 608, *609*
Chara 98, 470
Charaphytes 470
chelating agents 69
chemical and biochemical weathering 52–60
chemical and biochemical weathering products 57–60, 66, *67*
chemical divides 105–107, *106*
chemical divides: saline-lake water evolution 574–576
chemogenic sediments 85–86

chenier 494
chert 102, 104–105
chert nodules 104
chert: bedded 104
chevron (coronet) halite 576, *577*
Chezy equation 133
chicken-wire anhydrite 586
chlorite 65, *66*
chlorozoan (calcium carbonate shallow seas) 528
chute bar 377, *378–381*, 391–393, *392*
chute channel 377, *378–381*
chute cut-off *404*, 405
chute-and-pool cross strata 185, *186*
chutes-and-pools 183, *183*
cirque glacier 595
cirque *597*, 598
clasts 45
clay minerals 58, 62–67
clay minerals: cements *686*
clay minerals: reactions during diagenesis 694
clay minerals: structures 62–66, *66*
clay minerals: weathering 66, *67*
climate 19 *see* atmospheric circulation
climate: effects of plant evolution 732
climate: effects on paleosols 730
climate: effects on rivers 729–730
climate: effects on sedimentation 729–735
climate: variation over Earth's surface 729
climate: variation over time 731–735
climbing-ripple cross lamination 176, *177*, 434, *434*
clints 82, *83*
closed-basin lakes 463
closed basins 564, *565 see also* arid environments, Chapter 16
CO_2 *see* carbon dioxide
coal: climatic indicator 730
coarse ash (volcaniclastic) 281
coastal and shallow-marine deposits: economic aspects 562
coastal deposits: large-scale controls 707
coastal engineering and environmental concerns 560–562, *561*
coastal tufa 582, *584*
coasts and shallow seas Chapter 15, *see also* arid environments, Chapter 16
coasts and shallow seas: classification 476–477, *478*
coasts and shallow seas: effect of climate 474
coasts and shallow seas: effect of organic activity 475
coasts and shallow seas: effect of physical properties of water 475
coasts and shallow seas: effect of sea-level change 476
coasts and shallow seas: effect of terrigenous-sediment supply 474
coasts and shallow seas: effect of water chemistry 475
coasts and shallow seas: effect of water currents 476
coasts: spatial variation in water flow and sediment transport 477, *479*
coated grains 100

coccolithophores 91, *94*, 98
coccoliths *94*, 98
cohesion force 255
color mottles in soils 71
Colorado River delta 585
combined wind-shear and wave currents *224–226*, 226
comminution till 305
compaction 675, *675*, 676, 680–685, *681*
compaction curves 680
compaction: carbonate mud 683
compaction: differential 684, *684*
compaction: grain-to-grain contacts 682
compaction: over-pressure 685
compaction: terrigenous mud *682*, 683
compaction: terrigenous sand 680, *682*
complex dunes: aeolian 200
complex river confluences 391, *392*
compound bar 377, *378–381*
compound dunes: aeolian 200, *201*
compressional sedimentary basins 717–722, *718*
compressional sedimentary basins: coupled 3-D numerical
    models 718–722, *719*
compressional versus extensional sedimentary basins 722, *724*
compressive ice flow 295, *296*
computer models *see* numerical models
concretions 676, 685, 693
concretions: soils 70, *70*, 101
condensation surfaces (hardgrounds) 697
condensed sequences 738
congruent weathering reactions 56–57
conservation of ice mass 300
conservation of mass (continuity) equation 122, 127, 135, 137
conservation of momentum equation 127, 130, 135, 137
continental margin 13
continental rise *631*, 632
continental shelf 13, 473, 530, 654
continental-slope apron 632
continental-slope slumps 632, *633*, 637, *641*
continental-slope water (Kohout) convection 697, *697*
continental slope *631*, 632
continental slope: accretionary 637, *639*
continental slope: bypass 637, *639*
continental slope: carbonate versus siliciclastic 634, *638*, *639*
continental slope: erosional 637, *639*, *640*
continental slope: flow and sedimentary processes 653,
    654–656
contour currents 650, 659–661
contourite drift 632, *633*, 651, *651*
contourite 651, *651*
convolute lamination 270, *273*, 353
coprolites 311
corals 46, 95
coral animal 544, *545*
coral reef *see* reef
coral: zonation of reef growth forms 547, *547*

Coriolis force 17, 122, 222, 225, 240, 611, 642, 645, 650
Coulomb equation 255
crag-and-tail 598
crinoid skeletal ossicles 93, *94*
cross laminae and beds 173
cross strata formed by aeolian dunes and draas 205–210,
    207–211, 570, 572, *573*
cross strata formed by subaqueous ripples and dunes 170–179,
    171, *172*
cross strata formed by tidal dunes 247, *248*, 517, *518–520*, 519,
    526, *526*
cross strata formed by wave ripples 232, *235*, *486*, *487*, 488,
    499, 501, *518–520*, 521
cross strata: medium-scale 176, *176*, 407, *409*, 414, 435, *435*,
    485, 487, *487*
cross strata: small-scale 176, *176*, 270, *271*, 407, *409*, 414, 434,
    434, *435*, 485, 487, *487*, *518*, 521, 651
cross-bar channel 377, *378–381*, 391–393, *392*
cross-set definition 174, *176*
cross-set geometry: along-stream variation 178, *178*
cross-set thickness 174–179, *177*
Cruziana ichnofacies *321*, 322
cryptmicrobial laminite *see* microbial laminite
cryptobionts 315
crystals (volcaniclastic) 282, *283*
Cumberland Lake delta 483, *484*
cumulate textures 576, *577*
current crescents and sand shadows 192, *192*, 211, *211*
curved river channels: secondary (helical, spiral) flow 136,
    136, *396*, 396–397
curved river channels: theoretical water flow 402
curved river channels: water flow 136, *136*, *396*, 396–397
cutans 70, *70*
cyanobacteria 32, 438
cyanobacterial mats 525
cycles: Bond 622
cycles: carbonate tidal flat 549, *see also* sedimentary sequences
cycles: Dansgaard–Oeschler 622
cycles: Milankovitch 628

darcy 678
Darcy's law 678
Darcy–Weisbach equation 133, 135
Dead Sea 463, 572, 575
dead-ice moraine 308
Death Valley 468, 592, *593*, *594*
debris-flow deposits 264, *264*, 603, *610*, *612*, *616*
debris-flow geometry 262, *264*
debris flows *257*, *259*, 262–264, *262*, *263*, *264*
Debye–Huckel theory 48
deep-sea basin plain: flow and sedimentary processes 659–661
deep-sea biogenic ooze 98, 648, *649*
deep-sea brown clay 648, *649*
deep-sea burrows 650, *see also* Chapter 11
deep-sea-canyon deposits 653, *656*

deep-sea canyon 632, *633*

deep-sea channel-bar and fill deposits 655, *657–659*

deep-sea channel migration *634*, 655, *657*

deep-sea channel-levee system *658*, 658, *662, 663*, 664

deep-sea crevasse-splay deposits 657, *660, 661*

deep-sea deposits: large-scale controls 707

deep-sea environments: anoxic and hypersaline deposits 652

deep-sea environments: chemically precipitated sediments 652, 663, *664*

deep-sea environments: deep-water (thermohaline) circulation 644–646, *645*

deep-sea environments: geometry of basin margins *631*, 632–637, *633*

deep-sea environments: geometry of basin plains *631*, 638

deep-sea environments: geometry of volcanic hills and mid-oceanic ridges *631*, 638–640

deep-sea environments: geometry *631*, 632–640

deep-sea environments: iceberg sediment transport and deposition 647, *see also* Chapter 17

deep-sea environments: sediment gravity flows, slumps, and slides 647, 653–661, 669, *see also* Chapters 8 and 12

deep-sea environments: sediment supply 646

deep-sea environments: sediment transport by deep-water (thermohaline) currents 650–652

deep-sea environments: settling of fine-grained sediment 647–650

deep-sea environments: surface-water circulation 641–644, *642–644*

deep-sea environments: water flow and sediment transport 640–653

deep-sea fan channel 632, *633, 634, 635*

deep-sea fan crevasse splay 633, *633, 636*

deep-sea fan levee 633, *633, 636*

deep-sea fan sequences and glacio-eustasy 665

deep-sea fan terminal (frontal) lobe 633, *633, 635, 636*

deep-sea fan 632, *633*

deep-sea fans: flow and sedimentary processes 653–659

deep-sea fans: large-scale depositional models 664–666, *665, 666, 667*

deep-sea levee deposits 657, *658, 660, 661*

deep-sea mass-transport complexes 653, *654*, 655, *655, 661*, 664

deep-sea sediment (mud) waves 633, *633, 636, 637*, 651, 657

deep-sea terminal lobe deposits *635, 636*, 658

deep-sea volcanic hills and mid-ocean ridges: flow and sedimentary processes 661, *664*

deep-water waves 213, *215*, 215–217, *216, 218*

degree of channel splitting (braiding) 380, 382, 447

delta floodplain 482, *482*

delta front *482*, 483, *484*

delta lobe 488, *490*

delta plain 482, *482*, 485

delta *482*, 482–491

delta-front growth faults 488, *489*

delta-front mud diapirs *487*, 488, *489*

delta-front rift valleys 488, *489*

delta-front slump 483, 488, *489*

deltas: geometry 482–484

deltas: water flow and sediment transport 482–484

dendritic drainage 366, *369*

denitrification: microbial 32, 694

densiometric Froude number 268, 479

deposit feeding 315

depth-averaged flow velocity 132–133

desert roses (gypsum cements) 579, 689

desert soils *see* Aridisols

desert varnish 567

deserts *see* arid environments

desiccation cracks 359, *359, 360*, 438, *439*

Desulfovibrio 700

detritus 45

devitrification of volcanic glass 282

diagenesis: aqueous solutions 107, 677–678

diagenesis: calcium carbonate sediments 694–700, *695, 696, 697, 699*

diagenesis: cementation *675*, 676, 685–689, *686, 688–690*

diagenesis: compaction 675, *675*, 676, 680–685, *681, 682, 684*

diagenesis: definition 673

diagenesis: evaporites 700

diagenesis: general conditions and controls 673–675, *674*, 676–680

diagenesis: organic matter 676, 693–694

diagenesis: peat and coal 700

diagenesis: processes and products 675–676

diagenesis: replacement 676, 691, *692, 699*

diagenesis: soils 74–76

diagenesis: starting materials 677

diagenesis: terrigenous muds 693–694

diagenesis: terrigenous sands and gravels 691–693

diamicton 304, 602, *602*

diatom skeletal grains *102*, 103

diffusion equation 457

diffusivity of sediment 710

diffusivity of suspended sediment 150

dilatent flow *259*

dimensionless bed shear stress 143, *143*, 186, 227, 229

dish structures 273, *273*, 353, *355*

dissolution (secondary porosity) 676, 685

distributary channel 482, *482*

distributary channel mouth bar *480, 482*, 483, *484*, 488

diurnal inequality of tides 238

dolomite 86, 697–700, *699*

Dolomite Mountains (Italy) 551, *553, 739*, 741

dolomite problem 699

dolomite rhombohedra 699, *699*

dolomite: bacterial 700

dolomite: mixing zone (dorag) 700

dolomite: reflux 699

dolomite: sucrosic texture 699

dolomitization 697–700, *698*

downslope slumping of sediment 356

downstream decrease in grain size 154, *154*, 446
draas 570
drag and lift on bed grains *142*, 142–145
drag coefficient 140, *140*, 143
drag force on sediment grains *140*, 140–145
drainage basin 366, *367*
drainage basin area 366
drainage basin: sediment yield 374
drainage density 366
drainage divide 366, *367*
drainage pattern 366, *367*, *368*, *369*
driekanter 197, *197*
dropstone 309, *309*, 610, *610*, *617*
drumlin 600, *601*, *602*, *616*
dry mudflats 581, *583*
dune lag 165, *166*, 247, 389
dune-migration modes 165, *165*
dune-migration rate 165
dune superposition 166, *167*, 200–210
dunes (and draas): aeolian 200–205, *201–205*
dunes: tide-formed *240*, *245*, 247, *248*, 511, 513, *514*, 517, 519,
     524, *525*, *526*
dunes: unidirectional water flow 158, *161*, *163*, 163–167, *164*,
     400, 401, *401*, 435, *436*, 483, 485, 487, 498, 502, 650
duricrusts 71
dust (volcaniclastic) 281
dynamic equilibrium of rivers 376
dynamic grain friction coefficient 149, 259
dynamic pressure 122, 123, 142
dynamic sedimentation models 709
dynamic similarity 126

early diagenesis 674
Earth's orbital eccentricity 23, *24*, *25*
Earth's orbital obliquity 24, *24*, *25*
Earth's orbital precession 23, *24*, *25*
ebb-tidal delta 511, 514, *513–515*, 519–521
ebb-tidal delta: swash bars 514, *514*, *515*, *518*
ebb-tidal delta: swash platform *514*, *515*
echinoid burrows 311, *313*
ecology 35
ecosystems 35
ecosystems: marine 35
ecosystems: terrestrial 36
eddy viscosity 125, 148
edge waves 221, *222*, 506
Ediacaran fauna 42
efflorescent salt crusts 577
Eh half reactions 51
Eh scale 51, see pe
Ekman depth 223
Ekman spiral 222, *223*
El Niño 643
elastic lithosphere 614, 717
eluviation 68

ephemeral saline lakes *see* saline pans, saline mudflats, dry
     mudflats
ephemeral streams 568, 568–570
ephemeral streams: deposits 569
ephemeral streams: flow and sediment transport 569
epilimnion 465, *465*, 610
epitaxial cements 687, *688–690*
equigranular mosaic cements *688–690*, 689
equilibrium (snow) line 295, *296*
equilibrium (thermodynamic) 48, 49
equilibrium constant 47, 49
equilibrium–disequilibrium river geometry 376, *376*
equivalent sand roughness 131, 133, 146
erosion and deposition by turbidity current
     270–272, *271*
erosion and deposition of sediment 153–157
erosional bed forms in alluvial river channels 401
erosional bed forms: aeolian 210, *211*
erosional structures in cohesionless sediments 191, *193*
erratic block 309
escape structures 311, *313*
esker deposits *605*, *606*, *616*
esker 604, *605*, *606*, *616*
estuary 510, 511, *512*, *513*
Etosha Pan 463
eukaryotes 33
eustasy: causes 735
eustasy: continental sedimentation models 735–737, *737*
eustasy: marine sedimentation models 737–741
eustasy: sedimentation effects 735–741
eutrophic lakes 466
evaporation of seawater 107, *108*
evaporites 105–109, 438, *see also* arid environments
evaporites: calcite versus aragonite seas 109, *109*, *110*
evaporites: climate indicators 731
evaporites: conditions for formation 105
evaporites: continental shelf deposition 590, *590*
evaporites: diagenesis 700
evaporites: marine basin deposition *577*, 590–591, *591*
evaporites: mineralogy 105
evaporites: recycling 107, 577
evapotranspiration 54
evolutionary faunas 42
exfoliation *see* spheroidal weathering
exhumation diagenesis 674
experimental studies of turbidity currents 265, *265*
extending ice flow 295, *296*
extensional sedimentary basins 722–729, *see also* rift basins
extensional sedimentary basins: models of sediment fill
     722–729, *723*, *725–728*
extensional versus compressional sedimentary basins
     722, *724*
extraclasts 100
extrinsic (extra-basinal) controls: large-scale sedimentary
     sequences 707

failed rift basins 722
fair-weather wave base 499, *499*
fan delta 448, 485
fan-head entrenchment 450, *450*
faults 46
fecal pellets 311, *312*, *314*, *318*, *also see* pellets
feeding strategies of marine invertebrates 35
feldspar in terrigenous sandstones *80*, 81
fenestrae *345*, 345–346, 536, *537*, *538*
ferricrete 111
ferroan dolomite 87
ferromanganese nodules 652, *664*
fetch 214
fibrous cements 687, *688–690*
fine ash (volcaniclastic) 281
Finger Lakes, New York State 463, 598, *599*
fiord 598
firmground 99, 315, 696
flame structures 353, *353*
flaser bedding 244, *246*
flat pebbles 99
flexural (peripheral) bulge 614, 717, *718*, 721
flexural rigidity 614, 717
flood frequency 373–374
flood-frequency distributions *373*
flood hydrograph 372, *372*
flood recurrence interval 373, *373*
floodbasin 430, *431*
floodbasin deposits 438, *439*
floodplain channel fills 437
floodplain crevasse channel *431*, 432, *435–437*, 486
floodplain crevasse splay *431*, 432, *435–437*, 482, *482*, 486
floodplain definition 430
floodplain deposition and flood frequency 433
floodplain deposits 434–438, *435–439*
floodplain drainage channels 432
floodplain geometry 430–432, *431*
floodplain lake 432, 482, *482*, 486
floodplain levee 430, *431*, *435*, 437, 482, *482*, 486
floodplain soil catena 439, *440*
floodplain soils 438–440, *see* Chapter 3
floodplain soils: drying-out sequence 439, *440*
floodplains: crevasse-channel deposits 435, *435–437*
floodplains: crevasse-splay deposits 435, *435–437*
floodplains: deposition and erosion 433
floodplains: lacustrine-delta deposits 436, *437*
floodplains: lacustrine deposits 438, *437*, *439*
floodplains: levee deposits 435, *435*
floodplains: sediment transport 433
floodplains: water flow *432*, 432–433
flood-tidal channel fill 518, *518*
flood-tidal delta 511, 512, *512*, *514*, *515*, 519–521
flood-tidal delta: ebb shield *515*
flood-tidal delta: flood ramp 514, *515*, *519*
flood-wave models 138, *138*

Florida Bay mudbanks 540, *541*, *542*
Florida: western continental shelf 530, *530*
Florida–Bahama Banks 531, *532*, *533*, 635, *639*, *641*, 669
Florida–Bahama Banks: calcium carbonate production 90–91, *92*
Florida–Bahama Banks: Joulter's Cay ooid complex *543*, 544
flow and sediment transport over ripples and dunes 167–170, *168*, *169*
flow and sediment transport over vortex wave ripples 230, *234*
flow-duration curve 373, *373*
flow resistance 132–133
flow separation 135, *136*, 140, *140*, *141*, 142, *168*, *169*, 182, 230, 397
flow till 309
flowstone 346 *see also* cave deposits
fluid drag on cross strata 357
fluid inclusions 109
fluid pressure 122
fluid shear stress 122
fluid stability analysis 126, *126*
flute marks 192, *193*, 270, *271*, 500
fluvial aggradation 452, 458, *459*, 734, *737*
fluvial burrows 438, *see also* Chapter 11
fluvial degradation 458, *458*, *459*, 734, *737*
Flysch 708
foraminifera *95*, 95, 98
foramol (calcium carbonate shallow seas) 530
forced regression 509
forces acting on fluids 122–123, *130*
fore reef 546
foredeep basins 717
foreland basins 717
foresets 485, 611
foreshore 492, *494*
fossils 33–34
fossils in fluvial deposits 440–441, *see also* Chapter 11
fossils: major invertebrate groups 34
fossils: major plant groups 34
fossils: paleosols 72
framboids 111, *112*
frazil ice 292, 298
freeze–thaw 45–46
fringing reef 638, *664*
Froude number 137, *137*, 181, 186
fully turbulent layer 128, *128*

gastropod skeletal grains 93, *94*
general drag equation 140, 143
geo 558, *559*
geological time scale *15*
geometric sedimentation models 709
geostrophic currents 225
geosynclines 708
geothermal gradients 673, *674*
ghost shrimp *see Callianassa*

Gibbs free energy 48
Gibbs free energy of formation 48
gibbsite 60, 65, *65*
GIF *see* granular iron formations
Gilbert-type delta *467, 468,* 484, *485,* 611
gilgae 70
glacial deposits: climatic indicators 730
glacial environments: evolution over glacials–interglacials 614–618, *615–618*
glacial lake Missoula 608, *609*
glacial lake 463, 598, 599, *599, 605,* 609–612, *616*
glacial period 595, 618–625
glacial striations and gouges 301, *302*
glacial varves 611, *611*
glacial–interglacial cycles: effects on sedimentary environments and deposits 157, 591, 732–733, 734, *734*
glacial–interglacial depositional sequences 616, *616, 618*
glacier ice 292
glacier surge 607
glacier surges 295
glacio-eustasy 614, *615,* 623, *626*
glacio-isostasy 614, *615,* 623, *624, 625*
glaciolacustrine deposits 609–612, *610, 616*
glaciomarine deposits 609–612, *610, 617*
glacitectonite 305
glaebules 70
glass shards *see* shards
glauconite peloids 111
Glen's law 293
global ocean water conveyor *645,* 646
Goldrich weathering series 61, *61*
gradually varied non-uniform flow 135
grain flow deposits *260,* 261
grain flow geometry 260, *260*
grain flows *257, 259,* 259–262, *260*
grain flows on ripples and dunes 172–173
grain Reynolds number 140, *140, 142, 143,* 186, *196,* 227
grain shear stress 258, *259*
granular iron formations 114, *116*
granule ridges: aeolian 199
grapestone (calcium carbonate sediment grain) 99, 530, 696
gravitational (pendant) cements *688,* 689
gravity force 122, 255
Great Bear Lake 463
Great Salt Lake 462, 463, *467,* 572
Great Slave Lake 463, 584
green algae skeletal grains 92, *94,* 539, *540*
Green Lake 471
Green River Formation 106, *471,* 472, 575
greenhouse climate 732
greenhouse effect 38
greensands 112
green-stone belts 37
groundwater 54–56, *56, 371*

groundwater composition 45, *also see* weathering and groundwater composition
groundwater flow 56, 371, *371*
groundwater: diagenetic flow models 679
groundwater: hydrology of meteoric water *56, 371*
groundwater: hydrology of sedimentary basins 678–680, *679*
grykes *see* clints
Gulf Stream 642, *644*
gullied continental slope 632, 637, *641*
gutter casts 192, *192,* 266, 500
guyot 546, 638
gypsum: crystal mush 585, *588*
gypsum: desert roses 579
gypsum: lenticular crystals 579

Hadley cells 21
Halimeda 539, *540*
halite: bottom growth 576, *578*
halite: chevron (coronet) 576, *578*
halite: cumulate textures 576, *577, 578*
halite: rafts 576
halite: skeletal (hopper) crystals 576
halite: surface crusts 579, *580*
halite: surface crystal impressions 578, *581*
halite: surface growth on brines 576
halite: tepee structures *578,* 579
hanging valley 598, *599*
hardgrounds 99, 695, *696*
hardgrounds: conditions for development 696
hazards of mass movements 277
heavy-mineral sorting 154, 170
heavy minerals in terrigenous sandstones 82
Heinrich-event deposits 309
hemipelagic mud 647, 653, 659–661, 669
herringbone cross strata 244, *246, 247, 248,* 517
heterolithic bedding 244, *246,* 521, *522, 524*
high-magnesium calcite 87
high pore-water pressure 255, 352, 403, 488, 685
high-concentration turbidity-current deposits *271,* 272
high-density turbidity current 266
highstand systems tract 737, 740
hindered settling of sediment suspensions 141
homopycnal flow 479, *480,* 483, 611
hopper crystals (halite) 576
horn *597,* 598
hummocks and 3D wave ripples 230, *231, 233,* 500, 502
hummocky cross strata 232, *236, 238,* 487, 488, *499,* 501, *519,* 521
humpback (transitional) dunes 179, *179*
humus 69
hydraulic conductivity 154
hydraulic jump 137, 183, *183,* 184
hydraulic stability of equilibrium bed states: experiments 158, 186–187, *187–190*
hydraulic stability of equilibrium bed states: theory 188–191

hydraulic stability of equilibrium combined-flow bed states *233*, 237

hydraulic stability of equilibrium wave-formed bed states *227*, *233*, 237

hydraulically rough boundary 132, 133, 144

hydraulically smooth boundary 131, 133, 144

hydraulically transitional boundary 132

hydrograph 371–373, *372*

hydrological cycle *26*, 29

hydrological time series 373

hydrolysis, *see* incongruent weathering reactions

hydrosphere 24–30

hydrosphere: composition 25–27

hydrosphere: evolution 39–41, *see also* aragonite seas and calcite seas

hydrostatic gradient 673, *674*

hydrostatic pressure 122

hydrothermal brines 109

hyperaridity 563

hyperpycnal flow 480, 483, 611

hypolimnion *465*, *465*, 610

hypopycnal flow 481, *481*, 486, 611

ice ablation zone 295, *296*

ice abrasion 301

ice-accumulation zone 295, *296*

ice ages 38, 595, *596*, 618–625

ice cores 621, *622*, 732

ice entrainment of basal sediment 302, *303*

ice erosion 301–303

ice-flow mechanics *293*, 293–301

ice formation and properties 292

ice fractures, faults, and folds 295, *297*

ice plucking 301

ice segregation 46

ice sheet 292, 595, *596*, *612*

ice shelf 595

ice stratification 295, *296*

ice stream 295, 595, 606, *612*

ice viscosity 294

ice-volume history: recent 39, *39*

ice wedging 301

iceberg scour *303*, 303, 610

iceberg 298–300, 610, *610*, *617*

ice-contact delta 604, *605*

ice-contact environment 595

icehouse climates 731

ice-longitudinal crevasse 295, *297*

ice-marginal crevasse 295, *297*

ice-melt hydrograph *608*

ice-transverse crevasse 295, *297*

ice-wedge polygon 360, 612, *613*

ichnocoenosis 314, *315*, *325–328*

ichnocoenosis: terrestrial 324, *325–328*

ichnofabric 311

ichnofacies 321–324

ichnofossils *see* trace fossils

ichnogenera 321

ignimbrites 282

ikaite 582

illite 65

illuviation 68

imbricated sediment grains 154, 162, *163*, 183, *184*

immature mineralogy: terrigenous sediment grains 78

immersed weight of sediment 140

incised valley 457, *458*, *459*, 736

incongruent weathering reactions 57–60

inertial grain flow 259, *259*

infiltration-excess overland flow 371

initial flooding surface 737

initiation of sediment motion on slopes 255–256, *256*

inselbergs 82

INSTAAR 712

inter-avulsion period 441

interdistributary bay 482, *482*, 486

interdune areas 205

interdune deposits 210

interfering wave ripples 230, *231*

interglacial period 595, 618–625

intermediate-water waves 213, *215*, 215–217, *216*, *218*

intermittently suspended load 147, *152*

internal gravity waves 468, 479, *481*

intertidal meandering channels *510*, 511, *513*, 515, *523*, *524*

intertidal zone 510, *513*

intertropical convergence zone 21

intraclasts *97*, 99–100

intraformational mud-chip breccia 194, *194*, 359

intrinsic (intra-basinal) controls: large-scale sedimentary sequences 707

inverse density gradients in sediment–fluid mixtures 352–353, 488

ion activity product 51

ion-exchange capacity of clays 65

iron formations 110, *see also* banded iron formations and granular iron formations

iron formations: origins 116–117

iron mineralogy 109

iron mineralogy: mid-ocean-ridge hydrothermal vents 112–113, *113*, *114*, 652, *664*

iron mineralogy: soils *58*, *59*, *60*, 111

iron oxides 60

iron reduction (microbial) 32, 694

iron sulfides 60

iron-bearing sedimentary rocks 114–117

iron-bearing sediments 109–117

ironstones 110, 117, *117*

joints 46

jokulhlaup 298, 607, *608*

K see equilibrium constant
kame delta 605, *605*, *616*
kame terrace 604, *605*, *616*
kame 604, *605*
kankar 71, *also see* calcrete
kaolinite 65, *66*,
Kármán-vortex street 140, *140*
Kármán's constant 131, 148
karst topography 82–84
Kelvin–Helmholtz vortex 126, *126*, *265*, 266, *267*, *398*, 399, 478, *644*
kettle lake *605*, 607
Key Largo Formation (reef deposit) 551, *557*
keystone vugs 355 *see* bubble sands
kinematic eddy viscosity 125, 131
kinematic viscosity 124, 148
kinetic sieving 172, 205, 261, 614
knickpoint 457, *458*

lacustrine delta 432, 482, 486
lacustrine *see* lakes
lag deposit 153, 192
lagoon deposits 508
lagoon 482, *493*, *495*, 508, 511, 514, *514*
Laguna Madre 587
lahars 256, 264, 287, *see also* volcanic mudflows
Lake Baikal 462, 463
Lake Balkhash 572
Lake Bonneville 468
Lake Chad 462, 463, 565, *566*, 572
Lake Eyre 463
Lake Lahonton 468
Lake MacLeod 589, *589*
Lake Magadi 104, 463
Lake Natron 463
Lake Superior 462
Lake Tanganyika 5, 463
Lake Titicaca 463
Lake Unuyi 463, 577
Lake Victoria 463
lakes 462–472, *see also* arid environments, Chapter 16
lakes: ancient deposits 472
lakes: biogenic sediment 464, 470–471
lakes: biological productivity 465
lakes: chemogenic sediment 464, 470–471
lakes: circulation 466–468
lakes: classification by origin 463
lakes: classification by salinity 462
lakes: classifications 462–463
lakes: climate records 592–594, *593*, 730
lakes: closed basins 463
lakes: coastal deposition *467*, 468, *469*
lakes: deltas *467*, 468, *485*
lakes: floodplain 432, 482, *482*, 486
lakes: glacial 463, 598, 599, *599*, *605*, 609–612, *616*, *627*

lakes: laminae *439*, *471*, 471–472, 575, 611, *611*
lakes: organic matter 464
lakes: rift basins 463
lakes: sag basins 463
lakes: saline *see* arid environments, perennial saline lakes, *and* ephemeral saline lakes
lakes: salinity density stratification 466
lakes: sediment gravity flows 469, *470*
lakes: sediment supply 463–464
lakes: terrigenous sediment 464, *467*, 468–470, *469*, *470*
lakes: thermal density stratification *465*, 465–466
lakes: tufas and travertines *467*, 471, *584*
lakes: water composition 464–465
laminae: biogenic and chemogenic 332, *also see* stromatolite laminae
laminar flow 121, *122*, 123–124, 258, 293
laminated mud 193, *194*, 343, *343*, *439*, *470*, *471*, 498, *499*, 508, 521, *522*, 523, *525*, 536, *537*, 549, *550*, 585, 586, *588*, 611, *611*
landforms formed by weathering *see* weathering and landforms
Langmuir circulation 643
lapilli 281
lapillistone 282
large-scale erosion of rivers and floodplains 457–458, *458*, *459*
large-scale inclined strata 407, *409*, 411, 518, 523, 526, 655, *659*
large-scale inclined stratasets (storeys) 407, *409*, 411
Last Glacial Maximum 39, 595, 601, 608, 614, 623
laterite 71
Laurentide ice sheet *620*
law of mass action 49
law of the wall 131, *132*, 133, 167, 195
leaching 68
lenticular bedding 244, *246*, 519
lenticular gypsum crystals 579
LGM *see* Last Glacial Maximum
lifespan of turbidity current 266
lift coefficient 143
lift force on sediment grains *142*, 143, 146
linear (seif) dunes: aeolian *201*, 203, *203*, *204*
linear shelf sand ridge 492, 501, *501*
linear tidal sand ridge 511, 516, 525, *526*, 542, *543*
liquefaction 257, 352
liquefied sediment flow *257*, 272, *273*
lithic fragments, *see* terrigenous sandstones: rock fragments
lithoclasts 99
lithosphere: composition 10–11
lithosphere: elastic rheology 717
lithosphere: evolution 36–38
lithosphere: flexure and onlap of sediment 722, *723*
lithosphere: flexural rigidity 717
lithosphere: viscosity 717
lithosphere: viscoelastic rheology 717
lithosphere: physical properties 11–12
lithostatic gradient 673, *674*
load casts 352, *353*, *354*, 487, 488

Lockatong Formation 472, 593
lodgement-deformation till *304*, 305, *307*, *308*, 600, *604*, *610*, *617*
loess 198, *198*, 563, 570, 608, *616*
lonestone 309, *309*, 610
long profile of glacier 295, *296*
longitudinal ridges and grooves 192, *193*
longshore bar (ridge) *237*, 492, *494*, 497, *497*, 502, *502*, *503*
longshore-bar cross strata *237*, *502*, *503*, 504
longshore currents 220, *221*, *495*, 496, 505, 517
longshore sediment transport 228, 495, *495*, 506
longshore trough (runnel) *494*, 502, *503*
loose ground 315
low- and high-speed streaks 128, *128*
low-magnesium calcite 87
low-angle (zibar) laminae 205, *206*
low-density turbidity current 266
lowstand systems tract 736, 740
lugworm burrow 312, *314*
Lulworth Cove 558, *559*
luster mottling 689
lutocline 517
lysocline 648

macroform 158
macrotidal range 242, 477, 510
magadiite 104
Magnus force on sediment grains 141, 146
manganese modules and crusts: lakes 465
manganese reduction: microbial 32, 694
mangrove swamp 510, 515, *535*
Manning equation 133
marine regression 509, *510*, 527, *527*, 736
marine transgression 509, *509*, 527, *527*, 736
marsh 482, 486
mastication pellets 99
matrix: terrigenous sandstones 78–80
maturation (organic matter) 676, 694
mature mineralogy: terrigenous sediment grains 78
maximum annual discharge distribution 373, *373*
maximum flooding surface 737
maximum near-bed orbital velocity 216, 227, *227*, *233*, 505
mean annual flood 373, *373*
mean meridional circulation *20*, 21
meander bend expansion *404*, 405
meander bend model 395, *395*
meander bend symmetry–asymmetry 391, *391*
meander bend translation *404*, 405
meander bend: simple–compound 391, *391*
meander bend: sine-generated 394, *394*
meander belt width *385*, 391
meandering river *378*, 380
measurement and analysis of wind waves 214
Mediterranean Sea desiccation 565, *577*, 590–591

melt out till *304*, 305, *307*, 308, *308*, 602, *604*, 610, *610*, *616*, *617*
meltwater-channel deposits *604*, 604–606, *610*, *616*
meltwater channel 297, 298, *299*, 598, *605*
meltwater channel: sediment yield 607
meltwater discharge 607, *608*
meltwater erosion 598
meltwater megaflood 606, 607
meltwater 294, 296–298, *299*
meniscus cements 687, *688*, *690*
meniscus structure *313*, 318
meromictic lakes 466
mesoform 158
mesotidal range 241, 477, 510
metalimnion 465, *465*
metasomatism 676
methanogenesis: microbial 32, 694
Miami Oolite 551
micrite 96
micrite envelopes *95*, 96, *690*, *691*
micritization *see* micrite envelopes
microbial bioherm, *see* thrombolite
microbial buildup, *see* thrombolite
microbial laminites 342–345, *343*, 534–536, *537*, *538*, *also see* laminae, biogenic, *and* chemogenic
microbial mat 33, 333
microbial mat: zonation on Andros Island tidal flats 533, *535*
Microcodium 350
microcrystalline quartz 102, *102*
microfabrics in soils 71
microform 158
microspar 691
microtidal range 241, 477
mid ocean ridge hot springs 112–113, *113*, *114*
MIDAS 710
Milankovitch orbital cycles 23–24, *24*, *25*, 628
Milankovitch orbital variations: fluvial depositional sequences 735
Milankovitch orbital variations: sedimentary cycles 551, *553*, 732, 735
Mississippi delta 483, *484*, *490*
mixed layer clays 66, 694
modeling hillslope sediment transport and erosion 275–277
molality 48
Molasse 708, 717
molecular viscosity 123, 124, *124*, 148
Mono Lake 463, 582, *584*
monsoon 23
Montastrea growth rates 90, *91*
Monterey Formation 104
moraine: end 308, 602–604, *603*, *604*, 610, *610*, *616*
moraine: fluted 600
moraine: ground 306, 600, *601*, *605*, 610, *610*, *616*, *617*
moraine: lateral 602, *603*, *605*
moraine: linear 600

moraine: medial 602

moraine: terminal 602, *603*

motion of sediment gravity flows 258–259, *259*

moulins 297

Mt. Pinatubo 281, *287*

Mt. Pinatubo: eruption climatic effects 289, *291*

Mt. Saint Helens 285, *286*, *288*, 289, *290*

mud deposition by tidal currents 517, 521

mud lumps 355, 488, *489*

mudflats (playas) 580–581, *see* saline mudflats *and* dry
    mudflats

mychorrhizae 72

natural arches 82, *82*, 558, *559*

Navier-Stokes equations 127, 265

neap tides 238, *239*

neck cut-off *404*, 405

Neogene ice age 618–625

Neogene ice age: changes in land biota and soils 625

Neogene ice age: changes in water and sediment supply
    623, *627*

Neogene ice age: development 618, *620*

Neogene ice age: effect on fluvial environments 732, *734*

Neogene ice age: effect on lake environments 732

Neogene ice age: effect on marine environments 732

Neogene ice age: evidence from deep-sea cores *39*, 621, *622*

Neogene ice age: evidence from ice cores 621, *622*

Neogene ice age: evidence 619

Neogene ice age: glacio-eustasy 623, *626*

Neogene ice age: glacio-isostasy 623, *624*, *625*

neomorphism 676

nepheloid layer 650

Nereites ichnofacies 323, *324*

Newtonian fluid 124, *259*

nodal avulsion 441, 654, 658

nodular anhydrite 580, *582*, 700

nodular chert *see* chert nodules

nodular limestone: modern 696

nodules in soils *see* soil concretions

nodules *see* concretions

non-uniform flow 121, *122*

numerical models: alluvial *453–456*, 710–711, *712*, *715*, *716*,
    719–720

numerical models: carbonates 711

numerical models: coastal and marine 713–716, *719–720*

numerical models: drainage-system evolution 709–710

numerical models: glaciers and ice sheets 300–301

numerical models: large-scale sedimentary sequences
    709–712, *713–716*, *719*, *720*

Nyos crater lake 466

O soil horizon *67*, 68, *69*, *70*

obstacle to turbidity-current flow 269, *269*, 659

ocean anoxic events 731

ocean surface water gyres 641, *642*, *643*

ocean: ecosystems 35

ocean: salinity structure at depth 28, *30*

ocean: surface salinity 28, *29*

ocean: surface temperature 27, *28*

ocean: temperature structure at depth 27, *29*

oceanic mixed layer 27

oceanic trench *631*, 634

oceanic upwelling 643

ogive 295, *305*

oil field brines 677

oil shale 462, 465, *471*, 472, 575

oil window 694

omission surfaces (hardgrounds) 697

oncoids 100, 335

oncoids in lakes 470

ooids *97*, 100–101, 530

ooids: dropped nuclei (replacement) 691, *692*

ooids: halite 574

ooids: Joulter's Cay barrier island *543*

ooids: modern marine *97*, 100, 530

ooids: modern non marine 101

ooids: tidal sand ridges 542, *543*

opal-A 101

opal-CT 102, 103

open-framework gravel 162, *162*, 170, 173, *174*, 178

ophiolites 663

orbital water motions under waves 215–217, *216*

organic acids 55

organic features of soils *see* soils: organic features

organic matter: microbial diagenesis 693–694

oscillatory boundary layer under waves 217

oscillatory zoning (volcaniclastic crystals) *283*

ostracod skeletal grains 93, *94*

outwash plain deposits 607–608, *609*, *616*

outwash plain *603*, 607–608, *608*

over-filled compressional basins 718

overland flow 371, *371*

overturned cross strata 357, *357*

Owens Lake 463

oxidation – reduction reactions 50–51

oxygen in Earth's atmosphere 38

oxygen isotopes in paleosols 76

oxygen-isotopic composition of microfossils *39*, 621, *622*

oxygenated haloes around burrows 317

oyster-shell beach *503*

ozone layer 16

pack ice 292

paleoclimate: effects on rivers 729–730

paleoclimate: evidence in sedimentary deposits 729–731

paleohydrology 426

paleosol fossils 72

paleosols 76–77

paleosols: effects of climate 730

parabolic dunes: aeolian 202

parasequence 706
passive continental margin *631*, 632, 722, *723*
patch reefs 541, *541*, 546
path line 122
pe–pH diagrams *see* pH–pe diagrams
pe 50
peat 438, 483, 486, 488
pebble and cobble clusters 162, *162*, *163*
ped structures 69, *70*
pediments 567
pedotubules 71
pedotypes 74, *75*
pelagic mud 98, 647, *649*, 659–661, 669
pelagic ocean ecosystems 35
pelagic sediment: variation with climate 650
pellets 98–99
peloids *97*, 100
Penicillus 90, 539, *540*
perennial saline lakes 572–576
perennial saline lakes: biota 574
perennial saline lakes: brine evolution 574–576
perennial saline lakes: climate change 592–594, *593*
perennial saline lakes: density stratification 574
perennial saline lakes: deposits 575, *576–578*
perennial saline lakes: evaporative concentration 574–576
periglacial environment 612–614, *613*
periphyton 98, 470, 536, *537*
periplatform mud 647, 669
permafrost soils 84
permafrost 292, 612
permeability 154, *154*, *155*, 185, 422, 678
permeability: carbonate sediments 96
Persian Gulf 585
pH–pe diagram: iron minerals *58*
pH–pe diagrams *50*, 51
pH 49–50
phi grain-size scale 139
phosphate nodules 652
phosphate sediment grains *117*, 118
phreatic zone, *see* saturated zone
phreatophytes 569
phyla 34
physical properties of water 121
physical weathering 45–47
piedmont glacier 595, *596*
piled load-casted ripples *354*
pillar structures 353, *355*
pingo 612, *613*
pisoids *97*, 100–101
Pitzer theory 48
planar cross strata 173, *175*
planar strata (laminae) *162*, 163, *178*, 180, *180*, 236, *237*, 270,
    271, 407, *409*, 414, 434, *434*, *435*, 485, 487, *487*, *499*, 501,
    503, 507, *508*, *518*, 521, 651
plane bed: lower-stage 158, 160–163, *162*, *163*, *400*, 401

plane bed: upper-stage 158, 179–181, *180*, 270, *271*, *400*, 401,
    434, 485, 487, 507, *508*
plane bed: wave-formed *227*, *233*, 236, *237*, 500,
    502, *503*
plate tectonics *11*, 12–14, *13*
playa (mudflat) 565, 577, 580–581
plug flow *259*, 262, 264
plunging breaker *218*, 219
pluvial lakes 468, 472, 582, *583*, 592
pocket beach 559, *560*
poikilitic cement *689*, 689
point bar 377, *379*
polyframboids 111, *see also* framboids
porosity versus depth curves 680
porosity 154, *154*, *155*, 185, 422, 678
porosity: carbonate sediments 96, 678
porosity: terrigenous sands 678
postglacial sea-level rise 614, 623, *626*
postglacial uplift 614, 623, *624*, *625*
postvortex (anorbital) wave ripples 230, *233*
pre-cement porosity 682
precipitation: average annual 16, *19*
pre-Flysch 708
pre-Neogene glaciations 625
pressure (form) drag on sediment grains 140, *140*, 142
pressure release 46–47
pressure solution *675*, 676, 680, *681*, 682
pressure-melting of ice *293*, 294
primary current lineations 158, *160*, 180, *180*
prodelta *482*, 483
products 46, 47
proglacial environment 595
prokaryotes 31–33
prokaryotes: depth zonation in muddy sediment 694
prokaryotes: early diagenesis of organic matter 693–694
prokaryotes: metabolic diversity 31–33
Proterozoic 37
protodolomite 87, 534, 699
provenance 45
pseudo fecal pellets 99
pseudoanticlines 70
pseudo-imbricated sediment grains *174*
pseudonodules 353, *354*
pseudoplastic *259*, 293
pumice 282
push moraine 308, 603, *604*
pyramidal (star) dunes: aeolian *202*, 204, *205*
pyrite 60, 111, *112*
pyritized plants 483, 486
pyroclastic density currents 285–287, *286*, *287*, *288*
pyroclastic fall 282–285, *283–285*
pyroclastic flows 286, *286–290*
pyroclastic sediment 278, *also see* volcaniclastic
    sediments
pyroclastic surges 285, *286*, *290*

quartz: cements *675*, 686, 687, *690*

quartz: dust rims and syntaxial overgrowths *675*, 687, *690*

quartz: pressure solution 680

quartz: terrigenous sandstones *79*, 80

Racetrack Playa 581

radial drainage 366, *369*

radial fibrous (drusy) cement 687, *688*

radiolarian skeletal grains *102*, 103

rain-drop imprints 359, *360*, 438

rain-shadow (lee-side) deserts 564

rainwater composition 52–54, *53*, *55*

rainwater pH 53–54, *55*

raised peat mires 483

rapidly varied non-uniform flow 135

rates of chemical and biochemical weathering 60–62, *61*

rates of uplift and subsidence 713

reactants 47

reactivation surface 178, *179*, 247

recrystallization 676, 691

rectilinear drainage 366, *369*

recurved spit *493*, 495

red-algae skeletal grains 92, *94*, 539

Red Sea hydrothermal deposits 112, *115*

reef crest 546

reef diagenesis 695

reef flat 546

reef front 546, *548*

reef sub environments 546–549, *547*

reef: definition 544

reef: ecological requirements 545

reefs 544–549, *546–548*

reefs: atoll 545, *546*

reefs: deposits *550*, 551–555, *552*, *555–558*

reefs: evolution of reef building organisms 554, *558*

reefs: patch reefs 541, *541*, 546, *550*, *552*

reefs: processes of deposition and growth 547, *547*

reefs: productivity 544

reefs: shelf edge 545

reflected turbidity currents 269, 272

reflux: subsurface 574, 585

reg (stony desert) 570

regelation *293*, 294

regressive ripples *172*, 173, *520*

relative roughness 133, *520*

replacement 676, 691, *692*, 698

replacement: ghost 691, 699

resonant amplification of tidal waves 240

retroarc basins 717

reverse grading in grain and debris flows *260*, *264*

Reynolds equations *126*, 127

Reynolds number 123, 126, *126*, 259, *264*, 479

Reynolds stress 126

rhizoconcretions 72

rhizoliths, *see* rhizoconcretions

rhizosphere 72

rhomboid ripples 184, *184*, *237*, *503*

Rhynie Chert 104

rhythmites *243*, *499*, 611

Richardson number 148, 479

rift-basin lakes 463

rift-basin lakes: eastern US Mesozoic 593, 722

rip currents and nearshore circulation 220–221, *221*, 496, *504*

ripples formed by combined currents 230, *233*

ripples: aeolian *199*, 199–200

ripples: tide-formed 244–247, *245*, 514, 517, 519, 521

ripples: unidirectional water flow 158–160, *160*, *161*, *400*, 401, 434, *436*, *438*, 483, 485, 487, 502, 650

ripples: wave-formed 227, 230–232, *231*, *232*, *233*, 434, 438, *438*, 486, *499*, 500, 501, 502, 514, 517, 519, 521, *524*

river avulsion (diversion) 441–446, *442*, *444*, 448, 449, *450*, 452, 482, 485, 488, *490*, 607

river avulsion and anastomosis 441, *442*

river avulsion and base-level change 442

river avulsion and climate change 443

river avulsion and sedimentation rate 441

river avulsion and tectonism 443, *444*

river-avulsion frequency 441

river avulsion: theoretical models *442*, 443–445, *444*, 454

river avulsions: character 441, *442*, *444*

river avulsions: floodplain erosion and deposition 445

river-bank erosion and bar deposition *395*, 402, *403*, *404*, *406*

river-bank erosion rate 403

river-bank slumping 402, *404*

river channel bars 377, *378–381*, 483

river channel belt 407, *409*, 430, *430*, *452*, *453*, *454*

river channel cutting–filling *404*, 405

river channel talweg (thalweg) 385

river channel-bar deposits *409–421*, 411–417

river channel-fill deposits *412–421*, 417–419

river channels and bars: depositional models 407–422, *409–418*

river channels and bars: qualitative depositional models *409*, 410, *412–418*

river channels and bars: quantitative depositional models 409–410, *410*, *411*

river channels: hydraulic geometry 388–390, *389*, 446

river channels: origin and evolution 377–379

river channels: riparian vegetation 406

river confluence angle 391, *392*

river confluence geometry 391–393, *392*

river confluence migration 405

river confluence mixing zone 397, *398*

river confluence scour zone 391, *392*

river confluences: relative discharge 391

river confluences: water flow *396*, *398*, 397–399

river crossover 386

river deposits: controls on deposition 707

river deposits: falling stage features 419, *421*

river deposits: paleocurrent orientations 420

river deposits: porosity and permeability 422

river diffluence (bifurcation) zone geometry 393

river diffluence (bifurcation): water flow 399

river engineering 407, *408*

river long profiles 366, *369*, 446

river mouths: water flow and sedimentary processes 478–481, *480, 481*

river restoration 407

river system geometry 366–368, *367–369*

river systems: controls on geometry, water flow, and sediment transport *375*, 375–377

river systems: dissolved material 374

river systems: origin and evolution 368–371, *370*

river systems: sediment supply 374–375

river systems: theoretical evolution models 370, *370*

river systems: water supply *371*, 371–374

river terrace 457, *458, 459*, 734, *737*

river-dominated marine deltas: processes and deposits 485–490, *486, 487, 489, 490, 491, 492*

river-mouth (middle-ground) bar 478, *480, 481, 482*, 483, *484*, 486

rivers: effects of dams 446

roches moutonnées 597, *597*

rock drumlin 598

rock falls, slides and slumps 256–258, *257*

rock fragments: terrigenous sandstones *80, 81*

rock fragments: volcaniclastic 282

rock glacier 614

rocky coasts 557–560, *558, 559, 560*

rocky coasts: evolution 560, *560*

rocky coasts: rates of cliff erosion 557

Rocky Mountain compressional basins 717

Rogen (ribbed) moraine 600, *601*

rolling-grain wave ripples 230

roll-up structures *344*, 344–345, 580, *582*

root casts 72, 438

rotational slide 256, *257*

roughness height 131, 132, 148

Rouse equation 150

sabkha: coastal 585–587, *586*

sabkha: hydrology 585

sabkha: progradation and deposition 585, *587, 588*

saddle (baroque) dolomite 699, *699*

sag-basin lakes 463

sag basins 564, *565*, *see also* arid environments Chapter 16

Saint–Venant equations 137

saline deposits *see* evaporites

saline groundwater 107

saline lagoons: deposits 589, 590

saline mud flat 580–581, *582*

saline mud flat: gypsum (anhydrite) growth from groundwater brine 580, *582*

saline mud flat: halite growth from groundwater brine 580

saline-pan cycle 577–580, *579, 580, 581*

saline-pan deposits *579*, 580, *580, 581*

saline pans (salars, salinas, salt pans, chotts) 565, 577, *578*

Saline Valley, California 565, *565*

saline wedge in tidal channels 516

saltmarsh 510, *512, 513*, 515, *525*

sand ribbons 162, *163*, 511, 525, *526*

sand shoal 488, *491*, 543

sandur plain 607

sapropels 464

saturated solution 48

saturated zone 54

saturation state: solution 51–52

saturation-excess overland flow 371

scales of deep-sea channel deposits 655

scales of river deposits 407, *409*

Scandinavian ice sheet *620*

Scoyenia ichnofacies 324

scree 46, *46*

scroll bar 377, *378–381*

sea-salt aerosols 53, *53*

sea cave 558

sea grass 539, *540*

sea-grass banks 539, 540, *540, 541*

sea ice 292

sea monkeys *see* brine shrimp

sea stack 558, *558, 559*

sea waves 213

sea-bed sediment-grain size 496

sea-bed slope 497

seasonal turnover of lakes 465

seawater density 28, *30*

seawater: composition 25–27

seawater: evaporation mineral sequence 107, *108*

seawater: secular changes in composition *109, 110*, *see also* calcite seas, aragonite seas

secondary currents in straight channel flow 134, *134*

secondary porosity (diagenetic dissolution) 685

sediment abrasion during transport 152, *152*, 197, *197*

sediment continuity equation 153, 169, 275, 402, 457

sediment deposition from flowing ice *304*, 304–309, *307, 308*

sediment grain creep 196

sediment grain orientation 154

sediment grain properties 138, *139*

sediment grain reptation 196

sediment grain rolling and sliding 146

sediment grain saltation 146, 196

sediment grain shape 139, *139*, 152, *197*

sediment grain size 138, *139*

sediment grain sorting and orientation in debris flows 263, *263*

sediment grain sorting and orientation in grain flows 260, 261

sediment grain sorting during erosion and deposition 153–157, *154*

sediment grain sorting during transport 151, *152*

sediment grain sorting during wave-current transport 228

sediment grain sorting over ripples and dunes 170

sediment grain-size distribution 139, *139*, *152*, 155, *156*

sediment grain-size percentiles 139, *139*

sediment gravity flows *256*, 258–277, 449, *449*, 483, 603, 610, *610*, 611, *612*, *616*, *633*, *641*, 647, 653–661, 669

sediment preservation potential 153

sediment-routing models 155, *156*, 454

sediment shrinkage 358–361

sediment transport and deposition from icebergs *304*, 309–310, *610*

sediment transport by tidal currents *240*, 242, *242*

sediment transport by wave currents and combined currents 228, *229*

sediment-transport effect on flow 147–148

sediment-transport erosion and deposition from meltwater 309

sediment transport in flowing ice 303–304, *304–306*

sediment transport in turbidity current *268*, 270

sediment-transport rate 148–151

sediment-transport rate over ripples and dunes 165, 168–170

sediment transport, erosion, and deposition by tsunami 252–254, *253*

sediment volcano *273*, 353, *356*

sediment volume concentration 150, 258

sedimentary basins 13, *13*, *see also* Chapter 20

sedimentary dykes and necks 353, *356*

sedimentary facies: vertical changes 703, *704*

sedimentary sequences of unsteady, non-uniform wave currents 237, *238*

sedimentary sequences: accommodation 707

sedimentary sequences: calcium carbonate 549–555, *550*, *552*, *553*, *554*

sedimentary sequences: climate 729–735

sedimentary sequences: eustasy 735–741

sedimentary sequences: large-scale architecture in basins 703–707

sedimentary sequences: reefs *555*, 740, *740*

sedimentary sequences: scales 704

sedimentary sequences: sediment-supply controls 707

sedimentary sequences: Sloss 708

sedimentary sequences: tectonic effects 712–729

sedimentary sequences: types 704

sediment-cement fills of tepee structures 361, *361*

sediment–fluid density 147, 352

sediment–fluid viscosity 147, 258, 352

SEDSIM 710

seiche 213, 467, *467*

seismic stratigraphy 705, *706*

seismic stratigraphy: eustasy 735

seismic stratigraphy: sequences 706, *706*

seismic stratigraphy: unconformities 705

selenite 575, *576*

semi-arid environments *see* arid environments, Chapter 16

semi-Aridity 563

semidiurnal tides 238

septaria 71

sequence 706

sequence stratigraphy 705, *706*

sequence stratigraphy: deep-sea-fan model 740

sequence stratigraphy: eustasy 735

sequence stratigraphy: fluvial models 735–737, *737*

sequence stratigraphy: rimmed-shelf (carbonate) model *738*, 740

sequence stratigraphy: shallow-marine models 739

serpentinites 59

settling of sediment grains in air 195

settling of sediment grains in water *140*, 140–142, *141*, *142*

settling velocity of sediment grains 140

shadow dune: aeolian *201*, 205

shallow-sea burrows and trails 498, *498*, *499*, 501, *see also* Chapter 11

shallow-sea epifauna 498, *499*

shallow-sea infauna 498, *499*

shallow-water waves 213, *215*, 215–217, *216*, *218*

shards 282, *283*

Shark Bay, Western Australia 336, *337*, 537, *540*, 585, 587

shear strength 255, 352, 402

shear stress in ice 293, *293*, 294

shear velocity 128

sheet cracks 346

sheet joints *46*

shoreface channels 504

shoreface 492

shrinkage cracks 358, *358–360*

side bar 377, *378–380*, *392*, 397

siderite nodule 483, 486

silica: geochemistry 47, 102–103

silica: mineralogy 101–102

siliceous sediments and sedimentary rocks 101–105

siliceous sediments: modern 103–104

siliceous sinter 103, *104*

silicic acid 57

silcrete 71, *104*

sinkhole *83*, 84

sinter *see* siliceous sinter

Siwaliks compressional basin 717

skeletal-grain mineralogy *88*

skin-friction line 122

Skolithos ichnofacies *321*, 322

slope retreat 275, *276*

slopes of mass wasting 273–277, *276*

slump folds 356, *357*

smectite 65, *66*

snowball Earth 625

snowline 292, 295, *296*

sodium carbonate minerals 575

soft ground 315

soft-sediment deformation 352–362, *471*, 488, 604, 606

soil catena 73

soil creep *257*, 273, *275*

soil gas CO_2 55

soil horizons 67, 67–69, 69, 70
soil 45, 67–74, 73
soil: classifications 74, 75
soil: developmental stages 73
soil: diagenesis 74–76
soil: formation factors 67
soil: microfabrics 71
soil: organic features 72–73
soil: structure 69–72
solar insolation 23, 23
sole marks 192, 193, 194
solifluction 257, 273
solifluction deposits 273, 274, 614
solution chemistry 47–52
sorted stones (patterned ground) 613, 613
soup ground 315
source rocks 679
spar 686
spatial variation in turbidity-current deposits 271, 272
speleothems 349see also cave deposits
spheroidal weathering 78, 78
spicules see sponge spicules
spilling breaker 217, 218
spit 491, 493, 495
sponge spicules 103
spreiten 318
spring tides 238, 239
spring tufas 582, 584
spur and groove (reef) 546, 548
stagnation point on sediment grain 142
stalactites 83, 83
stalagmites 83, 83
standing wave 219, 220
static grain friction cofficient 149, 255
steady flow 121
stiffground 45, 45, 315
Stokes fluid drift 217, 217
Stokes' law of settling 140, 141
Stokes wave theory 217
stony desert 570
storm wave base 498, 499
stormflow 372, 372
storm-generated stratasets 237, 238, 499, 500
straight river channels: water flow 134, 134, 396
straight river 380
STRATSIM 711
stream channel segments 366, 368
stream magnitude 366, 368
stream morphometric laws 366, 368
stream order 366, 368
stream piracy 370
stream power 383, 384, more from Chapter 5
streamline 122, 122
stromatactis 553
stromatolite laminae 333, 334–337

stromatolites 333–338, 334–337, 544
structure grumeleuse 684
stylolites 676, 680, 681, 682
subaqueous levee 479, 480, 482, 483, 486
subcritical flow 137, 137
subglacial environment 595
subglacial meltwater lake 298
subglacial traction till 303, 304, 305, 306, 307, 308, 600, 602, 602, 604, 610, 610, 616, 617, 618
submarine fan/delta 610, 610, 617
subsurface channel-belt width 427, 428, 429
subsurface paleochannel depth 426, 427
subsurface superimposed channel-belt width 429, 430
subtidal zone 510, 513
sulfate concentration in rainwater 54, 55
sulfate-reducing bacteria 32, 60, 111, 694, 700
summer (berm-type) beach profile 497, 497, 502
sunspot cycles 627
supercritical flow 137, 137, 181
superimposed channel-bar and -fill deposits 422, 423, 424
Superior-type iron formations 115
supersaturated solution 49
support mechanisms for sediment gravity flows 258
supratidal (saltmarsh) deposits 494, 513, 523, 525
supratidal zone 510, 513
surf waves 213
surf zone 218, 219, 494, 504, 505
surging breaker 218, 219
suspended bed-material load 147
suspended sediment plume 481, 481, 484, 611
suspended-load sediment 143, 147, 150, 152
suspended-load transport in wind 197–198, 198
suspended-load transport rate 150, 150–151
suspension feeding 315
suspension-threshold criteria in water flows 147, 149
suspension-threshold criteria in wind 197
swaley cross strata 232, 236, 487, 488, 501, 511, 519, 521
swamp 482, 486
swash marks 494
swash 219, 505
swell waves 213
swelling clays 65
symmetrical–asymmetrical confluences 391, 392
synaeresis 358, 360
syntaxial bottom growth: gypsum 575, 576
syntaxial bottom growth: halite 576, 578
syntaxial cements 675, 686, 689, 690
syntectonic 710, 719
systems tract

Talapia 574
talus see scree
Taylor–Gortler vortex 128, 128, 266, 267, 643
tectonic activity: evidence in sedimentary basins 717

tectonic activity: rivers and floodplains: *438*, 443, *444*, 446–448, *447*, *448*, 719, 721, 725, *726*

tectonic activity: uplift and climatic change 719, *719*

tectonic activity: uplift and river diversion 721, *721*

tepee structures in halite, *578*, 579

tepee structures *360*, 361–362, *578*

tephra 278

terminal fan 448

terra rosa 350

terrestrial ecosystems 36

terrigenous sandstone: classifications 78–80

terrigenous sandstone: composition and tectonism 709, *709*

terrigenous-sediment composition 78

terrigenous-sediment grains 45, 80–82

texture: terrigenous-sediment grains 77–78

theoretical modeling of turbidity currents 265, *267*

thermocline: lakes 465, *465*

thermocline: oceans 27, 610

thermokarst 613, *613*

thickness of cross strata 171, 173

thinolite (ikaite) tufa 582, *584*

three-layer clays 65, *66*

threshold of mixed-size sediment transport in water flows 145

threshold of mud transport in water flows 146

threshold of sediment transport in water flows *143*, 143–146, *144*

threshold of sediment transport in wind 195, *196*

threshold of sediment transport under water waves 227, *227*

thrombolites 338–342, *339–341*

throughflow 371

tidal bore 239

tidal bundle in cross strata *240*, 247, *248*, 517, *518–519*, 519

tidal-channel migration 517

tidal channel 482, 491, *491*, 511, *512–515*, 517–521, *524*, 533

tidal currents in shallow seas 525, *see also* Chapter 7

tidal currents 239–242, 516

tidal currents: time–velocity asymmetry 239, *240*, 516

tidal flat 482, *482*, 491, *493*, 511, *512–515*, 514, 521–525, *522–524*, *531*, 532–539, *534*, *535*, 536, *537*, *538*, *539*

tidal-flat burrows 521, *522*

tidal-flat infauna 521, *552*

tidal inlet 511, *512–515*

tidal laminites 243, *243*, 521

tidal sand waves 516, 526, *526*

tidal sediment-transport paths 242, *242*, 516, 525

tidal wave convergence 240, *241*

tidal waves 213

tide generation 238, *239*

tide-dominated coasts and shallow seas 510

tide-dominated coasts and shallow seas: effects of marine transgressions and regressions 526, *527–529*

tide-dominated coasts and shallow seas: generalized water flow and sediment transport 516–517

tide-dominated coasts and shallow seas: geometry and origin 510–516, *512–515*, *524*, *526*, *534*, *535*, *538*, *540*, *541*, *543*

tide-dominated coasts: flow and sedimentary processes in tidal channels *512*, *515*, 517–521, *518–519*

tide-dominated coasts: flow and sedimentary processes of tidal flats 521–525, *522–524*, 536, 537, *537–542*, 549, *550*

tide-dominated coasts: tidal-channel deposits 517, 518, *518–520*, 536, *538*

tide-dominated coasts: tidal-delta deposits 519–521, *519*

tide-dominated delta front 491, *492*

tide-dominated delta 510

tide-dominated shallow seas: flow and sedimentary processes 525–526, *526*, *540*, *547*

till 304

Tivoli travertine 583

tombolo 491, *493*, 495

tool marks 192, *193*, *194*, 270, *271*, 500

topsets 485, 611

tors 82

towers 82

trace fossils 311–332, *312–315*, *318–321*, *323*, *330*

trace fossils: behavioral classification 319–321

trace fossils: controls 314–317, *316*

trace fossils: evolution *331*, 331–332

trace fossils: preservation 318–319, *319*

trace fossils: terrestrial *324–330*, 324–331

trace fossils: tiering 314, *315*, *331*

trackways *see* trace fossils

trails *see* trace fossils

transcurrent (planar) stratification 163, 178, *178*, 205, *206*

transgressive systems tract 737, 740

translational slide 256, *257*

transport-limited slopes 62, 273, *276*

transverse dunes: aeolian *201*, 202, *203*

transverse ribs 183, *184*

transverse ridge marks 192, *193*

travertine *see* tufa

tributary-mouth bar 377, *378–381*, 391–393, *392*, 397

trilobite skeletal grains *94*, 95

trough cross strata 173, *175*

truncated spurs 598, *599*

tsunami dynamics *250*, 250–252, *252*

tsunami generation 249, 249–250, *250*, *251*

tsunami wave geometry 249

tsunami 213, 247–254

tufa 346–349, *347*, *348*, 468, 471, 581–584, *584*, *see also* periphyton

tuff 282

turbidites 270, *271*

turbidity-current body 266, *268*, 270

turbidity-current body velocity 268

turbidity-current channel 269, 270

turbidity-current flow stripping 269, 656

turbidity-current flow *265*, 266–270, *267–269*

turbidity-current head velocity 266

turbidity-current head *265*, 266, *267*, 270
turbidity-current lobes and clefts *265*, 266, *267*
turbidity-current lofting 269
turbidity-current tail 270
turbidity current *256*, 264–272, *265*, *267*, *268*, *269*, *271*
turbulence anisotropy 133, 134, *134*, 147
turbulence intensity of turbidity current 268
turbulence intensity 124, 133, 167
turbulence vertical profiles 133
turbulent air flow 195
turbulent boils (kolks) 147, 168
turbulent boundary layer in air 195
turbulent boundary layer in water *125*, 127–132, *128*
turbulent eddy 121, 124, 125, 126, 128
turbulent ejections and sweeps 129, *129*, 135, 146, 167
turbulent macroscale 125
turbulent mixing length *125*, 131, 148
turbulent shear stress 125
turbulent wake 140, *140*, *141*
turbulent water flow 121–138, *122*, *124*
turnover: lakes 466
two-layer clays 65, *66*

under-filled compressional basins 718
undersaturated solution 48
uniform flow 121, *122*
unit bar 377, *378–381*
unit-bar deposits *409*, 412, *413–420*, 414
United Arab Emirates (Trucial Coast) 585, *586*, 588
unsaturated zone *56*
unsteady flow 121, 137–138, *138*
uplifted beaches and floodplains *459*, *583*, 623, *625*
U-shaped valley 598, *599*

vadose zone, *see* unsaturated zone
valley glacier 292, 595, *596*
varves: glacial lake 611, *611*
varves: saline lake 575
velocity distribution in ice 293, *293*, 294, *294*
velocity vertical profiles 130–132, *132*
velocity-defect law 131, *132*
ventifact 197, *197*
vermiculite 65
viscoelastic lithosphere 614, 717
viscosity of lithosphere 614, *718*
viscous (surface) drag on sediment grains 140, *140*, 142
viscous grain flow 259, *259*
viscous shear stress 123, 124, 125, 142
viscous sublayer 127, *128*, 144
void lining (isopachous) cement 687
volcanic-debris flows 287, *286–290*
volcanic eruptions 278, 281, *281*, 282, 286
volcanic eruptions: climatic effects *281*, 282, 289, *291*
volcanic glass *see* shards
volcanic mudflows 287, *288*, 290

volcaniclastic sediment 278–282, *283*, *284*, *285*, 661
volcaniclastic sediment: composition classification 281
volcaniclastic sediment: grain-size classification 281, 282
vortex wave ripples 230, *231–233*

wake region 128, 131
Walker Lake 463, 582
wall region 128, 131
Walsortian mounds 553
Walther's law of facies 708
wash load 147
washover (fan) *491*, 492, *493*, 507, *508*, *515*
washover deposits 507, *508, 515*
washover-lagoon burrows 508
water discharge 371, *372*
water *see* hydrosphere and seawater
water: chemical properties 24–25
water: density of pure 27, *27*
water: physical properties 27–29
water–sediment escape structures 273, *273*, 353–355, *355*, *356*
wave base 216
wave breaking 217, *218*, 219, 502
wave definitions 213, *214*
wave diffraction *219*, 220
wave dispersion 214
wave energy 217
wave orbital diameter 215, *216*, 227, *227*
wave reflection 219, *219*
wave refraction 219, *219*
wave set-up *218*, 219, 220
wave theories 214–217, *215*
wave-cut notch 557, *558*
wave-cut platform 557, *558*, *559*
wave-dominated coasts and shallow seas 491–510
wave-dominated coasts and shallow seas: effects of marine
    transgression and regression *509*, *510*, 509–510
wave-dominated coasts and shallow seas: generalized water
    flow and sediment transport 496–498, *see also* Chapter 7
wave-dominated coasts and shallow seas: geometry and origin
    491–496, *493–495*
wave-dominated coasts: flow and sedimentary processes on
    beaches *502*, 502–507, *503*
wave-dominated coasts: plan geometry *493*, 495, *495*
wave-dominated delta front 490, *492*
wave-dominated shallow seas: water flow and sedimentary
    processes 498–502, *499–501*
wavy bedding 244, *246*
weather 18
weathering 45, *also see* congruent weathering reactions *and*
    incongruent weathering reactions
weathering-limited slopes 62, 273, *276*
weathering residues 60
weathering: calcite 57
weathering: climate change 84
weathering: dolomite 57

weathering: feldspars 57, *59*
weathering: groundwater composition 45, 62
weathering: gypsum 57
weathering: halite 56
weathering: landforms *82*, 82–84
weathering: olivine 58, 59
weathering: silica 57, *57*
welded volcaniclastic sediment 282, *283*
well-to-well correlation of fluvial sandstone bodies 427, *428*
white-smoker tufa chimneys 112, *114*, 151, 652
whitings: lakes 465
whitings: shallow calcium carbonate seas 96, 530
wind set-down 225, *225*
wind set-up 224, *224, 226*, 466, *467*
wind waves 213

wind waves in shoaling water 217–221, *218*
wind-shear currents 221–227, *222–226*, 496
wind-wave currents, *see* chapter 7, 496, 516
wind-wave generation 213
winter (bar-type) beach profile 497, *497, 502*
wrinkle marks 353, *355*

yardang 211, *211*
Younger Dryas 619, *622*, 623

zebra mussel 469
Zechstein evaporites 590, *590*
zibar *201*, 205
zonal mean wind 19, *20*
Zoophycus ichnofacies 322, *324*
Zooxanthellae 545